ORGANIC REACTION MECHANISMS · 1987

ORGANIC REACTION MECHANISMS · 1987

An annual survey covering the literature dated December 1986 to November 1987

Edited by

A. C. Knipe and W. E. Watts
University of Ulster,
Northern Ireland

An Interscience® Publication

JOHN WILEY & SONS
Chichester · New York · Brisbane · Toronto · Singapore

Library of Congress Catalog Card Number 66–23143

British Library Cataloguing in Publication Data:

Organic reaction mechanisms.
1. Organic compounds. Chemical reactions.
Mechanisms—Serials
547.13′9

ISBN 0 471 92078 9

Phototypeset by Thomson Press (India) Ltd.
Printed and bound in Great Britain by
the Anchor Press Ltd, Tiptree, Essex

Contributors

R. A. AITKEN	Department of Chemistry, University of St. Andrews, Purdie Building, St. Andrews, Fife KY16 9ST, Scotland
R. A. COX	Department of Chemistry, University of Toronto, 80 George Street, Toronto, Ontario M5S 1A1, Canada
M. R. CRAMPTON	Department of Chemistry, Durham University, Durham DH1 3LE, UK
G. W. J. FLEET	Dyson Perrins Laboratory, Oxford University, South Parks Road, Oxford OX1 3QT, UK
A. FRY	Department of Chemistry and Biochemistry, University of Arkansas, Fayetteville, Arkansas 72701, USA
P. HANSON	Department of Chemistry, University of York, Heslington, York YO1 5DD, UK
A. C. KNIPE	Department of Chemistry, University of Ulster at Coleraine, Coleraine, Co. Londonderry BT52 1SA, Northern Ireland
R. B. MOODIE	Department of Chemistry, The University, Exeter EX4 4QD, UK
R. A. MORE O'FERRALL	Department of Chemistry, University College, Belfield, Dublin 4, Ireland
A. W. MURRAY	Department of Chemistry, The University, Dundee DD1 4HN, Scotland
D. C. NONHEBEL	Department of Pure and Applied Chemistry, University of Strathclyde, Thomas Graham Building, Glasgow G1 1XL, Scotland
M. I. PAGE	Department of Chemical Sciences, The Polytechnic, Queensgate, Huddersfield, West Yorkshire HD1 3DH, UK
R. M. PATON	Department of Chemistry, Edinburgh University, West Mains Road, Edinburgh EH9 3JJ, Scotland
J. SHORTER	Department of Chemistry, The University, Hull HU6 7RX, UK
W. J. SPILLANE	Department of Chemistry, University College, Galway, Ireland

Preface

The present volume, the twenty-third in the series, surveys research on organic reaction mechanisms described in the literature dated December 1986 to November 1987. In order to limit the size of the volume, we must necessarily exclude or restrict overlap with other publications which review specialist areas (e.g. photochemical reactions, biosynthesis, electrochemistry, organometallic chemistry, surface chemistry and heterogeneous catalysis). In order to minimize duplication, while ensuring a comprehensive coverage, the editors conduct a survey of all relevant literature and allocate publications to appropriate chapters. While a particular reference may be allocated to more than one chapter, we do assume that readers will be aware of the alternative chapters to which a border-line topic of interest may have been preferentially assigned.

One change of author has taken place since last year and we particularly welcome Dr. Derek Nonhebel on his return to the fold. He replaces Dr. David Cowley who has made a major contribution to the series over the past nine years by contributing authoritative reviews of Radical Reactions, initially as a co-author with Dr. Nonhebel.

Once again we wish to thank the publication and production staff of John Wiley & Sons and our team of experienced contributors for their efforts to ensure that the standards of this series are sustained. We are also indebted to Dr. N. Cully, who compiled the subject index.

A.C.K.
W.E.W.

Contents

Organic Reaction Mechanisms 1987
Edited by A. C. Knipe and W. E. Watts

CHAPTER 1

Reactions of Aldehydes and Ketones and their Derivatives

M. I. Page

Department of Chemical and Physical Sciences, Huddersfield Polytechnic

Formation and Reactions of Acetals, Ketals, and Orthoesters

Weaker than usual anomeric effects are observed in 2,3- and 2,5-*t*-butoxy and -trimethylsiloxy derivatives of 1,4-dioxane. Calculations indicate that this is not a steric but an electronic effect resulting from electron donation by the substituent.[1]

An analysis of *X*-ray data of COCOC fragments shows that there is no correlation between CO bond lengths or COC bond angle and torsion angles whilst the bond angle OCO at the anomeric centre is very sensitive to conformational variation.[2]

The pH-rate profile for the hydrolysis of the cyclic acetal of *trans*-1,3-cyclohexanediol (**1**) shows seven inflections, only one of which is due to the ionization of the dimethylamino group. The other inflections are attributed to changes in rate-limiting step and mechanism. Because of strain in the acetal, ring-opening is rapid and reversible. At $pH > 7$ the rate-limiting step is water or hydroxide ion attack on the oxocarbocation intermediate but at higher pH it becomes pH-independent ring-opening.[3]

There have been further reports on the general acid-catalysed and uncatalysed hydrolysis of acetaldehyde methyl phenyl acetals.[4]

The acid-catalysed reaction of alkyl β-tetrahydropyranyl acetals containing an enamine (**2**) form products in dry methanol consistent with exo- and endo-cyclic C—C bond cleavage.[5]

Acetals of 2-hydroxybenzaldehyde may be prepared from acylated salicylaldehyde under basic conditions such as sodium methoxide in methanol. This is attributed to hemiacetal ion formation and subsequent intramolecular acyl transfer (**3**). Loss of acetate from the acylal would generate a highly resonance-stabilized intermediate (**4**) which would be trapped by alcohol to give the acetal product.[6]

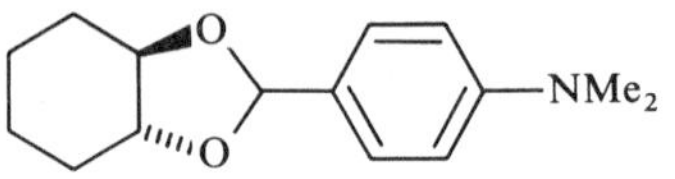

(**1**)

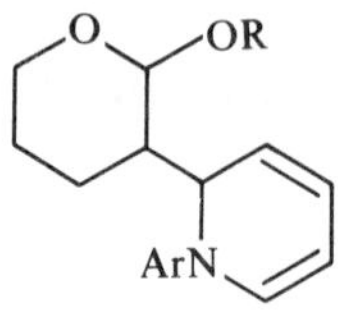

(**2**)

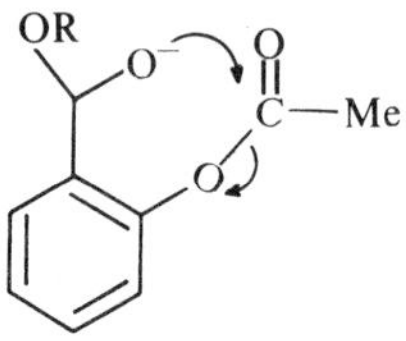

(**3**)

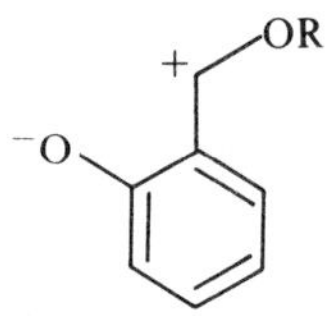

(**4**)

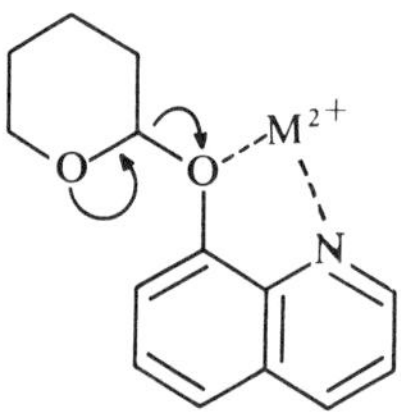

(**5**)

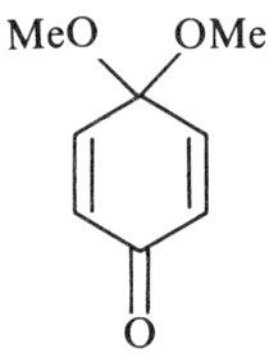

(**6**)

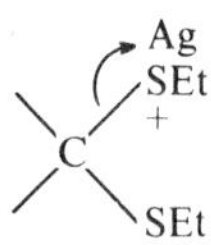

(**7**)

Divalent metal ions are very effective catalysts for the hydrolysis of acetals containing suitably placed pyridine and carboxylate residues, even though metal-ion binding to the acetals is weak. Catalysis is due to metal-ion stabilization of the leaving-group in the unimolecular decomposition such as in (**5**). Metal-ion catalysis does not occur when the leaving-group is an aliphatic alcohol even when the oxocarbocation intermediate is quite stable.[7]

The mechanism for the thermal disproportionation of quinone monoketals (**6**) involves reversible unimolecular dissociation into a radical pair followed by rate-limiting hydrogen transfer to give formaldehyde and 4-methoxyphenol.[8]

The sulphuric acid-catalysed hydrolysis of dihydropyranal acetals gives cyclenones whereas the use of pyridinium *p*-toluensulphonate gives δ-keto-aldehydes.[9]

As expected, the intermolecular addition of thiols to acyl nitriles shows much more negative volumes of reaction and activation compared with the intramolecular cyclization of a hydroxy-ketone to a hemiketal.[10]

Substituent effects on the silver ion-promoted hydrolysis of thioacetals suggest that the rate-limiting step involves unimolecular cleavage of the metal-bound substrate (**7**).[11]

A set of rules to describe selectivity in acetal formation has been prepared.[12]

The reductive cleavage of homo-chiral acetals using Lewis acid-hydride systems gives the opposite stereochemical outcome to that achieved using aluminium hydride.[13]

Oxocarbenium ions can be generated from acetals by electrolysis and trapped by silicon-mediated nucleophiles.[14] The kinetics of homolytic fission of acetals in solution at high pressure have been reviewed.[15] The mechanism of the liquid phase homolytic halogenation of acetals has also been reviewed.[16]

Hydrolysis and Formation of Glucosides, Nucleosides, Oxazines, and Related Compounds

Non-enzymic Reactions

Under high pressure, the specificity of glycosylation is increased which is attributed to a shift in the equilibrium from the monocyclic glycosyl cation towards the bicyclic acyloxonium cation with increasing pressure.[17]

Heptamolybdate ion is, thermodynamically, the most preferred species in aqueous solution among the numerous polyanions of molybdic acid. This ion catalyses the epimerization of D-glucose and racemization of D-glyceraldehyde. It is suggested that catalysis arises not simply from oxo-complexation of the monosaccharide and molybdate ion but from stabilization of a 3-oxa-allylic cationic species.[18]

The alkaline degradation of cytosine nucleosides involves predominantly deamination to the corresponding uracil derivatives. Substituent effects indicate that this occurs by direct displacement of the amino group by hydroxide ion attack at C(4). Uracil nucleosides are cleaved by complete fragmentation of the pyrimidine ring initiated by hydroxide ion attack on C(5) of the base.[19] The reaction with 6-substituted purine nucleosides consists of three consecutive reactions: attack of hydroxide ion on C(8) followed by ring-opening of the imidazole to an imine-aldehyde (**8**) and anomerization of the aglycone;[20,21] deformylation of (**8**); and finally cleavage of the *N*-glycosidic bond.

Carbon-bridged purine cyclonucleosides such as (**9**) show a slower rate of acid-catalysed hydrolysis, compared with that of 3-methyladenosine by a factor of over 10^4. This increased stability is attributed to a combination of the positively charged nitrogen retarding protonation and strain in the oxocarbocation intermediate.[22]

The effect of substituents in the purine has been studied in the acid-catalysed hydrolysis of 2′-deoxyadenosines. Sites of protonation were determined using ^{15}N NMR spectroscopy.[23]

Ring-chain tautomerism in 1,3-oxazines (**10**) shows equilibrium constants which may be correlated with substituents in the aromatic residue by a Hammett ρ^+-value of 0.76.[24]

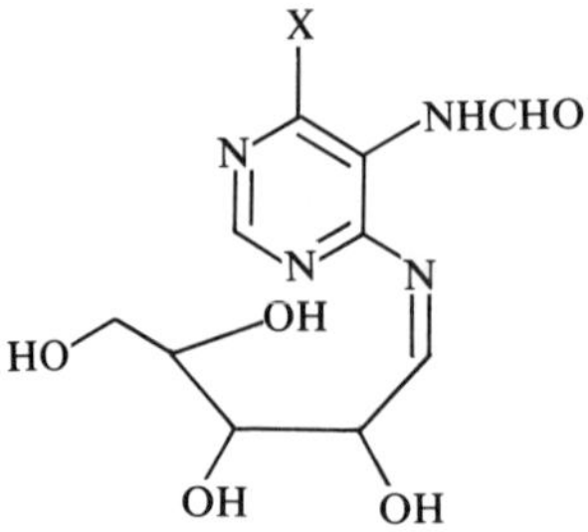

(8)

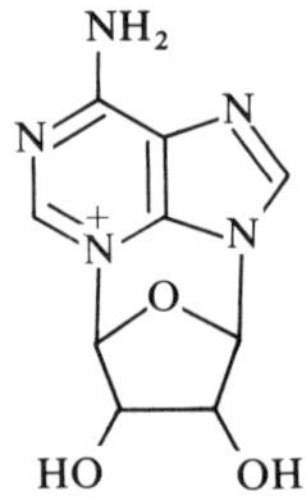

(9)

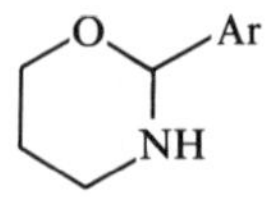

(10)

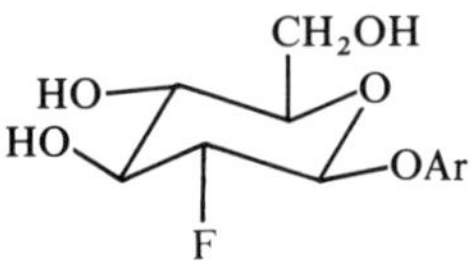

(11)

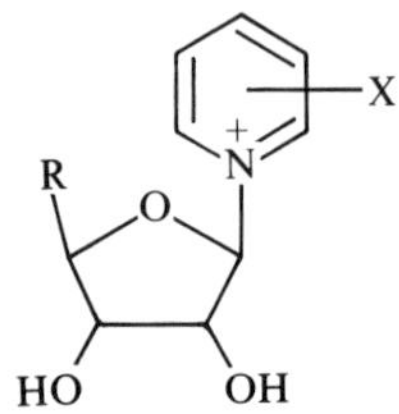

(**12**)

Enzymic Reactions

The incubation of 2-deoxy-2-fluoro-β-D-glucopyranoside (**11**) with β-glucosidase results in a time-dependent loss of enzyme activity. This is attributed to rapid loss of the leaving-group giving a relatively stable glucosyl enzyme intermediate.[25]

Solvent isotope effects and trapping experiments indicate that the mechanism of human liver α-L-fucosidase involves rate-limiting proton transfer and regeneration of the α-stereochemistry at the anomeric carbon. Initial-burst kinetics are consistent with either the formation of a glycosyl-enzyme intermediate or a tight ion-pair.[26]

Isotopic labelling experiments indicate that sucrose synthetase does not catalyse the cleavage of the scissile carbon oxygen bond of the glucosyl substrate in the absence of fructose.[27]

Enzymatic hydrolysis has been used for the stereospecific synthesis of enantiomerically pure pyranosides by hydrolysis of epoxide precursors.[28]

The NAD^+-glycohydrolase-catalysed hydrolysis of pyridine dinucleotides lacking the normal amide residue at C(3) of the pyridine residue (**12**) proceeds with rate-limiting pyridinium-ribose bond fission. The maximal rates of the enzyme-catalysed reaction, k_{cat}, exhibit a large dependence on the leaving-group pyridine with a Brønsted of -0.9. The Brønsted value for the pH-independent uncatalysed unimolecular reaction is -1.1. These results are consistent with the previously proposed late transition state with considerable oxocarbocation character.[29]

Reactions and Formation of Nitrogen Derivatives, Schiff Bases, Hydrazones, Oximes, and Related Species

Both increasing pressure and solvent polarity decrease the rate of isomerization about carbon-nitrogen double bonds. This is not expected for a rotational mechanism involving charge separation in the transition state. The mechanism of isomerization, even for strongly electron-withdrawing and -releasing substituents, appears to proceed by an inversion pathway (**13**).[30]

An inversion mechanism has been proposed for *E*/*Z*-isomerization of sulphonylimines on the basis of small substituent effects.[31]

1,3-Diazolidines (**14**) in trifluoroacetic acid are in equilibrium with the open-chain iminium ion (**15**) and, with no 2-substituent, 5–10% exists in the chain form. Aryl substitution in the 2-position increases this proportion to about 30% whereas *N*-alkyl substitution has little influence on the ring-chain equilibrium.[32]

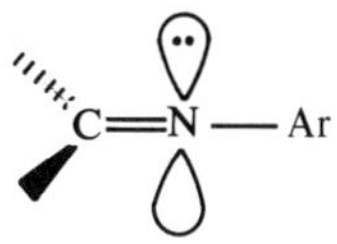

(**13**)

(**14**)

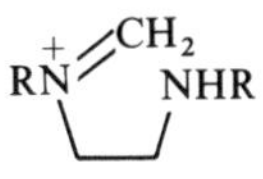

(**15**)

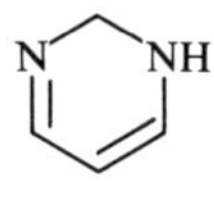

(**16**)

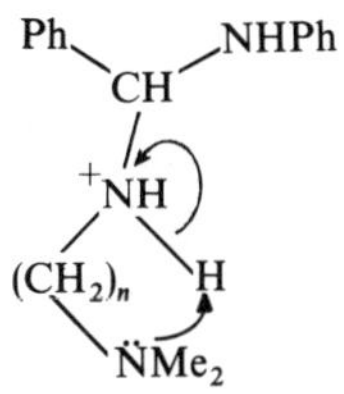

(**17**)

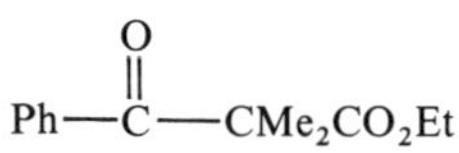

(**18**)

PhCONHN=CHAr

(**19**)

Imine-enamine tautomerism occurs in 1,2- (**16**) and 2,5-dihydropyrimidines and the position of equilibrium is strongly dependent on substituents at positions 4 and 6.[33]

Theoretical calculations predict that the nucleophilic addition of ammonia to formaldehyde is catalysed by one or two water molecules participating bifunctionally in cyclic activated complexes. Catalysis arises from the increased strength of hydrogen-bonding between water and the zwitterionic tetrahedral intermediate.[34]

The terminal tertiary amino group in (dimethylamino)alkylamines acts as an intramolecular general base catalyst in the transamination reaction of the diamines with *N*-benzylideneaniline (**17**).[35]

The aminolysis of the α,α-dimethyl-substituted keto-ester (**18**) proceeds by initial attack on the ketone carbon.[36] The kinetics and effect of substituents on the reaction of amines with *N*-sulphonylbenzimidates in chlorobenzene have been reported.[37]

The rate constants for the first five steps of the condensation of glucose with valine have been determined as well as the identification of the many products formed from the acid degradation of the Amadori intermediate.[38] The mechanisms of heterocyclic ring-closures to give five- and six-membered rings have been reviewed.[39]

Visual pigment rhodopsins contain a Schiff base which is hydrolysed at some stage during the photochemical cycle after *cis-trans* photo-isomerization and then reformed with 11-*cis*-retinal. The intermediate carbinolamine formed during the hydrolysis of *N*-retinylidine-*n*-butylamine, a model for the rhodopsin chromophore, has been detected using stopped-flow techniques. Hydrolysis in neutral micelles exhibits general base catalysis and spectroscopic changes in micelles are related to those observed in the natural system.[40]

An extensive study of substituent effects on the hydrolysis of Schiff bases to substituted benzaldehydes and 4-aminobenzoic has resolved carbinolamine formation and breakdown into Hammett ρ-values of -1.54 and -0.63, respectively. Some interesting, even if speculative, intermediates have been proposed.[41]

The effect of substituents on the acid-catalysed hydrolysis of the imines (**19**) has been investigated.[42] Carbinolamines are postulated intermediates in the base-catalysed hydrolysis of pyrrole-substituted imines.[43] Thermodynamic activation parameters for Schiff base formation from pyridoxal phosphate and hexylamine have been reported.[44] The kinetics and mechanism of the condensation of 2-hexosamines with pyridoxal phosphate has been investigated and changes in rate-limiting step with reaction conditions observed.[45]

The transamination reaction between pyridoxal and pyridoxamine Schiff bases has been shown to involve uncomplexed rather than metal-ion-complexed carbanions, as suggested previously.[46]

The reaction of chalcone with phenylhydrazine gives two products, differentiated by which nitrogen initially attacks the α,β-unsaturated ketone (**20**). The reaction at the NH_2 position, but not that at NH, is base-catalysed. Although not considered by the authors, it is conceivable that addition at NH is catalysed intramolecularly by the oxygen of the enolate anion acting as a general base (**21**).[47]

The formation of indoles from the reaction between arylhydrazones and phosphorus trichloride occurs by the formation of a diazaphospholine intermediate (**22**). It is proposed that acid-catalysed P—N fission occurs in (**22**) to give intermediates similar to those suggested for the Fischer indole reaction.[48]

Intermediates formed in the reaction of hydrazine with β-keto-esters have been studied by ^{13}C NMR spectroscopy. Nucleophilic attack occurs initially at the keto carbon followed by ring-closure of the hydrazones or enamine tautomers.[49]

Pyrazoles are formed from the reaction of α-aroyl-α-bromoketene dithioacetals with hydrazine. Rearrangement occurs in the episulphonium ion formed from the initial hydrazine adduct by displacement of bromide by the thioether.[50] Activation parameters and Hammett ρ-values have been reported for phenylhydrazone formation from substituted benzaldehydes.[51]

Tosylhydrazones react with peroxy-sulphur compounds to give the corresponding carbonyl derivatives by, it is assumed, base-catalysed fragmentation of the oxaziridine intermediate (**23**).[52]

Sulphonyl isocyanates react with aldehyde hydrazones at either the carbon or nitrogen of the azomethine. The position of electrophilic attack is determined by the nature of substituents.[53] The effect of substituents and solvent on the azo-hydrazone tautomeric equilibrium has been discussed.[54]

The intramolecular addition of an amino group to a carbonyl group generates the same carbinolamine tetrahedral intermediate **(24)** as that formed from the addition of hydroxide ion to a cyclic iminium ion. Unexpectedly, the cation **(25)** is in equilibrium with the protonated amino-ketone below pH 9.5. For electron-withdrawing substituents in the aromatic residue, the ring-opened form constitutes up to 14% of the mixture. Furthermore, in basic solution the predominant species is not the pseudo-base **(24)** but the ring-opened deprotonated amino-ketone **(26)**.[55]

During the biosynthesis of nicotinamide from tryptophan, the aldehyde **(27)** ring-closes to pyridine-2,3-dicarboxylate by a non-enzyme-catalysed pathway. At low buffer concentrations, the dehydration of the carbinolamine intermediate is rate-limiting whilst at high concentrations ring-closure becomes the slow step.[56]

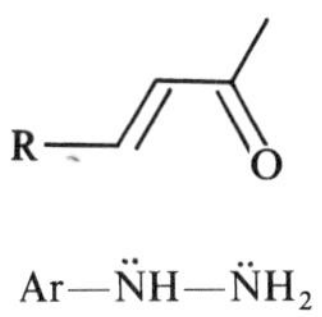

(20)

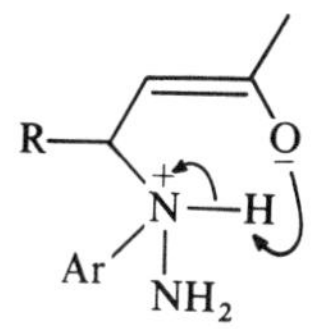

(21)

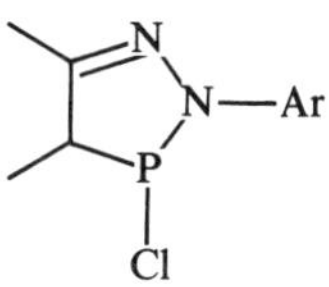

(22)

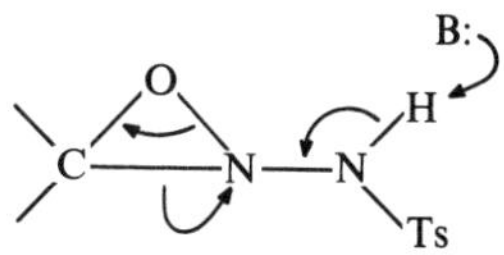

(23)

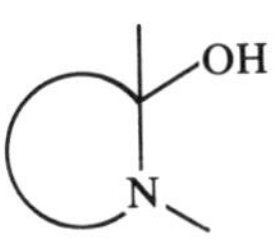

(24)

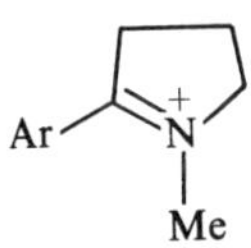

(25)

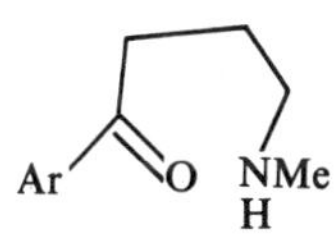

(26)

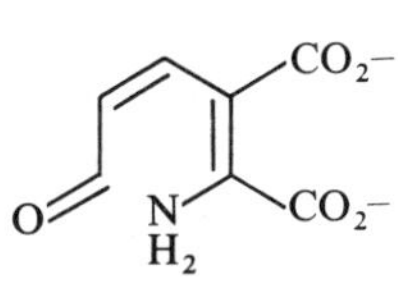

(27)

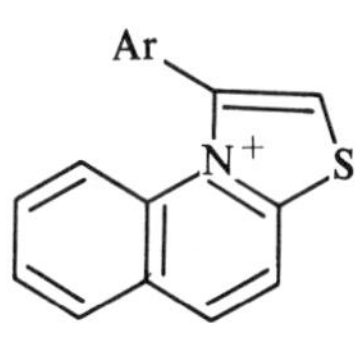

(28)

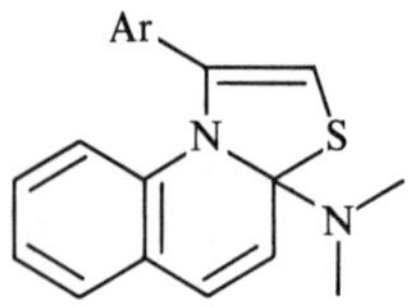

(29)

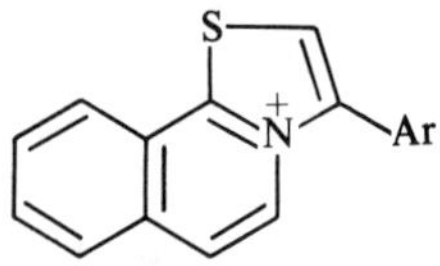

(30)

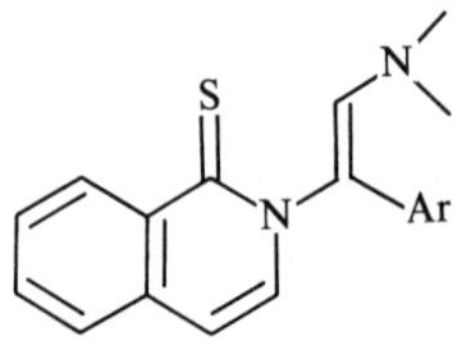

(31)

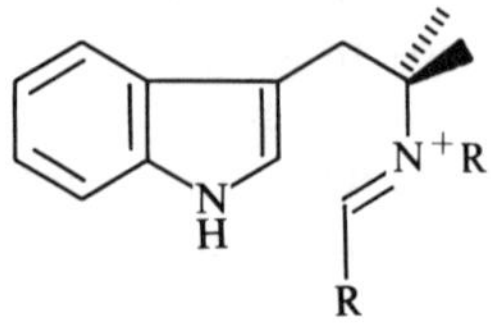

(32)

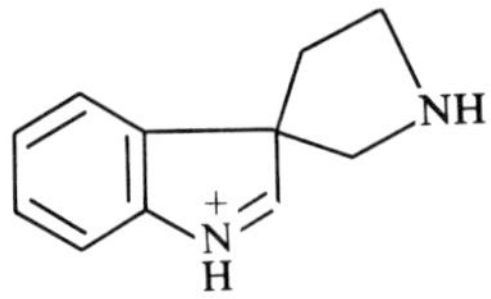

(33)

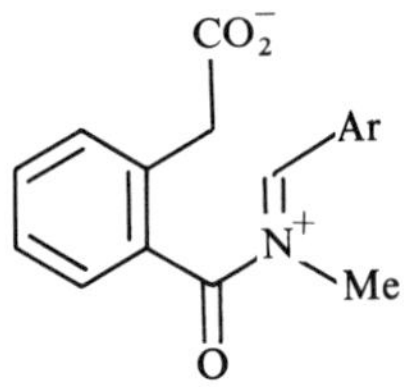

(34)

Amines add to the fused thiazolium salt (**28**) to form stable pseudo-bases (**29**) whereas the reaction with (**30**) leads to ring-opening giving the enamine (**31**). These, and other, annelation effects are rationalized by calculations of the charge distribution in the reactant thiazolium salts.[57]

The iminium ion (**32**) is the presumed intermediate in the ring-closure reaction of amino-indoles with aldehydes. A spiro intermediate (**33**) is then thought to be formed from enamine-type carbanion addition to the electrophilic carbon in (**32**). The stereochemistry of this addition step and the subsequent 1,2-shift controls the stereochemistry of the product.[58] This attack at the indole 3-position in preference to the 2-position is yet another example of the supposedly unfavourable 5-*endo-trig* ring-closure.[59]

The condensation of benzylidenemethylamine with homophthalic anhydride gives substituted piperidinones. Substituent effects indicate that the reaction occurs by initial iminolysis (**34**) in which *E*- and *Z*-imines ring-close to *cis*- and *trans*-isoquinolines, respectively, by different rate-limiting steps.[60]

Deamination of 1-aminopurinium salts with methanolic ammonia occurs by a nucleophilic attack of ammonia on the cyclic iminium ion followed by a ring-opening and -closing process (**35**) as indicated by isotopic labelling experiments.[61]

The reaction of methylamine with 1,4-dialdehydes to give intermediate imines which ring-close to give iminium salts (**36**) may be related to the 'hot taste' of the dialdehydes.[62]

The reaction of 1,3-diones with chloromethylene-iminium salts probably proceeds by the intermediate formation of iminium cations.[63] The kinetics of the cyclocondensation of *O*-phenylenediamine with benzil to give 2,3-diphenylquinoxaline have been reported.[64]

Iminium salts generated from primary amines and formaldehyde react with allylstannanes in protic media to give bishomoallylamines by a rapid aminomethano-destannylation.[65]

The hydrolysis of the imine of the 1,4-benzodiazepine (**37**) psychotropic drugs is thought to involve reversible ring-opening to the amino-ketone (**38**)[66] and the kinetics of this process in acid solution to give the triazole benzophenone derivative have been reported.[67]

The reaction of hydroxylamine with 4-phenyl-1,5-benzodiazepin 2-thiones and diazepinones (**39**) gives the 3-substituted amino-isoxazole (**40**). The proposed mechanism involves hydroxylamine attack at both unsaturated carbons followed by ring-closure.[68]

Analogous to the photo-hydration reactions of unsaturated carbon–carbon bonds, photo-hydrolysis of oximes occurs readily. Electronic excitation to the lowest singlet state (**41**) leads to a highly polarized state with an increase in the basicity of the oxime nitrogen. Oxazirane intermediates are formed from oximes in competition with the addition of water but not from oxime ethers, the reaction of which is not acid- or base-catalysed. Photolysis of *o*-hydroxy-substituted aromatic oximes gives substituted benzoxazoles in competition with acid-catalysed hydrolysis.[69]

The reaction of the carbinolamine (**42**) with hydroxylamine and other carbonyl reagents gives the expected ring-opened oxime and other products, respectively. This is contrary to an earlier report.[70]

Dialkyl acetals of *N,N*-dialkylformamide react irreversibly with amines to form amidines and although Hammett σ-values have been reported there are few mechanistic data available for this reaction.[71]

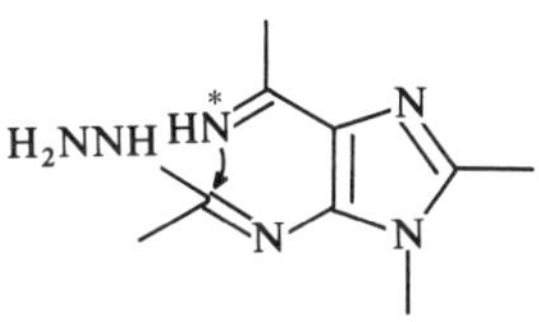

(**35**)

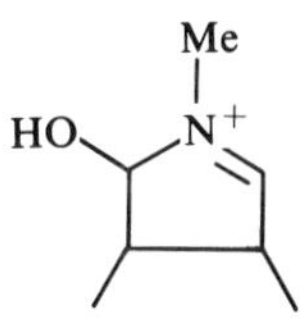

(**36**)

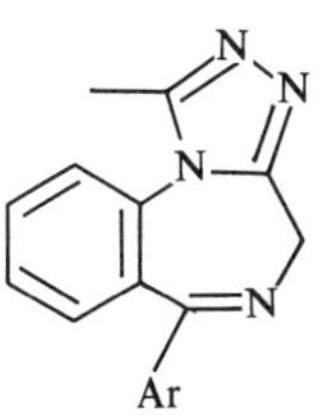

(**37**)

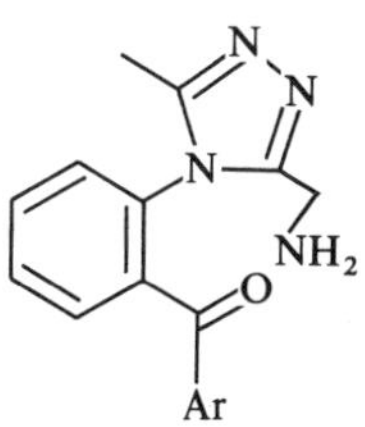

(**38**)

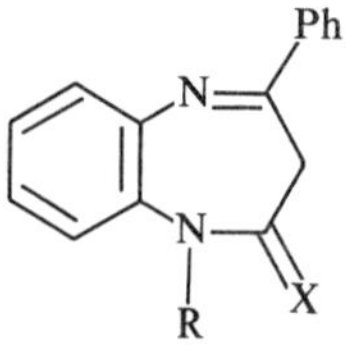

(39)

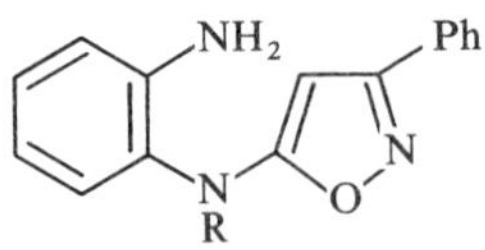

(40

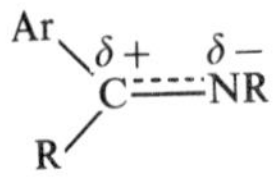

(41)

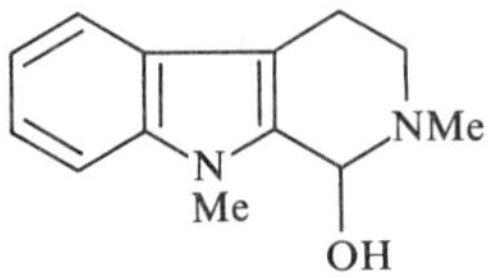

(42)

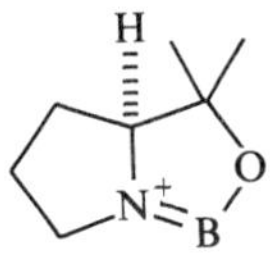

(43)

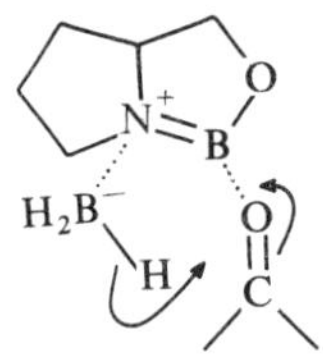

(44)

The nitrosation of ketones gives α-nitroso-ketones, usually existing as the oxime tautomers, and the acid-catalysed reaction proceeds by rate-limiting enol formation or nitrosation of the enol.[72]

Substituted nitrosobenzenes react with formaldehyde in an acid-catalysed reaction giving *N*-phenylhydroxamic acids by initial nucleophilic addition to the aldehyde conjugate acid.[73]

There is no observed intramolecular general base catalysis by a pyridine residue in the metal-ion-catalysed hydrolysis of acetylpyridine ketoxime pyridine-carboxylates.[74]

The reaction of chalcone with hydroxylamine in the presence of activated barium hydroxide as catalyst gives predominantly isoxazolines. It is suggested that the metal ion controls the cyclization process.[75] The same catalyst is also effective in the Claisen–Schmidt condensation. The reaction appears to be an inter-facial solid-liquid process between the adsorbed carbanion and the benzaldehyde.[76]

α-Bromoacetophenone oxime undergoes halogen substitution with cyanide to form an isoxazole by intramolecular attack of the oxime hydroxyl at the nitrile carbon.[77] Benzamidoximes react with aldehydes to form 4,5-dihydro-1,2,4-oxadiazoles, presumably by ring-closure of the intermediate amidine.[78]

Borane reacts with chiral vicinal amino-alcohols to form oxazaborolidines (**43**) which do not reduce ketones by themselves. However, in the presence of a molar equivalent of borane-tetrahydrofuran, reduction is rapid and enantioselective even with a catalytic amount of (**43**). The selectivity is attributed to reduction occurring in the coordinated complex (**44**).[79]

C—C Bond Formation and Fission; Aldol and Related Reactions

The reaction of α,α-dichloroacetoacetate with aldehydes in the presence of ethoxide ion gives epoxides. This novel reaction presumably involves a retro-acetoacetate condensation and subsequent dichlorocarbanion attack on the aldehyde. The tetrahedral intermediate anion (**45**) must then ring-close to the epoxide by displacing chloride.[80]

The general stereochemical outcome of aldol reactions that *E*- and *Z*-enolates give preferentially *anti* and *syn* aldols, respectively, is usually rationalized by the chair-like transition state. Assuming that only directionality and frontier-orbital interactions are important in these reactions under kinetic control, it has been shown that the stereochemistry of the product may be deduced.[81]

Geometrical equilibration of acyclic enolates from the often kinetically favoured *E*-isomer to the more stable *Z*-isomer (**46**) is a common method for controlling enolate stereochemistry. The effect of alkyl substituents on the equilibrium has been measured and compared with force-field calculations.[82]

The pyridine catalysed *Z*-*E* isomerization of arylideneflavanones occurs by initial addition of pyridine to the α,β-unsaturated ketone.[83] The diastereoselectivity of the aldol condensation of chiral glycol enolates is metal-tunable to give either *syn* or *anti* adducts.[84] The diastereofacial selectivity of enolates in aldol condensation can be varied by modification of a distant substituent.[85]

The enolate derived from the keto-pyranoside (**47**) undergoes aldol condensations in which the pyranoside core acts as a nucleophilic centre without β-elimination of the glycosidic methoxyl group and assists stereoselectivity.[86]

Sonication affects the stereochemical outcome of the Barbier reaction between 2-*S*-octyl halides, lithium and cyclohexanone. Ultrasonic waves are thought to affect single-electron-transfer at the solid-liquid interface.[87]

The Michael reaction of ascorbic acid with acrolein yields the spiran (**48**) which on distillation loses water to give the previously described product. This is the ring-closed hemiacetal form of the expected hydroxy-aldehyde adduct.[88]

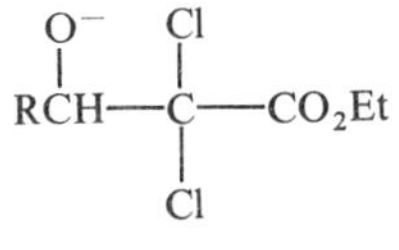

(**45**)

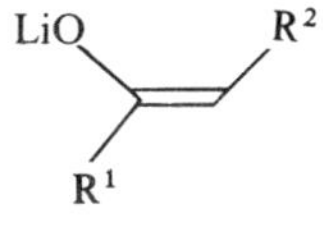

(**46**)

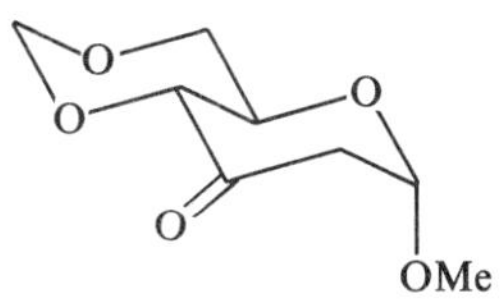

(**47**)

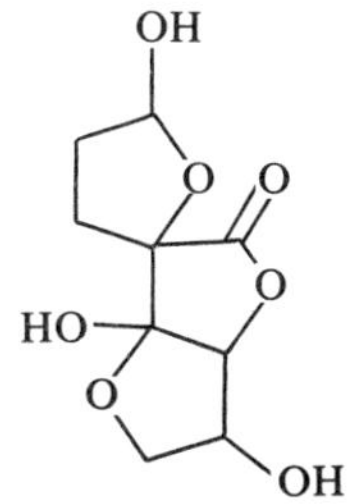

(**48**)

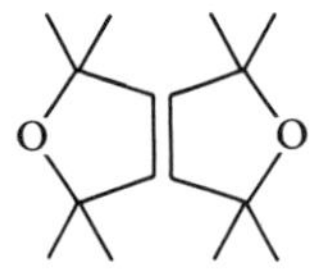

(49)

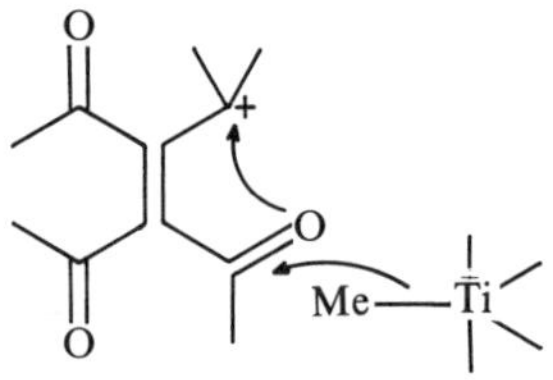

(50)

The first direct evidence of the intermediate Cram chelate of organometallics with α-alkoxy-ketones has been obtained using ^{13}C NMR spectroscopy.[89] Enantioselectivity is observed in pinacol formation from camphor which is attributed to metal-ion coordination.[90]

A model of the intermediate complex formed during the enantioselective addition of diethylzinc to benzaldehyde with chiral tridentate lithium complexes as catalysts is able to predict the observed direction of enantioselectivity.[91]

Steric effects in the transition state have been used to account for *anti*-selectivity in the aldol reaction using organoboron compounds.[92] The hydrogenation of β-keto-esters to β-hydroxy-esters in high enantiomeric purity can be achieved using substituted phosphines coordinated to ruthenium(II) as a catalyst.[93]

The synthesis of secondary alcohols in enantiomeric excess can be achieved by converting aldehydes into asymmetric dioxolanones, imidazolidinones, and oxazolidinones followed by nucleophilic addition. Using chiral diols coordinated to titanium, enantioselective addition of nucleophiles occurs to aromatic aldehydes.[94]

Using chiral diamine ligands, asymmetric addition of arylmagnesium bromide to aldehydes gives carbinols in enantiomeric excess.[95] The addition of *E*- and *Z*-crotylboronates to aldoximes shows unexpected diastereoselectivity with the *E*-isomer giving the *anti* and the *Z*-isomer giving the *syn* hydroxylamine product.[96] The addition of dialkylzincs to aldehydes catalysed by chiral pyrolidinylmethanols gives products in up to 100% enantiomeric excess.[97]

As usual, there have been many other reports on the stereochemical outcome of carbon-carbon bond-forming reactions.[98] The reaction of tetraacetylethene with dimethyltitanium dichloride gives **(49)** presumably via ring-closure of the intermediate carbocation **(50)**.[99]

The equilibrium between the intermediates formed in the Reformatsky reaction of benzaldehyde and cyclohexanone with the bromo-zinc salt derived from γ-bromo-senecionic acid has been investigated.[100]

The relative rates of addition of methyltitanium reagents to cyclic ketones follow the unexpected order for ring sizes 5, 6 and 7; aldehydes react about 500-fold faster than ketones.[101] Aldol condensations initiated electrochemically are assisted by the presence of lanthanide(III) ions.[102]

Silyl enol ethers add to acetals, aldehydes, and α,β-unsaturated ketones in the presence of trimethylsilyl chloride and tin(II) chloride. It is assumed that the tin(II) polarizes the silicon, improving coordination to the carbonyl oxygen.[103]

The reaction between silyl-ketene acetals and benzaldehyde in the presence of titanium(IV) chloride is thought to proceed by the formation of an octahedral, six-coordinate metal complex.[104]

The rapid addition of a tertiary amine to tin(IV) chloride-selenoacetal mixtures gives the relatively stable corresponding vinyl selenides, whereas the slow addition of the base gives aldol condensation products. The difference is attributed to the geometry of the vinyl-selenide formed.[105]

The rate of the intramolecular aldol reaction of (**51**) is second-order in *S*-proline but the enantiomeric excess increases non-linearly with the concentration of the base.[106]

Intramolecular cyclization of the C(3) carbon of dihydropyridine on an acetal carbon is catalysed by titanium tetrachloride and hydrogen chloride. The intermediate carbocation is presumably trapped by the enamine-type residue.[107]

Azoles react with ethylbromopyruvate oxime to give *N*-alkylated products of the azoles. Basic azoles, with pK_a values above 4, react by an elimination-addition mechanism.[108]

The condensation of salicylaldehyde and the piperidine dienamine of citral gives a stable hemiacetal (**52**) arising from initial δ-attack on the enamine.[109]

The regio- and stereo-selective *exo*-alkylation of Δ^2-isoxazolines is thought to be due to selective *Z*-aza-enolate formation followed by alkylation at the least hindered site.[110]

The high diastereoselectivity of the Grignard addition to triazole ketones (**53**) is rationalized in terms of azole coordination to magnesium.[111]

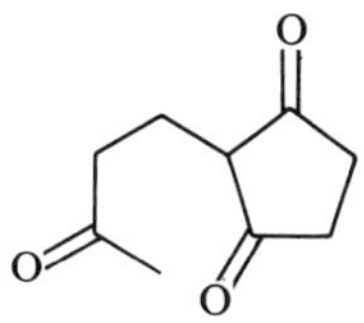

(**51**)

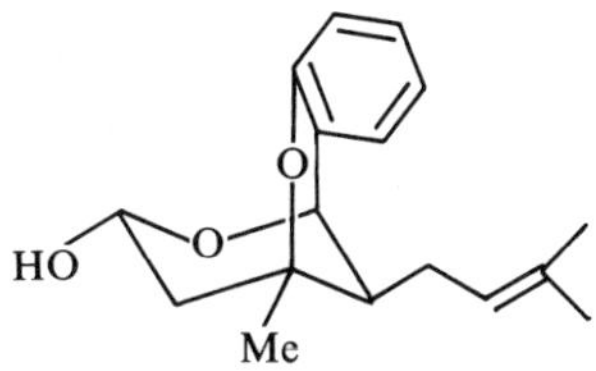

(**52**)

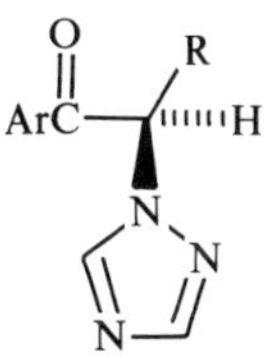

(**53**)

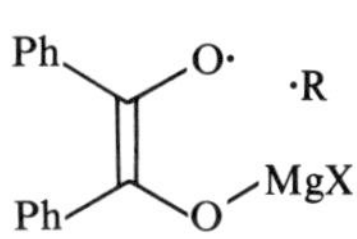

(**54**)

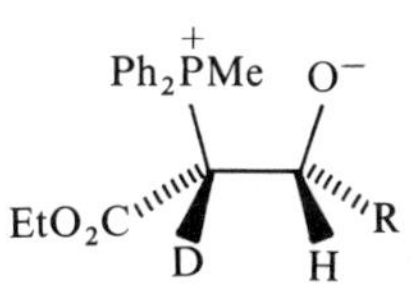

(**55**)

$$EtO_2C—\overset{-}{C}H—\overset{+}{P}Me_2Ph$$

(**56**)

There has been a further report that the reaction of Grignard reagents with benzil and benzophenone proceeds by an electron-transfer mechanism involving a radical cation of the Grignard reagent.[112] Although this is disputed,[113] the radical ion-pair dimer gives well-resolved ESR signals, the stability of which is related to the reduction potential of the ketone and the oxidation potential of the Grignard reagent. As well as the reduction product (benzoin), the reaction of benzil with Grignard reagents gives the *C*-alkylation (1,2-addition) and *O*-alkylation (1,4-addition) products. An electron-transfer mechanism is suggested involving a six-membered transition state (**54**). The *C*-alkylated product can result from recombination of the radicals after their separation from the solvent cage.[113]

The addition of methyllithium and methyl-Grignard reagents to α,β-epoxy-aldehydes occurs stereoselectively to give products which correspond to 'non-chelation-controlled' addition to the aldehyde group.[114]

Dithioacetals give the corresponding coupled alkene by treatment with Grignard reagents in the presence of a nickel-phosphine complex. It is suggested that the Grignard reagent traps the intermediate formed from carbon-sulphur bond-cleavage.[115]

Unsolvated dibutylmagnesium reacts with aldehydes in the presence of lithium alkoxide in methylcyclohexane to give not only the expected alcohol but also the ketone formed by oxidation.[116]

Although phosphonium ylides react with carbonyl groups to form alkenes and a phosphine oxide (Wittig reaction), sulphonium ylides form the thermodynamically more stable oxirane and sulphide. The sulphonium ylides are more reactive and add more readily to carbonyl groups and theoretical calculations indicate that the major difference is the fate of the betaine and cyclic intermediates.[117]

The stereochemistry and mechanism of the Wittig reaction have been reviewed.[118] Linear-free-energy relationships for the Wittig reaction between substituted benzaldehydes and triphenylbenzylidenephosphines in methanol have been reported.[119] The negative volumes of activation for the Wittig reaction have guided the application of high pressure to the Wittig synthesis of tri- and tetra-substituted alkenes.[120]

The stereoselectivity of the Wittig synthesis of β-unsaturated esters from carbohydrate-derived alkoxy-aldehydes depends on the substrate structure and the nature of the solvent. It is suggested that the alkoxy groups are involved in solvation of the phosphonium group.[121]

Replacement of triphenylphosphine by β-diphenylphosphinopropranoic acid in the Wittig reaction increases the *E*-stereoselectivity with semi-stabilized ylides.[122]

Magnesium and zinc oxides are efficient catalysts in the Wittig and Knoevenegel reactions because, it is suggested, the metal ion acts as an electrophile to the carbonyl oxygen and the oxide anion acts as a base catalyst in carbanion formation.[123]

α-Hydroxy-ketones undergo an accelerated Wittig reaction with stabilized phosphonium ylides to give *E*-trisubstituted alkenes; this is attributed to either intramolecular hydrogen-bonding to the carbonyl oxygen or to an interaction between the hydroxyl oxygen and the ylide phosphorus.[124]

The betaine (**55**), generated from its precursor alcohol, decomposes stereospecifically to the *Z*-alkene. The corresponding irreversible Wittig reaction of the ylide (**56**) with cyclohexanecarboxaldehyde gives a 95:5 *E*/*Z*-alkene ratio. This kinetic preference for *E*-alkenes indicates that equilibrium arguments used to explain high *E*-selectivity for related Wittig reactions of stabilized ylides with aldehydes are not general.[125]

The addition of sulphur ylides to aldehydes under heterogeneous conditions in the presence of small quantities of water gives epoxides, presumably by ring-closure of the intermediate (**57**).[126]

Base-catalysed ring-opening of hydroxythietanes (**58**) to give carbonyl compounds occurs in competition with nucleophilic substitution. The sulphane reacts 4×10^4-fold faster than the corresponding acyclic derivative. These reactions are thought to be concerted (**59**) rather than generating the free carbanion.[127]

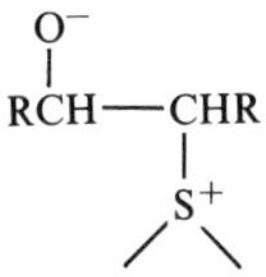

(**57**)

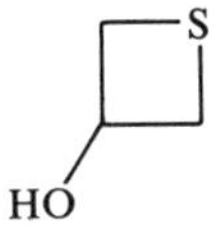

(**58**)

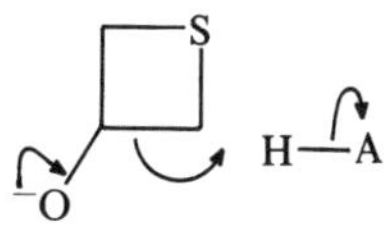

(**59**)

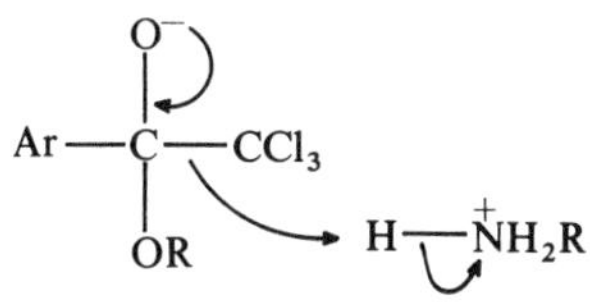

(**60**)

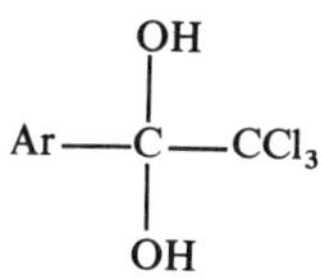

(**61**)

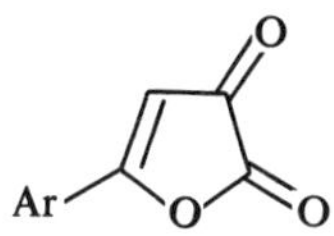

(**62**)

The alcoholysis of 2,2,2-trichloro-1-arylethanols is catalysed by amines which is thought to correspond to general acid-catalysed breakdown of the tetrahedral intermediate (**60**), although there must be doubts about the availability of electron density for the protonation step.[128]

The mono- and di-anions of the ketone hydrate (**61**) both expel the trichloromethyl anion.[129]

Nucleophilic addition to highly branched perfluoro-ketones invariably leads to carbonyl carbon–carbon bond cleavage, the leaving-group carbanion being stabilized by the electron-withdrawing fluorines. Another driving force may be steric hindrance in the tetrahedral intermediate since there is no evidence of hydrate formation of the fluorinated ketone in water.[130]

Substituent effects in the decomposition of benzoins in sodium methoxide-methanol indicate that the rate-limiting step in this carbon–carbon bond-cleavage reaction is the addition of methoxide ion to the carbonyl group.[131]

A suitably placed carboxyl group facilitates carbon–carbon bond-cleavage in a cyclopropane activated by a carbonyl group.[132] The C—C bond cleavage of non-enolizable α-phenyl ketones with sodamide replaces the benzoyl residue with hydrogen with retention of configuration.[133]

The ring-cleavage of the furandione (**62**) by amines is general base-catalysed in dioxane. Solvation effects on the rate of proton transfer have been considered in determining the rate-limiting step.[134]

The one-carbon ring-expansion reactions of bridged bicyclic ketones have been reviewed.[135]

Other Addition Reactions

The rate of oxygen exchange in α-methyl-substituted ketones (**63**) lends support to the suggestion that the interaction between the electron pair on an incoming nucleophile and the antibonding orbital of the antiperiplanar C- bond is greater for C—C compared with C—H bonds.[136]

Orbital arguments and torsional effects[137] suggest that nucleophilic attack on cyclohexanone occurs preferentially from the axial direction but this is frequently overwhelmed by steric effects. Support for this claim comes from the addition of small nucleophiles such as the acetonitrile carbanion which shows a preference for axial attack.[138]

Theoretical calculations on the trajectory of nucleophiles attacking chiral aldehydes and ketones (**64**) support the idea that diastereofacial selectivity arises from steric effects; when R is methyl or ethyl the trajectory is more on the R side of the normal plane than on the phenylethyl side; when R is *t*-butyl the approach trajectory of the attacking nucleophile is on the phenylethyl side of the normal plane.[139]

The suggestion that nucleophiles attack carbonyl groups along a trajectory that is displaced from the normal plane in the direction of the sterically less-demanding substituents attached to the ketone appears to be applicable to additions to chiral thionium ions.[140]

It has been suggested that the α-effect, the enhanced nucleophilicity of a base possessing an adjacent heteroatom with an unshared electron pair, arises from a stereoelectronic effect. In the addition of phosphite nucleophiles to carbonyl groups there can be more oxygen lone pairs antiperiplanar to the newly formed phosphorus-electrophile bond in acyclic compared with cyclic phosphites.[141]

The theoretical treatment of the stereoselective nucleophilic addition to carbonyl compounds has been reviewed,[142] as have steric effects[143] and product stability as a factor in kinetically-controlled addition reactions.[144]

The hydration equilibrium of aliphatic aldehydes is reduced when the aldehyde is bound to cationic and anionic micelles because of a decrease in the water activity at the micellar surface. The acid-catalysed hydration rate is increased by the presence of anionic micelles but decreased by cationic ones.[145]

The hydration and acid equilibria for 5-deoxypyridoxal have been reported.[146] Dehydration of formaldehyde hydrate is important in the kinetics of formation of the sulphonic acid adduct.[147]

The kinetic hydrogen isotope effect for the intramolecular hydride transfer in the glyoxal-glycollic acid rearrangement is calculated to be larger than that for analogous intermolecular hydride transfer. This is contrary to the general expectation of non-linear hydrogen transfer and, it is suggested, proton and hydride transfer may differ radically.[148]

Intramolecular hydride transfer in the hydroxy ketone (**65**) shows a primary kinetic isotope effect k_H/k_D of 3.4 for the migrating hydrogen which agrees well with the calculated value.[149]

The borohydride reduction of the propellane carbonyl group in an aqueous solution of a cationic surfactant occurs not only with the expected rate enhancement but with a stereochemical preference which is attributed to the mode of binding the substrate to the micelle.[150]

The stereoselective reduction of enol acetates to alcohols by lithium aluminium hydride is attributed to metal-ion coordination to the ketone intermediate.[151]

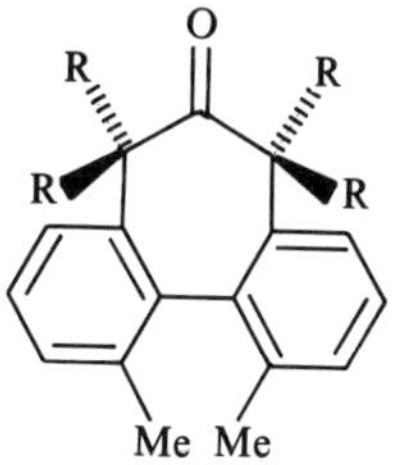

(63)

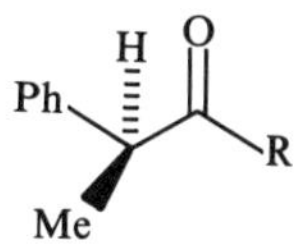

(64)

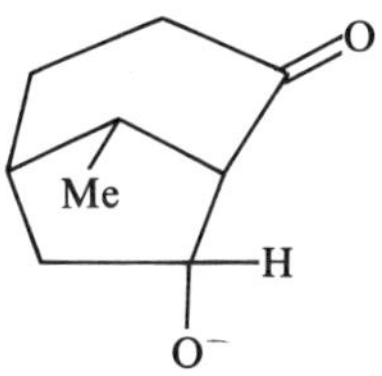

(65)

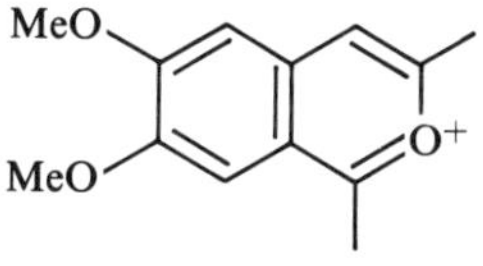

(66)

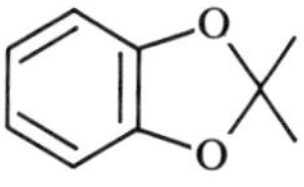

(67)

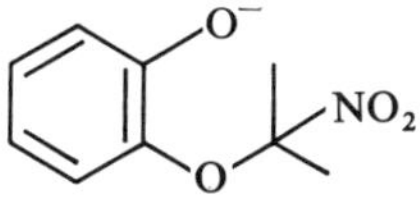

(68)

The enzyme mandelonitrile lyase catalyses the enantioselective formation of cyanohydrins from aldehydes in organic solvents because the rate of the normal chemical reaction is suppressed.[152]

Cyanide ion catalyses the dioxygen oxidation of α-keto-aldehydes by intermediate formation of the cyanohydrin anion which transfers an electron to oxygen.[153]

Nucleophilic addition of cyanide anion to 2-benzopyrylium salts (**66**) gives stable adducts.[154]

A limited kinetic study of the conversion of pyrylium cations into pyridinium derivatives with primary amines has led to speculation about the rate-limiting step, with little supporting evidence.[155]

The reaction between nitroalkanes and *o*-quinones in carbonate buffers gives cyclic acetals (**67**). The nitroalkane anion is thought to add to the carbonyl oxygen of the *o*-quinone to give (**68**) which subsequently ring-closes to the product. There are several other pathways by which the intermediate (**68**) could be generated.[156]

Enolization and Related Reactions

There have been more reports of stable enols derived from sterically-hindered ketones.[157] The photochemical generation of vinyl alcohol in aqueous solution has allowed the determination of the pK_a (10.5) for the enol and the pK_a (16.7) of acetaldehyde. Kinetic solvent isotope effects on both enolization and ketonization suggest that the reaction occurs by a stepwise rather than a concerted mechanism. Combined with other data for base-catalysed enolization, there appears to be an imbalance in the transition state with charge delocalization in the enolate anion as indicated by Brønsted α value, lagging behind proton transfer to the base, as indicated by the Brønsted β value.[158]

There is a remarkable linear correlation between the pK_{enol} values for sterically hindered ketones in hexane and those for substituted ketones (**69**) in water. Although the data cover a limited range of equilibrium constants, the linearity of the correlation must reflect a proportionality between substituent effects in both systems.[159]

The enol and hydrate are the dominant forms of 9-formylfluorene (**70**) in aqueous solution and the unstable aldehyde is generated from the reaction of its ethanethiol hemiacetal with iodine. The stability of the enol is indicated by the pK_a of 6.2 for the aldehyde (**70**).[160]

The equilibrium constants for the equilibrium between enols/alcohols and enol

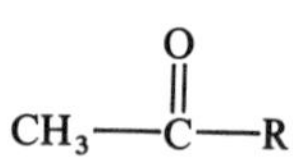

(**69**)

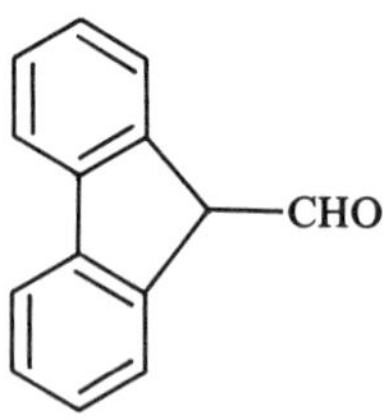

(**70**)

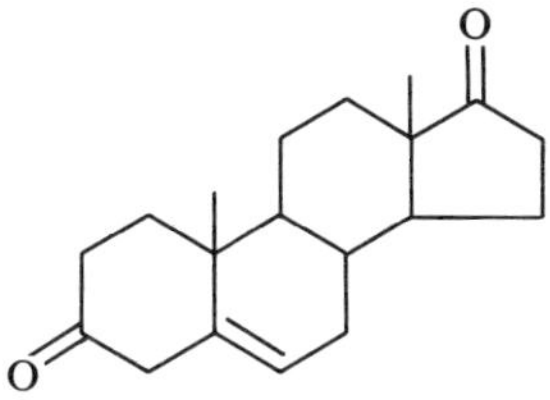

(71)

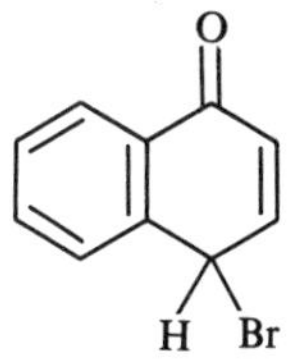

(72)

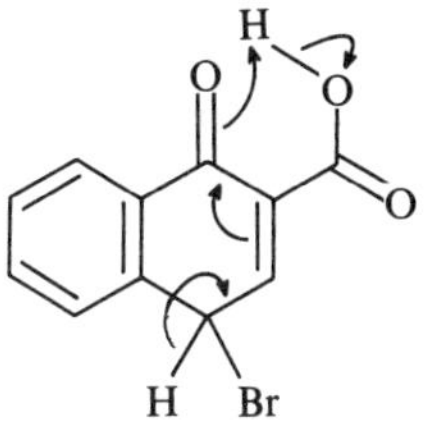

(73)

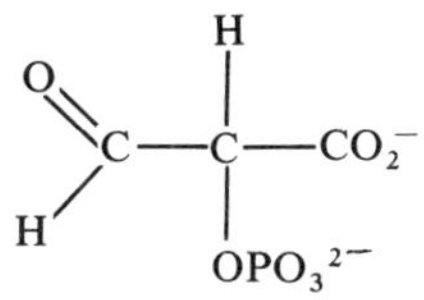

(74)

ethers/water have been calculated from known data. The values are not very substituent-dependent.[161]

The rate of the hydroxide ion-catalysed isomerization of the β,γ-unsaturated ketone (**71**) to its conjugated isomer shows saturation with increasing base concentration. This is consistent with complete conversion of the ketone of pK_a 12.7 into the dienolate anion with a partitioning ratio for backward and forward re-protonation of 25. The acidity of the ketone is thus 5 orders of magnitude more than that of comparable saturated derivatives.[162]

Metallo-porphyrins with a non-coordinating counter-ion catalyse the enolization of ketones. The enolate complex formed either undergoes metalation or aldol reaction with free ketone.[163]

The dienone intermediate (**72**), formed during the bromination of 1-naphthol, undergoes a general acid-catalysed enolization which shows a Brønsted α value of approximately 0. This is compatible with no proton transfer to the quinone oxygen in the transition state—perhaps diffusion-controlled encounter of the phenoxide ion and the general acid. Intramolecular general acid catalysis is observed in (**73**) with an effective molarity of 110 M.[164]

Enolase catalyses the enolization of (**74**) and, based on low fractionation factors on equilibrium isotope effects, it is suggested that the base on the enzyme which removes the proton is a sulphydryl group of a cysteine residue.[165]

The reaction of enolizable ketones with trichloroisocyanuric acid in the presence of chloride ion is suggested to involve a reaction of the conjugate acid of the ketone with a chlorinating species.[166]

The high reactivity of α-substituted carbonyl systems towards bimolecular nucleophilic substitution may be rationalized by a contributing enolate structure to the transition state (**75**) when the leaving-group is an arenesulphonate; Hammett ρ-values for rates and equilibria suggest 0.5 negative charge on the enolate residue in the transition state. The ρ-values for a variety of attacking- and leaving-groups indicate the enolate contribution is greatest when these groups readily accept a single electron rather than unshared pairs.[167]

The diffusion-controlled reaction of enolate ions with hypochlorous acid gives rise to estimated pK_a values for substituted acetophenones and a ρ-value of -1.95. At high hydroxide and low hypochlorite concentrations, the major products from the chlorination of acetophenones are the corresponding (α-hydroxy)arylacetic acids, whereas bromination gives substituted benzoic acids.[168]

The acid-catalysed deuteriation at position 3 of 5-acetoxy-2-pentanone occurs by the intermediate formation of the enol.[169]

The ratio of *O*- to *C*-methylation of acetophenone enolates by dimethyl sulphate is correlated with solvent polarity, Lewis basicity parameters, and the ionic radius of the metal. The results are interpreted in terms of negative-charge versus frontier-orbital control.[170]

O- and *C*-alkylation of the acetone enolate anion by acetone has been studied in the gas phase under high pressure.[171]

The stereochemistry of ketonization has been reviewed.[172] The water-catalysed ketonization of 1,3-cyclohexadienol (**76**), which cannot undergo a 1,5-sigmatropic rearrangement, is relatively slow, with a half-life of 48 s at 25 °C. The product is the β,γ-unsaturated ketone and ketonization is also catalysed by general bases.[173]

The non-catalysed ketonization of Z-dienols of α,β-unsaturated ketones proceeds by a 1,5-sigmatropic hydrogen shift. The lifetimes of dienols appear to be shorter than those of simple enols, which are relatively stable in the absence of acidic and basic catalysts.[174]

The small primary and cumulative secondary isotope effects observed in the thermolysis of β-hydroxy-ketones are consistent with the previously proposed cyclic transition state (**77**).[175]

A theoretical treatment of the 1,5-hydrogen shift isomerization of the enol form of malonaldehyde (**78**) has analysed the trajectory of the hydrogen.[176]

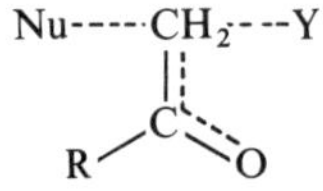

(**75**)

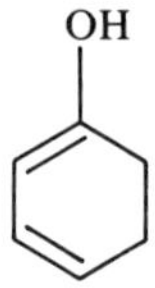

(**76**)

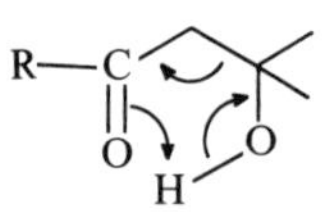

(**77**)

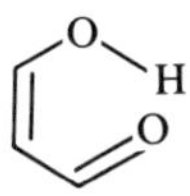

(**78**)

Hydrolysis and Reactions of Enol Ethers and Related Compounds

A solvent isotope effect k_H/k_D of 3.08 and the lack of deuterium exchange in the acid-catalysed hydrolysis of the heterocyclic enol methyl ether, 3-methoxybenzofuran, is indicative of rate-limiting *C*-protonation. By contrast, the hydrolysis of 3-methoxyfuran (**79**) proceeds by reversible protonation and rate-limiting addition of water.[177]

Contrary to an earlier claim[178] that the hydrolysis of the mixed vinyl ether acetal (**80**) proceeds by reversible vinyl ether carbon protonation in acetate buffers, it has been shown that the acetal function is hydrolysed first to give the enol intermediate (**81**). Consequently the saturation phenomenon previously observed is due to a change in rate-limiting step from buffer-catalysed ketonization of (**81**) to acid-catalysed acetal hydrolysis of (**80**) at high buffer concentration.[179]

The rate of the acid-catalysed hydrolysis of the divinyl ether (**82**) to the mono derivative (**83**) is 10^4-fold faster than the conversion of the intermediate to the diketone. Both stages show substantial solvent isotope effects and are compatible with the conventional mechanism involving rate-limiting proton transfer to carbon.[180]

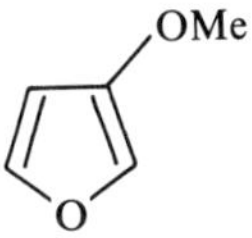

(79)

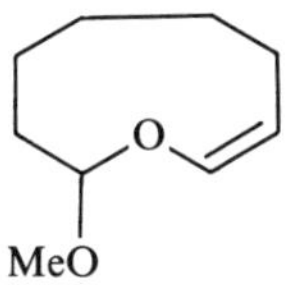

(80)

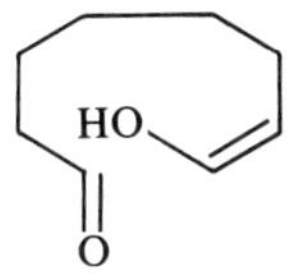

(81)

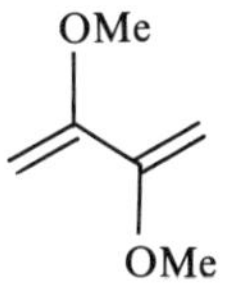

(82)

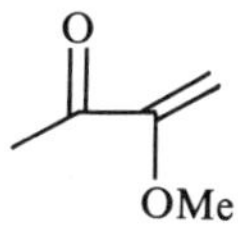

(83)

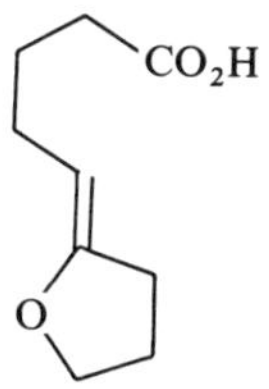

(84)

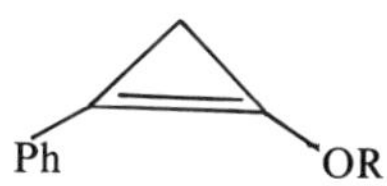

(85)

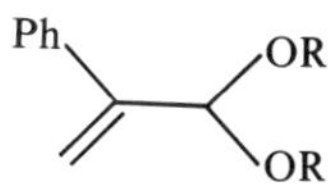

(86)

The mechanism of the hydrolysis of prostacyclin has been reviewed[181] and full details of the preliminary account of intramolecular carboxyl group catalysis in the hydrolysis of the vinyl ether (**84**) have been reported.[182]

Treatment of the strained vinyl ethers (**85**) with alcohols gives the ring-opened acetals (**86**). The reaction is not catalysed by alkoxide ion although the latter causes alcohol addition across the double bond to become competitive with ring-opening.[183]

A review of the mechanism of the acid-catalysed hydrolysis of ketene dithioacetals has emphasized the importance of reversible carbon protonation.[184]

The intermediate carbocation formed in the acid-catalysed hydrolysis of the ketene dithioacetal (**87**) is trapped by the neighbouring hydroxyl group but does not provide anchimeric assistance in water. The product distribution changes with pH in a similar way to that of hydrolysis of the cyclic intermediate (**88**).[185]

Acid-catalysed hydrolysis of the ketene dithioacetal proceeds by formation of the tetrahedral intermediate (**89**), the breakdown of which is rate-limiting at high pH, by an acid- and a base-catalysed step which is diffusion-controlled.[186]

The metal-ion-promoted hydrolysis of 2-phenyl-1,3-dithiane in aqueous dioxane solution proceeds by rate-limiting ring-opening of the metal-bound thioketal.[187]

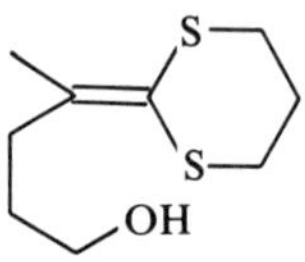

(**87**)

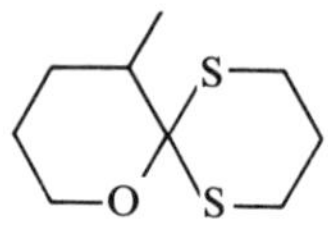

(**88**)

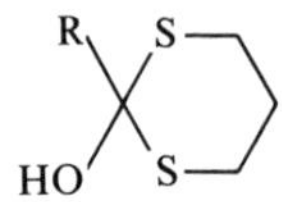

(**89**)

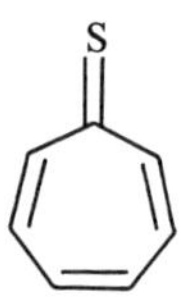

(**90**)

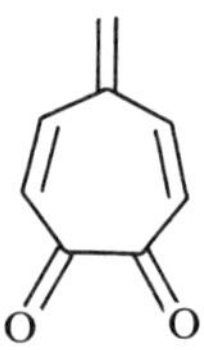

(**91**)

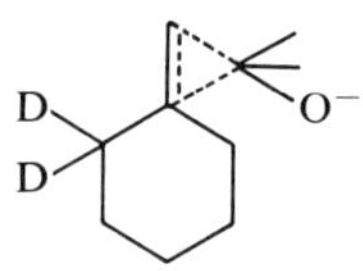

(**92**)

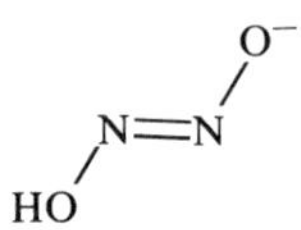

(**93**)

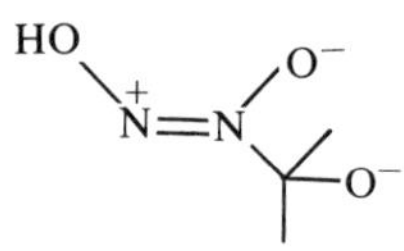

(**94**)

Other Reactions

Diketones show enhanced gas-phase basicities compared with their monofunctional counterparts.[188] Tropothione (**90**) has been synthesized and shows little aromatic character compared with its oxygen analogue.[189]

Seven-membered-ring analogues of *p*-benzoquinone (**91**) are unknown but a derivative has been synthesized which has a high electron affinity and a low barrier to rotation about the exocyclic methylene double bond.[190]

The base-catalysed oxidation of substituted benzaldoximes by periodatoargentate(III) is thought to occur by a two-electron-transfer mechanism involving the formation and dimerization of α-hydroxy-nitroso intermediates.[191]

The oxidation of acetone by molecular oxygen may proceed by the reaction of the enol as an intermediate since the rate of enolization is faster than that of oxidation under the same pressure conditions.[192]

The rate of oxidation of substituted benzaldehydes by nickel(III) is second-order in hydronium ion and shows a Hammett ρ-value of 1.9. The rate-limiting step is suggested to be carbon–hydrogen homolytic bond fission.[193]

Apparently anomalous product hydrogen-deuterium isotope effects in ene reactions in which the eneophile is made more electrophilic are attributed to the formation of the intermediate (**92**) which can partition between the steps of reversal, geometrical equilibration, and conversion into product.[194]

Intramolecular ene reactions of δ,ε-unsaturated aldehydes proceed with some stereoselectivity.[195] Chemiluminescent stable 1,2-dioxetanes containing a phosphate ester may be dephosphorylated using alkaline phosphalase to give an unstable dioxetane which rapidly fragments to the corresponding two ketones.[196]

The breakdown of *trans*-hyponitrite monoanion (**93**) into N_2O is catalysed by aldehydes and ketones and of the dianion by carbon dioxide. The mechanism is thought to involve nucleophilic addition to the carbonyl group followed by rate-limiting isomerization to the *cis* isomer (**94**) which regenerates the hyponitrite in the *cis* form which expels hydroxide ion by a *trans*-periplanar elimination.[197]

References

1 Aped, P., Apeloig, Y., Ellencweig, A., Fuchs, B., Goldberg, I., Karni, M., and Tartakovsky, E., *J. Am. Chem. Soc.*, **109**, 1486 (1987).
2 Cosse-Barbi, A. and Dubois, J.-E., *J. Am. Chem. Soc.*, **109**, 1503 (1987).
3 Fife, T. H. and Natarajan, R., *J. Am. Chem. Soc.*, **108**, 8050 (1986).
4 Lahti, M. and Kauppi, K., *Acta. Chem. Scand.*, **40A**, 533 (1986).
5 Gupta, R. B. and Franck, R. W., *J. Am. Chem. Soc.*, **109**, 6554 (1987).
6 Liepa, A. J. and Morton, T. C., *Aust. J. Chem.*, **39**, 1747 (1986).
7 Fife, T. H. and Przystas, T. J., *J. Chem. Soc., Perkin Trans. 2*, **1987**, 143.
8 Capparelli, M. P. and Swenton, J. S., *J. Org. Chem.*, **52**, 5360 (1987).
9 Thuy, V. M. and Maitte, P., *J. Heterocycl. Chem.*, **24**, 497 (1987).
10 Isaacs, N., Rzepa, H. S., Sheppard, R. N., Lobo, A. M., Prabhakar, S., and Merbach, A. E., *J. Chem. Soc., Perkin Trans. 2*, **1987**, 1477.
11 Penn, D., Saidi, Z., Satchell, D. P. N., and Satchell, R. S., *J. Chem. Res. (S)*, **1987**, 200.
12 Noe, C. R., Knollmuller, M., Gostl, G., Oberhauser, B., and Vollenkle, H., *Angew. Chem. Int. Edn*, **26**, 442 (1987).
13 Mori, A., Ishihara, K., Arai, I., and Yamamoto, H., *Tetrahedron*, **43**, 755 (1987).
14 Torii, S., Inokuchi, T., Takagishi, S., Horike, H., Kuroda, H., and Uneyama, K., *Bull. Chem. Soc. Jpn*, **60**, 2173 (1987).
15 Zhulin, V. M., Zorin, V. V., Zlotskii, S. S., and Rakhmankulov, D. L., *Izv. Vyssh. Ucheln. Zaved., Khim. Khim. Tekhnol.*, **29**, 3 (1986); *Chem. Abs.*, **106**, 175418 (1987).
16 Rakhmankulov, D. L., Zlotskii, S. S., Pastushenko, E. V., Rol'nik, L. Z., Karakhanov, R. A., Bartok, M., and Molnar, A., *Acta Phys. Chem.*, **31**, 743 (1985); *Chem. Abs.*, **106**, 213124 (1987).
17 Kochetkov, N. K., Zhulin, V. H., Klimov, E. M., Malysheva, N. N., Makarova, Z. G., and Ott, A. Ya., *Carbohydr. Res.*, **164**, 241 (1987).
18 Klaic, B., Raza, Z., Sankovic, M., and Sunjic, V., *Helv. Chim. Acta*, **70**, 59 (1987).
19 Lonnberg, H., Suokas, P., Kappi, R., and Darzynkiewicz, E., *Acta Chem. Scand.*, **40B**, 798 (1986).
20 Lonnberg, H., Kappi, R., Lehikoinen, P., and Oivanen, M., *Acta Chem. Scand.*, **41B**, 34 (1987).
21 Kappi, R. and Lonnberg, H., *Acta. Chem. Scand.*, **40B**, 768 (1986).
22 Lin, G. G., Bakthavachalam, V., Cherian, X. M., and Czarnik, A. W., *J. Org. Chem.*, **52**, 3113 (1987).
23 Remaud, G., Zhou, X.-X., Chattopadhyaya, J., Oivanen, M., and Lonnberg, H., *Tetrahedron*, **43**, 4453 (1987).

[24] Fulop, F., Pihlaja, K., Mattinen, J., and Bernath, G., *J. Org. Chem.*, **52**, 3821 (1987).
[25] Withers, S. G., Street, I. P., Bird, P., and Dolphin, D. H., *J. Am. Chem. Soc.*, **109**, 7530 (1987).
[26] White, W. J., Schray, K. J., Legler, G., and Alhadeff, J. A., *Biochim. Biophys. Acta*, **912**, 132 (1987).
[27] Singh, A. N., Hester, L. S., and Raushel, F. M., *J. Biol. Chem.*, **262**, 2554 (1987).
[28] Barili, P. L., Berti, G., Catelani, G., Colonna, F., and Mastrorilli, E., *J. Org. Chem.*, **52**, 2886 (1987).
[29] Tarnus, C. and Schuber, F., *Bioorg. Chem.*, **15**, 31 (1987).
[30] Asano, T. and Okada, T., *Chem. Lett.*, **1987**, 691.
[31] Prosyanik, A. K., Kol'tsov, N. Yu., Romanchenko, V. A., Belov, V. V., Brumistrov, K. S., and Loban, S. V., *Zh. Org. Khim.*, **23**, 375 (1987); *Chem. Abs.*, **107**, 197259 (1987).
[32] Lambert, J. B., Wang, G., Huseland, D. E., and Takiff, L. C., *J. Org. Chem.*, **52**, 68 (1987).
[33] Weis, A. L. and Vishkautsan, R., *J. Org. Chem.*, **51**, 4623 (1986).
[34] Williams, I. H., *J. Am. Chem. Soc.*, **109**, 6299 (1987).
[35] Toullec, J. and Razafindralambo, R., *J. Org. Chem.*, **52**, 1647 (1987).
[36] Henning, H. G. and Koppe, A. *Z. Chem.*, **27**, 367 (1987).
[37] Drizhd, L. P., Skrypka, A. V., and Savelova, V. A., *Ukr. Khim. Zh. (Russ. Edn)*, **52**, 1072 (1986); *Chem. Abs.*, **107**, 133658 (1987).
[38] Vernin, G., Boniface, C., Metzger, J., Obretenov, T., Kantasubrata, J., Siouffi, A. M., Larice, J. L., and Fraisse, D., *Bull. Soc. Chim. Fr. II*, **1987**, 681.
[39] Katritzky, A. R., Ostercamp, D. L., and Yousaf, T. I., *Tetrahedron*, **43**, 5171 (1987).
[40] Cooper, A., Dixon, S. F., Nutley, M. A., and Robb, J. L., *J. Am. Chem. Soc.*, **109**, 7254 (1987).
[41] Mitzner, R. and Seidel, S., *Z. Phys. Chem. (Leipzig)*, **267**, 1179 (1986).
[42] Kamal, M. M., Temerk, Y. M., Ahmed, M. E., and Ahmed, Z. A., *J. Prakt. Chem.*, **328**, 841 (1986); *Chem. Abs.*, **107**, 197221 (1987).
[43] Capitan, F., Capitan-Vallvey, L. F., Espinosa, P., and Molina, F., *An. Quim., Ser. B*, **82**, 224 (1986); *Chem. Abs.*, **107**, 197205 (1987).
[44] del Vado, M. A. G., Donoso, J., Munoz, F., Echevarria, G. R., and Blanco, F. G., *J. Chem. Soc., Perkin Trans. 2*, **1987**, 445.
[45] Pishchugin, F. V., Sharshenalieva, Z. Sh., and Dakenova, N. S., *Izv. Akad. Nauk Kirg. SSR*, **1986**, 65; *Chem. Abs.*, **106**, 155673 (1987).
[46] Bruce, M. R., *Inorg. Chem.*, **26**, 2197 (1987).
[47] Kenny, P. W. and Robinson, M. J. T., *Tetrahedron*, **43**, 4043 (1987).
[48] Baccolini, G., Dalpozzo, R., and Errani, E., *Tetrahedron*, **43**, 2755 (1987).
[49] Katritzky, A. R., Barczynski, P., and Ostercamp, D. L., *J. Chem. Soc., Perkin Trans. 2*, **1987**, 969.
[50] Singh, G., Ila, H., and Junjappa, H., *J. Chem. Soc., Perkin Trans. 2.*, **1987**, 1945.
[51] Gutierrez-Martin, F., Rodriguez-Pastor, A., and Mata-Perez, F., *Rev. Roum. Chim.*, **32**, 305 (1987).
[52] Kim, Y. H., Lee, H. K., and Chang, H. S., *Tetrahedron Lett.*, **28**, 4285 (1987).
[53] Brehme, R. and Klemann, A., *Tetrahedron*, **43**, 4113 (1987).
[54] Mustroph, H., *Z. Chem.*, **27**, 281 (1987).
[55] McClelland, R. A. and Seaman, N. E., *Can. J. Chem.*, **65**, 1689 (1987).
[56] Keys, L. D. and Hamilton, G. A., *J. Am. Chem. Soc.*, **109**, 2156 (1987).
[57] Hajos, G., Messmer, A., and Koritsanskyi, T., *J. Org. Chem.*, **52**, 2015 (1987).
[58] Bailey, P. D., *Tetrahedron Lett.*, **28**, 5181 (1987); Bailey, P. D., Hollinshead, S. P. and McLay, N. R., *ibid.*, p. 5177.
[59] Bailey, P. D., *J. Chem. Res. (S)*, **1987**, 202.
[60] Cushman, M. and Madaj, E. J., *J. Org. Chem.*, **52**, 907 (1987).
[61] Kos, N. J., Jongejan, H., and van der Plas, H. C., *Tetrahedron*, **43**, 4841 (1987).
[62] Cimino, G., Sodano, G., and Spinella, A., *Tetrahedron*, **43**, 5401 (1987).
[63] Katritzky, A. R. and Marson, C. M., *J. Org. Chem.*, **52**, 2726 (1987); Katritzky, A. R., Marson, C. M., and Wang, Z., *ibid.*, p. 2730.
[64] Lokshin, B. V., Siling, M. I., Belomoina, N. M., Krongauz, E. S., and Korshak, V. V., *Dokl. Akad. Nauk SSSR*, **289**, 370 (1986); *Chem. Abs.*, **106**, 137746 (1987).
[65] Grieco, P. A. and Bahsas, A., *J. Org. Chem.*, **52**, 1378 (1987).
[66] Jimenez, R. M., Dominguez, E., Badia, D., Alonso, R. M., Vicente, F., and Hernandez, L., *J. Heterocycl. Chem.*, **24**, 421 (1987).
[67] Jimenez, R. M., Alonso, R. M., and Vicente, F., *Bull. Soc. Chim. Belg.*, **96**, 265 (1987).
[68] Essassi, El. M., Salem, M., and Viallefont, P., *Bull. Soc. Chim. Fr. II*, **1987**, 890.
[69] Haley, M. F. and Yates, K., *J. Org. Chem.*, **52**, 1817, 1825 (1987).
[70] Mohrle, H. and Schillings, P., *Arch. Pharm. (Weinheim, Ger.)*, **320**, 16 (1987).
[71] Osek, J., Oszczapowicz, J., and Drzewinski, W., *J. Chem. Soc., Perkin Trans. 2*, **1986**, 1961.

[72] Leis, J. R., Pena, M. E., and Williams, D. L. H., *J. Chem. Soc., Chem. Commun.*, **1987**, 45.
[73] Kronja, O., Matijevic-Sosam, J., and Ursic, S., *J. Chem. Soc., Chem. Commun.*, **1987**, 463.
[74] Suh, J., Kwon, B. N., Lee, W. Y., and Chang, S. H., *Inorg. Chem.*, **26**, 805 (1987).
[75] Sinisterra, J. V. and Marinas, J. M., *Bull. Soc. Chim. Belg.*, **96**, 293 (1987).
[76] Aguilera, A., Alcantara, A. R., Marinas, J. M., and Sinisterra, J. V., *Can. J. Chem.*, **65**, 1165 (1987).
[77] Mohareb, R. M., Habashi, A., Hafez, E. A. A., and Sherif, S. M., *Arch. Pharm. (Weinheim, Ger.)*, **320**, 776 (1987).
[78] Srivastava, R. M., Freire, M. V. S., Chaves, A. S. S. C., and Beltrao, T. M., *J. Heterocycl. Chem.*, **24**, 101 (1987).
[79] Corey, E. J., Bakshi, R. K., and Shibata, S., *J. Am. Chem. Soc.*, **109**, 5551 (1987).
[80] Tsuboi, S., Furutani, H., Takeda, A., Kawazoe, K., and Sato, S., *Bull. Chem. Soc. Jpn*, **60**, 2475 (1987).
[81] Anh, N. T. and Thanh, B. T., *Nouveau. J. Chim.*, **10**, 681 (1986).
[82] Spears, G. W., Caufield, C. E., and Still, W. C., *J. Org. Chem.*, **52**, 1226 (1987).
[83] Dhavale, D. D., Joshi, P., and Marathe, K. G., *J. Chem. Soc., Perkin Trans. 2*, **1987**, 449.
[84] Pearson, W. H. and Cheng, M. C., *J. Org. Chem.*, **52**, 3176 (1987).
[85] McCarthy, P. A. and Kageyama, M., *J. Org. Chem.*, **52**, 4681 (1987).
[86] Handa, S., Tsang, R., McPhail, A. T., and Fraser-Reid, B., *J. Org. Chem.*, **52**, 3491 (1987).
[87] de Souza-Barboza, J. C., Luche, J. L., and Petrier, C., *Tetrahedron Lett.*, **28**, 2013 (1987).
[88] Eger, K., Folkers, G., Zimmerman, W., Schmidt, R., and Hiller, W., *J. Chem. Res. (S)*, **1987**, 277.
[89] Reetz, M. T., Hullmann, M., and Seitz, T., *Angew. Chem. Int. Edn*, **26**, 477 (1987).
[90] Pradhan, S. K., Thakker, K. R., and McPhail, A. T., *Tetrahedron Lett.*, **28**, 1813 (1987).
[91] Corey, E. J. and Hannon, F. J., *Tetrahedron Lett.*, **28**, 5233, 5237 (1987).
[92] Masamune, S., Sato, T., Kim, B. M., and Wollmann, T. A., *J. Am. Chem. Soc.*, **108**, 8279 (1986).
[93] Noyori, R., Ohkuma, T., Kitamura, M., Takaya, H., Sayo, N., Kumobayashi, H., and Akutagawa, S., *J. Am. Chem. Soc.*, **109**, 5856 (1987).
[94] von Seebach, D., Juaristi, E., Miller, D. D., Schickli, C., and Weber, T., *Helv. Chim. Acta*, **70**, 237 (1987); von Seebach, D., Imwinkelried, R., and Stucky, G., *ibid.*, p. 448; von Seebach, D., Beck, A. K., Imwinkelried, R., Roggo, S., and Wonnacott, A., *ibid.*, p. 954.
[95] Tomioka, K., Nakajima, M., and Koga, K., *Chem. Lett.*, **1987**, 65.
[96] Hoffmann, R. W. and Endesfelder, A., *Liebigs Ann. Chem.*, **1987**, 215.
[97] Soai, K., Ookawa, A., Kaba, T., and Ogawa, K., *J. Am. Chem. Soc.*, **109**, 7111 (1987).
[98] Yamamoto, T., *Acc. Chem. Res.*, **20**, 243 (1987); Claremon, D. A., Lumma, P. K., and Phillips, B. T., *J. Am. Chem. Soc.*, **108**, 8265 (1986); Epifani, E., Florio, S., and Ingrosso, G., *Tetrahedron*, **43**, 1937 (1987); Paquette, L. A., Learn, K. S., and Romine, J. L., *ibid.*, p. 4989; Garcia, J., Kim, B. M., and Masamune, S., *J. Org. Chem.*, **52**, 4831 (1987); Brown, H. C., Bhat, K. S., and Randad, R. S., *ibid.*, p. 3701; Hwang, K. J., Logusch, E. W., Brannigan, L. H., and Thompson, M. R., *ibid.*, p. 3435; Smaardijk, A. A. and Wynberg, H., *ibid.*, p. 135; Brown, D. W., Campbell, M. M., Taylor, A. P., and Zhang, X., *Tetrahedron Lett.*, **28**, 985 (1987); Murphy, P. J., Procter, G., and Russell, A. T., *ibid.*, p. 2037; Handa, Y. and Inanaga, J., *ibid.*, p. 5717; Arjona, O., de la Pradilla, R. F., Manzano, C., Perez, S., and Plumet, J., *ibid.*, p. 5547; Kira, M., Kobayashi, M., and Sakurai, H., *ibid.*, p. 4081; Jefford, C. W., Jaggi, D., and Boukouvalas, J., *ibid.*, p. 4037; Hull, C., Mortlock, S. V., and Thomas, E. J., *ibid.*, p. 5343; Dumartin, G., Pereyre, M., and Quintard, J. P., *ibid.*, p. 3935; Iida, H., Yamazaki, N., and Kibayashi, C., *J. Chem. Soc., Chem. Commun.*, **1987**, 746; Baldoli, C., Del Buttero, P., Licandro, E., Maiorana, S., and Papagni, A., *ibid.*, p. 762; Yamamoto, Y. and Matsuoka, K., *ibid.*, p. 923; Yamamoto, Y., Hatsuya, S., and Yamada, J., *ibid.*, p. 501; Pelter, A., Buss, D., and Colclough, E., *ibid.*, p. 297; Ando, A. and Shioiri, T., *ibid.*, p. 1620; Yamamoto, Y. and Yamada, J., *ibid.*, p. 1218; Braun, M., *Angew. Chem.*, **99**, 24 (1987); *Chem. Abs.*, **106**, 175370 (1987); Winterfeldt, E., *Kontakte (Darmstadt)*, **1986**, 37; *Chem. Abs.*, **107**, 76981 (1987); Winterfeldt, E., *Kontakte (Darmstadt)*, **1986**, 16; *Chem. Abs.*, **106**, 17576 (1987); Bartnick, R., Eddaif, A., Laurent, A., and Lesniak, S., *Bull. Pol. Acad. Sci., Chem.*, **34**, 27 (1986); *Chem. Abs.*, **106**, 4328 (1987).
[99] Dannheim, J., Grahn, W., Hopf, H., and Parrodi, C., *Chem. Ber.*, **120**, 871 (1987).
[100] Bellassoued, M., Habbsachi, F., and Gaudemar, M., *Tetrahedron*, **43**, 1785 (1987).
[101] Reetz, M. T. and Maus, S., *Tetrahedron*, **43**, 101 (1987); Reetz, M. T., Hugel, H., and Dresely, K., *ibid.*, p. 109.
[102] Fry, A. J., Susla, M., and Weltz, M., *J. Org. Chem.*, **52**, 2496 (1987).
[103] Iwasawa, N. and Mukaiyama, T., *Chem. Lett.*, **1987**, 463; Mukaiyama, T., Kobayashi, S., Tamura, M., and Sagawa, Y., *ibid.*, p. 491.
[104] Gennari, C., Columbo, L., Bertolini, G., and Schimperna, G., *J. Org. Chem.*, **52**, 2754 (1987).
[105] Nsunda, K. M. and Hevesi, L., *J. Chem. Soc., Chem. Commun.*, **1987**, 1518.

[106] Agami, C. and Puchot, C., *J. Mol. Catal.*, **38**, 341 (1986).
[107] Hartman, G. D., Phillips, B. T., Springer, J. P., and Hirshfield, J., *J. Heterocycl. Chem.*, **24**, 1397 (1987).
[108] Gilchrist, T. L., Stretch, W., and Chrystal, E. J. T., *J. Chem. Soc., Perkin Trans. 2*, **1987**, 2235.
[109] Rao, T. V., Rele, D. N., and Trivedi, G. K., *J. Chem. Res. (S)*, **1987**, 196.
[110] Curran, D. P. and Chao, J. C., *J. Am. Chem. Soc.*, **109**, 3036 (1987).
[111] Buschmann, E., Zeeh, B. Gotz, N., Sproesser, L., and Walker, N., *Liebigs Ann. Chem.*, **1987**, 349.
[112] Maruyama, K. and Katagiri, T., *Chem. Lett.*, **1987**, 735.
[113] Holm, T., *Acta Chem. Scand.*, **41B**, 278 (1987).
[114] Howe, G. P., Wang, S., and Procter, G., *Tetrahedron Lett.*, **28**, 2629 (1987).
[115] Ni, Z. J. and Luh, T. Y., *J. Chem. Soc., Chem. Commun.*, **1987**, 1515.
[116] Screttas, C. G. and Steele, B. R., *J. Organomet. Chem.*, **317**, 137 (1986).
[117] Volatron, F. and Eisenstein, O., *J. Am. Chem. Soc.*, **109**, 1 (1987).
[118] Huang, W., *Huaxue Tongbao*, **1986**, 1 64; *Chem. Abs.*, **106**, 155587 (1987).
[119] Li, D., *Sichuan Daxue Xuebao, Ziran Kexueban*, **1985**, 176; *Chem. Abs.*, **106**, 137717 (1987).
[120] Isaacs, N. S. and El-Din, G. N., *Tetrahedron Lett.*, **28**, 2191 (1987).
[121] Valverde, S., Martin-Lomas, M., Herradon, B., and Garcia-Ochoa, S., *Tetrahedron*, **43**, 1895 (1987).
[122] Daniel, H. and Le Corre, M., *Tetrahedron Lett.*, **28**, 1165 (1987).
[123] Moison, H., Texier-Boullet, F., and Foucaud, A., *Tetrahedron*, **43**, 537 (1987).
[124] Garner, P. and Ramakanth, S., *J. Org. Chem.*, **52**, 2629 (1987).
[125] Vedejs, E., Fleck, T. and Hara, S., *J. Org. Chem.*, **52**, 4637 (1987).
[126] Borredon, M. E., Delmas, M., and Gaset, A., *Tetrahedron*, **43**, 3945 (1987).
[127] Young, D. J. and Stirling, C. J. M., *J. Chem. Soc., Chem. Commun.*, **1987**, 552.
[128] Uieara, M., Zucco, C., Zanette, D., Rezende, M. C., and Nome, F., *J. Chem. Soc., Perkin Trans. 2*, **1987**, 175.
[129] Zucco, C., Lima, C. F., Rezende, M. C., Vianna, J. F., and Nome, F., *J. Org. Chem.*, **52**, 5356 (1987).
[130] Coe, P. L. and Owen, I. R., *Bull. Soc. Chim. Fr. II*, **1986**, 971.
[131] Morosov, A. A., *Dokl. Akad. Nauk SSSR*, **289**, 131 (1986); *Chem. Abs.*, **106**, 32221 (1987).
[132] Imanishi, T., Matsui, M., Yamashita, M., and Iwata, C., *J. Chem. Soc., Chem. Commun.*, **1987**, 1802.
[133] Paquette, L. A., Gilday, J. P., and Ra, C. S., *J. Am. Chem. Soc.*, **109**, 6858 (1987).
[134] Kozlov, A. P., Sychev, D. I., and Andreichikov, Yu. S., *Zh. Org. Khim.*, **22**, 1756 (1986); *Chem. Abs.*, **106**, 195664 (1987).
[135] Krow, G. R., *Tetrahedron*, **43**, 3 (1987).
[136] Fraser, R. R. and Stanciulescu, M., *J. Am. Chem. Soc.*, **109**, 1580 (1987).
[137] Wu, Y. D. and Houk, K. N., *J. Am. Chem. Soc.*, **109**, 908 (1987).
[138] Trost, B. M., Florez, J., and Jebaratnam, D. J., *J. Am. Chem. Soc.*, **109**, 613 (1987).
[139] Lodge, E. P. and Heathcock, C. H., *J. Am. Chem. Soc.*, **109**, 2819, 3353 (1987).
[140] Mori, I., Bartlett, P. A., and Heathcock, C. H., *J. Am. Chem. Soc.*, **109**, 7199 (1987).
[141] Chang, J. W. A., Taira, K., Urano, S., and Gorenstein, D. G., *Tetrahedron*, **43**, 3865 (1987).
[142] Houk, K. N. and Wu, Y. D., *Stereochem. Org. Bioorg. Transform., Proc. Workshop Conf. Hoechst, 17th*, 1986, 247; *Chem. Abs.*, **107**, 197185 (1987).
[143] Kuz'min, V. E., Kamalov, G. L., Lozitskaya, R. N., Siritsov, A. I., Alvares, J. R., and Stel'makh, I. B., *Teor. Eksp. Khim.*, **22**, 610 (1986); *Chem. Abs.*, **107**, 77049 (1987).
[144] Chandrasekhar, S., *Chem. Soc. Rev.*, **16**, 313 (1987).
[145] Hanke, V. R., Knoche, W., and Dutkiewicz, E., *J. Chem. Soc., Faraday Trans. 1*, **83**, 2847 (1987).
[146] Sanchez-Ruiz, J. M., Asensio, S., Llor, J., and Cortijio, M., *J. Chem. Soc., Perkin Trans. 2*, **1987**, 169.
[147] Olson, T. M., and Hoffmann, M. R., *Atmos. Environ.*, **20**, 2277 (1986); *Chem. Abs.*, **106**, 195652 (1987).
[148] Rajyaguru, I. H. and Rzepa, H. S., *J. Chem. Soc., Chem. Commun.*, **1987**, 998.
[149] Field, M. J., Hillier, I. H., Smith, S., Vincent, M. A., Mason, S. C., Whittleton, S. N., Watt, I. F., and Guest, M. F., *J. Chem. Soc., Chem. Commun.*, **1987**, 84.
[150] Natrajan, A., Ferrara, J. D., Youngs, W. J., and Sukenik, C. N., *J. Am. Chem. Soc.*, **109**, 7477 (1987).
[151] Orellana, G., Perez-Ossorio, R., Quiroga, M. L., and Sanchez-Asiain, M. P., *J. Chem. Soc., Perkin Trans. 2*, **1987**, 679.
[152] Effenberger, F., Ziegler, T., and Forster, S., *Angew. Chem. Int. Edn*, **26**, 458 (1987).
[153] Mashino, T. and Fridovich, I., *Arch. Biochem. Biophys.*, **252**, 163 (1987).
[154] Balaban, A. T., Gheorghiu, M. D., Shcherbakova, I. V., Kuznetsov, E. V., and Yudilevich, I. A., *Tetrahedron*, **43**, 409 (1987).
[155] Elshafie, S. M. M., *Z. Phys. Chem. (Leipzig)*, **268**, 1065 (1987).
[156] Itoh, S., Nii, K., Mure, M., and Ohshiro, Y., *Tetrahedron Lett.*, **28**; 3957 (1987).

[157] Nadler, E. B. and Rappoport, Z., *J. Am. Chem. Soc.*, **109**, 2112 (1987).
[158] Chiang, Y., Hojatti, M., Keefe, J. R., Kresge, A. J., Schepp, N. P., and Wirz, J., *J. Am. Chem. Soc.*, **109**, 4000 (1987).
[159] Rappoport, Z., *J. Am. Chem. Soc.*, **109**, 4730 (1987).
[160] Harcourt, M. P. and More O'Ferrall, R. A., *J. Chem. Soc., Chem. Commun.*, **1987**, 822.
[161] Toullec, J., El-Alaoui, M., and Kleffert, P., *Tetrahedron Lett.*, **28**, 1889 (1987).
[162] Pollack, R. M., Mack, J. P. G., and Eldin, S., *J. Am. Chem. Soc.*, **109**, 5048 (1987).
[163] Aoyama, Y., Tanaka, Y., Yoshida, H. T., and Ogoshi, H., *J. Organomet. Chem.*, **329**, 251 (1987).
[164] Tee, O. S. and Iyengar, N. R., *Can. J. Chem.*, **65**, 1714 (1987).
[165] Weiss, P. M., Boerner, R. J., and Cleland, W. W., *J. Am. Chem. Soc.*, **109**, 7201 (1987).
[166] Radhakrishnamurti, P. S., Rath, N. K., and Panda, R. K., *J. Chem. Soc., Perkin Trans. 2*, **1987**, 517.
[167] Yousaf, T. I. and Lewis, E. S., *J. Am. Chem. Soc.*, **109**, 6137 (1987).
[168] Guthrie, J. P., Cossar, J., and Klym, A., *Can. J. Chem.*, **65**, 2154 (1987).
[169] Staroverov, V. M., Popovich, V. I., and Shapet'ko, N. N., *Zh. Org. Khim.*, **22**, 1406 (1986); *Chem. Abs.*, **106**, 118990 (1987).
[170] Lazana, M. C. R. L. R., Franco, M. L. T. M. B., and Herold, B. J., *J. Chem. Soc., Perkin Trans. 2*, **1987**, 1399.
[171] Bouchoux, G. and Hoppilliard, Y., *Tetrahedron Lett.*, **28**, 4537 (1987).
[172] Zimmerman, H., *Acc. Chem. Res.*, **20**, 263 (1987).
[173] Pollack, R. M., Mack, J. P. G., and Blotny, G., *J. Am. Chem. Soc.*, **109**, 3138 (1987).
[174] Duhaime, R. M. and Weedon, A. C., *Can. J. Chem.*, **65**, 1867 (1987).
[175] Quijano, J., Rodriguez, M. M., Yepes, M. del S., and Gallego, L. H., *Tetrahedron Lett.*, **28**, 3555 (1987).
[176] Hutchinson, J. S., *J. Phys. Chem.*, **91**, 4495 (1987).
[177] Capon, B. and Kwok, F.-C., *Tetrahedron*, **43**, 69 (1987).
[178] See *Org. React. Mech.*, **1987**, 409.
[179] Burt, R. A., Chiang, Y., Chwang, W. K., Kresge, A. J., Okuyama, T., Tang, Y. S., and Yin, Y., *J. Am. Chem. Soc.*, **109**, 3787 (1987).
[180] Kresge, A. J. and Yin, Y., *Can. J. Chem.*, **65**, 1753 (1987).
[181] Kresge, A. J., *Acc. Chem. Res.*, **20**, 364 (1987).
[182] Bergman, N.-A., Jansson, M., Chiang, Y., Kresge, A. J., and Yin, Y., *J. Org. Chem.*, **52**, 449 (1987).
[183] Crossland, I., *Acta Chem. Scand.*, **41B**, 310 (1987).
[184] Okuyama, T., *Acc. Chem. Res.*, **19**, 370 (1986).
[185] Okuyama, T. and Fueno, T., *J. Org. Chem.*, **52**, 1405 (1987).
[186] Okuyama, T., Inaoka, T., and Fueno, T., *J. Org. Chem.*, **51**, 4988 (1986).
[187] Stachell, D. P. N. and Satchell, R. S., *J. Chem. Soc., Perkin Trans. 2*, **1987**, 513.
[188] Bouchoux, G. and Hoppilliard, Y., *Nouveau J. Chim.*, **11**, 225 (1987).
[189] Machiguchi, T., Otani, H., Ishii, Y., and Hasegawa, T., *Tetrahedron Lett.*, **28**, 203 (1987).
[190] Takahashi, K., Namekata, N., Takase, K., and Suzuki, T., *J. Chem. Soc., Chem. Commun.*, **1987**, 935.
[191] Rao, P. J. P., Sethuram, B., and Rao, T. N., *Oxid. Commun.*, **9**, 155 (1986); *Chem. Abs.*, **107**, 153765 (1987).
[192] Thomas, J. W. and Taylor, J. E., *Can. J. Chem.*, **64**, 2235 (1986).
[193] Reddy, T. R., Murthy, G. S. S., and Jagannadham, V., *Oxid. Commun.*, **9**, 83 (1986); *Chem. Abs.*, **106**, 101593 (1987).
[194] Song, Z., Chrisope, D. R., and Beak, P., *J. Org. Chem.*, **52**, 3938 (1987).
[195] Johnston, M. I., Kwass, J. A., Beal, R. B., and Snider, B. B., *J. Org. Chem.*, **52**, 5419 (1987).
[196] Schaap, A. P., Sandison, M. D., and Handley, R. S., *Tetrahedron Lett.*, **28**, 1159 (1987).
[197] Loechler, E. L., Schneider, A. M., Schwartz, D. B., and Hollocher, T. C., *J. Am. Chem. Soc.*, **109**, 3076 (1987).

Organic Reaction Mechanisms 1987
Edited by A. C. Knipe and W. E. Watts

CHAPTER 2

Reactions of Acids and their Derivatives

W. J. SPILLANE

Chemistry Department, University College, Galway, Ireland

CARBOXYLIC ACIDS

Tetrahedral Intermediates

A useful review of orthoesters (**1**) and dialkoxycarbenium ions (**2**) which considers reactivity, stability, structure, and new synthetic applications has appeared very recently.[1] It will be recalled[2] that these ions (**2**) are the precursors of hemiorthoesters (**3**), which are in turn the tetrahedral intermediates of ester-interchange reactions.

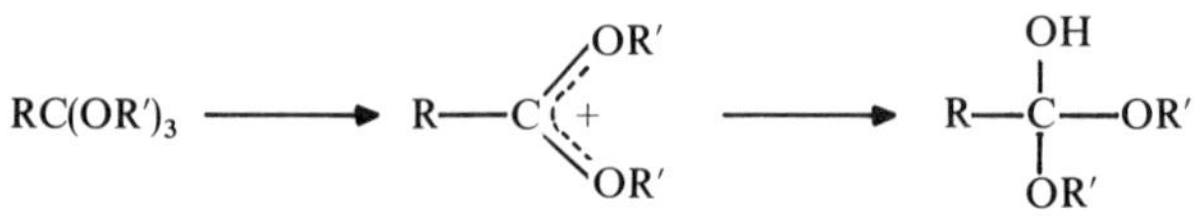

(**1**) (**2**) (**3**)

In another review[3] of direct relevance to tetrahedral intermediates, Okuyama considers the various stages in the acid-catalysed hydrolysis of ketene *S,S*-acetals (dithioacetals), typically the cyclic *S,S*-acetal (**4**), which proceeds with partial reversible carbon protonation to give the carbocation intermediate (**5**) and subsequently the tetrahedral intermediate (**6**) (Scheme 1).

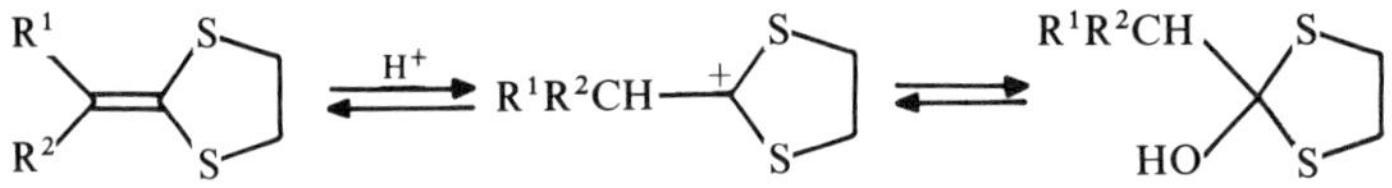

(**4**) (**5**) (**6**)

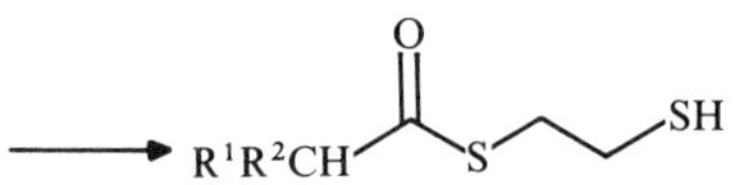

SCHEME 1

In other interesting work from the same laboratory, the hydrolyses of the four orthoesters (**7**)–(**10**), giving the final products ArCOOMe [from (**7**)] and ArCOSMe [from (**10**)] and both of these products from the mixed *O,S*-orthoesters, (**8**) and (**9**), has been interpreted in terms of different partitioning in acid-catalysed and spontaneous breakdowns of the tetrahedral intermediate hemiorthoesters, (**11**)–(**13**),

$ArC(OMe)_3$ (**7**) $ArC(OMe)_2(SMe)$ (**8**) $ArC(OMe)(SMe)_2$ (**9**) $ArC(SMe)_3$ (**10**)

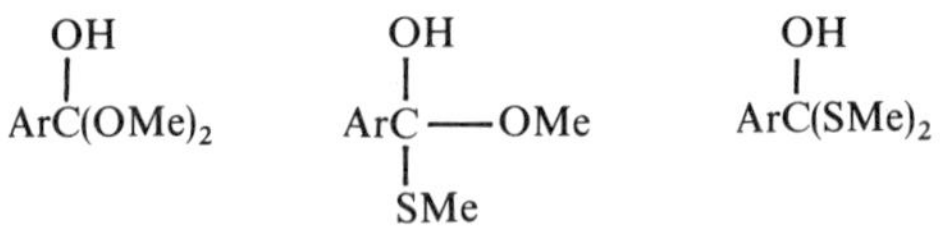

(**11**) (**12**) (**13**)

Ar = *p*-$MeOC_6H_4$

and the *O,S*-orthoesters.[4] The effect of dimethyl substitution on the reactivity of orthoesters has been probed by studying the kinetics of the acid-catalysed hydrolysis of the orthoisobutyrates (**14**; R = Me_2CH) and (**15**) and using the orthoesters (**14**; R = Me, Me_3C, *p*-$MeOC_6H_4$) as standards.[5] The alcoholysis (MeOH, EtOH, Pr^nOH,

$RC(OMe)_3$

(14)

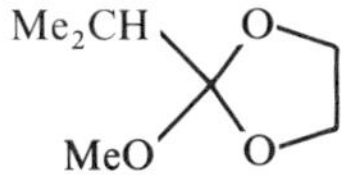

(15)

and Bu^nOH) of the 2,2,2-trichloro-1-arylethanones (**16**) stops at the hemiacetal stage (**17**) unless ethylamine is present in which case the final ester product (**18**) is formed (Scheme 2).[6] Further study of this interesting reaction is taking place. The

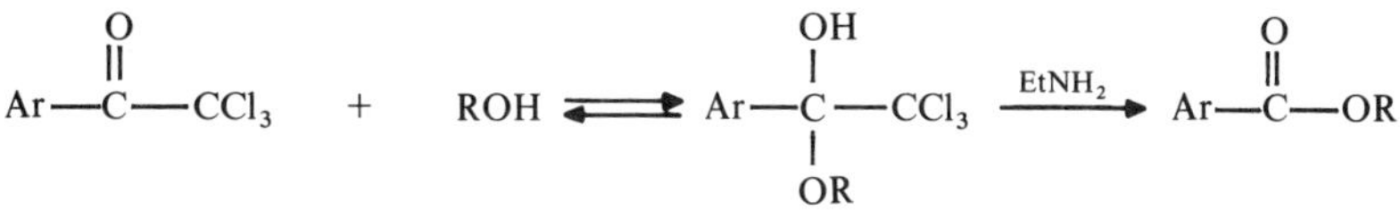

(16) **(17)** **(18)**

SCHEME 2

intramolecular hydroxy-ketone (**19**) ⇄ hemiacetal (**20**) interconversion shows a much less negative volume change than the intermolecular addition of $PhCH_2SH$ to acetyl cyanide (Scheme 3).[7] The changes have been studied by NMR and equilibrium constants measured as a function of pressure to obtain volumes of reaction which relate to the formation of the tetrahedral intermediates of ester hydrolysis.

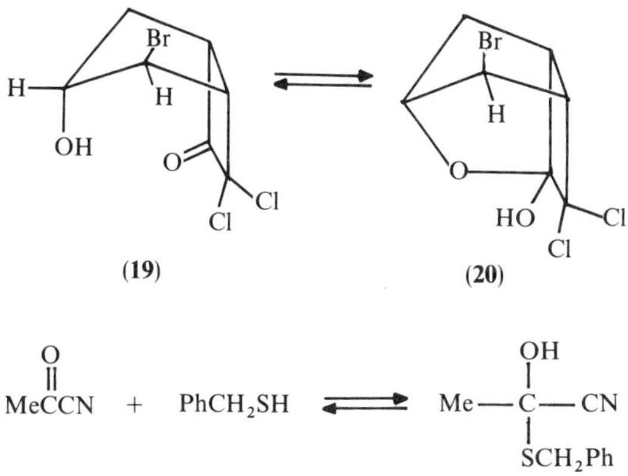

SCHEME 3

Williams, group[8] has reported a study of the transfer of the methoxycarbonyl group between isoquinoline and a series of pyridine nucleophiles (Scheme 4). A concerted rather than a stepwise process is supported by the linearity of the Brønsted plot ($\beta = 0.58$) over a wide range of pyridine nucleophilicity. The large electrostatic repulsion and large steric effects militate against the involvement of the tetrahedral intermediate (**22**) in the transfer of the methoxycarbonyl group. Hydrolysis of the *N*-methoxycarbonylisoquinolinium ion (**21**) gives the pseudobase (**23**).

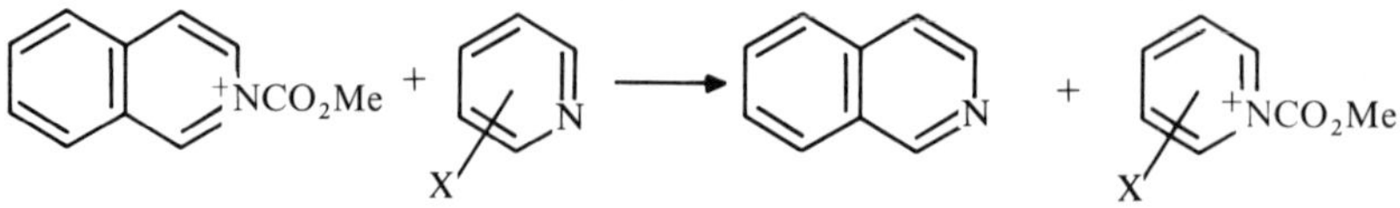

(**21**)

SCHEME 4

A change in the rate-determining step from breakdown to formation of the tetrahedral intermediate as the pyridine becomes more basic has been supported from kinetic studies (curved Brønsted plots) in the pyridinolysis of 2,4-dinitrophenyl acetate and carbonate in aqueous ethanol.[9] The same group,[10] from a study of the pyridinolysis of methyl chloroformate in aqueous ethanol, has shown that the nucleofugacity of pyridines is greater than that of chloride ion from the tetrahedral intermediate (**24**). The tetrahedral intermediate (**25**) in the reaction of pyridines and imidazoles with phenyl acetates and trifluoroacetates plays an important role in the mechanism.[11] Unsymmetrical acetamidines hydrolyse via the tetrahedral intermediates (**26**) and (**27**) and mixtures of two acetamides and two amines are obtained. The direction of cleavage is now found to be determined not by the relative basicities

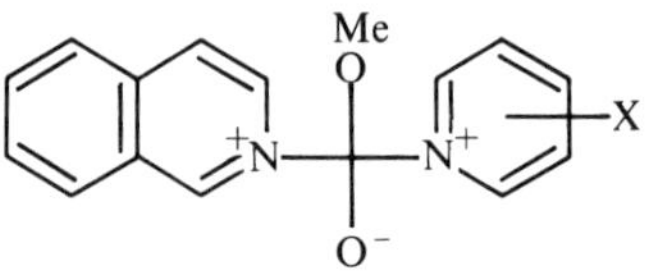

(**22**)

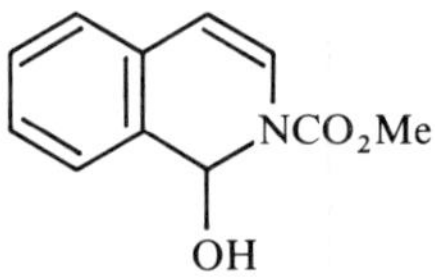

(**23**)

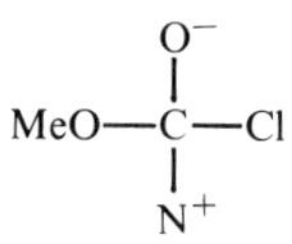

(**24**)

N^+ = Substituted *N*-pyridinyl

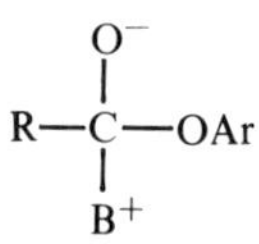

(**25**)

B^+ = Substituted *N*-pyridinyl or *N*-imidazolyl

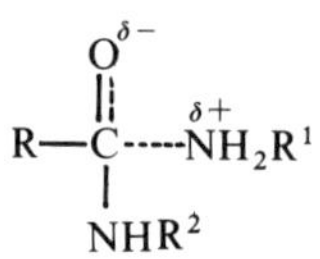

(**26**)

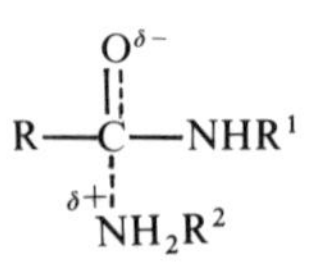

(**27**)

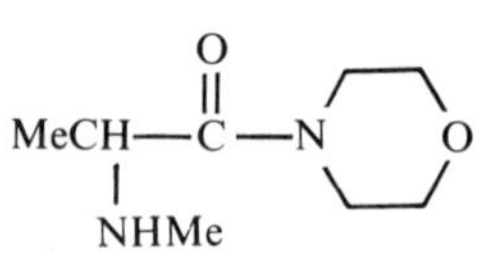

(**28**)

of the nitrogens of (**26**)/(**27**) but by the relative basicities of the two amines.[12] This work has afforded a means of testing the stereoelectronic theory. Two mechanisms are considered for the phosphate buffer catalysis of 4-(*N*-methylalanyl)morpholine (**28**) in formaldehyde solution—they involve either formation or breakdown of a tetrahedral intermediate in the slow step.[13] (See also Intramolecular Catalysis section).

The solvolysis of substituted phenyl *N*-phenylbenzimido-esters of type (**29**)[14,15] is of interest because the same tetrahedral intermediates are involved as those in the aminolysis of the corresponding esters or the hydrolysis of the amides. Both anionic (**30**) and zwitterionic (**31**) intermediates have been recognized from kinetics studies of the aminolysis of maleimide. Brønsted coefficients have been explained by assuming a pre-association stepwise mechanism.[16] The same group has studied the alkaline hydrolysis of *N*-substituted phthalimides of general type (**32**) and the involvement of anionic tetrahedral intermediates is supported.[17] Direct addition of water or hydroxyl to the carbonyl or to the imino nitrogen of nitroso-ureas (**33**) gives rise to different types of tetrahedral intermediates in the hydrolysis of these compounds,[18] some of which are important as drugs (e.g. the 2-chloro-ethylnitroso-ureas are anti-cancer agents). Anionic tetrahedral intermediates (**34**) are involved in the general acid-catalysed hydrolysis of azetidin-2-ylidene-ammonium salts.[19] Competitive endo- and exo-cyclic C—N bond cleavage occurs, the product ratio of β-lactone/amino-amide depending on pH and buffer.

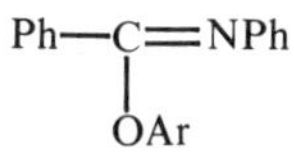

(**29**)

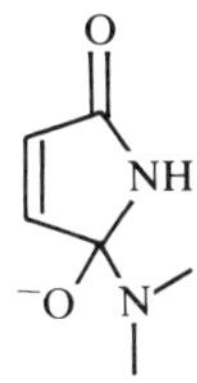

(**30**)

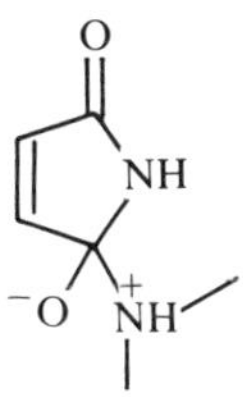

(**31**)

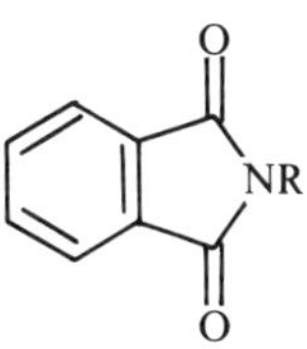

(**32**)

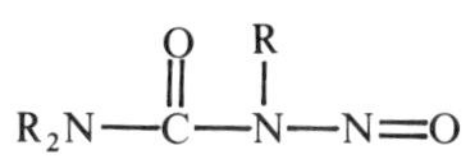

(**33**)

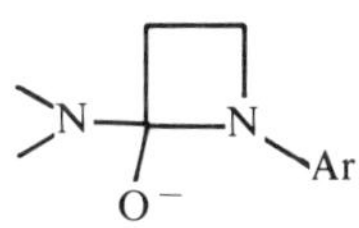

(**34**)

Formation and breakdown of tetrahedral intermediates of type (**35**) from diazoquinones can be rate-determining[20] during nucleophilic addition (at the carbonyl group). Hydrolysis (H_2O), azidolysis (HN_3), and perhydrolysis (H_2O_2) have been examined. The latter process often involves chemiluminescence and measurement of emitted light for this reaction has been used to study the breakdown of the tetrahedral intermediate in that case.

Guthrie's group[21] has measured heats of hydrolysis for the three acylimidazole acetals (**36**)–(**38**) and has calculated free energies of formation in aqueous solution for these and the corresponding tetrahedral intermediates. This has allowed a detailed analysis of the energetics of both the nucleophilic reaction of imidazole with ethyl and *p*-nitrophenyl acetates. The latter reaction may follow a concerted pathway.

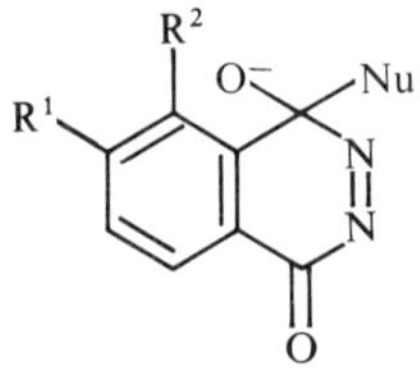

(35)

$MeC(OEt)_2Im$

(36)

$HC(OEt)_2Im$

(37)

$PhC(OMe)_2Im$

(38)

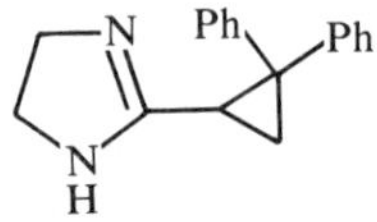

(39)

$R^1R^2R^2CCO_2H$

(40)

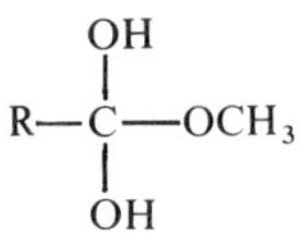

(41)

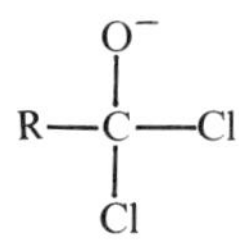

(42)

Examination of ^{13}C NMR spectra during the hydrolysis of the new antiarrhythmic agent, 2-(2′,2′-diphenylcycloprop-1-yl)-4,5-dihydro-1H-imidazole (**39**) suggested that a pseudo-base tetrahedral intermediate was not involved.[22]

Some theoretical studies have appeared concerning tetrahedral intermediates.[23–25] MNDO calculations on the A2 hydrolysis of acetamide show that the tetrahedral species are not at the energy minima in the reaction profile but at or near the saddle points.[23] Account has been taken of both *O*-protonated (the predominant form) and *N*-protonated tautomers in the calculations. A MINDO/3 semi-empirical method[24] has been used to study the kinetic stability of the tetrahedral intermediate formed after HO^- ion attack at methyl monoxalate.

DeTar's group[25] has calculated rates of esterification and of acid-catalysed hydrolysis of a set of esters of acetic acids (**40**; R groups = H or alkyl) using molecular mechanics and employing the tetrahedral intermediate (**41**) as a model of the transition state on the way to the tetrahedral intermediate. Gas-phase studies using a pulsed Fourier transform ion cyclotron resonance spectrometer suggest that the tetravalent adduct (**42**) must be either a transition state or an unstable intermediate.[26]

Tetrahedral intermediates are also involved in other work dealt with in this review but discussion of this work is deemed to be more appropriate in other sections (see references 41, 42, 89, 141, 145, 164, 200, 213, 240, and 310).

Intermolecular Catalysis and Reactions

Reactions in Hydroxylic Solvents

(a) *General*

A book on the subject of nucleophilicity has appeared recently.[27] The major subject headings give an indication of the coverage; they cover an introduction to nucleophilic reactivity, followed by (a) Marcus theory, alkyl transfer, and gas-phase reactions, (b) the Brønsted equation, HSAB theory, and factors in nucleophilicity, (c) LFERs for solvent nucleophilicity, (d) novel nucleophilic reactions, and (e) enhancement of nucleophilicity. Stereoelectronic effects in biomolecules have been reviewed.[28] The review covers a broad canvas and, after examining the basis for the stereoelectronic effect, it looks at the operation of this effect on reactivity of acetals, sugars, amides, phosphate esters, and enzymatic reactions (217 references). Reaction mechanisms in the radiolysis of peptides, polypeptides, and proteins have been comprehensively reviewed.[29] The review brings together more than 200 references and correlates a wide variety of studies of this important area.

MINDO/3 has been used to study the attack of hydroxide ion on methyl oxalate[24,30] and on methyl formate.[30] Three competitive mechanisms have been identified for reaction of the neutral esters in the gas phase. MINDO calculations have been made on the reaction of methanol and fulminic acid, HCNO and acetonitrile oxide, MeCNO to give the complex RC(OMe)=NOH. The formation of the H-bonded complex is rate-determining.[31] The force-field model was used to calculate steric factors of reactivity for the alkaline hydrolysis of substituted ethyl acetates and malonates, as examples.[32] The relation between steric indices of reactivity with steric substituent constants and rate constants was analysed. The same workers[33] have used the atom-atom potential-function method to analyse the steric effects in the hydride reduction of oxo-esters and alkylcyclohexanones and again the alkaline hydrolysis of malonates.

Both *ab initio* Gaussian 82 and ion cyclotron resonance and flowing-afterglow methods have been employed to study the mechanism of addition of alcohols to protonated acetone in the gas phase.[34] A stabilized adduct, possibly $Me_2C{=}\overset{+}{O}H\cdots O(H)Pr^i$ for the isopropyl alcohol reaction is observed.

(b) *Esters*

(i) *Formation*

Three Chinese papers[35–37] deal with esterification of carboxylic acids. An isokinetic relationship ($\beta = 372.3\,K$) for the ethanolysis of the benzoic acids $XC_6H_4CO_2H$ (X = H, 4-NO_2, 3-NO_2, 4-Cl, 3-Cl, 3-Me, 4-OMe, 4-Me, 4-OH, 3-OH, 4-Br) in the presence of *p*-toluenesulphonic acid has been found.[35] Methanolysis and ethanolysis rates for acetic acid have been determined at various temperatures by flow microcalorimetry.[36] Ethanolysis rates for the esterification of *n*-butyric acid have been reported in the range 25–90°C.[37] The kinetics of methanolysis of metacrylic acid in aqueous H_2SO_4 have been studied by a Russian group.[38] Another

Russian group[39] has measured equilibrium constants of esterification for the system $AcOH/HO(CH_2)_nOH$ ($n = 2, 3, 4, 6$). Higher values for the diester constants compared to those for the monoesters were attributed to the lower rates of hydrolysis of the diesters.

Calculation of the rates of esterification of a series of acetic acids (**40**) has been noted previously.[25] Acid-catalysed hydrolysis of 2,2-dimethoxy-3,4-dihydropyrans (**43**) yields mixtures of δ-keto-esters (**44**) and 3,4-dihydro-α-pyrones (**45**);[40] (**45**) is formed directly from (**43**) but (**44**) can form directly from (**43**) and also by methanolysis of (**45**).

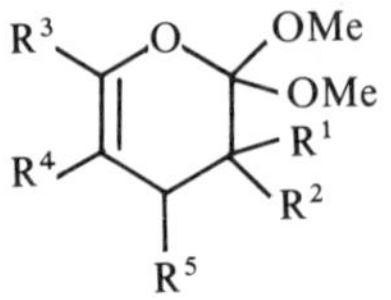

(**43**)

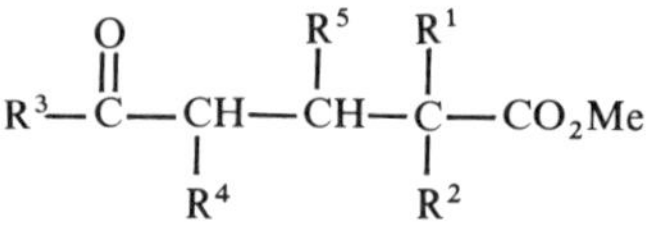

(**44**)

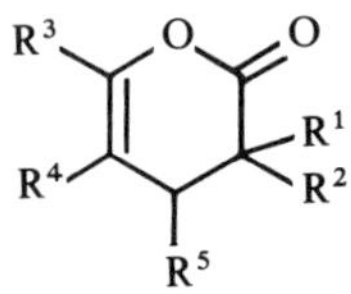

(**45**)

(ii) *Reactions*

Several papers have appeared in which the aminolysis of esters has been the main theme. The reaction of phenyl acetate with pyrazole is catalysed by general bases and by bifunctional general acids. Brønsted data are not consistent with trapping or stepwise pre-association mechanisms and the reaction is seen as involving concerted general-base catalysis of proton abstraction from pyrazole as it attacks the acetate.[41] The zwitterionic tetrahedral intermediate (**46**) plays a key role in the process. Both the aminolysis and hydrolysis of methyl salicylate (**47**) have been investigated and the involvement of tetrahedral intermediates is indicated.[42] Sodium cyanide was found to be the best (compared to imidazole and various pyridines) for the aminolysis of ethyl (*S*)-1-ethyl-2-pyrrolidinecarboxylate (**48**; R = Et).[43] The reaction was completely stereo-conservative, less than 1% racemization occurring. Cyanide was also efficient for *trans*-esterification with methanol solvent giving (**48**; R = Me). The order of reactivity for a variety of esters was $MeNH_2 > NH_3 > Me_2NH$. The aminolysis (PhNHMe, $MeNHCH_2COOEt$, $MeNHCH_2CN$, and piperidine) of α,α-dimethyl(benzoyl)acetate ethyl ester (**49**) has been studied.[44] The mechanism of reaction of phenyl acetates and trifluoracetates with pyridines and imidazoles[11] has been dealt with earlier (see Tetrahedral Intermediates).

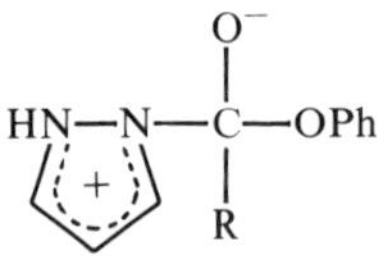

(**46**)

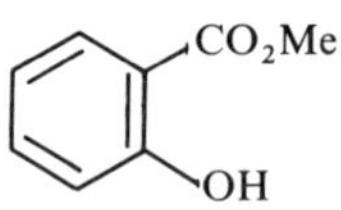

(**47**)

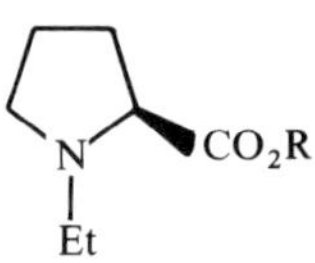

(**48**)

Numerous papers on the reactions of aryl esters have appeared. The solvolysis of phenyl *N*-phenylbenzimido-esters (**29**) has been mentioned previously[14] as has the hydrolysis of the imidazole acetals (**36**)–(**38**),[21] the resulting calculations throwing

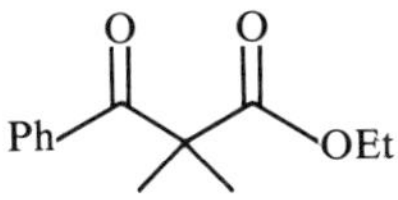

(49)

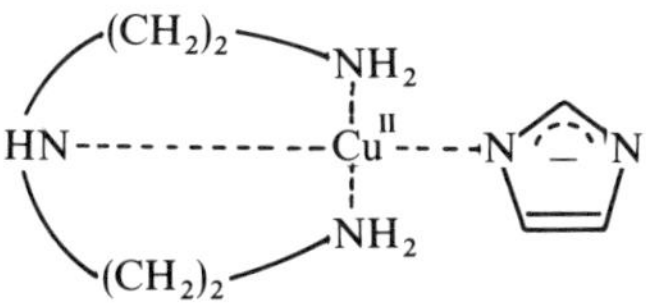

(50)

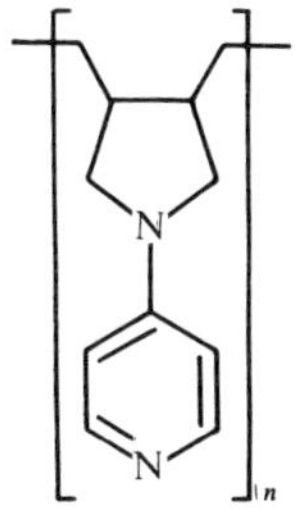

(51)

$Me(CH_2)_nCO_2C_6H_4NO_2$-*p*

(52)

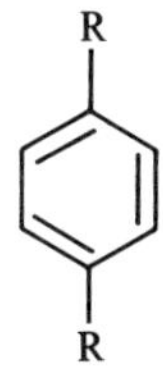

(53)

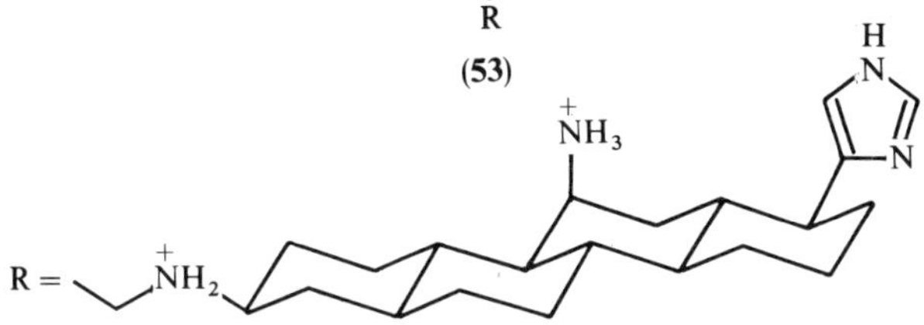

light on the reactions of imidazole with ethyl acetate and *p*-nitrophenyl acetate. The imidazolate-Cu(II) complex with diethylenetriamine (**50**) has a high nucleophilic reactivity in the cleavage of *p*-nitrophenyl benzoate in dimethyl sulphoxide.[45] *Ortho*-Iodosylbenzoate displays a significant α-effect in the cleavage of *p*-nitrophenyl acetate.[46] The poly(*N*,*N*-diallylaminopyridine) (**51**) is a super-nucleophilic catalyst for the hydrolysis of the activated esters (**52**). Rate-determining formation of an acylpyridinium *p*-nitrophenoxide moiety occurs.[47] The kinetics of the alkaline hydrolysis of dimethylphthalate to monomethyl phthalate (k_1) and thence to phthalic acid (k_2) have been reported.[48] Data on the effects of different organic co-solvents on k_1 and k_2 can be correlated by the Grunwald–Winstein equation. The effect of transfer from H_2O to H_2O-Me_2SO mixtures on the Brønsted, basicity and nucleophilic reactivity of the anions phenoxide, alkoxide, oximate, and hydroxamate towards *p*-nitrophenyl acetate has been presented.[49]

Guthrie has found rate enhancements of up to 5.5×10^2 relative to imidazole using the dimeric steroid (**53**) as catalyst in the hydrolysis of aryl propanoate esters ($RCH_2CH_2CO_2Ar$).[50] Hydrophobic binding plays a key role in the reactions. The catalytic effect of a series of heterocycles containing a thiol function towards *p*-nitrophenyl acetate has been assessed. Examples of bifunctional catalysis are reported for benzimidazolylmethanethiols.[51] Kinetics of hydrolysis of 2,4-dinitro-6-*sec*-butyl phenylisopropyl carbonate in Pr^iOH-H_2O have been reported.[52]

There are many reports of the hydrolysis of aliphatic esters. Acid-catalysed hydrolysis of esters of the acetic acids (**40**) has been referred to earlier.[25] The hydrolysis rate dependence on dielectric constant of the medium and temperature in the cleavage of neo- and iso-amyl acetates has been investigated.[53] The standard free energies of transfer of the 2-carbomethoxypropionate ion, $^{-}OCO(CH_2)_2CO_2CH_3$, have been determined from solubility studies and the values obtained have been used in an examination of the basic hydrolysis of this ion.[54] The first and second steps in the acid-catalysed hydrolysis of diethyl malonate proceed via $A_{AL}1$ and $A_{AC}2$ mechanisms concurrently.[55] The results are compared with those for di-*tert*-butyl succinate.

Salt effects on the kinetics of hydrolysis of monoethyl malonate, monomethyl glutarate and monoethyl adipate in alkaline media have also been reported.[56] The hydrolysis of the pharmacologically important diethyl monothioloxalate (**54**) and its attack by nucleophiles was examined to determine if oxalylthiol esters, $RSCOCO_2^{-}$, might be generated intracellularly and to see if they would be suitable as physiologically active metabolites of ascorbic acid.[57] From data collected, it has been estimated that compounds of type (**54**) would transfer the alkyl oxalyl group ($COCO_2R$) *in vivo* to other thiols many thousands of times faster ($t_{1/2}$ m sec) than hydrolysis ($t_{1/2}$ min) to less active derivatives would result. Thus compounds (**54**) could be metabolites for the vitamin. The kinetics of hydrolysis of the dibasic unsaturated aliphatic acid peresters (**55**) and (**56**) have been examined.[58] In acid media, (**55**) was more reactive than (**56**) but, in alkaline media, (**56**) was the more reactive.

The reaction of methyl acetate with acetyl iodide giving Ac_2O and MeI is acid-catalysed and an $A_{AC}2$ mechanism is proposed.[59] Reaction of the spiro-esters (**57**) with H_2O, HCl, and MeOH involves protonation at an O atom of the dioxolane ring and addition of the counter-ion to the α-carbon atom of the tetrahydrofuran ring.[60] A review in Russian on orthoesters has appeared (91 references).[61] Two further Russian reviews[62,63] have looked at the oxidation of esters[62] (31 references) and the liquid-phase homolytic halogenation of orthoesters and acetals[63] (36 references).

Acetoacetic acid methyl ester (**58**), chloracetic acid ethyl ester (**59**), acetonedicarboxylic acid diethyl ester (**60**), and *N*-methylacetoacetamide (**61**) have been found to generate dual-frequency oscillations in the Belousov–Zhabotinskii reaction.[64] Hitherto compound (**58**) was the only one known to produce this behaviour. The alkylation of the enolates of methyl bis(pentamethylphenyl)acetate (**62**; Ar $= C_6Me_5$, R = Me) and of isopropyl bis(pentachlorophenyl)acetate (**62**; Ar $= C_6Cl_5$; R = i-Pr) occurs exclusively on oxygen to yield diarylketene acetals such as (**63**) from (**62**; Ar $= C_6Cl_5$; R = i-Pr).[65]

$EtSCOCO_2Et$ (**54**)

(*Z*)–$HO_2CCH{=}CHCO_2OCMe_3$ (**55**)

$HO_2CC(=CH_2)CH_2CO_2OBu^t$ (**56**)

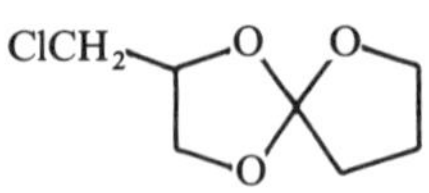

(**57**)

$MeCOCH_2CO_2Me$ (**58**)

$ClCH_2COCH_2CO_2Et$ (**59**)

$CO(CH_2CO_2Et)_2$	$MeCOCH_2CONHMe$	Ar_2CHCO_2R
(60)	**(61)**	**(62)**

$(C_6Cl_5)_2C{=}C(OMe)OPr^i$	Cl_3CCO_2Et
(63)	**(64)**

Several papers have appeared on studies (generally hydrolysis) of haloacetates. From a detailed recalculation of Taft σ^* values (mostly for halo-containing groups), Kanerva's group[66] has concluded that an $A_{AC}2$ mechanism operates during the acid-catalysed esterification of trichloroacetic acid in alcohol. However, the acid-catalysed hydrolysis of ethyl trichloroacetate (**64**) in aqueous solutions proceeds by an exceptional A-$B_{AC}3$ mechanism. The same group[67] concludes that the hydrolysis of chloromethyl fluoroacetate ($FCH_2CO_2CH_2Cl$) and of 1,2-dichloroethyl chloroacetate ($ClCH_2CO_2CHClCH_2Cl$) in acid involves in addition to a neutral $B_{AC}3$ mechanism a normal $A_{AC}2$ type mechanism occurring at high acid concentrations. In less concentrated acid an A-$B_{AC}3$ mechanism operates. Finally, these workers are reported studies on the neutral methanolysis and ethanolysis of chloromethyl trichloroacetate in aqueous MeOH and aqueous EtOH solutions.[68] Engeberts and Blandamer[69,70] have reported two collaborative studies on haloacetates. In the first paper,[69] they studied the effects of a 1:1 range of salts on the hydrolysis of phenyl and *p*-methoxyphenyl dichloroacetate. The importance of the role of overlap between hydration co-spheres of ions in solutions is noted. Added urea and tetramethyl-, 1,3-dimethyl-, and 1,3-diethyl-urea affects the rate constants for the neutral hydrolysis of *p*-methoxyphenyl dichloroacetate.[70] These effects have been examined using Gibbs function parameters and the Savage–Wood additivity principle.

(c) *Lactones and derivatives*

Alkoxycarbonyl-substituted heptalenecarboxylic acids (**65**) give 'ortho' anhydrides (**66**) and (**67**) when treated with the iminium salt formed from DMF and $(COCl)_2$ in the presence of MeOH.[71] The β-butyrolactone, (*S*)-4-methyloxetan-2-one (**68**), gives enantiomerically pure derivatives by S_N2 ring-opening of the β-lactone with various *C*, *N*, *O*, and *S* nucleophiles. Thus, (*S*)-2-hydroxy-4-octanone (**69**), (*R*)-3-aminobutanoic acid (**70**) and its *N*-benzyl derivative (**71**), and (*R*)-3-azidobutanoic acid (**72**) among others have been prepared.[72] A computational study of the acid-

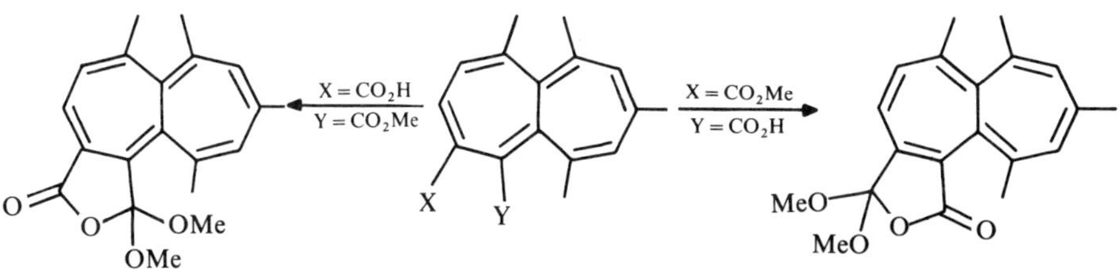

(66) **(65)** **(67)**

SCHEME 5

catalysed lactonizations of hydroxy-acids using Allinger's MM2 force-field method has been carried out.[73] The calculations indicate that there is no relationship between the angle of nucleophilic attack and the reactivity of a hydroxy-acid or the distance between reacting atoms in the starting material and the rate of reactions. Opening of the lactone ring in the alkaline hydrolysis of 4-methoxycoumarin **(73)** occurs to give finally the hydroxyacetophenone **(74)** and this process has recently been studied.[74]

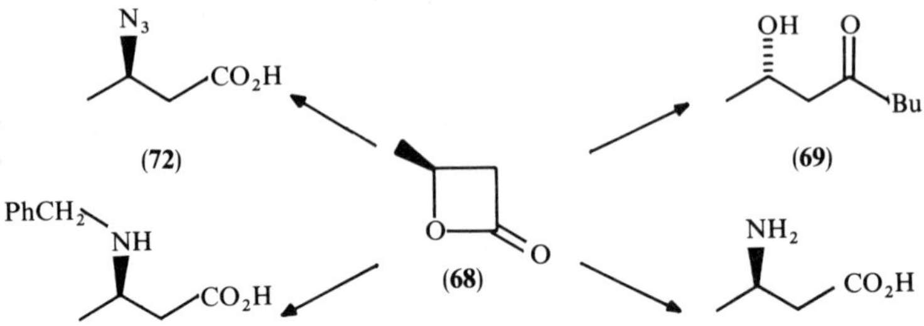

SCHEME 6

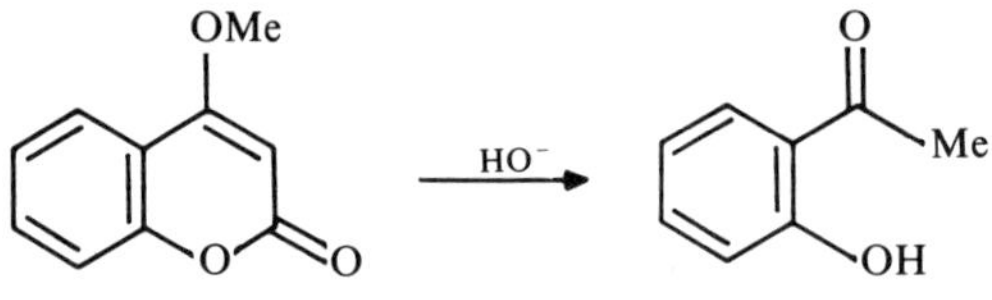

(73) **(74)**

SCHEME 7

The reactions of lactones and cyclic anhydrides with primary-secondary bis-Grignard reagents of type **(75)** have led to *cis*- to *trans*-diastereoisomers.[75] Formation of two interesting tricyclic trimers **(76)** and **(77)** of diketene **(78)** has been observed in Me_3SiCl/NaI solutions.[76]

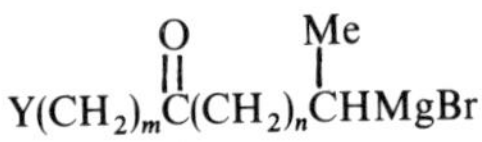

(75)

$Y = OMgBr, \quad CO_2MgBr$

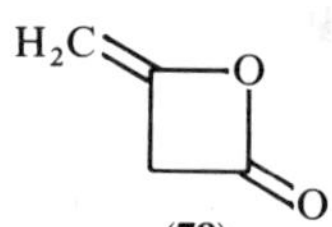

(78)

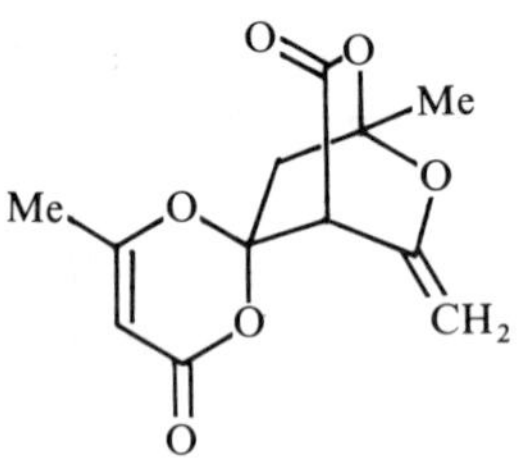

(76)

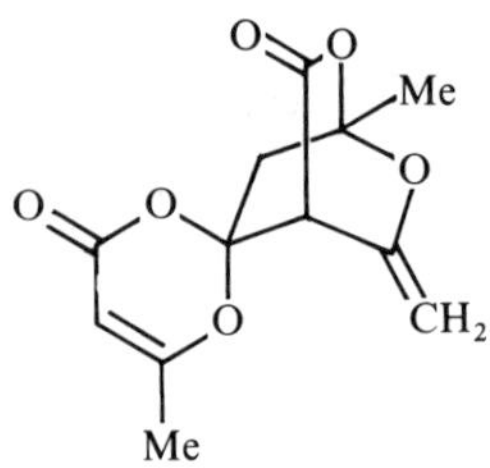

(77)

(d) *Acids, anhydrides, anilides*

Perpendicular effects on transition states have been examined in the context of free-energy profiles, and a two-dimensional energy surface for describing substituent effects in organic reactivity has been developed.[77] The various models are compared with the experimental free energy of activation of proton-transfer reactions between benzoic acids and MeOH and of the Cope and Claisen rearrangements.

A series of papers from Bowden's group[78–80] deals with the transmission of the polar effect. They have found[78] that the Kirkwood–Westheimer theory qualitatively and (in three cases) quantitatively predicts the p*K* values for [2.2]paracyclophane-4-carboxylic acid (**79**; R = H) and its four derivatives with a bromo substituent in the non-carboxylated ring; p*K* values were determined in 80% (w/w) 2-methoxyethanol-water at 25°C. Esterification with diazodiphenylmethane to give (**79**; R = Me) and the alkaline hydrolysis of the latter have been studied. In a second paper,[79] they have again accounted (both qualitatively and quantitatively) for the ionization constants of three 8-substituted 1-naphthoic acids (**80**; R = H) measured in the same solvent. The Kirkwood–Westheimer model also allows for the 'reversed' substituent effects so common in these special systems and observed here for the chloro- and bromo-substituted acids. The alkaline hydrolysis of six methyl esters (**80**; R = Me) has also been studied. In a further study, they have investigated a series of 5-substituted triptycene-7-carboxylic acids (**81**) (and esters) and again find that the Kirkwood–Westheimer treatment accounts for the ionization constants obtained.[80]

Some differences in *cis–trans* photo-isomerization have been noted for the *o*-protonated derivatives (acylium ions) of some carboxylic acids and derivatives, e.g. (**82**)–(**85**).[81] A mechanism of the A-2 type with pre-equilibrium protonation obeying H_β (the ketone acidity function) is favoured for the hydrolysis in H_2SO_4 of 2,2-dimethyl- and phenyl-acetohydroxamic acids (**86**).[82] The unsaturated intermediate (**87**) is a derivative of the hypothetical metasilicous acid (**88**), and is an analogue of

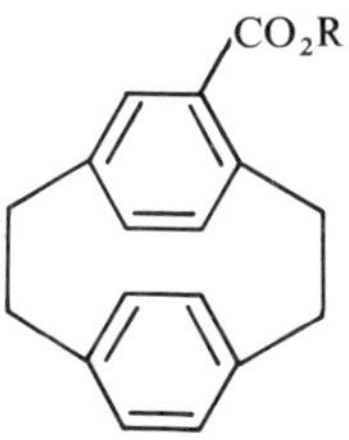

(**79**)

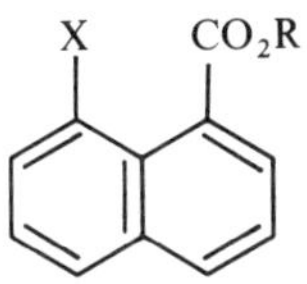

(**80**)

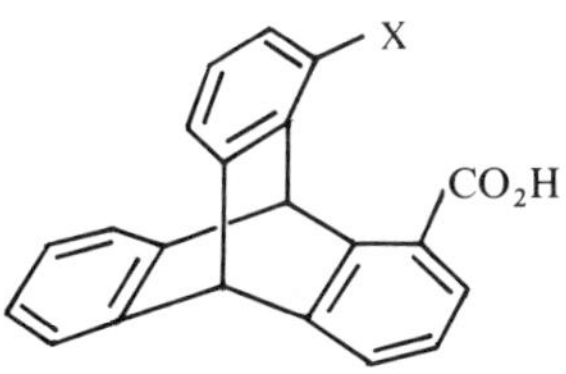

(**81**)

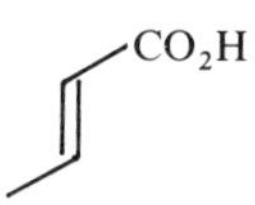

(**82**)

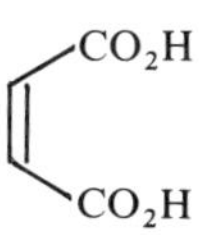

(**83**)

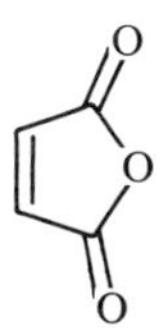

(**84**)

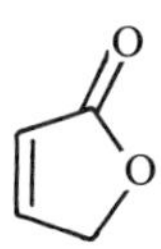

(**85**)

(86)

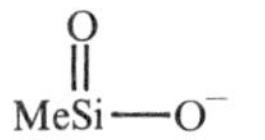

(87)

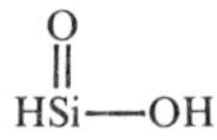

(88)

acetate ion. It is postulated to arise in the cleavage of *m*-$ClC_6H_4CH_2SiMe(OH)_2$ by NaOH in DMSO-H_2O to give initially the dianion *m*-$ClC_6H_4CH_2SiMe(O^-)_2$ and subsequently (**87**).[83]

Two papers report on reactions of phthalic anhydride.[84,85] The esterification of the anhydride with 2-ethylhexanol in the presence of the catalysts tetra-*n*-butyl-titanate and -zirconate and *p*-toluenesulphonic acid has been studied kinetically.[84] Aminolysis (1-naphthylamine, *p*-toluidine, aniline) of the anhydride gives imides. A relationship: $E_a = 27.0 + 7.51 \log A$, and an isokinetic temperature 392 K have been found.[85] The stereoselectivities of reactions of the dibenzoyloxysuccinic anhydrides (**89**) and (**90**) with chiral amines and alcohols have been examined by ^1H NMR analysis of the mixtures of diastereomers obtained. Stereoselectivity for (**89**) was higher in all cases.[86]

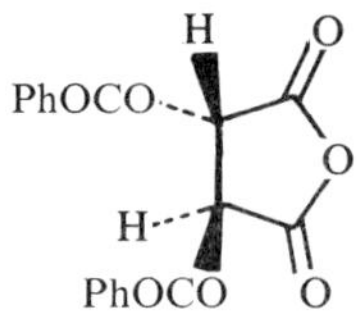

(89)

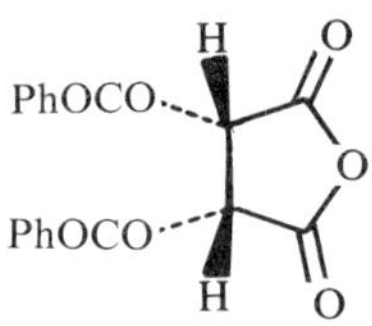

(90)

The p*K* values of substituted *β*-keto- (**91**) and thioketo- (**92**) anilides have been determined spectroscopically.[87] Acidic hydrolysis using ^{18}O-enriched acetanilide and *N*-cyclohexylacetamide results in loss of label from both starting substrates. Two reversibly formed *O*- and *N*-protonated tetrahedral intermediates are postulated.[88]

Hydrolysis in base with H_2O or D_2O of *N*-methyl-2,4-dinitroacetanilides (**93**) proceeds via the slow formation of a tetrahedral intermediate which decomposes with H_2O or hydroxide-ion catalysis. Breakdown of the intermediate proceeds by two pathways.[89]

Two papers have appeared on *N*-acetoxyanilides, whose derivatives are important carcinogenic metabolites of amines.[90,91] Both are the latest results of on-going programmes in the respective laboratories.[92] In addition to the finding of a base-catalysed hydrolysis, product distributions and rate constants show that there are two major pathways of decomposition of a series of *N*-acetoxyanilides at neutral pH, viz. N—O bond cleavage to give nitrenium ion intermediates and acyl-oxygen cleavage to give hydroxamic acid.[90] *N*-(3-Bromophenyl)hydroxylamine-*O*-sulphonate (**94**) and *N*-(3-bromophenyl-*O*-pivaloylhydroxylamine (**95**) have been identified as key models for certain suspected carcinogenic metabolites of polycyclic aromatic amines and amides.[91]

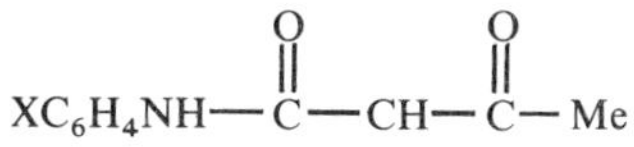

(91)

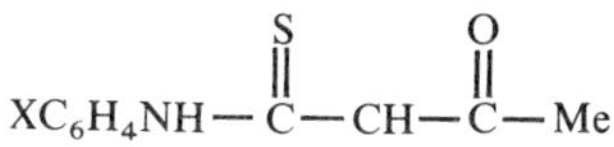

(92)

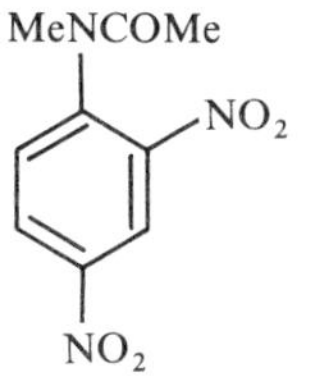

(93)

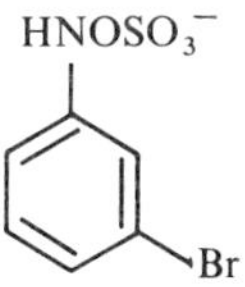

(94)

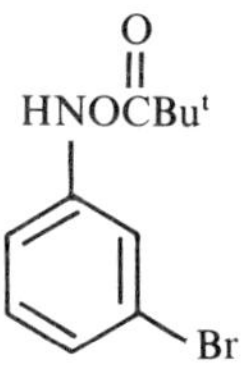

(95)

(e) *Acid halides*

Ab initio MO calculations have been used to study the optimal reaction paths for the displacement reactions of Cl^- with RCOCl (R = H, Me).[93] Some comparative studies have also been done with F^- and HCOF.

Six papers[94–99] deal with the solvolysis of various acid halides. The hydrolysis of thirty-nine cyclo-aliphatic and aromatic acid chlorides has been studied in 80% (v/v) aqueous acetone. The first-order hydrolysis rate for aliphatic acid chlorides decreased with increasing chain length, branching, and unsaturation, but increased in the presence of electron-donating substituents.[94] Methanolysis and phenolysis of six acyl chlorides (**96**) have been examined;[95] methanolysis is considered to proceed via a loose S_N2-type transition state subject to general base catalysis; phenolysis involves a retentive phenolysis mechanism in which electrophilic assistance followed by bonding at the phenolic oxygen in the ion-pair occurs. A study of the esterification of *n*-PrCOX (X = Cl, F) with *n*-BuOH in the presence and absence of $(PhO)_2P(O)OH$ and with *n*-BuOD in the presence and absence of $(PhO)_2P(O)OD$ showed no primary kinetic isotope effect with *n*-PrCOCl but gave a value of 2.42 with *n*-PrCOF.[96] A linear Taft correlation ($\rho^* = -0.86$) for the catalytic effects of the carboxylic acids (**97**) on the *n*-butanolysis of Cl_3COCl was found.[97] The hydrolysis of benzoyl chloride in aqueous acetone from 15°C to 30°C and at 1 bar to 2500 bars has been reported. An S_N1 mechanism is favoured at high pressures, higher temperatures, and high H_2O solvent content; an S_N2 process predominates when these variables are low.[98]

There is a dramatic difference in the effects of added potassium fluoride in the hydrolysis of p-XC_6H_4COF (X = Me_2N, MeO, and H);[99] for the fluoride (X = Me_2N) the rate is inhibited but KF increases the rates of hydrolysis of the other two. These differences are most likely attributable to the involvement of an acylium ion intermediate (**98**) for the compound (X = Me_2N) and the involvement of F^- and solvent both as general base catalysts for nucleophilic attack of H_2O on the other compounds (X = MeO and H).

RCOCl

(96)

R = Cl_2CH, CH_2Cl, Me, Et, Pr^i, Bu^t

RCO_2H

(97)

R = Me, Bu^t, $ClCH_2$, Cl_3C

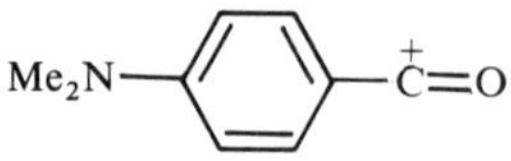

(98)

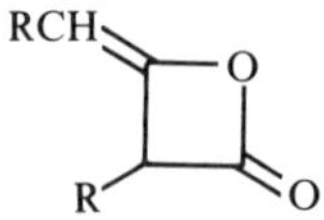

(99)

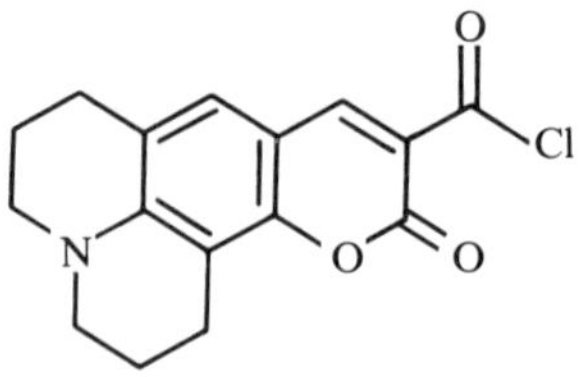

(100)

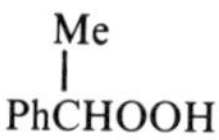

(101)

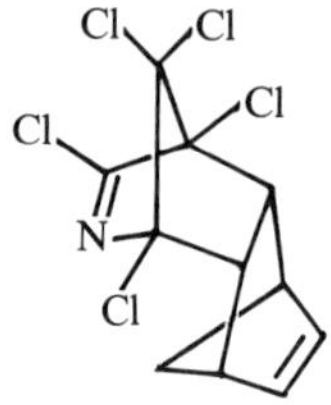

(102)

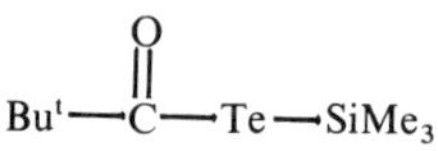

(103)

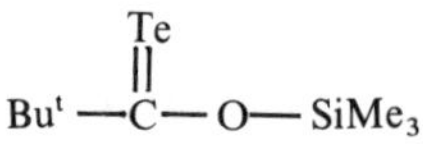

(104)

There are a number of reports of aminolysis of acyl halides. The kinetics of benzoylation (BzCl, BzBr) of a series of anilines in C_6H_6 in the presence of n-Bu_4NCl showed that BzBr was more reactive than BzCl.[100] The kinetics of the reaction of ArCOBr (Ar = 3,5-$Me_2C_6H_3$, Ph, 3-ClC_6H_4, 3-$O_2NC_6H_4$) with various substituted anilines suggested that the mechanism involves a 4-membered cyclic transition state.[101] The same Russian group has studied the kinetics of the reaction of a series of ring-substituted benzoyl chlorides with various anilines in chlorobenzene-Me_3COH mixtures. Both Hammett and Taft correlations were carried out.[102] Aldo-diketenes (**99**) form as by-products in the reaction of fatty acid chlorides with amines in the presence of a tertiary base.[103]

Acyl chlorides and nitriles react with ethyl α-(bromomethyl)acrylate to give α-methylene-γ-lactones and -γ-lactams;[104] these are 'Reformatsky-type' reactions. The reaction of CH_3COI with methyl acetate in an $A_{AC}2$ mechanism giving Ac_2O and MeI has an activation energy of 57 kJ mol^{-1} in $HClO_4$.[59] Treatment of the aminocoumarincarboxylic acid chloride (**100**) with 1-phenylethyl hydroperoxide (**101**) produced a strong chemiluminescence and gave acetophenone.[105] A mechanism involving intramolecular electron exchange, modelled on the firefly mechanism, and a secondary perester intermediate are proposed. The substrate aza-aldrin (**102**) containing an imidoyl moiety is very sensitive for determining the scope and limitation of the addition-elimination pathway of substitution of these halides;[106] (**102**) fails to react (under various conditions) with a variety of common nucleophiles, but reaction could be obtained with F^-, EtO^-, HO^-, PhS^-, and piperidine in alcohol. These results have been explained on the basis of Pearson's HSB classification of the nucleophiles. Reaction of pivaloyl chloride with bis (trimethylsilyl)telluride, $(Me_3Si)_2Te$, gives t-$BuCOTeSiMe_3$ which exists as a mixture of C=O and C=Te bonded isomers related by a 1,3-silyl shift: i.e. (**103**) ⇄ (**104**).[107] Acylation (RCOCl; R = Me, Ph) of 2,4-disubstituted 4(*R*)-alkoxycarbonyl thiazolidines and oxaxolidines gives the (2*R*,4*R*)-3-acyl isomers as the only products without any racemization. The mechanism involves concomitant epimerization at C(2).[108]

(f) *Ureas, carbamates and derivatives*

The hydrolysis of urea under open and closed conditions has been studied.[109] The mechanisms of hydrolysis of the *N*-nitroso-ureas has been discussed previously.[18] The reaction of the hydrochloride (**105**) with dimethylformamide dimethyl acetal (**106**) proceeds in two steps giving 2-amino-4-methoxy-1,3,5-triazine (**107**) as the final product.[110]

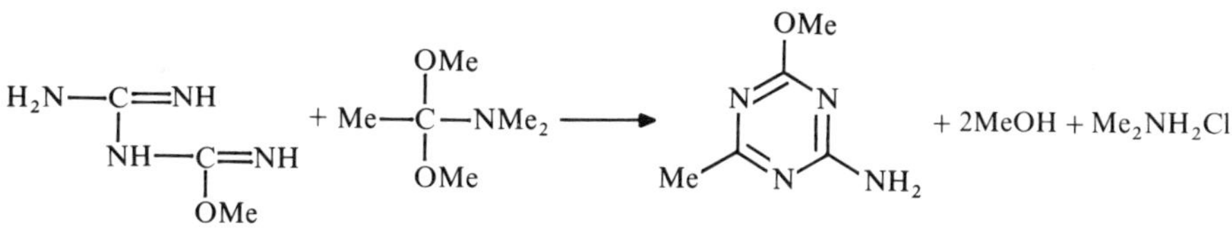

(**105**) (**106**) (**107**)

SCHEME 8

Three papers dealing with hydantoins have appeared.[111-113] The rates of base-catalysed cyclizations of the hydantoic acid amide (**108**) and hydantoic acid nitrile (**109**) derivatives to give (**110**) has been reported;[111] the same group has also studied the acid-catalysed cyclization of compounds (**108**; R^1, R^2, R^3 = Me and/or H) to give hydantoins; protonation occurs at the urea carbonyl.[112] In his latest studies of relevance to the carboxylation of biotin, Blagoeva[113] has compared the reactivity of the methylureido anion in (**111**) with the phenylureido anion in (**112**) towards CO_2^-. The latter, despite its lower basicity, is the better nucleophile. The strain introduced by the two methyl groups in (**112**) makes its cyclization much faster (10^3–10^4) than that of the parent compound 5-phenylhydantoic acid (**113**).

$R^3NHCONR^2CH_2CONHR^1$

(**108**)

$R^3NHCONR^2CHR^1CN$

(**109**)

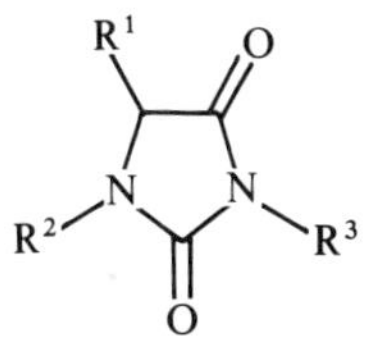

(**110**)

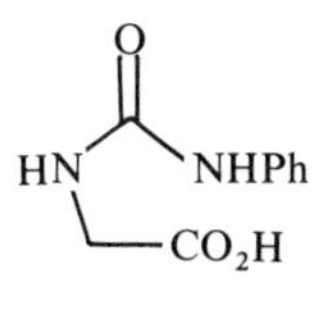

(**113**)

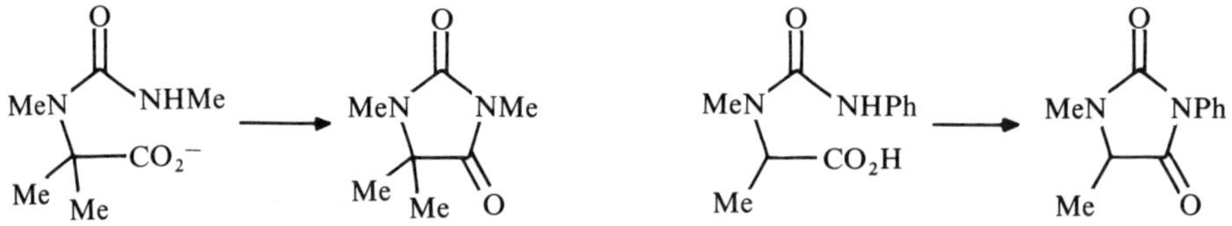

(111) (112)

SCHEME 9

Hydrolysis of flavin (**114**) is important biologically because of the flavin co-factors which are involved with *in vivo* redox processes; (**114**) gave the 4a-spiro-hydantoin (**115**) as the major product of such hydrolysis. Nucleophilic attack of hydroxide ion appears to occur preferentially at the 10a-position of (**114**).[114] Detailed studies have been reported[115,116] on the alkaline hydrolysis of 5-bromocytidine (**116**), 5-bromouridine (**117**), and related pyrimidine nucleosides, some of which exhibit antiviral properties. This work also helps to lead to a better understanding of the degradation of oligonucleotides and to develop methods for the removal of protecting groups during oligonucleotide synthesis.

The decarboxylation of a series of 4-phenylallophanate anions ($XC_6H_4HNCONHCO_2^-$) in H_2O at 25°C has been studied as a function of buffer and pH;[117] a concerted mechanism in acid and neutral media is proposed in which protonation (at the urea carbonyl) and cleavage of the C—N bond occur. In basic media, a spontaneous decomposition or a cyclic reaction with H_2O may be involved. The energetics of tautomerism in alloxans (**118**)–(**122**) have been investigated using *ab initio* STO-3G, 3-21G, and 6-31G calculations.[118] The 10-

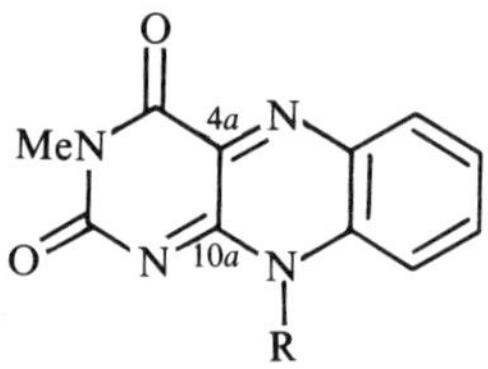

(114)

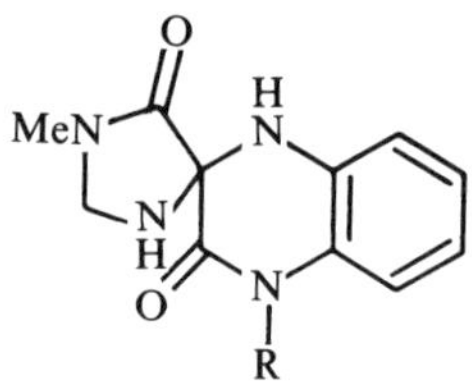

(115)

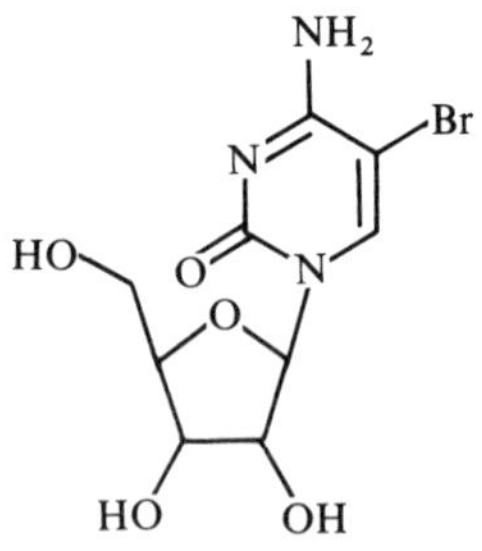

(116)

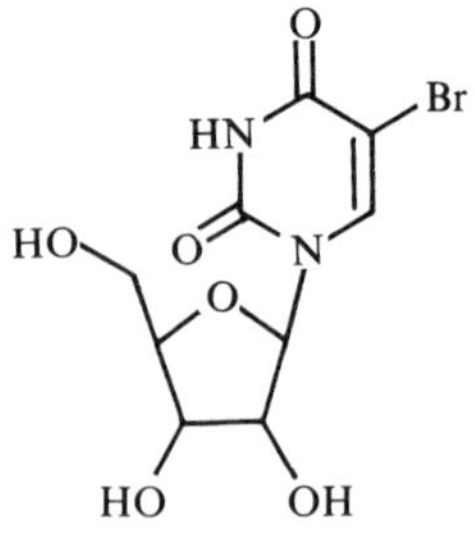

(117)

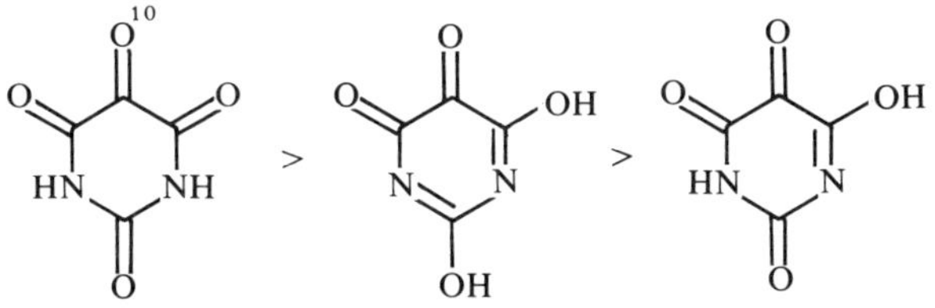

(118) **(119)** **(120)**

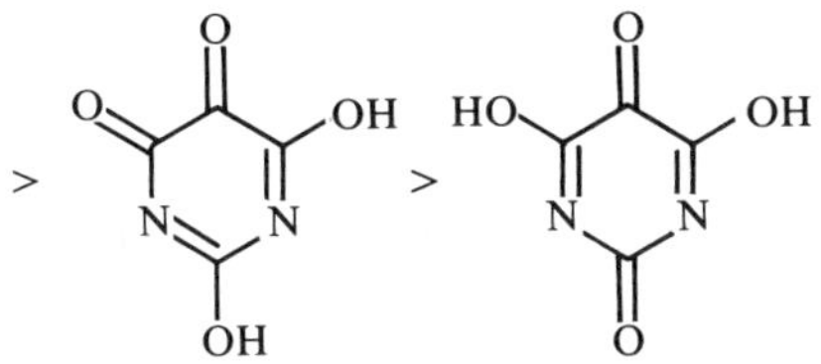

(121) **(122)**

SCHEME 10

oxygen atom in (**118**) is the preferred site of protonation and tautomerism does not occur in the vapour phase; the order of decreasing stability follows the order shown in Scheme 10.

(g) *Amides, β-lactams*

A theoretical study using the MNDO method has been carried out on the acid-catalysed hydrolysis of acetamide.[23] The excess acidity and target testing factor analysis methods for determining pK_{BH^+} (basicity constants) and slope parameters m* of amides have been compared. The latter method gave m* values about 0.08 units lower than those found by the excess acidity method.[119] The kinetics of the alkaline hydrolysis of formamide in acetone-water mixtures of varying composition at temperatures from 25 to 50°C have been investigated.[120] The rate was fastest in pure H_2O and addition of acetone appeared to cause a change in the solvent structure. The hydrolysis of the *N*-substituted amides (**123**) in aqueous NaOH in dioxane occurs by unusual CH_2—N bond cleavage to give (**124**). The amides (**125**) and (**126**) under identical conditions do not undergo C—N bond cleavage; (**124**) is formed from (**123**; R = Et) via (**123**; R = H).[121]

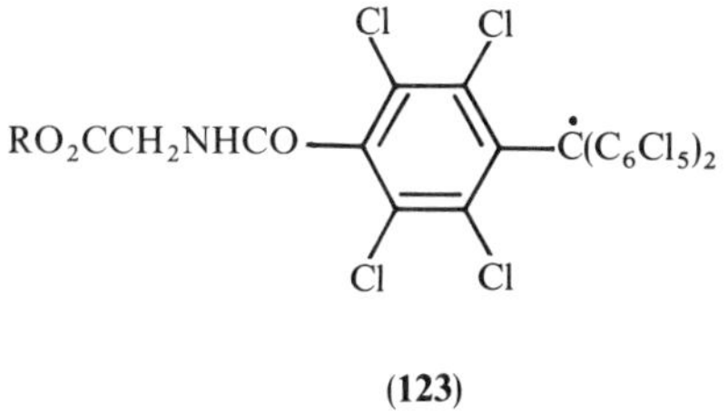

(123)

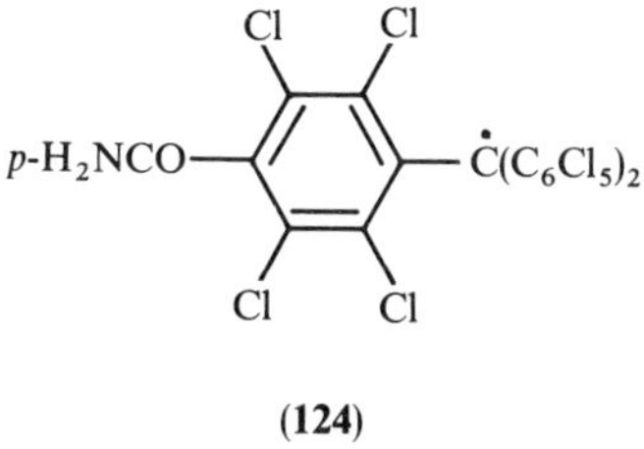

(124)

$C_6Cl_5CONHCH_2CO_2Et$

(125)

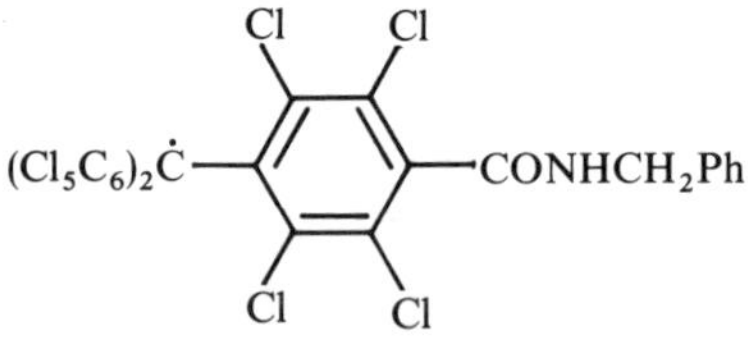

(126)

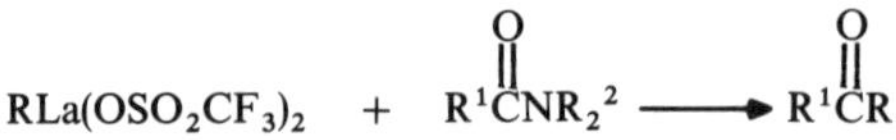

(127) **(128)** **(129)**

Alkyl- and aryl-lanthanum triflates (**127**) generated *in situ* from organolithium compounds and anhydrous lanthanum(III) triflate react with tertiary amides (**128**) to give 90% yields of the corresponding alkyl or aryl ketones (**129**).[122] An $A_{AL}1$ hydrolytic mechanism may be involved in another amide conversion in which bridgehead adamantyl or homoadamantyl amides such as (**130**) are converted in high yield to hydrocarbons on refluxing in ethanol and 50% H_2SO_4 (1:1 v/v).[123]

The effect of distortion on the hydrolytic reactivity of amides (*N*-pyramidalization) has been assessed from studies of the decomposition of the *N*-benzoylaziridines (**131**) in buffered aqueous media at 25°C as a function of pH.[124] *t*-Butoxycarbonyl derivatives of carboxamides (**132**) are cleaved on aminolysis with complete selectivity to give the corresponding *t*-butylcarbamates.[125] 2-(Diethylamino)ethylamine was the most suitable for preparative scale work. Reaction between *N*,*N*-dialkyl-2-(hydroxyimino)-3-oxoalkanamides (**133**) and meso-substituted *β*-diketones (**134**) on treatment with zinc in acetic acid gives *N*,*N*-dialkyl-2-pyrrolecarboxamides (**135**). Surprisingly, the analogous reaction using the 3-oxoalkanoate esters differs.[126]

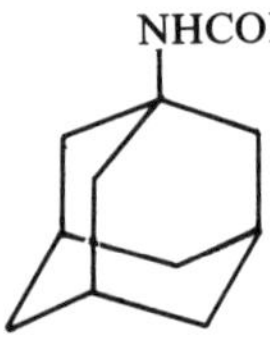

(130)

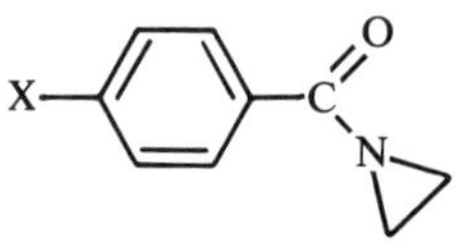

(131)

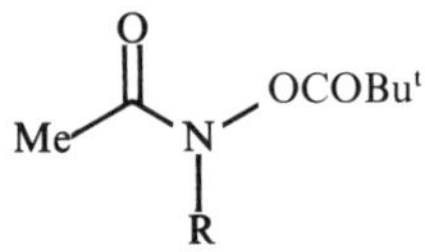

(132)

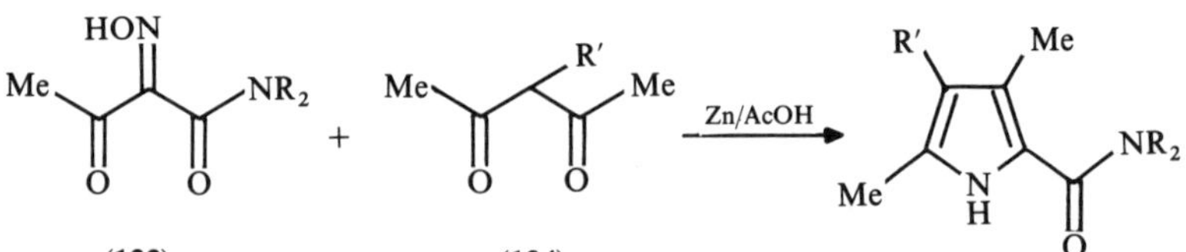

(133) **(134)** **(135)**

SCHEME 11

A simultaneous deprotection/acylation methodology using Boc or 2-(trimethylsilyl)ethoxycarbonyl-protected amino-acid pentafluorophenyl or *N*-hydroxysuccinimidyl active esters gives protected tripeptides in good yield.[127] Iodolactamization (I_2 in THF) of the γ,δ-unsaturated thioimidates (**136**) proceeds regio- and stereo-selectively giving the highly functionalized γ-lactams (**137**).[128] Formation of a β-lactam via (**34**) in the hydrolysis of azetidin-2-ylideneammonium salts has been dealt with earlier.[19]

Three papers[129–131] deal with reactions of 6-diazopenicillanate (**138**). Treatment of (**138**) with aromatic aldehydes in the presence of Lewis acids gave spiro-epoxides, oxazinones, and thiazepines, and 6-acylpenicillanate intermediates have been either isolated or trapped in these reactions.[129] With BF_3-Et_2O, (**138**) reacts with aromatic imines, ArCH=NAr^1 to give aziridines, thiazolo-pyrimidinones and, in one instance, an enaminopyrrolothiazole;[130] (**138**) in BF_3-Et_2O reacts with ketones to give mixtures of 6α-alkyl-6β-acylpenicillanates (**139**) and thiazolooxazinones (**140**).[131] Reaction of 6-cyano(or 6-nitro)-3,4 epoxy3,4-dihydro-2,2-dimethyl-2H-1-benzopyran with azacyclotridecanone gave the pyran ring-cleaved product (**141**) and not the expected cyclic amido-3-hydroxy benzopyrans.[132] MINDO/3 and MNDO have been used to explore the reactivity of 3-pyrrolin-2-one tautomers and their derivatives.[133]

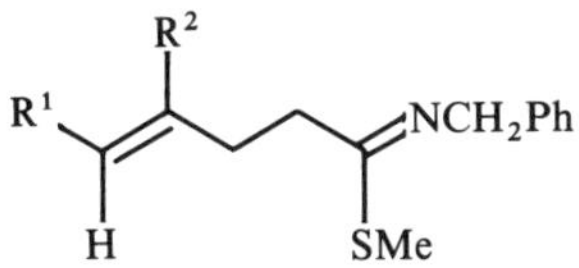

(**136**)

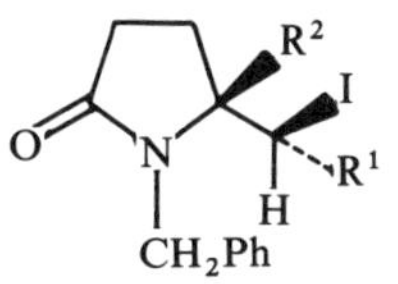

(**137**)

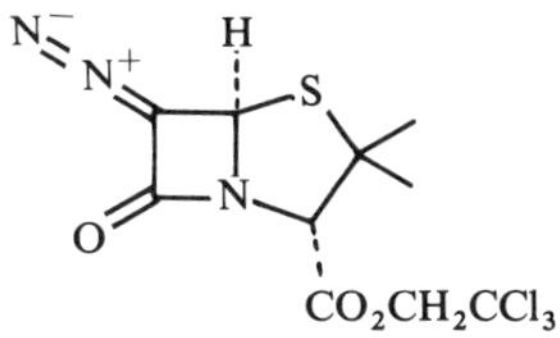

(**138**)

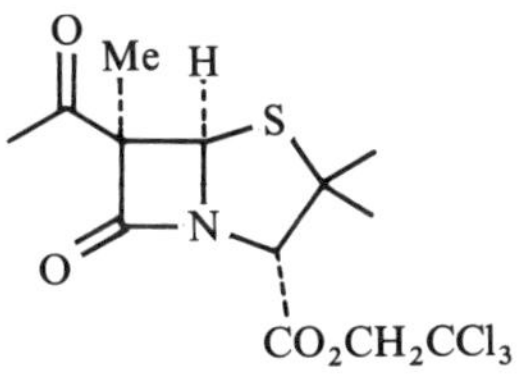

(**139**)

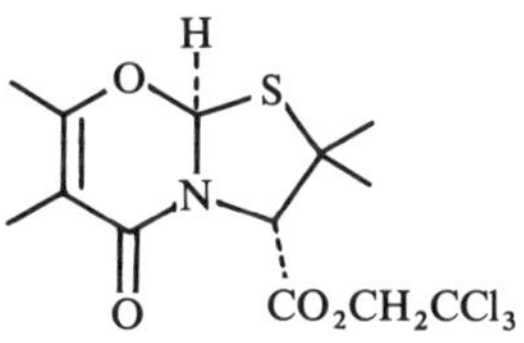

(**140**)

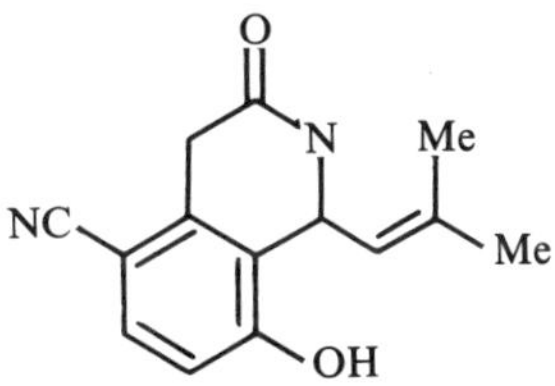

(**141**)

The reactivity of the 6-methyl-1,3-oxazine-2,4(3H)-diones (**142**) with alcohols shows that when R = H reaction generally occurs at C(2) but compounds (**142**; R = Me, $PhCH_2$) do not react.[134] Fragmentation of 3,5-dinitro-2-pyridones (**143**) gave nitroacetamides (**144**) and nitromalonaldehyde diimine (**145**).[135] Benzamide oxime (**146**) and *N,N*-dicyclohexylcarbodiimide (DDC) (**147**) gave the products[136] 5-cyclohexylamino-3-phenyl-1,2,4-oxadiazole (**148**) *N,N′,N″*-tricyclohexylguanidine (**149**) and *N,N′*-dicyclohexylurea.

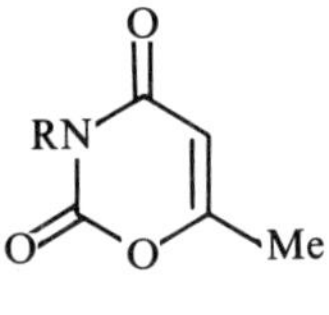

(142)

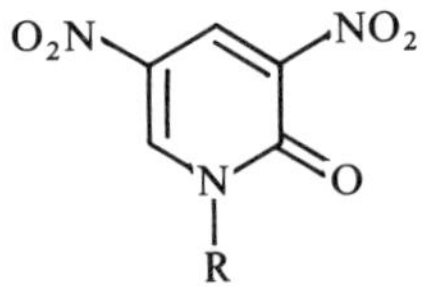

(143)

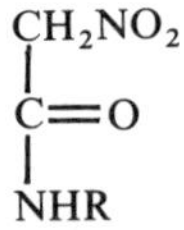

(144)

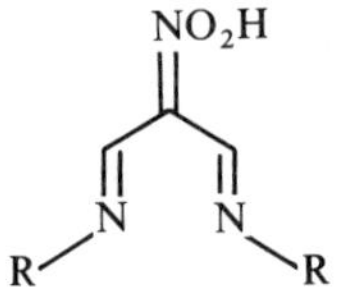

(145)

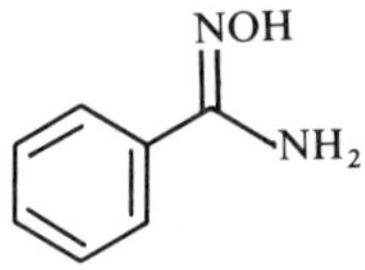

(146)

cyc-$C_6H_{11}N{=}C{=}NC_6H_{11}$-*cyc*

(147)

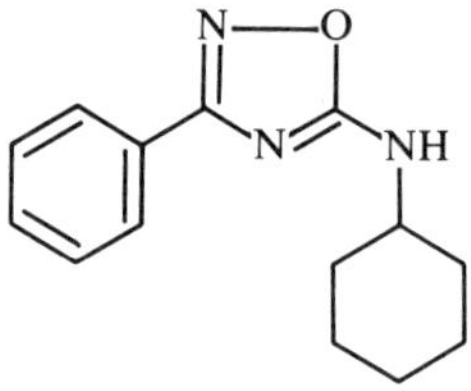

(148)

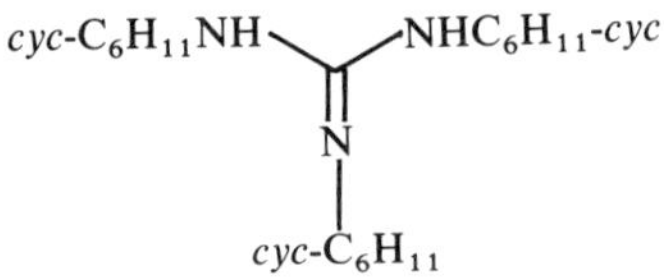

(149)

(h) *Nitriles*

Rate constants and activation parameters have been reported for the alkaline hydrolysis of some (benzylidene)phenylacetonitriles (**150**).[137,138] Phase-transfer-catalysed hydrations of aromatic benzyl and aliphatic nitriles to amides using $Mn(CO)_5Br$ as catalyst have been reported.[139] The reaction of $Me_2C(NH_2)CN$ with CO_2 gave the hydantoin (**151**) via (**152**). Only a little hydantoin (**153**) was formed from (**152**) or the salt (**154**); (**152**) and (**154**) are considered to be unlikely intermediates in the formation of (**153**).[140]

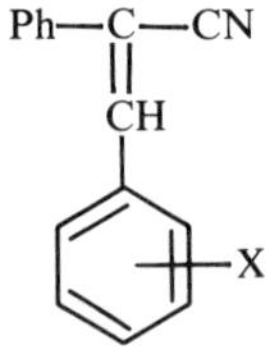

(150)

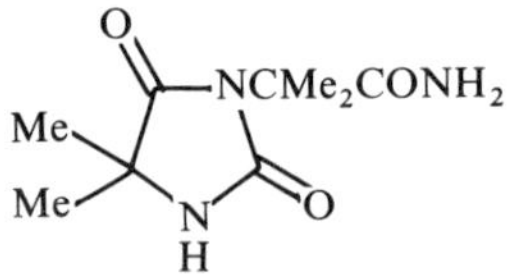

(151)

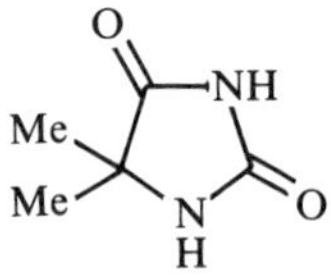

(153)

$Me_2C(CN)NHCONHCMe_2CONH_2$

(152)

$Me_2C(CN)\overset{+}{N}H_3\ {}^{-}O_2CNHCMe_2CN$

(154)

(i) *Other nitrogen centres*

The rates of alkaline hydrolysis of 4-nitrophenyl diarylimidates in 1:1 dioxane-H_2O do not depend on added salt (NaCl, $NaClO_4$, $LiClO_4$) and decrease with increasing salt concentration. An addition-elimination mechanism with a slow formation of a tetrahedral intermediate is proposed.[141]

Ab initio MO calculations on the rearrangement of the fulminate anion (CNO^-) to the cyanate anion (OCN^-) indicate that a metastable cyclic oxazirinyl anion intermediate (**155**) is involved.[142] A substantial review in German (317 references) on the insertion reactions of isocyanates (into various bonds) has appeared (see Scheme 12).[143]

(155)

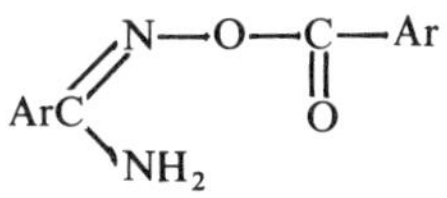

(156)

$RNHC(X)NHCH_2CO_2Et$

(157)

X = O, S

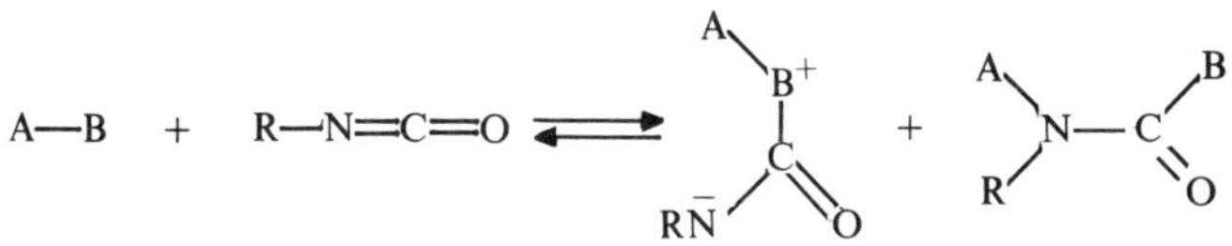

A—B = C—N, C—O, C—S, Si—N, Si—S, P—O, P—N, B—O, B—N, etc.

SCHEME 12

The kinetics of formation of 1,2,4-oxadiazoles from *O*-benzoylbenzamidoximes (**156**) have been studied in H_2SO_4 and aqueous EtOH media.[144] Concomitantly with oxadiazole formation, hydrolysis to benzamidoxime and benzoic acid occurs. Cyclization rate constants have been reported for substituted ethyl *N*-(phenylaminocarbonyl)-, -*N*-(alkylaminocarbonyl)-, and -*N*-(phenylaminothiocarbonyl)-glycinates (**157**).[145] The logarithms of the rate constants increases with decreasing amine basicity down to $pK_a(RNH_2)$ of 5.5; formation of a tetrahedral intermediate is rate-limiting.

The water-catalysed hydrolysis of the activated amide bond in 1-acyl-1,2,4-triazoles (**158**) has been studied in dilute aqueous solutions of carbohydrates.[146] These solutions show almost ideal behaviour and the kinetic medium effects observed (particularly $\Delta S^{\ddagger}$) are largely determined by carbohydrate-induced alterations in the 3-dimensional H-bond network of H_2O. The unstable 5-aryl-2-(3-benzylidene-2-phenylcarbazoyl)-1,2,3,4-tetrazoles (**159**) have been synthesized. On heating they undergo an interesting ring-change giving 5-aryl-2-(2-benzylidene-1-phenylhydrazino)-1,3,4-oxadiazoles (**160**); further heating eliminates a molecule of PhCN to give 2-anilino-5-aryl-1,3,4-oxadiazoles (**161**).[147]

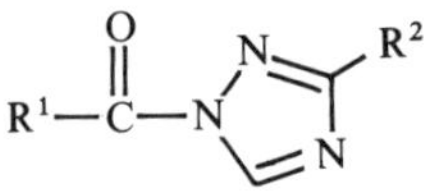

(158)

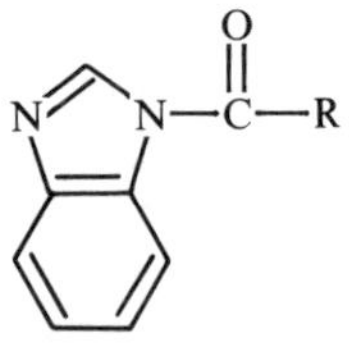

(162)

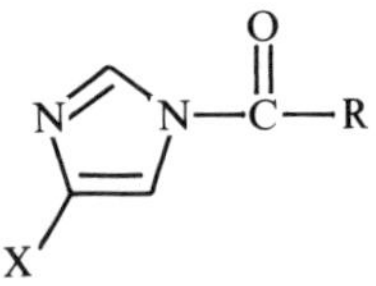

(163)

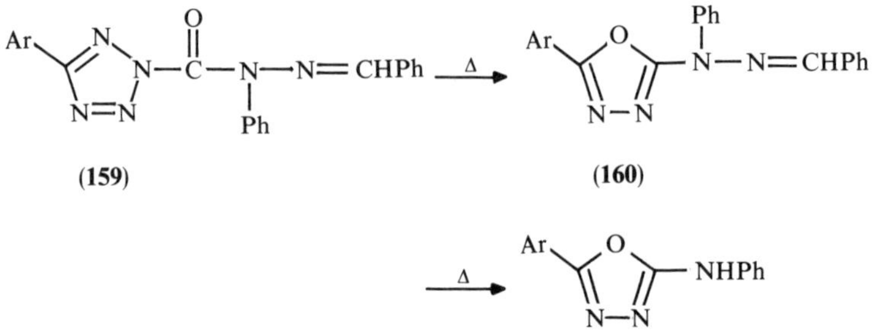

(159) **(160)** **(161)**

SCHEME 13

Fife's group[148] has studied the hydrolysis of a series of *N*-acylbenzimidazoles (**162**) and *N*-acyl-4(5)-substituted imidazoles (**163**); pH-independent hydrolysis of (**162**) probably involves concerted nucleophilic attack with protonation of the leaving-group by H_2O. An assessment of the structural requirements for a H_2O-catalysed reaction in the hydrolysis of an amide was possible for compound (**163**; R $= CH_2CMe_3$, $X = NO_2$).

Acetolysis (acetic anhydride, 10°C) of [$^{15}N_4$]-hexamethylenetetramine (hexamine) (**164**) to 3,7-diacetyl-1,3,5,7-tetrazabicyclo[3.3.1]nonane (DAPT) (**165**) gave relatively little isotopic mixing. However, more severe conditions (98°C) used to form, 1,3,5-triacetyl-1,3,5-triazacyclohexane (TRAT) (**166**) gave considerable scrambling of the nitrogen label.[149]

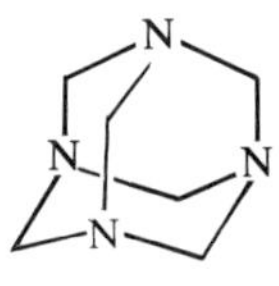

(164)

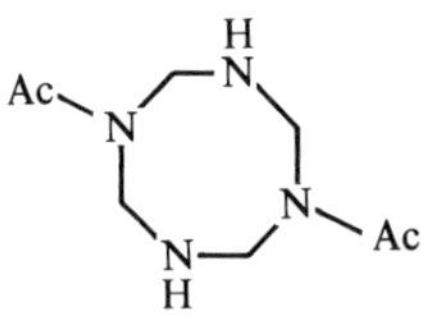

(165)

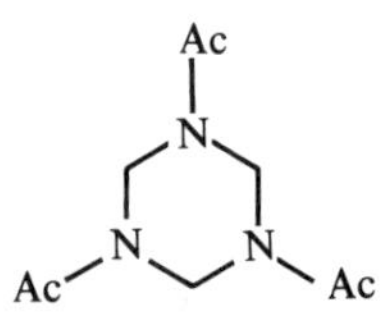

(166)

7-Oxo-3-(tetrahydropyran-2-yl)-7H-pyrido [2.1-i] purines (**167**) are highly fluorescent. These compounds react readily with primary and secondary amines to give (**168**) which are non-fluorescent.[150] The mechanism of the change resembles that of ester aminolysis. Acid hydrolysis of 9-(1-alkoxyethyl)purines (**169**) occurs via rate-limiting departure of the protonated base moiety and simultaneous formation of an alkoxyethyl carbocation (RO$\stackrel{+}{=}$CHMe).[151] The same mechanism is seen as applying to the hydrolysis of the 6-substituted 9-(2-deoxy-β-D-erythro-pentafuranosyl) purines.

3-Phenacyl- and 3-alkoxycarbonylmethyl-1-methylimidazolium bromides (**170**) react with two molecules of diethyl azodicarboxylate (DEAD) in the presence of triethylamine with the extrusion of the imidazole ring and the concomitant migration of an ethoxycarbonyl group to yield 3-substituted tetraethyl 1,2,4,5-tetraazapent-3-ene-1,1,2,5-tetracarboxylates (**171**).[152]

Several papers have appeared on triazenes[153–156]. The kinetics of solvolysis of 3-acetyl-1,3-bis(substituted phenyl)triazenes (**172**) in 40% EtOH and H_2SO_4 at 25°C have been measured.[153] A general acid-catalysed process occurs. A kinetic acidity function, H_K has been developed from the data and compared with H_0. The Hammett ρ-value for variation of X in (**172**) is ca. −1.4. In further work from the same laboratory,[154] an A-S_E2 mechanism is favoured for the decomposition of 3-substituted and 1,3-disubstituted 1,3-diphenyl-3-methyltriazenes, $X^1C_6H_4N{=}N{-}N(Me)C_6H_4X^2$, in 40% (v/v) aqueous EtOH buffers at 25°C.

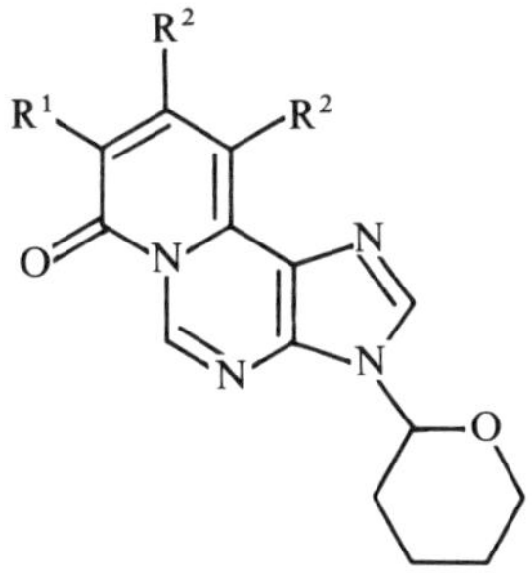

(**167**)

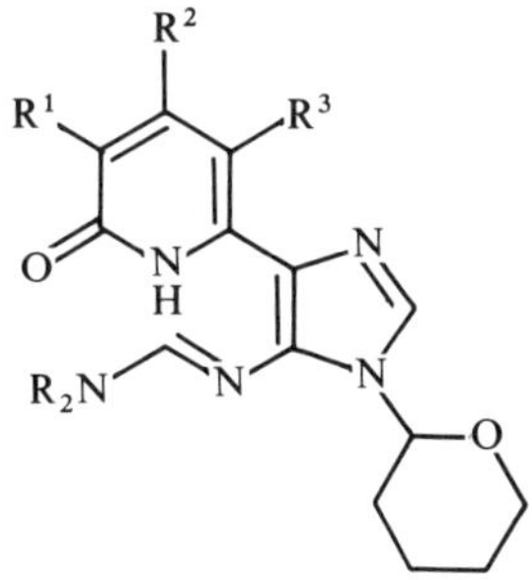

(**168**)

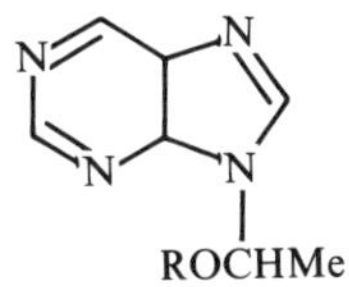

(**169**)

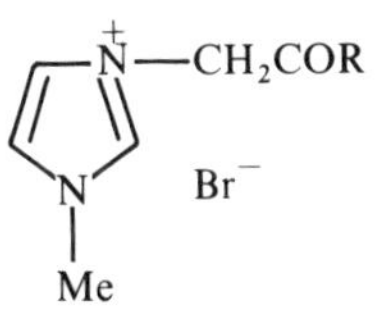

(**170**)

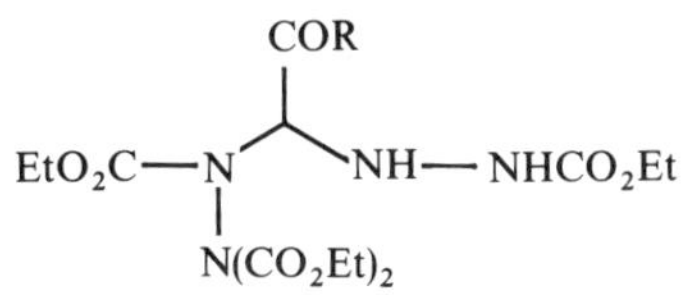

(**171**)

(**172**)

Scheme 14

The kinetics of hydrolysis of the medicinally important 1-aryl-3-benzoyloxymethyl-3-methyltriazenes (**173**) to 1-aryl-3-hydroxymethyl-3-methyltriazenes (**174**) in mixed aqueous organic solvents have been reported.[155] A battery of diagnostic methods (pH-rate, Hammett and Grunwald–Winstein plots, solvent isotope effect and thermodynamic parameters) are interpreted in terms of a unimolecular ionization of (**173**) to form an iminium cation (**175**) and a benzoate anion, which competes with H_2O as a trap for (**175**). In further work,[156] (**174**) has been converted into 3-alkoxymethyl- and 3-alkylthiomethyl-1-aryl-3-alkyltriazenes by reaction with alcohols (yields 80–90%) and thiols (yields ca. 40%), respectively.

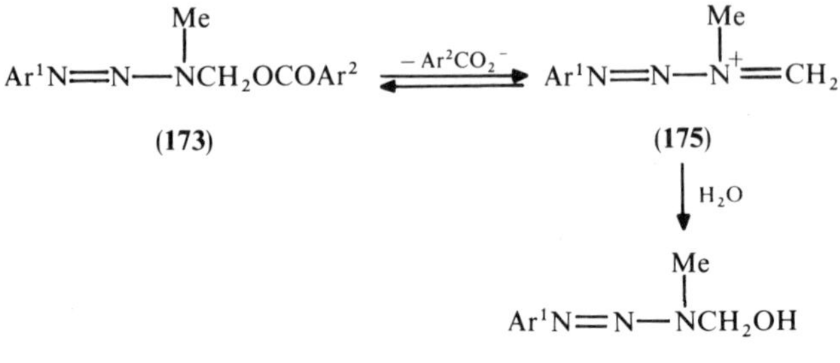

SCHEME 15

Reactions in Aprotic Solvents

(a) *Ester reactions*

The carbonylation of methanol to give methyl formate has been carried out with CO at 80–110°C/1–4MPa in the presence of sodium methoxide. The reaction was first-order in MeOH, methoxide, and CO. No deuterium (MeOD) isotope effect was observed.[157] Diazomethane in ether reacts readily in the dark at room temperature with phenyl trifluoracetate, to give 2-phenoxy-2-trifluoromethyloxirane (**176**).[158] Primary kinetic isotope effects (k_H/k_D) up to about 2 have been observed in the thermal decomposition of β-hydroxy-ketones and -esters (**177**; R^1 = Me, OEt; R^2, R^3 = H, Me); this supports the involvement of a cyclic transition state in the decomposition process.[159] An unusual mechanistic path is followed in the reaction of the dilithio derivative of dimethyl *cis*-4-cyclohexene-1,2-dicarboxylate (**178**) with phenyl 3-*t*-butylpropiolate (**179**) to give the fused bicyclic 2-cyclopentenones (**180**) in a single step.[160] The mechanism of the annulation is thought to involve the ketene intermediate (**181**). Another type of dilithio diester (**182**) reacts with dihalomethanes (**183**; X = Y = Br, I; X = Cl, Y = Br) to give the cyclopropanated products (**184**).[161]

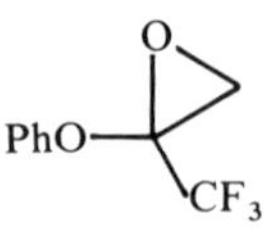

(**176**)

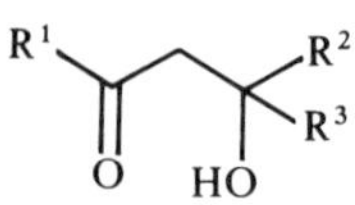

(**177**)

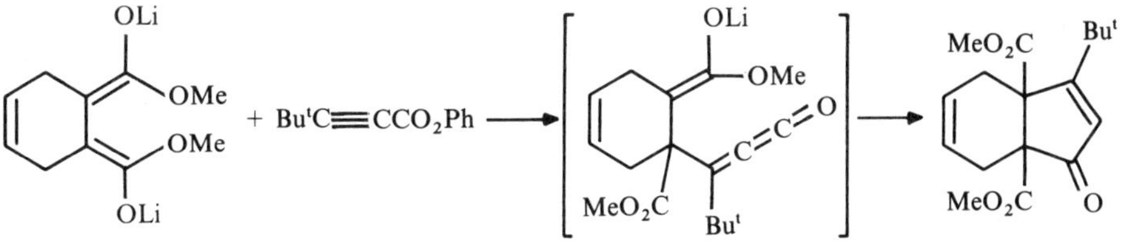

(178) **(179)** **(181)** **(180)**

SCHEME 16

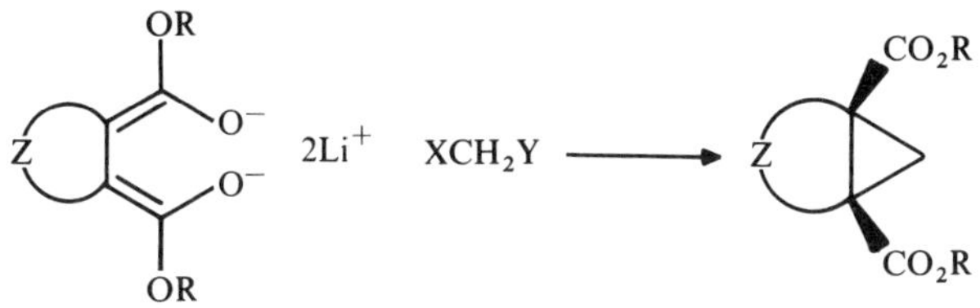

(182) **(183)** **(184)**

R = (*L*)-menthyl; Z = $(CH_2)_4$, $(CH_2)_3$, $(CH_2)_2$, CH=CH, $CH_2CH{=}CHCH_2$

SCHEME 17

Cesium alkanoates (mainly the propionate) have been examined for solubility, nucleophilic reactivity, and degree of ion-pairing in DMF and DMSO solutions;[162] ^{133}Cs NMR spectroscopy has been used and shows the Cs^+ ion to be virtually completely solvated, while the alkanoates are essentially free and highly reactive.

A new and interesting generation of ketene has been reported by gas-phase pyrolysis of alkyl (trimethylsilyl)acetates (**185**) giving only trimethylsilyl alkyl ethers and ketene (Scheme 18).[163]

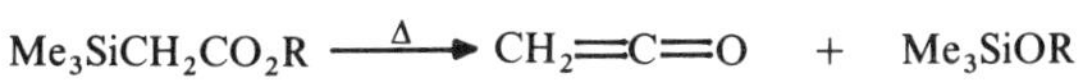

(185)

SCHEME 18

The aminolysis (aniline, piperidine, diethylamine, ring-substituted benzylamines, etc.) of aroyl-*N*-hydroxysuccinimide esters (**186**) in anhydrous dioxane displayed a 10^4-fold reactivity variation for (**186**; X = 4-NO_2, Y = H) over an amine basicity range of 6.5 p*K* units.[164] Reversible formation of a tetrahedral intermediate (**187**) followed by a slow breakdown to products is postulated.

(b) *Other reactions*

Dehydration and decarboxylation reactions of formic acid in the gas phase have been the subject of refined *ab initio* calculations.[165] A review (70 references) has appeared on the utility of the 9-fluorenylmethyloxycarbonyl (FMOC) amino-protecting group.[166] Further studies have been reported on another protecting group, the *t*-butoxycarbonyl (Boc) function, in regard to its introduction into amides.[167]

The kinetics of the reaction of 4-$Et_2NC_6H_4NH_2$ with PhCOCHXCONHPh (X = H, F) have been studied.[168] Aminolysis of benzoyl chloride in benzene in the presence of *n*-Bu_4NCl by 3,5-$(O_2N)_2C_6H_3NHR$ (R = H, Me)[169] and of XC_6H_4COCl (X = 4-OMe, H, 3-Cl, 3-NO_2, etc.) in a cyclohexane-chlorobenzene mixture (1:1) by $YC_6H_4NH_2$ (Y = H, 3-F, 3-NO_2, 4-OMe, etc.) has been reported.[170] The non-additivity of the effects of X and Y for the latter system was demonstrated. The aminolysis of the *N*-methylbenzimidoyl chlorides (**188**) with anilines leading to amidines has been studied.[171]

Ab initio MO calculations on the unimolecular decomposition of oxalic acid have been carried out using the Hartree–Fock method with LCAO approximation mainly using the 3–21G basis set.[172] Three conformers and eight possible transition states have been considered in the calculations. The lowest energy path was that shown in Scheme 19. Some experimental work on the decomposition was consistent with the results of the calculations.

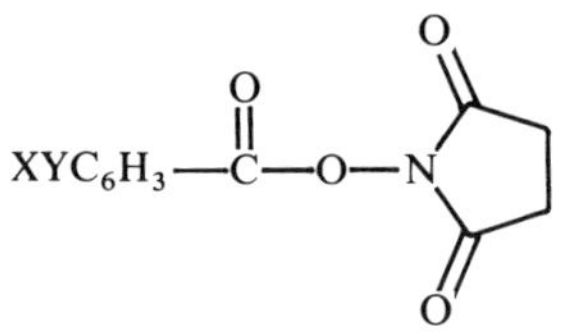

(**186**)

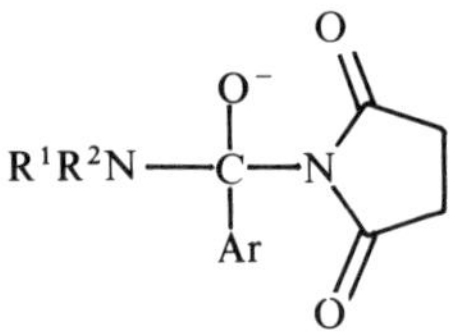

(**187**)

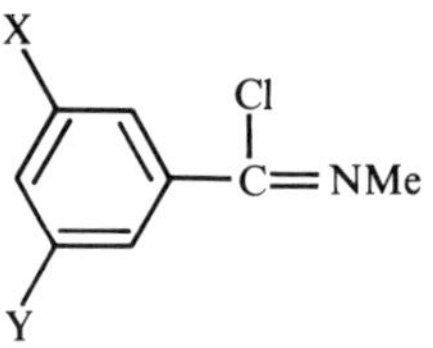

(**188**)

$$(COOH)_2 \longrightarrow CO_2 + CO + H_2O$$

SCHEME 19

Intramolecular Catalysis and Neighbouring-group Participation

The carbinolamine adduct (**189**) of (**28**), formed in equilibrium with (**28**), reacts via participation of the hydroxy group to give the tetrahedral intermediate (**190**) en route to the 5-oxazolidinone intermediate (**191**).[13] Latest results have been given[113] on the ureido systems (**111**) and (**112**), one of which (**111**) has previously been shown[173] to involve intramolecular nucleophilic attack by ureido anion on carboxylate. See also section (f) above; p. 45.

Neighbouring CO_2H catalyses the hydrolysis of the 1-arylethyl ethers of salicylic acid (**192**).[174] At pH values above the pK_a (3.95), the hydrolysis of (**192**; X = Y = H) is 900 times faster than that of its methyl ester. The reaction gives a ρ^+-value of -4.9 ± 0.5 for substitution in the benzylic ring (**192**; X = H, Y = 4-Me, H, 3-Br). For the other ring (**192**; Y = H = 4-OMe, 5-Me, H, 5-Cl, 5-NO_2), ρ is 1.25 ± 0.08 and further analysis gives ρ (PhOH) of 0.96 and ρ (CO_2H) of 0.23. Thus, the acid strength of the CO_2H has virtually no effect on the rate of the reaction that it is catalysing so effectively.

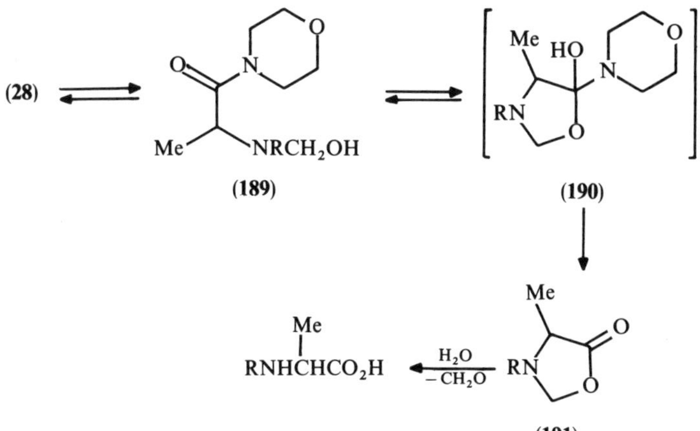

(189) (190) (191)

SCHEME 20

Methanolysis and ethanolysis of phenyl salicylate involves a pre-equilibrium formation of monomeric ROH from polymeric $(ROH)_n$, followed by an intramolecular general base-catalysed nucleophilic attack by monomeric ROH on the carbonyl.[175] Participation of the *O*-carbomethoxy group in the acid hydrolysis of (**193**) to give the oxo-lactones (**194**) lowered the kinetic isotope effect (k_D/k_H). Thus, for compounds (**193**; X = Y = NO_2, CO_2Me) $k_D/k_H > 1$ but all other substrates gave a value less than unity.[176]

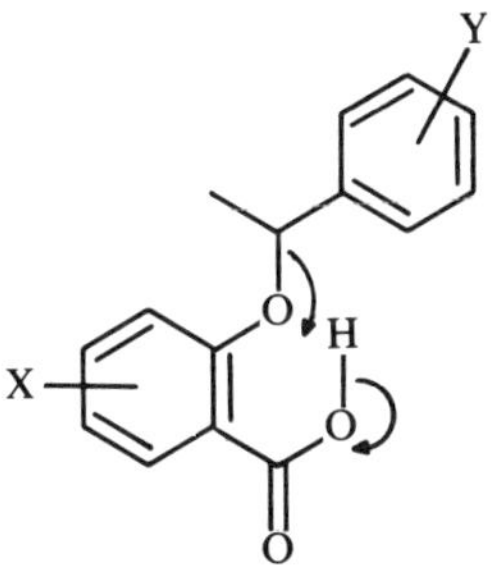

(192)

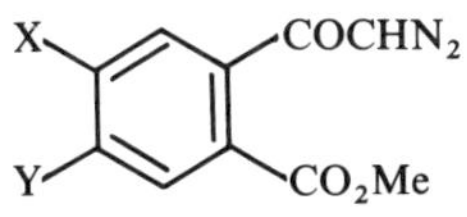

(193)

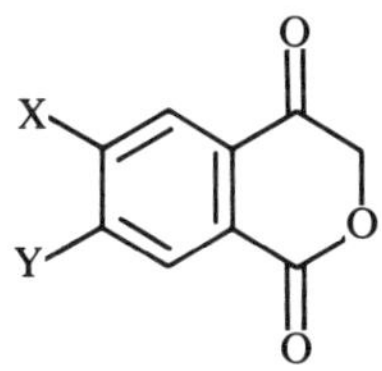

(194)

A first example of anchimeric assistance by the α-benzamido group in nitrile hydrolysis has been reported.[177] Hydrolysis of the 2-aminoadamantane-2-carbonitrile (**196**) to the amino-acid (**200**) is impossible to achieve in mineral acid or alkali; however, on addition of benzaldehyde and KOH in aqueous ethanol, the conversion via the crucial intermediate (**195**) occurs following the sequence shown (Scheme 21) to give the Schiff base (**198**), which can be converted into (**199**) with dilute acid. The final conversion into (**200**) proved difficult and a slightly modified scheme was used. The reaction of photoprotoporphyrin IX dimethyl ester (**201**) involves hydrolysis of the cyano-adduct (formed from added potassium cyanide) assisted by the neighbouring hydroxy group of the porphyrin (Scheme 22)[178] giving the porphyrin acetamide as final product.

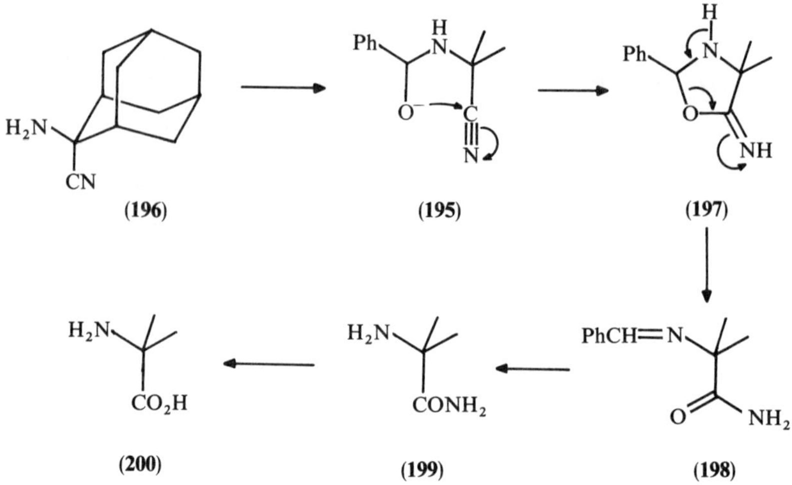

SCHEME 21

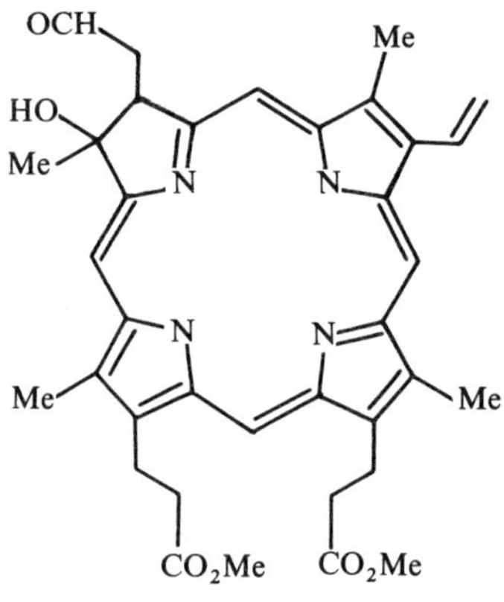

(201)

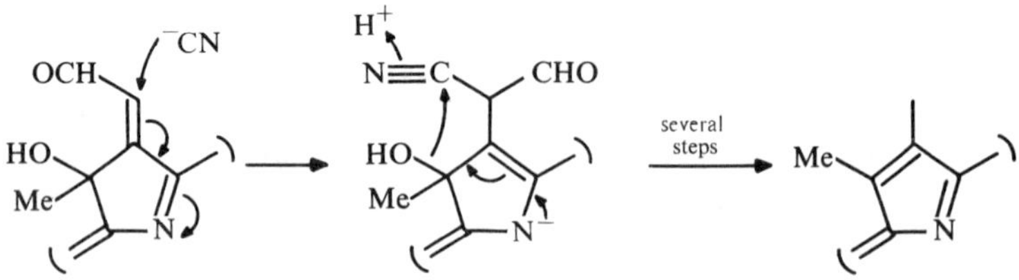

SCHEME 22

Hydrolysis of 2-(imidazol-2-ylmethyl)-*N*-acetylimidazole (**202**) exhibits three major domains in the pH-log(rate) profile corresponding to H_2O attack on protonated (**202**), HO^- attack on the neutral material, and an intramolecular general base catalysis of attack by H_2O on the neutral species.[179] The latter is supported by the observation of a plateau in the pH profile at neutrality and a $k_{H_2O}/k_{D_2O} = 2.8$ in that region.

The reactions of glycine with the β-isothiocyano-ketones (**203**) give via cyclodehydration various thiourea adducts such as (**204**), which undergoes intramolecular nucleophilic addition to give (**205**).[180] The influence of neighbouring groups in the racemization of dipeptides is supported by Smith's latest results;[181,182] (**206**) and (**207**) illustrate amino-terminal intramolecular assistance which may explain the high racemization rate of Gly-Ala.

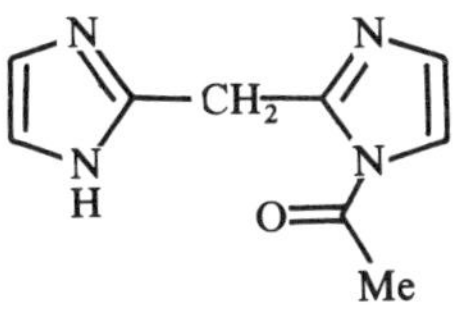

(202)

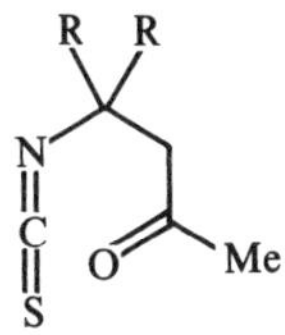

(203)

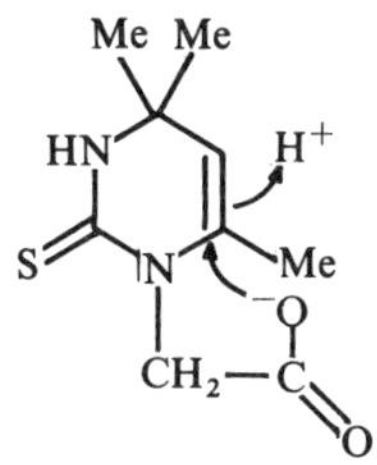

(204)

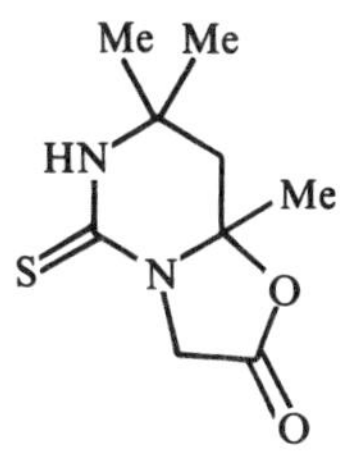

(205)

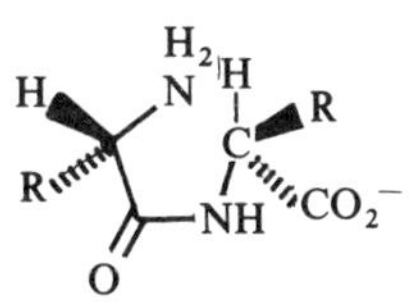

(206)

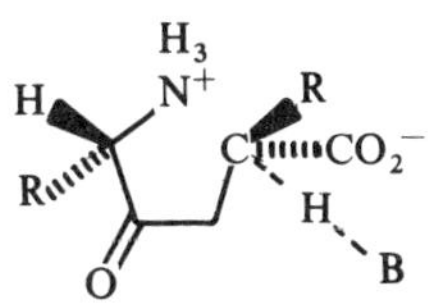

(207)

Association-prefaced Catalysis

Two reviews in this area have appeared. Bunton discusses the design of micelles as dephosphorylating agents[183] and a second review (in Japanese) by Odashima[184] deals with the control of binding and the reaction of guests by artificial hosts having well-defined hydrophobic cavities.

The hydrolysis of aspirin and two derivatives with hydrophobic chains (**208**) has been studied in the hope that the orientation of these substrates in the micelle would vary as the site of the hydrophobic chain was varied.[185] The CO_2^- groups of compounds (**208**; X = H, R = Me and X = H, R = *n*-C_7H_{15}) are aligned radially away from the micelle-H_2O surface but the CO_2^- group of (**208**; X = *n*-C_8H_{17}, R = Me) is tilted towards the micellar surface and is not as exposed to H_2O molecules as the CO_2^- groups of the first two compounds. Intramolecular general base-catalysed hydrolysis is therefore less favourable for (**208**; X = *n*-C_8H_{17}, R = Me).

Two new amphiphilic catalysts (**209**) and (**210**) have been proposed for ester hydrolysis and have been tested towards *p*-nitrophenyl *n*-hexanoate and *p*-nitrophenyl diphenyl phosphate, respectively, and found to be less effective than their 'parent' (**211**).[186]

A study of the hydrolysis rates of a series of *N*-methylnicotinyl and *N*-methylisonicotinyl compounds (**212a, b**) and (**213a, b**) in the presence of sodium dodecyl sulphate (SDS) anionic surfactant has been reported.[187] Rates for the hydrolysis of *p*-nitrophenyl acetate, *n*-hexanoate, and *n*-dodecanate catalysed by bifunctional catalysts, (**214**) and (**215**), containing imidazolyl and hydroxyl groups are reported in the presence of hexadecyltrimethylammonium bromide micelles at 25°C.[188] Analysis of the kinetics suggests the formation of an acylated intermediate which is deacylated at the imidazolyl group. The ester aminolysis of *p*-nitrophenyl *p*-nitrobenzoate (and some other esters) in the presence of dodecylammonium carboxylate reversed micelles shows that the catalytic role of the surfactant head-anion was concentration-dependent.[189]

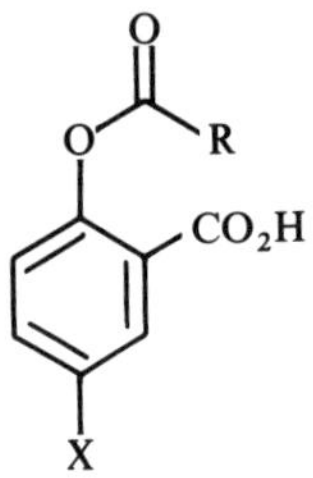

(**208**)

X = H, R = Me, *n*-C_7H_{15};
X = *n*-C_8H_{17}, R = Me

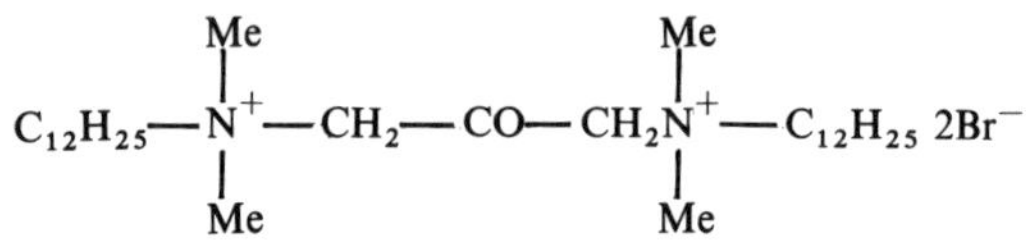

(**209**)

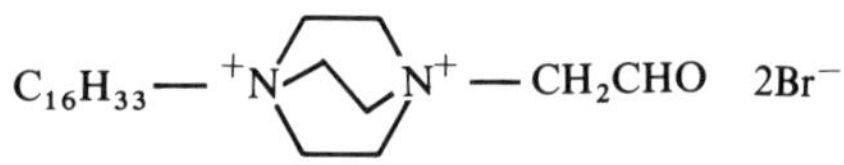

(**210**)

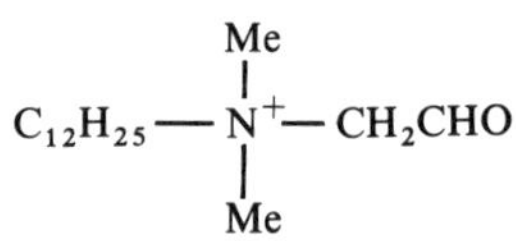

(**211**)

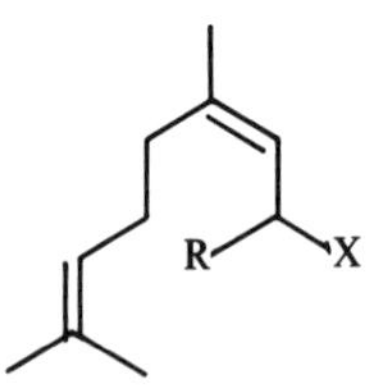

(**212**)

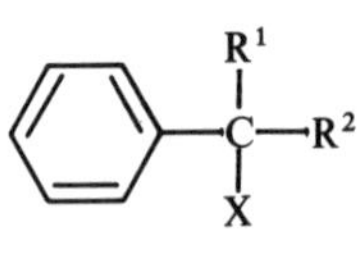

(**213**)

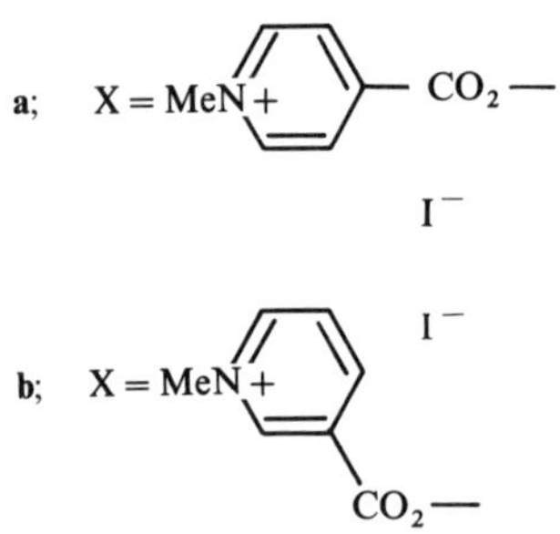

(214)

$R^1 = CH_2Ph$, $R^2 = CH_2Im$, etc.

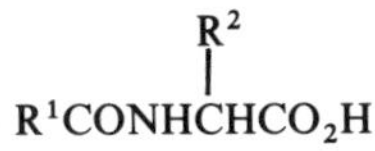

(215)

$R^1 = n$-nonyl,
$R^2 = CH_2C_6H_4OH$-p, etc.

$$R\overset{+}{N}Me_2\text{—}CH_2CO_2Me$$

(216)

$R = C_{10}H_{21}, C_{12}H_{25}, C_{14}H_{29}$

The reaction of hydroxide with methyl *N*-alkyl-*N,N*-dimethylglycinate (**216**) is catalysed by CTACl micelles and the data fit the pseudophase ion-exchange model, except when the surfactant counterion is HO^-. In this latter case, the kinetic data can be rationalized in a model in which the distribution of the HO^- between the aqueous and the micellar pseudophase depends upon $[HO^-]$.[190]

The imidazole-catalysed hydrolysis of enantiomeric pairs of three amino-acid esters (**217**)–(**219**) in reversed micelles, prepared from a combination of both racemic or chiral surfactants of type (**220**) and of both racemic or chiral (*S*)-2-octanol as co-surfactants, has been investigated.[191] The enantioselectivity observed is dependent on a number of factors. The results obtained indicate that the reaction takes place in a chiral micro-environment.

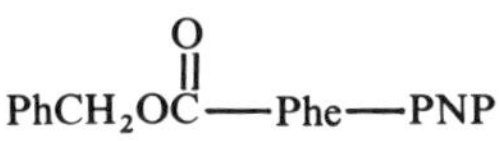

(217)

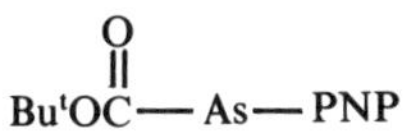

(218)

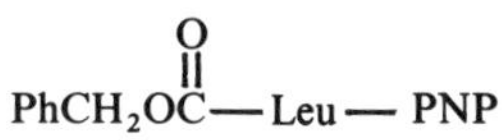

(219)

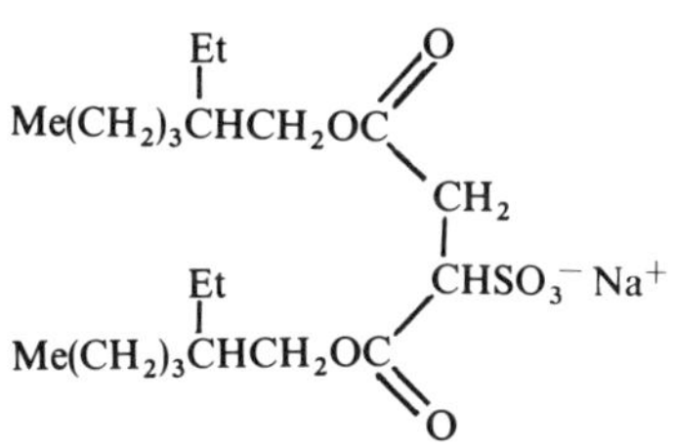

(220)

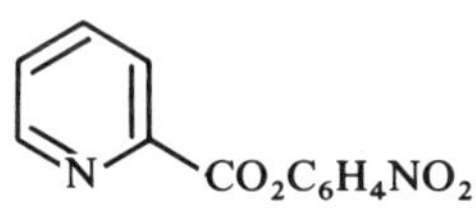

(221)

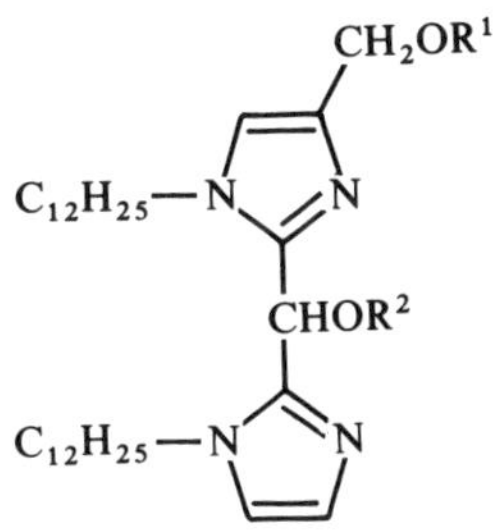

(222)

p-Nitrophenyl picolinate (**221**) has been hydrolysed by co-micelles of cetyltrimethylammonium bromide and lipophilic bis-imidazole ligands (**222**) in the presence of Zn^{2+} and Cu^{2+} at 25°C.[192] The results strongly suggest that the site-selective acylation of the ligand hydroxyl groups occurs via the formation of a tetrahedral Zn^{2+} ion and a square planar Cu^{2+} ion.

Typical micellar catalysis using perfluorooctanoic acid, a reactive counter-ion surfactant system, was observed for the hydroxamic acids (**223**; R = alkyl) but not for the two aryl-substituted acids (**223**; R = Ph, 4-BrC_6H_4).[193]

The reaction of HO^- with *N*-alkyl-2-bromopyridinium bromide (alkyl = *n*-$C_{12}H_{25}$, *n*-$C_{14}H_{29}$, *n*-$C_{16}H_{33}$) is catalysed by CTACl micelles. The pseudophase model accommodates the rate data but fails when CTAOH is used in place of CTACl.[194] The rate of reaction of HO^- with *p*-nitrophenyl diphenyl phosphate goes through minima with increasing concentrations of dodecyldimethylamine oxide, or *N*-hexadecyl-*N*,*N*-dimethyl-3-ammonio-1-propanesulphonate (SBS–16) (**224**).[195] The same phosphate and bis(*p*-nitrophenyl)phosphate are hydrolysed by 'metallomicelles' which bind these substrates very efficiently. Rate accelerations approach the 10^5-fold level and show turnover behaviour. The cupric ion complexes used were compounds (**225**).[196] *p*-Nitrophenyl diphenyl phosphate hydrolysis is effectively catalysed also by *o*-iodosobenzoate in an oil-in-water microemulsion stabilized by CTAB or CTACl and *n*-butanol.[197] The *o*-iodosobenzoate is shown to be a good turnover catalyst. A variety of cationic micelles (e.g. CBDAC, cetrimide C_{16}, benzalkonium chloride C_{14}, *N*-cetylpyridinium chloride, etc.) has been employed to catalyse the basic hydrolysis of diphenyl [1-(*N*-benzyloxycarbonylalanylamino-3-methyl]isobutyl phosphonate (**226**).[198]

There is a ten-fold retardation in the rate of the hydrolysis of acetylcholine (**227**) to choline (**228**) and acetic acid in the presence of the macrocyclic polyphenolate host (**229**).[199].

$RCH_2C(=O)NHOH$

(223)

n-$C_{16}H_{33}\overset{+}{N}Me_2(CH_2)_3SO_3^-$

(224)

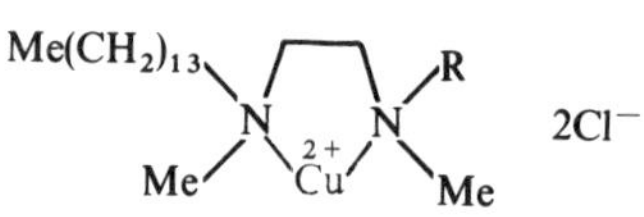

(225)

$PhCH_2OC(=O)NHCH(Me)C(=O)NHCHP(=O)(OC_6H_5)_2$

(226)

$Me_3\overset{+}{N}CH_2CH_2OCOMe\ Cl^-$

(227)

$Me_3\overset{+}{N}CH_2CH_2OH\ Cl^-$

(228)

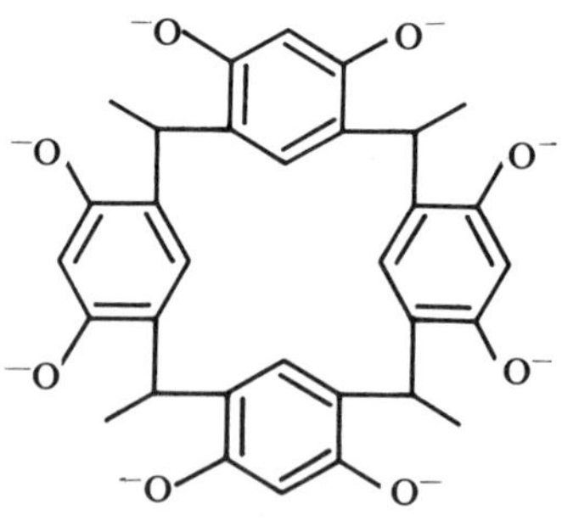

(229)

Some computer-modelling and experimental studies of micellar diastereoselectivity have been reported.[200] The substrates used were dipeptide *p*-nitrophenyl esters which were cleaved at pH 8 in buffer, non-functional CTA ion micelles, and in thiolate- and iodosobenzoate-functionalized micelles. The theoretical calculations involved combinations of molecular mechanics, MO calculations, and computer graphics methods to determine the preferred conformations and energies for the ground-state dipeptide substrates and for the derived tetrahedral intermediates.

Several papers have appeared on the use of vesicles.[201–203] Vesicles of the cationic non-phosphorolipid diester carbonate (**230**) have been prepared at pH 4; they support surface-specific exovesicular cleavage of *p*-nitrophenoxide from (**230**) by HO^- at pH 8.[201] It is of interest to note that vesicles from the analogous (**231**) fail to support surface-specific hydrolysis under similar conditions. The same group[202] has prepared the imidazole-derivatized phosphatidylcholine surfactant (**232**) and the reactivity of this compound is strongly increased in co-vesicles with non-functional cationic surfactants such as CTABr. The co-vesicular cleavage of *p*-nitrophenyl *n*-hexanoate was increased by a factor of 217 over native (**232**). These co-vesicles were also reactive towards other esters, 4-acetoxy-3-nitrobenzenesulphonate, and *p*-nitrophenyl acetate. Stereoselective hydrolysis of enantiomeric amino-acid esters (**233**) by the bilayer vesicular systems of histidine catalysts (**234**) and a double-chain surfactant, $[Me(CH_2)_{11}]_2NMe_2Br$, was carried out at 4–25°C in Tris buffer (pH 7.68) in 3% (v/v) $MeCN\text{-}H_2O$.[203]

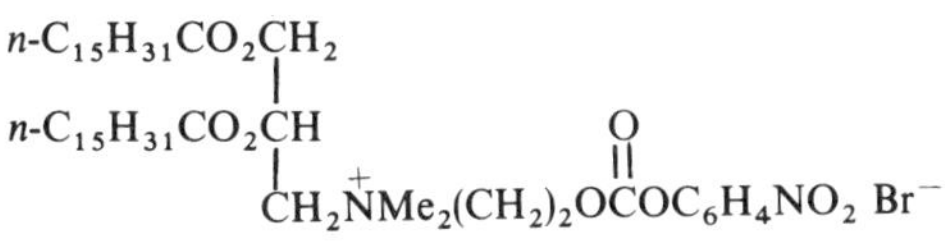

(**230**)

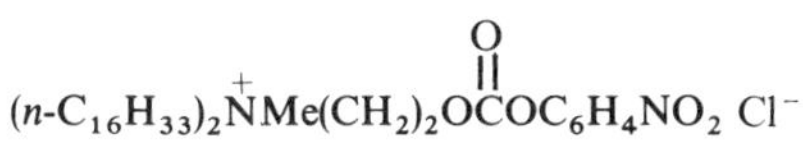

(**231**)

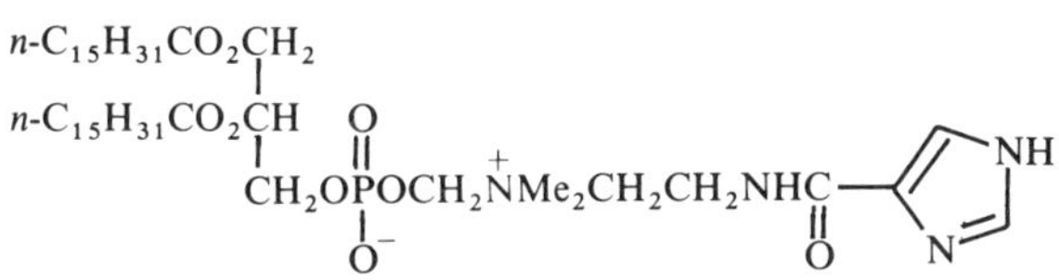

(**232**)

(**233**)

$R = PhCH_2, Me_2CHCH_2, Me; n = 0,8,12$

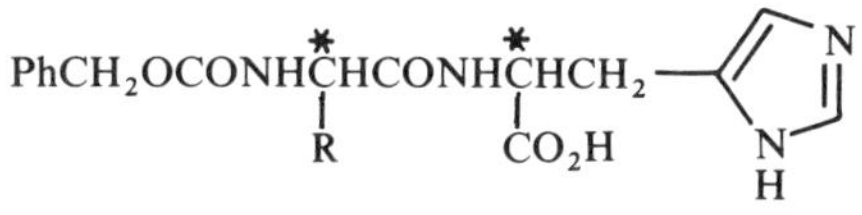

(**234**)

$R = PhCH_2, Me_2CHCH_2$

Hydrolysis of ATP, ADP, and pyrophosphate by macrocyclic polyamines, typically (**235**), has been studied;[204] these bind anionic substrates and perform reactions on the bound species. **235** ($X = (CH_2CH_2)_2O$, $Z = H$), a 24-membered macrocyclic, catalysed the hydrolysis of ATP at pH 3 and 7 with entropies of activation of -11 and -8.7 e.u., respectively. A large number of polyamines displayed catalytic ability. Such macrocycles are useful for probing the mechanisms of molecular catalysis in phosphate, and are models for the processes that may occur in enzymes utilizing ATP and related species as substrates.

Dimethyl β-cyclodextrin (**236**; R = H) inhibited the hydrolysis of *p*-nitrophenyl acetate whereas its 'imidazole' derivative (**236**; R = CH_2CH_2Im) accelerated the hydrolysis considerably.[205] The association constants of α-cyclodextrin with ionic 1-alkanesulphonate surfactants ($C_nH_{2n+1}SO_3Na$, SDS, SDBS, DAC, DTAC, DPC) have been determined conductometrically[206,207]. Italian workers have determined the kinetics of hydrolytic cleavage of the enantiomers of *p*- and *m*-nitrophenylacetate esters (**237**) and carbonates (**238**) in the presence of α- and β-cyclodextrins.[208] Rates of intra-complex cleavage were larger for the *R*- than for the *S*-enantiomers in most cases. The enantioselectivity factors vary from unity to ca. 19. The same group has achieved good rate enhancements and only slightly smaller enantioselectivities using some modified β-cyclodextrins, shown schematically in (**239**)–(**241**);[209] see also ref. 224.

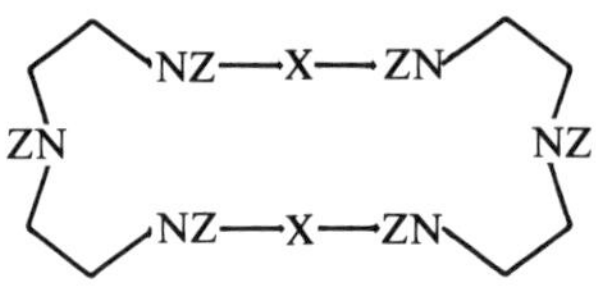

(**235**)

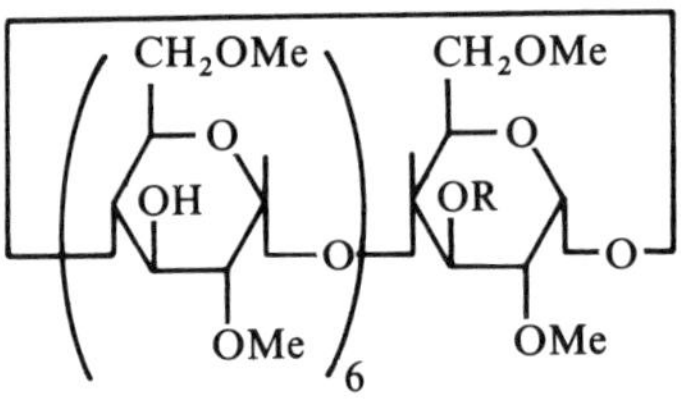

(**236**)

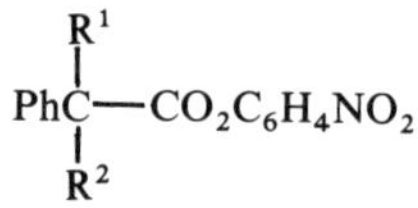

(**237**)

(**238**)

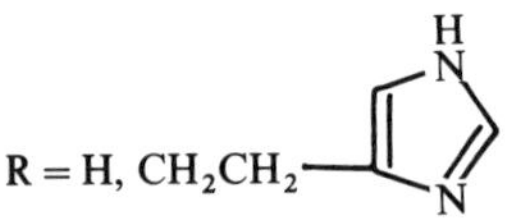

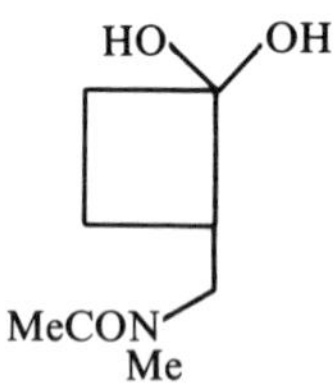

(**239**)

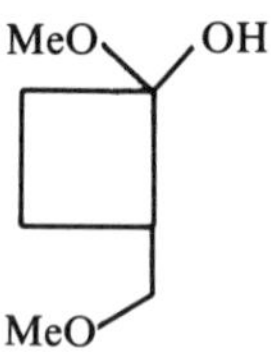

(**240**)

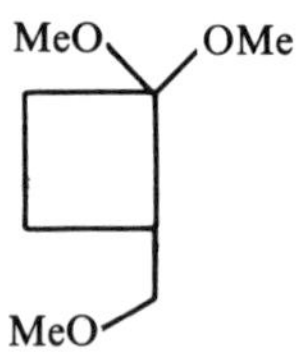

(**241**)

Metal-ion Catalysis

The imidazolate-Cu(II) complex (**50**) cleaves *p*-nitrophenylbenzoate.[45] The Cu(II) complexes (**225**) catalyse the cleavage of several phosphates.[196] The Cu(I)-catalysed decarboxylation of malonic acids (and hemi-esters) (**242**) has been achieved in ca. 96% yield; the decarboxylation can also be used to prepare optically-active products (**243**).[210] Cu(II) and Zn(II) play a vital role in the co-micellar catalysis of (**221**) by the bis-imidazole ligands (**222**).[192] A likely mechanism for the Zn(II)-catalysed hydrolysis of 2-acetylpyridine oxime phenylsulphite is illustrated (**244**). The increase in the leaving ability of the oxime anion is seen as the sole catalytic factor.[211]

The Co(III)-promoted hydrolysis of amino-acid esters and peptides and the synthesis of small peptides has been reviewed recently.[212] The hydrolysis of amides in a carboxypeptidase model containing Co(III) and phenolic and carboxylic functional groups has been reported.[213] The rate of amide hydrolysis was enhanced by two orders of magnitude above that due to the metal alone. No intramolecular catalysis was observed. A bifunctional mechanism for tetrahedral intermediate breakdown by phenol and acid (from the buffer) is proposed and it is suggested that a similar mechanism operates in the enzyme-catalysed hydrolysis of peptides. Co(II) chloride catalyses the acylation of alcohols and amines with anhydrides in ca. 90% yield.[214] Co(III), as the complex (**245**), catalyses the hydrolysis of adenosine 3′, 5′-monophosphate by a factor of about one hundred million![215] $ZnCl_2$ and $[Co(en)_2(NH_3)(OH)]Cl_2$ have no observable effect on the hydrolysis. The rate-determining step is the breakdown of (**246**).

$FeCl_3{\cdot}6H_2O$ catalyses the esterification of acetic acids by EtOH[216] and the esterification of phthalic anhydride with 2-ethylhexanol is catalysed by tetra-*n*-butyl-titanate and -zirconate.[84] Phenyltellurinyl trifluoroacetate (**247**) with ethyl carbamate and BF_3-Et_2O reacts regio- and stereo-selectively with alkenes in 1,2-dichloroethane via the carbamate intermediate (**248**) to give high yields of oxazolidin-2-ones (**249**).[217] The palladium complex (**250**) catalyses both the reaction

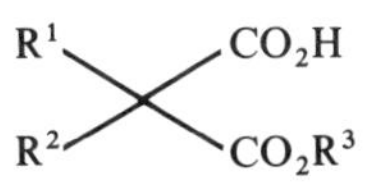

(**242**)

(**243**)

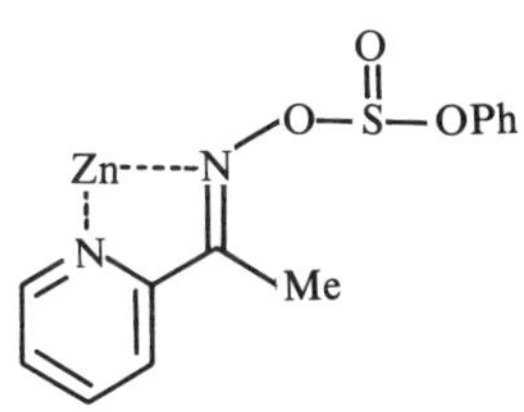

(**244**)

$Co(trien)(OH)(OH_2)Cl_2$

(**245**)

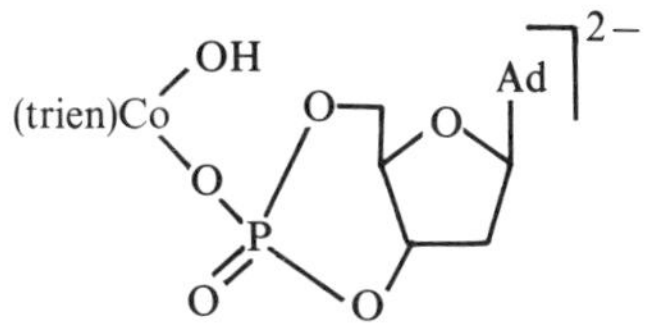

(**246**)

(**247**)

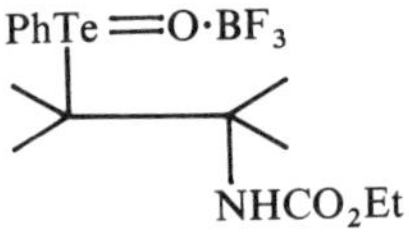

(248)

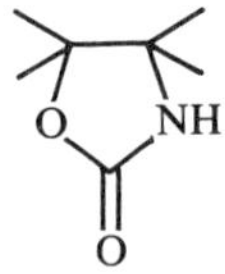

(249)

$(Ph_3P)_2Pd(Br)Ph$

(250)

of alkylphenylketene $Ph(RCH_2)C{=}C{=}O$ with benzoyl chlorides and the decarbonylative cross-condensation reaction of alkylphenylacetyl chlorides and benzoyl chlorides, giving in each case α,β-unsaturated ketones in high yields.[218]

(Dichloromethyl)lithium, is involved in the chiral synthesis of amino-acids (92–98% ee) from boronic esters.[219] The Ni(II)-catalysed hydrolysis of *O*-acetyl-2-pyridinecarboxaldoxime and related *O*-acetyl-2-pyridineketoxime and 6-carboxy-2-pyridinecarboxaldoxime has been studied; k_{cat} parameters have been calculated for one-metal-ion and two-metal-ion reaction pathways.[220] Comparisons have been made with previous findings.[221] Cu(II)-, Ni(II)-, Zn(II)-, Co(II)-, and Na-catalysed reactions of acrylonitrile, acrylamide, and ethyl acrylate have been reported.[222]

Decarboxylation

Refined *ab initio* calculations on the dehydration and decarboxylation reactions of formic acid in the gas phase have been made.[165] *Ab initio* methods using the STO-3G basis set and fully optimized geometries have been used to probe the decarboxylation of salicylic and *p*-aminobenzoic acids in aqueous solution.[223] In the proposed mechanism, there is a pseudo-unimolecular concerted decomposition of the free acid by interaction with a chain of H_2O molecules and the α-protonated form of the carboxylate anion represents the activated complex.

The decarboxylation of a series of 4-phenylallophanate anions, $XC_6H_4NHCONHCO_2^-$, in H_2O at 25°C has shown a changeover in mechanism, revealed from Hammett and Brønsted correlations, entropies, and solvent isotope effects.[117]

The effect of CTAB micelles and mixed DMSO and dioxane solvents in the presence of HCl on the decarboxylation of the *N*-alkyl-*N*-aryl-substituted 5-amino-1,3,4-oxadiazole-2-carboxylic acids (**251**) has been studied.[224,225] A unimolecular decarboxylation mechanism is favoured. The rates of decarboxylation and the dissocation constants of the methyl esters have also been investigated.[225] The

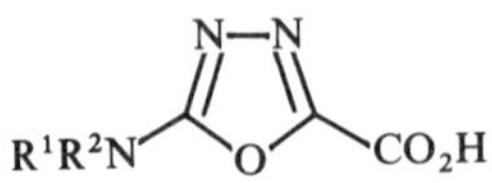

(251)

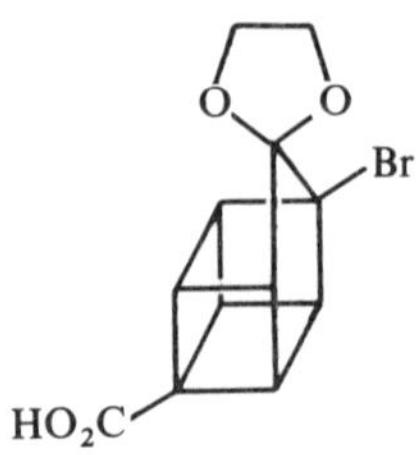

(252)

Me_3SiO_2CH

(253)

hypervalent iodine oxidative decarboxylation of homocubane (**252**) and cubane mono- and di-carboxylic acids has been achieved in 80–90% yields with $PhI(OAc)_2$-CCl_4-I_2 on irradiation with a 125W tungsten-filament lamp.[226] Trimethylsilyl formate (**253**) disproportionates to give methyl formate, carbon dioxide, and hexamethyldisiloxane, Me_6Si_2O. The process is reversible and has an equilibrium constant of 8.9×10^6 atm^2 M^{-1} at 250°C in dioxane. Methanol is formed in a further slower decarbonylation of methyl formate.[227]

The catalysis by thiamin diphosphate-dependent enzymes of the decarboxylation of pyruvate has been reviewed;[228] see phosphorus-containing acids section. The stereochemical course of the non-enzymatic decarboxylation of the adduct of pyruvate and thiamin has been determined[229] (see later).

Enzymic Catalysis

General

Several reviews of varying importance to this area have appeared.[28,230–235] Scott[230] has discussed his own and other work in a review (70 references) entitled NMR Studies of Biosynthesis and Enzyme Mechanism. Among the topics covered are aspects of the biosynthesis of vitamin B_{12} (timing of the methylation steps between urog'en III and cobyrinic acid), direct observation of enzyme-substrate complexes by ^{3}H NMR, use of ^{13}C NMR to study enzyme mechanisms, cryoenzymology and solid-state NMR as a probe for enzyme-substrate structure. Bender and D'Souza[231] have written a useful account of miniature organic models of enzymes—much of the account focuses on chymotrypsin. Molecular recognition by energy dissipation, a new enzymatic principle, has been reviewed (50 references).[232] The design of biospecific compounds which stimulate enzyme-substrate interaction has been reviewed, covering the following areas: classification of enzyme-specific compounds, design of binding site-interacting substances, design of specific substances involving both binding and catalytic site interactions, and the design of 'inverse' trypsin substrates.[233] The use of isotope effects in the detailed analysis of the catalytic mechanisms of enzymes has been reviewed (70 references) by Cleland.[234] Recent developments in the field of biologically active peptides have been reviewed.[235] A review of stereoelectronic effects in biomolecules looks at stereoelectronic effects in enzymatic reactions at some length (100 references).[28]

Candida cylindracea lipase-catalysed enantiospecific esterifications in H_2O-organic solvent mixtures using *dl*-menthol and *dl*-2-(*p*-chlorophenoxy)propionic acids (**254**) have been studied and some new quantitative expressions have been developed[236] which relate the kinetic and thermodynamic parameters that govern stereospecificity of enzyme-catalysed resolution of enantiomers in biphasic systems.

Klibanov's group has illustrated the furter use of enzymes in dry organic solvents to achieve some useful results.[237,238] Four commercially available lipases can catalyse trans-esterification reactions between 2,2,2-trichloroethyl butyrate and the glucose, galactose, and mannose, all blocked at the C(6) hydroxy position.[237] The lipases exhibit a remarkable regioselectivity by discriminating among the four available secondary hydroxyl groups in the C(6)-protected monosaccharides. Lipases also catalyse stereoselectively polycondensation reactions between racemic diesters and a chiral diols (or vice versa).[238]

Breslow's observation some years ago[239] that there is a 3.3×10^5-fold acceleration in hydrolysis rate for *p*-nitrophenyl 3-ferrocenylacrylate bound to *β*-cyclodextrin prompted Menger to probe the reason(s) for the slow hydrolysis ($t_{1/2}$ ca. 2 years!) of the ethyl ester under identical conditions.[240] The answer lies not in differences in degree of binding (all seven esters examined > 6% bound) nor that the complexes possess substantially different structures, but rather that the '*p*-nitrophenyl' tetrahedral intermediate partitions itself almost exclusively towards products whereas the 'ethyl' intermediate partitions towards reactants.

Non-enzymatic rates of hydrolytic deamination of adenosine and cytidine by acids and bases analogous to side-chains of naturally occurring amino-acids are compared with the rates of uncatalysed deamination in H_2O and with the rates of HO^- and H^+-catalysed reactions.[241] The catalysts used with both included $HS(CH_2)_2OH$, imidazole, dimethylamine, phosphate dianion, and (with adenosine only) acetate and selenide.

It has been possible to quantify the contributions of steric, electrical, and polarizability effects to the enantiomeric excesses of alcohols formed in the hydrolysis of esters of the general type ArCH(X)OAc by the mould *Rhizopus nigricans*.[242] From the equations set up it should be possible to predict the optical purity and the absolute stereochemistry of an alcohol prepared using this microorganism.

Serine Proteinases

Six papers, including a review, have appeared on acetylcholinesterase.[243–248] Quinn,[243] in his review (140 references), considers acetylcholinesterase (AChE) structure, reaction dynamics, and the virtual transition states for anilide and ester hydrolyses; looking ahead, he considers that the cloning of the AChE gene and determination of the X-ray structure of the enzyme are not too distant. In three papers in series, the latest results from this group on AChE-catalysed hydrolyses are presented.[244–246] The AChE-catalysed hydrolyses of the esters (**255**) and (**256**) involve a virtual transition state that is a weighted average of acylation and deacylation transition states that are each stabilized by single-proton transfers.[244] The second paper deals with the hydrolysis of the substrates *p*-methoxyphenyl formate and the anilides, *o*-nitrochloroacetanilide, *o*-nitroacetatanilide, and *o*-nitroformanilide. The slow stage of these enzyme-catalysed hydrolyses is an induced-fit conformation change that precedes chemical catalysis.[245] The same three anilides are considered further in the final paper.[246]

Electric eel AChE-catalysed hydrolysis of 3,7-diacetoxycycloheptene (**257**) gives optically pure (3*R*,7*S*)-3-hydroxy-7-acetoxycycloheptene but *Candida cylindracea* lipase-catalysed hydrolysis of (**257**) gives the same hydroxy-acetate but in only 44% ee.[247]

Alkali metal cations or NH_4^+ bind to the anionic site of AChE and reduce the hydration of the nucleophilic serine-OH in the esteratic site thus enhancing the nucleophilicity the OH group. This is the conclusion of a study of the role of H_2O in the reaction between AChE and methanesulphonyl fluoride.[248]

p-Amidinophenols (**258**) are 'inverse', that is to say, the site-specific group for the enzyme (trypsin), the amidinium ion, is included not in the acyl moiety but in the leaving-group. This topic of inverse substrates for trypsin and trypsin-like enzymes has recently been reviewed[249] and the same workers have used the method of 'inverse' substrates to carry out a topographical analysis of the trypsin active site.[250] The techniques involved the use of 'inverse' substrates carrying the fluorescent dimethylaminonaphthalene group; e.g. R in (**258**) is given by (**259**). The fluorescent spectra of the acyl-trypsins subsequently prepared gave information on the trypsin active site.

Four papers concerned with α-chymotrypsin have appeared.[251–254] The kinetics of the esterification of *N*-acetyl-L-tyrosine with EtOH catalysed by a hydrophobic derivative of α-chymotrypsin have been reported.[251] The energetics of the α-chymotrypsin-catalysed hydrolysis of *o*-hydroxytoluene-α-sulphonic acid sultone (**260**) have been reported.[252] The results suggest that there is a facile reversion of the sulphonyl enzyme species to the Michaelis complex and that this takes place in preference to hydrolysis of the covalent intermediate. *N*-Nitroso-propane and -butane sultams (**261**) are potent inhibitors of α-chymotrypsin.[253] A model for the acylation step of serine proteases, such as chymotrypsin, has been examined.[254] This is the reaction of β-amino-alcohols such as 2-(hydroxymethyl)imidazole and *N*,*N*-dimethylethanolamine with the strained amide, 2,3,4,5-tetrahydro-2-oxo-1,5-ethanobenzazepine (**262**).

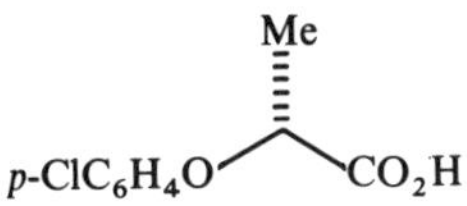

(**254**)

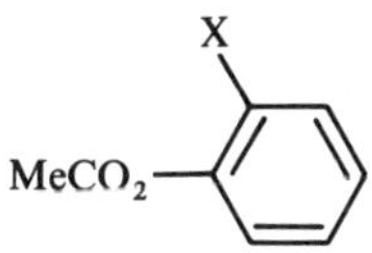

(**255**)

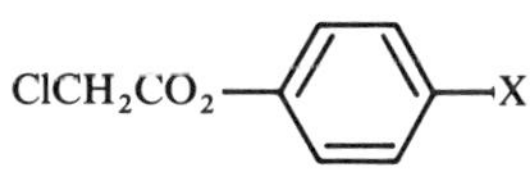

(**256**)
X = H, MeO

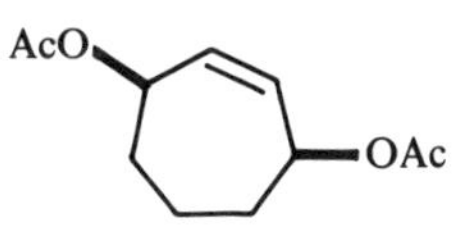

(**257**)

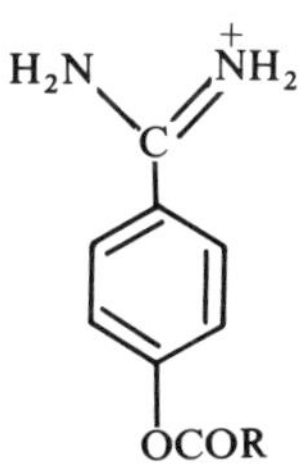

(**258**)

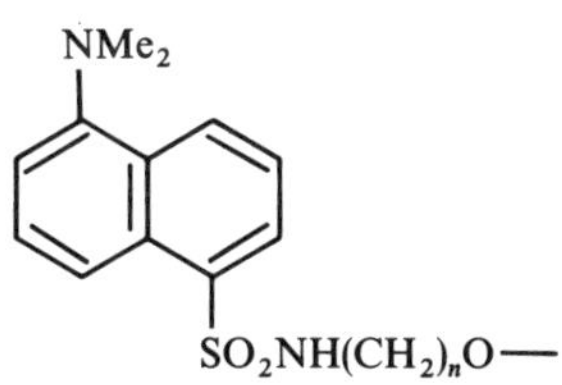

(**259**)
$n = 3$–6

Carboxypeptidase has received some attention from a Japanese group during the period covered by this review. At 150 MPa the yield of peptide was five-fold greater than at 0.1 MPa in the reaction of an *N*-acylamino-acid ester and an amino-acid amide using carboxypeptidase Y as a catalyst.[225] The activities of two carboxypeptidases, one from yeast and the other from wheat bran, on five endodipeptidase substrates have been explored.[256] In the accompanying paper,[257] further work on the reaction catalysed by the wheat bran carboxypeptidase is reported; see also under metallo-proteinases.

A pH-dependent change in the slow step of the human leukocyte elestase-catalysed hydrolysis of *p*-nitroanilide substrates has been found.[258] The change is from acylation at pH < 5.5 to deacylation at pH > 6.0.

Steady-state kinetic parameters have been obtained for the serine β-lactamase-catalysed hydrolysis of a series acyclic depsipeptides (i.e. ester analogues of acyl-D-alanyl-D-alanine peptides) (**263**).[259] Kinetic parameters for the action of bovine Factor Xa (E.C. 3.4.21.22) on 25 commercially available peptide *p*-nitroanilides have been determined.[260]

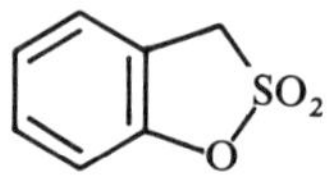

(**260**)

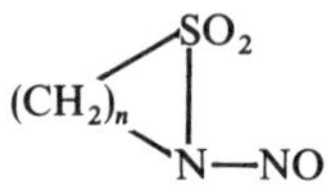

(**261**)
$n = 3,4$

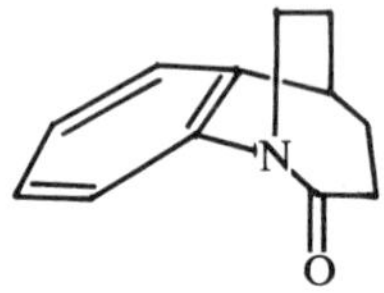

(**262**)

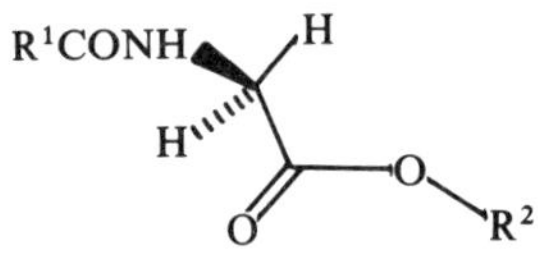

(**263**)

$R^1 = PhCH_2$; $R^2 =$ *o*-, *m*-, and *p*-carboxyphenyl; $OR^2 =$ D-lactate, D-mandelate

Metallo-proteinases

Carboxypeptidase A appears to react by the general-base mechanism for peptide hydrolysis but may react by a (anhydride) nucleophilic mechanism for some ester substrates.[261]

For the three substrates aspartic, 3-methyl-, and 3-ethylaspartic acids (**264**), C(3)—H bond cleavage is respectively, not rate-limiting, partially rate-limiting, and much less rate-limiting in the 3-methylaspartase-catalysed elimination of ammonia from these substrates.[262]

Conjugate addition of acyl groups to C=C by a vitamin B_{12}-catalysed electrochemical reduction of acetic anhydride in the presence of Michael olefins under visible light is shown to involve sequential formation and cleavage of the Co—C bond in acetylcobalamine.[263] The hydrolysis of amide in a carboxypeptidase A model using Co(III) and bifunctional catalysts has been mentioned earlier.[213]

Thiol and Acid Proteinases

Papain (E.C. 3.4.22.2)-catalysed hydrolysis kinetics of *N*-α-carbobenzoxyglycine *p*-nitrophenyl ester have been studied in the pH range 3.0–9.5 at 21°C.[264] The kinetics are pH-dependent and a minimum three-step mechanism involving the acyl-enzyme intermediate (E.P) has been proposed:

$$E + S \rightleftarrows E.S \overset{k_2}{\rightleftarrows} E.P \overset{k_3}{\rightleftarrows} E + P_2$$
$$+$$
$$P_i$$

The pH dependence of kinetic parameters indicates that at pH $\leqslant 3.5$, k_2 is rate-limiting, and at pH $\geqslant 3.5$, k_3 becomes rate-limiting. The results reported may be of general significance for cysteine proteinase-catalysed reactions.

One-step and two-step activation pathways have been recognized in the reactions of monkey and porcine pepsinogens A.[265] In both pepsinogens A, intramolecular reaction occurs first yielding pepsin A or the intermediate form, which then acts intermolecularly on the remaining pepsinogen or the intermediate. Hydrolysis of the distorted amide (**262**) by certain dicarboxylic acids may provide an accurate model for the enzymatic pathway for the action of aspartate proteinases.[266]

Other Enzymes

Porcine pancreatic lipase catalyses hydrolysis of racemic dimethyl succinates, aspartates, and glutamates giving (*R*)- and (*S*)-methyl esters enantioselectively.[267] The same enzyme is used efficiently to resolve (±)-sulcatol, an important pheromone.[268] Pig pancreatic lipase catalyses ester interchange of vinyl esters very efficiently (Scheme 23).[269]

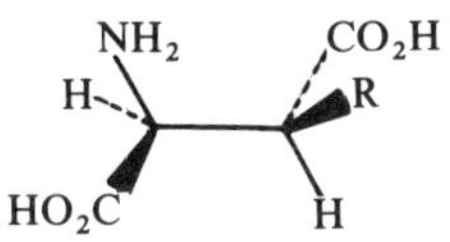

(**264**)

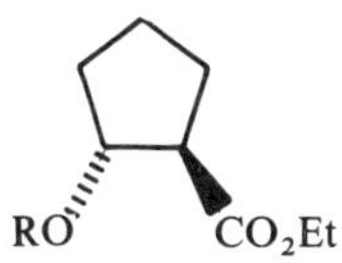

(**265**)

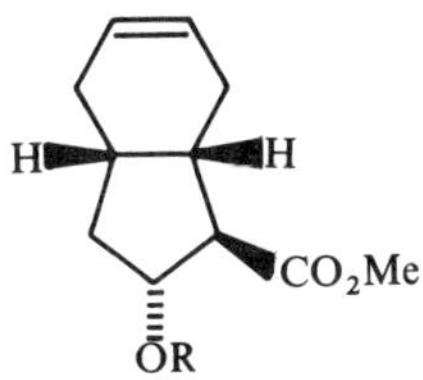

(**266**)

$$RCO_2CH{=}CH_2 \quad + \quad R'OH \longrightarrow RCO_2R' \quad + \quad [CH_2{=}CHOH]$$

SCHEME 23

Pseudomonas fluorescens lipase catalyses the enantioselective hydrolysis of five-membered racemic ring acetates such as (**265**) and (**266**) to the corresponding optically active alcohols.[270] *Candida cylindracea* lipase catalyses enantioselectively the esterification of racemic primary allenic alcohols such as (**267**), (**268**), and (**269**) with lauric acid in *n*-hexane.[271]

p-Nitrophenyl- (**270**) and cholesteryl-*N*-alkyl (**271**) carbamates are good inhibitors of porcine pancreatic cholesterol esterase-catalysed hydrolysis of *p*-nitrophenyl butyrate.[272] The racemization of phenylglycine, catalysed by pyridoxol and phosphate ion, has been investigated.[273] The flavoprotein *p*-hydroxybenzoate hydroxylase from *Pseudomonas fluorescens* was modified by several arginine specific reagents.[274] Modification by 2,3-butanedione led to loss of activity of the enzyme and modification by 4-hydroxy-3-nitrophenylglyoxal reduced the affinity of the enzyme for the substrate and NADPH.

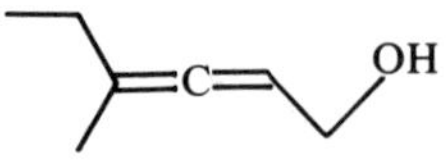

(267)

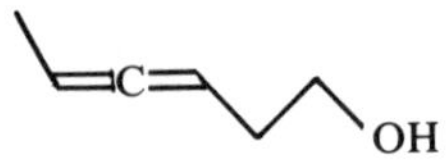

(268)

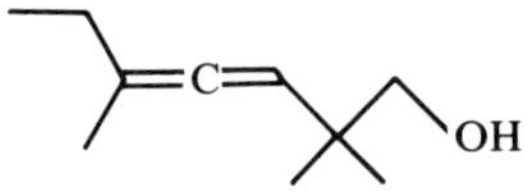

(269)

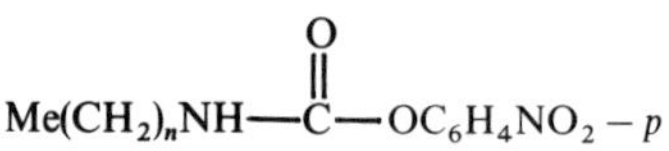

(270)
$n = 3,7$

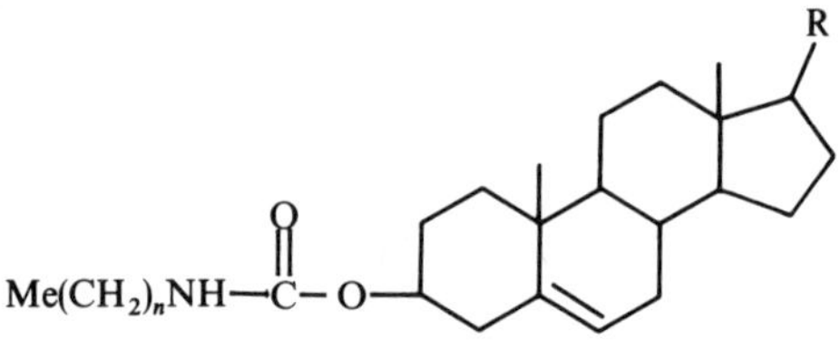

(271)
$n = 3,7$

Evidence for a branched pathway in cephalosporin biosynthesis comes from observation of a deuterium isotope effect when [3-^{2}H] penicillin N was incubated with preparations of deacetoxycephalosporin C/deacetylcephalosporin C synthetase.[275] Further experiments using the same preparation showed that there is a stepwise process for hydrogen removed during the ring-expansion of penicillin N to deacetoxycephalosporin C.[276] Evidence from experiments using the enzyme isopenicillin N synthetase suggests that a carbon-centred free radical or an equivalent iron-carbon-bonded intermediate is involved in C—S bond formation in the biosynthesis of penicillin.[277]

Pig liver esterase catalysed hydrolysis of *cis*-*N*-benzyl-2,5-bis(methoxycarbonyl)pyrrolidine in an aqueous medium and gave optically pure mono-ester which could be converted to the (*R*)-proline-related ester (**272**) by radical decarboxylation.[278] The enzymes in Baker's yeast catalyse the enantioselective hydrolysis of *N*-acetylamino-acid ester racemates.[279] Baker's yeast has again been employed to achieve the dual kinetic resolution of a keto-ester.[280]

Alkaline protease 'Alcolase' selectively hydrolyses at the α-carboxy position the tribenzyl ester-protected aspartyl and glutamyl dipeptides (**273**; R = Bz) to give the α-carboxy-free dipeptides (**273**; R = OH).[281]

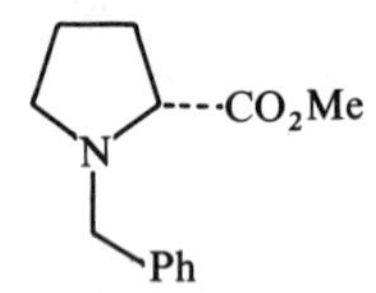

(272)

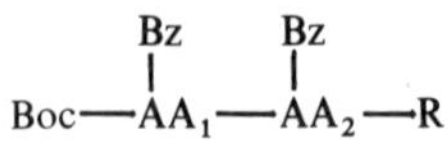

(273)

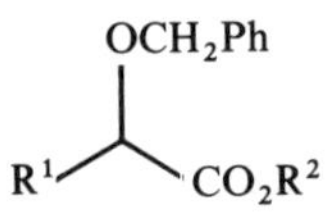

(274)

Asymmetric hydrolysis with the bacterium, Corynebacterium equi IFO 3730, of *dl*-α-benzyloxyalkane and arylalkanecarboxylic esters (**274**) gave chiral esters of high optical purity.[282]

NON-CARBOXYLIC ACIDS

Phosphorus-containing Acids

Non-enzymic Reactions

Theoretical calculations are reported in three papers.[283-285] Last year the facile O $\rightarrow$ N migration of an acyl group in a mixed carboxylic-phosphoric anhydride (**275**) was reported and it was noted that little charge developed in the transition state.[286] In a follow-up paper by the same group, a MNDO SCF-MO study supports the hypothesis that the reaction is concerted and that dipolar intermediates are not involved.[283]

An *ab initio* study of the diadic prototropic tautomerism: $H_3PX \rightleftarrows H_2PXH$ (X $= O, NH, CH_2$) has been made and assessed.[284] *Ab initio* electronic-structure calculations on the following molecules/anions $(HO)_3PS$, H_3PO_4, $H_2PO_3S^-$, $H_2PO_4^-$, HPO_3S^{2-}, HPO_4^{2-}, PO_3S^{3-} and PO_4^{3-} have been carried out using 3-21G and 3-21+G bases and a Gaussian 82 program.[285] Computed and experimentally determined bond lengths are compared. There is appreciable double-bond character in P—O bonds but an absence of this in P—S bonds; hence the title of the paper 'Sulfur Does Not Form Double Bonds in Phosphorothioate Anions'.

Hammett ρ-values for the first and second dissociations of arylphosphonic acids $(ArPO_3H_2)$ in water at 25°C are 0.923 and 1.140, respectively;[287] PO_3H^- appears to be electron-donating. The pK values of the *meta*- and *para*-substituted compounds have been correlated using σ^n, indicating little conjugation between the phosphonic group and the aromatic ring. These acids and their dianions $(ArPO_3^{2-})$ catalyse the enolization of acetone and a good Brønsted correlation ($\beta = 0.72$) has been found for catalysis by the latter.[288] For the acids a poorer plot ($\alpha = 0.37$) is obtained.

The design of micelles as dephosphorylating agents has been reviewed.[183] Zwitterionic micellar dephosphorylation of *p*-nitrophenyl phosphate using surfactants such as (**224**) has been mentioned.[195] The first report of an α-dephosphonylation mediated by pyridoxal occurs in the reaction of the latter with α-aminophosphonic acids which undergo a pyridoxal-promoted cleavage of the P—C bond in the presence of Cu(II).[289] The hydrolysis of the diphenylphosphonate (**226**) by cationic micelles has been mentioned[198] earlier in this review. Also, the synthesis of the imidazole-derivatized phosphatidylcholine (**232**) and the remarkable kinetic behaviour of vesicles made from this new surfactant have been referred to.[202] Hydrolysis of bismetallic complexes of linalyldiphosphate (**276**) with Mg^{2+} and Mn^{2+} has been studied.[290] The rate enhancements observed are proportional to the concentrations of the bimetallic complexes. This work has implications for cyclic terpene hydrocarbon biosynthesis catalysed by carbocyclase from *Citrus* limonum.

The hydrolysis of ATP by the macrocyclic polyamine (**235**) and related species was discussed earlier.[204]

Phosphate ester hydrolysis by the metallo-micelles (**225**), which bind substrates extremely well ($K > 10^5$M), give rate acceleration of 10^5 with turnover behaviour.[196] Hydrolysis of 4-nitrophenyl diphenyl phosphate in a CTAB micro-emulsion in the presence of *o*-iodosobenzoate has been discussed.[197] Interestingly, 2-iodoxybenzoic acids (*o*-$O_2IC_6H_4CO_2H$) are about 60–110% as efficient as their *o*-iodoso analogues in catalysing the hydrolysis of *p*-nitrophenyl diphenyl phosphate and *p*-nitrophenyl isopropyl phenylphosphinate.[291]

A useful review by Gorenstein on the stereoelectronic effect in bimolecules has a section on the operation of this effect in the hydrolysis of phosphate esters.[28] In this, the author has discussed earlier work[292] and his latest results[293,294] particularly in regard to a controversy that has arisen. Epimeric cyclic 6-membered-ring phosphonate phenyl esters, *cis*- and *trans*-2-phenoxy-2-oxo-*trans*-4,5-tetramethylene-1,2-oxaphosphorinane (**277**) have been synthesized and shown to exist in chair conformations for the phosphorinane ring.[295] Epimerization gave an equilibrium ratio of 89.5:10.5 (*trans*:*cis*) and methanolysis (NaOMe) proceeded with 100% inversion for both esters, consistent with placing the 6-membered ring in a diequatorial position of a trigonal bipyramidal pentacovalent intermediate.

Several groups[296–299] focus attention on the involvement of free monomeric metaphosphate and thiometaphosphate. The pH dependence of the reaction of anions of acetylphosphate and orthophosphate giving pyrophosphate in the presence of 0.25M orthophosphate monoanion and 6.4M $NaClO_4$ at 54°C gives different yields of pyrophosphate from acetylphosphate mono- and di-anions. This result is not consistent with trapping of free intermediate metaphosphate by orthophosphate and H_2O but, instead, points towards a concerted mechanism.[296] Evidence for the intermediacy of a symmetrically solvent metaphosphate in the phospho group-transfer to a hindered nucleophile in a protic solvent (*t*-butyl alcohol, 30°C) has been obtained from the determination of the stereochemical course of the solvolysis of the bis(tetra-*n*-butylammonium) salt of *p*-nitrophenyl(R_p)-[^{16}O, ^{17}O, ^{18}O]phosphate (0.25M) (**278**) to give product (**279**)[297] (Scheme 24). Analysis of (**278**) and (**279**) after about one half-life showed that the phospho group had suffered complete racemization. Good evidence has also emerged for a monomeric thiometaphosphate intermediate, PO_2S^-.[298,299] Cullis reports that the monoanion of R_p-4-nitrophenyl[^{18}O]thiophosphate (**280**) undergoes ethanolysis

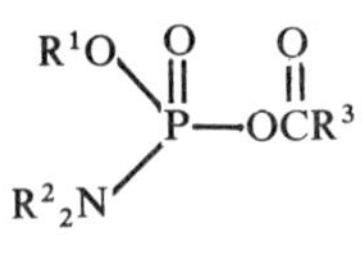

(**275**)

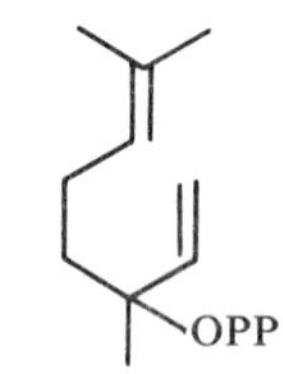

(**276**)

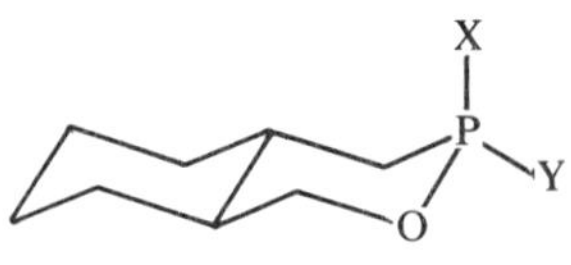

(**277**)

X = OR, Y = O;
X = O, Y = OR

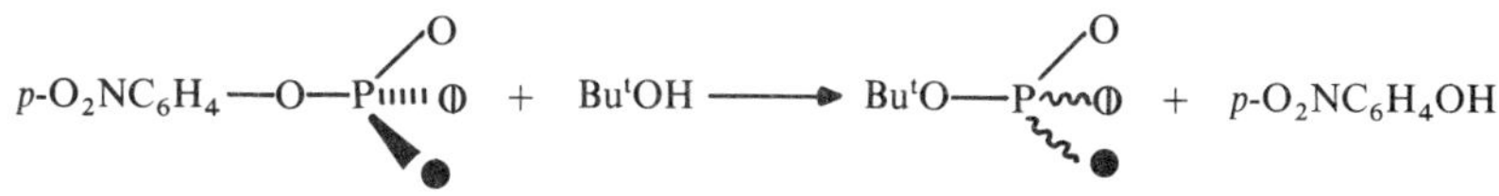
(278) (279)

SCHEME 24

with about 80% racemization and the dianion with complete racemization.[298] Lowe[299] finds further support for this intermediate in his studies of the hydrolysis of (R_p)- and (S_p)-deoxyadenosine-5′-[β-^{17}O]-β-thiodiphosphates (**281**).

p-Methoxyphenacyl esters of substituted phosphates (**282**), readily prepared from α-diazo-*p*-methoxyacetophenone, are photosensitive protecting groups which could be useful in biologically important phosphate synthesis.[300] The phosphate-phosphonate rearrangement has been studied with the diethylphosphates (**283**) rearranging to give the 1-hydroxyphosphonates (**284**) with retention of configuration at the carbon.[301] Reaction of diethyl benzoylphosphonate (**285**) with diethyl (methoxycarbonylmethyl)phosphonate (**286**) gives diethyl 2-methoxycarbonyl-1-styrylphosphonate (**287**) as a mixture of geometrical isomers. This is a new synthetic approach to vinylphosphonates.[302] Oxygen isotopes and high field ^{31}P NMR have been used to establish that dioxadiphosphetane intermediates are not involved in reaction of activated phorphoryl compounds with phosphoryl anions.[303] Some displacements in *cis*-2-chloro-3,4-dimethyl-5-phenyl-1,3,2-oxazaphospholidin-2-

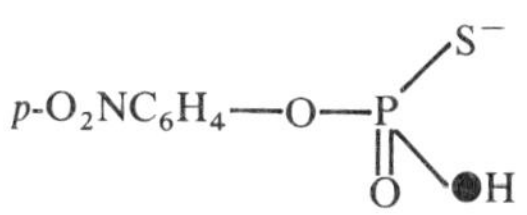
(280)

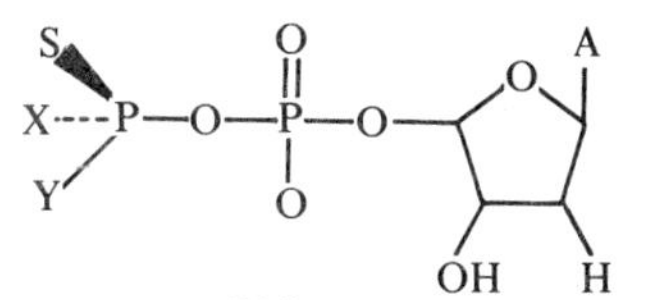
(281)

R_p: X = O, Y = O; S_p: X = Ⓞ, Y = O

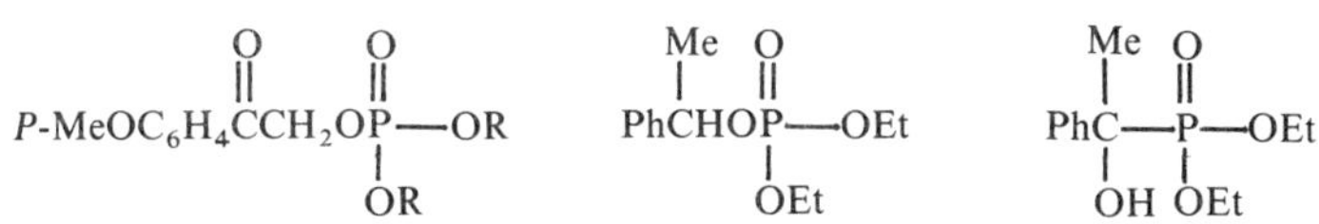
(282) (283) (284)

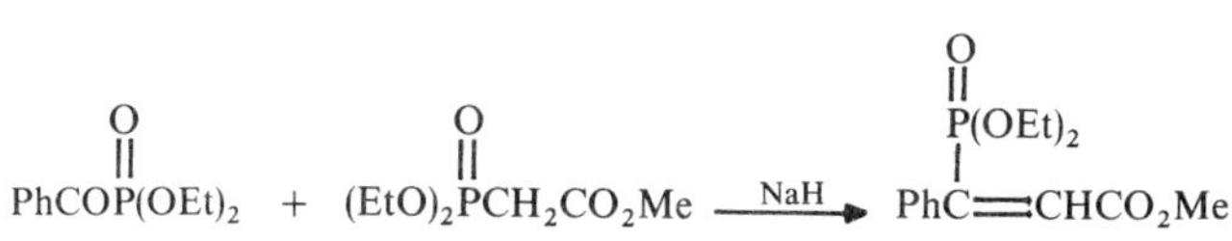
(285) (286) (287)

SCHEME 25

one (**288**; X = O; Y = Cl) and the corresponding -2-thione (**288**; X = S; Y = Cl) require an unexpected in-line exocyclic displacement at a phosphorus centre in a 5-membered ring.[304] Acid-catalysed fragmentation of monoalkyl α-oxyiminophosphonates (**289**) gives mixed phospho-diesters MeOP(=O)(OH)(OR) and nitriles in high yields.[305]

The synthesis of the cyclic (**290**), (**291**), and acyclic (**292**) phosphonylurea esters has been reported and their reactions with HO^- in MeCN probed.[306] Compounds (**290**; R = Ph, Me, Et) hydrolyse about 10^6 times more rapidly than do the exocyclic analogues (**291**). The reactivity patterns of these compounds in alkaline hydrolysis are interpreted in terms of ring strain and electronegativity of ligands. The acylation of phosphoramidates (**293**) under neutral conditions by acid chlorides and anhydrides has been studied to give carboxamides, by P—N bond-cleavage, and *N*-acylphosphoramidates.[307] The highly reactive phosphoramidates (**294**) have been prepared.[308] Both undergo acid-catalysed methanolysis very rapidly, which proceeds with 100% inversion at phosphorus. A completely free intermediate phosphacylium ion is unlikely.

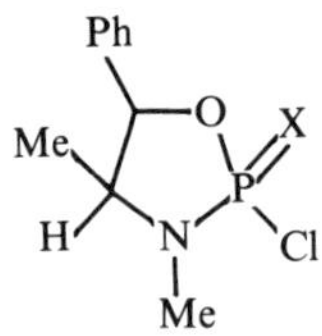

(**288**)
X = O, S

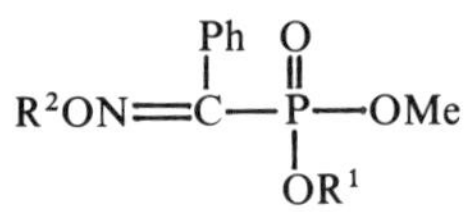

(**289**)

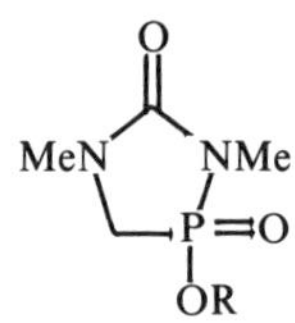

(**290**)

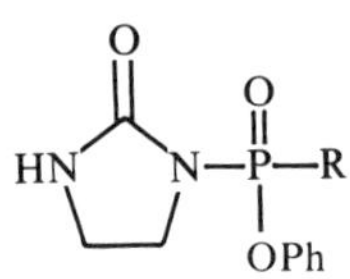

(**291**)

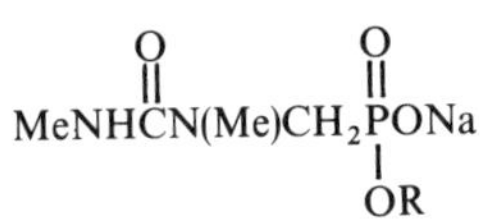

(**292**)

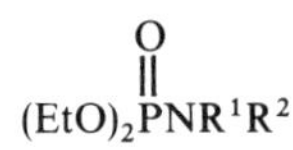

(**293**)

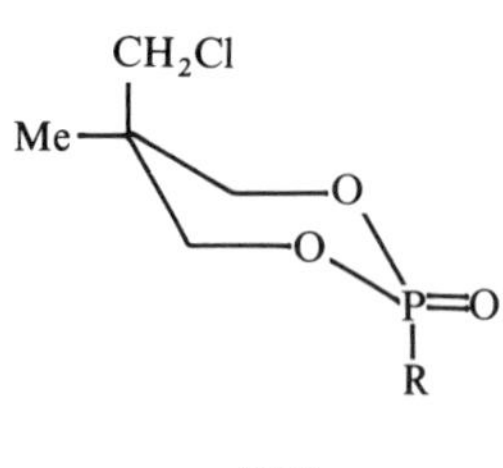

(**294**)
R = 1-imidazolyl, 1-benzimidazolyl

Enzymic Reactions

The use of isotope effects in the detailed analysis of catalytic mechanisms in enzymes has been reviewed.[234] Stereoelectronic effects in biomolecules including those in enzymatic reactions have been reviewed.[28] Thiamin diphosphate-enzymatic and non-enzymatic catalysis has been reviewed.[228]

The stereochemical course of the non-enzymatic decarboxylation of the adduct of pyruvate and thiamin has been studied. Reaction proceeds via an achiral intermediate or rapidly-equilibrating achirally-solvated enantiomeric intermediates. The mechanism of the enzymic reaction is also discussed.[229] The catalysis by the cobalt complex (**245**) of the hydrolysis of adenosine 3′,5′-monophosphate has been noted elsewhere in this review;[215] see Metal-ion Catalysis.

Ribonuclease TI specifically hydrolysed the phospha-diester linkages of guanosine 3′-phosphate of single-stranded RNA and the mechanism is two-step.[309] 5-Oxoprolinase catalyses a reaction in which the cleavage of ATP to ADP and P_i and the decyclization of 5-oxoproline to form glutamate are coupled.[310] Evidence is presented for enzyme-bound phosphorylated intermediates and a phosphorylated tetrahedral intermediate (**295**). Using $MgATP^{-2}$ and 3-P-glycerate as substrates, the effects of various anions on the kinetics of yeast phosphoglycerate kinase were studied. The anions were classified according to their effects on the reactions.[311] The kinetics of orotate phosphoribosyltransferase and orotidylate decarboxylase activities are found to fit a non-interacting sites model from computer-modelling analysis.[312]

Only (S_p)-adenosine 5′-*O*-phosphorothioate *O*-methyl ester (**296**) is a substrate for bovine intestinal mucosa 5′-nucleotide phosphodiesterase. The product adenosine 5′-*O*-phosphorothioate also had the S_p configuration. Thus there is retention of configuration at phosphorus presumably via a double-displacement mechanism. A nucleotidyl-enzyme intermediate is very likely on the reaction pathway.[313] The mechanism of catalysis of *E.coli* succinyl coenzyme A synthetase has been examined using positional oxygen isotope exchange. The catalysis involves phosphoryl transfer to a second site in a concerted fashion with transfer from the first.[314]

A mechanism involving a single general base and a binding group that helps orient the peptide for phosphoryl transfer is suggested from studies of the pH dependence of kinetic parameters and inhibition dissociation constants for the cyclic adenosine 3′,5-monophosphate-dependent protein kinase reaction.[315] Various isotope effects indicate that the transition-state complex of AMP nucleosidase is

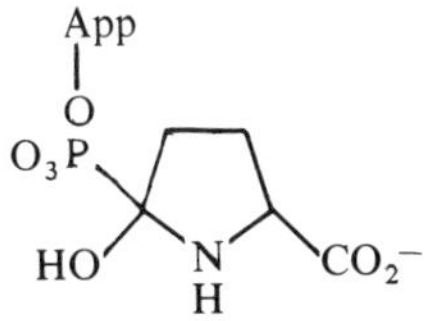

(**295**)

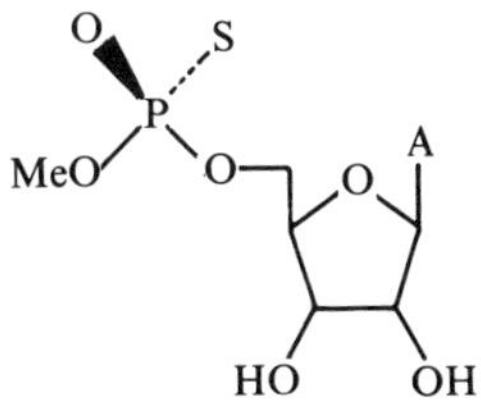

(**296**)

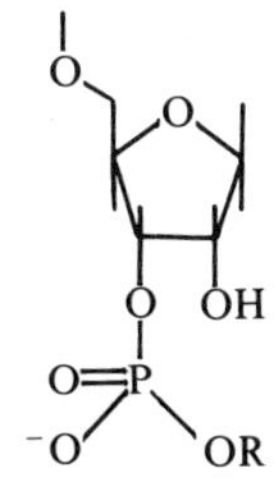

(**297**)

oxycarbocation-like. This is the conclusion of a study in which primary, β-secondary, and multiple-heavy-atom kinetic isotope effects were assessed using a variety of suitably labelled derivatives of AMP.[316] The results of Breslow and Labelle[317] on the cleavage of poly(uridylic acid) (**297**) by imidazole are fully in accord with the requirements of the stereoelectronic theory.[318]

Sulphur-containing Acids

The proceedings of the 12th International Symposium on Organic Sulphur, held at Nijmegen (Netherlands) have been published as 'Perspectives in the Organic Chemistry of Sulfur'.[319] A review (20 references) on nucleophilic substitution at sulphinyl and sulphonyl centres has appeared.[320] Sulphonyl transfers have been reviewed by King *et al.*[321] An account has appeared of the mechanism of acid-catalysed hydrolysis of ketene dithioacetates.[3]

The pK_a values of a series of arylsulphonic acids (**298**) have been estimated from their ^{33}S chemical shifts which are linearly related to pK_a.[322] An excellent correlation was also obtained between the chemical-shift data and a dual-parameter equation involving σ_I and σ_R. The hydrolysis in aqueous acid and aqueous acid-salt systems of some methyl-substituted benzenesulphonic acids has been reported.[323] Two kinetically indistinguishable mechanisms were proposed.

Sulphonylation by $XC_6H_4SO_2Cl$ (X = 3-NO_2, 4-Br, 4-Cl, H, 4-Me, etc.) of a series of substituted phenols in the presence of β-picoline or 3,5-lutidine as catalyst has been investigated;[324] ρ_x was 1.06 and substituent variation in the phenol gave inductive and resonance ρ-values of 0.39 and -0.45, respectively. The reactivity of the carboxylic acid chloride group was 900 times greater than that of the sulphonic acid chloride group in the aminolysis of 3-$ClO_2SC_6H_4COCl$ by aniline.[325] The kinetics of hydrolysis of several 4-substituted 1-naphthalenesulphonyl chlorides have been studied in aqueous acetone and an S_N2 or S_AN mechanism is favoured.[326]

Ten benzenesulphonyl chlorides, $XC_6H_4SO_2Cl$, and seven benzenedisulphonyl chlorides (**299**) have been hydrolysed in aqueous acetic acid containing $MeCO_2Na$ and the reaction has been followed by polarography. Pseudo-first-order rate constants for the two consecutive reactions of compounds (**299**) have been computed and the influence of the SO_2Cl and SO_3^- groups on the reactivity of the second SO_2Cl group have been discussed.[327] Methanolysis rates: methane- > ethane- > propane-2-sulphonyl chlorides in MeOH-MeCN mixtures are reported and an S_N2 mechanism is proposed.[328]

Nucleophilic substitution reactions of Y-substituted phenylmethanesulphonyl halides (**300**) with X-substituted anilines in MeOH-MeCN has been investigated.[329,330] An S_N2 mechanism is supported for both the fluorides and the chlorides because of the values of the cross-interaction constants, ρ_{XY}, which are said to reflect the distance between X and Y substituents in the transition state of the reaction and are calculated from the equation:

$$\log(k_{XH}/k_{HH}) = \rho_X\sigma_X + \rho_Y\sigma_Y + \rho_{XY}\sigma_X\sigma_Y$$

In the present instance, the two ρ_{XY} values are about -0.50 for (**300**; Z = Cl) and

− 1.02 for (**300**; Z = F) in MeOH at 55°C.[329] This analysis for the reactions of compounds (**300**) has been sharply criticized recently[331] on the grounds that the reactions of these sulphonyl halides do *not* involve nucleophilic displacements by the anilines at sulphur (as assumed[329,330]) but rather occur for the most part via a sulphene intermediate (**301**). On the basis of careful product analysis, it was possible to reach this conclusion. Further, it is contended that only about 5% reaction of (**300**) occurs by direct aniline displacement, the remaining 95% occurs via the sulphene route.[331]

2-Formylbenzenesulphonyl chloride (**302**) turns out to be a useful protecting group for phenols.[332] The 2-formylbenzenesulphonate esters of phenols are easily cleaved by alkali, which leaves carboxylate esters untouched. Reaction of thionyl chloride with 5-oxo-4-oxahomoadamantane (**303**) gives rise to a mixture of equatorial and axial (3:2) isomers of 4-chloro-2-adamantone. 2-Adamantanone reacts with thionyl chloride to give 2,2-dichloroadamantane.[333]

$XC_6H_4SO_3H$ (**298**)

$C_6H_{4-n}(Me)_n(SO_2Cl)_2$ (**299**)

$YC_6H_4CH_2SO_2Z$ (**300**)

$YC_6H_4CH{=}SO_2$ (**301**)

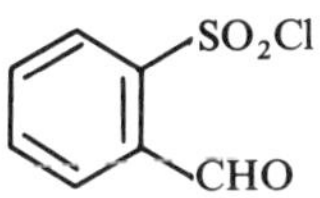

(**302**)

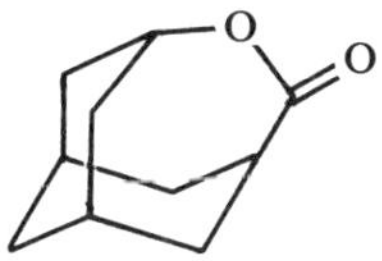

(**303**)

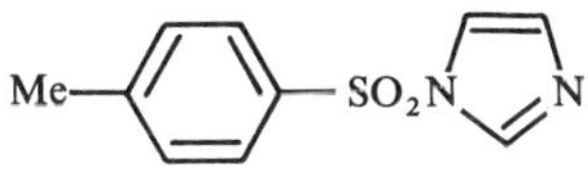

(**304**)

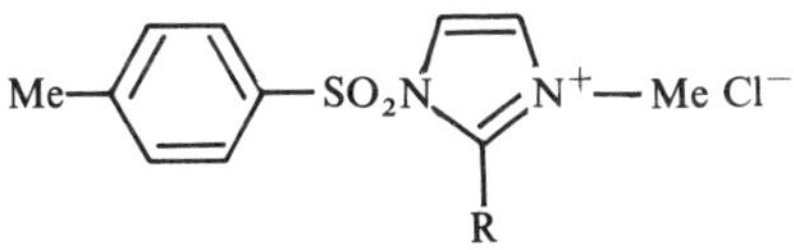

(**305**)

Hydrolysis of the sulphonylimidazole (**304**) and of the imidazolium chlorides (**305**) was in general base-catalysed.[334] Aminolysis (primary and secondary amines) rates give Brønsted plots for $pK_{N\mu}$ from 5 to 11, below and above those of the leaving-groups. The aminolysis rates are linearly related also to Ritchie N_+ and Swain–Scott *n* parameters. The *N*-imidazolylsulphonate derivatives of a number of manno- and galacto-pyranosides have been prepared and reacted with azide.[335] An acylated sulphite intermediate ($RCO_2SO_2C_2H_5$) in the reaction of SO_2, pyridine, and acyl chlorides in EtOH can undergo attack either at the carbonyl or sulphonyl group by EtOH to give $RCO_2C_2H_5$ or RCO_2H and disulphite, $(C_2H_5O)_2SO$, respectively.[336]

A study of equilibria in reactions of MeOH and EtOH with H_2SO_4 or of the hydrolysis of alkyl sulphates gives the following pK_a values for solution initially containing 20% (w/w) alcohol:[337] $pK_{Me\overset{+}{O}H_2} = -4.2$, $pK_{Et\overset{+}{O}H_2} = -3.7$, $pK_{Me\overset{+}{O}HSO_3^-} = -3.3$, and $pK_{Et\overset{+}{O}HSO_3^-} = -2.7$. The formation and hydrolysis of methyl, ethyl, and 2,4-dinitrobenzyl hydrogen sulphates in D_2O containing 9.24–13.28 M D_2SO_4 have been studied.[338] The rate-determining step of the hydrolysis was decomposition of a complex formed from $ROSO_3D$ and $D_2O_5^+$. Rates of alcoholysis (16 alcohols) of 3-phenylpropyl *p*-toluenesulphonate appear to take place via a bimolecular mechanism.[339] A combined Taft–Pavelich and Koppel–Palm equation correlated the rates. Rates of hydrolysis of a series of alkyl *p*-toluenesulphonates and methanesulphonates, including cyclohexyl- and 2-*endo*-norbornyl, in 0–70% w/w aqueous H_2SO_4 have been analysed using the equation:

$$\log(k/k_0)_{ROTS} = lN_{OTS} + mY_{OTS}$$

where the various parameters have the usual meaning. Values of N_{OTS} and Y_{OTS} have been correlated with H_0.[340] The rates of alkaline hydrolysis of the substituted phenyl tosylates (**306**) in the presence of *n*-Bu_4NBr have been determined together with activation parameters.[341a] The reactivity of 4-nitrophenol 4′-toluenesulphonate with amines, F^-, HO^-, HO_2^-, ClO^-, and BrO^- in water at 25°C and ionic strength of 0.1 M has been examined.[341b] The Brønsted β-value is 0.79 and a mechanism involving direct nucleophilic attack at the sulphur with S—O breakage is proposed.

Formation of benzylmethylsulphide in the reaction of benzylmercaptan with 2,2,2-trifluoroethyl methanesulphonate (**307**) is found to involve nucleophilic attack at sulphonyl sulphur.[342] The Zn(II)-catalysed hydrolysis of 2-acetylpyridine oxime phenyl sulphite involving the mechanism shown in (**244**) has been mentioned earlier.[211] The α-chymotrypsin-catalysed hydrolysis of the sultone (**260**) was also dealt with previously in this review.[252] Dissociation constants have been measured potentiometrically for arenesulphonamides of general formula (**308**) in a variety of organic solvents and in H_2O-EtOH mixtures.[343] Alkaline hydrolysis of the *N*-acylareneiminosulphonic acids (**309**) occurs by an S_N2 mechanism in aqueous acetonitrile. The aminolysis of (**309**) is complex, with non-linear dependence on [amine].[344]

There are three papers from a Czechoslovakian group on thioureas.[145,345,346] Base-catalysed cyclization of the ethyl *N*-(substituted aminothiocarbonyl)glycinates (**157**; X = S) was considered earlier.[145] The reactions of 1,3-diacylthioureas (**310**) and of benzoylthiourea with MeO^- and with amines have been reported and the dissociation constants of the thioureas measured in MeOH.[345] In a follow-up paper, the anomalous effect of *N*-methyl groups on thiourea dissociation constants has been examined.[346] Both HO^- and RO^- add competitively to the carbon atom of isothiocyanates and the products formed from alkoxides are the monothiourethanes (**311**).[347] Dissociation constants of the β-thioketo-amides (**92**) have been determined.[87]

p-$MeC_6H_4SO_2OC_6H_4X$ $F_3CCH_2OSO_2Me$ $ArSO_2NH_2$

(306) **(307)** **(308)**

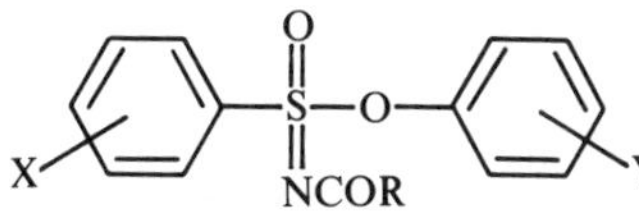

(309)

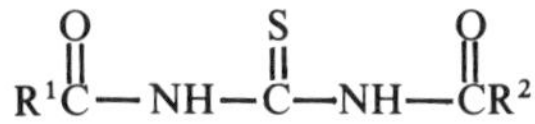

(310)

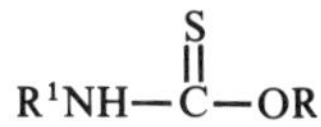

(311)

The mechanism of the reaction of diphenyl disulphide and toluene-*p*-sulphinate ions (**312**) in aqueous ethanol catalysed by silver ions giving phenyl toluene-*p*-thiolsulphonate (**313**) involves transition states containing one and two silver ions.[348] Sulphite reduction of phenyltellurium trichloride in dioxane gives the corresponding ditelluride (Ph_2Te_2) quantitatively. The mechanism of the conversion involves the intermediacy of the phenyl tellurate, $PhTeO_2^-$, $PhTeO^-$, and possibly the dimer Ph(OH)TeTe(OH)Ph.[349] The kinetics of the alkaline hydrolysis of the mono- and di-methyl esters of 2,3- and 3,4-thiophenedicarboxylic acids and their dihydro-analogues have been examined.[350] Iodine-induced lactamization of the thioimidates (**136**)–(**137**),[128] synthesis of sulphides from (**174**; O = S) and the effect of heterocycles with a thiol function on the reaction of *p*-nitrophenyl acetate were mentioned earlier.[51]

Reaction of sodium ethanethiolate at the 2-position of 1,2-dithiol-3-thiones (**314**) leads to various products depending on the substituents at C(4) and C(5).[351] Evidence for a highly coupled concerted transition state in the thiolate (5-nitro-5-thio- and 3-thio-benzoic acids) reduction of *S*-methyl-*S*-(substituted phenyl)sulphilimmonium salts has been found.[352] Reaction of Grignard and organolithium compounds with 2-sulphonyloxaziridines (**315**) involves a hemiaminal intermediate, $PhSO_2\bar{N}CH(OR)Ar\ M^+$.[353] The oxaziridines (**317**) have been obtained recently and their stereochemistry examined.[354] Acylation of 2,4-disubstituted 4(*R*)-alkoxycarbonylthiazolidines (**316**) with acyl chlorides gives the (2*R*,4*R*)-3-acyl isomer as the only product without any racemization.[108] Treatment of cyclic thiocarbamates with MeI/propylene oxide at 80°C leads smoothly to the carbonate.[355]

p-$MeC_6H_4SO_2^-$ $PhSSO_2C_6H_4Me$-p

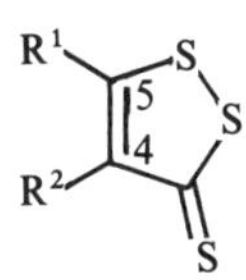

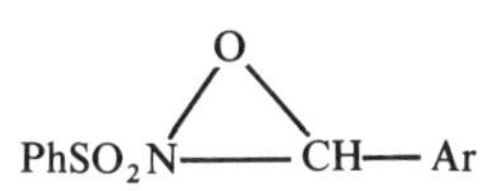

(312) **(313)** **(314)** **(315)**

The mechanisms of two reactions which give isomeric thiazol-5-yl ketones (**318**) have been elucidated.[356] The rates of hydrolysis of 2-alkoxycarbonylmethylthio-3-alkylbenzothiazolium salts in H_2O was pH-dependent and dependent on the alkoxy and alkyl substituents.[357] The rates of formation of dithizone (**319**) from phenylhydrazinium phenylhydrazine dithiocarbamate via phenylthiocarbazide have been studied.[358] Replacement of the enolic oxygen of diosphenols (**320**) by halogen is carried out via thiocarbonylation.[359] The reactions of 3-benzenesulphonyloxyalloxazine and its methyl analogue (**321**) with nucleophiles have been studied.[360]

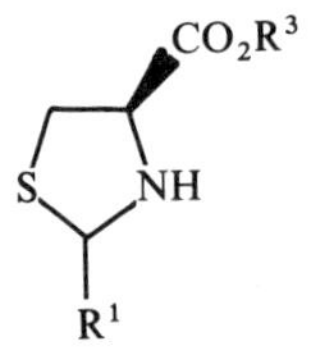

(**316**)

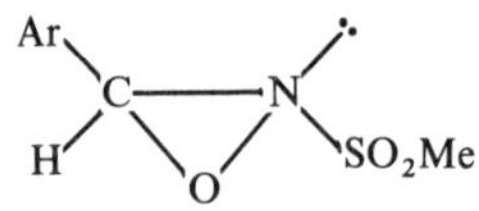

(**317**)

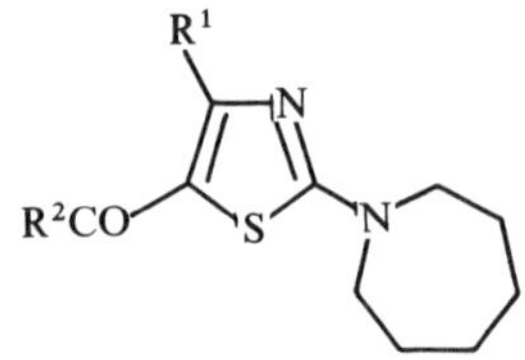

(**318**)

(**319**)

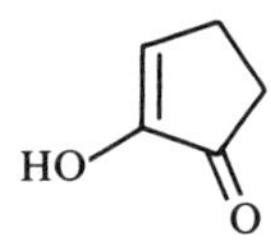

(**320**)

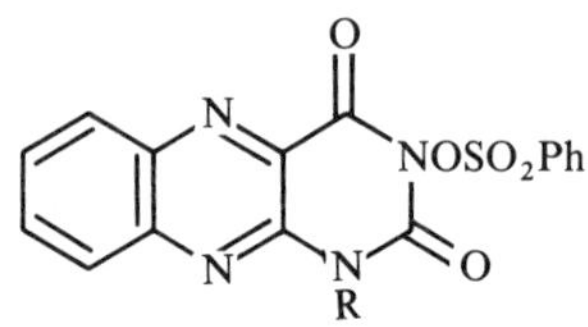

(**321**)

References

1 Pindur, U., Müller, J., Flo, C., and Witzel, H., *Chem. Soc. Rev.*, **16**, 75 (1987).
2 See *Org. Reaction Mech.*, **1983**, 34.
3 Okuyama, T., *Acc. Chem. Res.*, **19**, 370 (1986).
4 Okuyama, T., Kitano, M., and Fueno, T., *J. Org. Chem.*, **52**, 2657 (1987).
5 Okuyama, T. and Fueno, T., *J. Org. Chem.*, **51**, 5433 (1986).
6 Uieara, M., Zucco, C., Zanette, D., Rezende, M. C., and Nome, F., *J. Chem. Soc., Perkin Trans. 2*, **1987**, 175.
7 Isaacs, N. S., Rzepa, H. S., Sheppard, R. N., Lobo, A. M., Prabhakar, S., and Merbach, A. E., *J. Chem. Soc., Perkin Trans. 2*, **1987**, 1477.
8 Chrystiuk, E. and Williams, A., *J. Am. Chem. Soc.*, **109**, 3040 (1987).
9 Castro, E. A., Bórquez, M. T., and Parada, P. M., *J. Org. Chem.*, **51**, 5072 (1986).
10 Castro, E. A. and Ureta, C., *J. Chem. Res. (S)*, **1987**, 358.
11 Neuvonen, H., *J. Chem. Soc., Perkin Trans. 2*, **1987**, 159.
12 Perrin, C. L. and Nuñez, O., *J. Am. Chem. Soc.*, **109**, 522 (1987).
13 Pascal, R., Laspéras, M., Taillades, J., and Commeyras, A., *Nouveau J. Chim.*, **11**, 235 (1987).
14 Kaválek, J., Hejtmánkova, L., Štěrba, V., *Collect. Czech. Chem. Commun.*, **52**, 1285 (1987).
15 Prudchenko, A. P., Drizhd, L. P. and Savelova, V. A., *Zh. Org. Khim.*, **22**, 1909 (1986); *Chem. Abs.*, **106**, 155659 (1987).
16 Khan, M. N., *J. Chem. Soc., Perkin Trans. 2*, **1987**, 819.
17 Khan, M. N., *Int. J. Chem. Kinet.*, **19**, 143 (1987).
18 Buckley, N., *J. Org. Chem.*, **52**, 484 (1987).
19 Page, M. I., Webster, P., Bogdan, S., Tremerie, B., and Ghosez, L., *J. Chem. Soc., Chem. Commun.*, **1987**, 318.
20 Merényi, G., Lind, J., and Eriksen, T. E., *J. Am. Chem. Soc.*, **108**, 7716 (1986).
21 Guthrie, J. P. and Pike, D. C., *Can. J. Chem.*, **65**, 1951 (1987).
22 Ross, A. J., Go, M.-V., C., Casey, D. L. and Palling, D. T., *J. Pharm. Sci.*, **76**, 306 (1987).
23 Lee, I., Kim, C. K., and Seo, H. S., *Tetrahedron*, **42**, 6627 (1986).
24 Maraver, J. J., Marcos, E., and Bertran, J., THEOCHEM, **30**, 293 (1986); *Chem. Abs.*, **106**, 118938 (1987).

[25] DeTar, D. F., Binzet, S., and Darba, P., *J. Org. Chem.*, **52**, 2074 (1987).
[26] Han, C. C. and Brauman, J. I., *J. Am. Chem. Soc.*, **109**, 589 (1987).
[27] Harris, J. M. and McManus, S. P. (eds), *Adv. Chem. Ser.*, **1**, 215 (1987) (Nucleophilicity).
[28] Gorenstein, D. G., *Chem. Rev.*, **87**, 1047 (1987).
[29] Garrison, W. M., *Chem. Rev.*, **87**, 381 (1987).
[30] Maraver, J. J., Marcos, E., and Bertran, J., THEOCHEM, **35**, 251 (1987); *Chem. Abs.*, **107**, 197229 (1987).
[31] Sharma, K. K. and Aggarwal, A. K., *Int. J. Quantum Chem.*, **30**, 213 (1986); *Chem. Abs.*, **106**, 155594 (1987).
[32] Kuz'min, V. E., Pozigun, D. V., Lozitskaya, R. N., and Kamalov, G. L., *Teor. Eksp. Khim.*, **22**, 366 (1986); *Chem. Abs.*, **106**, 175489 (1987).
[33] Kuz'min, V. E., Kamalov, G. L., Lozitskaya, R. N., Siritsov, A. I., Alvares, J. R., and Stel'makh, I. B., *Teor, Eksp. Khim*, **22**, 610 (1986); *Chem. Abs.*, **107**, 77049 (1987).
[34] Sheldon, J. C., Currie, G. J., and Bowie, J. H., *Aust. J. Chem.*, **39**, 839 (1986).
[35] Huang, H., Xu, Z., and Liu, A., *Jilin Daxue Ziran Kexue Xuebao*, **1986**, 99; *Chem. Abs.*, **107**, 153713 (1987).
[36] Xie, C., Xu, Q., and Qu, S., *Wuli Huaxue Xuebao*, **2**, 363 (1986); *Chem. Abs.*, **106**, 195638 (1987).
[37] Tang, Y. and Jie, R., *Xiangtan Daxue Ziran Kexue Xuebao*, **1986**, 86; *Chem. Abs.*, **107**, 175422 (1987).
[38] Chubarov, G. A., Danov, S. M., and Balashov, A. L., *Izv. Vyssh. Uchebn. Zaved. Khim. Khim. Tekhnol.*, **29**, 41 (1986); *Chem. Abs.*, **107**, 38869 (1987).
[39] Silis, U., Alksnis, A., and Sile, A., *Latv. PSR Zinat. Akad. Vestis, Kim. Ser.*, **1986**, 303; *Chem. Abs.*, **107**, 38841 (1987).
[40] Scheeren, J. W., Bakker, C. G., Peperzak, R., and Nivard, R. J. F., *Tetrahedron Lett.*, **28**, 1821 (1987).
[41] Fishbein, J. C., Baum, H., Cox, M. M., and Jencks, W. P., *J. Am. Chem. Soc.*, **109**, 5790 (1987).
[42] Khan, M. N., *Int. J. Chem. Kinet.*, **19**, 415 (1987).
[43] Högberg, T., Ström, P., Ebner, M., and Ramsby, S., *J. Org. Chem.*, **52**, 2033 (1987).
[44] Henning, H.-G. and Koppe, A., *Z. Chem.*, **27**, 367 (1987).
[45] Sato, M., Ohmae, K., Nagae, S., and Nakaya, J., *J. Chem. Soc., Chem. Commun.*, **1987**, 839.
[46] Moss, R. A., Swarup, S., Ganguli, S., *J. Chem. Soc., Chem. Commun.*, **1987**, 860.
[47] Vaidya, R. A. and Mathias, L. J., *Polym. Prepr.*, **27**, 101 (1986); *Chem. Abs.*, **107**, 38843 (1987).
[48] Khan, M. N., *Indian J. Chem.*, **25A**, 831 (1986); *Chem. Abs.*, **106**, 155661 (1987).
[49] Terrier, F., Laloi-Diard, M., Degorre, F., and Kiffer, D., Abstr. RSC Intern. Symp. Organic Reactivity, Paris, **1987**, C23.
[50] Guthrie, J. P., Cassar, J., and Dawson, B. A., *Can. J. Chem.*, **64**, 2456 (1986).
[51] Brembilla, A., Roizard, D., Lochon, P., *J. Chim. Phys. Phys.-Chim. Biol.*, **83**, 577 (1986); *Chem. Abs.*, **107**, 96100 (1987).
[52] Putsykina, E. B., Vasil'ev, A. F., Belyakova, I. L., Karas, M. K., and Putsykin, Yu. G., *Khim. Prom.-st. (Moscow)*, **1986**, 333; *Chem. Abs.*, **106**, 32154 (1987).
[53] Martinez, P. and Rodriguez, A. F., *Z. Phys. Chem. (Leipzig)*, **267**, 1098 (1986).
[54] Biswas, P. K., Bag, B. C., and Das, M. N., *Can. J. Chem.*, **65**, 235 (1987).
[55] Khalil, F. Y., Hanna, M. T., and Assad, A. N., *Z. Naturforsch.*, **41A**, 1330 (1986); *Chem. Abs.*, **107**, 77053 (1987).
[56] Herrera, M. M., Puig, J. J. M., and Burgos, F. S., *Collect. Czech. Chem. Commun.*, **51**, 2781 (1986).
[57] Al-Arab, M. M. and Hamilton, G. A., *Biorg. Chem.*, **15**, 81 (1987).
[58] Fedorova, V. A., Gorin, Ya. I., and Chuiko, L. S., *Ukk. Khim. Zh. (Russ. edn)*, **52**, 966 (1986); *Chem. Abs.*, **107**, 77058 (1987).
[59] Shvets, V. F., Makarov, M. G., and Nevskaya, E. B., *Kinet. Katal.*, **27**, 996 (1986); *Chem. Abs.*, **106**, 155830 (1987).
[60] Zhidkova, N. V., Kobryanskii, V. M., and Turovskaya, L. N., *Izv. Akad. Nauk. SSSR, Ser. Khim.*, **1986**, 1612; *Chem. Abs.*, **106**, 155677 (1987).
[61] Pavlova, L. A., Davidovich, Yu. A., and Rogozhin, S. V., *Usp. Khim.*, **55**, 1803 (1986); *Chem. Abs.*, **106**, 101424 (1987).
[62] Martem'yanov, V. S., *Neftekhimiya*, **26**, 753 (1986); *Chem. Abs.*, **107**, 6533 (1987).
[63] Rakhmankulov, D. L., Zlotskii, S. S., Partushenko, E. V., Rol'nik, L. Z., Karakhanov, R. A., Bartok, M., and Molnar, A., *Acta Phys. Chem.*, **31**, 743 (1985); *Chem. Abs.*, **106**, 213124 (1987).
[64] Tsukada, M., *Chem. Lett.*, **87**, 1537 (1987).
[65] O'Neill, P. and Hegarty, A. F., *J. Org. Chem.*, **52**, 2113 (1987).
[66] Kanerva, L. T. and Euranto, E. K., *J. Chem. Soc., Perkin Trans. 2*, **1987**, 441.
[67] Euranto, E. K., Kanerva, L. T., and Nurmi, U., *Acta Chem. Scand.*, **40B**, 337 (1986).
[68] Kanerva, L. T., *Acta Chem. Scand.*, **40B**, 731 (1986).

[69] Engeberts, J. B. F. N., Blandamer, M. J., Burgess, J., Clarke, B., and Hakin, A. W., *J. Chem. Soc., Faraday Trans. 1*, **83**, 865 (1987).
[70] Blokzijl, W., Engeberts, J. B. F. N., Jager, J., and Blandamer, M. J., *J. Phys. Chem.*, **91**, 6022 (1987).
[71] Weber, R. H., Brügger, P., Arnold, W., Schönholzer, P., and Hansen, H.-J., *Helv. Chim. Acta*, **70**, 1439 (1987).
[72] Griesbeck, A. and Seebach, D., *Helv. Chim. Acta*, **70**, 1326 (1987).
[73] Dorigo, A. E. and Houk, K. N., *J. Am. Chem. Soc.*, **109**, 3698 (1987).
[74] Aldersey, M. F., Benson, J. M., Dean, F. M., El-Kadri, S., and Lythgoe, D. J., *Tetrahedron*, **43**, 5417 (1987).
[75] Canonne, P., Boulanger, R., and Bernatchez, M., *Tetrahedron Lett.*, **28**, 4997 (1987).
[76] Ernest, I., Fritz, H., and Rihs, G., *Helv. Chim. Acta*, **70**, 203 (1987).
[77] Agmon, N., *J. Org. Chem.*, **52**, 2192 (1987).
[78] Acevedo, S. and Bowden, K., *J. Chem. Soc., Perkin Trans. 2*, **1986**, 2045.
[79] Acevedo, S. and Bowden, K., *J. Chem. Soc., Perkin Trans. 2*, **1986**, 2049.
[80] Acevedo, S. and Bowden, K., *J. Chem. Soc., Perkin Trans. 2*, **1986**, 2051.
[81] Amat, A. M., Asensio, G., Castello, M. J., Miranda, M. A. and Simon-Fuentes, A., *Tetrahedron*, **43**, 905 (1987).
[82] Buglass, A. J., Dorr, M., and Juffkins, M., *Tetrahedron Lett.*, **28**, 3283 (1987).
[83] Chmielecka, J., Chojnowski, J., Eaborn, C., and Stanczyk, W. A., *J. Chem. Soc., Chem. Commun.*, **1987**, 1337.
[84] Bhutada, S. R. and Pangarkar, V. G., *J. Chem. Technol. Biotechnol.*, **36**, 61 (1986); *Chem. Abs.*, **105**, 225509 (1987).
[85] Popov, A. F., Maslosh, V. Z., Kudyukov, Yu. P., and Grebenyuk, Yu. V., *Izv. Vyssh. Uchebn. Zaved., Khim. Khim. Tekhnol.*, **29**, 30 (1986); *Chem. Abs.*, **106**, 155642 (1987).
[86] Bell, K. H., *Aust. J. Chem.*, **40**, 399 (1987).
[87] Nashid, B., Salama, S. B., and Ibrahim, O., *Z. Phys. Chem.* (*Leipzig*), **268**, 347 (1987).
[88] Slebocka-Tilk, H. and Brown, R. S., Abstr, RSC Intern. Symp. Organic Reactivity, Paris, **1987**, C5.
[89] Kijima, A. and Sekiguchi, S., *J. Chem. Soc., Perkin Trans. 2*, **1987**, 1203.
[90] Underwood, G. R. and Callahan, R. J., *Tetrahedron Lett.*, **28**, 5427 (1987).
[91] Novak, M., Rovin, L. H., Pelecanou, M., Mulero, J. J., and Lagerman, R. K., *J. Org. Chem.*, **52**, 2002 (1987).
[92] See *Org. Reaction Mech.*, **1985**, 44; **1986**, 42.
[93] Blake, J. F. and Jorgensen, W. L., *J. Am. Chem. Soc.*, **109**, 3856 (1987).
[94] Macho, V. and Rusina, M., *Petrochemia*, **25**, 127 (1985); *Chem. Abs.*, **107**, 22761 (1987).
[95] Kevill, D. N. and Kim, C.-B., Abstr. Intern. Symp. Organic Reactivity, Paris, **1987**, C17.
[96] Litvinenko, L. M., Zhil'tsov, N. P., Semenyuk, G. V., and Shubkina, N. I., *Ukr. Khim. Zh.* (*Russ. edn*), **52**, 854 (1986); *Chem. Abs.*, **107**, 115053 (1987).
[97] Semenyuk, G. V., Razumova, N. G., and Oleinik, N. M., *Zh. Org. Khim.*, **23**, 30 (1987); *Chem. Abs.*, **107**, 197214 (1987).
[98] Jee, J. G. and Ree, T., *Bull. Korean Chem. Soc.*, **8**, **31**, (1987); *Chem. Abs.*, **107**, 133670 (1987).
[99] Song, B. D. and Jencks, W. P., *J. Am. Chem. Soc.*, **109**, 3160 (1987).
[100] Titskii, G. D. and Turovskaya, M. K., *Zh. Org. Khim.*, **22**, 2132 (1986); *Chem. Abs.*, **107**, 96095 (1987).
[101] Shpan'ko, I. V., Goncharova, A. N., Chetverova, E. V., and Likhomanenko, E. E., *Zh. Org. Khim.*, **22**, 1432 (1986); *Chem. Abs.*, **106**, 155645 (1987).
[102] Shpan'ko, I. V. and Likhomanenko, E. E., *Zh. Org. Khim.*, **22**, 1896 (1986); *Chem. Abs.*, **106**, 175500 (1987).
[103] Metzger, J. and Jung, G., *Liebigs Ann. Chem.*, **1987**, 895.
[104] El Alami, N., Beland, C., and Villieras, J., *J. Organomet. Chem.*, **319**, 303 (1987).
[105] Van Gompel, J. and Schuster, G. B., *J. Org. Chem.*, **52**, 1465 (1987).
[106] Rammash, B. K. and Wong, J. L., *J. Org. Chem.*, **52**, 64 (1987).
[107] Severengiz, T. and du Mont, W.-W., *J. Chem. Soc., Chem. Commun.*, **1987**, 820.
[108] Ando, W., Igarashi, Y., and Huang, L., *Chem. Lett.*, **87**, 1361 (1987).
[107] Wolf, B. and Horsch, W., *Pharmazie*, **41**, 485 (1986); *Chem. Abs.*, **107**, 133651 (1987).
[110] Kaválek, J., Panchartek, J., Potěšil, T., and Štěrba, V., *Collect. Czech. Chem. Commun.*, **51**, 1964 (1986).
[111] Macháček, V., Svobodová, G., and Štěrba, V., *Collect. Czech. Chem. Commun.*, **52**, 140 (1987).
[112] Kaválek, J., Macháček, V., Svobodová, G., and Štěrba, V., *Collect. Czech. Chem. Commun.*, **52**, 1999 (1987).
[113] Blagoeva, I. B., *J. Chem. Soc., Perkin Trans. 2*, **1987**, 127.
[114] Harayama, T., Tezuka, Y., Taga, T., and Yoneda, F., *J. Chem. Soc., Perkin Trans. 1*, **1987**, 75.

[115] Käppi, R. and Lönnberg, H., *Acta Chem. Scand.*, **40B**, 768 (1986).
[116] Lönnberg, H., Suokas, P., Käppi, R., and Darzynkiewicz, E., *Acta Chem. Scand.*, **40B**, 798 (1986).
[117] Al Sabbagh, M. M., Calmon, M., Calmon, J. P., *J. Chem. Soc., Perkin Trans. 2*, **1987**, 929.
[118] Millefiori, S. and Millefiori, A., *J. Heterocycl. Chem.*, **24**, 525 (1987).
[119] Ebber, A., Haldna, U., and Murshak, A., *Org. React. (Tartu)*, **23**, 40 (1986); *Chem. Abs.*, **107**, 6686 (1987).
[120] Salem, S. M. and Sidahmed, I. M., *J. Chin. Chem. Soc. (Taipei)*, **32**, 451 (1985); *Chem. Abs.*, **106**, 66547 (1987).
[121] Ballaster, M., Riera, J., Castaner, J., Onrubia, C., and Ubierna, J. J., *An. Quim. Ser. C.*, **81**, 262 (1985); *Chem. Abs.*, **106**, 49405 (1987).
[122] Collins, S. and Hong, Y., *Tetrahedron Lett.*, **28**, 4391 (1987).
[123] Bishop, R. and Burgess, G., *Tetrahedron Lett.*, **28**, 1585 (1987).
[124] Slebocka-Tilk, H. and Brown, R. S., *J. Org. Chem.*, **52**, 805 (1987).
[125] Grehn, L., Gunnarsson, K., and Ragnarsson, U., *Acta Chem. Scand.*, **41B**, 18 (1987).
[126] Paine, J. B., Brough, J. R., Buller, K. K., and Erikson, E. E., *J. Org. Chem.*, **52**, 3993 (1987).
[127] Shute, R. E. and Rich, D. H., *J. Chem. Soc., Chem. Commun.*, **1987**, 1155.
[128] Takahata, H., Takamatsu, T., Mozumi, M., Chen, Y.-S., Yamazaki, T., and Aoe, K., *J. Chem. Soc., Chem. Commun.*, **1987**, 1627.
[129] Jephcote, V. J., Jowett, I. C., John, D. I., Edwards, P. D., Luk, K., Slawin, A. M. Z., and Williams, D. J., *J. Chem. Soc., Perkin Trans. 1*, **1986**, 2187.
[130] Jephcote, V. J., John, D. I., and Williams, D. J., *J. Chem. Soc., Perkin Trans. 1*, **1986**, 2195.
[131] Jephcote, V. J. and John, D. I., *J. Chem. Soc., Perkin Trans. 1*, **1986**, 2203.
[132] Cassidy, F., Evans, J. M., Smith, D. M., and Stamp, G., *Tetrahedron Lett.*, **28**, 2403 (1987).
[133] Ribó, J. M. and Valles, A., *J. Heterocycl. Chem.*, **24**, 457 (1987).
[134] Singh, H. and Kumar, S., *J. Chem. Res. (S)*, **1987**, 390.
[135] Tohda, Y., Ariga, M., Kawashima, T., and Matsumura, E., *Bull. Chem. Soc. Jpn*, **60**, 201 (1987).
[136] Kawashima, E. and Tabei, K., *J. Heterocycl. Chem.*, **23**, 1657 (1986).
[137] Zavoinau, D., Moga-Gheorghe, S., and Cuza, O., *Rev. Chim. (Bucharest)*, **37**, 461 (1986); *Chem. Abs.*, **106**, 213205 (1987).
[138] Zavoianu, D., Moga-Gheorghe, S., Cuza, O., and Zavoianu, R., *Rev. Chim. (Bucharest)*, **37**, 120 (1986); *Chem. Abs.*, **106**, 66543 (1987).
[139] Wang, J.-X. and Alper, H., *J. Chem. Res. (S)*, **1986**, 456.
[140] Worman, J. J., Uhrich, K., Olson, E., Diehl, J., Farnum, S., and Hawthorne, S., *Prepr. Pap.-ACS Div. Fuel Chem.*, **31**, 273 (1986); *Chem. Abs.*, **106**, 137744 (1987).
[141] Prudchenko, A. P., Drizhd, L. P., and Savelova, V. A., *Zh. Org. Khim.*, **22**, 1029 (1986); *Chem. Abs.*, **106**, 4351 (1987).
[142] Li, W.-K., Baker, J., and Radom, L., *Aust. J. Chem.*, **39**, 913 (1986).
[143] Noack, R. and Schwetlick, K., *Z. Chem.*, **27**, 77 (1987).
[144] Grambal, F. and Lasovsky, J., *Collect. Czech. Chem. Commun.*, **51**, 2786 (1986).
[145] Mindl, J. and Štěrba, V., *Collect. Czech. Chem. Commun.*, **52**, 156 (1987).
[146] Nusselder, J. J. H. and Engeberts, J. B. F. N., *J. Org. Chem.*, **52**, 3159 (1987).
[147] Milcent, R. and Barbier, G., *J. Heterocycl. Chem.*, **24**, 1233 (1987).
[148] Fife, T. H., Natarajan, R., and Werner, M. H., *J. Org. Chem.*, **52**, 740 (1987).
[149] Cooney, A. P., Crampton, M. R., Jones, M., and Golding, P., *J. Heterocycl. Chem.*, **24**, 1163 (1987).
[150] Odijk, W. M. and Koomen, G.-J., *J. Chem. Soc., Perkin Trans. 2*, **1987**, 733.
[151] Oivanen, M., Lönnberg, H., Zhou, X.-X., and Chattopadhyaya, J., *Tetrahedron*, **43**, 1133 (1987).
[152] Jones, A. R., Arques, J. S., Garcia, E. Z., Bates, P. A., Hursthouse, M. B., *J. Chem. Soc., Chem. Commun.*, **1986**, 1745.
[153] Pytela, O., Kaska, M., Ludwig, M., and Večerǎ, M., *Collect. Czech. Chem. Commun.*, **52**, 2212 (1987).
[154] Svoboda, P., Pytela, O. and Večerǎ, M., *Collect. Czech. Chem. Commun.*, **52**, 2217 (1987).
[155] Iley, J. N., Moreira, R., and Rosa, E., *J. Chem. Soc., Perkin Trans. 2*, **1987**, 1503.
[156] Iley, J. N., Rosa, E., and Fernandes, L., *J. Chem. Res. (S)*, **1987**, 264.
[157] Tagaev, O. A., Pazderskii, Yu. A., Gutor, I. M., and Moiseev, I. I., *Kinet. Katal*, **27**, 1122 (1986); *Chem. Abs.*, **106**, 175720 (1987).
[158] Verardo, G., Strazzolini, P., and Giumanini, A. G., *Tetrahedron Lett.*, **28**, 3011 (1987).
[159] Quijano, J., Rodriguez, M. M., Yepes, M. del S., and Gallego, L. H., *Tetrahedron Lett.*, **28**, 3555 (1987).
[160] Corey, E. J. and Su, W., *Tetrahedron Lett.*, **28**, 5241 (1987).
[161] Garratt, P. J. and Porter, J. R., *J. Chem. Res. (S)*, **1987**, 108.
[162] Dijkstra, G., Kruizinga, W. H., and Kellogg, R. M., *J. Org. Chem.*, **52**, 4230 (1987).

[163] Taylor, R., *J. Chem. Soc., Chem. Commun.*, **1987**, 741.
[164] Cline, G. W. and Hanna, S. B., *J. Am. Chem. Soc.*, **109**, 3087 (1987).
[165] Ruelle, P., *J. Am. Chem. Soc.*, **109**, 1722 (1987).
[166] Caprino, L. A., *Acc. Chem. Res.*, **20**, 401 (1987).
[167] Grehn, L., Gunnarsson, K., Ragnarsson, U., *Acta Chem. Scand.*, **40B**, 745 (1986).
[168] Bogolyubskii, A. V., Il'chenko, A. Ya., Kul'chitskii, M. M., Alperovich, M. A., and Yagupolskii, L. M., *Ukr. Khim. Zh.* (*Russ. edn*), **52**, 746 (1986); *Chem. Abs.*, **107**, 153699 (1987).
[169] Titskii, G. D. and Turovskaya, M. K., *Zh. Org. Khim.*, **22**, 2302 (1986); *Chem. Abs.*, **107**, 58226 (1987).
[170] Goncharova, A. N., Shpan'ko, I. V., Chetverova, E. V., and Likhomanenko, E. E., *Ukr. Khim. Zh.* (*Russ. edn*), **52**, 1077 (1986); *Chem. Abs.*, **107**, 133659 (1987).
[171] Litvinenko, L. M., Mikhailov, V. A., Savelova, V. A., and Drizhd, L. P., *Zh. Org. Khim.*, **22**, 2343 (1986); *Chem. Abs.*, **107**, 58227 (1987).
[172] Kakumoto, T., Saito, K., and Imamura, A., *J. Phys. Chem.*, **91**, 2366 (1987).
[173] See *Org. React. Mech.*, **1984**, 56.
[174] Barber, S. E. and Kirby, A. J., *J. Chem. Soc., Chem. Commun.*, **1987**, 1775.
[175] Khan, M. N., *Int. J. Chem. Kinet.*, **19**, 757 (1987).
[176] Kirkovskii, L. P. and Chapyshev, S. V., *Dokl. Akad. Nauk. SSSR*, **291**, 847 (1986); *Chem. Abs.*, **107**, 77081 (1987).
[177] Paventi, M., Chubb, F. L., and Edward, J. T., *Can. J. Chem.*, **65**, 2114 (1987).
[178] Clezy, P. S. and Tonkin, K., *Aust. J. Chem.*, **40**, 413 (1987).
[179] Brown, R. S. and Clewley, R. G., *J. Org. Chem.*, **52**, 1216 (1987).
[180] Singh, H. and Kumar, S., *Tetrahedron*, **43**, 2177 (1987).
[181] Smith, G. G. and Baum, R., *J. Org. Chem.*, **52**, 2248 (1987).
[182] See *Org. Reaction Mech.*, **1986**, 54.
[183] Bunton, C. A., Report 1985, ARO-18290.12-CH Order No. AD A155589; Avail, NTIS; *Chem. Abs.*, **106**, 49363 (1987).
[184] Odashima, K., *Yakugaku Kenkyu no Shinpo*, **2**, 163 (1986); *Chem. Abs.*, **106**, 83675 (1987).
[185] Broxton, T. J., Christie, J. R., and Sango, X., *J. Org. Chem.*, **52**, 4814 (1987).
[186] Menger, F. M. and Persichetti, R. A., *J. Org. Chem.*, **52**, 3451 (1987).
[187] Juršič, B., Ladika, M., and Sunko, D. E., *Tetrahedron*, **43**, 1955 (1987).
[188] Ihara, Y., Nango, M., and Koga, J., *Bull. Chem. Soc. Jpn*, **60**, 815 (1987).
[189] Al Seoud, O., Vieira, R. C., and Novaki, L. P., *Bull. Chem. Soc. Jpn*, **60**, 1163 (1987).
[190] Al-Lohedan, H. A., *Tetrahedron*, **43**, 345 (1987).
[191] Andriamanampisoa, R., Boyer, B., Lamaty, G., and Roque, J. P., *Tetrahedron*, **43**, 77 (1987).
[192] Ogino, K., Kashihara, N., Fujita, T., Ueda, T., Isaka, T., and Tagaki, W., *Chem. Lett.*, **87**, 1303 (1987).
[193] Berndt, D. C., Ayoub, M. E., and Akhavan-Tafti, M. H., *Int. J. Chem. Kinet.*, **19**, 513 (1987).
[194] Al-Lohedan, H. A., *J. Phys. Chem.*, **91**, 4524 (1987).
[195] Bunton, C. A., Mhala, M. M., and Moffatt, J. R., *J. Org. Chem.*, **52**, 3832 (1987).
[196] Menger, F. M., Gan, L. H., Johnson, E., and Durst, D. H., *J. Am. Chem. Soc.*, **109**, 2800 (1987).
[197] Mackay, R. A., Longo, F. R., Knier, B. L., and Durst, H. D., *J. Phys. Chem.*, **91**, 861 (1987).
[198] Kafarski, P., Lejczak, B., and Plucinski, P., *Tetrahedron*, **43**, 799 (1987).
[199] Schneider, H.-J. and Schneider, U., *J. Org. Chem.*, **52**, 1613 (1987).
[200] Moss, R. A., Hendrickson, T. F., Ueoka, R., Kim, K. Y., and Weiner, P. K., *J. Am. Chem. Soc.*, **109**, 4363 (1987).
[201] Moss, R. A., Bhattacharya, S., Scrimin, P., and Swarup, S., *J. Am. Chem. Soc.*, **109**, 5740 (1987).
[202] Moss, R. A., Scrimin, P., Bhattacharya, S., and Swarup, S., *J. Am. Chem. Soc.*, **109**, 6209 (1987).
[203] Ohkubo, K. and Miyake, S., *J. Chem. Soc., Perkin Trans. 2*, **1987**, 995.
[204] Hosseini, M. W., Lehn, J.-M., Maggiora, L., Mertes, K. B., and Mertes, M. P., *J. Am. Chem. Soc.*, **109**, 537 (1987).
[205] Ikeda, H., Kojin, R., Yoon, C.-J., Ikeda, T., and Toda, F., *Chem. Lett.*, **87**, 1495 (1987).
[206] Satake, I., Ikenoue, T., Takeshita, T., Hayakawa, K., and Maeda, T., *Bull. Chem. Soc. Jpn*, **58**, 2746 (1985).
[207] Satake, I., Yoshida, S., Hayakawa, K., Maeda, T., and Kusumuto, Y., *Bull. Chem. Soc. Jpn*, **59**, 3991 (1986).
[208] Fornasier, R., Reniero, F., Scrimin, P., and Tonellato, U., *J. Chem. Soc., Perkin Trans. 2*, **1987**, 193.
[209] Fornasier, R., Reniero, F., Scrimin, P., and Tonellato, U., *J. Chem. Soc., Perkin Trans. 2*, **1987**, 1121.
[210] Toussaint, O., Capdevielle, P., and Maumy, M., Abstr. RSC Intern. Symp. Organic Reactivity, Paris, **1987**, P35.
[211] Suh, J. and Koh, D., *J. Org. Chem.*, **52**, 3446 (1987).
[212] Sutton, P. A. and Buckingham, D. A., *Acc. Chem. Res.*, **20**, 357 (1987).

[213] Schepartz, A. and Breslow, R., *J. Am. Chem. Soc.*, **109**, 1814 (1987).
[214] Ahmad, S. and Iqbal, J., *J. Chem. Soc., Chem. Commun.*, **1987**, 114.
[215] Chin, J. and Zou, X., *Can. J. Chem.*, **65**, 1882 (1987).
[216] Liu, F., Biao, J., Yang, M., and Huang, H., *Jilin, Daxue Ziran Kexue Xuebao*, **1986**, 101; *Chem. Abs.*, **107**, 96111 (1987).
[217] Hu, N. X., Aso, Y., Otsubo, T., and Ogura, F., *J. Chem. Soc., Chem. Commun.*, **1987**, 1447.
[218] Mitsudo, T., Kadokura, M., and Watanabe, Y., *J. Org. Chem.*, **52**, 3186 (1987).
[219] Matteson, D. S. and Beedle, E. C., *Tetrahedron Lett.*, **28**, 4499 (1987).
[220] Suh, J. and Chang, B., *Bioorg. Chem.*, **15**, 167 (1987).
[221] See *Org. Reaction Mech.*, **1986**, 59; **1985**, 59.
[222] Nanjappan, P. and Czarnik, A. W., *J. Am. Chem. Soc.*, **109**, 1826 (1987).
[223] Ruelle, P., *J. Chem. Soc., Perkin Trans. 2*, **1986**, 1953.
[224] Noto, R., Buscemi, S., Werber, G., and Spinelli, D., *J. Heterocycl. Chem.*, **24**, 1449 (1987).
[225] Noto, R., Werber, G., Buccheri, F., and Arnone, C., *J. Heterocycl. Chem.*, **24**, 1457 (1987).
[226] Moriarty, R. M., Khosrowshahi, J. S., and Dalecki, T. M., *J. Chem. Soc., Chem. Commun.*, **1987**, 675.
[227] Klingler, R. J. and Rathke, J. W., *Organometallics*, **5**, 2568 (1986).
[228] Kluger, R., *Chem. Rev.*, **87**, 863 (1987).
[229] Kluger, R., Karimian, K., and Kitamura, K., *J. Am. Chem. Soc.*, **109**, 6368 (1987).
[230] Scott, A. I., *Pure Appl. Chem.*, **58**, 753 (1986).
[231] D'Souza, V. T. and Bender, M. L., *Acc. Chem. Res.*, **20**, 146 (1987).
[232] Cramer, F. and Freist, W., *Acc. Chem. Res.*, **20**, 79 (1987).
[233] Tanizawa, K. and Kanaoka, Y., *Top. Curr. Chem.*, **136**, 81 (1987).
[234] Cleland, W. W., *Bioorg. Chem.*, **15**, 283 (1987).
[235] Schmidt, G., *Top. Curr. Chem.*, **136**, 109 (1987).
[236] Ching-Shih, C., Shih-Hsiung, W., Girdaukas, G., and Sih, C. J., *J. Am. Chem. Soc.*, **109**, 2812 (1987).
[237] Therisod, M. and Klibanov, A. M., *J. Am. Chem. Soc.*, **109**, 3977 (1987).
[238] Margolin, A. L., Crenne, J.-Y., and Klibanov, A. M., *Tetrahedron Lett.*, **28**, 1607 (1987).
[239] See *Org. Reaction Mech.*, **1983**, 57.
[240] Menger, F. M. and Ladika, M., *J. Am. Chem. Soc.*, **109**, 3145 (1987).
[241] Frick, L., MacNeela, J. P., and Wolfenden, R., *Bioorg. Chem.*, **15**, 100 (1987).
[242] Charton, M. and Ziffer, H., *J. Org. Chem.*, **52**, 2400 (1987).
[243] Quinn, D. M., *Chem. Rev.*, **87**, 955 (1987).
[244] Acheson, S. A., Dedopoulou, D., and Quinn, D. M., *J. Am. Chem. Soc.*, **109**, 239 (1987).
[245] Acheson, S. A., Barlow, P. N., Lee, G. C., Swanson, M. L., and Quinn, D. M., *J. Am. Chem. Soc.*, **109**, 246 (1987).
[246] Barlow, P. N., Acheson, S. A., Swanson, M. L., and Quinn, D. M., *J. Am. Chem. Soc.*, **109**, 253 (1987).
[247] Pearson, A. J., Bansal, H. S., and Lai, Y.-S., *J. Chem. Soc., Chem. Commun.*, **1987**, 519.
[248] Pavlič, M. R., *Arch. Biochem. Biophys.*, **253**, 446 (1987).
[249] Tanizawa, K., Kanaoka, Y., and Lawson, W. B., *Acc. Chem. Res.*, **20**, 337 (1987).
[250] Tanizawa, K., Nakano, M., and Kanaoka, Y., *Bioorg. Chem.*, **15**, 50 (1987).
[251] Remy, M.-H., Bacou, P., Bourdillon, C., and Thomas, D., *Biochim. Biophys. Acta*, **911**, 252 (1987).
[252] Bolen, D. W., Kimura, T., and Nitta, Y., *Biochemistry*, **26**, 146 (1987).
[253] White, E. H. and Lim, H. M., *J. Org. Chem.*, **52**, 2162 (1987).
[254] Somayaji, V., Skorey, K. I., and Brown, R. S., *J. Org. Chem.*, **51**, 4866 (1986).
[255] Kunugi, S., Tanabe, K., Fukuda, M., Makimoto, S., and Taniguchi, Y., *J. Chem. Soc., Chem. Commun.*, **1987**, 1335.
[256] Kunugi, S., Tanabe, K., Yamashita, K., and Fukuda, M., *Bull. Chem. Soc. Jpn*, **60**, 1399 (1987).
[257] Shima, H., Fukuda, M., Tanabe, K., Ito, T., and Kunugi, S., *Bull. Chem. Soc. Jpn*, **60**, 1403 (1987).
[258] Stein, R. L. and Strimpler, A. M., *J. Am. Chem. Soc.*, **109**, 6538 (1987).
[259] Govardhan, C. P. and Pratt, R. F., *Biochemistry*, **26**, 3378 (1987).
[260] Lottenberg, R., Hall, J. A., Pautler, E., Zupan, A., Christensen, U., and Jackson, C. M., *Biochim. Biophys. Acta*, **874**, 326 (1986).
[261] Breslow, R. and Schepartz, A., *Chem. Lett.*, **1987**, 1.
[262] Botting, N. P., Akhtar, M., Cohen, M. A., and Gani, D., *J. Chem. Soc., Chem. Commun.*, **1987**, 1371.
[263] Walder, L. and Orlinski, R., *Organometallics*, **6**, 1606 (1987).
[264] Ascenzi, P., Aducci, P., Torroni, A., Amiconi, G., Ballio, A., Menegatti, E., and Guarneri, M., *Biochim. Biophys. Acta*, **912**, 203 (1987).
[265] Kageyama, T. and Talahashi, K., *Eur. J. Biochem.*, **165**, 483 (1987).
[266] Somayaji, V. and Brown, R. S., *J. Am. Chem. Soc.*, **109**, 4738 (1987).
[267] Guibe-Jampel, E., Rousseau, G., and Salaün, J., *J. Chem. Soc., Chem. Commun.*, **1987**, 1080.

[268] Stokes, T. M. and Oehlschlager, A. C., *Tetrahedron Lett.*, **28**, 2091 (1987).
[269] Degueil-Castaing, M., De Jeso, B., Drouillard, S., and Maillard, B., *Tetrahedron Lett.*, **28**, 953 (1987).
[270] Xie, Z.-F., Suemune, H., and Sakai, K., *J. Chem. Soc., Chem. Commun.*, **1987**, 838.
[271] Gil, G., Ferre, E., Meou, A., Le Petit, J., and Triantaphylides, C., *Tetrahedron Lett.*, **28**, 1647 (1987).
[272] Hosie, L., Sutton, L. D., and Quinn, D. M., *J. Biol. Chem.*, **262**, 260 (1987).
[273] Tabushi, I., Kuroda, Y., and Yamada, M., *Tetrahedron Lett.*, **28**, 5695 (1987).
[274] Wijnands, R. A., Müller, F., and Visser, A. J. W. G., *Eur. J. Biochem.*, **163**, 535 (1987).
[275] Baldwin, J. E., Adlington, R. M., Aplin, R. T., Crouch, N. P., Knight, G., and Schofield, C. J., *J. Chem. Soc., Chem. Commun.*, **1987**, 1651.
[276] Baldwin, J. E., Adlington, R. M., Aplin, R. T., Crouch, N. P., Schofield, C. J., and Ting, H.-H., *J. Chem. Soc., Chem. Commun.*, **1987**, 1654.
[277] Baldwin, J. E., Adlington, R. M., Domayne-Hayman, B. P., Knight, G., and Ting, H.-H., *J. Chem. Soc., Chem. Commun.*, **1987**, 1661.
[278] Björkling, F., Boutelje, J., Hjalmarsson, M., Hult, K., and Norin, T., *J. Chem. Soc., Chem. Commun.*, **1987**, 1041.
[279] Glanzer, B. I., Faber, K., and Griengl, H., *Tetrahedron*, **43**, 771 (1987).
[280] Brooks, D. W., Wilson, M., and Webb, M., *J. Org. Chem.*, **52**, 2244 (1987).
[281] Chen, S.-T. and Wang, K.-T., *J. Chem. Res. (S)*, **1987**, 308.
[282] Kato, Y., Ohta, H., and Tsuchihashi, G., *Tetrahedron Lett.*, **28**, 1303 (1987).
[283] Modro, T. A., Rzepa, H. S., and Symes, J., *J. Chem. Soc., Perkin Trans. 2*, **1987**, 767.
[284] Nguyen, M. T. and Hegarty, A. F., *J. Chem. Soc., Perkin Trans. 2*, **1987**, 47.
[285] Liang, C. and Allen, L. C., *J. Am. Chem. Soc.*, **109**, 6449 (1987).
[286] See *Org. Reaction Mech.*, **1986**, 71.
[287] Nagarajan, K., Shelly, K. P., Perkins, R. R., and Stewart, R., *Can. J. Chem.*, **65**, 1729 (1987).
[288] Shelly, K. P., Nagarajan, K., and Stewart, R., *Can. J. Chem.*, **65**, 1734 (1987).
[289] Calvo, K. C., *J. Org. Chem.*, **52**, 3654 (1987).
[290] Portilla, G., Rojas, M. C., Chiong, E., and Cori, O., *J. Chem. Soc., Perkin Trans. 1*, **1987**, 1479.
[291] Katritzky, A. R., Duell, B. L., Durst, H. D., and Knier, B., *Tetrahedron Lett.*, **28**, 3899 (1987).
[292] See *Org. Reaction Mech.*, **1986**, 70; **1985**, 67; **1984**, 68.
[293] Gorenstein, D. G., Chang, A., and Yang, J.-C., *Tetrahedron*, **43**, 469 (1987).
[294] Yang, J.-C. and Gorenstein, D. G., *Tetrahedron*, **43**, 479 (1987).
[295] Chang, J. W. A. and Gorenstein, D. G., *Tetrahedron*, **43**, 5187 (1987).
[296] Herschlag, D. and Jencks, W. P., *J. Am. Chem. Soc.*, **108**, 7938 (1986).
[297] Freeman, S., Friedman, J. M., and Knowles, J. R., *J. Am. Chem. Soc.*, **109**, 3166 (1987).
[298] Cullis, P. M., Misra, R., and Wilkins, D. J., *J. Chem. Soc., Chem. Commun.*, **1987**, 1594.
[299] Harnett, S. P. and Lowe, G., *J. Chem. Soc., Chem. Commun.*, **1987**, 1416.
[300] Epstein, W. W. and Garrossian, M., *J. Chem. Soc., Chem. Commun.*, **1987**, 532.
[301] Hammerschmidt, F. and Völlenkle, H., *Liebigs Ann. Chem.*, **1986**, 2053.
[302] Breuer, E. and Moshe, R., *Isr. J. Chem.*, **27**, 45 (1987).
[303] Cullis, P. M., Kaye, A. D., and Trippett, S., *J. Chem. Soc., Chem. Commun.*, **1987**, 1464.
[304] Cullis, P. M., Iagrossi, A., Rous, A. J., and Schilling, M. B., *J. Chem. Soc., Chem. Commun.*, **1987**, 996.
[305] Breuer, E., Karaman, R., Leader, H., and Goldblum, A., *J. Chem. Soc., Chem. Commun.*, **1987**, 671.
[306] Kluger, R. and Thatcher, G. R. J., *Can. J. Chem.*, **65**, 1838 (1987).
[307] Challis, B. C. and Iley, J. N., *J. Chem. Soc., Perkin Trans. 2*, **1987**, 1489.
[308] Wadsworth, W. S., *J. Org. Chem.*, **52**, 1748 (1987).
[309] Ikehara, M., Nishikawa, S., Morioka, H., Kim, H. J., Fuchimura, K., Kimura, T., Adiwinata, J., Tanaka, T., Ohtsuka, E., and Uesugi, S., *Pure Appl. Chem.*, **59**, 565 (1987).
[310] Li, L., Seddon, A. P., and Meister, A., *J. Biol. Chem.*, **262**, 11020 (1987).
[311] Khamis, M. M. and Larsson-Raźnikiewicz, M., *Acta Chem. Scand.*, **41B**, 348 (1987).
[312] McClard, R. W. and Shokat, K. M., *Biochemistry*, **26**, 3385 (1987).
[313] Cummins, J. H. and Potter, B. V. L., *Eur. J. Biochem.*, **162**, 123 (1987).
[314] Williams, S. P. and Bridger, W. A., *Biochemistry*, **26**, 4483 (1987).
[315] Yoon, M.-Y. and Cook, P. F., *Biochemistry*, **26**, 4118 (1987).
[316] Parkin, D. W. and Schramm, V. L., *Biochemistry*, **26**, 913 (1987).
[317] See *Org. Reaction Mech.*, **1986**, 73.
[318] Taira, K., *Bull. Chem. Soc. Jpn.*, **60**, 1903 (1987).
[319] *Perspectives in Org. Sulfur Chemistry*, Studies in Organic Chemistry, Vol. 28, B. Zwanenburg and A. J. H. Klunder (Eds.), Elsevier (Amsterdam), 1987.
[320] Mikolajczyk, M., *Phosphorus Sulfur*, **27**, 31 (1986); *Chem. Abs.*, **106**, 118928 (1987).

[321] King, J. F., Skonieczny, S., Khemani, K. C., and Lock, J. D., *Adv. Chem. Ser.*, **215** (Nucleophilicity), 385 (1987).
[322] Crumrine, D. S., Shankweiler, J. M., and Hoffman, R. V., *J. Org. Chem.*, **51**, 5013 (1986).
[323] Kozlov, V. A. and Bagrovskaya, N. A., *Zh. Org. Khim.*, **22**, 1228 (1986); *Chem. Abs.*, **106**, 17651 (1987).
[324] Vizgert, R. V., Maksimenko, N. N., Rubleva, L. I., and Derevyanko, S. V., *Zh. Org. Khim.*, **22**, 1438 (1986); *Chem. Abs.*, **106**, 155646 (1987).
[325] Timoshenko, G. N., Grigorichev, A. K., Moskviche, Yu. A., and Miromov, G. S., *Izv. Vyssh. Uchebn., Zaved., Khim. Khim. Tekhnol.*, **29**, 23 (1986); *Chem. Abs.*, **106**, 195628 (1987).
[326] Perumal, S., Selvaru, S., Viswanathan, T. K., and Arumugam, N., *Indian J. Chem.*, **25A**, 436 (1986); *Chem. Abs.*, **105**, 225514 (1987).
[327] Sanecki, P. and Rokaszewski, E., *Can. J. Chem.*, **65**, 2263 (1987).
[328] Lee, I., Kang, H. K., and Park, B. S., *Kich'o, Kwahak Yonguso Nonmunjip* (*Inha Taehakkyo*), **1986**, 83; *Chem. Abs.*, **106**, 195621 (1987).
[329] Lee, I., Kang, H. K., and Lee, H. W., *J. Am. Chem. Soc.*, **109**, 7472 (1987).
[330] Lee, I. and Kang, H. K., *Tetrahedron Lett.*, **28**, 1183 (1987).
[331] King, J. F. and Skonieczny, S., *Tetrahedron Lett.*, **28**, 5001 (1987).
[332] Shashidhar, M. S. and Bhatt, M. V., *J. Chem. Soc., Chem. Commun.*, **1987**, 654.
[333] Janku, J., Burkhard, J., and Vodička, L., *Collect. Czech. Chem. Commun.*, **52**, 752 (1987).
[334] Monjoint, P. and Ruasse, M. F., Abstr. RSC. Intern. Symp. Organic Reactivity, Paris, **1987**, P65.
[335] El Sayed Ahmed, F. M., David, S., and Vatele, J.-M., *Carbohydr. Res.*, **155**, 19 (1986).
[336] Kirzecky, N. D., Thomson, M. L., and Wayne, R. S., *J. Org. Chem.*, **52**, 3452 (1987).
[337] Almstead, N., Christ, W., Miller, G., Reilly-Packard, S., Vargas, K., and Zuman, P., *Tetrahedron Lett.*, **28**, 1627 (1987).
[338] Vinnik, M. I., Kislina, I. S., Kitaigorodskii, and Nikitaev, A. T., *Izv. Akad. Nauk, SSSR, Ser. Khim.*, **1986**, 2671 *Chem. Abs.*, **107**, 77057 (1987).
[339] Sendega, R. V., Makitra, R. G., and Pirig, Ya. N., *Kinet. Katal.*, **27**, 789 (1986); *Chem. Abs.*, **106**, 155657 (1987).
[340] Bentley, T. W., Jurczyk, S., Roberts, K., and Williams, D. J., *J. Chem. Soc., Perkin Trans. 2*, **1987**, 293.
[341] (a) Nummert, V., Ojassalu, K., and Piirsalu, M., *Org. React.* (*Tartu*), **23**, 83 (1986); *Chem. Abs.*, **106**, 213200 (1987); (b) Simanenko, Yu. S., Prokop'eva, T. M., Savelova, V. A., Popov, A. F., Sakulin, G. S., and Belousova, I. A., *Org. React.* (*Tartu*), **23**, 209 (1986); *Chem. Abs.*, **107**, 197230 (1987).
[342] Langler, R. F. and Morrison, N. A., *Can. J. Chem.*, **65**, 2385 (1987).
[343] Ludwig, M., Pytela, O. and Večeřa, M., *Collect. Czech. Chem. Commun.*, **51**, 1948 (1986).
[344] Said, F. M. and Douglas, K. T., *Phosphorus Sulfur*, **27**, 361 (1986); *Chem. Abs.*, **106**, 101524 (1987).
[345] Kaválek, J., Jirman, J., and Štěrba, V., *Collect. Czech. Chem. Commun.*, **52**, 120 (1987).
[346] Kaválek, J., Jirman, J., Macháček, V., and Štěrba, V., *Collect. Czech. Chem. Commun.*, **52**, 1992 (1987).
[347] Augustin, J. and Baláž, S., *Collect. Czech. Chem. Commun.*, **52**, 443 (1987).
[348] Castle, L. A., Satchell, D. P. N., and Satchell, R. S., *J. Chem. Res. (S)*, **1987**, 114.
[349] Miller, J. D. and Tahir, T. A., *J. Chem. Soc., Dalton Trans.*, **1987**, 985.
[350] Struhárik, M., Hrnčiar, P., and Loos, D., *Collect. Czech. Chem. Commun.*, **52**, 2005 (1987).
[351] Largeron, M., Martens, T., and Fleury, M. B., *Tetrahedron*, **43**, 3421 (1987).
[352] Young, P. R. and Reid, K. J., *J. Org. Chem.*, **52**, 2695 (1987).
[353] Davis, F. A., Wei, J., Sheppard, A. C., and Gubernick, S., *Tetrahedron Lett.*, **28**, 5115 (1987).
[354] Forni, A., Moretti, I., Torre, G., Brückner, S., and Malpezzi, L., *J. Chem. Soc., Perkin Trans. 2*, **1987**, 699.
[355] Patroni, J. J. and Stick, R. V., *Aust. J. Chem.*, **40**, 795 (1987).
[356] Challacombe, K., Plackett, S. J., and Meakins, G. D., *Tetrahedron Lett.*, **28**, 5767 (1987).
[357] Gaplovsky, A. and Sutoris, V., *Chem. Pap.*, **40**, 115 (1986); *Chem. Abs.*, **106**, 155626 (1987).
[358] Sarangi, B., Misra, B. K., and Behera, G. B., *J. Indian Chem. Soc.*, **62**, 979 (1985); *Chem. Abs.*, **107**, 197183 (1987).
[359] Ponaras, A. A. and Zaim, O., *J. Org. Chem.*, **51**, 4741 (1986).
[360] Hamby, J. M. and Bauer, L., *J. Heterocycl. Chem.*, **24**, 1013 (1987).

Organic Reaction Mechanisms 1987
Edited by A. C. Knipe and W. E. Watts

CHAPTER 3

Radical Reactions: Part 1

P. HANSON

Department of Chemistry, University of York

Introduction

A considerable number of short reviews of radical chemistry has appeared this year. Several are published together as the proceedings of a NATO Advanced Science Institutes workshop which took place in Louvain-la-Neuve in January 1986.[1] Leading figures in free radical research have given general accounts of their approaches to the study of fundamental processes in the gas[2] and liquid[3] phases.

Specialized reviews on kinetic ESR and CIDEP applied to the monomer-dimer equilibria of pyridinyl radicals[4] and on the positive muon as a kinetic probe for radical reactions have also appeared.[5] The hydroxyl radical in aqueous media has been reviewed[6] as have recent advances in the radical chemistry of organic sulphides.[7] Three surveys of the chemistry of peroxy radicals[8–10] and two of the homolytic reactions of acetals and orthoesters[11,12] have been published in the Soviet literature. The burgeoning applications of radical chemistry in synthesis have been surveyed in short reviews,[13,14] at greater length,[15] and in book form.[16]

Structure, Stereochemistry and Stability

Carbon-centred Radicals

The formation of radicals by the thermolysis of bonds in diamagnetic precursors has been the subject of a number of experimental and theoretical studies. The initiation step in the pyrolysis of acetonitrile, studied behind reflected shocks in a single-pulse shock tube over the temperature range 1350–1950 K, was found to be:

$$CH_3CN \rightarrow \cdot CH_2CN + H\cdot$$

for which an Arrhenius expression was quoted; the major reaction products are CH_4 and HCN, but several others were identified and their formation accounted for.[17] The thermolysis of azoalkanes has been the subject of two investigations.[18,19] Activation parameters for the homolytic decomposition and the *cis*/*trans* isomerization of a large series of bridgehead azo-compounds have been found, together with enthalpies of isomerization.[18] It was concluded that there is no change in mechanism across the range of *trans*-azo decompositions examined and that the observations are consistent with a concerted two-bond decomposition pathway. Similarly, a single isomerization mechanism, an inversion process, was inferred in which the transition states resemble the *cis*-isomers. The second study concerned the investigation of the kinetics of thermolysis of diastereoisomeric 1,1′-disubstituted *trans*-azoneopentanes, e.g. (**1**), by differential scanning calorimetry.[19] The rate of homolysis of (**1**; R = alkyl) is sterically accelerated by increasing the size of R but is inhibited for aryl R. These effects result from the change is strain during the reaction (**1**) → 2(**2**) + N_2. The increase in strain for aryl R arises on account of the repulsion between Bu^t and an aryl ring coplanar with the radical centre in (**2**).

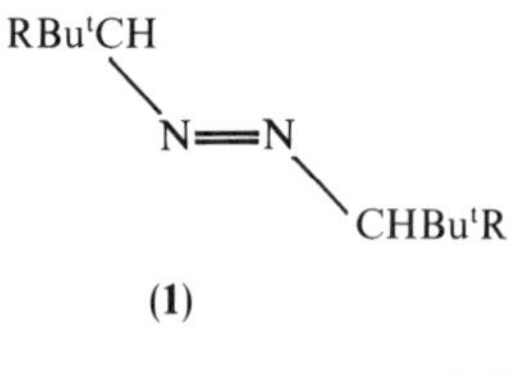

(**1**)

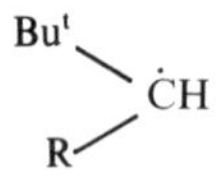

(**2**)

Whereas the decomposition of a series of diazonium ions in 1,2-dichloroethane is heterolytic, their complexes with 18-crown-6 ether decompose by mixed heterolytic and homolytic pathways.[20] The heterolytic process was shown to be amenable to multiple regression analysis using various substituent constants but the homolytic dediazoniation was not. Various ground-state decomposition pathways for MeSH and EtSH have been investigated by *ab initio* calculations of electronic structure; C—S bond homolysis was found to be preferable to α-elimination of H_2S in each case.[21]

Energies and structures have been calculated for a large number of substituted methyl radicals and methanes, at the geometry-optimized UHF and ROHF 4-31G levels, and the results used to calculate radical stabilization energies from the isodesmic reaction:

$$X_n\dot{C}H_{(3-n)} + CH_4 \rightarrow X_nCH_{(4-n)} + {}^{\cdot}CH_3$$

The stabilization energies found were compared with other radical stability scales and with experimental data.[22] Theoretical calculations on the structure and barriers to internal rotation of the 2-propyl radical[23] and on the ground- and low-lying doublet excited states of the propargyl radical have been reported.[24] Structures have been computed for $Ph\dot{C}H_2$ and $Ph^{\cdot}$ by the UHF method for comparison with ROHF calculations reported by others.[25] The application of the MINDO/3 method to methyl-substituted allyl and to vinyl and ethynyl radicals has been described in the German literature.[26,27]

Muonic radicals have featured in three studies: a theoretical study has indicated that the factors which determine the most stable conformation of a β-substituted ethyl radical are dominated by the zero-point energy;[28] the muon-electron hyperfine splitting constants in the α- and β-phosphonylated radicals $(EtO)_2P(O)\dot{C}HCH_2Mu$ and $H_2\dot{C}CH(Mu)P(O)(OEt)_2$ show that the muonium 'atom' prefers the out-of-plane position in the α-phosphonylated radical but is constrained towards the in-plane position in the β-phosphonylated radical.[29] The addition of muons to mono- and bis-(trimethylsilyl)alkynes produces vinylic radicals, the magnitudes of the hyperfine splittings of which exclude linear structures or equilibrating bent structures but the regioselectivity of muon addition to the monosubstituted alkyne was ambiguous.[30]

A series of medium-ring cycloalkylmethyl radicals has been studied by ESR.[31] From the 9-membered to the 15-membered rings (excluding the 14-membered which was not investigated) the radicals occur in at least two conformers which place the radical centre in either quasi-equatorial or quasi-axial positions, distinguished by characteristic β-proton hyperfine splittings. The preferred ring conformations were identified and the energetics of conformational change measured from the exchange broadening of the ESR spectra. A number of 1-substituted cyclopropyl radicals (**3**) has been studied by ESR at low temperatures.[32] The radicals (**3a**–**c**) exhibit hyperfine splittings which are larger for the ring protons that are *syn* with respect to the SOMO, than for those that are *anti*. The *syn*/*anti* ratios of splittings were smaller than that previously recorded by others for (**3d**). The out-of-plane bend of the 1-substituent, calculated from the α–^{13}C hyperfine splitting, for (**3a**), (**3b**), (**3e**) and (**3f**) was found to be smaller than values indicated by *ab initio* calculation. Variable-temperature studies permitted the evaluation of Arrhenius parameters for the inversion of (**3a**) and for rotation about the C(1)—OEt bond in (**3b**). The radicals (**4**) and (**5**), where the alkene functions are held relatively fixed with respect to the

SOMO, were investigated in a search for evidence of homoallylic conjugation.[33] The hyperfine splittings found for respective δ- and γ-protons are so small that such conjugation may be ruled out. Arrhenius parameters for the inversion of the isomeric cyclohexenyl radicals (**6**) and (**7**) have been reported.[34]

$$\text{For } (\mathbf{6}),\ \log(k/s^{-1}) = (13.1 \pm 0.8) - (7.0 \pm 0.9)/\theta$$
$$\text{and for } (\mathbf{7}),\ \log(k/s^{-1}) = (12.6 \pm 0.8) - (2.4 \pm 0.5)/\theta$$

where $\theta = 2.3RT$ kcal mol^{-1}. The isomeric bridgehead decalin radicals have been the subject of a study by both MM2 and MNDO methods. Both radicals are predicted to be non-planar and the calculations show no evidence of a planar transition state. The detailed structure of the *cis*-radical and the nature of the interconversion process require more demanding theoretical work.[35] Carbohydrate-derived radicals have been generated in non-aqueous solutions in the temperature range 303–373 K; the tetraacetylglucosyl radical exhibits hyperfine splittings in its ESR spectrum consistent with a preferred boat conformation (**8**).[36] However, this same radical, when formed radiolytically in the solid phase at 77 K, retains the chair conformation of its precursor but annealing induces partial conversion to the boat form observed in the liquid phase.[37]

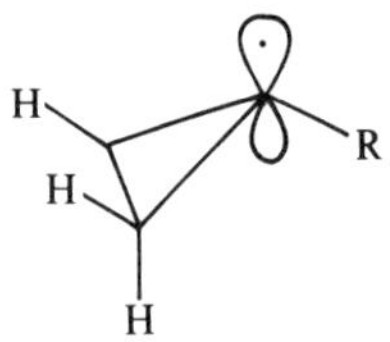

(**3**) **a**; R = Me
b; R = EtO
c; R = Cl
d; R = F
e; R = Me_3Si
f; R = H

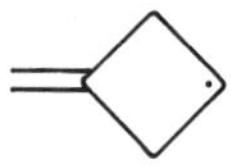

(**4**)

(**5**)

(**6**)

(**7**)

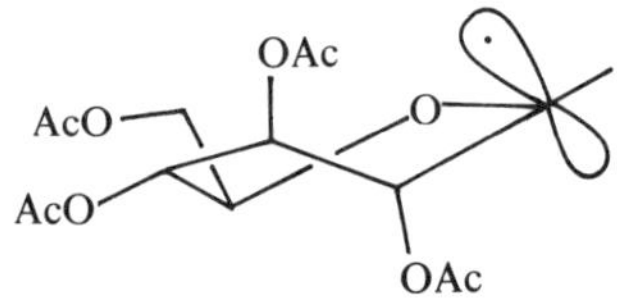

(**8**)

The captodative effect (merostabilization) has continued to attract attention.[1] As a test for synergetic captodative radical stabilization, the homolytic equilibria yielding the radicals (**9**) were examined from the point of view of strain, enthalpy and entropy of dissociation, and spin distribution.[38] It was concluded that pairs of captodative substituents do stabilize the radicals in an additive way in contrast to the geminal OMe groups in (**9d**); the spin distribution in the radicals, sampled as the hyperfine splitting of the *para* position of the phenyl rings, responded to the substituents differently from the bond enthalpy. This serves as a caution against adopting the magnitudes of hyperfine splittings as indicators of radical stability. On the basis of just such an assumption, others have also claimed a synergetic effect for captodative substituents and an 'antagonistic' effect for like substituents.[39] The dimerization equilibria of **(9f)** and of **(10)** have been examined in various solvents.[40] The results permit the refutation of the recent suggestion that solvent polarity plays an important role in the captodative effect. The relation of bond dissociation energies to the charge distribution in the dissociating molecule has been examined in an approach which makes interesting comparison with that of the Rüchardt school.[41] Captodative substitution in series of amino- and hydroxy-propynyl radicals resulted in extra delocalization, as inferred from hyperfine splittings; in one amino series the barriers to C—N bond rotation were measured and used to estimate radical stabilization energies.[42]

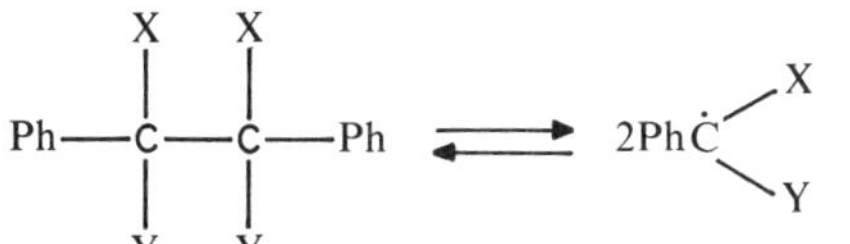

(9)

	X	Y
a;	Me	Me
b;	Et	Me
c;	Et	OMe
d;	OMe	OMe
e;	CN	Me
f;	CN	OMe
g;	CN	CN

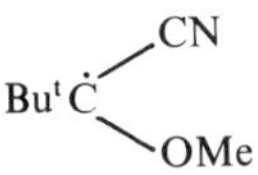

(10)

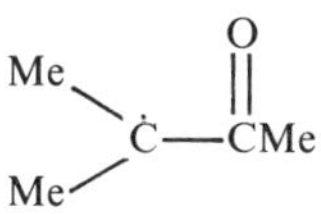

(11)

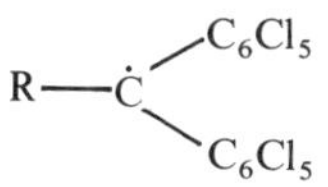

(12) **a**; R = CO_2Pr^i

b; R = CN

c; R = C(O)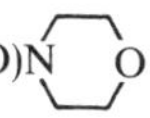

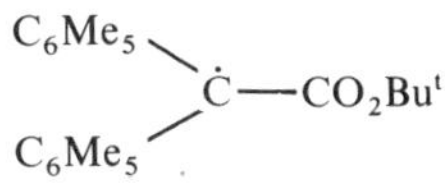

(13)

The capacity of β-C—Si bonds to function as the donor component in captodative substitution has been demonstrated[43] and the stabilization afforded alkyl radicals by both α- and β-silyl substituents has been estimated as *ca* 2.6 kcal mol^{-1}.[44] The stabilization afforded the radical **(11)** by the acetyl group has been found comparable with that afforded by the cyano group by comparison of the decomposition rates of the appropriately substituted azoalkanes.[45]

The vibrational spectrum of the nitromethyl radical has been analysed: the force constant for the N—O bond stretch was evaluated as 8.07×10^2 N m^{-1} rather smaller than that observed for the corresponding bond in nitromethane; the value observed for the C—N bond stretch, 4.75×10^2 N m^{-1}, is in the range appropriate for a single bond.[46] A number of new sterically stabilized radicals has been synthesized, e.g. **(12)** and **(13)** by oxidation of the corresponding anionic precursors.[47,48] Further chemistry of the well-established perchlorotriphenylmethyl radical has also been reported: its behaviour as a clathrate host[49] and its photolysis;[50] the synthesis, structure and spin distribution in related polychlorotriphenylmethyl radicals have also been described.[51]

Nitrogen-centred Radicals

State-of-the-art *ab initio* molecular orbital techniques have been used in accurate calculations of the structures and energies of the succinimidyl π- and σ-radicals.[52] The results were found to be extremely sensitive both to the size of the basis set used and to the amount of electron correlation included, but the ground state appears to be the π-radical. Unlike other diarylhydrazyls, radical (**14**) exhibits the larger nitrogen hyperfine splitting from the trivalent N(2); this is in accord with the stabilization afforded by the trifluoroacetyl group arising via pair delocalization [canonical structure (**14a**)] rather than by spin withdrawal [canonical structure (**14b**)].[53] Semi-empirical calculations have been performed for the phenyldiazenyl radical, PhN=Ṅ: viz. MINDO/3 calculations relating heat of formation to the CNN bond angle and the CCNN dihedral angle, and INDO-UHF calculations of hyperfine splitting constants.[54] A series of iminyl radicals (**15**) has been generated photolytically and shown to be π-radicals.[55] Various tritylthioaminyl radicals, $R\dot{N}SCPh_3$ (R = Me, Et, $PhCH_2$, Ph), have been obtained by hydrogen abstraction from the corresponding thioamines and their ESR parameters recorded.[56] A complete evaluation of ESR parameters, including hyperfine splitting constants for ^{33}S and ^{17}O has been given for the acylated thioaminyls RC(O)ṄSPh (R = Bu^t, Ph); it is estimated that about 47% of the π-spin density resides on S (a higher proportion than in simple thioaminyls) and only about 5% on O, consistent with little withdrawal of spin by the acyl group, as noted previously for the comparable case of (**14**).[57] Homolytic fission of the S—N bond has been shown to be one of the reaction

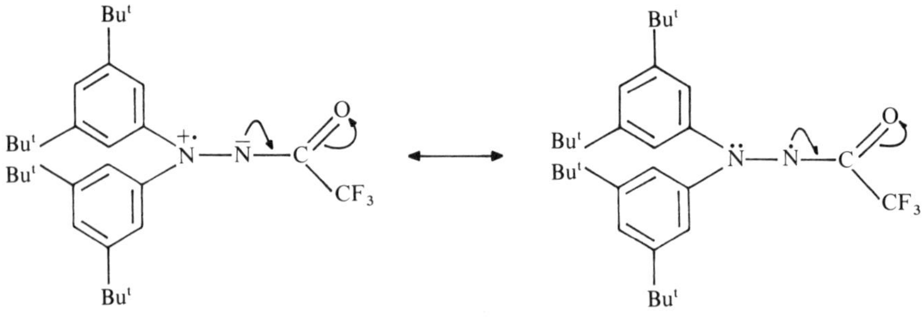

(**14a**) *major* (**14b**) *minor*

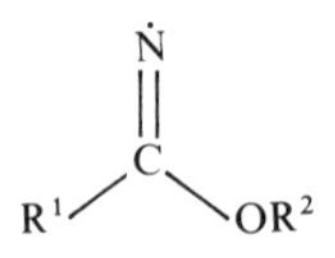

(**15**) R^1 = alkyl;

R^2 = Et, SiMe$_3$

paths followed during the thermal reaction of arylsulphenanilides in benzene or furan. The homolytic path is favoured by electron-donating substituents in the aniline moiety and by higher temperatures of reaction.[58]

Oxygen-centred Radicals

Theoretical calculations have been reported for $PhCO_2^{\bullet}$[59] and for HC(O•)=CHOR (R = H, Li, BeH, BH_2 and Me).[60] Laser excitation spectra, all showing well-resolved rotational structure, have been recorded at low temperature in a supersonic free jet expansion for several alkoxyl radicals.[61] The reactions of $^{\bullet}CF_3$ with O_2 and O_3 to give $CF_3O^{\bullet}$, $CF_3O_2^{\bullet}$ and $CF_3O_3^{\bullet}$ have been studied and thermochemical estimates given; calculations were used to predict the structure and vibrational spectrum of $CF_3O^{\bullet}$.[62] The results were used in discussion of the ozone-destroying catalytic cycles of the atmosphere. The effect of semiquinone radical formation on the protolytic equilibria and kinetics of (**16**) and (**17**) has been the subject of a pulse-radiolytic study; a considerable effect is found.[63] On the other hand, progressive protonation of the oxygens of benzo-1,4-semiquinone radical anion has relatively little effect on the electronic structure of the semiquinone system, monitored by resonance Raman Spectroscopy.[64] ESR and ENDOR studies have been performed on phenoxyl radicals (**18**) and (**19**) and activation parameters for rotations about the CH_2—C bond found.[65]

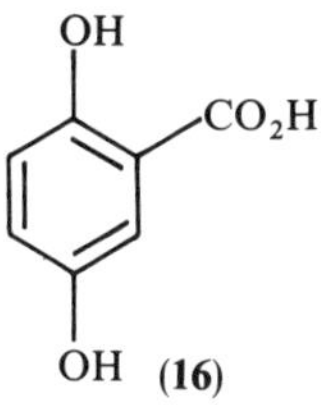

(**16**)

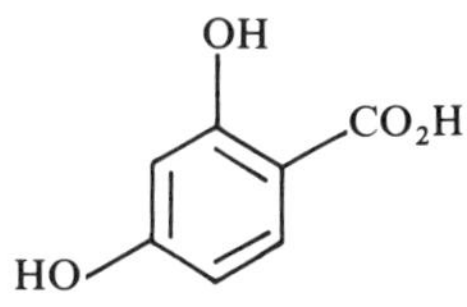

(**17**)

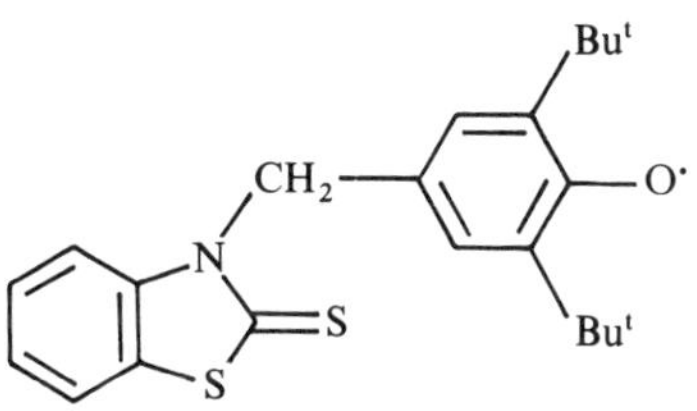

(**18**)

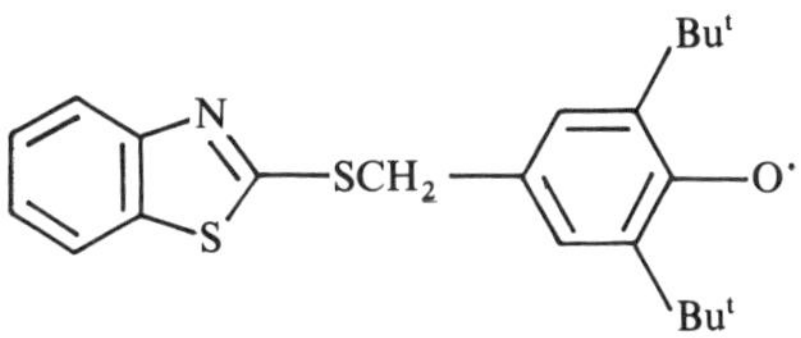

(**19**)

Miscellaneous Radicals

The germanium-centred radical $(PhMe_2CCH_2)_3Ge^{\bullet}$ has been the subject of an ESR spectroscopic study.[66] At temperatures greater than 60°C the spectra show the six methylene protons to be equivalent but, as the temperature is lowered to − 120°C an alternating linewidth effect is observed. The associated inequivalence of the methylene couplings arises from the adoption by the radical of a pyramidal structure with concomitant hindered rotation about the C—Ge bonds. An activation energy for the bond rotation of $5.4 \pm 0.8\ \text{kcal mol}^{-1}$ was determined. The thermally stable radical (**20**) has been isolated. In the liquid phase it occurs 65% as the monomeric radical and 35% as a diamagnetic dimer. A gas-phase structure has been determined together with a crystal structure of (**21**) for comparison.[67]

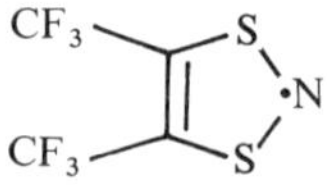

(**20**)

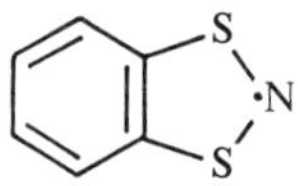

(**21**)

Rearrangements

Group Migration

A 1,2-migration of fluorine in 1,1,2-trifluoroethyl radical has been proposed following the observation that $CF_2{=}CFT$ is the major tritiated product from the addition of energetic tritium atoms to $CF_2{=}CFH$. The initial adduct, excited $^{\bullet}CF_2CFHT$ would be expected to decompose to $CF_2{=}CFT$, on the basis of relative bond strengths, and the rearrangement provides a step in the route to the unexpected observed product.[68] ESR studies have shown that $CF_2Cl\dot{C}FCl$, produced in γ-irradiated $CF_2ClCFCl_2$ solid solutions, can rearrange to give $CF_3\dot{C}Cl_2$ in the presence of amines of low ionization potential. A mechanism via carbanions produced by electron transfer has been suggested:[69]

$$Me_3N + CF_3Cl\dot{C}FCl \rightleftarrows Me_3N^{+\bullet} + CF_2Cl\bar{C}FCl$$
$$CF_2Cl\bar{C}FCl \rightleftarrows CF_2{=}CFCl + Cl^- \rightleftarrows \bar{C}F_2CFCl_2$$
$$\rightleftarrows CF_2{=}CCl_2 + F^- \rightleftarrows CF_3\bar{C}Cl_2$$
$$CF_3\bar{C}Cl_2 + Me_3N^{+\bullet} \rightleftarrows CF_3\dot{C}Cl_2 + Me_3N$$

Polychlorobenzenes have been found to undergo a photo-rearrangement in which a crowded chlorine migrates to a *meta* position.[70]

The tetraacetylgalactosyl radical (**22**) undergoes 1,2-migration of an acetyl group to give the rearranged radical (**23**). The movement of the radical centre away from the stabilized position adjacent to the ring oxygen atom is apparently offset by the gain in stabilization from the anomeric effect which is possible in (**23**).[36]

Intramolecular Hydrogen Atom Transfer

The minimum-energy path for the rearrangement of $Me\dot{C}{=}CH_2$ to $CH_2{=}CH\dot{C}H_2$ has been simulated by MINDO/3 calculations which suggest a barrier to rearrangement of $154\ \text{kJ mol}^{-1}$.[71] Vinyl radicals generated in aqueous solution by addition of thiyl radicals to alkynes have been shown to rearrange by intramolecular 1,4- and 1,5-hydrogen shifts.[72] The 1,4-shift is rapid if the resulting radical is stabilized captodatively by α-S and α-CO_2H substituents, otherwise 1,5-shift predominates (Scheme 1). Highly excited vinyl radicals, generated in the gas phase by direct photolysis of 2-methyl- and 2,3-dimethyl-but-2-ene at 147 nm have been found to isomerize to allylic radicals before the occurrence of fragmentation.[73] Rate constants have been measured for the 1,4- and 1,5-hydrogen transfers which occur during the isomerization of primary hexyl and secondary octyl radicals in the gas phase. The reactions were studied in the presence of high concentrations of methyl radicals which trap the rearranged alkyl radicals by combination and disproportionation. The conversion of 1-hexyl to 2-hexyl was found to have rate coefficients given by the Arrhenius expression:

$$\log(k/\mathrm{s}^{-1}) = (9.5 \pm 0.3) - (11.6 \pm 0.3)/\theta$$

where $\theta = 2.3RT$ kcal mol^{-1}. The study of the octyl isomerizations was complicated by the rapid interconversion of several secondary octyl radicals but, nevertheless, estimates of rate coefficients were made.[74] Six- and seven-membered transition states for intramolecular hydrogen abstraction in alkoxyl radicals have been modelled by both molecular-orbital and force-field calculations.[75] The calculations imply that the experimentally preferred six-membered transition state is enthalpically less stable than the seven-membered alternative on account of a non-linear C---H---O arrangement; the preferred 1,5-shifts therefore take place for entropic reasons. It is predicted that 1,6-hydrogen shifts via seven-membered transition states should become preferred at temperatures below about -190°C. The intramolecular hydrogen abstractions of carbamyl[76] and sulphonamidyl[77] radicals have been investigated and the selectivities in abstraction from chains on either side of the amide functions discussed.

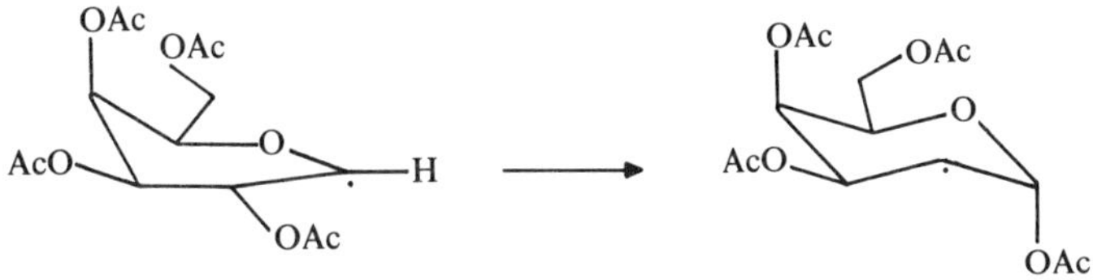

(22) **(23)**

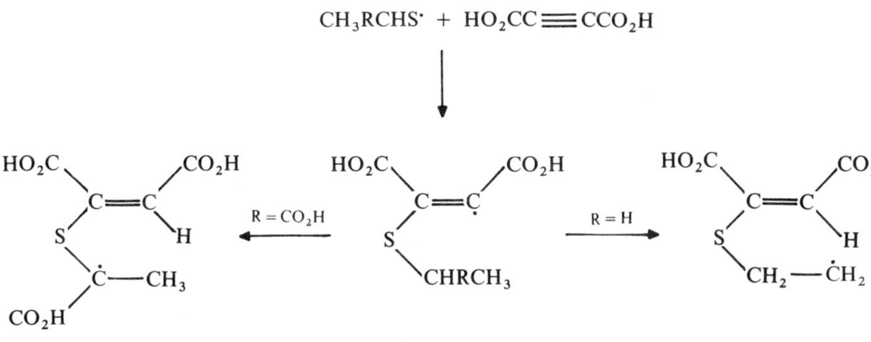

SCHEME 1

Ring Fragmentation

Rate constants for β-scission of the four-membered ring in the 3-methylenecyclobutylmethyl radical (**24**) have been found to be given by the Arrhenius expression:

$$\log(k/s^{-1}) = (12.9 \pm 0.3) - (11.5 \pm 0.3)/\theta$$

where $\theta = 2.3\,RT$ kcal mol^{-1}; this represents a very small lowering of activation energy relative to the analogous scission of cyclobutylmethyl radical, despite the radical produced being allylic, a result taken to imply development of the resonance stabilization late along the reaction coordinate.[33] β-Scission of a four-membered ring has been applied to provide a regiospecific positioning of the double bond in a synthesis of ($\pm$)-silphinene (**25**) (Scheme 2).[78]

(24)

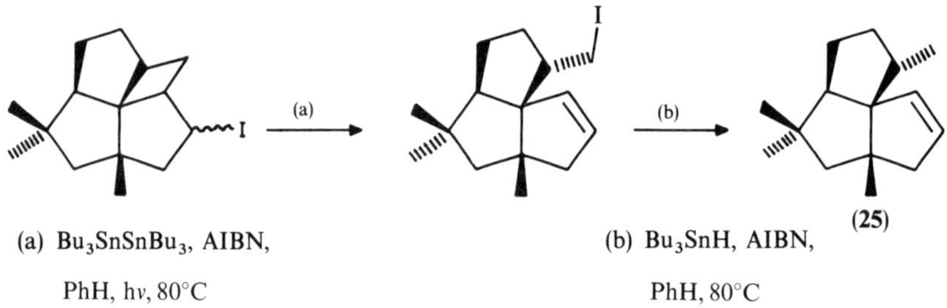

SCHEME 2

The homobenzvalenyl radical (**26**) has been generated in both solid and liquid solutions and characterized by ESR; it rearranges via progressive fragmentations of the minor rings to the tropylium radical (Scheme 3).[79] β-Scission of the C—C bond of oxiranes is diagnostic of adjacent homolysis and has been shown to occur faster than the initial radical could cyclize (Scheme 4);[80] the absence of concerted ring-opening was also demonstrated for similar systems. The possibility that radical, or radicaloid, intermediates play a role in penicillin biosynthesis and in penam-to-cepham ring-expansion has been probed using both radical reaction sequences and enzymically mediated processes involving the homolytic fragmentations of rings.[81,82]

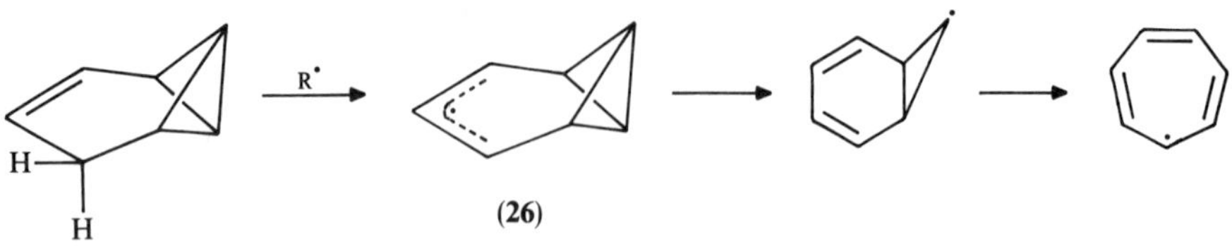

SCHEME 3

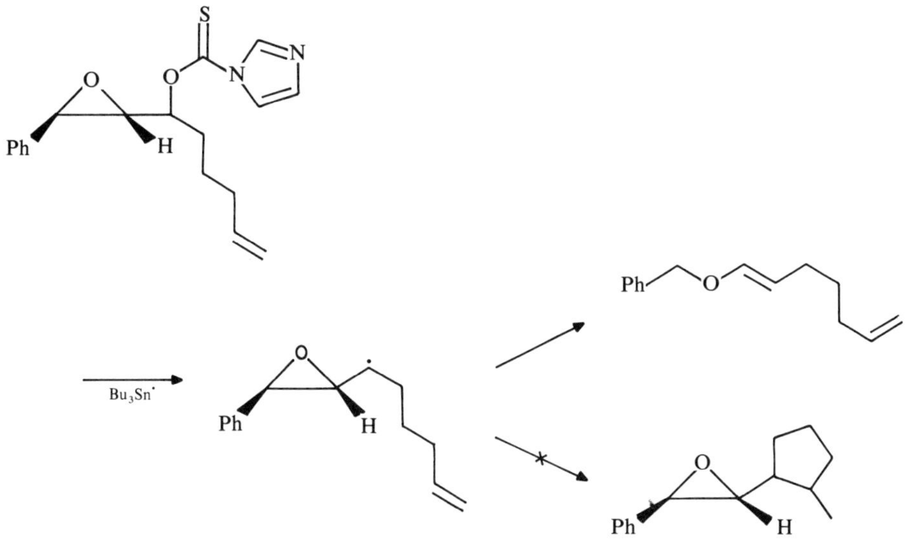

SCHEME 4

Evidence, both experimental and theoretical, has been presented that the 1,2-migration of vinyl and formyl groups in radicals;

$$\text{e.g. } O{=}CH{-}CMe_2{-}\dot{C}H_2 \rightarrow \dot{C}Me_2{-}CH_2{-}CH{=}O$$

occurs via closure and alternative β-scission of transient cyclopropane intermediates.[83] An analogous mechanism has been invoked in a synthetically promising ring-expanding adjunct to the Dieckmann reaction (Scheme 5),[84] and the reaction has been adapted for ring-expansion by 3 and 4 carbons also.[85] Comparable processes underlie a number of other procedures which produce medium-ring products.[86-88] Transient ring-closure and scission surely account for the acyl-group migration in the carbohydrate derivative (**27**) which has been proved by isotopic labelling experiments to be mediated homolytically and not heterolytically by anions (Scheme 6).[89]

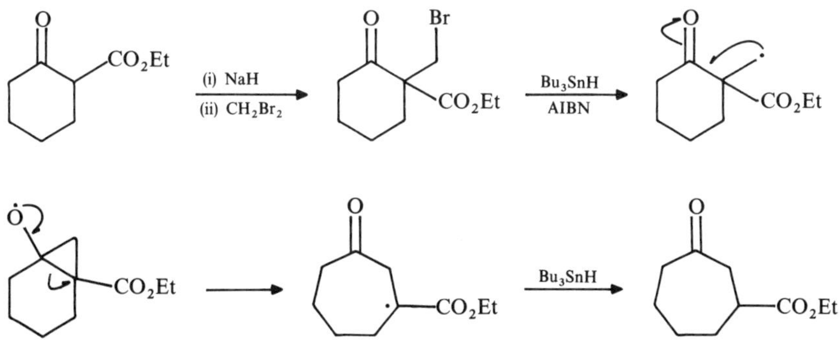

SCHEME 5

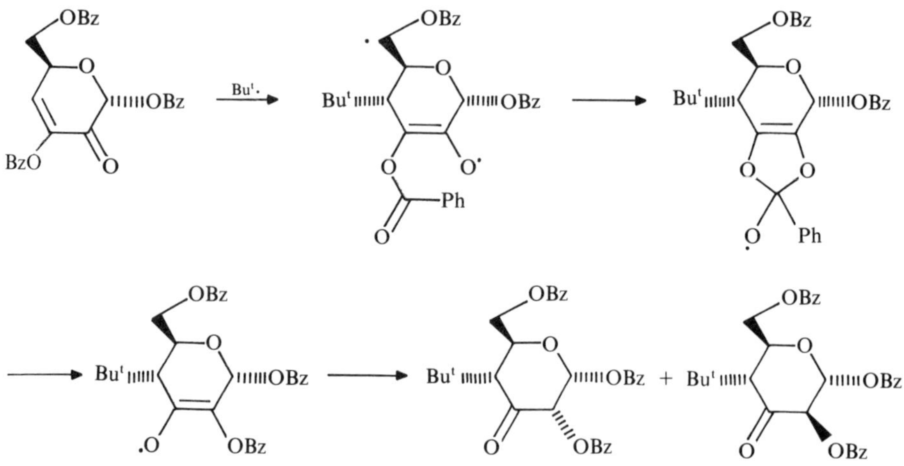

SCHEME 6

Intramolecular Addition

A new force-field model for intramolecular radical addition has appeared.[90] It is based on Allinger's MM2 force field and is adapted for use with that program. A combination of MINDO/3 SCF MO calculations of molecular properties and RRKM calculations of forward and backward rate constants for the ring-chain isomerism of Ph$^{\bullet}$ and HC≡C—CH=CH—CH=CH$^{\bullet}$ has been given;[91] the reaction is of importance for soot nucleation. The ring-chain isomerism of O=C=N—C(O)$CH_2\dot{C}H_2$ and the succinimidyl radical has been studied experimentally;[92] the 2-(isocyanatocarbonyl)ethyl radical cyclizes rapidly to succinimidyl as both form a common adduct with Bu^t_2C=CH_2. The isomerization is forbidden for a ground-state π-configuration of the succinimidyl radical although the π-ground state is that indicated by calculation (cf. ref. 52).

'Best' Arrhenius parameters for radical clocks based on alkenyl radical cyclizations have been brought together from new data and data in the literature;[93] the data for a tertiary radical are less reliable than those for primary and secondary radicals. Similar data have also been given for oxygen-containing radicals.[94]

A number of papers have elaborated on the Baldwin–Beckwith rules for cyclization. Whereas the 5-(enol-*endo*)-*exo*-ring-closure (Scheme 7a) is inhibited on account of a lack of coincidence in the approach vectors of the potentially reactive trigonal carbons, the analogous radical cyclization does occur (Scheme 7b).[95] The reason is the lesser degree of C=C double-bond character in a ketonyl radical than in an enolate anion. The cyclization of ketonyl radicals to bicyclic cyclopentanones has been developed synthetically and the dependence of the ring-junction stereochemistry upon the size of ring closed has been investigated for related radical closures. For the ring-closure of Scheme 8, *trans* ring-junctions are disfavoured for $n = 1$ or 2 but are allowed for $n = 3$.[96] Certain ring-closures by the intramolecular addition of acyl and vinyl radicals onto alkynes have been found to take place readily when the

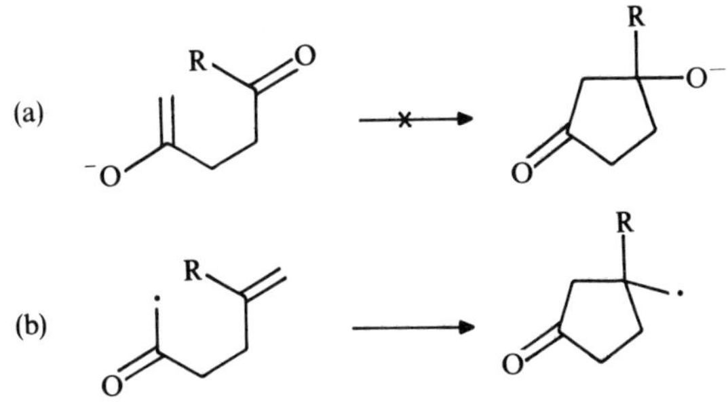

SCHEME 7

SCHEME 8

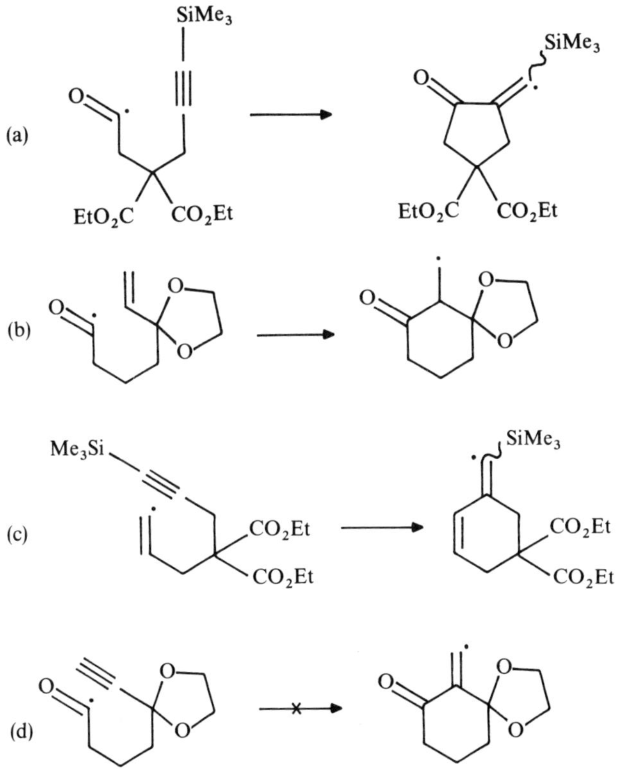

SCHEME 9

ring produced is five-membered but not when it is six-membered. Thus, whereas cyclic products derived from the 5-(π-*exo*)-*exo*-dig, 6-(π-*exo*)-*exo*-trig, and 6-(π-*endo*)-*exo*-dig ring-closures of Scheme 9a–c, respectively, are observed, cyclic products arising by 6-(π-*exo*)-*exo*-dig closure, Scheme 9d, are not observed.[97] The rationale suggests that the geometry of the transition state requires (i) the two π-systems to be nearly coplanar in order to facilitate the development of conjugation in the product, and (ii) the attacking radical centre to approach the alkyne at an acute angle of 60° to the C≡C triple bond. (It is interesting to note that this alkyne approach vector, originally given in the Baldwin rules for ring-closure, has been criticized on the ground that it requires the attacking radical to 'over-fly' one alkyne carbon atom and pass through the nodal plane of the alkyne LUMO before attacking the second carbon atom.)

Developments in synthetic methodology have involved the application of radical cyclizations: a combination of Michael addition with 5-*exo*-dig radical ring-closure provides a means of annelating cyclohexanone by five-membered carbocycles;[98] radical isomerization of unsaturated α-iodocarbonyl compounds gives γ-iodocarbonyl cycloalkanes;[99] acetonitrile(trichloro)bis(triphenylphosphine)-rhenium(III) initiates the radical addition of CCl_4 or $CBrCl_3$ to 1,6-dienes yielding mixed *cis*- and *trans*-1-halomethyl-2-(2, 2, 2-trichloroethyl)cyclopentanes in which the *cis*-isomer predominates;[100] the cyclization of enynes to cyclohexylidene thioethers is initiated by the addition of thiyl radicals to the alkyne function;[101] the stereoselectivity of radical cyclization of ω-substituted α,β-unsaturated esters is greatest for *exo*-closure on a *Z*-alkene function;[102] manganese(III) is of value in effecting oxidative radical polycyclization reactions although with some systems there is evidence that intermediate tertiary radicals may be oxidized to carbocations before later cyclization steps.[103,104]

The competitive addition of radicals to alkenes and carbonyl functions has shown that addition to carbonyl may often be preferred (Scheme 10);[105,106] such reactions suggest potential synthetic applications as yet unexplored.

The capacity of medium-ring cycloalkenyl radicals for rearrangement to bicyclic radicals by transannular reaction has been investigated. Radicals such as cyclooct-4-

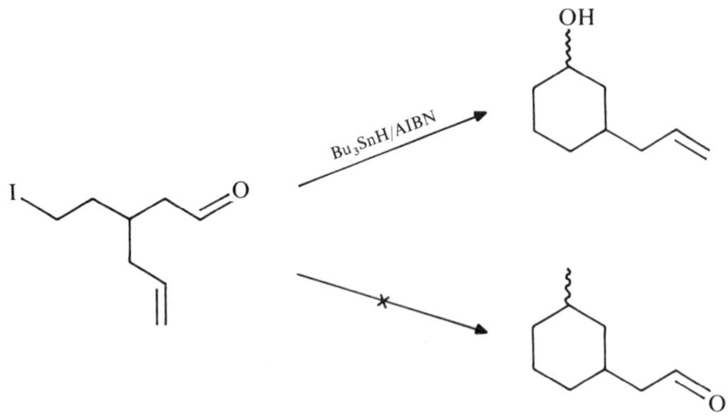

SCHEME 10

enylmethyl (**27**) react readily to give bicyclic products, although more slowly than the simple cyclization of hex-5-enyl.[107] By contrast, the endocyclic radical (**28**) is reluctant to rearrange and may be trapped efficiently by various reagents.[108] Other substituents in the ring may accelerate or inhibit transannular rearrangement depending on their nature and position.

(**27**) (**28**)

The mechanism of iododediazoniation of arenediazonium ions has been elucidated and a new scheme which supersedes previous variants has been advanced as follows:[109]

$$ArN_2^+ + I^- \rightarrow Ar^\bullet + N_2 + I^\bullet$$
$$ArN_2^+ + I_2^{\bullet -} \rightarrow Ar^\bullet + N_2 + I_2$$
$$Ar^\bullet + I_2 \rightarrow ArI + I^\bullet$$
$$Ar^\bullet + I_3^- \rightarrow ArI + I_2^{\bullet -}$$
$$I^\bullet + I^- \rightarrow I_2^{\bullet -}$$
$$I^\bullet + I^\bullet \rightarrow I_2$$
$$I_2 + I^- \rightarrow I_3^-$$

In appropriately substituted diazonium ions, the intermediate aryl radicals rearrange by intramolecular cyclization (Scheme 11). The synthetic potential of this and comparable cyclizations has been indicated,[110] and the occurrence of isomerization of the initial radical-cyclization products, by neophyl rearrangement to thermodynamically stabler isomers, has been discussed.[111] Directly comparable reactions of aryl iodides mediated by Co(I)-salophen reagents have been described.[112,113] Aryl radicals will rearrange by intramolecular addition to appropriately located azo-groups; both *exo* and *endo* cyclizations take place rapidly, a difference from alkenic and alkynic analogues which was ascribed to differences in geometry.[114]

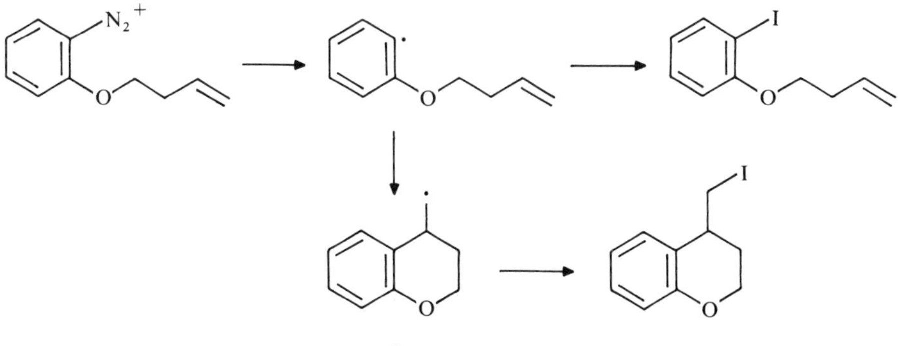

SCHEME 11

Several tetrahydrofuran-forming radical rearrangements have been reported,[115–118] as have comparable pyrrolidine-producing reactions,[119,120] and cyclizations leading to sulphur-containing heterocycles.[121–123] Intramolecular cyclization of aminyl radicals (**29**), produced by flash vacuum pyrolysis, leads to products via spirodienyl radicals (**30**), indicating a preference for the 5-*exo*-trig mode in radical attack at aromatic carbon which is comparable to that in nitrene-mediated rearrangements.[124] The same preference for 5-*exo*-trig ring-closure upon aromatic carbon is demonstrated by the rearrangement of the carbamyl radical (**31**).[125] Macrolide formation by radical cyclization has been successfully accomplished:[126] the lactone (**32**), amongst others, was synthesized in 78% isolated yield by radical cyclization of the ω-iodoacrylate precursor. It was correctly reasoned that, since the effective molarity for the cyclization of a chain of greater than twelve members commonly lies in the range 0.01 to 0.1 M, radical reactions which propagate intermolecularly at such concentration should also occur intramolecularly. The macrocyclization of acrylates is exclusively *endo*-cyclic; for fumarates both modes of closure occur but the *endo*-mode predominates.

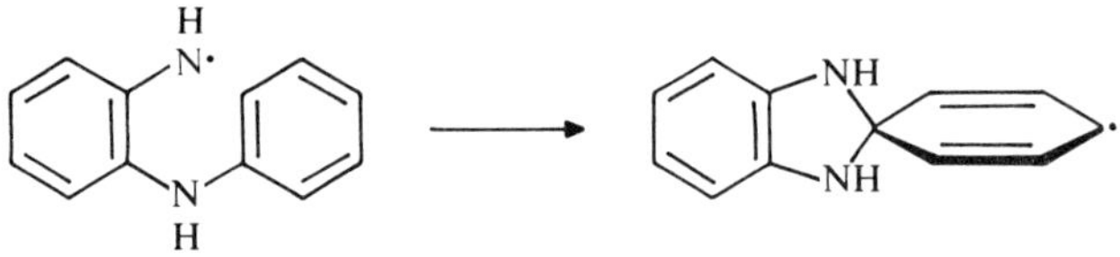

(**29**) (**30**)

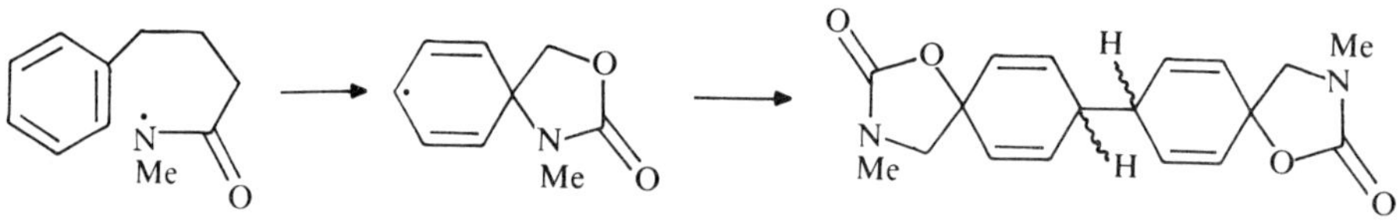

(**31**)

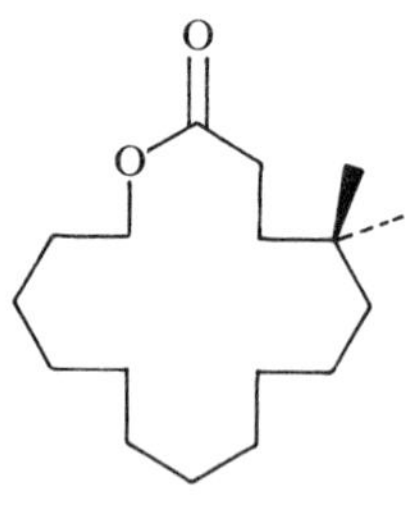

(**32**)

Radical-cyclization steps have been incorporated into various natural-product syntheses: viz fredericamycin A,[127] (±)silphiperfolene,[128] forskolin,[129] and (±)-dihydrocorynantheol.[130]

Miscellaneous Reactions

Evidence has been presented that the rearrangement $CH_3O\dot{C}H_2 \rightarrow CH_3CH_2O^{\cdot}$ is not significant for the formation of C—C bonds during the Zeolite ZSM-5 catalysed methanol conversion reaction.[131] The formation of methylcyclopentane during the reduction of 1-bromo-5-hexene by organotin hydrides plus lithium diisopropylamide or methyllithium has implicated homolytic steps in the reaction mechanism.[132] The paramagnetic liquid (**33**) has been isolated in the dark and quantitatively photochemically isomerized to (**34**).[133] Photo-CIDNP studies of calix(4)arenes (**35**) and related species show that, while photo-excited flavin may abstract any of the phenolic hydrogens, the phenoxyl radicals so formed rapidly rearrange by a reshuffling of hydrogens so that the radical centre is stabilized by being located in a cresol unit.[134]

(**33**) (**34**)

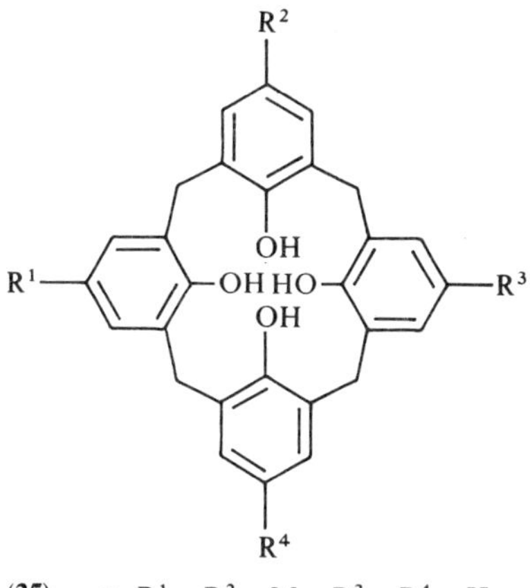

(**35**) **a**; $R^1 = R^2 = Me$; $R^3 = R^4 = H$

b; $R^1 = R^3 = Me$; $R^2 = R^4 = H$

Fragmentations

A one-parameter analytical potential-energy function for β-bonds in free radicals has been derived which accounts quantitatively for their observed stretching frequencies and for the activation barriers for their fission.[135] The β-fragmentation of $Bu^tO^{\cdot}$ has been investigated experimentally in the temperature range 402–443 K, by competition with the addition to NO; the pressure-dependent process:

$$Bu^tO^{\cdot} + M \rightarrow Me^{\cdot} + Me_2CO + M$$

was examined for several inert gases M. A value for the rate constant under atmospheric conditions was interpolated.[136] The ready β-scission of dialkylsulphide-boryl ($R_2S \rightarrow \dot{B}H_2 \rightarrow R^{\bullet} + R\overset{+}{S}{=}\overset{-}{B}H_2$) is one of several types of reaction of these radicals which has been observed in solution by ESR spectroscopy.[137] β-Nitro-sulphones react stereospecifically with tributyltin radicals to give alkenes; the mechanism indicated in Scheme 12 was suggested.[138] Such involvement of alkyl(trialkylstannyloxy)-nitroxyl radicals as intermediates in the denitration of aliphatic nitro-compounds with trialkylstannyl radicals has been confirmed in an ESR study.[139] The decay kinetics of the nitroxyl radicals were investigated in the temperature range 248–326 K; for the radicals from tertiary nitro compounds, a first-order process was found; primary and secondary nitro-compounds decayed by processes of mixed order. After prolonged reaction times the presence of dialkyl-nitroxyl radicals could be detected. A mechanism supported by product studies was presented.

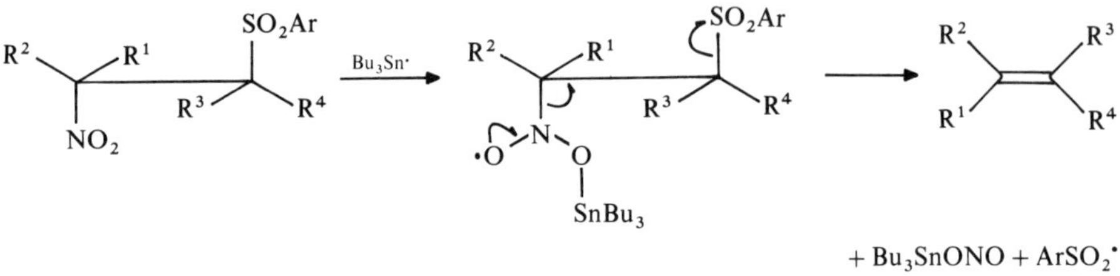

$R^1 = R^3 = Me$, $R^2 = Et$, $R^4 = CN$, CO_2Et

SCHEME 12

The formation of ethene from β-halopropanoic acid by reaction with peroxydisulphate and silver ion has been suggested to involve silver-induced fragmentation of 2-chloroethyl radical.[140] The decay of diazenyl radicals, $ArN{=}\dot{N}$, in cyclopropane solution has been investigated using a time-resolved ESR method.[141]

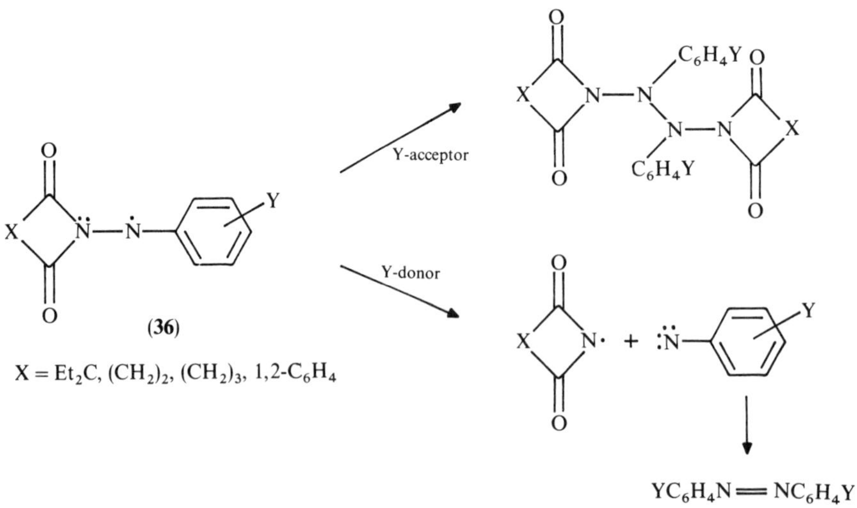

(36)

$X = Et_2C, (CH_2)_2, (CH_2)_3, 1,2\text{-}C_6H_4$

A first-order process was measured in the temperature range − 117 to − 48°C. At − 96°C, the rate constant for first-order decomposition of $C_6D_5N_2^{\bullet}$ was $34\,s^{-1}$; at the same temperature, the first-order rate constants for an extensive series of substituted diazenyl radicals were correlated in a Hammett plot with $\rho = 1.53$, $r = 0.91$. Second-order constants were also measured during the initial part of the decay; for $C_6D_5N_2^{\bullet}$ the second-order rate constant at − 93°C was $9.2 \times 10^5\,M^{-1}\,s^{-1}$. Arrhenius parameters were estimated for both processes. A new α-scission of 2,2-diacylhydrazyl radicals (**36**) has been discovered as one of the reaction paths followed by radicals with electron-donating substituents in the phenyl ring.[142] The mediation of nitrenes is inferred from the nature of the products observed.

Combination and Disproportionation Reactions

Overall cross-reaction rate constants have been measured by molecular modulation spectrometry for the reaction of Me˙ with Et˙, Pr$^{i\bullet}$ and Bu$^{t\bullet}$; the overall self-reaction rate constants and absorption cross-sections were also measured for the four alkyl radicals.[143] Bond strength-bond length (BSBL) calculations have been made for the disproportionation and cross-disproportionation reactions of these alkyl radicals and of $CH_2{=}\dot{C}H$.[144] An Arrhenius expression for the disappearance of Me˙ in a shock tube in the temperature range 1700–2700 K and pressure range 0.3–3 atm. has been given in the Russian literature.[145] The combination of Me˙ with PhO˙ has been examined by means of RRKM theory and kinetic modelling.[146] Both high-temperature shock-tube data for the production of cresols and low-temperature branching ratios for the production of cresols against methylcyclohexadienones were accounted for in terms of a single mechanism. Canonical variational transition-state theory has been used to calculate bimolecular rate constants for the reaction of H˙ and D˙ with Me˙;[147] the values obtained are in good agreement with experimental values. The kinetics of the reaction of H˙ with $H{-}\dot{C}{=}O$ have been studied as a function of temperature at 296 K, 350 K and 418 K;[148] the measured rate constants are in agreement with values calculated by others. The rate constants for the reaction of $H{-}\dot{C}{=}O$ with alkyl radicals have been measured at 297 K following exciplex laser flash photolyses of aldehydes in the gas phase.[149] Combination of time-resolved measurements of the $H\dot{C}{=}O$ concentration with end-product analysis showed that the course of reaction was exclusively disproportionation to alkane and CO, except for Me˙ + $H\dot{C}{=}O$ where a small contribution from recombination is likely. Other acyl/alkyl radical reactions have been studied following steady-state and pulsed photolyses of cyclobutyl methyl ketone.[150] A bimolecular QRRK analysis of reactions of Me˙ has been given including addition to O_2 and to ˙OH.[151] The reaction of Me˙ with excited oxygen ($^1\Delta g$) has been described[152] as have the reactions of alkyl radicals with ˙OH[153] and of Pr^{i}O˙ with O_2, NO and NO_2.[154]

The self-reactions of $Pr^iO_2^{\bullet}$ in the liquid phase at various temperatures within the range 135–373 K and in several solvents have been investigated for comparison with those in the gas phase.[155–157] At temperatures above 363 K, there is evidence that reaction with hydrocarbon solvents becomes increasingly important; below 210 K, the dimerization of the radical to the tetroxide must be considered;

the variation in the ratio of the major products from the reaction, propanone and 2-propanol, with temperature is explained by a reaction in which the dimer fragments to propanone and H_2O_2. The recombination of cyclohexylperoxy radicals in the presence of lauryl peroxide has been investigated.[158] Although allylperoxyl radicals were observed when the pyrolysate of hexa-1,5-diene and oxygen mixtures was condensed in an adamantane matrix at 77 K, no mass spectrometric evidence for them would be found in the gas-phase pyrolysate.[159] The observations were explained in terms of a reversible equilibrium between allylperoxyl radicals on the one hand and allyl radicals and molecular oxygen on the other.

The disproportionation-combination rate constant ratios k_D/k_C have been determined for the reactions $R^{\bullet} + RCH_2\dot{C}HCl$ and for the self-reactions of $RCH_2\dot{C}HCl$ where $R = CF_3$, C_2F_5, and C_3F_7; the finding that, for $R = CF_3$ and C_2F_5, the ratios were very sensitive to the surface/volume ratio of the reaction vessel suggested a heterogeneneous component for disproportionation.[160] The equilibrium constant and kinetics of the reactions of $^{\bullet}C_2F_5$ and $I^{\bullet}$ have been determined.[161] The self-reaction of $Me_3Si\dot{S}iMe_2$ has been investigated in solution at 298 K. The ratio of products due to disproportionation and combination was found in a ratio less than or equal to 0.45, but there was no evidence of silylene formation.[162] It was concluded that silylenes formed during photolysis of polysilanes arise by a photo-extrusion mechanism and not by the self-reaction of polysilyl radicals. ESR spectroscopy has been used to determine bimolecular rate constants for the decay of the trimethylsiloxy iminyl radical (**37**) in the temperature range 213–263 K[55] and to measure the dimerization equilibrium constants of 1,3,2-dithiazolidin-2-yl radicals e.g. (**38**).[163] ESR spectroscopy has also been used to investigate the nature, distribution and dimerization of radicals produced during the solid-phase and aqueous-phase radiolysis of amides.[164] The rate of loss of alkylsuphonyl radicals in alkane solvents has been shown to be diffusion-controlled,[165] and the chemiluminescence which accompanies the sulphoxidation of alkanes by $SO_2/O_2/Ar$ mixtures in the presence of dicyclohexyl peroxide initiator has been ascribed to various

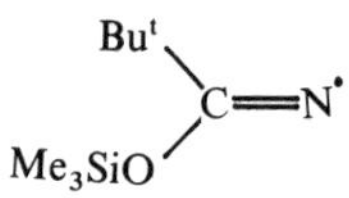

(**37**)

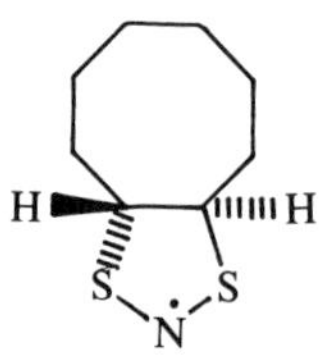

(**38**)

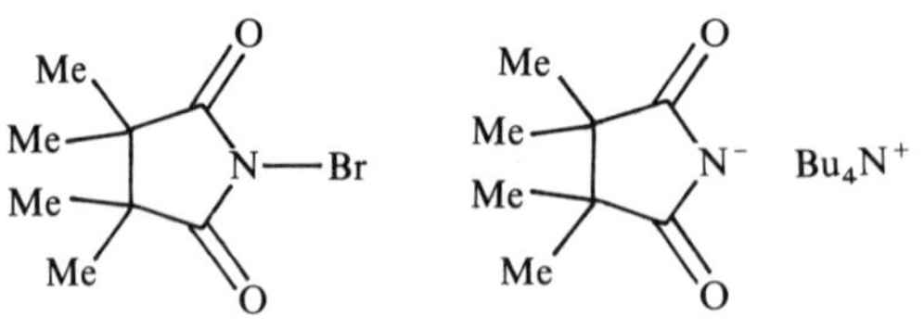

(**39**)

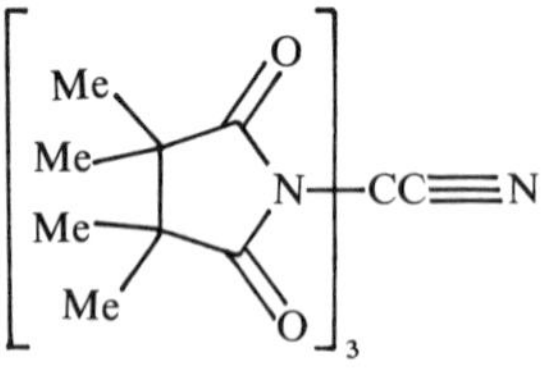

(**40**)

radical-recombination processes.[166] Radical-recombination processes have also been implicated in the solvent dependence of the activation enthalpies for homolysis of Co—C bonds[167] and in the pyrolysis mechanism for *n*-hexane in the temperature range 723 to 823 K.[168]

The consequences of radical-pair interactions in micelles have been reported in a number of papers: a model for polarized EPR spectra generated by radical-pair reactions in micelles has been proposed;[169] the effects of additives on the recombination efficiences and other properties of benzyl radicals[170] and of phenoxyl and semiquinone radicals[171] have been investigated. CIDNP of ^{13}C in the products and reactants has been observed during the nitration of phenols by nitrous acid; the evidence implicates phenoxyl radicals in the radical-pairs which generate the polarization.[172] CIDNP has also provided evidence of a new type of reaction between two radicals in a solvent cage: the primary geminate pair, $Bu^tSO_2^{\bullet}$ and $Ph^{\bullet}$, produced by thermolysis of $Bu^tSO_2N{=}N{-}Ph$, exchange SO_2 to produce $Bu^{t\bullet}$ and $PhSO_2^{\bullet}$.[173] The cage recombination reactions of $PhS^{\bullet}$ in complex hydrocarbon solvents and other reactions have been also reported.[174] An intriguing reaction has been reported following the thermal decomposition in acetonitrile of the complex (**39**) formed by *N*-bromotetramethylsuccinimide and tetrabutylammonium tetramethylsuccinimide: the nitrile (**40**) is produced.[175] The tentative rationale suggests the formation of solvent-caged pairs of tetramethylsuccinimidyl radicals which react with the solvent, the one by hydrogen abstraction, the other by combination with the cyanomethyl radicals so formed to give ultimately (**40**).

Atom Abstraction Reactions

Hydrogen Atom Abstraction by Carbon-centred Radicals

Rate coefficients for the high-temperature reaction of $Me^{\bullet}$ with H_2 have been determined directly using shock-tube techniques and time-resolved measurements of the absorption of $Me^{\bullet}$ at 216.5 nm; the derived Arrhenius parameters agree with measurements indirectly obtained.[176] Two symposium papers concerned with hydrogen-atom tunnelling have used the $Me^{\bullet}/CH_4$ hydrogen atom exchange as the basis of their respective considerations, the one comparing two theoretical approaches to the tunnelling phenomenon,[177] the other concerned with solvent effects on tunnelling reactions.[178] Other theoretical approaches to hydrogen abstraction reactions have also appeared.[179–181]

Rate constants for hydrogen abstraction, relative to that for its termination, have been measured for $Ph\dot{C}HMe$ abstracting from Ph(Me)CHOH and $PhCH_2OH$ in the liquid phase at 120°C.[182] The rates of benzylic hydrogen abstraction have also been determined for $Ph_2\dot{C}H$ and fluorenyl radical abstracting from 1,2,3,4-tetrahydronaphthalene, 9,10-dihydroanthracene, and toluene at temperatures around 500 K; the equilibrium constant for H-atom exchange between $Ph_2\dot{C}H$ and fluorenyl radical was also evaluated.[183] The stereochemistry of abstraction of deuterium from stannanes and germanes by various cyclohexyl radicals has been investigated: 4-*t*-butylcyclohexyl radical reacts to give predominantly *cis*-4-*t*-butyl-

1-^{2}H-cyclohexane, and 1,4-dimethylcyclohexyl radical gives mixed *cis* and *trans* products.[184] The rate at which a secondary alkyl radical abstracts hydrogen from $Bu^n{}_3GeH$ has been shown to be some eighty-fold slower than the parallel abstraction from $Bu^n{}_3SnH$ at 298 K.[93] The transfer of hydrogen to alkenes from transition metal hydrides has been shown to have homolytic character, and its rate measured, by incorporation of a radical clock.[185] Hydrogen atom abstraction from CH_3CN has been found to be the main reaction path when 1-adamantyl radicals react with the solvent; minor reaction paths arise by addition.[186]

Bond strength-bond length calculations have been carried out for hydrogen atom abstractions by $CH_2{=}\dot{C}H$ from a series of hydrogen-donor molecules, including H_2, alkanes, and carbonyl compounds.[187] Intermolecular hydrogen abstraction reactions by substituted vinyl radicals have been shown to be of sufficiently high rates to compete with intramolecular hydrogen transfer and additions to alkenes.[188] Variously generated 2-methylphenyl radical has been shown to abstract hydrogen from the methyl groups of toluenes in preference to addition to the aromatic rings, in contrast with its 3- and 4-isomers. A steric cause was considered probable.[189]

Hydrogen Atom Abstraction by Oxygen-centred Radicals

Absolute rate constants for hydrogen abstraction by $HO^{\bullet}$ from various alkanes have been reported. Thus, concordant results at 295 K have been obtained by pulse radiolysis and by laser photolysis for ethane[190,191] and for cyclohexane;[190,192] values are also given for cyclopentane and for perdeuteriated derivatives of cyclopentane and cyclohexane and fitted by modified Arrhenius expressions.[192] Arrhenius parameters for hydrogen abstraction by $HO^{\bullet}$ from methanol in the temperature range 260–283 K and from ethanol in the range 255–459 K have been presented.[193] Those for MeOH are in excellent agreement with precedent but those for EtOH differ markedly. The formation of radicals from polyhydric alcohols and carbohydrates by radiolysis in N_2O-saturated aqueous solution has been measured and their reactions with O_2 and with 1,4-dithiothreitol in various pH conditions investigated.[194] Absolute rate coefficients for the reactions of $HO^{\bullet}$ with MeOMe and EtOEt over the temperature range 295–442 K have been measured and fitted by Arrhenius equations.[195] Transition-state-theory calculations previously used for modelling the abstraction of hydrogen by $HO^{\bullet}$ from alkanes have been adapted for haloalkanes.[196] For each haloalkane $\Delta S^{\ddagger}$ is calculated and the result used with experimental values of the rate coefficients at 298 K to estimate $k(T)$ at higher temperatures. For the temperature range up to 500 K good agreement was found between experimental and calculated rate coefficients for 10 halomethanes and 18 haloethanes with the notable exception of $CHCl_3$. Since it is unlikely that for this compound the dynamics of the reaction are unique within the whole family, it is suggested that the experimental data are erroneous. The calculated $\Delta S^{\ddagger}(298)$ values given by the calculations for α-abstraction are fitted, with a maximum error of 1.4 and an average error of 0.3 cal K^{-1} by the expression:

$$\Delta S^{\ddagger}(298) = -2.2 \ln M - 18.0 + R \ln n(\mathrm{H})$$

where M is the relative molecular mass and n(H) is the number of abstractable α-H atoms.[197] The activation energy $E(298)$ can be fitted with a maximum error of 1.0 and an average error of 0.3 kcal mol^{-1} by the expression:

$$E(298)/R = 2100 - 85n(\mathrm{F}) - 515n(\mathrm{Cl}) - 950n(\mathrm{CH_3}) \\ - 600n(\mathrm{CH_2X}) - 650n(\mathrm{CHX_2}) - 250n(\mathrm{CX_3})$$

where n(i) indicates the number of H atoms on CH_4 replaced by indicated substituents and X is Cl or F. These empirical expressions were combined to yield a universal rate expression which depends only on the relative molecular mass and the number of extractable hydrogen atoms.

Hydrogen abstraction by HO$^{\bullet}$ from trimethylamine in aqueous solution has been reported to give rise to a series of interconvertible radicals: $Me_3N^{+\bullet}$, $Me_2\overset{+}{N}H\dot{C}H_2$ benzophenone have been found to correlate with the ionization potentials of the radicals were discussed.[198] When HO$^{\bullet}$, generated from Ti^{3+} and H_2O_2, abstracts hydrogen from certain amino-acid derivatives, the expected α-abstraction products are observed; thus abstraction from *N*-acetylglycine, for example, gives $CH_3CONH\dot{C}HCO_2H$ which is captodatively stabilized.[199] However, abstraction from other compounds, e.g. *N*-acetylvaline, gave predominantly the γ-abstraction product $^{\bullet}CH_2CH(Me)CH(CO_2H)NHCOCH_3$. The authors accounted for this behaviour in terms of a complexation of the amide group by Ti rather than by a selective destruction of the α-radicals.

The rate constants for HO$^{\bullet}$ reacting with isomeric butenes near 1200 K have been found to increase with the number of available allylic hydrogens.[200] Hydrogen abstraction is the dominant reaction mode for HO$^{\bullet}$ and $CH_2{=}CH_2$ at elevated temperatures (748–1173 K).[201] Likewise, H-abstraction is important for benzene derivatives at elevated temperatures where the importance of addition reactions is reduced; substituent effects have been investigated for H-abstraction reactions of substituted benzenes.[202] A generalized approach to the empirical prediction of room temperature rate constants of HO$^{\bullet}$ with organic substances, including H-abstraction reactions, has been presented.[203] This has been tested against the available data base of about 300 compounds, and for only 18 of these do the calculated and experimental rate constants fail to agree within a factor of 2.

Relative rates of H-abstraction by Bu^{t}O$^{\bullet}$ from a series of 13 homoaryl- and heteroaryl-methanes at 70°C have been determined.[204] Although the range of relative reactivities was small, significant variations between isomers were found. Fairly satisfactory correlations were obtained between the relative reactivities and various computed energies; it was found that a carbocation model described the transition state better than a radical model. Stereochemical effects have been found in the abstraction of hydrogen from diastereoisomeric 2,3-dichlorobutanes by Bu^{t}O$^{\bullet}$: different apparent relative reactivities were observed at C(1) and C(2) between *meso* and racemic diasteoreoisomers, *apparent* because other possible more complex mechanisms have not yet been specifically excluded.[205] Reaction rates for H-abstraction by *cyclo*-$C_6H_{11}O^{\bullet}$ from 1,3-dioxolanes (**41**) in benzene at 50°C have been measured.[206]

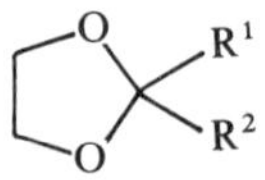

(41)

R^1	R^2	$10^{-6}k/M^{-1}\,s^{-1}$
H	H	2.9
H	Me	7.2
H	Pr^i	6.0
Me	Me	0.9

(42a) **(42b)**

Hydrogen abstraction reactions have featured in two studies of aroyloxyl radicals $ArCO_2^{\bullet}$. Reactions of $MeOC_6H_4CO_2^{\bullet}$ have been monitored via the broad featureless absorption at 500–800 nm which is ascribed to the radical,[207] and H-abstraction from thiophenols by $PhCO_2^{\bullet}$ has been investigated by quantitative product studies.[208] The rates of abstraction of hydrogen from various RH by triplet benzophenone have been found to correlate with the ionization potentials of the radicals $R^{\bullet}$.[209] The transition state for the hydrogen transfer may be represented by (**42**). The difference in energy between the reaction products ($Ph_2\dot{C}$—OH + $R^{\bullet}$) and (**42b**) is approximated by:

$$I_P(R^{\bullet}) + D(OH) - EA(O^{\bullet}) - C$$

where $I_P(R^{\bullet})$ is the ionization potential of $R^{\bullet}$, $D(OH)$ is the O—H bond strength in the ketyl radical, $EA(O^{\bullet})$ is the electron affinity of the oxy-radical obtained by removing $H^{\bullet}$ from the ketyl, and C is the coulombic energy evolved in bringing $Ph_2\dot{C}$—O^- and R^+ from infinity to their separation in (**42**). For a series of substrates R—H, only $I_P(R^{\bullet})$ varies, hence the correlation. Methanol is approximately 2×10^3 times more reactive than acetonitrile, for similar C—H bond dissociation energies, because of the low ionization potential of $^{\bullet}CH_2OH$ by comparison with that of $^{\bullet}CH_2CN$. The complex reaction of nitrogen dioxide with alkenes involves both dimeric N_2O_4 and monomeric NO_2; the reaction of the latter proceeds, in part, by abstraction of allylic hydrogens.[210] The formation of gas-phase radicals by the heterogeneous reaction of hydrocarbons with metal oxides appears, from the nature of the radicals, to involve hydrogen abstraction at the oxide surface.[211,212]

Hydrogen Atom Abstraction by Halogen Atoms

The equilibrium:

$$Cl^{\bullet} + CH_4 \rightleftarrows HCl + {}^{\bullet}CH_3$$

has been used in an analysis of the flow dynamics of a new, very-low-pressure reactor with the outcome that new accurate determinations of kinetic and thermodynamic parameters of the reaction have been made: the forward rate constant $k_f = (0.993 \pm 0.013) \times 10^{13}\ cm^3\ mol^{-1}\ s^{-1}$ at 25°C; the equilibrium constant $K = 1.406 \pm 0.034$; the reaction enthalpy is $\Delta H^0 = 1.89 \pm 0.10\ kcal\ mol^{-1}$ and the heat of formation of the methyl radical is $\Delta H_f^0(Me) = 35.06 \pm 0.10\ kcal\ mol^{-1}$.[213] The new reactor led also to a reappraisal of the rate constant for abstraction of hydrogen from isobutane by Br ;[214] this latter reaction additionally formed a case history discussed by Benson at the 1986 Polanyi Medal presentation.[215] Gas-phase kinetic measurements of the reaction of Br˙ with CH_3CH_2Br in the temperature interval 40–100°C have yielded a hydrogen abstraction rate constant of

$$k/\text{cm}^3\,\text{mol}^{-1}\,\text{s}^{-1} = 10^{12.645 \pm 0.099} \exp[-(10583 \pm 158)/RT],$$

a bond dissociation energy $D^0(CH_3CHBr—H) = 96.4 \pm 1.1\,\text{kcal mol}^{-1}$ and a heat of formation of the 1-bromoethyl radical $\Delta H_f{}^0(CH_3\dot{C}HBr) = 29.6 \pm 1.1\,\text{kcal mol}^{-1}$.[216] Soviet workers have measured a rate constant for the gas-phase bromination of $ClCF_2H$ and have estimated a bond energy of 412 kJ mol^{-1} for the C—H bond in this compound.[217] The importance of solvent viscosity for the kinetics of abstraction of benzylic hydrogen by Br˙ has been emphasized. It was shown that viscosity dependence of the relative rates of bromination of toluene and 4-chlorotoluene ran counter to the temperature dependence. The work thus shows the need to recognize the importance of viscosity effects in any consideration of anomalous Arrhenius plots. The kinetic isotope effect in benzylic bromination was analysed for its viscosity dependence by reference to the gas-phase reaction.[218] Kinetic isotope effects for benzylic chlorination of toluene in a two-phase aqueous hypochlorite/CH_2Cl_2 solvent with a phase-transfer agent and for bromination by *N*-bromosuccinimide in CCl_4 have been evaluated. From the relative magnitudes of the primary and secondary effects it was concluded that extensive C—H bond breaking has occurred in the transition states for hydrogen abstraction, but that little rehybridization of the carbon centre has occurred.[219]

Hydrogen Atom Abstraction by Miscellaneous Radicals

Kinetic isotope effects for hydrogen abstraction from various saturated alkanes and their perdeuteriated analogues by D˙ have been determined. The results, in the temperature range 363–473 K, may be expressed by an Arrhenius expression:

$$k_H/k_D = A_H/A_D \exp(E_D - E_H)/RT$$

The variation in the value of the ratio of pre-exponential factors was discussed, and that in the activation energy difference was related to the strength of the C—H/D bond broken. Theoretical calculations of the isotope effects were carried out which gave good agreement with experiment when tunnelling effects were allowed for.[220] Soviet work has investigated the relative reactivities of different C—H bonds in alkenes towards abstraction by H˙.[221]

The hydrogen abstraction reactions of $^{\bullet}NH_2\,(\tilde{X}^2B_1)$ with neopentane, isobutane, toluene and acetaldehyde have been investigated in an isothermal-discharge-flow system over given temperature and pressure ranges. Arrhenius expressions for the rate coefficients were quoted and correlations of activation energies with bond energies of the broken C—H bonds considered.[222] Synthetic virtues of nitrogen-centred radicals (1,8-naphthalenedicarboximidyl and phthalidimidyl radicals) have been assessed[223] and the synthetic application of *N*-cyanamide radicals, $R\dot{N}—CN$ (for steroidal and similar R) described.[224]

Polarity reversal catalysis in H-atom abstractions by amine-boryl radicals has been recognized. Thus, H-abstraction from CH_3CN by $Bu^tO^\bullet$ is inefficient and gives only weak ESR signals from $^\bullet CH_2CN$ when effected in cyclopropane and similar solvents. However, in the presence of the amine-borane $Me_3N \rightarrow BH_2$(*t*-hexyl) a strong signal from $^\bullet CH_2CN$ was observed. The difference in behaviour was explained in terms of a catalytic cycle in which H-abstraction from the amine borane by $Bu^tO^\bullet$ takes place readily and the nucleophilic radical so formed, $Me_3N \rightarrow \dot{B}HHx^t$, then abstracts hydrogen from CH_3CN more readily than the electrophilic $Bu^tO^\bullet$.[225]

Halogen Atom Abstraction

The MINDO semi-empirical SCF MO method has been applied in calculations to locate transition states for the reactions of $H^\bullet$ with methyl halides. Calculations for CH_3F failed, giving anomalous Arrhenius parameters; for the remaining methyl halides, the calculations indicated that atom abstraction reactions should occur in preference to substitutions at thermal energies and that halogen abstraction is the most favoured. The increase in reactivity along the series Cl < Br < I was predicted in accord with observation.[226] Gas-phase chlorine atom abstractions by $^\bullet CH_3$ and $^\bullet CF_3$ from CF_2Cl_2 and $CFCl_3$ have been measured in the 500–600 K region, relative to the methyl radical combination rates. Arrhenius pre-exponential factors for seven chlorine atom transfers of the two radicals were calculated using transition-state theory.[227] Second-order rate constants have been found for the abstraction of halogen atoms by 1-octyl radicals from a variety of alkyl halides in benzene at 50°C.[228] Two methods were used: competition for 1-Oc$^\bullet$ by alkyl iodides and Bu_3SnH and competition for 1-Oc$^\bullet$ by alkyl halide and (**43a**). The values for halogen transfer were deduced from measurements of the product distribution, the reactant ratios, and a knowledge of the rates of reaction of 1-Oc$^\bullet$ with Bu_3SnH and (**43a**); the problems caused by unwanted heterolytic reactions of alkyl halide probes of homolytic mechanisms were also discussed. The rate constant for abstraction of chlorine from CCl_4 by hex-5-enyl has been determined using the cyclization of the radical as the clock.[229] For the temperature range 274–353 K the rate constant is given by:

$$\log(k/\text{M}^{-1}/\text{s}^{-1}) = (8.4 \pm 0.3) - (6.2 \pm 0.4)/\theta$$

where $\theta = 2.3RT$ kcal mol^{-1}. The rate coefficient evaluated for 298 K using this expression, 7.2×10^3 M^{-1} s^{-1}, is significantly smaller than other estimates for primary radicals.

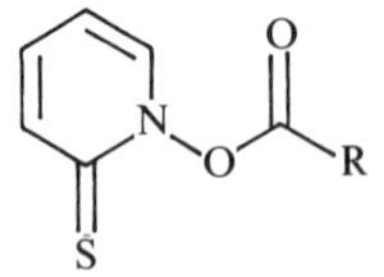

(**43**) **a**; R = C_8H_{17}

b; R = aryl

The replacement of the carboxyl group in aromatic carboxylic acids by Br or I occurs cleanly on treatment of (**43b**) with CCl_3Br, CH_3I or CH_2I_2 in the presence of an initiator. Homolysis of (**43b**) produces Ar$^\bullet$ which abstracts the halogen from its donor.[230]

The abstraction of halogen from toluenesulphonyl halides by a number of radicals has been reported and interpretations made on the nature of the transition states in the light of observed relative reactivities.[231,232] Absolute rate constants have been found for the abstraction of chlorine from ring-substituted benzyl halides by $Et_3\dot{S}i$ radicals. The Hammett ρ value found for the reaction (0.64) is one of the largest positive values to be reported for a radical reaction.[233]

The rate constants for the reactions of Me$^\bullet$, Et$^\bullet$, $CH_2{=}\dot{C}H$, $CH_2{=}CH{-}\dot{C}H_2$ and $CH{\equiv}C{-}\dot{C}H_2$ with Cl_2 have been measured over various ranges of temperature, depending on the radical, and Arrhenius expressions found.[234,235]

Halogenation

New results for the selective chlorination of steroids by attached nicotinic ester templates have been reported and earlier results retracted.[236] The chlorination of mixed ethene and dichloroethane has been investigated in a dark system using an agitated vessel with a flat gas-liquid interface. The reaction kinetics were elucidated via a theoretical treatment of simultaneous absorption and reaction for gas-liquid systems. The order of reaction for the substitution of the dichloroethane differed according to whether ethene was present or not. An explanation in terms of an enhanced production of Cl$^\bullet$ in the presence of ethene was offered.[237] Soviet workers have investigated the chlorination of $CCl_2{=}CCl_2$ in the temperature range 473–773 K. A non-branched-chain mechanism was inferred whose rate at lower temperatures, under irreversible chlorination conditions, approximated that of the dissociation of Cl_2.[238] Other Soviet work has been concerned with the exhaustive chlorination of toluene. Rate-constant ratios (k_1/k_2, k_2/k_3) were obtained for the three consecutive substitutions of the side-chain in the temperature interval 25 to 90°C and were found to decrease with temperature.[239]

The selectivity of the photo-chlorination of tertiary and primary sites in 2,3-dimethylbutane varies in the presence of fluorinated benzenes, decreasing as the extent of fluorination increases. As usual in such systems of variable selectivity which contain aromatic substrates, the formation of σ-complexes between Cl$^\bullet$ and the aromatic has been invoked. The relationship of selectivity to extent of fluorination was explained in terms of slower rates of σ-complex formation, higher rates of σ-complex dissociation, and a lower intrinsic selectivity of chlorination by the complex, as the degree of fluorination of the aromatic is increased.[240] Reaction of CH_4 with low concentrations (1% by volume) of Cl_2 at high temperatures leads to the formation of C_2H_6, C_2H_4, and C_3H_6 as well as CH_3Cl. The percentage of methane that is converted to non-chlorinated hydrocarbons rises with temperature reaching 77% at 1000°C. The radical mechanisms which divert chlorination to hydrocarbon products have been elucidated.[241] Side-chain acetoxylation of toluenes is effected by the decomposition of persulphate in acetic acid, induced by catalytic amounts of bromide ion.[242] The mechanism which has been proposed involves an intermediate radical bromination with heterolytic conversion of the intermediate benzyl bromide to the benzyl acetate.

$$S_2O_8^{2-} + 2Br^- \rightarrow 2SO_4^{2-} + 2Br^\bullet$$
$$Br^\bullet + ArCH_3 \rightarrow Ar\dot{C}H_2 + HBr$$
$$2Br^\bullet \rightarrow Br_2$$
$$Ar\dot{C}H_2 + Br_2 \rightarrow ArCH_2Br + Br^\bullet$$
$$HBr + AcO^- \rightarrow AcOH + Br^-$$
$$ArCH_2Br + AcO^- \rightarrow ArCH_2OAc + Br_2$$

Overall:

$$ArCH_3 + S_2O_8^{2-} + 2AcO^- \rightarrow ArCH_2OAc + 2SO_4^{2-} + AcOH$$

Various by-products and other ramifications of the mechanism were also discussed.

Addition Reactions

Theoretical work has been carried out to model the addition of $Me^\bullet$ to $CH_2{=}CH_2$ complexed by Li^+; a significant acceleration over addition to uncomplexed alkene is indicated. The simple rationale offered is that most radical additions to alkenes involve a predominant radical SOMO/alkene π^*-LUMO interaction; complexation lowers the LUMO energy hence increasing the interaction. It was suggested that perhaps Ag^+ complexes of alkenes, which are most stable for those alkenes which are least reactive towards radicals, might usefully extend the range of alkene substrates suitable for radical additions.[243] Papers have appeared in the Soviet literature which consider the idea of radical polarity and its importance for addition reactions to alkenes and allenes.[244–246]

Carbon-centred Radicals

The kinetics of the reaction of $Me^\bullet$ with propene have been analysed in some detail. Rate constants have been determined for overall addition and for specific terminal and non-terminal additions; combination with data for lower temperatures permitted the evaluation of Arrhenius parameters for the different addition modes. The selectivity of addition relates to differences in activition energy rather than in pre-exponential factor. Comparison of the activation parameters with those for allylic H-abstraction shows that they differ in the pre-exponential factors; addition is the more important primary process but at high temperatures its reversal reduces its importance and allows abstraction to predominate.[247] The addition of $^\bullet C{\equiv}N$ to ethene and ethyne has been studied and Arrhenius parameters found for reaction in the temperature range ambient to 700 K and at 50 torr total pressure.[248] The thermal gas-phase addition of $CF_3O_3CF_3$ to $CF_2{=}CF_2$ has been investigated. Several products of the form $CF_3O(CF_2CH_2)_nO_2CF_3$ ($n = 1$–12) are produced. The reaction is a free-radical telomerization, initiated by addition of $CF_3O^\bullet$ to alkene monomer; it becomes explosive at a particular temperature. A value for the rate constant for addition of telomeric radicals to alkenes has been estimated and also the corresponding activation energy.[249]

The regioselectivity of radical addition of $BrCH(CN)_2$ to alkenes has been examined. Relative rates of addition to eleven alkenes were measured with the finding that steric effects do not influence the addition rates to *n*-alkyl-substituted alkenes. A correlation was found of $\ln(k_{rel})$ values with Fukui's superdelocalizability index, $S_r^{(R)}$. The values of the index showed significant differences for attack by $^{\bullet}CH(CN)_2$ at the different alkene carbons, implying a regioselectivity determined by polar effects.[250] Others have also addressed the problem of regioselectivity of radical additions to alkenes by investigating the competition of pairs of alkenes for $^{\bullet}CR^1R^2CO_2Et$ (R^1, R^2 = H and alkyl). No simple conclusions could be drawn.[251] A mechanism has been proposed to account for the regioselectivity of lactone formation by radical addition of 2-bromoalkanoic acids to terminal alkenes (Scheme 13).[252] The photo-stimulated radical-chain addition of Bu^{t}HgCl to a series of alkenes and alkynes giving products of particular regiochemistry has also been described.[253]

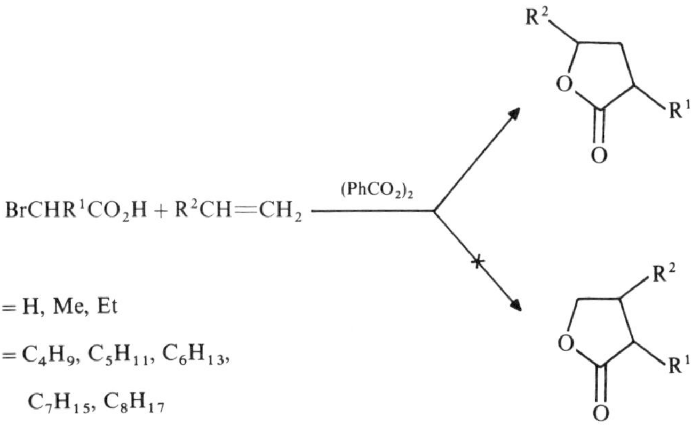

SCHEME 13

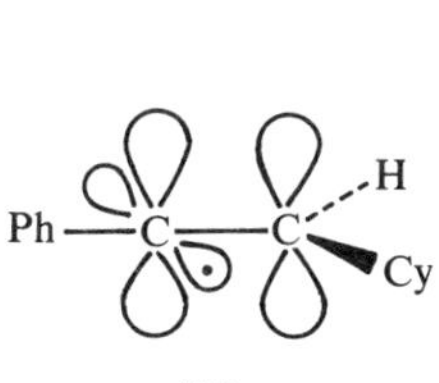

(44)

The stereoselectivity of radical addition reactions has aroused a degree of interest. It has been shown that, for the radical addition of R—H to alkenes and alkynes, the H-donor and the temperature can be of considerable importance for the stereochemistry observed.[254] Thus, in the addition of cyclohexane to PhC≡CH, the *Z*/*E* ratio of the product varied with the H-donor used and with temperature. In general, the *Z*-product is favoured by the lower enthalpy of activation: approach of the H-donor to the intermediate radical (**44**) is less hindered when *anti* to cyclohexyl (Cy) than when *syn*; but the difference in energy of the diastereoisomeric transition states, and hence the predominance of the *Z*-product, varies with the reactivity of the H-donor. Entropic effects may, however, counter enthalpic effects to the extent of inverting the product ratio at high temperatures. Stereoselective addition of $BrCCl_3$ to an acyclic double bond has been observed; addition to either dimethyl fumarate or dimethyl maleate gives a predominance of the *erythro* diastereoisomer of the adduct whereas addition to maleic anhydride gives the *threo* adduct.[255] The thiohydroxamic ester method has been used in the development of stereoselective radical additions for synthetic purposes. An achiral radical was added to the acrylate ester of a chiral alcohol resulting in a moderate enantiomeric excess at the new asymmetric centre created (Scheme 14a);[256] a chiral radical was added to methyl acrylate with essentially complete retention of configuration, although in this case there was no enantiomeric excess at the new asymmetric centre (Scheme 14b).[257] Similar chiral syntheses of amino-acids have been reported.[258]

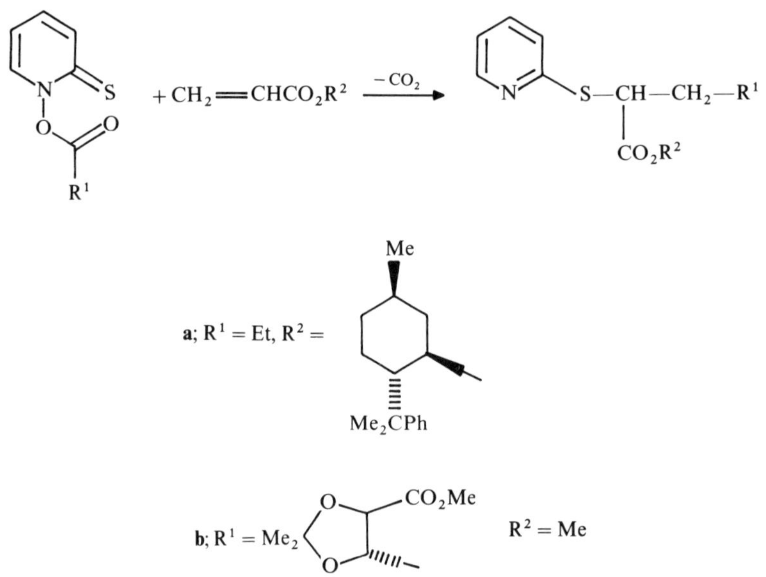

SCHEME 14

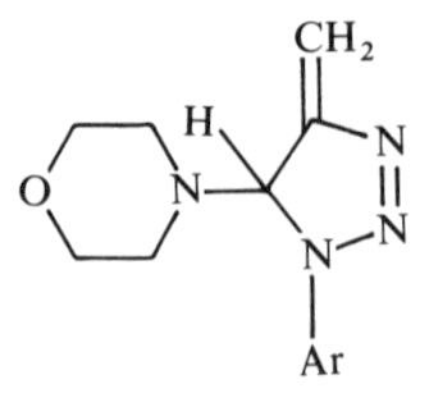

(**45**)

Stereoselective functionalization of a penicillanate via an addition reaction of a chiral radical has also been described.[259] The addition of radicals, generated by the thiohydroxamic ester method, to quinones has been investigated;[260] other radical additions to quinones have been described in Russian.[261] A complex array of reactions which follows upon the addition of trihalomethyl radicals to 1-aryl-4-methylene-5-morpholino-1,2,3-triazolines (**45**) has been, in part, elucidated.[262] A process for the reductive coupling of two methanal molecules to form ethane-1,2-diol, which involves a radical addition step, has been described (Scheme 15)[263] and a closely similar reaction, the alkylation of aldoximes by radical addition to the double bond, has also been reported.[264]

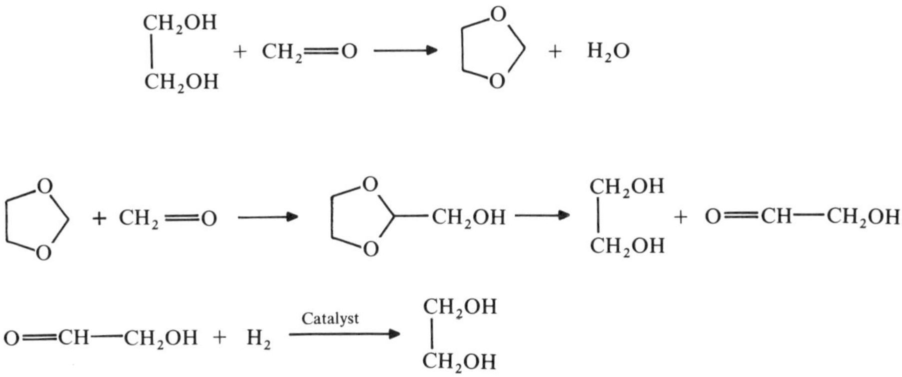

SCHEME 15

Oxygen-centred Radicals

Rate constants and activation parameters have been determined for the reaction of $HO_2^{\bullet}$ with a series of alkenes in the temperature interval 400–520°C. The results show that addition is characterized by high activation energies which correlate well with the ionization potentials of the alkenes. Arrhenius parameters for the reversal of the addition, calcualted thermochemically, indicate this process to be slow in comparison with the alternative $S_{H}i$ decomposition to oxirane plus $^{\bullet}OH$.[265] Reactions of $HO^{\bullet}$ with alkynes have been studied by a discharge-flow resonance fluorescence technique in the temperature range 253–343 K. New rate constants for pent-1-yne and hex-1-yne fit into a general pattern with earlier values for other alkynes and are consistent with electrophilic addition as the first reaction step; the relative reactivities of alkynes with $HO^{\bullet}$ were compared with those with $NO_3^{\bullet}$.[266] Several new rate constants have been determined for the addition of the latter radical at ambient temperatures to a series of alkynes, chloroalkenes, and unsaturated aldehydes; they are quoted relative to the additions to ethene and propene.[267] Addition dominates the reaction of $HO^{\bullet}$ with $CH_2{=}CH_2$ in the temperature interval 343–563 K under gas-phase pulse radiolytic conditions; at higher temperatures reversal of addition and hydrogen abstraction supersede addition.[201] The synthetic utility of addition of $HO^{\bullet}$, from photo-decomposition of

H_2O_2 in CH_3CN, to cycloalkenes as a route to secondary alcohols has been explored.[268]

Aromatic substrates undergo addition of HO˙ in the gas phase at relatively low temperatures, and rate constants have been determined for benzene and a series of halogenated benzenes at 296 K. Hexafluorobenzene remained well behaved at higher temperatures, but H-bearing compounds exhibited H-abstraction reactions at elevated temperatures. The temperature of onset of H-abstraction increased with the size of the halogen indicating an increase in stability of the adducts with size of the halogen. The data were used to estimate stabilization energies for the adducts of HO˙ with PhX (X = H, F, Cl, Br and I).[191] Addition of HO˙ to 6- and 9-substituted purines occurs at two, perhaps more, ring positions (Scheme 16). Attack at position 4 gives an adduct which readily dehydrates producing a cation-radical; this process is especially susceptible to the influence of the substituent at C(6). Attack at position 8 gives an adduct which undergoes rupture of the imidazole ring.[269,270]

The radical $Bu_2{}^tC{=}NO^•$ oxidizes phenols in various ways. Dihydric phenols give corresponding quinones whereas simple phenols form adducts such as (**46**) and (**47**). The rate constants for some of these reactions have been measured by kinetic ESR spectroscopy.[271]

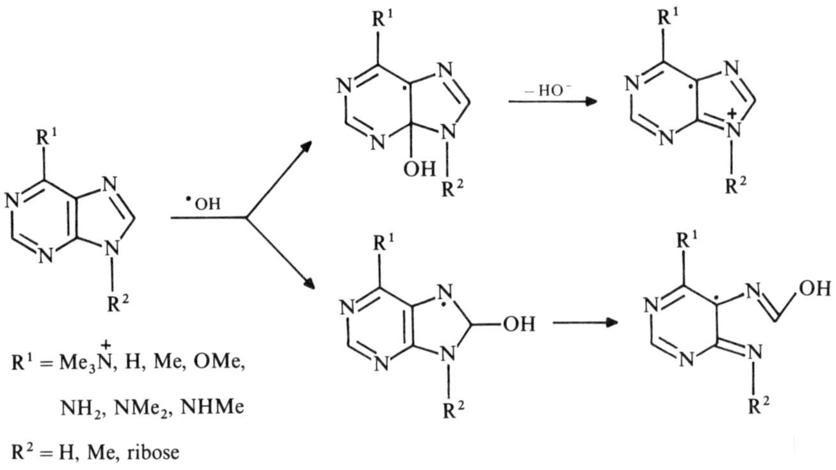

SCHEME 16

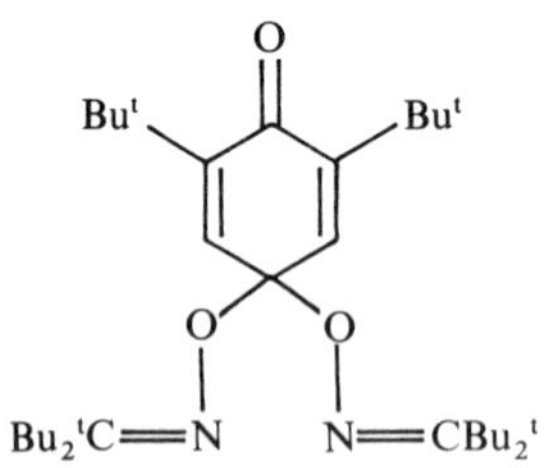

(**46**)

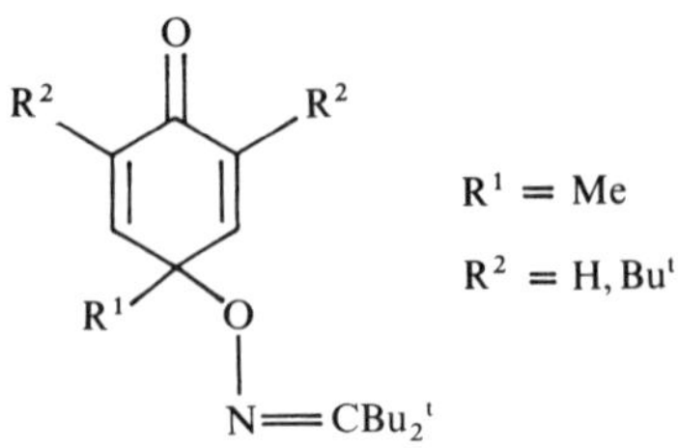

(**47**)

Miscellaneous Radicals

Matrix-isolation experiments have enabled the observation of two products resulting from the addition of $H^{\bullet}$ to Ph≡CH. These were identified as $Ph\dot{C}{=}CH_2$ and as the 1-ethynylcyclohexa-2,4-dienyl radical and UV/visible spectroscopic absorptions were assigned to each; 1-phenylpropyne gave comparable results. The relative stabilities of the various adducts were discussed.[272]

The kinetics and mechanism of the reaction of $^{\bullet}NF_2$ with alkenes have been reconsidered. A logical flaw in the interpretation of measurements in 1965 was noticed and the original data were reinterpreted. The assumed mechanism was tested by modelling with RRKM theory and the experimental results were combined with recent thermochemical data to give estimates of C—N bond dissociation energies in fluoroamines.[273] Addition of nitrogen-centred radicals to an alkene features in the photochemically initiated addition of *N*-chloroamides (RCONHCl) to $Bu^tCH{=}CH_2$, but the quantum yield has been shown to be governed by the electronegativity of R and its influence on the chlorine-atom-transfer step.[274]

The photo-addition of SF_5Cl to alkenes has been shown to occur via a radical mechanism in which $^{\bullet}SF_5$ initiates attack on the double bond. The regiochemistry of addition is not influenced by steric effects as sterically strained adducts were observed.[275] The radical addition of thiyl radicals to alkynes is induced by the presence of Et_3B. Aliphatic thiyl radicals add less efficiently than aromatic but their reaction is facilitated by the presence of MeOH in the solvent. These phenomena have been reported and applied but not accounted for mechanistically.[276] Other reactions of thiyl radicals with alkenes, alkynes, and hydrogen-donor molecules at high temperatures have been reported,[277] as have acid-induced additions to vinyl ethers.[278]

Toluenesuphonyl cyanide has been added to various alkenes, dienes, and hex-1-yne by a radical mechanism. The reactions occur in a regioselective and stereoselective manner and involve an initial attack of $MeC_6H_4SO_2^{\bullet}$ upon the double bond.[279] Sulphonyl radicals have also been implicated in the asymmetric addition of sulphonyl chlorides to alkenes within the coordination sphere of a chirally ligated Ru(II)-phosphine complex.[280] Thiyl, sulphonyl, selenyl and $Bu^tO^{\bullet}$ radicals have all been reported to be trapped by thioketones by addition at S. The adduct radicals exhibit conformations which vary with the nature of the groups attached to the thiocarbonyl carbon and to the radical. This was inferred from the variable *g* values measured for the adducts.[281] Thiyl and phosphinyl radicals have been reported to add to the C=C double bond of diketene yielding new oxetan-2-ones. The adducts of thiols are unstable but the resultant sulphides can be oxidized to stable sulphoxides and sulphones; the phosphine adducts are, in general, stable.[282] The additions of various phosphonyl and phosphinyl radicals to 4-substituted pyridines has been monitored by ESR spectroscopy. Phosphonyl radicals, $R_2\dot{P}(O)$, add at nitrogen giving 1-substituted-1-hydropyridinyl radicals e.g. (**48**); $Ph_2P^{\bullet}$ added giving a product with an exchangeable proton. The nature of this product, possibly a 1-hydropyridinyl radical, and its mode of formation remain unclear.[283]

An addition of Ph_3GeH to alkynes induced by Et_3B has been recorded (cf. ref.

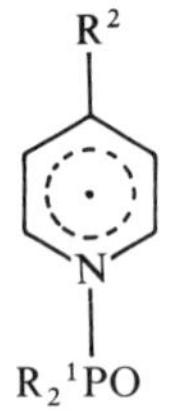

$R^1 = EtO, Ph$

$R^2 = CN, CO_2H, CO_2^-$

(48)

276). The reaction was found to occur with high regio- and stereo-selectivity. It was also found that the addition of Ph_3GeH to alkenes is facilitated by Et_3B; the *Z*-adduct is the kinetic product and the *E*-adduct is thermodynamically preferred; *Z*-to-*E* isomerization of alkenes is thus catalysed by the presence of small quantities of Ph_3GeH and Et_3B.[284] Radical adducts formed between group IV atom-centred radicals and various hydroxyquinones have been described. The consequences for the spin distribution in the adducts of the regioselectivity of addition, the capacity of the adduct for hydrogen-bonding and the nature of the group IV added, were discussed.[285]

Homolytic Aromatic Substitution

Previous work on the Gomberg–Bachmann reaction of purines has been retracted in view of new results: the refluxing of adenine derivatives with isoamyl nitrite in anisole for two days results in the formation of 6-anisyl derivatives of the purine in which the 6-(2-anisyl) isomer predominates. It has therefore been inferred that the reaction gives rise to a 6-purinyl radical which behaves as a normal aryl radical and substitutes the anisole in a typical manner. This corrects the conclusions from the earlier work in which the major product had been lost during work-up.[286]

The homolytic substitutions of the conjugate acids of heteroaromatic bases by nucleophilic C-centred radicals have been further investigated. The nature of the solvent was found to have an influence on the regioselectivity and relative rates. The influence of solvent and the nucleophilicity of the attacking radical on the deuterium kinetic isotope effect in the heterocycle were also studied. The results indicated the importance of reversibility and of polar effects for the mechanism.[287] Analogous substitutions of protonated quinolines by hydroxyalkyl radicals, formed by the γ-radiolysis of alcohols[288] and by alkyl radicals from the γ-radiolysis or the photolysis, promoted by $Fe_2(SO_4)_3$, of alkanoic acids,[289] have been reported. The substitutions of pyrazine-2,3-dicarbonitrile and of 1,3-dimethyllumazine by alkyl radicals produced by the oxidative decarboxylation of alkanoic acids with $AgNO_3/(NH_4)_2S_2O_8$ have also been reported. It was found that the reactivity of the alkyl radicals decreased in the order tertiary > secondary > primary and that the pyrazine derivative was more reactive than the lumazine.[290]

The homolytic substitution of electron-rich aromatics by electrophilic radicals, the mechanistic converse of alkyl substitution of protonated heteroaromatic bases, has been described. Arylamines were converted into trifluoroalkylated derivatives by reaction with CF_3Br under slight pressure in the presence of Zn and SO_2. The mechanism is exemplified in Scheme 17.[291]

$$Zn + SO_2 \longrightarrow Zn^+ + SO_2^{\overline{\cdot}}$$

$$SO_2^{\overline{\cdot}} + CF_3Br \longrightarrow SO_2 + CF_3Br^{\overline{\cdot}}$$

$$CF_3Br^{\overline{\cdot}} \longrightarrow Br^{\overline{\cdot}} + CF_3$$

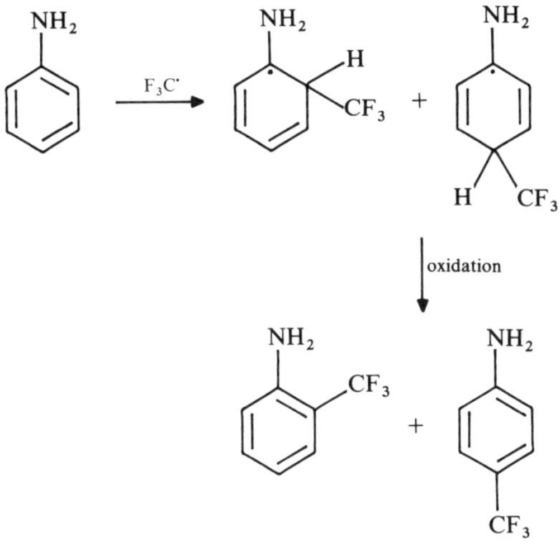

Also $\quad CF_3^{\cdot} + SO_2^{\overline{\cdot}} \longrightarrow CF_3SO_2^-$

SCHEME 17

S_H2 and S_Hi Reactions

The stereochemical outcome of the S_H2 reaction of ^{18}F atoms (generated by the cyclotron reaction: $^{20}Ne(d,\alpha)^{18}F$) with *S*-(+)-$CH_3CHClCOCl$ has been shown to depend upon the kinetic energy of the F atoms; only the most energetic can bring about front-side attack and give the halogen-exchanged product with retention of configuration. The results imply that the activation energy for front-side attack is considerably larger than that for the alternative back-side mode.[292] Bromine atoms have been found to react with bicyclo[2.2.0]hexane at a bridgehead in an S_H2 mode resulting in scission of the inter-ring bond and the formation of 1,4-dibromocyclohexanes (Scheme 18).[293]

Visible-light photolysis of primary and secondary alkyl-cobaloximes produces alkyl radicals which displace upon PhSSPh, PhSeSePh and $BrCCl_3$ to produce alkyl aryl sulphides and selenides and alkyl bromide; if a hex-5-enyl-cobaloxime is used, cyclization of the radical supersedes the S_H2 displacement at high dilution but, nevertheless, other side-reactions limit the yield of rearranged product. The nature of these side-reactions has been investigated.[294] But-3-enyl- and allyl-cobaloximes

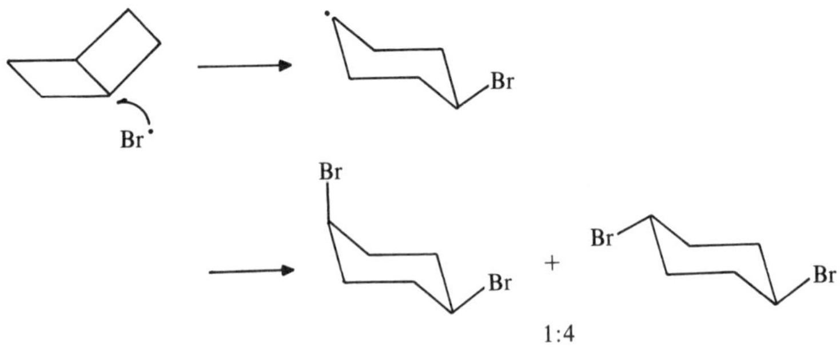

SCHEME 18

react with *N*-chloro-*N*-methylsulphonamides to give *N*-cyclopropylcarbinyl- or rearranged *N*-allyl-*N*-methyl sulphonamides, respectively. *N*-halogenocarboxamides, by contrast, react to give cyclopropylcarbinyl or rearranged allyl halide. It was concluded that the sulphonamide reaction is initiated by the attack of $MeSO_2\dot{N}Me$ on the terminal carbon of the alkene ligand but the nature of the initial step in the reaction of *N*-halogenocarboxamides was less clear. In either case the incipient adduct radical can then rearrange to observed products by S_Hi reaction which displaces cobaloxime(II) (Scheme 19).[295] Substitutions at carbon by S_H2 and S_Hi mechanisms have been reported for compounds of the types $CH_2{=}CHCH_2MR_n$ and $PhCH{=}CHMR_n$ (M = Sn, Ge, Pb, Hg) when attacked by a variety of radicals: PhS•, PhSe•, PhTe•, $Pr^{i\bullet}$, $PhSO_2^\bullet$ $(EtO)_2\dot{P}O$; relative reactivities were found for the different M and for different attacking radicals.[296,297]

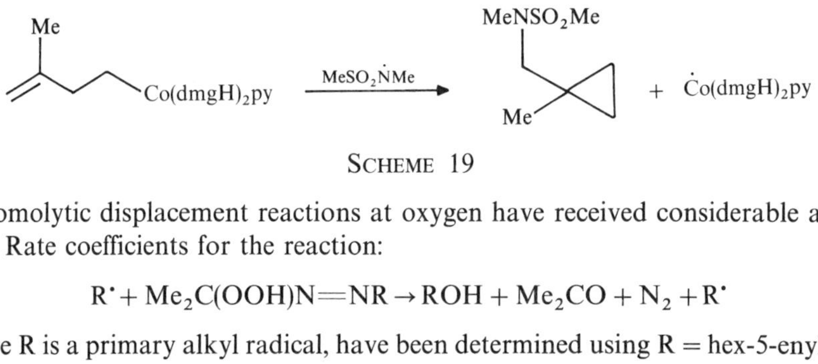

SCHEME 19

Homolytic displacement reactions at oxygen have received considerable attention. Rate coefficients for the reaction:

$$R^\bullet + Me_2C(OOH)N{=}NR \rightarrow ROH + Me_2CO + N_2 + R^\bullet$$

where R is a primary alkyl radical, have been determined using R = hex-5-enyl and comparing the S_H2 process with the known rate of cyclization of hex-5-enyl. The rate constant for the S_H2 process in hexafluorobenzene at 50°C was found to be dependent on the concentration of the substrate hydroperoxide, a fact ascribed to intermolecular hydrogen-bonding. The high rate of decomposition ($1.8 \times 10^7\ \text{M}^{-1}\ \text{s}^{-1}$ at a concentration of 0.1 M) was explained in terms of a concerted process in which displacement is coupled with scission of the adjacent C—N bond.[298] It has already been noted that cyclization to oxirane by S_Hi reaction is the preferred mode of decomposition of the adducts of $HO_2^\bullet$ with alkenes.[265] Similar oxirane-forming reactions have been reported during the thermolysis of $CH_2{=}CHCH_2O_2Bu^t$[299] and homologous alkenyl peroxides have been used to synthesize oxetanes and tetrahydrofurans.[300] The induced decomposition of *t*-butyl perpent-4-enoate yielding lactones could conceivably occur via either an intermediate radical adduct or a concerted addition-elimination (Scheme 20a); evidence for the two-step process has been obtained by decomposition of the perester (**49**): the intermediate radical is diverted to the cyclopentane perester (Scheme 20b).[301]

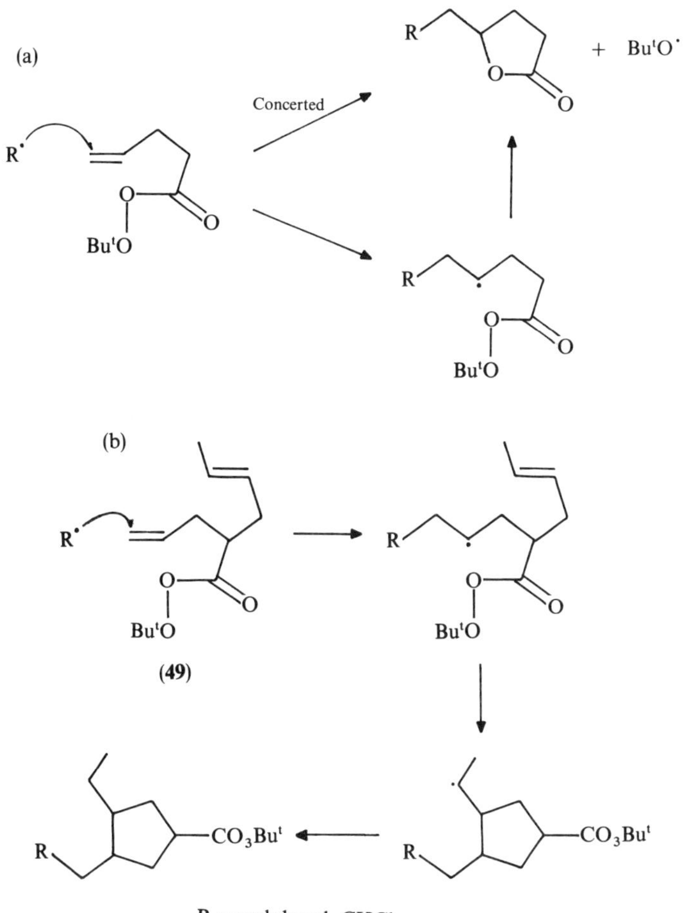

(**49**)

R = cyclohexyl, $CHCl_2$

SCHEME 20

Arrhenius parameters have been found for the intramolecular displacement reactions of $RS(CH_2)_3\dot{C}H_2$ to give tetrahydrothiophene and R$^\cdot$ for R = $PhCH_2$, Bu^t, and Pr^n, by competition against H-abstraction from Bu_3SnH. The values of the parameters were judged to be more consistent with a transition state in which the departing alkyl radical is well developed than with one corresponding to the rate-determining formation of an intermediate sulphuranyl radical. INDO calculations for sulphuranyl radicals and MM2 calculations for transition-state models were also reported as aids to interpretation of the results.[302] Rate constants for the S_H2'

scavenging of octyl radicals by (**43a**) have been determined by competition with H-abstraction from Bu_3SnH; the Arrhenius expression found was:

$$\log (K/\mathrm{M}^{-1}\mathrm{s}^{-1}) = (9.04 \pm 0.3) - (3.95 \pm 0.4)/\theta$$

where $\theta = 2.3RT$ kcal mol^{-1}.[303] An unusual S_H2 displacement at S by the phosphoranyl radical (**50**) has been brought to light; the reaction is of synthetic utility for reactions with dialkyl disulphides, RSSR, with R = Me, Bu^n, Bu^tCH_2, Bu^s, but not for R = Bu^t. The different dialkyl disulphides reacted at widely different relative rates (Bu^t:Me = 1:578), evidently reflecting the importance of steric effects, though competition experiments using mixed disulphides showed S—S bond strength was also relevant.[304]

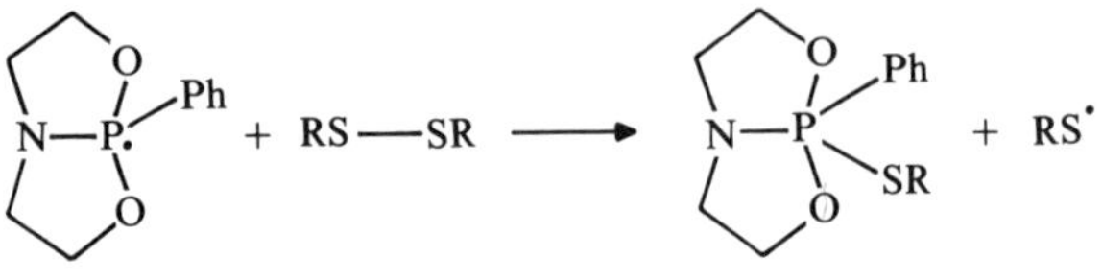

(**50**) (**51**)

The induced decomposition of *S-t*-butyl benzenethioseleninate, $PhSe(O)SBu^t$, has been investigated, the thioseleninate function perhaps being important for the enzyme glutathione peroxidase.[305] The ester was stable in aqueous dioxan (10^{-4} M), but at 0.5 M at room temperature it underwent rapid decomposition. On the basis of the products observed and initiation by the ready homolysis of benzeneseleninic anhydride, present as an impurity, the mechanism tentatively proposed is:
Initiation:

$$\mathrm{PhSeOSePh} \rightarrow \mathrm{PhSeO^{\bullet}} + \mathrm{PhSe^{\bullet}}$$

Propagation:

$$\mathrm{PhSeO^{\bullet}} + \mathrm{PhSe(O)SBu^t} \rightarrow \mathrm{PhSeOSe(O)Ph} + \mathrm{Bu^tS^{\bullet}}$$
$$\mathrm{Bu^tS^{\bullet}} + \mathrm{PhSe(O)SBu^t} \rightarrow \mathrm{Bu^tSSBu^t} + \mathrm{PhSeO^{\bullet}}$$

Nitroxides and Spin-trapping

ESR spectra of the nitroxyl (**52**) have been measured for solutions in various solvents: the hyperfine splitting for N(1) was found to increase markedly in polar solvents, showing a correlation with their E_T values. The work reinterprets earlier work of others.[306] Substituent effects on the ESR spectra of the nitroxyl (**53**) have been investigated and Hammett correlations of the hyperfine splittings obtained.[307]

A complex suite of reactions has been found following the reaction of 2,2-diphenylindolinone *N*-oxyl (**54a**) with aminyl radicals generated *in situ* by oxidation of amines by PbO_2. Aminyls from secondary amines couple with (**54a**) at C(5) and the adduct is then oxidized to a new aminated nitroxyl (**55b**). If aminyls from primary amines were scavenged, reaction proceeded further to (**55**). New radicals were characterized by ESR and an X-ray crystal-structure analysis was performed for (**55**; Ar = 4-$MeOC_6H_4$).[308]

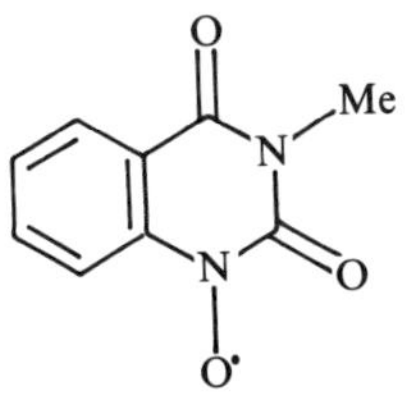

(52)

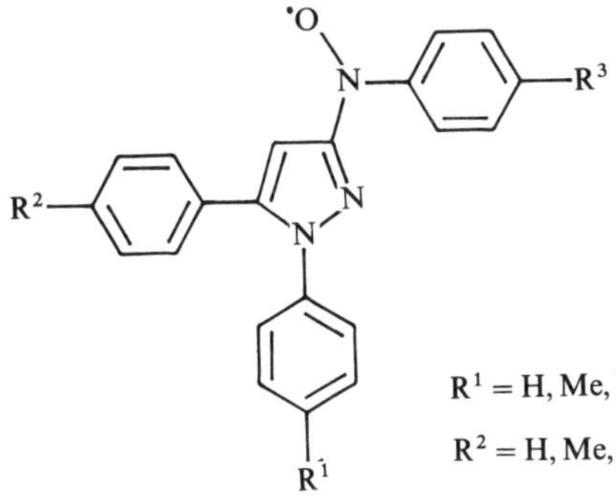

$R^1 = H, Me, Br, NO_2$

$R^2 = H, Me, Br, Cl$

$R^3 = H, Cl, Me$

(53)

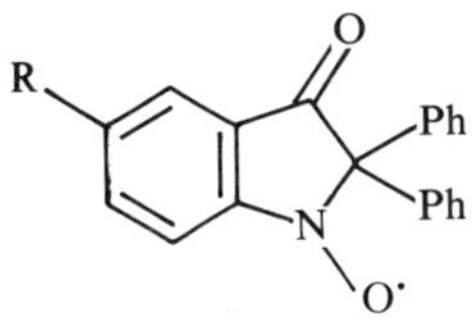

(54) a; R = H

b; R = Ph_2N, $(4\text{-}MeOC_6H_4)_2N$

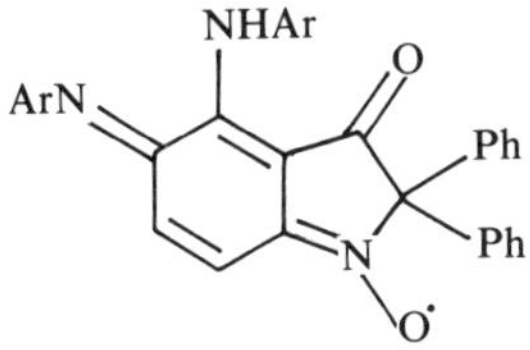

(55)

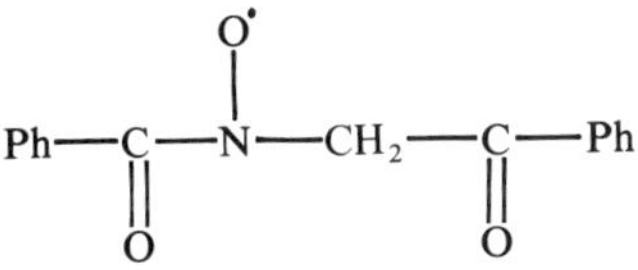

(56)

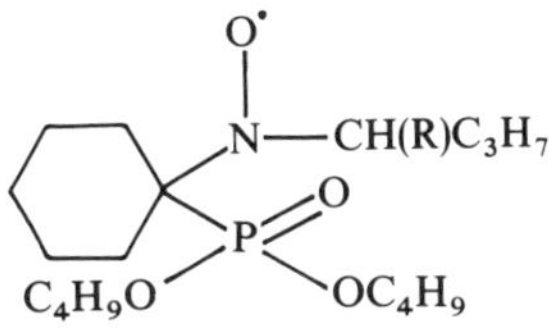

(57)

The p*H* and p*K* dependence of the formation of nitroxides by a series of drugs on oxidation by H_2O_2 has been examined by ESR spectroscopy. All the drugs were secondary amines which react only in the free base form. Given the p*K* values of the drugs, they would be expected to exist to the extent of *ca* 1% in the reactive form at physiological p*H*.[309] Photo-decomposition of α-nitratomethylbenzoin has been shown by an ESR study to give rise to a single observable radical identified as (**56**); a tentative mechanism for its formation was proposed.[310] A stable nitroxyl (**57**) has also been reported as a product of oxidation of a pesticide; the nature of R in (**57**) is uncertain but it is possibly an alkyl group trapped by an intermediate nitrone produced in the degradation of the pesticide precursor.[311]

A new water-soluble spin-trap, 3,5-dibromo-4-nitrosobenzenesulphonate, has been described and used to trap alkyl radicals produced oxidatively by the Ti^{3+}/H_2O_2 or Fe^{2+}/H_2O_2 system; $HO^{\bullet}$ produced by these systems is not trapped.[312] The kinetics of the reactions of $HO^{\bullet}$ with nitrone spin-traps have been investigated by fast optical kinetic techniques.[313] The spin-trapping of radicals in liquid CCl_4 following radiolysis with X-rays of different wavelength has been described. Both the unstable adduct of $PhCH{=}N(O)Bu^t$ and $Cl^{\bullet}$ and the stabler adduct with $^{\bullet}CCl_3$ were detected. The trapping of radicals depended on the frequency of the X-rays: energies above the *K* absorption edge for chlorine led to enhanced radical production as a consequence of the Auger effect which follows inner-shell photo-absorption. Results were also described for the trapping of radicals generated by the decay of tritiated methanol in CCl_4.[314] Spin-trapping by nitrosodurene has indicated the mediation of $^{\bullet}CCl_3$, $Cl_3CCH_2\dot{C}HBu$, and $Cl\dot{C}r(CO)_5$ as intermediates during the addition of CCl_4 to hex-1-ene in the presence of $Cr(CO)_6$. Details of kinetics and mechanism were also reported.[315] Other Soviet work has examined the oxidative lactonization of $R^1CHR^2CHR^3CH_2CO_2H$ (R^1, R^2, R^3 = H, H, H; Me, H, H; Me, H, Me; Bu, Et, H) in the presence of Bu^tNO and $PhCH{=}N(O)Bu^t$ as spin-traps.[316] Work has been reported in Chinese of the trapping, by 2,4,6-$Bu^t{}_3C_6H_2NO$, of radicals produced by abstraction of hydrogen from aliphatic hydrocarbons[317] and from ethers and phenols[318] by photo-excited benzophenone.

References

1 Viehe, H. G., Janousek, Z., and Merenyi, R., (eds) *Substituent Effects in Radical Chemistry*, NATO ASI Ser., Ser. C, vol. **189**, D. Reidel Publ. Co., Dordrecht, Neth., 1986.
2 Benson, S. W., *Acc. Chem. Res.*, **19**, 335 (1986).
3 Fischer, H. and Paul, H., *Acc. Chem. Res.*, **20**, 200 (1987).
4 Ikegami, Y., *Rev. Chem. Intermed.*, **7**, 91 (1987); *Chem. Abs.*, **106**, 175417 (1987).
5 Roduner, E., *Progr. React. Kinet.*, **14**, 1 (1986); *Chem. Abs.*, **106**, 195563 (1987).
6 Barnes, A. R. and Sugden, J. K., *Pharm. Acta Helv.*, **61**, 218 (1986); *Chem. Abs.*, **107**, 6525 (1987).
7 Asmus, K. D., *Prepr.-Am. Chem. Soc., Div. Pet. Chem.*, **31**, 870 (1986); *Chem. Abs.*, **106**, 213144 (1987).
8 Denisov, E. T., *Neftekhimiya*, **26**, 723 (1986); *Chem. Abs.*, **107**, 58152 (1987).
9 Opeida, I. A., *Neftekhimiya*, **26**, 736 (1986); *Chem. Abs.*, **107**, 6531 (1987).
10 Opeida, I. A., *Neftekhimiya*, **26**, 745 (1986); *Chem. Abs.*, **107**, 6532 (1987).
11 Zhulin, V. M., Zorin, V. V., Zlotskii, S. S., and Rakhmankulov, D. L., *Izv. Vyssh. Uchebn. Zaved., Khim. Khim. Tekhnol.*, **29**, 3 (1986); *Chem. Abs.*, **106**, 175418 (1987).
12 Rakhmankulov, D. L., Zlotskii, S. S., Pastushenko, E. V., Rol'nik, L. Z., Karakhanov, R. A., Bartok, M., and Molnar, A., *Acta Phys. Chem.*, **31**, 743 (1985); *Chem. Abs.*, **106**, 213124 (1987).
13 Griller, D., *Annu. Rep. Prog. Chem., Sect. B*, **82**, 75 (1985) (Publ. 1986); *Chem. Abs.*, **106**, 195579 (1987).
14 Barton, D. H. R. and Zard, S. Z., *Janssen Chim. Acta*, **4**, 3 (1986); *Chem. Abs.*, **106**, 213128 (1987).
15 Ramaiah, M., *Tetrahedron*, **43**, 3541 (1987).
16 Giese, B., *Radicals in Organic Synthesis: Formation of Carbon-Carbon Bonds*, Pergamon Press, Oxford (1986); *Chem. Abs.*, **105**, 225458 (1986).
17 Lifshitz, A., Moran, A., and Bidani, S., *Int. J. Chem. Kinet.*, **19**, 61 (1987).
18 Schmittel, M. and Rüchardt, C., *J. Am. Chem. Soc.*, **109**, 2750 (1987).
19 Peyman, A., Hickl, E., and Beckhaus, H.-D., *Chem. Ber.*, **120**, 713 (1987).
20 Nakazumi, H., Kitao, T., and Zollinger, H., *J. Org. Chem.*, **52**, 2825 (1987).
21 Baldridge, K. K., Gordon, M. S., and Johnson, D. E., *J. Phys. Chem.*, **91**, 4145 (1987).
22 Pasto, D. J., Krasnansky, R., and Zercher, C., *J. Org. Chem.*, **52**, 3062 (1987).
23 Pacansky, J. and Yoshimine, M., *J. Phys. Chem.*, **91**, 1024 (1987).
24 Honjou, H., Yoshimine, M., and Pacansky, J., *J. Phys. Chem.*, **91**, 4455 (1987).
25 Hameka, H. H., *J. Org. Chem.*, **52**, 5025 (1987).
26 Ziegler, U., Ondruschka, B., and Koehler, H. J., *J. Prakt. Chem.*, **328**, 886 (1986); *Chem. Abs.*, **107**, 77026 (1987).
27 Kuehnel, W., Gey, E., and Ondruschka, B., *Z. Phys. Chem.*, **268**, 23 (1987); *Chem. Abs.*, **106**, 195599 (1987).
28 Claxton, T. A. and Graham, A. M., *J. Chem. Soc., Chem. Commun.*, **1987**, 1167.

[29] Rhodes, C. J., Symons, M. C. R., Scott, C. A., Roduner, E., and Heming, M., *J. Chem. Soc., Chem. Commun.*, **1987**, 448.
[30] Rhodes, C. J., Symons, M. C. R., Scott, C. A., Roduner, E., and Heming, M., *J. Chem. Soc., Chem. Commun.*, **1987**, 447.
[31] Ingold, K. U. and Walton, J. C., *J. Am. Chem. Soc.*, **109**, 6937 (1987).
[32] Deycard, S., Hughes, L., Lusztyk, J., and Ingold, K. U., *J. Am. Chem. Soc.*, **109**, 4954 (1987).
[33] Walton, J. C., *J. Chem. Soc., Perkin Trans. 2*, **1987**, 231.
[34] Berson, J. A., Griller, D., Owens, K., and Wayner, D. D. M., *J. Org. Chem.*, **52**, 3316 (1987).
[35] Kurtz, H. A., Lloyd, R. V., and Vaughan Williams, R., *J. Org. Chem.*, **52**, 302 (1987).
[36] Sustmann, R. and Korth, H.-G., *J. Chem. Soc., Faraday Trans. 1*, **83**, 95 (1987).
[37] Chandra, H. and Symons, M. C. R., *Tetrahedron Lett.*, **28**, 1455 (1987).
[38] Birkhofer, H., Hädrich, J., Beckhaus, H.-D., and Rüchardt, C., *Angew. Chem. Int. Edn*, **26**, 573 (1987).
[39] Korth, H. G., Lommes, P., Sustmann, R., Sylvander, L., and Stella, L., *Nouveau J. Chim.*, **11**, 365 (1987).
[40] Beckhaus, H.-D. and Rüchardt, C., *Angew. Chem. Int. Edn*, **26**, 770 (1987).
[41] Fliszár, S. and Minichino, C., *Can. J. Chem.*, **65**, 2495 (1987).
[42] MacInnes, I. and Walton, J. C., *J. Chem. Soc., Perkin Trans. 2*, **1987**, 1077.
[43] Sakurai, H., Kyushin, S., and Nakadaira, Y., *Chem. Lett.*, **1987**, 297.
[44] Davidson, I. M. T., Barton, T. J., Hughes, K. J., Ijadi-Maghsoodi, S., Revis, A., and Paul, G. C., *Organometallics*, **6**, 644 (1987).
[45] Luedtke, A., Meng, K., and Timberlake, J. W., *Tetrahedron Lett.*, **28**, 4255 (1987).
[46] Jacox, M. E., *J. Phys. Chem.*, **91**, 5038 (1987).
[47] Hegarty, A. F. and O'Neill, P., *Tetrahedron Lett.*, **28**, 901 (1987).
[48] O'Neill, P. and Hegarty, A. F., *J. Chem. Soc., Chem. Commun.*, **1987**, 744.
[49] Veciana, J., Carilla, J., Miravitlles, C., and Molins, E., *J. Chem. Soc., Chem. Commun.*, **1987**, 812.
[50] Fox, M. A., Gaillard, E., and Chen, C.-C., *J. Am. Chem. Soc.*, **109**, 7088 (1987).
[51] Armat, O., Veciana, J., Roviva, C., Riera, J., Castañer, J., Molins, E., Rius, J., Miravitlles, C., Olivella, S., and Brichfeus, J., *J. Phys. Chem.*, **91**, 5608 (1987).
[52] Field, M. J., Hillier, I. H., and Guest, M. F., *J. Chem. Soc., Perkin Trans. 2*, **1987**, 1311.
[53] Bologa, U., Balaban, A. T., Grecu, N., Negoita, N., Capriou, M. T., and Walter, R. I., *J. Org. Chem.*, **52**, 4176 (1987).
[54] Gey, E., Radeglia, R., Becker, H. G., and Pfeifer, D., *Z. Chem.*, **26**, 405 (1986).
[55] Glover, S. A. and Beckwith, A. L. J., *Aust. J. Chem.*, **40**, 701 (1987).
[56] Miura, U., Isogai, M., and Kinoshita, M., *Bull. Chem. Soc. Jpn*, **60**, 3065 (1987).
[57] Miura, Y., Shibata, Y., and Kinoshita, M., *Bull. Chem. Soc. Jpn*, **59**, 3291 (1986).
[58] Benati, L., Montevecchi, P. C., and Spagnolo, P., *J. Chem. Soc., Perkin Trans. 1*, **1987**, 99.
[59] Sakai, S., Imoto, M., Ouchi, T., and Oiwa, M., *Nippon Kagaku Kaishi*, **1986**, 739.
[60] Raevskii, N. I. and Borisov, Yu. A., *Dokl. Akad. Nauk SSSR.*, **288**, 664 (1986); *Chem. Abs.*, **106**, 17677 (1987).
[61] Foster, S. C., Hsu, Y.-C., Damo, C. P., Liu, X., Kung, C.-Y., and Miller, T. A., *J. Phys. Chem.*, **90**, 6766 (1986).
[62] Franciso, J. S., Steinfeld, J. I., and Williams, I. H., *NBS Spec. Publ. (U.S.)*, **716**, 250 (1986).
[63] Sun, Q. and Schuler, R. H., *J. Phys. Chem.*, **91**, 4591 (1987).
[64] Tripathi, G. N. R. and Schuler, R. H., *J. Phys. Chem.*, **91**, 5881 (1987).
[65] Kadirov, M. K., Il'yasov, A. V., Kadirova, V. Kh., Mukmeneva, N. A., and Kirpichnikov, P. A., *Izv. Akad. Nauk SSSR., Ser. Khim.*, **1986**, 1773.
[66] Lehnig, M., Neumann, W. P., and Wallis, E., *J. Organomet. Chem.*, **333**, 17 (1987).
[67] Awere, E. G., Burford, N., Mailer, C., Passmore, J., Schriver, M. J., White, P. S., Banister, A. J., Oberhammer, H., and Sutcliffe, L. H., *J. Chem. Soc., Chem. Commun.*, **1987**, 66.
[68] Kotaka, M. and Sato, S., *J. Chem. Soc., Chem. Commun.*, **1986**, 1783.
[69] Qin, X.-Z., Guo, Q.-X., Wang, J. T., and Williams, F., *J. Chem. Soc., Chem. Commun.*, **1987**, 1553.
[70] Morisaki, K., Miura, Y., Abe, Y., Hirota, M., and Nakada, M., *Chem. Lett.*, **87**, 1589 (1987).
[71] Ondruschka, B., Ziegler, U., and Zimmerman, G., *Z. Phys. Chem. (Leipzig)*, **267**, 1127 (1986).
[72] Gilbert, B. C., Parry, D. J., and Grossi, L., *J. Chem. Soc., Faraday Trans. 1*, **83**, 77 (1987).
[73] Collin, G. J., Deslauriers, H., and De Mare, G. R., *Can. J. Chem.*, **65**, 391 (1987).
[74] Dóbé, S., Bérces, T., Réti, F., and Márta, F., *Int. J. Chem. Kinet.*, **19**, 895 (1987).
[75] Dorigo, A. E. and Houk, K. N., *J. Am. Chem. Soc.*, **109**, 2195 (1987).
[76] Dicks, P. F., Glover, S. A., Goosen, A., and McCleland, C. W., *Tetrahedron*, **43**, 923 (1987).
[77] Troyanskii, E. I., Lazareva, M. I., and Nikishin, G. I., *Izv. Akad. Nauk SSSR., Ser. Khim.*, **1986**, 1575 (1986); *Chem. Abs.*, **106**, 155690 (1987).

[78] Crimmins, M. T. and Mascarella, S. W., *Tetrahedron Lett.*, **28**, 5063 (1987).
[79] Korth, H.-G., Müller, W., Sustmann, R., and Christl, M., *Chem. Ber.*, **120**, 1257 (1987).
[80] Johns, A., Murphy, J. A., Patterson, C. W., and Wooster, N. F., *J. Chem. Soc., Chem. Commun.*, **1987**, 1238.
[81] Baldwin, J. E., Adlington, R. M., Kang, T. W., Lee, E., and Schofield, C. J., *J. Chem. Soc., Chem. Commun.*, **1987**, 104.
[82] Baldwin, J. E., Adlington, R. M., Domayne-Hayman, B. P., Knight, G., and Ting, H.-H., *J. Chem. Soc., Chem. Commun.*, **1987**, 1661.
[83] Giese, B., Heinrich, N., Horler, H., Koch, W., and Schwarz, H., *Chem. Ber.*, **119**, 3528 (1986).
[84] Dowd, P. and Choi, S.-C., *J. Am. Chem. Soc.*, **109**, 3493 (1987).
[85] Dowd, P. and Choi, S.-C., *J. Am. Chem. Soc.*, **109**, 6548 (1987).
[86] Suginome, H. and Yamada, S., *Tetrahedron*, **43**, 3371 (1987).
[87] Suginome, H., Washiyama, H., and Yamada, S., *Bull. Chem. Soc. Jpn*, **60**, 1071 (1987).
[88] Beckwith, A. L. J., O'Shea, D. M., Gerba, S., and Westwood, S. W., *J. Chem. Soc., Chem. Commun.*, **1987**, 666.
[89] Giese, B. and Witzel, T., *Tetrahedron Lett.*, **28**, 2571 (1987).
[90] Spellmeyer, D. C. and Houk, K. N., *J. Org. Chem.*, **52**, 959 (1987).
[91] Dewar, M. J. S., Gardiner, W. C., Frenklach, M., and Oref, I., *J. Am. Chem. Soc.*, **109**, 4456 (1987).
[92] Kaushal, P., Roberts, B. P., and Ryan, E. J., *J. Chem. Soc., Chem. Commun.*, **1987**, 1587.
[93] Lusztyk, J., Maillard, B., Deycard, S., Lindsay, D. A., and Ingold, K. U., *J. Org. Chem.*, **52**, 3509 (1987).
[94] Beckwith, A. L. J. and Glover, S. A., *Aust. J. Chem.*, **40**, 157 (1987).
[95] Clive, D. L. J. and Cheshire, D. R., *J. Chem. Soc., Chem. Commun.*, **1987**, 1520.
[96] Clive, D. L. J., Cheshire, D. R., and Set, L., *J. Chem. Soc., Chem. Commun.*, **1987**, 353.
[97] Crich, D. and Fortt, S. M., *Tetrahedron Lett.*, **28**, 2895 (1987).
[98] Clive, D. L. J., Boivin, T. L. B., and Angoh, A. G., *J. Org. Chem.*, **52**, 4943 (1987).
[99] Curran, D. P. and Chang, C.-T., *Tetrahedron Lett.*, **28**, 2477 (1987).
[100] Grigg, R., Devlin, J., Ramasubbu, A., Scott, R. M., and Stevenson, P., *J. Chem. Soc., Perkin Trans. 1*, **1987**, 1515.
[101] Broka, C. A. and Reichert, D. E. C., *Tetrahedron Lett.*, **28**, 1503 (1987).
[102] Hanessian, S., Dhanoa, D. S., and Beaulieu, P. L., *Can. J. Chem.*, **65**, 1859 (1987).
[103] Mohan, R., Kates, S. A., Dombroski, M. A., and Snider, B. B., *Tetrahedron Lett.*, **28**, 845 (1987).
[104] Snider, B. B., Mohan, R., and Kates, S. A., *Tetrahedron Lett.*, **28**, 841 (1987).
[105] Tsang, R. and Fraser-Reid, B., *J. Am. Chem. Soc.*, **108**, 8102 (1986).
[106] Tsang, R., Dickson, J. K., Pak, H., Walton, R., and Fraser-Reid, B., *J. Am. Chem. Soc.*, **109**, 4384 (1987).
[107] MacCorquodale, F. and Walton, J. C., *J. Chem. Soc., Chem. Commun.*, **1987**, 1456.
[108] Bloodworth, A. J., Crich, D., and Melvin, T., *J. Chem. Soc., Chem. Commun.*, **1987**, 786.
[109] Abeywickrema, A. N. and Beckwith, A. L. J., *J. Org. Chem.*, **52**, 2568 (1987).
[110] Beckwith, A. L. J. and Meijs, G. F., *J. Org. Chem.*, **52**, 1922 (1987).
[111] Abeywickrema, A. N., Beckwith, A. L. J, and Gerba, S., *J. Org. Chem.*, **52**, 4072 (1987).
[112] Patel, V. F. and Pattenden, G., *Tetrahedron Lett.*, **28**, 1451 (1987).
[113] Patel, V. F. and Pattenden, G., *J. Chem. Soc., Chem. Commun.*, **1987**, 871.
[114] Beckwith, A. L. J., Wang, S., and Warkentin, J., *J. Am. Chem. Soc.*, **109**, 5289 (1987).
[115] Morikawa, T., Nishiwaki, T., Iitaka, Y., and Kobayashi, Y., *Tetrahedron Lett.*, **28**, 671 (1987).
[116] Srikrishna, A. and Pullaiah, K. C *Tetrahedron Lett.*, **28**, 5203 (1987).
[117] Begley, M. J., Bhandal, H., Hutchinson, J. H., and Pattenden, G., *Tetrahedron Lett.*, **28**, 1317 (1987).
[118] Kraus, G. A. and Thurston, J., *Tetrahedron Lett.*, **28**, 4011 (1987).
[119] Watanabe, Y., Ueno, Y., Tanaka, C., Okawara, M., and Endo, T., *Tetrahedron Lett.*, **28**, 3953 (1987).
[120] Tokuda, M., Yamada, Y., Takagi, T., and Suginome, H., *Tetrahedron*, **43**, 281 (1987).
[121] Kiesewetter, R. and Margaretha, P., *Helv. Chim. Acta*, **70**, 121 (1987).
[122] Kiesewetter, R. and Margaretha, P., *Helv. Chim. Acta*, **70**, 125 (1987).
[123] Kaafarani, M., Crozet, M. P., and Surzur, J.-M., *Bull. Soc. Chim. Fr. II*, **1987**, 885.
[124] Cadogan, J. I. G., Hitchison, H. S., and McNab, H., *J. Chem. Soc., Perkin Trans. 1*, **1987**, 1407.
[125] Dicks, P. F., Glover, S. A., Goosen, A., and McCleland, C. W., *J. Chem. Soc., Perkin Trans. 1*, **1987**, 1243.
[126] Porter, N. A. and Chang, V. H.-T., *J. Am. Chem. Soc.*, **109**, 4976 (1987).
[127] Clive, D. L. J., Angoh, A. G., and Bennett, S. M., *J. Org. Chem.*, **52**, 1339 (1987).
[128] Curvan, D. P. and Kuo, S.-C., *Tetrahedron*, **43**, 5653 (1987).
[129] Hutchinson, J. H., Pattenden, G., and Myers, P. L., *Tetrahedron Lett.*, **28**, 1313 (1987).

[130] Ihara, M., Taniguchi, N., Fukumoto, K., and Kametani, T., *J. Chem. Soc., Chem. Commun.*, **1987**, 1438.
[131] Hunter, R., Hutchings, G. J., and Pickl, W., *J. Chem. Soc., Chem. Commun.*, **1987**, 843.
[132] Reimann, W., Kuivila, H. G., Farah, D., and Apoussidis, T., *Organometallics*, **6**, 557 (1987).
[133] Brooks, W. V. F., Burford, N., Passmore, J., Schriver, M. J., and Sutcliffe, L. H., *J. Chem. Soc., Chem. Commun.*, **1987**, 69.
[134] Bohner, V., Goldmann, H., Kaptein, R., and Zetta, L., *J. Chem. Soc., Chem. Commun.*, **1987**, 1358.
[135] Grela, M. A. and Colussi, A. J., *Int. J. Chem. Kinet.*, **19**, 869 (1987).
[136] Batt, L. and Robinson, G. N., *Int. J. Chem. Kinet.*, **19**, 391 (1987).
[137] Baban, J. A. and Roberts, B. P., *J. Chem. Soc., Perkin Trans. 2*, **1987**, 497.
[138] Ono, N., Kamimura, A., and Kaji, A., *J. Org. Chem.*, **52**, 5111 (1987).
[139] Korth, H.-G., Sustmann, R., Dupuis, J. and Giese, B., *Chem. Ber.*, **120**, 1197 (1987).
[140] Eberhardt, M. K. and Soto, K., *Tetrahedron Lett.*, **28**, 3439 (1987).
[141] Suehiro, T., Masuda, S., Nakausa, R., Taguchi, M., Mori, A., Koike, A., and Date, M., *Bull. Chem. Soc. Jpn*, **60**, 3322 (1987).
[142] Schulz, M., Kluge, R., and Willscher, S., *Liebigs Ann. Chem.*, **1987**, 671.
[143] Anastasi, C. and Arthur, N. L., *J. Chem. Soc., Faraday Trans. 2*, **83**, 277 (1981).
[144] Claus, P., Scherzer, K., and Loeser, U., *Z. Chem.*, **27**, 370 (1987).
[145] Zborovskii, A. G., Petrov, Yu. P., Tereza, A. M., and Tyurin, A. N., *Khim. Fiz.*, **5**, 1528 (1986); *Chem. Abs.*, **106**, 119209 (1987).
[146] Lin, C. Y. and Lin, M. C., *Aust. J. Chem.*, **39**, 723 (1986).
[147] Hase, W. L., Mondo, S. L., Duchovic, R. J., and Hirst, D. M., *J. Am. Chem. Soc.*, **109**, 2916 (1987).
[148] Timonen, R. S., Ratajczak, E., and Gutman, D., *J. Phys. Chem.*, **91**, 692 (1987).
[149] Baggott, J. E., Frey, H. M., Lightfoot, P. D., and Walsh, R., *J. Phys Chem.*, **91**, 3386 (1987).
[150] Baldwin, P. J., Canosa-Mas, C. E., Monty Frey, H., and Walsh, R., *J. Chem. Soc., Faraday Trans. 2*, **83**, 1049 (1987).
[151] Dean, A. M. and Westmoreland, P. R., *Int. J. Chem. Kinet.*, **19**, 207 (1987).
[152] Fiedler, E., *Ber. Max-Plank-Inst. Stroemungs-forsch.*, **15**, 97 (1986); *Chem. Abs.*, **107**, 77134 (1987).
[153] Anderson, L. G. and Bonilla, J. L., *NBS Spec. Publ. (U.S.)*, **716**, 19 (1986); *Chem. Abs.*, **106**, 4478 (1987).
[154] Balla, R., Nelson, H. H., and MacDonald, J. R., *NBS Spec. Publ. (U.S.)*, **716**, 30 (1986); *Chem. Abs.*, **106**, 195631 (1987).
[155] Bennett, J. E., *J. Chem. Soc., Faraday Trans. 1*, **83**, 1805 (1987).
[156] Bennett, J. E., Brunton, G., Lindsay Smith, J. R., Salmon, T. M. F., and Waddington, D. J., *J. Chem. Soc., Faraday Trans. 1*, **83**, 2421 (1987).
[157] Bennett, J. E., Brunton, G., Lindsay Smith, J. R., Salmon, T. M. F., and Waddington, D. J., *J. Chem. Soc., Faraday Trans. 1*, **83**, 2433 (1987).
[158] Kosyak, Z. M. and Chernyak, B. I., *Dopov. Akad. Nauk Ukr. RSR., Ser. B: Geol. Biol. Nauki*, **1986**, 33; *Chem. Abs.*, **107**, 197212 (1987).
[159] Mach, K., Novakova, J., Hanus, V., and Dolejsek, Z., *Collect. Czech. Chem. Commun.*, **51**, 2675 (1986).
[160] Holmes, B. E. and Rakestraw, D. J., *Int. J. Chem. Kinet.*, **19**, 401 (1987).
[161] Skorobogatov, G. A., Dymov, B. P., Lebedev, V. N., and Khripun, V. K., *Zh. Obshch. Khim.*, **56**, 2632 (1986); *Chem. Abs.*, **107**, 77148 (1987).
[162] Hawari, J. A., Griller, D., Weber, W. P., and Gaspar, P. P., *J. Organomet. Chem.*, **326**, 335 (1987).
[163] Fairhurst, S. A., Stewart, N. J., and Sutcliffe, L. H., *Magn. Reson. Chem.*, **25**, 60 (1987).
[164] Dusance, A.-C., De Doncker, J., Couillard, C., De Laet, M., and Tilquin, B., *J. Chem. Soc., Faraday Trans. 1*, **83**, 125 (1987).
[165] Safiullin, R. L., Enikeeva, L. R., and Komissarov, V. D., *Kinet. Katal.*, **27**, 742 (1986); *Chem. Abs.*, **107**, 58414 (1987).
[166] Safiullin, R. L., Zaripov, R. N., and Komissarov, V. D., *Izv. Akad. Nauk SSSR., Ser. Khim.*, **1986**, 1447 (1986); *Chem. Abs.*, **106**, 17714 (1987).
[167] Geno, M. K. and Halpern, J., *J. Chem. Soc., Chem. Commun.*, **1987**, 1052.
[168] Imbert, F. E. and Marshall, R. M., *Int. J. Chem. Kinet.*, **19**, 81 (1987).
[169] Closs, G. L., Forbes, M. D. E., and Norris, J. R., *J. Phys. Chem.*, **91**, 3592 (1987).
[170] Turro, N. J., Zimmt, M. B., Lei, X. G., Gould, I. R., Nitschke, K. S., and Cha, Y., *J. Phys. Chem.*, **91**, 4544 (1987).
[171] Kuz'min, V. A. and Levin, P. P., *Izv. Akad. Nauk SSSR., Ser. Khim.*, **1986**, 1421; *Chem. Abs.*, **106**, 119203 (1987).
[172] Ali, M., Ridd, J. H., Sandall, J. P. B., and Trevellick, S., *J. Chem. Soc., Chem. Commun.*, **1987**, 1168.
[173] Kobayashi, M., Gotoh, M., and Yoshida, M., *Bull. Chem. Soc. Jpn*, **60**, 295 (1987).

174 Scott, T. W. and Liu, S. N., *Springer Ser. Chem. Phys.*, **46** (*Ultrafast Phenom.*, 5), 338 (1986); *Chem. Abs.*, **107**, 115061 (1987).
175 Finkelstein, M., Hart, S. A., Moore, W. M., Ross, S. D., and Eberson, L., *Acta. Chem. Scand.*, **41B**, 232 (1987).
176 Moeller, W., Mozzhukhin, E., and Wagner, H. G., *Ber. Bunsen-Ges. Phys. Chem.*, **90**, 854 (1986); *Chem. Abs.*, **106**, 119003 (1987).
177 Pacey, P. D., Siebrand, W., and Wildman, T. A., *Jerusalem Symp. Quantum Chem. Biochem.*, **19** (*Tunnelling*), 117 (1986); *Chem. Abs.*, **107**, 96060 (1987).
178 Ali, D. P. and Hynes, J. T., *Jerusalem Symp. Quantum Chem. Biochem.*, **19** (*Tunnelling*), 203 (1986); *Chem. Abs.*, **107**, 115030 (1987).
179 Ponec, R., Krupicka, S., and Hajek, M., *Collect. Czech. Chem. Commun.*, **52**, 547 (1987).
180 Wünsch, E., Lluch, J. M., Oliva, A., and Bertran, J., *J. Chem. Soc., Perkin Trans. 2*, **1987**, 211.
181 Lee, I., Song, C. H., and Park, B. S., *Taehan Hwahakhoe Chi*, **30**, 166 (1986); *Chem. Abs.*, **106**, 4312 (1987).
182 Gedra, A., Lukacs, J., Vidoczy, T., and Sumegi, L., *Magy. Kem. Foly.*, **92**, 12 (1986); *Chem. Abs.*, **105**, 225573 (1986).
183 Manka, M. J., Brown, R. L., and Stein, S. E., *Int. J. Chem. Kinet.*, **19**, 943 (1987).
184 Dneprovski, A. S., Pertsikov, B. Z., and Chertkov, V. A., *Zh. Org. Khim.*, **23**, 327 (1987); *Chem. Abs.*, **107**, 197275 (1987).
185 Bullock, R. M. and Samsel, E. G., *J. Am. Chem. Soc.*, **109**, 6542 (1987).
186 Engel, P. S., Lee, W.-K., Marschke, G. E., and Shine, H. J., *J. Org. Chem.*, **52**, 2813 (1987).
187 Scherzer, K., Loeser, U., and Stiller. W., *Z. Chem.*, **27**, 300 (1987).
188 Gilbert, B. C., McLay, N. R., and Parry, D. J., *J. Chem. Soc., Perkin Trans. 2*, **1987**, 329.
189 Battaglia, L. P., Nardelli, M., Pelizzi, C., Predieri, G., and Chiusoli, G. P., *J. Organomet. Chem.*, **317**, 19 (1986).
190 Nielsen, O. J., Munk, J., Pagsberg, P., and Sillesen, A., *Chem. Phys. Lett.*, **128**, 168 (1986); *Chem. Abs.*, **106**, 101525 (1987).
191 Wallington, T. J., Neuman, D. M., and Kurylo, M. J., *Int. J. Chem. Kinet.*, **19**, 725 (1987).
192 Droege, A. T. and Tully, F. P., *J. Phys. Chem.*, **91**, 1222 (1987).
193 Greenhill, P. G. and O'Grady, B. V., *Aust. J. Chem.*, **39**, 1775 (1986).
194 Akhlaq, M. S., Al-Baghdadi, S., and Von Sonntag, C., *Carbohydr. Res.*, **164**, 71 (1987).
195 Tully, F. P. and Droege, A. T., *Int. J. Chem. Kinet.*, **19**, 251 (1987).
196 Cohen, N. and Benson, S. W., *J. Phys Chem.*, **91**, 162 (1987).
197 Cohen, N. and Benson, S. W., *J. Phys. Chem.*, **91**, 171 (1987).
198 Das, S. and Von Sonntag, C., *Z. Naturforsch.*, **41B**, 505 (1986); *Chem. Abs.*, **106**, 49522 (1987).
199 Burgess, V. A. and Easton, C. J., *Tetrahedron Lett.*, **28**, 2747 (1987).
200 Smith, G. P., *Int. J. Chem. Kinet.*, **19**, 269 (1987).
201 Liu, A.-D., Mulac, W. A., and Jonah, C. D., *Int. J. Chem. Kinet.*, **19**, 25 (1987).
202 Mulder, P. and Louw, R., *J. Chem. Soc., Perkin Trans. 2*, **1987**, 1167.
203 Atkinson, R., *Int. J. Chem. Kinet.*, **19**, 799 (1987).
204 Mahiou, B. and Gleicher, G. J., *J. Org. Chem.*, **52**, 1555 (1987).
205 Bunce, N. J., Cheung, H. K. Y., and Langshaw, J. A., *J. Org. Chem.*, **51**, 5421 (1986).
206 Germash, A. V., Yakupov, I. S., Zlotskii, S. S., and Rakhmankulov, D. L., *Khim. Fiz.*, **5**, 1534 (1986); *Chem. Abs.*, **106**, 137884 (1987).
207 Chateauneuf, J., Lusztyk, J., and Ingold, K. U., *J. Am. Chem. Soc.*, **109**, 897 (1987).
208 Kim., S. S., Seo, J. S., and Yoon, M. H., *J. Org. Chem.*, **52**, 3691 (1987).
209 Naguid, Y. M. A., Steel, C., Cohen, S. G., and Young, M. A., *J. Phys. Chem.*, **91**, 3033 (1987).
210 Giamalva, D. H., Kenion, G. B., Church, D. F., and Pryor, W. A., *J. Am. Chem. Soc.*, **109**, 7059 (1987).
211 Driscoll, D. J., Martir, W., and Lunsford, J. H., *J. Phys. Chem.*, **91**, 3585 (1987).
212 Labinger, J. A. and Ott, K. C., *J. Phys. Chem.*, **91**, 2682 (1987).
213 Dobis, O. and Benson, S. W., *Int. J. Chem. Kinet.*, **19**, 691 (1987).
214 Benson, S. W., Kondo, O., and Marshall, R. M., *Int. J. Chem. Kinet.*, **19**, 828 (1987).
215 Benson, S. W., *J. Chem. Soc., Faraday Trans. 2*, **83**, 791 (1987).
216 Tschnikow-Roux, E., Salomon, D. R., and Paddison, S., *J. Phys. Chem.*, **91**, 3037 (1987).
217 Zhdanov, G. F., Poluektov, B. A., Belevtsev, E. G., and Orlov, G. D., *Zh. Fiz. Khim.*, **61**, 77 (1987); *Chem. Abs.*, **107**, 197219 (1987).
218 Tanner, D. D. and Nguyen, H. N., *J. Org. Chem.*, **52**, 4686 (1987).
219 Hanzlik, R. P., Schaefer, A. R., Moon, J. B., and Judson, C. M., *J. Am. Chem. Soc.*, **109**, 4926 (1987).

[220] Fujisaki, N., Gaeumann, T., and Ruf, A., *Ber. Bunsen-Ges. Phys. Chem.*, **90**, 375 (1986); *Chem. Abs.*, **106**, 66546 (1987).
[221] Shevel'kova, L. V., Vedeneeva, L. M., and Nametkin, N. S., *Dokl. Akad. Nauk SSSR.*, **288**, 1164 (1986); *Chem. Abs.*, **106**, 32291 (1987).
[222] Hack, W., Kurzke, H., Rouveirolles, P., and Wagner, H. G., *Ber. Bunsen-Ges. Phys. Chem.*, **90**, 1210 (1986); *Chem. Abs.*, **107**, 38872 (1987).
[223] Day, A. C., Covindaraj, N., McBain, D. S., Skell, P. S., and Tanko, J. M., *J. Org. Chem.*, **51**, 4959 (1986).
[224] Carrau, R., Hernandez, R., and Suarez, E., *J. Chem. Soc., Perkin Trans. 1*, **1987**, 937.
[225] Paul, V. and Roberts, B. P., *J. Chem. Soc., Chem. Commun.*, **1987**, 1322.
[226] Gonzalez, M. and Sayos, R., *Chem. Phys.*, **104**, 49 (1986); *Chem. Abs.*, **106**, 4313 (1987).
[227] Taccone, R. A., Salinovich, O., and Staricco, E. H., *Int. J. Chem. Kinet.*, **19**, 627 (1987).
[228] Newcomb, M., Sanchez, R. M., and Kaplan, J., *J. Am. Chem. Soc.*, **109**, 1195 (1987).
[229] Jewell, D. R., Mathew, L., and Warkentin, J., *Can. J. Chem.*, **65**, 311 (1987).
[230] Barton, D. H. R., Lacher, B., and Zard, S. Z., *Tetrahedron*, **43**, 4321 (1987).
[231] Da Silva Corrêa, C. M. M. and Oliviera, M. A. B. C. S., *J. Chem. Soc., Perkin Trans. 2*, **1987**, 811.
[232] Chatgilialoglu, C., *NATO ASI Ser., Ser. C*, **184** (*Adv. Chem. React. Dyn.*), 571 (1986); *Chem. Abs.*, **107**, 153703 (1987).
[233] Chatgilialoglu, C., Ingold, K. U. and Scaiano, J. C., *J. Org. Chem.*, **52**, 938 (1987).
[234] Timonen, R., Kalliorinne, K., and Koskikallio, J., *Acta Chem. Scand.*, **40A**, 459 (1986); *Chem. Abs.*, **107**, 22774 (1987).
[235] Timonen, R., Russell, J. J., Sarzyński, D., and Gutman, D., *J. Phys. Chem.*, **91**, 1873 (1987).
[236] Breslow, R., Brandl, M., Hunger, J., and Adams, A. D., *J. Am. Chem. Soc.*, **109**, 3799 (1987).
[237] Wachi, S. and Morikawa, H., *J. Chem. Eng. Jpn*, **19**, 437 (1986); *Chem. Abs.*, **107**, 58219 (1987).
[238] Mal'tseva, A. S. and Rozlovskii, A. I., *Dokl. Akad. Nauk SSSR.*, **290**, 1156 (1986); *Chem. Abs.*, **106**, 175502 (1987).
[239] Aver'yanov, V. A., Treger, Yu. A., Kirichenko, S. E., *Khim. Prom-st.* (*Moscow*), **1986**, 659; *Chem. Abs.*, **107**, 22778 (1987).
[240] Bunce, N. J., Joy, R. B., Landers, J. P., and Nakai, J. S., *J. Org. Chem.*, **52**, 1155 (1987).
[241] Chambon, M., Marquaira, P.-M., and Côme, G. M., *Can. J. Chem.*, **65**, 1491 (1987).
[242] Citterio, A., Santi, R., and Pagani, A., *J. Org. Chem.*, **52**, 4925 (1987).
[243] Clark, T., *J. Chem. Soc., Chem. Commun.*, **1986**, 1774.
[244] Volovik, S. V., Dyadyusha, G. G., and Staninets, V. I., *Teor, Eksp. Khim.*, **22**, 562 (1986); *Chem. Abs.*, **107**, 77048 (1987).
[245] Volovik, S. V., Dyadyusha, G. G., and Staninets, V. I., *Zh. Org. Khim.*, **22**, 1359 (1986); *Chem. Abs.*, **106**, 83748 (1987).
[246] Volovik, S. V., Dyadyusha, G. G., and Staninets, V. I., *Dopov. Akad. Nauk Ukr. RSR., Ser. B: Geol. Khim. Biol. Nauki*, **1986**, 33; *Chem. Abs.*, **106**, 213210 (1987).
[247] Baldwin, R. R., Keen, A., and Walker, R. W., *J. Chem. Soc., Faraday Trans. 2*, **83**, 759 (1987).
[248] Lichtin, D. A. and Lin, M. C., *Chem. Phys.*, **104**, 325 (1986); *Chem. Abs.*, **106**, 4346 (1987).
[249] Czarnowski, J., *J. Chem. Soc., Faraday Trans. 2*, **83**, 579 (1987).
[250] Riemenschneider, K., Bartels, H. M., Dornow, R., Drechsel-Grau, E., Eichel, W., Luthe, H., Matter, Y. M., Michaelis, W., and Boldt, P., *J. Org. Chem.*, **52**, 205 (1987).
[251] Ghodoussi, V., Gleicher, G. J., and Kravetz, M., *J. Org. Chem.*, **51**, 5007 (1986).
[252] Nakano, T., Kayama, M., and Nagai, Y., *Bull. Chem. Soc. Jpn*, **60**, 1049 (1987).
[253] Russell, G. A., Jiang, W., Hu, S. S., and Khanna, R. K., *J. Org. Chem.*, **51**, 5498 (1986).
[254] Giese, B., González-Gómez, J. A., Lachhein, S., and Metzger, J. O., *Angew. Chem. Int. Edn*, **26**, 479 (1987).
[255] Nedelec, J.-Y., Blanchet, D., Lefort, D., and Guilhem, J., *J. Chem. Res. (S)*, **1987**, 315.
[256] Crich, D. and Davies, J. W., *Tetrahedron Lett.*, **28**, 4205 (1987).
[257] Barton, D. H. R., Gateau-Olesker, A., Gero, S. D., Lacher, B., Tachdjian, C., and Zard, S. Z., *J. Chem. Soc., Chem. Commun.*, **1987**, 1790.
[258] Barton, D. H. R., Hervé, Y., Potier, P., and Thierry, J., *Tetrahedron*, **43**, 4297 (1987).
[259] Sacripante, G. and Just, G., *J. Org. Chem.*, **52**, 3659 (1987).
[260] Barton, D. H. R., Bridon, D., and Zard, S. Z., *Tetrahedron*, **43**, 5307 (1987).
[261] Golubev, V. B., Mun, G. A., and Zubov, V. P., *Zh. Fiz. Khim.*, **60**, 585 (1986); *Chem. Abs.*, **105**, 225506 (1986).
[262] Almirante, N., Cicardi, S., Napoletano, C., and Serravalle, M., *Tetrahedron*, **43**, 625 (1987).
[263] Sanderson, J. R., Yeakey, E. L., Lin, J. J., Duranlean, R., and Marquis, E. T., *J. Org. Chem.*, **52**, 3243 (1987).

264 Citterio, A. and Filippini, L., *Synthesis*, **1986**, 473; *Chem. Abs.*, **106**, 101434 (1987).
265 Gulati, S. K., Mather, S., and Walker, R. W., *J. Chem. Soc., Faraday Trans. 2*, **83**, 2171 (1987).
266 Boodaghians, R. B., Hall, I. W., Toby, F. S., and Wayne, R. P., *J. Chem. Soc., Faraday Trans. 2*, **83**, 2073 (1987).
267 Atkinson, R., Aschmann, S. M., and Goodman, M. A., *Int. J. Chem. Kinet.*, **19**, 299 (1987).
268 Sonawane, H. R., Nanjundiah, B. S., and Kelkar, R. G., *Tetrahedron*, **42**, 6673 (1986).
269 Vieira, A. J. S. C. and Steenken, S., *J. Am. Chem. Soc.*, **109**, 7441 (1987).
270 Vieira, A. J. S. C. and Steenken, S., *J. Phys. Chem.*, **91**, 4138 (1987).
271 Ngo, M., Larson, K. R., and Mendenhall, G. D., *J. Org. Chem.*, **51**, 5390 (1986).
272 Andews, L. and Kelsall, B. J., *J. Phys. Chem.*, **91**, 1435 (1987).
273 Sanders, W. A. and Lin, M. C., *J. Chem. Soc., Faraday Trans. 2*, **83**, 905 (1987).
274 Tuaillon, J., Couture, Y., and Lessard, J., *Can. J. Chem.*, **65**, 2194 (1987).
275 Grelbig, T., Kruegerke, T., and Seppelt, K., *Z. Anorg. Allg. Chem.*, **544**, 74 (1987); *Chem. Abs.*, **107**, 77067 (1987).
276 Ichinose, Y., Wakamatsu, K., Nozaki, K., Birnbaum, J.-L., Oshima, K., and Utimoto, K., *Chem. Lett.*, **87**, 1647 (1987).
277 Alnajjar, M. S. and Franz, J. A., *Prepr. Pap.-Am. Chem. Soc., Div. Fuel Chem.*, **31**, 287 (1986); *Chem. Abs.*, **106**, 4418 (1987).
278 Trofimov, B. A., Nedolya, N. A., and Komel'kova, V. I., *Sulfur Lett.*, **4**, 139 (1986); *Chem. Abs.*, **106**, 175479 (1987).
279 Fang, J.-M. and Chen, M.-Y., *Tetrahedron Lett.*, **28**, 2853 (1987).
280 Kameyama, M., Kamigata, N., and Kobayashi, M., *J. Org. Chem.*, **52**, 3312 (1987).
281 Alberti, A., Bonini, B. F., and Pedulli, G. F., *Tetrahedron Lett.*, **28**, 3737 (1987).
282 Dingwall, J. G. and Tuck, B., *J. Chem. Soc., Perkin Trans. 1*, **1986**, 2081.
283 Alberti, A., Hudson, A., and Pedulli, G. F., *J. Chem. Soc., Faraday Trans. 1*, **83**, 91 (1987).
284 Ichinose, Y., Nozaki, K., Wakamatsu, K., Oshima, K., and Utimoto, K., *Tetrahedron Lett.*, **28**, 3709 (1987).
285 Ahmed, I. A., Hudson, A., and Alberti, A., *J. Organomet. Chem.*, **333**, 9 (1987).
286 McKenzie, T. C. and Rolfes, S. M., *J. Heterocycl. Chem.*, **24**, 859 (1987).
287 Minisci, F., Vismara, E., Fontana, F., Morini, G., Serravalle, M., and Giordano, C., *J. Org. Chem.*, **52**, 730 (1987).
288 Sugimori, A., Yamada, T., Ishida, H., Nose, M., Terashima, K., and Oohata, N., *Bull. Chem. Soc. Jpn*, **59**, 3905 (1986).
289 Sugimori, A. and Yamada, T., *Bull. Chem. Soc. Jpn*, **59**, 3911 (1986).
290 Tada, M., Ito, M., and Ohshima, K., *J. Heterocycl. Chem.*, **23**, 1893 (1986).
291 Wakselman, C. and Tordeux, M., *J. Chem. Soc., Chem. Commun.*, **1987**, 1701.
292 Firouzbakht, M. L., Ferrieri, R. A., and Wolf, A. P., *J. Am. Chem. Soc.*, **109**, 2213 (1987).
293 Walton, J. C., *J. Chem. Soc., Chem. Commun.*, **1987**, 1252.
294 Branchaud, B. P., Meier, M. S., and Malekzadeh, M. N., *J. Org. Chem.*, **52**, 212 (1987).
295 Johnson, M. D., Lampman, G. M., Koops, R. W., and Gupta, B. D., *J. Organomet. Chem.*, **326**, 281 (1987).
296 Russell, G. A., Ngoviwatchai, P., Tashtoush, H., and Hershberger, J., *Organometallics*, **6**, 1414 (1987).
297 Light, J. P., Ridenour, M., Beard, L., and Hershberger, J. W., *J. Organomet. Chem.*, **326**, 17 (1987).
298 Lakhani, C., Mathew, L., and Warkentin, J., *Can. J. Chem.*, **65**, 1748 (1987).
299 Montaudon, E., Rakotomanana, F., Agorrody, M., and Maillard, B., *Bull. Soc. Chim. Fr. II*, **1987**, 375.
300 Agorrody, M., Campagnole, M., Montaudon, E., and Maillard, B., *Tetrahedron*, **43**, 3429 (1987).
301 Maillard, B., Kharrat, A., and Gardrat, C., *Nouveau J. Chim.*, **11**, 7 (1987).
302 Franz, J. A., Roberts, D. H., and Ferris, K. F., *J. Org. Chem.*, **52**, 2256 (1987).
303 Newcomb, M. and Kaplan, J., *Tetrahedron Lett.*, **28**, 1615 (1987).
304 Bentrude, W. G., Kawashima, T., Keys, B. A., Garroussian, M., Heide, W., and Wedegaertner, D. A., *J. Am. Chem. Soc.*, **109**, 1227 (1987).
305 Kice, J. L. and Purkiss, D. W., *J. Org. Chem.*, **52**, 3448 (1987).
306 Sueishi, Y. and Nishimura, N., *Bull. Chem. Soc. Jpn*, **59**, 3305 (1986).
307 Schulz, M., Moegel, L., Omelka, L, and Tkac, A., *J. Prakt. Chem.*, **328**, 222 (1986); *Chem. Abs.*, **106**, 101657 (1987).
308 Alberti, A., Greci, L., Stipa, P., Sgarabotto, P., and Ugozzoli, F., *Tetrahedron*, **43**, 3031 (1987).
309 Lagercrantz, C., *Acta Chem. Scand.*, **41B**, 526 (1987).
310 Batelaan, J. G., Hageman, H. J., and Verbeek, *Tetrahedron Lett.*, **28**, 2163 (1987).
311 Linsel, G., Lassmann, G., and Damerau, W., *Z. Chem.*, **27**, 340 (1987).

[312] Ozawa, T. and Hanaki, A., *Bull. Chem. Soc. Jpn*, **60**, 2304 (1987).
[313] Sridhar, R., Beaumont, P. C., and Powers, E. L., *J. Radioanal. Nucl. Chem.*, **101**, 227 (1986); *Chem. Abs.*, **107**, 39048 (1987).
[314] Halpern, A., *J. Chem. Soc., Faraday Trans. 1*, **83**, 219 (1987).
[315] Sadykhov, E. G. and Gasanov, R. G., *Kinet. Katal.*, **27**, 984 (1986); *Chem. Abs.*, **106**, 175495 (1987).
[316] Zubarev, V. E., Troyansky, E. I., and Nikishin, G. I., *Izv. Akad. Nauk SSSR., Ser. Khim.*, **1986**, 2630; *Chem. Abs.*, **107**, 197245 (1987).
[317] Ji, H. and Xu, G., *Bopuxue Zashi*, **3**, 125 (1986); *Chem. Abs.*, **106**, 195770 (1987).
[318] Ji, H. and Xu, G., *Bopuxue Zazhi*, **3**, 165 (1986); *Chem. Abs.*, **106**, 195771 (1987).

Organic Reaction Mechanisms 1987
Edited by A. C. Knipe and W. E. Watts

CHAPTER 4

Radical Reactions: Part 2

D. C. NONHEBEL

Department of Pure and Applied Chemistry, University of Strathclyde

Homolytic Oxidation and Reduction

Reaction of cytochrome P-450 with methylcyclopropane gives exclusively cyclopropanemethanol, but under similar conditions the more strained bicyclo[2.1.0]pentane gives a mixture of 2-hydroxybicyclo[2.1.0]pentane and cyclopent-3-en-1-ol:[1] formation of the latter product is consistent with the intermediacy of the bicyclo[2.1.0]pentyl radical which ring-opens to the cyclopent-3-en-1-yl radical. The absence of any ring-opening in the reaction of methylcyclopropane indicates that the rate constant (k) for collapse of the intermediate radical pair must be at least $1 \times 10^9 \, l \, mol^{-1} \, s^{-1}$.

In hemin-catalysed epoxidations of alkenes the initial step involves electron transfer to a high-valent iron intermediate[2] (Scheme 1). Pinol, carveol and sobrerol are all obtained in the anodic oxidation of α-pinene: the four-membered ring in the α-pinene radical cation undergoes ring-opening in the key step of the reaction.[3]

Side-chain oxidations of 5-substituted 1,2,3-trimethylbenzenes with CAN and under anodic conditions give the same proportions of reaction in the 1- and 2-methyl groups consistent with a common radical cation intermediate.[4] Different

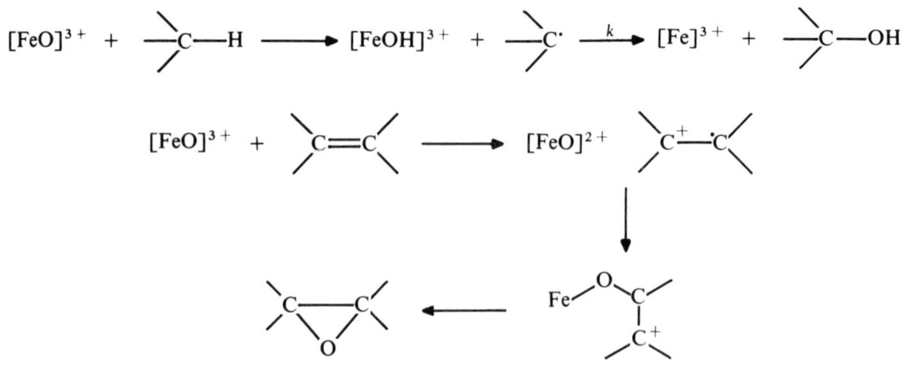

SCHEME 1

product ratios were obtained with cobalt(III) acetate and NBS indicating different pathways. Studies on the anodic oxidation of the anions of isomeric benzylpyridines indicate that, whereas the nitrogen stabilizes the anion, it has little influence on the stabilization of the desired radicals.[5] Peroxydisulphate and lithium bromide have been shown to induce side-chain bromination of alkyl aromatics, e.g. *p*-nitrotoluene.[6] The reaction involves oxidation of bromide to bromine atoms which bring about the reaction: *p*-nitrotoluene is not oxidized to its radical cation under the reaction conditions.

$$SO_4^{\bar{\cdot}} + Br^- \rightarrow SO_4^{2-} + Br^{\cdot}$$
$$ArCH_3 + Br^{\cdot} \rightarrow ArCH_2^{\cdot} + HBr$$
$$ArCH_2^{\cdot} + Br_2 \rightarrow ArCH_2Br + Br^{\cdot}$$

Enzymic hydroxylation (with mono-oxygenases) of ethylbenzene gives optically active 1-phenylethanol: the products are formed in a cage combination of $Ph\dot{C}HCH_3$ and $HO^{\cdot}$.[7]

Reaction of indole with copper(II) chloride results in chlorination of the heterocyclic ring with high molar ratio of $CuCl_2$: indole and dimeric products with low molar ratios: both products are derived from the indole radical cation.[8] 1-Alkyl-3-(trimethylsilyl)cyclohexanols undergo oxidative fragmentation with CAN to give keto-alkenes.[9]

Anodic oxidation of methyl ethers of *p*-alkylphenols in methanol leads to *p*-quinol ether ketals via coupling of the radical cation of the substrate and a methoxyl radical. Side-chain oxidation products are also formed via benzylic radicals.[10,11] This competing process becomes very much less important in reactions of 2-hydroxyethyl ethers of *p*-alkylphenols in which intramolecular coupling can occur.

A study of the kinetics of oxidation of phenoxide by hexacyanoferrate suggests that phenoxyl radicals are generated reversibly and that they are subsequently oxidized to the phenoxonium ion.[12] Oxidations of phenols unsubstituted at the 4-position with Fremy's salt give *p*-quinones: MNDO calculations indicate that the spin density in phenoxyl radicals is highest at the 4-position favouring reaction at this site.[13] 5-Hydroxytryptamine has been oxidized electrochemically to the phenoxyl radical: this either dimerizes or undergoes further oxidation to give initially a quinone imine.[14]

An interesting development has been the study of the oxidation of dianions of *p*-(4-nitrobutyl)phenol and phenolic *β*-diketones. In both instances the phenoxide is oxidized to the radical which undergoes intramolecular coupling at the carbanionic centre (Scheme 2).[15,16]

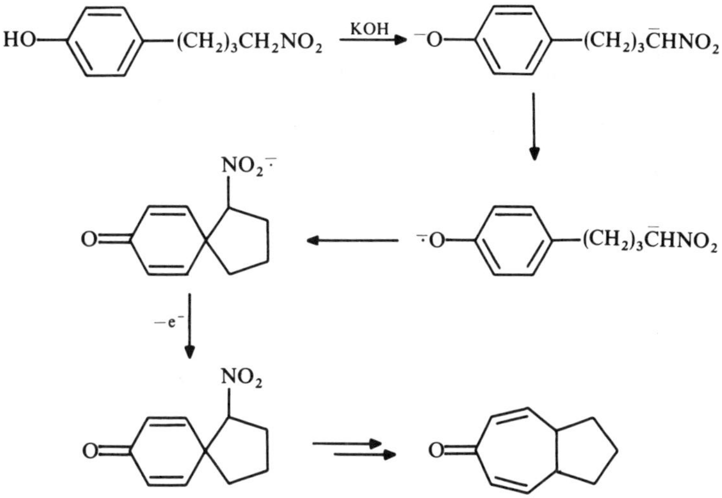

SCHEME 2

The oxidative cyclization of $CH_2{=}CHCH_2CH_2COCH_2COR$ (R = OMe, Me) with manganese(III) acetate/copper(II) acetate leads to excellent yields of methyl salicylate and *o*-hydroxyacetophenone respectively. The initial radical adds reversibly to the double bond to give the thermodynamically more stable cyclohexyl radical. When the double bond in the starting material has terminal substituents, increasing amounts of cyclopentanones are formed.[17] Under similar conditions (**1**; R = Me) is oxidized to (**4**): the initial cyclization of (**2**; R = Me) gives the tertiary cyclohexyl radical (**3**), whereas (**2**; R = H) cyclizes to give a mixture of the secondary cyclohexyl radical (**3**; R = H) and the cyclopentylmethyl radical (**5**; R = H) leading to both (**4**; R = H) and (**6**).[18] In another example the initial cyclized cyclohexyl radical is oxidized to the cation prior to further cyclization when it is tertiary, whereas the second cyclization occurs prior to oxidation when the initial cyclized cyclohexyl radical is secondary.[19] Manganese(III) acetate promotes the addition of enolizable *β*-dicarbonyl compounds to enol ethers and enol esters to give 2,3-dihydrofurans.[20] But-2-enamides and but-2-en-4-olides are formed in reactions of 1,1-diarylethenes with *α*-cyanoacetamide and manganese(III) acetate.[21] Oxidation of 3-phenylpropenoic acids with manganese(III) acetate and lithium chloride gives $PhCHClCHCl_2$ and $PhCH(OAc)CHCl_2$.[22] The electrochemical oxidation of 1,3-diketones in presence of alkenes leads to [3 + 2]-cycloaddition products.[23]

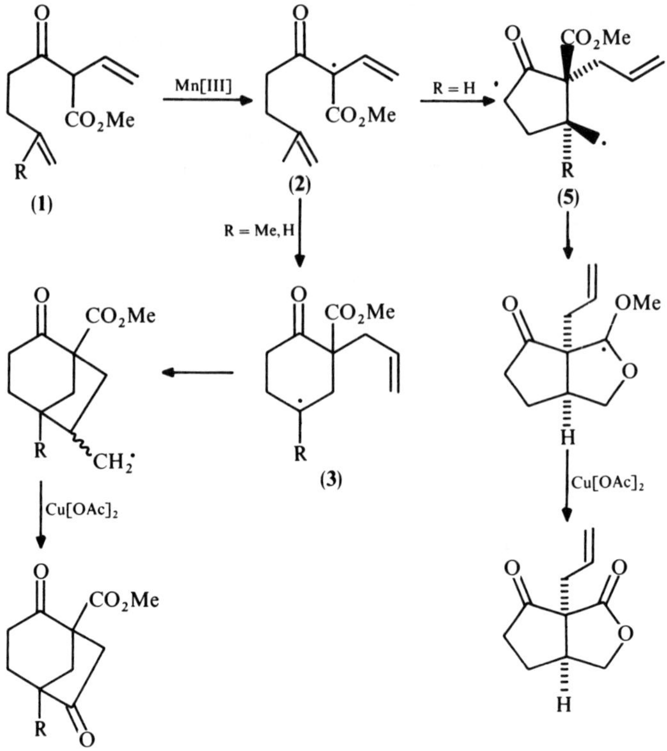

Cobalt(II) chloride promotes the coupling of acetic anhydride and aldehydes to give 1,2-diketones.[24]

$$\begin{aligned} Co^{II} + Ac_2O &\rightarrow Ac_2O^{\cdot -} + Co^{III} \\ Ac_2O^{\cdot -} &\rightarrow CH_3\dot{C}O + AcO^- \\ RCHO + Co^{III} &\rightarrow R\dot{C}O + H^+ + Co^{II} \\ R\dot{C}O + CH_3\dot{C}O &\rightarrow RCOCOCH_3 + RCOCOR \\ &\quad + CH_3COCOCH_3 \end{aligned}$$

Long-lived radicals are generated in the oxidative desilyation of ketene acetals.[25] The anodic oxidation of β-ionone proceeds via its radical anion, acetoxylation and oxidation of which leads to an allylic cation.[26]

The rates of peroxydisulphate-promoted decarboxylations of both arylacetic acids and aliphatic carboxylic acids have been measured: the rates of decarboxylation correlate with the stabilities of the radicals obtained after decarboxylation. The reactions are postulated to proceed by electron transfer from HOMO of the carboxylate anion to $SO_4^{\cdot -}$.[27] A different conclusion was reached on the basis of an ESR study of the oxidation of phenylalkanoic acids, $Ph(CH_2)_nCOOH$, by $SO_4^{\cdot -}$ or $HO^{\cdot}$. In the pH range 6–9, oxidation is said to give the radical zwitterions,

$\overset{+\bullet}{\mathrm{P}}\mathrm{h(CH_2)_{\mathit{n}}COO^-}$, which undergo decarboxylation via intramolecular electron transfer.[28] In acid solution oxidation gives the corresponding radical cations which undergo a variety of reactions including hydration, decarboxylation, and proton loss (to give benzylic radicals).

Anodic oxidation of alkylthioacetonitriles in methanol leads to the α-methoxylated products via initial formation of $\mathrm{R\overset{+\bullet}{S}CH_2CN}$.[29]

Evidence for a SET mechanism for the oxidative *N*-dealkylation of $PhNHCH_2CH_2NH_2$ by dopamine β-mono-oxygenase has been published.[30] The product obtained in the oxidative dimerization of diarylamines is controlled by the spin-density distribution in the radical cation intermediates.[31,32] Oxidation of primary aromatic amines by 2,4,6-tri-*t*-butylphenoxyl leads to a mixture of C—N and C—C coupled products.[33] An ECE process is involved in the electrochemical oxidation of 2,4-dibromo-6-nitroaniline to 2-bromo-6-nitrobenzoquinone.[34]

The reaction of hydroxyl radicals with ethane-1,2-diol gives $\mathrm{HOCH_2\dot{C}HOH}$ which undergoes acid-catalysed dehydration to $\mathrm{\dot{C}H_2CHO}$.[35] The oxidizing radicals, $HO^\bullet$, $O^{\bar{\bullet}}$, $SO_4^{\bar{\bullet}}$, and $Br_2^{\bar{\bullet}}$ all undergo reversible addition to 4,6-dihydroxypyrimidine at C(5). At pH > 8 the adduct radicals undergo base-catalysed dehydration.[36] The initial step in the reaction of indole with $HO^\bullet$ involves electron transfer.[37] The rate constants for reactions of $HO^\bullet$ with aliphatic ketones have been determined.[38] Addition of $HO^\bullet$ to pyridinedicarboxylic acids occurs exclusively *ortho* or *meta* to the heterocyclic nitrogen.[39]

Haloarenes are reduced by primary and secondary alcohols in an electro-catalytic process.[40] Initiation involves electrochemical reduction of the haloarene followed by loss of halide to give an aryl radical.

$$\mathrm{ArX \xrightarrow{e^-} ArX^{\bar{\bullet}} \rightarrow Ar^\bullet + X^-}$$
$$\mathrm{Ar^\bullet + R_2CHO^- \rightarrow ArH + R_2\dot{C}O^-}$$
$$\mathrm{ArX + R_2\dot{C}O^- \rightarrow ArX^{\bar{\bullet}} + R_2CO}$$

The rate-determining step in the reduction of monohalobiphenyls by electrochemically-generated (phenanthridine)$^-$ involves electron transfer.[41] The electrochemical reduction of *trans*-1,2-dibromocyclohexane is subject to homogeneous catalysis by aromatic electron acceptors, e.g. *p*-dicyanobenzene.[42]

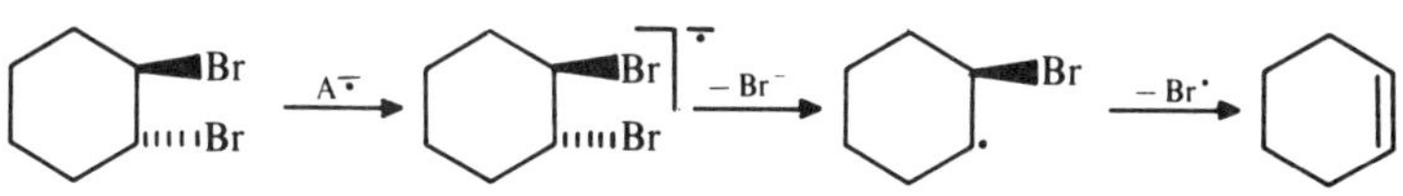

The electrochemical reduction of $\mathrm{PhCHBrBu^t}$ gives dimeric products from coupling of $\mathrm{Ph\dot{C}HBu^t}$.[43] Reductive electrochemical dimerization of ethyl α-bromo-*p*-halophenylacetates gives a mixture of racemic and *meso* substituted succinates.[44] The cathodic reduction of 1-iododec-5-yne yields *inter alia* pentylidenecyclopentane as a result of cyclization of the dec-5-yn-1-yl radical.[45]

Samarium(II) iodide effects the reduction of a wide range of alkyl and aryl halides: the reactions proceed via the intermediate alkyl or aryl radical.[46]

$$ArX \xrightarrow{SmI_2} ArX^{\cdot -} \rightarrow Ar^{\cdot} + X^-$$

Reductive coupling of 1,3-dioxolane[47] and THF[48] with carbonyl compounds has been promoted by reduction of iodobenzene with samarium(II) iodide (Scheme 3). Samarium(II) iodide has also been employed to induce the stereoselective cyclizations of 2-(ω(iodoalkyl)-β-keto-esters and 2-(ω-iodoalkyl)-β-keto-amides.[49] Tricarbonyl(arene)chromium complexes are more readily reduced than the corresponding arenes: the initial radical anions couple to give hydrodimeric products.[50] Cathodic reduction of dicyanoethenes similarly results in

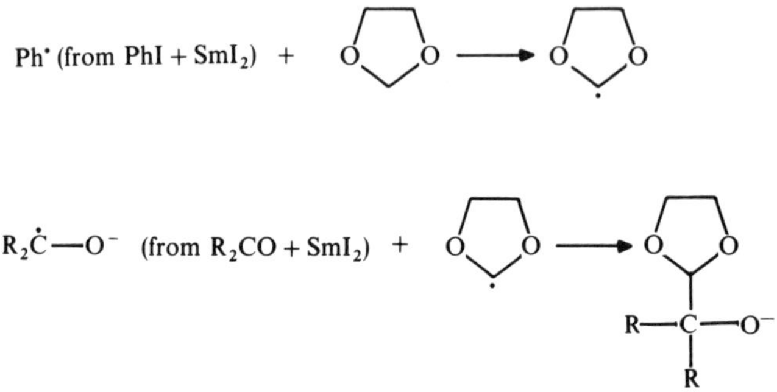

SCHEME 3

hydrodimerization.[51] Reductive dimerization of tricarbonyl(1-methoxycarbonylcyclohexadienyl)iron hexafluorophosphate proceeds on reaction with zinc dust or electrochemically via radical intermediates. The corresponding *N,N*-diallylamide complex gives cyclic products.[52]

A product study of the reduction of (+)-camphor by alkali metals in THF or liquid ammonia led to the conclusion that the reactions proceed via a ketyl radical anion and that, contrary to earlier suggestions, dianions are not intermediates.[53] Electrochemical reductive alkylations of tricarbonyl(benzophenone)-chromium[54] and thiopivalophenone[55] proceed via the ketyl radical anion. The reductions of daunomycin, daunomycinone, and 7-deoxydaunomycin by xanthine oxidase to the corresponding semiquinones involve a one-electron-transfer reaction.[56]

Radical Ions

Radical Cations—Structure, Generation, and Properties

A review of the structure, stability, and methods of generation of hydrocarbon radical cations has been published.[57] Reviews of the formation and reactions of radical cations in the gas phase,[58] and of the structures and rearrangements of radical cations in solution have been published.[59] *Ab initio* calculations have been carried out to investigate odd-electron σ and π interactions in radicals and radical ions.[60]

Ab initio calculations on the ground states, triplet states, radical cations, and radical anions of propene, ethanal, ethanalimine, dimethylbutadiene, and biacetyl indicate that, whereas in the ground state and in radical cations a C—H bond in the methyl group is eclipsed with the C=X bond, in the excited state and in radical anions a staggered conformation is preferred due to a more favourable overlap of the π_{C-X} and $\pi^*_{CH_3}$ orbitals.[61] Alkane radical cations have been generated by X-irradiation of alkanes adsorbed on zeolites at 4K: dimeric species are formed with increasing alkane concentration.[62] Radical cations of *n*-alkanes at 77K have an extended rather than a gauche structure: the two in-plane C—H bonds of the terminal methyl groups preferentially participate in the σ-delocalized MO.[63] At 100 K proton loss to give 1-alkyl radicals occurs: in CF_2ClCF_2Cl the isomeric 2-alkyl radicals are formed via the gauche conformation of the radical cations as the electron density is higher in the in-plane C—H at C(2) rather than at the terminal methyl group. Rotation of the methyl groups in $C_4H_{10}^{+\bullet}$ does not occur at 4.2 K.[64] The radical cation of [1.1.1.1]pagodane is remarkably long-lived.[65] Radical cations derived from Bredt's rule-protected alkenes are long-lived in solution: the C_β—H bonds lie close to the nodal plane of the π-SOMO.[66] An ESR spectroscopic study of tetra(1-adamantyl)cyclobutadiene showed that the adamantyl groups cause considerable distortion of the central ring from planarity but that there is no restriction of rotation about the C-adamantyl bonds.[67] The ESR spectrum of $CH_3CH_2Br^{+\bullet}$ has been obtained.[68] The *trans*-stilbene radical cation has been generated by photolysis of a mixture of either *cis*- or *trans*-stilbene and mercury(II) trifluoroacetate in TFA: there is no evidence of restricted rotation about the C-phenyl bonds, nor of cylization to the phenanthrene radical cation.[69] The kinetics of decay of $Ph_2C{=}CPh_2^{+\bullet}$ have been determined.[70] Diphenyliodinium salts catalyse the formation of arene radical cations by pulse radiolysis.[71] Arene radical cations are generated on reaction with DDQ in TFA/CH_2Cl_2.[72] ESR studies indicate that $PhCF_3^{+\bullet}$ is a π^* species whereas INDO and CNDO/2 calculations suggest it is a σ^* species.[73] The radical cations of furans, thiophens, and pyrroles have been generated by photolysis in TFA in the presence of mercury(II) trifluoroacetate: the ESR spectra indicate that the unpaired electron is in a $^1a_2$MO with the node passing through the heteroatom.[74] 1,4-Dimethoxybenzene and 1,2,4,5-tetramethoxybenzene radical cations have been generated by pulse radiolysis.[75] The cyclic trimers of aromatic *o*-diethers are readily converted to the corresponding radical cations.[76] In the flavin radical cation, electron loss occurs from the main π-system rather than from a specific site.[77] ESR, ENDOR, and TRIPLE resonance studies of *N*,*N'*-diaryl-4,4'-bipyridylium radical cations have been carried out.[78] The radical cations of 3-alkyl- and 3-acyl-2-oxa-3-azabicyclo[2.2.2]octanes (**7**) are unusually long-lived as the bridgehead C—H bond is not aligned with the SOMO: approximately 75% of the unpaired spin is associated with the N—O bond.[79] The rates of electron transfer between tetraalkylhydrazines and tetraalkylhydrazine radical cation varies between 18 and 0.03×10^{-11} cm^3 mol^{-1} s^{-1} at 550 K.[80] The low efficiencies of the reaction

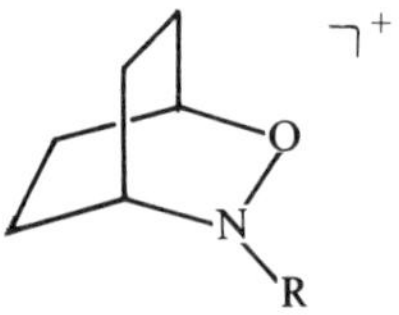

(7)

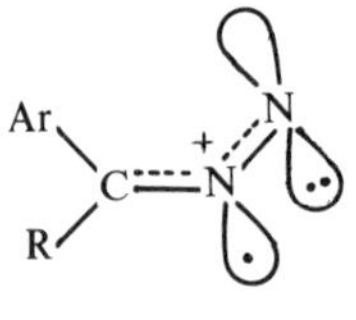

(8)

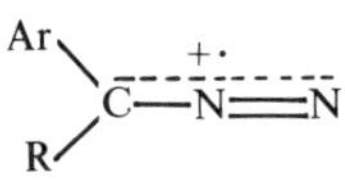

(9)

are associated with the large geometrical change required in the reactions. The double nitrogen inversion in sesquibicyclic hydrazine radical cations has been studied.[81] Photolyses of charge-transfer complexes of azoalkanes with electron acceptors give azoalkane radical cations.[82] Electrochemical oxidations of aryldiazomethanes lead to either σ- or π-radical cations (**8**) or (**9**).[83] ESR studies have shown that the tetramethylurea radical cation (**10**) has an orthogonal structure, and that it does not decompose to $Me_2N^{\cdot}$.[84–86] The influence of phosphorus substituents on the half-wave reduction potentials of a series of aromatic phosphorus compounds has been determined.[87] Electron transfer occurs from a benzene ring in $Ph_3P^{+\cdot}$ and $PhOP(O)(OPh)_2{}^{+\cdot}$.[88] The SOMO in $HMPA^{+\cdot}$ is on one of the dimethylamine groups. The radical cations from dinuclear molybdenum carbonyl complexes, e.g. $[(\eta^5\text{-}C_5H_5)_2Mo_2(CO)_4(\mu\text{-}SR)]$ have been obtained by anodic oxidation.[89] Simple oxirane radical cations have never been observed, as facile C—C bond cleavage occurs: this is avoided in polycyclic oxiranes where steric constraints prevent opening of the oxirane ring.[90] Ylid ions $\dot{C}H_2\overset{+}{X}H$ (X = OH, NH_2, F, Ph_2, SH, Cl) are much more stable than the prototype ylids CH_2XH.[91]

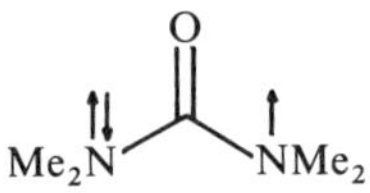

(**10**)

A mass spectrometric study of diisopropyl sulphide has provided evidence for the dimeric radical cation $Pr^{i}{}_{2}S\overset{+\cdot}{-}SPr^{i}{}_{2}$ with a three-electron S—S bond.[92] Time-resolved pulse radiolysis studies have been employed to study the reaction $R_2S^{+\cdot} + R_2S \rightleftarrows R_2S\overset{+\cdot}{-}SR_2$.[93] The dimeric radical cation from thiirane breaks down to give ethylene and *c*-C_2H_4S—$S^{\cdot}$: the extrusion of ethylene is depicted as a concerted reaction in which the driving force is supplied by the transfer of the unpaired electron from the σ^* (S—S) orbital of the dimeric radical cation to a vacant *p* orbital on the terminal sulphur.[94] New examples of species (**11**) and (**12**) with $2\sigma/1\sigma^*$ three-electron bonds have been detected in oxidations of 1,ω-diiodoalkanes[95] and 1-iodo-3-

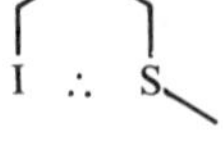

(**11**)

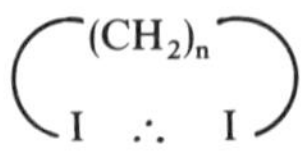

(**12**)

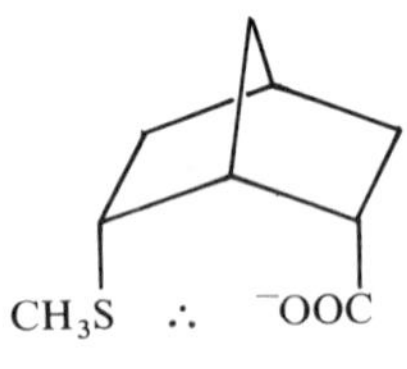

(**13**)

methylthiopropane.[96] Neighbouring-group participation by carboxylate and alcohol groups results in stabilization of an oxidized sulphur atom, e.g. (**13**) in which the $S^{+\cdot} \leftrightarrow {}^{-}OOC$ interaction is estimated to be 50 kJ mol^{-1}.[97]

Radical Cation Reactions

The role of radical cations in the dehydro-coupling of aromatic systems by catalyst-oxidant systems has been reviewed.[98] The rate of dimerization of anthracene radical cations has been studied.[99] Disubstituted 2-nitrobenzenesulphenamide radical cations undergo dimerization at the 4-position.[100] The iron(III) chloride-promoted polymerization of pyrrole proceeds via a Radical-cation mechanism.[101] Dimers are obtained from the oxidation of 5,12-dimethoxynaphth[2,3-*a*]azulene using iron(III) chloride as the oxidant: substitution products are formed with tropylium tetrafluoroborate as oxidant.[102] Diene radical cations, formed on photolysis of dienes in presence of TiO_2, give Diels–Alder products or, in the presence of oxygen, oxygenated products.[103] Oxidative sulphurization of acetylene with S_2Cl_2 leads successively to 1,2-dithiete and 1,4-dithiane radical cations.[104]

Radical cation-nucleophile reactions have been shown to be extremely fast.[105] This contrasts with predictions that suggest that such reactions are high-energy processes. The reaction of arene radical cations with chloride is inhibited by non-nucleophilic counter-ions which form ion-pairs with $ArH^{+\cdot}$.[106] Electro-generated $Ph_3P^{+\cdot}$ reacts with allyltrialkylstannane to give allyltriphenylphosphonium.[107]

The anodic oxidation of diaryldiazoalkanes leads to the corresponding radical cations which undergo loss of nitrogen to give $Ar_2\overset{+}{C}H$: the reaction is inhibited by 2,6-lutidine, suggesting that protons may act as chain carriers.[108,109] The radical cation $CF_3COCH_3^{+\cdot}$ undergoes cleavage to $\dot{C}F_3$ and $CH_3C{=}O^{+}$.[110] The trimethylsilylallyl radical cation, an important intermediate in photo-induced electron-transfer reactions of silylalkenyl iminium salts, fragments to give allyl radicals on reaction with alcohols.[111] Radical cations of *β*-amino-alcohols (**14**), generated by photolysis in presence of an electron acceptor, fragment to give the corresponding aldehydes.[112,113]

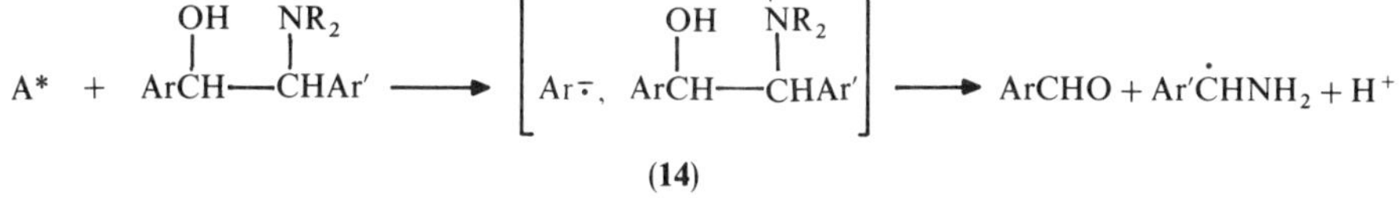

(14)

Ab initio MO calculations of 1,3-hydrogen shifts in radical cations suggest that they should occur readily in a suprafacial sense.[114] *Ab initio* calculations predict that $HOCH_2CH_2OH^{+\cdot}$ should undergo a barrier-free 1,4-hydrogen shift to give $H_2\overset{+}{O}CH_2CH_2O^{\cdot}$: this is consistent with mass spectrometric observations.[115] Another *ab initio* study on the barriers to 1,*n*-hydrogen transfer in alkylamine radical cations indicates a regular decrease in barrier height from 1,2- to 1,5-hydrogen migrations associated with a decrease in ring strain and increasing linearity of the C---H---N angle in the transition state.[116] Cyclic derivatives of 1,2-diphenylcyclobutene rearrange to the corresponding 2,3-diphenylcyclobutenes: this rearrangement constitutes the first example of a suprafacial [1,3]-sigmatropic hydrogen shift via a radical cation.[117]

Ab initio calculations indicate that the ring-opening of the methylenecyclopropane radical cation to the trimethylenemethane radical cation is exothermic and should proceed in a disrotatory fashion.[118] Evidence for the intervention of cyclopropylamine radical cations in the deactivation of monoamine oxidase and cytochrome P450 by cyclopropylamine is seen in the formation of ring-opening to $H_2\overset{+}{N}{=}CHCH_2\dot{C}H_2$, which attacks the enzyme site.[119] *cis*-3,4-Diarylcyclobutene radical cations, formed in photo-induced electron-transfer reactions, undergo stereospecific conrotatory ring-opening to give *cis,trans*-1,4-diarylbuta-1,3-dienes (Scheme 4).[120] Different products are obtained in the CAN-catalysed single-electron-transfer oxidations of (**15**) and (**16**) indicating that the derived radical cations are different but undergo a slow valence isomerization (Scheme 5).[121]

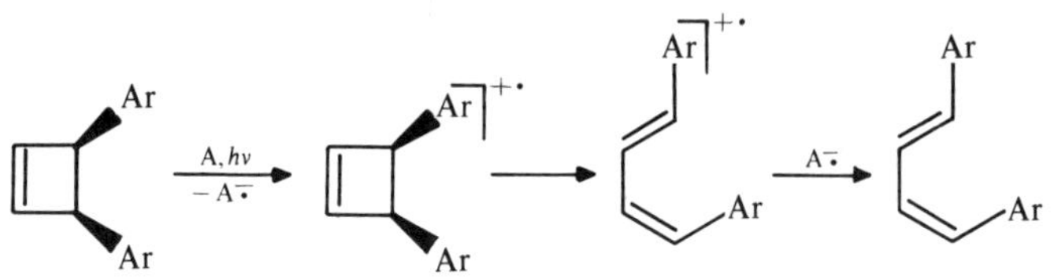

SCHEME 4

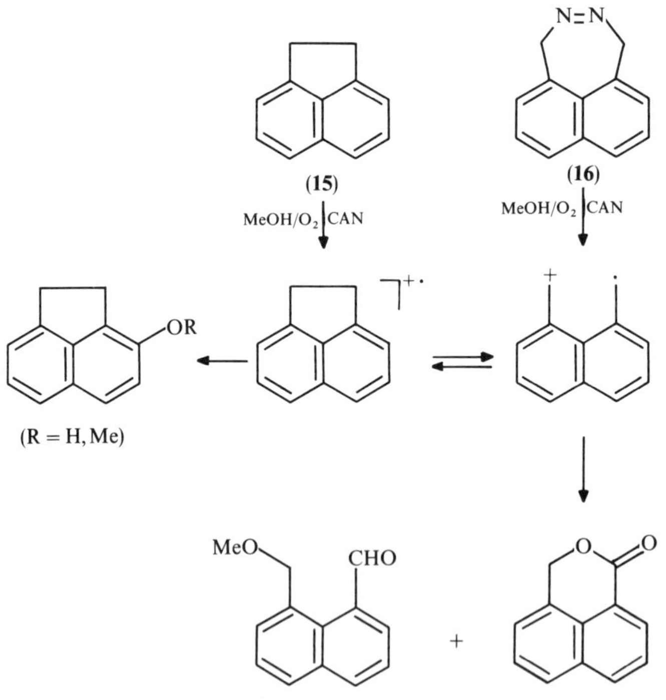

SCHEME 5

Quadricyclene undergoes valence isomerization to norbornadiene in presence of photoexcited *n*-type semiconductors: the key step in the reaction is the oxidation of quadricyclene to its radical cation by photo-generated holes that migrate to the semiconductor surface.[122,123]

N-Hydroxypyridine-2-thione carbamates have been used as sources of aminyl radicals, e.g. (**17**): in acid solution these are protonated to the radical cation (**18**).[124] The rate of cyclization of (**18**) is much greater than that of (**17**) as seen by the exclusive formation of the acyclic product (**19**) from (**17**), and the cyclic product (**20**) from (**18**) in the presence of a good hydrogen donor, e.g. *t*-butyl mercaptan.

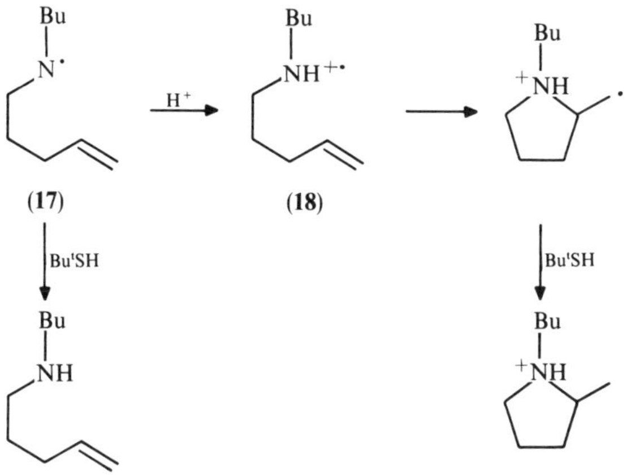

(**17**) (**18**) (**19**) (**20**)

Radical Anions—Structure, Generation, and Properties

Methods of generation and characterization of radical trianions have been reviewed.[125] NMR spectroscopy has been used to show that polycyclic dianions undergo a photo-ejection process to give radical anions.[126] The reduction of a number of electron acceptors to their radical anions has been achieved by reaction with hydroxide in DMSO.[127] Photolyses of 2-(*p*-nitrophenyl)-1-phenylethanol, and *p*-nitrophenylacetate at pH 12 give the corresponding radical anions and subsequently *p*-nitrobenzyl radicals.[128,129] The sulphite radical anion adds to alkenes to give the adduct radical anions: the direction of addition is said to be determined by steric factors.[130] The structure and reactivity of $PhMe^{\cdot -}$ has been reviewed.[131] Calculations on the distribution of electron density in $PhCN^{\cdot -}$ have been published.[132] ESR studies on radical anions of chloronitrobenzenes,[133] 2,5-diarylfurans,[134] and anthraquinone and 9,10-anthraquinol[135] have been carried out. The 1,1-diphenylcyclopropabenzene radical anion is stable.[136] The cyclohexane ring in 1,2-diphenylcyclohexenes is in a half-chair conformation.[137] The 2,3,4,5-tetraphenyldiazocyclopentadiene radical anion loses nitrogen to give the corresponding carbene radical anion: the rate of loss of nitrogen indicates that the extent of spin delocalization in the carbene radical anion is not the sole controlling factor in the reaction.[138] The SOMO in the radical anion of 1,2:9,10-dibenzo[2.2]-paracyclophane lies in the plane of the aromatic rings.[139]

Radical anions of aromatic ketones and hydrocarbons are generated by reaction with $EtOCH_2CH_2OLi$.[140] The camphorquinone radical anion is produced by photo-reduction in propan-2-ol,[141] and the 1,4-naphthoquinone radical anions by alkali metal reduction in DME.[142] The spin density on the carbons in alkali metal ion-pairs of 1-aminoanthraquinone radical anions increases with increasing effective ionic charge of M^+.[143] IR studies indicate that $PhCOCOPh^{\cdot -}$ exists in the *s-trans* conformation.[144] ESR studies show that conformation is also favoured in mono- and di-thiosemidione radical anions and 1,1-dithio-oxalic esters and amides.[145] The spin density in the monothiosemidiones is largely associated with the C—S bond. The spin density distribution in 2*H*-thiapyran-2-thione radical anion has been determined.[146] Radical anions of phenyl benzenecarbodithioates have been studied by polarography.[147] ENDOR studies of lumiflavin radical anions in micelles have been reported.[148] The reaction of lithium with sp^3-halides can be interpreted to involve the formation of $R^{\cdot}$---X^-, Li^+: sp^2 halides react by electron transfer with the ρ^* C—X orbitals or into the π-system of the double bond.[149] The radical anion of di-*t*-butylsilanone, $Bu^t{}_2SiO^{\cdot -}$, is generated by reaction of $Bu^t{}_2Si(OH)H$ with di-*t*-butyl peroxide in alkaline solution.[150]

Radical Anion Reactions

The rate constants for loss of chloride from *o*-, *m*-, and *p*-chloroacetophenones are 5.3×10^6, 9.6×10^4 and $2.8 \times 10^5\ s^{-1}$, respectively: the negative charge is initially in the delocalized π-orbital of the acetophenone moiety and undergoes intramolecular electron transfer and loss of chloride.[151] Halogenobenzonitrile radical anions also undergo intramolecular electron transfer and synchronous loss of halide: the rates of dehalogenation are dependent on the ease of electron transfer and the level of the LUMO of the C—X bond.[152]

The anthraquinone radical anion undergoes preferential disproportionation rather than further one-electron reduction in hydrogen-bonding solvents.[153] The protonation of the anthracene radical anion in aprotic solvents has been studied.[154,155] The protonation of naphthalene radical anions by diethyl malonate is catalysed by base: this catalyses the conversion of the malonate into its enol form.[156] The radical anion $Ph_2C^{\cdot -}$ is protonated to give $Ph_2\dot{C}H$ in alcohols or carbonyl compounds possessing C—H bonds, reacts with alkyl halides to give $Ph_2\dot{C}R$, and adds to carbonyl compounds not possessing C—H bonds: the adduct radicals undergo subsequent fragmentation.[157]

$$Ph_2C^{\cdot -} + Bu^tCHO \rightarrow Ph_2\dot{C}{-}CH(O^-)Bu^t \rightarrow Ph_2\bar{C}CHO + Bu^{t\cdot}$$

The key step in the photo-induced electron-transfer deprotection of carbohydrate tosylates involves the fragmentation of the tosylate radical anion.[158]

$$ArSO_2OR^{\cdot -} \rightarrow ArSO_2^{\cdot} + RO^-$$

2-Trifluoroacetoxymethylanthraquinone radical anion undergoes facile cleavage of trifluoroacetate.[159] Radical anions of 9,9-diarylfluorenes fragment to give 9-arylfluorenyl radicals and aryl anions, which subsequently undergo electron transfer to give the more stable 9-arylfluorenyl anion and aryl radical.[160] The bis(*p*-dimethylaminophenyl)phenylphosphine radical anion, unlike $Ph_3P^{\bullet -}$, does not fragment.[161] The homoazulene (1,5-methano[10]annulene) radical anion rearranges to give secondary reduction products with the bicyclo[5.4.0]- and bicyclo[6.3.0]-1,3,5,7,9-pentaenyl carbon frameworks.[162] Esters of oxalic acid, squaric acid and oxamic acid on cathodic reduction in aprotic solvents yield the corresponding alkanes: the efficiency of the reaction depends on the direction of fragmentation of the intermediate radical anion, which in turn depends on the stability of the radical fragment.[163,164] Thus dibenzyl oxalate undergoes very efficient reduction to give toluene.

$$PhCH_2OCOCOOCH_2Ph^{\bullet -} \rightarrow Ph\dot{C}H_2 + {}^{-}O_2COOCH_2Ph$$
$$Ph\dot{C}H_2 + SH \rightarrow PhCH_3 + S^{\bullet}$$

The $PhCH_2SPh^{\bullet -}$ radical anion undergoes rapid fragmentation to benzyl radicals and benzenethiolate ions.[165] The fuschone radical anion has been employed as a base in Wittig reactions: it is generated at a low cathodic potential.[166]

Electron-transfer Reactions

Electron-transfer reactions are one of the major growth points in organic chemistry. This is recognized by the publication of two books which survey the use of electron-transfer processes in synthesis.[167,168] Theoretical aspects of electron-transfer reactions have also been reviewed.[167,169] The role of electron tunnelling in solid-state electron-transfer reactions has been surveyed.[170] A review has also emphasized the parallels between inorganic and organic chemistry in electron-transfer-induced reactions.[171] Models have been developed to describe the kinetics of concerted electron-transfer bonding-breaking reactions for the electrochemical and homogeneous reduction of alkyl halides[172] and for the reaction of arenes with solvated electrons.[173]

Interest in $S_{RN}1$ reactions of aromatic substrates remains high. The rate constants for the reaction of phenyl, 1-naphthyl and 3-pyridyl radicals with PhS^-, CN^-, and $MeCOCH_2^-$ are close to the diffusion-controlled limit: the more negative the standard redox potential of the ArX/$ArX^{\bullet -}$ couple, the faster the rate of addition of the nucleophile to the aryl radical.[174] In competitive studies of the reactions of *p*-anisyl and adamantyl radicals with Ph_2P^- and PhS^-, the adamantyl radicals are more selective: this is attributed to their greater stability.[175] However, 1-bromoadamantane and 4-bromoanisole have very similar reactivities towards Ph_2P^-, suggesting that the radical anion intermediates undergo electron transfer to both substrates at comparable rates.[175] The isomeric chalcogeno benzophenones have been synthesized in an electrochemical $S_{RN}1$ reaction from the appropriate benzophenone and either $PhSe^-$ or $PhTe^-$ in the presence of azobenzene as a redox catalyst.[176] The photo-stimulated reactions of 2-halothiophenes and PhS^- give phenyl thienyl sulphides with diphenyl sulphide as a by-product: this arises from fragmentation of $ThSPh^{\bullet -}$ to give phenyl radicals and thence diphenyl sulphide.[177] Yields of the phenyl thienyl sulphides are improved in presence of weak electron acceptors, e.g. PhCN or $PhNO_2$, which quench $ThSPh^{\cdot -}$ without complete inhibition of the reaction. 1-Aryl-2-naphthylamines are formed in the photo-stimulated reactions of iodoarenes with 2-naphthylamide ions.[178] The first example of an electrochemically-induced $S_{RN}1$ reaction using a phenoxide as the nucleophile has been published: exclusive *C*-arylation occurs *ortho* or *para* to the oxygen.[179] Thus reactions of *o*-, *m*- and *p*-chlorobenzonitriles with 2,6-di-*t*-butylphenoxide give the appropriate biaryl in yields of 68–89%. 2-Haloacetophenones and 2′-acetonaphthone enolates react by an $S_{RN}1$ route to give hydroxylated 2,2′-binaphthyls.[180] *o*-Diiodo- and *o*-bromochloro-benzenes react with 3,4-toluenedithiolate to give 2-methylthianthrene.[181] An intramolecular $S_{RN}1$ reaction occurs in the electrochemically-induced formation of dibenzosulphone from *o*-bis(phenylsulphonyl)benzene: in the presence of benzenethiolate the intermediate 2-(phenylsulphonyl)phenyl radical is trapped.[182] Other reactions which have been postulated to proceed by the $S_{RN}1$ route are the cobalt-catalysed carbonylation of haloarenes to give the corresponding benzoic acid,[183] and the copper bronze-catalysed formation of 2-phenylbenzofuran from *o*-bromophenylacetylbenzene.[184]

Free-radical chain processes in aliphatic systems involving an electron-transfer reaction have been reviewed.[185] Further studies on $S_{RN}1$ reactions of *p*-nitrobenzylic compounds and related substrates have been reported. Facile substitution at the tertiary carbon atom of *p*-nitrocumyl chloride and α,*p*-dinitrocumene occurs with a wide range of nucleophiles [PhS^-, $PhSO_2^-$, $Me_2\bar{C}NO_2$, $R\bar{C}(COOEt)_2$, NO_2^-, N_3^-, CN^-, 1-methyl-2-naphthoxide, pyrrolidine, piperidine, quinuclidine and DABCO].[186] The reactions invariably proceed with yields in excess of 80% in mild conditions. A feature of these reactions is their relative insensitivity to steric effects in contrast to S_N2 reactions. Steric effects are, however, evident in reactions of 2-chloro-1,1-dialkyl-6-nitroacenaphthenes (**21**) which proceed with N_3^- and *p*-$MeC_6H_4S^-$ but not with more sterically hindered nucleophiles, e.g. $Me_2\dot{C}NO_2$.[187] Quantum yields of up to 6000 have been recorded for the reaction of α,*p*-dinitrocumene with azide, substantiating the chain character of the reaction.[188] Consistent with a radical intermediate, $S_{RN}1$ reactions of optically active 2-(*p*-nitrophenyl)-2-nitrobutane (**24**) with a range of nucleophiles proceed with complete racemization giving (**25**),[189] while both (**21a**) and (**21b**) give identical mixtures of (**22**)

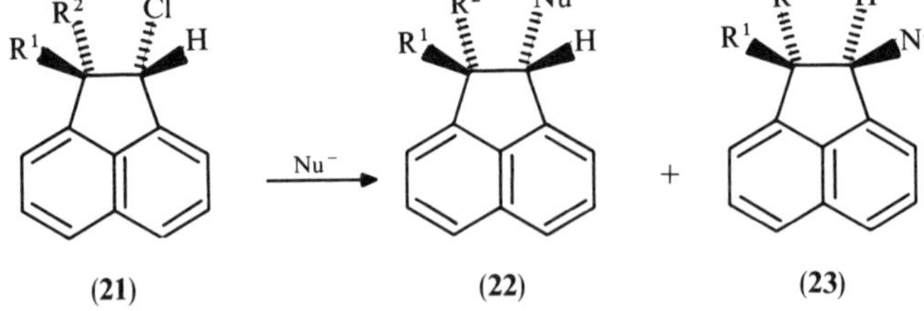

(**21**) (**22**) (**23**)

a; $R^1 = Me$, $R^2 = Et$
b; $R^1 = Et$, $R^2 = Me$

$Nu^- = N_3^-$, *p*-$MeC_6H_4S^-$

and (**23**).[187] The 5-nitrofurans (**26**) and (**27**) react with *p*-toluenethiolate by an $S_{RN}1$ mechanism.[190,191] ESR studies have been carried out on the α-ray irradiation of

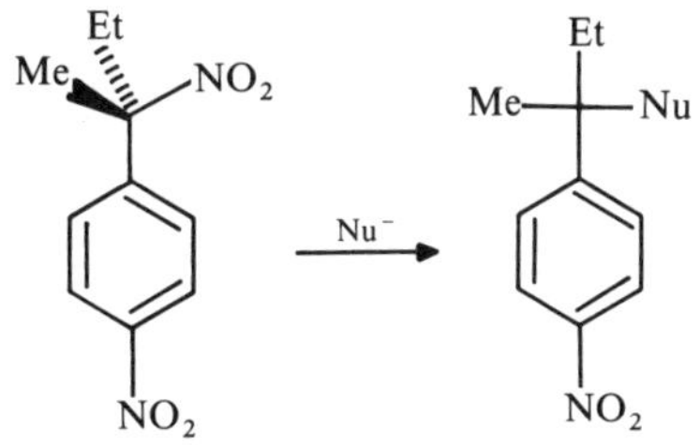

(24) **(25)**

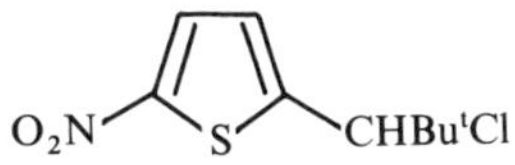

(26)

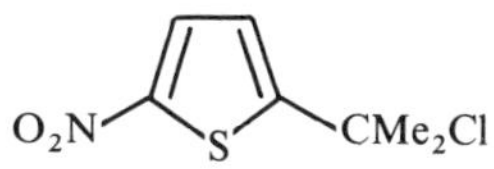

(27)

substituted 2-methyl-5-nitrofurans at 77 K.[192] Both the initial radical anion and the derived radical were unambiguously identified. There is considerable overlap in the radical anions between the C—X σ^* orbitals and the nitroaromatic π^* orbitals and, contrary to earlier suggestions, it seems that dissociation involves smooth reorganization of the MOs to the required transition state for loss of X^-. The reactions of α-halomethyl-5-nitrofurans with ArS^- unlike those with $Me_2\bar{C}NO_2$ are not catalysed by light nor are they subject to inhibition by radical traps, suggesting an SET mechanism (Scheme 6) rather than an $S_{RN}1$ mechanism. The reaction of *N*-

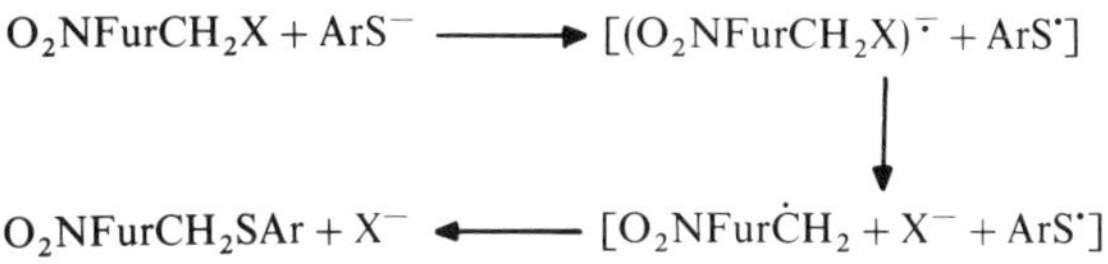

SCHEME 6

alkylpyridinium ions with nitronate anions also proceeds by a non-chain radical mechanism.[193] The indirect electrochemical reduction of benzylic halides by the 1,4-dihydro-4-methoxycarbonyl-1-methylpyridine anion proceeds by an electron-transfer process giving intermediate benzylic radicals.[194] The reaction of 2-halomethylanthaquinones with the 2-nitropropane anion proceeds by an $S_{RN}1$ mechanism.[195] Imidazole, benzimidazole and pyrazole have been employed as nucleophiles in $S_{RN}1$ reactions with *p*-nitrobenzyl chloride and α-bromo-α-nitroalkanes.[196] High yields of fluoroalkylimidazoles are obtained regioselectively from $S_{RN}1$ reactions of imidazole with fluoroalkyl bromides or iodides.[197] α-Bromo-α-nitroalkanes react with *t*-butylthiolate to give coupled products via either both an $S_{RN}1$ process and a radical-coupling reaction, or exclusively via the radical-coupling route.[198]

A seminal review[199] of the relationship between nucleophilic substitution reactions and single-electron-transfer reactions has suggested that the following factors favour SET:

(a) strong donor-acceptor pair ability of the nucleophile and electrophile;

(b) steric interactions in the transition state;
(c) delocalization of electrons in the nucleophile-substrate bond;
(d) the nucleophile-substrate bond strength.

The absence of complete inversion of configuration in reactions of nucleophiles with optically active 2-halooctanes has been used to infer the incursion of a competing SET pathway: iodoalkanes consistently gave products with less inversion than bromo- and especially chloro-alkanes.[200] A SET mechanism is deduced for reactions of *t*-butyllithium with *endo*-5-(2-bromoethyl)-2-norbornene and other halo-alkenes as seen in the formation of cyclized product.[201,202] 1-Phenylalkyllithium reacts with *t*-alkyl halide by two alternative pathways, viz. a polar pathway leading to coupling at C(1) and an SET pathway leading to C—C bond formation at C(3).[203] This is supported by the formation of cyclized products at C(3) and acyclic coupled products at C(1) with the radical probe 6-bromo-6-methylhept-1-ene (Scheme 7). The low deuterium content in cyclized products

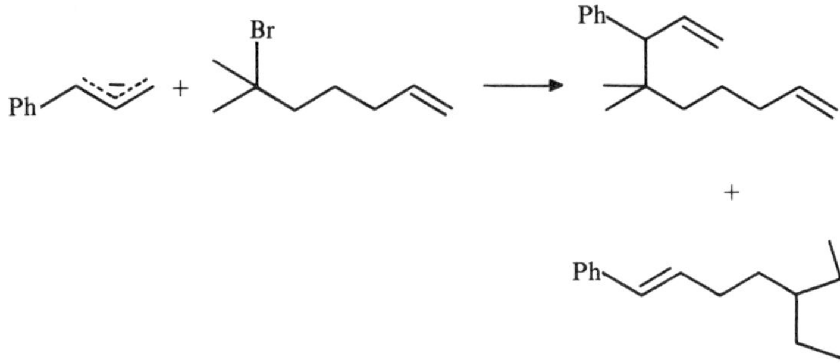

SCHEME 7

obtained from the reduction with $LiAlD_4$ of four cyclizable probes indicates that the product is derived from a cyclized radical which abstracts hydrogen from the solvent (Scheme 8).[204] Reactions with 5-iodocyclooct-1-ene with lithium dimethylcuprate

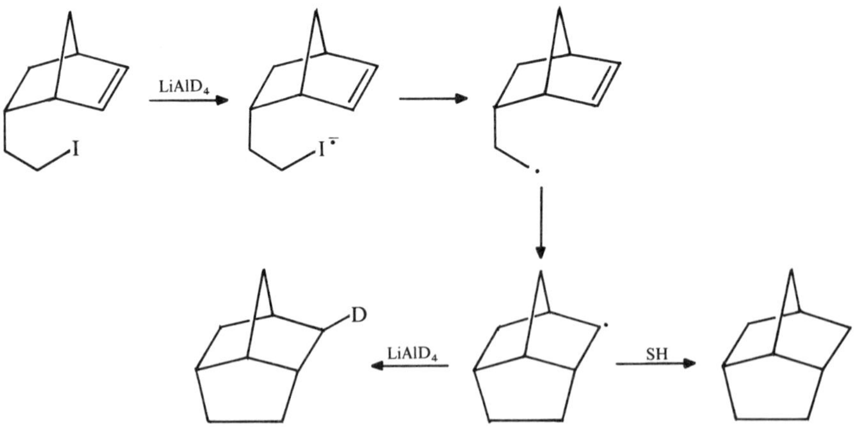

SCHEME 8

give considerable amounts of bicyclo[3.3.0]octane indicating an ET reaction leading to the formation of the cyclooct-1-en-5-yl radical which cyclizes to the bicyclo[3.3.0]octyl radical.[205] The corresponding tosylate gives no bicyclic products in accordance with simple nucleophilic substitution. The validity of using a cyclizable iodoalkane as a probe for an SET mechanism in their reactions with nucleophiles has been questioned. Rearranged products can be formed in the reactions of haloalkane radical probes with nucleophiles via a sequence involving radical-chain isomerization that converts the halide into a rearranged halide[206,207] followed by nucleophilic attack on the isomerized halide (Scheme 9). The rates of the various steps in this sequence indicate that it proceeds four orders of magnitude faster than simple nucleophilic substitution of the unrearranged halide. Reductions of 7-bromo- and 7-iodo-2-methoxyhept-2-enenitriles with a variety of metal hydrides have been shown not to proceed by an SET route as previously claimed.[208] The importance of ET processes in reduction of alkyl bromide with anions of transition metal hydrides has been evaluated.[209]

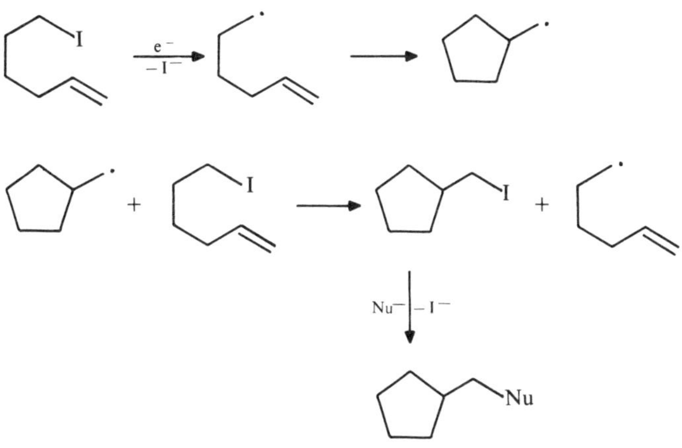

SCHEME 9

Reactions of 2-substituted 9-phenylfluorenide anions proceed by a simple S_N2 mechanism with simple alkyl halides as their reduction potentials are not positive enough to elicit electron transfer from the fluorenide anion.[210] Alkyl halides with lower reduction potentials, e.g. c-$C_6H_{10}(NO_2)X$ ($X = NO_2$, CN, Ts), however, proceed by the SET route. 9-Dialkylaminofluorenide anions react with benzyl chloride and 1-bromobutane by a merged S_N2-SET mechanism but with CF_3CH_2I solely by an SET pathway.[211] 9-Substituted fluorenide anions (9-MeFl$^-$, 9-MeOFl$^-$, and 9-Me_2NFl$^-$) which have similar basicities but become progressively easier to oxidize have been used to probe for the intervention of SET in reactions of alkyl halides: the relative rates of reactions with CF_3CH_2I, $PhSO_2CH_2Cl$, and c-$C_6H_{10}(NO_2)_2$ support an SET route.[212] The rates of electron transfer between methylviologen and some sterically hindered alkyl halides are consistent with an SET process.[213] Electron transfer occurs in the reactions of Me_2CCl_2 and

R_3SnLi,[214] *p*-nitrobenzyl bromide and the copper(I) complex, $Cu(dap)_2^+$,[215] CF_3Br and SO_2^- (generated *in situ* from zinc and SO_2),[216] acyl halides and 1,4-diaza-1,3-dienes,[217] and benzoyl chloride and *t*-nitrosobutane.[218] Several electron-transfer reactions of tetrahalomethanes have been reported.[219–221]

Single-electron-transfer steps are involved in both the initiation and propagation steps in reductions of α-haloacetophenones by triphenyltin hydride[222] and 1,4-dihydronicotinamides.[223]

$$PhCOCH_2X + Ph_3SnH \rightarrow PhCOCH_2X^{\cdot-}, Ph_3SnH^{\cdot+}$$
$$PhCOCH_2X^{\cdot-} \rightarrow PhCO\dot{C}H_2 + X^-$$
$$PhCO\dot{C}H_2 + Ph_3SnH \rightarrow PhCOCH_3 + Ph_3Sn^{\cdot}$$
$$Ph_3Sn^{\cdot} + PhCOCH_2X \rightarrow Ph_3Sn^+ + PhCOCH_2X^{\cdot-}$$

An electron-transfer mechanism is also involved in the reduction of α,α-dihalogenocarbonyl compounds with $LiAlH_4$.[224] The indirect electrochemical reduction of *meso*- and (±)-1,2-dichloro-1,2-diphenylmethanes involving radical anions as mediators is an electron-transfer process.[225,226]

$$A^{\cdot-} + PhCHClCHClPh \rightarrow A + PhCHClCHClPh^{\cdot-}$$
$$PhCHClCHClPh^{\cdot-} \rightarrow Ph\dot{C}HCHClPh + Cl^{\cdot}$$
$$A^{\cdot-} + Ph\dot{C}HCHClPh \rightarrow A + Ph\bar{C}HCHClPh$$
$$Ph\bar{C}HCHClPh \rightarrow PhCH{=}CHPh + Cl^-$$

Reactions of ketones with organolithium reagents proceed via an initial ET, which is followed by a C—C bond-forming step. The rate-determing step is dependent on the structure of the ketone and the reagent: the reaction of benzophenone with MeLi involves rate-limiting electron transfer, while with the Me_2CuLi C—C bond-forming is the slow step.[227] Direct evidence for electron transfer has been observed using ESR in the reaction of 2-methylbenzophenone with EtMgBr.[228] Treatment of diphenylmethylcyclopropenylium perchlorate with Grignard reagents gives both 3-substituted 3-methyl-1,2-diphenylcyclopropenes and 3-substituted 1-methyl-2,3-diphenylcyclopropenes: the proportion of the latter is greater in reactions with allylic and benzylic Grignard reagents and is interpreted in terms of an ET mechanism.[229] The formation of cyclized products in the reaction of thianthrene radical cation with hex-5-enylmagnesium bromide indicates an ET process.[230] Radicals are implicated in reactions of α-diketones and Grignard reagents[231,232] and with the potassium-graphite intercalation compound C_8K.[233] The ease of electron transfer with aromatic carbonyl and nitro compounds is in the order: ArMgBr > ArSMgBr > ArNHMgBr > ArOMgBr, i.e. in the order of decreasing electron-donating ability.[234]

Amidyl radicals are generated in an ET reaction between amidates and TCNE.[235] *N*-Acryloylaziridines undergo an SET reaction with hydrocarbon radical anions to give pyrrolidones: the reactions proceed via ring-opening of the ketyl followed by intramolecular trapping by the C=C bond of the acryloyl moiety.[236]

Benzaldehyde radical anions have been detected by ESR spectroscopy in basic conditions which give high yields of products of the Cannizzaro reaction.[237] There is also evidence for ET in sonochemical Cannizzaro[238] and Wittig–Horner reactions,[239] and in the reaction of benzaldehyde with isopropoxide.[240]

The denitration of aliphatic nitro-compounds by tributyltin radicals involves an ET process: the intermediate nitroxide is readily detected by ESR.[241]

$$RNO_2 + Bu_3Sn \rightarrow [RNO_2^{\cdot-}, Bu_3Sn^{+\cdot}] \rightarrow R{-}N(O^{\cdot}){-}OSnBu_3$$

$$R{-}N(O^{\cdot}){-}OSnBu_3 \rightarrow R^{\cdot} + Bu_3SnONO$$

$$R^{\cdot} + SH \rightarrow RH + S^{\cdot}$$

Trialkyltin anions react with $ArSnR_3$ in an ET process to give the corresponding radical anions which fragment to give either aryl radicals or anions.[242]

The role of electron-transfer processes in electrophilic aromatic substitution has been reviewed.[243] Theoretical studies using INDO-SCR and SCF-CI calculations indicate that, whereas nitration of nitrobenzene appears to proceed via the classical electrophilic mechanism, nitrations of toluene and xylenes proceed by radical-pair recombination and only this mechanism can successfully predict the product distribution.[244] Aromatic nitrations of a series of aromatic ethers using conventional conditions (HNO_3) and charge-transfer conditions [$C(NO_2)_4/h\nu$] gave the same yields and product distributions, pointing to a common mechanism.[245] Time-resolved spectroscopy established that charge-transfer nitration proceeds via [$ArH^{+\cdot}, NO_2^{\cdot}$]: the intermediate radical cations from *p*-$CH_3OC_6H_4OCH_2COOH$ can be trapped by the pendant nucleophile in benzene. 1,3-Dimethoxybenzene gives biaryls as well as the nitrated product as a result of coupling of the radical cations. Nitration of fluoroanthene with NO_2 in CCl_4 gives predominantly the 2- and 3-nitrofluoranthenes by an ET route, but in CH_2Cl_2 an electrophilic mechanism gives the 3- and 8-nitro-compounds.[246] A laser flash photolysis study indicates that ET occurs in reaction of $NO_3^{\cdot}$ (generated by photolysis of CAN) with aromatic and alkylaromatic compounds.[247] Nitration of alkylaromatics with nitric acid in acetic anhydride leads to formation of the side-chain nitro-compound via a homolytic route, whereas in CCl_4 the corresponding nitrate is formed by an electrophilic pathway.[248]

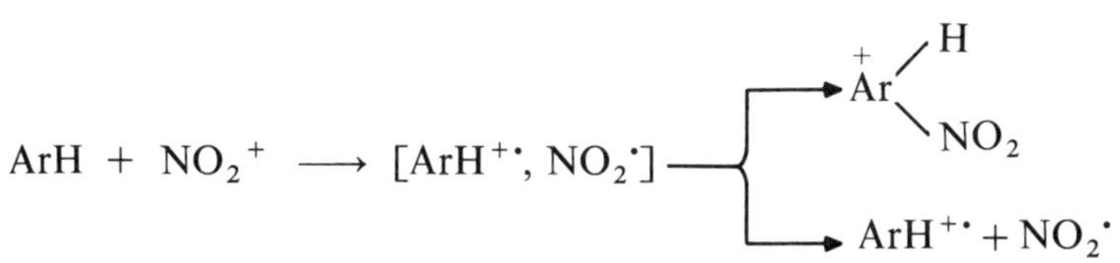

Cation radical pericyclic reactions induced by triarylaminium salts or by photosensitized electron transfer have been reviewed.[249] The kinetics of aminium salt-initiated cycloadditions are consistent with the Scheme below (M = monomer, D = dimer).

$$\textit{Initiation}\ \ M + Ar_3N^{+\bullet} \rightarrow M^{+\bullet} + Ar_3N$$
$$\textit{Propagation}\ \ M^{+\bullet} + M \rightarrow D^{+\bullet}$$
$$D^{+\bullet} + M \rightarrow D + M^{+\bullet}$$
$$\textit{Termination}\ \ 2M^{+\bullet} \rightarrow \text{Dication (for cyclohexa-1,3-diene)}$$
$$M^{+\bullet} \rightarrow \text{Radical} + H^{+}$$
$$\text{(for } p\text{-methoxyphenylprop-1-ene)}$$

The reaction is inhibited by triarylamine which causes back electron-transfer: this is a characteristic feature of the cation-radical-chain mechanism.[250] Triarylaminium salts promote the cyclization of 2,4-dimethylpenta-1,3-diene to (**28**) by a mechanism involving Brønsted acid catalysis: a different product (**29**) is obtained using PET catalysis or by the use of triarylaminium salts in presence of a hindered base. This latter product is formed by a radical-cation process.[251] Other criteria for the occurrence of the radical-cation mechanism include high stereospecificity (see Scheme 10).[251,252] There is disagreement as to whether the first propagation step in

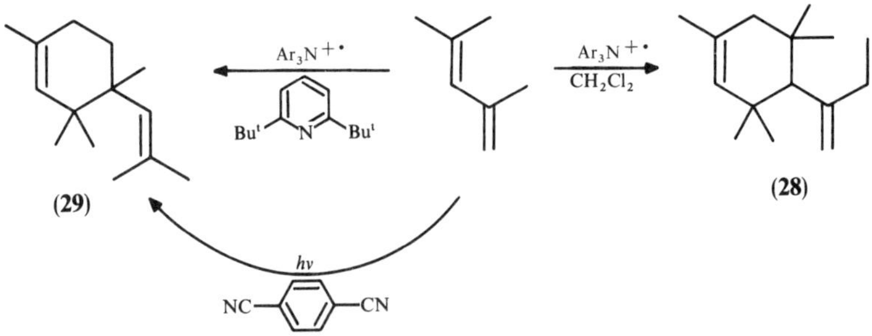

(**29**) (**28**)

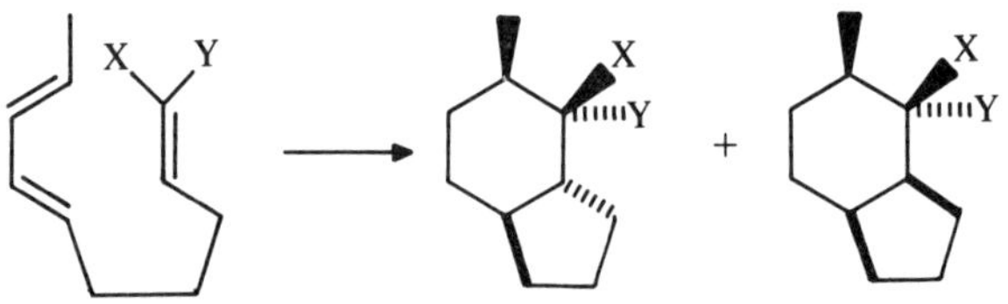

X = *p*-MeOC$_6$H$_4$; Y = H	98	:	2
X = H; Y = *p*-MeOC$_6$H$_4$	4	:	96

SCHEME 10

the cation radical Diels–Alder reaction involves reaction of the diene cation radical and the dienophile[253] or the diene and the dienophile radical cation.[254] Evidence in favour of the former route is that, in the electrochemically-induced Diels–Alder reaction between cyclohexa-1,3-diene and *p*-chlorostyrene, reaction occurs at a potential 0.5V more positive than the oxidation potential of the styrene.[253] Electron-donor acceptor complexes of 3-membered rings, e.g. 1,1,2,2-tetraarylcyclopropanes, with TCNE lead to oxygenation and/or [3+2]-cycloaddition with TCNE: oxygenation occurs via solvent-separated radical cations, and cycloaddition via coupling of the radical ion-pairs.[255a] 1,2-Dioxetanes have been synthesized by oxygenation of enol ethers in presence of triarylaminium salts: the enol ether radical cation reacts with triplet oxygen.[255b] Biphenyl-2-ylethenes undergo electrocyclization with triarylaminium salts giving 9-methyl-9-phenylfluorene in protic solvents and 9-phenylphenanthrene in presence of base.[256] The photo-cyclization of quadricyclene to norbornadiene is initiated by triarylaminium salts.[122,123,257] The PET-catalysed cyclization of cage ketones with the 1,2-diarylbicyclo[2.2.0]hexane moiety has been reported.[258]

Triarylaminium salts react with Br^- and I^- but not Cl^- or CN^- in an ET reaction.[259] Diazomethane radical cations are implicated in the formation of dimeric olefins in the oxidation of diphenyldiazomethane with $Ar_3N^{+\cdot}$.[260] *cis*-1-*p*-Anisyl-2-vinylcyclopropane is rapidly isomerized to its *trans*-isomer by triarylaminium salts at −90°C: labelling studies suggest that the reaction proceeds via a ring-opened intermediate.[261] Triarylaminium salts have been employed as redox catalysts in the indirect electrochemical α-methoxylation of aliphatic ethers and acetals.[262]

Bromination of 3-substituted bicyclobutanecarbonitriles in methanol results in stereospecific *cis*-addition in an electron-transfer reaction involving the bicyclobutane radical cation. *Ab initio* calculations support this mechanism.[263]

Electron transfer occurs in the reaction of the 4,4′-bis(dimethylamino)trityl cation and the 4,4′,4″-trinitrotrityl anion.[264]

$$Ar_2\dot{C}Ph + Ar'_3C^- \rightarrow Ar_2\bar{C}Ph + Ar'_3C^\cdot$$

The kinetics and equilibria of ET reactions have been reviewed.[265] Electron transfer occurs in the reaction of trityl cations and rhenium-alkyl complexes.[266] Sterically hindered nitro-compounds, e.g. 2,4,6-tri-*t*-butylnitrobenzene, react with 9-methoxyfluorenide anions in the ET reaction to give 9,9′-dimethoxy-9,9′-bifluorenyl.[267] Tetrahydroquinolines are formed via metallo-porphyrin-catalysed radical cyclo additions of alkenes and *p*-cyano-*N*,*N*-dimethylaniline in an ET reaction.[268] The reduction of diphenylacetylene by tetraphenylborate occurs by an ET route.[269] 2-Cumyladamantan-2-ol has been synthesized from the reaction of adamantanone and cumyl chloride initiated by lithium 4,4′-di-*t*-butylbiphenyl.[270] The reactions of *O*-cholesteryl chalcogenoformates with 2,4,6-tri-*t*-butylphenyllithium also proceed via an SET pathway.[271] There is evidence for SET activation in the cleavage of Co—C bonds of alkylcobalt(III) complexes with iodine.[272] Birch reduction of thiophene derivatives gives, in part, ring-opened products.[273] CIDNP was observed on mixing solutions of *cis*-6-hydroperoxy-5-hydroxy-5,6-dihydrothymine and ascorbic acid.[274] Other ET reactions reported include the reactions of superoxide with the α-tocopherol anion,[275] adriamycin,[276] and diacyl peroxides,[277] ozonate anions and nitrosobenzene,[278] cyanate anions with phenols and anilines,[279] dithionite and acridines in presence of an iron-sulphur cluster,[280] potassium and Δ^2-oxazoline-5-

ones[281] and nitrones,[282] and halogens with 2,2′-azinobis(3-ethylbenzothiazole-6-sulphonates).[283]

Photo-induced Electron Transfers

Reviews of the role of radical ions and electron-transfer processes in photochemical reactions[284] and of photo-additions of aromatic compounds proceeding via an ET mechanism have been published.[285]

The fluorescence of 9,10-dicyanoanthracene is quenched by alkenes and dienes: the rate of quenching is consistent with an ET process.[286] A photo-induced ET reaction occurs on irradiation of solid solutions of 9,10-dicyanoanthracene and 1,4-dimethylnaphthalene.[287] Charge transfer is involved in the photo-cycloaddition reaction between 9-cyanoanthracene and anthracene or naphthalene.[288] Polycyclic aromatic hydrocarbon, e.g. phenenthrene undergo photo-amination in the presence of electron acceptors, e.g. *m*-dicyanobenzene, to give aminated dihydroarenes.[289]

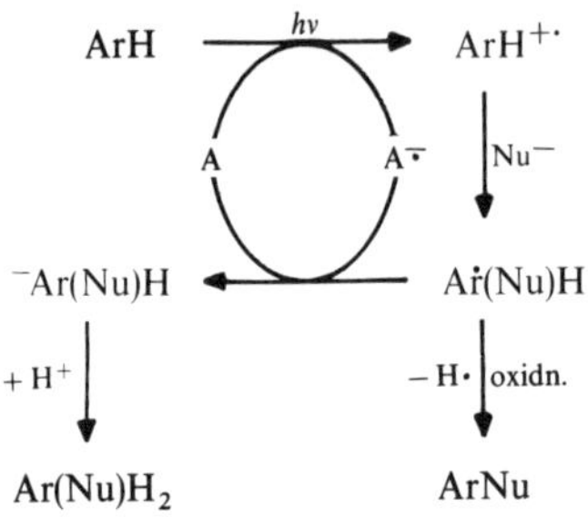

Methylarenes undergo photoinduced oxidation in presence of 1,4-dicyanonaphthalene to give benzylic alcohols.[290] 1,1-Diphenylalk-1-enes undergo a photo-rearrangement in presence of an electron acceptor to give the non-conjugated isomer.[291] 9,10-Dicyanoanthracene acts as a photosensitizer in the cycloaddition of 1-butyl-*cis*-2,3-diphenylaziridine and TCNE to give the stereoisomeric 1-butyl-3,3,4,4-tetracyano-2,5-diphenylpyrrolidines.[292] 1-Cyanonaphthalene acts as a photosensitizer in the anti-Markovnikov addition of alcohols to 1-methylcyclohexene.[292] 2-Hydroxychalcones are converted into the isomeric flavanoids on photolysis in presence of 9,10-dicyanoanthracene.[292] γ,ε-Unsaturated acids undergo a photo-induced ET reaction to γ-lactones using 1-cyanonaphthalene and biphenyl as an electron-acceptor electron-donor pair (Scheme 11).[295]

The 1,5-radical cations, formed as intermediates in the photosensitized reactions of five-membered cyclic ethers, undergo cleavage and nucleophilic attack by the solvent as with acyclic analogues. The bond-cleavage process requires orbital overlap with the SOMO. When the SOMO is held orthogonal to the adjacent bond as in the radical cations of 3-phenyl-2,3-dihydrobenzofurans and 2-methoxy-1-phenylindanes, deprotonation[296] occurs. The importance of back electron-transfer within geminate radical ions has been demonstrated from reactions between electron acceptors and electron donors using laser flash photolysis.[297] The rate of diffusive separation of the geminate radical ion pair is $5 \times 10^8\ s^{-1}$. The radical cations can be scavenged by low concentrations of 4,4′-dimethoxystilbene.

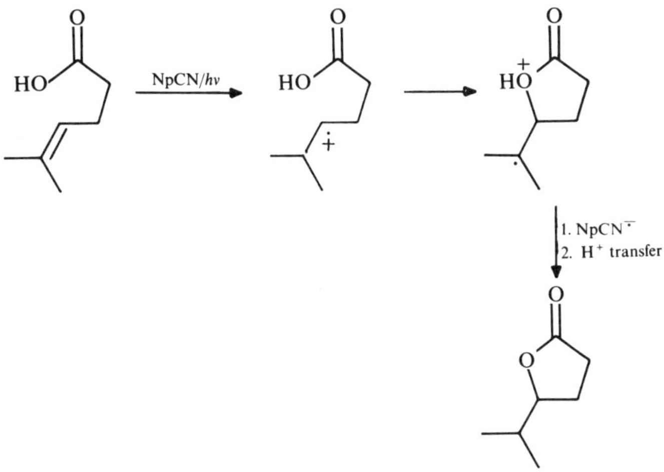

SCHEME 11

Photochemical reactions which proceed via sequential electron transfer and proton transfer have been reviewed.[298] The preference for this mechanism rather than simple hydrogen abstraction is based on kinetics rather than thermodynamics. Exergonic ET is sufficiently rapid to occur within the lifetime of the excited-state acceptor whereas hydrogen transfer is not. Direct evidence for ET has been provided by the photo-induced reactions of biacetyl with electron-rich alkenes: both the intermediate radical ions (**30**) and (**31**) have been detected by ESR.[299] The radical ion-pair undergoes hydrogen abstraction and subsequent radical coupling to give reduction products (**32**) and/or collapses to give oxetanes (**33**) (Scheme 12). Radical ions from both reactants have been observed by flash photolysis ESR in the photo-reaction of DABCO and xanthone.[300] The benzophenone ketyl radical anion has been observed by laser flash photolysis of benzophenone and isobutylamine.[301] Photolysis of nitrobenzene in presence of amines leads to $PhNO_2^{\cdot -}$.[302]

The amine-catalysed photo-reductions of polyfluorobenzene,[303] 4-chlorobiphenyl,[304,305] and 2-(4-chlorophenyl)benzoxazole[306] proceed via ET. Photolysis of the 1:1 DABCO:CBr_4 complex leads to electron transfer.[307] An ET process occurs in the reaction of 1-phenyl-2,3-diazabicyclo[2.2.2]oct-2-ene and $BrCCl_3$.[308]

The triethylamine-catalysed photolysis of *N*,*N*-diallyl-2-oxocycloalkane-carboxamides results in cyclization to give 6-membered ketones.[309] The triethylamine-catalysed photo-reduction of a δ,ζ-unsaturated ketone has been employed as a key step in the synthesis of hirsutene.[310] Photo-reduced ET from amino-acids and proteins to triplet 4-nitroquinoline-*N*-oxide has been reported.[311] Photo-reductions of carbonyl compounds by $Ru(bpy)_3{}^{2+}$ proceed by a series of ET reactions.[312,313] The rate constants for the reduction of benzophenone with a series of hydrogen donors indicate the reactions involve initial ET.[314] The photosensitized reductions of $NAD(P)^+$ models with Ph_4B^- and alcohols using a rhodium(III)-porphyrin complex as photosensitizer have been reported.[315] The photo-reduction

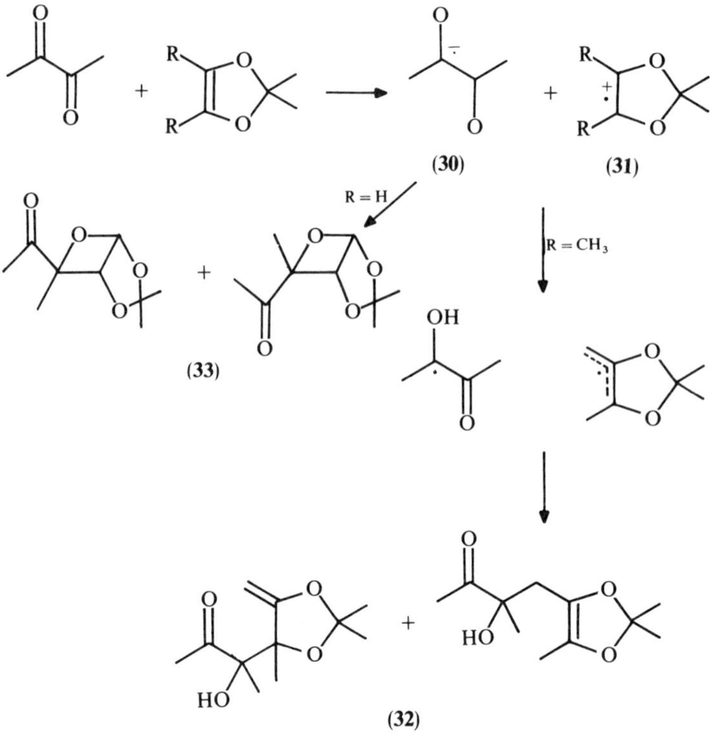

SCHEME 12

of 10-methylacridinium chloride in methanol proceeds via initial formation of 9-methoxy-9,10-dihydro-10-methylacridine.[316] The 10-methylacridinium cation has been used as a model NAD^+ compound to catalyse substrate-selective photo-oxidation of benzyl alcohols with oxygen: the reaction is initiated by ET to singlet photo-excited $AcrH^+$ from the benzyl alcohol.[317] Phe $Ru(bpy\ _3{}^{2+}$-sensitized photolyses of 1-benzyl-4-alkyl-4-(1-hydroxyalkyl)-1,4-dihydronicotinamides proceed via initial formation of the radical cation which fragments to $R\overset{+}{C}(OH)R_2$ and BNA.[318] $Ru(bpy)_3{}^{2+}$ catalyses the oxidation of aromatic alcohols in presence of diazonium salts.[319]

There is spectroscopic evidence for the radical anion of 1-methoxy-4-nitronaphthalene in its photo-substitution reaction with diethylamine to give 1-diethylamino-4-nitronaphthalene.[320] The photo-substitution of 2,3-dichloronaphtho-1,4-quinone with thiophene proceeds via an ET route[321] as does the photo-rearrangement and substitution of 7,8-dichloro-10-methoxy-2,3:5,6-dibenzobicyclo[2.2.2]octa-2,5-dienes.[322] Photolysis of phenanthrene in the presence of an amine and carbon dioxide results in reductive carboxylation.[323] ET is involved in the reaction of $RC{\equiv}CIPh^+$ with triphenylphosphine in sunlight.[324] Haloarenes, e.g. 1-cyano-4-bromonaphthalene, are photo-reduced by R_4B^- or

$CH_3BP_3^-$ via an ET pathway.[325]

$$ArX + R_4B^- \xrightarrow{h\nu} ArX^{\cdot -} + R_4B^{\cdot}$$

$$ArX^{\cdot -} \longrightarrow Ar^{\cdot} + X^-$$

$$Ar^{\cdot} \xrightarrow{BH_4^- \text{ or } CH_3CN} ArH$$

Photo-induced ET allylation of aromatic carbonyl compounds, and furyl and thienyl dicyanoethenes by allylstannanes[326,327] and of *N*-methylarenedicarboximides by allylsilanes[328] have been reported: the initial step involves formation of the radical cations $CH_2{=}CHCH_2MR_3^{+\cdot}$, which undergo loss of the electrofugal group to give allyl radicals.

Further studies of photo-induced ET reactions of iminium salts have been reported. Thus isotope-labelling studies have demonstrated dual intramolecular diradical [route (*a*)] and intramolecular cation diradical [route (*b*)] cyclization pathways in photo-cyclizations of silylalkenyliminium salts. (Scheme 13).[329] Photo-

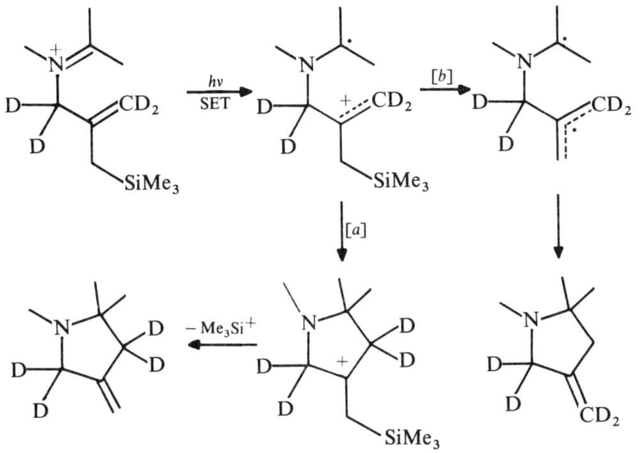

SCHEME 13

lyses of 2-phenyl-1-pyrrolium perchlorate with allenes leads to photo-addition: the allenes are converted into their radical cations which either undergo deprotonation or undergo nucleophilic capture by solvent (MeOH) prior to radical coupling.[330] Cyclohex-3-enones and $Et_2NCH_2SiMe_3$ participate in ET photo-additions: $Et_2\overset{+\cdot}{N}CH_2SiMe_3$ undergoes proton loss to $Et_2N\dot{C}HSiMe_3$ and also, in polar solvents, desilyation to give $Et_2NCH_2^{\cdot}$ prior to radical coupling.[331] Photo-induced ET occurs in photo-reactions of arenes, e.g. $PhCH_2SiMe_3$, and 2-phenyl-1-pyrrolium salts (**34**):[332] $PhCH_2SiMe_3^{+\cdot}$ undergoes ready desilyation to the benzyl radical which couples with the radical derived from the iminium salt to give 2-benzyl-2-phenylpyrrolidine (**35**) (Scheme 14). The analogous reaction with toluene gives only a trace of (**35**) together with some of the arylation products (**36**; Ar = *o*-Me_6H_4, *p*-MeC_6H_4) as a result of radical–radical cation coupling before deproton-

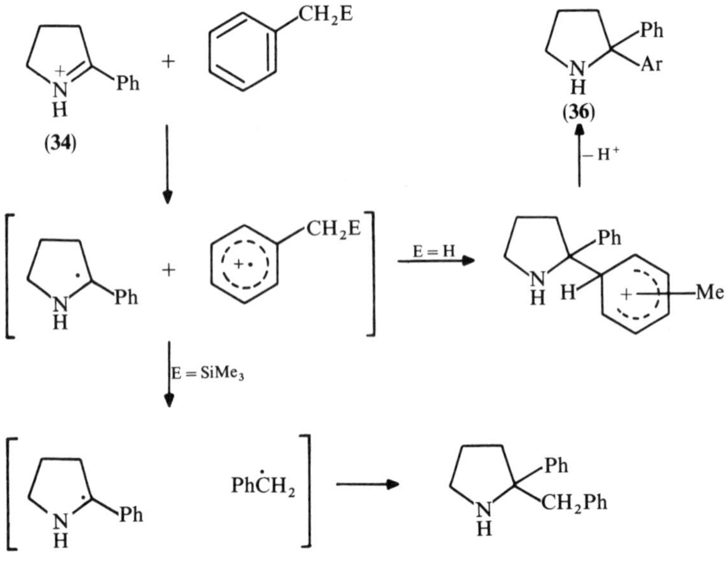

SCHEME 14

ation of the toluene radical cation: $SiMe_3$ is a better electrofugal group than is hydrogen. An intramolecular version of the above reactions has been used in the syntheses of benzoindolizines (**37**) and benzopyrrolizidines (**38**) (Scheme 15).[333] Further developments of this general procedure using benzyltrialkylsilyl groups have been employed in the syntheses of a number of alkaloids, e.g. xylopinine and

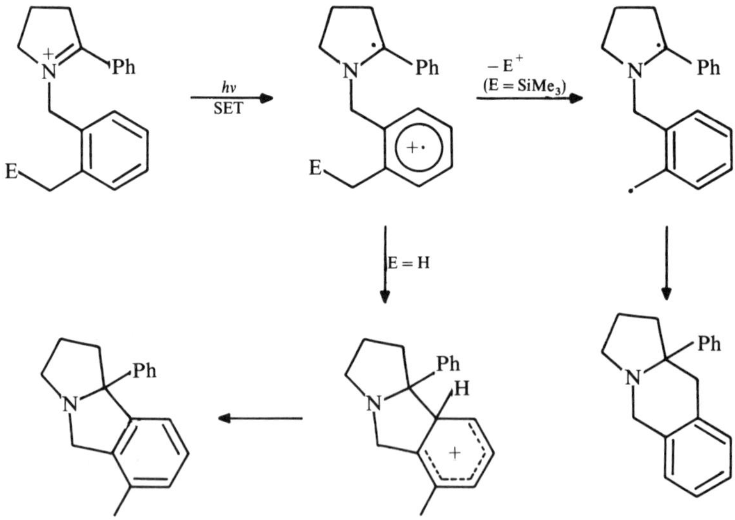

SCHEME 15

stylopine,[334] and protoberberines.[335] Erythina alkaloids have been synthesized using an electron-transfer-induced photo-cyclization of a silylalkenyl-3,4-dihydroisoquinolinium perchlorate.[336]

The photo-oxygenations of adamantylideneadamantane (AA) and related compounds[337–339] are induced by photosensitizers, e.g. triphenylpyrylium (TPP^+) perchlorate or 9,10-dicyanoanthracene, and also by triarylaminium salts.

$$AA + TPP^+ \xrightarrow{h\nu} AA^{+\bullet} + TPP^\bullet$$
$$AA^{+\bullet} + O_2 \longrightarrow {}^{+}AA\text{-}O_2$$
$${}^{+}AA\text{-}O_2 + TPP^\bullet \longrightarrow \text{Dioxetane} + TPP^+$$

3,3-Diaryl-1,1-dideuteriomethylenecyclobutanes undergo a degenerate photo-induced rearrangement in presence of electron acceptors: the intermediate radical cations can be trapped by oxygen to give 3,3-diaryl-5-methylene-1,2-dioxacyclohexanes.[340]

The photo-decomposition of triarylsulphonium salts yields phenylthiobiphenyls by a cage fragmentation-combination process in addition to biphenyl.[341]

$$Ph_3S^+X^- \xrightarrow{h\nu} [Ph_2S^{+\bullet}\,Ph^\bullet\,X^-] \longrightarrow \text{phenylthiobiphenyls}$$

p-Cyanobenzylthianthrenium trifluorosulphinate undergoes a similar photo-induced rearrangement to give 1- and 2-*p*-cyanobenzylthianthrenes.[342] Previous proposals that photolysis of diphenyliodinium iodide proceeds via radical pathway have now been discounted.[343]

A review of photo-induced redox chemistry on irradiated semiconductor surfaces has been published.[344] The CdS-photomediated reaction of 1,1′-di-*p*-anisylethene leads to the formation of two cyclized [2 + 4] and two open-chain dimers which are derived from reaction of the substrate with the substrate radical cation.[345] 1-*p*-Anisyl-2-phenylcyclopropane and 1,2-di-*p*-anisylcyclopropane undergo rapid photo-induced *cis-trans* isomerizations on a CdS surface, whereas 1,2-diphenylcyclopropane is unaffected.[346] Electrodes modified by a poly(pyrrole-phenothiazine) film act as photo-electrodes via a charge-transfer complex with an electron-acceptor in solution.[347]

Pyrimidine dimers, which occur in DNA exposed to UV light as a result of photo-cycloaddition between two stacked pyrimidines in the double helix, are split by a photo-induced ET reaction which produces dimeric radical anions which fragment.[348] An unusual photo-activated ET occurs on irradiation of 9-diazo-3,6-diazafluorene in methanol: this involves proton transfer to the excited species with simultaneous electron transfer.[349] Further examples of photo-induced ET reactions, involving the reaction of *N*-methyl-1,8-naphthalimide with α-methylstyrene or stilbene to give tetracyclic imides, have been reported.[350] The photo-stimulated mercuration of arenes with mercury(II) trifluoroacetate proceeds via an ET pathway.[351]

$$ArH + HgX_2 \xrightarrow{h\nu} [Ar^{+\bullet}\,HgX_2^{-\bullet}] \rightarrow [ArH^{+\bullet}\,HgX^\bullet\,X^-] \rightarrow ArHgX + HX$$

Trialkyltin iodides undergo homolytic substitution by photochemically-generated iodine atoms by an ET pathway.[352]

Photolysis

Semi-empirical SCFMO methods indicate that the activation barrier in the photolysis of acetaldehyde leading to $CH_3\dot{C}O$ and $H^{\bullet}$ is about $25\,kJ\,mol^{-1}$ less than that leading to $CH_3^{\bullet}$ and $^{\bullet}CHO$.[353] The primary cleavage in the photolysis of unsymmetrically substituted dibenzyl ketones results in the formation of the more stable benzylic radical.[354] Dialkoxyacetophenones, $[PhCOCH(OR^1)(OR^2)]$ (and also α,α-diphenylacetophenones), are useful photochemical initiators of polymerization: initiation is mainly effected by the $[Ph\dot{C}O\ \dot{C}H(OR^1)(OR^2)]$ radical pair.[355] 2,4,6-Trimethylbenzoyl diphenylphosphine oxide, $ArCOP(O)Ph_2$, on photolysis gives the radical pair $[Ar\dot{C}O\ ^{\bullet}P(O)Ph_2]$.[356] The 2,4,6-trimethylbenzoyl radical can be oxidized to the acyl cation by diphenyliodinium salts.

$$Ar\dot{C}O + Ph_2I^+ \rightarrow Ar\overset{+}{C}O + PhI + Ph^{\bullet}$$

The decay of the radical pair in sodium dodecyl sulphate is influenced by magnetic fields.[357] Photolysis of 2,*n*-diphenylcycloalkanones (*n*-membered ring) gives products that are significantly enriched in ^{13}C: the extent of enrichment depends on the electron-nuclear hyperfine coupling mechanism of inter-system crossing.[358]

Triplet keto-sulphides derived from $PhCO(CH_2)_4SBu$ undergo very rapid internal quenching as a result of intramolecular charge transfer: the efficiency of this is in the order: $n = 2 > 3 > 1 \gg 4 > 5$ indicating the ease of attaining overlap between the lone pair on the sulphur and the half-filled oxygen *p*-orbital.[359] δ-Substituted ketones undergo β-elimination of sulphur-centred radicals: the rates of β-cleavage are in the order: $SNC > PhS \sim BuS(O) \gg BuS, BuSO_2, CH_3COS$ and range from 10^9 to $3 \times 10^5\ s^{-1}$.

$$PhCOCH_2CH_2CH_2CH_2X \xrightarrow{h\nu} Ph\dot{C}(OH)CH_2CH_2\dot{C}HCH_2X$$

$$\downarrow$$

$$PhCOCH_2CH_2CH{=}CH_2 + HX \longleftarrow [Ph\dot{C}(OH)CH_2CH_2CH{=}CH_2\ X^{\bullet}]$$

Photolyses of $PhCOCH_2SR$, $PhCOCH_2S(O)R$, and $PhCOCH_2SO_2R$ all give acetophenone as the major product is presence of benzenethiol: this traps the $PhCO\dot{C}H_2$ radical formed by β-cleavage.[360] The rates of β-cleavage are in the order: $PhS > MeS(O) > BuS > BuSO_2$ and range from 10^{10}–10^{11} to $10^7\ s^{-1}$.

Photolysis of 2,4-di-*t*-butylacetophenone (**39**) gives the rearranged compound (**44**), formed via (**43**), as the major product with smaller amounts of the cyclic compound (**42**), whereas 2,4-di-*t*-butylbenzophenone affords only the analogous cyclic compound.[361] The preferred conformation of the 1,5-biradical of (**40**) is (**41**): rotation about the $Ar{-}\dot{C}(OH)CH_3$ bond would have the cage in the reaction of the benzophenone as the phenyl group can conjugate with benzylic radical centre. α,β-

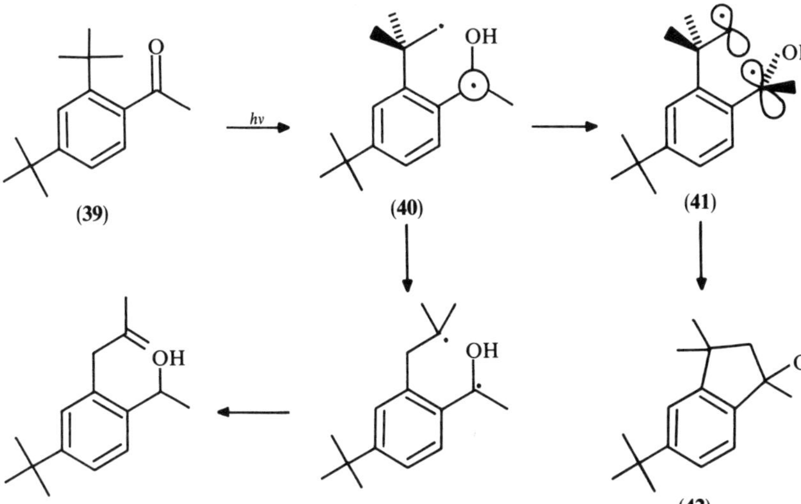

(39) (40) (41) (44) (43) (42)

Cyclopropyl ketones, e.g. 2,3-dihydro-2,3-methano-1,4-naphthoquinone, rearrange on photolysis: the intermediate cyclopropylcarbinyl radical ring-opens prior to recyclization of the biradical.[362] Photolysis of the ketone (**45**) results in the unusual 1,7-hydrogen abstraction via (**46**), through a nine-membered transition state, to give the stereoisomeric tetrahydrobenzoxepinols (**47**).[363] Photolyses of β,γ-unsaturated amides give cyclization products via 1,6-biradical intermediates.[364] 3-

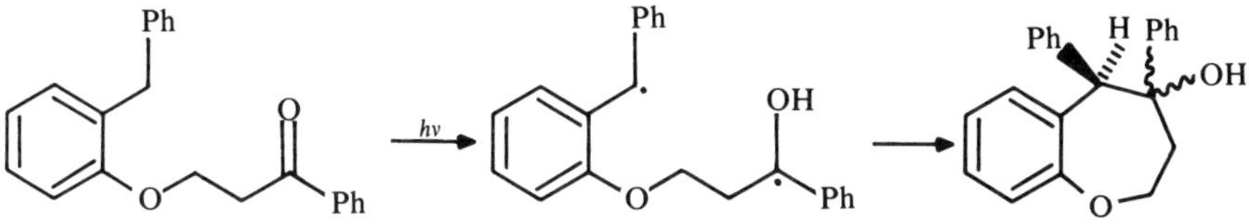

(45) (46) (47)

Alkoxy-1,2-naphthoquinones undergo photo-induced hydrogen abstraction to give methylenedioxy compounds.[365] Cyclobutanones are formed on photolysis of allylnaphthalene-2(1*H*)-ones as a result of ring-closure of the ambident benzylic radical at the endocyclic carbon.[366] Irradiation of *trans*-4,5-dibenzoyl-1,2-dimethylcyclohexane gives a mixture of 5,6-dibenzoyl-2,3-dimethylcyclohexene and 4,5-dibenzoyl-2-methyl-1-methylenecyclohexane: the reaction proceeds via an intermediate 1,4-allylic exocyclic biradical.[367] Benzophenone undergoes photo-addition to haloketene acetals to give oxetanes and tetrahydrofurancarboxylates via a 1,4-biradical.[368] Radical processes are involved in the photochemical reaction between ketones and dibenzodisilacyclohexadienes.[369]

Solid-state photochemical reactions have been reviewed.[370,371] The solid-state photochemistry of methyl-substituted benzophenones depends on their crystal packing: reaction occurs when the distance between the C=O···H angle is also not too large.[372] Cyclodextrin has been shown to modify the photochemical behaviour of alkyl aryl ketones: the results were interpreted on the basis of steric constraints imposed by the cyclodextrin cavity on rotational motions of the 1,4-biradical.[373] The photolysis products obtained from long-chain benzoin alkyl ethers differ from those from the analogous short-chain compounds when complexed to cyclodextrin due to the influence of packing of the substrate in the cyclodextrin cavity.[374] Solid-state photo-reactions of inclusion molecular complexes of *N,N*-dialkylpyruvamides and cyclodextrin or deoxycholic acid gave the corresponding β-lactams: the chemo- and regio-selectivities were different from solution reactions.[375] Asymmetric induction due to the chirality of the host was also observed. The photochemistry of ketones absorbed on silica is influenced by the nature of the porous surface of the silica which provides geometrical effects.[376]

Photolyses of the epimeric 2-acetyl-3,3-dimethylnorbornanes in isotropic liquid media proceed in distinct pathways: the conformation of the intermediate biradicals are immobilized by the medium.[377] The elimination/cyclization ratios from photolyses of alkanones in heneicosane[378] and other liquid-crystalline solvents[379] are very sensitive to the solvent phase, providing evidence that solvent order can control the dynamics of Norrish II reactions. The quantum yields for Norrish II fragmentation of 4-methoxyvalerophenone in liquid-crystalline solvents are very sensitive to the phase type.[380]

Photo-additions of 1-acyl-3-oxahexa-1,5-dienes[381] and 2-acyl-4-oxahexa-1,5-dienes[382] give products resulting from both 1,5- and 1,6-ring-closure: this is the first reported example of 1,6-ring-closure in the 2-acyl series. Photolyses of 1- and 2-acylhexa-1,5-dienes give freely reverting biradicals which also afford products resulting from 1,5- and 1,6-ring-closure.[383]

The use of IR laser photochemistry as a means of radical generation has been reviewed.[384] Both benzil[385] and 1,3-bis(1-naphthyl)propan-2-one[386] are stable to conventional photolysis but give rise to radicals by a two-photon process effected by laser photolysis. Laser flash photolysis of *p*-acetylphenyl 2-(*p*-hydroxyphenyl)ethyl ether leads to a short-lived triplet which decays via hydrogen abstraction from the remote OH group: the resultant 1,13-biradical has a lifetime of 118 ns in acetonitrile.[387] Laser flash photolysis of di-*t*-butyl peroxide in presence of arylamines provides a clean source of the $Ar\dot{N}H$ radical: the results indicate that in $Ph_2N^{\cdot}$ there is effective overlap of the SOMO and only one benzene ring.[388] Laser flash photolysis of $(CF_3)_3CI$ gives high concentrations of $(CF_3)_3C^{\cdot}$.[389]

Photolysis of $(CF_3)_2CFI$ leads to $(CF_3)_2\dot{C}F$ radicals.[390] 2,2′-Bithienyl-2-carboxaldehyde has been synthesized by the irradiation of 2-iodothiophene-5-carboxaldehyde in thiophene.[391] Photo-cyclization of 3-chloro-*N*-phenylbenzo[b]thiophene-2-carboxamide gives [1]benzothieno[2,3-*c*]quinolin-6(5*H*)-2-one.[392] Photolysis of pentachlorophenol in acetonitrile gives 2-methyl-4,5,6,7-tetrachlorobenzoxazole.[393] The photo-substitution of 1-chloro-2-naphthol has been reported.[394] Irradiation of 5- and 6-iodouracils in presence of allylsilanes leads to the corresponding allyluracils: photolysis leads to C—X homolysis with the resultant radical reacting with the allylsilane in an S_H2'-process.[395] Photolyses of substituted (1-naphthylmethyl)trimethylammonium chlorides lead to both 1-methoxymethylnaphthalene via a heterolytic pathway and 1-methylnaphthalene as a result of homolysis of the C—N bond; the latter route is favoured by electron-withdrawing substituents in the naphthalene nucleus.[396] A detailed study of the photo-decomposition of *N*-bromosuccinimide showed that more β-bromopropanoyl isocyanate is formed from direct photolysis than in the presence of bromine: direct photolysis yields vibrationally excited succinimidyl radicals which undergo ring-opening more rapidly than the ground-state species which is formed from the reaction of bromine atoms with NBS.[397] It was concluded that both succinimidyl radicals and bromine atoms can act as the chain-propagating species.

Irradiation of the mixed anhydrides derived from *N*-hydroxy-2-thiopyridone and alkanecarboxylic acids gives rise to alkyl radicals: these add to quinones,[398] and protonated hetero-aromatic bases to give alkylated products,[399] and they also react with disulphides, diselenides or ditellurides to give the corresponding chalcogenide.[400] An undesirable by-product in this reaction is the thioether formed in the decarboxylative rearrangement (Scheme 16).[401] This chain-reaction can be repressed by carrying out the photolyses at very low temperature (−64°C): under these conditions $k_2 > k_{-1}$. The resultant alkyl radicals couple to give good yields of dimeric products. The photo-reductive decarboxylation of carboxylic acids has been achieved in good yield by irradiation of their benzophenone oxime esters, $Ph_2C{=}NOCOR$, in presence of *t*-butyl mercaptan: yields are lower with esters of aromatic acids.[402] The alkylation of pyridines and quinolines has been carried out by irradiation of a solution of the protonated base and carboxylic acid: the reaction is accelerated by iron(III) sulphate.[403]

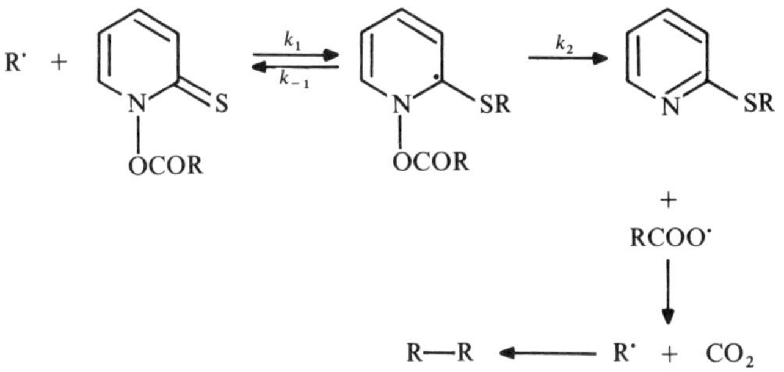

SCHEME 16

Photolysis of benzylstannanes results in efficient C—Sn bond cleavage: the resultant benzyl and stannyl radicals couple to give bibenzyls and distannanes.[404] The 2,4′- and 4,4′-dimers of 1-benzyl-4,4-dihydronicotinamides undergo photosensitized isomerization to give mixtures of the two compounds.[405] Photolysis of isopropyl nitrate leads to isopropxyl radicals: the reactions of PrO˙ with O_2, NO, and NO_2 have been studied.[406] Photolyses of hypoiodites of steroidal cyclic homoallylic alcohols in presence of $I_2/Pb(OAc)_4$ give lactol acetates.[407] A radical-pair mechanism is operative in the photo-rearrangements of the sodium salt of phenylsulphamic acid to the isomeric sodium anilinesulphonate,[408] of *N*-aroylcarbazoles to 1- and 3-aroylcarbazoles,[409] of 2,4-dioxofuro[2,3-*d*]-

pyrimidines to 5,7-diazaspiro[2.5]octanes,[410] and of dithioacetal *S*-oxide, RCH(SR′)(SOAr), to the aldehyde, RCHO, and disulphide, R′SSAr.[411] The phosphoranyl biradical (**49**) has been shown to be an intermediate in the benzophenone-sensitized photo-rearrangement of allyloxydimethoxyphosphite (**48**) to the isomer (**50**).[412] Alkyl-, allyl-, vinyl- and alkylidene-cyclopropanes rearrange via biradical intermediates on irradiation in the far ultraviolet.[413] Photolysis of 2-(alkylthio)carbonyl-2-(arylethyl)propylcobaloxime results in cleavage of the Co—C bond to give $CH_3C(X)(COSR)\dot{C}H_2$ which rearranges to $CH_3\dot{C}HXCH_2COSR$.[414] Photo-addition of alkenes to quinoline, isoquinoline, and phthalazin-thiones proceeds via a radical route.[415] Alkanes have been successfully functionalized by the mercury-photosensitized cross-dimerization with alcohols, ethers, and silanes.[416] Thus cyclohexane reacts with methanol to give cyclohexylmethanol with a yield of 30%. The red-light photolysis of 10-bromo-2-chloro-2-nitrosocamphene gives two nitroxides.[417]

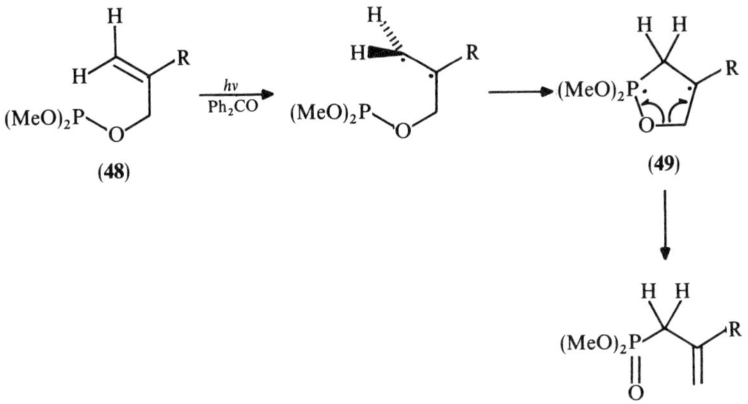

(**48**) (**49**) (**50**)

Biradicals

Spin-orbit coupling constants for the trimethylene biradical and a pair of interacting methyl radicals have been calculated: the introduction of a CH_2 group increases the spin-orbit coupling, the magnitude of which is related to the activation energy for singlet-triplet crossing.[418] 2,2-Dimethylcyclopentane-1,3-diyl and other 2,2-dimethyl-1,3-diyl radicals have triplet lifetimes at least one hundred times shorter than that of the parent cyclopentane-1,3-diyl: this is interpreted on the basis of a lower singlet-triplet energy gap because of perturbation of 'through-bond' and 'through-space' couplings.[419] Triplet 1-phenylcyclohexane-1,4-diyl is several thousand times longer-lived than cyclohexane-1,4-diyl: stabilization by phenyl group favours the twist-boat conformation (**52**) while the unsubstituted biradical adopts the boat conformation (**51**) which favours inter-system crossing through favourable interaction of the orbital lobes for spin-orbit coupling.[420] Triplet cyclopentane-1,3-diyl (**53**), in which the orbital lobes are parallel, is also relatively

(51) (52) (53)

long-lived as is, to a lesser extent, triplet cyclohexane-1,3-diyl in which the inter-orbital angle is 45°.[420] In contrast, bicyclo[2.2.1]heptane-2,7-diyl with an inter-orbital angle of 60° is very short-lived. The cyclohexane-1,3-diyl collapses to give 33% of bicyclo[2.2.0]hexane whereas 1-phenylcyclohexane-1,3-diyl gives exclusively 1-phenylhexa-1,5-diene.[421] The biradicals [MeĊ(R)$(CH_2)_4$Ċ(R)Me; R = H,Me] have been observed by time-resolved ESR spectroscopy: they have unexpectedly long lifetimes ($>1\ \mu s$).[422] The lifetimes of substituted 1,*n*-diphenyl-1,*n*-diyls are reduced by *p*-bromo substituents: the influence of substituents and chain length on τ were interpreted in terms of spin-orbit coupling.[423] Photolyses of cyclomers of the 1,1′-ethylenebis[4-(methoxycarbonyl)pyridinyl] biradical[424] and 1,1′-ethylenebis(pyridinyl) biradical[425] lead to the monomeric biradicals.

Ab initio calculations predict that the planar tetramethylene-ethane biradical (**54**) has a singlet ground state in contrast to experimental results.[426] Both related biradicals (**55**)[427] and (**56**),[428] however, have triplet ground states. The deviation from planarity in (**53**) is less than 5° and so it is possible that tetramethylene-ethane is very twisted and is possibly a perpendicular species. Contrary to theoretical calculations 2,3,5,6-tetramethylenecyclohexane-1,4-diyl has a triplet ground state.[429] Matrix-isolation NMR spectroscopy indicates that the ground state 3,4-dimethylenefuran (**57**; X = O) is a singlet.[430] 3,4-Dimethylenethiophene (**57**; X = S) has been detected spectroscopically in the thermolysis of ($CH_2{=}C{=}CH_2S$).[431] The tetra-radical, tetrakisgalvinol, has been synthesized and studied by ENDOR spectroscopy.[432]

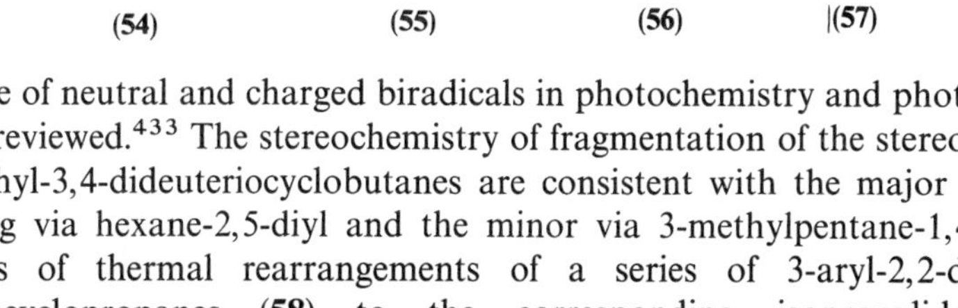

(54) (55) (56) (57)

The role of neutral and charged biradicals in photochemistry and photophysics has been reviewed.[433] The stereochemistry of fragmentation of the stereoisomeric 1,2-dimethyl-3,4-dideuteriocyclobutanes are consistent with the major pathway proceeding via hexane-2,5-diyl and the minor via 3-methylpentane-1,4-diyl.[434] The rates of thermal rearrangements of a series of 3-aryl-2,2-dimethyl-methylenecyclopropanes (**58**) to the corresponding isopropylidenecyclopropanes (**60**) are consistent with the reaction proceeding via a biradical intermediate (**59**).[435] All substituents in the *para*-position exert a rate-enhancing effect which is interpreted in terms of increased stabilization of the radical intermediate: the magnitude of the effect has enabled a series of σ values to be determined.

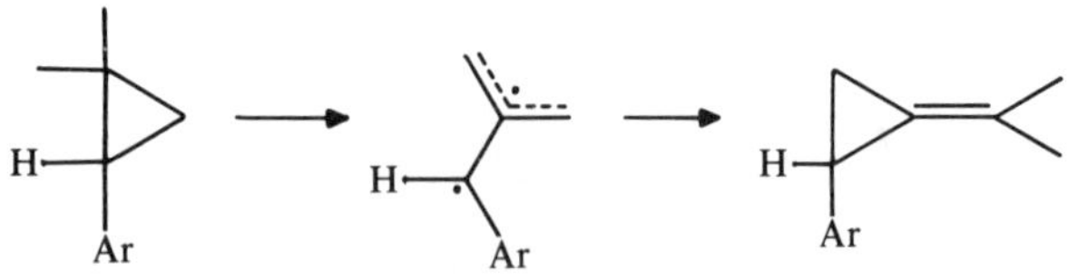

(58) (59) (60)

A tetramethylene biradical is an intermediate in the reaction of *p*-methoxystyrene and 3,3-dicyanoacrylate.[436] In non-polar solvents the reaction products include a 1:1 alternating copolymer and a cyclobutane adduct, whereas only the latter is formed in dipolar aprotic solvents: in these solvents the biradical has considerable polar character, and the resultant coulombic interaction between the termini favours a gauche conformation of the biradical leading to the cyclobutane. [1.1.1]Propellane reacts with TCNE and dimethyl acetylenedicarboxylate to give 1:1 and 2:1 adducts, respectively, via a dipolar biradical intermediate or an ET route.[437] Bicyclo[3.3.1]non-1-en-3-one decomposes above −40°C to give three dimeric products arising from combination and disproportionation of biradical intermediates.[438] An extremely short-lived 1,4-biradical has been detected in the laser flash-initiated cycloaddition of aromatic thiones and alkenes: the short lifetime is attributed to delocalization of unpaired spin to sulphur from both unpaired electron.[439] Biradicals are implicated in thermolyses of 2-methylene-2,3,4-thiadiazolines to allene episulphides, benzothiophenes, and divinyl sulphides.[440] The intramolecular 1,3-diyl-trapping reaction has been used as a key step in the total synthesis of hypnophilin and coriolin.[441] A key step in the DNA cleavage by neocarzinostatin involves a nucleophile-induced formation of a biradical intermediate.[442]

Perinaphtha-1,3-diyl undergoes a 1,2-hydrogen migration below 100 K to give phenalene: the rate of rearrangement is independent of temperature indicative of quantum-mechanical tunnelling.[443] Reactions of O(^{3}P) atoms with styrenes yield mainly epoxides and carbonyl compounds as a consequence of 1,2-hydrogen migration in the intermediate biradical.[444] Photolyses of hydrobenzoin carbonates lead to the PhĊHCHPhO• biradical: this undergoes a 1,2-phenyl shift to give diphenylacetaldehyde and a 1,2-hydrogen shift to give 1,2-diphenylethanone.[445]

Pyrolysis

The compound (**61**) undergoes facile homolysis to the captodative radical (**62**): the rate constants at 25°C are $1.4 \times 10^{-5}\ s^{-1}$ in CH_3CN and $7 \times 10^{-3}\ s^{-1}$ in H_2O.[446]

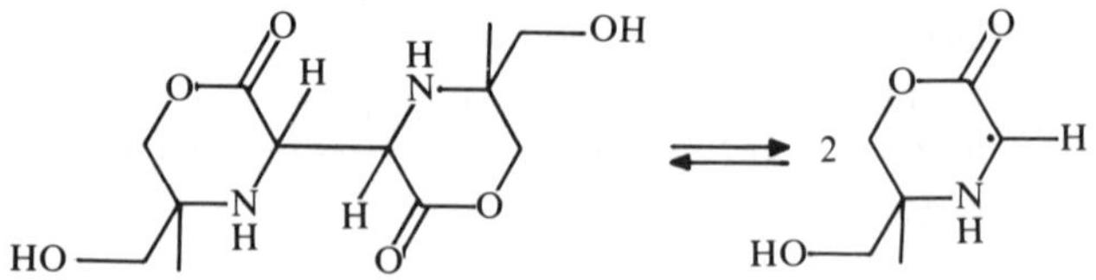

(61) (62)

Evidence that the thermal degradation of poly(methyl methacrylate) comes from studies of the thermolyses of model compounds for an unsaturated end-group (**63**)[447] and for a head-to-head unit (**64**)[448] which indicates that both (**63**) and (**64**) break down to give radicals at 230–300°C and 210°C, respectively. The ratio of

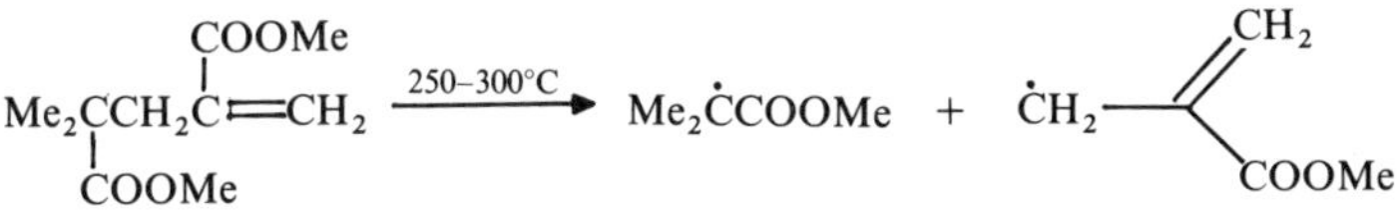

(**63**)

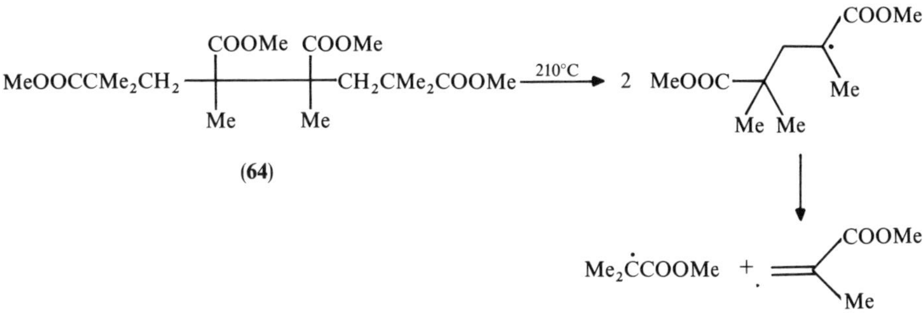

(**64**)

products obtained from thermolyses of sterically-hindered tertiary alcohols, $RCBu^t{}_2OH$ gives a measure of the relative strain energies of $R^{\bullet}$ and $Bu^{t\bullet}$.[449] Thermolysis of *trans*-3,4-di-*t*-butylcyclobutane-1,2-dione results in cleavage of the CO—CO bond followed by decarbonylation to give *trans*-di-*t*-butylethene.[450] Bicyclic cyclobutane-1,2-diones possessing a double bond in conjugation with the cyclobutanedione ring undergo C—CO bond cleavage to give an allylic acyl biradical which may rearrange or undergo mono- or bis-decarbonylation.

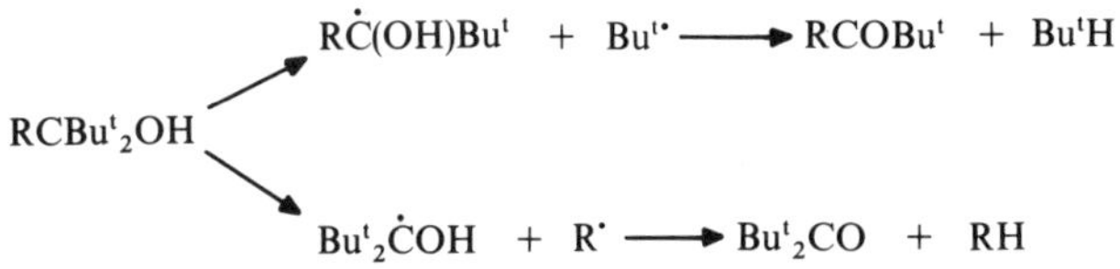

Ab initio calculations indicate that the lowest-energy pathway for the thermal decomposition of ethane involves C—C homolysis.[451] The pyrolysis of butane at 660°C gives twice as many secondary as primary radicals.[452] The termination step in the pyrolysis of isobutane involves methyl and *t*-butyl radicals.[453] Pyrolysis of *n*-pentane at 660°C results in C—H bond cleavage to give 1-pentyl radicals which rearrange to 2-pentyl radicals.[454] Pyrolyses of butane and hexane, in which the termination step involves combination and disproportionation of ethyl radicals, are not inhibited by alkenes.[455] In contrast, the pyrolysis of neopentane, which gives methyl but not ethyl radicals, is strongly inhibited by alkenes: the more reactive methyl radical abstracts an allylic hydrogen from the alkene. Pyrolyses of mixtures of cyclohexane and isomeric alk-1-enes at 940°C give allyl radicals.[456] A radical mechanism is operative in the pyrolysis of 1-phenyldodecane.[457]

Cyclopropylacetylene rearranges at 600–900°C: the initial step involves ring opening to give $\dot{C}H_2CH_2\dot{C}HC{\equiv}CH$.[458] 8-Vinylbicyclo[4.2.0]oct-1-ene on thermolysis yields 1,2,3,5,8,8*a*-hexahydronaphthalene via a biradical intermediate.[459] The flash vacuum pyrolyses of dispiro[2.2.*n*.2]alkadienes proceed via biradical intermediates: the first step involves homolysis of a spiro bond.[460] The formation of pyrocarbon from the pyrolysis of benzene proceeds via a radical pathway.[461] Radicals are involved in the pyro-condensation reactions of anthracenes with other aromatic compounds.[462]

The pyrolyses of 1,2-dichloroethane,[463] fluoropropenes,[464] and chloromethylsilanes[465] proceed at least in part by a radical mechanism. Flash vacuum pyrolysis of alkoxyvinylsilanes proceeds via initial intramolecular hydrogen transfer to give a 1,4-biradical.[466] RRKM theory indicates that pyrolysis of ethanol should proceed via $\dot{C}H_3$ and $\dot{C}H_2OH$.[467] The kinetic-chain length in the pyrolysis of butanone is relatively short: the results indicate that $CH_3\dot{C}HCOCH_3$ is less reactive than $CH_3CH_2CO\dot{C}H_2$ or $\dot{C}H_2CH_2COCH_3$.[468] 1,4,3,6-Dianhydromannopyranose has been prepared by the flash vacuum pyrolysis of isomannide dinitrate: the reaction involves homolysis of the O—N bond and subsequent 1,5-hydrogen migration followed by loss of NO_2.[469] The thermal decomposition of arylaminoanthraquinone dyes involves C—N homolysis.[470]

Radiolysis

Exposure of acetobromoglucose to γ-rays results in dissociative electron-capture with loss of bromide: the resultant tetraacetylglucosyl radical has a boat structure.[471] γ-Ray irradiations of CCl_4, MeOH, and Me_2CHOH give $\dot{C}Cl_3$, $\dot{C}H_2OH$, and $Me_2\dot{C}OH$, respectively.[472] High yields of carboxylic acids (and HCl) are obtained from the radical-mediated degradation of chlorinated ethanes in an aqueous environment generating the radical by radiation: the reactions proceed via peroxyl radicals, e.g. $RCCl_2OO^\bullet$.[473] γ-Radiolysis of acetone in oxygenated aqueous solutions gives $MeCOCH_2OO^\bullet$ radicals which decay by second-order kinetics: the rate of reaction of acetonyl radicals with oxygen is $3 \times 10^9\ mol^{-1}\ s^{-1}$.[474] Radiolysis and ESR spectroscopic studies establish that the radical cation $Me_2C{\overset{+\bullet}{-}}CMe_2$ is formed by acid-catalysed elimination of HO^- from $HOCMe_2\dot{C}Me_2$: the radical cation deprotonates to give $Me_2C{=}CMe\dot{C}H_2$ with $k = 3.9 \times 10^5\ s^{-1}$.[475] Radiolysis of trimethylamine in H_2O/O_2 under alkaline conditions gives $O_2^{\bullet -}$.[476]

$$Me_3N \xrightarrow{HO^\bullet} Me_2N\dot{C}H_2 \xrightarrow{O_2} Me_2NCH_2^+ + O_2^{\bullet -}$$

Pulse-radiolysis studies indicate that the fractions of attack by $HO^\bullet$ on 4-hydroxybenzoic acid at the 1-, 2-, 3- and 4-positions are 0.16:0.04:0.65:0.15, respectively, thus confirming the electrophilic character of $HO^\bullet$.[477] *Ipso* attack on the phenolic group gives rise to the phenoxyl radical through rapid loss of water. Radiolysis of N-heterocyclic systems gives the corresponding radical cations in an addition-elimination reaction.[478]

Azo Compounds and Diazonium Salts

The parameters for the decomposition of a series of both symmetrical and unsymmetrical azoalkanes support one-bond scission.[479] The gas-phase photodissociation of MeN=NCMe$_2$CH=CH$_2$ has been shown, by time-resolved anti-Stokes Raman spectroscopy, to proceed in a two-step process: the 1,1-dimethylallyl radical is formed within 2 ns of excitation whereas the methyl radical and nitrogen are formed through the decay of a reaction intermediate with a liftime of 12 ns.[480] Irradiation of *cis,anti*-[5,6-d_2]-2,3-diazacyclo[2.2.2]oct-2-ene leads to bicyclo-[2.2.0]hexane in which inversion of configuration has occurred: the inversion is explained in terms of a one-bond cleavage followed by conformational changes in the diazenyl radical.[481] Conclusive evidence for the intermediacy of diazenyl radicals in photochemical denitrogenations of cyclic azo compounds comes from the detection of the rearranged compound **(66)** from the photolysis of **(65)**;[482] thermolysis did not lead to rearrangement; photolysis (n,π^*-excitation) gave π-type diazenyl radicals while thermolysis gave σ-type radicals which would be expected to lose nitrogen very rapidly. Thermolysis of *trans*-1-phenyl-2-

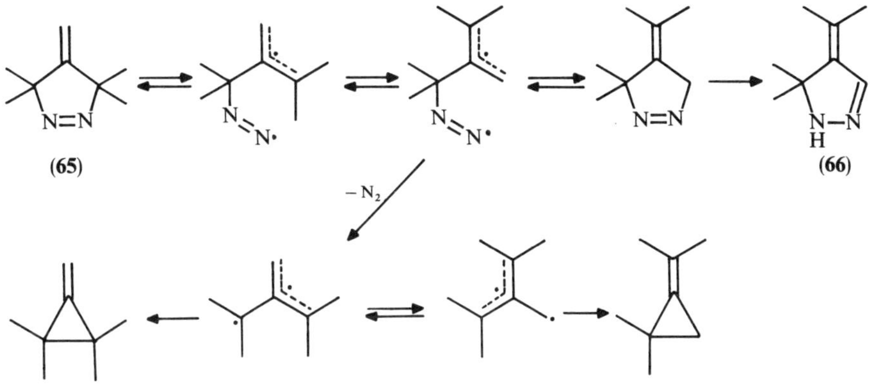

(65) **(66)**

(methylazo)cyclopropane resulted in its isomerization and also led to the formation of 1-methyl-5-phenyl-2-pyrazoline via the relatively stable α-azo(hydrazonyl) radical, PhĊHCH$_2$ĊHN=NMe.[483] No evidence for the formation of a tetrayl was obtained from the thermolysis or photolysis of a bis-azoalkane: nitrogen extrusion occurs in a stepwise fashion via diazenyl radicals.[484] Studies on the thermolysis of bridgehead azo compounds indicate, in contrast to the results just discussed, that a concerted two-bond scission occurs.[485] Studies of the rates of decomposition of azoalkanes XCMe$_2$N=NCMe$_2$X indicate the lack of polar contributions in azoalkane thermolyses.[486] The rate of decomposition of a series of *trans*-azoalkanes, R^{1}R^{2}CHN=NCHR^{1}R^2, increases with increasing bulk of the alkyl groups due to back-strain. In contrast PhCHMeN=NCHMePh decomposes one hundred times faster than does PhCHBu^{t}N=NCHBu^{t}Ph: this is due to the steric repulsion between the benzene ring and the alkyl group in the radical which requires that the π-system can overalp with the SOMO.[487]

Photo-denitrogenation of appropriate azoalkanes gave the same products as were obtained in related di-π-methane rearrangements, suggesting a common intermediate.[488] Photolysis of a pyrazolopyridine gave cyclopropapyridine and 4-isopropenylpyridine via a biradical intermediate.[489] Direct or photosensitized photolysis of (**67**) gave the expected tricyclic hydrocarbon (**68**), whereas electron-transfer photolysis using triphenylpyrylium perchlorate gave norbornene as a consequence of rearrangement of the intermediate radical cation (**69**).[490] The photolysis of sterically-protected 1,2,3-selenadiazoles in presence of alkenes gives rise to cycloadducts: the carbon terminus of the critical biradical adds to the alkene and the resultant biradical gives the product.[491] Photolyses and thermolyses of 1,4,4-trialkyl-4,5-dihydromethylene-1*H*-1,2,3-triazoles give products derived from the biradical intermediates formed by loss of nitrogen.[492]

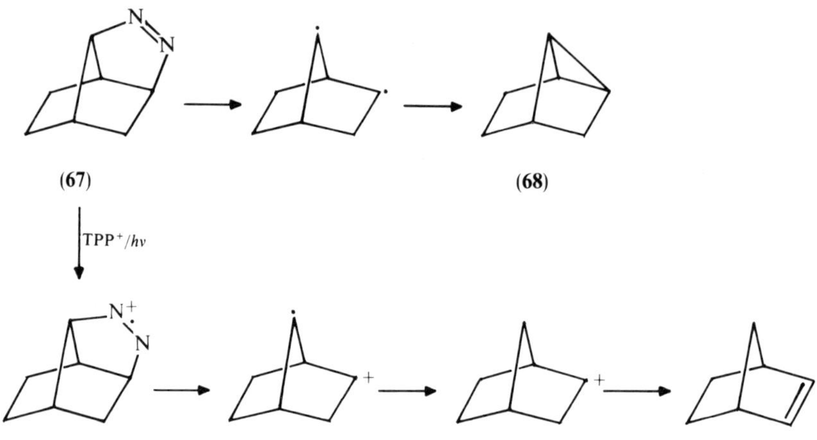

(**67**) (**68**) (**69**)

The rate constants for thermolyses of azides are consistent with the rate-limiting step involving N—N bond cleavage.[493]

The rate of denitrogenation of the phenyldiazenyl radical is 4.8×10^5 s^{-1}.[494] An unusual radical switch within the solvent cage in the decomposition of the diazosulphone, PhN=NSO$_2$Bu^t, is postulated to explain the formation of products containing the PhSO$_2$ group.[495]

$$PhN{=}NSO_2Bu^t \rightarrow [PhN_2^{\bullet}\ {}^{\bullet}SO_2Bu^t] \rightarrow [Ph^{\bullet}\ {}^{\bullet}SO_2Bu^t] \rightarrow [PhSO_2^{\bullet}\ {}^{\bullet}Bu^t]$$

Arenediazonium salts undergo facile reduction by ferrocyanide and decamethylferrocene in an outer-sphere process.[496] The intermediacy of aryl radicals in the reactions of *o*-but-3-enyloxybenzene and related diazonium salts (**70**) with iodide[497] and with benzenethiolate[498] has been demonstrated by the formation of cyclic products. The variation in relative yields of the simple substitution product (**73**) and the cyclic product (**74**) indicates the reaction with iodide proceeds via an electron-transfer mechanism (Scheme 17), and with the thiolate by an $S_{RN}1$ mechanism (Scheme 18) via the intermediate radical (**71**) which cyclizes to (**72**).

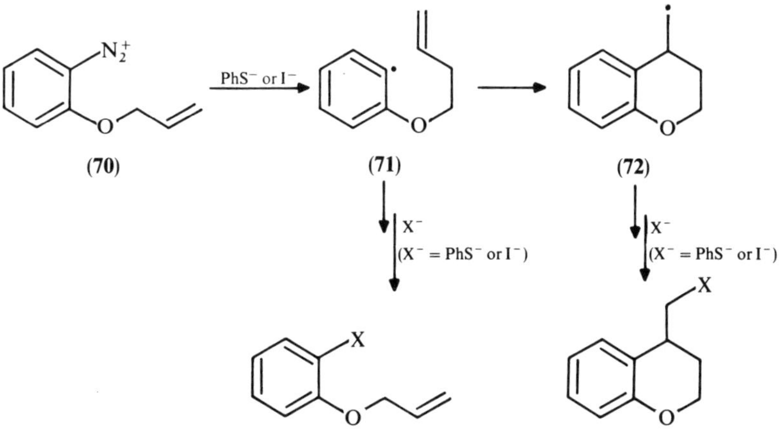

$$ArN_2^{+} + I^{-} \text{ (or } I_2^{\cdot -}) \rightarrow Ar^{\cdot} + N_2 + I^{\cdot} \text{ (or } I_2)$$
$$Ar^{\cdot} + I_2 \text{ (or } I_3^{\cdot -}) \rightarrow ArI + I^{\cdot} \text{ (or } I_2^{\cdot -})$$

SCHEME 17

$$ArN_2^{+} + PhS^{-} \rightarrow ArN{=}NSPh \rightarrow Ar^{\cdot} + N_2 + PhS^{\cdot}$$
$$Ar^{\cdot} + PhS^{-} \rightarrow ArSPh^{\cdot -}$$
$$ArSPh^{\cdot -} + ArN{=}NNSPh \rightarrow ArSPh + [ArN{=}NSPh]^{\cdot -}$$
$$[ArN{=}NSPh]^{\cdot -} \rightarrow Ar^{\cdot} + N_2 + PhS^{-}$$

SCHEME 18

The reaction of diazonium salts with thiolate ions yields diaryl sulphides in an $S_{RN}1$ reaction which is inhibited by *m*-dinitrobenzene.[499] Diazo-sulphides react with cyanide in DMSO under $S_{RN}1$ conditions (photo-activation or cathodic stimulation) to give the corresponding nitriles.[500]

$$ArN{=}NSPh + e^{-} \rightarrow [ArN{=}NSPh]^{\cdot -}$$
$$[ArN{=}NSPh]^{\cdot -} \rightarrow Ar^{\cdot} + N_2 + PhS^{-}$$
$$Ar^{\cdot} + CN^{-} \rightarrow ArCN^{\cdot -}$$
$$ArCN^{\cdot -} + ArN{=}NSPh \rightarrow ArCN + [ArN{=}NSPh]^{\cdot -}$$

Bromo- and chloro-substituted diazo-sulphides give dinitriles: the intermediate $XC_6H_4CN^{\cdot -}$ undergoes loss of halide to give the corresponding cyanophenyl radical.[501] *p*-Bromobenzenethiolate undergoes polymerization in DMSO using arenediazonium salts as initiators.[502]

$$Ar^{\cdot} \text{ (from } ArN_2^{+}) + p\text{-}BrC_6H_4S^{\cdot} \rightarrow [ArSC_6H_4Br]^{\cdot -} \rightarrow ArSC_6H_4^{\cdot} + Br^{-}$$
$$ArSC_6H_4^{\cdot} \rightarrow \text{polymer}$$

Peroxides

Carbonyl inversion in the decompositions of $(Bu^tCH_2CMe_2CO_2)_2$ and $[Me(CH_2)_4CO_2]_2$ has been shown to proceed by a radical mechanism.[503] Both radical and ion-pair routes are operative in decompositions of $(Me_2CHCO_2)_2$ in supercritical CO_2, $CHCl_3$ and CCl_4.[504] The formation of rearranged products, e.g. PhCH=CHD, in the decomposition of $PhCH_2CD_2CO_2—O_2CC_6H_4Cl$-*m* is taken to indicate that the initial radical-pair $[PhCH_2\dot{C}D_2\,{}^{\bullet}O_2CC_6H_4Cl\text{-}m]$ undergoes internal electron-transfer to give $PhCH_2\overset{+}{C}D_2$ which in turn rearranges to $Ph\overset{+}{C}HCHD_2$.[505] The gas-phase decomposition of dibenzoyl peroxide proceeds by a radical mechanism.[506] Both aryl and arylcyclohexadienyl radicals are trapped in the decomposition of diaroyl peroxides in presence of a spin trap.[507] The decomposition of the peracetylated dodecanoyl peroxides of tartaric and D-gluconic acids in the solid state at −60°C proceeds with a diastereoselectivity of 90–95% ee.[508] The thermolysis of enanthoyl peroxide has been examined.[509]

Di-*t*-butyl nitroxide causes induced homolysis of perfluorodiacyl peroxides.[510] Chemiluminescence is observed in reactions initiated by AIBN and lauryl peroxide.[511] The decomposition dibenzoyl peroxide in presence of tetramethyl-*p*-phenylenediamine proceeds via an ET mechanism.[512] Bisheptafluorobutyryl peroxide undergoes induced decomposition in presence of Grignard reagents by an ET process.[513] The acid-catalysed cleavage of the pinacolone (An_3CCOAn; An = *p*-anisyl) with diaroyl peroxides proceeds via an ET route.[514]

$$An_3CCOAn + (ArCO)_2 \xrightarrow{H^+} An_3C^+ + An\dot{C}O + ArCO_2H + ArCO_2{}^{\bullet}$$

$$An\dot{C}O + ArCO_2{}^{\bullet} \longrightarrow AnCO^+ + ArCO_2{}^-$$

$$AnCO^+ + CF_3COOH \longrightarrow AnCOCOOCF_3 + H^+$$

^{13}C CIDNP Studies indicate that *t*-butyl phenylperacetate decomposes by a non-concerted pathway with the observed rate of decomposition dependent on the O—O bond strength, the rate of decarboxylation, and the rate of internal return.[515] The decomposition of *t*-butyl 2-ethylperoxyhexanoate has been studied.[516] The decomposition of peresters in hydrogen-donor solvents results in the conversion of aldoximes into ketoximes.[517] The rate constants for the initiation step in the oxidation of cumene in presence of peresters have been determined.[518]

The influence of metal salts on the rate of decomposition of peroxy acids decreases in the order: Mn(II) > Co(II) > Ce(IV) > Ce(III).[519]

The rate of decomposition of cyclohexyl peroxide has been measured:[520] the decomposition in benzene is catalysed by vanadyl dibenzoylmethane and inhibited by cyclohexanol.[521] The photo-decomposition of tetralin-1-hydroperoxide is catalysed by iron(III) and cobalt(III) acetylacetonates.[522] 2,2,6,6-Tetramethylpiperidine-*N*-oxyl induces the decomposition of hydroperoxides in presence of traces of strong acids.[523] The nitroxide disproportionates to give 1-hydroxy-2,2,6,6-tetramethylpiperidine and the 1-oxo-2,2,6,6-tetramethylpiperidinium cation: this catalyses the decomposition of the peroxide. Different products are obtained in the thermal and metal-catalysed decompositions of methyl linolenate hydroperoxides: in the thermal decomposition, homolysis occurs either side of the hydroperoxy group.[524] The rearrangement products from the allylic hydroperoxides derived from oleic acid are consistent with a concerted rearrangement of allylperoxyl radicals via a five-membered transition state (Scheme 19).[525] The intermediate alkoxyl radical obtained from the iron(II)-copper(II)-catalysed decomposition of hydroperoxyeudesmanolides undergoes fragmentation to give 1,10-*sec*-eudesmanolides.[526] A study of the induced decomposition of alkyl and acyl hydroperoxides indicates that the reactivity of the O—H bond in alkyl hydroperoxides is 50–100 times greater than in acyl hydroxides, but that of the O—O bond in acyl hydroperoxides is 150–200 times greater than in alkyl hydroperoxides.[527]

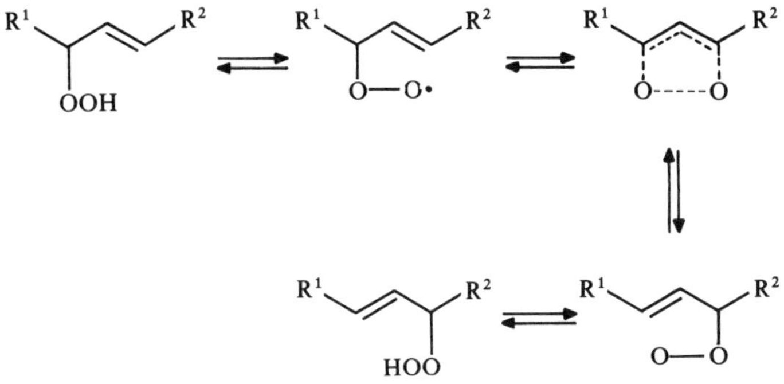

SCHEME 19

Thermolysis of the α-azohydroperoxide, $Me_2C(OOH)N{=}N(CH_2)_4CH{=}CH_2$, leads to hex-5-enyl radicals which react with the OOH group in competition with their cyclization to cyclopentylmethyl radicals.[528] From the known rate of cyclization of hex-5-enyl radicals, the rate of induced decomposition of the azohydroperoxide was determined and shown to be concentration-dependent, reflecting the effects of intermolecular hydrogen-bonding. The rate of induced decomposition at 0.1 M concentration in C_6F_6 is 1.8×10^7 mol^{-1} s^{-1}: the high value is taken to indicate that there is concerted breaking of the O—O bond and the nearest C—N bond. The stability of α-azohydroperoxides ArCH(COOH)N=NAr′ is increased by electron-withdrawing groups in Ar and by electron-releasing groups in Ar′.[529]

Further studies on radical addition of ZH hydrogen donors (e.g. CH_2Cl_2, CH_3CH_2COOMe, $c\text{-}C_6H_{12}$, THF) to unsaturated peroxides have been reported. Allyl *t*-butyl peroxide and also $CH_2{=}CMeCH_2OOBu^t$ and $CH_2{=}CHCMe_2OOBu^t$ give epoxides while $Me_2C{=}CHCH_2OOBu^t$ yields oxetanes (Scheme 20).[530,531] In a similar manner, γ- and δ-unsaturated peroxides lead to four- or five-membered rings,[532] and five- or six-membered rings, respectively.[533]

The prostaglandin endoperoxide model compound (**75**) was converted into the thromboxane B ring system (**76**) on treatment with iron(II), providing evidence for the biosynthetic pathway for the formation of the thromboxane skeleton (Scheme 21).[534]

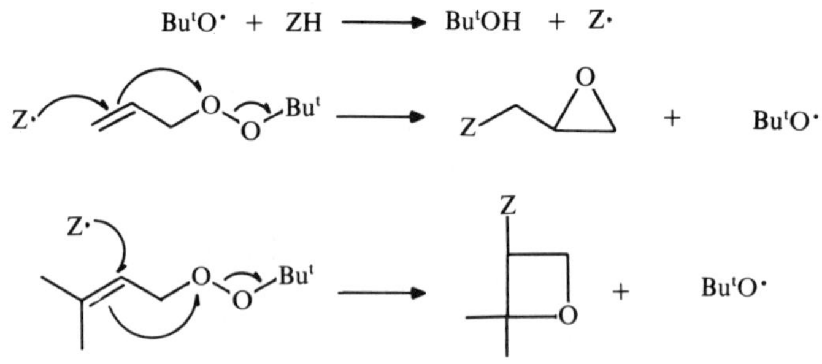

SCHEME 20

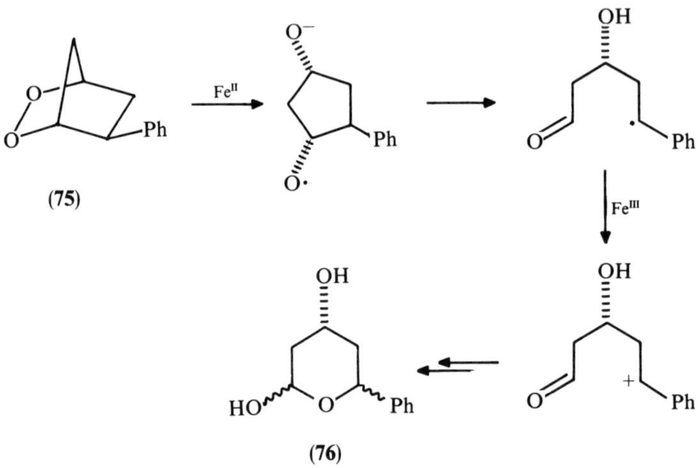

SCHEME 21

The relative rates of decomposition of a series of substituted dioxetanes are consistent with a biradical mechanism of decomposition.[535,536] The kinetics of decomposition of *trans*-3,6-dimethyl-3,6-diphenyl-1,2,4,5-tetraoxane have been determined.[537]

Intramolecular CIEEL occurs in the reaction of an aminocoumarincarboxylic acid chloride with 1-phenylethyl hydroperoxide: the aminocoumarin functions as the electron-acceptor and the peroxide as the electron-donor.[538] The reaction serves as a model for light production in bioluminescence. An ET mechanism may be operative in the chemiluminescence of phenyl acridine-9-percarboxylate in the reaction of phenyl acridine-9-carboxylate with HO_2^-.[539] The thermal decomposition of hydrotrioxides in presence of adamantylideneadamantane occurs with intense IR chemiluminescence: addition of oxygen gives the dioxetane.[540] The thermolysis of secondary alkaneperoxysulphonic acids is accompanied by chemiluminescence.[541]

Autoxidation

A book on the autoxidation of unsaturated lipids has been published.[542] The reactivities of unsaturated compounds and alkylbenzenes towards peroxy radicals have been reviewed.[543]

The autoxidation of alkanes can be induced by radiation and inhibited by BHT and diphenylamine.[544] The influence of methyl radicals on the combustion of methane in argon-oxygen mixtures has been studied.[545]

The rate of autoxidation of PUFA and simple esters is proportional to the number of doubly allylic positions: the kinetics of autoxidation of triglycerides are more complex.[546] The enzymic lipoxygenation of PUFA proceeds via an organo-iron intermediate but does not involve radical intermediates. The conclusion was reached from a study of the autoxidation of a cyclopropane-containing PUFA: no ring-opening was observed.[547] A related compound in peroxide-catalysed autoxidation gave ring-opened products. There is little evidence that peroxidation is initiated by carbon tetrachloride.[548] The phenoxyl radicals generated from phenolic antioxidants, during the autoxidation of polyunsaturated lipids, are rapidly destroyed.[549] The results support the theory that vitamins E and C act as chain-breaking antioxidants in the *in vivo* autoxidation of polyunsaturated lipids of cellular membranes. The radical generated from vitamin E is intercepted by vitamin C to give a radical which is subsequently oxidized. Radicals have been shown to be key intermediates in oxidative stress involving damage to lipids, proteins, carbohydrates and DNA as a result of action by $HO^{\bullet}$, $O_2^{\bullet-}$, $ROO^{\bullet}$ or $RO^{\bullet}$.[550] The rate constants for reaction of *t*-butoxyl radicals with PUFA are about 10 times greater in water than in non-polar solvents.[551] *t*-Butoxyl radicals are reactive towards lipids, nucleosides, and other compounds involved in antioxidative defence, but not glutathione. Autoxidation of **(77)** occurs at the 11- and 14-positions: the other allylic positions are more sterically hindered.[552]

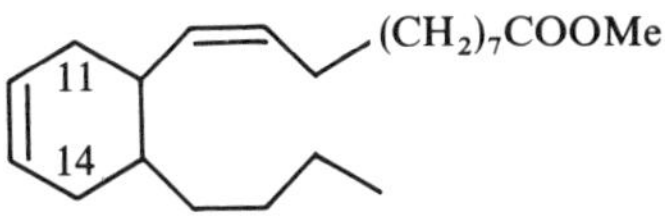

(77)

p-Xylene is oxidized by air to terephthalic acid using cobalt(II) bromide as catalyst.[553] The role of zinc compounds in the autoxidation of cumene has been investigated.[554] The autoxidations of 4,6,8-trimethylazulene and 1,4-dimethyl-7-isopropylazulene give complex product mixtures: the reactions probably involve initial electron-transfer to give the azulene radical cation and superoxide.[555] Superoxide is also formed in the autoxidation of 1,8-dihydroxy-9-anthrone.[556] The autoxidation of adrenalone has been examined.[557]

The low-temperature autoxidation of organoboranes led to diperoxyboranes, which subsequently can be converted by reaction with hydrogen peroxide into the corresponding alkyl hydroperoxides: only two of the three alkyl groups in the trialkylborane can be converted into the hydroperoxides.[558] The reactions are inhibited by radical scavengers and the following mechanism has been proposed:

$$R_3B + O_2 \rightarrow R_2BOO^{\bullet} + R^{\bullet}$$
$$R^{\bullet} + O_2 \rightarrow ROO^{\bullet}$$
$$ROO^{\bullet} + R_3B \rightarrow ROOBR_2 + R^{\bullet}$$
$$ROOBR_2 + O_2 \rightarrow (ROO)_2BR$$

The photo-oxidation of cholesterol in erythrocyte membranes gives the 7α-hydroperoxide due to reaction with $^1O_2$: in contrast, oxidation with xanthine oxidase gives both 7α- and 7β-hydroperoxides and, unlike the photo-oxidation, the reaction is inhibited by catalase, superoxide dismutase, and phenolic antioxidants implicating the involvement of $O_2^{\bullet -}$.[559] The autoxidation of cholesterol in phospholipid bilayers also gives the 7-hydroperoxycholesterol.[560]

Products obtained from the autoxidation of diethyl ether include MeCH(OOH)OEt, MeCH(OH)OCH(OH)Me, and MeCH(OOH)OCH(OOH)-Me.[561] Bis(dimethoxymethyl) peroxide and bis(1,1-dimethoxyethyl) peroxide are obtained in the benzophenone-sensitized photo-oxygenations of trimethoxymethane and trimethoxyethane, respectively.[562]

$$(CH_3O)_3CH \rightarrow (CH_3O)_2CHO\dot{C}H_2 \rightarrow (CH_3O)_2\dot{C}H + CH_2O$$

$$(CH_3O)\dot{C}H + O_2 \rightarrow CH_3O_2CHOO^{\bullet} \xrightarrow{(CH_3O)_2\dot{C}H} (CH_3O)_2CHOCH(OCH_3)_2$$

The autoxidation of diethylene glycol has been studied.[563] There is evidence for electron-transfer in autoxidations of alkenes and enzymes promoted by thiophenol (Scheme 22).[564–567] The aerobic and peroxide-induced coupling of thiols has been reported.[568]

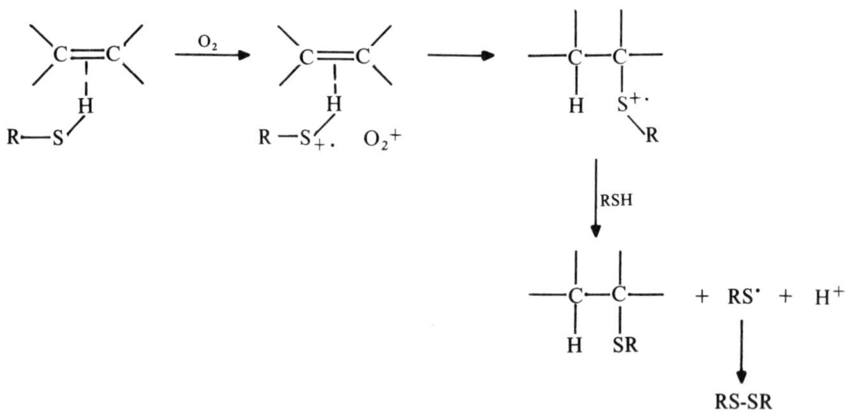

SCHEME 22

The rate constants for reaction of *t*-butylperoxyl radicals with benzaldehyde are little influenced by solvent whereas the reactions with phenol proceed less rapidly in solvents with strong donor-acceptor properties.[569] The absolute rate constants for reactions of $Cl_3COO^\bullet$ with cyclohexane, cyclohexene, and hexamethylbenzene are 1×10^3, 1×10^5, and 7.5×10^4 $l\,mol^{-1}\,s^{-1}$, respectively.[570]

The photo-oxidation of alkanes in oxygen matrices at 10 K proceeds via $R_2\dot{C}OO^\bullet$ biradicals.[571] Radical intermediates are also involved in the photo-oxidation of tetrabromofuran: in the solid state this gives tetrabromo-γ-lactone, and in solution dibromomaleic anhydride.[572]

α-Aminoperoxyl radicals are involved in the principal propagation steps in the autoxidation of tertiary amines.[573] The autoxidation of trialkylamines in basic media results in oxidative dealkylation with the intervention of secondary nitroxyl radicals.[574] The kinetics of decomposition of tetraphenylhydrazine in presence of BHT have been reported.[575]

Kinetic studies of antioxidant activity of tocopherol model compounds reveal that the rate constants increase with increasing electron-donating ability of alkyl substituents.[576] *Ab initio* MO calculations have shown that the neutral and anion radicals generated on oxidation of ascorbic acid have similar geometries but different spin-density distributions: the spin density is primarily at C(1), C(2), C(7), and O(8) in the neutral radical and at O(3) and O(4) in the radical anion.[577] The 5- and 6- but not the 2-esters of ascorbic acid are effective antioxidants.[578] The mechanisms of action and reactivities of radicals derived from amine and phenolic inhibitors have been reviewed.[579]

The recombination of tertiary peroxyl radicals proceeds via a transoid conformation of the tetroxide, but primary and secondary peroxyl radicals react via a cisoid transition state.[580] The rate of recombination of cyclohexylperoxyl radicals has been investigated.[581]

The rate of reaction of acetonyl with oxygen is $3 \times 10^9\,l\,mol^{-1}\,s^{-1}$: the resultant acetonylperoxyl radicals decay by second-order kinetics to give *inter alia* pyruvaldehyde, pyruvic acid, and formaldehyde ($2k_t = 8 \times 10^8\,l\,mol^{-1}\,s^{-1}$).[582] The rate of reaction of acetylperoxyl radicals with NO_2 is $2.29 \times 10^9\,l\,mol^{-1}\,s^{-1}$.[583] The rate of one-electron oxidation of chloropromazine by halogenated alkylperoxyl radicals is greater the higher the oxidation potential of the radical, i.e. the rates are in the order: $CCl_3OO^\bullet > CHCl_2OO^\bullet > CH_2ClOO^\bullet$.[584] The reaction of excited vinylidene radicals with oxygen has been reported.[585] Arylnitrenes react with oxygen to yield nitroarenes: the reaction possibly proceeds via arylnitroso oxides which behave as electrophilic radicals.[586] Acetate has been shown to exert a stabilizing effect on aqueous solutions of ozone due to its ability to scavenge hydroxyl radicals:[587] the resultant acetate radical reacts with oxygen to give the relatively inert peroxyacetate radical.

$$HO^\bullet + CH_3COO^- \rightarrow \dot{C}H_2COO^- + H_2O$$
$$O_2 + \dot{C}H_2COO^- \rightarrow {}^\bullet OOCH_2COO^-$$

Muon-containing Radicals

Reviews on muons as a kinetic probe for radical reactions including ET reactions and rearrangements[588] and in μSR (muon spin rotation) spectroscopy[589] have been published. Muons serve as a passive probe and have been used to determine absolute rate constants which are difficult to obtain by conventional methods. Nuclear attack occurs to a greater extent in the reactions of muons with $PhCH_2CH{=}CH_2$ than $PhCH{=}CH_2$.[590] The rate constants for the reactions of muons with aromatic compounds[591] and ethene[592] have been determined. There is evidence for quantum-mechanical tunnelling in reactions of muons in solution.[593] Additions of muons to 1,2-bis(trimethylsilyl)acetylene give a rapidly equilibrating mixture of *cis*- and *trans*-$Me_3SiC(Mu){=}\dot{C}SiMe_3$ with the unpaired electron in an sp^2-orbital.[594]

References

1 Ortiz de Montellano, P. R. and Stearns, R. A., *J. Am. Chem. Soc.*,**109**, 3415 (1987).
2 Traylor, T. G. and Miksztal, A. R., *J. Am. Chem. Soc.*, **109**, 2770 (1987).
3 Monteil, V., Segura, M. L., Aldaz, A., and Barba, F., *J. Chem. Res. (S)*, **1987**, 27.
4 Baciocchi, E., Cort, A. D., Eberson, L., Mandolini, L., and Rol, C., *J. Org. Chem.*, **51**, 4544 (1986).
5 Bank, S. and Gernon, M., *J. Org. Chem.*, **52**, 5105 (1987).
6 Citterio, A., Santi, R., and Pagani, A., *J. Org. Chem.*, **52**, 4925 (1987).
7 Holland, H. L., Bergen, E. J., Chenchaiah, P. C., Khan, S. H., Munoz, B., Ninniss, R. W., and Richards, D., *Can. J. Chem.*, **65**, 502 (1987).
8 Balogh-Hergovich, E. and Speier, G., *J. Chem. Soc., Perkin Trans. 1*, **1986**, 2305.
9 Hwu, J. R. and Gilbert, B. A., *J. Organomet. Chem.*, **332**, 53 (1987).
10 Capparelli, M. P., DeSchepper, R. S., and Swenton, J. S., *J. Org. Chem.*, **52**, 4953 (1987).
11 Capparelli, M. P., DeSchepper, R. S., and Swenton, J. S., *J. Chem. Soc., Chem. Commun.*, **1987**, 610.
12 Panda, P. P., Panda, R. K., and Murti, P. S. R., *Int. J. Chem. Kinet.*, **19**, 155 (1987).
13 Deya, P. M., Dopico, M., Raso, A. G., Morey, J., and Saa, J. M., *Tetrahedron*, **43**, 3523 (1987).
14 Wrona, M. L. and Dryhurst, G., *J. Org. Chem.*, **52**, 2817 (1987).
15 Kende, A. S. and Koch, K., *Tetrahedron Lett.*, **27**, 6051 (1986).
16 Leboff, A., Carbonnelle, A.-C., Alazard, J.-P., Thal, C., and Kende, A. S., *Tetrahedron Lett.*, **28**, 4163 (1987).
17 Peterson, J. R., Egler, R. S., Horsley, D. B., and Winter, T. J., *Tetrahedron Lett.*, **28**, 6109 (1987).
18 Snider, B. B. and Dombroski, M. A., *J. Org. Chem.*, **52**, 5487 (1987).
19 Snider, B. B., Mohan, R., and Kates, S. A., *Tetrahedron Lett.*, **28**, 841 (1987); Mohan, R., Kates, S. A., Dombroski, M. A., and Snider, B. B., *Tetrahedron Lett.*, **28**, 845 (1987).
20 Corey, E. J. and Ghosh, A. K., *Tetrahedron Lett.*, **28**, 175 (1987); *Chem. Lett.*, **1987**, 223.
21 Yonemura, H., Nishino, H., and Kuvosawe, E., *Bull. Chem. Soc. Jpn*, **59**, 3153 (1986).
22 Sato, H., Nishino, H., and Kurosawa, K., *Bull Chem. Soc. Jpn*, **60**, 1753 (1987).
23 Yoshida, J., Sakaguchi, K., and Isoe, S., *Tetrahedron Lett.*, **27**, 6075 (1986).
24 Ahmed, S. and Iqbal, J., *J. Chem. Soc., Chem. Commun.*, **1987**, 692.
25 O'Neill, P. and Hegarty, A. F., *J. Chem. Soc., Chem. Commun.*, **1987**, 744.
26 Guirado, A., Moss, G. P., and Utley, J. H. P., *J. Chem. Soc., Chem. Commun.*, **1987**, 41.
27 Tanner, D. D. and Osman, S. A. A., *J. Org. Chem.*, **52**, 4689 (1987).
28 Gilbert, B. C., Scarratt, C. J., Thomas, C. B., and Young, J., *J. Chem. Soc., Perkin Trans. 2*, **1987**, 371.
29 Kimura, M., Koie, K., Matsubara, S., Sawaki, Y., and Iwamura, H., *J. Chem. Soc., Chem. Commun.*, **1987**, 122.
30 Wimalasena, K. and May, S. W., *J. Am. Chem. Soc.*, **109**, 4036 (1987).
31 Pankratov, A. N., Mushtakova, S. P., and Gribov, L. A., *Zh. Anal. Khim.*, **42**, 352 (1987); *Chem. Abs.*, **107**, 197432 (1987).
32 Pankratov, A. N., Shmakov, S. L., Mushtakova, S. P., and Gribov, L. A., *Zh. Anal. Khim.*, **41**, 810 (1986); *Chem. Abs.*, **107**, 58288 (1987).
33 Hedayatullah, M. and Thevenet, F., *Bull Soc. Chim. Belg.*, **96**, 311 (1987).
34 Arias, S., Brillas, E., and Corta, J. M., *J. Chem. Soc., Faraday Trans. 1*, **83**, 2619 (1987).
35 Gilbert, B. C. and Whitwood, A. C., *J. Chem. Res. (S)*, **1987**, 416.
36 Novais, H. M. and Steenken, S., *J. Phys. Chem.*, **91**, 426 (1987).
37 Shen, X., Lind, J., and Merényi, G., *J. Phys. Chem.*, **91**, 4403 (1987).
38 Wallington, T. J. and Kurylo, M. J., *J. Phys. Chem.*, **91**, 5050 (1987).
39 Taniguchi, H., *J. Phys. Chem.*, **91**, 5213 (1987).

[40] Andrieux, C. P., Badoz-Lambling, J., Combellas, C., Lacombe, D., Savéant, J.-M., Thiébault, A., and Zann, D., *J. Am. Chem. Soc.*, **109**, 1518 (1987).
[41] Arena, J. V. and Rushing, J. F., *J. Phys. Chem.*, **91**, 3368 (1987).
[42] Lexa, D., Savéant, J.-M., Su, K. B., and Wang, D. L., *J. Am. Chem. Soc.*, **109**, 6464 (1987).
[43] Fry, A. J. and Powers, T. A., *J. Org. Chem.*, **52**, 2498 (1987).
[44] de Luca, C., Inesi, A., and Rampazzo, L., *J. Chem. Soc., Perkin Trans. 2*, **1987**, 847.
[45] Shao, R. and Peters, D. G., *J. Org. Chem.*, **52**, 652 (1987).
[46] Inanaga, J., Ishikawa, M., and Yamaguchi, M., *Chem. Lett.*, **1987**, 1485.
[47] Matsukawa, M., Inanaga, J., and Yamaguchi, M., *Tetrahedron Lett.*, **28**, 5877 (1987).
[48] Otsubo, K., Kawamura, K., Inanaga, J., and Yamaguchi, M., *Chem. Lett.*, **1987**, 1487.
[49] Molander, G. A., Etter, J. A., and Zinke, P. W., *J. Am. Chem. Soc.*, **109**, 453 (1987); Molander, G. A., and Kenny, C., *Tetrahedron Lett.*, **28**, 4367 (1987).
[50] Chiu, Y. L., Sant'ana, A. E. G., and Utley, J. H. P., *Tetrahedron Lett.*, **28**, 1349 (1987).
[51] Clarke, N. C., Runciman, P. J. I., Utley, J. H. P., and Landquist, J. K., *J. Chem. Soc., Perkin Trans. 2*, **1987**, 435.
[52] Pearson, A. J., Chen, Y.-S., Daroux, M. L., Tanaka, A. A., and Zettler, M., *J. Chem. Soc., Chem. Commun.*, **1987**, 155.
[53] Huffman, J. W., Liao, W.-P., and Wallace, R. H., *Tetrahedron Lett.*, **28**, 3315 (1987).
[54] Gautheron, B., Degrand, C., and Bikrani, M., *J. Organomet. Chem.*, **329**, 349 (1987).
[55] Langner, K., Tesch-Schmidtke, S., and Voss, J., *Chem. Ber.*, **120**, 67 (1987).
[56] Schreiber, J., Mottley, C., Sinha, B. K., Kalyanaraman, B., and Mason, R. P., *J. Am. Chem. Soc.*, **109**, 348 (1987).
[57] Courtneidge, J. L. and Davies, A. G., *Acc. Chem. Res.*, **20**, 90 (1987).
[58] Tureček, F., *Collect. Czech. Chem. Commun.*, **52**, 1928 (1987).
[59] Roth, H. D., *Acc. Chem. Res.*, **20**, 343 (1987).
[60] Clark, T., *NATO ASI Ser., Ser. C.*, **1986**, 189; *Chem. Abs.*, **107**, 96048 (1987).
[61] Doriso, A., Pratt, D. W., and Houk, K. N., *J. Am. Chem. Soc.*, **109**, 6591 (1987).
[62] Toriyama, K., Nunome, K., and Iwasaki, M., *J. Am. Chem. Soc.*, **109**, 4496 (1987).
[63] Toriyama, K., Nunome, K., and Iwasaki, M., *J. Phys. Chem.*, **90**, 6836 (1986).
[64] Lindgren, M. and Lund, A., *J. Chem. Soc., Faraday Trans. 1*, **83**, 1815 (1987).
[65] Prinzbach, H., Murty, B. A. R. C., Fessner, W.-D., Mortensen, J., Heinze, J., Gescheidt, G., and Gerson, F., *Angew. Chem. Int. Edn*, **26**, 457 (1987).
[66] Clark, T., Teasley, M. F., Nelsen, S. F., and Wynberg, H., *J. Am. Chem. Soc.*, **109**, 5719 (1987).
[67] Chan, W., Courtneidge, J. L., Davies, A. G., Neville, A. G., Evans, J. C., and Rowlands, C. C., *J. Chem. Soc., Perkin Trans. 2*, **1986**, 2017.
[68] Symons, M. C. R., Muto, H., Toriyama, K., Nunome, K., and Iwasaki, M., *J. Chem. Soc., Perkin Trans. 2*, **1986**, 2011.
[69] Courtneidge, J. L., Davies, A. G., and Gregory, P. S., *J. Chem. Soc., Perkin Trans. 2*, **1987**, 1527.
[70] Brede, O., Mueller, E., and Naumann, W., *J. Radioanal. Nucl. Chem.*, **105**, 37 (1986); *Chem. Abs.*, **106**, 155652 (1987).
[71] Yamamoto, Y., Ma, X.-H., and Hayashi, K., *J. Phys. Chem.*, **91**, 5343 (1987).
[72] Handoo, K. L. and Gadru, K., *Curr. Sci.*, **55**, 920 (1986); *Chem. Abs.*, **106**, 195698 (1987).
[73] Chandra, H., Symons, M. C. R., and Hasegawa, A., *J. Chem. Soc., Faraday Trans. 1*, **83**, 759 (1987).
[74] Davies, A. G., Julia, L., and Yazdi, S. N., *J. Chem. Soc., Chem. Commun.*, **1987**, 929.
[75] Brandy, S. M., Sassoon, R. E., and Rabani, J., *J. Phys. Chem.*, **91**, 953 (1987).
[76] Le Berre, V., Angély, L., Simonet-Guéguen, N., and Simonet, J., *J. Chem. Soc., Chem. Commun.*, **1987**, 984.
[77] Heelis, P. F., Parsons, B. J., Phillips, G. O., and Swallow, A. J., *J. Phys. Chem.*, **90**, 6833 (1986).
[77] Evans, J. C., Morris, C. R., and Rowlands, C. C., *J. Chem. Soc., Perkin Trans. 2*, **1987**, 1375.
[79] Nelsen, S. F., Thompson-Colon, J. A., Kirste, B., Rosenhouse, A., and Manahem, K., *J. Am. Chem. Soc.*, **109**, 7128 (1987).
[80] Nelsen, S. F., Rumack, D. T., and Meot-Nev, M., *J. Am. Chem. Soc.*, **109**, 1373 (1987).
[81] Nelsen, S. F., Frigo, T. B., Kim, Y., and Thompson-Colón, J. A., *J. Am. Chem. Soc.*, **108**, 7926 (1986).
[82] Blackstock, S. C. and Kochi, J. K., *J. Am. Chem. Soc.*, **109**, 2484 (1987).
[83] Ishiguro, K., Sawaki, Y., Izuoka, A., Sugawara, T., and Iwamura, H., *J. Am. Chem. Soc.*, **109**, 2531 (1987).
[84] Symons, M. C. R., *J. Chem. Soc., Chem. Commun.*, **1987**, 866.
[85] Qin, X.-Z., Pentecost, T. C., Wang, J. T., and Williams, F., *J. Chem. Soc., Chem. Commun.*, **1987**, 450.
[86] Rhodes, C. J., Podmore, I. D., and Symons, M. C. R., *J. Chem. Soc., Chem. Commun.*, **1987**, 1638.
[87] Bock, H., Lechner-Knoblauch, U., and Hänel, P., *Chem. Ber.*, **119**, 3749 (1987).

88 Symons, M. C. R. and Janes, R., *J. Chem. Soc., Faraday Trans. 1*, **83**, 383 (1987).
89 Guéguen, M., Guerchais, J. E., Pétillon, F. Y., and Talarmin, J., *J. Chem. Soc., Chem. Commun.*, **1987**, 557.
90 Williams, F., Dai, S., Snow, L. D., Qin, X.-Z., Bally, T., Nitsche, S., Haselbach, E., Nelsen, S. F., and Teasley, M. F., *J. Am. Chem. Soc.*, **109**, 7526 (1987).
91 Yates, B. F., Bouma, W. J., and Radom, L., *J. Am. Chem. Soc.*, **109**, 2250 (1987).
92 Drewello, T., Lebrilla, C. B., Schwarz, H., de Koning, L. J., Fokkens, R. H., Nibbering, N. M. M., Anklam, E., and Asmus, K.-D., *J. Chem. Soc., Chem. Commun.*, **1987**, 1381.
93 Moenig, J. Goslich, R., and Asmus, K.-D., *Ber. Bunsen-Ges. Phys. Chem.*, **90**, 115 (1986); *Chem. Abs.*, **106**, 155753 (1987).
94 Qin, X.-Z., Meng, Q.-C., and Williams, F., *J. Am. Chem. Soc.*, **109**, 6778 (1987).
95 Mohan, H. and Asmus, K.-D., *J. Am. Chem. Soc.*, **109**, 4745 (1987).
96 Anklam, E., Mohan, H., and Asmus, K.-D., *J. Chem. Soc., Chem. Commun.*, **1987**, 629.
97 Mahling, S., Asmus, K.-D., Glass, R. S., Hojjatie, M., and Wilson, G. S., *J. Org. Chem.*, **52**, 3717 (1987).
98 Kovacic, P. and Jones, M. B., *Chem. Rev.*, **87**, 357 (1987).
99 Masnovi, J. M. and Kochi, J. K., *J. Phys. Chem.*, **91**, 1878 (1987).
100 Suyo, H., Michida, T., and Hatsumura, H., *Chem. Pharm. Bull.*, **34**, 558 (1986); *Chem. Abs.*, **106**, 66654 (1987).
101 Bjorklund, R. B., *J. Chem. Soc., Faraday Trans. 1*, **83**, 1507 (1987).
102 Bindl, J., Seitz, P., Seitz, U., Salbeck, E., Selbeck, J., and Daub, J., *Chem. Ber.*, **120**, 1747 (1987).
103 Fox, M. A., Sackett, D. D., and Younathan, J. N., *Tetrahedron*, **43**, 1643 (1987).
104 Bock, H., Rittmeyer, P., and Stein, U., *Chem. Ber.*, **119**, 3766 (1986).
105 Parker, V. D. and Tilset, M., *J. Am. Chem. Soc.*, **109**, 2521 (1987).
106 Yamamoto, Y., Nishida, S., and Hayashi, K., *J. Chem. Soc., Faraday Trans. 1*, **83**, 1795 (1987).
107 Takanami, T., Suda, K., and Ohmori, H., *Chem. Lett.*, **1987**, 1335.
108 Parker, V. D., and Bethell, D., *J. Am. Chem. Soc.*, **109**, 5066 (1987).
109 Bakke, J. M., Bethell, D., and Parker, V. D., *Acta Chem. Scand.*, **41B**, 253 (1987).
110 Rhodes, C. J., and Symons, M. C. R., *J. Chem. Soc., Chem. Commun.*, **1987**, 258.
111 J. Yoshida, and S. Isoe, *Chem. Lett.*, **1987**, 631.
112 Ci, X., Lee, L. Y. C., and Whitten, D. G., *J. Am. Chem. Soc.*, **109**, 2536 (1987).
113 Ci, X., and Whitten, D. G., *J. Am. Chem. Soc.*, **109**, 7215 (1987).
114 Clark, T., *J. Am. Chem. Soc.*, **109**, 6838 (1987).
115 Yates, B. F., Bouma, W. J., MacLeod, J. K., and Radom, L., *J. Chem. Soc., Chem. Commun.*, **1987**, 204.
116 Yates, B. F., and Radom, L., *J. Am. Chem. Soc.*, **109**, 2910 (1987).
117 de Mayo, P., and Wenska, G., *Tetrahedron*, **43**, 1661 (1987).
118 Du, P., and Borden, W. T., *J. Am. Chem. Soc.*, **109**, 5330 (1987).
119 Qin, X.-Z., and Williams, F., *J. Am. Chem. Soc.*, **109**, 595 (1987).
120 Miyashi, T., Wakamatsu, K., Akiya, T., Kikuchi, K., and Mukai, T., *J. Am. Chem. Soc.*, **109**, 5270 (1987).
121 Adam, W., Casado, A., and Misanda, M. A., *Angew. Chem. Int. Edn*, **26**, 797 (1987).
122 Draper, A. M., and de Mayo, P., *Tetrahedron Lett.*, **27**, 6157 (1986).
123 Ikezawa, H., and Kutal, C., *J. Org. Chem.*, **52**, 3299 (1987)
124 Newcomb, M., and Deeb, T. M., *J. Am. Chem. Soc.*, **109**, 3163 (1987).
125 Gerson, F., and Huber, W., *Acc. Chem. Res.*, **20**, 85 (1987).
126 Frim, R., Rabinovitz, M., and Muszkat, K. A., *J. Chem. Soc., Chem. Commun.*, **1987**, 770.
127 Umemoto, K. and Okamura, N., *Bull. Chem. Soc. Jpn*, **59**, 3047 (1986).
128 Craig, B. B. and Pace, M. D., *J. Chem. Soc., Chem. Commun.*, **1987**, 1144.
129 Muralidharan, S. and Wan, P., *J. Chem. Soc., Chem. Commun.*, **1987**, 1142.
130 Ozawa, T. and Kwan, T., *Polyhedron*, **5**, 1531 (1986); *Chem. Abs.*, **107**, 115052 (1987).
131 Ohsawa, T. and Oishi, T., *Yuki Gosei Kagaku Kyokaishi*, **44**, 953 (1986); *Chem. Abs.*, **107**, 6510 (1987).
132 Villar, H. O. and Castro, E. A., *Z. Naturforsch.*, **41A**, 1339 (1986); *Chem. Abs.*, **107**, 22740 (1987).
133 Starichenko, V. F., Selivanova, G. A., and Shteingarts, V. D., *Zh. Org. Khim.*, **22**, 1255 (1986); *Chem. Abs.*, **106**, 17777 (1987).
134 Franco, M. L. T. M. B., Lazana, M. C. R. L. R., and Herold, B. J., *J. Chem. Soc., Perkin Trans. 2*, **1987**, 1407.
135 Vuolle, M. and Mäkelä, R., *J. Chem. Soc., Faraday Trans. 1*, **83**, 51 (1987).
136 Ashley, K., Sarafarazi, F., Buckland, S. J., Foley, J. K., Mei, Q., Halton, B., Stang, P. J., and Pons, S., *Can. J. Chem.*, **65**, 2062 (1987).

[137] Gerson, F. and Martin, W. B., *Helv. Chim. Acta*, **70**, 1558 (1987).
[138] Bethell, D. and Parker, V. D., *J. Chem. Res. (S)*, **1987**, 116.
[139] Gerson, F., Martin, W. B., Wong, H. N. C., and Chan, C. W., *Helv. Chim. Acta*, **70**, 79 (1987).
[140] Screttas, C. G. and Micha-Screttas, M., *J. Am. Chem. Soc.*, **109**, 7573 (1987).
[141] Depew, M. C. and Wan, J. K. S., *J. Phys. Chem.*, **90**, 6597 (1986).
[142] Flint, N. J. and Tabner, B. J., *J. Chem. Soc., Faraday Trans. 1*, **83**, 167 (1987).
[143] Bock, H., Hierholzer, B., and Schmolz, P., *Angew. Chem. Int. Edn*, **26**, 791 (1987).
[144] Juchnovski, I. and Kolev, T., *Spectrosc. Lett.*, **19**, 529 (1986); *Chem. Abs.*, **106**, 83866 (1987).
[145] Buddensiek, D., Köpke, B., and Voss, J., *Chem. Ber.*, **120**, 575 (1987); Günther, H. and Voss, J., *J. Chem. Res. (S)*, **1987**, 68.
[146] Roeske, R. and Voss, J., *Phosphorus Sulphur*, **26**, 257 (1987); *Chem. Abs.*, **106**, 66656 (1987).
[147] Prangova, L., Böge, A., Wollny, B., and Voss, J., *J. Chem. Res. (S)*, **1987**, 182.
[148] Bretz, N., Mastalsky, I., Elsner, M., and Kurreck, H., *Angew. Chem. Int. Edn*, **26**, 345 (1987).
[149] Reynolds, J. L., Doshi, D., and Schechter, H., *J. Am. Chem. Soc.*, **109**, 8032 (1987).
[150] Davies, A. G. and Neville, A. G., *J. Chem. Soc., Chem. Commun.*, **1987**, 16.
[151] Kimura, N. and Takamuku, S., *Bull. Chem. Soc. Jpn*, **59**, 3653 (1986).
[152] Ye, S. and Dai, S., *Gaodeng Xuexiao Huaxue Xuebao*, **7**, 1127 (1986); *Chem. Abs.*, **107**, 175427 (1987).
[153] Wipf, D. O., Wehmeyer, K. R., and Wightman, R. M., *J. Org. Chem.*, **51**, 4760 (1986).
[154] Nielsen, M. F., Hammerich, O., and Parker, V. D., *Acta Chem. Scand.*, **41B**, 50 (1987).
[155] Nielsen, M. F., Hammerich, O., and Parker, V. D., *Acta Chem. Scand.*, **41B**, 64 (1987).
[156] Nielsen, M. F., Porat, Z., Eggert, H., and Hammerich, O., *Acta Chem. Scand.*, **40B**, 652 (1986).
[157] McDonald, R. N. and Gung, W. Y., *J. Am. Chem. Soc.*, **109**, 7328 (1987).
[158] Masnovi, J., Koholic, D. J., Berki, R. J., and Binkley, R. W., *J. Am. Chem. Soc.*, **109**, 2851 (1987).
[159] Blankespoor, R. L., Schutt, D. L., Tubergen, M. B., and De Jong, R. L., *J. Org. Chem.*, **52**, 2059(1987).
[160] Walsh, T. D., *J. Am. Chem. Soc.*, **109**, 1511 (1987).
[161] van Doorn, J. A. and Meijboom, N., *J. Chem. Soc., Chem. Commun.*, **1987**, 1700.
[162] Gerson, F., Knöbel, J., Metzger, A., Scott, L. T., Kirms, M. A., Oda, M., and Sumpter, C. A., *J. Am. Chem. Soc.*, **108**, 7920 (1986).
[163] Islam, N., Sopher, D. W., and Utley, J. H. P., *Tetrahedron*, **43**, 959 (1987).
[164] Islam, N., Sopher, D. W., and Utley, J. H. P., *Tetrahedron*, **43**, 2741 (1987).
[165] Severin, M. G., Arévalo, M. C., Farnia, G., and Vianello, E., *J. Phys. Chem.*, **91**, 466 (1987).
[166] Goulart, M. O. F., Ling-Chung, S. K., and Utley, J. H. P., *Tetrahedron Lett.*, **28**, 6081 (1987).
[167] Eberson, L., *Electron-transfer in Organic Chemistry*, Springer-Verlag, Berlin, 1987.
[168] E. Steckhan, (ed.), *Topics in Current Chemistry*, no. 142, Springer-Verlag, Berlin, 1987.
[169] Marcus, R. A., *Nouveau J. Chim.*, **11**, 79 (1987).
[170] Mikkelsen, K. V. and Ratner, M. A., *Chem. Rev.*, **87**, 113 (1987).
[171] Chanon, M., *Acc. Chem. Res.*, **20**, 214 (1987).
[172] Savéant, J.-M., *J. Am. Chem. Soc.*, **109**, 6788 (1987).
[173] Grimshaw, J. and Grimshaw, E. K., *J. Chem. Soc., Perkin Trans. 2*, **1987**, 1345.
[174] Amatore, C., *Prepr.-Am. Chem. Soc., Div. Pet. Chem.*, **31**, 886 (1986); *Chem. Abs.*, **109**, 22903 (1987).
[175] Bornanci, E. R. N., Alonso, R. A., and Rossi, R. A., *J. Org. Chem.*, **52**, 2166 (1987).
[176] Degrand, C., Prest, R., and Compagnon, P.-L., *J. Org. Chem.*, **52**, 5229 (1984).
[177] Novi, M., Garbarino, G., Petrillo, G., and Dell'Erba, C., *J. Org. Chem.*, **52**, 5382(1987).
[178] Pierini, A. B., Baumgartner, M. T., and Rossi, R. A., *Tetrahedron Lett.*, **28**, 4653 (1987).
[179] Alam, N., Amatore, C., Combellas, C., Thiebault, A., and Verpeaux, J. N., *Tetrahedron Lett.*, **28**, 6171 (1987).
[180] Beugelmans, R., Bois-Choussy, M., and Ting, Q., *J. Org. Chem.*, **52**, 3880 (1987).
[181] Novi, M., Garbarino, G., Petrillo, G., and Dell'Erba, C., *J. Chem. Soc., Perkin Trans. 2*, **1987**, 623.
[182] Pierini, A. B., Baumgartner, M. T., and Rossi, R. A., *J. Org. Chem.*, **52**, 1089 (1987).
[183] Kudo, K., Shibata, T., Kashimura, T., Mori, S., and Sugita, N., *Chem. Lett.*, **1987**, 577.
[184] Grimshaw, J. and Thompson, N., *J. Chem. Soc., Chem. Commun.*, **1987**, 240.
[185] Russell, G. A., *Adv. Phys. Org. Chem.*, **23**, 271 (1987).
[186] Kornblum, N., Cheng, L., Davies, T. M., Earl, G. W., Holy, N. L., Kerber, R. C., Kestner, M. M., Manthey, J. W., Musser, M. T., Pinnick, H. W., Snow, D. H., Stuchal, F. W., and Swiger, R. T., *J. Org. Chem.*, **52**, 196 (1987).
[187] McLure, F. I. and Norris, R. K., *Aust. J. Chem.*, **40**, 523 (1987).
[188] Wade, P. A., Morrison, H. A., and Kornblum, N., *J. Org. Chem.*, **52**, 3102 (1987).
[189] Kornblum, N. and Wade, P. A., *J. Org. Chem.*, **52**, 5301 (1987).
[190] Lee, M. S. K., Newcombe, P. J., Norris, R. K., and Wilson, K., *J. Org. Chem.*, **52**, 2796 (1987).
[191] McLure, F. I., Norris, R. K., and Wilson, K., *Aust. J. Chem.*, **40**, 49 (1987).

[192] Symons, M. C. R. and Bowman, W. R., *J. Chem. Soc., Perkin Trans. 2*, **1987**, 1133.
[193] Katritzky, A. R., *NATO ASI Ser., Ser. C*, **1986**, 189; *Chem. Abs.*, **107**, 133686 (1987).
[194] Lund, T. and Lund, H., *Acta Chem. Scand.*, **41B**, 93 (1987).
[195] Crozet, M. P., Jentzer, O., and Vanello, P., *Tetrahedron Lett.*, **28**, 5531 (1987).
[196] Beugelmans, R., Lechevallier, A., Kiffer, D., and Maillous, P., *Tetrahedron Lett.*, **27**, 6209 (1986).
[197] Chen, Q.-Y. and Qiu, Z.-M., *J. Chem. Soc., Chem. Commun.*, **1987**, 1240.
[198] Amrollah-Madjdabadi, A., Beugelmans, A., and Lechevallier, A., *Tetrahedron Lett.*, **28**, 4525 (1987).
[199] Pross, A., *Adv. Chem. Ser.*, **1987**, 215; *Chem. Abs.*, **107**, 153720 (1987).
[200] Ashby, E. C. and Pham, T. N., *Tetrahedron Lett.*, **28**, 3183 (1987).
[201] Ashby, E. C. and Pham, T. N., *J. Org. Chem.*, **52**, 1291 (1987).
[202] Cooke, M. P. and Widener, R. K., *J. Org. Chem.*, **52**, 1381 (1987).
[203] Tanaka, J., Nojima, M., and Kusabayashi, S., *J. Am. Chem. Soc.*, **109**, 3391 (1987).
[204] Ashby, E. C. and Pham, T. N., *Tetrahedron Lett.*, **28**, 3197 (1987).
[205] Ashby, E. C. and Coleman, D., *J. Org. Chem.*, **52**, 4554 (1987).
[206] Newcomb, M., Sanchez, R. M., and Kaplan, J., *J. Am. Chem. Soc.*, **109**, 1195 (1987).
[207] Curran, D. P. and Kim, K., *Tetrahedron Lett.*, **27**, 5821 (1986).
[208] Park, S.-U., Chung, S.-K., and Newcomb, M., *J. Org. Chem.*, **52**, 3275 (1987).
[209] Ash, C. E., Hurd, P. W., Darensbourg, M. Y., and Newcomb, M., *J. Am. Chem. Soc.*, **109**, 3313 (1987).
[210] Bordwell, F. G., Clemens, A. H., and Cheng, J.-P., *J. Am. Chem. Soc.*, **109**, 1773 (1987).
[211] Bordwell, F. G. and Harrelson, J. A., *J. Am. Chem. Soc.*, **109**, 8112 (1987).
[212] Bordwell, F. G. and Wilson, C. A., *J. Am. Chem. Soc.*, **109**, 5470 (1987).
[213] Fuhlendorff, R., Lund, T., Lund, H., and Pedersen, J. A., *Tetrahedron Lett.*, **28**, 5335 (1987).
[214] Prezzavento, B. A. and Kuivila, H. G., *J. Org. Chem.*, **52**, 929 (1987).
[215] Kern, J.-M. and Sauvage, J.-P., *J. Chem. Soc., Chem. Commun.*, **1987**, 546.
[216] Wakselman, C. and Tordeux, P., *J. Chem. Soc., Chem. Commun.*, **1987**, 1701.
[217] Armesto, D., Bosch, P., Horspool, W. M., and Ortiz, M. J., *Tetrahedron Lett.*, **28**, 4605 (1987).
[218] Smith, J. and Tedder, J. M., *J. Chem. Soc., Perkin Trans. 2*, **1987**, 895.
[219] Löfås, S. and Ahlberg, P., *J. Chem. Soc., Chem. Commun.*, **1987**, 216.
[220] Eberson, L. and Ekström, M., *Acta Chem. Scand.*, **41B**, 41 (1987).
[221] Bakkas, S., Juillard, M., and Chanon, M., *Tetrahedron*, **43**, 501 (1987).
[222] Tanner, D. D. and Singh, H. K., *J. Org. Chem.*, **51**, 5182 (1986).
[223] Tanner, D. D., Singh, H. K., Kharrat, A., and Stein, A. R., *J. Org. Chem.*, **52**, 2142 (1987).
[224] Chandresekar, S. and Murthy, T., *J. Chem. Res. (S)*, **1987**, 414.
[225] Lund, T., Pederson, S. U., Lund, H., Cheung, K. M., and Utley, J. H. P., *Acta Chem. Scand.*, **41B**, 285 (1987).
[226] Gatti, N., Jugelt, W., and Lund, H., *Acta Chem. Scand.*, **41B**, 646 (1987).
[227] Yamataka, H., Fujimura, N., Kawafuji, Y., and Hanafusa, T., *J. Am. Chem. Soc.*, **109**, 4305 (1987).
[228] Maruyama, K. and Katagiri, T., *Chem. Lett.*, **87**, 731 (1987).
[229] Padwa, A., Goldstein, S. I., and Rosenthal, R. J., *J. Org. Chem.*, **52**, 3278 (1987).
[230] Shine, H. J. and Soroka, M., *Prepr.-Am. Chem. Soc., Div. Pet. Chem.*, **31**, 814 (1986); *Chem. Abs.*, **107**, 6696 (1987).
[231] Maruyama, K. and Katagiri, T., *Chem. Lett.*, **87**, 735 (1987).
[232] Holm, T., *Acta Chem. Scand.*, **41B**, 278 (1987).
[233] Tamarkin, D. and Rabinovitz, M., *J. Org. Chem.*, **52**, 3472 (1987).
[234] Ōkubo, M., Tsutsumi, T., and Matsuo, K., *Bull. Chem. Soc. Jpn*, **60**, 2085 (1987).
[235] Asensío, G., Gonzalez-Nuñez, E., and Rodrigo, M. J., *J. Chem. Soc., Chem. Commun.*, **1987**, 263.
[236] Bentz, G., Besbes, N., Laurent, A., and Stamm, H., *Tetrahedron Lett.*, **28**, 2511 (1987).
[237] Ashby, E. C., Coleman, D., and Gamasa, M., *J. Org. Chem.*, **52**, 4079 (1987).
[238] Fuentes, A., Marinas, J. M., and Sinisterra, J. V., *Tetrahedron Lett.*, **28**, 2947 (1987).
[239] Sinisterra, J. V., Fuentes, A., and Marinas, J. M., *J. Org. Chem.*, **52**, 3875 (1987).
[240] Yamataka, H. and Hanafusa, T., *Chem. Lett.*, **1987**, 643.
[241] Korth, H.-G., Sustmann, R., Dupuis, J., and Giese, B., *Chem. Ber.*, **120**, 1197 (1987).
[242] Mochida, K., *Bull. Chem. Soc. Jpn*, **60**, 3299 (1987).
[243] Eberson, L. and Radner, F., *Acc. Chem. Res.*, **20**, 53 (1987).
[244] Feng, J., Zheng, X., and Zerner, M. C., *J. Org. Chem.*, **51**, 4531 (1986).
[245] Sankararaman, S., Haney, W. A., and Kochi, J. K., *J. Am. Chem. Soc.*, **87**, 5235 (1987).
[246] Squadrito, G. L., Church, D. L., and Pryor, W. A., *J. Am. Chem. Soc.*, **109**, 6535 (1987).
[247] Baciocchi, E., Del Giacco, T., Murgia, S. M., and Sebastiani, G. V., *J. Chem. Soc., Chem. Commun.*, **1987**, 1246.
[248] Suzuki, H., Koide, H., Taki, Y., Ohbayashi, E., and Ogawa, T., *Chem. Lett.*, **87**, 891 (1987).

[249] Bauld, N. L., Bellville, D. J., Harirchian, G., Lorenz, K. T., Pabon, R. A., Reynolds, D. W., Wirth, D. D., Chiou, H.-S., and Marsh, B. K., *Acc. Chem. Res.*, **20**, 371 (1987).
[250] Lorenz, K. T. and Bauld, N. L., *J. Am. Chem. Soc.*, **109**, 1157 (1987).
[251] Reynolds, D. W., Lorenz, K. T., Chiou, H.-S., Belville, D. J., Pabon, R. A., and Bauld, N. L., *J. Am. Chem. Soc.*, **109**, 4960 (1987).
[252] Harirchian, B. and Bauld, N. L., *Tetrahedron Lett.*, **28**, 927 (1987).
[253] Mlcoch, J. and Steckhan, E., *Tetrahedron Lett.*, **28**, 1081 (1987).
[254] Todres, Z. V., *Tetrahedron*, **43**, 3839 (1987).
[255] [(a)]Miyashi, T., Kamata, M., and Mukai, T., *J. Am. Chem. Soc.*, **109**, 2780 (1987); [(b)]Curci, R., Lopez, L., Troisi, L., Rashid, S. M. K., and Schaap, A. P., *Tetrahedron Lett.*, **28**, 5319 (1987).
[256] Lapouyade, R., Villeneuve, P., Nourmamode, A., and Morand, J.-P., *J. Chem. Soc., Chem. Commun.*, **1987**, 776.
[257] Gassman, P. G. and Hershberger, J. W., *J. Org. Chem.*, **52**, 1337 (1987).
[258] Yamashita, Y., Ikeda, H., and Mukai, T., *J. Am. Chem. Soc.*, **109**, 6682 (1987).
[259] Eberson, L. and Larsson, B., *Acta Chem. Scand.*, **41B**, 367 (1987).
[260] Ishiguro, K., Sawaki, Y., and Iwamara, H., *Chem. Lett.*, **1987**, 1853.
[261] Dinnocenzo, J. P. and Schmittel, M., *J. Am. Chem. Soc.*, **109**, 1561 (1987).
[262] Ginzel, K.-D. and Steckhan, E., *Tetrahedron*, **43**, 5797 (1987).
[263] Hoz, S., Livneh, M., and Cohen, D., *J. Am. Chem. Soc.*, **109**, 5149 (1987); Hoz, S., Basch, H., and Cohen, D., *J. Am. Chem. Soc.*, **109**, 6891 (1987).
[264] Arnett, E. M. Molter, K. E., Marchot, E. C., Donovan, W. H., and Smith, P., *J. Am. Chem. Soc.*, **109**, 3788 (1987).
[265] Chowdhury, S. and Kebarle, P., *NATO ASI Ser., Ser. C.*, **1987**, 193; *Chem. Abs.*, **107**, 133727 (1987).
[266] Bodner, G. S., Galdysz, J. A., Nielsen, M. F., and Parker, V. D., *J. Am. Chem. Soc.*, **109**, 1757 (1987).
[267] Guthrie, R. D., Hartmann, C., Neill, R., and Nutter, D. E., *J. Org. Chem.*, **52**, 736 (1987).
[268] Dicken, C. M., Lu, F.-L., and Bruice, T. C., *Tetrahedron Lett.*, **27**, 5967 (1986).
[269] Wilkey, J. D. and Schuster, G. B., *J. Org. Chem.*, **52**, 2117 (1987).
[270] Choi, H., Pinkerton, A. A., and Fry, J. L., *J. Chem. Soc., Chem. Commun.*, **1987**, 225.
[271] Ishii, A., Okazaki, R., and Inamoto, N., *Bull. Chem. Soc. Jpn*, **60**, 1037 (1987).
[272] Fukuzumi, S., Ishikawa, K., and Tanaka, T., *Chem. Lett.*, **1986**, 1801.
[273] Kumamoto, T., Hosoya, K., Kanzaki, S., Masuko, K., Watanabe, M., and Shirai, K., *Bull. Chem. Soc. Jpn*, **59**, 3097 (1986).
[274] Itahara, T., *Tetrahedron Lett.*, **28**, 4079 (1987).
[275] Afanas'ev, I. B., Kupriyanova, N. S., and Grabovetskii, V. V., *Zh. Obshch. Khim.*, **56**, 1326 (1986); *Chem. Abs.*, **106**, 17708 (1987).
[276] Afanas'ev, I. B. and Polozova, N. I., *J. Chem. Soc., Perkin Trans. 2*, **1987**, 835.
[277] Forrester, A. R. and Purushotham, V., *J. Chem. Soc., Perkin Trans. 1*, **1987**, 945.
[278] Forrester, A. R. and Purushotham, V., *J. Chem. Soc., Faraday Trans. 1*, **83**, 211 (1987).
[279] Alfassi, Z. B., Huie, R. E., Mosseri, S., and Neta, P., *J. Phys. Chem.*, **91**, 3888 (1987).
[280] Tabushi, I., Kuroda, Y., and Sasaki, Y., *J. Chem. Soc., Chem. Commun.*, **1987**, 1622.
[281] Muneer, M., Tikare, R. K., Kamat, P. V., and George, M. V., *Can. J. Chem.*, **65**, 1624 (1987).
[282] Ashok, K., Scaria, P. M., Kamat, P. V., and George, M. V., *Can. J. Chem.*, **65**, 2039 (1987).
[283] Maruthamuthu, P., Venkatasubramanian, L., and Dharmalingham, P., *Proc. Indian Acad. Sci., Chem. Sci.*, **97**, 213 (1986); *Chem. Abs.*, **106**, 155730 (1987).
[284] Mattay, J., *Angew. Chem. Int. Edn*, **26**, 825 (1987).
[285] McCullough, J. J., *Chem. Rev.*, **87**, 811 (1987).
[286] Chang, S. L. P. and Schuster, D. I., *J. Phys. Chem.*, **91**, 3644 (1987).
[287] Truong, T. B. and Santamaria, J., *J. Chem. Soc., Perkin Trans. 2*, **1987**, 1.
[288] Albini, A., Fasani, E., and Faiardi, D., *J. Org. Chem.*, **52**, 155 (1987).
[289] Yasuda, M., Yamashita, T., Shima, K., and Pac, C., *J. Org. Chem.*, **52**, 753 (1987).
[290] Albini, A. and Spreti, S., *J. Chem. Soc., Perkin Trans. 2*, **1987**, 1175.
[291] Arnold, D. R. and Mines, S. A., *Can. J. Chem.*, **65**, 2312 (1987).
[292] Caër, Y., Laurent, A., Laurent, E., Tardivel, R., Celbulska, Z., and Bartnik, R., *Nouveau J. Chim.*, **11**, 351 (1987).
[293] Gassman, P. G. and Bottorff, K. J., *Tetrahedron Lett.*, **28**, 5449 (1987).
[294] Pandey, G., Krishna, A., and Kumaraswamy, G., *Tetrahedron Lett.*, **28**, 4615 (1987).
[295] Gassman, P. G. and Bottorff, K. J., *J. Am. Chem. Soc.*, **109**, 7547 (1987).
[296] Arnold, D. R., Fahie, B. J., Lamont, L. J., Wierzchowski, J., and Young, K. M., *Can. J. Chem.*, **65**, 2734 (1987).

[297] Gould, I. R., Ege, D., Mattes, S. L., and Farid, S., *J. Am. Chem. Soc.*, **109**, 3794 (1987).
[298] Lewis, F. D., *Acc. Chem. Res.*, **19**, 401 (1986).
[299] Mattay, J., Gersdorf, J., and Buchkremer, K., *Chem. Ber.*, **120**, 307 (1987).
[300] Kaise, M. and Someno, K., *Chem. Lett.*, **1987**, 1295.
[301] Hoshino, M. and Shizuka, H., *J. Phys. Chem.*, **91**, 714 (1987).
[302] Akiyama, K., Ikegami, Y., Ikenoue, T., and Tero-Kubota, S., *Bull. Chem. Soc. Jpn*, **59**, 3269 (1986).
[303] Freeman, P. K. and Srinivasa, R., *J. Org. Chem.*, **52**, 252 (1987).
[304] Tanaka, Y., Tsujimoto, K., and Ohashi, M., *Bull. Chem. Soc. Jpn*, **60**, 788 (1987).
[305] Tanaka, Y., Uryu, T., Ohashi, M., and Tsujimoto, K., *J. Chem. Soc., Chem. Commun.*, **1987**, 1703.
[306] Fery-Forgues, S., Lavabre, D., and Paillous, N., *J. Org. Chem.*, **52**, 3381 (1987).
[307] Blackstock, S. C., Lorand, J. P., and Kochi, J. K., *J. Org. Chem.*, **52**, 1451 (1987).
[308] Engel, P. S., Kitamura, A., and Keys, D. E., *J. Org. Chem.*, **52**, 5015 (1987).
[309] Cossy, J., Belotti, D., and Pete, J. P., *Tetrahedron Lett.*, **28**, 4545 (1987).
[310] Cossy, J., Belotti, D., and Pete, J. P., *Tetrahedron Lett.*, **28**, 4547 (1987).
[311] Seki, H., Takematsu, A., and Arai, S., *J. Phys. Chem.*, **91**, 176 (1987).
[312] Ishitani, O., Yanagida, S., Takamuku, S., and Pac, T., *J. Org. Chem.*, **52**, 2790 (1987).
[313] Fukuzumi, S., Ishikawa, K., Hironaka, K., and Tanaka, T., *J. Chem. Soc., Perkin Trans. 2*, **1987**, 751.
[314] Naguib, Y. M. A., Steel, C., Cohen, S. G., and Young, M. A., *J. Phys. Chem.*, **91**, 3033 (1987).
[315] Aoyama, Y., Midorikawa, K., Toi, H., and Ogoshi, H., *Chem. Lett.*, **1987**, 1651.
[316] Kano, K., Zhou, B., and Hashimoto, S., *Bull. Chem. Soc. Jpn*, **60**, 1041 (1987).
[317] Fukuzumi, S., Kuroda, S., and Tanaka, T., *J. Chem. Soc., Chem. Commun.*, **1987**, 120.
[318] Ishitani, O., Yanagida, S., Takamuku, S., and Pac, C., *Bull. Chem. Soc. Jpn*, **60**, 1801 (1987).
[319] Cano-Yelo, H. and Deronzier, A., *Nouveau J. Chim.*, **11**, 479 (1987).
[320] Bunce, N. J., Cater, S. R., Scaiano, J. C., and Johnston, L. J., *J. Org. Chem.*, **52**, 4214 (1987).
[321] Marayuma, K. and Tamiaki, H., *Bull. Chem. Soc. Jpn*, **60**, 1847 (1987).
[322] Cristol, S. J., Aeling, E. O., and Heng, R., *J. Am. Chem. Soc.*, **109**, 830 (1987).
[323] Tazuke, S., Kazama, S., and Kitamura, N., *J. Org. Chem.*, **51**, 4548 (1986).
[324] Ochiai, M., Kunishima, M., Nagao, Y., Fuji, K., and Fujita, E., *J. Chem. Soc., Chem. Commun.*, **1987**, 1708.
[325] Kropp, M. and Schuster, G. B., *Tetrahedron Lett.*, **28**, 5295 (1987).
[326] Takuwa, A., Tagawa, H., Iwamoto, H., Soga, O., and Maruyama, K., *Chem. Lett.*, **1987**, 1091.
[327] Mizuno, K., Toda, S., and Otsuji, Y., *Chem. Lett.*, **1987**, 203.
[328] Kubo, Y., Imaoka, T., Shiragami, T., and Araki, T., *Chem. Lett.*, **1986**, 1749.
[329] Tu, C.-T. and Mariano, P. S., *J. Am. Chem. Soc.*, **109**, 5287 (1987).
[330] Haddaway, K., Somekawa, K., Fleming, P., Tossell, J. A., and Mariano, P. S., *J. Org. Chem.*, **52**, 4239 (1987).
[331] Yoon, U.-C., Kim, J.-U., Hasegawa, E., and Mariano, P. S., *J. Am. Chem. Soc.*, **109**, 4421 (1987).
[332] Borg, R. M., Heuckeroth, R. O., Lan, A. J. Y., Quillen, S. L., and Mariano, P. S., *J. Am. Chem. Soc.*, **109**, 2728 (1987).
[333] Lan, A. J. Y., Heuckeroth, R. O., and Mariano, P. S., *J. Am. Chem. Soc.*, **109**, 2738 (1987).
[334] Dai-Ho, G. and Mariano, P. S., *J. Org. Chem.*, **52**, 704 (1987).
[335] Dai-Ho, G. and Mariano, P. S., *J. Org. Chem.*, **52**, 706 (1987).
[336] Ahmed-Schofield, R. and Mariano, P. S., *J. Org. Chem.*, **52**, 1478 (1987).
[337] Akaba, R., Aihara, S., Sakuragi, H., and Tokumaru, K., *J. Chem., Chem. Commun.*, **1987**, 1262.
[338] Nelsen, S. F., *Acc. Chem. Res.*, **20**, 269 (1987).
[339] Akasaka, T. and Ando, W., *J. Am. Chem. Soc.*, **109**, 1260 (1987).
[340] Miyashi, T., Takahashi, Y., Yokogawa, K., and Mukai, T., *J. Chem., Chem. Commun.*, **1987**, 175.
[341] Dektar, J. L. and Hacker, J. L., *J. Chem. Soc., Chem. Commun.*, **1987**, 1591.
[342] Saeva, F. D., *J. Chem. Soc., Chem. Commun.*, **1987**, 37.
[343] Devoe, R. J., Sahyun, M. R. V., Serpone, N., and Sharma, D. K., *Can. J. Chem.*, **65**, 2342 (1987).
[344] Fox, M. A., *Nouveau J. Chim.*, **11**, 129 (1987).
[345] Al-Ekabi, H. and de Mayo, P., *J. Org. Chem.*, **52**, 4756 (1987).
[346] Carson, P. A. and de Mayo, P., *Can. J. Chem.*, **65**, 976 (1987).
[347] Deronzier, A., Essakalli, M., and Moutet, J.-C., *J. Chem. Soc., Chem. Commun.*, **1987**, 773.
[348] Hartman, R. F., Van Camp, J. R., and Rose, S. D., *J. Org. Chem.*, **52**, 2684 (1987).
[349] Li, Y.-Z. and Schuster, G. B., *J. Org. Chem.*, **52**, 3975 (1987).
[350] Somich, C., Mazzochi, P. H., and Ammon, H. L., *J. Org. Chem.*, **52**, 3614 (1987).
[351] Courtneidge, J. L., Davies, A. G., Gregory, P. S., McGuchan, D., and Yazdi, S. N., *J. Chem. Soc., Chem. Commun.*, **1987**, 1192.
[352] De Ryck, P. H., Verdonck, L., Van Hooste, H., and Van der Kelen, G. P., *Bull. Soc. Chim. Belg.*, **95**, 1033 (1986).

[353] Malar, E. J. P. and Chandra, A. K., *J. Phys. Chem.*, **91**, 5111 (1987).
[354] Gould, I. R., Baretz, B. H., and Turro, N. J., *J. Phys. Chem.*, **91**, 925 (1987).
[355] Fouassier, J. P. and Lougnot, D. J., *J. Chem. Soc., Faraday Trans. 1*, **83**, 2935 (1987).
[356] Yagci, Y. and Schnabel, W., *Macromol. Chem. Rapid Commun.*, **8**, 209 (1987).
[357] Hayashi, H., Sakaguchi, Y., Kamachi, M., and Schnabel, W., *J. Phys. Chem.*, **91**, 3936 (1987).
[358] Turro, N. J., Doubleday, C., Hwang, K.-C., Cheng, C.-C., and Fehlner, J. R., *Tetrahedron Lett.*, **28**, 2929 (1987).
[359] Wagner, P. J. and Lindstrom, M. J., *J. Am. Chem. Soc.*, **109**, 3057 (1987).
[360] Wagner, P. J. and Lindstrom, M. J., *J. Am. Chem. Soc.*, **109**, 3062 (1987).
[361] Wagner, P. J., Giri, B. P., Pabon, R., and Singh, S. B., *J. Am. Chem. Soc.*, **109**, 8104 (1987).
[362] Osuka, A., Shimuzu, H., Suzuki, H., and Maruyama, K., *Chem. Lett.*, **1987**, 1061.
[363] Carless, H. A. J. and Mwesigye-Kibende, S., *J. Chem. Soc., Chem. Commun.*, **1987**, 1673.
[364] Aoyama, H., Arata, Y., and Omote, Y., *J. Org. Chem.*, **52**, 4639 (1987).
[365] Takuwa, A., Soga, O., Itoh, M., and Maruyama, K., *Chem. Lett.*, **1987**, 727.
[366] Geraghty, N. W. A., Monaghan, M. J., and Hanley, N. W., *Tetrahedron Lett.*, **28**, 4729 (1987).
[367] Vulpian, R., Bouteillfer, J.-C., Lauricella, R., and Aycard, J.-P., *Bull. Soc. Chim. Fr. II*, **1987**, 842.
[368] Ooms, P. and Hartmann, W., *Tetrahedron Lett.*, **28**, 4271 (1987).
[369] Sakurai, H., Sakamoto, K., and Kira, M., *Chem. Lett.*, **1987**, 983.
[370] Cohen, M. D., *Tetrahedron*, **43**, 1211 (1987).
[371] Ramamurthy, V. and Venkatesan, K., *Chem. Rev.*, **87**, 433 (1987).
[372] Ito, Y., Matsuura, T., Tabata, K., and Ji-Ben, T., *Tetrahedron*, **43**, 1307 (1987).
[373] Reddy, G. D., Jayasree, B., and Ramamurthy, V., *J. Org. Chem.*, **52**, 3107 (1987).
[374] Reddy, G. D. and Ramamurthy, V., *J. Org. Chem.*, **52**, 3952 (1987).
[375] Kaftory, M., *Tetrahedron*, **43**, 1513 (1987).
[376] Turro, N. J., *Tetrahedron*, **43**, 1589 (1987).
[377] Sonawane, H. R., Nanjundiah, B. S., Rajput, S. I., and Kumar, M. U., *Tetrahedron Lett.*, **27**, 6125 (1986).
[378] Nuñez, A. and Weiss, R. G., *J. Am. Chem. Soc.*, **109**, 6215 (1987).
[379] Treanor, R. L. and Weiss, R. G., *Tetrahedron*, **43**, 1371 (1987).
[380] Leigh, W. J. and Jakobs, S., *Tetrahedron*, **43**, 1393 (1987).
[381] Matlin, A. R. and McGarvey, D. J., *Tetrahedron Lett.*, **28**, 5087 (1987).
[382] Matlin, A. R., Leckta, T. C., McGarvey, D. J., Jacob, P. W., and Picken, H. A., *Tetrahedron Lett.*, **28**, 5083 (1987).
[383] Schröder, C., Wolff, S., and Agosta, W. C., *J. Am. Chem. Soc.*, **109**, 5491 (1987).
[384] Lupo, D. W. and Quack, M., *Chem. Rev.*, **87**, 181 (1987).
[385] McGimpsey, W. G. and Scaiano, J. C., *J. Am. Chem. Soc.*, **109**, 2179 (1987).
[386] Johnston, L. J. and Scaiano, J. C., *J. Am. Chem. Soc.*, **109**, 5487 (1987).
[387] Scaiano, J. C., McGimpsey, W. G., Leigh, W. J., and Jakobs, S., *J. Org. Chem.*, **52**, 4540 (1987).
[388] Leyva, E., Platz, M. S., Niu, B., and Wirz, J., *J. Phys. Chem.*, **91**, 2293 (1987).
[389] Sviridov, A. P. and Turovets, I. M., *Vestn. Mosk. Univ., Ser. 2: Khim.*, **27**, 470 (1986); *Chem. Abs.*, **106**, 155778 (1987).
[390] Skorobogatov, G. A., Slesar, O. N., and Torbin, N. D., *Kinet. Katal.*, **27**, 1055 (1986); *Chem. Abs.*, **106**, 119135 (1987).
[391] D'Auria, M., De Mico, A., D'Onofrio, F., and Piancatelli, G., *J. Chem. Soc., Perkin Trans. 1*, **1987**, 1777.
[392] McKenney, J. D. and Castle, R. N., *J. Heterocycl. Chem.*, **24**, 1525 (1987).
[393] Choudhry, G. G., Graham, N. J., and Webster, G. R. B., *Can. J. Chem.*, **65**, 2223 (1987).
[394] Ivanov, V. L., Eggert, L., and Kuzmin, M. G., *Zh. Org. Khim.*, **22**, 1493 (1986); *Chem. Abs.*, **106**, 118991 (1987).
[395] Saito, I., Ikehira, H., and Matsuura, T., *J. Org. Chem.*, **51**, 5148 (1986).
[396] Foster, B., Gaillard, B., Mathur, N., Pincock, A. L., Pincock, J. A., and Sehmbey, C., *Can. J. Chem.*, **65**, 1599 (1987).
[397] Chow, Y. L. and Zhao, D.-C., *J. Org. Chem.*, **52**, 1931 (1987).
[398] Barton, D. H. R., Bridon, D., and Zard, S. Z., *Tetrahedron*, **43**, 5307 (1987).
[399] Castagnino, E., Corsano, S., Barton, D. H. R., and Zard, S. Z., *Tetrahedron Lett.*, **27**, 6337 (1986).
[400] Barton, D. H. R., Bridon, D., and Zard, S. Z., *Heterocycles*, **25**, 449 (1987).
[401] Barton, D. H. R., Bridon, D., Fernandez-Picot, I., and Zard, S. Z., *Tetrahedron*, **43**, 2733 (1987).
[402] Hasebe, M. and Tsuchiya, T., *Tetrahedron Lett.*, **28**, 6207 (1987).
[403] Sugimori, A. and Yamada, T., *Bull. Chem. Soc. Jpn*, **59**, 3911 (1986).
[404] Soundararajan, N. and Platz, M. J., *Tetrahedron Lett.*, **28**, 2813 (1987).
[405] Ishitani, O., Yanagida, S., Takamuku, S., and Pac, C., *Bull. Chem. Soc. Jpn*, **60**, 2419 (1987).

406 Balla, R. J., Nelson, H. H., and McDonald, J. R., *NBS Spec. Publ. (U.S.)*, **716**, 30 (1986); *Chem. Abs.*, **106**, 195631 (1987).
407 Suginome, H., Washiyama, H., and Yamada, S., *Bull. Chem. Soc. Jpn*, **60**, 1071 (1987).
408 Lally, J. M. and Spillane, W. J., *J. Chem. Soc., Chem. Commun.*, **1987**, 8.
409 Ghosh, S., Das, T. K., Datta, D. B., and Mehta, S., *Tetrahedron Lett.*, **28**, 4611 (1987).
410 Ming, Y. and Wamhoff, H., *Chem. Ber.*, **120**, 1433 (1987).
411 Ogura, K., Itoh, S., Takahashi, K., and Ida, H., *Tetrahedron Lett.*, **27**, 6381 (1987).
412 Bentrude, W. G., Lee, S.-G., Akutagawa, K., Ye, W., and Charbonnel, Y., *J. Am. Chem. Soc.*, **109**, 1577 (1987).
413 Leigh, W. J. and Srinivasan, R., *Acc. Chem. Res.*, **20**, 107 (1987).
414 Tada, M., Nakamura, T., and Matsumoto, M., *Chem. Lett.*, **1987**, 409.
415 Sato, E., Ikeda, Y., and Kanaoka, Y., *Chem. Lett.*, **1987**, 273.
416 Brown, S. H. and Crabtree, R. H., *J. Chem. Soc., Chem. Commun.*, **1987**, 970.
417 Majeed, N. N. and Porte, A. L., *J. Chem. Soc., Perkin Trans. 2*, **1987**, 1139.
418 Carlacci, L., Doubleday, C., Furlani, T. R., King, H. F., and McIver, J. W., *J. Am. Chem. Soc.*, **109**, 5323 (1987).
419 Adam, W., Günther, E., Hössel, P., and Platsch, H., *Tetrahedron Lett.*, **28**, 4407 (1987).
420 Adam, W., Grabowski, S., Marshall, R. M., Hannemann, K., and Wirz, J., *J. Am. Chem. Soc.*, **109**, 7572 (1987).
421 Adam, W., Hössel, P., Hümmer, W., Platsch, H., and Wilson, R. M., *J. Am. Chem. Soc.*, **109**, 7570 (1987).
422 Closs, G. L. and Forbes, M. D. E., *J. Am. Chem. Soc.*, **109**, 6185 (1987).
423 Zimmt, M. B., Doubleday, C., and Turro, N. J., *Chem. Phys. Lett.*, **134**, 549 (1987); *Chem. Abs.*, **107**, 58376 (1987).
424 Ikegami, Y., Muramatsu, T., Hanaya, K., Onodera, S., Nakayama, N., and Kosower, E. M., *J. Am. Chem. Soc.*, **109**, 2876 (1987).
425 Muramatsu, T., Hanaya, K., and Ikegami, Y., *Chem. Lett.*, **1986**, 2139.
426 Du, P. and Borden, W. T., *J. Am. Chem. Soc.*, **109**, 930 (1987).
427 Dowd, P., Chang, W., and Paik, Y. H., *J. Am. Chem. Soc.*, **109**, 5284 (1987).
428 Roth, W. R., Kowalczik, U., Maier, G., Reisenauer, H. P., Sustmann, R., and Müller, W., *Angew. Chem. Int. Edn*, **26**, 1285 (1987).
429 Roth, W. R., Langer, R., Bartmann, M., Stevermann, B., Maier, G., Reisenauer, H. P., Sustmann, R., and Müller, W., *Angew. Chem. Int. Edn*, **26**, 256 (1987).
430 Zilm, K. W., Merrill, R. A., Greenberg, M. M., and Berson, J. A., *J. Am. Chem. Soc.*, **109**, 1567 (1987).
431 Greenberg, M. M., Blackstock, S. C., and Berson, J. A., *Tetrahedron Lett.*, **28**, 4263 (1987).
432 Grimm, M., Kirste, B., and Kurreck, H., *Angew. Chem. Int. Edn*, **25**, 1097 (1987).
433 Bonačić-Koutecký, V., Koutecký, J., and Mich, J., *Angew. Chem. Int. Edn*, **26**, 170 (1987).
434 Wang, Y.-S. and Chickos, J. S., *J. Org. Chem.*, **52**, 4776 (1987).
435 Creary, X., Mehrsheikh-Mohammadi, M. E., and McDonald, S., *J. Org. Chem.*, **52**, 3254 (1987).
436 Padias, A. P. and Hall, H. K., *J. Org. Chem.*, **52**, 4536 (1987).
437 Wiberg, K. G. and Waddell, S. T., *Tetrahedron Lett.*, **28**, 151 (1987).
438 Campbell, K. A., House, H. O., Surber, B. W., and Trahanovsky, W. S., *J. Org. Chem.*, **52**, 2474 (1987).
439 Kamphuis, J., Bos, H. J. T., Visser, R. J., Huizer, B. H., and Varma, C. A. G. O., *J. Chem. Soc., Perkin Trans. 2*, **1986**, 1867.
440 Furuhata, T. and Ando, T., *Tetrahedron Lett.*, **28**, 1179 (1987).
441 Hijfte, L. V., Little, R. D., Petersen, J. L., and Moeller, K. D., *J. Org. Chem.*, **52**, 4647 (1987).
442 Myers, A. G., *Tetrahedron Lett.*, **28**, 4493 (1987).
443 Fisher, J. J. and Michl, J., *J. Am. Chem. Soc.*, **109**, 583 (1987).
444 Zadok, E., Rubinraut, S., and Mazur, Y., *J. Org. Chem.*, **52**, 385 (1987).
445 White, R. C. and Rix, T., *J. Org. Chem.*, **52**, 2309 (1987).
446 Gaudino, G. and Koch, T. H., *J. Org. Chem.*, **52**, 3073 (1987).
447 Morrow, G. R. and Rae, I. D., *Aust. J. Chem.*, **40**, 1477 (1987).
448 Holland, K. A. and Rae, I. D., *Aust. J. Chem.*, **40**, 687 (1987).
449 J. S. Lomas, *J. Org. Chem.*, **52**, 2627 (1987).
450 Rubin, M. B. and Harel, Y., *Tetrahedron Lett.*, **28**, 5373 (1987).
451 Gordon, M. S., Truong, T. N., and Pople, J. A., *Chem. Phys. Lett.*, **130**, 245 (1986); *Chem. Abs.*, **106**, 101631 (1987).
452 Rumyantsev, A. N. and Oganesova, E. Yu., *Khim. Tekhnol. Topl. Masel*, **1987**, 29; *Chem. Abs.*, **107**, 77155 (1987).

453 Billaud, F. and Baronnet, F., *J. Anal. Appl. Pyrolysis*, **10**, 181 (1987); *Chem. Abs.*, **107**, 133751 (1987).
454 Rumyantsev, A. N. and Oganesova, E. Yu., *Khim. Tekhnol. Topl. Masel*, **1987**, 26; *Chem. Abs.*, **107**, 197318 (1987).
455 Marshall, I. G., *Int. J. Chem. Kinet.*, **19**, 649 (1987).
456 Gusel'nikov, L. E., Volkova, V. V., Zigler, U., Zimmermann, G., Ondruska, B., Ivanov, P. E., and Shevel'kova, L. V., *Izv. Akad. Nauk SSSR, Ser. Khim.*, **1986**, 2152; *Chem. Abs.*, **106**, 155774 (1987).
457 Savage, P. E. and Klein, M. T., *Ind. Eng. Chem. Res.*, **26**, 374 (1987); *Chem. Abs.*, **106**, 101634 (1987).
458 Hopf, H. and Wachholz, G., *Chem. Ber.*, **120**, 1259 (1987).
459 Christl, M. and Schreck, M., *Angew. Chem. Int. Edn*, **26**, 449 (1987).
460 Kraakman, P. A., Nibbering, E. T. J., De Wolf, W. H., and Bickelhaupt, F., *Tetrahedron*, **43**, 5109 (1987).
461 Aref'eva, E. F. and Tesner, P. A., *Kinet. Katal.*, **28**, 184 (1987); *Chem. Abs.*, **107**, 39006 (1987).
462 Stein, S. E., Griffith, L. L., Billmers, R., and Chen, R. H., *J. Org. Chem.*, **52**, 1582 (1987).
463 Schneider, M. and Wolfrum, J., *Ber. Bunsen-Ges. Phys. Chem.*, **90**, 1058 (1986); *Chem. Abs.*, **106**, 213216 (1987).
464 Cifuentes, L. O., Casas, F., and Aguilera, F. A., *Bol. Soc. Quim. Peru*, **51**, 224 (1985); *Chem. Abs.*, **107**, 38995 (1987).
465 Davidson, I. M. T. and Dean, C. E., *Organometallics*, **6**, 966 (1987).
466. Bain, S., Ijadi-Maghsoodi, S., and Barton, T. J., *Organometallics*, **6**, 1136 (1987).
467 Al'minderov, V. V., Milikh, G. M., and Trakhtenberg, L. I., *Zh. Fiz. Khim.*, **61**, 833 (1987); *Chem. Abs.*, **107**, 197332 (1987).
468 LeRoux, E. and Baronnet, F., *J. Chim. Phys., Phys.-Chim. Biol.*, **83**, 569 (1986); *Chem. Abs.*, **107**, 22853 (1987).
469 Batelaan, J. G., Weber, A. J. M., and Wiersum, U. E., *J. Chem. Soc., Chem. Commun.*, **1987**, 1397.
470 Kharitonova, I. A., *Tr. Inst.-Mosk. Khim.-Tekhnol. Inst. im. D. I. Mendeleeva*, **138**, 26 (1985); *Chem. Abs.*, **107**, 133744 (1987).
471 Chandra, H., Symons, M. C. R., Korth, H.-G., and Sustmann, R., *Tetrahedron Lett.*, **28**, 1455 (1987).
472 Bryantseva, N. V., Lugovoi, Yu. M., Shostenko, A. G., and Glebov, A. S., *Khim. Vyv. Energ.*, **20**, 553 (1986); *Chem. Abs.*, **106**, 155663 (1987).
473 Lal, M., Mönig, J. and Asmus, K.-D., *J. Chem. Soc., Perkin Trans. 2*, **1987**, 1639.
474 Asmus, K.-D., Williams, P. S., Gilbert, B. C., and Winter, J. N., *J. Chem. Soc., Chem. Commun.*, **1987**, 208.
475 Zegota, H., Schuchmann, M. N., Schulz, D., and von Sonntag, C., *Z. Naturforsch.*, **41B**, 1015 (1986); *Chem. Abs.*, **107**, 22904 (1987).
476 Das, S., Schuchmann, M. N., Schuchmann, H.-P., and von Sonntag, C., *Chem. Ber.*, **120**, 319 (1987).
477 Anderson, R. F., Patel, K. B., and Stratford, M. L. R., *J. Chem. Soc., Faraday Trans. 1*, **83**, 3177 (1987).
478 Steenken, S., *J. Chem. Soc., Faraday Trans. 1*, **83**, 113 (1987).
479 Ács, G. and Péter, A., *Int. J. Chem. Kinet.*, **19**, 929 (1987).
480 Adams, J. S., Burton, K. A., Andrews, B. K., Weirman, R. B., and Engel, P. S., *J. Am. Chem. Soc.*, **108**, 7935 (1986).
481 Edmunds, A. J. F. and Samuel, C. J., *J. Chem. Soc., Chem. Commun.*, **1987**, 1179.
482 Adam, W. and Dörr, M., *J. Am. Chem. Soc.*, **109**, 1240 (1987).
483 Engel, P. S. and Bodager, G. A., *J. Org. Chem.*, **51**, 4792 (1986).
484 Adam, W., Hannemann, K., Peters, E.-M., Peters, K., von Schnering, H. G., and Wilson, R. M., *J. Am. Chem. Soc.*, **109**, 5250 (1987).
485 Schmittel, M. and Rüchardt, C., *J. Am. Chem. Soc.*, **109**, 2750 (1987).
486 Luedtke, A., Meng, K., and Timberlake, J. W., *Tetrahedron Lett.*, **28**, 4255 (1987).
487 Peyman, A., Hickl, E., and Beckhaus, H.-D., *Chem. Ber.*, **120**, 713 (1987).
488 Adam, W., Dörr, M., Kron, J., and Rosenthal, R. J., *J. Am. Chem. Soc.*, **109**, 7074 (1987).
489 Bambal, R., Fritz, H., Rihs, G., Tschamber, T., and Streith, J., *Angew. Chem. Int. Edn*, **26**, 668 (1987).
490 Adam, W. and Dörr, M., *J. Am. Chem. Soc.*, **109**, 1570 (1987).
491 Ando, W., Kumamoto, Y., and Tokitoh, N., *Tetrahedron Lett.*, **27**, 6107 (1987).
492 Quast, H. and Meichsner, G., *Chem. Ber.*, **120**, 1049 (1987).
493 Stepanov, R. S., Kruglyakova, L. A., and Buka, E. S., *Kinet. Katal.*, **27**, 479 (1986); *Chem. Abs.*, **105**, 225617 (1987).
494 Suehiro, T., Masuda, S., Nakausa, R., Toguchi, M., Mori, A., Koike, A., and Date, M., *Bull. Chem. Soc. Jpn*, **60**, 3321 (1987).
495 Kobayashi, M., Gotoh, M., and Yoshida, M., *Bull. Chem. Soc. Jpn*, **60**, 295 (1987).

[496] Doyle, M. P., Guy, J. K., Brown, K. C., Mahapatro, S. N., Van Zyl, C. M., and Pladziewiecz, J. R., *J. Am. Chem. Soc.*, **109**, 1536 (1987).
[497] Abeywickrema, A. N. and Beckwith, A. L. J., *J. Org. Chem.*, **52**, 2568 (1987).
[498] Abeywickrema, A. N. and Beckwith, A. L. J., *J. Am. Chem. Soc.*, **108**, 8227 (1987).
[499] Petrillo, G., Novi, M., Garbarino, G., and Dell'Erba, C., *Tetrahedron*, **42**, 4007 (1986).
[500] Novi, M., Petrillo, G., and Dell'Erba, C., *Tetrahedron Lett.*, **28**, 1345 (1987).
[501] Petrillo, G., Novi, M., Garbarino, G., and Dell'Erba, C., *Tetrahedron*, **43**, 4625 (1987).
[502] Novi, M., Petrillo, G., and Sartirana, M. L., *Tetrahedron Lett.*, **27**, 6129 (1987).
[503] Liu, Y. and Guo, Q., *Huaxue Xuebao*, **43**, 1172 (1985); *Chem. Abs.*, **105**, 225558 (1987).
[504] Sigman, M. E., Barbas, J. T., and Leffler, J. E., *J. Org. Chem.*, **52**, 1754 (1987).
[505] Lee, S.-G., *J. Chem. Soc., Chem. Commun.*, **1987**, 1115.
[506] Mahmood, A. J., Islam, T. S. A., Quader, M. A., and Chowdhury, G. K., *J. Bangladesh Acad. Sci.*, **10**, 239 (1986); *Chem. Abs.*, **107**, 38998 (1987).
[507] Kobrina, L. S., Sever, A. V., and Yakobson, G. G., *Izv. Sib. Otd. Akad. Nauk SSSR, Ser. Khim. Nauk*, **1986**, 91; *Chem. Abs.*, **106**, 137761 (1987).
[508] Lomölder, R. and Schäfer, H. J., *Angew. Chem. Int. Edn*, **26**, 1253 (1987).
[509] Zhavnerko, G. K. and Agabekov, V. E., *Vesti Akad. Navuk SSSR, Ser. Khim. Navuk*, **1986**, 36; *Chem. Abs.*, **107**, 77153 (1987).
[510] Xhao, C., Chen, G., Wang, X., and Jiang, X., *J. Fluorine Chem.*, **31**, 417 (1986); *Chem. Abs.*, **106**, 175663 (1987).
[511] Chernyak, B. I. and Kosyak, Z. M., *Neftekhimiya*, **26**, 810 (1986); *Chem. Abs.*, **107**, 58313 (1987).
[512] Nikonov, A. L., Efremova, E. P., Chikhacheva, I. P., Kartsev, G. N., and Stavrova, S. D., *Zh. Obshch. Khim.*, **59**, 1320 (1986); *Chem. Abs.*, **106**, 4389 (1987).
[513] Yoshida, M., Shimokoshi, K., and Kobayashi, M., *Chem. Lett.*, **1987**, 433.
[514] Kitagawa, T., Takeuchi, K., Murai, O., Matsui, S., Inoue, T., Nishimara, M., and Okamoto, K., *J. Chem. Soc., Perkin Trans. 2*, **1986**, 1987.
[515] Kurhanewicz, J. and Jurch, G. R., *J. Am. Chem. Soc.*, **109**, 5038 (1987).
[516] Zhavnerko, G. K., Agabekov, V. E., Beresnevich, L. B., and Moiseichuk, K. L., *Dokl. Akad. Nauk BSSR*, **31**, 239 (1987); *Chem. Abs.*, **107**, 197327 (1987).
[517] Citterio, A. and Filippini, L., *Synthesis*, **1986**, 473; *Chem. Abs.*, **106**, 101434 (1987).
[518] Butovskaya, G. V., Supichenko, G. N., Beresnevich, L. B., Agabekov, V. E., Moiseichuk, K. L., Antonovskii, V. L., and Mitskevich, N. I., *Vestsi Akad. Nauk BSSR, Ser. Khim. Nauk*, **1986**, 26; *Chem. Abs.*, **106**, 155719 (1987).
[519] Mokrii, E. N., Suprun, V. Ya., Melnik, S. R., and Kucher, R. V., *Dopov. Akad. Nauk Ukr. RSR, Ser. B: Geol. Khim. Biol. Nauki*, **1986**, 37; *Chem. Abs.*, **106**, 155754 (1987).
[520] Lipes, V. V., Bakova, E. A., and Morozova, V. I., *Kinet. Katal.*, **27**, 570 (1986); *Chem. Abs.*, **106**, 83857 (1987).
[521] Zhan, Z. and Wan, X., *Gaodeng Xuexiao Huaxue Xuebao*, **7**, 156 (1986); *Chem. Abs.*, **107**, 77151 (1987).
[522] Macova, E. and Lederer, P., *Collect. Czech. Chem. Commun.*, **51**, 2830 (1986).
[523] Dulog, D. and Bleher, R., *Makromol. Chem.*, **187**, 2357 (1986); *Chem. Abs.*, **105**, 225550 (1987).
[524] Frankel, E. N., Neff, W. E., Selke, E., and Brooks, D. D., *Lipids*, **22**, 322 (1987).
[525] Porter, N. A. and Wujek, J. S., *J. Org. Chem.*, **52**, 5085 (1987).
[526] González, A., Galindo, A., Mansilla, H., and Trigos, A., *Tetrahedron Lett.*, **28**, 4203 (1987).
[527] Lipes, V. V., *Kinet. Katal.*, **27**, 1046 (1986); *Chem. Abs.*, **106**, 119134 (1987).
[528] Lakani, C., Mathew, L., and Warkentin, J., *Can. J. Chem.*, **65**, 1748 (1987).
[529] Tezuka, T. and Ando, S., *Chem. Lett.*, **1986**, 1671.
[530] Montaudon, E., Agorrody, M., Rakotomanana, F., and Maillard, B., *Bull. Soc. Chim. Belg.*, **96**, 769 (1987).
[531] Montaudon, E., Rakotomanana, F., Agorrody, M., and Maillard, B., *Bull. Soc. Chim. Fr. II*, **1987**, 842.
[532] Agorrody, M., Campagnole, M., Montaudon, E., and Maillard, B., *Bull. Soc. Chim. Belg.*, **96**, 3429 (1987).
[533] Agorrody, M., Montaudon, E., and Maillard, B., *Can. J. Chem.*, **65**, 2694 (1987).
[534] Takahashi, K. and Kishi, M., *J. Chem. Soc., Chem. Commun.*, **1987**, 722.
[535] Richardson, W. H., Stiggal-Estberg, D. L., Chen, Z., Baker, J. C., Burns, D. M., and Sherman, D. G., *J. Org. Chem.*, **52**, 3143 (1987).
[536] Baumstark, A. L., Retter, C. A., Tehrani, K., and Kellogg, C., *J. Org. Chem.*, **52**, 3308 (1987).
[537] Cafferata, L. F. R., *An. Asoc. Quim. Argent.*, **74**, 501 (1986); *Chem. Abs.*, **107**, 115146 (1987).
[538] Van Gompel, J. and Schuster, G. B., *J. Org. Chem.*, **52**, 1465 (1987).

539 White, E. H., Roswell, D. F., Dupont, A. C., and Wilson, A. A., *J. Am. Chem. Soc.*, **109**, 5189 (1987).
540 Shereshovets, V. V., Sharipov, G. L., Kabal'nova, N. N., Kazakov, V. P., Komissarov, V. D., and Tolstikov, G. A., *Zh. Org. Khim.*, **22**, 2549 (1986); *Chem. Abs.*, **107**, 153792 (1987).
541 Safiullin, R. L., Nikolaev, A. I. Enikeeva, L. R., and Komissarov, V. D., *Izv. Akad. Nauk SSSR, Ser. Khim.*, **1986**, 1690; *Chem. Abs*, **106**, 155766 (1987).
542 Chan, H. W.-S., (Ed.), *Autoxidation of Unsaturated Lipids*, Academic Press, 1987.
543 Kucher, R. V. and Timokhin, Y. I., Neftekhimiya, **26**, 737 (1986); *Chem. Abs.*, **107**, 6531 (1987); Opeida, I. A., *Neftekhimiya*, **26**, 745 (1987); *Chem. Abs.*, **107**, 6532 (1987).
544 Brede, O., Hermann, R., and Mehnert, R., *J. Chem. Soc., Faraday Trans. 1*, **83**, 2365 (1987).
545 Zborovskii, A. G., Petrov, Yu. P., Smirnov, V. N., Tereza, A. M., and Tyurin, A. N., *Khim. Fiz.*, **6**, 506 (1987); *Chem. Abs.*, **107**, 115155 (1987).
546 Cosgrove, J. P., Church, D. F., and Pryor, W. A., *Lipids*, **22**, 299 (1987).
547 Corey, E. J. and Nagata, R., *J. Am. Chem. Soc.*, **109**, 8107 (1987).
548 Frank, H., Wiegand, M., Strecker, M., and Thiel, D., *Lipids*, **22**, 689 (1987).
549 Lambelet, P., Ducret, F., Saucy, F., Savoy, M.-C., and Löliger, J., *J. Chem. Soc., Faraday Trans. 1*, **83**, 141 (1987).
550 Sies, H., *Angew. Chem. Int. Edn*, **25**, 1058 (1986).
551 Erben-Russ, M., Michel, C., Bors, W., and Saran, M., *J. Phys. Chem.*, **91**, 2362 (1987).
552 Awl, R. A., Frankel, E. N., and Weisleder, D., *Lipids*, **22**, 721 (1987).
553 Hronec, M., Masarovič, F., Cvengrošová, Z., and Ilavský, J., *Collect. Czech. Chem. Commu.*, **52**, 2241 (1987).
554 Kozlov, S. K. and Potekhin, V. M., *Zh. Prikl. Khim. (Leningrad)*, **59**, 1633 (1986); *Chem. Abs.*, **106**, 119076 (1987).
555 Matsubara, Y., Takekuma, S., Yokoi, K., Yamamoto, H., and Nozoe, T., *Bull. Chem. Soc. Jpn*, **60**, 1415 (1987).
556 Bruce, J. M., Kerr, C. M., and Dodd, N. J. F., *J. Chem. Soc., Faraday Trans. 1*, **83**, 85 (1987).
557 Hartmut, B., Stolze, K., and Scheffler, K., *Z. Naturforsch.*, **41B**, 776 (1986); *Chem. Abs.*, **106**, 83890 (1987).
558 Brown, H. C. and Midland, M. M., *Tetrahedron*, **43**, 4059 (1987).
559 Girotti, A. W., Bachowski, G. J., and Jordan, J. E., *Lipids*, **22**, 401 (1987).
560 Sevaniau, A. and McLeod, L. L., *Lipids*, **22**, 627 (1987).
561 Anthoni, U., Larsen, C., Nielsen, P. H., and Christophersen, C., *Acta Chem. Scand.*, **41B**, 216 (1987).
562 Kopecky, K. R. and Molina, J., *Can. J. Chem.*, **65**, 2350 (1987).
563 Gordienok, N. I., Freidin, B. G., and Proskurina, L. S., *Zh. Prikl. Khim. (Leningrad)*, **59**, 1549 (1986); *Chem. Abs.*, **106**, 119075.
564 D'Souza, V. T., Nanjundiah, R., Baeza, H. J., and Szmant, H. H., *J. Org. Chem.*, **52**, 1729 (1987).
565 Broka, A. and Reichert, D. E. C., *Tetrahedron Lett.*, **28**, 1503 (1987).
566 D'Souza, V. T., Iyer, V. K., and Szmant, H. H., *J. Org. Chem.*, **52**, 1725 (1987).
567 Chung, M.-I., D'Souza, V. T., and Szmant, H. H., *J. Org. Chem.*, **52**, 1741 (1987).
568 Giles, D. W., Cha, J. A., and Lim, P. K., *Chem. Eng. Sci.*, **41**, 3129 (1986); *Chem. Abs.*, **107**, 197271 (1987).
569 Tavadyan, L. A., Mardoyan, V. A., and Nalbandyan, A. B., *Khim. Fiz.*, **5**, 1377 (1986); *Chem. Abs.*, **106**, 195804 (1987).
570 Mosseri, S., Alfassi, Z. B., and Neta, P., *Int. J. Chem. Kinet.*, **19**, 309 (1987).
571 Martinez, R. I., *J. Phys. Chem.*, **91**, 1345 (1987).
572 Shoppee, C. W. and Wu, W.-Y., *Aust. J. Chem.*, **40**, 1137 (1987).
573 Aleksandrov, A. L., *Izv. Akad. Nauk SSSR, Ser. Khim.*, **1986**, 1736; *Chem. Abs.*, **106**, 195700 (1987).
574 Grossi, L., *Tetrahedron Lett.*, **28**, 3387 (1987).
575 Varlamov, V. T. and Denisov, E. T., *Izv. Akad. Nauk SSSR, Ser. Khim.*, **1986**, 2417; *Chem. Abs.*, **107**, 58341 (1987).
576 Mukai, K., Yokoyama, S., Fukuda, K., and Uemoto, Y., *Bull. Chem., Soc. Jpn*, **60**, 2163 (1987).
577 Abe, Y., Okada, S., Horii, H., Taniguchi, S., and Yamabe, S., *J. Chem. Soc., Perkin Trans. 2*, **1987**, 715.
578 Takahashi, M., Niki, E., Kawakami, A., Kumasaka, A., Yamamoto, Y., Kamiya, Y., and Tanaka, K., *Bull. Chem. Soc. Jpn*, **59**, 3179 (1986).
579 Denisov, E. T. and Khudyakov, I. V., *Chem. Rev.*, **87**, 1313 (1987).
580 Dmitruk, A. F., Lobanov, V. V., and Kholoimova, L. I., *Theor. Eksp. Khim.*, **22**, 363 (1986); *Chem. Abs.*, **106**, 155828 (1987).
581 Kosyak, Z. M. and Chernyak, B. I., *Dopov. Akad. Nauk Ukr. RSR, Geol, Khim. Biol. Nauki*, **1986**, 33; *Chem. Abs.*, **107**, 197212 (1987).
582 Henryk, Z., Schuchmann, M. N., Schulz, D., and von Sonntag, C., *Z. Naturforsch.*, **41B**, 1015 (1986).

[583] Basco, N. and Parmar, S. S., *Int. J. Chem. Kinet.*, **19**, 115 (1987).
[584] Alfassi, Z. B., Mosseri, S., and Neta, P., *J. Phys. Chem.*, **91**, 3383 (1987).
[585] Fahr, A. and Laufer, A. H., *J. Am. Chem. Soc.*, **109**, 3843 (1987).
[586] Sawaki, Y. and Ishikawa, S., *J. Am. Chem. Soc.*, **109**, 584 (1987).
[587] Sehested, K., Holcman, J., Bjergbakke, E., and Hart, E. J., *J. Phys. Chem.*, **91**, 2359 (1987).
[588] Roduner, E., *Prog. React. Kinet.*, **14**, 1 (1986); *Chem. Abs.*, **106**, 195563 (1987).
[589] Wu, L., *Huaxue Tongbao.*, **1987**, 15; *Chem. Abs.*, **107**, 133610 (1987).
[590] Stadlbauer, J. M., Ng, B. W., and Walker, D. C., *Hyperfine Interact.*, **32**, 721 (1986); *Chem. Abs.*, **107**, 115059 (1987).
[591] Louwrier, P. W. F., Brinkman, G. A., and Roduner, E., *Hyperfine Interact.*, **32**, 831 (1986); *Chem. Abs.*, **107**, 22887 (1987).
[592] Jean, Y. C., Ganti, R. L., Cheng, K. L., Venkateswaran, K., and Walker, D. C., *Hyperfine Interact.*, **32**, 813 (1986); *Chem. Abs.*, **107**, 39046 (1987).
[593] Ganti, R. L. and Jean, Y. C., *Hyperfine Interact.*, **32**, 819 (1986); *Chem. Abs.*, **107**, 39047 (1987).
[594] Rhodes, C. J., Symons, M. C. R., Scott, C. A., Roduner, E., and Heming, M., *J. Chem. Soc., Chem. Commun.*, **1987**, 447,448.

Organic Reaction Mechanisms 1987
Edited by A. C. Knipe and W. E. Watts

CHAPTER 5

Oxidation and Reduction

G. W. J. FLEET

Dyson Perrins Laboratory, South Parks Road, Oxford

Oxidation by Metal Ions and Related Species

Chromium, Manganese, and Technetium

Oxidations of alcohols to carbonyl compounds by 3-carboxypyridinium[1] and imidazolium[2] dichromates and by potassium chlorochromate[3] have been investigated. A chromium(IV) species (**1**) is produced in the oxidations of alcohols[4] and of substituted toluenes (Scheme 1)[5] with pyridinium fluorochromate. Pyridinium dichromate induces oxidative rearrangement of acetylenic cyclohexenols (Scheme 2).[6] *tert*-Butylhydroperoxide with a catalytic amount of chromium(VI) anhydride causes efficient oxidations of alcohols[7] and benzylic methylene groups[8] to carbonyl compounds. A chromium(V) species has been identified as an intermediate in the oxidation of tartaric acid by dichromate;[9] the oxidation of α-hydroxy-acids by dichromate involves formation of a cyclic chromate ester which

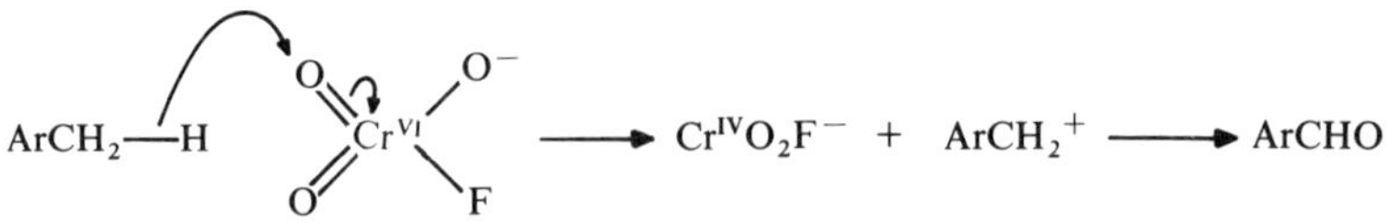

(1)

SCHEME 1

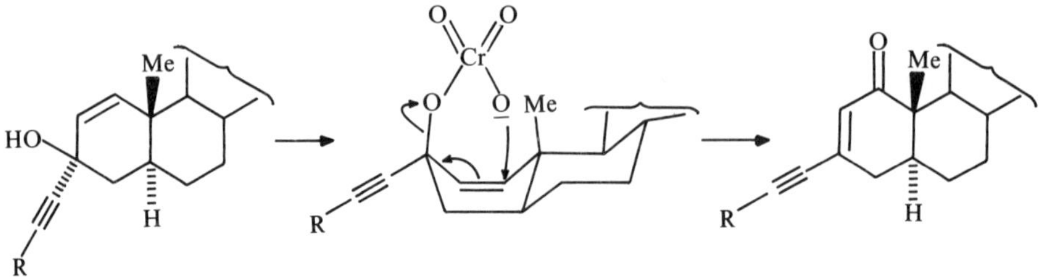

SCHEME 2

decomposes in the rate-determining step by oxidative decarboxylation.[10] The kinetics of the co-oxidation of oxalic acid and aryl sulphoxides with chromium(VI) have been studied.[11] An unusual and regiospecific oxidation was observed in the reactions of secododecahedrane (**2**) or of its monoepoxide (**3**) with chromate to give (**4**) (Scheme 3).[12] Mechanistic studies on the chromium(VI) oxidations of arylmethyl

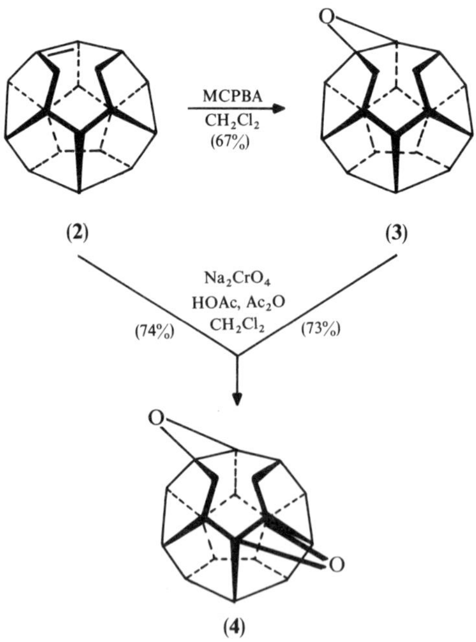

SCHEME 3

sulphides,[13] olefins,[14] and (phenylazo)dimethylaniline[15] have been reported. Solid supports with small surface areas and large pore diameter give potassium dichromate-supported reagents with higher activity than supports with high surface area and small pores.[16]

Kinetic studies on the permanganate oxidations of hydroxy-acids,[17,18] carboxylic acids,[19] and hydrocarbons[20] have been reported. Several competing reactions have been identified in the permanganate oxidation of formic acid.[21] Formation of permanganic acid in an anhydrous medium may be involved in the oxidation of acetic acid by methyltributylammonium permanganate in methylene chloride.[22] Radicals are intermediates in the oxidation of unsaturated carboxylic acids by methyltributylammonium permanganate in methylene chloride; a reactive manganese(V) cyclic diester undergoes a rapid free-radical reduction to manganese(III).[23] Permanganate oxidation of hindered homoannular dienes gives epoxy-diols as products, whereas unhindered substrates give rise to carbon–carbon bond cleavage.[24] A pathway has been proposed for the oxidative ring-cleavage of 3-(1-nitro-2-oxocyclohexyl)propanal by permanganate (Scheme 4).[25] Cetyltrimethylammonium permanganate has been used as a reagent for the selective oxidative cleavage of aryl-substituted olefins.[26] The oxidation of methylamine by permanganate is autocatalysed by colloidal manganese dioxide; the catalytic pathway takes place by reaction of permanganate ion from solution with methylammonium ions previously adsorbed on the colloid surface.[27] The autocatalytic

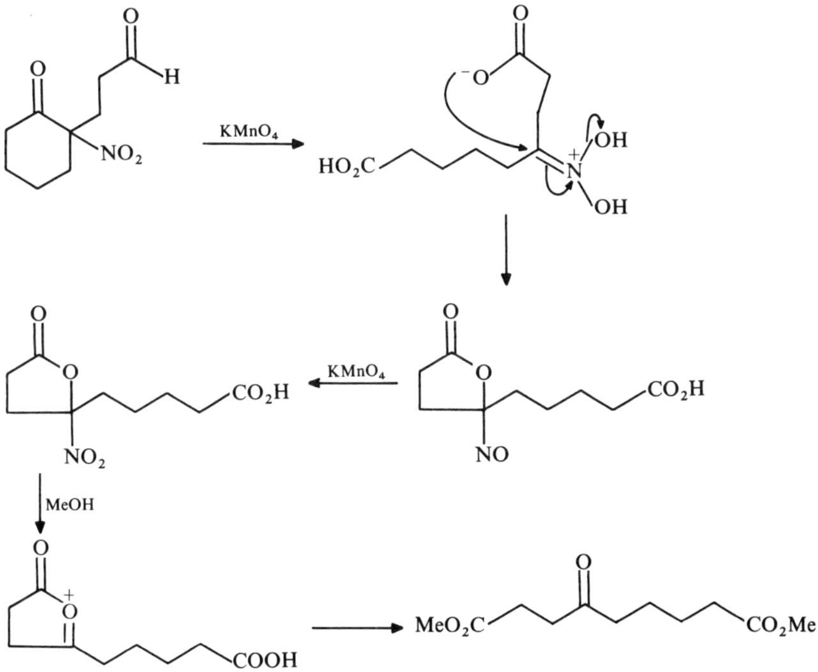

SCHEME 4

role of colloidal manganese dioxide in the permanganate oxidations of some α,β-unsaturated esters[28] and amino-acids[29,30] has been discussed.

The oxidations of organic compounds by active manganese dioxide[31] and by manganese(III) acetate[32] have been reviewed. The oxidation of α-hydroxy-acids with manganese(III) proceeds by initial complex formation followed by disproportionation with carbon-carbon bond cleavage.[33] The oxidation of propanol by manganese(III)[34] and the reaction of olefins with α-cyanoacetamide in the presence of manganese(III)[35] have been investigated. The manganese(III)-initiated oxidative free-radical polycyclization of β-keto-esters such as (**5**) may be terminated by copper(II) oxidative β-hydride elimination (Scheme 5).[36] Manganese(III) is the active

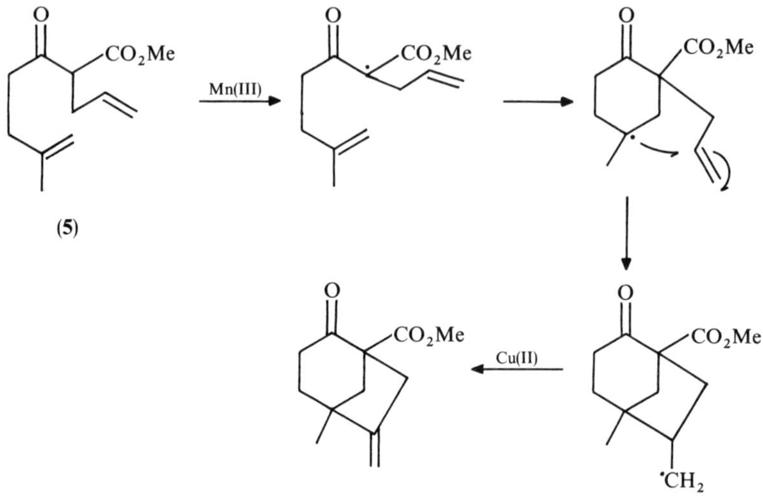

SCHEME 5

species in the manganese-catalysed oxidation of 2-hydroxynaphthaldehyde thiosemicarbazone, causing the generation of superoxide ion.[37] A dimeric manganese(II)-histidine complex is a catalyst for the oxidation of indigo carmine by hydrogen peroxide.[38] Manganese(II) catalysts are uniquely effective for the formation of ibuprofen (**6**) by oxidation of the corresponding aldehyde, possibly due to the efficacy of manganese in reducing peracids to acids thereby suppressing competing Baeyer–Villiger reactions.[39] Other manganese(II) catalysed oxidations have been studied.[40–42]

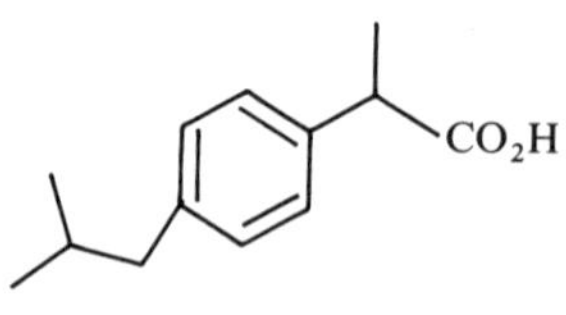

(**6**)

The epoxidation of alkenes catalysed by manganese porphyrins may provide suitable models for the mode of action of cytochrome P-450. Studies on the epoxidation of 2,3-dimethylbut-2-ene by *tert*-butylhydroperoxide catalysed by imidazole-ligated TPP [(*meso*-tetraphenylporphinato)manganese(III)] have demonstrated that both an intermediate imidazole-ligated higher-valent manganese oxoporphyrin species and another intermediate capable of transferring oxygen to the olefin are present.[43,44] The rate enhancement observed on addition of styrene in the epoxidation of aliphatic alkenes by the monooxygenase model (TPP)manganese(III) chloride–sodium hypochlorite is caused by phenylacetaldehyde produced under the reaction conditions.[45] Both iron(II) and manganese(III) porphyrins are effective catalysts for the oxidation of nitroso to nitro compounds.[46] A manganese(IV) porphyrin, capable of epoxidizing olefins in the absence of any auxiliary oxidant, has been characterized.[47]

The epoxidation of olefins by hydroperoxides catalysed by several technetium compounds has been studied.[48]

Copper, Silver, and Gold

Kinetic studies have appeared on copper(III) oxidations of 1,2-diols[49] and nitrophenols,[50] and on copper(II) catalysis of chloramine-T oxidation of aspartic acid.[51] Copper-catalysed allylic oxidation of *cis*- and *trans*-pent-2-ene by *tert*-butyl peroxyacetate affords *trans*-4-acetoxypent-2-ene as the major product; an initially formed allylic radical reacts with copper(II) carboxylate to form a copper(III) intermediate (**7**) in the direction forming the most stable alkene; (**7**) fragments in a pericyclic reaction to give the product.[52] Di-*tert*-butyl peroxide is an efficient oxidant for the copper(I)-catalysed carbonylation of alcohols to dialkyl carbonates or succinic esters;[53] copper(II) alkoxides may be intermediates in such processes.[54]

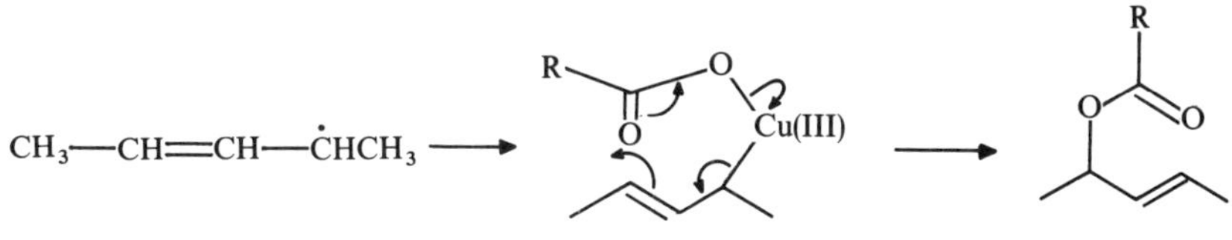

(**7**)

The oxidation of organic compounds by copper-amine complexes has been reviewed.[55] No ring-cleavage products are found in the dehydrogenation of indolines to indoles by chloro(pyridine)copper(I) complexes.[56] The biomimetic complex copper(II)-tetrapyridine, a non-enzymic analogue of galactose oxidase, catalyses the oxidation of α-keto-alcohols by oxygen; three different pathways have been identified for the oxidation of monohydroxyacetone.[57] In the study of a series of copper chelate catalysts, such as (**8**), for the oxidation of cyclohexene, the activity of the catalysts increased in going from a square planar to a distorted tetrahedral chelate; the chelates initiated the oxidation by accelerating the decomposition of cyclohexene hydroperoxide.[58] The effect of copper and other salts on the oxidation of ethylbenzene has been reported.[59]

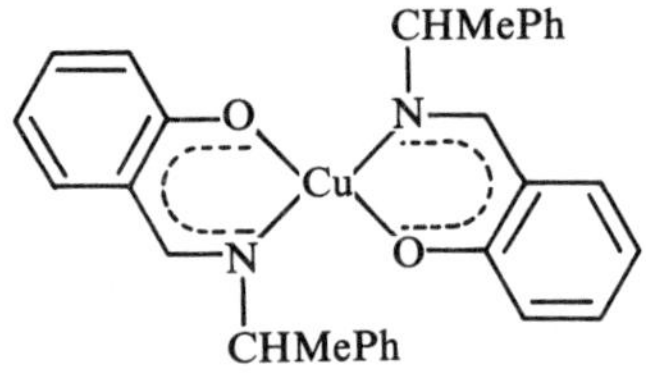

(8)

An extensive review of the use of Fetizon's reagent (silver carbonate on celite) has appeared.[60] The oxidation of allyl alcohol on silver(110), covered with a monolayer of atomic oxygen, involves an alkoxy intermediate which affords hydrogen, water, and acrolein.[61] A two-electron transfer mechanism involving the formation and dimerization of an α-hydroxy-nitroso compound has been proposed for the oxidative deoximation of benzaldoximes with diperiodatoargentate(III).[62] The oxidation of aldoses with tetrachloroaurate(III) proceeds via one-electron-transfer processes.[63]

Thallium, Lead, and Bismuth

The reactions of thallium salts with organic compounds have been classified as one- or two-electron oxidations.[64] Lead tetraacetate oxidations of organic compounds have been reviewed.[65] In studies on the pyridine-catalysed oxidation of substituted methyl mandelates by lead tetraacetate, a Hammett ρ-value of $+0.75$ indicates rate-limiting proton transfer of the benzylic C—H;[66] ρ-value of -1.00 was found for the lead tetraacetate oxidation of benzylic alcohols.[67] Reaction of 1-naphthol with lead tetraacetate gives 4-acetoxylation as the major pathway.[68] Oxidative intramolecular cyclization of germacrone **(9)** by lead(IV) gives products derived from **(10)**.[69] The lead tetraacetate oxidation of 3-aminoquinazolone **(11)** to give *O*-acylhydroxylamine **(12)** has been studied; **(12)** causes aziridination of alkenes by a mechanism that does not involve *N*-nitrene or *N*-nitrenium ion intermediates.[70]

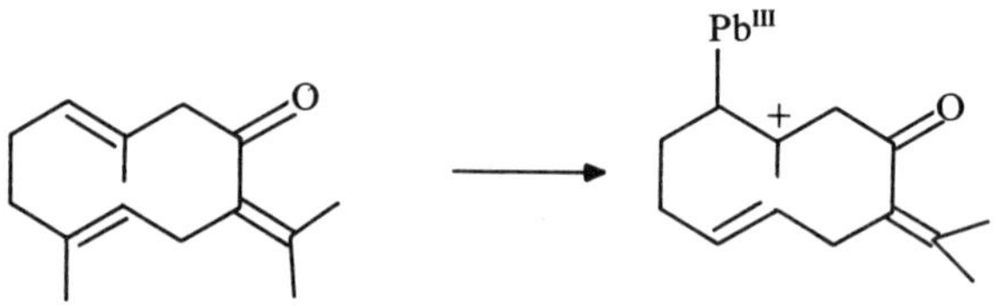

(9) **(10)**

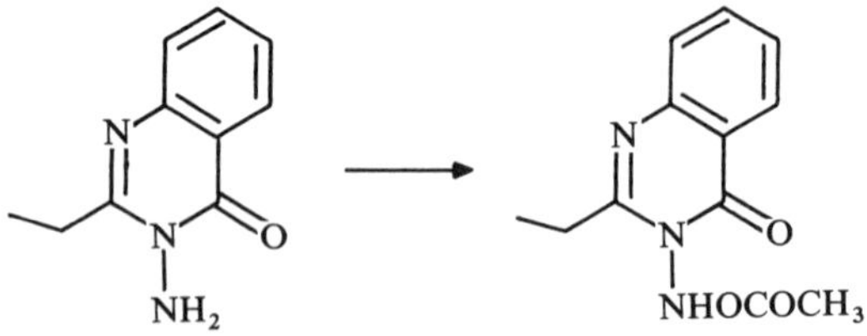

(11) **(12)**

The use of bismuth(v) in the oxidation of organic substrates has been discussed.[71,72] Several possible intermediates for the bismuth(v) oxidation of 2,6-dialkylphenols have been considered.[73]

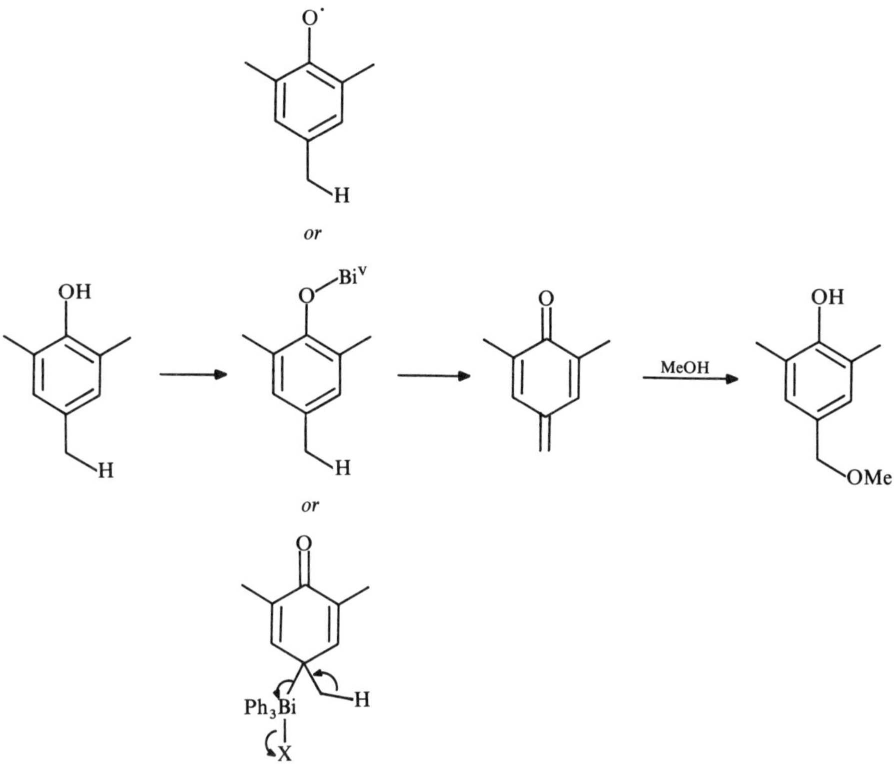

SCHEME 6

Cerium, Titanium, Vanadium, Molybdenum, and Tungsten

Cerium(IV) oxidations of organic compounds have been reviewed.[74] Mechanistic aspects of the oxidations of alkanes,[75] alkyl-aromatic compounds,[76] methylcyclohexanones,[77] and nitroanilines[78] with cerium(IV) have been investigated. Two complexes are formed in the cerium(IV) oxidation of monosaccharides; the first complex is formed in a pre-equilibrium reaction, the dissociation of which involves partly oxidation of the first complex and partly formation of a second more stable complex which undergoes slow oxidation.[79] The reactive species in the cerium(IV) oxidation of malonic acid and malonate ester are $Ce(SO_4)_2$ and $Ce(SO_4)_3^{2-}$.[80] Among other studies on the Belousov–Zhabotinskii reaction,[81,82] an oscillating reaction without the presence of strong acid was found in the potassium bromate oxidation of malonic acid with cerium(III) or manganese(II).[83]

Studies of the electronic and steric factors controlling the asymmetric epoxidation of allylic alcohols by titanium tartrate complexes by the frontier orbital approach have indicated that 2 two-electron interactions determine the orientation and reactivity of the alkene part of the allylic alcohol: (i) interaction of the peroxyoxygen lone pair with the π^* orbital of the alkene, and (ii) interaction of the titanium-peroxyoxygen antibonding orbital with the π orbital of the alkene.[84] A wide range of binding modes of tartrate esters to titanium(IV) has been demonstrated.[85] The titanium(IV)-pyridinediol complex (**13**) has been studied as a catalytic model for the asymmetric epoxidation reaction.[86] Use of 3A or 4A molecular sieves substantially increases the scope of titanium(IV)-catalysed asymmetric epoxidation of allylic alcohols; the role of molecular sieves and the effects of reaction stoichiometry, oxidant, and tartrate on the course of the reaction have been discussed.[87] Enantioselectivity in the asymmetric oxidation of sulphides by Sharpless reagents in enhanced by the use of cumene hydroperoxide rather than *tert*-butyl hydroperoxide.[88]

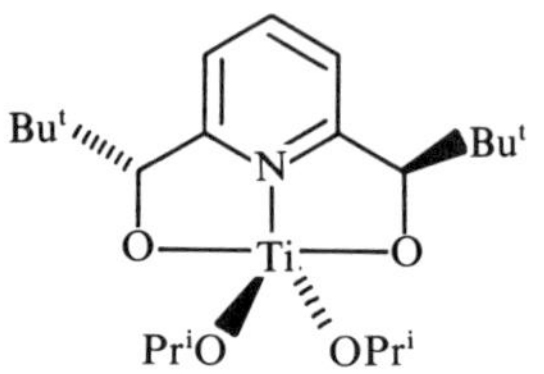

(**13**)

Mechanisms involving radical intermediates have been suggested for the vanadium(V) oxidations of pentanone[89] and galacturonic acid.[90] Studies on the catalytic action of vanadium(II) on the oxidation of cyclohexene[91] and of vanadium and molybdenum oxides on the oxidation of benzene[92] have been reported. Different types of catalysis by vanadium(V) and molybdenum(VI) have been identified in the bromate-iodide oxidation of ascorbate.[93] An anionic molybdenum-picolinate *N*-oxido-peroxo complex is an effective catalyst for the oxidation of alcohols in non-polar solvents.[94] The role of molybdenum catalysis in the hydrogen peroxide oxidations of sulphoxides to sulphones[95] and of acetylacetone[96] have been

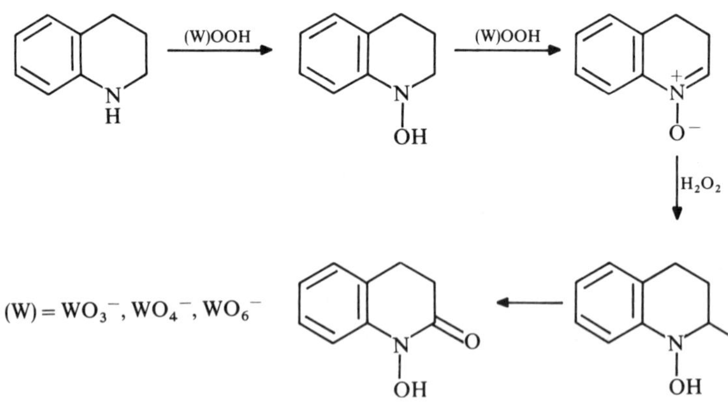

SCHEME 7

elucidated. Peroxomolybdenum(VI) complexes bearing mono- and bi-dentate ligands show asymmetric induction in the epoxidation of *trans*-2-octene.[97] The effect of ligand structure on the molybdenum-catalysed epoxidation of alkenes has been studied.[98-100] An identical mechanism is involved in the stoichiometric oxidation of olefins by peroxomolybdenum and peroxotungsten complexes.[101] The highest turnovers so far reported for the epoxidation of olefins with hydrogen peroxide under homogeneous catalysis were found using catalytic trioctylmethylammonium tungstate complexed by lipophilic phosphorus ligands containing a P=O unit.[102] The tungstate-catalysed oxidation of 1,2,3,4-tetrahydroquinolines with hydrogen peroxide gives 1-hydroxy-2-oxo-1,2,3,4-tetrahydroquinolines (Scheme 7).[103] Aspects of oxidations of organic compounds by peroxo complexes[104,105] have been reviewed.

Group VIII Metals

Superoxide is the key oxidant for iron in the selective oxidation of saturated hydrocarbons under Gif conditions (iron cluster-zinc-oxygen-acetic acid).[106] Kinetic studies have been reported on the ferrate oxidation of aldehydes,[107] and on the hexacyanoferrate oxidations of diphenylamines,[108] *p*-methoxyacetophenone,[109] and thioglycollic acid.[110] Various substituted ferrocenes accelerate the oxidation of cumene and the decomposition of cumeme hydroperoxide.[111]

Rearrangements known to proceed through alkene cation radical intermediates have been found to accompany hemin-catalysed olefin epoxidation; the radical cations are formed by one-electron transfer from the olefin to a high-valent iron intermediate.[172] An *N*-alkyl-hemin is formed during the epoxidation by addition of the hemin catalyst to the alkene; this transient species is a catalyst for epoxidation but is not an intermediate.[113] Many models studies for cytochrome P-450 have been reported.[114,115] A coherent theoretical model for the mechanism of stereoselective epoxidation of alkenes by oxo-iron porphyrins has been presented in which the first step is the binding of the alkene to the iron centre in an orientation where the alkene is perpendicular to Fe=O bond; from this complex, the epoxide is formed by a slipping displacement of the alkene towards the oxygen which leads to a favourable interaction between the alkene π^* orbital and a lone pair on the oxygen.[116] Anhydrous iron(III) chloride in acetonitrile catalyses demethylation of *N,N*-dimethylaniline, epoxidation of olefins, and cleavage of 1,2-diols by hydrogen peroxide, and the products from each reaction closely parallel those from enzymatic oxidation of the substrates by cytochrome P-450; the form of the reactive intermediate in this system may be the same as that in the enzyme-catalysed reactions since there is a close product similarity and acetonitrile provides a solvent analogous to the proteinaceous lipid matrix of the biomembrane. The mechanism of the oxidation proceeds by Lewis acid interaction of iron(III) chloride with hydrogen peroxide to give a highly electrophilic iron(III)-oxene as the reactive intermediate; this view is in contrast to the prevailing idea that cytochrome P-450 acts as a redox catalyst to generate an iron(V)-oxo species or an iron(IV)-oxo cation radical.[117] Ethylenediamine tetraacetate-iron(III) reacts with peroxycarboxylic acids and alkyl hydroperoxides resulting in transfer of oxygen to iron(III) by heterolytic oxygen–oxygen bond-cleavage.[118] The direct observation of oxygen–oxygen bond-cleavage in the formation of an oxoiron(IV) porphyrin radical cation has been reported.[119] Investigations[120,121] of the mechanism of oxygen transfer in the reaction of *p*-cyano-*N*,*N*-dimethylaniline *N*-oxide with metalloporphyrin salts have provided evidence for intermediate iron(IV)-oxo porphyrin π cation radicals.[122] Deuterium isotope studies have been reported on the hydroxylation of geraniol by cytochrome P-450[123] and on the one-electron reduction of diacetylcytochrome P-450.[124] Metal oxo-species may be involved in oxygen-transfer reactions mediated by iron-bleomycin complexes.[125]

The use of cobalt compounds, including peroxo and superoxo complexes, as oxidizing agents for organic compounds has been discussed.[126] Two pathways have been found for oxygen-atom transfer in cobalt-catalysed epoxidation of olefins; a radical chain mechanism occurs with *tert*-butyl hydroperoxide as oxidant, but a two-electron pathway involving an oxocobalt(IV) complex occurs with iodosylbenzene as oxidant.[127] The highly selective cobalt(III)-catalysed epoxidation of oleic acid with oxygen in the presence of benzaldehyde proceeds via a radical-chain pathway.[128] Magnetic field effects on the cobalt-catalysed oxidation of 2,6-di-*tert*-butylphenol have been investigated.[129] Mechanistic studies have been reported on cobalt-catalysed oxidations of hydroquinone,[130] succinic acid,[131] *N*-benzylamides,[132] ionones,[133] *p*-xylene,[134,135] and methylbenzenes.[136]

Ruthenium tetraoxide oxidations of organic substrates have been reviewed.[137] The ruthenium-catalysed transformation of alcohols to ketones and of diols to lactones is enhanced by addition of a hydrogen acceptor such as acetone (Scheme 8); the key step is the oxidative addition of ruthenium to the O—H bond of the alcohol and a subsequent β-elimination of ruthenium hydride to give the corresponding carbonyl species.[138] Dichlorotris(triphenylphosphine)ruthenium catalyses the oxidation of secondary alcohols by *N*-methylmorpholine *N*-oxide; the rate-determining step of the reaction is attack of a ruthenium(IV)-oxo complex on the substrate.[139] A catalytic amount of ruthenium(III) also causes oxidation of alcohols to ketones by *N*-methylmorpholine *N*-oxide; in this case, a ruthenium(V)-oxo complex may be involved.[140] Other methods of producing a ruthenium(V)-oxo complex for the oxidation of alcohols have been studied.[141] Ruthenium(III) catalyses the oxidation of allyl alcohol by oxygen.[142] A primary kinetic isotope of k_H/k_D of over 50 has been observed for the oxidation of alcohols by hydride transfer, catalysed by the ruthenium oxide complex (**14**).[143] The catalytic oxidation of vicinal diols to diacids by ruthenium pyrochlore oxides may involve oxoruthenium surface groups.[144] The regioselectivity of ruthenium tetraoxide oxidation of *N*-alkyl-lactams has been

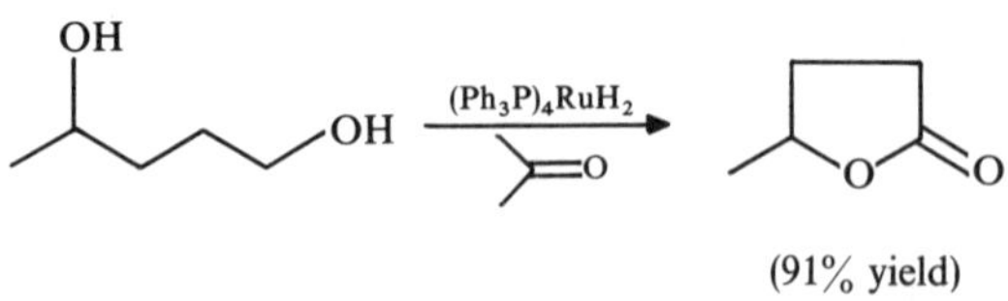

SCHEME 8

$[(bpy)_2(py)Ru(O)]^{2+}$

(14)

pby = 2,2′-bipyridyl py = pyridine

investigated.[145] The rate-determing step in the oxidation of 1,2-diols by hexacyanoferrate(III) catalysed by ruthenium(VIII) is fragmentation of a complex between the diol and ruthenium to the aldehydes and a ruthenium hydride species which is oxidized by iron(III).[146,147] Studies on the ruthenium-catalysed oxidations of sulphides and sulphoxides,[148] and α-hydroxy-acids[149] have been reported.

Oxidation of organic compounds with palladium reagents[150] and the role of palladium in liquid-phase oxidation[151] have been discussed. The biomimetic aerobic oxidation of 1,3-dienes by cobalt tetraphenylporphyrin-hydroquinone-palladium(II) provides an example of triple catalysis (Scheme 9).[152,153] Terminal alkenes are oxidized to methyl ketones by palladium(II)-hydroquinone catalysis (Scheme 10).[154] Palladium catalyses the dehydrogenation of alcohols via allyl carbonates (Scheme 11).[155] Mechanistic aspects of the palladium catalysis of oxidations of ethylene[156,157] and methylbenzenes[158] have been discussed.

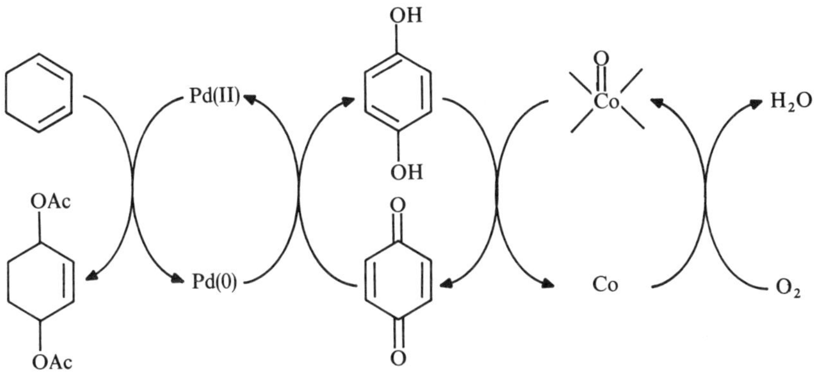

SCHEME 9

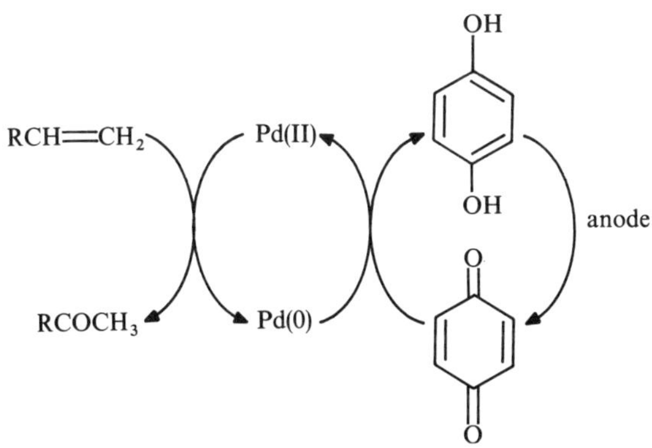

SCHEME 10

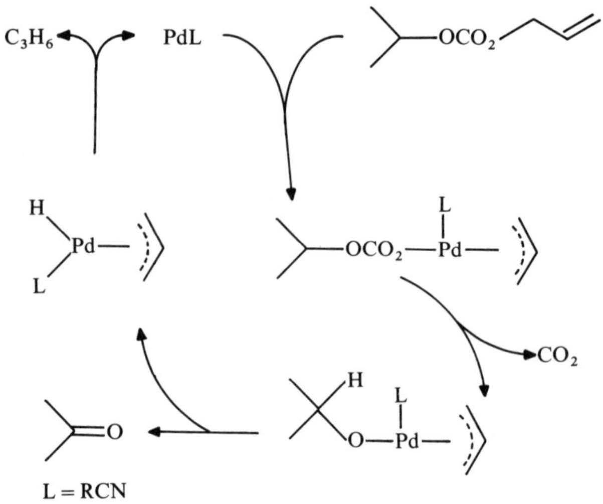

SCHEME 11

The oxidations of organic compounds with osmium tetraoxide have been reviewed.[159] Asymmetric dihydroxylation of β-substituted *N*-(α,β-enoyl)bornane-10,2-sultams (**15**) by osmium tetraoxide/*N*-methylmorpholine *N*-oxide proceeds in a highly stereospecific and π-face-selective manner.[160] The osmium tetraoxide

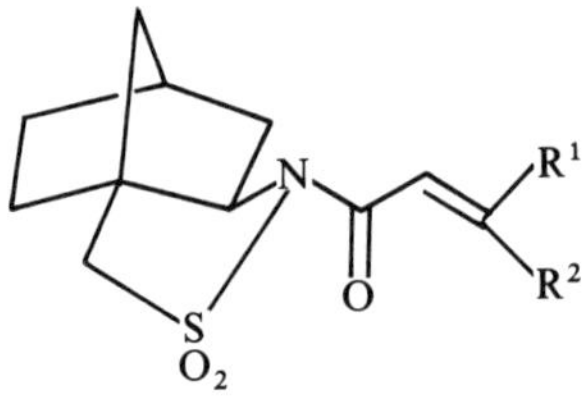

(**15**)

hydroxylation of chiral allylic β-hydroxy-sulphoxides can be highly stereoselective, depending on the nature of the substituent linked to the double bond and on the configurations of the double bond and the sulphoxide.[161] The reaction of osmium tetraoxide with a conjugated isomer of avermectin aglycone with a free hydroxyl group at C(13) occurs with high regio- and facial selectivity.[162] The oxidation of isopyrene (**16**) by osmium tetraoxide and other oxidizing agents has been

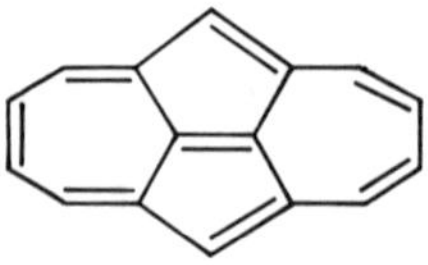

(**16**)

investigated.[163] Studies on the osmium(VIII) oxidations of ketones,[164] substituted glycolic acids,[165] styrenes,[166] and benzoin[167] have been described.

The use of nickel peroxide in organic chemistry[168] has been reviewed and the role of nickel and other metal stearates on the initial stages of the oxidation of *n*-pentadecane[169] has been discussed. The porphyrin-nickel(II) catalysis of olefin epoxidation has been investigated with iodosylbenzene as the terminal oxidant; the catalytic activity has been tentatively ascribed to an active oxo-nickel(IV) intermediate that is rather indiscriminate in oxygen-atom transfer and in homolytic attack on solvent and ligand.[170] The oxidations of alcohols by nickel(IV)[171] and of aldehydes by nickel(III)[172] have been studied. The mechanism of oxidation of murexide by periodate, catalysed by iridium(IV), has been investigated.[173]

Oxidation by Compounds of Non-metallic Elements

Nitrogen, Sulphur, and Selenium

The oxidation of propan-2-ol by nitric acid proceeds in two consecutive steps, the first giving acetone and the second giving acetic and formic acids.[174] The oxidation of the tertiary alcohol 2-phenylpropan-2-ol (**17**) by nitrous acid involves rate-determining dehydration to α-methylstyrene; the oxidation of 1-arylethanols (**18**) proceeds by two competing pathways: direct hydride abstraction by NO^+ and dehydration followed by oxidation of the substituted styrene.[175] Studies on the nitrobenzene oxidation of *m*- and *p*-hydroxybenzyl alcohols indicate that quinonemethides are not intermediates in these oxidations; accordingly, it may be unlikely that quinonemethides are involved in lignin oxidations.[176] The fast and selective oxidations of alcohols to carbonyl compounds by sodium hypochlorite catalysed by oxo-ammonium salts under two-phase conditions is co-catalysed by bromide.[177] MNDO calculations have been used to analyse the regioselectivity of Fremy's salt oxidations of phenols.[178] Mechanistic aspects of the photo-oxidation of alkanes with *N*-oxides[179] and of the oxidation of hydrazines by nitrosobenzene[180] have been investigated.

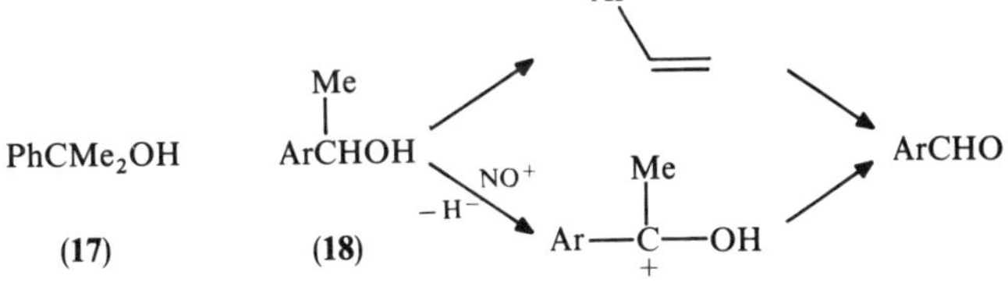

(17) (18)

A mechanism has been suggested for the sulphuryl chloride oxidation of several gibberellin epoxides to α,β-unsaturated aldehydes (Scheme 12).[181] Unlike the Swern oxidant, dimethyl sulphoxide activated by trifluoroacetic anhydride gives good yields of α-dicarbonyl compounds from vicinal diols, including substrates with halogen substituents.[182] Various thiocarbonyl compounds undergo desulphurization to the corresponding carbonyl compound with *tert*-butyl thionitrate (**19**) as

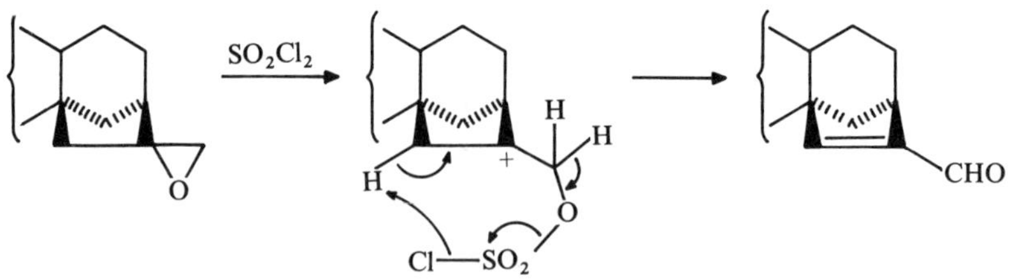

SCHEME 12

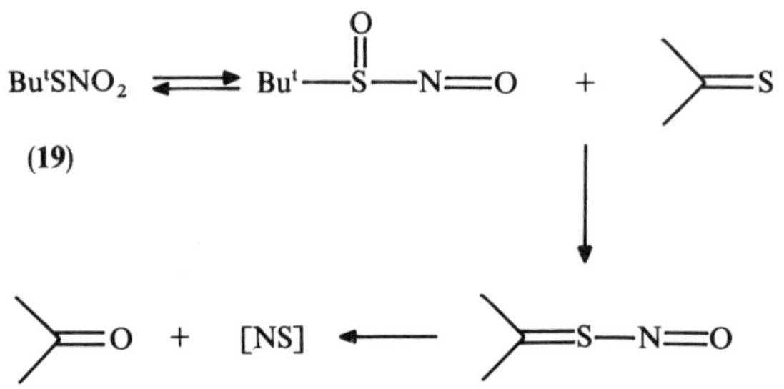

(19)

shown.[183] The mechanism of peroxydisulphate oxidation of phenyl-substituted carboxylic acids has been investigated.[184]

An initial [4 + 2] cycloaddition reaction between 2,4-alkadienoic esters and selenium dioxide gives 5-alkyl-2-furan carboxylic esters and 5-alkyl-2-selenophen carboxylic esters (Scheme 13).[185] The selenium dioxide oxidation of hydrazines such as (**20**) causes cleavage of the bicyclic system.[186] The major products of the reaction of α,β-unsaturated esters with hydrogen peroxide, catalysed by benzeneseleninic acid, are formyloxiranes which are subject to further oxidation.[187] The selenium dioxide-catalysed hydrogen peroxide oxidation of secondary amines affords nitrones.[188]

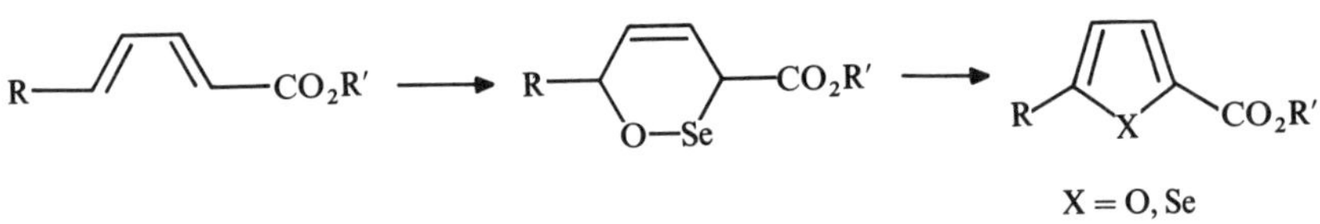

SCHEME 13

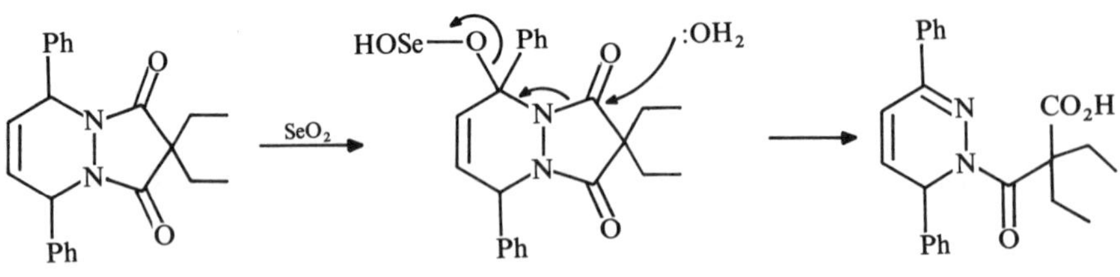

(**20**)

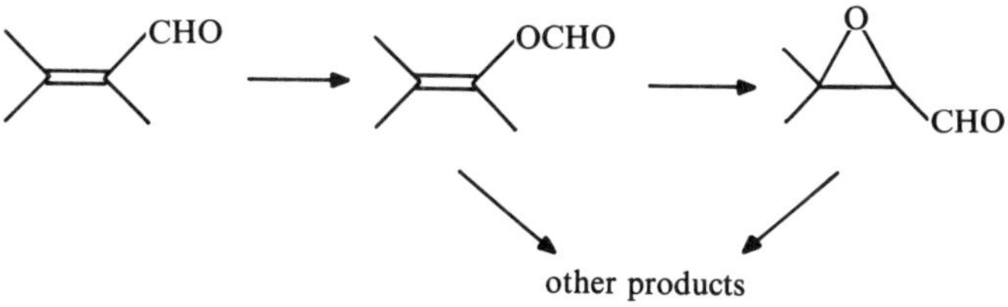

SCHEME 14

Halogens

The mechanism of the periodate oxidation of benzaldoxime has been studied.[189] The hypervalent iodine reagent, [hydroxy(tosyloxy)iodo]benzene (**21**), causes alkynes to undergo an oxidative rearrangement to form carboxylic esters (Scheme 15).[190] *m*-Chloroperbenzoic acid oxidation of a steroidal homoallylic iodide leads to rearranged products via a hypervalent iodine intermediate (Scheme 16).[191] Transannular oxidative cyclization of 1-methylcyclooct-4-en-1-ol

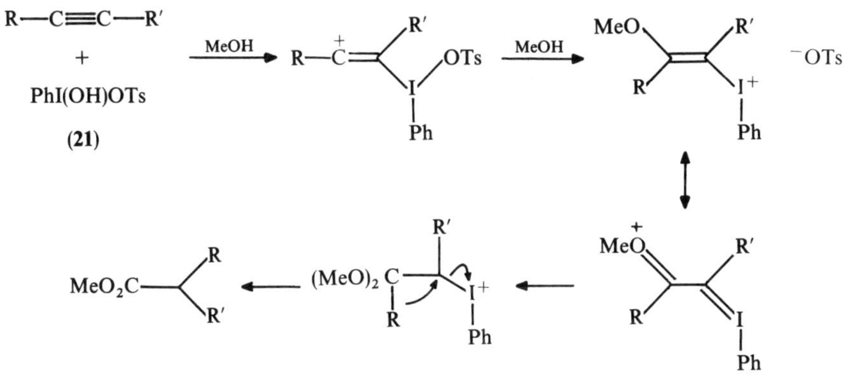

SCHEME 15

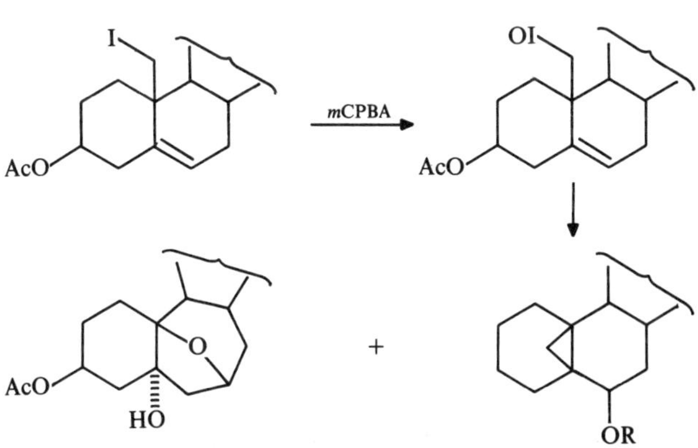

SCHEME 16

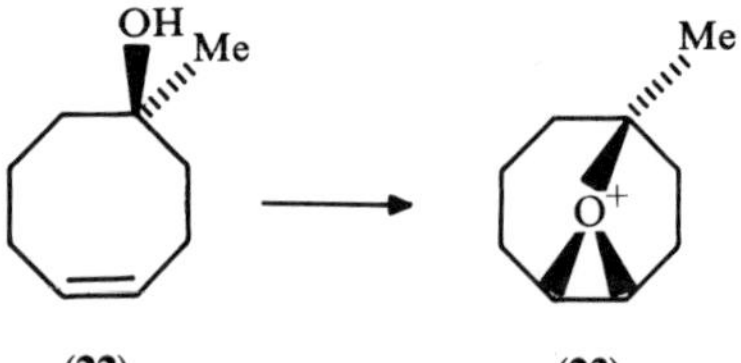

(22) **(23)**

(**22**) with hypervalent iodine proceeds via a tricyclic oxonium species (**23**).[192] Hypervalent iodine oxidations of silylenol ethers in the presence of Lewis acids can lead to carbon–carbon bond formation,[193] α-hydroxy-ketones,[194] or α-methoxy-

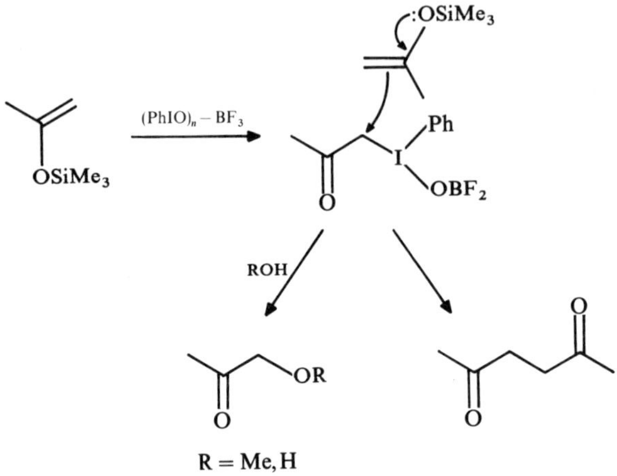

SCHEME 17

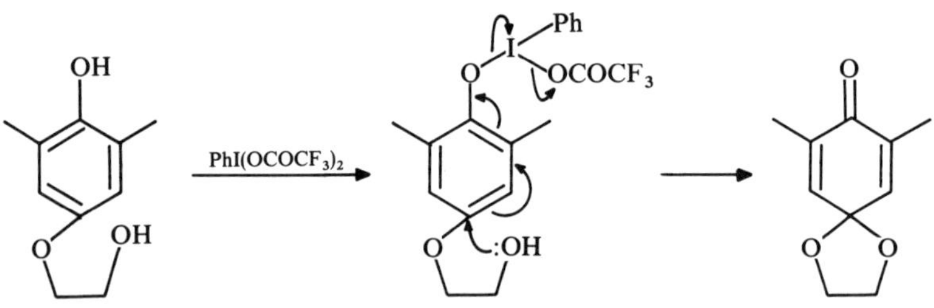

SCHEME 18

ketones[195] in high yield (Scheme 17). Hypervalent iodine oxidation of *p*-alkoxyphenols and related compounds provides a general route to *p*-benzoquinone monoacetals and spiro-lactones (Scheme 18).[196] (Diacetoxyiodo)benzene (**24**) in methanol causes oxidative cleavage of 5-substituted pyrazol-3-(2*H*)-ones (**25**) to 2-alkynoic methyl esters.[197] Hypervalent oxidation of chromium tricarbonyl complexes of benzocyclanones gives enantiomerically pure α-hydroxy dimethyl acetals.[198] Hypervalent oxidations of thiouracils to uracils, including

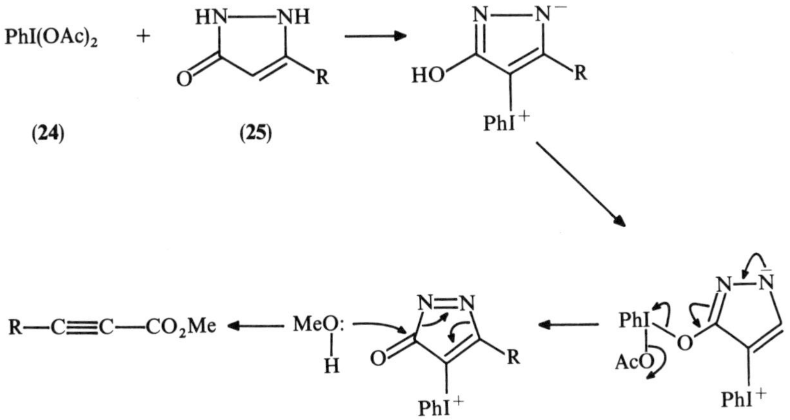

(24) (25)

the 4-thiouridine site in tRNA,[199] and of protected methionines[200] have been reported. A mechanism has been proposed for the oxidative decarboxylation of homocubane- and cubane-carboxylic acids with (diacetoxyiodo)benzene-carbon tetrachloride-iodine.[201] Studies of the reaction of iodine monoxide, generated from ozone and iodine, with dimethyl sulphide have indicated that this process may be an important sink of dimethyl sulphide in marine atmospheres.[202] Oxidation of aryliminodimagnesium reagents (**26**) with a number of iodine oxidizing agents affords *syn*-azobenzenes via formation of aryl nitrene-like species.[203] The bicyclic cation (**28**) is formed by neighbouring-group participation by the transannular amide during the oxidation of (**27**) by aqueous iodine.[204] The kinetics of the oxidation of cyclohexanol by aqueous iodine have been studied.[205]

$ArN(MgBr)_2$

(**26**)

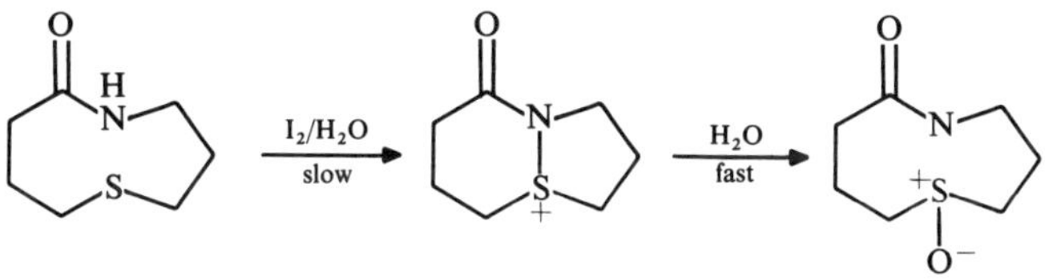

(**27**) (**28**)

Bis(quinuclidine)bromine(I) tetrafluoroborate (**29**) is a new Br^+ reagent for the oxidation of alcohols to carbonyl compounds.[206] The bromate oxidation of 2,4-dinitrophenylhydrazine involves initial fast conversion into the corresponding

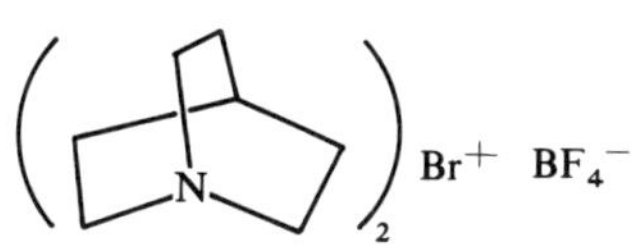

(**29**)

diazonium salt, followed by rate-determining decomposition of the salt to products such as 1,3-dinitrobenzene.[207] Hammett studies on the bromate oxidations of aryl sulphides and sulphoxides show that the transition states of both series have considerable charge separation.[208] Electron transfer is the rate-determining step in

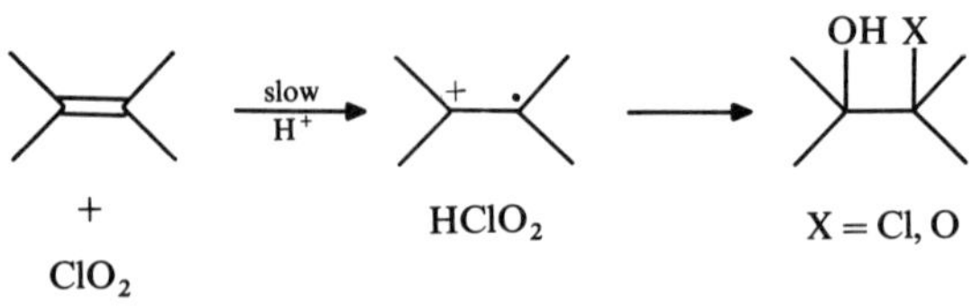

SCHEME 19

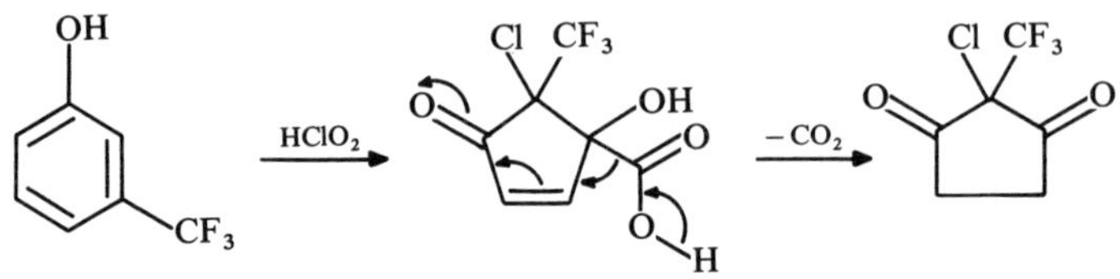

(30) **(31)** **(32)**

the oxidation of olefins by chlorine dioxide (Scheme 19).[209] Formation of the ring-contracted dione (**32**) in the chlorous acid oxidation of *m*-(trifluoromethyl)phenol (**30**) involves decarboxylation of the cyclopentenecarboxylic acid (**31**).[210] Highly regio- and stereo-selective oxidation sequences for *gem*-dimethyl olefins to give terminal *trans*-allylic alcohols have been described in which the initial halogenation by calcium hypochlorite is followed by amine displacement, oxidation, and [2,3]-shift of the corresponding *N*-oxide (Scheme 20).[211] Hypochlorite and chlorite oxidations of sulphides have been investigated.[212]

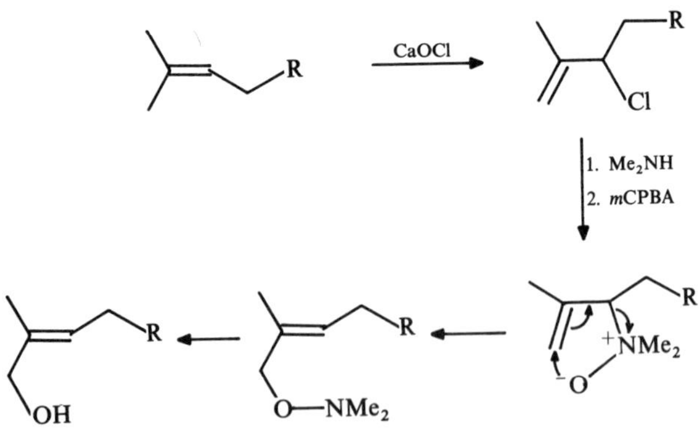

SCHEME 20

The oxidation of thiourea (Scheme 21) by aqueous bromine involves fast initial attack by bromine on thiourea, followed by a second slower oxidative step which is autocatalytic in bromide.[213] Kinetic studies on the oxidation of benzhydrol by

$$CS(NH_2)_2 + 4Br_2 + 6H_2O \longrightarrow 8Br^- + 2NH_4^+ + SO_4^{2-} + 8H^+ + CO_2$$

SCHEME 21

aqueous bromine and *N*-bromosuccinimide have been reported.[214,215] In the oxidations of *ortho*-substituted benzyl alcohols[216] and α-hydroxy-acids[217] by *N*-bromoacetamide in acid, the rate-determining step is attack of substrate on protonated hypobromous acid. The stereoselectivity observed in the ring-opening of 2-[(phenylthio)methyl]cyclohexanols mediated by *N*-chlorosuccinimide and triethylamine (Scheme 22) has been rationalized.[218] Rate-determining hydride transfer to oxidant has been found for the oxidations of alcohols by *N*-bromo-3,5-dinitrobenzamide,[219] *N*-bromosuccinimide,[220] and *N*-bromobenzenesulphonamide,[221] and for the oxidation of substituted benzaldehydes by *N*-bromobenzamide.[222] Extensive kinetic studies have continued on the oxidation by *N*-halogenated oxidants of a wide range of organic substrates including amino-acids,[223–225] ketones,[226] oximes,[227] diols,[228] secondary alcohols,[229] thiosemicarbazide,[230] and derivatives of mercaptoacetic acid.[231]

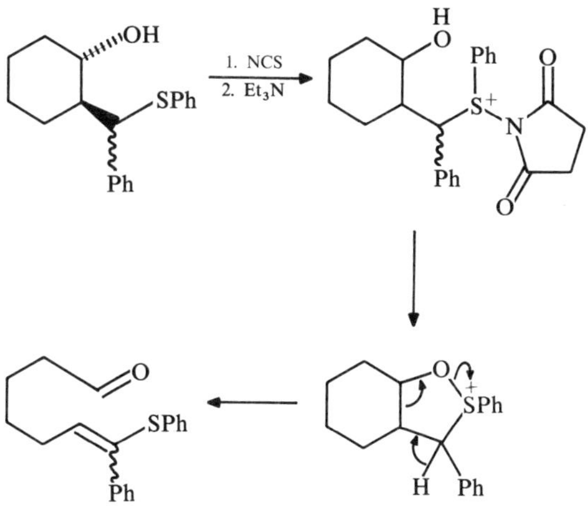

SCHEME 22

Miscellaneous Oxidations

The mechanism of asymmetric oxidation of chiral acyclic tetrasubstituted enolates with chiral oxaziridines such as (**33**) has been discussed.[232] The oxidation of thiones to thione *S*-oxides (Scheme 23) by sulphonylaziridines has been compared with the corresponding oxidation by perbenzoic acid.[233] The formation of aromatic radical cations in the oxidation of aromatic systems by dichlorodicyano-*p*-benzoquinone (DDQ) in trifluoroacetic acid-dichloromethane has been studied by spectroscopic techniques.[234] The DDQ dehydrogenation of flavanones involves a two-step ionic mechanism in which the rate-determining step is hydride abstraction by protonated DDQ.[235] The *cis*-selective two-step mechanism in hydrogen transfer from 1,4-

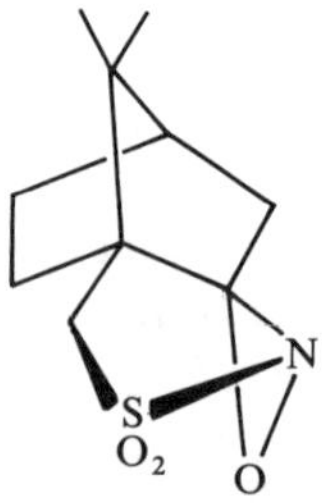

(33)

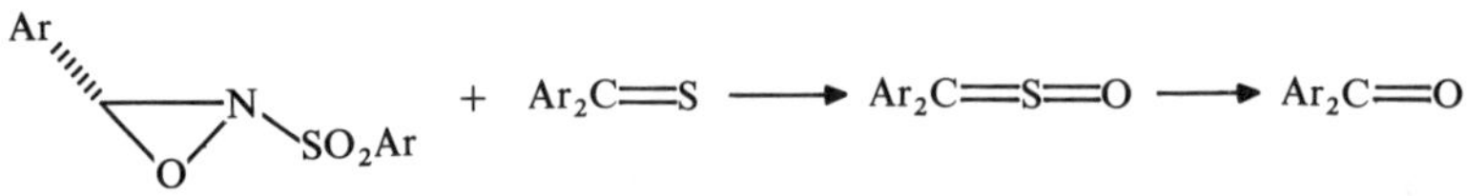

SCHEME 23

dihydroarenes to quinones has been studied using deuterium labelling; solvent and primary kinetic isotope effects exclude a symmetry-allowed synchronous transfer of both hydrogen atoms and are consistent only with a primary hydride abstraction to give an orientationally fixed ion-pair.[236] The irradiation with visible light of a mixture of an alkene in degassed aqueous acetonitrile, with tetraphenylporphyrinato-antimony(V) as sensitizer and methyl viologen as electron acceptor, induced epoxidation of the alkene accompanied by hydrogen evolution with water acting as the hydrogen donor.[237] A magnesium Oppenauer oxidation procedure gives high yields of ketones from secondary saturated or allylic alcohols.[238]

Ozonolysis and Ozonation

The relevance of laboratory studies on the photochemistry of ozone to interpretations of atmospheric chemistry has been surveyed.[239] The reactions of ozone with cyclopentene and cyclohexene under atmospheric conditions have been examined.[240] Ozonolysis of the bis(vinyl ether) (**34**) gave only (**35**) in carbon tetrachloride, although (**36**) was the major product in methanol; evidence was found for intramolecular oxygen transfer from the carbonyl oxide in (**37**) to the methoxyvinyl group.[241] In studies on the ozone cleavage of chloroalkenes, high selectivity for cleavage of the non-halogen-substituted double bond was observed in methanol (Scheme 24).[242] Diozonolysis of conjugated 1-chloro-dienes results in cleavage of both double and single bonds of the diene units (Scheme 25).[243] The couse of the reaction of ozone with tetramethylene has been investigated by stopped-flow and other techniques (Scheme 26).[244,245] The course of the reaction of ozone with five-membered-ring aromatic heterocycles has been correlated to the highest bond-order value in the ring, derived from HMO calculations.[246] Solvent effects on

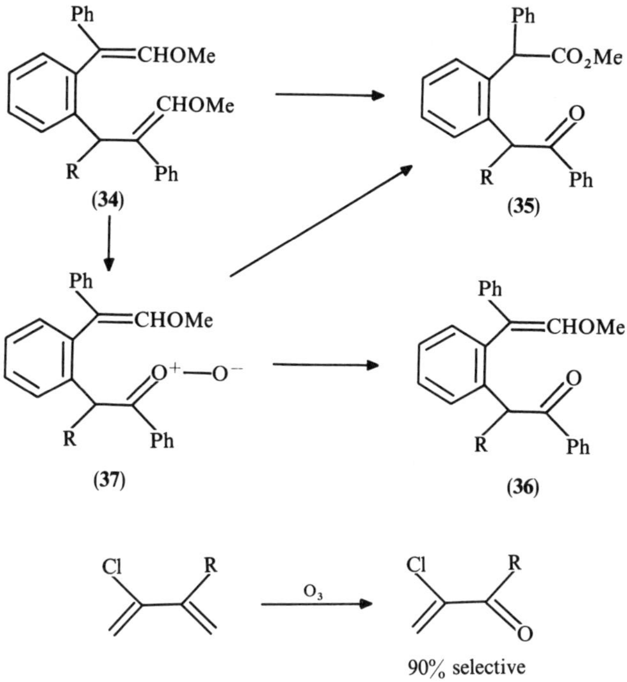

SCHEME 24

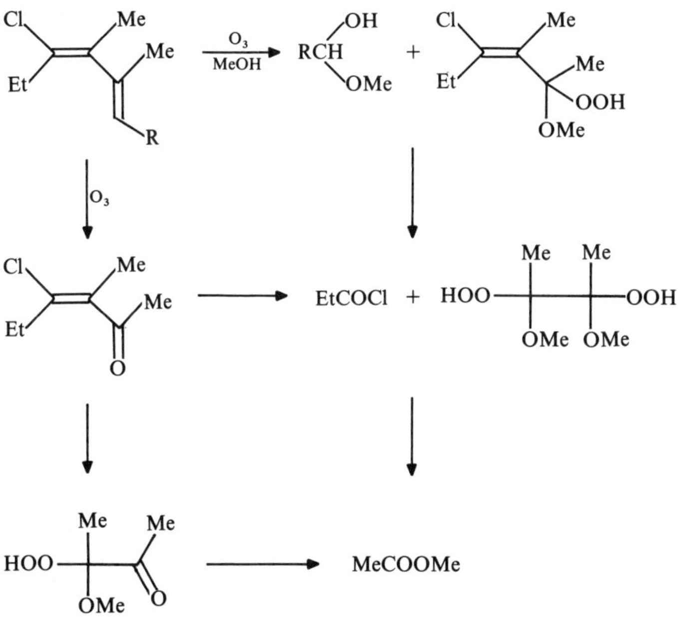

SCHEME 25

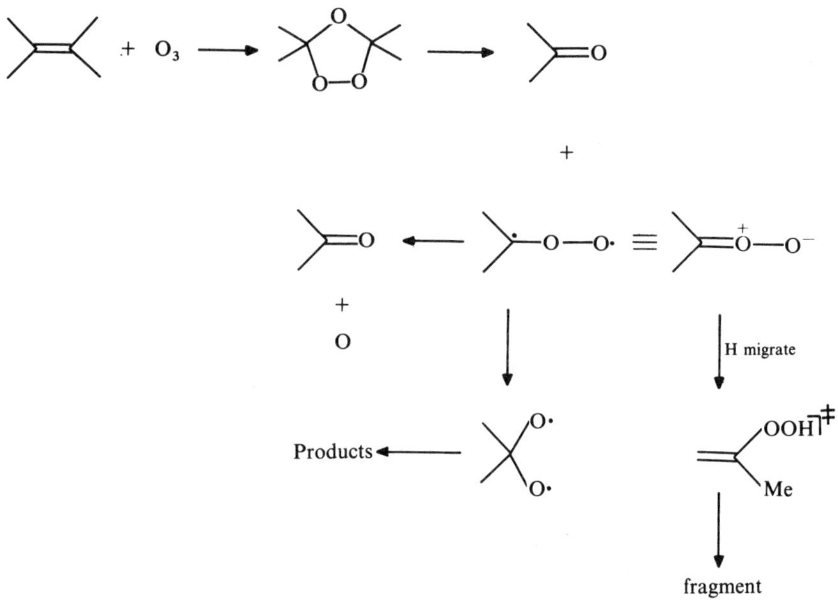

SCHEME 26

the ozonolysis of hetarenes have been studied.[247] Reactions of ozone with benzene derivatives,[248,249] including benzoic acid,[250] have been investigated. Hydroperoxides were the primary products of the reaction of oxygen and ozone with a series of crown ethers.[251]

Peracids and Peroxides

Isotopic labelleling studies of the oxidation of benzil with perbenzoic acid are not consistent with a Baeyer–Villiger oxidation; an alternative pathway is proposed

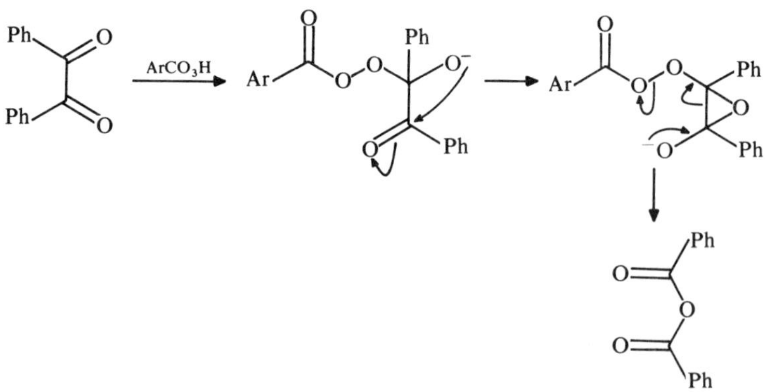

SCHEME 27

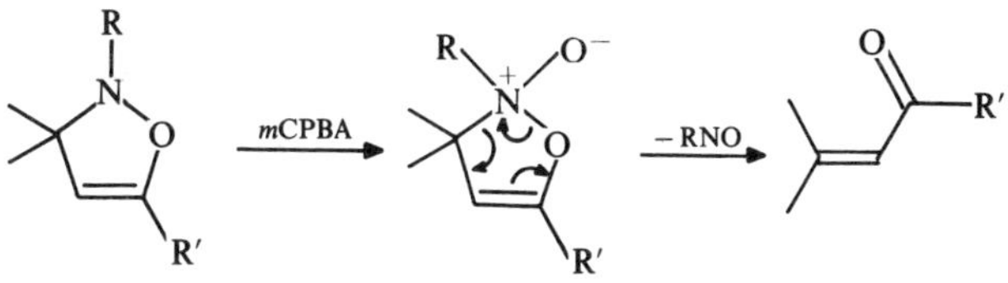

SCHEME 28

(Scheme 27).[252] Peracid oxidation of isoxazolines derived from dipolar cycloadditions of nitrones with alkynes or alkenes results in chelotropic elimination to α,β-unsaturated ketones (Scheme 28).[253] Peracid treatment of 3,6-di-*tert*-butyl-2,2,7,7-tetramethyl-3,4,5-octatriene (**38**) gives the stable methylenecyclopropanone (**39**);[254]

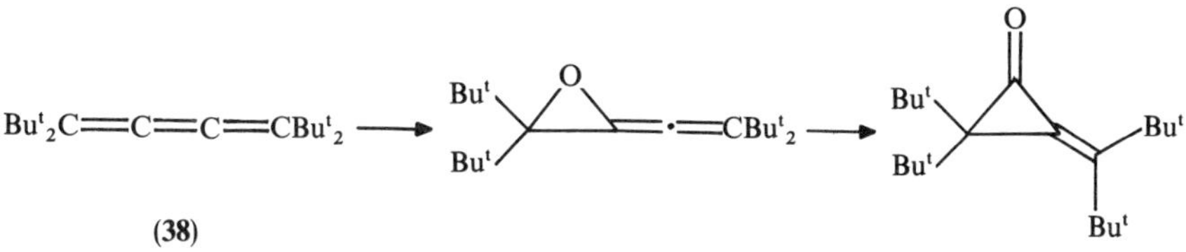

(**38**)

(**39**)

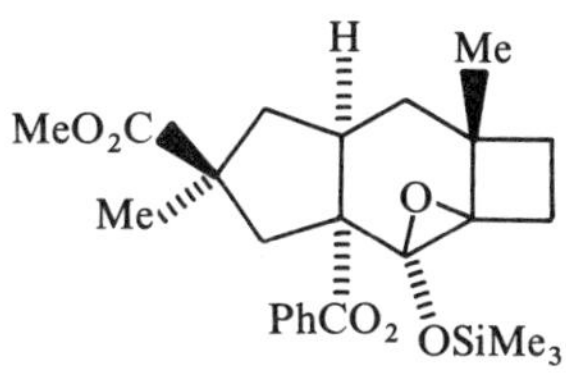

(**40**)

the first stable silyloxy epoxide (**40**) has been fully characterized.[255] The superior performance of amides as directing groups in the epoxidation of cyclic as opposed to acyclic olefins may be related to conformational factors.[256] The high diastereoselectivity observed in the epoxidation of *cis*-4-protected amino-allylic alcohols with peracid may be rationalized by a cooperative effect of the allylic hydroxyl group and the carbonyl oxygen of the protecting group forming hydrogen bonds with the peracid (Scheme 29).[257] Studies on the peracid oxidations of 1-octene[258] and pyrimidines[259] have been reported. The relative reactivity for the epoxidation of alkenes by dimethyldioxiran (**41**) showed that the reaction is unexpectedly sensitive to steric factors; the transition state for the oxygen transfer may be a spiro-complex between (**41**) and the alkene.[260]

An *ab initio* study of the S_N2 reactions of the hydroperoxide ion with chloromethane has been reported.[261] *Ab initio* calculations suggest that the unanticipated stability of (**41**) may be due to the *gem*-dimethyl group conferring thermodynamic as well as kinetic stability; transfer of oxygen to thianthrene 5-oxide causes mainly oxidation of the sulphoxide oxygen, revealing a dominating

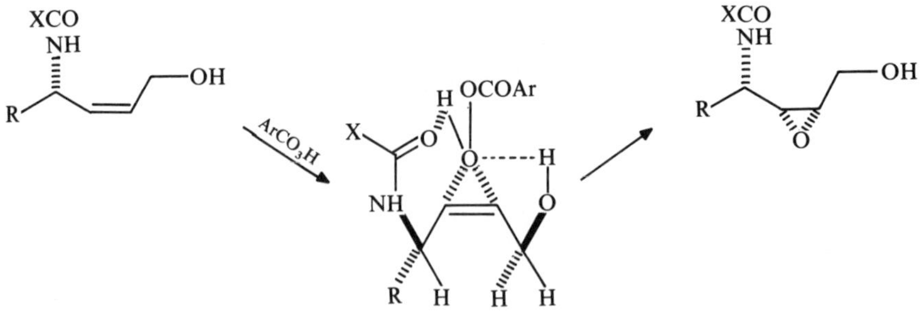

SCHEME 29

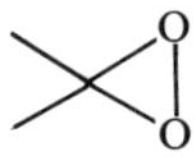

(41)

nucleophilic character for the oxidation.[262] Various features of the nucleophilic peroxide epoxidations of α,β-unsaturated aldehydes[263–265] and of *N*-methyl-4-styrylpyridinium iodides[266] have been investigated. The epoxidation of substituted naphthaquinones with *tert*-butyl hydroperoxide in the presence of bovine serum albumin as a chiral catalyst gives epoxynaphthaquinones in up to 80% enantiomeric excess.[267]

The mechanism for the transfer of oxygen from α-azo-hydroperoxides to give amine oxides, sulphoxides, and epoxides is similar to that for oxygen transfer from peracids; hydrogen-bonding is important in the transition state (Scheme 30).[268] The acid-induced oxidative cleavage of benzpinacolones by diaroyl peroxides is consistent with a single-electron-transfer pathway; the acid promotes the reaction by assisting oxygen–oxygen bond cleavage and by the formation of protonated peroxide as a more powerful oxidizing species.[269] Studies on the hydrogen peroxide oxidations of *o*-nitrobenzeneselenenic acid[270] and of formaldehyde,[271] and of the oxidation of benzidine by dibenzoyl peroxide[272] have been reported. The initiation of liquid-phase oxidations by diacyl peroxides has been investigated.[273]

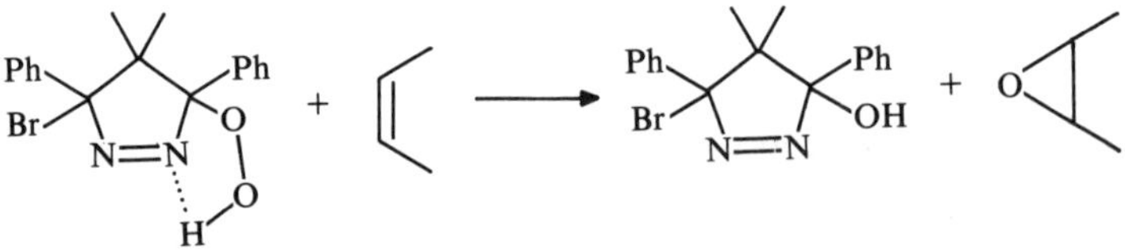

SCHEME 30

Photo-oxygenation, Singlet Oxygen, and Superoxide

The value of 3-adamantylidenetricyclo[3.2.1.0^{2,4}]octane as a suitable mechanistic probe for distinguishing between oxygenation by singlet oxygen or by electron transfer has been demonstrated.[274] The cation radical-catalysed chain oxygenation of olefins and dienes has been surveyed.[275] Earlier work on matrix photo-oxidation of alkenes by oxygen has been reinterpreted in terms of a generalized mechanism in which dioxymethylenes (**42**) are generated by a non-ozonolytic pathway.[276] Photo-

$$R_2C{=}\overset{+}{O}{-}O^-$$

(42)

induced electron transfer in excited contact charge-transfer pairs of 1-arylcyclohexenes and oxygen generates radical cations of the alkene and superoxide.[277] The reaction of adamantylideneadamantane (**43**) with singlet oxygen gives the dioxetane (**44**) as the major product with only traces of the epoxide (**45**); when Rose Bengal is added, the epoxide (**45**) becomes the major product. Charge transfer complexes between *N*-ethylcarbamate and fluorene with 2,4,7-trinitrofluoren-9-one behave

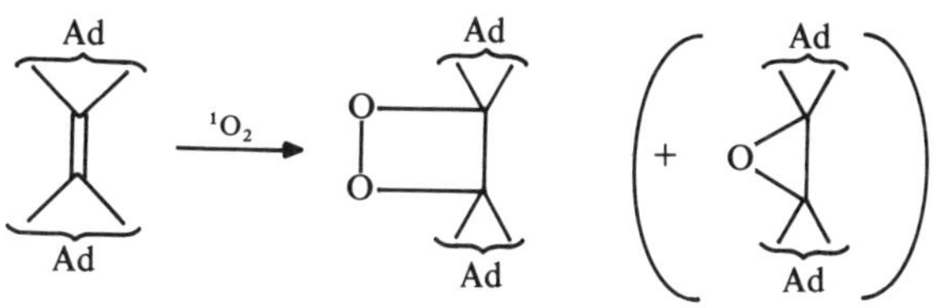

(43) **(44)** **(45)**

Ad = adamantylidene

like Rose Bengal; thus the charge-transfer complexes and Rose Bengal compete for singlet oxygen giving superoxide which is responsible for the formation of epoxide (**45**).[278] Rose Bengal-functionalized phase-transfer catalysts promote photo-oxygenations with singlet oxygen.[279] Dicyanoanthracene-sensitized photo-oxygenations of stilbenes[280] and conjugated dienes[281] have been studied. The regio- and stereo-selectivities of the photo-oxygenations of phenyl-substituted propenes and related compounds have been discussed in terms of the conformations of the intermediate peroxyoxiranes.[282] The product distribution from the low temperature photo-oxidation of substituted indenes is highly solvent-dependent.[283]

The radiationless deactivation of singlet oxygen by solvents[284] and the quenching of singlet oxygen by pyrrole in solution[285] have been studied; the interaction of singlet oxygen with aliphatic amines and with hydroxylamines results in different quenching mechanisms.[286]

The rearrangements of the two hydroperoxides formed by the singlet oxygenation of oleic acid have been investigated.[287] The course of the reaction of singlet oxygen with methylenecyclopropanes to give 4-methylene-1,2-dioxolanes or lactones has been rationalized in terms of initial formation of a perepoxide, followed by conversion into an allylic cation or carbonyl oxide (Scheme 31).[288] The relative stabilities of two possible perepoxide regioisomers may play a major role in determining the product distribution in the reaction of (*E*)- and (*Z*)-1-*tert*-butoxy-1,3-butadienes.[289] Singlet oxygen is more reactive as a Diels–Alder dienophile with

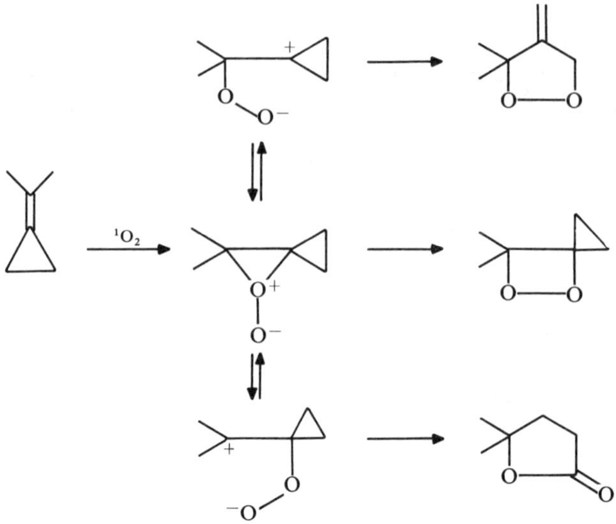

SCHEME 31

conjugated dienes with short,[290] rather than long,[291] separation of the terminal carbons of the diene unit. Photo-oxidation of 2,5-dimethyl-2,3,4-hexatriene allows the isolation of the trisdioxetane (**46**).[292] Thiirane reacts with singlet oxygen to give peroxythiirane (**47**) which can undergo trapping either by thiirane or by a nucleophilic solvent.[293]

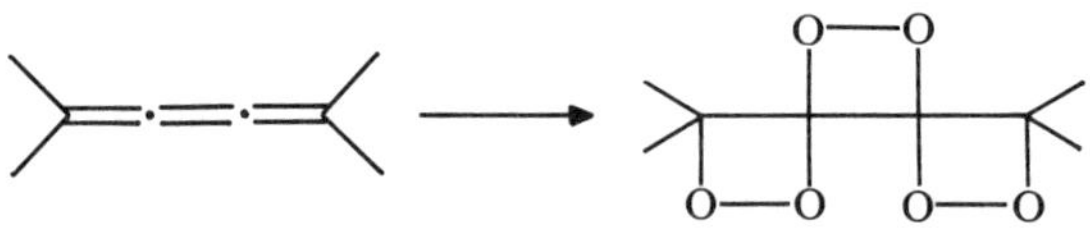

(**46**)

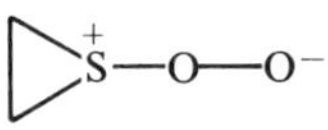

(**47**)

A peroxysulphur intermediate, generated *in situ* from 2-nitrobenzenesulphonyl chloride and superoxide, causes rapid oxidation of sulphoxides to sulphones (Scheme 32).[294] Superoxide ion is formed by hydroxyl-induced oxidation of trimethylamine in oxygenated aqueous solution.[295] The rate-determining step of the superoxide-mediated oxidation of diarylmethanes involves proton transfer from the substrate to superoxide ion.[296] In both aprotic and protic media, one-electron reduction of adriamycin by superoxide is favoured over deprotonation; the relevance of this to the reaction of superoxide with other anthraquinones has been considered.[297] Both α-tocopherol and ascorbic acid react with superoxide by deprotonation to form the corresponding anions; in subsequent steps these anions

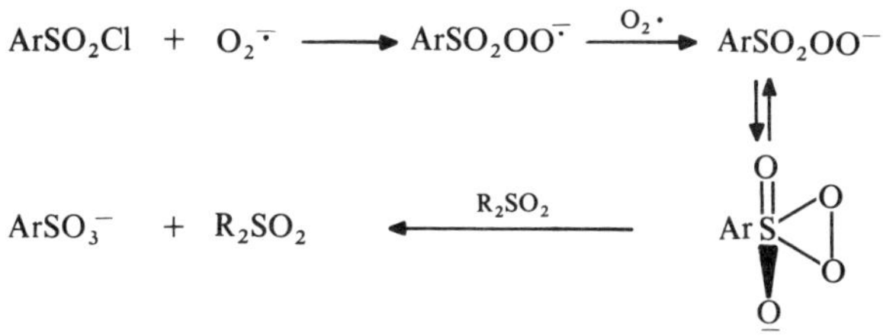

SCHEME 32

are oxidized by oxygen to form superoxide.[298] Superoxide initiates the autoxidation of dihydroxyacetone.[299] The reactions of superoxide and ozonate radical anions with 9,10-dichloro-9,10-dihydroanthracene give products derived from nucleophilic substitution and reductive dehalogenation.[300]

Atomic Oxygen, Autoxidation, and Other Reactions of Oxygen

The oxidation of phenylethylenes with atomic oxygen $O(^3P)$ in the condensed phase gives mainly epoxides and carbonyl compounds formed by 1,2-hydride shifts; a diradical intermediate is involved and the effect of temperature on the distribution of products may be explained by the interconversion of different electronic states of the diradical.[301] Kinetic studies on the reactions of atomic oxygen with ethylene,[302,303] tetrafluoroethene,[304] acetylene,[305] and diacetylene[306] have been reported.

An argument that all autoxidations are metal-catalysed has been presented.[307] Reviews have appeared on various aspects of autoxidation, including metal ion and bromide catalysis,[308] and catalysis by variable-valence metal ions;[309,310] the oxidations of hydrocarbons,[311] naphthalene,[312] and esters[313] have been surveyed. Mechanistic investigations have been reported on the autoxidations of aromatic aldehydes,[314,315] aliphatic aldehydes,[316,317] alkylazulenes,[318,319] methylbenzenes,[320–322] ethylbenzene,[323] isopropylbenzenes,[324,325] tetrahydroanthraquinones,[326] phenol,[327] amines,[328–330] adamantane,[331] trichloroethane,[332] butan-2-ol,[333] friedlin,[334] and diethyl ether.[335] The antioxidant activity of α-tocopherol has been investigated.[336] The differential reactivity patterns observed in inner-sphere and outer-sphere reductions of various oxidants by neutral ascorbic acid and its monoanion have been discussed in terms of mechanistic factors such as redox potential, electrostatic considerations, ion-pair formation, hydrogen-bonding, and proton rearrangement in the transition state of the oxidation.[337] *Ab initio* calculations indicate that the five-ring lactone of ascorbic acid is planar during the oxidation process.[338] The facile oxidation of vinyl ethers to esters may proceed via a fragmentation of a dioxetane (Scheme 33).[339] The base-catalysed autoxidation of 2,3,4-trimethylnaphth-1-ols has been studied (Scheme 34); naphthalenes with additional substituents at C(5) are more sensitive to oxygen, possibly resulting from 4,5-periinteractions.[340] Selectivity in the base-catalysed oxygenation of a pair of associating thiols has been studied.[341] Product distribution[342] and other features, including solvent effects[343] and temperature,[344] of the thiol-olefin co-oxidation reaction have been investigated. Studies on the cyanide-catalysed oxidations of

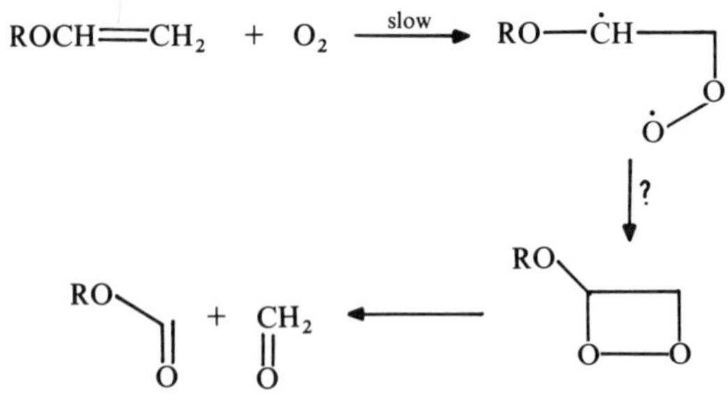

SCHEME 33

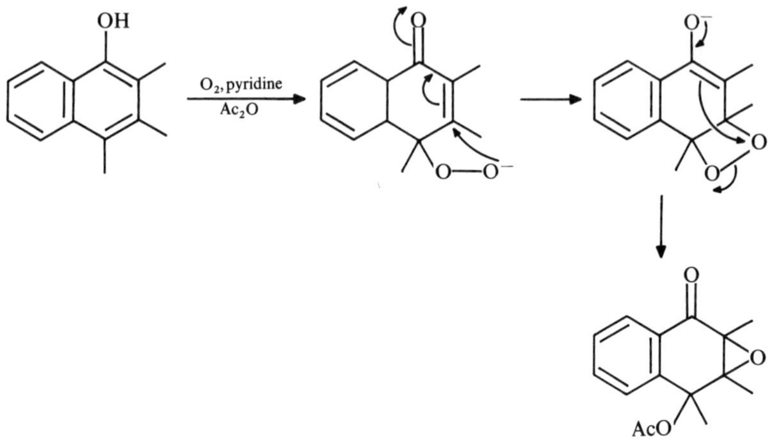

SCHEME 34

α-keto-aldehydes and α-keto-alcohols have been reported,[345] and dehydro-coupling of aromatic nuclei to produce poly(p-phenylene) has been reviewed.[346]

Reduction by Complex Metal Hydrides

The reactions of thexylchloroborane-methyl sulphide in methylene chloride have been compared with other boranes.[347] Many investigations of stereochemical features of hydride reductions have been reported. The effectiveness of various hydride reagents for the asymmetric reduction of different classes of ketones has been critically assessed.[348] The use of the optically active forms of the diastereomers of 2-(1-phenylethyl)amino-1-phenylethanol as chiral auxiliaries in the asymmetric reduction of acetophenone with modified derivatives of lithium aluminium hydride has been studied.[349] The degree of asymmetric induction in the lithium aluminium hydride reduction of chiral benzoylformamides derived from chiral amino-alcohols is dependent largely on the substituents of the amino-alcohol moiety.[350] A metallated intermediate (**48**) is formed during the reduction of enol acetates with lithium aluminium hydride.[351] The reduction of α-fluoro-α-methyl-β-keto-esters with hydrosilanes takes place with a very high degree of stereoselectivity.[352]

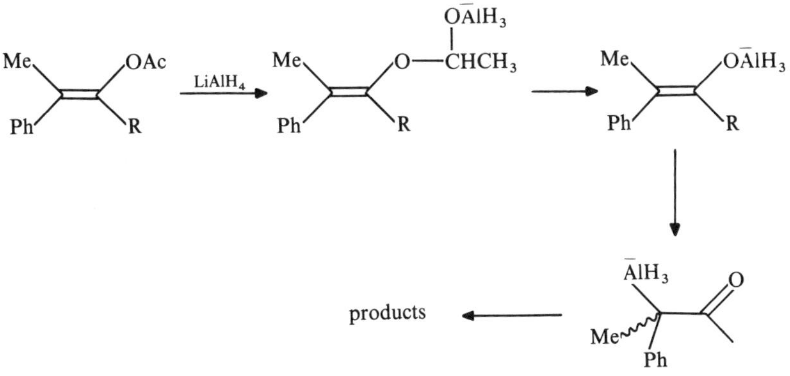

(48)

Chelation- and non-chelation-controlled reductions of β-dicarbonyl compounds to 1,3-diols with three adjacent chiral centres by modified aluminium hydride reagnets have been investigated; the stereoselectivity depends largely on the solvent and the bulk of the hydride reagent.[353] Addition of the magnesium cation enhances the stereoselectivity of 1,5-asymmetric reduction of the C_2-symmetric acetal ketones

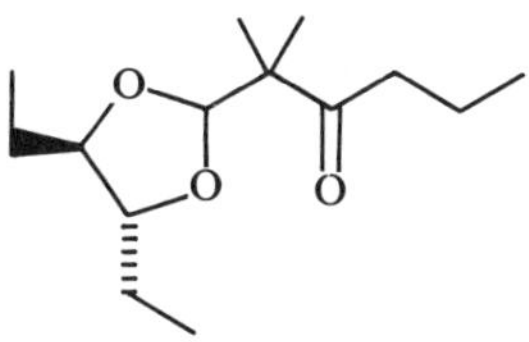

(49)

(49).[354] Very high selectivity for the formation of *erythro*-4-hydroxy-3-methylbenzenebutanoic acids is found in the reduction of the corresponding ketones by diisobutylaluminium hydride in the presence of zinc chloride (Scheme 35).[355] A

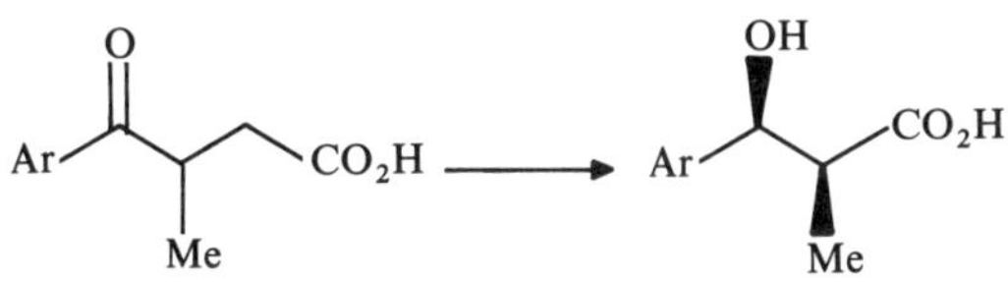

SCHEME 35

possible transition state has been proposed to account for the diastereoselectivity in the reduction of chiral α-keto-unsaturated acetals with lithium aluminium hydride in the presence of lithium bromide.[356] The borohydride reduction of propellanediones in an aqueous solution of a cationic surfactant provides clear evidence that micellar binding can control diastereoselectivity; in addition to the expected rate enhancement that cationic micelles provide for borohydride reduction, they direct the approach of the reducing agent to one face of the substrate.[357] Remote electronic control by a sulphonyl group is found in the regio- and stereo-specific reduction of **(50)** to **(51)** by lithium aluminium hydride.[358] Stereoselectivity in

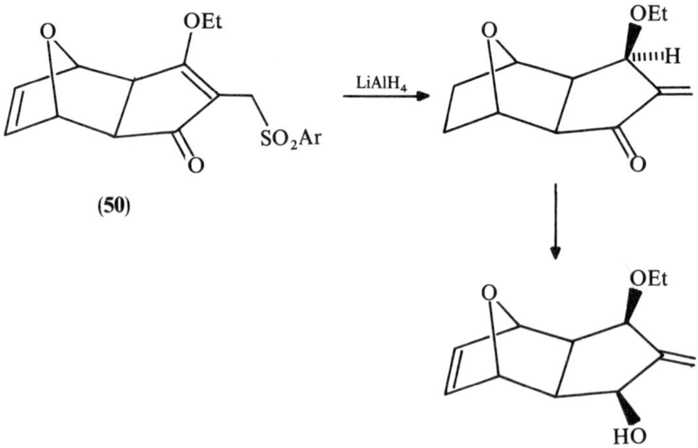

(50)

(51)

hydride reductions of carbonyl groups in imino-ketones,[359,360] amino-ketones,[361] indolo[2, 1-b]quinazoline-6, 12-dione,[362] quassin,[363] and cyclohexanones with α-sulphur substituents[364] has been studied. The stereoselectivity in the borohydride reduction of chiral iminium ions derived from α-phenylethylamine may be rationalized by approach to the less-hindered face of the carbon-nitrogen double bond.[365] *anti*-Selectivity in the α-chelation-controlled hydride addition to acylic alkoxy-ketone oximes with hydride reagents affords chiral primary *anti*-amines.[366] The reduction of *N*-diphenylphosphinyl-imines of substituted alkylcyclohexanones with lithium tri-*sec*-butylborohydride provides highly diastereoselective conversion into protected axial primary amines (Scheme 36).[367]

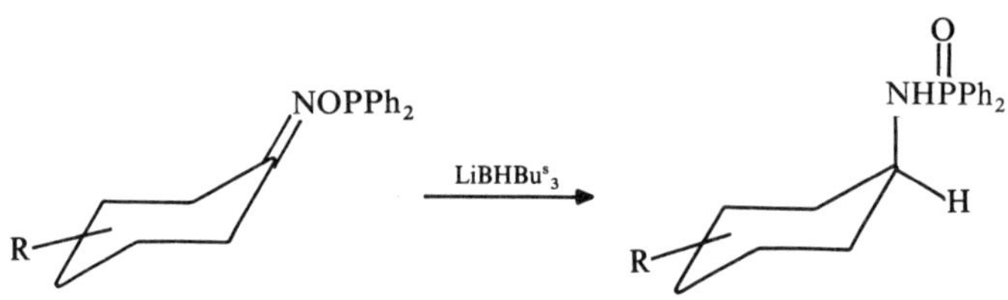

SCHEME 36

The major reduction products of 2,4,6-trisubstituted pyrylium salts with sodium triacetoxyborohydride are *cis*-2,6-substituted dihydropyrans (Scheme 37).[368] The different roles of acetoxyborohydride species of formula $NaBH_{4-n}(OAc)_n$ in the

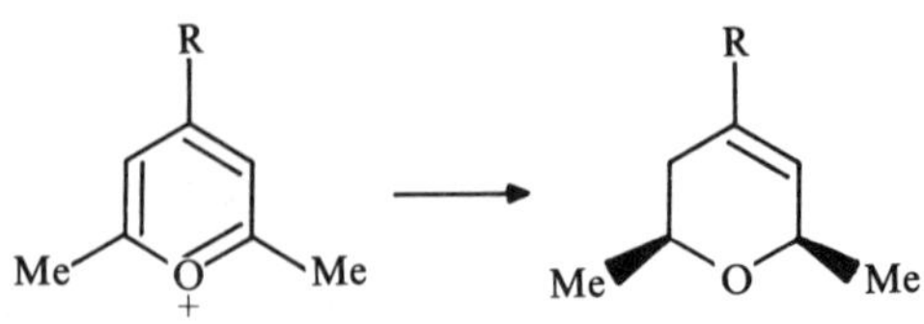

SCHEME 37

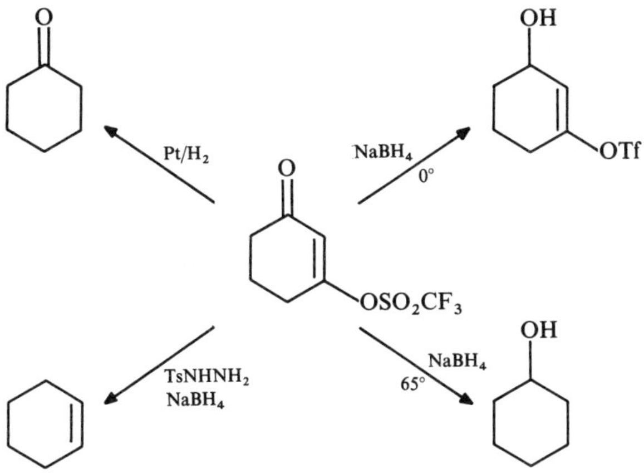

SCHEME 38

reduction of ketones have been discussed.[369] The reduction of triflates derived from cyclic 1,3-diketones by sodium borohydride and other reagents has been investigated (Scheme 38).[370] Hydride reduction of the aldehyde-anhydride, fujenal (**52**), proceeds first at C(7); the C(7) alkoxide thus generated then participates in reaction of the C(6) carbonyl group.[371] A hexavalent silicate (**53**) is the active hydride species

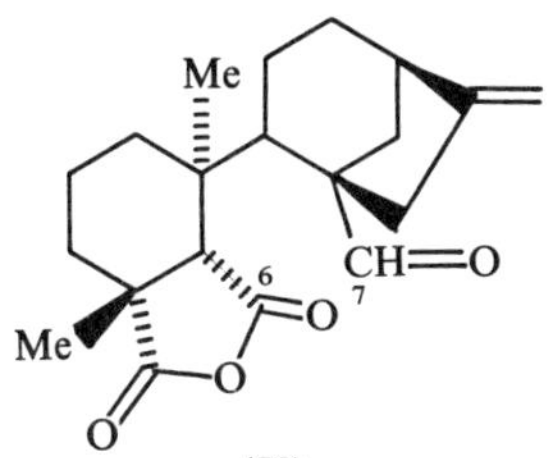

(**52**)

$$HSiR_3 \xrightarrow[\text{HMPA}]{F^-} [HSiR_3F(HMPA)]^-$$

(**53**)

in the fluoride-catalysed reduction of ketones with hydrosilanes in HMPA; no evidence was found for the interaction of the carbonyl group with silicon.[372] Mono- and *gem*-di-substituted epoxides are highly regioselectively reduced by sodium borohydride in mixed solvents containing methanol to the more substituted alcohols.[373] An exceptionally facile reduction of carboxylic esters to aldehydes may be achieved with lithium aluminium hydride in the presence of diethylamine; in the presence of diethylamine, only one hydride equivalent is available since a stable alane-amine complex precipitates from the ether.[374] The borohydride reduction of non-symmetrical *cis*-hexahydrophthalimides is very much more regioselective than the reduction of the corresponding tetrahydrophthalimides; the different behaviour is explained on the basis of conformational preference of the imide and the antiperiplanar effect.[375] The regioselectivity of metal hydride reductions of 3-substituted phthalic anhydrides has been re-examined.[376] Kinetic studies of the reduction of a tertiary amide by lithium aluminium hydride in the presence of aluminium chloride have been reported;[377] the reductive cleavage of homochiral acetals with Lewis acid-catalysed metal hydrides has been investigated.[378]

Catalytic amounts of tri-*n*-butyltin chloride and a polyether phase-transfer catalyst added to a suspension of sodium borohydride in toluene provide a suitable reducing system for alkyl halides.[379] Lithium aluminium hydride reduction of 4,8-dihalo-2-thiaadamantane (**54**) occurs with complete retention of configuration,

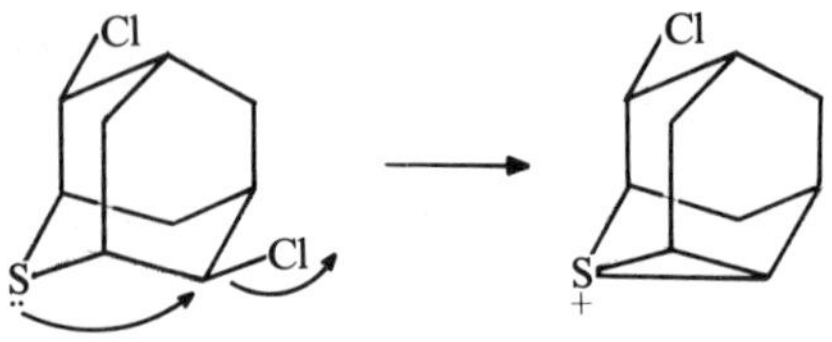

(**54**) (**55**)

probably via intermediate formation of the thiiranium ion (**55**).[380] Transannular cyclizations occur during the sodium borohydride reduction of 1,2-dithia-5,8-diazacyclodeca-4,8-dienes to give bicyclic products, rather than the ten-membered-ring monocyclic compounds previously claimed.[381] A one-pot reductive alkylation of nitroarenes has been devised, based on *in situ* generation of hydrides from lithium aluminium hydride in the presence of palladium (Scheme 39).[382] The mechanism of

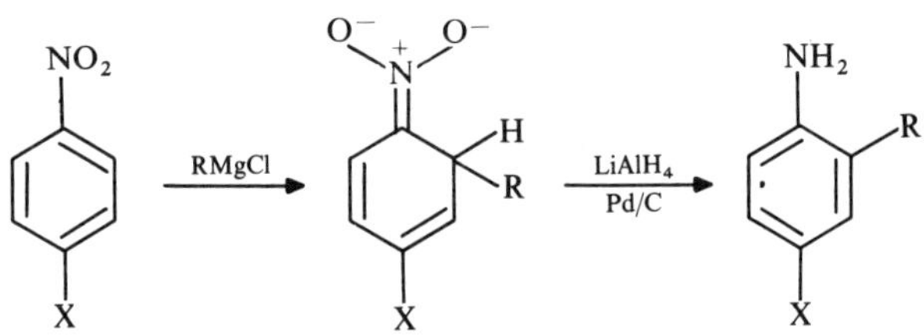

SCHEME 39

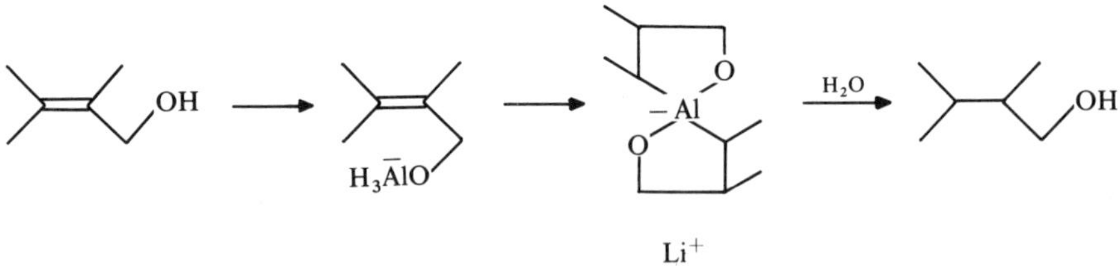

SCHEME 40

the reduction of the double bond in allylic alcohols by lithium aluminium hydride has been studied by NMR techniques (Scheme 40).[383] The conjugate reductions of α,β-acetylenic ketones and esters by diisobutylaluminium hydride-HMPA have provided the first examples of alkenylaluminium intermediates with a carbonyl functionality (Scheme 41).[384]

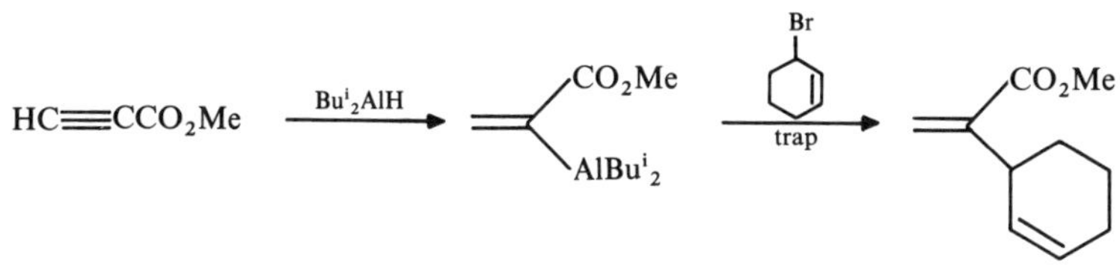

SCHEME 41

Reduction by Metals, Metal Ions, and Metal Complexes

The mechanism and stereochemistry of the alkali metal reduction of cyclic saturated and unsaturated ketones in protic solvents has been reviewed.[385] The reduction of camphor with alkali metals in tetrahydrofuran gives the same results as those obtained in liquid ammonia.[386] Some features of the steric effects of substituents on the reduction of 1,4-lactones by sodium in liquid ammonia have been identified.[387] In contrast to literature suggestions, benzonitriles and benzamides are good substrates for Birch reductions; benzonitriles give 1,4-cyclohexadienes, although the product of HCN elimination is obtained unless ammonium chloride is added prior to work-up.[388] Trimethylsilyl substituents can be used to control the regiochemistry of metal reductions of naphthalenes, allowing the formation of misoriented Birch

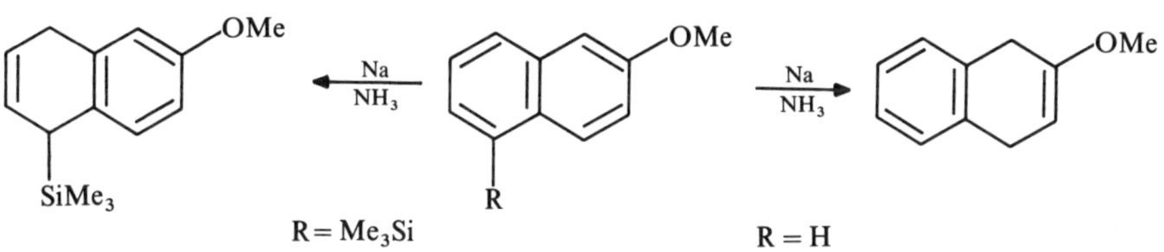

SCHEME 42

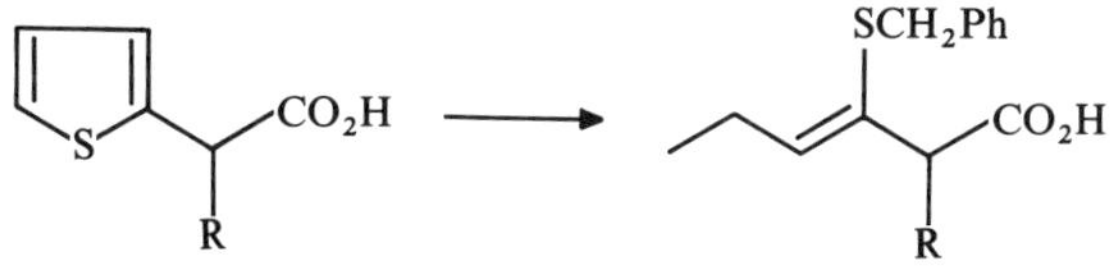

(56) **(57)**

products (Scheme 42).[389] the ring-opening by Birch reduction of 2,3-dialkylthiophens and subsequent alkylation gave a tetrasubstituted olefin formed by selective fragmentation of the S—C(5) bond; in the contrast, reduction of the thiophencarboxylic acid **(56)** and subsequent benzylation gave **(57)** with cleavage of the S—C(2) bond.[390] Unlike the carbon-phosphorus bond fragmentation observed

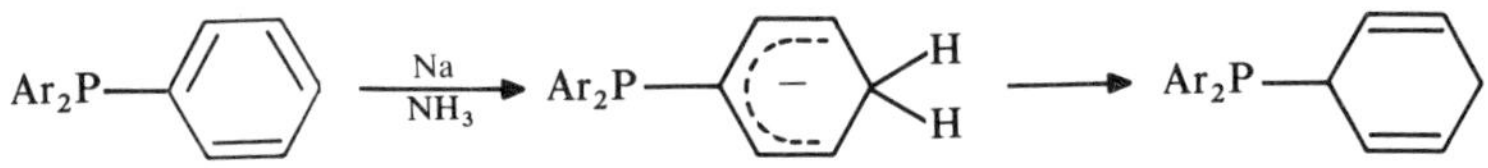

(58) **(59)**

Ar = *p*-$Me_2NC_6H_4$

with triphenylphosphine, bis(4-dimethylaminophenyl)phenylphosphine (**58**) is not cleaved by sodium in liquid ammonia; the intermediate phosphinocyclohexadienyl anion gives the Birch reduction product (**59**).[391] Calcium in liquid ammonia causes chemoselective cleavage of benzyl ethers;[392] aluminium in liquid ammonia reduces disulphides to sulphides.[393] The Clemmensen reduction of α,β-unsaturated ketones gives mainly olefins, showing no reduction of the carbon–carbon double bond.[394] The deoxygenation of aryl and α,β-unsaturated aldehydes and ketones with zinc and trimethylsilyl chloride does not involve pinacolic coupling prior to deoxygenation (Scheme 43).[395] The selective reduction of α,β-unsaturated esters to saturated esters without affecting unsaturation elsewhere in the molecule is achieved by use of magnesium in methanol.[396]

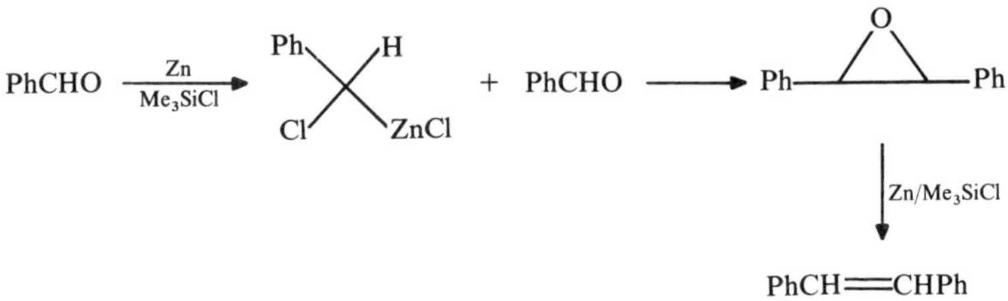

SCHEME 43

The reduction of a series of diketo-selenides with a low-valent titanium reagent generated from titanium(IV) chloride and zinc affords 2,5-dihydroselenophenes and 1,3-dienes (Scheme 44).[397] In the reduction of α-keto-esters by titanium(III), it is possible by choice of solvent to choose between dimeric or monomeric products; dimerization increases with increasing dielectric constant whereas alcohol formation increases with decreasing proton concentration.[398] Studies on iron-cluster-mediated reductions[399] of nitroarenes[400] and other organic substrates have been reported. Arenediazonium tetrafluoroborate salts undergo reduction with iron(II), and the second-order rate constants for the ferrocyanide reductions of *p*-nitro- to *p*-methoxy-substituted arenes span five orders of magnitude.[401] The mechanism of the coupling of aryl iodides to biphenyls by palladium(II) complexes has been

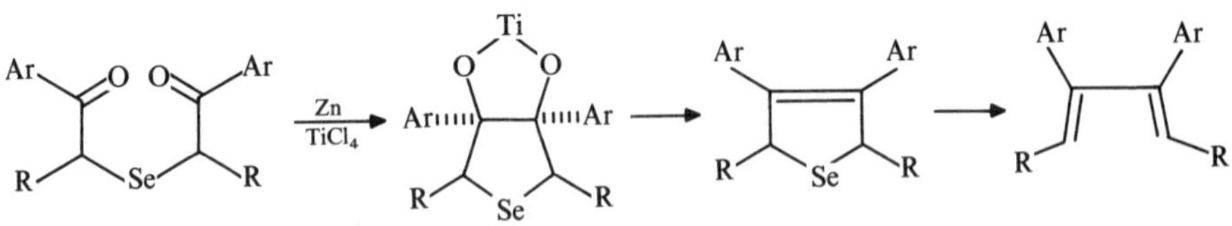

SCHEME 44

studied.[402] The reductions of nitro-compounds and thioketones by sodium hydrogen telluride involve nucleophilic attack by telluride.[403]

Miscellaneous Reductions

The structure-activity relationships for thiol-disulphide interchange between 36 di- and tri-thiols and the disulphides from 2-mercaptoethanol and dithreitol have been used to rank the thiols in terms of reduction potential; there is a general correlation between the reducing ability of the dithiols and the size of ring disulphide formed—dithiols which produce six-membered-ring disulphides are about one order of magnitude stronger reducing agents than dithiols which produce five- or seven-ring disulphides.[404] Aliphatic nitro-compounds react with carbon disulphide in the

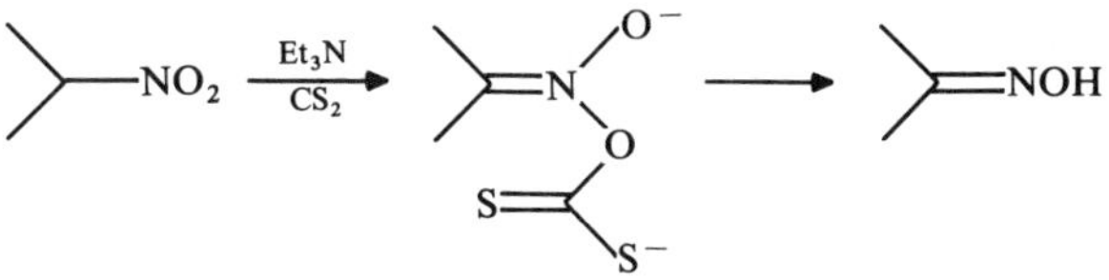

SCHEME 45

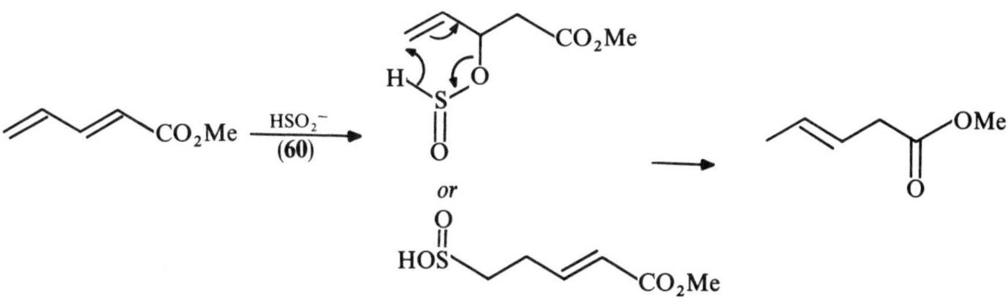

SCHEME 46

presence of triethylamine to give the corresponding oximes (Scheme 45);[405] both nitro-compounds and oximes are reduced to imines by tributylphosphine-diphenyl disulphide.[406] The sulphoxylate anion (**60**) may be the reactive species in the reduction of alkadienoates with sodium dithionite (Scheme 46).[407] the sodium dithionite reductions of 5-iminodaunomycin[408] and of dihexadecylphosphate vesicle-bound viologens[409] have been studied.

Lewis acid-promoted reductions of α,β-unsaturated carbonyl compounds by 2-phenylimidazoline (**61**)[410] and 2-phenylimidazothiazoline (**62**)[411] may both

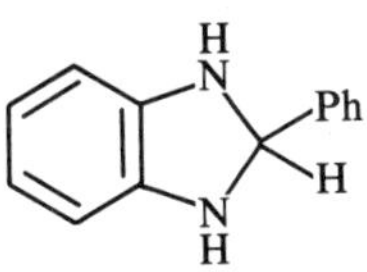

(61)

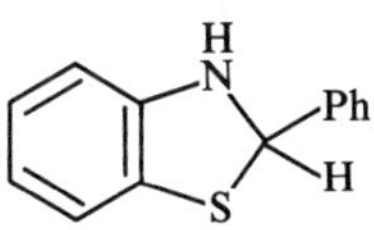

(62)

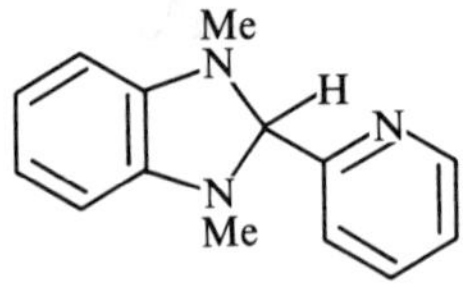

(63)

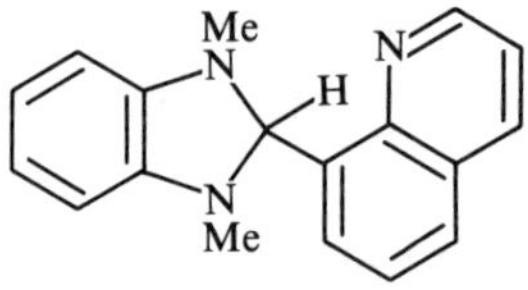

(64)

involve synchronous transfer of two hydrogens. The benzimidazolines (**63**)[412] and (**64**)[413] are both designed to bind electrophilic substrates and reduce them by internal source of hydride. The Huang–Minlon reduction of (**65**) gave (**66**) as a result of a transannular 1,6-hydride shift enforced by relief of steric strain.[414] Several studied on the reduction of aromatic nitro groups by alkoxides have been reported.[415,416] Competing stepwise and concerted pathways have been identified for the iodide reduction of sulphilimines;[417] catalysis of the decomposition of an intermediate sulphurane by hydrogen-bonding has been demonstrated.[418]

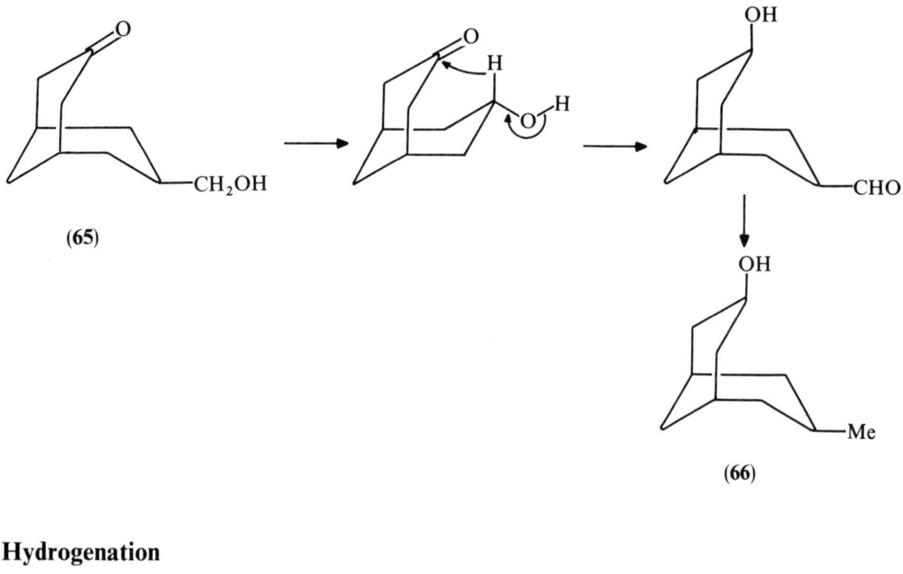

(65)

(66)

Hydrogenation

The isodesmic enthalpies of hydrogenation of all 29 monoalkenes containing fewer than seven carbon atoms are nearly the same, regardless of branching of the carbon skeleton.[419] Semi-empirical SCF-MO calculations for a variety of reaction involving dihydrogen transfer to a π bond predict that the energy difference between the concerted pericyclic and the stepwise radical-pair pathways is relatively insensitive to structural variation in regard to the π bond; the concerted route is favoured more by the AM1 than by the MNDO procedure—thus, dihydrogen transfer reactions may provide one means of calibrating semi-empirical against *ab initio* calculations of pericyclic mechanisms.[420]

The power of BINAP as a chiral ligand for asymmetric hydrogenation is further demonstrated in the use of ruthenium complexes for the asymmetric hydrogenation of β-keto-carboxylic esters to give access to β-hydroxy-esters in high enantiomeric purity,[421] and for the enantioselective hydrogenation of allylic and homoallylic alcohols.[422] The predominant product enantiomer in the asymmetric hydrogenation of methyl Z-α-acetamidocinnamate catalysed by (DIPAMP)rhodium(I) is derived from the less stable rhodium(I)-olefin adduct.[423] Studies of the asymmetric hydrogenation of N,N'-dimethyl-5-alklidenehydantoins catalysed by an achiral base coordinated to bis(dimethylglyoximato)cobalt(II), with a chiral quinuclidine co-catalyst as a proton donor, have been reported.[424,425] An unusual stereoselectivity is found in the hydrogenation of the methylenecyclohexane (**67**).[426]

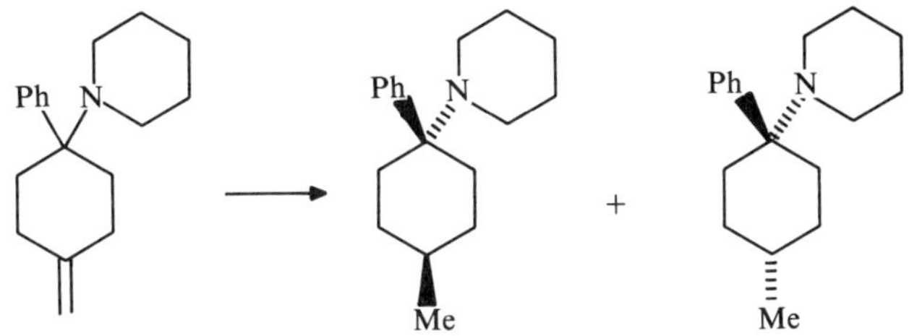

(**67**) 24 : 1

Stereochemical features of the hydrogenation of 2,3-dihydrofurans[427] and 1,3,5-trimethylcyclohexene[428] have been discussed. Kinetic studies on the homogeneous rhodium-catalysed hydrogenations of olefins have been reported.[429–431] Catalytic hydrogenation of the allene (**68**) occurs selectively at a terminal double bond.[432] Catalytic hydrogenation of [7]paracyclophane results in absorption of only two moles of hydrogen to give (**69**); the resistance to further hydrogenation and MM2 calculations shows that (**69**) is a member of the class of hyperstable olefins.[433]

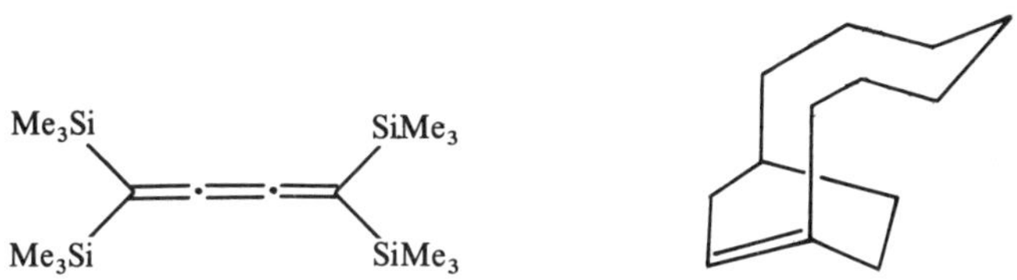

(**68**) (**69**)

Regioselective hydrogenations in the presence of palladium of quinolines and indoles,[434] and pyrànonaphthaquinones[435] have been studied. The reaction of an anionic ruthenium cluster with hydrogen gives reversible carbon-hydrogen bond formation.[436] Mechanistic aspects of rhodium-[437] and ruthenium[438]-catalysed hydrogenation of arenes, and of the catalytic hydrogenation of trinitrobenzenes[439,440] have been investigated. Aspects of the design of catalysts for the selective reduction of alkynes[441,442] and of the photo-hydrogenation of acetylenes in a water-oil two-phase system[443] have been discussed. In the hydrogenolysis of epoxides catalysed by platinum or palladium, the cleavage of the carbon-oxygen bond is rate-determining.[444] In the hydrogenolysis of aryloxytetrazolyl ethers, dissociation of a heterogeneous complex between the catalysts and the tetrazolone product is rate-limiting.[445] Selectivity in the hydrogenolysis of cyclopropane derivatives has been studied.[446,447]

Reductions and Oxidations of Biological Interest

The reduction by 1,4-dihydropyridines as a model for NADH reactions has been reviewed.[448] The mechanism for the reduction of α-haloacetophenones by NADH model compounds proceeds by a radical chain pathway in which both the initiation and propagation steps contain single-electron-transfer reactions; a critical assessment is given of the use of dihydronicotinamides as a model for the role of NADH in enzyme-mediated reactions.[449] The first non-enzymatic example of a stereospecific

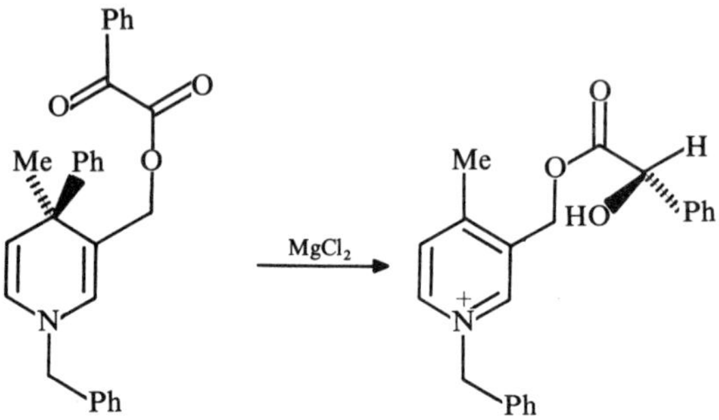

SCHEME 47

intramolecular reduction by an NADH model containing a covalently bound carbonyl moiety has been reported.[450] Intramolecular hydride transfer from the dihydropyridine group to the side-chain α-keto-ester in (**70**) is readily observed in water and is *not* subject to general acid catalysis.[451] A comparison has been made between the energetics of photo-induced electron transfer from NADH models to organic oxidants, and hydride transfers from NADH models to *p*-benzoquinones.[452] The photo-oxidation of benzyl alcohols with oxygen catalysed by the NADH model, 10-methylacridinium ion, is initiated by an electron-transfer reaction from the benzyl alcohol to the singlet excited state of the acridinium ion.[453] Other studies on photosensitized NADH reactions have appeared.[454]

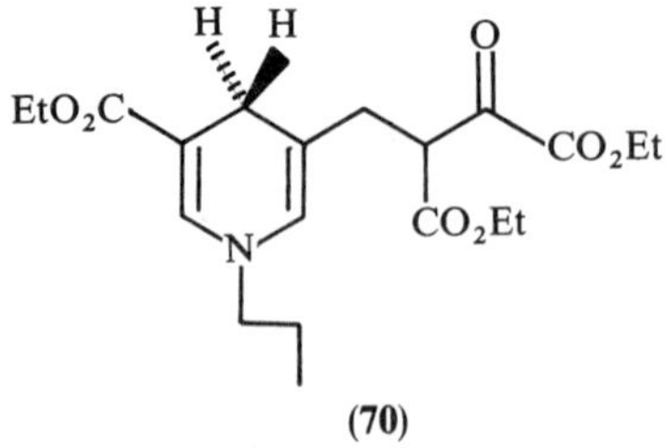

(**70**)

The magnesium ion-catalysed reductions of imines[455] and the silica gel-catalysed reductions of α,β-unsaturated carbonyl compounds[456,457] by Hantsch ester (**71**)

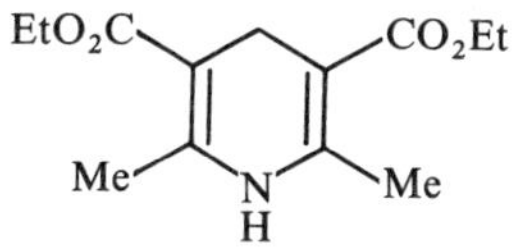

(71)

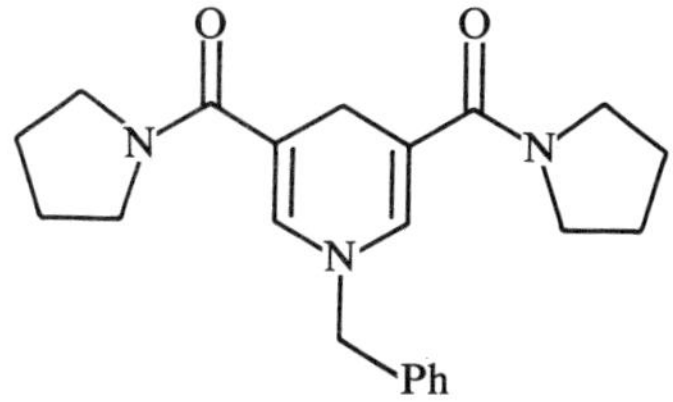

(72)

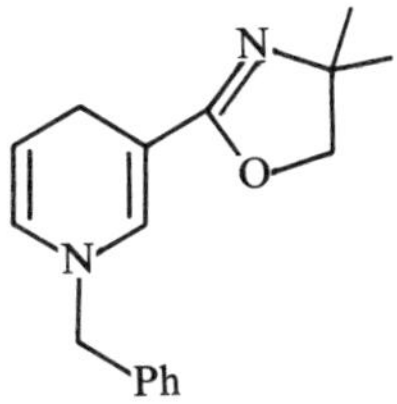

(73)

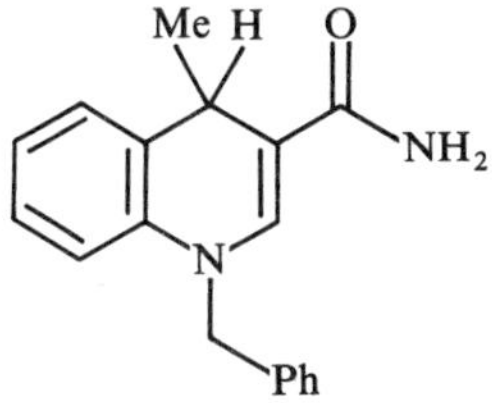

(74)

may provide examples of general acid catalysis. Metal ion[458] and carboxylic acid[459] catalysis of the reduction of nitrosoarenes by the NADH model (**72**) indicates the reduction is subject to general acid catalysis; the reduction of aryl nitroso-compounds by NADH under physiological conditions has been studied.[460] Other NADH model studies have been reported with (**73**)[461] and (**74**).[462] The addition of cyanide to NAD has been correlated with addition of hydride.[463] Slow conformational change controlling NADH release from the enzyme is rate-limiting for the oxidation of propionaldehyde by aldehyde dehydrogenase.[464] The hydride transfer to NAD by several forms of rat liver aldehyde dehydrogenase from the aldehyde substrate was A(pro-*R*)-stereospecific.[465] The mechanism of transfer of NADH among hydrogenases has been studied by transient stopped-flow techniques.[466]

α-Bromoacetophenone functions as an affinity label for human liver aldehyde dehydrogenase and provides the first identification of the active site of aldehyde dehydrogenase.[467]

References

1 Cossio, F. P., Lopez, M. C., and Palomo, C., *Tetrahedron*, **43**, 3963 (1987).
2 Kim, S. and Lhim, D. C., *Bull. Chem. Soc. Jpn*, **59**, 3297 (1986).
3 Carlsen, P. H. J. and Braeden, J. E., *Acta Chem. Scand.*, **41B**, 313 (1987).
4 Bhattacharjee, M. N., Chaudhuri, M. K., and Purkayastha, S., *Tetrahedron*, **43**, 5389 (1987).
5 Bhattacharjee, B., Bhattacharjee, M. N., Bhattacharjee, M., and Bhattacharjee, A. K., *Bull. Chem. Soc. Jpn*, **59**, 3217 (1986).
6 Liotta, D., Brown, D., Hoekstra, W., and Monahan, R., *Tetrahedron Lett.*, **28**, 1069 (1987).
7 Muzart, J., *Tetrahedron Lett.*, **28**, 2133 (1987).
8 Muzart, J., *Tetrahedron Lett.*, **28**, 2131 (1987).
9 Alvarez, M. M. P. and Diez, H. O., *An. Quim.*, **82A**, 231 (1986); *Chem. Abs.*, **107**, 6631 (1987).
10 Saran, N. K., Acharya, R. C., and Rao, S. R., *J. Indian Chem. Soc.*, **62**, 747 (1985); *Chem. Abs.*, **106**, 49462 (1987).
11 Koda, S., Endo, Y., Hirota, E., and Tsuchiya, S., *J. Phys. Chem.*, **91**, 5840 (1987); *Chem. Abs.*, **107**, 197312 (1987).

[12] Paquette, L. A. and Kobayashi, T., *Tetrahedron Lett.*, **28**, 3531 (1987).
[13] Perumal, S., Srinivasan, C., and Arumugam, N., *Indian J. Chem.*, **25A**, 367 (1986); *Chem. Abs.*, **106**, 32192 (1987).
[14] Radhakrishnamurti, P. S., Panda, H. P., and Pradhan, D. C., *React. Kinet. Catal. Lett.*, **31**, 79 (1986); *Chem. Abs.*, **107**, 38939 (1987).
[15] Rao, P. V. S., Srinivas, K., and Murty, K. S., *J. Indian Chem. Soc.*, **63**, 481 (1986); *Chem. Abs.*, **107**, 153778 (1987).
[16] Climent, M. S., Marinas, J. M., and Sinisterra, J. V., *Tetrahedron*, **43**, 3303 (1987).
[17] Mahmood, A. J., Fouzia, B., and Nawab, M. A., *Dhaka Univ., Stud., Part B*, **34**, 145 (1986); *Chem. Abs.*, **106**, 49488 (1987).
[18] Rao, M. A., Sethuram, B., Rao, T. N., and Jagannadham, V., *Oxid. Commun.*, **9**, 247 (1986); *Chem. Abs.*, **107**, 133726 (1987).
[19] Zielinski, M., *Nukleonika*, **31**, 81 (1986); *Chem. Abs.*, **107**, 197297 (1987).
[20] Rudakov, E. S., Tishchenko, N. A., and Volkova, L. K., *Kinet. Katal.*, **27**, 1101 (1986); *Chem. Abs.*, **106**, 119091 (1987).
[21] Monzo, I. S., Palou, J., Penalver, J., and Valero, R., *Int. J. Chem. Kinet.*, **19**, 741 (1987).
[22] Perez-Benito, J. F., *J. Chem. Soc., Chem. Commun.*, **1987**, 831.
[23] Perez-Benito, J. F. and Lee, D. G., *J. Org. Chem.*, **52**, 3239 (1987).
[24] Boncza-Tomaszewkski, Z., *Can. J. Chem.*, **65**, 656 (1987).
[25] Zurcher, A. and Hesse, M., *Helv. Chim. Acta*, **70**, 1937 (1987).
[26] Rathore, R. and Chandrasekaran, S., *J. Chem. Res. (S)*, **1986**, 458.
[27] Mata-Perez, F. and Perez-Benito, J. F., *Can. J. Chem.*, **65**, 2373 (1987).
[28] Freeman, F., Chang, L. Y., Kappos, J. C., and Sumarta, L., *J. Org. Chem.*, **52**, 1460 (1987).
[29] Perez-Benito, J. F., Mata-Perez, F., and Brillas, E., *Can. J. Chem.*, **65**, 2329 (1987).
[30] Garrido, J. A., Perez-Benito, J. F., Rodriguez, R. M., de Andres, J., and Brillas, E., *J. Chem. Res. (S)*, **1987**, 380.
[31] Fatiadi, A. J., *Org. Synth. Oxid. Met. Compd.*, **1986**, 119; *Chem. Abs.*, **107**, 38740 (1987).
[32] De Klein, W. J., *Org. Synth. Oxid. Met. Compd.*, **1986**, 261; *Chem. Abs.*, **107**, 38741 (1987).
[33] Ghandi, P. K. and Rao, B. M., *Z. Phys. Chem. (Leipzig)*, **268**, 513 (1987).
[34] Mehta, M., Nagori, R. R., and Mehrotra, R. N., *Can. J. Chem.*, **65**, 277 (1987).
[35] Sato, H., Nishino, H., and Kurosawa, K., *Bull. Chem. Soc. Jpn*, **60**, 1753 (1987).
[36] Snider, B. B. and Dombroski, M. A., *J. Org. Chem.*, **52**, 5487 (1987).
[37] Sengar, S. K. S. and Yadav, B. S., *Bull. Assoc. Kinet. India*, **8**, 3 (1986); *Chem. Abs.*, **107**, 58329 (1987).
[38] Tran, V. N., Nguyen, V. X., Tran, V. N., and Ho, S. T., *Tap Chi Hoa Hoc*, **23**, 3 (1985); *Chem. Abs.*, **106**, 195692 (1987).
[39] Riley, D. P., Getman, D. P., Beck, G. R., and Heintz, R. M., *J. Org. Chem.*, **52**, 287 (1987).
[40] Agabekov, V. E., Mitskevich, N. I., Krylova, S. V., Fedorishcheva, M. N., and Nevdakh, M. N., *Izv. Khim.*, **18**, 624 (1986); *Chem. Abs.*, **106**, 175560 (1987).
[41] Lopez, B. F. J., Hernandez, H. H. F., Medina, E. J., and Marin, S. R., *Analyst (London)*, **111**, 1325 (1986); *Chem. Abs.*, **106**, 175573 (1987).
[42] Petkevich, T. S., Kovalenko, N. A., and Mitskevich, N. I., *Vestsi Akad. Navuk USSR, Ser. Khim. Navuk*, **1986**, 20; *Chem. Abs.*, **106**, 175574 (1987).
[43] Balasubramanian, P. N., Sinha, A., and Bruice, T. C., *J. Am. Chem. Soc.*, **109**, 1456 (1987).
[44] Wong, W. H., Ostovic, D., and Bruice, T. C., *J. Am. Chem. Soc.*, **109**, 3428 (1987).
[45] van der Made, A. W., van Gerwen, M. J. P., Drenth, W., and Nolte, R. J. M., *J. Chem. Soc., Chem. Commun.*, **1987**, 888.
[46] Jorgensen, K. A., *J. Chem. Soc., Chem. Commun.*, **1987**, 1405.
[47] Groves, J. T. and Stern, M. K., *J. Am. Chem. Soc.*, **109**, 3812 (1987).
[48] Rummel, S. and Lorenz, B., *Isotopenpraxis*, **22**, 36 (1986); *Chem. Abs.*, **106**, 119049 (1987).
[49] Reddy, K. B., Sethuram, B., and Rao, T. N., *Z. Phys. Chem. (Leipzig)*, **268**, 706 (1987).
[50] Pathy, C. A. and Panigrahi, G. P., *Indian J. Chem.*, **25A**, 147 (1986); *Chem. Abs.*, **105**, 225579 (1987).
[52] Beckwith, A. L. J. and Zavitsas, A. A., *J. Am. Chem. Soc.*, **108**, 8230 (1986).
[53] Morris, G. E., Oakley, D., Pippard, D. A., and Smith, D. J. H., *J. Chem. Soc., Chem. Commun.*, **1987**, 410.
[54] Morris, G. E., Oakley, D., Pippard, D. A., and Smith, D. J. H., *J. Chem. Soc., Chem. Commun.*, **1987**, 411.
[55] De Jonge, C. R. H. I., *Org. Synth. Oxid. Met. Compd.*, **1986**, 423; *Chem. Abs.*, **107**, 38744 (1987).
[56] Balogh-Hergovich, E. and Speier, G., *J. Mol. Catal.*, **37**, 309 (1986); *Chem. Abs.*, **106**, 213208 (1987).
[57] Driscoll, J. J. and Kosman, D. J., *J. Am. Chem. Soc.*, **109**, 1765 (1987).

[58] Solozhenko, E. G., Panaova, G. V., Potapov, V. M., and Garbar, A. V., *Zh. Obsch. Khim.*, **56**, 405 (1986); *Chem. Abs.*, **105**, 225582 (1987).
[59] Boneva, M. and Ivanov, S., *Izv. Khim.*, **19**, 115 (1986); *Chem. Abs.*, **107**, 58294 (1987).
[60] Fetizon, M., Golfier, M., Mourgues, P., and Louis, J. M., *Org. Synth. Oxid. Met. Compd.*, **1986**, 503; *Chem. Abs.*, **107**, 38746 (1987).
[61] Solomon, J. L. and Madix, R. J., *J. Phys. Chem.*, **91**, 6241 (1987).
[62] Rao, P. J. P., Sethurma, B., and Rao, T. N., *Oxid. Commun.*, **9**, 155 (1986); *Chem. Abs.*, **107**, 153765 (1987).
[63] Gupta, K. K. S., Tarafder, A., and Mahapatra, A., *J. Chem. Res. (S)*, **1987**, 306.
[64] McKillop, A., *Org. Synth. Oxid. Met. Compd.*, **1986**, 695; *Chem. Abs.*, **106**, 195510 (1987).
[65] Milhailovic, M. L., Cekovic, Z., and Lorenc, L., *Org. Synth. Oxid. Met. Compd.*, **1986**, 741; *Chem. Abs.*, **106**, 195511 (1987).
[66] Banerjee, S. K., Shanker, R., and Sachdeva, O. P., *Z. Naturforsch.*, **41B**, 467 (1986); *Chem. Abs.*, **106**, 83790 (1987).
[67] Banerjee, S. K. and Shanker, R., *Z. Phys. Chem. (Leipzig)*, **268**, 409 (1987).
[68] Greenland, H., Pinhey, J. T., and Sternhell, S., *Aust. J. Chem.*, **39**, 2067 (1986).
[69] Tsankova, E. and Enev, V., *Tetrahedron*, **43**, 4425 (1987).
[70] Atkinson, R. S. and Kelly, B. J., *J. Chem. Soc., Chem. Commun.*, **1987**, 1362.
[71] Kitchin, J. P., *Org. Synth. Oxid. Met. Compd.*, **1986**, 633; *Chem. Abs.*, **106**, 195512 (1987).
[72] Ho, T. L., *Org. Synth. Oxid. Met. Compd.*, **1986**, 569; *Chem. Abs.*, **107**, 38747 (1987).
[73] Barton, D. H. R., Yadav-Bhatnagar, N., Finet, J.-P., Khamsi, J., Motherwell, W. B., and Stanforth, S. P., *Tetrahedron*, **43**, 323 (1987).
[74] Venediktoc, E. A. and Balakina, B. D., *Kinet. Katal.*, **27**, 1465 (1986); *Chem. Abs.*, **107**, 39096 (1987).
[75] Baciocchi, E., Giacco, T. D., and Sebastiani, G. V., *Tetrahedron Lett.*, **28**, 1941 (1987).
[76] Kreh, R. P., Spotnitz, R. M., and Lundquist, J. T., *Tetrahedron Lett.*, **28**, 1067 (1987).
[77] Dziegiec, J., *Pol. J. Chem.*, **59**, 979 (1986); *Chem. Abs.*, **106**, 175579 (1987).
[78] Ignaczak, M., Dziegiec, J., and Leszczynski, L., *Pol. J. Chem.*, **59**, 1039 (1986); *Chem. Abs.*, **107**, 77110 (1987).
[79] Virtanen, P. O. I., Lindroos, R., Oikarinen, E., and Vasjkuri, J., *Carbohydr. Res.*, **167**, 29 (1987).
[80] Vaidya, V. K., Joshi, S. N., and Bakore, G. V., *Z. Phys. Chem. (Leipzig)*, **268**, 364 (1987).
[81] Gyorgi, L., Deutsch, T., and Koros, E., *Int. J. Chem. Kinet.*, **19**, 35 (1987).
[82] Gyorgi, L., Deutsch, T., and Koros, E., *Int. J. Chem. Kinet.*, **19**, 435 (1987).
[83] Tsukada, M., *Chem. Lett.*, **1987**, 1707.
[84] Jorgensen, K. A., Wheeler, R. A., and Hoffmann, R., *J. Am. Chem. Soc.*, **109**, 3240 (1987).
[85] Pedersen, S. F., Dewan, J. C., Eckman, R. R., and Sharpless, K. B., *J. Am. Chem. Soc.*, **109**, 1279 (1987).
[86] Hawkins, J. M. and Sharpless, K. B., *Tetrahedron Lett.*, **28**, 2825 (1987).
[87] Schmidt, R. and Brauer, H.-D., *J. Am. Chem. Soc.*, **109**, 6976 (1987).
[88] Hawkins, J. M. and Sharpless, K. B., *Tetrahedron Lett.*, **28**, 2825 (1987).
[89] Pandey, A., Sharma, K., and Sharma, V. K., *Natl. Acad. Sci. Lett. (India)*, **8**, 383 (1985); *Chem. Abs.*, **107**, 58290 (1987).
[90] Micera, G., Pusino, A., and Gessa, C., *Inorg. Chim. Acta*, **120**, 49 (1986); *Chem. Abs.*, **106**, 101607 (1987).
[91] Kholuiskaya, S. N., Kasparov, V. V., Kovalev, V. V., Konovalova, A. A., and Rubailo, V. L., *Izv. Akad. Bauk SSSR, Ser. Khim.*, **1987**, 44; *Chem. Abs.*, **107**, 153760 (1987).
[92] Belokopytov, Y. V., Pyatnitskii, Y. L., Glushchuk, N. V., and Astrelin, I. M., *Ukr. Khim. Zh. (Russ. edn)*, **52**, 1145 (1986); *Chem. Abs.*, **107**, 197288 (1987).
[93] Mazsa, G. and Fabian, I., *J. Chem. Soc., Dalton Trans.*, **1986**, 2675.
[94] Bortolini, O., Campestrini, S., Di Furia, F., and Modena, G., *J. Org. Chem.*, **52**, 5467 (1987).
[95] Bortolini, O., Campestrini, S., Di Furia, F., and Modena, G., *J. Org. Chem.*, **52**, 5093 (1987).
[96] Gekhman, A. E., Nekipelov, V. M., Sakharov, S. G., Zamaraev, K. I., and Moiseev, I. I., *Izv. Akad. Nauk SSSR, Ser. Khim.*, **1986**, 1242; *Chem. Abs.*, **106**, 101592 (1987).
[97] Bortolini, O., Di Furia, F., Modena, G., and Schionato, A., *J. Mol. Catal.*, **35**, 47 (1986); *Chem. Abs.*, **106**, 49435 (1987).
[98] Dawoodi, Z. and Kelly, R. L., *Polyhedron*, **5**, 271 (1986); *Chem. Abs.*, **106**, 119050 (1987).
[99] Tempesti, E., Giuffre, L., Di Renzo, F., Mazzocchia, C., and Airoldi, G., *Appl. Catal.*, **26**, 285 (1986); *Chem. Abs.*, **107**, 133818 (1987).
[100] Herbowski, A., Sonczak, J. M., and Ziolkowski, J. J., *Polyhedron*, **5**, 277 (1986); *Chem. Abs.*, **106**, 83843 (1987).

[101] Amato, G., Arcoria, A., Ballistreri, F. P., Tomaselli, G. A., Bortolini, O., Conte, V., Di Furia, F., Modena, G., and Valle, G., *J. Mol. Catal.*, **37**, 165 (1986); *Chem. Abs.*, **107**, 58292 (1987).
[102] Quenard, M., Bonmarin, V., and Gelbard, G., *Tetrahedron Lett.*, **28**, 2237 (1987).
[103] Murahashi, S.-I., Oda, T., Sugahara, T., and Masui, Y., *J. Chem. Soc., Chem. Commun.*, **1987**, 1471.
[104] Khan, M. M. T., *Oxid. Commun.*, **9**, 105 (1986); *Chem. Abs.*, **106**, 17604 (1987).
[105] Ogata, Y. and Sawaki, Y., *Org. Synth. Oxid. Met. Compd.*, **1986**, 63; *Chem. Abs.*, **106**, 195513 (1987).
[106] Barton, D. H. R., Boivin, J., Motherwell, W. B., Ozbalik, N., Schwartzentruber, K. M., and Jankowski, K., *Nouveau J. Chim.*, **10**, 387 (1986).
[107] Bartzatt, B. and Carr, J., *Transition Met. Chem. (Weinheim, Ger.)*, **11**, 414 (1986); *Chem. Abs.*, **106**, 49489 (1987).
[108] Dasgupta, G. and Mahanti, M. K., *Indian J. Chem.*, **25A**, 958 (1986); *Chem. Abs.*, **107**, 77117 (1987).
[109] Gupta, K. C., Gupta, R., and Saxena, U., *Natl. Acad. Sci. Lett. (India)*, **8**, 323 (1985); *Chem. Abs.*, **107**, 58289 (1987).
[110] Singh, S. H., Dasgupta, G., and Mahanti, M. K., *Int. Congr. Catal.* [*Proc.*], *8th*, **5**, V463 (1984); *Chem. Abs.*, **106**, 83810 (1987).
[111] Kosmacheva, T. G., Vodo'pyanova, A. E., Agabekov, V. E., and Mitskevich, N. I., *Vestsi Akad. Nauk BSSR, Ser. Khim. Navuk*, **1986**, 28; *Chem. Abs.*, **106**, 175575 (1987).
[112] Traylor, T. G. and Miksztal, A. R., *J. Am. Chem. Soc.*, **109**, 2770 (1987).
[113] Traylor, T. G., Nakano, T., Miksztal, A. R., and Dunlap, B. E., *J. Am. Chem. Soc.*, **109**, 3625 (1987).
[114] Mansuy, D., *Pure Appl. Chem.*, **59**, 759 (1987).
[115] Nagiev, T. M., Nagieva, Z. M., and Tagieva, S. A., *Vestn. Mosk. Univ., Ser. 2: Khim.*, **28**, 42 (1987); *Chem. Abs.*, **107**, 197292 (1987).
[116] Jorgensen, K. A., *J. Am. Chem. Soc.*, **109**, 698 (1987).
[117] Sugimoto, H., Spencer, L., and Sawyer, D. T., *Proc. Natl. Acad. Sci. U.S.A.*, **84**, 1731 (1987).
[118] Balasubramanian, P. N. and Bruice, T. C., *Proc. Natl. Acad. Sci. U.S.A.*, **84**, 1734 (1987).
[119] Groves, J. T. and Watanabe, Y., *J. Am. Chem. Soc.*, **108**, 7834 (1987).
[120] Bruice, T. C., Dicken, C. M., Balasubramanian, P. N., Woon, T. C., and Lu, F.-L., *J. Am. Chem. Soc.*, **109**, 3436 (1987).
[121] Ostovic, D., Knobler, C. B., and Bruice, T. C., *J. Am. Chem. Soc.*, **109**, 3444 (1987).
[122] Woon, T. C., Dicken, C. M., and Bruice, T. C., *J. Am. Chem. Soc.*, **108**, 7990 (1986).
[123] Fretz, H. and Woggon, W.-D., *Helv. Chim. Acta*, **69**, 1959 (1986).
[124] Kobayashi, K., Amano, M., Kanbara, Y., and Hayashi, K., *J. Biol. Chem.*, **262**, 5445 (1987).
[125] Girardet, M. and Meunier, B., *Tetrahedron Lett.*, **28**, 2955 (1987).
[126] Freeman, F., *Org. Synth. Oxid. Met. Compd.*, **1986**, 315; *Chem. Abs.*, **107**, 38742 (1987).
[127] Koola, J. D. and Kochi, J. K., *J. Org. Chem.*, **52**, 4545 (1987).
[128] Kuo, M. C. and Chou, T. C., *Ind. Eng. Chem. Res.*, **26**, 277 (1987); *Chem. Abs.*, **106**, 83838 (1987).
[129] Perito, R. P. and Corden, B. B., *J. Am. Chem. Soc.*, **109**, 4418 (1987).
[130] Bratushko, Y. I., Ermokhina, N. I., and Yatsimirskii, K. B., *Zh. Neorg. Khim.*, **32**, 403 (1987); *Chem. Abs.*, **107**, 153762 (1987).
[131] Bratushko, Y. I. and Ermokhina, N. I., *Zh. Neorg. Khim.*, **32**, 398 (1987); *Chem. Abs.*, **107**, 153761 (1987).
[132] Shcherbina, F. F., *Ukr. Khim. Zh. (Russ. edn)*, **52**, 1279 (1986); *Chem. Abs.*, **107**, 197295 (1987).
[133] Guirado, A., Moss, G. P., and Utley, J. H. P., *J. Chem. Soc., Chem. Commun.*, **1987**, 41.
[134] Zaidi, S. A. H., *Appl. Catal.*, **27**, 99 (1986); *Chem. Abs.*, **106**, 175576 (1987).
[135] Hronec, M. and Ilavsky, J., *Oxid. Commun.*, **9**, 227 (1986); *Chem. Abs.*, **107**, 197278 (1987).
[136] Shcherbina, F. F., *Neftekhimiya*, **26**, 409 (1986); *Chem. Abs.*, **105**, 225587 (1987).
[137] Courtney, J. L., *Org. Synth. Oxid. Met. Compd.*, **1986**, 633; *Chem. Abs.*, **106**, 195514 (1987).
[138] Murahashi, S.-I., Naota, T., Ito, K., Maeda, T., and Taki, H., *J. Org. Chem.*, **52**, 4319 (1987).
[139] Vijayasri, K., Rajaram, J., and Kuriacose, J. C., *Proc. Indian Acad. Sci., Chem. Sci.*, **97**, 125 (1986); *Chem. Abs.*, **106**, 137781 (1987).
[140] Vijayasri, K., Rajaram, J., and Kuriacose, J. C., *J. Mol. Catal.*, **39**, 203 (1987); *Chem. Abs.*, **107**, 133718 (1987).
[141] Wong, K. Y., Che, C. M., and Anson, F., *Inorg. Chem.*, **26**, 737 (1986); *Chem. Abs.*, **106**, 119111 (1987).
[142] Khan, M. M. T. and Rao, A. P., *J. Mol. Catal.*, **35**, 237 (1986); *Chem. Abs.*, **106**, 17697 (1987).
[143] Roecker, L. and Meyer, T. J., *J. Am. Chem. Soc.*, **109**, 746 (1987).
[144] Felthouse, T. R., *J. Am. Chem. Soc.*, **109**, 7566 (1987).
[145] Yoshifuji, S., Arakawa, Y., and Nitta, Y., *Chem. Pharm. Bull.*, **35**, 357 (1987); *Chem. Abs.*, **107**, 197274 (1987).
[146] Verma, M. K., Tandon, P. K., Singh, H. S., and Singh, M. P., *Z. Phys. Chem. (Leipzig)*, **267**, 1219 (1986).

[147] Verma, M. K., Tandon, P. K., and Singh, M. P., *Z. Phys. Chem.* (*Leipzig*), **268**, 565 (1987).
[148] Roecker, L., Dobson, J. C., Vining, W. J., and Meyer, T. J., *Inorg. Chem.*, **26**, 779 (1987); *Chem. Abs.*, **106**, 119110 (1987).
[149] Gupta, K. C., Kumar, P., and Behari, K., *Bull. Soc. Chim. Fr. II*, **1987**, 232.
[150] Maitlis, P. M., *Org. Synth. Oxid. Met. Compd.*, **1986**, 469; *Chem. Abs.*, **107**, 38745 (1987).
[151] Lyons, J. E., Suld, G., and Hsu, C. Y., *Mezhdunar. Simp. Svyazi Gomogennym Geterog. Katal., 5th*, **1**, 107 (1986); *Chem. Abs.*, **107**, 115021 (1987).
[152] Backvall, J.-E., Awasthi, A. K., and Renko, Z. D., *J. Am. Chem. Soc.*, **109**, 4750 (1987).
[153] Backvall, J.-E. and Gogoll, A., *J. Chem. Soc., Chem. Commun.*, **1987**, 1236.
[154] Tsujii, J. and Minato, M., *Tetrahedron Lett.*, **28**, 3683 (1987).
[155] Minami, I. and Tsuji, J., *Tetrahedron*, **43**, 3903 (1987).
[156] Grover, G. S. and Chaudhari, R. V., *Chem. Eng. J.* (*Lausanne*), **32**, 93 (1986); *Chem. Abs.*, **106**, 83796 (1987).
[157] Gusevskaya, E. V., Beck, I. E., Stepanov, A. G., Likholobov, V. A., Nekipelov, V. M., and Yermakov, Y. I., *J. Mol. Catal.*, **37**, 177 (1986); *Chem. Abs.*, **106**, 137780 (1987).
[158] Kozhevnikov, I. V., Kim, V. I., and Sidel'nikov, V. N., *Izv. Akad. Nauk, SSSR, Ser. Khim.*, **1986**, 1352; *Chem. Abs.*, **106**, 83809 (1987).
[159] Singh, H. S., *Org. Synth. Oxid. Met. Compd.*, **1986**, 633; *Chem. Abs.*, **106**, 195509 (1987).
[160] Oppolzer, W. and Barras, J.-P., *Helv. Chim. Acta*, **70**, 1666 (1987).
[161] Solladie, G., Frechou, C., Hutt, J., and Demailly, G., *Bull. Soc. Chim. Fr. II*, **1987**, 827.
[162] Selnick, H. G. and Danishefsky, S. J., *Tetrahedron Lett.*, **28**, 4955 (1987).
[163] Vogel, E., Schmalstieg, L., Weyer, H.-J., and Gleiter, R., *Chem. Lett.*, **1987**, 33.
[164] Singh, A. K., Tewari, A., and Sisodia, A. K., *Natl. Acad. Sci. Lett.* (*India*), **8**, 209 (1985); *Chem. Abs.*, **106**, 175564 (1987).
[165] Mishra, P. and Khandaul, N. C., *Indian J. Chem.*, **25A**, 902 (1986); *Chem. Abs.*, **107**, 77115 (1987).
[166] Tripathy, K. S., *Indian J. Chem.*, **25A**, 762 (1986); *Chem. Abs.*, **106**, 66611 (1987).
[167] Misra, P., Mohapatra, R. C., and Khandaul, N. C., *J. Indian Chem. Soc.*, **63**, 291 (1986); *Chem. Abs.*, **106**, 119078 (1987).
[168] George, M. V., *Org. Synth. Oxid. Met. Compd.*, **1986**, 373; *Chem. Abs.*, **107**, 38743 (1987).
[169] Krylova, S. V., Agabekov, V. E., Fedorishcheva, M. N., Butovskaya, G. V., and Mitskevich, N. I., *Neftekhimya*, **27**, 125 (1987); *Chem. Abs.*, **107**, 153768 (1987).
[170] Koola, J. D. and Kochi, J. K., *Inorg. Chem.*, **26**, 908 (1987); *Chem. Abs.*, **106**, 155744 (1987).
[171] Murthy, C. P., Sethuram, B., and Rao, T. N., *Z. Phys. Chem.* (*Leipzig*), **267**, 1212 (1986).
[172] Reddy, T. R., Murthy, G. S. S., and Jagannadham, V., *Oxid. Commun.*, **9**, 83 (1986); *Chem. Abs.*, **106**, 101593 (1987).
[173] Kalinina, V. E., Zhukov, Y. A., Velasko, P. G., *Izv. Vyssh. Uchebn. Zaved., Khim. Khim. Tekhnol.*, **30**, 51 (1987); *Chem. Abs.*, **107**, 197298 (1987).
[174] El Shayeb, H. A., El Wahab, F. M. A., and Sultan, D. M., *Thermochim. Acta*, **109**, 367 (1987); *Chem. Abs.*, **107**, 58318 (1987).
[175] Moodie, R. B., Richards, S. N., and Thomas, M. P., *J. Chem. Soc., Chem. Commun.*, **1987**, 870.
[176] Schultz, T. P., *J. Org. Chem.*, **52**, 279 (1987).
[177] Anelli, P. L., Biffi, C., Montanari, F., and Quici, S., *J. Org. Chem.*, **52**, 2559 (1987).
[178] Deya, P. M., Dopico, M., Raso, A. G., Morey, J., and Saa, J. M., *Tetrahedron*, **43**, 3523 (1987).
[179] Schneider, H.-J. and Sauerbrey, R., *J. Chem. Res.* (*S*), **1987**, 14.
[180] Kano, K., Koga, M., and Anselme, J. P., *Bull. Soc. Chim. Belg.*, **96**, 137 (1987).
[181] Willis, C. L., *Tetrahedron Lett.*, **28**, 2175 (1987).
[182] Amon, C. M., Banwell, M. G., and Gravatt, G. L., *J. Org. Chem.*, **52**, 4851 (1987).
[183] Kim, H. J. and Kim, Y. H., *Tetrahedron Lett.*, **28**, 1669 (1987).
[184] Gilbert, B. C., Scarratt, C. J., Thomas, C. B., and Young, J., *J. Chem. Soc., Perkin Trans. 2*, **1987**, 371.
[185] Tsuboi, S., Mimura, S., Ono, S., Watanabe, K., and Takeda, A., *Bull. Chem. Soc. Jpn*, **60**, 1807 (1987).
[186] Sheradsky, T. and Moshenberg, R., *J. Org. Chem.*, **52**, 101 (1987).
[187] Syper, L., *Tetrahedron*, **43**, 2853 (1987).
[188] Murahashi, S.-I. and Shiota, T., *Tetrahedron Lett.*, **28**, 2383 (1987).
[189] Rao, M. P., Sethuram, B., and Rao, T. N., *Z. Phys. Chem.* (*Leipzig*), **268**, 578 (1987).
[190] Moriarty, R. M., Vaid, R. K., Duncan, M. P., and Vaid, K. V., *Tetrahedron Lett.*, **28**, 2845 (1987).
[191] Sicinski, R. R. and Szczepek, W. J., *Tetrahedron Lett.*, **28**, 5729 (1987).
[192] Kim, H. and Schlechi, M. F., *Tetrahedron Lett.*, **28**, 5229 (1987).
[193] Moriarty, R. M., Prakash, O., and Duncan, M. P., *J. Chem. Soc., Perkin Trans. 1*, **1987**, 559.
[194] Moriarty, R. M., Duncan, M. P., and Prakash, O., *J. Chem. Soc., Perkin Trans. 1*, **1987**, 1781.

[195] Moriarty, R. M., Prakash, O., Duncan, M. P., Vaid, R. K., and Musallam, H. A., *J. Org. Chem.*, **52**, 150 (1987).
[196] Tamura, Y., Yakura, T., Haruta, J., and Kita, Y., *J. Org. Chem.*, **52**, 3927 (1987).
[197] Moriarty, R. M., Vaid, R. K., and Farid, P., *J. Chem. Soc., Chem. Commun.*, **1987**, 711.
[198] Moriarty, R. M., Engerer, S. C., Prakash, O., Prakash, I., Gill, U. S., and Freeman, W. A., *J. Org. Chem.*, **52**, 153 (1987).
[199] Moriarty, R. M., Prakash, I., Clarisse, D. E., Penmasta, R., and Awasthi, A., *J. Chem. Soc., Chem. Commun.*, **1987**, 1209.
[200] Ranganathan, S., Ranganathan, D., Singh, S. K., Bhattacharyya, D., Shanthy, S., and Singh, G. P., *Tetrahedron*, **43**, 5363 (1987).
[201] Moriarty, R. M., Khosrowshahi, J. S., and Dalecki, T. M., *J. Chem. Soc., Chem. Commun.*, **1987**, 675.
[202] Martin, D., Jourdan, J. L., Laverdet, G., and Le Bras, G., *Int. J. Chem. Kinet.*, **19**, 503 (1987).
[203] Okubo, M., Yoshida, S., Egami, Y., Matsuo, K., and Nagase, S., *Bull. Chem. Soc. Jpn*, **60**, 1741 (1987).
[204] Doi, J. T., Bharadwaj, P. K., and Musker, W. K., *J. Org. Chem.*, **52**, 2581 (1987).
[205] Sharma, S. K. and Manglik, A., *J. Iraqi Chem. Soc.*, **10**, 75 (1985); *Chem. Abs.*, **106**, 119066 (1987).
[206] Blair, L. K., Parris, K. D., Lee, O. F. D., Jenkins, K. F., Feese, R. C., Belcher, T., Badger, D., Morris, D., and Kuhn, C., *J. Org. Chem.*, **51**, 5454 (1986).
[207] Jonnalagadda, S. B. and Muthakia, G. K., *J. Chem. Soc., Perkin Trans. 2*, **1987**, 1539.
[208] Srinivasan, C., Jegatheesan, P. P., and Arumugam, N., *Indian J. Chem.*, **25A**, 678 (1986); *Chem. Abs.*, **106**, 175548 (1987).
[209] Rav-Acha, C., Choshen, E., and Sarel, S., *Helv. Chim. Acta*, **69**, 1728 (1986).
[210] Blazejewski, J.-C., Dorme, R., and Wakselman, C., *J. Chem. Soc., Perkin Trans. 1*, **1987**, 1861.
[211] Inoue, S., Iwase, N., Miyamoto, O., and Sato, K., *Chem. Lett.*, **1986**, 2035.
[212] Weber, J. V., Schneider, M., Salami, B., and Paquer, D., *Recl. Trav. Chim. Pays-Bas*, **105**, 99 (1986); *Chem. Abs.*, **105**, 225575 (1986).
[213] Simoyi, R. H. and Epstein, I. R., *J. Phys. Chem.*, **91**, 5124 (1987).
[214] Ameta, R. C., Suresh, K. S., and Shanker, R., *Z. Phys. Chem. (Leipzig)*, **267**, 1227 (1986).
[215] Ameta, R. C., Suresh, K. S., and Shanker, R., *Z. Phys. Chem. (Leipzig)*, **268**, 502 (1987).
[216] Agarwal, A., Mathur, S., and Banerji, K. K., *J. Chem. Res. (S)*, **1987**, 176.
[217] Bishnoi, M. L., Negi, S. C., and Banerji, K. K., *Indian J. Chem.*, **25A**, 660 (1986); *Chem. Abs.*, **106**, 155709 (1987).
[218] Takaki, K., Yasumura, M., Tamura, T., and Negoro, K., *J. Org. Chem.*, **52**, 1256 (1987).
[219] Kothari, S., Agarwal, A., and Banerji, K. K., *Indian J. Chem.*, **25A**, 722 (1986); *Chem. Abs.*, **106**, 175572 (1987).
[220] Jambulingam, M., Nanjappan, P., Natarajan, K., and Ramalingam, K., *Indian J. Chem.*, **25B**, 390 (1986); *Chem. Abs.*, **106**, 101581 (1987).
[221] Mathur, A. and Banerji, K. K., *J. Chem. Soc., Perkin Trans. 2*, **1987**, 1645.
[222] Banerji, K. K., *J. Org. Chem.*, **51**, 4764 (1986).
[223] Singh, M. B., Tandon, P. K., and Singh, B., *Z. Phys. Chem. (Leipzig)*, **268**, 787 (1987).
[224] Hiremath, R. C., Mayanna, S. M., and Venkatasubramanian, N., *J. Chem. Soc., Perkin Trans. 2*, **1987**, 1569.
[225] Abe, Y., Okada, S., Horii, H., Taniguchi, S., and Yamabe, Y., *J. Chem. Soc., Perkin Trans. 2*, **1987**, 715.
[226] Singh, B., Sisodia, A. K., and Chand, R., *Oxid. Commun.*, **9**, 219 (1986); *Chem. Abs.*, **107**, 197277 (1987).
[227] Sundari, V. and Devi, R. C., *Curr. Sci.*, **55**, 850 (1986); *Chem. Abs.*, **106**, 195691 (1987).
[228] Beebe, T. R., Adkins, R., Baldridge, R., Hensley, V., McMillen, D., Morris, D., Noe, R., Ng, F. W., Powell, G., Spielberger, C., and Stolte, M., *J. Org. Chem.*, **52**, 5472 (1987).
[229] Zielinski, M., *Nukleonika*, **31**, 81 (1986); *Chem. Abs.*, **107**, 197297 (1987).
[230] Gowda, B. T. and Bhat, J. I., *Tetrahedron*, **43**, 2119 (1987).
[231] Ganapathy, K. and Kabilan, S., *Indian J. Chem.*, **25A**, 681 (1986); *Chem. Abs.*, **106**, 175549 (1987).
[232] Davis, F. A., Ulatowski, T. G., and Haque, M. S., *J. Org. Chem.*, **52**, 5288 (1987).
[233] Maccagnani, G., Innocenti, A., Zani, P., and Bhattaglia, A., *J. Chem. Soc., Perkin Trans. 2*, **1987**, 1113.
[234] Handoo, K. L. and Gadru, K., *Curr. Sci.*, **55**, 920 (1986); *Chem. Abs.*, **106**, 195698 (1987).
[235] Hoshino, Y. and Takeno, N., *Nippon Kagaku Kaishi*, **1986**, 157; *Chem. Abs.*, **106**, 4379 (1987).
[236] Brock, M., Hintze, H., and Heesing, A., *Chem. Ber.*, **119**, 3727 (1986).
[237] Inoue, H., Sumitani, M., Sekita, A., and Hida, M., *J. Chem. Soc., Chem. Commun.*, **1987**, 1681.
[238] Byrne, B. and Karras, M., *Tetrahedron Lett.*, **28**, 769 (1987).
[239] Wayne, R. P., *Atmospheric Environment*, **21**, 1683 (1987).
[240] Bennett, P. J., Harris, S. J., and Kerr, J. A., *Int. J. Chem. Kinet.*, **19**, 609 (1987).

241 Nakamura, N., Nojima, M., and Kusabayashi, S., *J. Am. Chem. Soc.*, **109**, 4969 (1987).
242 Griesbaum, K. and Meister, M., *Chem. Ber.*, **120**, 1573 (1987).
243 Greisbaum, K. and Bandyopadhyay, A. R., *Can. J. Chem.*, **65**, 487 (1987).
244 Niki, H., Maker, P. D., Savage, C. M., Breitenbach, L. P., and Hurley, M. D., *J. Phys. Chem.*, **91**, 941 (1987).
245 Martinez, I. and Herron, J. T., *J. Phys. Chem.*, **91**, 946 (1987).
246 Kashima, C., Hibi, S., Maruyama, T., Harada, K., and Omote, Y., *J. Heterocycl. Chem.*, **24**, 637 (1987).
247 Tyupalo, N. F. and Dmitruk, A. F., *Dopov. Akad. Nauk Ukr. RSR, Ser. B: Geol., Khim. Biol. Nauki*, **1986**, 50; *Chem. Abs.*, **107**, 197269 (1987).
248 Yakobi, V. A., Shpak, L. P., Ivanova, N. N., Ponomarev, V. A., and Fersyuk, Z. I., *Izv. Khim.*, **19**, 206 (1986); *Chem. Abs.*, **107**, 6623 (1987).
249 Galstyan, T. M., Yakobi, V. A., and Galstyan, G. A., *Izv. Khim.*, **19**, 215 (1986); *Chem. Abs.*, **107**, 38928 (1987).
250 Yakobi, V. A. and Fersyuk, Z. I., *Khim. Tekhnol. Vody*, **8**, 29 (1986); *Chem. Abs.*, **106**, 175577 (1987).
251 Kuramshin, E. M., Kochinashvili, M. V., Kulak, L. G., Kotlyar, S. A., Zlotskii, S. S., and Rakhmankulov, D. L., *Dokl. Akad. Nauk SSSR*, **288**, 1120 (1986); *Chem. Abs.*, **106**, 175542 (1987).
252 Cullis, P. M., Arnold, J. R. P., Clarke, M., Howell, R., DeMira, M., Naylor, M., and Nicholls, D., *J. Chem. Soc., Chem. Commun.*, **1987**, 1088.
253 Padwa, A., Kline, D. N., and Perumattam, J., *Tetrahedron Lett.*, **28**, 913 (1987).
254 Crandall, J. K., Salazar, G. E., and Watkins, R. J., *J. Am. Chem. Soc.*, **109**, 4338 (1987).
255 Paquette, L. A., Lin, H.-O., and Gallucci, J. C., *Tetrahedron Lett.*, **28**, 1363 (1987).
256 Rousch, W. R., Straub, J. A., and Brown, R. J., *J. Org. Chem.*, **52**, 5127 (1987).
257 Kogen, H. and Nishi, T., *J. Chem. Soc., Chem. Commun.*, **1987**, 311.
258 Pazderskaya, I. Y., Pikh, Z. G., Samarik, V. Y., and Tolopko, D. K., *Zh. Fiz. Khim.*, **60**, 1872 (1986); *Chem. Abs.*, **106**, 175565 (1987).
259 Harayama, T., Kotoji, K., Yanada, R., Yoneda, F., Taga, T., Osaki, K., and Nagamatsu, T., *Chem. Pharm. Bull.*, **34**, 2354 (1986); *Chem. Abs.*, **106**, 137752 (1987).
260 Baumstark, A. L. and McCloskey, C. J., *Tetrahedron Lett.*, **28**, 3311 (1987).
261 Evanseck, J. D., Blake, J. F., and Jorgensen, W. L., *J. Am. Chem. Soc.*, **109**, 2349 (1987).
262 Adam, W., Chan, Y.-Y., Cremer, D., Gauss, J., Scheitzow, D., and Schindler, M., *J. Org. Chem.*, **52**, 2800 (1987).
263 Tishchenko, I. G., Revinskii, I. F., and Burd, V. N., *Dokl. Akad. Nauk BSSR*, **30**, 632 (1986); *Chem. Abs.*, **106**, 17713 (1987).
264 Revinskii, I. F., Tishchenko, I. G., and Burd, V. N., *Zh. Org. Khim.*, **22**, 1070 (1986); *Chem. Abs.*, **106**, 195677 (1987).
265 Tishchenko, I. G., Burd, V. N., and Revinskii, I. F., *Dokl. Akad. Nauk BSSR*, **30**, 826 (1986); *Chem. Abs.*, **106**, 195694 (1987).
266 Shunmugasundaram, A., Chellamani, A., and Raj, M. J., *Indian J. Chem.*, **25A**, 899 (1986); *Chem. Abs.*, **107**, 77114 (1987).
267 Colonna, S., Manfredi, A., Annunziata, R., and Spadoni, M., *Tetrahedron*, **43**, 2157 (1987).
268 Baumstark, A. L. and Vasquez, P. C., *J. Org. Chem.*, **52**, 1939 (1987).
269 Kitagawa, T., Takeuchi, K., Murai, O., Matsui, S., Inoue, T., Nishimura, M., and Okamoto, K., *J. Chem. Soc., Perkin Trans. 2*, **1986**, 1987.
270 Kice, J. L. and Manek, M. B., *J. Org. Chem.*, **52**, 2953 (1987).
271 Bizjak, M., Benner, H., and Novakov, T., *Vestn. Slov. Kem. Drus.*, **33**, 207 (1986); *Chem. Abs.*, **106**, 155751 (1987).
272 Amira, M. F., Sallam, S. A., Isklander, M. F., and Morsi, S. E., *Indian J. Chem.*, **25A**, 442 (1986); *Chem. Abs.*, **106**, 32193 (1987).
273 Butovskaya, V. V., Supichenko, G. N., Zhavnerko, G. K., Agabekov, V. E., Antonovskii, V. L., Vandrovskii, V. A., and Mitskevich, N. I., *Vestsi Akad. Navuk BSSR, Ser. Khim. Navuk*, **1986**, 14; *Chem. Abs.*, **107**, 38923 (1987).
274 Akasaka, T. and Ando, W., *J. Am. Chem. Soc.*, **109**, 1260 (1987).
275 Nelsen, S. F., *Acc. Chem. Res.*, **20**, 269 (1987).
276 Martinez, I., *J. Phys. Chem.*, **91**, 1345 (1987).
277 Kojima, M., Sakauragi, H., and Tokomaro, K., *Bull. Chem. Soc. Jpn*, **60**, 3331 (1987).
278 Jefford, C. W., Estrada, M. J., and Barchietto, G., *Tetrahedron*, **43**, 1737 (1987).
279 Guarini, A. and Tundo, P., *J. Org. Chem.*, **52**, 3501 (1987).
280 Gollnick, K. and Schnatterer, A., *Photochem. Photobiol.*, **43**, 265 (1986); *Chem. Abs.*, **106**, 119052 (1987).

281 Chen, J., Cao, Y., Zhang, B., and Ming, Y., *Huaxue Xuebao*, **44**, 192 (1986); *Chem. Abs.*, **106**, 101578 (1987).
282 Kropf, H. and Reichwaldt, R., *J. Chem. Res. (S)*, **1987**, 412.
283 Jiancheng, Z. and Foote, C. S., *Tetrahedron Lett.*, **27**, 6153 (1986).
284 Schmidt, R. and Brauer, H.-D., *J. Am. Chem. Soc.*, **109**, 6976 (1987).
285 Venediktoc, E. A. and Balakina, B. D., *Kinet. Katal.*, **27**, 1465 (1986); *Chem. Abs.*, **107**, 39096 (1987).
286 Encinas, M. V., Lemp, E., and Lissi, E. A., *J. Chem. Soc., Perkin Trans. 2*, **1987**, 1125.
287 Porter, N. A. and Wujek, J. S., *J. Org. Chem.*, **52**, 5085 (1987).
288 Akasaka, T. and Ando, W., *Tetrahedron Lett.*, **28**, 217 (1987).
289 Clennan, E. L. and Lewis, K. K., *J. Am. Chem. Soc.*, **109**, 2475 (1987).
290 Clennan, E. L. and Matusuno, K., *J. Org. Chem.*, **52**, 3483 (1987).
291 Clennan, E. L. and Nagraba, K., *J. Org. Chem.*, **52**, 294 (1987).
292 Sander, W. and Patyk, A., *Angew. Chem. Int. Edn*, **26**, 475 (1987).
293 Jensen, F. and Foote, C. S., *J. Am. Chem. Soc.*, **109**, 1478 (1987).
294 Kim, Y. H. and Lee, H. K., *Chem. Lett.*, **1987**, 1499.
295 Das, S., Schuchmann, M. N., Schuchmann, H.-P., and Von Sonntag, C., *Chem. Ber.*, **120**, 319 (1987).
296 Frimer, A. A. and Farkash-Solomon, T., *Superoxide Dismutase Chem. Biol. Med., Proc. Int. Conf., 4th*, **1985**, 41; *Chem. Abs.*, **107**, 6624 (1987).
297 Afanas'ev, I. B. and Polozova, N. I., *J. Chem. Soc., Perkin Trans. 2*, **1987**, 835.
298 Afanas'ev, I. B., Grabovetskii, V. V., and Kuprianova, N. S., *J. Chem. Soc., Perkin Trans. 2*, **1987**, 281.
299 Mashino, T. and Fridovich, I., *Arch. Biochem. Biophys.*, **254**, 547 (1987).
300 Forrester, A. R. and Purushotham, V., *Tetrahedron Lett.*, **28**, 3279 (1987).
301 Abriola, D. P., Fields, R., Stein, S., MacKerell, A. D., and Pietruszko, R., *Biochemistry*, **26**, 5679 (1987).
302 Klemm, R. B., Nesbitt, F. L., Skolnik, E. G., Lee, J. H., and Smalley, J. F., *J. Phys. Chem.*, **91**, 1574 (1987).
303 Mahmud, K., Marshall, P., and Fontjin, A., *J. Phys. Chem.*, **91**, 1568 (1987).
304 Zhan, M., Zhou, S., Qiu, Y., Liu, S., Shi, J., Li, F., and Yao, J., *Huaxue Xuebao*, **44**, 1149 (1986); *Chem. Abs.*, **107**, 133719 (1987).
305 Mahmud, K. and Fontjin, A., *J. Phys. Chem.*, **91**, 1918 (1987).
306 Mitchell, M. B., Nava, D. F., and Stief, L. J., *J. Chem. Phys.*, **85**, 3300 (1986); *Chem. Abs.*, **106**, 175563 (1987).
307 Veprek-Siska, J., *Oxid. Commun.*, **8**, 301 (1985); *Chem. Abs.*, **106**, 175537 (1987).
308 Zakharov, I. V. and Geletii, Y. V., *Neftekhimiya*, **26**, 776 (1986); *Chem. Abs.*, **107**, 6635 (1987).
309 Solyanikov, V. M., Petrov, L. V., and Drozdova, T. I., *Neftekhimiya*, **26**, 766 (1986); *Chem. Abs.*, **107**, 6634 (1987).
310 Digurov, N. G., Bukharkina, T. V., and Lebedev, N. N., *Neftekhimiya*, **26**, 787 (1986); *Chem. Abs.*, **107**, 6636 (1987).
311 Kharitonov, V. V., Psikha, B. L., and Krashakov, S. A., *Khim. Fiz.*, **6**, 218 (1987); *Chem. Abs.*, **107**, 153764 (1987).
312 Zhong, Y., *Huaxue Fanying Gongcheng Yu Gongyi*, **2**, 60 (1986); *Chem. Abs.*, **107**, 153653 (1987).
313 Martem'yanov, V. S., *Neftekhimiya*, **26**, 753 (1986); *Chem. Abs.*, **107**, 6533 (1987).
314 Taketomo, K., *Nagasaki Daigaku Kyoikugakubu Shizen Kagaku Kenkyu Hokoku*, **1986**, 39; *Chem. Abs.*, **106**, 101604 (1987).
315 Lissi, E. A. and Alvarado, J. J., *Bol. Soc. Chil. Quim.*, **30**, 171 (1985); *Chem. Abs.*, **106**, 175535 (1987).
316 Borisov, A. A., Zamanskii, V. M., Konnov, A. A., and Skachkov, G. L., *Khim. Fiz.*, **6**, 407 (1987); *Chem. Abs.*, **107**, 197289 (1987).
317 Kaiser, E. W., *Int. J. Chem. Kinet.*, **19**, 457 (1987).
318 Matsubara, Y., Takekuma, S., Yokoi, K., Yamamoto, H., and Nozoe, T., *Bull. Chem. Soc. Jpn*, **60**, 1415 (1987).
319 Matsubara, Y., Takezuma, S., Yamamoto, H., and Nozoe, T., *Chem. Lett.*, **1987**, 455.
320 Valechrachova, G., Hronec, M., and Vesely, V., *Collect. Czech. Chem. Commun.*, **51**, 2574 (1986).
321 Hronec, M., Masarovic, F., Cvengrosova, Z., and Ilavsky, J., *Collect. Czech. Chem. Commun.*, **52**, 2241 (1987).
322 Kutuev, A. A., *Khim. Tekhnol. Pererab. Nefti Gaza*, **1985**, 70; *Chem. Abs.*, **107**, 153775 (1987).
323 Kharlampidi, K. E., Chirkunov, E. V., Kornilova, L. I., Miroshkin, N. P., Batyrshin, N. N., and Nigmatullina, F. I., *Khim. Tekhnol. Pererab. Nefti Gaza*, **1985**, 10; *Chem. Abs.*, **107**, 153774 (1987).
324 Kozlov, S. K., Tovstokhat'ko, F. I., and Potekhin, V. M., *Zh. Prikl. Khim. (Leningrad)*, **59**, 1381 (1986); *Chem. Abs.*, **106**, 83805 (1987).

325 Kozlov, S. K. and Potekhin, V. M., *Zh. Prikl. Khim. (Leningrad)*, **59**, 1397 (1986); *Chem. Abs.*, **106**, 83806 (1987).
326 Santacesaria, E., Ferro, R., Ricci, S., and Carra, S., *Ind. Eng. Chem. Res.*, **26**, 155 (1987); *Chem. Abs.*, **106**, 49480 (1987).
327 Willms, R. S., Balinsky, A. M., Reible, D. D., Wetzel, D. M., and Harrison, D. P., *Ind. Eng. Chem. Res.*, **26**, 148 (1987); *Chem. Abs.*, **106**, 49479 (1987).
328 Aleksandrov, A. L., *Izv. Akad. Nauk SSSR, Ser. Khim.*, **1986**, 1736; *Chem. Abs.*, **106**, 195700 (1987).
329 Aleksandrov, A. L. and Sapozhkova, V. M., *Izv. Akad. Nauk SSSR, Ser. Khim.*, **1986**, 1742; *Chem. Abs.*, **106**, 195701 (1987).
330 Pankratov, A. N., Shmakov, S. L., Mushtakova, S. P., and Gribov, L. A., *Zh. Anal. Khim.*, **41**, 810 (1986); *Chem. Abs.*, **107**, 58288 (1987).
331 Belayakov, V. A., Zalevskaya, N. M., Kachurin, I. O., and Opeida, I. A., *Neftekhimya*, **26**, 670 (1986); *Chem. Abs.*, **107**, 153749 (1987).
332 Chang, W. D., Karra, S. B., and Senkan, S. M., *Combust. Sci. Technol.*, **49**, 107 (1986); *Chem. Abs.*, **107**, 6627 (1987).
333 Diaz, R. F., Alvarez, D. M., and Moreno, P. J., *An. Quim.*, **82A**, 579 (1986); *Chem. Abs.*, **107**, 153767 (1987).
334 Nishihama, T. and Takahashi, T., *Bull. Chem. Soc. Jpn*, **60**, 2117 (1987).
335 Anthoni, U., Larsen, C., Nielsen, P. H., and Christopherson, C., *Acta Chem. Scand.*, **41B**, 216 (1987).
336 Mukai, K., Yokoyama, S., Fukuda, K., and Uemoto, Y., *Bull. Chem. Soc. Jpn*, **60**, 2163 (1987).
337 Acharya, S., Neogi, G., and Panda, R. K., *Z. Phys. Chem. (Leipzig)*, **267**, 1189 (1986).
338 Abe, Y., Okada, S., Horii, H., Taniguchi, S., and Yamabe, Y., *J. Chem. Soc., Perkin Trans. 2*, **1987**, 715.
339 Taylor, R., *J. Chem. Res. (S)*, **1987**, 178.
340 Greenland, H., Pinhey, J. T., and Sternhell, S., *Aust. J. Chem.*, **40**, 325 (1987).
341 Endo, T., Ueda, T., Isago, T., Kato, J., Suzuki, T., Misaka, M., and Ito, M. M., *Bull. Chem. Soc. Jpn*, **60**, 2298 (1987).
342 D'Souza, V. T., Iyer, V. K., and Szmant, H. H., *J. Org. Chem.*, **52**, 1725 (1987).
343 D'Souza, V. T., Nanjundiah, R., Jaime, B. H., and Szmant, H. H., *J. Org. Chem.*, **52**, 1729 (1987).
344 Chung, M.-I., D'Souza, V. T., and Szmant, H. H., *J. Org. Chem.*, **52**, 1741 (1987).
345 Mashino, T. and Fridovich, I., *Arch. Biochem. Biophys.*, **252**, 163 (1987).
346 Kovacic, K. and Jones, M. B., *Chem. Rev.*, **87**, 357 (1987).
347 Brown, H. C., Nazer, B., Cha, J. S., and Sikorski, J., A., *J. Org. Chem.*, **51**, 5264 (1986).
348 Brown, H. C., Park, W. S., Cho, B. T., and Ramachandran, P. V., *J. Org. Chem.*, **52**, 5406 (1987).
349 Garry, S. W. and Neilson, D. G., *J. Chem. Soc., Perkin Trans. 1*, **1987**, 601.
350 Kashima, C., Harada, K., and Omote, Y., *J. Chem. Soc., Perkin Trans. 1*, **1987**, 1495.
351 Orellana, G., Perez-Ossorio, R., Quiroga, M. L., and Sanchez-Asiain, M. P., *J. Chem. Soc., Perkin Trans. 2*, **1987**, 679.
352 Kitazume, T., Kobayashi, T., Yamamoto, T., and Yamazaki, T., *J. Org. Chem.*, **52**, 3218 (1987).
353 Barluenga, J., Resa, J. G., Olano, B., and Fustero, S., *J. Org. Chem.*, **52**, 1425 (1987).
354 Hasegawa, F., Yanagiya, M., and Matsumoto, T., *Tetrahedron Lett.*, **28**, 1671 (1987).
355 Frenette, R., Kakushima, M., Zamboni, R., Young, R. N., and Verhoeven, T. R., *J. Org. Chem.*, **52**, 304 (1987).
356 Tamura, Y., Annoura, H., and Fujioka, H., *Tetrahedron Lett.*, **28**, 5681 (1987).
357 Natrajan, A., Ferrara, J. D., Youngs, W. J., and Sukenik, C. N., *J. Am. Chem. Soc.*, **109**, 7477 (1987).
358 Girardet, M. and Meunier, B., *Tetrahedron Lett.*, **28**, 2955 (1987).
359 Armesto, D., Perez-Ossorio, R., Plumet, J., and Ramos, A., *J. Chem. Res. (S)*, **1987**, 80.
360 Alcaide, B., Lopez-Mardomingo, C., and Plumet, J., *J. Chem. Res. (S)*, **1987**, 270.
361 Bartnik, R., Eddaif, A., Laurent, A., and Lesniak, S., *Bull. Pol. Acad. Sci., Chem.*, **34**, 27 (1986); *Chem. Abs.*, **106**, 4328 (1987).
362 Bergman, J., Tilstam, U., Tornroos, K.-W., *J. Chem. Soc., Perkin Trans. 1*, **1987**, 519.
363 Ceccherelli, P., Curini, M., Diamantini, M., Marcotullio, M. C., and Grandolini, G., *Tetrahedron*, **43**, 243 (1987).
364 Carreno, M. C., Dominguez, E., Garcia-Ruano, J. L., and Rubio, A., *J. Org. Chem.*, **52**, 3619 (1987).
365 Polniaszek, R. P. and McKee, J. A., *Tetrahedron Lett.*, **28**, 4511 (1987).
366 Iida, H., Yamazaki, N., and Kibayashi, C., *J. Chem. Soc., Chem. Commun.*, **1987**, 746.
367 Hutchins, R. O. and Rutledge, M. C., *Tetrahedron Lett.*, **28**, 5619 (1987).
368 Balaban, T.-S. and Balaban, A. T., *Tetrahedron Lett.*, **28**, 1341 (1987).
369 Nieminen, T. E. A. and Hase, T. A., *Tetrahedron Lett.*, **28**, 4725 (1987).
370 Martinez, A. G., Alvarez, R. M., Casado, M. M., Subramanian, L. R., and Hanack, M., *Tetrahedron*, **43**, 275 (1987).

[371] Baynham, M. K., Dickinson, J. M., Hanson, J. R., and Hitchcock, P. B., *J. Chem. Soc., Perkin Trans. 1*, **1987**, 1987.
[372] Fujita, M. and Hiyama, T., *Tetrahedron Lett.*, **28**, 2263 (1987).
[373] Ookawa, A., Hiratsuka, H., and Soai, K., *Bull. Chem. Soc. Jpn*, **60**, 1813 (1987).
[374] Cha, J. S. and Kwon, S. S., *J. Org. Chem.*, **52**, 5486 (1987).
[375] Vijn, R. J., Hiemstra, H., Kok, J. J., Knotter, M., and Speckamp, W. N., *Tetrahedron*, **43**, 5019 (1987).
[376] Soucy, C., Favreau, D., and Kayser, M. M., *J. Org. Chem.*, **52**, 129 (1987).
[377] Ismailov, F. K., Avanesova, S. S., Shakhtakhtinskii, T. N., Alieva, S. A., and Samedova, D. Z., *Azerb. Khim. Zh.*, **1985**, 23; *Chem. Abs.*, **106**, 32196 (1987).
[378] Mori, A., Ishihara, K., Arai, I., and Yamamoto, H., *Tetrahedron*, **43**, 755 (1987).
[379] Blanton, J. R., *J. Org. Chem.*, **52**, 472 (1987).
[380] Fort, R. C., Stahl, M. H., and Sky, A. F., *J. Org. Chem.*, **52**, 2396 (1987).
[381] Joshua, A. V., Scott, J. R., Sondhi, S. M., Ball, R. G., and Lown, J. W., *J. Org. Chem.*, **52**, 2447 (1987).
[382] Bartoli, G., Bosco, M., Pozzo, R. D., Petrini, M., *Tetrahedron*, **43**, 4221 (1987).
[383] Vinvens, M., Fadel, R., and Vidal, M., *Bull. Soc. Chim. Fr. II*, **1987**, 462.
[384] Tsuda, T., Yoshida, T., Kawamoto, T., and Saegusa, T., *J. Org. Chem.*, **52**, 1624 (1987).
[385] Pradhan, S. K., *Tetrahedron*, **42**, 6351 (1986).
[386] Huffman, J. W., Liao, W.-P., and Wallace, R. H., *Tetrahedron Lett.*, **28**, 3315 (1987).
[387] Chakraborti, A. K., Saha, B., Ray, C., and Ghatak, U. R., *Tetrahedron*, **43**, 4433 (1987).
[388] Schultz, A. G. and Macielag, M., *J. Org. Chem.*, **51**, 4983 (1986).
[389] Rabideau, P. W. and Karrick, G. L., *Tetrahedron Lett.*, **28**, 2481 (1987).
[390] Kumamoto, T., Hosoya, K., Kanzaki, S., Masuko, K., Watanabe, M., and Shirai, K., *Bull. Chem. Soc. Jpn*, **59**, 3097 (1986).
[391] van Doorn, J. A. and Meijboom, N., *J. Chem. Soc., Chem. Commun.*, **1987**, 1700.
[392] Hwu, R. R., Chua, V., Schroeder, J. E., Barrans, R. E., Khoudary, K. P., Wang, N., Wetzel, J. M., *J. Org. Chem.*, **51**, 4731 (1986).
[393] Sato, R., Akaishi, R., Goto, T. and Saito, M., *Bull. Chem. Soc. Jpn*, **60**, 773 (1987).
[394] Banerjee, A. K., Alvarez, A. G., Santana, M., and Carrasco, M. C., *Tetrahedron*, **42**, 6615 (1986).
[395] Banerjee, A. K., de Carrasco, M. C. S., Frydrych-Houge, C. S. V., and Motherwell, W. B., *J. Chem. Soc., Chem. Commun.*, **1986**, 1803.
[396] Hudlicky, T., Sinai-Zingde, G., and Natchus, M. G., *Tetrahedron Lett.*, **28**, 5287 (1987).
[397] Nakayama, J., Ikuina, Y., Murai, F., and Hoshino, M., *J. Chem. Soc., Chem. Commun.*, **1987**, 1072.
[398] Clerici, A. and Porta, O., *J. Org. Chem.*, **52**, 5099 (1987).
[399] Itoh, T., Nagano, T., and Hirobe, M., *Chem. Pharm. Bull.*, **34**, 2013 (1986); *Chem. Abs.*, **106**, 155707 (1987).
[400] Inoue, H., Shirai, M., and Haruki, E., *J. Chem. Soc., Chem. Commun.*, **1987**, 674.
[401] Dolye, M. P., Guy, J. K., Brown, K. C., Mahapatro, S. N., Van Zyl, C. M., and Pladziewicz, J. R., *J. Am. Chem. Soc.*, **109**, 1536 (1987).
[402] Ozawa, F., Hidaka, T., Yamamoto, T., and Yamamoto, A., *J. Organomet. Chem.*, **330**, 253 (1987).
[403] Suzuki, H., Manabe, H., Kawaguchi, T., and Inouye, M., *Bull. Chem. Soc. Jpn*, **60**, 771 (1987).
[404] Houk, J. and Whitesides, G. M., *J. Am. Chem. Soc.*, **109**, 6825 (1987).
[405] Barton, D. H. R., Fernandez, I., Richard, C. S., and Zard, S. Z., *Tetrahedron*, **43**, 551 (1987).
[406] Barton, D. H. R., Motherwell, W. B., Simon, E. S., and Zard, S. Z., *J. Chem. Soc., Perkin Trans. 1*, **1986**, 2243.
[407] Camps, F., Coll, J., and Guitart, J., *Tetrahedron*, **43**, 2329 (1987).
[408] Bird, D. M., Boldt, M., and Kock, T. H., *J. Am. Chem. Soc.*, **109**, 4046 (1987).
[409] Thompson, D. H. P., Barrette, W. C., and Hurst, J. K., *J. Am. Chem. Soc.*, **109**, 2003 (1987).
[410] Chikashita, H., Nishida, S., Miyazaki, M., Morita, Y., and Itoh, K., *Bull. Chem. Soc. Jpn*, **60**, 737 (1987).
[411] Chikashita, H., Miyazaki, M., and Itoh, K., *J. Chem. Soc., Perkin Trans. 1*, **1987**, 699.
[412] Ramos, S. M., Tarazi, M., and Wuest, J. D., *J. Org. Chem.*, **52**, 5437 (1987).
[413] Bachand, B., Ramos, S. M., and Wuest, J. D., *J. Org. Chem.*, **52**, 5443 (1987).
[414] Momose, T., Itooka, T., Nishi, T., Uchimoto, M., Ohnishi, K., and Muraoka, O., *Tetrahedron*, **43**, 3713 (1987).
[415] Prato, M., Quintily, U., Scapol, L., and Scorrano, G., *Bull. Soc. Chim. Fr. II*, **1987**, 99.
[416] Machacek, V., Hassanien, M. M. M., Sterba, V., and Lycka, A., *J. Chem. Soc., Perkin Trans. 2*, **1987**, 867.
[417] Young, P. R. and Huang, H. C., *J. Am. Chem. Soc.*, **109**, 1805 (1987).
[418] Young, P. R. and Huang, H. C., *J. Am. Chem. Soc.*,**109**, 1810 (1987).
[419] Rogers, D. W., *Tetrahedron Lett.*, **28**, 1967 (1987).
[420] Agrafiotis, D. K. and Rzepa, H. S., *J. Chem. Soc., Chem. Commun.*, **1987**, 902.

[421] Noyori, R., Ohkuma, T., Kitamura, M., Takaya, H., Sayo, N., Kumobayashi, H., and Akutagawa, S., *J. Am. Chem. Soc.*, **109**, 5856 (1987).
[422] Takaya, H., Ohta, T., Sayo, N., Kumobayashi, H., Akutagawa, S., Inoue, S., Kasahara, I., and Noyori, R., *J. Am. Chem. Soc.*, **109**, 1596 (1987).
[423] Landis, C. R. and Halpern, J., *J. Am. Chem. Soc.*, **109**, 1746 (1987).
[424] Takeuchi, S. and Ohgo, Y., *Bull. Chem. Soc. Jpn*, **60**, 1449 (1987).
[425] Ohgo, Y., Tashiro, Y., and Takeuchi, S., *Bull. Chem. Soc. Jpn*, **60**, 1549 (1987).
[426] Vinick, F. J., Nowakowska, J., and Jung, S., *Tetrahedron Lett.*, **28**, 741 (1987).
[427] Touboul, E., Dana, G., and Convert, O., *Tetrahedron*, **43**, 543 (1987).
[428] Augustine, R. L. and Yaghmaie, F., *J. Org. Chem.*, **52**, 1862 (1987).
[429] Khan, M. M. T., Khan, B. T., and Safia, M., *React. Kinet. Catal. Lett.*, **29**, 63 (1985); *Chem. Abs.*, **106**, 4380 (1987).
[430] Leelamani, E. G., Gayathri, V., Reddy, G. K. N., and Venkatasubramanian, N., *Indian J. Chem.*, **25A**, 719 (1986); *Chem. Abs.*, **107**, 38930 (1987).
[431] Zuber, M., Szuberia, W. A., and Pruchnik, F. P., *J. Mol. Catal.*, **38**, 309 (1986); *Chem. Abs.*, **107**, 58230 (1987).
[432] Kusumoto, T. and Hiyama, T., *Tetrahedron Lett.*, **28**, 1811 (1987).
[433] Li, Z.-H. and Jones, M., *Tetrahedron Lett.*, **28**, 753 (1987).
[434] Shaw, J. E. and Stapp, P. R., *J. Heterocycl. Chem.*, **24**, 1477 (1987).
[435] Ferreira, V. F., Pinto, A. V., Pinto, M. C. F. R., and Silva, M. M., *J. Chem. Res. (S)*, **1987**, 26.
[436] Sailor, M. J., Sabat, M., and Shriver, D. F., *Organometallics*, **6**, 728 (1987).
[437] Blum, J., Amerm, I., Vollhardt, K. P. C., Schwarz, H., and Hohne, G., *J. Org. Chem.*, **52**, 2804 (1987).
[438] Sokol'skii, D. V., Uaklikhanova, A., and Temirbulayova, A. E., *Zh. Prikl. Khim. (Leningrad)*, **59**, 1643 (1986); *Chem. Abs.*, **106**, 119077 (1987).
[439] Hadhoud, M. K., Shokry, S. A., El Morsi, A. K., and Ragab, S., *Int. Jahrestag.-Fraunhofer-Inst. Treib-Explosivst.*, **1986**, 54/1; *Chem. Abs.*, **107**, 22768 (1987).
[440] Varnikova, G. V. and Shchel'tsyn, V. K., *Monomery dlya Termstoik. Polimerov, M.*, **1985**, 56; *Chem. Abs.*, **107**, 133714 (1987).
[441] Ulan, J. G., Kuo, E., Maier, W. F., Rai, R. S., and Thomas, G., *J. Org. Chem.*, **52**, 3126 (1987).
[442] Ulan, J. G., Maier, W. F., and Smith, D. A., *J. Org. Chem.*, **52**, 3132 (1987).
[443] Mandler, D. and Willner, I., *J. Phys. Chem.*, **91**, 3600 (1987).
[444] Notheisz, F., Zsigmond, A. G., Bartok, M., and Smith, G. V., *J. Chem. Soc., Faraday Trans. 1*, **83**, 2359 (1987).
[445] Johnstone, R. A. W. and Price, P. J., *J. Chem. Soc., Perkin Trans. 1*, **1987**, 1069.
[446] Bruckner, C. and Reissig, H.-U., *Chem. Ber.*, **120**, 617 (1987).
[447] Bruckner, C. and Reissig, H.-U., *Chem. Ber.*, **120**, 627 (1987).
[448] Zehani, S. and Gelbard, G., *Nouveau J. Chim.*, **10**, 511 (1986).
[449] Fort, R. C., Stahl, M. H., and Sky, A. F., *J. Org. Chem.*, **52**, 2396 (1987).
[450] Meyers, A. I. and Brown, J. D., *J. Am. Chem. Soc.*, **109**, 3155 (1987).
[451] Kirby, A. J. and Walwyn, D. R., *Tetrahedron Lett.*, **28**, 2421 (1987).
[452] Fukuzumi, S., Koumitsu, S., Hironaka, K., and Tanaka, *T., J. Am. Chem. Soc.*, **109**, 305 (1987).
[453] Fukuzumi, S., Kuroda, S., and Tanaka, T., *J. Chem. Soc., Chem. Commun.*, **1987**, 120.
[454] Xu, H.-J., Deng, G., and Yu, Q., *J. Chem. Soc., Chem. Commun.*, **1987**, 916.
[455] Van Niel, J. C. G., Kort, C. W. F., and Pandit, U. K., *Recl. Trav. Chim. Pays-Bas*, **195**, 262 (1986); *Chem. Abs.*, **106**, 195697 (1987).
[456] Yasui, S., Fujii, M., Nakamura, K., and Ohno, A., *Bull. Chem. Soc. Jpn*, **60**, 963 (1987).
[457] Fujii, M., Nakamura, K., Yasui, S., Oka, S., and Ohno, A., *Bull. Chem. Soc. Jpn*, **60**, 2423 (1987).
[458] Awano, H., Takemoto, K., Ohya, H., Tomio, M., Tamagaki, S., and Tagaki, W., *Bull. Chem. Soc. Jpn*, **60**, 1887 (1987).
[459] Awano, H. and Tagaki, W., *Bull. Chem. Soc. Jpn*, **59**, 3117 (1987).
[460] Leskovac, V., Trivic, S., and Hong, Z.-L., *J. Ser. Chem. Soc.*, **50**, 503 (1985); *Chem. Abs.*, **107**, 6616 (1987).
[461] Binay, P., Dupas, G., Bourguignon, J., and Queguiner, G., *Can. J. Chem.*, **65**, 648 (1987).
[462] Romoff, T. T., Sampson, N. S., and van Eikeren, P., *J. Org. Chem.*, **52**, 4454 (1987).
[463] Burgner, J. W., Oppenheimer, N. J., and Ray, W. J., *Biochemistry*, **26**, 91 (1987).
[464] Blackwell, L. F., Motion, R. L., MacGibbon, A. K., Hardman, M. J., and Buckley, P. D., *Biochem. J.*, **242**, 803 (1987).
[465] Jones, H. K., Lindahl, R., Baker, D. C., and Timkovich, R., *J. Biol. Chem.*, **262**, 10911 (1987).
[466] Srivastava, D. K. and Bernhard, S. A., *Biochemistry*, **26**, 1240 (1987).
[467] Abriola, D. P., Fields, R., Stein, S., MacKerell, A. D., and Pietruszko, R., *Biochemistry*, **26**, 5679 (1987).

Organic Reaction Mechanisms 1987
Edited by A. C. Knipe and W. E. Watts

CHAPTER 6

Carbenes and Nitrenes

R. A. Aitken

Department of Chemistry, University of St Andrews

Reviews have appeared on the role of metal-carbene intermediates in catalytic reactions[1] and the use of transition metal carbene complexes for cyclopropanation.[2] A review of organic photochemistry in the far UV includes many carbene reactions.[3] Reviews in the nitrene area cover new synthetic applications of oxycarbonylnitrenes[4] and the gas-phase pyrolysis of azides.[5]

Structure and Reactivity

The interplay of qualitative and quantitative theoretical methods as applied to the energetics, structure, and reactivity of carbenes has been reviewed.[6] Several studies on the structure and reactivity of methylene have been reported. Spin-coupled valence-bond calculations have been applied to the reaction of singlet $:CH_2$ with ethene.[7] The photo-decomposition of ketene to singlet $:CH_2$ and CO in the presence of argon, ethene and ethyne has been examined by pulsed laser techniques.[8] Laser magnetic resonance has been used to study the formation of triplet $:CH_2$ by intersystem crossing in the presence of hydrocarbons[9] and its abstraction of hydrogen from the latter.[10] A study of the reactivity of singlet $:CH_2$ in perfluorohexane[11] and in particular the CIDNP effect observed for the Cl_3CCH_2D

produced by insertion into added $CDCl_3$[12] have allowed estimation of the rate constant for intersystem crossing of singlet $:CH_2$ in solution as $\leqslant 8 \times 10^8\,s^{-1}$. Application of a new *ab initio* method to calculation of the singlet-triplet energy gap for several simple carbenes gives values in better agreement with experiment than previous methods.[13] This energy difference has also been calculated using MNDO and *ab initio* methods for :CHX and $:CX_2$ (X = Li, BeH, BH_2, NH_2, OH, F, and CN).[14] Ring-opening of bicyclic alkanes such as (**1**) with $:CF_2$ to give (**2**) is thought to occur by a two-bond cleavage rather than a diradical mechanism.[15] Frontier-orbital calculations for the insertion and addition of carbon monosulphide predict it to behave like a singlet carbene, S=C:.[16] *Ab initio* calculations on two carbene isomers of acetaldehyde, viz. methylhydroxycarbene and methoxycarbene, have been used to predict the energies, structure, vibrational spectra, and unimolecular reactivity of these species.[17]

Detailed kinetic and philicity studies for the addition of methoxy(phenyl)carbene to alkenes have shown it to be ambiphilic with a strong nucleophilic component attributed to the high-lying HOMO in the *Z* configuration.[18] Competition between addition to chloroacrylonitrile and acrylonitrile has been introduced as a useful device to detect ambiphilic behaviour in carbenes. Even the normally electrophilic phenylhalocarbenes react as nucleophiles.[19] Laser flash photolysis studies on methoxycarbonyl(phenyl)carbene and its derived carbonyl oxide have shown the carbene to be intermediate in stability between Ph_2C: and fluorenylidene.[20] For the photolytic or thermal decomposition of substituted 3-benzyl-3-chlorodiazirines in the presence of alkenes, addition to the alkene is unaffected by substituents on the benzyl group while 1,2-H shift is accelerated by OMe or Me and retarded by Cl. These results support the existence of an energy barrier for the 1,2-migration and suggest the involvement of carbene-alkene complexation.[21] Theoretical studies on the formation of cyclopentylidene carbene (**3**) from cyclohexyne predict an activation energy of 26.4 kcal mol^{-1}. The carbene is 19 kcal mol^{-1} less stable than cyclohexyne and rearranges to cyclohexa-1,3-diene via (**4**) with an energy barrier of 38.9 kcal mol^{-1}.[22] A highly significant paper has highlighted the importance of cycloheptatetraene (**5**) as the key intermediate in many phenylcarbene reactions which were previously thought to involve cycloheptatrienylidene (**6**).[23] Photochemical rearrangement of PhCH: at 10 K gives only (**5**) with no evidence of (**6**) or the isomer (**7**). Formation of heptafulvalene cannot be taken as evidence for the involvement of (**6**) since this product is rapidly formed by electrocyclic ring-opening of (**8**), the dimer of (**5**). *Ab initio* calculations show that the triplet state of (**5**) in its equilibrium geometry is 6.7 kcal mol^{-1} lower in energy than the singlet state.[24] The bicyclic carbene (**9**) readily ring-opens to (**10**) which has been trapped by both [2 + 2]- and [4 + 2]-cycloadditions.[25] A study of the addition of (**11**) to substituted styrenes shows it to be a weak ambiphile with evidence for reaction in singlet and triplet states.[26] No evidence was found for heavy-atom effects in the reaction of 2,7-dihalofluorenylidenes (**12**; X = Cl, Br, I) with MeOH. Kinetic and product studies give singlet-triplet energy gaps of $\sim$ 4.2 kcal mol^{-1} in each case, roughly twice the value for (**12**; X = H).[27] The reactivity of thioxanthylidene *S,S*-dioxide (**13**) has been reported. It adds to styrenes, inserts into the OH of alcohols, and forms an ylide with acetonitrile, all reactions being from the singlet state.[28] A comparison in reactivity between 1,8-diazafluorenylidene (**14**) and the 3,6-isomer (**15**) shows them to behave similarly but both differ from fluorenylidene, suggesting a through-bond rather than through-space effect of the nitrogens.[29] The formation of (**14**) by photolysis of the corresponding diazo-compound is wavelength dependent: irradiation at 310 nm causes rearrangement to the diazirine which gives the singlet carbene only, while at 420 nm the carbene is formed directly in the singlet and triplet states with no involvement of the diazirine.[30] Full details have appeared on the formation and reactivity of dimethylpentatetraenylidene (**16**): the structure has a significant contribution from the dipolar form (**17**) and with 2,3-dimethylbut-2-ene a dimeric adduct is formed which probably has structure (**18**).[31] Correlation of the ESR and optical spectra of triplet Ph_2C: and (**19**) in a benzophenone crystal gives a consistent description of the carbenes' properties.[32] Measurements of the magnetic properties and intramolecular interactions in dicarbenes (**20**) and (**21**),[33] as well as three different ('*o*, *m*, and *p*-') isomers of (**22**), have been reported.[34]

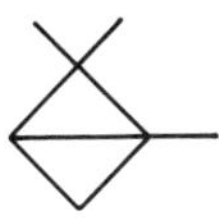

(**1**)

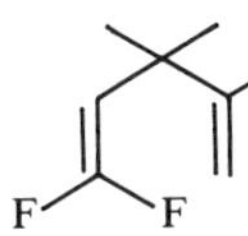

(**2**)

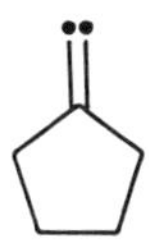

(**3**)

(**4**)

(**5**)

(**6**)

(**7**)

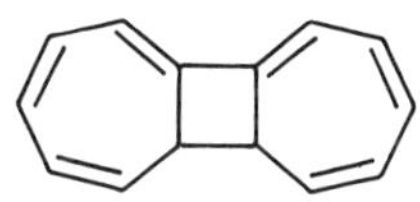

(**8**)

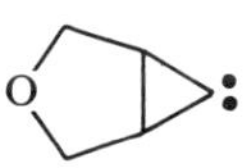

(**9**)

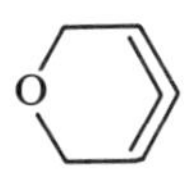

(**10**)

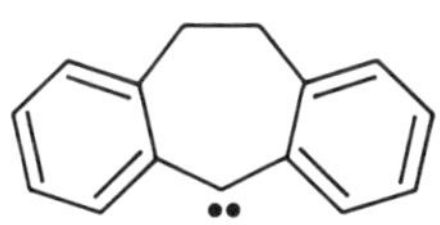

(**11**)

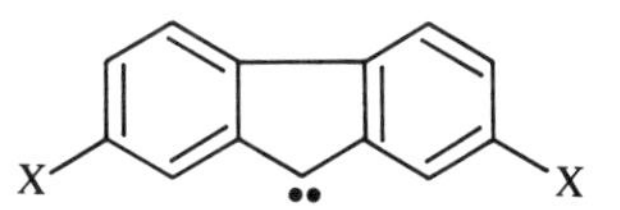

(**12**)

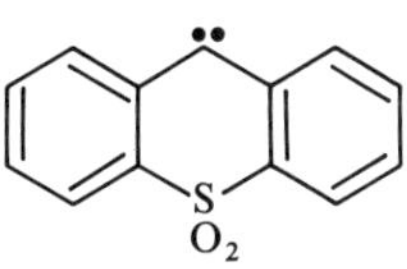

(**13**)

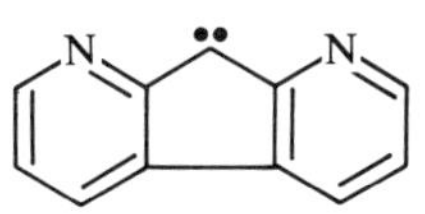

(**14**)

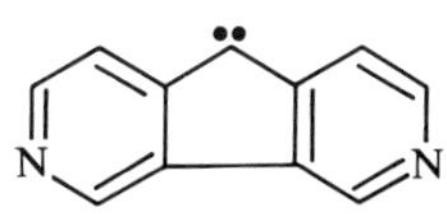

(**15**)

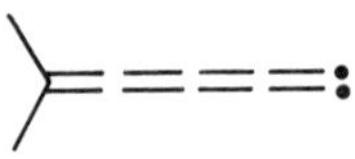

(16)

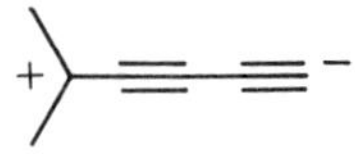

(17)

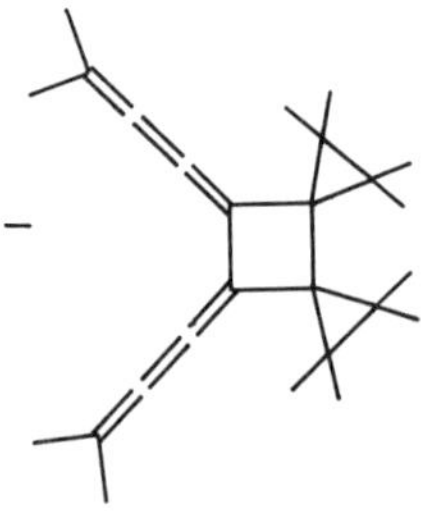

(18)

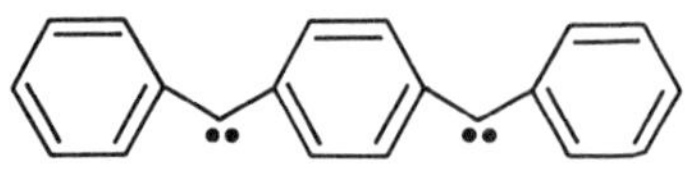

(19)

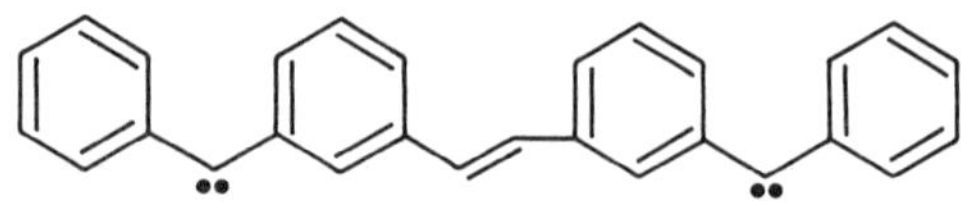

(20)

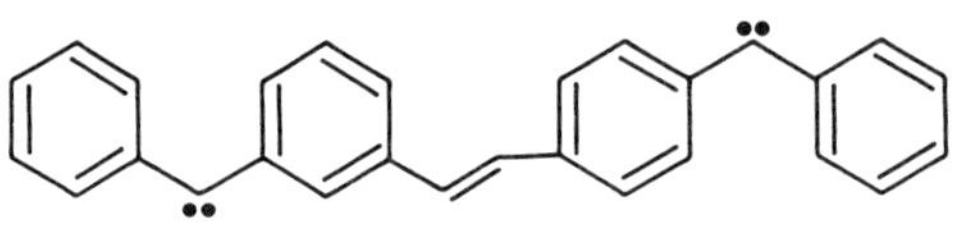

(21)

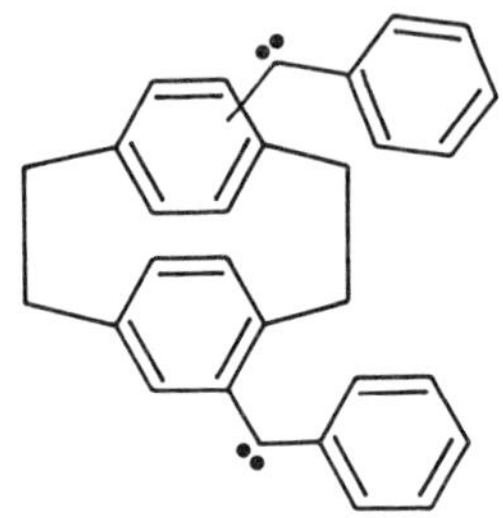

(22)

Several reports have appeared on carbene radical ions. The reactivity of methylene radical anion has been examined in a flowing afterglow/selected ion flow tube.[35] In an electrochemical study the rate of nitrogen loss from diazotetraphenylcyclopentadiene radical anion has been measured and the similarity to the rate for $PhCOC(N_2)Ph^{-\bullet}$ suggests that charge delocalization is not the only important effect in these reactions.[36] The heat of formation of $Ph_2C^{-\bullet}$ has been estimated and its reactivity with a variety of substrates investigated.[37] The heat of formation of methoxycarbene radical cation, which is involved in the mass spectrometric breakdown of ethylene oxide, has been estimated.[38] The carbene radical cations (**23**; $X = H_2$, CH_2CH_2 and CH=CH) show both electrophilic and radical reactions, interaction with methanol, for example, producing $Ar_2C(OMe)_2$, Ar_2CO, and Ar_2CHOMe.[39] The $C_4N^{+\bullet}$ cation, which can be drawn as (**24**) and occurs in interstellar gas clouds, has been generated in a selected ion flow tube and its reactions examined.[40]

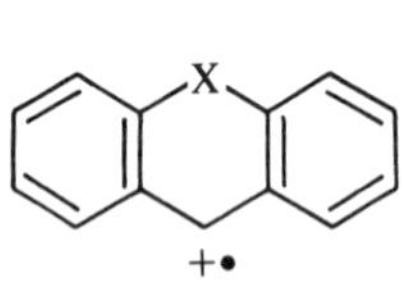

(23)

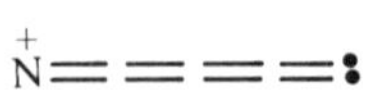

(24)

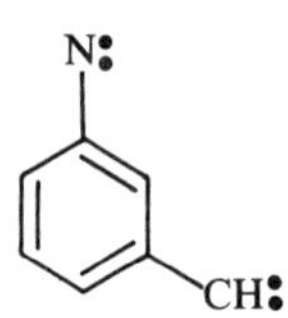

(25)

The carbene-nitrene (**25**) has been generated in a matrix at 18 K and shown by ESR to have a quintet ground state.[41] High-resolution gas-phase emission spectroscopy has allowed determination of the geometry and spin-spin splitting for the triplet ground state of methylnitrene.[42] A study of the photochemically initiated autocatalytic decomposition of aryl azides indicates that triplet ArN: acts as a chain propagator.[43] By carrying out the photolysis of 1-naphthyl azide with diethylamine at -60 to -78°C, moderate yields of products from ring-opening of azirines (**26**) and (**27**) by the amine are obtained.[44] On reducing the temperature further the triplet product 1-azonaphthalene predominates. Photolytic studies of 2-naphthoyl azide give only singlet products with no sign of the triplet nitrene indicating a singlet ground state for this nitrene, a conclusion supported by qualitative MO considerations.[45] The use of time-resolved IR spectroscopy in the photolysis of *N,N*-dimethyl-*p*-azidobenzamide shows the formation of didehydroazepine (**28**) rather than the azirine (**29**), and addition of diethylamine gives (**30**).[46] It is suggested that ring-expansion is also the predominant pathway in the photolysis of phenyl azide. The photo-decomposition of 2-azidophenazines (**31**) has been shown to involve singlet and triplet nitrenes, the azirine, and a didehydroazepine. For X = OMe, the last of these was intercepted by starting azide to form adduct (**32**) which then lost nitrogen.[47]

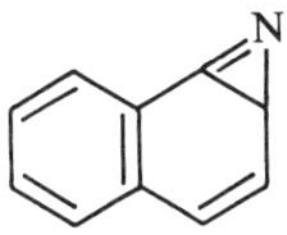

(26)

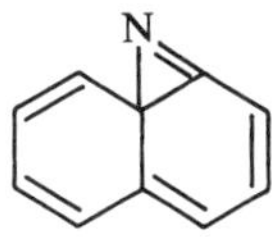

(27)

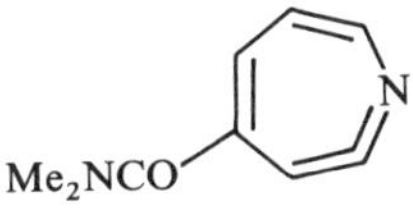

(28)

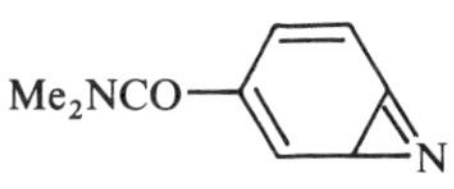

(29)

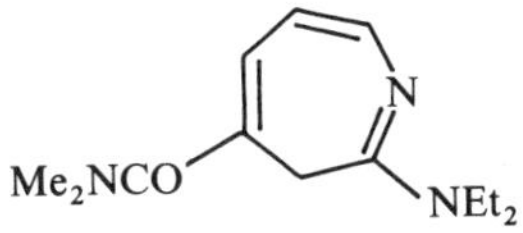

(30)

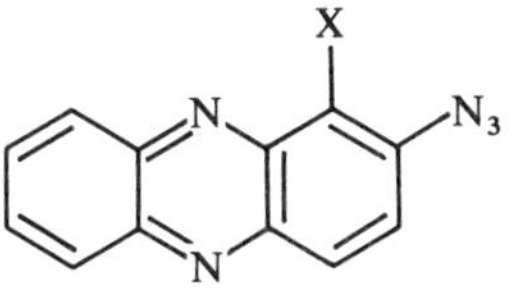

(31)

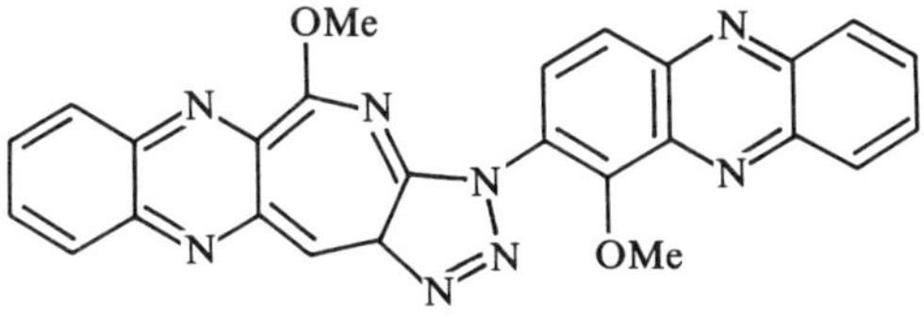

(32)

Generation

The involvement of methylene in oxidative coupling of CH_4 over lithium-promoted magnesium oxide has been demonstrated by the formation of cyclopropane from added ethene.[48] Formation and desorption of methylene from CH_2I_2 on an aluminium surface at 170 K has been directly observed by X-ray photoemission spectroscopy.[49] Further evidence has been obtained for thermal generation of vinylidene in the pyrolysis of acetylene[50] and vinylidene has also been characterized on a ruthenium surface exposed to ethene.[51] Reaction of 1,1-dibromo-2-methylpropene with Bu^nLi followed by $CuI.PBu^n_3$ at $-100°C$ gives a copper carbenoid-ate complex corresponding to dimethylvinylidene. On warming, this provides the first synthesis of decamethyl[5]radialene (**33**).[52] Quaternary ammonium chlorometallates such as $PhCH_2\overset{+}{N}Et_3\ FeCl_4^-$ have been reported as useful new catalysts in the phase-transfer generation of Cl_2C:.[53] The influence of the catalyst used on the reactivity of phase-transfer-generated Br_2C: with allyl bromide has been examined. While ammonium salts and crown ethers give mainly addition product (**34**), two new catalysts, $Ph_4\overset{+}{As}\ Cl^-$ and $Ph_3P{=}\overset{+}{N}{=}PPh_3\ Cl^-$ give mainly

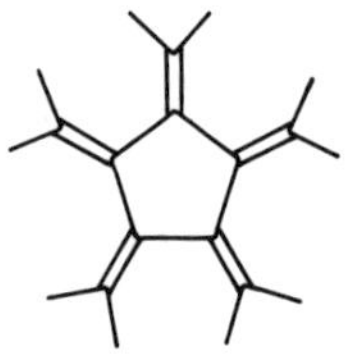
(**33**)

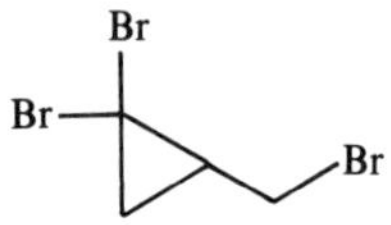

(**34**)

(**35**)

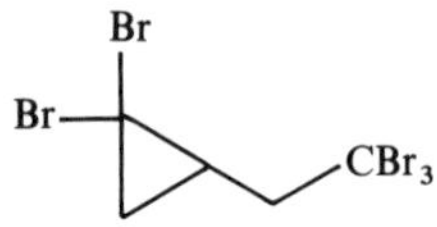

(**36**)

the insertion products (**35**) and (**36**).[54] The dissolution of copper metal in a mixture of Me_2SO and CCl_4 involves the generation of Cl_2C: according to the equation:

$$Cu + 2Me_2SO + CCl_4 \rightarrow CuCl_2(Me_2SO)_2 + Cl_2C:$$

the carbene is rapidly consumed by oxidation by Me_2SO to phosgene.[55] A number of oligohalopropenes react with sterically hindered bases to form halo(halovinyl)carbenes by α-elimination. As an example, (**37**) gives (**38**) which has been trapped by addition to alkenes.[56] Thermolysis of 3-chloro-3-phenyldiazirine in the presence of diazo-compounds R_2CN_2 gives only alkenes (**39**) by conversion into phenylchlorocarbene which then interacts with the diazo-compound. This implies

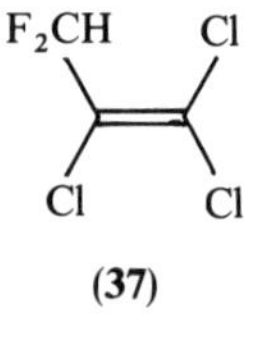

(**37**)

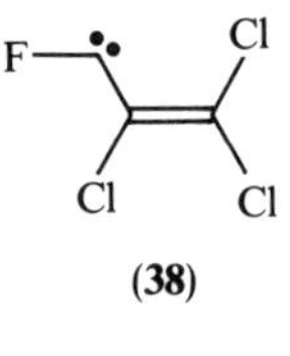

(**38**)

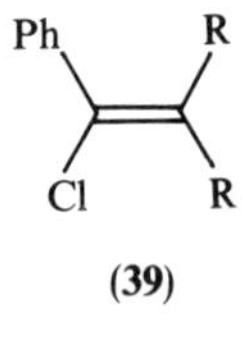

(**39**)

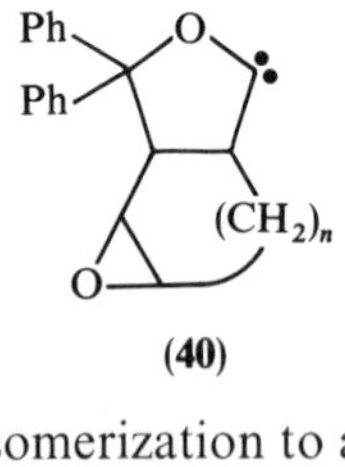

(**40**)

that formation of alkenes from diazirines relies on initial isomerization to a diazo-compound which then reacts with the carbene. In contrast, interaction of carbene with diazirine gives azine products.[57] 3-Methoxy-3-phenoxydiazirine has been obtained as the first diazirine precursor for a dioxacarbene.[58] Two further cases of the photochemical ring-expansion of cyclic ketones to oxacarbenes have been reported: (**40**) is trapped by insertion with methanol in a route to leukotrienes,[59] and (**41**) inserts into isopropanol as part of the total synthesis of pederol.[60] Treatment of $PhCOCBr_2Ph$ with samarium iodide generates an α-oxocarbenoid equivalent to benzoylphenylcarbene which undergoes regio- and stereo-selective cycloaddition with alkenes. For example, α-methylstyrene gives (**42**) and insertion product (**43**).[61]

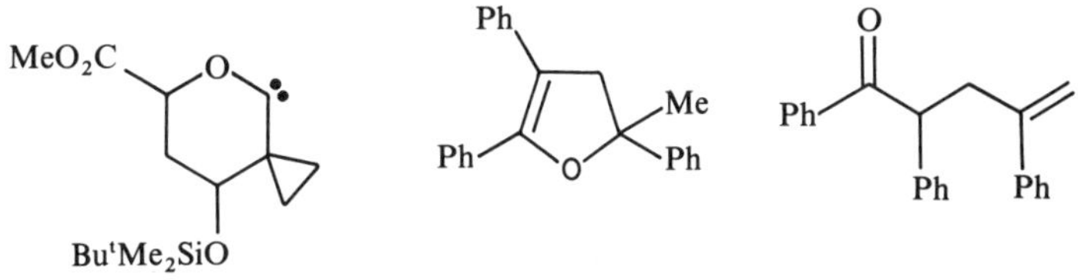

(**41**) (**42**) (**43**)

Dipolar cycloaddition is also observed for α-oxocarbene (**45**), formed by thermal decomposition of the Cu(II)complex of hydrazone (**44**). With dimethyl acetylenedicarboxylate the cycloadduct from (**46**) is obtained together with products from abstraction of a hydrogen atom or *p*-chlorophenyl radical by (**45**).[62] Flash vacuum

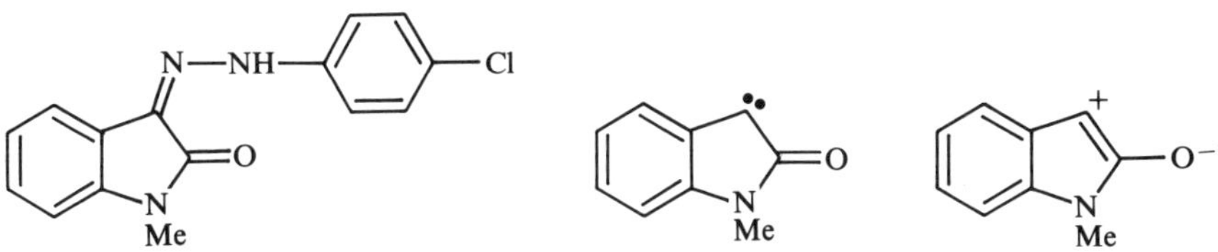

(**44**) (**45**) (**46**)

pyrolysis of anhydride (**47**) proceeds by loss of CO and CO_2, rearrangement of the resulting aryne to 7-methylindenylidene carbene, and intramolecular insertion to give (**48**). For (**49**), loss of 2CO is followed by a similar sequence to give (**48**) but there is competing ring-expansion to oxacarbene (**50**) which inserts into Me and then loses CO_2 to give acenaphthylene.[63] A new synthesis of the tosylhydrazones of α,β-

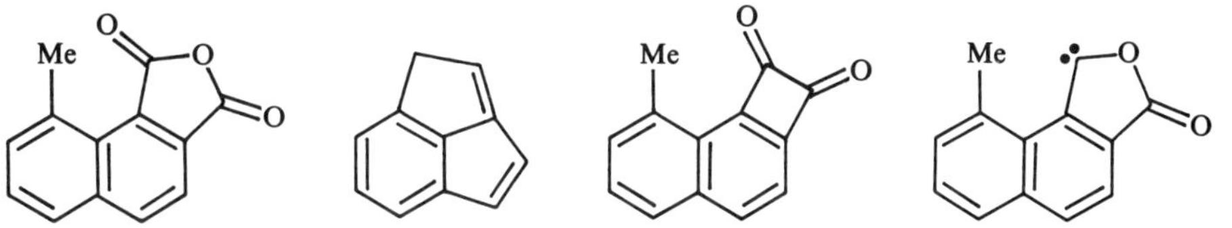

(**47**) (**48**) (**49**) (**50**)

unsaturated ketones has been reported which provides access to the corresponding diazocompounds. Photolysis of (**51**) and (**52**) prepared by this method in a matrix of 10 K gives cyclopentadiene and cycloheptatetraene (**5**), respectively.[64] Pyrolysis of norbornane-2-thiol at 400–500°C leads to loss of H_2S to give mainly (**53**) by intramolecular insertion of norbornan-2-ylidene together with some norbornene formed by a radical pathway.[65] Photolysis of a variety of 1-arylbenzotriazoles results in loss of N_2 to give carbenes (**54**) which, depending on the nature of Ar, may undergo addition, insertion, or nucleophilic attack.[66] The photochemical decarbonylation of (**55**) is accompanied by 1,5-ring closure of the resulting diradical to give products derived from vinylcarbene (**56**).[67] The slow dimerization of methyl

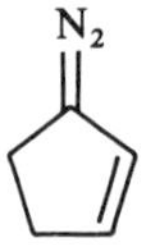

(51)

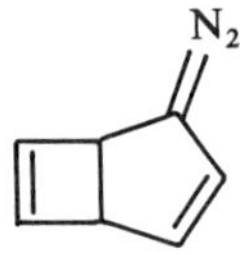

(52)

(53)

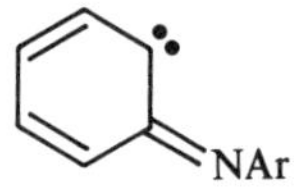

(54)

3,3-dimethylcyclopropene-1-carboxylate to (**57**) may involve ring-opening of the cyclopropene to a vinylcarbene although attempts to trap this were unsuccessful.[68]

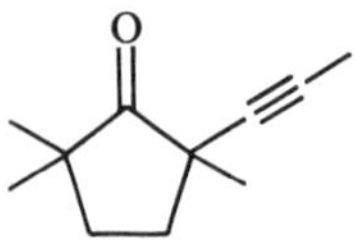

(55)

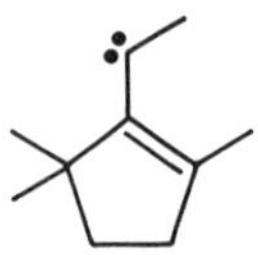

(56)

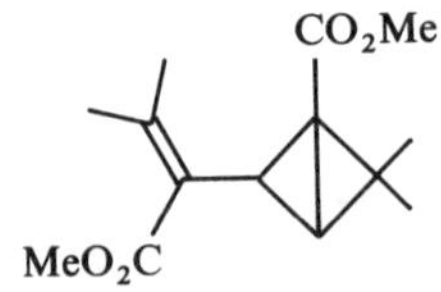

(57)

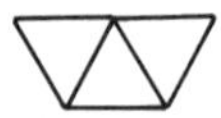

(58)

The highly strained hydrocarbon (**58**), whose formation is supported by NMR data at $-78°C$, decomposes above $-50°C$ by means of a retroaddition to give cyclopentene-4-ylidene which goes on to cyclopentadiene.[69] While thermolysis of azidocinnamates (**59**) gives products from an intramolecular ene reaction of the corresponding azirine and indoles formed by nitrene insertion into aromatic CH, photolysis gives the additional products (**60**) by ring-opening of the azirine to an iminocarbene and intramolecular addition.[70] Vinylcarbenes (**62**) have been gen-

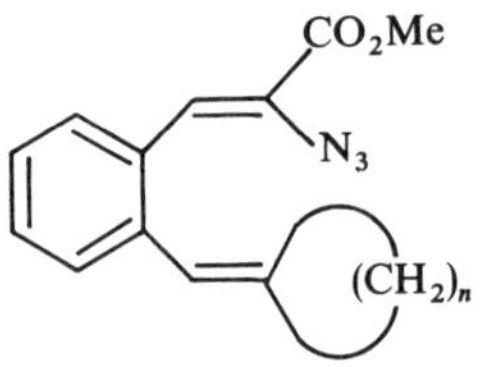

(59)

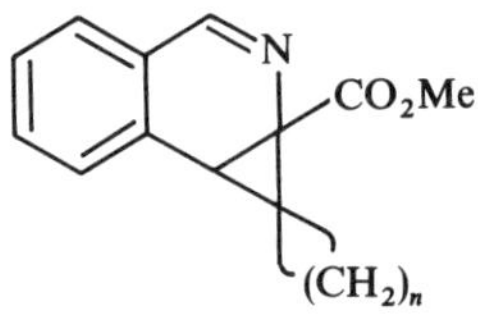

(60)

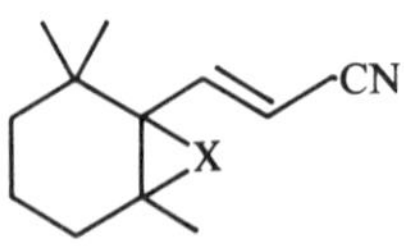

(61)

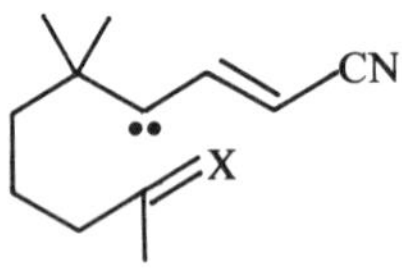

(62)

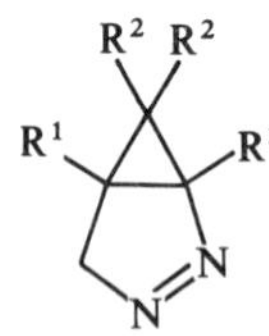

(63)

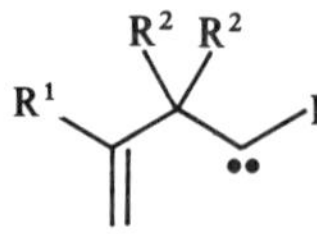

(64)

erated by irradiation of cyclopropane (**61**; $X = CH_2$) and epoxide (**61**; $X = 0$).[71] The decomposition of two types of ring-fused 1-pyrazolines has been shown to involve cycloreversion to an unsaturated diazo-compound and subsequent formation of carbene products.[72] Depending on the substituents present, the carbenes (**64**) formed from (**63**) may undergo intramolecular addition, or a 1,2-hydrogen shift, or a 1,2-vinyl shift. The isomeric mixture of 6,6- and 7,7-difluorocompounds (**65**) is reported

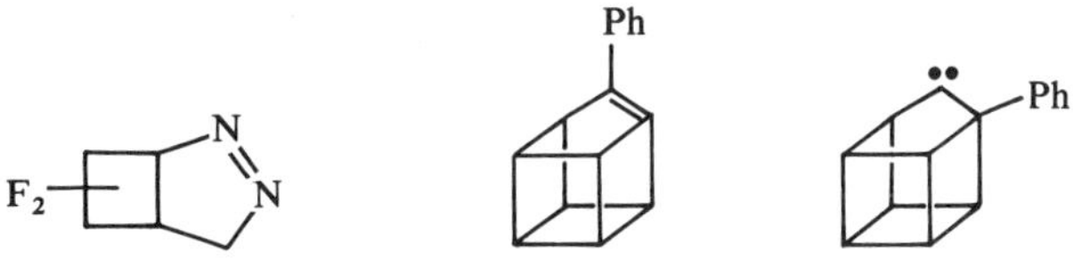

(65) (66) (67)

to decompose partly by cycloreversion to give carbene products as for the non-fluorinated analogue, but in contrast to the tetrafluoro derivative where only radical reactions occur.[73] Photolysis of cubylphenyldiazomethane initially gives the highly strained bridgehead alkene (**66**) by 1,2-alkyl migration in the initially formed carbene and this product can be trapped with ethanol. It further rearranges by a second 1,2-alkyl shift to give (**67**) which adds to but-2-ene and inserts into ethanol.[74] Thermal decomposition of α-diazo-α-trialkylsilyl ketones (**68**) proceeds by a 1,3-silyl shift from C to O followed by loss of nitrogen to give silyloxyvinylidenes which insert to give (**69**).[75] Flash vacuum pyrolysis of bis(trimethylsilyl)thioketene at 770°C leads

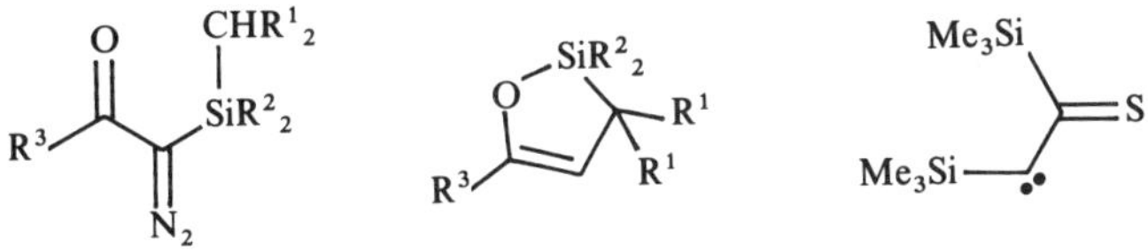

(68) (69) (70)

to a 1,2-shift of Me_3Si producing α-thiocarbene (**70**) which undergoes intramolecular addition, insertion, and rearrangement.[76] In addition to rearrangement, irradiation of pyridinium mesoionic compounds (**71**) in benzene leads to fragment-

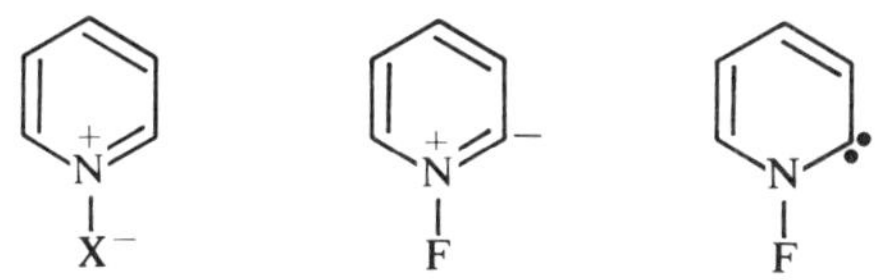

(71) (72) (73)

ation: for $X = C(CN)_2$ singlet and triplet dicyanocarbene are produced, for $X = NCO_2Et$ triplet ethoxycarbonyl nitrene, and for $X = O$ triplet oxene.[77] Treatment of *N*-fluoropyridinium salts with Et_3N leads to deprotonation to (**72**) which reacts as (**73**) and undergoes addition to benzene and furan and ylide formation with acetonitrile, acetone, and tetrahydrofuran.[78] Heating the selenium ylide (**74**) in an aprotic solvent generates the carbene (**75**) which dimerizes and trimerizes.[79] The copper(II)-catalysed decomposition of iodonium ylide (**76**) gives the corresponding dioxocarbene which undergoes cycloaddition reactions with A=B to give (**77**).[80] Nucleophilic attack on alkynyliodonium salts (**78**) is followed by loss of iodobenzene to give a vinylidene (**79**) which inserts into the δ-CH forming a cyclopentene.[81]

Comparison of the reactivity of the species generated by photolysis of aroyl azides, aryldioxazolones, and aromatic nitrile oxides suggests that the aroylnitrene

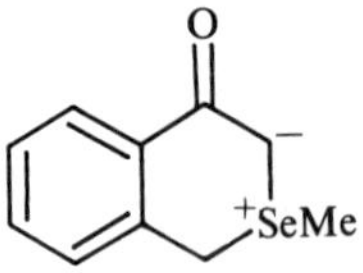

(74)

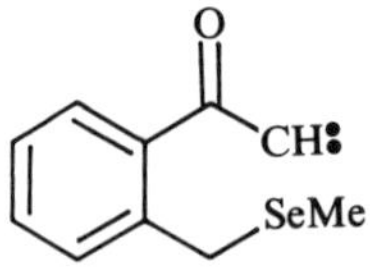

(75)

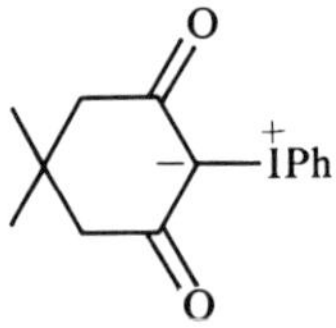

(76)

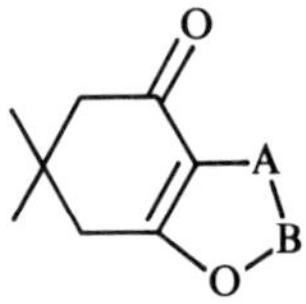

(77)

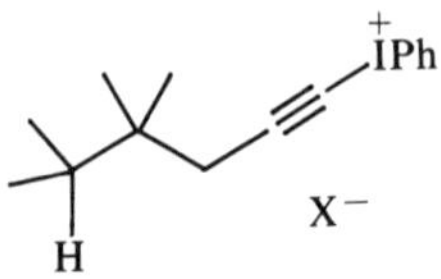

(78)

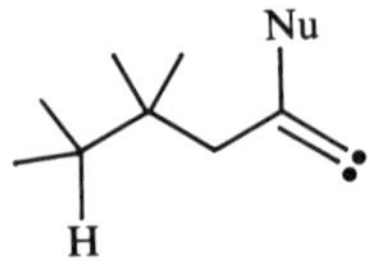

(79)

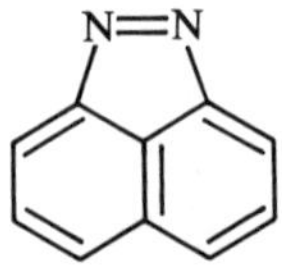

(80)

is involved as a common intermediate in each case.[82] The laser photolysis of 1,8-diazidonaphthalene has now been carried out at room temperature and the formation of (**80**) is shown to involve a two-photon process.[83] The *N*-acetoxy compound (**81**) has been isolated from the $Pb(OAc)_4$ oxidation of corresponding *N*-aminoquinazolone at −20°C and shown to react with alkenes, thus suggesting that nitrenes or nitrenium ions are not involved in this reaction or that of *N*-aminophthalimide with $Pb(OAc)_4$ at 80°C.[84]

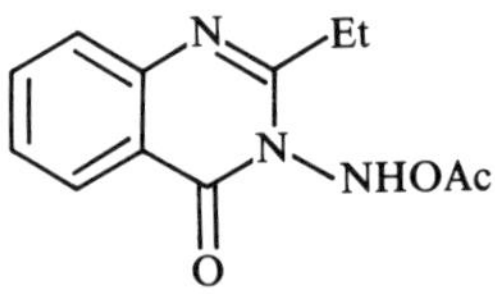

(81)

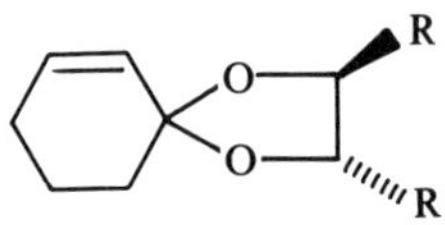

(82)

Addition

Carbenes

Diastereoselectivity in the cyclopropanation of ketals (**82**) has been examined as a function of R.[85] Good stereoselectivity is observed in the samarium-promoted cyclopropanation of allylic alcohols with $R^1R^2CI_2$.[86] For a series of enol ethers, Simmons–Smith cyclopropanation is accompanied by ZnI_2-promoted isomerization to allyl ethers, the reaction of (**83**), for example, giving (**84**) and (**85**).[87] The reaction of phase-transfer-generated Cl_2C: with azobenzenes has been reviewed,[88]

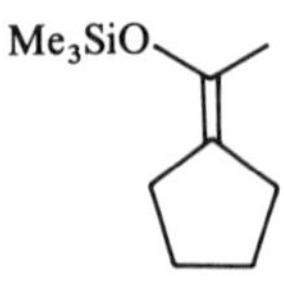

(83)

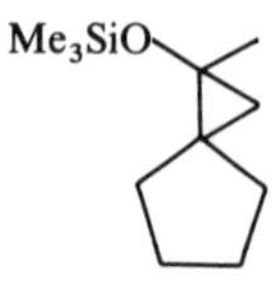

(84)

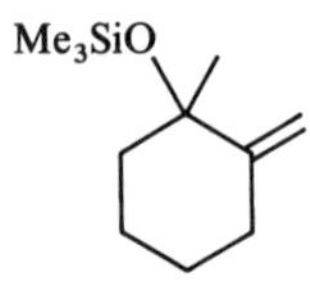

(85)

and a kinetic study of the addition of phase-transfer-generated dihalocarbenes to styrenes and 1-phenyl-2-methylenecyclopropanes has been reported.[89] The apparent non-stereospecific addition of Br_2C: to *E*-cyclooctene under phase-transfer conditions, reported previously, has been explained by reversible nucleophilic addition of Br_3C^- allowing isomerization of the double bond.[90] A multinuclear NMR study of α- and β-trialkylsilylstyrenes and their dichlorocarbene adducts has shown that electron-withdrawing alkyl groups cause increased polarization of the double bond leading to reduced reactivity towards the electrophilic Cl_2C:.[91] Phase-transfer-generated Cl_2C: and Br_2C: add to 3H-1,5-benzodiazepines to give (**86**), a new ring system.[92] While a series of 3-methyldihydrophosphole-1-oxides undergo

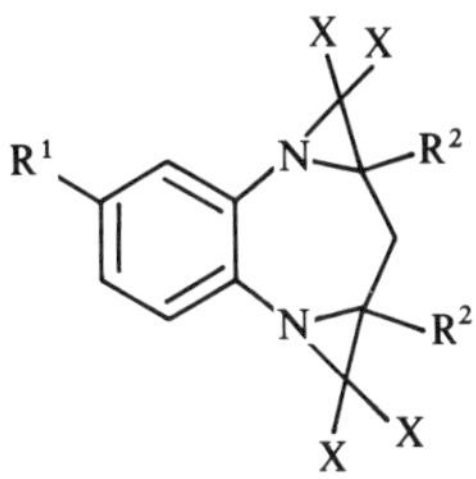

(86)

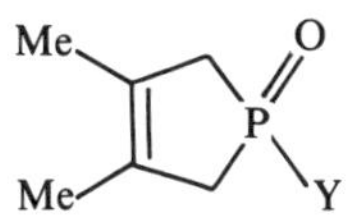

(87)

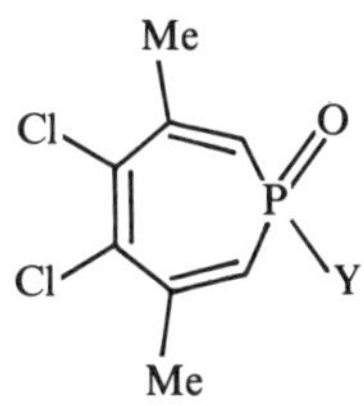

(88)

simple addition of Cl_2C: under phase-transfer conditions, the 3,4-dimethyl analogues (**87**; Y = Ph, OMe) give (**88**) by a sequence involving isomerization, loss of HCl, and addition of a second carbene.[93] A mechanism has been proposed for the unusual addition of Cl_2C: to cyclophane (**89**) to give (**90**) and (**91**) whose structures are unambiguously established by X-ray diffraction.[94] The absolute rate constants

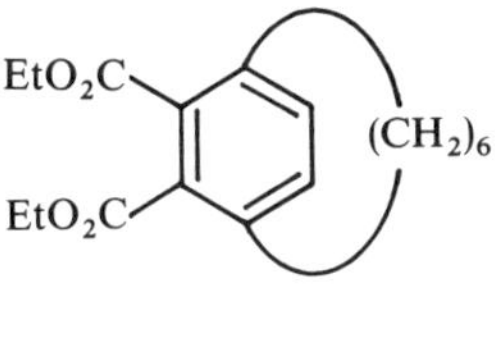

(89)

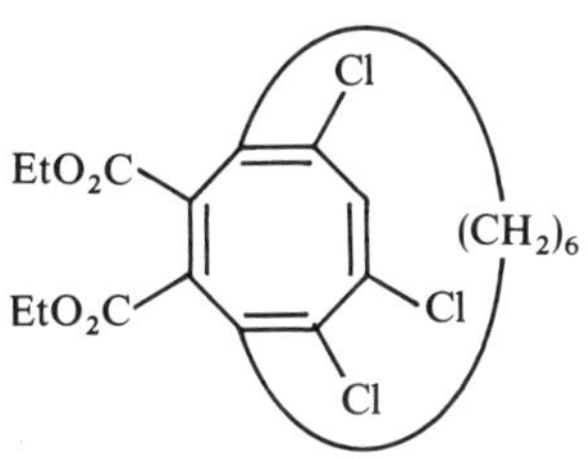

(90)

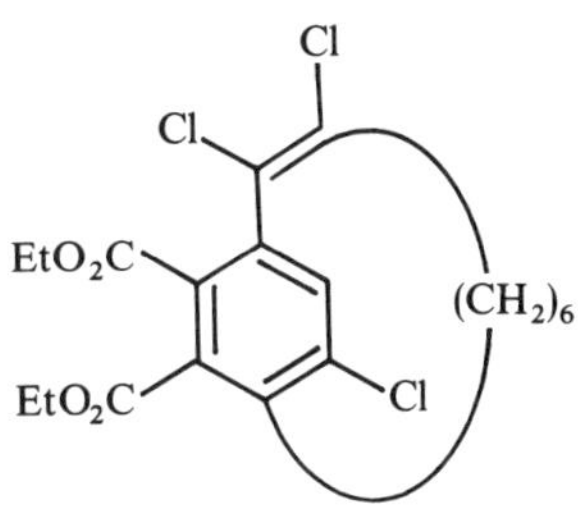

(91)

as a function of pressure for addition of phenylhalocarbenes to 2,3-dimethylbut-2-ene and pent-2-ene suggest formation of a reversible carbene-alkene complex or an early one- or two-bond transition state.[95] Addition of benzenesulphonylcarbene, from thermolysis of $PhSO_2CHN_2$ in benzene, to give the cycloheptatriene in low yield has been reported.[96] The $Rh_2(OAc)_4$-catalysed addition of diazoesters (**92**; X = H) to furan is thought to involve cyclopropanation to give (**93**; Y = O) followed by a Cope rearrangement to give (**94**) rather than a concerted addition.[97] This conclusion is supported by the isolation of (**93**; Y = CH_2) from addition of (**92**; X = OR) to cyclopentadiene.[98] In the addition of $Me_2C{=}C{=}C$: with a series of cycloalkenes, the rate is controlled by steric hindrance from the hydrogens α and β to

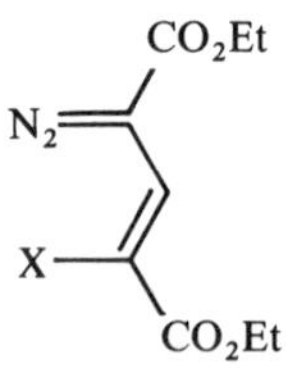

(92)

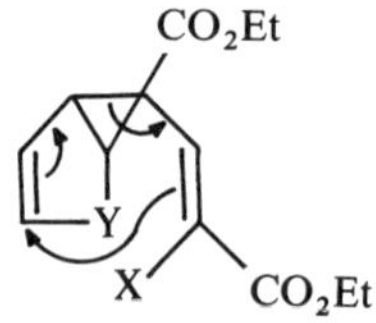

(93)

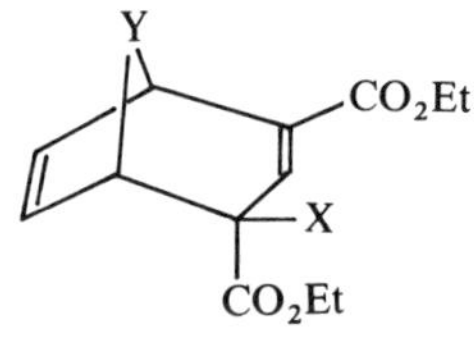

(94)

the double bond.[99] Reaction of this carbene with ring-methylated pyrroles is followed by ring-expansion of the initial adducts (**95**) to give vinylpyridines such as (**96**) in low yield.[100] Addition of phenylchlorocarbene to 1,8-

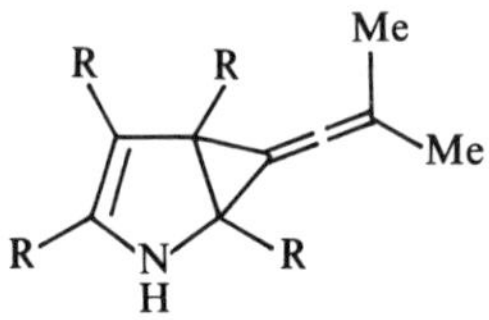

(95)

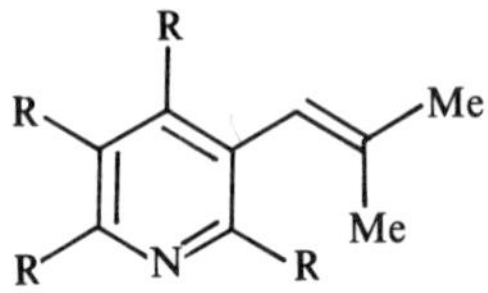

(96)

bis(phenylethynyl)naphthalene is followed by an ene-type reaction of the initial cyclopropene to give (**97**). Upon heating, this ring-opens to carbene (**98**) which undergoes electrocyclization followed by a 1,5-chlorine shift to give (**99**).[101] Intramolecular addition of (**100**) provides access to propellane (**101**)[102] and carbenoids corresponding to (**102**) and (**103**) similarly give [1,1,1]-propellanes.[103]

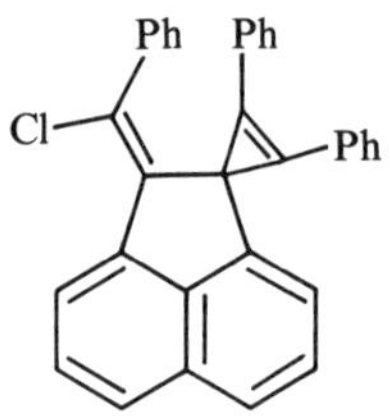

(97)

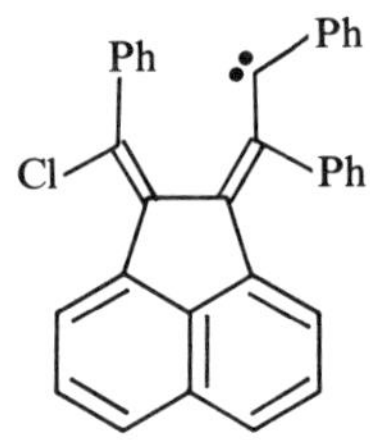

(98)

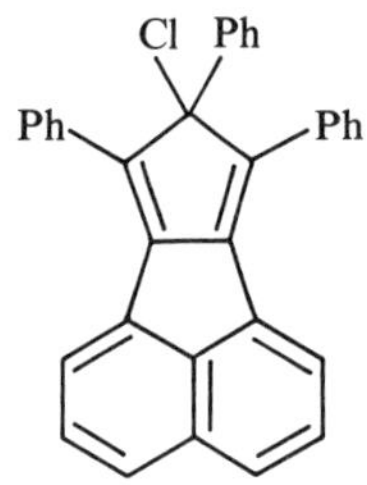

(99)

(100)

(101)

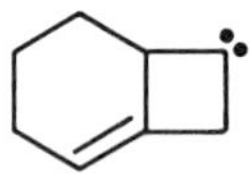

(102)

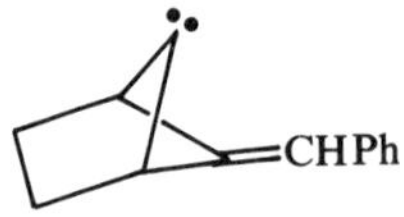

(103)

Thermolysis of $Ph_2Bu^tSiC(N_2)COBu^t$ results in 1,3-silyl migration and loss of N_2 to give (**104**) which adds intramolecularly, leading to heptafulvalene (**105**).[104] The competition between addition to form spiropentanes and rearrangement to allenes has been examined for a series of cyclopropylidenes (**106**).[105] $Rh_2(OAc)_4$-catalysed decomposition of (**107**) results in a novel 1,6-addition to give (**108**).[106]

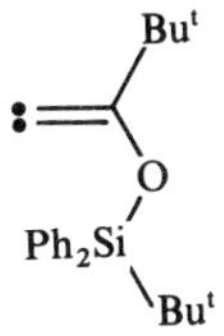

(104)

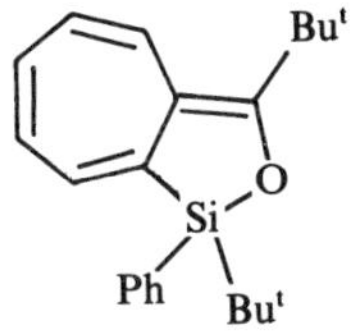

(105)

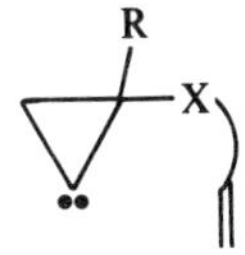

(106)

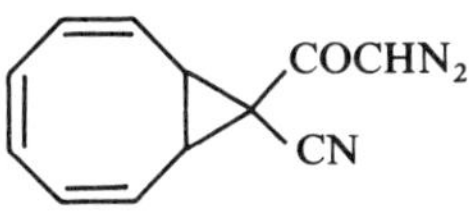

(107)

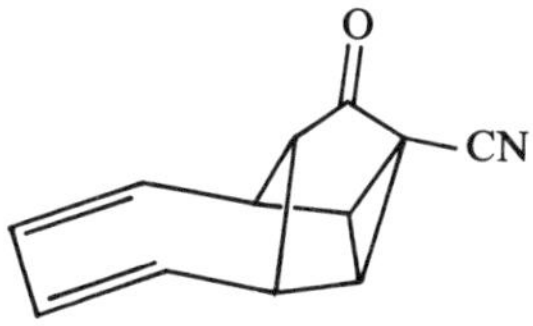

(108)

Nitrenes

The potential surface for addition of aminonitrene H_2N—N: to ethene has been calculated by *ab initio* methods.[107] Addition of methoxycarbonylnitrene to naphthalene gives 1H- and 3H- but no 2H-azepine products due to a 'walk' rearrangement of the initial naphtho[1,2]aziridine to the [2,3]- and [1,9]-isomers.[108] Asymmetric addition of EtO_2CN: to a series of proline-derived chiral enamines gives the corresponding α-ethoxycarbonylamino-ketones in up to 77% enantiomeric excess.[109] Full details have appeared of the diastereoselective addition of *N*-nitrenes (**109**) and (**110**) to a variety of alkenes.[110] By adding CF_3CO_2H, the addition of (**110**) to α,β-unsaturated esters can be carried out at −60°C giving improved selectivity and this is thought to involve the N(1)-protonated form of (**110**).[111] Examination of

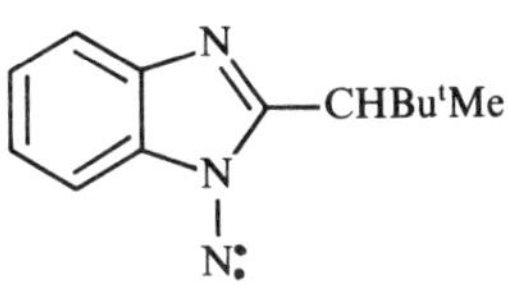

(109)

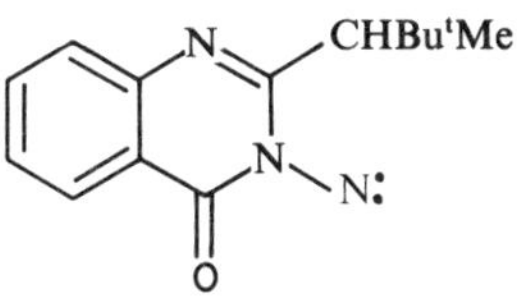

(110)

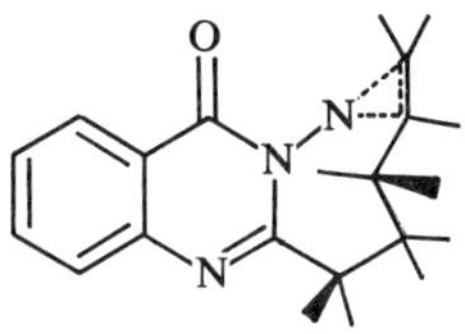

(111)

intramolecular addition in a series of model compounds suggests the preferred transition state geometry to be (**111**).[112] While thermolysis of (**112**; X = PhNH) results in addition to give an azirine which rearranges to (**113**), addition to the allylic double bond occurs in (**112**; X = $NHCH_2CH{=}CH_2$) producing (**114**) after re-

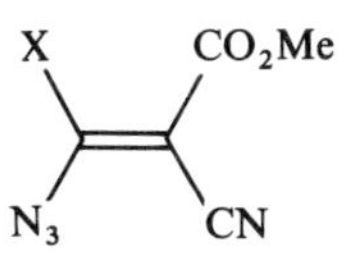

(112)

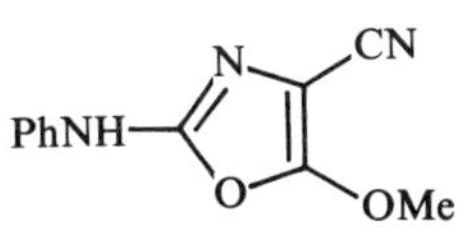

(113)

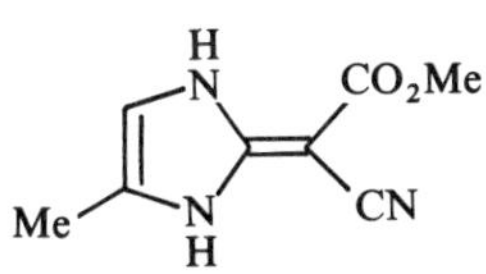

(114)

arrangement.[113] Thermolysis of (**112**; X = NR_2) produces both an azirine and a triazole.[114] A variety of 2,3-diazidobuta-1,3-dienes give biazirinyls thermally or photochemically.[115] Intramolecular addition occurs upon irradiation of (**115**; Y = N_3) to give hexacyclic triaziridines (**116**; X = CH, CMe, P, and PO). Reaction of (**116**; X = CH) with MeO_2CN: regenerates the azo-group to give (**115**; Y = N=NCO_2Me).[116] The related azonitrene (**117**) undergoes both addition to form the triaziridine and rearrangement to the imine (**118**).[117]

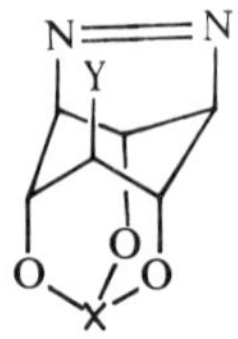

(**115**)

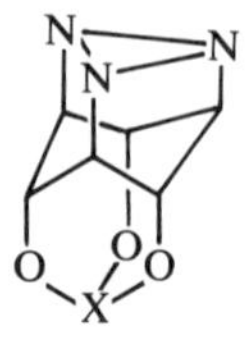

(**116**)

(**117**)

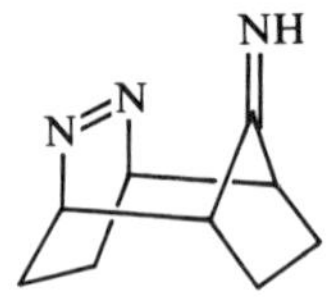

(**118**)

Insertion and Abstraction

Carbenes

A short review of selective alkylation at the α-position of alcohols by carbenes has appeared.[118] *Ab initio* calculations on the insertion of methylene into H_2[119] and into CH and CC of ethane as well as CC of cyclopropane have been reported.[120] Gas-phase reaction of :CH_2 with $Pr^i{}_2O$ proceeds by insertion, abstraction and displacement.[121] A kinetic and mechanistic study of the insertion of :CF_2 into Br_2 gives an activation energy of 4.3–5.1 kJ mol^{-1} for the process.[122] *Ab initio* calculations show that direct insertion of HN=C: into water or ammonia is unlikely but it occurs catalytically on hydrogen-bonded dimers accounting for the experimentally observed non-stereospecificity of the process.[123] Insertion of Me_2C=C: and Me_2C=C=C=C: into SiH of enantiomerically pure methylphenylnaphthylsilane occurs with complete retention of configuration at silicon supporting a concerted mechanism and ruling out an earlier suggestion of the involvement of ion-pairs.[124] The involvement of carbene insertion in the isomerization of but-1-yn-3-one and pent-1-yn-3-one to cyclobut-2-enone and cyclopent-2-enone, respectively, is supported by *ab initio* calculations.[125] Cyclobutylidene (**119**) undergoes competing intramolecular insertion and rearrangement to produce (**120**) and (**121**), respectively.[126] Formation of (**122**) by catalytic decomposition of the corresponding diazo-compound is followed by insertion into NH to afford 1,2-

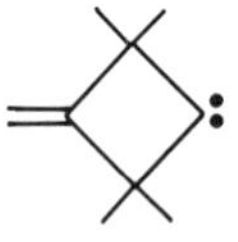

(**119**)

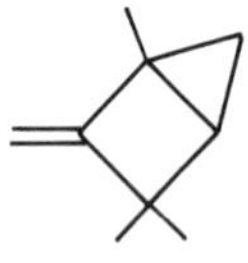

(**120**)

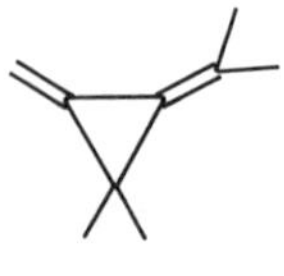

(**121**)

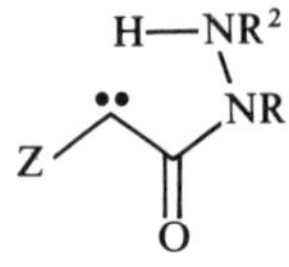

(**122**)

diazetidinones.[127] The reaction of α,α-dimethoxy-ketones with diethyl diazomethylphosphonate provides a new synthesis of furans by insertion of carbene (**123**) to give (**124**) which then loses methanol.[128] Aromatic α-diazo-β-oxo-esters (**125**) undergo intramolecular insertion to form lactones (**126**) in high yield on treatment

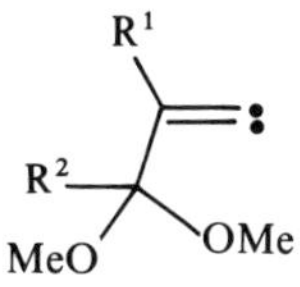

(**123**)

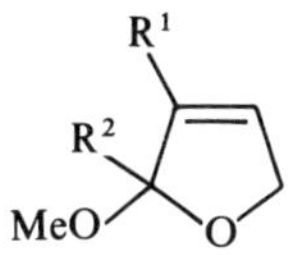

(**124**)

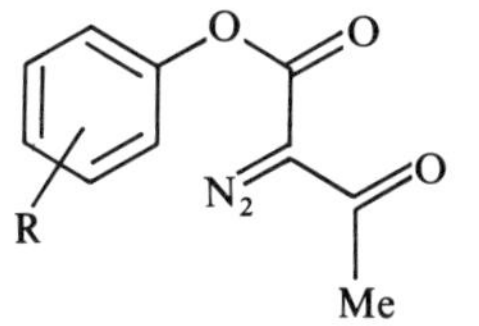

(**125**)

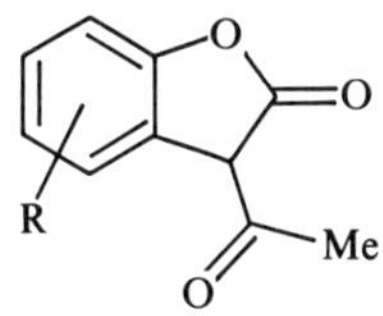

(**126**)

with $Rh_2(OAc)_4$.[129] The competition between intramolecular insertion into CH and addition has been examined for a series of acylcarbenes (**127**; R = H, Me).[130] Carbenes (**128**; $n = 3,4$) undergo both insertion into the δ-CH and isomerization to allenes.[131] The kinetics of hydrogen abstraction from Et_2O and toluene, and chlorine abstraction from CCl_4 have been studied for triplet Ph_2C: and dibenzocycloheptadienylidene (**11**). In the latter reaction triplet (**11**) reacts directly with CCl_4, but for Ph_2C: intersystem crossing to the singlet state is involved.[132]

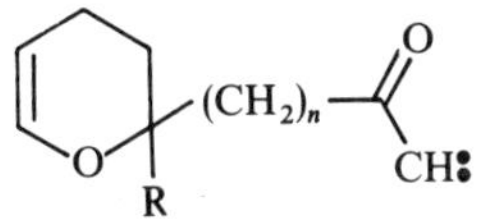

(**127**)

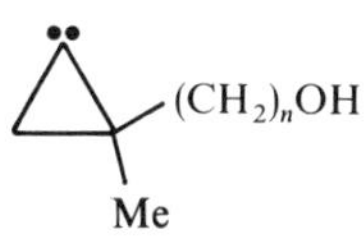

(**128**)

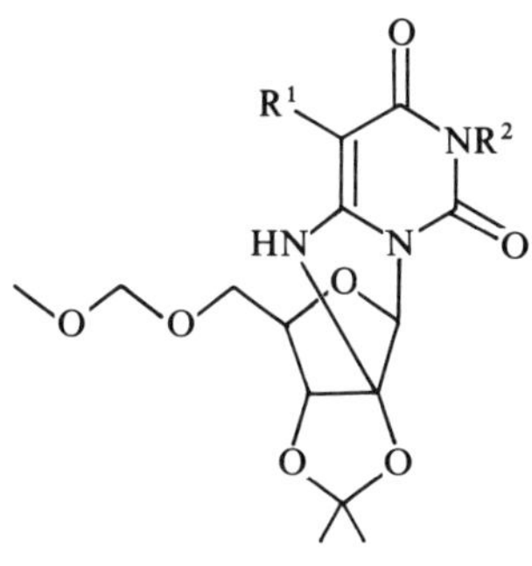

(**129**)

Nitrenes

Upon treatment of long-chain amines $RNMe_2$ ($R = C_8H_{17}, C_{12}H_{25}$) with EtO_2CN: in CH_2Cl_2 or $CHCl_3$, reaction only occurs when the concentration of amine is high enough for aggregation to form reverse micelles and then proceeds by selective insertion to give $RNMeCH{=}NCO_2Et$.[133] Intramolecular insertion as a function of conformational equilibrium has been studied for a series of cyclohexylnitrenoformates.[134] Irradiation of a series of 6-azido-uridine derivatives leads to efficient insertion to give (**129**).[135] Nitrene insertion into a peri- CH is observed on thermolysis of (**130**; X = OMe, SMe)[136] and (**131**).[137] Nitrenes (**132**) generally insert into aromatic CH to give indoles but for R = H there is competing insertion into the formyl CH to give an isoquinolone.[138] Full details have appeared of the double insertion sequence of aromatic vinylnitrenes to give pyrrolo[3.2-e]indole phosphodiesterase inhibitors.[139]

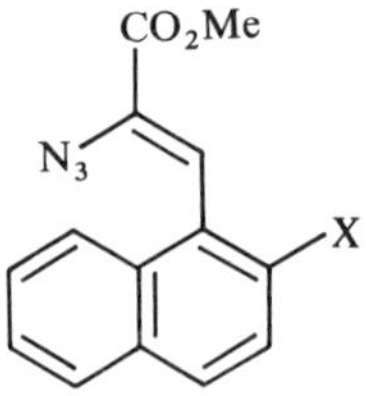

(130)

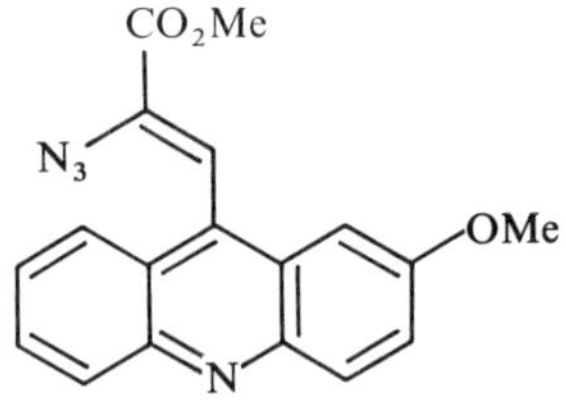

(131)

Rearrangement

A mass spectrometric study has allowed estimation of the relative stability of methylhydroxycarbene and its isomers acetaldehyde and vinyl alcohol.[140] Matrix isolation has allowed the direct observation of the rearrangement of *o*-tolylcarbene to *o*-quinodimethane by a 1,4-H shift and of phenylmethylcarbene to styrene by a 1,2-H shift.[141] In the thermal and photochemical decomposition of (**133**), rearrangement by the exclusive 1,2-migration of Me_3Si gives products derived from silene (**134**) for $R = CO_2Et$,[142] H and $SiMe_3$.[143] The first generation of the parent

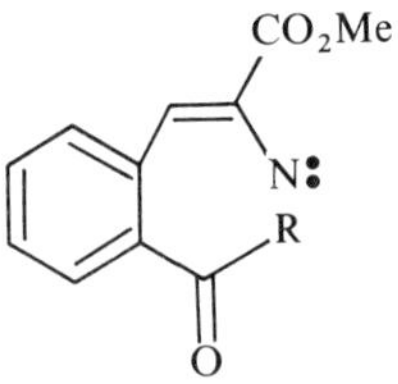

(132)

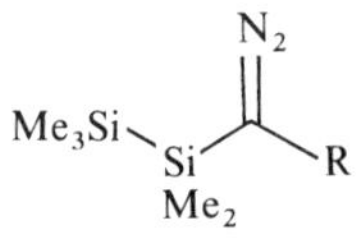

(133)

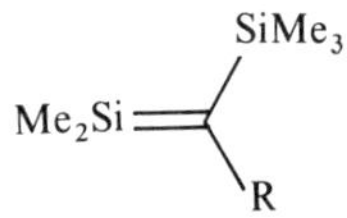

(134)

vinylidene carbene, $H_2C{=}C{=}C$: has been reported by photochemical rearrangement of matrix-isolated cyclopropenylidene at 12 K first to triplet propargylene and then to singlet $H_2C{=}C{=}C$: whose IR spectrum is in accord with *ab initio* predictions.[144] Thermal decomposition of 3-benzyloxy-3-chlorodiazirine at room temperature gives no carbene products, due to rapid fragmentation of the carbene to CO, benzyl cation, and Cl^-. In acetonitrile benzyl chloride is formed, accompanied, in the presence of methanol, by benzyl methyl ether.[145] Photochemical cleavage of tricyclic cyclopropyl alkoxycarbene (**135**) to (**136**) has been directly observed in a matrix at 20 K.[146] Isotopic labelling studies show that the transformation of (**137**) to an isomeric mixture of dihydropentalenes proceeds by rearrangement first to (**138**)

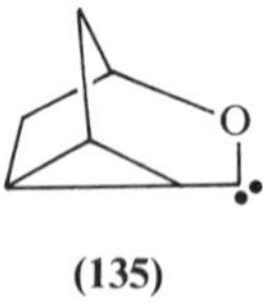

(135)

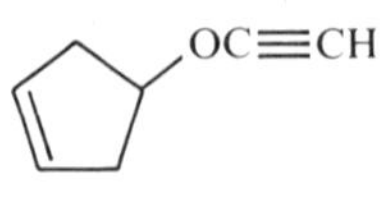

(136)

(137)

(138)

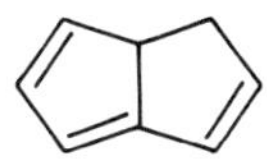

(139)

and then to (**139**).[147] The formation of cyclopentadiene in the pyrolysis of ethynylcyclopropane at 850°C is explained by radical reaction to give pent-3-en-1-yne followed by rearrangement to *Z*-propenylvinylidene which then inserts.[148] Several carbenes are involved in the complex rearrangement of epoxides (**140**) to substituted furans,[149] and nitrones (**141**) to pyrroles and pyridines.[150] While photochemically generated carbenes (**142**) generally undergo Wolff rearrangement

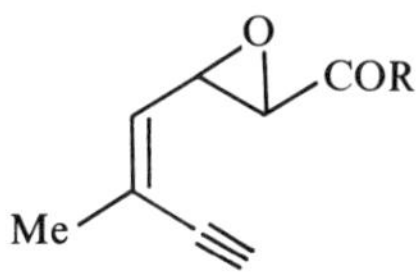

(140)

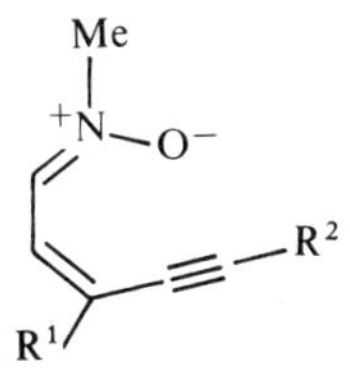

(141)

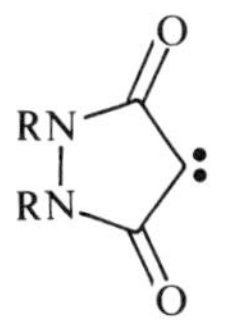

(142)

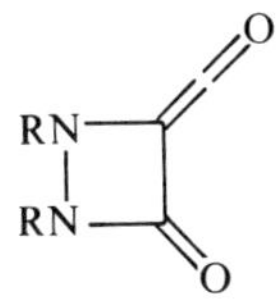

(143)

to (**143**), a cycloreversion to give RN=NR and carbon suboxide occurs for $R_2 = (CH_2)_4$.[151] A theoretical study related to the Wolff rearrangement indicates that singlet or triplet formylmethylene can rearrange to oxirene which then readily forms ketene.[152] Crossover is observed in the photochemical addition of unsymmetrically substituted perfluoro-α-diazo-ketones to hexafluorobut-2-yne suggesting equilibration of the isomeric α-oxocarbenes via an oxirene.[153] This type of equilibration also occurs in the photolysis of $PhC(N_2)CO$-*o*-C_6H_4Ph to give Wolff rearrangement and direct carbene products.[154] The interchange of groups R^1,R^2 in the indole products from photolysis of (**144**) is reported to occur by rearrangement of carbene (**145**) to a 1H-azirine which may open in the opposite sense.[155].

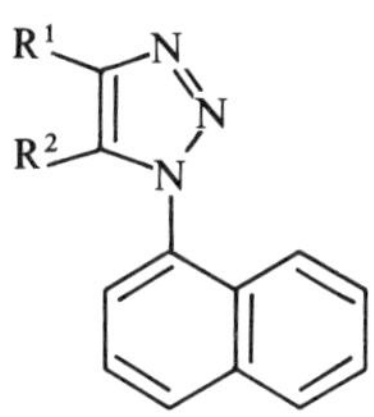

(144)

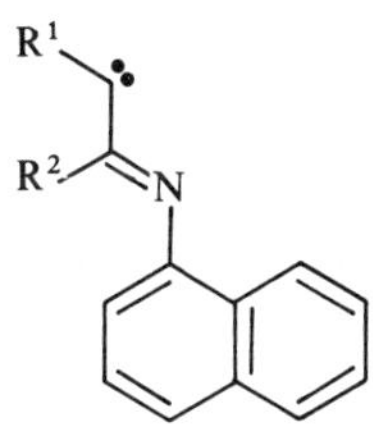

(145)

An *ab initio* study on the rearrangement of silylnitrene, H_3SiN:, to silanimine, $H_2Si{=}NH$, suggests that, while thermolysis or direct photolysis of H_3SiN_3 do not involve the nitrene, its singlet state may be involved in the photo-sensitized reaction.[156] An unusual mode of rearrangement has been reported for methylcubylnitrene (**146**) formed by photolysis of the corresponding azide; an apparent retro-carbene insertion gives (**147**).[157]

Nitrenium Ions

In a study related to the mechanism of carcinogenesis, solvolysis of *N*-acetoxyacetanilides at neutral p*H* has been shown to give arylnitrenium ions as well as hydroxamic acids.[158] Photolysis of substituted *N*-dialkylaminopyridinium salts (**148**) generates the pyridine and the singlet nitrenium ion $R^1R^2N^+$ which inserts into CH of benzene or toluene.[159] The BF_3-promoted decomposition of aryl azides in toluene proceeds by way of $Ar\overset{+}{N}{-}\overset{-}{B}F_3$ which inserts into aliphatic or aromatic CH of the solvent depending on Ar.[160] A detailed study of the reactivity of $R^1C({=}O)\overset{+}{N}OR^2$ formed by treatment of the corresponding *N*-chloro compound with $AgBF_4$ shows electrophilic substitution to occur readily onto a suitably placed aromatic ring in R^1 or R^2. Where both groups contain an aromatic ring, the direction of attack depends on the length of the connecting saturated chains.[161]

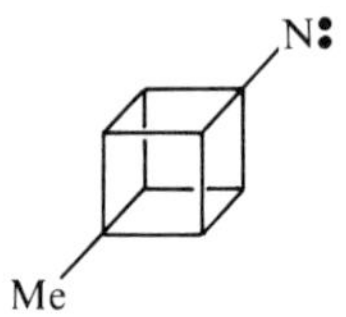

(146)

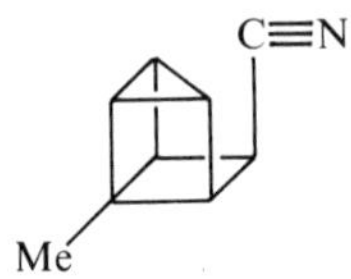

(147)

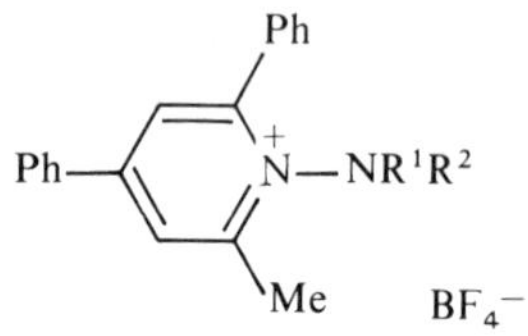

(148)

Nucleophiles and Electrophiles

The behaviour of the unusual boron compound (**149**) as carbene (**150**) has allowed preparation of the first stable stannaethene (**151**; $MR_2 = Sn[CH(SiMe_3)_2]_2$)[162] and germaethene (**151**; $MR_2 = Ge[N(SiMe_3)_2]_2$),[163] by coupling with the corresponding stannylene and germylene. The novel reaction of F_2C: with $Ph_2C{=}NCH_2CO_2Et$ to give the cyclopropane (**152**) involves initial electrophilic attack by the carbene on nitrogen followed by proton transfer and ring-closure.[164] Evidence for the formation of ylide $Me_2\overset{+}{C}O\overset{-}{C}H_2$ by photolysis of CH_2N_2 in acetone is provided by its cycloaddition reactions with a variety of dipolarophiles.[165] A kinetic study has shown that ylide formation between *p*-nitrophenylchlorocarbene and acetone is essentially irreversible.[166] Irradiation of 9-diazofluorene with epoxides is reported to give fluorenone and the alkene by initial electrophilic attack of singlet fluorenylidene to form ylide (**153**) which then fragments.[167] The competition between electrophilic attack on sulphur and 1,5-insertion into CH as a function of *n* has been examined for the $Rh_2(OAc)_4$-catalysed decomposition of a series of diazoesters (**154**).[168] Electrophilic attack of phase-transfer-generated Cl_2C: and Br_2C: on Br^- and I^- provides a useful synthesis of mixed trihalomethanes.[169]

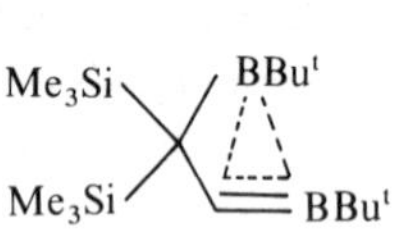

(149)

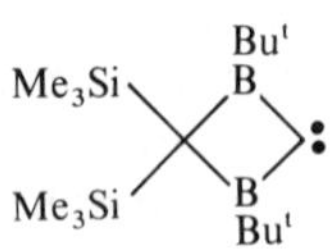

(150)

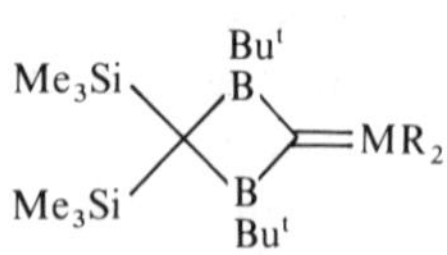

(151)

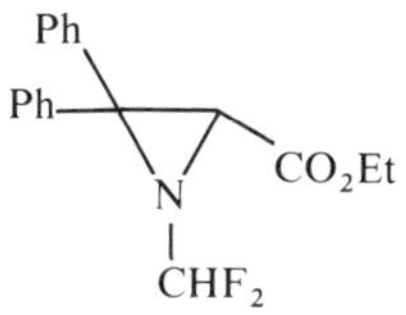

(152)

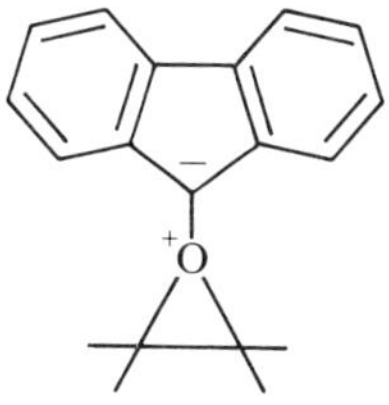

(153)

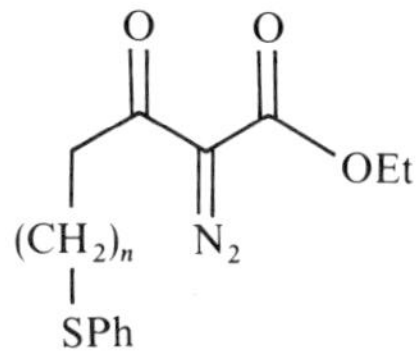

(154)

Silylenes and Germylenes

A mass spectrometric study has given a value of 69.0 kcal mol^{-1} for ΔH_f^0 of silylene, :SiH_2.[170] Kinetic studies on the reaction of this species with alkanes, alkenes, butadiene, and ethyne have been reported.[171,172] *Ab initio* calculations on Si_3H_6 isomers predict a ground-state singlet for disilylsilylene $(H_3Si)_2Si$: with a singlet-triplet energy gap of 6 kcal mol^{-1}.[173] A detailed kinetic and thermodynamic study shows the main process in thermolysis of $MeSiH_3$ to be loss of H_2 to give methylsilylene which then inserts into the silane to produce $(MeSiH_2)_2$ as the major product.[174] A similar process occurs for Bu^nSiH_3, but in this case the silylene can insert intramolecularly to give (**155**).[175] A deuterium-labelling study has shown the decomposition of silacyclobutane to silylene and propene to proceed by retro-insertion to give $Pr^n\ddot{S}iH$, which is in equilibrium with 2-methylsilacyclopropane, and decomposes to propene and :SiH_2.[176] Photochemical generation of a cyclopropenyl-silylene has been achieved by elimination of Si_2Me_6 from (**156**) and its insertion into the OH of ethanol has been observed.[177] Di-*t*-butylsilylene, generated by photolysis of hexa-*t*-butylcyclotrisilane adds to the triple bond of phosphaalkynes to afford the first phosphasilirenes.[178] Irradiation of hexaneopentyltrisilaoxetane results in elimination to give the disilaoxirane and dineopentylsilylene which inserts into EtOH.[179] The gamma-ray-induced formation of Me_2Si: from dodecamethylcyclohexasilane has been reported.[180]

(155)

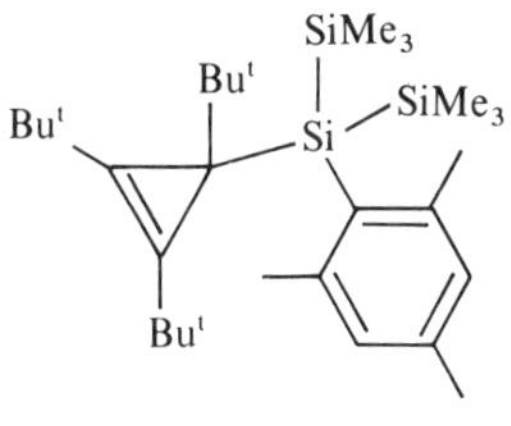

(156)

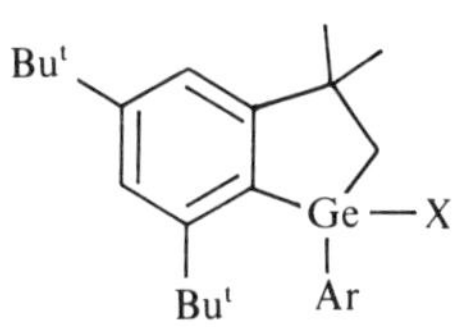

(157)

The reaction of various dichlorogermanes with Mg and $MgBr_2$ in THF leads to formation of cyclic trimers and tetramers of the corresponding germylenes.[181] Reaction of $GeCl_2$ with 2ArLi (Ar = 2,4,6-tri-*t*-butylphenyl) at -70°C to 10°C gives the corresponding germylene Ar_2Ge:. This reacts by intramolecular insertion at 20°C to give (**157**; X = H), while reaction with sulphur proceeds presumably by way of the thiagermanone to give (**157**; X = SH).[182]

References

[1] Doyle, M. P., *Acc. Chem. Res.*, **19**, 348 (1986).
[2] Brookhart, M. and Studabaker, W. B., *Chem. Rev.*, **87**, 411 (1987).
[3] Leigh, W. J. and Srinivasan, R., *Acc. Chem. Res.*, **20**, 107 (1987).
[4] Meth-Cohn, O., *Acc. Chem. Res.*, **20**, 18 (1987).
[5] Bock, H. and Dammel, R., *Angew. Chem. Int. Edn*, **26**, 504 (1987).
[6] Liebman, J. F. and Simons, J., *Mol. Struct. Energ.*, **1**, 51 (1986); *Chem. Abs.*, **107**, 58155 (1987).
[7] Sironi, M., Raimondi, M., Cooper, D. L., and Gerratt, J., *J. Chem. Soc., Faraday Trans. 2*, **83**, 1651 (1987).
[8] Becerra, R., Canosa-Mas, C. E., Frey, M., and Walsh, R., *J. Chem. Soc., Faraday Trans. 2*, **83**, 435 (1987).
[9] Boehland, T., Temps, F., and Wagner, H. G. G., *Report 1985 Max-Planck Inst. Stroemungsforsch*, **10** (1985); *Chem. Abs.*, **107**, 22720 (1987).
[10] Boehland, T., Dobe, S., Temps, F., and Wagner, H. G. G., *Report 1986, Max Planck Inst. Stroemungsforsch*, **11** (1985); *Chem. Abs.*, **107**, 58223 (1987).
[11] Turro, N. J., Cha, Y., and Gould, I. R., *J. Am. Chem. Soc.*, **109**, 2101 (1987).
[12] Turro, N. J. and Cha, Y., *Tetrahedron Lett.*, **27**, 6149 (1986).
[13] Carter, E. A. and Goddard, W. A., *J. Phys. Chem.*, **91**, 4651 (1987).
[14] Farras, J., Olivella, S., Sole, A., and Vilarrasa, J., *J. Comput. Chem.*, **7**, 428 (1986); *Chem. Abs.*, **105**, 225481 (1986).
[15] Jackson, J. E., Misslitz, U., Jones, M., and de Meijere, A., *Tetrahedron*, **43**, 653 (1987).
[16] Jorgensen, K. A., Moltzen, E. K., and Senning, A., *J. Org. Chem.*, **52**, 2505 (1987).
[17] Yadav, J. S. and Goddard, J. D., *J. Chem. Phys.*, **85**, 3975 (1986); *Chem. Abs.*, **106**, 101566 (1987).
[18] Moss, R. A., Shen, S., Hadel, L. M., Kmiecik-Lawrynowicz, G., Wlostowska, J., and Krogh-Jespersen, K., *J. Am. Chem. Soc.*, **109**, 4341 (1987).
[19] Moss, R. A., Fan, H., Hadel, L. M., Shen, S., Wlostowska, J., Wlostowski, M., and Krogh-Jespersen, K., *Tetrahedron Lett.*, **28**, 4779 (1987).
[20] Fujiwara, Y., Tanimoto, Y., Itoh, M., Hirai, K., and Tomioka, H., *J. Am. Chem. Soc.*, **109**, 1942 (1987).
[21] Liu, M. T. H., Soundararajan, N., Paike, N., and Subramanian, R., *J. Org. Chem.*, **52**, 4223 (1987).
[22] Tseng, J., McKee, M. L., and Shevlin, P. B., *J. Am. Chem. Soc.*, **109**, 5474 (1987).
[23] McMahon, R. J., Abelt, C. J., Chapman, O. L., Johnson, J. W., Kreil, C. L., LeRoux, J.-P., Mooring, A. M., and West, P. R., *J. Am. Chem. Soc.*, **109**, 2456 (1987).
[24] Janssen, C. L. and Schaefer, H. F., *J. Am. Chem. Soc.*, **109**, 5030 (1987).
[25] Schreck, M. and Christl, M., *Angew. Chem. Int. Edn*, **26**, 690 (1987).
[26] Doss, S. H., Abdel-Wahab, A. A., Frühof, E. M., Dürr, H., Gould, I. R., and Turro, N. J., *J. Org. Chem.*, **52**, 434 (1987).
[27] Rak, S. F., Lapin, S. C., Falvey, D. E., and Schuster, G. B., *J. Am. Chem. Soc.*, **109**, 5003 (1987).
[28] Abdel-Wahab, A. A., Doss, S. H., Dürr, H., Turro, N. J., and Gould, I. R., *J. Org. Chem.*, **52**, 429 (1987).
[29] Li, Y.-Z. and Schuster, G. B., *J. Org. Chem.*, **52**, 3975 (1987).
[30] Li, Y.-Z. and Schuster, G. B., *J. Org. Chem.*, **52**, 4460 (1987).
[31] Basak, S., Srivastava, S., and le Noble, W. J., *J. Org. Chem.*, **52**, 5095 (1987).
[32] Sixl, H., Mathes, R., Schaupp, A., Ulrich, K., and Huber, R., *Chem. Phys.*, **107**, 105 (1986); *Chem. Abs.*, **107**, 6678 (1987).
[33] Murata, S., Sugawara, T., and Iwamura, H., *J. Am. Chem. Soc.*, **109**, 1266 (1987).
[34] Izuoka, A. Murata, S., Sugawara, T., and Iwamura, H., *J. Am. Chem. Soc.*, **109**, 2631 (1987).
[35] DePuy, C. H., Barlow, S. E., Van Doren, J. M., Roberts, C. R., and Bierbaum, V. M., *J. Am. Chem. Soc.*, **109**, 4414 (1987).
[36] Bethell, D. and Parker, V. D., *J. Chem. Res. (S)*, **1987**, 116.
[37] McDonald, R. N. and Gung, W. Y., *J. Am. Chem. Soc.*, **109**, 7328 (1987).
[38] Buschek, J. M., Holmes, J. L., and Terlouw, J. K., *J. Am. Chem. Soc.*, **109**, 7321 (1987).
[39] Parker, V. D. and Bethell, D., *J. Am. Chem. Soc.*, **109**, 5066 (1987).
[40] Bohme, D. K., Wlodek, S., and Raksit, A. B., *Can. J. Chem.*, **65**, 1563 (1987).
[41] Tukada, H., Mutai, K., and Iwamura, H., *J. Chem. Soc., Chem. Commun.*, **1987**, 1159.
[42] Carrick, P. G., Brazier, C. R., Bernath, P. F., and Engelking, P. C., *J. Am. Chem. Soc.*, **109**, 5100 (1987).
[43] Jenkins, R. F., Waddell, W. H., and Richter, H. W., *J. Am. Chem. Soc.*, **109**, 1583 (1987).
[44] Leyva, E. and Platz, M. S., *Tetrahedron Lett.*, **28**, 11 (1987).
[45] Autrey, T. and Schuster, G. B., *J. Am. Chem. Soc.*, **109**, 5814 (1987).

[46] Shields, C. J., Chrisope, D. R., Schuster, G. B., Dixon, A. J., Poliakoff, M., and Turner, J. J., *J. Am. Chem. Soc.*, **109**, 4723 (1987).
[47] Albini, A., Bettinetti, G., and Minoli, G., *J. Org. Chem.*, **52**, 1245 (1987).
[48] Martin, G.-A. and Mirodatos, C., *J. Chem. Soc., Chem. Commun.*, **1987**, 1393.
[49] Domen, K. and Chuang, T. J., *J. Am. Chem. Soc.*, **109**, 5288 (1987).
[50] Duran, R. P., Amorebieta, V. T., and Colussi, A. J., *J. Am. Chem. Soc.*, **109**, 3154 (1987).
[51] Hills, M. M., Parmeter, J. E., and Weinberg, W. H., *J. Am. Chem. Soc.*, **109**, 597 (1987).
[52] Iyoda, M., Otani, H., Oda, M., Kai, Y., Boba, Y., and Kasai, N., *J. Chem. Soc., Chem. Commun.*, **1986**, 1794.
[53] Iovel, I. G., Goldberg, Yu. Sh., Shymanska, M. V., and Lukevics, E., *J. Chem. Soc., Chem. Commun.*, **1987**, 31.
[54] Dehmlow, E. V. and Wilkenloh, J., *Tetrahedron Lett.*, **28**, 5489 (1987).
[55] Tezuka, Y., Miya, M., Hashimoto, A., and Imai, K., *J. Chem. Soc., Chem. Commun.*, **1987**, 1642.
[56] Keyaniyan, S., Göthling, W., and de Meijere, A., *Chem. Ber.*, **120**, 395 (1987).
[57] Doyle, M. P., Devia, A. H., Bassett, K. E., Terpstra, J. W., and Mahapatro, S. N., *J. Org. Chem.*, **52**, 1619 (1987).
[58] Moss, R. A., Wlostowski, M., Terpinski, J., Kmiecik-Lawrynowicz, G., and Krogh-Jespersen, K., *J. Am. Chem. Soc.*, **109**, 3811 (1987).
[59] Davies, H. G., Rahman, S. S., Roberts, S. M., Wakefield, B. J., and Winders, J. A., *J. Chem. Soc., Perkin Trans. 1*, **1987**, 85.
[60] Pirrung, M. C. and Kenney, P. M., *J. Org. Chem.*, **52**, 2335 (1987).
[61] Fukuzawa, S., Fujinami, T., and Sakai, S., *J. Chem. Soc., Chem. Commun.*, **1987**, 919.
[62] Corsico Coda, A., Desimoni, G., Gamba Invernizzi, A., Quadrelli, P., Righetti, P. P. and Tacconi, G., *Tetrahedron*, **43**, 2843 (1987).
[63] Brown, R. F. C., Coulston, K. J., Eastwood, F. W., and Saminathan, S., *Aust. J. Chem.*, **40**, 107 (1987).
[64] Chapman, O. L. and Abelt, C. J., *J. Org. Chem.*, **52**, 1218 (1987).
[65] Johnson, D. E. and Dimian, A. F., *J. Chem. Soc., Chem. Commun.*, **1987**, 416.
[66] Mitchell, G. and Rees, C. W., *J. Chem. Soc., Perkin Trans. 1*, **1987**, 403.
[67] Rudolph, A., Margaretha, P., and Agosta, W. C., *Helv. Chim. Acta*, **70**, 339 (1987).
[68] Baird, M. S., Hussain, H. H., and Clegg, W., *J. Chem. Soc., Perkin Trans. 1*, **1987**, 1609.
[69] Wiberg, K. B. and McClusky, J. V., *Tetrahedron Lett.*, **28**, 5411 (1987).
[70] Moody, C. J. and Warrellow, G. J., *J. Chem. Soc., Perkin Trans. 1*, **1987**, 913.
[71] Ishii, K., Abe, M., and Sakamoto, M., *J. Chem. Soc., Perkin Trans. 1*, **1987**, 1937.
[72] Baird, M. S. and Hussain, H. H., *Tetrahedron*, **43**, 215 (1987).
[73] Dolbier, W. R. and Al-Fekri, D. M., *Tetrahedron*, **43**, 39 (1987).
[74] Eaton, P. E. and Hoffmann, K.-L., *J. Am. Chem. Soc.*, **109**, 5285 (1987).
[75] Brückmann, R. and Maas, G., *Chem. Ber.*, **120**, 635 (1987).
[76] Barton, T. J. and Paul, G. C., *J. Am. Chem. Soc.*, **109**, 5292 (1987).
[77] Strub, H., Strehler, C., and Streith, J., *Chem. Ber.*, **120**, 355 (1987).
[78] Umemoto, T. and Tomizawa, G., *Tetrahedron Lett.*, **28**, 2705 (1987).
[79] Hori, M., Kataoka, T., Shimizu, H., and Tsutsumi, K., *J. Org. Chem.*, **52**, 1397 (1987).
[80] Hadjiarapoglou, L. P., *Tetrahedron Lett.*, **28**, 4449 (1987).
[81] Ochiai, M., Kunishima, M., Nagao, Y., Fuji, K., Shiro, M., and Fujita, E., *J. Am. Chem. Soc.*, **108**, 8281 (1986).
[82] Eibler, E., Käsbauer, J., Pohl, H., and Sauer, J., *Tetrahedron Lett.*, **28**, 1097 (1987).
[83] Yabe, A., Ouchi, A., and Moriyama, H., *J. Chem. Soc., Chem. Commun.*, **1987**, 1744.
[84] Atkinson, R. S. and Kelly, B. J., *J. Chem. Soc., Chem. Commun.*, **1987**, 1362.
[85] Mash, E. A., Nelson, K. A., and Heidt, P. C., *Tetrahedron Lett.*, **28**, 1865 (1987).
[86] Molander, G. A. and Etter, J. B., *J. Org. Chem.*, **52**, 3942 (1987).
[87] Ryu, I., Aya, T., Otani, S., Murai, S., and Sonoda, N., *J. Organomet. Chem.*, **321**, 279 (1987).
[88] Sekiguchi, S., *Senryo to Yakuhin*, **31**, 87 (1986); *Chem. Abs.*, **106**, 83666 (1987).
[89] Sheng, H., Lin, S., Wang, Z., and Chen, Y., *Huaxue Xuebao*, **44**, 431 (1986); *Chem. Abs.*, **105**, 225552 (1986).
[90] Dehmlow, E. V. and Stiehm, T., *Angew. Chem. Int. Edn*, **26**, 329 (1987).
[91] Liepins, E., Goldberg, Yu., Iovel, I., and Lukevics, E., *J. Organomet. Chem.*, **335**, 301 (1987).
[92] Takahashi, M., Takada, T., and Sakagami, T., *J. Heterocycl. Chem.*, **24**, 797 (1987).
[93] Keglevich, G., Petneházy, I., Miklós, P., Almásy, A. Tóth, G., Töke, L., and Quin, L. D., *J. Org. Chem.*, **52**, 3983 (1987).
[94] Königstein, V., Tochtermann, W., Peters, E.-M., Peters, K. and von Schnering, H. G., *Tetrahedron Lett.*, **28**, 3483 (1987).

[95] Turro, N. J., Okamoto, M., Gould, I. R., Moss, R. A., Lawrynowicz, W., and Hadel, L. M., *J. Am. Chem. Soc.*, **109**, 4973 (1987).
[96] Bartnik, R. and Wawrynko, K., *J. Chem. Soc., Perkin Trans. 1*, **1987**, 715.
[97] Davies, H. M. L., Clark, D. M., Alligood, D. B., and Eiband, G. R., *Tetrahedron*, **43**, 4265 (1987).
[98] Davies, H. M. L., Smith, H. D., and Korkor, O., *Tetrahedron Lett.*, **28**, 1853 (1987).
[99] Okamoto, Y. and Iwamoto, T., *Kenkyu Kiyo-Wakayama Kogyo Koto Senmon Gakko*, **20**, 43 (1985); *Chem. Abs.*, **106**, 4368 (1987).
[100] Birtwistle, I. and Rogers, V., *J. Chem. Soc., Perkin Trans. 1*, **1987**, 1347.
[101] Komatsu, K., Arai, M., Hattori, Y., Fukuyama, K., Katsube, Y., and Okamoto, K., *J. Org. Chem.*, **52**, 2183 (1987).
[102] Majerski, Z. and Zuanić, M., *J. Am. Chem. Soc.*, **109**, 3496 (1987).
[103] Belzner, J. and Szeimies, G., *Tetrahedron Lett.*, **28**, 3099 (1987).
[104] Brückmann, R. and Maas, G., *J. Chem. Soc., Chem. Commun.*, **1986**, 1782.
[105] Skattebol, L., Nilsen, N. O., and Myhren, F., *Acta Chem. Scand.*, **40B**, 782 (1986).
[106] Birkhahn, M., Dehmlow, E. V., and Bögge, H., *Angew. Chem. Int. Edn.*, **26**, 72 (1987).
[107] Lazareva, M. A., *Teor, Eksp. Khim.*, **22**, 477 (1986); *Chem. Abs.*, **106**, 119027 (1987).
[108] Satake, K., Mizushima, H., Kimura, M., and Morosawa, S., *J. Chem. Soc., Chem. Commun.*, **1987**, 197.
[109] Fioravanti, S., Loreto, M. A., Pellacani, L., and Tardella, P. A., *J. Chem. Res. (S)*, **1987**, 310.
[110] Atkinson, R. S. and Tughan, G., *J. Chem. Soc., Perkin Trans. 1*, **1987**, 2787, 2797, 2803.
[111] Atkinson, R. S. and Tughan, G., *J. Chem. Soc., Chem. Commun.*, **1987**, 456.
[112] Atkinson, R. S. and Grimshire, M. J., *J. Chem. Soc., Perkin Trans. 1*, **1987**, 1135.
[113] Saalfrank, R. W., Fischer, M., Wirth, U., and Zimmermann, H., *Angew. Chem. Int. Edn*, **26**, 1160 (1987).
[114] Saalfrank, R. W., Ackermann, E., Fischer, M., and Wirth, U., *Chem. Ber.*, **120**, 2003 (1987).
[115] Banert, K., *Chem. Ber.*, **120**, 1891 (1987).
[116] Klingler, O. and Prinzbach, H., *Angew. Chem. Int. Edn*, **26**, 566 (1987).
[117] Marterer, W., Fritz, H., and Prinzbach, H., *Tetrahedron Lett.*, **28**, 5497 (1987).
[118] Oku, A. and Harada, T., *Kenkyu Hokoku–Asahi Garasu Kogyo Gijutsu Shoreikai*, **49**, 65 (1986); *Chem. Abs.*, **107**, 175187 (1987).
[119] Ortega, M., Lluch, J. M., Oliva, A., and Bertran, J., *Can. J. Chem.*, **65**, 1995 (1987).
[120] Gordon, M. S., Boatz, J. A., Gano, D. R., and Friederichs, M. G., *J. Am. Chem. Soc.*, **109**, 1323 (1987).
[121] Mehta, R. K. S., *Indian J. Chem.*, **25B**, 157 (1986); *Chem. Abs.*, **106**, 32149 (1987).
[122] Mel'nikovich, S. V. and Moin, F. B., *Kinet. Katal.*, **28**, 181 (1987); *Chem. Abs.*, **107**, 39098 (1987).
[123] Nguyen, M. T. and Hegarty, A. F., *J. Chem. Soc., Perkin Trans. 2*, **1987**, 1675.
[124] Stang, P. J. and Learned, A. E., *J. Am. Chem. Soc.*, **109**, 5019 (1987).
[125] Nguyen, M. T. and Hegarty, A. F., *J. Chem. Soc., Perkin Trans. 2*, **1987**, 55.
[126] Brinker, U. H. and Erdle, W., *Angew. Chem. Int. Edn*, **26**, 1260 (1987).
[127] Lawton, G., Moody, C. J., and Pearson, C. J., *J. Chem. Soc., Perkin Trans. 1*, **1987**, 899.
[128] Buxton, S. R., Holm, K. H., and Skattebol, L., *Tetrahedron Lett.*, **28**, 2167 (1987).
[129] Hrytsak, M. and Durst, T., *J. Chem. Soc., Chem. Commun.*, **1987**, 1150.
[130] Adams, J., Frenette, R., Belley, M., Chibante, F., and Springer, J. P., *J. Am. Chem. Soc.*, **109**, 5432 (1987).
[131] Harada, T., Yasunari, Y., and Oku, A., *Bull. Chem. Soc. Jpn*, **60**, 1715 (1987).
[132] Barcus, R. L., Platz, M. S., and Scaiano, J. C., *J. Phys. Chem.*, **91**, 695 (1987).
[133] Bertolaccini, R., Loreto, M. A., Pellacani, L., Tardella, P. A., and Cerichelli, G., *J. Org. Chem.*, **52**, 1859 (1987).
[134] Marais, P. C. and Meth-Cohn, O., *J. Chem. Soc., Perkin Trans. 1*, **1987**, 1553.
[135] Miyasaka, T., Tanaka, H., Satoh, K., Imahashi, M., Yamaguchi, K., and Iitaka, Y., *J. Heterocycl. Chem.*, **24**, 873 (1987).
[136] Isomuru, K., Sakurai, M., Komura, T., Saruwatari, M., and Taniguchi, H., *Chem. Lett.*, **87**, 883 (1987).
[137] Labarca, C. V., MacKenzie, A. R., Moody, C. J., Rees, C. W., and Vaquero, J. J., *J. Chem. Soc., Perkin Trans. 1*, **1987**, 927.
[138] Hickey, D. M. B., MacKenzie, A. R., Moody, C. J. and Rees, C. W., *J. Chem. Soc., Perkin Trans. 1*, **1987**, 921.
[139] Bolton, R. E., Moody, C. J., Rees, C. W., and Tojo, G., *J. Chem. Soc., Perkin Trans. 1*, **1987**, 931.
[140] Wesdemiotis, C. and McLafferty, F. W., *J. Am. Chem. Soc.*, **109**, 4760 (1987).
[141] McMahon, R. J. and Chapman, O. L., *J. Am. Chem. Soc.*, **109**, 683 (1987).
[142] Sekiguchi, A., Sato, T., and Ando, W., *Organometallics*, **6**, 2337 (1987).
[143] Sekiguchi, A. and Ando, W., *Organometallics*, **6**, 1857 (1987).

[144] Maier, G., Reisenauer, H. P., Schwab, W., Carsky, P., Hess, B. A., and Schaad, L. J., *J. Am. Chem. Soc.*, **109**, 5183 (1987).
[145] Moss, R. A., Wilk, B. K., and Hadel, L. M., *Tetrahedron Lett.*, **28**, 1969 (1987).
[146] Kesselmayer, M. A. and Sheridan, R. S., *J. Am. Chem. Soc.*, **109**, 5029 (1987).
[147] Fleischhauer, I. and Brinker, U. H., *Chem. Ber.*, **120**, 501 (1987).
[148] Hopf, H. and Wachholz, G., *Chem. Ber.*, **120**, 1259 (1987).
[149] Eberbach, W. and Roser, J., *Tetrahedron Lett.*, **28**, 2685 (1987).
[150] Eberbach, W. and Roser, J., *Tetrahedron Lett.*, **28**, 2689 (1987).
[151] Lawton, G., Moody, C. J., and Pearson, C. J., *J. Chem. Soc., Perkin Trans. 1*, **1987**, 877.
[152] Novoa, J. J., McDouall, J. J. W., and Robb, M. A., *J. Chem. Soc., Faraday Trans. 2*, **83**, 1629 (1987).
[153] Mahaffy, P. G., Visser, D., Torres, M., Bourdelande, J. L., and Strausz, O. P., *J. Org. Chem.*, **52**, 2680 (1987).
[154] Ruehl, W., Boelsing, F., Hofer, E., Speer, D., and Jansen, M., *Z. Naturforsch.*, **41B**, 772 (1986); *Chem. Abs.*, **106**, 101562 (1987).
[155] Mitchell, G. and Rees, C. W., *J. Chem. Soc., Perkin Trans. 1*, **1987**, 413.
[156] Nguyen, M. T., Faul, M., and Fitzpatrick, N. J., *J. Chem. Soc., Perkin Trans. 2*, **1987**, 1289.
[157] Eaton, P. E. and Hormann, R. E., *J. Am. Chem. Soc.*, **109**, 1268 (1987).
[158] Underwood, G. R. and Callahan, R. J., *Tetrahedron Lett.*, **28**, 5427 (1987).
[159] Takeuchi, H., *J. Chem. Soc., Chem. Commun.*, **1987**, 961.
[160] Spagnolo, P. and Zanirato, P., *Tetrahedron Lett.*, **28**, 961 (1987).
[161] Glover, S. A., Goosen, A., McCleland, C. W., and Schoonraad, J. L., *Tetrahedron*, **43**, 2577 (1987).
[162] Meyer, H., Baum, G., Massa, W., Berger, S., and Berndt, A., *Angew. Chem. Int. Edn*, **26**, 546 (1987).
[163] Meyer, H., Baum, G., Massa, W., and Berndt, A., *Angew Chem. Int. Edn*, **26**, 798 (1987).
[164] McCarthy, J. R., Barney, C. L., O'Donnell, M. J., and Huffman, J. C., *J. Chem. Soc., Chem. Commun.*, **1987**, 469.
[165] Turro, N. J. and Cha, Y., *Tetrahedron Lett.*, **28**, 1723 (1987).
[166] Liu, M. T. H., Soundararajan, N., Anand, S. M. and Ibata, T. *Tetrahedron Lett.*, **28**, 1011 (1987).
[167] Shields, C. J. and Schuster, G. B., *Tetrahedron Lett.*, **28**, 853 (1987).
[168] Davies, H. M. L. and Crisco, L. V. T., *Tetrahedron Lett.*, **28**, 371 (1987).
[169] Sülzle, D., *Tetrahedron Lett.*, **28**, 2769 (1987).
[170] Boo, B. H. and Armentrout, P. B., *J. Am. Chem. Soc.*, **109**, 3549 (1987).
[171] Rogers, D. S., Walker, K. L., Ring, M. A., and O'Neal, H. E., *Organometallics*, **6**, 2313 (1987).
[172] Chu, J. O., Beach, D. B., and Jasinski, J. M., *J. Phys. Chem.*, **91**, 5340 (1987).
[173] Gordon, M. S. and Bartol, D., *J. Am. Chem. Soc.*, **109**, 5948 (1987).
[174] Neudorfl, P. S., Lown, E. M., Safarik, I., Jodhan, A., and Strausz, O. P., *J. Am. Chem. Soc.*, **109**, 5780 (1987).
[175] Sawrey, B. A., O'Neal, H. E., and Ring, M. A., *Organometallics*, **6**, 720 (1987).
[176] Barton, T. J. and Tillman, N., *J. Am. Chem. Soc.*, **109**, 6711 (1987).
[177] Fink, M. J. and Puranik, D. B., *Organometallics*, **6**, 1809 (1987).
[178] Schäfer, A., Weidenbruch, M., Saak, W., and Pohl, S., *Angew. Chem. Int. Edn*, **26**, 776 (1987).
[179] Watanabe, H., Tabei, E., Goto, M., and Nagai, Y., *J. Chem. Soc., Chem. Commun.*, **1987**, 522.
[180] Oka, K., Nakao, R., Nagata, Y., and Takaaki, D., *J. Chem. Soc., Perkin Trans. 2*, **1987**, 337.
[181] Ando, W. and Tsumuraya, T., *J. Chem. Soc., Chem. Commun.*, **1987**, 1514.
[182] Lange, L., Meyer, B., and duMont, W.-W., *J. Organomet. Chem.*, **329**, C17 (1987).

Organic Reaction Mechanisms 1987
Edited by A. C. Knipe and W. E. Watts

CHAPTER 7

Nucleophilic Aromatic Substitution

MICHAEL R. CRAMPTON

Department of Chemistry, Durham University

General

There has been renewed interest in the binding of arenediazonium ions by macrocyclic compounds. A spherand, synthesized from cyclic urea and aryl ether units, has been found to be more effective than 18-crown-6 polyether in the complexation and stabilisation of diazonium ions.[1] It has also been shown that hexasulphonated calix[6]arenes can suppress dediazoniations in aqueous systems where 18-crown-6 is ineffective.[2] The dediazoniation of *para*-substituted benzenediazonium ions in 1,2-dichloroethane occurs by a heterolytic mechanism. However the presence of reduction products from the reaction of 18-crown-6 complexes of these diazonium ions indicates an additional homolytic pathway. The effects of substituents on rate constants for both reaction pathways have been correlated with dual substituent parameters.[3] ^{15}N NMR spectroscopy has shown that, in the absence of a reactant, solutions in acetonitrile of the diazonium ions (**1**; X = MeO, NO_2, SO_3H) do not undergo nitrogen rearrangement; it was suggested that the nitrogen scrambling which is found during dediazoniation reactions occurs within an addition complex in equilibrium with the reactants.[4]

The initial step in many transformations of diazonium ions involves electron transfer. Rate constants for one-electron, outer-sphere reduction of arenediazonium ions (Equation 1) have been determined and Marcus theory

$$ArN_2^+ + Red^- \rightarrow ArN_2^{\bullet} + Red^{\bullet} \qquad (1)$$

applied.[5] The reaction of the diazonium salt (**2**) with sodium benzenethiolate in dimethyl sulphoxide (DMSO) yields a mixture of the cyclized (**3**) and uncyclized (**4**) products. The mechanistic route for this thiodediazoniation involves initial formation of diazosulphide (ArN=NSAr′) followed by homolytic cleavage to give aryl radicals which may ring-close or couple with thiolate anions to give radical anions. Electron transfer from the latter to diazosulphide results in chain

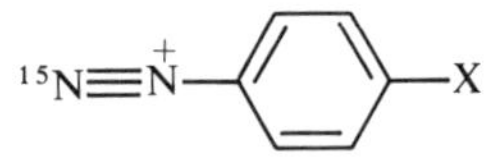

(**1**)

propagation.[6] Treatment of (**2**) and its analogues with sodium iodide in acetone may also yield cyclized and uncyclized products.[7,8] There has been a short review of the radical-chain $S_{RN}1$ mechanism.[9] Substitution of the (phenylthio)azo group in (**5**) by cyanide ions may be achieved by photo- or electron-stimulated $S_{RN}1$ reaction,[10,11] and the photo-stimulated reaction of aryl iodides with 2-naphthylamide ions gave 1-aryl-2-naphthylamines.[12] Double $S_{RN}1$ reaction of dithiolate ions with 1- or 2-halonaphthalenes yields the corresponding double sulphides,[13] while reaction of dithiolate ions with *ortho*-dihalo-aromatics gives disubstituted cyclic products.[14] Photo-stimulated reaction of halothiophenes with benzenethiolate ions in acetonitrile yields phenyl thienyl sulphides although the yields are reduced by fragmentation of the ThSPh$^-$ radical anion.[15] Photo-stimulated $S_{RN}1$ reactions have also proved useful in the syntheses of unsymmetrically hydroxylated 2,2′-binaphthalenes, by reaction between 2′-haloacetophenones and enolates derived from 2′-acetonaphthone,[16] and of unsymmetrically substituted hydroxybiphenyls, by reaction of 2,6- and 2,4-di-*tert*-butylphenoxides with chlorobenzonitriles.[17] Electrochemical reduction of halobenzonitriles in acetonitrile in the presence of benzeneselenate ions yields (phenylseleno)benzonitriles,[18] and reduction of bromobenzophenone in the presence of PhSe$^-$ or PhTe$^-$ yields, respectively, (phenylseleno)- or (phenyltelluro)-benzophenones.[19] Electrochemical reduction of substituted *ortho*-bis(phenylsulphonyl)benzenes results in the formation of a radical anion which expels $PhSO_2^-$; in the presence of arenethiolate ions, coupling of the resulting radical with the nucleophile gives diaryl sulphides by $S_{RN}1$ reaction.[20]

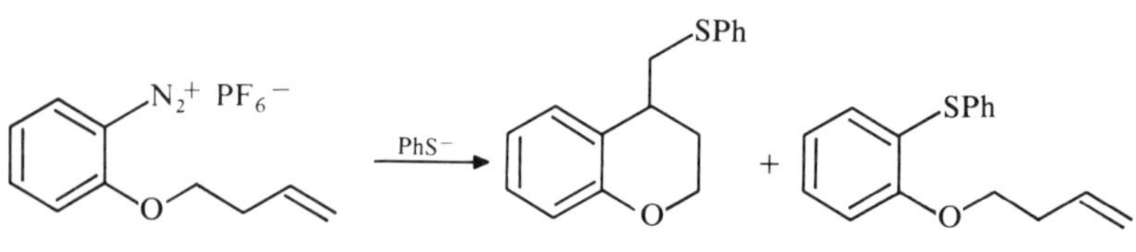

(**2**) (**3**) (**4**)

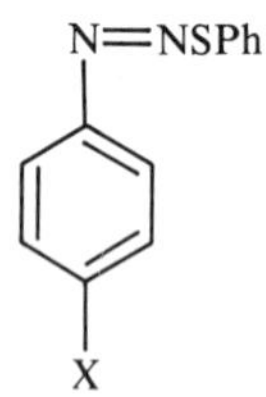

(**5**)

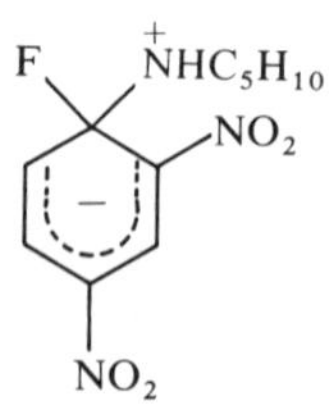

(**6**)

The S_NAr Mechanism

A book on the subject of nucleophilicity has been published. It includes interesting chapters on nucleophilic reactivity,[21] the relationship between reactivity and separation of nucleophile and electrophile,[22] reactions of ambident nucleophiles with nitro-aromatic compounds,[23] and competition between substitution and reduction in the reactions of chloronitrobenzenes with alkoxide ions.[24] An examination by stopped-flow spectrophotometry claims the observation of π-complex and radical intermediates on the pathways to nucleophilic substitution and σ-adduct formation.[25]

There have been further studies of base catalysis in substitutions by amine nucleophiles including a review of reactions in non-polar aprotic solvents.[26] Kinetic studies of the reaction of 1-fluoro-2,4-dinitrobenzene with piperidine, involving the intermediacy of (**6**), show that the nature of the rate-determining step is solvent-dependent; in solvents such as chloroform, acetonitrile, and nitromethane, which show hydrogen-bond donor properties, formation of (**6**) is rate-limiting whereas in other solvents the observation of base catalysis is interpreted in terms of rate-limiting expulsion of fluoride.[27] The intermediacy of charge-transfer complexes, prior to σ-bond formation, has been postulated in reactions of 1,2,3,5-tetranitrobenzene and fluoro-2,4,6-trinitrobenzene with aromatic amines in non-polar solvents,[28] and also in the reactions of 1,2-dinitrobenzene with aliphatic amines in *n*-hexane.[29,30]

The predominant reaction of (**7**) with aliphatic amines in DMSO and in acetonitrile is S_NAr displacement of methoxide although there is some S_N2 attack at alkyl carbon with secondary amine nucleophiles. Base catalysis of the S_NAr reaction was observed with both primary and secondary amines and the results were interpreted by the SB-GA mechanism in which general acid-catalysed leaving-group departure is rate-limiting.[31] The hydrolyses of 4-(4′-methoxyphenylazo)pyridine, (**8**), and its 3-isomer have been studied in aqueous sulphuric acid. These hydrolyses involve the diprotonated substrates and reaction may occur by an *A*-2 mechanism involving rate-limiting attack by water at aryl carbon or by an A-S_E2 mechanism where nucleophilic attack and proton transfer are rapid and the slow step is general acid-catalysed expulsion of methoxide.[32] The rate coefficients for the alkaline

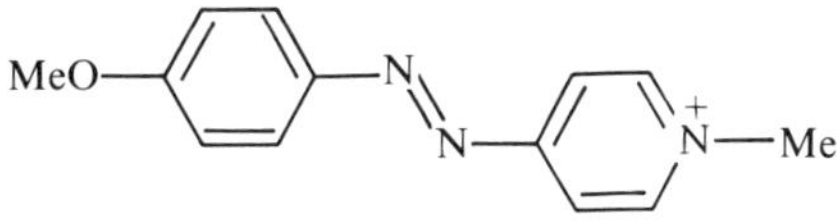

(**7**)

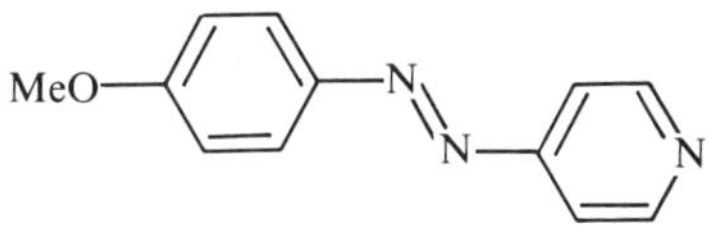

(**8**)

hydrolyses of 2,4-dinitroanisole and 2-methoxy-5-nitropyridine have been measured in mixtures of water with a series of dipolar aprotic solvents, and the results correlated with the H_- acidity function for these media. In water-pyridine mixtures there is evidence[33] for nucleophilic catalysis involving the intermediate (**9**). Rates of reaction with methoxide of 1-substituted 2,4-dinitrobenzenes and of 2-substituted 5-nitropyridines in methanol-DMSO mixtures have also been correlated[34] with an acidity function; in media rich in DMSO, attack by methoxide at unsubstituted ring positions is faster than at the substituted position leading to the observation of σ-adducts (**10**).

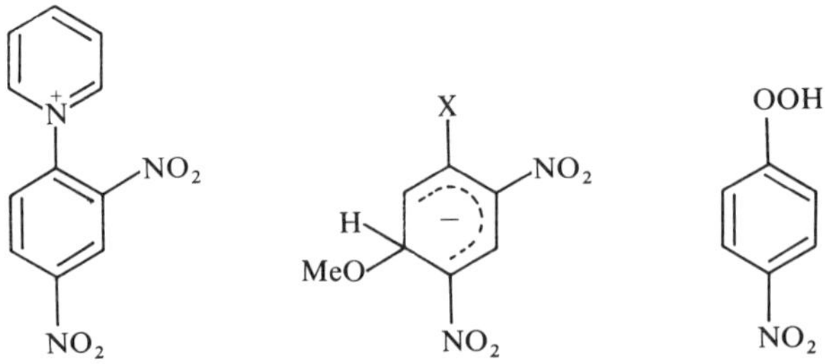

(9) **(10)** **(11)**

There have been short reviews of substitutions in nitrobenzenes and in chlorobenzenes.[35] Kinetic and NMR studies of the reaction of 1,4-dinitrobenzene with alkaline hydrogen peroxide show that 4-nitrophenyl hydroperoxide (**11**) is an observable intermediate in the conversion into 4-nitrophenol.[36] Nitro-group displacement also occurs in the reaction of 1,4-dinitrobenzene with sodium phenoxide and with substituted phenoxides in DMSO, although the mechanism here is likely to involve radicals.[37] Amino-dechlorinations have been reported in reactions of nitro-activated aromatic chlorides with primary and secondary amines in THF under high pressure; DABCO and quinuclidine gave *N*-aryl quaternary salts although acyclic tertiary amines were unreactive.[38] There have also been reports of the synthesis of picryl cellulose by S_NAr displacement of chloride from picryl chloride by sodium cellulosate,[39] and of the mono- and di-piperidinodechlorination of 2,3-dichloro-5,6-dicyano-1,4-benzoquinone (DDQ) in chloroform.[40]

There has been an interesting report of the direct amination of substituted nitrobenzenes by reaction with 4-amino-1,2,4-triazole in DMSO in the presence of potassium *tert*-butoxide. The mechanism (Scheme 1) represents an extension of vicarious substitution of hydrogen to a nitrogen nucleophile.[41] The formylation of nitroarenes has been reported by the vicarious substitution of hydrogen by the

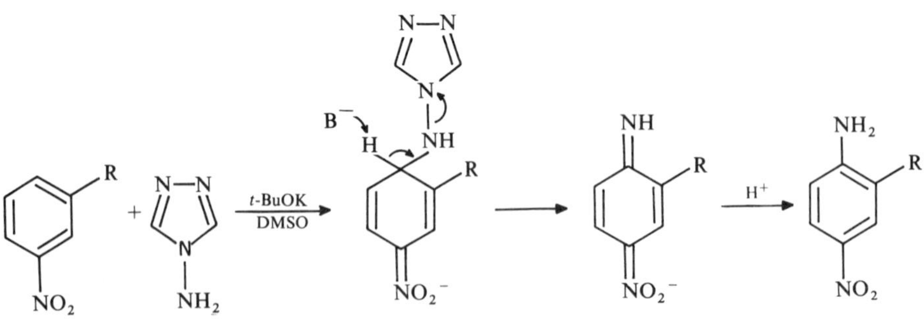

SCHEME 1

dichloromethyl carbanion followed by hydrolysis of the substituted product.[42] Reaction of the chloromethyl aryl sulphone carbanion with 1-nitronaphthalene leads to vicarious substitution of hydrogen[43] whereas reaction with 1-cyanonaphthalene gives bis-annulated addition products; the difference is ascribed to the lower electron-withdrawing effect of the cyano group which results in greater nucleophilicity of ring-carbon atoms in the cyano-substituted ring.[44] The initial step in the formation of diphosphorylated naphthalene derivatives from 4-substituted-1-nitronaphthalenes and dimethyl phosphite is thought to be nucleophilic addition of the dimethyl phosphite anion to the aromatic ring.[45]

The unexpected products from the reaction of 6-cyano- and 6-nitro-benzopyran derivatives with azacyclotridecanone are thought[46] to arise from initial intramolecular attack at the electron-deficient 5-position as shown in (**12**). The intramolecular rearrangement[47] of *N*-(2,4-dinitrophenyl)-piperidine and -morpholine *N*-oxides is likely to involve intermediates such as (**13**), and there is crystallographic evidence for strong intramolecular N---C interaction in 2-methoxycarbonyl-1-naphthyl *N*-(1-naphthyl)benzimidate which undergoes the Chapman rearrangement to give dibenz[*ch*]acridinone.[48]

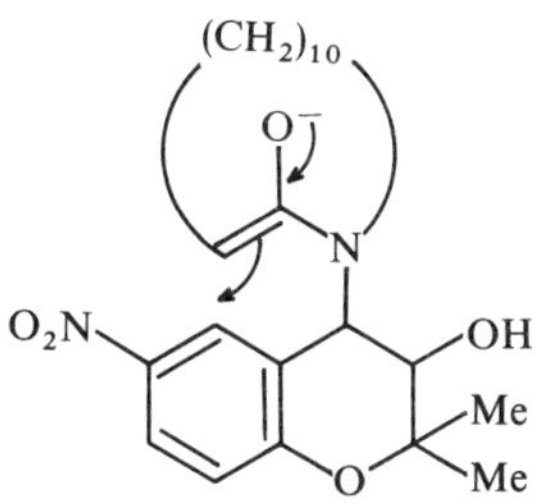

(**12**)

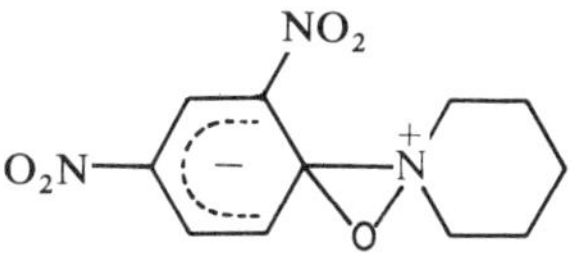

(**13**)

A study of the photochemical reactions of 1-methoxy-4-nitronaphthalene with amines has shown that secondary amines displace the methoxy substituent by an electron-transfer process while primary amines displace the nitro substituent, probably by an S_N2Ar^* process.[49] A duality of mechanisms is also observed in the photo-substitution of 4-nitroveratrole (**14**) by *n*-hexylamine resulting in isomeric products formed by substitution of either the 1- or 2-methoxyl groups.[50] The photochemical reactions of 4,5-dinitroveratrole (**15**) with amines lead to either substitution of methoxide or reduction of nitro groups.[51] UV photolyses of 3,4-dimethoxybenzonitrile and 3,4-dimethoxyacetophenone in the presence of nucleo-

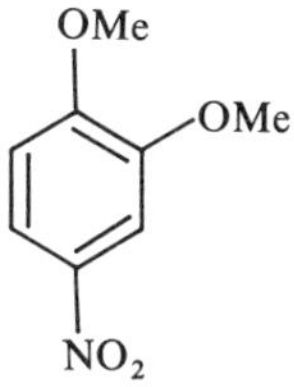

(**14**)

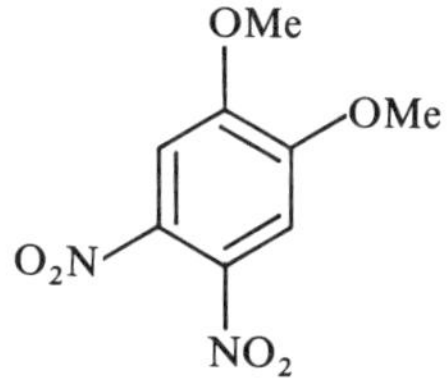

(**15**)

philes may lead to substitution of either the 3- or 4-substituent; reactions with hydroxide ions give 3-substituted products exclusively while reaction with cyanide ions give mixtures.[52]

The unexpectedly easy N—N exchange reactions (Scheme 2) of *N,N*-dimethyl-2,4-bis(trifluoroacetyl)-1-naphthylamine with ammonia, primary amines, and pyrrolidine have been reported.[53] The use of ^{36}Cl has allowed the measurement of rates of chlorine exchange of lithium chloride with various nitro-activated 1-chloronaphthalenes in sulpholane[54] and also with 4-substituted 1-chloro-2,6-dinitrobenzenes.[55] There have been reports of the catalysis by activated nickel powder of halide exchange reactions between aryl halides and alkali metal halides,[56] and of the use of alumina and charcoal supported copper(I) iodide in the conversion of bromo- and chloro-benzenes into iodobenzenes.[57] Tetrabutylammonium bifluoride is reported to be a stable source of fluoride ion for nucleophilic substitutions,[58] and it is reported that the reactivity of potassium fluoride is enhanced by the presence of tetraphenylphosphonium bromide.[59] The solvent dependence of the rate of reaction of potassium fluoride with 4-chloronitrobenzene has been examined.[60] Potassium fluoride has been found to form a stable hydrogen-bonded complex with 4-cyanophenol; in DMSO the complex acts as an oxygen nucleophile capable of forming biaryl ethers by denitration or dechlorination.[61] Other studies relevant to the reactivity of fluoride ions include a spectroscopic study of the solvation of fluorides,[62] and an examination of the basicity of fluoride-alumina reagents.[63]

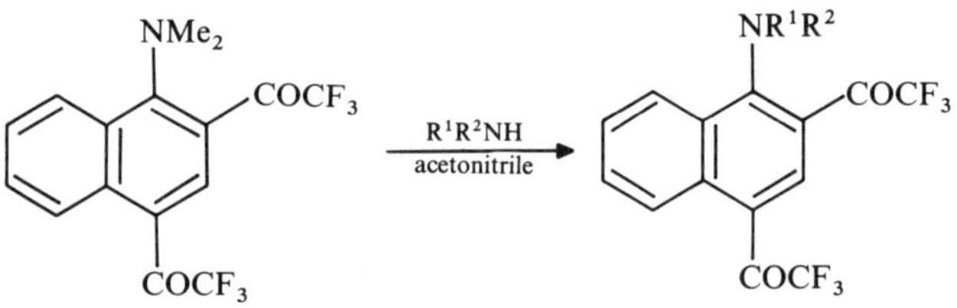

SCHEME 2

Rate constants for methoxydefluorination reactions in polyfluoroarene ring systems have been measured,[64] and kinetics of the reaction with sodium methoxide and with piperidine have been used to compare the reactivities of perfluorinated methylbenzenes and benzocycloalkenes.[65]

There is continued interest in obtaining rate enhancements by modification of the reaction medium. Quaternary ammonium salts have been found to increase the solubility of ammonia in organic media with resultant increases in rates of substitution reactions,[66] and catalysis by a macrocyclic quaternary ammonium salt of the reaction of 4-chloro-3,5-dinitrobenzenesulphonate with azide ions has been reported.[67] There have been further reports of the catalysis by β-cyclodextrin of amino-dehalogenation reactions.[68,69] Micellar catalysis of reactions of uni- and divalent ions has been reviewed,[70] and there have been reports of the effects on the rates of reaction of hydroxide ions with 1-chloro-2,4-dinitrobenzene and with

1-chloro-2,4-dinitronaphthalene produced by sub-micellar aggregates,[71] cationic bolaform surfactants,[72] and single- and twin-tailed surfactants.[73]

Much effort has been devoted to the study of catalysis of substitution reactions by metal ions and by metal complexes,[74] and there has been a review of cyanation reactions using copper cyanide and other metal cyanides reagents.[75] Copper(II) complexed with acetylacetone or with ethyl acetoacetate has been found to catalyse the ammonolyses of bromobenzoate ions,[76] but the effective catalytic species on the Ullman condensation of diarylamines with iodobenzenes has been shown to be cuprous ion in solution.[77] The presence of two donor atoms in (**16**) is thought to facilitate the cyanodebromination reaction by copper(I) complexes.[78] Intramolecular chelation of the metal is also thought to increase reactivity in the reactions of the oxazole derivative (**17**) with arylmagnesium bromide and aryllithium to give substituted biphenyls,[79] and in the nickel(II)-catalysed reaction of (**18**) with tertiary phosphines to give arylphosphonium salts.[80]

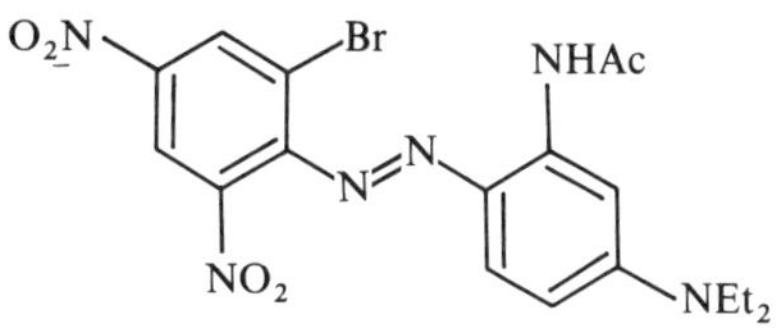

(**16**)

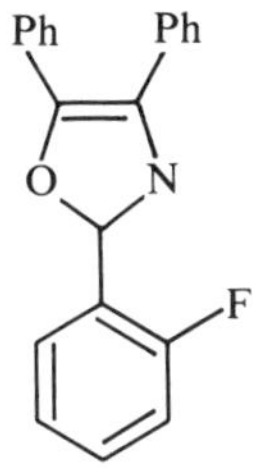

(**17**)

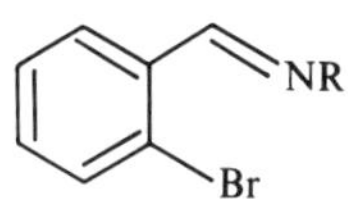

(**18**)

There have been reports of the effects of photo-stimulation[81] and of change of solvent[82] on the cobalt-catalysed carbonylations of aryl halides. The reaction of 1-amino-2-iodoarenes with thioamides is efficiently catalysed by palladium(0) to give 2-alkylbenzothiazoles.[83] Palladium has also been found to catalyse the carbon-carbon bond-forming reactions of halogenoarenes with buta-1,3-diene[84] and with enol ethers.[85] Other examples of palladium catalysis are the cyanation of aryl halides by trimethylsilyl cyanide[86] and the coupling of aryl halides with (trimethylstannyl)diphenylphosphine and with (trimethylsilyl)diphenylphosphine.[87]

Activation of aromatic compounds to nucleophilic substitution may be achieved by π-complexing with transition metal-ligand groups. There have been reports of substitutions in (cyclopentadienyl)iron complexes of arenes[88] and in (cyclopentadienyl)ruthenium complexes of 4- or 5-chloroindole (**19**; R = H, Me).[89] *Tele*-substitution of the chloro group of (1-chloro-2,6-dimethylbenzene)-

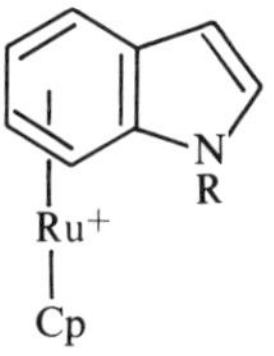

(**19**)

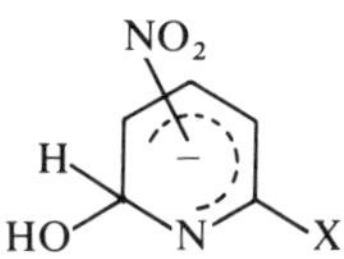

(**20**)

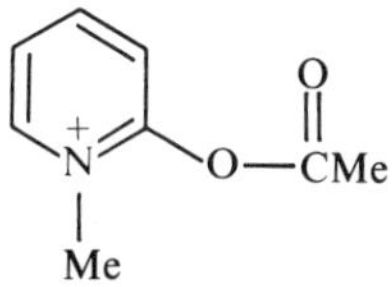

(**21**)

tricarbonylchromium by a carbon nucleophile is reported to occur at a position *para* to the leaving-group;[90] *tele*-substitution, at the *meta* position, occurs in substitutions of tricarbonylchromium complexes of diphenyl ethers.[91]

Heterocyclic Systems

The reactions of pyrazine-*N*-oxides have been reviewed.[92] The rates of alkaline hydrolysis of 2-substituted 3- and 5-nitropyridines, 4-nitropyridine *N*-oxide and 2- and 4-substituted pyridine methiodides have been measured in water-DMSO mixtures and the results related to the H_- acidity function. In the halogenonitropyridines, initial hydroxide attack occurs at the 6-position to give (**20**; X = halogen) and is followed by ring-opening.[93] ^{18}O studies of the buffer catalysis of the hydrolysis of 2-fluoro-1-methylpyridinium iodide are consistent[94] with nucleophilic catalysis involving the intermediate (**21**). There have been reports of abnormal micellar effects on the reactions of azide ions with *N*-alkyl-2-bromopyridinium ions,[95] and of the catalytic effects of tris(3,6-dioxaheptyl)amine on the alkoxydehalogenation reactions of 2- and 4-chloropyridine under phase-transfer catalysis.[96] It is reported that reaction of methyl thioglycolate with halopyridines carrying electron-withdrawing groups may lead to the formation of thienopyridines.[97] The regiochemistry of reaction of nitro-activated 4(5)-haloimidazoles with oxygen, nitrogen, and carbon nucleophiles has been examined;[98] reaction with arylthiolate ions yields products which are readily oxidized to sulphones.[99]

Nucleophilic substitution in pentafluoropyridine by the trifluoromethanethiolate ion (CF_3S^-) has been found[100] to yield mono-, di-, and tri-substituted products. Study of the effects of counter-ions, solvent, and solvating agents on the reactions of transition metal carbonylates with 2-(pentacarbonylmanganese)-4,6-difluoro-1,3,5-triazine indicates enhanced reactivity for contact ion-pairs compared to solvent-separated ion-pairs and to free ions.[101] The preparation of perfluoroalkyl-1,2,3-triazines, such as (**22**), has been reported; unusually, they are subject to nucleophilic attack at the 2-nitrogen atom.[102]

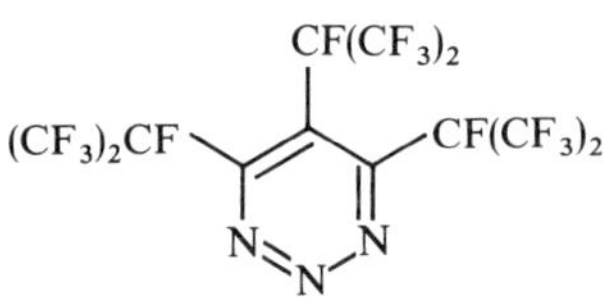

(**22**)

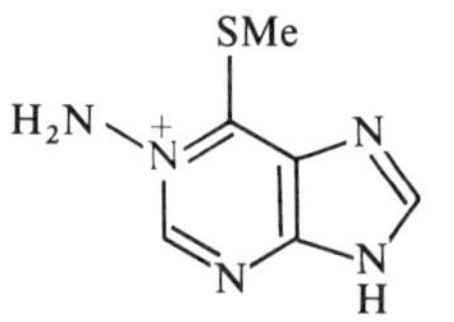

(**23**)

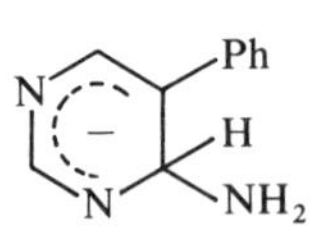

(**24**)

Ring transformations of 1,2,4-triazines into 1,2,4-triazoles and 1,3,5-triazines have been reported; ^{15}N labelling experiments indicate a mechanism involving initial attack of amide ions at unsubstituted ring positions and ring-opened intermediates.[103] Reaction with methanolic ammonia of 1-amino-6-(methylthio)-purinium mesitylenesulphonate (**23**) results in deamination and partial substitution of the methylthio group without ring-opening; however, reaction of the 2-(methylthio) isomer results in ring-opening followed by substitution.[104] The amide adduct (**24**) has been observed spectroscopically during the Chichibabin amination (replacement of H by NH_2) of 5-phenylpyrimidine; a mixture of 2- and 4-aminated products is formed, the former by a ring-opening – ring-closure sequence. Amination of phenylpyrazine yields 3-amino and 5-amino derivatives without opening of the pyrazine ring.[105] Treatment of 2-alkylpyrimidines with potassium amide and potassium permanganate in liquid ammonia yields aminated products. However, corresponding treatment of *N*-methylpyridinium salts gives imino derivatives (**25**; R = H, Me). The imino-compounds react[106] with base to give 6-methylaminopyrimidines (**26**) by an ANRORC mechanism as shown in Scheme 3.

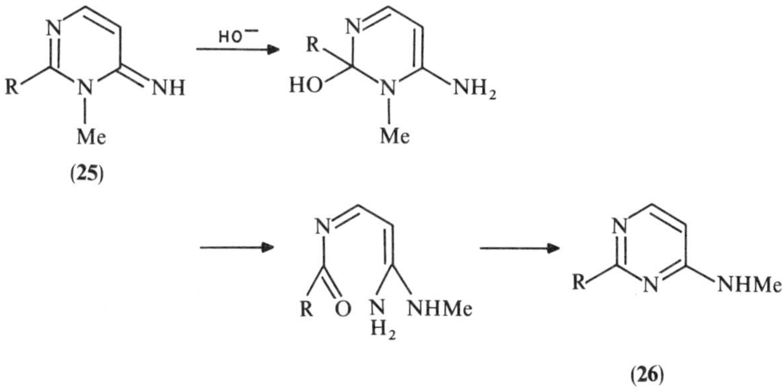

SCHEME 3

Meisenheimer and Related Adducts

Spectroscopic studies of the reactions of some amide and imide anions with 1,3,5-trinitrobenzene (TNB) in DMSO and in acetonitrile show the formation of *N*-bonded adducts (**27**) which may exist as *Z*- or *E*-isomers.[107] ^{13}C NMR measurements indicate that solutions in liquid ammonia of the TNB-methoxide adduct eventually yield 1:1 and 1:2 TNB:NH_2^- adducts, while similar solutions of the TNB-acetonate adduct result in cyclization to form a bicyclic *meta*-bridged complex.[108] The reactions of 2,4,6-trinitrocumene and 2,4,6-trinitroethylbenzene with 1,1,3,3-tetramethylguanidine in *N,N*-dimethylformamide may result in the formation of *N*-bonded adducts or transfer of a side-chain proton.[109] Theoretical calculations by the MNDO method have been reported for 2,4,6-trinitroanisole and its 1,1-dimethoxy adduct and for picryl fluoride and its 1,1-difluoro adduct; changes in electron density on adduct formation were compared with corresponding changes in ^1H and ^{13}C chemical shifts.[110] There have also been reports of the formation of 1,3- and 1,1-adducts in the reaction of 2,4,6-trinitroanisole with dimethyl malonate ions,[111] of the formation of methoxide adducts from picryl nitrocellulose,[40,112] and of complex formation from nitroarenes with tertiary alkylphosphines.[113]

There have been several reports of intramolecular reaction in trinitro-activated systems.[114] The spiro-adduct (**28**) has been identified[115] and there is spectroscopic evidence for the intermediacy of (**29**) during reaction of 2-(3,5-dinitrophenylthio)ethanol with base.[116] Kinetic and mechanistic studies have been reported for the interesting reactions of alkyl esters of *N*-(2,4,6-trinitrophenyl)-glycine and -α-alanine with methoxide which lead to 4,6-dinitro-2-nitrosoaniline.[117] The latter product is also formed by reaction of *N*-[(2,4,6]-trinitrophenyl)]aminoacetanilide with base, although its *N*-methyl derivative is reported[118] to yield the spiro-adduct (**30**).

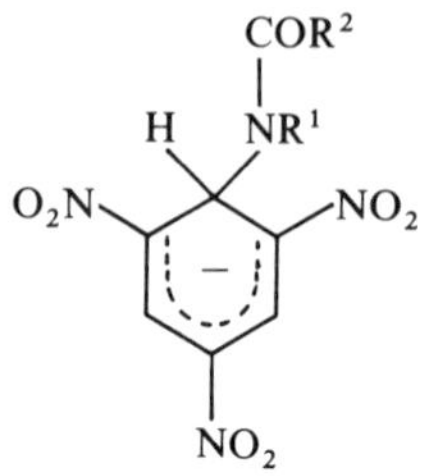

(**27**)

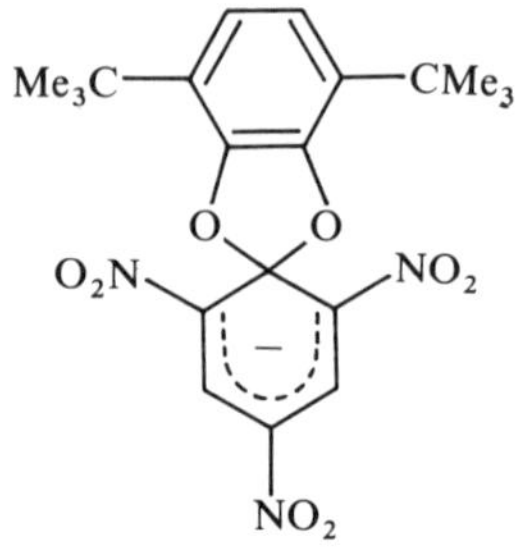

(**28**)

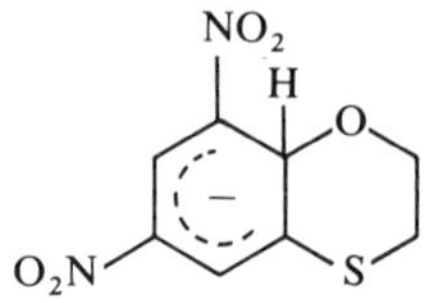

(**29**)

The Janovsky reaction involves the addition of a carbanion derived from a ketone to a ring-activated arene. Such adducts have been described resulting from reaction of the acetonate ion with polynitrodiphenyl sulphones[119] and from reaction of cardenolide with 1,3-dinitrobenzene.[120] There have been kinetic studies of the formation of a Janovsky adduct by acetonate addition to 2,4-dinitrophenyl 2-methoxy-4-formylphenyl ether,[121] and of the solvent dependence (in THF-water mixtures) of the decomposition of Janovsky adducts.[122] An interesting development has been the report of adducts, such as (**31**), formed between cyclopentadienyliron complexes of substituted benzenes and carbanions derived from ketones.[123]

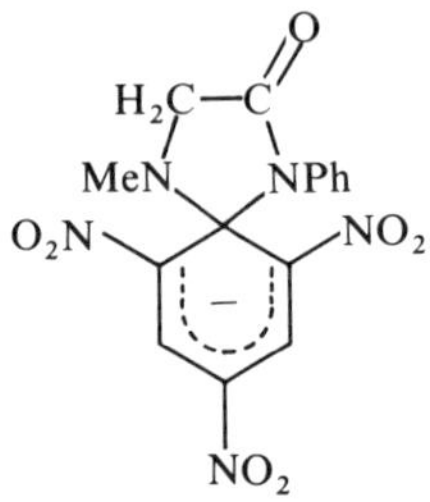

(**30**)

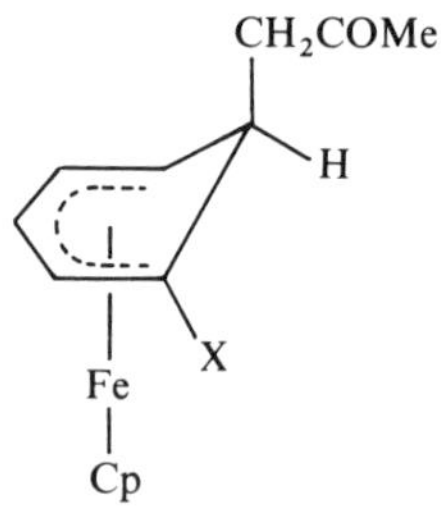

(**31**)

The strong electrophilic character of 4,6-dinitrobenzofuroxan and of 4,6-dinitro-2-(2′,4′,6′-trinitrophenyl)benzotriazole 1-oxide has been used to examine ambident nucleophilic reactivity in aryl oxide and arylamine nucleophiles; in general, the reactions involve kinetically preferred but reversible reaction via the heteroatom of the nucleophile followed by conversion into carbon-carbon bonded adducts which are thermodynamically preferred.[124] An NMR study of the adducts formed by addition of 4,6-dinitro-benzofuroxan and -benzofurazan to 1,8-

bis(dimethylamino)naphthalene has revealed the presence of two conformers due to restricted rotation about the new C—C bond.[125]

There is spectroscopic evidence for addition of cyanide ions at the 1-position of 2-benzopyrilium salts to yield[126] adducts (**32**; $R^1, R^2 = H$, alkyl), and for the addition of sulphite ions to a thiamin analogue[127] to yield (**33**). Kinetic and equilibrium data have been reported for methoxide ion addition to 4-substituted 2,6-diphenylthiopyrylium cations to yield 2*H*- and 4*H*-adducts; the results for the 4-methoxy derivative allow an estimate of 12 kJ mol^{-1} for the effect of *gem*-dimethoxy stabilization.[128] Equilibrium constants have been evaluated for pseudobase formation by hydroxide addition at the 4-position of 3-substituted-1-methylpyridinium ions; the value for 1-methylpyridinium ion was estimated using a linear-free-energy equation.[129] In the case of certain *N*-(hydroxyalkyl)- and *N*-(aminoalkyl)-isoquinolinium cations it has been shown[130] that intramolecular cyclization to give (**34**; X = O, NH; $n = 2,3$) can compete effectively with pseudobase formation; cyclization is more efficient for X = NH than for X = 0, and for $n = 3$ than for $n = 2$.

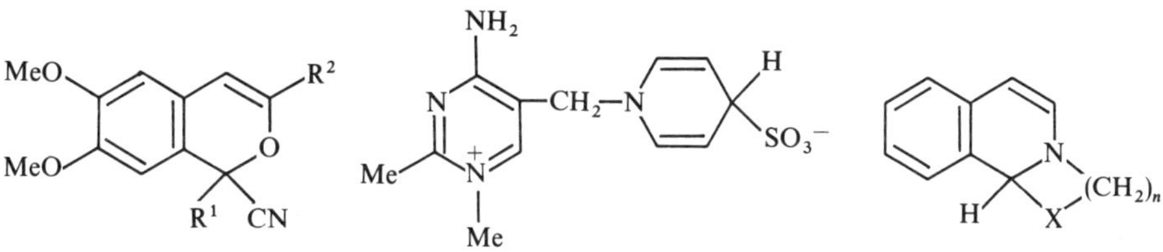

(**32**) (**33**) (**34**)

Benzyne and Related Intermediates

New *ab initio* MO calculations, including electron correlation, have been reported for benzyne and give a geometry which is closer to a cumulene-like structure than that calculated previously in the absence of correlation effects.[131]

The orientation of the reaction of methoxy-substituted arynes with amine and nitrile nucleophiles has been investigated,[132] and reaction with nucleophiles of arynes derived from 4-bromo-1,2-dimethylbenzene has been found to yield mainly the 4-substituted products.[133] The treatment of *o*-chloro- and *o*-bromo-phenyl phenyl sulphoxides and also (3-bromo-4-pyridyl) phenyl sulphoxide with Grignard reagents has been found to produce arynes; the reaction is thought to involve *o*-halogenoaryl Grignard reagents as intermediates.[134] Arynes are thought also to be involved in the reaction of polyhalobenzenes with alkenyl Grignard reagents which produces new aryl-alkenyl bonds.[135]

The X-ray structure of benzyne coordinated at tetra- and penta-ruthenium centres has been reported[136] and a complex involving coordination of benzyne at tantalum has been postulated[137] as an intermediate in the cyclometallation of 2,6-di-*tert*-butylphenoxide ligands by tantalum-aryl groups. There have been further studies of zirconocene complexes (**35**) of substituted benzynes. Trapping of the latter complexes with nitriles leads to the formation of azazirconacyclopentenes which can be converted[138,139] into functionalized aromatic compounds. The synthesis, structure and reactions of the zirconocene-benzdiyne complex (**36**) have also been reported.[140]

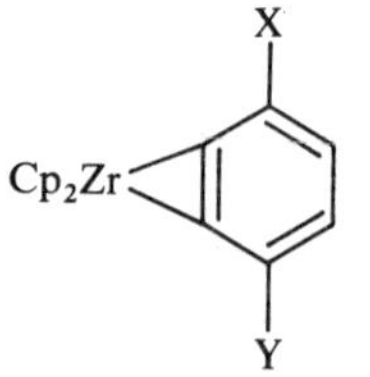

(**35**)

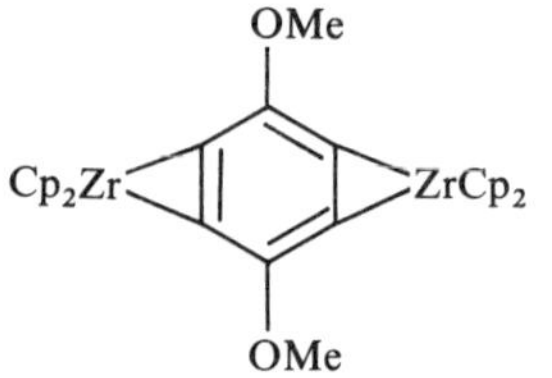

(**36**)

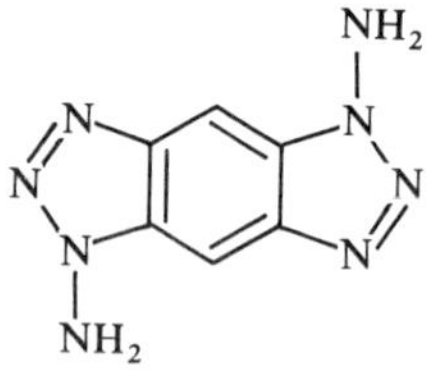

(**37**)

There have been reports of the addition of benzyne to azulene,[141] of the use of tetrahalobenzenes as di-aryne equivalents in the syntheses of polycyclic arenes,[142] and of the highly regioselective cycloaddition of the 1,4-benzadiyne equivalent (**37**) with hexamethyl-2,4-cyclohexadienone.[143]

References

1 Cram, D. J. and Doxsee, K. M., *J. Org. Chem.*, **51**, 5068 (1986).
2 Shinkai, S., Mori, S., Arimura, T., and Manabe, O., *J. Chem. Soc., Chem. Commun.*, **1987**, 238.
3 Nakazumi, H., Kitao, T., and Zollinger, H., *J. Org. Chem.*, **52**, 2825 (1987).
4 Anderson, L. M., Butler, A. R., and McIntosh, A. S., *J. Chem. Soc., Perkin Trans. 2*, **1987**, 1239.
5 Doyle, M. P., Guy, J. K., Brown, K. C., Mahapatro, S. N., Van Zyl, C. M., and Pladziewicz, J. R., *J. Am. Chem. Soc.*, **109**, 1536 (1987).
6 Abeywickrema, A. N., and Beckwith, A. L. J., *J. Am. Chem. Soc.*, **108**, 8227 (1986).
7 Beckwith, A. L. J. and Meijs, G. F., *J. Org. Chem.*, **52**, 1922 (1987).
8 Abeywickrema, A. N. and Beckwith, A. L. J., *J. Org. Chem.*, **52**, 2568 (1987).
9 Xia, C. and Chen, Z., *Huaxue Tongbao*, **1986**, 20; *Chem. Abs.*, **106**, 137665 (1987).
10 Novi, M., Petrillo, G., and Dell'Erba, C., *Tetrahedron Lett.*, **28**, 1345 (1987).
11 Petrillo, G., Novi, M., Garbarino, G., and Dell'Erba, C., *Tetrahedron*, **43**, 4625 (1987).
12 Pierini, A. B., Baumgartner, M. T., and Rossi, R. A., *Tetrahedron Lett.*, **28**, 4653 (1987).
13 Beugelmans, R. and Ginsburg, H., *Tetrahedron Lett.*, **28**, 413 (1987).
14 Pierini, A. B., Baumgartner, M. T., and Rossi, R. A., *J. Org. Chem.*, **52**, 1089 (1987).
15 Novi, M., Garbarino, G., Petrillo, G., and Dell'Erba, C., *J. Org. Chem.*, **52**, 5382 (1987).
16 Beugelmans, R., Bois-Choussy, M., and Tang, Q., *J. Org. Chem.*, **52**, 3880 (1987).
17 Alam, N., Amatore, C., Combellas, C., Thiébault, A., and Verpeaux, J. N., *Tetrahedron Lett.*, **28**, 6171 (1987).
18 Degrand, C., *J. Org. Chem.*, **52**, 1421 (1987).
19 Degrand, C., Prest, R., and Compagnon, P.-L., *J. Org. Chem.*, **52**, 5229 (1987).
20 Novi, M., Garbarino, G., Petrillo, G., and Dell'Erba, C., *J. Chem. Soc., Perkin Trans. 2*, **1987**, 623.
21 Harris, J. M. and McManus, S. P., *A.C.S. Adv. Chem. Ser.*, **215**, 1 (1987).
22 Menger, F. M., *A.C.S. Adv. Chem. Ser.*, **215**, 209 (1987).
23 Buncel, E., Dust, J. M., Park, K. T., Renfrow, R. A., and Strauss, M. J., *A.C.S. Adv. Chem. Ser.*, **215**, 369 (1987).
24 Paradisi, C. and Scorrano, G., *A.C.S. Adv. Chem. Ser.*, **215**, 339 (1987).
25 Bacaloglu, R., Bunton, C. A., and Cerichelli, G., *J. Am. Chem. Soc.*, **109**, 621 (1987).
26 Palleros, D. R., Cerdeira, S., and Nudelman, N. S., *Rev. Latinoam. Quim.*, **17**, 12 (1986); *Chem. Abs.*, **106**, 17603 (1987).
27 Nudelman, N. S., Manicini, P. M. E., Martinez, R. D., and Vottero, L. R., *J. Chem. Soc., Perkin Trans. 2*, **1987**, 951.
28 Hayami, J., Otani, S., Yamaguchi, F., and Nishikawa, Y., *Chem. Lett.*, **87**, 739 (1987).
29 Cattana, R. I., Singh, J. O., Anunziata, J. D., and Silber, J. J., *J. Chem. Soc., Perkin Trans. 2*, **1987**, 79.
30 Chiacchiera, S. M., Singh, J. O., Anunziata, J. D., and Silber, J. J., *J. Chem. Soc., Perkin Trans. 2*, **1987**, 987.

[31] Onyido, I. and Ubochi, C. I., *J. Chem. Soc., Perkin Trans. 2*, **1987**, 563.
[32] Buncel, E. and Onyido, I., *Can. J. Chem.*, **64**, 2115 (1986).
[33] Bowden, K. and Prasannan, S., *J. Chem. Soc., Perkin Trans. 2*, **1987**, 185.
[34] Bowden, K. and Nadvi, N. S., *J. Chem. Soc., Perkin Trans. 2*, **1987**, 189.
[35] Oda, R., *Kagaku (Kyoto)*, **41**, 177 and 482 (1986); *Chem. Abs.*, **107**, 77014 and 77015 (1987).
[36] Hiller, R. A. and Weiler, R., *Can. J. Chem.*, **65**, 251 (1987).
[37] Sammes, P. G., Thetford, D., and Voyle, M., *J. Chem. Soc., Chem. Commun.*, **1987**, 1373.
[38] Ibata, T., Isogami, Y., and Toyoda, J., *Chem. Lett.*, **87**, 1187 (1987).
[39] Strauss, M. J., Torres, R., Phelan, J., Craft, A., Pitner, B., Nason, D., Carignan, Y., Dust, J. M., and Buncel, E., *Can. J. Chem.*, **65**, 1891 (1987).
[40] Muralikrishna, U. and Krishnamurthy, M., *Indian J. Chem.*, **25A**, 949 (1986); *Chem. Abs.*, **107**, 58229 (1987).
[41] Katritzky, A. R. and Laurenzo, K. S., *J. Org. Chem.*, **51**, 5039 (1986).
[42] Makosza, M. and Owczarczyk, Z., *Tetrahedron Lett.*, **28**, 3021 (1987).
[43] Makosza, M., Danikiewicz, W., and Wojciechowski, K., *Liebigs Ann. Chem.*, **1987**, 711.
[44] Makosza, M., Glinka, T., Ostrowski, S., and Rykowski, A., *Chem. Lett.*, **1987**, 61.
[45] Danikiewicz, W. and Makosza, M., *Tetrahedron Lett.*, **28**, 1707 (1987).
[46] Cassidy, F., Evans, J. M., Smith, D. M., and Stemp, G., *Tetrahedron Lett.*, **28**, 2403 (1987).
[47] Khuthier, A.-H., Al-Mallah, K. Y., and Hanna, S. Y., *J. Chem. Soc., Perkin Trans. 2*, **1987**, 109.
[48] Kimura, M., *J. Chem. Soc., Perkin Trans. 2*, **1987**, 205.
[49] Bunce, N. J., Cater, S. R., Scaiano, J. C., and Johnston, L. J., *J. Org. Chem.*, **52**, 4214 (1987).
[50] Cantos, A., Marquet, J., and Moreno-Mañas, M., *Tetrahedron Lett.*, **28**, 4191 (1987).
[51] Marquet, J., Moreno-Mañas, M., Vallribera, A., Virgili, A., Bertran, J., Gonzalez-Lafont, A., and Lluch, J. M., *Tetrahedron*, **43**, 351 (1987).
[52] Kuzmic, P. and Soucek, M., *Collect. Czech. Chem. Commun.*, **52**, 980 (1987).
[53] Hojo, M., Masuda, R., and Okada, E., *Tetrahedron Lett.*, **28**, 6199 (1987).
[54] Attia, M., Gore, P. H., and Morris, D. F. C., *J. Chem. Res. (S)*, **1987**, 160.
[55] Gore, P. H., Ikejiani, A. O. O., and Morris, D. F. C., *J. Chem. Res. (S)*, **1987**, 210.
[56] Yang, S. H., Li, C. S., and Cheng, C. H., *J. Org. Chem.*, **52**, 691 (1987).
[57] Clark, J. H. and Jones, C. W., *J. Chem. Soc., Chem. Commun.*, **1987**, 1409.
[58] Bosch, P., Camps, F., Chamorro, E., Gasol, V., and Guerrero, A., *Tetrahedron Lett.*, **28**, 4733 (1987).
[59] Clark, J. H. and Macquarrie, D. J., *Tetrahedron Lett.*, **28**, 111 (1987).
[60] Svestka, S., Vecera, M., and Vetesnik, P., *Sb. Ved. Vys. Sk. Chemickotechnol. Pardubice*, **48**, 41 (1985); *Chem. Abs.*, **106**, 4344 (1987).
[61] Clark, J. H. and Owen, N. D. S., *Tetrahedron Lett.*, **28**, 3627 (1987).
[62] Carmona, C., Eaton, G., and Symons, M. C. R., *J. Chem. Soc., Chem. Commun.*, **1987**, 873.
[63] Ando, T., Clark, J. H., Cork, D. G., Hanafusa, T., Ichihara, J., and Kimura, T., *Tetrahedron Lett.*, **28**, 1421 (1987).
[64] Atwal, R. K. and Bolton, R., *Aust. J. Chem.*, **40**, 241 (1987).
[65] Rodionov, P. P., Osina, O. I., Platonov, V. E., and Yakobson, G. G., *Bull. Soc. Chim. Fr. II*, **1986**, 986.
[66] Barak, G. and Sasson, Y., *J. Chem. Soc., Chem. Commun.*, **1987**, 1267.
[67] Schmidtchen, F. P., *J. Mol. Catal.*, **38**, 273 (1986); *Chem. Abs.*, **107**, 77056 (1987).
[68] Barra, M., de Rossi, R. H., and de Vargas, E. B., *J. Org. Chem.*, **52**, 5004 (1987).
[69] Barra, M., de Vargas, E. B., and de Rossi, R. H., *Contrib. Cient. Technol.*, **1985**, 75; *Chem. Abs.*, **106**, 49406 (1987).
[70] Bunton, C. A. and Moffatt, J. R., *Ann. Chim. (Rome)*, **77**, 117 (1987).
[71] Bunton, C. A. and Bacaloglu, R., *J. Colloid Interface Sci.*, **115**, 288 (1987); *Chem. Abs.*, **107**, 115058 (1987).
[72] Cipiciani, A., Fracassini, M. C., Germani, R., Savelli, G., and Bunton, C. A., *J. Chem. Soc., Perkin Trans. 2*, **1987**, 547.
[73] Cipiciani, A., Germani, R., Savelli, G., Bunton, C. A., Mhala, M., and Moffatt, J. R., *J. Chem. Soc., Perkin Trans. 2*, **1987**, 541.
[74] Oda, R., *Kagaku (Kyoto)*, **42**, 362 (1987); *Chem. Abs.*, **107**, 197143 (1987).
[75] Ellis, G. P. and Romney-Alexander, T. M., *Chem. Rev.*, **87**, 779 (1987).
[76] Kondratov, S. A. and Shein, S. M., *Ukr. Khim. Zh. (Russ. edn)*, **52**, 858 (1986); *Chem. Abs.*, **107**, 77050 (1987).
[77] Paine, A. J., *J. Am. Chem. Soc.*, **109**, 1496 (1987).
[78] Connor, J. A., Gibson, D., and Price, R., *J. Chem. Soc., Perkin Trans. 1*, **1987**, 619.

[79] Cram, D. J., Bryant, J. A., and Doxsee, K. M., *Chem. Lett.*, **1987**, 19.
[80] Allen, D. W., Cropper, P. E., Smithurst, P. G., Ashton, P. R., and Taylor, B. F., *J. Chem. Soc., Perkin Trans. 1*, **1986**, 1989.
[81] Kudo, K., Shibata, T., Kashimura, T., Mori, S., and Sugita, N., *Chem. Lett.*, **1987**, 577.
[82] Miura, M., Akase, F., and Nomura, M., *J. Org. Chem.*, **52**, 2623 (1987).
[83] Takagi, K., Iwachido, T., and Hayama, N., *Chem. Lett.*, **1987**, 839.
[84] Uno, M., Takahashi, T., and Takahashi, S., *J. Chem. Soc., Chem. Commun.*, **1987**, 785.
[85] Andersson, C.-M., Hallberg, A., and Davies, G. D., *J. Org. Chem.*, **52**, 3529 (1987).
[86] Chatani, N. and Hanafusa, T., *J. Org. Chem.*, **51**, 4714 (1986).
[87] Tunney, S. E. and Stille, J. K., *J. Org. Chem.*, **52**, 748 (1987).
[88] Sutherland, R. G., Chowdhury, R. L., Piórko, A., and Lee, C. C., *J. Org. Chem.*, **52**, 4618 (1987).
[89] Moriarty, R. M., Ku, Y.-Y., and Gill, U. S., *J. Chem. Soc., Chem. Commun.*, **1987**, 1493.
[90] Rose-Munch, F., Rose, E., and Semra, A., *J. Chem. Soc., Chem. Commun.*, **1987**, 942.
[91] Bouttonet, J. C., Rose-Munch, F., Rose, E., and Semra, A., *Bull. Soc. Chim. Fr. II*, **1987**, 640.
[92] Sato, N., *Yokohama-shiritsu Daigaku Ronso, Shizen Kagaka Keiretsu*, **37**, 25 (1985); *Chem. Abs.*, **107**, 6530 (1987).
[93] Bowden, K., Prasannan, S., and Ranson, R. J., *J. Chem. Soc., Perkin Trans. 2*, **1987**, 181.
[94] Muscio, O. J. and Rutherford, D. R., *J. Org. Chem.*, **52**, 5194 (1987).
[95] Bunton, C. A. and Cuenca, A., *J. Org. Chem.*, **52**, 901 (1987).
[96] Ballesteros, P., Claramunt, R. M., and Elguero, J., *Tetrahedron*, **43**, 2557 (1987).
[97] Dunn, A. D. and Norrie, R., *J. Heterocycl. Chem.*, **24**, 85 (1987).
[98] Kulkarni, S., Grimmett, M. R., Hanton, L. R., and Simpson, J., *Aust. J. Chem.*, **40**, 1399 (1987).
[99] Kulkarni, S. and Grimmett, M. R., *Aust. J. Chem.*, **40**, 1415 (1987).
[100] Dmowski, N. and Haas, A., *J. Chem. Soc., Perkin Trans. 1*, **1987**, 2119.
[101] Artamkina, G. A., Mil'chenko, A. Yu., Beletskaya, I. P., and Reutov, O. A., *J. Organomet. Chem.*, **321**, 371 (1987).
[102] Chambers, R. D., Tamura, M., Howard, J. A. K., and Johnson, O., *J. Chem. Soc., Chem. Commun.*, **1987**, 1697.
[103] Rykowski, A. and van der Plas, H. C., *J. Org. Chem.*, **52**, 71 (1987).
[104] Kos, N. J., Jongejan, H., and van der Plas, H. C., *Tetrahedron*, **43**, 4841 (1987).
[105] Breuker, K., van der Plas, H. C., and van Veldhuizen, A., *Isr. J. Chem.*, **27**, 67 (1987).
[106] Buurman, D. J. and van der Plas, H. C., *J. Heterocycl. Chem.*, **24**, 1337 (1987).
[107] Hallé, J.-C., Zah-Letho, J., Simonnin, M.-P., and Terrier, F., *J. Chem. Soc., Perkin Trans. 2*, **1987**, 805.
[108] Chudek, J. A., Foster, R., and Marr, A. W., *J. Chem. Soc., Perkin Trans. 2*, **1987**, 1341.
[109] Biswas, M. K., Jarczewski, A., and Leffek, K. T., *Can. J. Chem.*, **65**, 1007 (1987).
[110] Nudelman, N. S. and MacCormack, P., *J. Chem. Soc., Perkin Trans. 2*, **1987**, 227.
[111] Macháček, V., Štěrba, V., and Lyčka, A., *Collect. Czech. Chem. Commun.*, **52**, 132 (1987).
[112] Strauss, M. J., Torres, R., Carignan, Y., and Buncel, E., *Tetrahedron Lett.*, **28**, 159 (1987).
[113] Beg, M. A. A. and Siddiqui, S. M., *Pak. J. Sci. Ind. Res.*, **29**, 315 (1986); *Chem. Abs.*, **107**, 58321 (1987).
[114] Mel'nikov, A. I., Gitis, S. S., and Kaminskii, A. Ya., *Zh. Org. Khim.*, **22**, 1504 (1986); *Chem. Abs.*, **106**, 101662 (1987).
[115] Knyazev, V. N., Lipovtsev, V. N., Belostofskaya, I. S., Vol'eva, V. B., Drozd, V. N., and Erstov, V. V., *Zh. Org. Khim.*, **22**, 1951 (1986); *Chem. Abs.*, **106**, 195983 (1986).
[116] Knyazev, V. N. and Drozd, V. N., *Zh. Org. Khim.*, **21**, 2590 (1985); *Chem. Abs.*, **106**, 32947 (1987).
[117] Macháček, V., Hassanien, M. M. M., Štěrba, V., and Lyčka, A., *J. Chem. Soc., Perkin Trans. 2*, **1987**, 867.
[118] Macháček, V., Hassanien, M. M. M., and Štěrba, V., *Collect. Czech. Chem. Commun.*, **52**, 2225 (1987).
[119] Alekhina, N. N., Gitis, S. S., Grudtsyn, Yu. D., and Kaminskii, A. Ya., *Zh. Org. Khim.*, **22**, 1263 (1986); *Chem. Abs.*, **106**, 17652 (1987).
[120] Megges, R., Weiland, J., Rollka, W., and Repke, K. R. H., *Z. Chim.*, **27**, 72 (1987).
[121] Ciurlizza, G. A. and Ramirez, A. J., *Acta Mex. Cienc. Tecnol.*, **3**, 35 (1985); *Chem. Abs.*, **107**, 38857 (1987).
[122] Gitis, S. S., Savinova, L. N., Glaz, A. I., Golopolosova, T. V., and Kaminskii, A. Ya., *Org. React. (Tartu)*, **23**, 73 (1986); *Chem. Abs.*, **107**, 38981 (1987).
[123] Sutherland, R. G., Chowdhury, R. L., Piórko, A., and Lee, C. C., *Can. J. Chem.*, **64**, 2031 (1986).
[124] Buncel, E., Renfrow, R. A., and Strauss, M. J., *J. Org. Chem.*, **52**, 488 (1987).
[125] Simonnin, M.-P., Halle, J.-C., Pouet, M.-J., Terrier, F., and Davoust, D., *Can. J. Chem.*, **65**, 404 (1987).
[126] Balaban, A. T., Gheorghiu, M. D., Shcherbakova, I. V., Kuznetsov, E. V., and Yudilevich, I. A., *Tetrahedron*, **43**, 409 (1987).

[127] Uray, G. and Kriessmann, I., *J. Org. Chem.*, **52**, 802 (1987).
[128] Doddi, G., Ercolani, G., and Nunziante, P., *J. Chem. Soc., Perkin Trans. 2*, **1987**, 1427.
[129] Bunting, J. W., *Tetrahedron*, **43**, 4277 (1987).
[130] Bunting, J. W. and Moors, R. G., *Tetrahedron Lett.*, **28**, 3915 (1987).
[131] Rigby, K., Hillier, I. H., and Vincent, M., *J. Chem. Soc., Perkin Trans. 2*, **1987**, 117.
[132] Razzuk, A. and Biehl, E. R., *J. Org. Chem.*, **52**, 2619 (1987).
[133] Biehl, E. R., Razzuk, A., Jovanovic, M. V., and Khanapure, S. P., *J. Org. Chem.*, **51**, 5157 (1986).
[134] Furukawa, N., Shibutani, T., and Fujihara, H., *Tetrahedron Lett.*, **28**, 2727 (1987).
[135] Du, C.-J. and Hart, H., *J. Org. Chem.*, **52**, 4311 (1987).
[136] Knox, S. A. R., Lloyd, B. R., Orpen, A. G., Viñas, J. M., and Weber, M., *J. Chem. Soc., Chem. Commun.*, **1987**, 1498.
[137] Chamberlain, L. R., Kerschner, J. L., Rothwell, A. P., Rothwell, I. P., and Huffman, J. C., *J. Am. Chem. Soc.*, **109**, 6471 (1987).
[138] Buchwald, S. L., Sayers, A., Watson, B. T., and Dewan, J. C., *Tetrahedron Lett.*, **28**, 3245 (1987).
[139] Buchwald, S. L., Watson, B. T., Lunn, R. T., and Nugent, W. A., *J. Am. Chem. Soc.*, **109**, 7137 (1987).
[140] Buchwald, S. L., Lucas, E. A., and Dewan, J. C., *J. Am. Chem. Soc.*, **109**, 4396 (1987).
[141] Cresp, T. M. and Wege, D., *Tetrahedron*, **42**, 6713 (1986).
[142] Hart, H., Lai, C., Nwokogu, G. C., and Shamouilian, S., *Tetrahedron*, **43**, 5203 (1987).
[143] Ok, D. and Hart, H., *J. Org. Chem.*, **52**, 3835 (1987).

Organic Reaction Mechanisms 1987
Edited by A. C. Knipe and W. E. Watts

CHAPTER 8

Electrophilic Aromatic Substitution

R. B. Moodie

Department of Chemistry, University of Exeter

General

An experimental and theoretical study of the structure of aromatic σ-complexes has been reported. The results of EH, MNDO, and *ab initio* 3-21G calculations are in agreement with crystal structures determined for (**1**), (**2**) and (**3**); the cyclohexadienylium ring of (**1**) is planar but those of (**2**) and (**3**) are bent with the larger substituent at the tetrahedral carbon in the axial position. In re-aromatization, it is the axial substituent which departs, so steric effects must in part determine whether σ-complex formation is rate-determining.[1] A discussion of the relative merits of the σ-complex and excited-charge-transfer-complex models of transition states for electrophilic aromatic substitutions comes down firmly in favour of the former, on the basis of comparisons of reactivity ratios with σ-basicities on the one hand and electrode potentials, ionization potentials, and charge-transfer transition energies on the other.[2]

An examination of the effect of the trifluoromethoxy substituent on reactivity and selectivity in electrophilic substitutions has been reported,[3] and there has been a review of the effect of the acid-base properties of heteroaromatic compounds on their electrophilic substitution reactions.[4] The replacement of one phenyl ring by tetrafluorophenyl in [2.2]paracyclophane caused a remarkable deactivation of the other ring towards electrophilic aromatic substitution, indicating some transfer of electron density between the rings.[5]

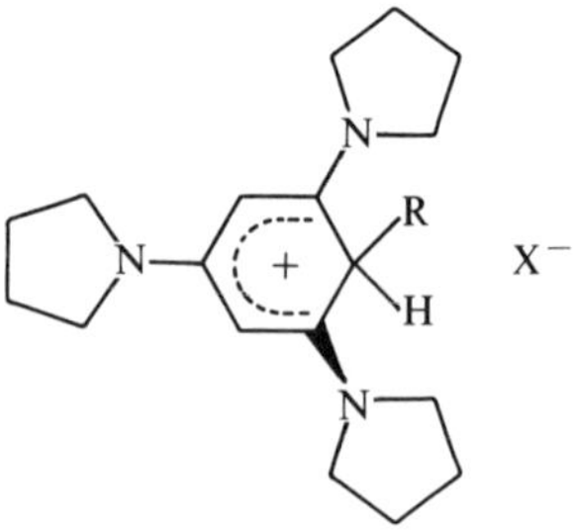

	R	X
(**1**)	H	ClO_4
(**2**)	alkyl	ClO_4
(**3**)	Br	Br

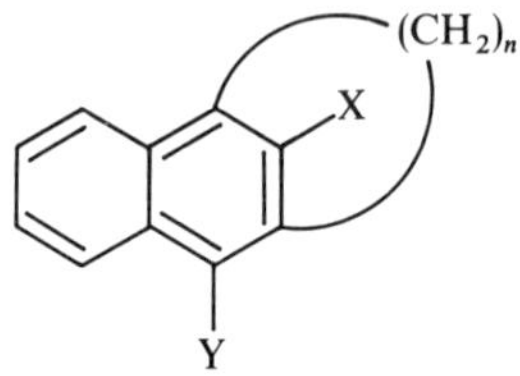

(**4**)
$n = 7, 8$, or 10
X or Y = ^3H

Hydrogen Exchange

Partial rate factors for hydrogen-deuterium exchange of some pyrroles bearing an electron-withdrawing substituent have been determined; the 1-substituted pyrroles exchange fastest at C(2) and the 2-substituted pyrroles at C(4).[6] Protodetritiation rates of compounds (**4**) have been measured and discussed.[7] Rate constants and activation parameters for reactions of muonium with aromatics indicate the possibility of tunnelling.[8] The extent of protonation of hexamethylbenzene has been used to examine solvent basicities.[9]

Halogenation

An approach to selective fluorination, in which aryl-metal bonds in aryltrialkylstannanes and aryl-mercurials are cleaved with F_2, CF_3OF, or $CsSO_4F$, is reported; $CsSO_4F$ is especially efficient.[10]

Kinetic evidence for substantial pre-equilibrium formation of a π-complex in the molecular chlorination of alkylbenzenes and naphthalene has been adduced.[11] Chlorination of *n*-alkyl phenyl ethers by chlorine water on reversed-phase h.p.l.c. columns is reported to give enhanced substrate selectivity.[12] The ratio of *ortho* to *para* chlorination of acetanilide and benzanilide is changed in favour of *para* substitution when the reactants are entrapped in cyclodextrins or amylose.[13] The chlorination of anisole and phenol with *N*-chlorodialkylamines in trifluoroacetic acid occurs almost entirely at C(4); the electrophile is probably the *N*-chlorodialkylammonium ion.[14]

Quaternary ammonium salts enhance the reactivity of bromine towards aromatic compounds in the presence of hydrogen peroxide under heterogeneous conditions, and may well act both as Lewis acid and as phase-transfer catalysts.[15] Compounds as deactivated as *m*-dinitrobenzene can be brominated at low temperatures in high yield by BrF.[16] Benzo-4-bromo-2,5-cyclohexadienone, an intermediate in the aqueous bromination of 1-naphthol, undergoes enolization by general acid- and general base-catalysed pathways; the former shows, puzzlingly, a negative value of α.[17] Bromination of thiazole by Br_2 in the presence of $AlCl_3$ occurs at C(2); an ylid mechanism is suggested (Scheme 1).[18]

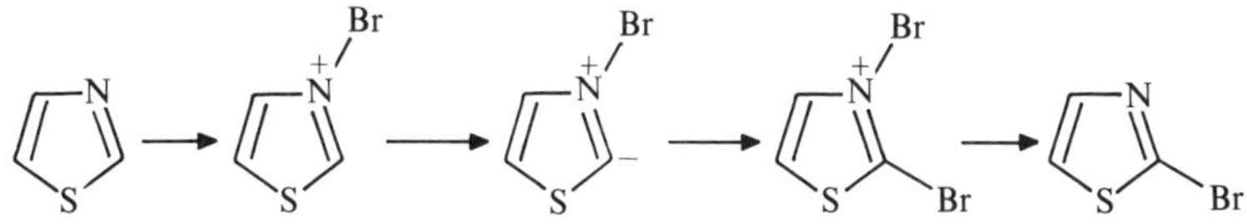

SCHEME 1

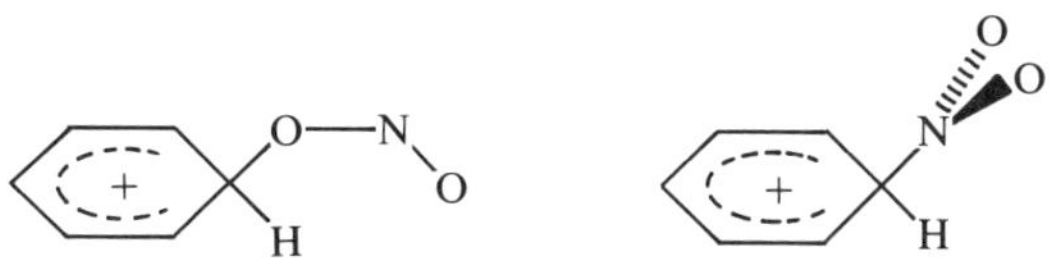

(5) (6)

Nitration

Ab initio (4-31G/ /STO-3G) calculations indicate that, in the interaction of nitronium ion with benzene, the nitrito complex (**5**) is slightly more stable than the nitro complex (**6**); semi-empirical (MINDO/3) calculations suggest that solvation may change this hierachy.[19]

Calculations by MNDO, INDO-SCF, and SCF-CI methods indicate that gas-phase nitration of substances more reactive than benzene should proceed through radical pairs.[20] An account of present views of electron-transfer mechanisms concludes that, in the absence of nitrous acid, aromatic radical cations are formed if at all by homolysis of a Wheland intermediate and not by 'outer sphere' electron transfer from aromatic to nitronium ion.[21] Gas-phase nitration at pressures near atmospheric (with protonated methyl nitrate as the electrophile) are typical electrophilic aromatic substitutions of rather low selectivity ($\rho = -3.87$). Highly activated substrates react at a limiting rate analogous to the encounter rate in solution.[22]

Nitration of dialkoxybenzenes with tetranitromethane may be effected photochemically by irradiation of the charge-transfer band of the donor-acceptor complex of the reactants.[23,24] Methoxytoluenes under the same conditions give the products of trinitromethylation. It is suggested that both reactions proceed through a solvent-caged triplet ($ArH^{+\cdot}$, $C(NO_2)_3^-$, $NO_2\cdot$).[23] For the nitration reaction, comparisons are made with the results of an avowedly normal electrophilic nitration, but in the latter reaction no steps were taken to avoid nitrous acid-catalysed nitration.[24] Nitrations of alkylbenzenes by HNO_3/H_2SO_4 and by $AcONO_2/CCl_4$ have been compared and contrasted; both systems show substrate reactivity to correlate better with the ionization potentials than with the basicities of the arenes. For nitration in $AcONO_2/CCl_4$, toluene gives an unusually high yield of the *meta* nitro product together with substantial amounts of phenylnitromethane and benzyl alcohol.[25]

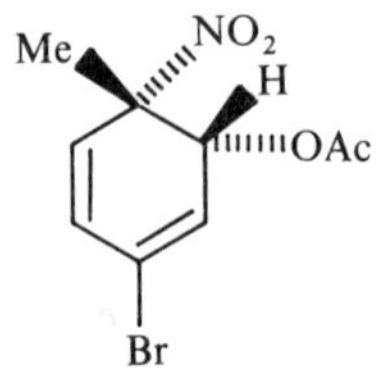

(7)

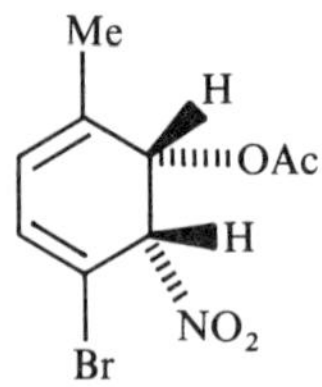

(8)

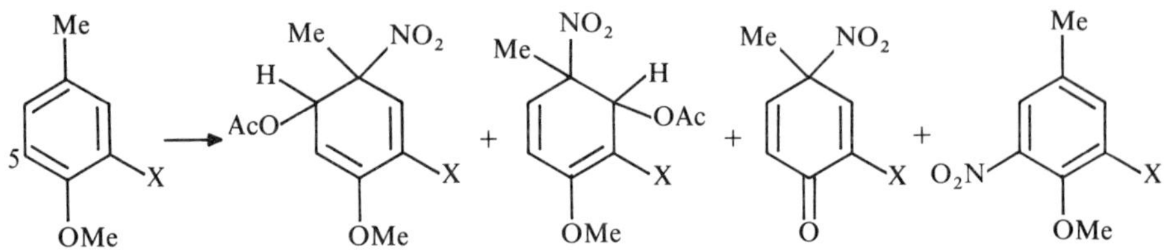

(9) **(10)** **(11)** **(12)** **(13)**

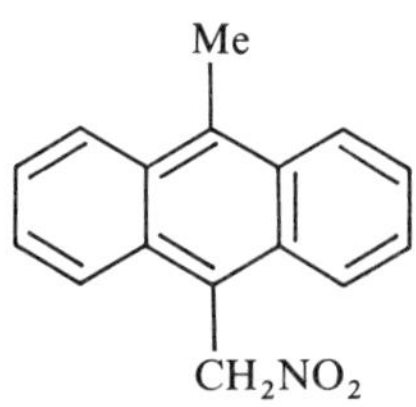

(14)

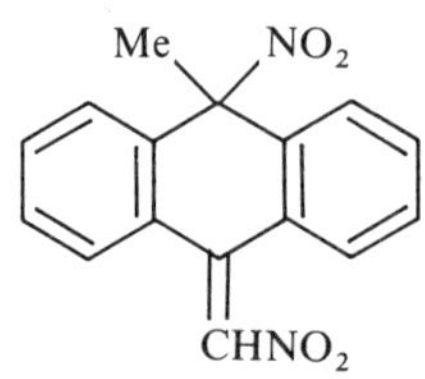

(15)

In the nitration of 4-bromotoluene in acetic anhydride, only one diastereomer of the 1,2-adduct (**7**) is obtained; one of three possible explanations advanced is that nitronium ion only becomes free of complexing acetic acid within the encounter pair and that the *ipso*-Wheland intermediate is captured by acetic acid before relaxation of the solvent shell. The stereoselective thermal rearrangement of (**7**) to (**8**) is suprafacial, as required if new C—NO_2 bond formation and old C—NO_2 bond fission are synchronous.[26] Yields of the products of nitration of (**9**; X = Cl, Br, NO_2) show that the ratio of initial *ipso* to non-*ipso* attack, measured as [(**10**) + (**11**) + (**12**)]/(**13**), is greater than expected from results for 4-methoxytoluene indicating that with (**9**) there is buttressed steric inhibition of attack at C(5); (**11**) is only observed with X = NO_2, perhaps because its formation (by nucleophilic capture of the *ipso*-Wheland intermediate) retains important resonance interation between the 3- and 4-substituents.[27] Nitration of (**14**) gives (**15**) as a major and stable product; it is presumably formed by *ipso*-attack at C(9) followed by proton loss from the nitromethyl group at C(10).[28] 2-Hydroxyacetophenones, treated with nitric acid in acetic acid, give 2-nitroaryl acetates; *ipso* attack at the acetylated carbon is implicated and there is evidence that the 1,3-shift of the acetyl group from carbon to oxygen is intramolecular (Scheme 2).[29]

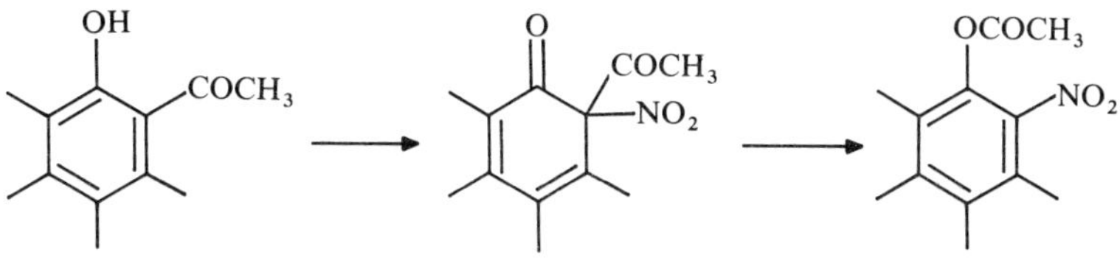

SCHEME 2

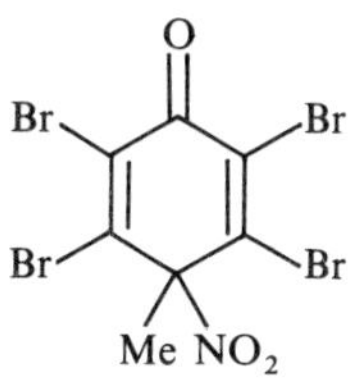

(16)

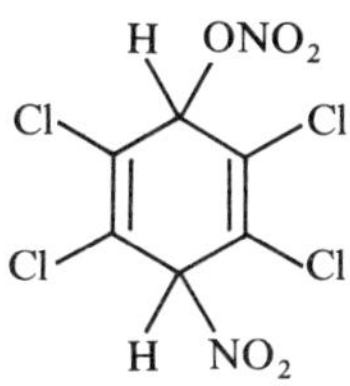

(17)

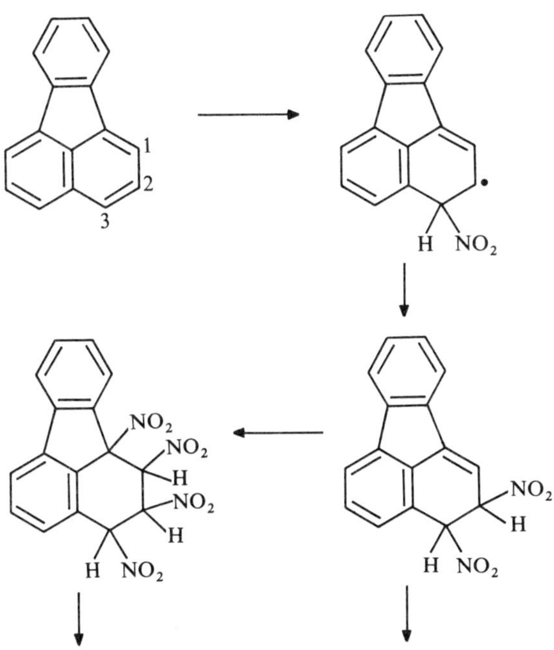

1,2-dinitrofluoroanthene + 1,3-dinitrofluoroanthene

2-nitrofluoroanthene + 3-nitrofluoroanthene

SCHEME 3

Nitrocyclohexadienones such as (**16**) have been used in non-polar solvents as mild nitrating agents for reactive compounds such as hydroxynaphthalenes; the mechanism was not investigated.[30] Kinetics and products of nitration of deactivated aromatic compounds in mixtures of dinitrogen pentaoxide and nitric acid have been examined; is some cases adducts like (**17**) are formed along with expected nitroaromatic products.[31]

Nitration of fluoroanthene with N_2O_4 in CCl_4 gives, abnormally, mononitration mainly at C(2) and products disubstituted in the same ring; the mechanism of Scheme 3 is suggested.[32] The presence of $HClO_4$ or metal nitrates in the nitration of naphthalene by N_2O_4 in sulpholane is claimed to affect the concentration of nitrosonium ion and thence the initial rate.[33]

Unusually high *para/ortho* ratios have been achieved in the nitration of halobenzenes, using cupric nitrate supported on clay in the presence of acetic anhydride and a non-polar solvent.[34]

Structures and energies of complexes between C_6H_6 and NO^+ of various geometries have been explored by *ab initio* molecular-orbital computations at several levels.[35]

Sulphonation

There has been a quantitative study of the kinetics and products of reactions of polymethylbenzenes in concentrated sulphuric acid; the complex reactions include sulphonation and desulphonation, *ipso*-protonation, and intermolecular methylation.[36] Initial sulphonation of (**18**) with SO_3 in dioxane occurs at the position indicated in accord with the results of semi-empirical calculations; further sulphonation steps have also been investigated.[37] Yields of aniline and of orthanilic and sulphanilic acids from the photolysis of metanilic acid are pH- and solvent-dependent and indicate the intermediacy of σ-complexes which interconvert by 1,2-shifts of the sulphonic acid group.[38] The kinetics of sulphonation of *N*-substituted arylamines in oleum containing tetrabisulphatoboric acid have been investigated.[39]

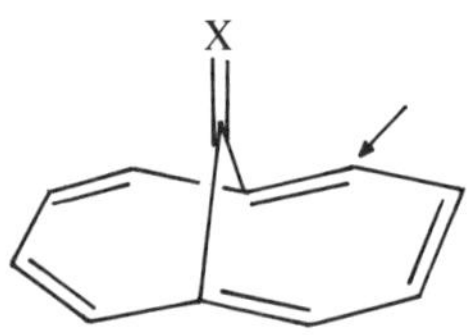

(**18**)
$X = CH_2, CMe_2, O$

Friedel–Crafts and Related Reactions

There have been two studies of gas-phase alkylations with isopropyl cations. With 5-membered heterocycles, yields reflect the competing rates of the isomerizations and deprotonations of Scheme 4 and indicate preferred sites of initial attack.[40] With dihalobenzenes, enhanced *ortho*-orientation is found in some cases, and the formation from 4-bromofluorobenzene of 2-bromo-4-fluorocumene suggests *ipso*-attack at C(4) followed by 1,2-migration of the bromine substituent.[41] The intramolecular rearrangement of 3-alkyl-3-methylindolium to give 2,3-disubstituted indoles, known to be a step in the electrophilic alkylation of 3-substituted indoles, has now been studied quantitatively and independently; relative migratory aptitudes are reported.[42] For the rearrangement of Scheme 5 $k_H/k_D = 1.08$, showing that the first step is not significantly reversible.[43]

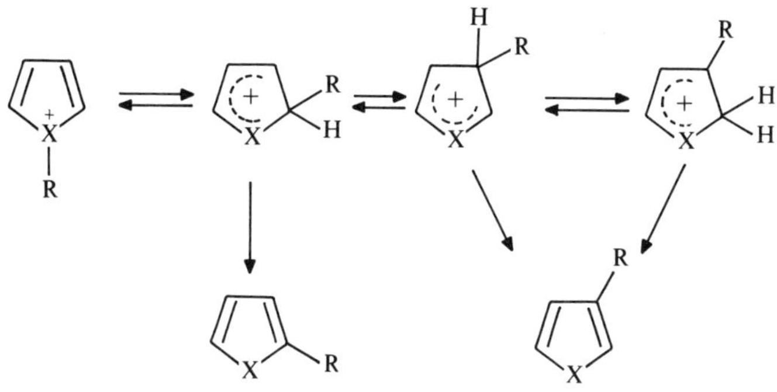

$R = Pr^i$; X = NH, NCH_3, O, or S

SCHEME 4

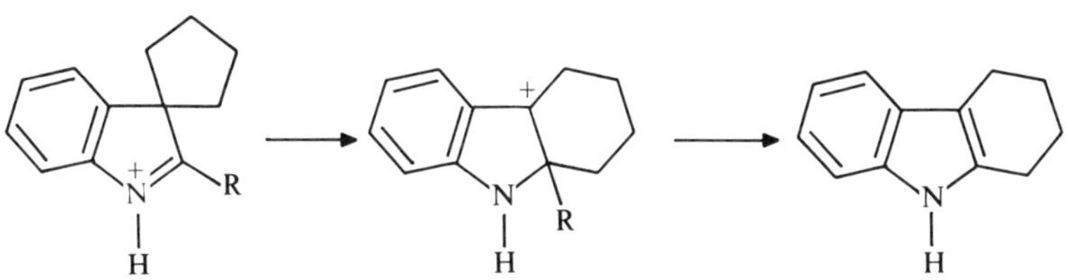

R = H or D

SCHEME 5

Triflic acid is a recoverable and particularly effective catalyst for the ethylation of benzene with ethene in a two-phase system; however, yields of ethylbenzene are far from quantitative because of a build-up of protonated polyethylbenzenes in the catalyst layer and because of competing formation of ethyl triflate.[44] Alkylation of phenols by primary alcohols can be achieved using triflic acid or Nafion-H as catalyst.[45] Aroylations with aromatic nitriles and triflic acid have been examined; electron-withdrawing substituents in the nitrile enhance the yield of diaryl ketone at the expense of the *s*-triazine by-product.[46]

The major initial reaction of aromatic aldehydes with arenes in the presence of Lewis and Brønsted acids is electrophilic substitution to give a diarylmethanol which reacts further to give a triarylmethane; transformylation does not occur.[47]

There has been a study of the acidity dependence of the rate of the chloromethylation of toluene in homogeneous mixtures of toluene, formaldehyde, hydrochloric and acetic acids, and zinc chloride.[48]

Cyclizations like that in Scheme 6 are rather insensitive to ring substituents; epoxide ring-opening may be at least in part rate-limiting.[49] Moderate yields are reported for the reaction of Scheme 7.[50] For the cyclodehydration of Scheme 8, $\rho = -8.54$.[51] There have been explorations of the preparative potential of the three types of acid-induced intramolecular amidoalkylations of Scheme 9[52] and of the annulation reactions of Scheme 10, in which there is electrophilic substitution by intermediate bis(phenylthio)carbocations.[53]

The Friedel–Crafts acetylation of arene(tricarbonyl)chromium has been investigated; the tricarbonylchromium group deactivates to a small extent the ring to which it is attached.[54] There is evidence for pre-equilibrium formation of acyl chloride in the acylation of benzene with phthalic anhydride in the presence of $AlCl_3$.[55] Fluorinated catalyst systems such as HF/BF_3 are recommended for acylation of trifluoromethoxy- and trifluoromethylthio-benzene because halogen exchange occurs if, for instance, $AlCl_3$ is used. Very high *para* regioselectivity is reported.[56] Pivaloylation of *N*-methylpyrrole takes place preferentially at C(3) and disubstitution follows, at a comparable rate, mainly at C(5); the contrast with acetylations is mainly a consequence of steric factors.[57] The competition between acyldesilylation (followed by acylation of anisole) and acylation in the reaction of *m*-trimethylsilylanisole with benzoyl or acetyl chloride in the presence of $AlCl_3$ has been examined.[58]

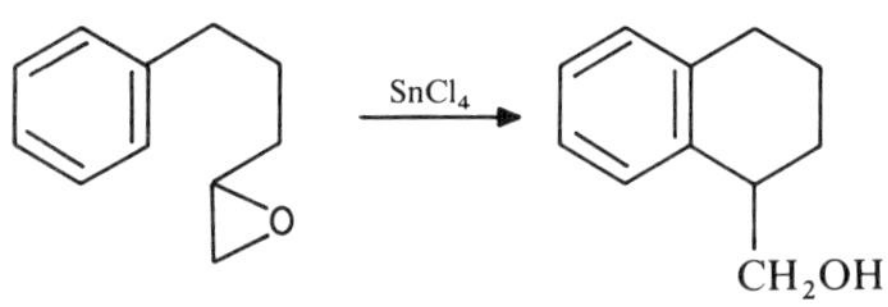

SCHEME 6

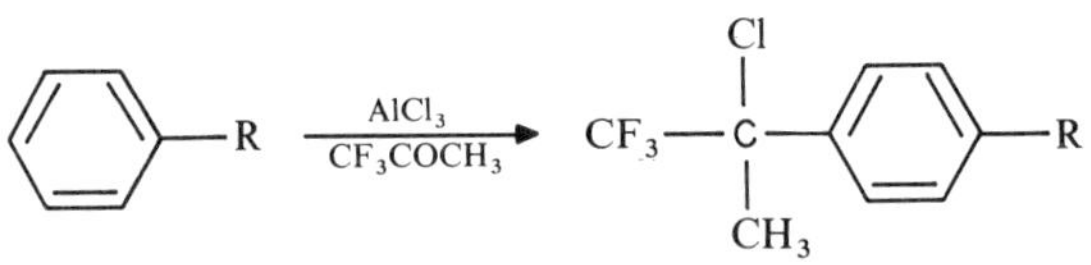

SCHEME 7

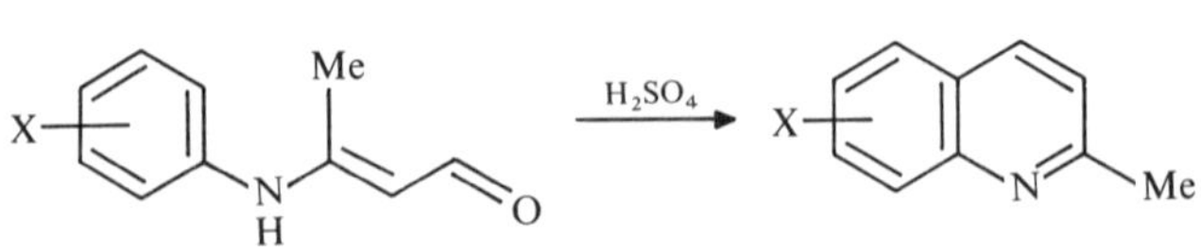

SCHEME 8

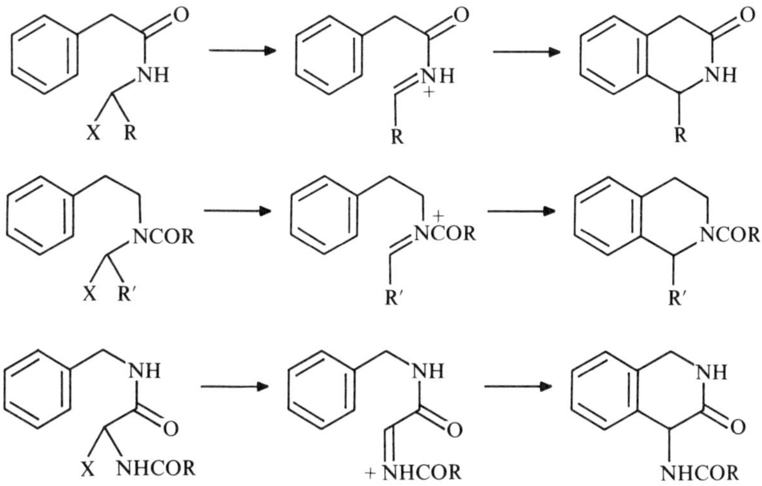

SCHEME 9

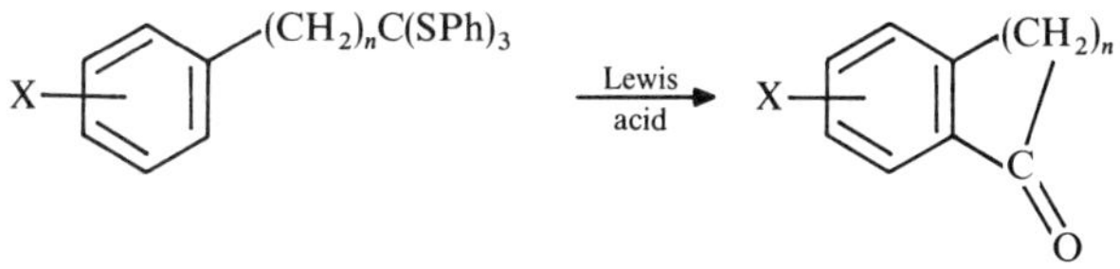

SCHEME 10

Phenylsulphonylation of *o*-xylene, indan, and tetralin by benzenesulphonyl chloride in the presence of $AlCl_3$ has been studied.[59] Lewis acid-catalysed reactions of *p*-chlorobenzenesulphenyl chloride with methylbenzenes gives mainly the products of mono- or di-sulphenylation; disulphides are also formed.[60]

Miscellaneous Reactions

Photolysis of 1-amino- or 1-alkylamino-2-methyl-4,6-diphenylpyridinium tetrafluoroborate in benzene or toluene produces the singlet nitrenium or alkylnitrenium ion, which partitions between intersystem crossing and electrophilic substitution; for the latter reaction $\rho = -2.1$.[61] *N*-alkoxy-*N*-acylnitrenium ions, generated by treatment of *N*-alkoxy-*N*-chloroamides with silver ions, can cyclize onto aromatic nuclei in alkoxy or acyl side-chains.[62] Phenyl azide reacts with aromatic compounds in the presence of $AlCl_3$ according to the mechanism of Scheme 11; the second step is rate-determining and, from competition experiments under heterogeneous conditions, a ρ-value of -6.0 has been deduced[63] for the final fast product-determining step.

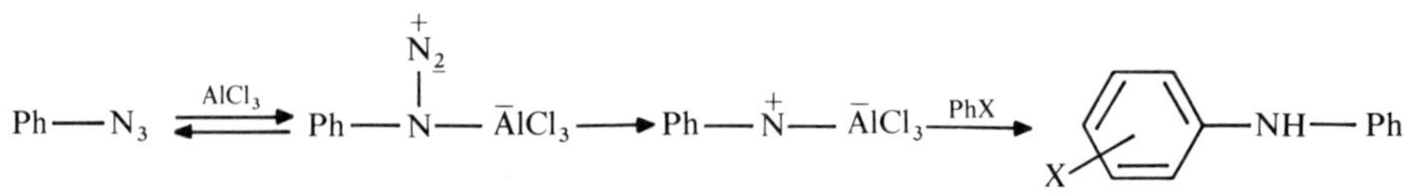

SCHEME 11

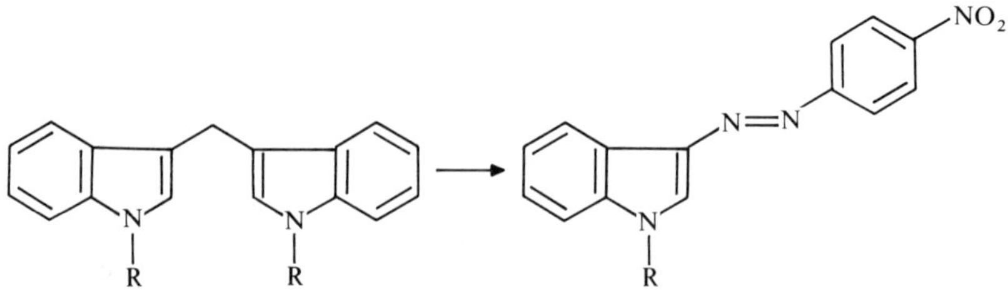

(19) (20)

+

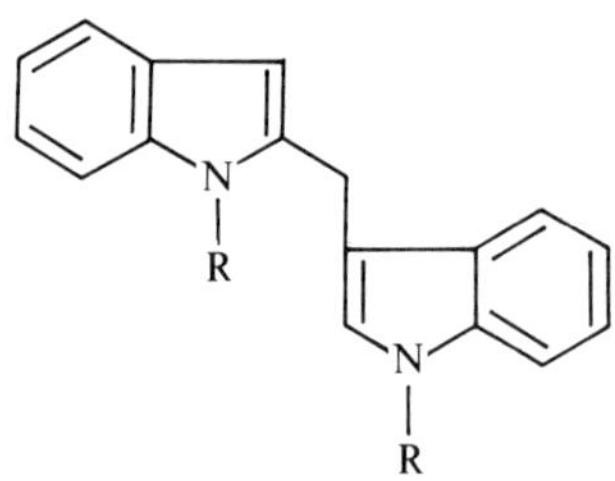

(21)

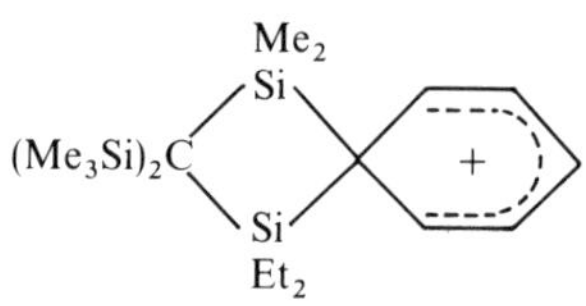

(22)

There has been a review of azo-coupling mechanisms in relation to other S_E2 reactions, together with new results on the reaction of *p*-nitro- and *p*-methoxydiazonium salts with 1-naphthylamine in polar aprotic media in the presence of alkylammonium halides.[64] Coupling at C(3) of 1- and 2-methylindoles with *p*-nitrobenzenediazonium ion shows no primary kinetic hydrogen isotope effect; in contrast and surprisingly 3-methylindole, which couples at C(2) following initial attack at C(3) and rearrangement, does show an isotope effect.[65] In the reaction of (**19**; R = H or Me) with *p*-nitrobenzenediazonium fluoroborate, (**20**) and (**21**) are formed indicating attack at C(3) or C(3′) followed by displacement or rearrangement; the cation formed by displacement gives 3-hydroxymethylindole which after attack by the diazonium ion, also at C(3), and loss of formaldehyde and a proton gives more (**20**).[66] Reaction of $(Me_3Si)_2C(SiMe_2Ph)(SiEt_2I)$ with $AgBF_4$ in Et_2O gives both unrearranged $(Me_3Si_2)C(SiMe_2Ph)(SiEt_2F)$ and rearranged $(Me_3Si)_2C(SiMe_2F)(SiEt_2Ph)$, probably via the cation (**22**).[67] The kinetics of protodesilylation of silylated derivatives of (**23**), (**24**), and (**25**) have been investigated.[68] Hydroxylation of anilides and phenyl esters with hydrogen peroxide in HF-SbF_5 gives important proportions of *meta*-substitution, which results from reaction of protonated hydrogen peroxide with the protonated substrate.[69]

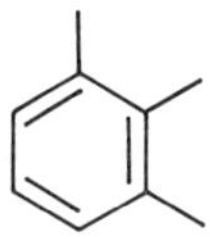
(**23**)

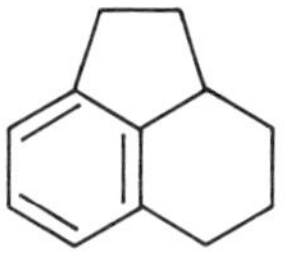
(**24**)

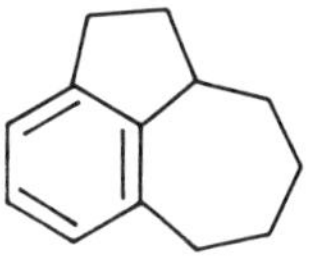
(**25**)

References

1 Effenberger, F., Reisinger, F., Schönwälder, K. H., Bäurle, P., Stezowski, J. J., Jogun, K. H., Schöllkopf, K., and Stohrer, W.-D., *J. Am. Chem. Soc.*, **109**, 882 (1987).
2 Baciocchi, E. and Mandolini, L., *Tetrahedron*, **43**, 4035 (1987).
3 Olah, G. A., Yamato, T., Hashimoto, T., Shih, J. G., Trivedi, N., Singh, B. P., Piteau, M., and Olah, J. M., *J. Am. Chem. Soc.*, **109**, 3708 (1987).
4 Belen'kii, L. I., *Khim. Get. Soedin*, **22**, 749 (1986); *Chem. Abs.*, **106**, 49361 (1987).
5 Filler, R., Cantrell, G. L., and Choe, E. W., *J. Org. Chem.*, **52**, 511 (1987).
6 Gilow, H. M., Hong, Y. H., Millirons, P. L., Snyder, R. C., and Casteel, W. J., *J. Heterocycl. Chem.*, **23**, 1475 (1986).
7 Laws, A. P., Neary, A. P., and Taylor, R., *J. Chem. Soc., Perkin Trans. 2*, **1987**, 1033.
8 Louwrier, P. W. F. and Brinkman, G. A., *Hyperfine Interactions*, **32**, 831 (1986); *Chem. Abs.*, **107**, 22887 (1987).
9 Fărcaşiu, D., *Adv. Chem. Ser.*, **215**, 285 (1987); *Chem. Abs.*, **107**, 197133 (1987).
10 Bryce, M. R., Chambers, R. D., Mullins, S. J., and Parkin, A., *Bull. Soc. Chim. Fr. II*, **1986**, 930.
11 El-Dusouqui, O. M. E., Mahmud, K. A. M., and Sulfab, Y., *Tetrahedron Lett.*, **28**, 2417 (1987).
12 Jaeger, D. A., Clennan, M. W., Leyden, D. E., and Murthy, R. S. S., *Tetrahedron Lett.*, **28**, 4805 (1987).
13 Chênevert, R. and Ampleman, G., *Can. J. Chem.*, **65**, 307 (1987).
14 Lindsay-Smith, J. R., McKeer, L. C., and Taylor, J. M., *J. Chem. Soc., Perkin Trans. 2*, **1987**, 1533.
15 Dakka, J. and Sasson, Y., *J. Chem. Soc., Chem. Commun.*, **1987**, 1421.
16 Rozen, S. and Brand, M., *J. Chem. Soc., Chem. Commun.*, **1987**, 752.
17 Tee, O. S. and Iyengar, N. R., *Can. J. Chem.*, **65**, 1714 (1987).
18 Gol'dfarb, Y. L., Gromova, G. P., and Belen'kii, L. I., *Khim. Get. Soedin*, **22**, 837 (1986); *Chem. Abs.*, **106**, 175478 (1987).
19 Gleghorn, J. T. and Torossian, G., *J. Chem. Soc., Perkin Trans. 2*, **1987**, 1303.
20 Feng, J., Zheng, X., and Zerner, M. C., *J. Org. Chem.*, **51**, 4531 (1986).
21 Eberson, L. and Radner, F., *Acc. Chem. Res.*, **20**, 53 (1987).
22 Attinà, M., Cacace, F., and Yañez, M., *J. Am. Chem. Soc.*, **109**, 5092 (1987).
23 Sankararaman, S. and Kochi, J. K., *Recl. Trav. Chim. Pays-Bas*, **105**, 278 (1986).
24 Sankararaman, S., Harney, W. A., and Kochi, J. K., *J. Am. Chem. Soc.*, **109**, 5235 (1987).
25 Rudakov, E. S., Lobachev, V. L., Ignatenko, V. M., and Velichko, L. I., *Kinet. Katal.*, **27**, 1315 (1986); *Chem. Abs.*, **107**, 115063 (1987).
26 Bushnell, G. W., Fischer, A., Henderson, G. N., and Mahasay, S. R., *Can. J. Chem.*, **64**, 2382 (1986).
27 Fischer, A., Henderson, G. N., and Mahasay, S. R., *Can. J. Chem.*, **65**, 1233 (1987).
28 Suzuki, H., Kimura, E., Katayama, T., and Ogawa, T., *Chem. Lett.*, **87**, 979 (1987).
29 Benedikt, G. M. and Traynor, L., *Tetrahedron Lett.*, **28**, 763 (1987).
30 Lemaire, M., Guy, A., Roussel, J., and Guette, J.-P., *Tetrahedron*, **43**, 835 (1987).

[31] Moodie, R. B. and Stephens, R. J., *J. Chem. Soc., Perkin Trans. 2*, **1987**, 1059.
[32] Squadrito, G. L., Church, D. F., and Pryor, W. A., *J. Am. Chem. Soc.*, **109**, 6535 (1987).
[33] Boughriet, A., Brémard, C., and Wartel, M., *Nouveau J. Chim.*, **11**, 245 (1987).
[34] Laszlo, P. and Pennetreau, P., *J. Org. Chem.*, **52**, 2407 (1987).
[35] Raghavachari, K., Reents, W. D., and Haddon, R. C., *J. Comput. Chem.*, **7**, 265 (1986); *Chem. Abs.*, **106**, 101472 (1987).
[36] Koeberg-Telder, A. and Cerfontain, H., *Recl. Trav. Chim. Pays-Bas*, **106**, 85 (1987).
[37] Cerfontain, H., Koeberg-Telder, A., Lambrechts, H. J. A., Lindner, W., Zhang, S.-Z., and Vogel, E., *J. Org. Chem.*, **52**, 3373 (1987).
[38] Lally, J. M. and Spillane, W. J., *J. Chem. Soc., Chem. Commun.*, **1987**, 1571.
[39] Khelevin, R. N., *Zhur. Org. Khim.*, **22**, 2315 (1986); *Chem. Abs.*, **107**, 77054 (1987).
[40] Laguzzi, G. and Speranza, M., *J. Chem. Soc., Perkin Trans. 2*, **1987**, 857.
[41] Aliprandi, B., Cacace, F., and Fornarini, S., *Tetrahedron*, **43**, 2831 (1987).
[42] Jackson, A. H. and Lynch, P. P., *J. Chem. Soc., Perkin Trans. 2*, **1987**, 1215.
[43] Ibaceta-Lizana, J. S. L., Jackson, A. H., Prasitpan, N., and Shannon, P. V. R., *J. Chem. Soc., Perkin Trans. 2*, **1987**, 1221.
[44] Booth, B. L., Al-Kinany, M., and Laali, K., *J. Chem. Soc., Perkin Trans. 1*, **1987**, 2049.
[45] Rajadhyaksha, R. A. and Chaudhari, D. D., *Ind. Eng. Chem. Res.*, **26**, 1276 (1987); *Chem. Abs.*, **107**, 58245 (1987).
[46] El-Khawaga, A. M., *Bull. Soc. Chim. Fr. II*, **1987**, 372.
[47] Roberts, R. M., El-Khawaga, A. M., Sweeney, K. M., and El-Zohry, M. F., *J. Org. Chem.*, **52**, 1591 (1987).
[48] Rey, M. I., Ortiz, M. I., and Irabien, A., *J. Mol. Catal.*, **39**, 105 (1987); *Chem. Abs.*, **107**, 96104 (1987).
[49] Taylor, S. K., Davisson, M. E., Hisson, B. R., Brown, S. L., Pristach, H. A., Schraam, S. B., and Harvey, S. M., *J. Org. Chem.*, **52**, 425 (1987).
[50] Bonnet-Delpon, D. and Charpentier-Morize, M., *Bull. Soc. Chim. Fr. II*, **1986**, 933.
[51] Ono, M., Todoriki, R., and Tamura, S., *Chem. Pharm. Bull.*, **34**, 1468 (1986); *Chem. Abs.*, **106**, 83737 (1987).
[52] Ben Ishai, D., Satati, I., Peled, N., and Goldshore, R., *Tetrahedron*, **43**, 439 (1987).
[53] bin-Manus, A. R. and Smith, R. A. J., *Tetrahedron*, **43**, 1847 (1987).
[54] von Rosenberg, J. L. and Pinder, A. R., *J. Chem. Soc., Perkin Trans. 1*, **1987**, 747.
[55] Hashimoto, I. and Hori, M., *Kenkyo Kiyo-Wakayama Kogyo Koto Senmon Gakko*, **21**, 22 (1986); *Chem. Abs.*, **107**, 235724 (1987).
[56] Desbois, M., *Bull. Soc. Chim. Fr. II*, **1986**, 885.
[57] Harsanyi, M. C. and Norris, R. K., *J. Org. Chem.*, **52**, 2209 (1987).
[58] Bennetau, B., Krempp, M., and Dunoguès, J., *J. Organomet. Chem.*, **334**, 263 (1987).
[59] Yoshii, Y., Ito, A., Tsuneaki, H., and Osamu, M., *Nippon Kagaku Kaishi*, **1986**, 1117; *Chem. Abs.*, **106**, 195637 (1987).
[60] Grant, D. W., Hogg, D. R., and Wardell, J. L., *J. Chem. Res. (S)*. **1987**, 392.
[61] Takeuchi, H., *J. Chem. Soc., Chem. Commun.*, **1987**, 961.
[62] Glover, S. A., Goosen, A., McCleland, C. W., and Schoonraad, J. L., *Tetrahedron*, **43**, 2577 (1987).
[63] Takeuchi, H., Maeda, M., Mituni, M., and Koyama, K., *J. Chem. Soc., Perkin Trans. 1*, **1987**, 57.
[64] Bagal, I. L., Stovpovai, P. A., and El'tsov, A. V., *Zhur. Obsch. Khim.*, **56**, 489 (1986); *Chem. Abs.*, **105**, 23761 (1986).
[65] Jackson, A. H. and Lynch, P. P., *J. Chem. Soc., Perkin Trans. 2*, **1987**, 1483.
[66] Jackson, A. H., Prasitpan, N., Shannon, P. V. R., and Tinker, A. C., *J. Chem. Soc., Perkin Trans. 1*, **1987**, 2543.
[67] Eaborn, C., Lickiss, P. D., Najim, S. T., and Stańczyk, W. A., *J. Chem. Soc., Chem. Commun.*, **1987**, 1461.
[68] Gruber, R., Kirsch, G., and Cagniant, D., *Bull. Soc. Chim. Fr. II*, **1987**, 498.
[69] Berrier, C., Jacquesy, J.-C., Jouannetaud, M.-P., and Morellet, G., *Bull. Soc. Chim. Fr. II*, **1987**, 158.

Organic Reaction Mechanisms 1987
Edited by A. C. Knipe and W. E. Watts

CHAPTER 9

Carbocations

ROBIN A. COX

Department of Chemistry, University of Toronto, Canada

The 2-Norbornyl Cation

Crystal structures for two substituted 2-norbornyl cations, (**1**) and (**2**), are reported.[1,2] Compound (**1**) has the more classical structure of the two, but even so the C(1)—C(6) distance is long at 1.604 Å;[1] in (**2**) this distance is 1.742 Å.[2] The C(2)—C(6) non-bonded distance is 2.35 Å in (**1**) and only 2.09 Å in (**2**). The C(1)—C(2) and C(2)—C(3) bonds are on the short side.[1,2] The unsubstituted 2-norbornyl cation has been characterized in the gas phase by means of its collisional activation and charge-stripping mass spectra,[3] and its gas-phase heat of formation reassessed at 767 ± 5 kJ mol^{-1}.[3]

Saunders and Johnson[4] have re-examined the question of heavy-atom tunnelling in the 2-norbornyl cation.[5] They find that this phenomenon can explain neither the 5 K NMR results[6] nor the isotopic perturbation data[7] obtained for this ion.[4] [2,3-^{13}C]Norborn-2-yl chloride in SbF_5 at -65°C gives an ion with completely scrambled ^{13}C labels; the small difference between the observed and a statistical distribution, due to the ^{12}C—^{13}C equilibrium isotope effect, enables isotopic equilibrium constants to be calculated with some accuracy.[8] Isotopic perturbation and isotopic scrambling in the nortricycyl-norbornenyl cations have also been studied.[9] The results were explained in terms of a single-ion structure, nortricyclyl with extensive charge delocalization.[9]

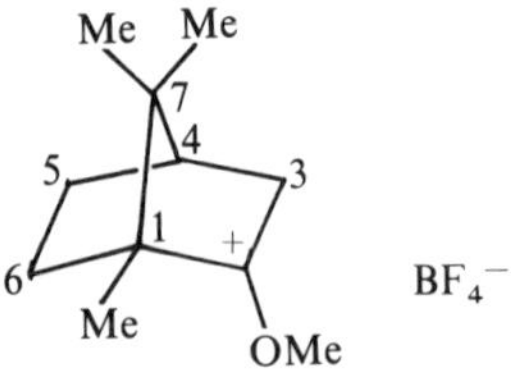

(1)

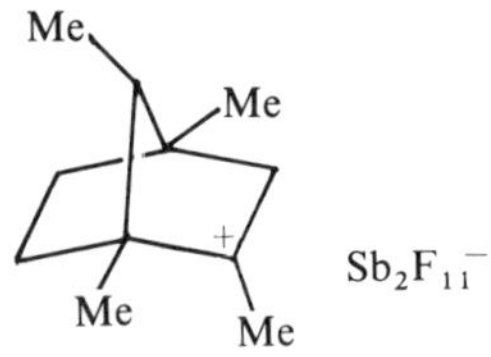

(2)

Deuterium isotope effects on the ^{13}C chemical shifts observed in 2-substituted 2-norbornyl cations are stated to be consistent with the onset of σ-bridging as electron demand increases.[10] Similar studies on the 2-methyl-2-norbornyl and 2-methylbicyclo[2.2.2]octyl cations associate the isotope shifts with bending motions along the bridging coordinate.[11] The individual gauge for localized molecular orbitals (IGLO) method has been used to calculate the NMR parameters of a number of carbocations.[12] Agreement with experiment is good for delocalized structures; it was found that a bicyclobutonium structure was consistent with the observed spectrum for $C_4H_7^+$, and a methylbicyclobutonium one for $C_4H_6CH_3^+$. For the 2-norbornyl cation: 'Its experimental NMR spectra are in good agreement with those computed for the non-classical structure and very different from that calculated for several classical geometries.'[12] However, MP4SDQ/6-31G* level calculations indicate an unsymmetrical π-complex structure for corner-protonated cyclopropane, which has implications for norbornyl cation structures.[13]

Treatment of (**3**) with HSO_3F at $-78°C$ results in the quantitative formation of the ring-expanded product (**4**); the mechanism was elucidated by deuterium-labelling experiments, as shown.[14]

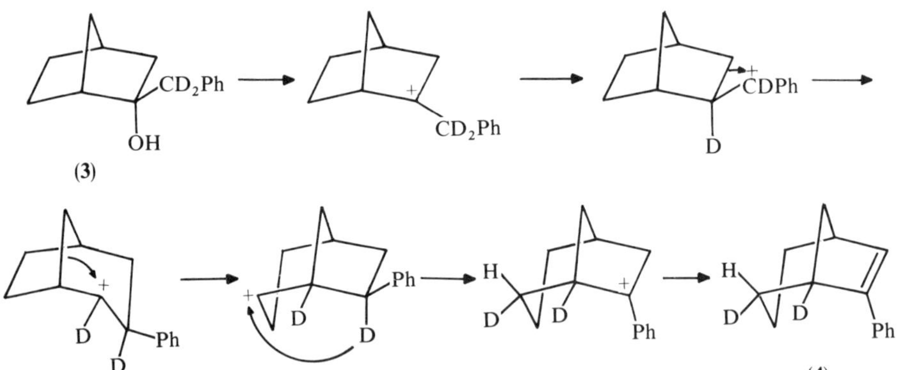

(3) (4)

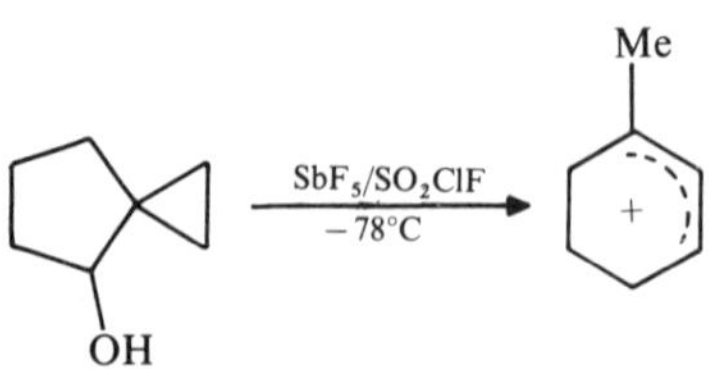

(5) (6)

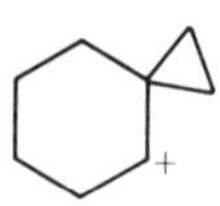

(7)

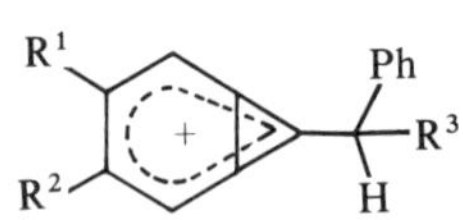

(8)

Polycyclic and Bridged Ions

4-Spiro[2.*n*]alkyl cations have been studied.[15] Typically they quickly rearrange; for instance (**5**) gives (**6**), but ion (**7**) was stable under the conditions, and could be prepared from three different precursors.[15] Attempts to prepare the bullvalenyl cation from fluorobullvalene in SbF_5/SO_2ClF resulted only in α-protonated naphthalene, even at $-110°C$.[16] Alkylidenecycloproparenes protonate only at the exocyclic cyclopropyl carbon, giving (**8**).[17] The sesquinorbornyl cation (**9**) rearranges irreversibly to (**10**) upon raising the temperature.[18]

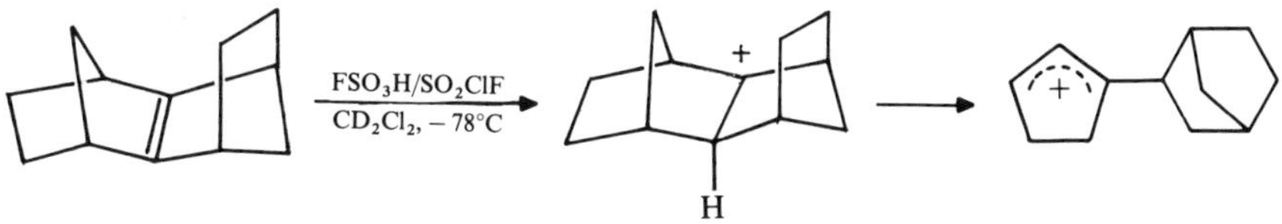

(**9**) (**10**)

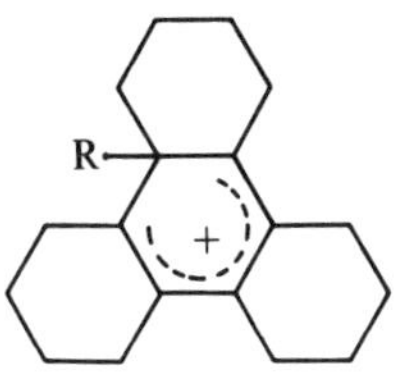

(**11**)
R = NO_2, H, Cl

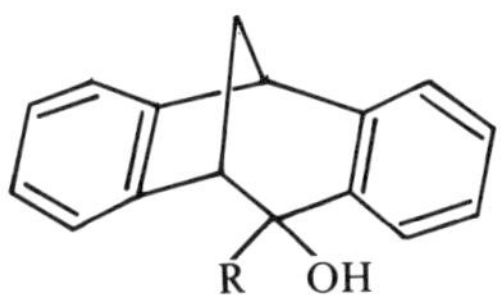

(**12**)

R = Me, Ph

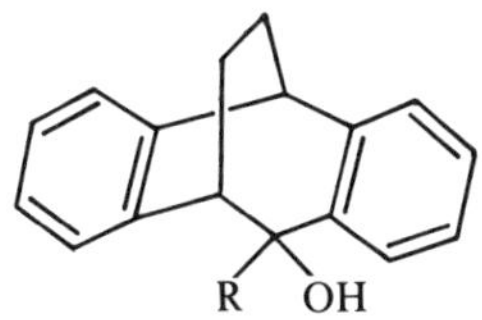

(**13**)

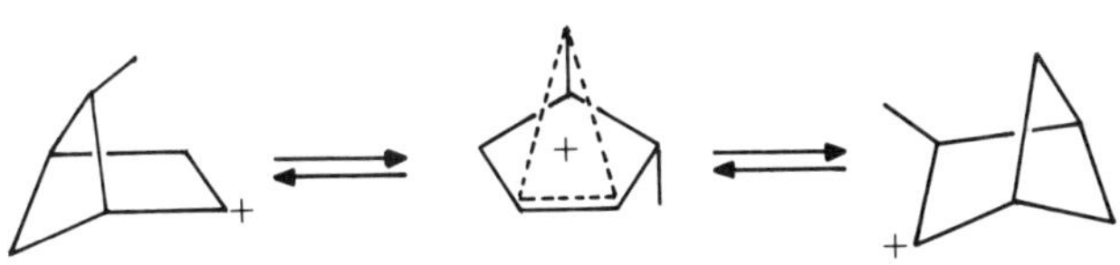

(**14**) (**16**) (**15**)

Degenerate 1,2-shifts in (**11**; R = NO_2) are subtly different from those in (**11**; R = H, Cl) in their regiochemistry.[19] Trifluoroacetolyses of (**12**) and (**13**) proceed with rearrangement, giving products with one more carbon in the saturated bridge.[20] Rearrangement and ring-cleavage reactions in the 3- and 5-methylbicyclo[2.1.1]hex-2-yl cations (**15**) and (**14**) may involve the intermediacy of (**16**).[21] Wagner–Meerwein rearrangements occurring in the radical cations generated during the triphenylpyrylium tetrafluoroborate-photosensitized electron transfer of a number of different azoalkanes have been studied.[22] The rearrangement pathways available to bicyclo[*n*.1.1]alkylium cations have been the subject of AM1, MNDO and MINDO/3 calculations; general agreement with experimental observations was found.[23]

It has been found that 1-bromobicyclo[3.2.1]heptane solvolyses in methanol eight times faster than does *t*-butyl bromide; it is suggested that the postulated bridgehead carbocation intermediate is stabilized by transannular orbital interactions.[24] Investigation of the protolytic cleavage of a number of bicyclic cyclopropane derivatives leads to the conclusion that the protonated cyclopropane and σ routes are equivalent in generating bridged carbocations, and that edge-protonated cyclopropanes play at best a minor role in product formation.[25]

Treatment of the 1-adamantyl cation with carbon monoxide and adamantane in CF_3SO_3H results in 3,4-homoadamantanediol, formed by addition of the 1-adamantanecarbonyl cation to 1-adamantanecarbaldehyde followed by Wagner–Meerwein rearrangement.[26] Extraction of 3-methyl-3-vinylmethylenecyclohexane into FSO_3H/SO_2ClF at −100°C followed by quenching with water results in cyclization to 2,4-dimethylbicyclo[2.2.2]octan-2-ol.[27] Molecular-mechanics calculations have given force-field parameters for a number of bridgehead and tertiary carbenium ions; the rate of solvolysis of bridgehead substrates with halide and tosylate leaving-groups can be expressed as strain changes between the substrate and the intermediate cation.[28]

Open Systems

A review of the reactivity, stability, and structure of orthoesters and diakoxycarbenium ions has appeared; new synthetic applications are also covered.[29] Structure-reactivity correlations are reported for general base catalysis of the addition of alcohols to (**17**), in the context of an attempt to determine how a reaction changes its mechanism.[30] It has been found that acid-catalysed isomerization of the 4-Me group in $CD_3CD(OH)CD_2CH_3$ between the 1- and 4-positions occurs faster than dehydration to 2-butene; based on this and other evidence it is inferred that oxygen exchange into 2-butanol and hydration of 1-butene do not involve a common carbocation intermediate.[31] A study of the vinyl cation $Tol(An)C{=}\overset{+}{C}Tol$ completes a study of the nine possible phenyl-, tolyl-, and anisyl-substituted vinyl cations, and the degenerate 1,2-aryl shifts that can take place in them.[32] The formation of carbocations from C_3 compounds in zeolites of different acidities has been investigated using UV-visible spectroscopy.[33]

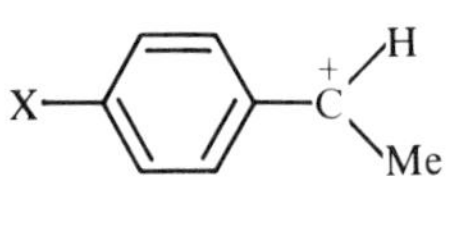

(**17**)
X = NMe_2, OMe

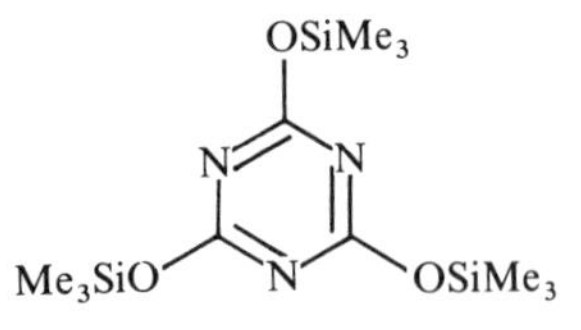

(**18**)

$$R^1R^2C{=}\overset{+}{N}{=}CR^3R^4$$

(**19**)

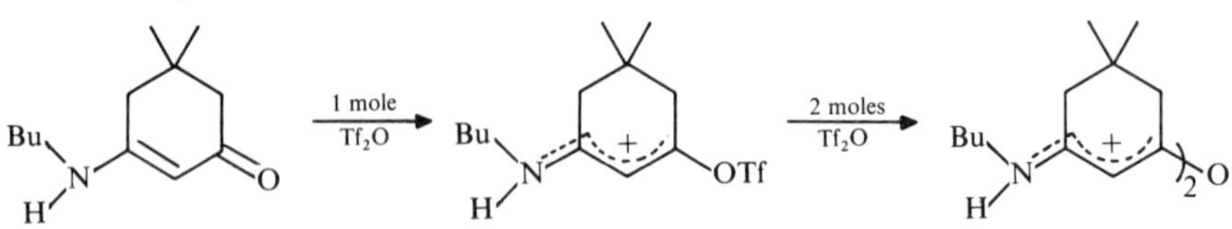

(**20**) (**21**) (**22**)

Chlorocarbenium ions $R^1R^2ClC^+$ react with (**18**) and ketones or tertiary carboxamides to give 2-azaallenium ions (**19**); one of the R groups in (**19**) may be Cl.[34] The ion (**19**) with all R = phenyl has been shown to be linear, and its electrophilic ring-closure to an isoindole has been studied.[35] In HF/SbF_5 or FSO_3H/SbF_5, thiocarbonyl fluorides XFC=S give $X—\overset{+}{C}=S$ species (X = F or SCF_3); these react as electrophiles with aromatic molecules.[36] Enaminones like (**20**) react with triflic anhydride to give cations like (**21**), and then dication ether species (**22**).[37] The ambifunctional reactivity of 3-trifloxypropeniminium salts towards nucleophiles has been investigated.[38]

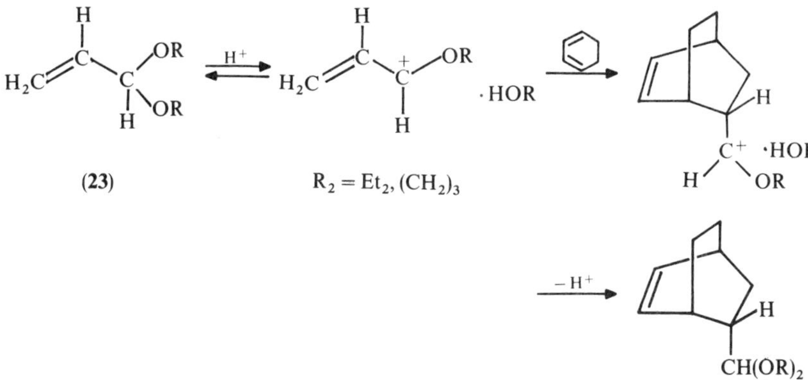

(**23**) $R_2 = Et_2, (CH_2)_3$

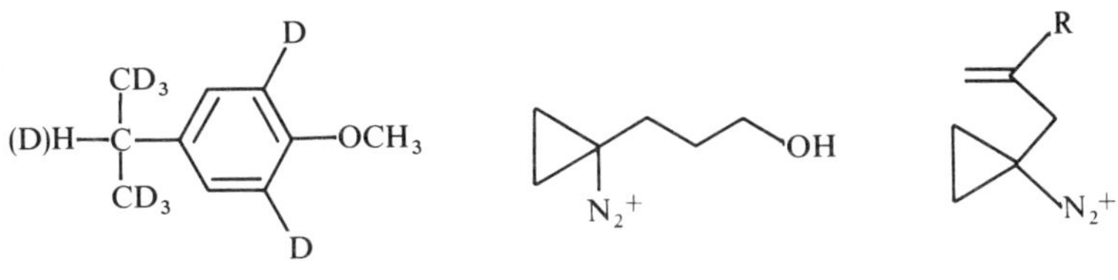

(**24**) (**25**) (**26**)

Acrolein acetals (**23**) have been used as allyl cation precursors in the ionic Diels–Alder reaction.[39] The effect of increasing acidity on the selective aliphatic hydrogen exchange of (**24**) in trifluoroacetic acid has been examined, in order to clarify results on the exchange of methyl hydrogens observed in the undeuteriated material in CF_3CO_2D.[40] The decompositions of (**25**) and (**26**) have been studied, along with the cyclizations and other reactions of the resulting carbocations.[41,42]

The polyfluorinated cations resulting from the reactions of $(CF_3)_2C=C(OMe)R$ ($R = F, C_2F_5$) with electrophiles have been characterized, along with a polyfluorinated acylium ion.[43] The reactions of a number of dienylic chlorides with isobutene and $ZnCl_2$/ether in CH_2Cl_2 have been studied; orientation effects in the observed acyclic products are explained on the basis of steric effects.[44] A convenient preparation of [3.*n*]metacyclophanes involves a cationic cyclo-codimerization.[45] The relative electrophilicity of alkylating agents can be controlled by variation of the Lewis acid strength and concentration; conditions leading to the highest selectivity have been explored.[46] CIDNP evidence for a 'hot' carbocation formed by electron transfer in the pair of radicals $PhCH_2\dot{C}D_2{}^{\bullet}O_2CAr$ is presented.[47]

Benzhydryl and Trityl Cations

The reactions of diarylcarbenium ions, p-XC_6H_4—$\overset{+}{C}H$—C_6H_4Y-p, with alkenes have been studied.[48,49] The 'isokinetic rate constant' for additions to 2-methyl-2-butene has been measured directly, the isokinetic substituent (no influence of temperature on rate) being found to be X = Y = Me.[48] Linear reactivity-selectivity correlations for additions to five different alkenes were obtained; selectivity was found to decrease, be constant, or increase with increasing reactivity, inconsistent with the predictions of the reactivity-selectivity principle.[49] The intermediacy of diarylcarbenium ions in the formation of a substituted benzyltoluene, formed during the acylation of 1-methylcyclohexene in CF_3SO_3H, has been proposed.[50]

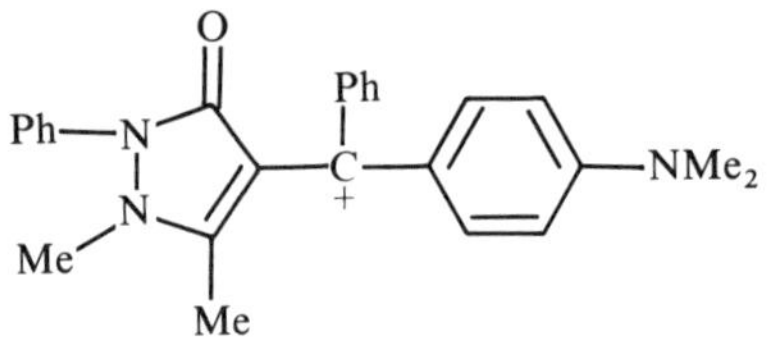

(**27**)

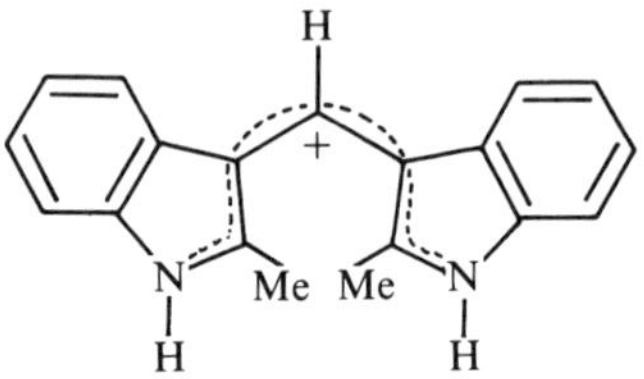

(**28**)

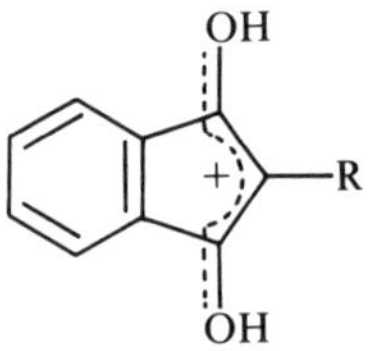

(**29**)

The effects of ionic strength and dielectric constant on the kinetics of the reaction between triarylmethyl cations and sulphite anions have been determined;[51,52] an N_+ scale for the $SO_3{}^{2-}/H_2O$ system is proposed.[51] In the case of (**27**) the dielectric-constant behaviour was in agreement with theory, but specific solvent effects were postulated for the reactions of Malachite Green and Crystal Violet.[52] Bisindolylcarbenium ions like (**28**) react regiospecifically with some *O*- and *C*-nucleophiles.[53] The kinetics of the reactions between trityl cations and nucleophiles, among other reactions, can be followed using photoacoustic detection of small absorbance changes.[54]

Monocyclic Systems

Enoloxonium ions (**29**), formed when 2-aryl-1,3-indanediones are treated with sulphuric acid, have been examined spectroscopically.[55] The fluorination and skeletal transformations which perfluorobenzocycloalkanes undergo when treated

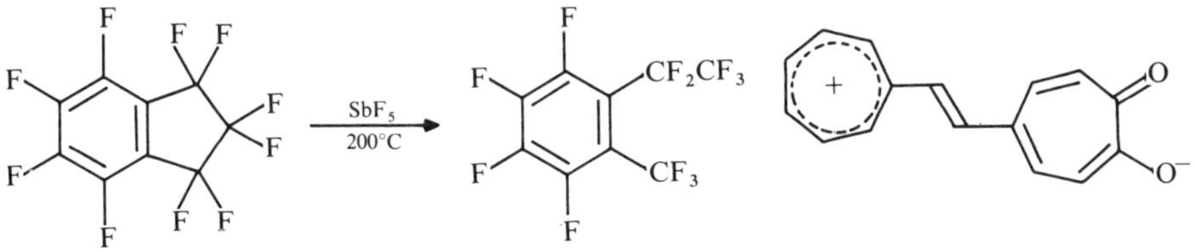

(30) (31) (32)

with SbF_5 at 200°C have been investigated; for instance (**30**) gives (**31**).[56] The molecule 8-(heptafulven-8-yl)-*p*-tropoquinone methide, one of whose resonance forms is (**32**), has been studied.[57] Benzenonium and phenanthrenonium ions with hydroxy and methoxy groups at the sp^3 centre have been generated in superacid and examined by ^{13}C-NMR spectroscopy; a linear correlation was found between 1,2-methyl shift rates and ^{13}C chemical shifts in these species.[58]

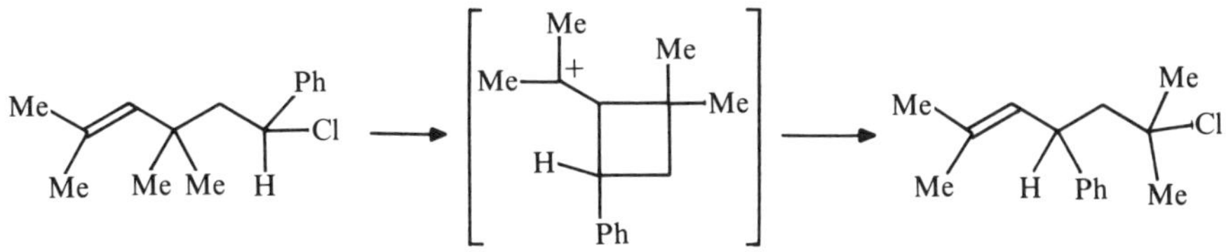

(33) (35) (34)

The Lewis acid-catalysed rearrangement of (**33**) to (**34**) is explained by postulating a carbocationic 1,3-alkenyl shift with (**35**) as an intermediate.[59] Evidence for a mechanism implying different benzylic carbocation species in the acid-catalysed solvolysis of rigid 1-arylcyclohexene oxides comes from a study of substituent effects on the stereoselectivity.[60] The reaction of 5-chloro-1,3-cycloheptadiene with silyl enol ethers in the presence of Lewis acids gives good yields of α-cyclohepta-dienyl-substituted carbonyl compounds, which opens a route to 3-substituted 1,4-cycloheptadienes by way of attack of the cycloheptadienyl cation at the 3-position.[61]

(36) (37) (38) (39)

High-quality calculations show that the most stable $C_5H_7^+$ species is (**36**), followed by (**37**) and (**38**); (**38**) is destabilized relative to (**39**) by the inductive effect of the double bond, which explains why the double bond provides no solvolysis rate enhancement in these species, despite stereochemical evidence for anchimeric assistance.[62]

Heterocyclic Systems

Suitably substituted isoquinolinium cations like (**40**) may cyclize, giving structures like (**41**) or (**42**).[63] A number of equilibrium constants associated with some of these

(40) (41)

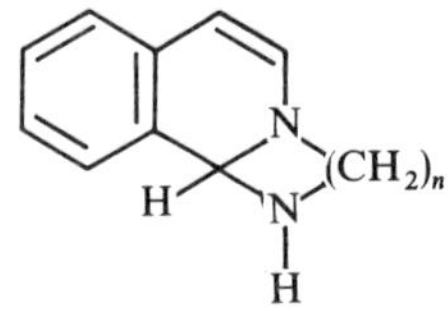

(42)

processes have been evaluated.[63] A linear-free-energy relationship between the equilibration of substituted 1-methylpyridinium cations with their C(4) hydroxide adducts (pseudo-bases), and σ_p^-, has been evaluated.[64] Equilibrium constants for the ionization of trityl bromide and 3-bromo-3-phenyldiazirine in acetonitrile are compared.[65]

Trityl cations catalyse the Michael reactions of ethylthioketene dithioacetals with silyl enol ethers[66] and stannylated nucleophiles[67] to give the corresponding adducts.[66,67] 1,3-Dioxolan-2-ylium cations derived from aldehyde ethylene acetals and trityl cation react with ketene silyl acetals and enamines, giving the corresponding β-keto-esters and 1,3-diketones.[68] Theoretical studies illuminate the formation, electronic structures, and electrophilic reactivity of the pyrazole cation system.[69]

Multiply Charged Systems

In triflic acid, β-nitrostyrenes protonate twice to give *N,N*-dihydroxyiminiumbenzyl dications (**43**); these species are electrophilic enough to react with benzene.[70]

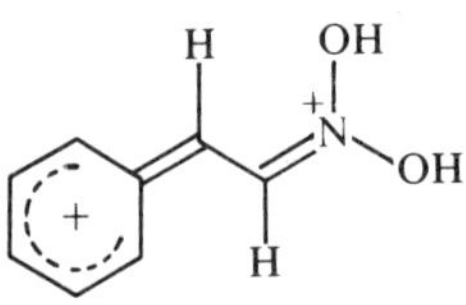

(43)

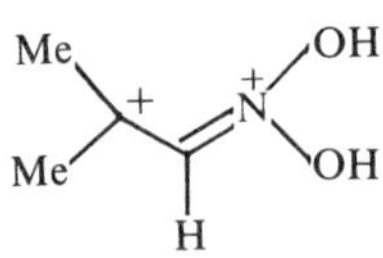

(44)

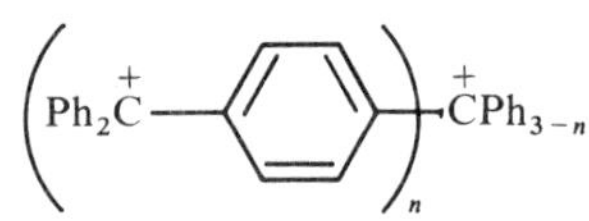

(45)

Similarly, nitroethylene is diprotonated to (**44**) in the medium.[71] This species is isoelectronic with guanidinium ion, and MO calculations confirm the stabilization conferred by Y-delocalization.[71] Multi-tritylium ions (**45**) can be obtained by treating the trityl ether precursors with CF_3SO_3H; the $n = 2$ and $n = 3$ ions are an intense red and red-violet, respectively.[72]

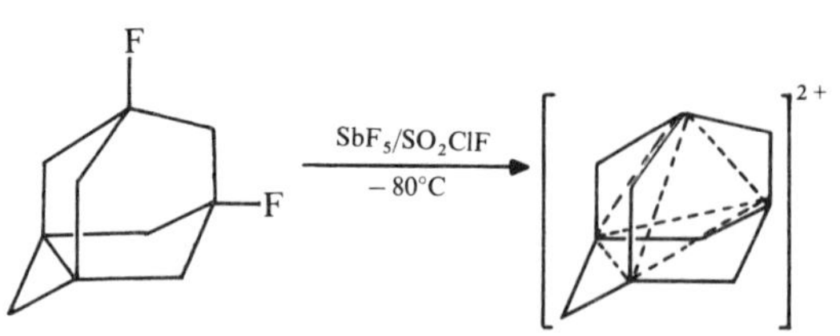

(46)

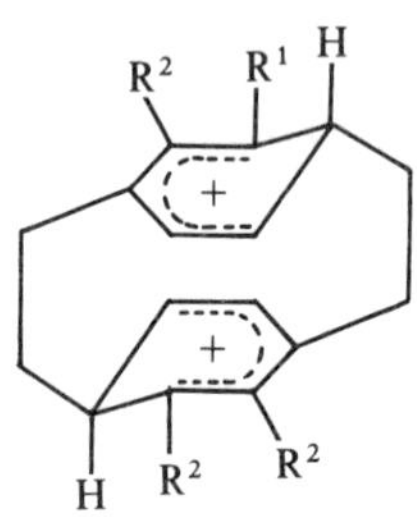

(47)

A three-dimensional homoaromatic structure is proposed for the dication (**46**), which is stable up to 0°C in solution; the non-classical structure shown is supported by ^{13}C-NMR data, the bridgehead C-atom signal being shifted upfield rather than downfield.[73] Treatment of [2.2]paracyclophanes with superacids at low temperatures results in a red monoprotonated species if FSO_3H/SO_2ClF is used, and pale yellow (**47**) with $FSO_3H/SbF_5/SO_2ClF$.[74]

Destabilized Carbocations

Treatment of the corresponding alcohols with $ClSO_3H$ in SO_2 or CD_2Cl_2 at −95°C gives α-aryl- and diaryl-thioformamidyl cations (**48**), which efficiently cyclize and deprotonate to benzothiophenes (**49**) when added to ice.[75] *N*-chloro-*N*-methoxyamides give nitrogen heterocycles in good yield when treated with anhydrous zinc acetate, probably by way of the destabilized nitrenium ion (**50**).[76]

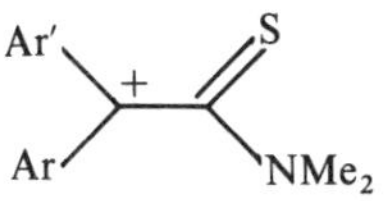

(**48**)

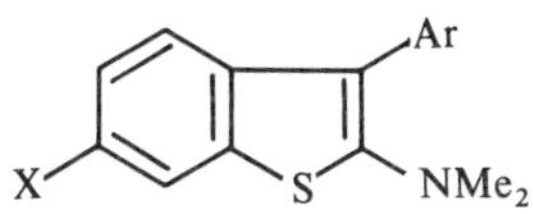

(**49**)

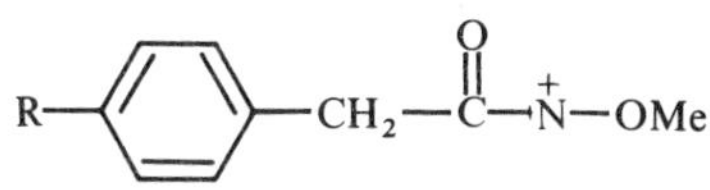

(**50**)

The rates of capture of the 1-(4-methoxythiophenyl)-2,2,2-trifluoroethyl cation by alkylamines decrease with increasing amine pK_a, and are up to 200 times slower than diffusion-limited capture by azide; the reaction is thought to be desolvation-limited.[77] The substituent effect of the pentafluorophenyl group has been investigated; $C_6F_5CH(OTs)CH_3$ is three to five powers of ten less reactive than is $PhCH(OTs)CH_3$ in solvolysis reactions, but the C_6F_5 group is still much more activating than hydrogen.[78]

A new method of preparing benzylic α-fluoro or α,α-difluoro ketones, esters or nitriles, by anodic oxidation of the unfluorinated precursors in $CH_3CN/(Et_3N,3HF)$ has been described; the reaction probably involves α-carbonyl or α-cyano carbocation intermediates.[79] The structures and stabilities of the various $C_2H_2F_3{}^+$ cations have been studied by *ab initio* MO calculations.[80] $CF_3CH_2{}^+$ is less stable than $FCH_2CF_2{}^+$, and has a low barrier for 1,2-F migration; the nett effect of F is highly destabilizing in the former. $FCH_2CF_2{}^+$ is also more stable than F_2CHCHF^+, but the barrier to hydrogen migration is quite high.[80]

Direct Carbocation/Nucleophile Reactions

Ritchie has summarized what is known of the rate and equilibrium constants for the reactions of carbocations with anionic and neutral nucleophiles in several solvents; the factors affecting kinetic nucleophile reactivities are discussed.[81] The reported correlation of nucleophilic reactivities with one-electron oxidation potentials of nucleophiles does not apply to hydrazine in aqueous solutions or to several nucleophiles in DMSO, but the data could be rationalized in other ways.[82]

Evidence is presented for the equilibration of resonance-stabilized carbocations and carbanions with the equivalent radicals by single-electron-transfer in solution, along with evidence that the transfer is reversible.[83] It is shown that nucleophilicity towards the pyronium cation in aqueous solution correlates with the vertical ionization potential of the nucleophile; implications for the physical significance of the N_+ scale are discussed.[84] Monte Carlo simulations are used to probe the point at which a contact ion-pair becomes a solvent-separated one; for $Me_3C^+.Cl^-$ the distances are 2.9 Å for contact, and 5.5 Å for solvent separation, between the ions.[85]

Silicon-containing Systems

The first demonstration of the solvolytic generation of R_3Si^+, the silicenium ion (**52**), comes from the methyl migration taking place in (**51**), as inferred from the composition of solvent-trapped products.[86] Increasing solvent nucleophilicity leads to increasing amounts of trapping product formed from (**52**). The methyl-migration route is supported by *ab initio* calculations, which predict that $R_3Si\overset{+}{C}R_2$ is less stable than $R_2\overset{+}{Si}CR_3$.[86] Unusual properties do not have to be ascribed to (**52**) provided that the solvolysis results are interpreted as product formation from solvent-separated ion-pairs.[87] In the gas phase, the observed dissociation products from several silicon-containing species include those formed from silicenium ions and from α- and β-silyl carbocations; isomeric $C_5H_{13}Si^+$ ions are readily interconverted by rearrangement.[88]

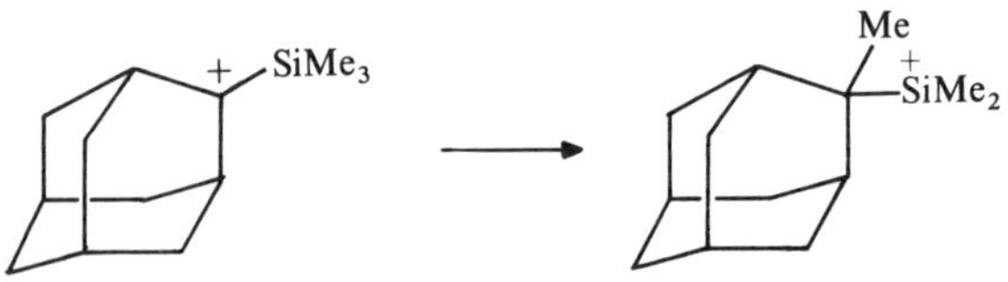

(**51**) (**52**)

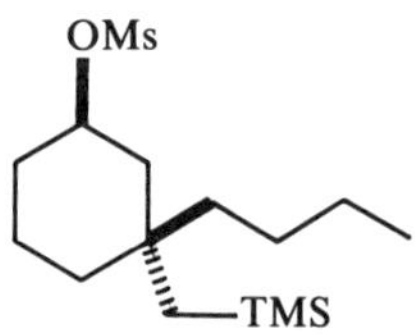

(**53**)

A γ-silicon substituent is found to promote solvolysis and stabilize the carbocation intermediate in at least one case.[89] The directing effect of a silyl group on the rearrangement of mesylates like (**53**) during reaction with dimethylaluminium triflate has been observed; stable β-silyl cationic species are formed preferentially on the way to the observed product olefins.[90] Reaction of silyl enol ethers with *N*-chlorosuccinimide leads to the formation of a siloxycarbinyl cation that can be trapped by azide.[91] A new synthesis of vinyl sulphides by the Lewis acid-promoted reaction of [chloro(arylthio)methyl]trimethylsilanes with trisubstituted alkenes is described.[92]

Organometallic Systems

A general carbon basicity scale (pK_C) is proposed, based on the equilibrium constants for the reversible addition of a wide range of pyridine and phosphorus nucleophiles to the ethene ligand of the cation $[CpFe(CO)_2(C_2H_4)]^+$.[93] The pK_{R^+} values for a series of carbocations (**54**) have been determined; the organometallic substituent is found to be more stabilizing than a benzyl group.[94] A detailed kinetic study of the reversible addition of 2-ethylpyridine to the dienyl portion of a series of tricarbonyl(1-5-η-dienyl)iron(II) cations has been performed; the dissociation process is found to be entropy-controlled.[95] The reaction of tin enolates with (**55**) has been studied,[96] as has the dynamic behaviour of (**56**) and its reaction with chiral nucleophiles.[97]

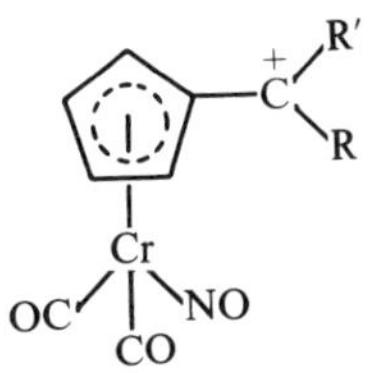

(**54**)

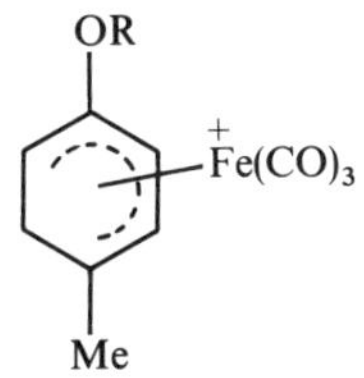

(**55**)

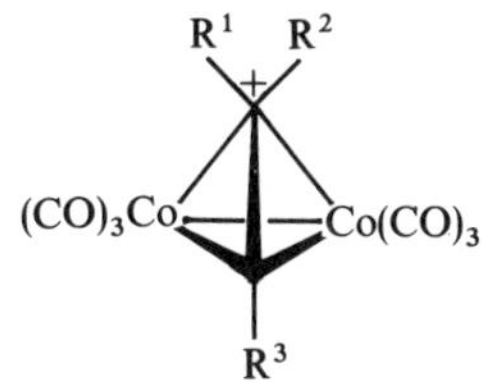

(**56**)

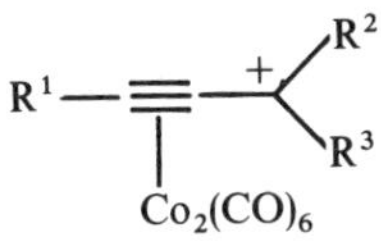

(**57**)

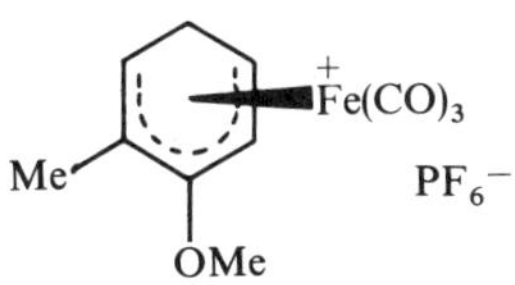

(**58**)

The chemistry and synthetic utility of cobalt-complexed propargyl cations (**57**) has been reviewed.[98] It is found that compounds like (**58**) can be resolved into (+) and (−) enantiomers;[99,100] the applications to organic enantiomer synthesis have been explored.[100]

The Gas Phase

Free butyl cations can be formed from $CH_3CH_2CH_2CH_2T$ in the gas phase by β-decay.[101] When these react with CO in the presence of EtOH only ethyl α-methylbutyrate is obtained, indicating complete isomerization of the primary cation to the secondary.[101] Similarly the free vinyl cations $Ph\overset{+}{C}{=}CHT$ and $PhCT{=}\overset{+}{C}H$, formed from $PhCT{=}CHT$, react with H_2O, EtOH, and Et_2O, giving mainly electrophilic addition products, with some deprotonation to phenylacetylene.[102] The simple vinyl cation $C_2X_3^+$ (X = H or T) inserts into the σ-bonds of hydrogen or methane, giving $C_2H_5^+$ and $Me_2\overset{+}{C}H$, respectively; the latter decomposes to allyl cation and H_2 at pressures below 200 torr.[103] No evidence was obtained for protonated cyclopropane structures.[103]

A kinetic and theoretical study of the ion/molecule reactions of methoxymethyl cation with ammonia has been carried out.[104] Three reactions are possible, the products being $CH_2{=}\overset{+}{N}H_2$ and CH_3OH, $CH_3\overset{+}{N}H_3$ and $CH_2{=}O$, or $\overset{+}{N}H_4$ and ethylene oxide; the latter is a specific gas-phase reaction.[104] Chloride-ion affinities of substituted benzyl cations in the gas phase are represented by a dual-parameter linear-free-energy relationship.[105] A study of gas-phase proton transfer in halotoluene mixtures leads to the suggestion of a 'dynamic' ion structure, in which a mobile hydrogen atom is shared by two or more unsubstituted ring carbon atoms in the protonated molecule.[106] Entropies well in excess of expected values have been measured for some additional aromatic protonations, of benzene, halobenzenes and xylenes, and a similar explanation is proposed.[107]

Theoretical Calculations

Calculations at the 6-31G* level show that the thioacetyl ion $H_3C—\overset{+}{C}=S$ is the global minimum on the $C_2H_3S^+$ surface, with the 1-thiovinyl cation $H_2C=C=\overset{+}{S}H$ as the next lowest; all rearrangement barriers were found to be lower than the dissociation barrier.[108] The decomposition of ethynyldiazonium ion has been examined, as a possible source of *sp*-centred carbenium ions.[109] A MINDO/3 study of the rearrangement of the 1-methylcyclohexyl cation to the 1,2-dimethylcyclopentyl cation shows that the two non-equivalent pathways, α and β scission, are relatively close in energy.[110] MINDO/3 and MNDO calculations of the heats of formation of 119 carbocations are reported.[111] Antiaromatic cyclopentadienyl cations have been studied.[112]

An MNDO study of some polyfluorinated allyl cations, viz. $[CF_2CR=CF_2]^+$ (R = H,F,Cl,and CF_3) and (*E*)- and (*Z*)-$[CF_2CR=CFCl]^+$ (R = F and CF_3), leads to a linear relation between $^4J_{FF}$ values and F····F distances.[113] An *ab initio* investigation of the fluorine-substituted ethylene dications $C_2H_nF_{4-n}{}^{2+}$ ($n = 0$-4) shows that $C_2F_4{}^{2+}$ and $C_2HF_3{}^{2+}$ are planar, but that the others are not.[114] Substituted methylene dications HCX^{2+}, with X = F, OH, NH_2, CH_3, Cl, SH, PH_2, and SiH_3, have had their geometries optimized at the MP2/6-31G* and higher levels; some extremely short C—X bonds were found.[115] The helium-containing species $CHe_3{}^{3+}$ and $CHe_4{}^{4+}$ were found to be remarkably stable, according to high-level *ab initio* calculations, and should be experimentally observable despite the availability of highly exothermic fragmentation processes.[116]

References

1 Montgomery, L. K., Grendze, M. P., and Huffman, J. C., *J. Am. Chem. Soc.*, **109**, 4749 (1987).
2 Laube, T., *Angew. Chem. Int. Edn*, **26**, 560 (1987).
3 Blanchette, M. C., Holmes, J. L., and Lossing, F. P., *J. Am. Chem. Soc.*, **109**, 1392 (1987).
4 Saunders, M. and Johnson, C. S., *J. Am. Chem. Soc.*, **109**, 4401 (1987).
5 Dewar, M. J. S. and Merz, K. M., *J. Am. Chem. Soc.*, **108**, 5634 (1986).
6 Yannoni, C. S., Macho, V., and Myhre, P. C., *J. Am. Chem. Soc.*, **104**, 7380 (1982).
7 Saunders, M. and Kates, M. R., *J. Am. Chem. Soc.*, **102**, 6867 (1980).
8 Jarret, R. M. and Saunders, M., *J. Am. Chem. Soc.*, **109**, 3366 (1987).
9 Saunders, M., Jarret, R. M., and Pramanik, P., *J. Am. Chem. Soc.*, **109**, 3735 (1987).
10 Servis, K. L., Domenick, R. L., Forsyth, D. A., and Pan, Y., *J. Am. Chem. Soc.*, **109**, 7263 (1987).
11 Forsyth, D. A., Botkin, J. H., Puckace, J. S., Servis, K. L., and Domenick, R. L., *J. Am. Chem. Soc.*, **109**, 7270 (1987).
12 Schindler, M., *J. Am. Chem. Soc.*, **109**, 1020 (1987).
13 Dewar, M. J. S., Healey, E. F., and Ruiz, J. M., *J. Chem. Soc., Chem. Commun.*, **1987**, 943.
14 Barrow, C. J., Bright, S. T., Coxon, J. M., and Steel, P. J., *J. Org. Chem.*, **52**, 5300 (1987).

[15] Surya Prakash, G. K., Fung, A. P., Olah, G. A., and Rawdah, T. N., *Proc. Natl. Acad. Sci. U.S.A.*, **84**, 5092 (1987).
[16] Saunders, M. and Jarret, R. M., *J. Org. Chem.*, **51**, 5427 (1986).
[17] Buckland, S. J., Halton, B., Mei, Q., and Stang, P. J., *Aust. J. Chem.*, **40**, 1375 (1987).
[18] Bartlett, P. D., Ravenscroft, M. D., and Roof, A. A. M., *J. Org. Chem.*, **52**, 1847 (1987).
[19] Loktev, V. F., Shakirov, M. M., and Shubin, V. G., *Zh. Org. Khim.*, **22**, 1110 (1986); *Chem. Abs.*, **105**, 225562 (1986).
[20] Stănescu, M., Elian, M., and Ciorănescu, E., *Rev. Roum. Chim.*, **31**, 1001 (1986).
[21] Kirmse, W., Kampmann, K.-H., and Zellmer, V., *Chem. Ber.*, **120**, 1301 (1987).
[22] Adam, W. and Dörr, M., *J. Am. Chem. Soc.*, **109**, 1570 (1987).
[23] Della, E. W. and Schiesser, C. H., *Tetrahedron Lett.*, **28**, 3869 (1987).
[24] Della, E. W., Pigou, P. E., and Tsanaktsidis, J., *J. Chem. Soc., Chem. Commun.*, **1987**, 833.
[25] Kirmse, W. and Streu, J., *J. Org. Chem.*, **52**, 515 (1987).
[26] Takeuchi, K., Akiyama, F., Miyazaki, T., Kitagawa, I., and Okamoto, K., *Tetrahedron*, **43**, 701 (1987).
[27] Coates, R. M. and Kang, H.-Y., *J. Chem. Soc., Chem. Commun.*, **1987**, 232.
[28] Müller, P. and Mareda, J., *Helv. Chim. Acta*, **70**, 1017 (1987).
[29] Pindur, U., Müller, J., Flo, C., and Witzel, H., *Chem. Soc. Rev.*, **16**, 75 (1987).
[30] Ta-Shma, R. and Jencks, W. P., *J. Am. Chem. Soc.*, **108**, 8040 (1986).
[31] Dietze, P. E. and Jencks, W. P., *J. Am. Chem. Soc.*, **109**, 2057 (1987).
[32] Lee, C. C. and Wanigasekera, D., *Can. J. Chem.*, **65**, 933 (1987).
[33] Förster, H., Seebode, J., Fejes, P., and Kiricsi, I., *J. Chem. Soc., Faraday Trans. 1*, **83**, 1109 (1987).
[34] Hamed, A., Müller, E., and Jochims, J. C., *Tetrahedron*, **42**, 6645 (1986).
[35] Frey, H., Mehlhorn, A., and Rühlmann, K., *Tetrahedron*, **43**, 2945 (1987).
[36] Haas, A. and Wanzke, W., *Chem. Ber.*, **120**, 429 (1987).
[37] Singer, B. and Maas, G., *Chem. Ber.*, **120**, 485 (1987).
[38] Singer, B. and Maas, G., *Chem. Ber.*, **120**, 1683 (1987).
[39] Gassman, P. G., Singleton, D. A., Wilwerding, J. J., and Chavan, S. P., *J. Am. Chem. Soc.*, **109**, 2182 (1987).
[40] Nilsson, Å., *Acta Chem. Scand.*, **41B**, 222 (1987).
[41] Kirmse, W. and Rode, K., *Chem. Ber.*, **120**, 839 (1987).
[42] Kirmse, W. and Rode, K., *Chem. Ber.*, **120**, 847 (1987).
[43] Snegirev, V. F., Galakhov, M. V., Petrov, V. A., Makarov, K. N., and Bakhmutov, V. I., *Izv. Akad. Nauk SSSR, Ser. Khim.*, **1986**, 1318; *Chem. Abs.*, **106**, 101659 (1987).
[44] Mayr, H. and Heilmann, W., *Tetrahedron*, **42**, 6657 (1986).
[45] Nishimura, J., Ohbayashi, A., Horiuchi, Y., Okada, Y., Yamanaka, S., and Oku, A., *J. Org. Chem.*, **52**, 1409 (1987).
[46] Mayr, H., Schade, C., Rubow, M., and Schneider, R., *Angew. Chem. Int. Edn*, **26**, 1029 (1987).
[47] Lee, S.-G., *J. Chem. Soc., Chem. Commun.*, **1987**, 1115.
[48] Schneider, R. and Mayr, H., *Angew. Chem. Int. Edn*, **25**, 1016 (1986).
[49] Mayr, H., Schneider, R., and Grabis, U., *Angew. Chem. Int. Edn*, **25**, 1017 (1986).
[50] Roussel, C. and Shaimi, L., *J. Chem. Soc., Chem. Commun.*, **1987**, 1315.
[51] Sinev, V. V. and Ivanov, Yu. E., *Org. React. (Tartu)*, **22**, 436 (1985); *Chem. Abs.*, **106**, 118977 (1987).
[52] Ivanov, Yu. E. and Sinev, V. V., *Org. React. (Tartu)*, **22**, 444 (1985); *Chem. Abs.*, **105**, 225519 (1986).
[53] Pindur, U. and Müller, J., *J. Heterocycl. Chem.*, **24**, 159 (1987).
[54] Anderson, V. E., Cheng, H. Z., Diebold, G. J., Mahmood, A., and Sweigart, D. A., *J. Am. Chem. Soc.*, **109**, 6191 (1987).
[55] Kampars, V., Bruvers, Z., Adamsone, B., Kampare, R., and Neilands, O., *Latv. PSR Zinat. Akad. Vestis, Kim. Ser.*, **1986**, 496; *Chem. Abs.*, **106**, 137828 (1987).
[56] Karpov, V. M., Mezhenkova, T. V., Platonov, V. E., and Yakobson, G. G., *Bull. Soc. Chim. Fr. II*, **1986**, 980.
[57] Takahashi, K., Namekata, N., Takase, K., and Suzuki, T., *J. Chem. Soc., Chem. Commun.*, **1987**, 935.
[58] Chikinev, A. V., Bushmelev, V. A., Shakirov, M. M., and Shubin, V. E., *Zh. Org. Khim.*, **22**, 1454 (1986); *Chem. Abs.*, **106**, 155692 (1987).
[59] Bäuml, E., Kolberg, G., and Mayr, H., *Tetrahedron Lett.*, **28**, 387 (1987).
[60] Crotti, P., Dell'Omodarme, G., Ferretti, M., and Macchia, F., *J. Am. Chem. Soc.*, **109**, 1463 (1987).
[61] Mahr, H., Heilmann, W., and Lammers, R., *Tetrahedron*, **42**, 6663 (1986).
[62] Schleyer, P. von R., Bentley, T. W., Koch, W., Kos, A. J., and Schwarz, H., *J. Am. Chem. Soc.*, **109**, 6953 (1987).

63 Bunting, J. W. and Moors, R. G., *Tetrahedron Lett.*, **28**, 3915 (1987).
64 Bunting, J. W., *Tetrahedron*, **43**, 4277 (1987).
65 Liu, M. T. H. and Paike, N., *Tetrahedron Lett.*, **28**, 3763 (1987).
66 Hashimoto, Y., Sugumi, H., Okauchi, T., and Mukaiyama, T., *Chem. Lett.*, **1987**, 1691.
67 Hashimoto, Y., Sugumi, H., Okauchi, T., and Mukaiyama, T., *Chem. Lett.*, **1987**, 1695.
68 Hayashi, Y., Wariishi, K., and Mukaiyama, T., *Chem. Lett.*, **1987**, 1243.
69 Pasto, D. J., Freeman, J. P., and Fields, M. K., *J. Org. Chem.*, **52**, 4058 (1987).
70 Ohwada, T., Ohta, T., and Shudo, K., *Tetrahedron*, **43**, 297 (1987).
71 Ohwada, T., Itai, A., Ohta, T., and Shudo, K., *J. Am. Chem. Soc.*, **109**, 7036 (1987).
72 Hellwinkel, D., Stahl, H., and Gaa, H. G., *Angew. Chem. Int. Edn*, **26**, 794 (1987).
73 Bremer, M., Schleyer, P. von R., Schötz, K., Kausch, M., and Schindler, M., *Angew. Chem. Int. Edn*, **26**, 761 (1987).
74 Hopf, H., Shin, J.-H., and Volz, H., *Angew. Chem. Int. Edn*, **26**, 564 (1987).
75 Ablenas, F. J., George, B. E., Maleki, M., Jain, R., Hopkinson, A. C., and Lee-Ruff, E., *Can. J. Chem.*, **65**, 1800 (1987).
76 Kikugawa, Y. and Shimada, M., *Chem. Lett.*, **1987**, 1771.
77 Richard, J. P., *J. Chem. Soc., Chem. Commun.*, **1987**, 1768.
78 Allen, A. D., Kwong-Chip, J. M., Mistry, J., Sawyer, J. F., and Tidwell, T. T., *J. Org. Chem.*, **52**, 4164 (1987).
79 Laurent, E., Marquet, B., Tardivel, R., and Thiebault, H., *Tetrahedron Lett.*, **28**, 2359 (1987).
80 Charpentier, M., Fossey, J., Tidwell, T. T., and Wolfe, S., *Can. J. Chem.*, **65**, 473 (1987).
81 Ritchie, C. D., *Can. J. Chem.*, **64**, 2239 (1986).
82 Ritchie, C. D., *Adv. Chem. Ser.*, **215** (Nucleophilicity), 169 (1987); *Chem. Abs.*, **107**, 153773 (1987).
83 Arnett, E. M., Molter, K. E., Marchot, E. C., Donovan, W. H., and Smith, P., *J. Am. Chem. Soc.*, **109**, 3788 (1987).
84 Shaik, S. S., *J. Org. Chem.*, **52**, 1563 (1987).
85 Jorgensen, W. L., Buckner, J. K., Huston, S. E., and Rossky, P. J., *J. Am. Chem. Soc.*, **109**, 1891 (1987).
86 Apeloig, Y. and Stanger, A., *J. Am. Chem. Soc.*, **109**, 272 (1987).
87 Kevill, D. N., *J. Chem. Res. (S)*, **1987**, 272.
88 Apeloig, Y., Karni, M., Stanger, A., Schwarz, H., Drewello, T., and Czekay, G., *J. Chem. Soc., Chem. Commun.*, **1987**, 989.
89 Shiner, V. J., Ensinger, M. W., and Rutkowske, R. D., *J. Am. Chem. Soc.*, **109**, 804 (1987).
90 Tanino, K., Hatanaka, T., and Kuwajima, I., *Chem. Lett.*, **1987**, 385.
91 Ohkata, K., Mase, M., and Akiba, K., *J. Chem. Soc., Chem. Commun.*, **1987**, 1727.
92 Ishibashi, H., Nakatani, H., Sakashita, H., and Ikeda, M., *J. Chem. Soc., Chem. Commun.*, **1987**, 338.
93 Chapman, S., Kane-Maguire, L. A. P., and Kanitz, R., *J. Organomet. Chem.*, **329**, C11 (1987).
94 England, K. and Reeves, P. C., *J. Organomet. Chem.*, **319**, 355 (1987).
95 Odiaka, T. I., *J. Chem. Soc., Dalton Trans.*, **1986**, 2707.
96 Pearson, A. J. and O'Brien, M. K., *J. Chem. Soc., Chem. Commun.*, **1987**, 1445.
97 Schreiber, S. L., Klimas, M. T., and Sammakia, T., *J. Am. Chem. Soc.*, **109**, 5749 (1987).
98 Nicholas, K. M., *Acc. Chem. Res.*, **20**, 207 (1987).
99 Alexander, R. P. and Stephenson, G. R., *J. Chem. Soc., Dalton Trans.*, **1987**, 885.
100 Palotai, I. M., Stephenson, G. R., and Kane-Maguire, L. A. P., *J. Organomet. Chem.*, **319**, C5 (1987).
101 Sinotova, E. N., Nefedov, V. D., Skorokhodov, S. S., and Arkhipov, Yu. M., *Radiokhimiya*, **29**, 69 (1987); *Chem. Abs.*, **107**, 153745 (1987).
102 Nefedov, V. D., Sinotova, E. N., Kuzhelev, L. P., and Lebedev, V. P., *Radiokhimiya*, **29**, 73 (1987); *Chem. Abs.*, **107**, 153858 (1987).
103 Fornarini, S. and Speranza, M., *J. Phys. Chem.*, **91**, 2154 (1987).
104 Okada, S., Abe, Y., Taniguchi, S., and Yamabe, S., *J. Am. Chem. Soc.*, **109**, 295 (1987).
105 Mishima, M., Arima, K., Usui, S., Fujio, M., and Tsuno, Y., *Chem. Lett.*, **1987**, 1047.
106 Fernandez, M. T., Jennings, K. R., and Mason, R. S., *J. Chem. Soc., Faraday Trans. 2*, **83**, 159 (1987).
107 Mason, R. S., Fernandez, M. T., and Jennings, K. R., *J. Chem. Soc., Faraday Trans. 2*, **83**, 89 (1987).
108 Rodriquez, C. F., and Hopkinson, A. C., *Can. J. Chem.*, **65**, 1209 (1987).
109 Glaser, R., *J. Am. Chem. Soc.*, **109**, 4237 (1987).
110 Viruela-Martin, P. M., Nebot-Gil, I., Tomas, F., and Viruela-Martin, R., *J. Comput. Chem.*, **7**, 417 (1986); *Chem. Abs.*, **106**, 101564 (1987).
111 Halim, H., Heinrich, N., Koch, W., Schmidt, J., and Frenking, G., *J. Comput. Chem.*, **7**, 93 (1986); *Chem. Abs.*, **105**, 225475 (1986).
112 Mehlhorn, A., Fabian, J., Tyutyulkov, N., Olbrich, G., and Dietz, F., *J. Prakt. Chem.*, **328**, 603 (1986); *Chem. Abs.*, **106**, 195598 (1987).

[113] Borisov, Yu. A., Galakhov, M. V., and Bakhmutov, V. I., *Izv. Akad. Nauk SSR, Ser. Khim.*, **1986**, 1498; *Chem. Abs.*, **106**, 155597 (1987).
[114] Frenking, G., Koch, W., and Schwarz, H., *J. Comput. Chem.*, **7**, 406 (1986); *Chem. Abs.*, **106**, 101474 (1987).
[115] Wong, M. W., Yates, B. F., Nobes, R. H., and Radom, L., *J. Am. Chem. Soc.*, **109**, 3181 (1987).
[116] Wong, M. W., Nobes, R. H., and Radom, L., *J. Chem. Soc., Chem. Commun.*, **1987**, 233.

Organic Reaction Mechanisms 1987
Edited by A. C. Knipe and W. E. Watts

CHAPTER 10

Nucleophilic Aliphatic Substitution

J. SHORTER

Department of Chemistry, The University, Hull

Vinylic Systems

Bimolecular nucleophilic substitution at a vinylic centre has been reviewed, including quantum-chemical aspects, stereochemistry, and structure-activity relations.[1]

Rappoport's extensive studies of vinylic substitution have continued with a study of the [monochloro]/[monobromo] substitution product ratios for the attack of 9-(bromochloromethylene)fluorene (**1**) and β,β-bis(*p*-nitrophenyl)-α-bromo-α-chloroethylene (**2**) by *p*-toluenethiolate and *p*-cresolate ions in CD_3CN-DMSO-d_6 and CD_3OD-DMSO-d_6 mixtures.[2] The results were interpreted in terms of a multistep nucleophilic vinylic substitution proceeding via an intermediate carbanion, the product ratios thus being identified as ratios of rate constants for expulsion of Br^- and Cl^- (the 'intramolecular element effect') from the carbanion. In a further study, the stereochemistry of substitution of methyl trideuteriomethyl (*E*)-

(α-bromo-*p*-methyl- and -*p*-nitrobenzylidene)malonates (**3** and **4** respectively) with various nucleophiles was examined in a variety of solvents by NMR spectroscopy.[3] It was found that all reactions proceeded with a high preference for retention of configuration.

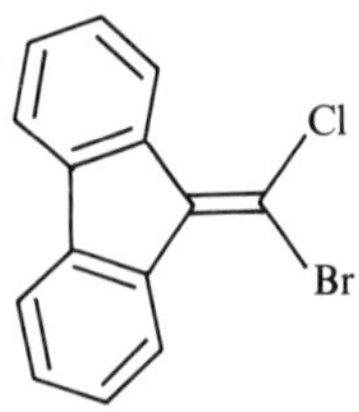

(**1**)

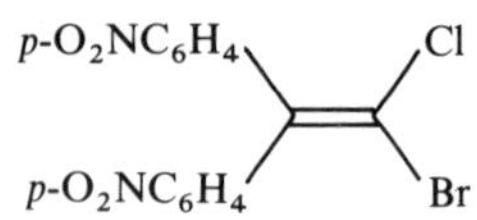

(**2**)

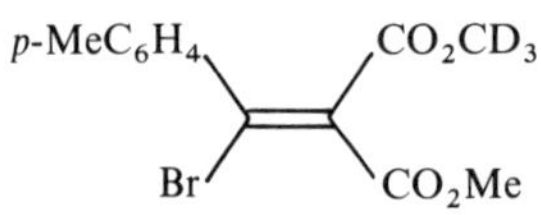

(**3**)

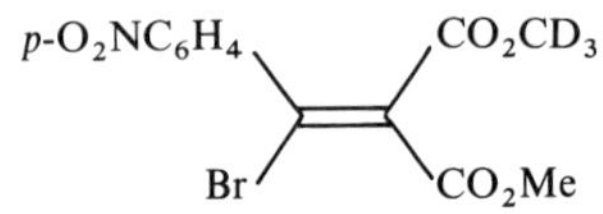

(**4**)

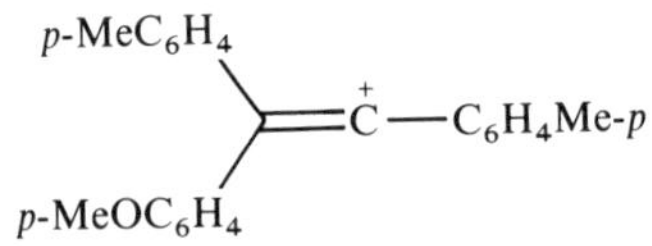

(**5**)

C. C. Lee's long series of studies of solvolytic rearrangements of ^{14}C- or ^{13}C-labelled triarylvinyl halides has continued with work on (*E*)- and (*Z*)-2-anisyl-1,2-ditolyl[2-^{13}C]vinyl bromides.[4] This has provided information on the behaviour of (**5**), the last to be studied of the nine triarylvinyl cations with various combinations of phenyl, tolyl, and anisyl groups capable of giving rise to degenerate 1,2-aryl shifts across the double bond.

Further details have now appeared of the synthesis of the cyclic dienynyl trifluoromethanesulphonates (**6**) and (**7**), and their solvolyses to cyclized aromatic products via the aryl cations (**8**) and (**9**).[5]

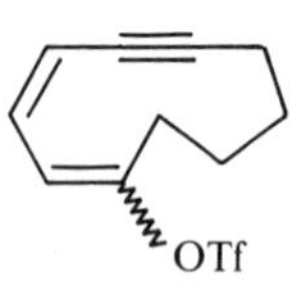

(**6**)

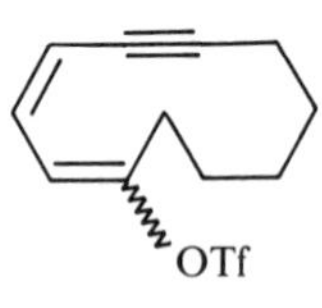

(**7**)

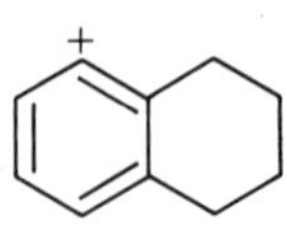

(**8**)

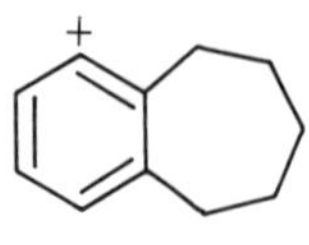

(**9**)

Rate constants and activation parameters have been determined for the reactions of 2-(β,β-dihalovinyl)-5-nitrofurans with piperidine.[6]

Allylic and Various Unsaturated Systems

Kinetic and product studies have been carried out for the reactions of 9-substituted fluorenide carbanions with allylic chlorides in DMSO.[7] The rate measurements, interpreted through Brønsted plots, yielded information on the steric effects of the 9-substituent. The product studies indicated that $CH_2{=}CHCH_2Cl$ reacts by an S_N2 mechanism, $CH_2{=}CHCHMeCl$ gives both S_N2 and S_N2' products, and $CH_2{=}CHCMe_2Cl$ appears to give only S_N2' products.

2-(phenylsulphonyl)-3-bromo-1-phenyl-1-propene (**10**) and 3-bromo-2-(methylsulphonyl)-1-phenyl-1-propene (**11**) react with various amines to give mixtures of S_N2' and S_N2 products in ratio depending on the amine.[8] For example, the S_N2'/S_N2 ratio ranged from approximately 20:1 for reaction of (**10**) with piperidine to 1:9 for reaction of (**10**) with diisopropylamine.[8a] Alkoxide, phenoxide, and thiolate nucleophiles attack 3-chloro-1-methoxypropene exclusively at C(1) in the presence of *N*,*N*-diisopropylethylamine.[9] The reaction is formally a highly regioselective S_N2' process. Evidence has been obtained for *syn* stereochemistry in an intramolecular S_N' reaction of stereoisomeric allylic acetates.[10]

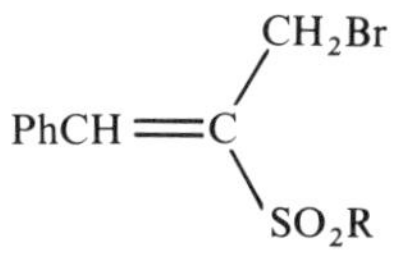

(**10**) R = Ph

(**11**) R = Me

The role of metallic compounds in allylic reactions continues to arouse some mechanistic interest. The reaction of *n*-butylmagnesium bromide-copper(I) iodide with 12-, 14-, and 16-membered optically active (*R*)-2-methylenecycloalkylidene epoxides in THF-dimethyl sulphide yielded S_N2' products, with the *trans* isomer predominating.[11] Related work on $LiMe_2Cu$ reacting with vinyloxiranes found (*E*)-allylic alcohol products formed by *anti* S_N2'.[12] Nucleophiles have been classified under two headings according to whether they attack Pd(0)-allylic complexes at the allylic ligand or the metal.[13] The reaction between Cu(II) perchlorate and metallic copper to form Cu(I) perchlorate is accelerated by allylic chlorides and acetates, and the Cu(I) perchlorate is in turn a very good catalyst for nucleophilic substitution of these substrates.[14]

The sulphonyl group in allylic sulphones, particularly tertiary sulphones, is easily displaced by various nucleophiles and the reaction is catalysed by Lewis acids.[15] Nucleophilic displacement of the 4-arylamino group of a number of 2,4-di(arylamino)cyclopent-2-enones has been studied.[16]

Norbornyl and Closely Related Systems

Molecular-mechanics calculations of strain in carbenium ions have been tested on Bentley's unified reactivity scale for bridgehead solvolysis.[17] Good correlations were obtained for various force-fields and the approach was extended to several rigid but planar cations. The behaviour of the cation derived from the 2-methyl-2-*endo*-norbornyl derivative (**12**) was adequately correlated but the *exo* isomer (**13**) showed enhanced reactivity by a factor of *ca.* 10^2 to 10^3.

The contributions from C. A. Grob's group have continued. Log *k* values for the solvolysis (in 80% v/v ethanol) of 4- and 5-*exo*-substituted 2-*exo*- and 2-*endo*-norbornyl tosylates, (**14**) to (**17**), correlate rectilinearly with the inductive constants σ_I^q for the substituents.[18] Comparison of the reaction constants ρ_I with those obtained previously for the corresponding 6- and 7-substituted compounds shows that inductivity is greater for 2-*exo* ionization than for 2-*endo* ionization in all pairs. This was attributed to the more favourable alignment of neighbouring carbon atoms for dorsal participation in *exo* ionization. In a further study, two methyl groups at C(6) of 2-*exo*- and 2-*endo*-norbornyl tosylate are shown to reduce the solvolysis rates in 80% ethanol by factors of 28 and 16, respectively.[19] The corresponding figures for inclusion of C(6) in a spirocyclopropyl group are 250 and 8, respectively. These findings are explained in terms of the inductive model.

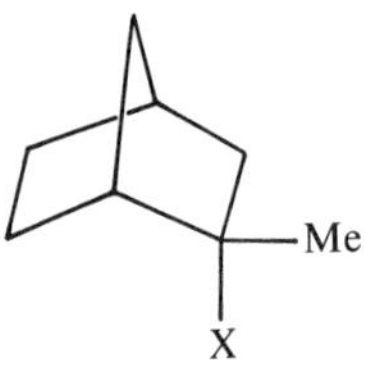

(**12**)

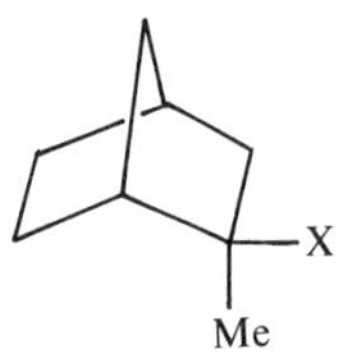

(**13**)

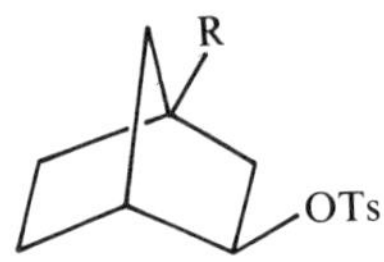

(**14**)

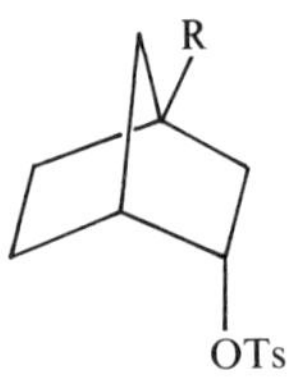

(**15**)

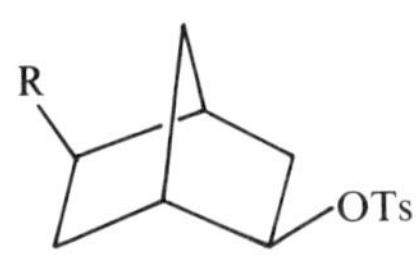

(**16**)

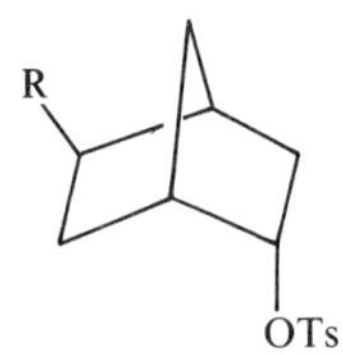

(**17**)

endo-2,3-Epoxy-1,7,7-trimethylbicyclo[2.2.1]heptane reacts with Me_3SiCN and zinc iodide to give 72% of *anti*-7-trimethylsiloxy-*endo*-2,3,3-trimethyl-*exo*-2-isocyanobicyclo[2.2.1]heptane and eight minor products.[20] All could be rationalized in terms of the initial generation of a carbocation. In contrast to the bridgehead derivative 1-bromobicyclo[2.2.1]heptane, it has been found that 1-bromobicyclo[3.1.1]heptane is very reactive even under mild solvolytic conditions.[21] The bridgehead carbenium ion appears to be stabilized by transannular orbital interactions. Secondary and tertiary alkyl and alicyclic chlorides, including various norbornyl and adamantyl derivatives, may be converted into the corresponding bromides by HBr in the presence of small amounts of anhydrous iron(III) bromide.[22]

endo- and *exo*-Benzobicyclo[3.2.1]octen-3-yl chlorides, (**18**) and (**19**), may be photolysed (254 nm light in methanol) to give a variety of products derived from carbocationic intermediates.[23] The *endo* isomer is more reactive than the *exo*, in contrast to the behaviour of the [2.2.1] and [2.2.2] series. Further studies have been made of the solvolysis products of the tetracyclononane (**20**) and the tetracyclodecane (**21**) in order to elucidate the effects of the fused cyclopropane rings.[24] The application of 2D NMR and molecular mechanics was involved.

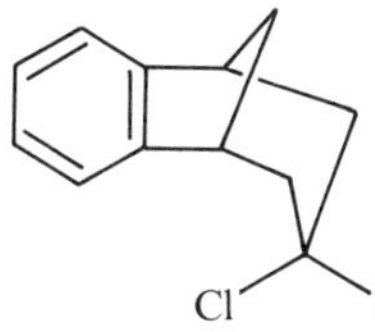

(**18**)

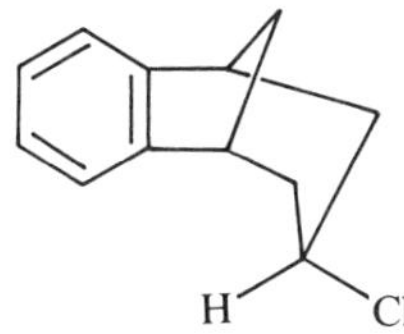

(**19**)

(**20**)

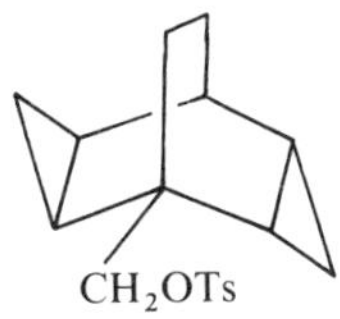

(**21**)

Rates and products of solvolysis in the presence of silver ion have been determined for four tropenium ion analogues of bromobenzonorbornenes (as their hexafluoroantimonates) (**22**) to (**25**).[25] The results were held to demonstrate anchimeric assistance by the tropenium moiety and this was said to be the first example of participation by this aromatic system. Studies of acid-catalysed rearrangements of 5,6-epoxy-7-oxabicyclo[2.2.1]hept-2-yl derivatives have shed light on the migratory aptitudes of acyl versus alkyl groups in Wagner–Meerwein transpositions.[26] The study of solvolyses of derivatives of 5,6,7,8-tetrahydro-5,8-methanoquinoline has complemented the findings of previous work on related systems in respect of participation by pyridine and benzene rings.[27]

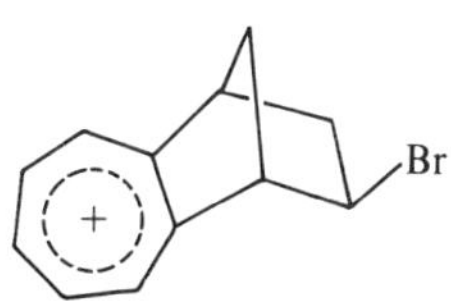

(**22**)

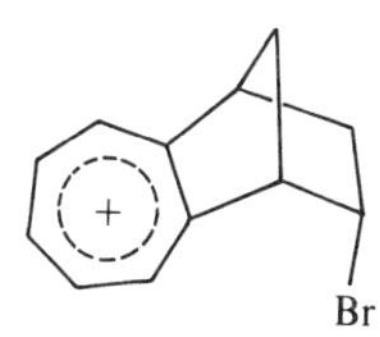

(**23**)

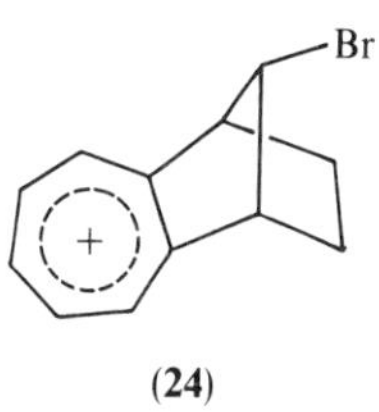

(**24**)

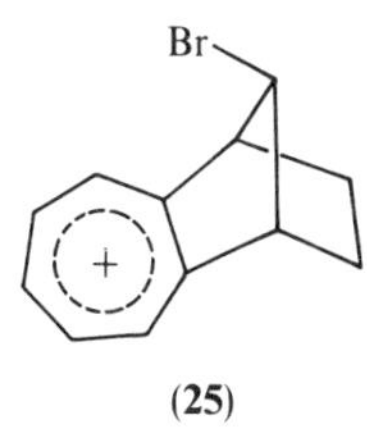

(**25**)

Miscellaneous Polycyclic Systems

One γ-trimethylsilyl substituent in 1-bromoadamantane enhances the rate of solvolysis in 80% ethanol by a factor of 8.6; a second such substituent by a further factor of 3.8 (see **26a** and **b**).[28] Log k values for the solvolysis of the δ-substituted 1-adamantyl tosylate (**27**) and 2-adamantyl p-nitrobenzenesulphonate (**28**) in 80% ethanol are well correlated by Grob's σ_I^q values, giving ρ_I values of -0.97 and -1.02, respectively.[29] (These are considered identical within the error limits.) The inductivity for the 5-position in the 2-adamantyl system is thus greater than for the less remote 4-position, for which ρ_I is -0.80.

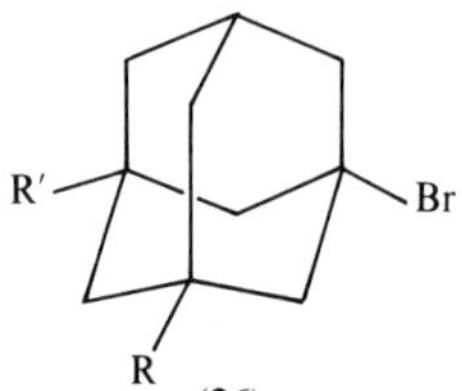

(26)

a; R = Me_3Si, R′ = H

b; R = R′ = Me_3Si

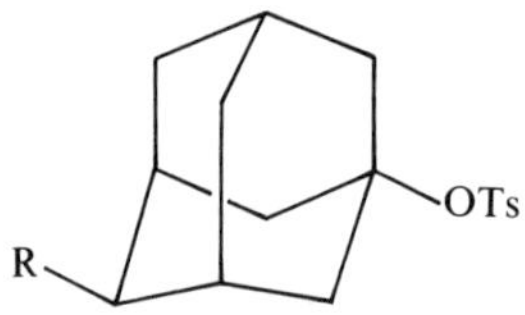

(27)

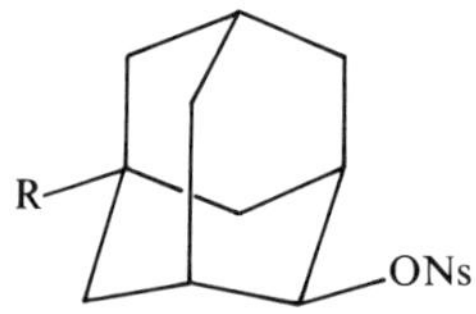

(28)

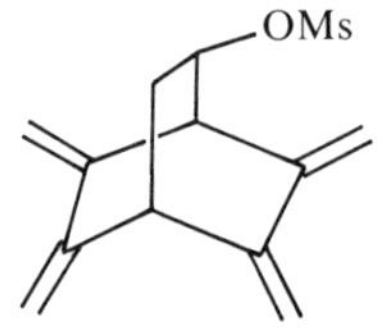

(29)

The treatment of the epimeric 4,4-dimethyl-2-adamantyl tosylates with tributylhexadecylphosphonium azide in aprotic solvents yielded the displacement products with a high degree of inversion, contrasting with the behaviour in polar protic solvents in which retention was favoured.[30] This and other evidence indicates the possibility of S_N2 processes for 2-adamantyl derivatives. The solvolyses of 5,6,7,8-tetramethylidenebicyclo[2.2.2]oct-2-yl methanesulphonate (**29**) and its tricarbonyliron mono- and di-nuclear complexes have been studied.[31] The reactions of the complexes can be highly chemo- and stereo-selective, depending on configuration and medium. The transient formation of a hemiacetal is involved in the cyclization of 3,3;20,20-bis(ethylenedioxy)-C-nor-9,11-seco-5α-pregnane-9β,11-diol to 11-oxa-5α,17α-pregnane-3,20-dione by refluxing in acetone with a catalytic amount of toluene-p-sulphonic acid.[32]

Epoxide Reactions

The effect of substituents on the stereoselectivity of acid-catalysed solvolysis of the conformationally rigid diastereoisomeric 4-*tert*-butyl-1-arylcyclohexene oxides has been studied.[33] The electron-donating *p*-methyl substituent increases the *syn/anti* adduct ratios in ring-opening, while the electron-withdrawing *m*-chloro substituent reduces it. Hammett–Brown σ^+ correlations were applied and the nature of benzylic carbocationic intermediates was discussed. The same research group has found anomalous stereochemistry for the gas-phase acid-induced ring-opening of 1-phenylcyclohexene oxide.[34]

The ring-opening of 1,2-epoxycyclohexane with sodium 1,3-diphenyl-2-azapropenide led to a variety of products, which indicated that the reagent behaved 'both as an α-benzylamine carbanion equivalent and as a benzoyl anion equivalent'.[35]

The ring-opening of *erythro*- and *threo*-2,3-epoxy-alcohols has been studied for a variety of nucleophiles, both in the presence and absence of titanium tetraisopropoxide.[36] In general, opening of the *threo* compounds proceeded faster and with higher selectivity than the same reaction with the corresponding *erythro* isomers. The opening of 2,3-epoxy alcohols with organocuprates shows enhanced C(2) selectivity when carried out in coordinating solvents, such as mixtures of THF and tetramethylethylenediamine.[37]

Complexes of early transition metals (Ti, V, etc.) catalyse the ring-opening of epoxides by trimethylsilyl azide.[38] Palladium and ruthenium complexes catalyse the ring-opening of oxiranes by aromatic acyl chlorides.[39] When 1,2-epoxy-1-vinylcyclopentane reacts with methanol, Hg(II) nitrate (as a Lewis acid) promotes the opening of the oxirane ring, while Hg(II) acetate favours oxymercuration of the double bond.[40] Orthomolybdate and orthotungstate ions (MO_4^{2-}, M = Mo, W) are selective catalysts for the hydrolysis of ethylene oxide.[41]

The transition-state structure for the ring-opening of ethylene oxide by *trans* attack of NH_2^- has been studied by *ab initio* MO methods.[42] MNDO studies have been made on the gas-phase reaction of dypnone oxide with chloride ion.[43]

Kinetic studies involving epoxides have included a modelling of the kinetics of the non-catalytic hydration of ethylene oxide,[44] the reaction of propylene oxide with propionic acid,[45] the reactions of mercaptide ions in the β-hydroxyethyl mercaptan-triethylamine system with olefin oxides,[46] the reactions of piperazine with 1-substituted 2,3-epoxypropanes,[47,48] the reactions of oxiranes with bis(2-hydroxyalkyl) disulphides and of bis(2-hydroxyalkyl) sulphides in the presence of *N,N*-dimethylcyclohexylamine,[49] and the reactions of some acid esters with epichlorohydrin.[50]

Functionalized bicyclic dioxopiperazines have been synthesized via intramolecular epoxide opening.[51] The stereochemistry of ring-opening has been examined for chiral epoxides derived from allylic alcohols having two substituted phenyl groups.[52] Dimethylboron bromide effects the cleavage of cyclic ethers (3- to 7-membered rings) under mild conditions to give the corresponding bromo-alcohols.[53] In particular, 2-substituted tetrahydrofurans were cleaved regioselectively by a predominantly S_N2 mechanism favouring the formation of primary rather than secondary bromides.

Enantiomorphous tri-*o*-thymotide clathrates of prochiral oxiranes were treated with gaseous hydrogen halides and showed ring-opening reactions which differed from those occurring in homogeneous phase.[54]

In the area of carcinogens, bifunctional catalysis has been found in the nucleotide catalysed hydrolysis of (±)-7β,8α-dihydroxy-9α,10α-epoxy-7,8,9,10-tetrahydrobenzo[a]pyrene.[55] The same research group has also examined the pH dependence of the mechanism of hydrolysis of benzo[a]pyrene-*cis*-7,8-diol 9,10-epoxide catalysed by DNA, Poly(G), and Poly(A).[56]

4′-Phenylchalcones, chalcone oxides, and related compounds have been tested as inhibitors of cytosolic epoxide hydrolase, microsomal epoxide hydrolase, and glutathione *S*-transferases from mouse and rat liver.[57]

Other Small Rings

A long series of papers on deamination reactions has continued with a study of the decomposition of 1-alkylcyclopropanediazonium ions.[58] For 1-methyl, 1,2-dimethyl- (both isomers), and 1-cyclopropyl compounds in methanol, cyclopropyl-allyl transformation and nucleophilic displacement are the only reactions, but for 1-propyl and 1-isopropyl compounds there is competition from elimination and 1,2α-hydrogen shifts. The decomposition of 1-arylcyclopropanediazonium ions has been interpreted in terms of competing k_s, k_c, and k_Δ reactions.[59]

The roles of aminocyclopropyl-substituted CH acids as substrates[60] and of 2-cyclopropylidene-1,3-cycloalkanediones as intermediates[61] in nucleophilic substitution at a three-membered ring have been examined. In Ag^+-assisted solvolysis of the trichlorocyclopropane (**30**), the intermediate cyclopropyl cation is stabilized by the electron-donating aromatic ring.[62] CASSCF and contracted CCI MO calculations have been performed on the ring-opening of cyclopropane by various Pd(II) complexes.[63] Asymmetric induction has been studied in the nucleophilic cyclopropane ring-cleavage by vitamin B_{12s}.[64]

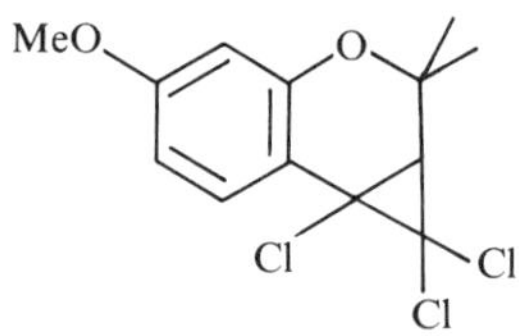

(**30**)

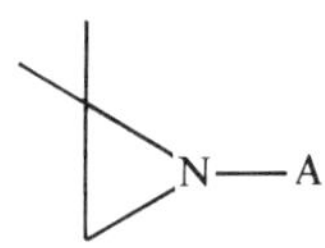

(**31**)

A long series of papers on the reactions of aziridines has continued with a study of alcoholyses of activated aziridines with respect to site selectivity and competing side-reactions as influenced by the basicity of the leaving-group and the size of ROH.[65] In a further study, the ring-opening of 2,2-dimethylaziridines (**31**; various groups A) by Grignard reagents RMgHal (Hal = Br or I; various groups R) has been examined.[66] Product studies suggest that the ring is opened by Lewis acid-assisted attack of Hal^- on the tertiary carbon atom. The fluorene anion opens sulphonyl-activated 2,2-dimethylaziridine (**31**; A = Ts) at the unsubstituted carbon atom, but the anions of fluorene and of carbazole attack the carbonyl-activated substrate (**31**; A = ArCO) at the tertiary carbon atom.[67] A further study of ring-opening of 1-acyl-2,2-

dimethylaziridines involved carbanions of the β-dicarbonyl type.[68] The influence of the size of the nucleophile on the regioselectivity was examined. Differences in the behaviour of *cis-trans* pairs have been shown for the ring-opening of stilbene oxides and activated stilbene imines.[69]

Another series of studies also involves aziridines. 1-(2-Aminoethyl)aziridine reacts with ethylenediamine, piperazine, and ammonia in the presence of their respective hydrochlorides by a mechanism in which the substrate is protonated on the ring N.[70] The kinetics of aziridine ammonolysis have been investigated.[71] The role of solvation in this process has been subjected to quantum-chemical study through non-empirical MO calculations (STO-3G)[72] and through the MINDO/3 method.[73] Comparison was made with the behaviour of oxirane.

7-Methoxyindano[1,2-b]aziridine (**32a**) undergoes hydrolysis at pH 8.88 approximately 250 times faster than the isomeric 6-methoxy compound (**32b**).[74] Protonation of the aziridine precedes the rate-limiting ring-opening at C(1) to generate a benzylic cation which is conjugatively stabilized by a 7-OMe group. Reactions of the nitrogen mustard 2-chloro-*N*-(2-chloroethyl)-*N*-methylethanamine with nucleophiles in aqueous solution have been shown to proceed via the *N*-2-chloroethyl-*N*-methylaziridium ion.[75]

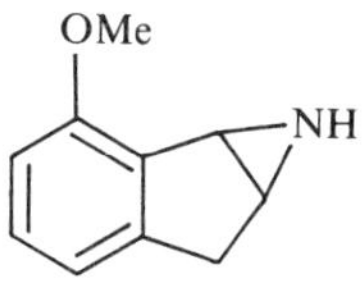

(**32a**)

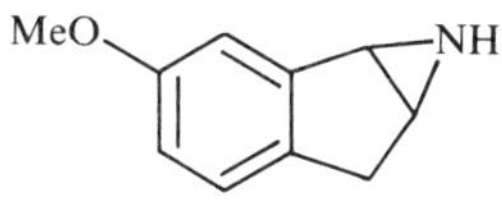

(**32b**)

The attack of ammonia on thiirane has been examined by *ab initio* MO calculations (3.21 G*) and the concerted reaction was found to be favoured.[76] A thiiranium ion formed in two different ways reacts with competing nucleophiles under identical conditions to form the same products but in different proportions.[77] Possibly different paths involving different kinds of ion-pairs are involved. The conversion of *S*-(2-hydroxyethyl)-L-cysteine into the corresponding chloro-, bromo-, or iodo-ethyl compound in strong acid proceeds via an intermediate thiiranium ion.[78]

Substitution at Elements Other than Carbon

Minimum-strain-energy conformations from molecular mechanics have been used in an MO treatment designed to model inversion and retention pathways for nucleophilic substitution at tetracoordinate silicon, with fluorine or chlorine as leaving-group.[79] Attacking nucleophile prefers axial entry and leaving-group prefers axial departure, compared with equatorial entry and departure, respectively. Also, retention via pseudorotation is of lower energy than retention via equatorial attack and axial departure. In a further paper, MO calculations are presented for regular trigonal-bipyramidal and square-pyramidal conformations of $[SiH_3XY]^-$, a proposed five-coordinate transition state, with X = H and Y = OH, SH, Cl, F;

X = Cl, F, OH, SH and Y = Cl, F.[80] The results of the calculations were correlated with experimental data on retention and inversion at silicon.

MNDO and *ab initio* MO calculations have been made for gas-phase alkoxide exchange at C and Si centres.[81] Facile reaction at silicon compared with carbon is due to easy valence-shell expansion of the former.

The rates of methanolysis of $(Me_3Si)_3CSiH(C_6H_4Y\text{-}p)I$ (Y = OMe, Me, H, Cl, or F) in 1:1 v/v methanol/dioxan are increased by electron-withdrawing Y and decreased by electron-releasing Y substituents.[82] 'This is taken to imply that the methanol is covalently involved in the transition state, and thus that, contrary to an earlier suggestion, the reaction cannot have an S_N2 (intermediate) mechanism.' The kinetics of the acid-catalysed hydrolysis of 1-chloro-1-cyclohexyl-3,3-dimethyl-1,3-diphenyl-disiloxane and -disilamethylene have been investigated.[83]

The mechanism and stereochemistry of nucleophilic substitution at sulphinyl and sulphonyl centres have been reviewed and some new kinetic studies of isotopic chloride-exchange in sulphonyl chlorides have also been presented.[84] Sulphonyl-transfer mechanisms have also been briefly reviewed and two recent studies described in greater detail.[85] The latter include the hydrolysis of 2-hydroxyethanesulphonyl chloride, which proceeds by way of β-sultone. Kinetic studies of the reactions of Y-substituted phenylmethanesulphonyl halides with X-substituted anilines in MeOH-MeCN mixtures have been made.[86] The fluorides had markedly lower rates and smaller magnitudes of ρ_X and ρ_Y compared with the chlorides, but the magnitude of the cross-interaction constant $|\rho_{XY}|$ for the former was the greater. The mechanism is considered to be associative S_N2.

Equilibrium constants were determined for thiol-disulphide interchange between 36 di- and tri-thiols and the disulphides derived from 2-mercaptoethanol or dithiothreitol.[87] Various structure-reactivity relationships were developed. By use of the highly hindered sulphonates (**33**), S—O scission in S_N2 reactions can be greatly reduced as compared with C—O scission.[88]

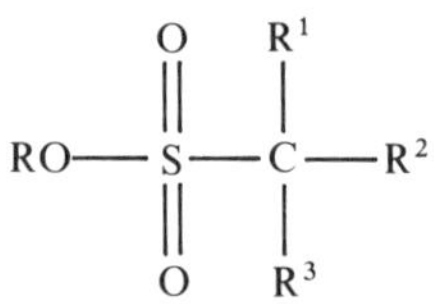

(**33**)

	R^1	R^2	R^3
a;	Me	Me	Me
b;	CF_3	Ph	Ph

The reaction of hydroxy compounds ROH with acids HX in the presence of triphenylphosphine and diethyl azodicarboxylate (Mitsunobu reaction) is widely used for the functionalization of alcohols and related compounds.[89] A revised mechanism for this has been proposed, involving alternative pathways for formation of the ultimate intermediate $Ph_3\overset{+}{P}(OR)$ by nucleophilic attack of ROH on phosphorus in earlier intermediates.

The reaction between (α-halomethyl)triphenylphosphonium salts and some sulphur nucleophiles giving $Ph_3\overset{+}{P}CH_2SR\ X^-$ is not a normal S_N2 reaction.[90] It is a three-step chain nucleophilic substitution in which attack of RS^- on $Ph_3\overset{+}{P}CH_2X$ (giving RSX) and on R*S*X (giving RSSR) is involved. Perhalofluoroalkanes react readily with many kinds of nucleophiles.[91] Most of the reactions proceed via anionic chain processes initiated by different types of halophilic attacks. Solvent effects have been studied for the reactions of various perfluoroalkanes with PhS^-.[92]

Intramolecular Substitution

Diethoxytriphenylphosphorane initiates stereospecific cyclodehydrations of bis(2-hydroxyethyl) sulphides (**34**) to the corresponding 1,4-oxathianes (**36**).[93] Mechanistic details for various examples were probed by specific deuterium labelling in combination with ^1H and ^{13}C NMR. There is strong support for initial phosphoranylation of the primary OH, followed by loss of ethanol to form the betaine (**35**), and subsequent intramolecular displacement of triphenylphosphine oxide by secondary alkoxide ion. The use of $Ph_3P—CCl_4$ as a reagent was also examined.

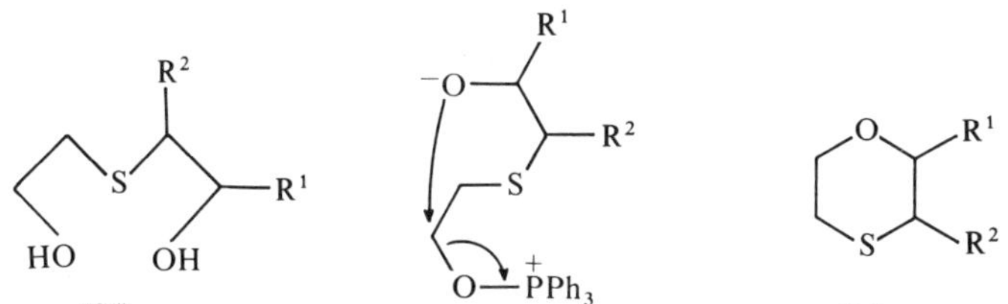

(**34**) (**35**) (**36**)

Tertiary (2-hydroxyethyl)dialkylamines (**37**) react with epichlorohydrin to yield equal amounts of 2-(hydroxymethyl)-4,4-dialkylmorpholinium chlorides (**38**) and perhydro-6-hydroxy-4,4-dialkyl-1,4-oxazepinium chlorides (**39**).[94] The corresponding reactions of secondary amines ($R^1 = H$) give a 9:1 ratio of the 6- and 7-membered ring products, in accordance with the predictions of Baldwin's rules. For the ring-closure of 1-(benzylamino)-3-chloro-2-(trialkylsiloxy)propanes to 1-benzyl-3-(trialkylsiloxy)azetidines, there is little or no advantage in using substituents larger than trimethylsilyl.[95]

$HOCH_2CH_2NR^1R^2$

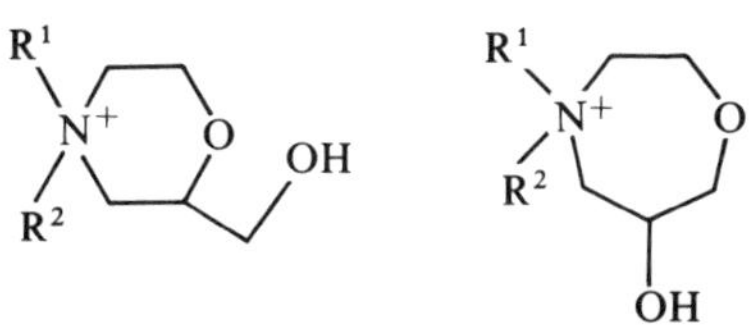

(**37**) (**38**) (**39**)

The base-catalysed cyclization of ω-epoxy 1,1-bis-sulphones (**40**) can give two different types of product, (**41**) and (**42**), depending on the value of *n*.[96] The rates of ring-closure, yielding 3- to 7-membered rings, span about three orders of magnitude. Kinetic studies have shown that haloalkylphosphines $Ph_2P(CH_2)_{n-1}Cl$ undergo anchimerically-assisted cyclization to Ph_2P^+ $(CH_2)_{n-1}$ with ring-size preference $5>6>3>4$.[97] Phenylthio and diphenylphosphino groups have similar intramolecular nucleophilicities. A study of the leaving-group effect (k_{Br}/k_{Cl}) has helped to shed light on the question: 'Why are strained small rings so easily formed in intramolecular nucleophilic substitution reactions?'[98] It is tentatively suggested that 'the adverse effect of ring strain is partially offset by a significant reduction of non-bonded interactions in the transition states'.

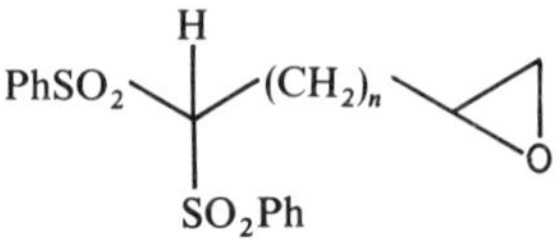

(40)

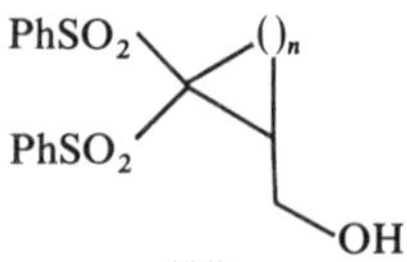

(41)

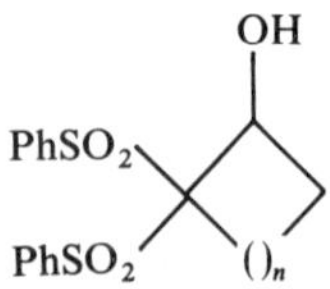

(42)

$n=1$ to 4

The kinetics of base-catalysed cyclization of 1,5-dianilino-2,4-diphenylpenta-1,4-dien-3-one to substituted 4-pyridone and aniline have been rationalized in terms of a fast proton abstraction followed by a rate-determining ring-closure.[99] The kinetics of cyclization and muscarinic actions of a (2-bromoethyl)methylamino analogue of oxotremorine have been investigated.[100]

Whereas some nucleophiles such as I^- and SCN^- attack 2-isothiocyanatobenzyl bromide at CH_2Br, primary or secondary amines, or phenolates react exclusively with the NCS group to form unstable thioureas, which then cyclize to the corresponding 4*H*-benzo[*d*][1,3]thiazines.[101] Pyridine, however, reacts to give 2-isothiocyanatobenzylpyridinium bromide;[102] this will, in turn, react with amines at NCS to give products which undergo ring-closure, pyridine being the leaving-group.

Anchimeric Assistance

Aryl-assisted halogen-exchange and rearrangements of 2-bromo-1-chloro-3-arylpropanes and 1-bromo-3-chloro-2-arylpropanes have been investigated in great detail.[103] In the case of 1,2-dibromo-3-arylpropanes, there is evidence for a competitive bromine-assisted pathway, i.e. there are bromine-bridged bromonium-ion intermediates as well as aryl-bridged benzenium-ion intermediates. Comparative studies of the solvolysis of neryl (**43**) and geranyl (**44**) 2,4-dinitrophenolates have shown the participation of the remote double bond in the neryl system.[104] The solvolyses of medium-ring 2-cycloalkyn-1-ylmethyl sulphonates (**45**) in solvents of various ionizing powers proceed exclusively with participation of the triple bond.[105]

Kinetic and product studies of the pyrolysis of 4-chloro-1-butanol in the gas phase indicate the participation of the OH group in stabilizing a carbocation in an intimate ion-pair.[106]

A series of studies of possible neighbouring-group participation in allyl derivatives has continued with kinetic and product investigation of the solvolysis of 2-(ω-alkoxyalkyl)-3-methyl-2-cyclohexenyl *p*-nitrobenzoates.[107] Rates, secondary

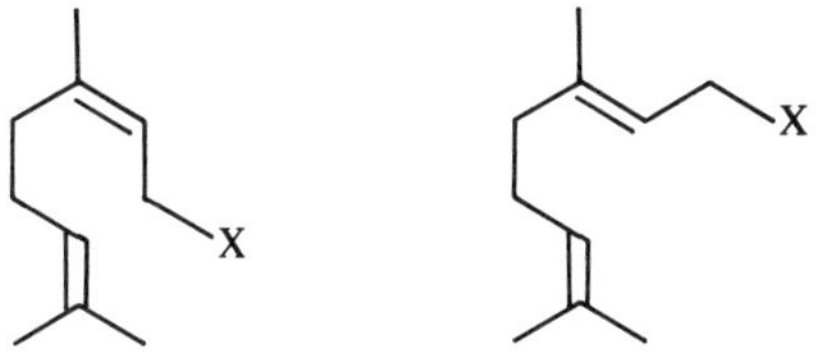

(43) **(44)**

X = $OC_6H_3(NO_2)_2$

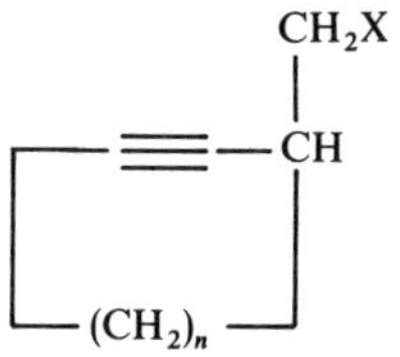

(45)

X = ONos or OTf

n = 6 or 7

α-deuterium isotope effects, and product composition for solvolyses in 80% ethanol support oxygen participation to a certain degree when a six-membered oxygen-containing ring can be formed, but in all other cases, and also when 97% trifluoroethanol was used as solvent, no oxygen participation was found. Various lines of evidence show that the sulphonate esters of 2-methoxyethanol undergo solvent-assisted displacement in a variety of solvents but neighbouring-group participation by the ether oxygen does not occur.[108] However, the branched compound $MeOCMe_2CH_2OSO_2Ph$ shows participation by the ether oxygen and gives completely rearranged product.

The hydrolysis of chloro-ketones $ArCO(CH_2)_nCl$ is very dependent on intramolecular assistance by the carbonyl group and is sensitive to the pattern of substitution in Ar and to the value of n.[109] When $n = 3$, a 4-OH is necessary for facile hydrolysis via the stabilized cation (**46**), but if $n = 4$ the hydrolysis even of the 4-OH derivative is difficult. Adjacent methyl groups reduce the favourable influence of 4-OH and this is attributed to steric hindrance preventing the OH being coplanar with the aryl ring in (**46**). For the displacement of an acetate group from 1,4-butanediol diacetate by heating with anhydrous aluminium chloride at *ca.* 100°C, an ^{18}O labelling study has provided direct evidence for the intervention of a seven-membered acetoxonium ion (**47**).[110] Previously five-membered and six-membered acetoxonium intermediates have been postulated in various reactions. Evidence has been obtained from product analyses in the solvolyses of certain β-halo-sulphones that the weakly nucleophilic character of the sulphonylic oxygens may permit the sulphone group to provide anchimeric assistance for the departure of good leaving-groups.[111]

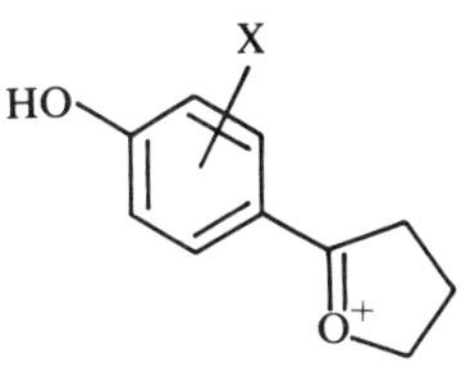

(46)

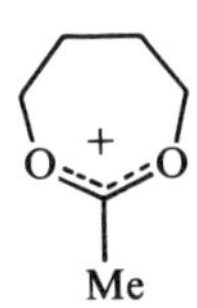

(47)

Solvolysis rate constants for various substrates $CH_3SCH_2CH_2Y$ (Y = OTs, I, Br, Cl, ODNB, OPNB) show that participation by sulphur reduces such rate ratios as

Br/Cl, Cl/OPNB, etc. by comparison with substrates without MeS.[112] However, the authors warn against taking rate ratios as criteria to determine the degree of participation. Further, it has subsequently been found that rate constants determined for the solvolysis of 2-(methylthio)ethyl tosylate have been rendered erroneous by the ready dimerization of the substrate.[113] New OTs/Cl rate ratios were reported, rather higher than those published previously, but still apparently deemed 'low' because of participation by sulphur.

The role of neighbouring-group participation by sulphur in the acetolysis of substrates $PhS(CH_2)_nCD_2OTs$ ($n = 1$ to 5) has been examined.[114] It has been confirmed that participation decreases with increasing ring size of the intermediate cyclic sulphonium salt. For the displacement reactions of sulphonate esters of 5-thio-D-allose, 5-thio-D-altrose, and 5-thio-D-glucose derivatives, participation by the ring sulphur atom and formation of intermediate episulphonium ions have been proposed.[115]

Ambident Nucleophiles

The silver salts of arylnitroacetonitriles $ArCH(CN)NO_2$ react with 9-bromofluorene to give both *C*- and *O*-alkylated products (**48**) and (**49**), respectively.[116] The *C/O* product ratio varied from 1.16 for $Ar = C_6H_5$ to 0.10 for $Ar = o\text{-}MeC_6H_4$.

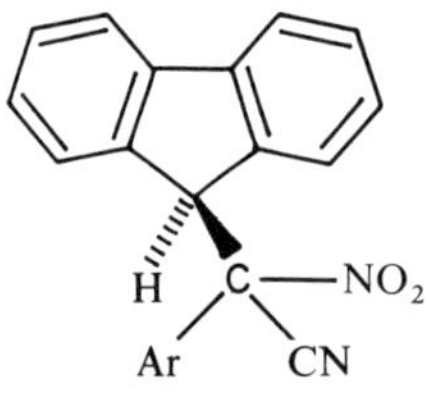

(**48**)

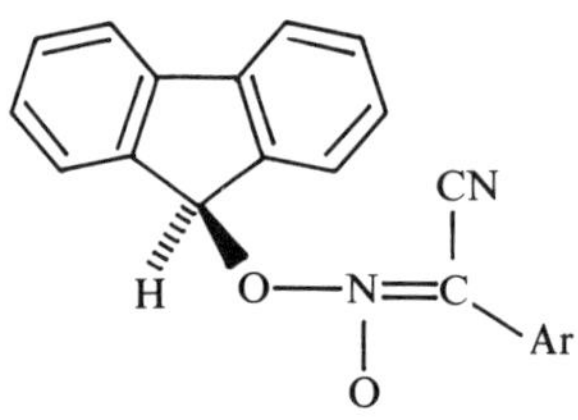

(**49**)

The reaction of benzyl bromides RBr with trimethylsilyl isothiocyanate (TMSTC) gave both RSCN and RNCS, in proportions varying with the substituent in R.[117] For benzyl bromide itself in HMPA, RSCN was overwhelmingly the predominant product but the introduction into benzyl bromide of the electron-releasing *p*-OMe or *p*-Me on the one hand or the electron-attracting *p*-CN or *p*-NO_2 on the other produced somewhat more RNCS than RSCN. The formation of RNCS was considered to involve a competitive *N*-attack of TMSTC on benzyl bromide and not a rearrangement of RSCN to RNCS.

Lithium, sodium, and potassium enolates of acetophenone have been methylated with dimethyl sulphate in 14 aprotic solvents.[118] Log (molar ratio) of *O*-methylation versus *C*-methylation was correlated by multiple regression with solvent polarity and Lewis basicity parameters and with functions of the ionic radius.

Isotope Effects

The radionuclide ^{11}C (half-life 20.34 m) has been used for the first time in the study of kinetic isotope effects (KIE).[119] The combined use of ^{11}C and ^{14}C in such studies was claimed to have several advantages. The ${}^{11}k/{}^{14}k$ ratio for C^*H_3I reacting with *N,N*-dimethyl-*p*-toluidine in methanol at 30°C was 1.23.

The mechanism of the *S*-adenosylmethionine synthetase reaction from *E. coli* has been characterized by KIE (involving ^{14}C and ^{3}H) and substrate-trapping measurements.[120] An S_N2-like transition state is said to be involved.

Gas-phase Reactions

The gas-phase negative-ion chemistry of organosilanes has been reviewed;[121] 'the anion chemistry of organosilicon compounds in all its rich diversity can be studied in the gas phase as well as (and often more easily than) in solution... In addition, the reactions of neutral organosilicon compounds, including highly unstable species like silenes and silylenes, with anions of all types, can be studied.' The same research group has studied the reactions of HO^- and MeO^- with Me_4Si (and related systems) by *ab initio* MO calculations (6-21G, Gaussian 82) and both ion-cyclotron-resonance and flowing-afterglow experiments.[122]

When monosolvated alkoxide ions $(R^1O\text{---}H\text{---}OR^2)^-$ react with the ethers of B, C, Si, or Ti, the major displacement reaction is always that in which the smaller alkoxide R^1O^- reacts with the central atom, while the larger alcohol R^2OH becomes part of the solvated alkoxide product:[123]

e.g. $$(R^1OH\text{---}H\text{---}OR^2)^- + Me_3SiOR^3 \rightarrow (R^2O\text{---}H\text{---}OR^3)^- + Me_3SiOR^1$$

these reactions are exothermic and very fast.

The comparative behaviour of nucleophiles, in gas and solution phase, with respect to acylation, alkylation, and phosphorylation has been reviewed.[124]

Cis- and *trans*-indanediols react with the NH_3/NH_4^+ system under chemical ionization conditions via nucleophilic substitution.[125] Competition between S_N2 and S_Ni mechanisms depends on the stereochemistry of the diol.

Theoretical studies of gas-phase processes have included the effect of electron correlation in the S_N2 reaction of HO^- with CH_3F,[126] quantum-dynamic calculations on the S_N2 reactions of MeF with H^-, F^-, and HO^-,[127] and non-empirical calculation of potential surfaces for the same reactions.[128]

Radical Processes

The relationship between nucleophilic reactions and single-electron transfer (SET) has been reviewed.[129] Such processes often compete and the *configuration mixing model* may be used to predict the factors likely to encourage one pathway rather than the other. In this model both nucleophilic and SET processes involve a single-electron shift.

The reactions of optically active 2-halooctanes with a variety of nucleophiles in DMF have been used to detect the occurrence of an electron-transfer pathway.[130] This is the latest instalment of extensive work which has produced evidence for SET in the reactions of alkyl halides with nucleophiles. In the present case evidence for inversion being less than 100% is obtained, loss of optical activity from RX increasing in the order $RCl < RBr < RI$.

9-Substituted fluorenide carbanions 9-MeFl$^-$, 9-MeOFl$^-$, and 9-Me$_2$NFl$^-$, which have nearly the same basicities but are progressively easier to oxidize, have been used to probe for electron-transfer components of reactions.[131] In reactions with PhCH$_2$Cl, Bu^{i}Br, or Bu^{i}I, the above nucleophiles showed a decrease in rate in the order stated, corresponding to steric effects on an S_N2 process. However, the order was reversed for reactions with *c*-C$_6$H$_{10}$(NO$_2$)Ts, PhSO$_2$CH$_2$Cl, or F$_3$CCH$_2$I, corresponding to the order of ease of oxidation of the carbanions, and products derived from radicals were found.

The factors influencing the regiochemistry of the reactions of phenyl-substituted allyllithiums with tertiary alkyl bromides have been investigated systematically.[132] The interpretation of the rather complicated results was in terms of two alternative pathways, a polar one and SET.

The reactions of several Group VI metal carbonyl hydrides with 6-bromo-1-hexene and 4-bromo-3,3-dimethyl-1-butene in THF have been studied.[133] The latter neopentyl-like substrate was reduced by a radical-chain process, while for the former unhindered substrate an S_N2 pathway and a radical-chain process were found to be in competition. In the reactions of unhindered alkyl bromides and iodides with boron and aluminium hydride reducing agents, however, SET was shown to be of minor occurrence.[134]

Medium Effects

The contributions of M. J. Kamlet, R. W. Taft, and their associates, to the study of solvent effects have continued.[135–137] The solvolysis/dehydrohalogenation rates of *t*-butyl chloride in 21 hydrogen-bond donor (HBD) and non-HBD solvents have been well correlated by multiple regression on the solvent parameters δ_H^2, π^*, α, and β.[135] The same procedure has been applied to analogous data for other substrates and the respective regression coefficients have been compared and interpreted. In a further contribution, arguments are advanced to show that at least four solvent parameters are needed in any general equation for solvent effects on rate constants and on Gibbs energies of individual solutes.[136] The authors argue for the merits of their favourite solvent parameters as listed above and show that others such as Y and N values are 'blends' of these. In a third paper, the authors present a 'double-difference' method for estimating electrophilic solvent assistance effects in solvolysis reactions.[137] The method requires the solvolysis rates to be measured in four solvents: methanol, ethanol, trifluoroethanol, and hexafluoro-2-propanol.

Kinetic studies of the solvent effect on the quaternization of trimethylamine by dithiophosphoric acid esters have been interpreted in terms of linear-free-energy relationships (LFER) with empirical solvent parameters[138] and by means of a multiparameter LFER.[139] The predominant effect of solvent on the rates of reactions of ω-halogenoacetophenones with piperidine is solvent polarizability.[140] A theoretical analysis has been presented of the activation parameters for the reaction of ethyl iodide with bromide ion in *N*-methylacetamide–acetonitrile and *N*-methylacetamide – *N*,*N*-dimethylacetamide mixtures.[141]

The development of quantitative scales of nucleophilicities of aqueous, alcoholic, and acidic media based on solvolysis reactions has been reviewed extensively.[142] The rates of solvolysis of the methyldiphenylsulphonium ion in 21 (mainly mixed) solvents have been correlated with solvent nucleophilicity $N_{Et_3O^+}$ values.[143] The solvolyses of *t*-butyl halides and of 1-adamantyl tosylate in methanol-acetone mixtures have been treated through the Grunwald–Winstein equation.[144] *Y* values based on 1-adamantyl tosylate are considered superior to others and the order of electrophilic assistance to the leaving groups is $Ts^- > Cl^- > Br^- > I^-$. Solvolysis rates for 1-phenyl-1-cyano-2,2,2-trifluoroethyl brosylate were measured in various solvents and the results were analysed through the extended Grunwald–Winstein equation, giving $m = 0.88$ and $l = 0.49$.[145] The solvolysis of 3-chloro-1,2-benzisothiazole-1,1-dioxide (pseudosaccharin chloride) has been studied in various solvent mixtures.[146] Treatment of the results in various ways led to the conclusion that the reaction proceeds via an S_N2 mechanism.

Rate constants for the solvolysis of *p*-chlorobenzyl chloride in ethanol-water mixtures have been determined at 30°C and 40°C and up to 1600 bar.[147] Values of ln *k* were fitted to a second-order function of *P* and values of $\Delta V^\ddagger$ and $\Delta\beta^\ddagger$ were obtained. These exhibit extremum behaviour at about 0.20 mol fraction of ethanol. In a related study of *p*-methoxybenzyl chloride, the extremum behaviour was shown at about 0.30 mol fraction of ethanol.[148]

There have been several theoretical studies of the role of the solvent. Molecular dynamics were computed for the exchange reaction of Cl^- with CH_3Cl in water and they depend strongly on the instantaneous local configuration of the solvent at the transition state barrier.[149] A simple 'frozen solvent theory involving non-adiabatic solvation' was presented. The origin has been determined of the decrease in rate of about 20 orders of magnitude for S_N2 reactions as between gas phase and solution.[150] The participation of more than 60 molecules of water is necessary to explain this rate difference. A previously developed non-adiabatic model has been used to calculate rate constants and pre-exponential factors for

$$Y^- + MeX \rightarrow YMe + X^- \quad (X, Y = F, Cl, Br, I)$$

and the calculated results agreed with experiment.[151] The potential-energy surface for the attack of F^- on MeF was calculated by the MNDO/CI method with solvation included.[152] Free-energy profiles were computed for the separation of the *t*-butyl cation and Cl^- in dilute aqueous solution at 25°C and 1 atm.[153] The approach involved extensive Monte Carlo (MC) simulations and integral-equation calculations using extended RISM methodology. MC found a contact ion-pair at a C—Cl distance of 2.9 Å and the onset of a solvent-separated ion-pair at 5.5 Å, but solvent-separated ion-pairs and 'free' ions were not found to be energetically distinct species.

Study of the solvolytic reactivities of various nicotinates found that in hexafluoroisopropanol (97%) hydrogen bonding was particularly important in helping the departure of the nicotinate leaving-group.[154] The roles of basicity, nucleophilicity, and electrophilicity have been examined in relation to the protonation of aromatics.[155]

In a further study of the mechanism of solvolysis of benzyl azoxytosylate, rates and products of reaction in 1:1 v/v aqueous trifluoroethanol have been measured in the presence of various added bases and nucleophiles.[156] Non-basic solutes give only small rate effects but if nucleophilic they are incorporated in the products, 'the classic evidence of an S_N1 mechanism'. (There are many other pertinent detailed findings.) In kinetic studies of the decomposition of benzyltriphenylphosphonium chloride by sodium hydroxide in binary aqueous solvent mixtures, the addition of amines was found to increase the rate to a maximum at 0.1 M amine and thereafter to decrease it.[157]

The addition of thiourea has been developed as a probe for nucleophilic solvent assistance.[158] Thiourea accelerates solvolysis reactions known to proceed by a k_s mechanism but k_c or k_Δ reactions show slight retardation if there is no external ion return. If the latter does occur, however, the effect is similar to that observed for k_s substrates.

In the hydrolysis of primary alkyl halides, metal-ion electrophiles as additives may have a double function: as Lewis acids they enhance heterolysis of RX, while as charged particles they affect hydrolysis selectivity.[159] Salt effects have been studied for the arylaminolysis of *N*-(phenylsulphonyl)benzimidoyl chloride in acetonitrile.[160]

Phase-transfer Catalysis and Other Intermolecular Effects

A long series on phase-transfer catalysis has continued with the discovery that sodium trichloroacetate and alkyl halides in the presence of tetrabutylammonium bromide yield alkyl 2,2-dichloro-4-pentenoates.[161] Alkyl trichloroacetates, the primary products, are apparently dechlorinated by Cl_3C^- to give the anions of alkyl dichloroacetates as intermediates. In a further study, the influence of phase-transfer catalyst on a branching of reaction paths has been examined, and an almost complete change-over of paths was demonstrated for a reaction of allyl bromide.[162]

Concentrated aqueous solutions of sodium and potassium hydroxides affect the hydration and hence the reactivity of anions, such as Cl^-, Br^-, I^-, SCN^-, or N_3^-, in aliphatic nucleophilic substitutions under phase-transfer conditions.[163] Systems involving quaternary salts, crown ethers, and cryptands show a somewhat different behaviour in this respect. Pseudo-crown ethers based on polyphosphazene are catalysts for nucleophilic substitution.[164] The solubility of ammonia in organic media is enhanced by quaternary ammonium salts and its reactions with some alkyl and aryl halides are accelerated.[165]

The conversion of benzyl bromides into oxime ethers by passing nitric oxide through a solution of the bromide in *t*-pentyl alcohol–3 M aqueous sodium hydroxide is catalysed by a Pd or Ru complex, with a quaternary ammonium salt as phase-transfer agent.[166] For the phase-transfer-catalysed exchange reaction of alkyl chloride with metal bromide salts, the selectivity constant $K^{sel}_{Cl/Br}$ depends on the nature of the metal cation and the inorganic phase composition.[167] The phase-transfer-catalysed etherifications of pentaerythritol [2,2-bis(hydroxymethyl)-1,3-propanediol] and ethylenetetramethanol [2,3-bis(hydroxymethyl)-2-butene-1,4-diol] have been compared with respect to rate and selectivity.[168] The results clearly indicate that the facile alkylation of the former substrate is due to its tetrahedral structure.

The manipulation of nucleophilic displacement reactions by host-guest complexes has been extensively reviewed.[169] Most of the examples involve encapsulation of an organic substrate in the cavity of a macrocyclic ammonium ion in aqueous solution. Details have now been published concerning a new type of phase-transfer catalysis, in which substrates in an inner organic phase react with anions in an outer aqueous phase, the phase-transfer catalyst being grafted onto an ultrathin nylon capsule membrane separating the phases.[170] Copolymers of β-cyclodextrin (β-CD) with epichlorohydrin act as more efficient phase-transfer catalysts than β-CD itself in liquid-liquid two-phase nucleophilic displacement reactions.[171]

The selectivity of CsF as a base in solid-liquid systems is improved in the presence of a phase-transfer catalyst.[172] The reactivity of KF in nucleophilic substitution may be enhanced by tetraphenylphosphonium bromide, particularly in non-dipolar aprotic solvents.[173]

Structural Effects

An extensive review of nucleophilic substitution, with respect to mechanism and reactivity, has appeared.[174] Topics covered include quantum-chemical calculations, gas-phase reactions, potential-energy surfaces, substituent effects, and correlations. Nucleophilic substitution in β-hetero-functional α,β-unsaturated iminium salts has also been reviewed at length.[175] The computer program CAMEO (Computer-Assisted Mechanistic Evaluation of Organic Reactions) has been used to develop and implement a mechanistic model of nucleophilic chemistry.[176] It is thereby possible to predict the outcome of nucleophilic processes by mechanistic reasoning and by applying rules governing competing reactions that are based on precedents in the literature.

A series on the ion-pair mechanism and bimolecular displacement at saturated carbon has continued with a study of the racemization and radio-bromide exchange for substituted 1-phenylbromoethanes in acetonitrile, nitromethane, and acetone.[177] Changing substituents from 4-nitro to 3,4-Me_2 and solvent from acetone to the other more polar solvents shifts the transition state from an early to a late stage of the scheme: substrate $\rightleftarrows$ intimate ion-pair $\rightleftarrows$ various solvated ion-pairs $\rightleftarrows$ free or dissociated ions.

In presenting a study of the substituent effect of the pentafluorophenyl group, the authors address the question: 'Is there a correlation between X-ray crystal structures and reactivity in carbocation formation?'[178] Caution is advised in applying such correlations.

Katritzky's series on kinetics and mechanism of nucleophilic displacements with heterocycles as leaving-groups has continued.[179,180] Study of further substrates (various *N*-substituents and pyridine leaving-groups) reacting with a variety of nucleophiles finds first- and second-order rate components and 'the conclusion that these nucleophilic displacements can proceed by five independent mechanistic pathways is strengthened'.[179] X-ray crystal structures were determined for seven

N-benzylpyridinium salts.[180] The molecules show steric overcrowding and various measures of this in terms of bond angles and lengths parallel nucleophilic displacement rates.

The epimeric *trans,trans,trans*-tricyclo[7.3.1.0^{5,13}]tridecanyl-3-tosylates (**50a** and **b**) have restricted conformational freedom but are free from substituents near the functional group and are without the strain associated with ground-state deformations of bridged systems.[181] They are thus very suitable for investigating unambiguously the true reactivity of the *ax* and *eq* nucleofuge in solvolysis. For reaction in buffered acetic acid at 50°C, the *ax*:*eq* rate ratio is 3.4:1, while in 97% HFIP at 25°C it is *ca.* 1:1.

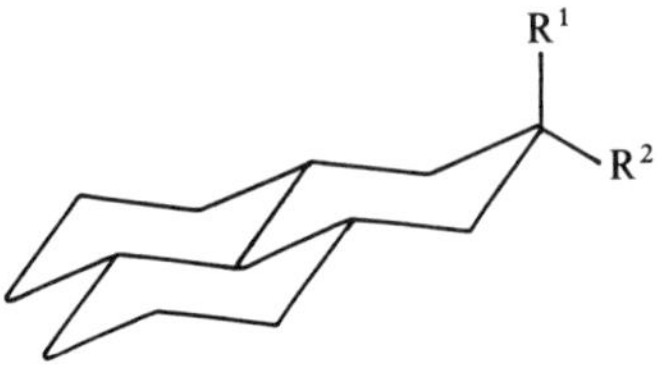

(50)

a; $R^1 = OTs$, $R^2 = H$

b; $R^1 = H$, $R^2 = OTs$

Correlation Analysis by Hammett, Brønsted, or Taft Equations

The second-order reactions of propyl, propargyl, and allyl bromides with substituted anilines in methanol or DMF have Hammett ρ, values of *ca.* -1, suggesting reaction via a loose S_N2 transition state.[182] The second-order reactions of ethyl bromoacetate with substituted phenoxide ions in 90% v/v acetone-water mixtures gave a Hammett ρ value of -2.069 (40°C), Brønsted α value of 0.88 (25°C), and an isokinetic temperature β of 510.6K.[183] Deviations from second-order behaviour at higher concentrations of sodium phenoxide were attributed to ion-pair formation. The rate constants for reactions of 4- and 6-substituted 1-naphthoate ions and of 6-substituted 2-naphthoate ions with ethyl bromoacetate in 90% acetone-water at 30°C correlate well with Hammett σ_p values, yielding ρ values of -0.54, -0.19, and -0.25, respectively.[184] The reactivity-selectivity principle is not obeyed.

In a series of papers on dynamic NMR as a non-destructive method for determining rates of dissociation, those of the C—N bond in substituted 1-(1,3-dithiolan-2-yl)pyridinium perchlorates in CD_3NO_2 or CD_3CN have been measured.[185] Correlation of $\Delta G^{\ddagger}$ with pK_a values of the substituted pyridines was found and an S_N2 mechanism was suggested, with the perchlorate ion acting as nucleophile.

Rate constants for the substitution reactions of (*E*)-4-$RC_6H_4SO_2CH{=}CH\overset{+}{N}Et_3Cl^-$ with piperidine were correlated with σ^0 constants for 4-R and values of ρ were found to be 2.05 (in MeCN) and 1.76 (in DMF).[186] A zwitterionic intermediate was proposed.

1-Arylethyl ethers of salicylic acid (**51**) are hydrolysed more rapidly than expected for simple 1-arylethyl ethers, cf. the corresponding methyl esters.[187] This indicates intramolecular proton transfer catalysis by CO_2H (**51, 52**). The effects of substituents Y on rate were correlated by σ^+, as would be expected for a reaction generating a carbocationic centre adjacent to the ring, ρ^+ being *ca.* -4.9, while the effects of substituents Z were correlated by σ, ρ being 1.25.

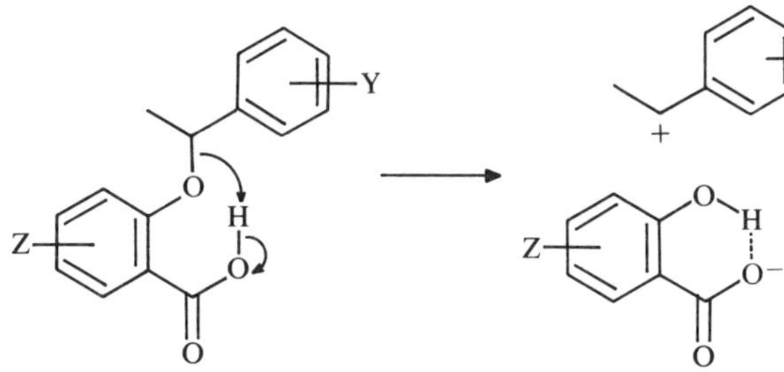

(51) **(52)**

E. S. Lewis's work on methyl transfers has continued with a study of reaction rates and equilibria for the methylation of the phenyl selenide anion by aryl methyl selenides.[188] For reaction at 90°C in sulpholane the Hammett ρ value for the rates of the forward reactions is 1.1 and for the equilibria is 2.9. The charge on the methyl group in the transition state was calculated to be -0.24 and the significance of 'this unusual negative charge' was discussed.

Experimental evidence has now been obtained to support the theoretical conclusion that the high S_N2 reactivity of α-halocarbonyl compounds is due to 'lowering of the intrinsic barrier by a major contribution of enolate structure to the transition state'.[189] The evidence largely involves a comparison of Hammett ρ values for various pertinent systems.

In the correlation analysis of the rates of S_N2 reactions of X-substituted-anilines with Y-substituted-1-arylethyl Z-substituted-arenesulphonates in methanol it is necessary to include cross-interaction terms $\rho_{XY}\sigma_X\sigma_Y$, $\sigma_{XZ}\sigma_X\sigma_Z$, and $\rho_{YZ}\sigma_Y\sigma_Z$, in addition to the ordinary Hammett terms $\rho_X\sigma_X$, $\rho_Y\sigma_Y$, and $\rho_Z\sigma_Z$.[190] Cross-interaction between X and Z is unexpectedly high and this is rationalized in terms of a four-centre transition state.

Hammett σ_p^+ values for arylazo and *t*-butylazo groups as determined from measurements on the solvolysis of cumyl chlorides were considerably more positive than expected and differed from earlier estimates based on electrophilic substitutions.[191]

The rates of reactions of 2-phenylethyl arenesulphonates with pyridine in acetonitrile have been measured at various temperatures and pressures.[192,193] Hammett ρ values were decreased by raising the pressure, 'indicating that the reaction is probably closer to tight S_N2 character at higher pressures'. In a related study on 2-phenylethyl brosylate reacting with substituted pyridines in acetonitrile, it was found that the Hammett ρ and Brønsted β values were increased at higher pressures, 'indicating that the reaction has increasing S_N2 character with increasing pressure'.[194] A further study from the same research group concerns the application of the Hammett equation to the reactions of Z-substituted-2-phenylethyl *m*-nitrobenzenesulphonates with Y-substituted pyridines. The values of ρ_Z and ρ_Y were interrelated.[195]

The Hammett and Brønsted equations have been applied to the second-order rate constants for the reactions of substituted anilines with methyl iodide and tosylate in MeOH–MeCN mixtures.[196] The S_N2 reactions of benzyl benzenesulphonates with anilines have been treated by all the main dual-substituent parameter equations.[197] It was claimed that the Swain–Lupton equation was the most reliable.

Rates of acetolysis of an extended series of 2-aryl-2-methylpropyl brosylates have been treated by the LArSR relationship, i.e. Yukawa–Tsuno equation, giving an *r* value of 0.57.[198] This was considered reasonable for a process involving an aryl-assisted transition state. The validity of the LArSR treatment and the significance of *r* as indicating resonance demand were defended. A related study of the acetolysis rates of 2-arylethyl tosylates found a non-linear substituent effect, which was accounted for in terms of two linear LArSR relationships, one for the aryl-assisted and one for the unassisted process.[199] A comparative study of the ratios of aryl-assisted to unassisted processes through deuterium scrambling in α-D_2-labelled substrates confirmed the validity of the interpretation in terms of two linear LArSR relationships.[200]

An inverse order of reactivities, $k_p/k_m > 1$, was found for certain electron-attracting α-aryl substituents in the solvolysis of certain highly crowded toluene *p*-sulphonates and *p*-nitrobenzoates.[201] It was claimed that such findings can be used to probe the absence of a resonance contribution of the aryl ring in the cationic transition state and that Hammett–Brown and Yukawa–Tsuno treatments of such systems are liable to give a misleading impression.

The reactivities of alkylamines in nucleophilic reactions at sp^2 and sp^3 carbon atoms may be correlated successfully by an equation involving both the steric constant E_s and n_H, the number of hydrogen atoms on N.[202] The contributions of steric, electrical, and polarizability effects in enantioselective hydrolyses with the mould *Rhizopus nigricans* have been assessed quantitatively.[203]

Nucleophilicity and Leaving-group Effects

A substantial volume on 'Nucleophilicity' has appeared in the *Advances in Chemistry* series, and some references to this have already been given in this chapter. An introduction by the editors of the volume provides a general review of nucleophilic reactivity including Marcus theory, methyl transfers, gas-phase reactions, Brønsted equation, hard and soft acids and bases, LFER for solvent nucleophilicity, etc.[204] The study of gas-phase S_N2 reactions of anions with neutral alkyl halides is reviewed as an approach to a measure of intrinsic nucleophilicity.[205] The effect of solvation on nucleophilic reactivity in hydroxylic solvents is examined.[206] Desolvation is often required before the nucleophile can enter into reaction. Electrophile-nucleophile combination reactions are examined from several points of view.[207] A 'quasi-thermodynamic theory of nucleophilic reactivity' has been presented.[208] In this theory 'nucleophilic reactivity is related by a semiempirical equation to solvation energy and electron affinity of the nucleophile and the dissociation energy of the bond formed'. The α-effect was also discussed. The relationship between the reactivity of a nucleophile towards an electrophile, and the distance separating the two species has been discussed.[209] Rates and stereochemistry have been determined for solvolyses of 1-arylethyl tosylates containing electron-withdrawing substituents on the α-carbon or the aryl ring.[210] These reactions proceed via electron-deficient carbocations and show large Hammett–Brown ρ^+ values but modest values of the Grunwald–Winstein *m*.

1-Azabicyclo[2.2.1]heptane should be less hindered than quinuclidine.[211] However, in plots of log *k versus* pK_a for substituted quinuclidines and 1-azabicyclo[2.2.1]heptane entering into various reactions as bases or nucleophiles, the supposedly less-hindered compound is substantially *less* reactive than would correspond to its pK_a values.

Equilibrium constants, oxidation potentials, and nucleophilicity in S_N2 reactions have been discussed.[212] There is only very poor correlation between one-electron oxidation potentials and reactivity towards methyl iodide for a wide range of nucleophiles. Desolvation energies are important in governing the nucleophilicities of anions. A thermodynamic scale of leaving-group abilities was presented.

The application of hard and soft acid-base theory to nucleophilic substitution has been reviewed in Chinese.[213]

Tetrabutylammonium bifluoride dissolved in 1:1 THF:HMPT has been advocated as a convenient source of the 'naked' fluoride ion, which will convert various alkyl chlorides, bromides, etc., into the corresponding fluorides in good yield.[214]

Theoretical Treatments

In a long series on the determination of reactivity by MO theory, geometries and energies of S_N2 transition states were examined by MNDO calculations for

$$CH_3X + Y^- \rightarrow CH_3Y + X^-$$

where X = F, Cl, CN and Y = CN, OH, F, Cl.[215] Leaving-group ability had a more marked effect than nucleophilicity on the activation barrier. In a further study, two contributions to the activation barrier of the S_N2 reaction, intrinsic and thermodynamic, were discussed in connection with various rate-equilibrium relationships.[216] Substituent effects on S_N2 transition states have also been treated theoretically.[217]

MNDO calculations on extremely 'loose' S_N2 transition states have found some unexpected relationships between bond order of the forming and that of the breaking bond.[218] The MNDO method has also been used to study the alkylation of nucleophiles by $ClCH_2CH_2NR^1R^2$ (R^1,R^2 = H or an electron-donating or -withdrawing substituent) via a cyclic ethylenimmonium ion.[219]

A general intersecting-state model has been applied to calculate the energy barriers of methyl nucleophilic substitutions in the vapour and in solution.[220] A theoretical study has addressed the question: 'Is the $PhCH_2(N_2O)^+$ cation likely to be a reaction intermediate?'[221] *Ab initio* calculations on $H(N_2O)^+$ and $CH_3(N_2O)^+$ as model systems apparently support the possible existence of $PhCH_2(N_2O)^+$, while has been postulated as an intermediate in the solvolysis of benzyl azoxytoluene-*p*-sulphonate.

The formation, electronic structures, and electrophilic reactivity of the pyrazole cation system have been treated by MO methods.[222] This cation undergoes nucleophilic attack only at the 3 (or 5) position and this has been explained in terms of HOMO–LUMO interactions. The limitations of geometry-reactivity relationships have been examined in connection with the solvolyses of cycloalkane tosylates.[223]

Kinetic Studies (Miscellaneous)

These studies, at various levels of thoroughness, have included the following: nucleophilic substitution of thienylneopentyl substrates;[224] isomerization in the hydrolysis of isobutyl halides;[225] reactions of amines with β-bromopropiophenone;[226] reaction of *N*-(β-hydroxyethyl)diethylamine with allyl bromide in acetone;[227] effect of pressure on a bimolecular reaction involving Ph_3P as nucleophile;[228] reactions of sodium monochloroacetate with some mercapto-acids;[229] reactions of tribenzylamine with Br_2 and Br_3^- (re-investigation leading to modified findings);[230] an anion chain Arbuzov reaction;[231] the hydrolysis of bis(2-chloroethyl) sulphide[232] and the formation of sulphonium chlorides in this reaction;[233] the esterification of butanol with hydrobromic acid;[234] the acid-catalysed cleavage of 9-anthryl ethers in aqueous solution;[235] the decomposition of trimethylphenylammonium iodide in propanol.[236]

References

[1] Shainyan, B. A., *Usp. Khim.*, **55**, 942 (1986); *Russ. Chem. Rev.*, **55**, 511 (1986); *Chem. Abs.*, **106**, 49362 (1987).
[2] Avramovitch, B., Weyerstahl, P., and Rappoport, Z., *J. Am. Chem. Soc.*, **109**, 6687 (1987).
[3] Rappoport, Z., and Gazit, A., *J. Am. Chem. Soc.*, **109**, 6698 (1987).
[4] Lee, C. C. and Wanigasekera, D., *Can. J. Chem.*, **65**, 933 (1987).
[5] Hanack, M. and Rieth, R., *Chem. Ber.*, **120**, 1659 (1987).
[6] Kravchenko, V. V., Popov, A. F., Kostenko, L. I., Kotenko, A. A., Vegh, D., and Kovac, J., *Zh. Org. Khim.*, **22**, 1520 (1986); *Chem. Abs.*, **106**, 83749 (1987).
[7] Bordwell, F. G., Clemens, A. H., and Cheng, J.-P., *J. Am. Chem. Soc.*, **109**, 1773 (1987).
[8a] Domes, E., Clarke, U., and Neitzel, J. J., *J. Org. Chem.*, **52**, 1540 (1987).
[8b] Doomes, E. and Overton, B. M., *ibid.*, p. 1544.
[9] Hagen, J. P., Harris, J. J., and Lakin, D., *J. Org. Chem.*, **52**, 782 (1987).
[10] Bäckvall, J. E., Vågberg, J. O., and Genêt, J. P., *J. Chem. Soc., Chem. Commun.*, **1987**, 159.
[11] Marshall, J. A. and Audia, V. H., *J. Org. Chem.*, **52**, 1106 (1987).
[12] Marshall, J. A. and Trometer, J. D., *Tetrahedron Lett.*, **28**, 4985 (1987).
[13] Flaud, J.-C. and Legros, J.-Y., *J. Org. Chem.*, **52**, 1907 (1987).
[14] Baruah, J. B. and Samuelson, A. G., *J. Chem. Soc., Chem. Commun.*, **1987**, 36.
[15] Julia, M., Nel, M., and Uguen, D., *Bull. Soc. Chim. Fr. II*, **1987**, 487.
[16] D'Arcy, B. R., Lewis, K. G., and Mulquiney, C. E., *Aust. J. Chem.*, **40**, 509 (1987).
[17] Müller, P. and Mareda, J., *Helv. Chim. Acta*, **70**, 1017 (1987).
[18] Fuso, F., Grob, C. A., Sawlewicz, P., and Yao, G. W., *Helv. Chim. Acta*, **69**, 2098 (1986).
[19] Altmann-Schaffner, E. and Grob, C. A., *Helv. Chim. Acta*, **70**, 43 (1987).
[20] Gassman, P. G., Okuma, K., Lindbeck, A., and Allen, R., *Tetrahedron Lett.*, **27**, 6307 (1986).
[21] Della, E. W., Pigou, P. E., and Tsanaktsidis, J., *J. Chem. Soc., Chem. Commun.*, **1987**, 833.
[22] Yoon, K. B. and Kochi, J. K., *J. Chem. Soc., Chem. Commun.*, **1987**, 1013.
[23] Morrison, H., Muthuramu, K., and Severance, D., *J. Org. Chem.*, **51**, 4681 (1986).
[24] Adams, B. R. and Chenier, P. J., *J. Org. Chem.*, **51**, 5016 (1986).
[25] Wilt, J. W., George, C., and Peeran, M., *J. Org. Chem.*, **52**, 3739 (1987).
[26] Le Drian, C. and Vogel, P., *Helv. Chim. Acta*, **70**, 1703 (1987).

[27] Tanida, H. and Irie, T., *J. Org. Chem.*, **52**, 5218 (1987).
[28] Grob, C. A. and Sawlewicz, P., *Tetrahedron Lett.*, **28**, 951 (1987).
[29] Grob, C. A., Wang, G., and Yang, C., *Tetrahedron Lett.*, **28**, 1247 (1987).
[30] Banert, K. and Kurnianto, A., *Chem. Ber.*, **119**, 3826 (1986).
[31] Zwick, J.-C., Gabioud, R., and Vogel, P., *Helv. Chim. Acta*, **70**, 1079 (1987).
[32] Nagano, H., Iwadare, T., and Shiota, M., *J. Chem. Soc., Perkin Trans. 1*, **1986**, 2291.
[33] Crotti, P., Dell'Omodarme, G., Ferretti, M., and Macchia, F., *J. Am. Chem. Soc.*, **109**, 1463 (1987).
[34] Crotti, P., Macchia, F., Pizzabiocca, A., Renzi, G., and Speranza, M., *Tetrahedron Lett.*, **28**, 3393 (1987).
[35] Bradamante, S., Ferraccioli, R., and Pagani, G. A., *J. Chem. Soc., Perkin Trans. 1*, **1987**, 515.
[36] Kirshenbaum, K. S. and Sharpless, K. B., *Chem. Lett.*, **1987**, 11.
[37] Chong, J. M., Cyr, D. R., and Mar, E. K., *Tetrahedron Lett.*, **28**, 5009 (1987).
[38] Blandy, C., Gervais, D., and Cardenas, M. S., *J. Mol. Catal.*, **34**, 39 (1986); *Chem. Abs.*, **106**, 4367 (1987).
[39] Roloff, A., *Chimia*, **39**, 392 (1985); *Chem. Abs.*, **107**, 197244 (1987).
[40] Kagel, J. R. and Mertes, M. P., *J. Org. Chem.*, **52**, 2950 (1987).
[41] Briggs, J. R., Harrison, A. M., and Robson, J. H., *Polyhedron*, **5**, 281 (1986); *Chem. Abs.*, **106**, 119022 (1987).
[42] Yu. J. and Liu, R., *Huaxue Xuebao*, **44**, 755 (1986); *Chem. Abs.*, **106**, 175515 (1987).
[43] Kim, W. K., Sohn, C. K., and Lee, I., *Bull. Korean Chem. Soc.*, **7**, 279 (1986); *Chem. Abs.*, **106**, 195663 (1987).
[44] Shargorodskii, M. A., Gordeev, L. S., Groshev, G. L., and Labutin, A. N., *Izv. Vyssh. Uchebn. Zaved., Khim. Khim. Tekhnol.*, **29**, 136 (1986); *Chem. Abs.*, **107**, 38902 (1987).
[45] Sarancha, V. N., Guzechak, O. Ya., Romanyuk, I. M., Il'nitskaya, L. A., and Prisyazhnyuk, S. M., *Zh. Prikl. Khim. (Leningrad)*, **60**, 566 (1987); *Chem. Abs.*, **107**, 153735 (1987).
[46] Malievskii, A. D. and Gorbunova, O. I., *Izv. Akad. Nauk SSSR, Ser. Khim.*, **1986**, 2678; *Chem. Abs.*, **107**, 77079 (1987).
[47] Mislyuk, O. A., Shibaev, V. I., and V'yunov, K. A., *Zh. Org. Khim.*, **22**, 2502 (1986); *Chem. Abs.*, **107**, 115094 (1987).
[48] Mislyuk, O. A., Shibaev, V. I., V'yunov, K. A., and Ginak, A. I., *Zh. Org. Khim.*, **22**, 2227 (1986); *Chem. Abs.*, **107**, 175271 (1987).
[49] Wilk, K. A., Sokolowski, A., Burczyk, B., and Chlebicki, J., *Int. J. Chem. Kinet.*, **19**, 615 (1987).
[50] Karat, L. D., Shologon, I. M., and Klebanov, M. S., *Ukr. Khim. Zh. (Russ. Ed.)*, **52**, 532 (1986); *Chem. Abs.*, **106**, 155675 (1987).
[51] Williams, R. M. and Maruyama, L. K., *J. Org. Chem.*, **52**, 4044 (1987).
[52] Takahashi, K. and Ogata, M., *J. Org. Chem.*, **52**, 1877 (1987).
[53] Guindon, Y., Therien, M., Girard, Y., and Yoakim, C., *J. Org. Chem.*, **52**, 1680 (1987).
[54] Gerdil, R. and Barchietto, G., *Tetrahedron Lett.*, **28**, 4685 (1987).
[55] Gupta, S. C., Islam, N. B., Whalen, D. L., Yagi, H., and Jerina, D. M., *J. Org. Chem.*, **52**, 3812 (1987).
[56] Islam, N. B., Whalen, D. L., Yagi, H., and Jerina, D. M., *J. Am. Chem. Soc.*, **109**, 2108 (1987).
[57] Miyamoto, T., Silva, M., and Hammock, B. D., *Arch. Biochem. Biophys.*, **254**, 203 (1987).
[58] Kirmse, W., Rode, J., and Rode, K., *Chem. Ber.*, **119**, 3672 (1986).
[59] Kirmse, W. and Rode, J., *Chem. Ber.*, **119**, 3694 (1986).
[60] Weidner, J. and Vilsmaier, E., *Monatsh. Chem.*, **118**, 1057 (1987).
[61] Weidner, J., Vilsmaier, E., and Fries, R., *Monatsh. Chem.*, **118**, 1039 (1987).
[62] Brown, P. E. and Islam, Q., *Tetrahedron Lett.*, **28**, 3047 (1987).
[63] Blomberg, M. R. A., Siegbahn, P. E. M., and Bäckvall, J.-E., *J. Am. Chem. Soc.*, **109**, 4450 (1987).
[64] Ogoshi, H., Kikuchi, Y., Yamaguchi, T., Toi, H., and Aoyama, Y., *Organometallics*, **6**, 2175 (1987).
[65] Buchholz, B. and Stamm, H., *Isr. J. Chem.*, **27**, 17 (1986).
[66] Onistschenko, A., Buchholz, B., and Stamm, H., *Tetrahedron*, **43**, 565 (1987).
[67] Assithianakis, P., Onistschenko, A., and Stamm, H., *Arch. Pharm. (Weinheim, Ger.)*, **320**, 604 (1987).
[68] Buchholz, B. and Stamm, H., *Chem. Ber.*, **120**, 1239 (1987).
[69] Mall, T. and Stamm, H., *J. Org. Chem.*, **52**, 4812 (1987).
[70] Borisenko, V. S., Bobylev, V. A., Koldobskii, S. G., and Tereshchenko, G. F., *Zh. Obshch. Khim.*, **56**, 2768 (1986); *Chem. Abs.*, **107**, 77083 (1987).
[71] Dalin, A. R., Bobylev, V. A., Suslikov, V. F., Koldobskii, S. G., and Tereshchenko, G. F., *Zh. Obshch. Khim.*, **56**, 2774 (1986); *Chem. Abs.*, **107**, 58258 (1987).
[72] Koldobskii, S. G., Tereshchenko, G. F., Bobylev, V. A., and Dalin, A. R., *Khim. Geterotsikl. Soedin.*, **1986**, 1036; *Chem. Abs.*, **106**, 155679 (1987).

[73] Koldobskii, S. G., Tereshchenko, G. F., Bobylev, G. F., and Dalin, A. R., *Khim. Geterosikl. Soedin.*, **1986**, 1042; *Chem. Abs*, **106**, 175516 (1987).
[74] Nguy, N. M., Chiu, I.-C., and Kohn, H., *J. Org. Chem.*, **52**, 1649 (1987).
[75] Golding, B. T., Kebbell, M. J., and Lockhart, I. M., *J. Chem. Soc., Perkin Trans. 2*, **1987**, 705.
[76] Guimon, C., Pfister-Guillouzo, G., Tozzolino, P., and Labat, Y., *Nouveau J. Chim.*, **11**, 259 (1987).
[77] Schmid, G. H., Strukelj, M., and Dalipi, S., *Can. J. Chem.*, **65**, 1945 (1987).
[78] Dohn, D. R. and Casida, J. E., *Bioorg. Chem.*, **15**, 115 (1987).
[79] Deiters, J. A. and Holmes, R. R., *J. Am. Chem. Soc.*, **109**, 1686 (1987).
[80] Deiters, J. A. and Holmes, R. R., *J. Am. Chem. Soc.*, **109**, 1692 (1987).
[81] Lee, I., Yang, K., Park, B. S., and Lee, K. S., *Bull. Korean Chem. Soc.*, **7**, 231 (1986); *Chem. Abs.*, **106**, 155595 (1987).
[82] Azarian, D. B., Eaborn, C., and Lickiss, P. D., *J. Organomet. Chem.*, **328**, 255 (1987).
[83] Brumme, J., Scheim, U., Porzel, A., Rühlmann, K., and Grosse-Ruyken, H., *J. Organomet. Chem.*, **334**, 277 (1987).
[84] Mikolajczyk, M., *Phosphorus Sulfur*, **27**, 31 (1986); *Chem. Abs.*, **106**, 118928 (1987).
[85] King, J. F., Skonieczny, S., Khemani, K. C., and Lock, J. D., *Adv. Chem. Ser.*, **215**, 385 (1987); *Chem. Abs.*, **107**, 153643 (1987).
[86] Lee, I., Kang, H. K., and Lee, H. W., *J. Am. Chem. Soc.*, **109**, 7472 (1987).
[87] Houk, J. and Whitesides, G. M., *J. Am. Chem. Soc.*, **109**, 6825 (1987).
[88] Netscher, T., Schwesinger, R., Trupp, B., and Prinzbach, H., *Tetrahedron Lett.*, **28**, 2115 (1987).
[89] Varasi, M., Walker, K. A. M., and Maddox, M. L., *J. Org. Chem.*, **52**, 4235 (1987).
[90] Galli, R., *J. Org. Chem.*, **52**, 5349 (1987).
[91] Li, X.-Y., Jiang, X.-K., Pan, H.-Q., Hu, J.-S., and Fu, W.-M., *Pure Appl. Chem.*, **59**, 1015 (1987).
[92] Li, X.-Y., Pan, H.-Q., Fu, W.-M., and Jiang, X.-K., *J. Fluorine Chem.*, **31**, 213 (1986); *Chem. Abs.*, **106**, 17641 (1987).
[93] Murray, W. T., Kelly, J. W., and Evans, S. A., *J. Org. Chem.*, **52**, 525 (1987).
[94] Buriks, R. S. and Lovett, E. G., *J. Org. Chem.*, **52**, 5247 (1987).
[95] Higgins, R. H., *J. Heterocycl. Chem.*, **24**, 1489 (1987).
[96] Benedetti, F., Fabrissin, S., Gianferrara, T., and Risaliti, A., *J. Chem. Soc., Chem. Commun.*, **1987**, 406.
[97] Neamati-Mazraeh, N. and McManus, S. P., *Tetrahedron Lett.*, **28**, 837 (1987).
[98] Casadei, M. A., Di Martino, A., Galli, C., and Mandolini, L., *Gazz. Chim. Ital.*, **116**, 659 (1986); *Chem. Abs.*, **107**, 133683 (1987).
[99] El-Kholy, A. E. and Soliman, A. A., *Indian J. Chem.*, **25A**, 1038 (1986); *Chem. Abs.*, **107**, 77084 (1987).
[100] Ringdahl, B. and Jenden, D. J., *J. Med. Chem.*, **30**, 852 (1987); *Chem. Abs.*, **106**, 195656 (1987).
[101] Gonda, J. and Kristian, P., *Collect. Czech. Chem. Commun.*, **51**, 2802 (1986).
[102] Gonda, J. and Kristian, P., *Collect. Czech. Chem. Commun.*, **51**, 2810 (1986).
[103] Faïn, D. and Dubois, J.-E., *J. Org. Chem.*, **52**, 260 (1987).
[104] Astin, K. B. and Musaad, N. S., *J. Org. Chem.*, **52**, 2106 (1987).
[105] Hanack, M. and Wächtler, A. E. F., *Chem. Ber.*, **120**, 727 (1987).
[106] Chuchani, G. and Martin, I., *Int. J. Chem. Kinet.*, **19**, 183 (1987).
[107] Ladika, M., Juršić, B., and Sunko, D. E., *Tetrahedron*, **43**, 179 (1987).
[108] McManus, S. P., Karaman, R. M., Sedaghat-Herati, R., Neamati-Mazraeh, N., Hovanes, B. A., Paley, M. S., and Harris, J. M., *J. Org. Chem.*, **52**, 2518 (1987).
[109] Kalyanan, N. and Likhate, M. A., *J. Chem. Soc., Chem. Commun.*, **1987**, 1028.
[110] Wilen, S. H., Delguzzo, L., and Saferstein, R., *Tetrahedron*, **43**, 5089 (1987).
[111] Carretero, J. C., Ruano, J. L. G., Martinez, M. C., and Rodriguez, J. H., *Tetrahedron*, **43**, 4417 (1987).
[112] Sunko, D. E., Juršić, B., and Ladika, M., *J. Org. Chem.*, **52**, 2299 (1987).
[113] McManus, S. P., Sedaghat, M. R., and Harris, J. M., *Tetrahedron Lett.*, **28**, 5299 (1987).
[114] Rosnati, V., Saba, A., Angius, A., and Casarini, D., *J. Org. Chem.*, **52**, 4094 (1987).
[115] Al-Masoudi, N. A. L. and Hughes, N. A., *J. Chem. Soc., Perkin Trans. 1*, **1987**, 2061.
[116] Lianis, P. S., Rodios, N. A., and Alexandrou, N. E., *Liebigs, Ann. Chem.*, **1987**, 537.
[117] Nishiyama, K. and Oba, M., *Bull. Chem. Soc. Jpn*, **60**, 2692 (1987).
[118] Lazana, M. C. R. L. R., Franco, L. T. M. B., and Herold, B. J., *J. Chem. Soc., Perkin Trans. 2*, **1987**, 1399.
[119] Axelsson, B. S., Långström, B., and Matsson, O., *J. Am. Chem. Soc.*, **109**, 7233 (1987).
[120] Markham, G. D., Parkin, D. W., Mentch, F., and Schramm, V. L., *J. Biol. Chem.*, **262**, 5609 (1987).
[121] DePuy, C. H., Damrauer, R., Bowie, J. H., and Sheldon, J. C., *Acc. Chem. Res.*, **20**, 127 (1987).
[122] Sheldon, J. C., Hayes, R. N., Bowie, J. H., and DePuy, C. H., *J. Chem. Soc., Perkin Trans. 2*, **1987**, 275.

[123] van der Wel, H., Nibbering, N. M. M., Sheldon, J. C., Hayes, R. N., and Bowie, J. H., *J. Am. Chem. Soc.*, **109**, 5823 (1987).
[124] Caserio, M. C. and Kim, J. K., *Adv. Chem. Ser.*, **215**, 65 (1987); *Chem. Abs.*, **107**, 175178 (1987).
[125] Tabet, J. C., Audier, H. E., and Denhez, J. P., *Tetrahedron*, **43**, 873 (1987).
[126] Cernusak, I., Diercksen, G. H. F., and Urban, M., *Chem. Phys. Lett.*, **128**, 538 (1986); *Chem. Abs.*, **106**, 195635 (1987).
[127] Ryaboi, V. M., *Teor. Eksp. Khim.*, **22**, 435 (1986); *Chem. Abs.*, **106**, 118999 (1987).
[128] Ryaboi, V. M., *Teor. Eksp. Khim.*, **22**, 271 (1986); *Chem. Abs.*, **106**, 155650 (1987).
[129] Pross, A., *Adv. Chem. Ser.*, **215**, 331 (1987); *Chem. Abs.*, **107**, 153720 (1987).
[130] Ashby, E. C. and Pham, T. N., *Tetrahedron Lett.*, **28**, 3183 (1987).
[131] Bordwell, F. G. and Wilson, C. A., *J. Am. Chem. Soc.*, **109**, 5470 (1987).
[132] Tanaka, J., Nojima, N., and Kusabayashi, S., *J. Am. Chem. Soc.*, **109**, 3391 (1987).
[133] Ash, C. E., Hurd, P. W., Darensbourg, M. Y., and Newcomb, M., *J. Am. Chem. Soc.*, **109**, 3313 (1987).
[134] Park, S.-U., Chung, S.-K., and Newcomb, M., *J. Org. Chem.*, **52**, 3275 (1987).
[135] Abraham, M. H., Doherty, R. M., Kamlet, M. J., Harris, J. M., and Taft, R. W., *J. Chem. Soc., Perkin Trans. 2*, **1987**, 913.
[136] Abraham, M. H., Doherty, R. M., Kamlet, M. J., Harris, J. M., and Taft, R. W., *J. Chem. Soc., Perkin Trans. 2*, **1987**, 1097.
[137] Doherty, R. M., Abraham, M. H., Harris, J. M., Taft, R. W., and Kamlet, M. J., *J. Org. Chem.*, **51**, 4872 (1986).
[138] Fuellbier, B., Foerster, J., Klemmer, K., Zahn, K., and Haberland, D., *Wiss. Z. Ernst-Moritz-Arndt-Univ. Greifsw., Math.-Naturwiss. Reihe*, **35**, 58 (1986); *Chem. Abs.*, **106**, 195657 (1987).
[139] Fuellbier, B. and Haberland, D., *Wiss. Z. Ernst-Moritz-Arndt-Univ. Greifsw., Math.-Naturwiss. Reihe*, **35**, 63 (1986); *Chem. Abs.*, **106**, 195658 (1987).
[140] Litvinenko, L. M., Popov, A. F., and Anikeev, A. V., *Org. React.* (*Tartu*), **23**, 64 (1986); *Chem. Abs.*, **106**, 213199 (1987).
[141] Kondo, Y. and Kusabayashi, S., *J. Chem. Soc., Faraday Trans. 1*, **83**, 1089 (1987).
[142] Bentley, T. W., *Adv. Chem. Ser.*, **215**, 255 (1987); *Chem. Abs.*, **107**, 197132 (1987).
[143] Kevill, D. N. and Anderson, S. W., *J. Org. Chem.*, **51**, 5029 (1986).
[144] Lee, I., La, S. M., Kong, B. H., Lee, B. S., and Sohn, S. C., *Taehan Hwahakhoe Chi*, **31**, 25 (1987); *Chem. Abs.*, **107**, 153718 (1987).
[145] Fujiyama, R., Kamimura, I., Yoshida, M., Suzuki, K., Kiyooka, S., Murata, A., Goto, M., Fujio, M., and Tsuno, Y., *Chem. Lett.*, **87**, 1043 (1987).
[146] Koo, I. S., Kim, J. J., Yoon, Y. J., and Lee, I., *Taehan Hwahakhoe Chi*, **30**, 296 (1986); *Chem. Abs.*, **106**, 175473 (1987).
[147] Kwun, O. C. and Kyong, J. B., *Taehan Hwahakhoe Chi*, **30**, 188 (1986); *Chem. Abs.*, **106**, 101509 (1987).
[148] Kwun, O. C., Kyong, J. B., and Choi, K. J., *Taehan Hwahakhoe Chi*, **30**, 301 (1986); *Chem. Abs.*, **106**, 175474 (1987).
[149] Bergsma, J. P., Gertner, B. J., Wilson, K. R., and Hynes, J. T., *J. Chem. Phys.*, **86**, 1356 (1987); *Chem. Abs.*, **107**, 115071 (1987).
[150] Kong, Y. S. and Jhon, M. S., *Theor. Chim. Acta*, **70**, 123 (1986); *Chem. Abs.*, **106**, 175488 (1987).
[151] German, E. D. and Tikhomirov, V. A., *Izv. Akad. Nauk SSSR, Ser. Khim.*, **1986**, 1904; *Chem. Abs.*, **106**, 195644 (1987).
[152] Burstein, K. Ya. and Isaev, A. N., *Zh. Strukt. Khim.*, **27**, 8 (1986); *Chem. Abs.*, **106**, 195630 (1987).
[153] Jorgensen, W. L., Buckner, J. K., Huston, S. E., and Rossky, P. J., *J. Am. Chem. Soc.*, **109**, 1891 (1987).
[154] Juršić, B., Sunko, D. E., and Ladika, M., *Tetrahedron*, **43**, 4367 (1987).
[155] Farcasiu, D., *Adv. Chem. Ser.*, **215**, 285 (1987); *Chem. Abs.*, **107**, 197133 (1987).
[156] Maskill, H. and Jencks, W. P., *J. Am. Chem. Soc.*, **109**, 2062 (1987).
[157] Hanna, M. T., Khalil, F. Y., and Beder, S. M., *Bull. Soc. Chim. Belg.*, **96**, 27 (1987).
[158] McManus, S. P., Neamati-Mazraeh, N., Karaman, R. M., and Harris, J. M., *J. Org. Chem.*, **51**, 4876 (1986).
[159] Zamashchikov, V. V., Bezbozhnaya, T. V., and Kovach, N. A., *React. Kinet. Catal. Lett.*, **32**, 233 (1986); *Chem. Abs.*, **107**, 153744 (1987).
[160] Mikhailov, V. A., Drizhd, L. P., and Savelova, V. A., *Zh. Org. Khim.*, **22**, 2319 (1986); *Chem. Abs.*, **107**, 38867 (1987).
[161] Raths, H.-C. and Dehmlow, E. V., *Chem. Ber.*, **120**, 647 (1987).
[162] Dehmlow, E. V. and Wilkenloh, J., *Tetrahedron Lett.*, **28**, 5489 (1987).
[163] Landini, D., Maia, A., and Montanari, F., *Isr. J. Chem.*, **26**, 263 (1985).

[164] Janout, V., Cefelin, P., Tur, D. R., and Vinogradova, S. V., *Polym. Bull. (Berlin)*, **15**, 303 (1986); *Chem. Abs.*, **106**, 137718 (1987).
[165] Barak, G. and Sasson, Y., *J. Chem. Soc., Chem. Commun.*, **1987**, 1267.
[166] Falicky, S. and Alper, H., *J. Chem. Soc., Chem. Commun.*, **1987**, 1039.
[167] Yonovich-Weiss, M. and Sasson, Y., *Isr. J. Chem.*, **26**, 243 (1985).
[168] Nouguier, R. M. and Mchich, M., *J. Org. Chem.*, **52**, 2995 (1987).
[169] Schneider, H. J., Busch, R., Kramer, R., Schneider, U., and Theis, I., *Adv. Chem. Ser.*, **215**, 457 (1987); *Chem. Abs.*, **107**, 197135 (1987).
[170] Okahata, Y. and Ariga, K., *J. Org. Chem.*, **51**, 5064 (1986).
[171] Tanaka, N., Yamaguchi, A., Araki, Y., and Araki, M., *Chem. Lett.*, **87**, 715 (1987).
[172] Sebti, S. and Foucaud, A., *J. Chem. Res. (S)*, **1987**, 72.
[173] Clark, J. H. and Macquarrie, D. J., *Tetrahedron Lett.*, **28**, 111 (1987).
[174] Bazilevskii, M. V., Koldobskii, S. G., and Tikhomirov, V. A., *Usp. Khim.*, **55**, 1667 (1986); *Russ. Chem. Rev.*, **55**, 948 (1986); *Chem. Abs.*, **107**, 175172 (1987).
[175] Usov, V. A., Timokhina, L. V., and Voronkov, M. G., *Usp. Khim.*, **55**, 1761 (1986); *Russ. Chem. Rev.*, **55**, 1003 (1986); *Chem. Abs.*, **106**, 101422 (1987).
[176] Metivier, P., Gushurst, A. J., and Jorgensen, W. L., *J. Org. Chem.*, **52**, 3724 (1987).
[177] Stein, A. R., *Can. J. Chem.*, **65**, 363 (1987).
[178] Allen, A. D., Kwong-Chip, J. M., Mistry, J., Sawyer, J. F., and Tidwell, T. T., *J. Org. Chem.*, **52**, 4164 (1987).
[179] Katritzky, A. R., Schultz, H., Lopez-Rodriguez, M. L., Musumarra, G., and Cirma, G., *J. Chem. Soc., Perkin Trans. 2*, **1987**, 73.
[180] Katritzky, A. R., Lamba, D., Spagna, R., Vaciago, A., Prewo, R., Bieri, J. H., Stezowski, J. J., and Musumarra, G., *J. Chem. Soc., Perkin Trans. 2*, **1987**, 1391.
[181] Coles, C. J. and Maskill, H., *J. Chem. Soc., Perkin Trans. 2*, **1987**, 1083.
[182] Rao, T. J., Punnaiah, G., and Sundaram, E. V., *Proc. Indian Acad. Sci., Chem. Sci.*, **97**, 55 (1986); *Chem. Abs.*, **106**, 49413 (1987).
[183] Raghavan, R. S., Kalavathy, K., and Gandhirajan, C. K., *J. Inst. Chem. (India)*, **57**, 237 (1985); *Chem. Abs.*, **106**, 175472 (1987).
[184] Rajasekaran, K. and Gnanasekaran, C., *Proc. Indian Acad. Sci., Chem. Sci.*, **97**, 47 (1986); *Chem. Abs.*, **106**, 66561 (1987).
[185] Morita, T. and Ōki, M., *Bull. Chem. Soc. Jpn*, **59**, 3605 (1986).
[186] Kravchenko, V. V., Kostenko, L. I., Popov, A. F., and Kotenko, A. A., *Zh. Org. Khim.*, **22**, 2322 (1986); *Chem. Abs.*, **107**, 197207 (1987).
[187] Barber, S. E. and Kirby, A. J., *J. Chem. Soc., Chem. Commun.*, **1987**, 1775.
[188] Lewis, E. S., Yousaf, T. I., and Douglas, T. A., *J. Am. Chem. Soc.*, **109**, 2152 (1987).
[189] Yousaf, T. I. and Lewis, E. S., *J. Am. Chem. Soc.*, **109**, 6137 (1987).
[190] Lee, I., Kim, H. Y., and Kang, H. K., *J. Chem. Soc., Chem. Commun.*, **1987**, 1216.
[191] Byrne, C. J., Happer, D. A. R., Hartshorn, M. P., and Powell, H. K. J., *J. Chem. Soc., Perkin Trans. 2*, **1987**, 1649.
[192] Yoh, S.-D., Park, J.-H., Lee, K.-A., and Han, I.-S., *Bull. Chem. Soc. Jpn*, **60**, 1149 (1987).
[193] Yoh, S. D. and Park, J. H., *Taehan Hwahakhoe Chi*, **30**, 383 (1986); *Chem. Abs.*, **106**, 155638 (1987).
[194] Yoh, S.-D., Kim, S.-H., and Park, J.-H., *J. Chem. Soc., Perkin Trans. 2*, **1987**, 1439.
[195] Yoh, S.-D., Yang, K.-H., and Han, I.-S., *Tetrahedron*, **43**, 4089 (1987).
[196] Lee, I., Lee, B. C., Oh, Y. J., and Sohn, S. C., *Kich'o Kwahak Yonguso Nonmunjip (Inha Taehakkyo)*, **7**, 77 (1986); *Chem. Abs.*, **106**, 175475 (1987).
[197] Lee, I. and Sohn, S. C., *Bull. Korean Chem. Soc.*, **7**, 321 (1986); *Chem. Abs.*, **107**, 6687 (1987).
[198] Fujio, M., Funatsu, K., Goto, M., Mishima, M., and Tsuno, Y., *Tetrahedron*, **43**, 307 (1987).
[199] Fujio, M., Funatsu, K., Goto, M., Seki, Y., Mishima, M., and Tsuno, Y., *Bull. Chem. Soc. Jpn*, **60**, 1091 (1987).
[200] Fujio, M., Goto, M., Seki, Y., Mishima, M., Tsuno, Y., Sawada, M., and Takai, Y., *Bull. Chem. Soc. Jpn*, **60**, 1097 (1987).
[201] Liu, K.-T. and Kuo, M.-Y., *J. Chem. Soc., Chem. Commun.*, **1987**, 640.
[202] Akshentsev, V. V., Dragalov, V. V., and Chimishkyan, A. L., *Zh. Org. Khim.*, **23**, 305 (1987); *Chem. Abs.*, **107**, 153717 (1987).
[203] Charton, M. and Ziffer, H., *J. Org. Chem.*, **52**, 2400 (1987).
[204] Harris, J. M., and McManus, S. P., *Adv. Chem. Ser.*, **215**, 1 (1987); *Chem. Abs.*, **107**, 153642 (1987).
[205] Brauman, J. I., Dodd, J. A., and Han, C. C., *Adv. Chem. Ser.*, **215**, 23 (1987); *Chem. Abs.*, **107**, 115019 (1987).

[206] Jencks, W. P., *Adv. Chem. Ser.*, **215**, 155 (1987); *Chem. Abs.*, **107**, 175180 (1987).
[207] Ritchie, C. D., *Adv. Chem. Ser.*, **215**, 169 (1987); *Chem. Abs.*, **107**, 153773 (1987).
[208] Hudson, R. F., *Adv. Chem. Ser.*, **215**, 195 (1987); *Chem. Abs.*, **107**, 197150 (1987).
[209] Menger, F. M., *Adv. Chem. Ser.*, **215**, 209 (1987); *Chem. Abs.*, **107**, 153658 (1987).
[210] Allen, A. D., Kanagsabapathy, V. M., and Tidwell, T. T., *Adv. Chem. Ser.*, **215**, 315 (1987); *Chem. Abs.*, **107**, 153719 (1987).
[211] Hine, J. and Chen, Y.-J., *J. Org. Chem.*, **52**, 2091 (1987).
[212] Pearson, R. G., *J. Org. Chem.*, **52**, 2131 (1987).
[213] Song, X. and Han, C., *Huaxue Tongbao*, **1986**, 46; *Chem. Abs.*, **107**, 22711 (1987).
[214] Bosch, P., Camps, F., Chamorro, E., Gasol, V., and Guerrero, A., *Tetrahedron Lett.*, **28**, 4733 (1987).
[215] Lee, I., Kim, C. K., and Song, C. H., *Bull. Korean Chem. Soc.*, **7**, 391 (1987); *Chem. Abs.*, **107**, 38864 (1987).
[216] Lee, I. and Seo, H. S., *Bull. Korean Chem. Soc.*, **7**, 448 (1986); *Chem. Abs.*, **107**, 96058 (1987).
[217] Lee, I. and Song, C. H., *Bull. Korean Chem. Soc.*, **7**, 186 (1986); *Chem. Abs.*, **106**, 118942 (1987).
[218] Ford, G. P., and Smith, C. T., *J. Chem. Soc., Chem. Commun.*, **1987**, 44.
[219] Voityuk, A. A. and Bliznyuk, A. A., *Zh. Strukt. Khim.*, **27**, 25 (1986); *Chem. Abs.*, **106**, 83750 (1987).
[220] Formosinho, S. J., *Tetrahedron*, **43**, 1109 (1987).
[221] Gill, P. M. W., Maskill, H., Poppinger, D., and Radom, L., *J. Chem. Res. (S)*, **1987**, 54.
[222] Pasto, D. J., Freeman, J. P., and Fields, M. K., *J. Org. Chem.*, **52**, 4058 (1987).
[223] Schneider, H.-J., Bucheit, U., and Schmidt, G., *J. Chem. Res. (S)*, **1987**, 92.
[224] McLure, F. I., Norris, R. K., and Wilson, K., *Aust. J. Chem.*, **40**, 49 (1987).
[225] Zamashchikov, V. V., Bezbozhnaya, T. V., and Chanysheva, I. R., *Zh. Org. Khim.*, **22**, 1145 (1986); *Chem. Abs.*, **106**, 32183 (1987).
[226] Popov, A. F., Piskunova, Zh. P., and Matvienko, V. N., *Zh. Org. Khim.*, **22**, 1442 (1986); *Chem. Abs.*, **106**, 155647 (1987).
[227] Alekperov, N. M., Azizov, A. M., Movsumzade, E. M., and Babaev, N. M., *Zh. Prikl. Khim. (Leningrad)*, **59**, 1814 (1986); *Chem. Abs.*, **106**, 155832 (1987).
[228] Dauben, W. G., Gerdes, J. M., and Look, G. C., *Synthesis*, **1986**, 532; *Chem. Abs.*, **106**, 137673 (1987).
[229] Tsirul'nikova, N. V., Danielyan, D. G., and Temkina, V. Ya., *Zh. Obshch. Khim.*, **56**, 1377 (1986); *Chem. Abs.*, **106**, 83743 (1987).
[230] Bellucci, G., Bianchini, R., and Ambrosetti, R., *J. Chem. Soc., Perkin Trans. 2*, **1987**, 39.
[231] Spears, L. G., Liao, A., Minsek, D., and Lewis, E. S., *J. Org. Chem.*, **52**, 61 (1987).
[232] Yang, Y. C., Ward, J. R., Wilson, R. B., Burrows, W., and Winterle, J. S., *Thermochim. Acta*, **114**, 313 (1987); *Chem. Abs.*, **107**, 38979 (1987).
[233] Yang, Y.-C., Szafraniec, L. L., Beaudry, W. T., and Ward, J. R., *J. Org. Chem.*, **52**, 1637 (1987).
[234] Larranga, J. M., Irabien, J. A., and Ortiz, M. I., *Elhuyar*, **12** (Lehen Alea), 5 (1986); Chem. Abs., **106**, 17660 (1987).
[235] Powell, M. F., *J. Org. Chem.*, **52**, 56 (1987).
[236] Gavrilov, V. I., Gumerov, N. S., and Rakhmatulin, R. R., *Zh. Obshch. Khim.*, **57**, 206 (1987); *Chem. Abs.*, **107**, 58355 (1987).

Organic Reaction Mechanisms 1987
Edited by A. C. Knipe and W. E. Watts

CHAPTER 11

Carbanions and Electrophilic Aliphatic Substitution

A. C. KNIPE

Department of Chemistry, University of Ulster, Coleraine

Carbanion Structure and Stability

MO Calculations

Structural conclusions, based on measurements of photoelectron spectra of $(CH_2CN)^-$ and $(CD_2CN)^-$, have been reinforced by *ab initio* Hartree–Fock and Moller–Plesset calculations on both the cyanomethyl radical and cyanomethide ion. The radical is found to be a planar C_{2v} species whereas the ion is almost described as $CH_2{=}C{=}N^-$ but with the hydrogens bent slightly out of the molecular plane.[1] Likewise, the isocyanomethide anion ($^-CH_2NC$) has been found to be a pyramidal species and a localized ion.[2] These studies have permitted determination of the bond dissociation energy of CH_3CN, the electron affinities of the radicals (CH_2CN and CH_2NC), and the isomerization energies for isocyano/cyano isomerizations of the carbanions and the radicals.

Structures of C^-, CH^-, CH_2^-, and CH_3^- have been investigated by a multiconfigurational SCF + CI approach which was also applied to the parent neutral systems for which electron affinities have been determined; there is reason to question the supposed non-planarity of CH_3^-.[3]

Ab initio MO calculations of the structures of ethyl, ethenyl, and ethynyl carbanions have been performed and various rearrangement processes involving these species have also been explored. These included 1,2-hydrogen shifts for all three systems, inversion of ethenyl and ethyl anions, and rotation and inversion-rotation for the ethyl anion.[4] For the ethenyl anion, 1,2-hydrogen shift may occur with migration of *syn* or *anti* hydrogens (with C_2 and C_s transition-state symmetries, respectively) while for ethynyl and ethyl anions the 1,2-hydrogen shifts occur with C_{2v} and C_2 symmetry, respectively.

Complete Active Space SCF (CAS-SCF) and externally contracted CI (CCI) *ab initio* methods and optimized basis sets have been used to study the stability of the molecular ions CH_3F^- and CH_3Cl^-; neither of the parent molecules has a positive electron affinity and the extra electron occupies a very diffuse molecular orbital extending out from the methyl group.[5]

MNDO and *ab initio* SCF-MO methods have been applied to determine the barriers to inversion to configuration at the centre X and to ring-opening in a series of isoelectronic three-membered-ring systems (**1**)–(**4**). Loss of configuration in (**1a**)–(**3a**) is predicted to occur via ring-opening to give an enolate anion, rather than via direct inversion at C^- which has the lowest barrier for (**4**). Oxirane and aziridine anions are expected to be configurationally stable whereas thiirane carbanions are predicted to ring-open rapidly to give a thioenol.[6]

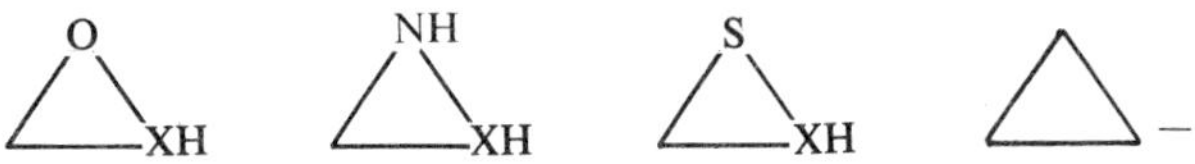

(**1**) (**2**) (**3**) (**4**)

a; $X = C^-$; **b**; $X = N$; **c**; $X = O^+$

The computational realization of planar tetra-coordinate carbon has been confirmed for 3,3-dilithio-1,2-diboracyclopropane, 1,1-dilithiocyclopropane, and 3,3-dilithiocyclopropene, with allowance for effects of polarization functions and correlation energy;[7] however, planar 3,3-dilithiocyclopropene is found to have no barrier to decomposition to cyclopropenylidene and dilithium.

Ab initio exploration (at the RHF/3-21G level) of the potential-energy surface for 2,3-dilithiopropene has revealed two local minima and one transition structure; the bonding to lithium in this highly fluctional molecule is essentially ionic.[8] The structure of the homoantiaromatic carbanion 7-norbornadienyllithium in its monomer, dimer, and solvent dimer forms has been investigated by the semi-empirical MNDO MO method.[9]

Organolithiums

In addition to theoretical studies of organolithium structures, reviewed in the previous section, there have been reports of their direct observation by NMR spectroscopy.[10–12] ^1H and ^{13}C chemical shifts for anions of 2-, 3-, and 4-benzylpyridine as a function of cation, temperature, and solvent have been determined.[10] Calculated charge distributions reveal a shift of charge to the pyridine ring in the order 4- > 2- > 3-benzylpyridine with concurrent charge decrease in the phenyl rings. The lithium salts form tight ion-pairs in THF with cation coordination to nitrogen in each case. The effects of the lithium ion on charge distribution and on the barriers to phenyl and pyridyl ring rotation have been reported.

Controversy surrounding the relative merits of principal components (PC) analysis versus linear solvation energy relationships, for correlation and prediction of properties depending on solute-solvent interactions, has continued with conflicting interpretations of the ^{13}C NMR shifts of the planar anion indenyllithium.[11,12]

Previous attempts to prepare the parent vicinal dilithioalkenes, *cis*- and *trans*-1,2-dilithioethylene (**9** and **10**), have been unsuccessful. This has now been achieved by a double metal–metal exchange sequence, from the corresponding 1,2-bis-(tributylstannyl)ethylenes (**5** and **6**) via less sterically demanding two-coordinate mercury derivatives (**7** and **8**).[13] Experiments using *cis-trans* mixtures of [D_2]-(**7**) + (**8**) and of (**7**) + [D_2]-(**8**) with *t*-BuLi established that (**9**) and (**10**) do not interconvert under the reaction conditions.

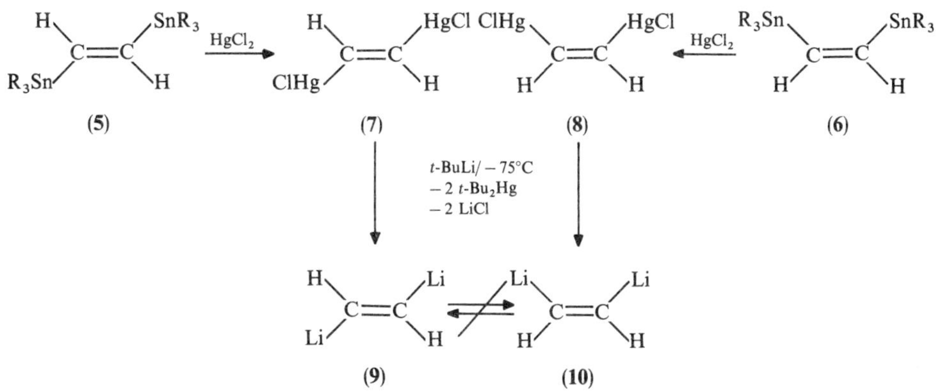

Treatment of 1,1,3,3-tetraphenylallene with lithium at low temperatures in etheral solvents gives a dilithium allenide which is believed to exist in two equilibrating 2-lithioallyllithium forms (**11**).[14] An analogous, less stable, species is obtained on reaction with sodium but with potassium the dipotassium allenide could not be detected since its protonation by solvent forms 1,1,3,3-tetraphenylallylpotassium which is reduced by excess potassium to give the trianion (**12**). The allyl geometry of the hydrocarbon framework is maintained in the 1,1,3,3-tetraphenylallyl trianion. Isotopic perturbation studies using variable temperature

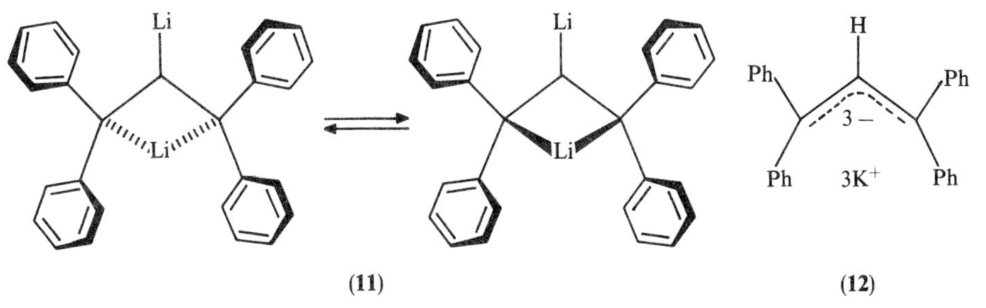

^{13}C NMR spectroscopy have established that allylsodium and allylpotassium have symmetrical structures whereas allyllithium is an unsymmetrical dimer in THF.[15] The ^{13}C NMR spectrum of 3-(phenylsulphonyl)tricyclo[4.1.0.0^{2,7}]hept-4-en-3-yllithium reveals a considerable interaction between the allyl anion and bicyclobutane components.[16]

The crystal structure of α-(phenylsulphinyl)-α-methylbenzyllithiumtetramethylethylenediamine in its dimeric form (bonded through a Li_2O_2 four-membered ring) has been reported; the configuration is similar to that proposed previously to account for diastereofacial differentiation in addition of electrophiles to α-sulphinyl carbanions.[17] The crystal structure of the tetramethyl lithium-coordinated α-sulphonimidoyl carbanion $[(Me_3Si)CH\{S(O)(NSiMe_3)\}Li]_4$ has been discussed in the context of asymmetric synthesis using chiral lithium alkyl sulphoxamides.[18]

Aromatic and Other Delocalized Ions

Structures of the cyclononatetraenyl anion (CNT) and of its benzo (BCNT) and 1,2:3,4-dibenzo (DBCNT) derivatives have been investigated by semiempirical MNDO and AM1 methods.[19] The results support the observed interconversions of the all-*cis* and single-*trans* isomers of CNT; the latter is substantially delocalized although it is calculated to be significantly non-planar and *ca* 10–20 kcal mol^{-1} less stable than the former in the gas phase. The *trans,cis,trans,cis*-isomer is not an important contributor to CNT. In contrast, the single-*trans* isomer (**13**) of BCNT was calculated to be 4-8 kcal mol^{-1} more stable than the all-*cis* isomer, in keeping with their observed interconversion on warming. The preferred highly non-planar structure (**14**) of all-*cis* DBCNT is unlikely to isomerize to a planar form, or to single- or double-*trans* isomers, whereas conversion into a dihydroindane-type analogue might well explain its known transformation.

(13)

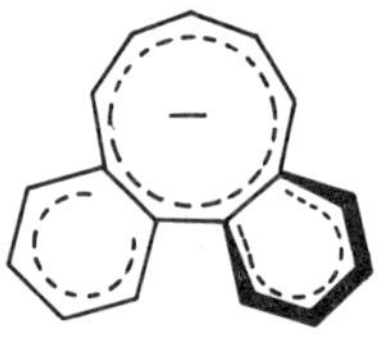

(14)

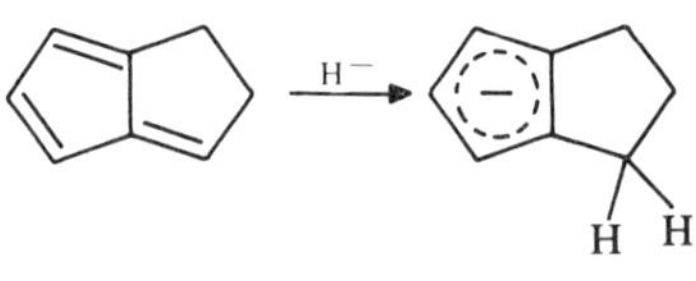

(15) **(16)**

Semiempirical MNDO calculations suggest that 9- or 10-anions derived from 9-R-9,10-dihydroanthracenes (R = H, Me, Et, *i*-Pr, Ph) are planar or almost planar structures in which the anionic centre is close to sp^2-hybridized even under the coulombic influence of a co-cation. However, 9-lithio-9,10-dihydroanthracenes are predicted to be more puckered than the anionic species.[20]

Equilibrium acidity constants of 22.7 and 22.1 have been reported for 6,6-dimethylfulvene and 6-phenyl-6-methylfulvene, respectively, in DMSO.[21] The reaction of 1,2-dihydropentalene (**15**) with $LiAlH_4$, to give 1,2,3,4-tetrahydropentalene and 1,2,3,5-tetrahydropentalene via (**16**), has been discussed.[22]

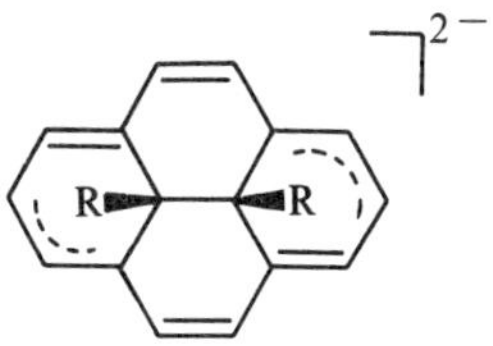

(17)

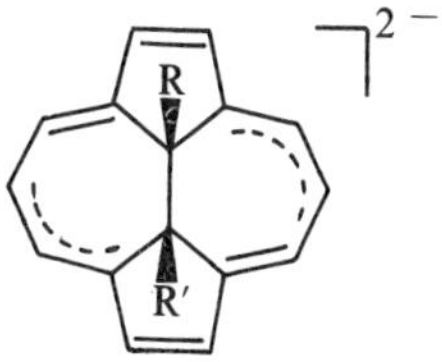

(18)

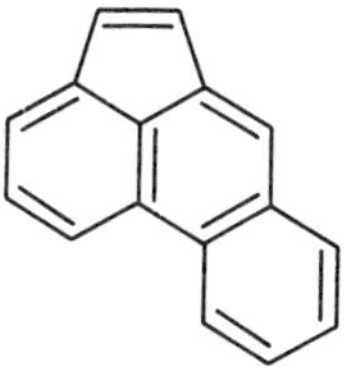

(19)

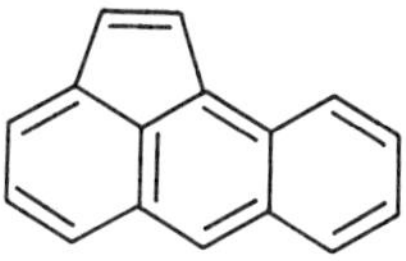

(20)

The connection between the aromaticity of annulenes and annulene ions with $4n + 2$ π-electrons and the stability of the closed-shell solutions for these systems has been analysed; estimates are given of the critical ring size at which the Hückel rule breaks down.[23] The geometries of dianions (**17**) and (**18**) and their precursor bridged [14]annulenes have been studied by a modified π-SCF force-field approach.[24] The diatropic neutral molecules are almost planar aromatic structures whereas the dianions are paratropic and of reduced symmetry as a consequence of the sign inversion of the ring-current contribution. An investigation of the dianions and dications derived from a wide range of bridged $[4n + 2]$annulenes has revealed that the observed paratropism of the $4n$ π-ions is markedly affected by structural variations such as size, conformation, and configuration of the perimeter; the results have been discussed in terms of the HOMO/LUMO gap and the degree of bond fixation.[25]

^{13}C and ^{1}H NMR techniques have also been used to study charge delocalization paratropicity and structure in the dianions of acephenanthrylene (**19**) and aceanthrylene (**20**).[26] The paths of electron delocalization of the dianions differ from those of the neutral systems in that the bridge double bond makes a significant contribution to delocalization. In each case the dominant delocalization path is that which has the minimum paratropic contribution. The first paratropic three-dimensional $4n\pi$ system (**21**) has also been studied by NMR techniques; this double-deck anion involves a charged polycyclic layer interacting with a neutral benzene ring which therefore serves as a useful probe of the through-space interaction. The unusually pronounced low-field shift of the benzene ring protons is attributed to the anisotropy of the $4n\pi$ doubly charged polycyclic moiety.[27] The synthesis of nonafulvene has been realized by oxidative coupling of cyclononatetraenide.[28]

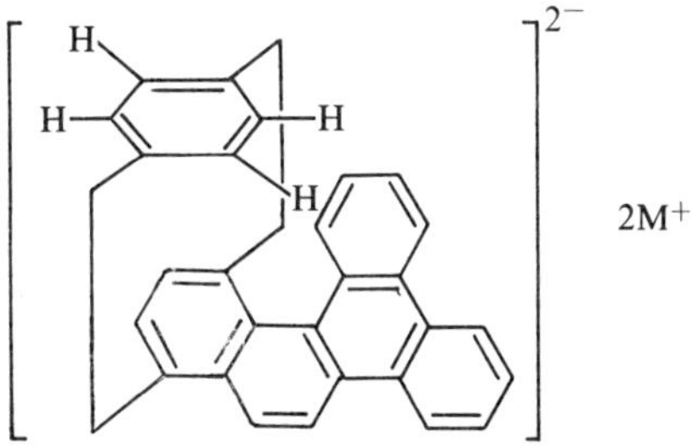

(21)

Evidence for the formation of π-dianions of 9-fluorenone and 10,11-dihydro-5H-dibenzo[a,d]cyclohepten-5-one has been gained from proton NMR and CNDO charge distributions.[29]

The ($\pi2+\sigma2+\sigma2$)-cycloreversion route to 1,4-diheterocines from *cis*-aza/oxa-bis-σ-homobenzenes has been applied successfully[30] to synthesis of the monohetero-analogues (**22**) from which planar 10π-diatropic anions (**23**) may be obtained,[31] regardless of the nature of the hetero-ring-elements.

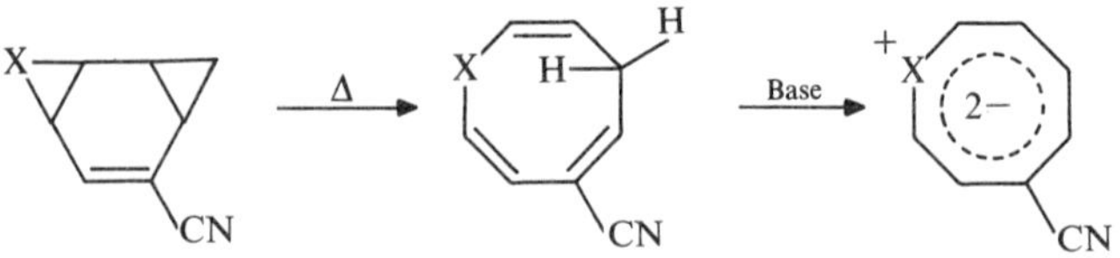

(**22**) (**23**)

X = N-Bu^t, N-Tos, O, S

A high stability and unusual charge distribution have been reported for the 1,3-diphenylbenzo[*c*]thiophene dianion, which is a rare example of an heterocyclic 4$n\pi$ charged system.[32]

Homoaromaticity does not play an important role in stabilization of the 2,4-diphenylbicyclo[3.2.1]oct-6-en-3-one dianion which, although displaying some ^{13}C NMR features suggestive of long-range π-interactions, is not much more stable than its 6,7-dihydro analogue.[33]

^{13}C NMR data reveal that the fraction of π-charge withdrawn by electronegative groups (X) attached to carbanion centres is subject to saturation effects in the case of $PhCHX^-$ and Ph_2CX^-.[34] The ^{13}C NMR spectra of 1,4-diphenyl- and 1,1,4,4-tetraphenyl-1,3-butadiene dianions have been reported.[35]

Carbanion Reactions

Enolates and Related Species

The regioselectivity of alkylation of acetophenone enolate by dimethyl sulphate has been studied as a function of the cation (lithium, sodium, or potassium) in fourteen aprotic solvents.[36] Linear correlations of the logarithm of the molar ratio of *O*-methylation versus *C*-methylation products with solvent polarity and Lewis basicity parameters and either the ionic radius or the reciprocal of the ionic volume of the metal were found. The results have been interpreted by consideration of the HOMO energies of the enolates, their polarity index (excess of negative charge densities on O and α-C), and frontier orbital index, as a function of the correcting parameter in HMO calculations which characterizes the effect of the cation field upon the electron distribution.

Alkylation of ester enolates has been reviewed.[37] Exclusive *O*-alkylation, with formation of the corresponding diarylketene acetals, has been reported for reaction of the enolates of methyl bis(pentamethylphenyl)acetate and isopropyl bis(pentachlorophenyl)acetate; the steric hindrance which suppresses *C*-alkylation is also responsible for the stability of the ketene acetals in acid.[38] Regiospecific alkylation of 3*a*,4,7,7*a*-tetrahydroindanone (**24**; X = H) has been achieved by using a phenylthio group (X = PhS) either to direct formation of enolate ion (**25**) by dissolving metal reduction or to protect the tertiary carbon during generation and alkylation of anion (**26**); the stereochemical outcome has been determined for alkylation with methyl iodide, allyl bromide, and methyl bromoacetate.[39]

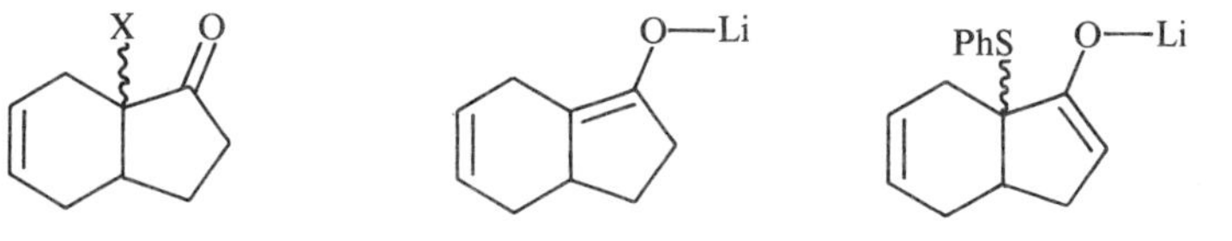

(**24**) (**25**) (**26**)

Phase-transfer 2-methylation of 6,7-dichloro-5-methoxy-2-phenyl-1-indanone by methyl chloride has been effected in 94% enantiomeric excess (ee) using *N*-benzylcinchoninium halides in 50% NaOH/toluene;[40] a Hammett plot of log ee/ee_0 vs σ (for substitution of the catalyst) gave a reaction constant $\rho = 0.21$ with an ee range of 60–94%.

A novel approach to synthesis of non-proteinogenic amino-acids with high optical purities has involved stereoselective alkylation of *cis*-disubstituted β-lactam ester enolates followed by reductive cleavage of the lactam ring.[41] The enantioselectivity is a consequence of approach of the electrophile from the least hindered side of the enolate which is chelated with the β-lactam oxygen. Enolate chelation is also believed to account for the enantioface differentiation (0–76% ee in formation of **28**) of the prochiral enolates (**27a**) and (**27b**) by a series of mixed sulphates derived from D-(+)-glucose; the electronic, steric, and conformational requirements of the chiral group (A*) on the sulphate have been studied and its *O*-coordination to lithium, which acts as a pivot during methyl-group transfer, has been proposed.[42]

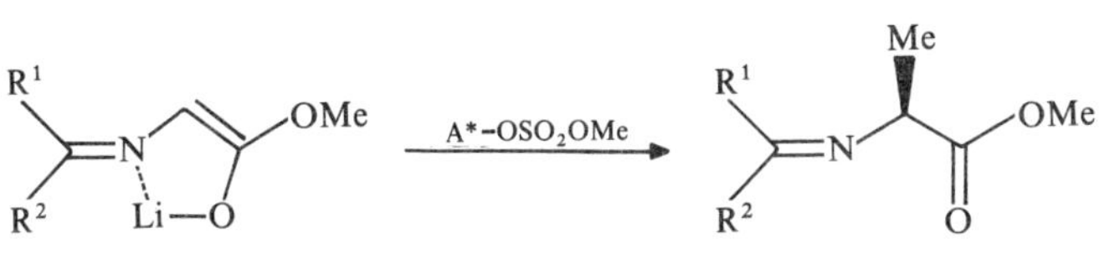

(**27**) (**28**)

a; $R^1 = R^2 = Ph$; **b**; $R^1 = p\text{-}Me_2NC_6H_4$, $R^2 = Bu^t$

Formation of an eight-membered chelate has been implicated in the 1,4-chiral induction, directed by a hydroxyl group, found for alkylation of lithium enolates generated from *t*-butyl esters of δ-hydroxy-carboxylic acids.[43] The conformation of the enolate derived from the 14-membered lactone, 13-tetradecanolide, has been discussed in an attempt to rationalize its predominant conversion into the 2*R**,13*S**-dimethyltridecanolide on alkylation with methyl iodide.[44]

The site-selectivity for arolylation of aza-allyl anions [derived from *N*-(diphenylmethyl)arylmethanimines] has been achieved.[45] Products of acylation of α-fluoromethanephosphonate carbanions have been studied.[46]

1,3-asymmetric induction achieved by promoting a Reformatsky-type reaction of (**29**) by samarium diiodide apparently proceeds via a rigid cyclic transition state (**30**), enforced by chelation.[47]

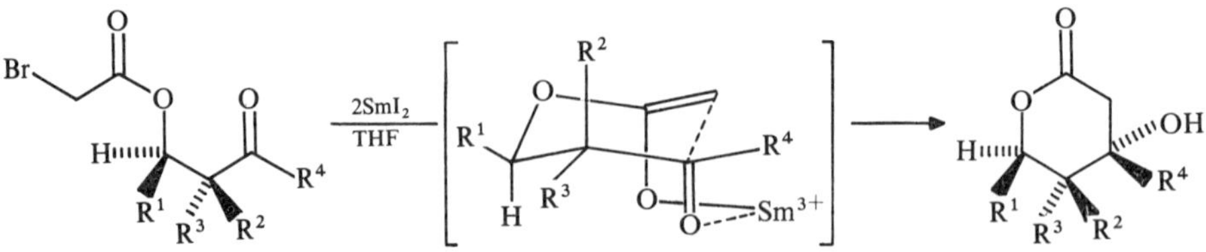

(**29**) (**30**) (**31**)

Results of a detailed MNDO study of the Reformatsky reaction of α-bromo-esters suggest that the reaction proceeds through an aromatic transition state by a [3,3]-sigmatropic shift of a zinc enolate; an antiaromatic transition state disfavours a 1,4-shift of the isomeric and more stable organozinc derivative.[48]

Reaction between monosubstituted cyclohexanones and the lithium dienolate derived from crotonic acid proceeds by nucleophilic addition of the α and γ carbon atoms in the cold and on heating, respectively;[49] the kinetic preference is for equatorial approach whereas equilibration favours the products from axial attack. Directionality and stereochemistry of aldol addition have been investigated.[50]

Equilibrium caesium ion-pair acidities of acetophenone, propiophenone, and related ketones decrease as the equilibrium enolate concentration is increased; this is a consequence of aggregation which is influenced by both electronic and steric factors.[51] Studies of the kinetics of protonation of enols, enolates and their analogues, and the stereochemistry of ketonization have been reviewed.[52] The enolate ion $(CH_3COCHCOCH_3)^-$ has been generated by high-pressure gas-phase condensation between acetone and its enolate anion.[53] Collision-induced dissociations of enolate anions in the gas phase have been reviewed[54] and further studies have featured eliminative fragmentation reactions of enolate negative ions derived from dimethyl succinates,[55] hexane-2,5-dione,[55] acetylacetone,[56] and 3,3-dimethylheptan-4-one.[57]

The ring transformation which occurs upon reaction of 1,3-dinitroquinolizin-4-one with enolate anions $(R^1CH_2COCHCOR^2)^-$, to give the corresponding 2-COR^2-6-R^1-4-nitrobenzene-1,3-diols, has been explained.[58] Mono- and bi-functional compounds have been prepared by reaction of the masked propenal homoenolate, 3,3-diethoxypropyllithium, with a range of electrophilic reagents.[59] Homoenolization in 7-substituted camphenilones has been probed by ^{13}C NMR spectroscopy.[60]

Ring-opening of 1-benzoyl-2,2-dimethylaziridine by nitrile anions $(R^1R^2CHCN)^-$ has been found to occur in the abnormal mode with high regio-selectivity.[61] Ring-opening of 1,1-diactivated 2-vinylcyclopropanes with concomitant addition of a carbon nucleophile has been found to occur with high regio- and stereo-selectivity when catalysed by palladium(0) complexes.[62] In [*n*.1.0]bicyclic systems (**32**) the alkylation occurs *syn* to the cyclopropane bond being cleaved.

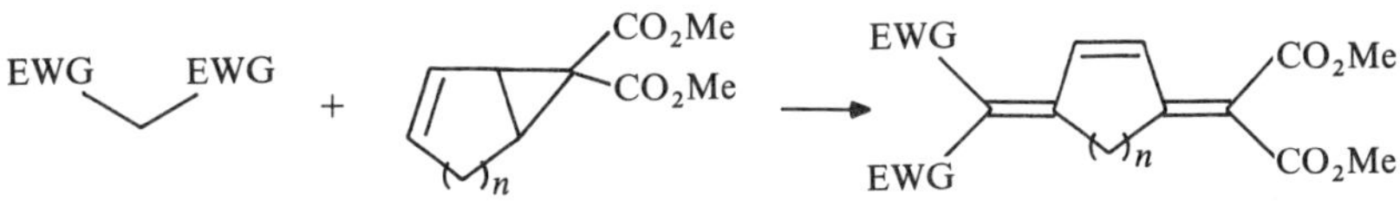

EWG = COMe, SO_2Ph (**32**) $n = 1,2$

Heteroatom-stabilized Species

A review of twentyfive years of use of dimethylsulphoxonium methylide (Corey's reagent) has highlighted the versatile chemistry of this nucleophile.[63]

Diastereoselective α-alkylation of the dianions of β-hydroxy-sulphoxides, to give *threo*-α-alkyl-β-hydroxy-sulphoxides with high 1,2-asymmetric induction, has been applied in the synthesis of (+)- and (−)-disparlure.[64] The salient features of the structure of α-sulphonyl carbanions have been discussed in the light of X-ray crystal structures.[65] The SO_2—C^- bond distance is significantly shorter than for the free sulphone; the S—O bond distances are essentially unchanged; the coordination about carbon is planar, and the counterion interacts with the sulphonyl oxygen atoms but not with the carbanionic carbon. These observations are consistent with an ylide-like structure featuring an electron-rich carbon, a barrier to rotation about the SO_2—C^- bond, and little change in electron density between S and O.

The rates of exchange of α-hydrogens of cyclopropyl isopropyl sulphone indicate that carbanion formation is not inhibited by enforced pyramidalization; the cyclopropyl: isopropyl exchange ratio is 47:1.[66] Base-catalysed exchange of 2-H and 4-H in 3-thiatricyclo[2.2.1.0^{2,6}]heptane 3,3-dioxide (**33**) is even more rapid, apparently because the rigid skeleton provides the optimum gauche configuration for stabilization of the incipient carbanion.

Metallation of 1,3-oxathiane 3,3-dioxides occurs at C(4) and not C(2) since the ring oxygen destabilizes the sulphonyl carbanion.[67] Mono- and di-anions of 3-(4-methoxyphenylsulphonyl) furan have been generated regioselectively and trapped by electrophiles to give regioisomers in high yield;[68] the bidentate carbanion stabilization of (**34**) is believed to account for *o*-lithiation of the phenyl ring, rather than metallation at C(5) of the furan, to give the furan dianion. The sulphonyl group

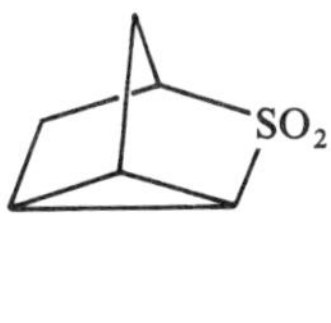

(**33**)

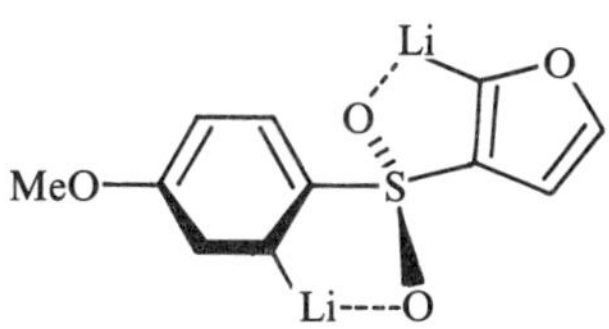

(**34**)

also directs formation of (**36**) by stereospecific elimination of methanol followed by regiospecific metallation to give (**37**) upon reaction between BuLi and the acetal (**35**).[69]

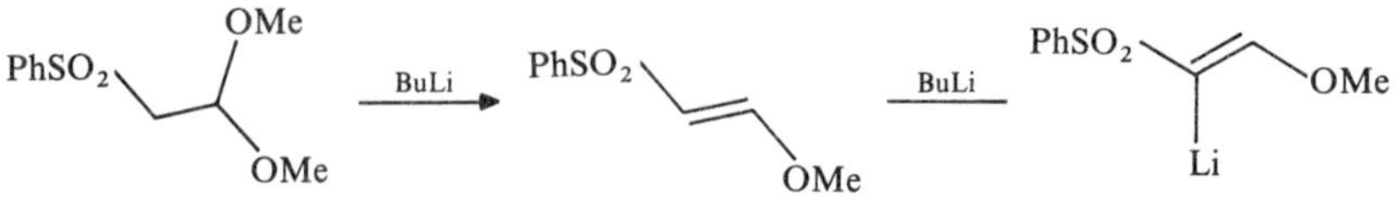

(**35**) (**36**) (**37**)

The crown ether ring of *N*-phenylaza-15-crown-5 and of *N*-phenyl-13-aza-1,4,7,10-tetraoxacyclopentadecane is believed to promote nucleophilic reaction of DMSO with the aromatic amine by complexation through the sulphur atom.[70] The dimethyl sulphoxide anion has been found to be a non-transferable ligand in a wide variety of reactions of mixed homocuprates ($MeSOCH_2CuR$)Li,[71] and transannular addition of α-sulphinyl carbanions to unactivated double bonds has been reported.[72]

Factors which influence stabilization of a carbanionic centre by adjacent OH and SH groups have been analysed in an *ab initio* SCF—MO study (at an extended level) of the Y and W conformations of H_2C—OH and H_2C—SH.[73] The SH group is found to stabilize the anionic centre more than the OH group, as a consequence of σ effects rather than *d*-orbital overlap which is relatively insignificant; the effect of non-bonded interaction favours OH over SH.

The various factors which cause carbanions to undergo carbanion addition to α,β-unsaturated carbonyl compounds have been discussed with regard to the product proportions found for aprotic conjugate addition reactions of the carbanions of oct-2-enyl sulphides and thiocarbamates with γ-crotonolactone; reaction through C(1), relative to C(3), of the carbanion is favoured in the presence of hexamethylphosphoric triamide (hmpa) but this cannot be explained easily.[74] Reactions of a similar range of anions with 4-*t*-butylcyclopent-2-enone in the presence of hmpa in THF at −70°C give, as predominant products, diastereomeric mixtures of *syn* and *anti* allylic sulphides arising from reaction through C(1) of the octenyl carbanions.[75] The configurations of vinyl sulphides obtained through reaction at C(3), apparently under orbital control, reflect the configurations of carbanions from which they are derived.

The HSAB principle has been used to explain the regiochemistry of reaction of the dithiosubstituted crotyl lithium (**38**) with α,β-unsaturated ketones at either the α- or γ-site of the carbanion, depending on the nature of the ketone. Reaction of (**38**) with

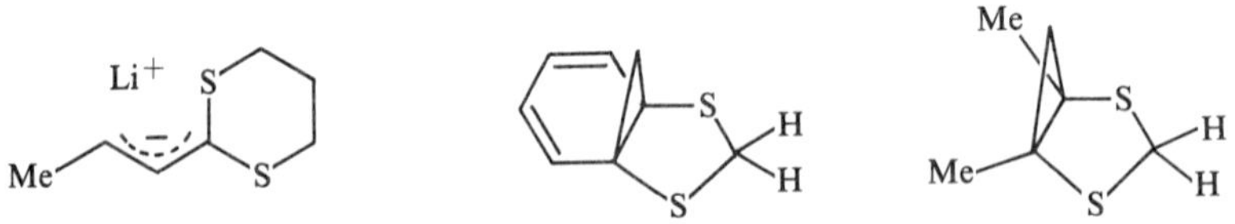

(**38**) (**39**) (**40**)

an α,β-unsaturated aldehyde gives γ-products in favour of the *anti* isomer;[76] and γ-(1,2)-adducts obtained from reaction with enals undergo alkoxy-Cope rearrangement on treatment with KH to give virtually complete α-(1,4)-addition products.

Carbanions of dithiabicyclo[3.1.0]hexane derivatives (**39**) and (**40**) are highly stereoselective in their reaction with electrophiles, there being a 50:1 preference for electrophile approach at the side *anti* to the cyclopropane ring.[77] A ninefold kinetic stereoselectivity (with an isotope effect of 2.8) has been determined for the lithiation reaction by study of the two diasteriomeric 8-deuterio derivatives of (**39**). Thus, the *trans* stereoselectivity is mainly thermodynamically controlled and, in the light of *ab initio* MO calculations, explicable in terms of stereoelectronic effects.

Reactions of carbanions generated from dialkylamino(methylthio)acetonitriles have been studied.[78]

A previously unrecognised directing effect of the neighbouring hydroxyl group has been proposed to account for the enhanced rate and *E*-selectivity observed for Wittig reactions of both acyclic and cyclic α-hydroxy-ketones with stabilized phosphonium ylides.[79] The preparation and characterization of pure phospholyl anions and the corresponding phospholes have been reported for the first time;[80] the former exhibit huge ^{1}J(C-P) NMR couplings. The dianion obtained by deprotonation of sodium cyanotriphenylphosphoranylidenemethanide has been characterized and used to synthetic advantage.[81]

Formation of perfluorinated carbanions by reaction of perfluorocycloalkenes with CsF has been studied by ^{13}C and ^{19}F NMR spectroscopy.[82]

Organometallic Species

The formation of benzylpotassium upon reaction between alkyllithium, potassium alkoxide, and toluene has been studied.[83]

ω-Phenyl-ω-(trimethylsilyl)alkyllithium compounds undergo a range of intramolecular rearrangements. Transmetallation of a methyl group at silicon by a 1,$(n+2)$ proton transfer occurs for all values of the chain length n whether intramolecular 1,n shift of the benzylic proton takes place only for $n \geqslant 4$. For $n = 3$ and 4 the predominant reaction is a 1,n trimethylsilyl shift via a cyclic ate complex.[84]

Stereo- and regio-control by complex-induced proximity effects have been reviewed for reactions of organolithium compounds.[85] The efficient asymmetric alkylation of secondary amines by using formamidines as chiral auxiliaries has been shown to proceed with inversion of the C—Li bond following a metallation (deprotonation) reaction which is not rate-determining.[86] The proposed reaction mechanism, for which k_c is believed to be rate-determining, is illustrated in Scheme 1. A related asymmetric alkylation of chiral aminooxazolines proceeds via a thermodynamic mixture of lithiated species (dipole-stabilized) which, as before, is quenched with retention or inversion of configuration on reaction with d_6-DMSO or RX, respectively.[87] The different deprotonation behaviour may be a consequence of bidentate versus monodentate coordination of lithium.

The cyclopropyl lithiate derived from *trans*-1-methoxy-2-(*p*-tolylsulphonyl)-3,3-dimethylcyclopropane reacts with a variety of electrophiles to give only a single

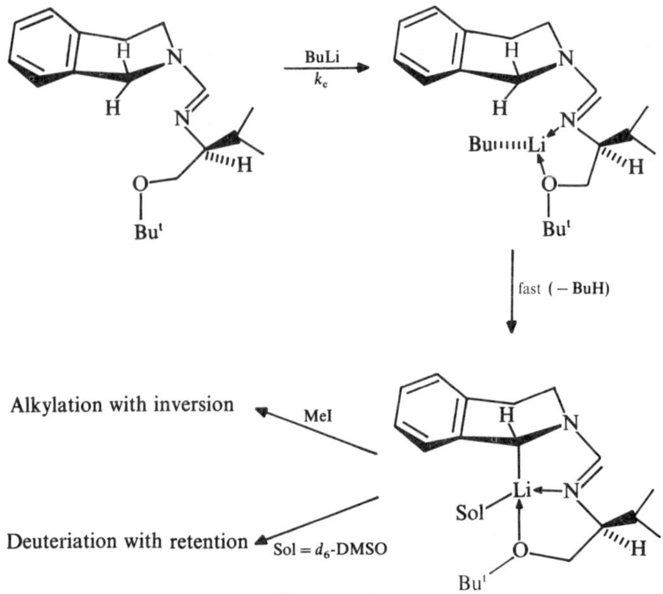

SCHEME 1

stereoisomer; this stereoselectivity has been attributed to intramolecular chelation of the cyclopropyl lithiate with the neighbouring methoxy group.[88]

The well-known ability of amides to promote lithiation at adjacent sites has been applied to effect *ortho*-lithiation of 1-methyl-1(*N,N*-diisopropylcarbamoyl)cyclopropane;[89] reaction with lithium tetramethylpiperidine/$HgCl_2$ gives complete conversion into the organomercury derivative (**41b**) via the intermediate (**41a**). The previously reported[90] *ortho*-directing effect in lithiation of cubanecarboxamide has been studied by SCF MO calculations at the STO-3G level.[91] The results demonstrate that there is an enhancement of the acidity of the adjacent hydrogen which is therefore susceptible to formation of the 2-lithio derivative, which in turn is stabilized by interaction between the lithium and the amide oxygen.

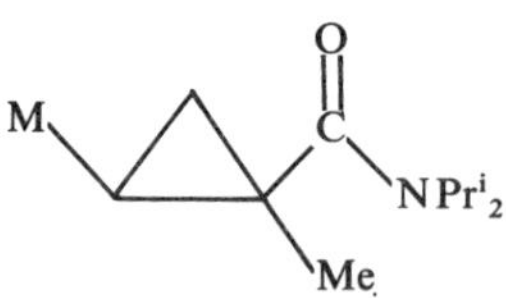

(41)

a; M = Li; **b**; M = HgCl

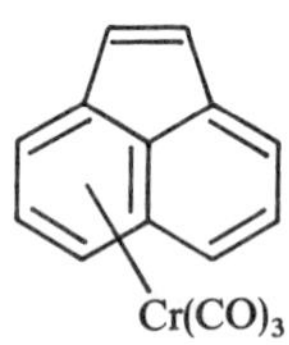

(42)

A range of carboxamides has been shown to undergo complex-induced β-lithiation upon reaction with *s*-butyllithium, even though α-lithiation is thermodynamically favoured;[92] the regio- and stereo-chemistry of reaction of the β-lithiated amides and of the corresponding Grignard and aluminium derivatives with a variety of electrophiles have been interpreted. Chelation-directed deprotonation of 3,5-dimethylisoxazole-4-(*S*-2′-hydroxymethyl-*N*-pyrrolidino)carboxamide is dependent upon the substitution on the 2′-hydroxymethyl moiety and metallation (BuLi/THF) at both methyl groups has been observed.[93]

Intramolecular coordination of lithium within the dimer of 3-lithio-1,5-dimethoxypentane in pentane has been probed by ^{1}H and ^{13}C NMR spectroscopy,[94] and ether cleavage with organo-alkali-metal compounds and alkali metals has been reviewed.[95] The total rates of deprotonation plus nucleophilic substitution at C(2) of 3-(methoxymethyl)- and 3-(dimethylaminomethyl)-oxetanes by butyllithium are *ca* 25–4000 times higher than for 3,3-diethyloxetane; this has been ascribed to pre-coordination of BuLi to the substrate.[96]

Chelation-controlled alkylation (*s*-BuLi/MeI in THF) of optically active 1-alkoxy-1,4-cyclohexadienes with 1,4-chirality transfer has been found to occur with regio- and diastereo-selectivities dependent on the nature of the chiral auxiliary.[97]

Treatment of *cis*- and *trans*-cinnamonitrile with excess LDA in aprotic media forms one preferred isomer of PhC(Li)=C(CN)Li in high yield. While *trans*-PhCD=CDCN is obtained quantitatively upon reaction of the α,β-dilithio derivative with MeOD, both α-substituted and α,β-disubstituted derivatives of *cis*- and *trans*-cinnamonitrile are obtained upon reaction with various electrophiles (MeI, RCHO, MeSSMe).[98] Substitution is believed to proceed via an equilibrium mixture of the β-lithiated intermediates, *cis*-PhC(Li)=C(E)CN and *trans*-PhC(Li)=C(CN)E, the products obtained being dependent upon the nucleophilic reactivity and configurational stability of the C_{β}—Li bonds.

The ready availability of distannylalkenes $RR'C{=}C(SnMe_3)_2$ ($R' = H$, alkyl, aryl), which form $RR'C{=}C(Li)SnMe_3$ on reaction with MeI, has prompted an investigation of the stability, isomerization, and thermal decomposition of the α-stannylvinyllithiums;[99] intramolecular coordination via a five-membered ring is an important influence. The more stable α-plumbylvinyllithiums $RR'C{=}C(Li)PbMe_3$ and the mixed bis(trimethylmetal)alkenes $RR'C{=}C(SnMe_3)MMe_3$ (M = Si, Ge, Pb) have also been prepared.[100]

Results of a systematic study of the effects of solvent and electrophile on the halodemetallation of tetraalkyltin compounds have established that cleavage products with predominant inversion and retention of configuration are obtained in polar solvents (especially aprotic) and non-polar solvents, respectively. The nature of the electrophile is also important, as revealed by comparison of Br_2, I_2, ICl, and IBr.[101]

Metallation of (fluorene)$Cr(CO)_2L$ complexes [L = CO, $P(OPh)_3$, PPh_3, PBu_3] with KH in THF gives the corresponding η^6-anions which equilibrate with the η^5-hapto isomers; the dependence on L has also been studied.[102]

The lithiation of tricarbonyl(naphthalene)chromium with *n*-BuLi or tetramethylpiperidyllithium is kinetically controlled, with a high regioselectivity for the 2-position in the latter case; in the presence of *i*-$PrNH_2$ an equilibrium mixture (*ca* 1:1) of 1- and 2-substitution products is formed.[103] Results of a related study of lithiation by lithium diisopropylamide indicate a kinetic preference for lithiation at C(2) and a thermodynamic preference for lithiation at C(1);[104] 5-methoxy-, 2,6- and 2,7-dimethoxy-naphthalene complexes were also studied.

Tricarbonyl(acenaphthylene)chromium (**42**) can be lithiated at the 3-position, on reaction with lithium diisopropylamide, and subsequently converted into 3-substituted derivatives by reaction with a range of electrophiles.[105] Reaction of (**42**) with lithium salts (RLi) of tertiary carbanions results in addition of the carbanion to the coordinated hydrocarbon.

In *t*-BuOK/THF, aromatic hydrocarbons complexed with $Cr(CO)_3$ deprotonate at the benzylic position and react with benzaldehyde and furfural to give condensation products.[106] However, the regiospecificity of benzylic attack cannot be explained by preferential attack at the carbon which is eclipsed by the Cr—CO bond; electronic effects predominate.[107]

Cyano-substituted carbanions generated α to the (η^3-allyl)$Mo(CO)_2$(Cp) system (Cp = η^5-C_5H_5) have been found to undergo a novel decyanation reaction.[108]

Highly stereoselective alkylation, deuteriation, and aldol reactions of the enolate derived from (η^5-$C_5H_5Fe(CO)(PPh_3)COCH_2OCH_2Ph$ have been reported.[109] The coordinated isoprene anion obtained on deprotonation of (η^4-isoprene)$Fe(CO)_3$ by lithium diisopropylamide forms simple adducts with electrophiles at low temperatures but rearranges to an isomeric (η^4-trimethylenemethane)$Fe(CO)_3$ anion derivative on warming.[110]

A double-well potential has been found to govern the motion of the proton in the C—H—C hydrogen bond of the carbanion of [1.1]ferrocenophane, for which a symmetrical structure must be discounted.[111]

Ab initio theoretical study of the organolithium-induced ring-opening of ethene episulphoxide has been reported. There is a low barrier for ring-opening of the *syn* carbanion whereas for the *anti* carbanion inversion competes with ring-opening and results in partial loss of stereospecificity.[112]

[^{13}C, $^2H_7$]bicyclo[3.2.2]nonatrienyl-lithium and -potassium and their degenerate rearrangements have been studied by NMR methods. A symmetrical delocalized structure has been proposed for the anion and it has been argued that its controversial stabilization is caused by an important interaction with the metal ion rather than by delocalization and homoaromaticity.[113] The observed scrambling of ^{13}C labels is consistent with a degenerate rearrangement via the 9-barbaralyl anion.

Reactions of (2-carbamoyl)lithium reagents obtained by directed β-lithiations of α,β-unsaturated amides have been reported.[114]

The reductive σ-bond cleavage of barbaralane, induced upon metallation by alkali metals, has been studied.[115]

Rate-determining carbon–carbon bond formation is a feature of the Grignard reaction of methyl iodide with benzophenone.[116] Direct evidence supporting a dimer structure of the radical intermediate in the Grignard reaction has been reported[117] and rates of electron transfer and of successive R transfer to ketones have been determined.[118]

Proton-transfer Reactions

Intrinsic barriers for reaction and the principle of non-perfect synchronization have been discussed with particular reference to carbanion-forming reactions.[119] High intrinsic barriers are generally associated with a lack of synchronization between concurrent reaction events such as bond formation and cleavage, solvation and desolvation, or changes in resonance interaction, etc. Increased DMSO concentrations have been shown to have little effect on the intrinsic rate constants (k_0) for reversible deprotonation of 9-cyanofluorene and 9-(carbomethoxy)fluorene by primary amines, piperidine and morpholine.[120] This is in contrast with the large solvent-induced increase in k_0 found previously for (i) reaction of oxyanions with the latter substrate (upon change from methanol to DMSO) or (ii) reaction of amines with acetylacetone, 1,3-indanone, nitromethane, and phenylnitromethane. The results have been interpreted in terms of late solvation of the developing carbanion combined with either later solvation of the developing ammonium ion or early desolvation of the oxy-anion. However, an additional dynamic solvent effect is also believed to be important.

The principle of imperfect synchronization has also been applied in interpretation of solvent effects on the rates of deprotonation of 2,2′,4,4′-tetranitrodiphenylmethane by general bases B^-, and reprotonation of the resulting carbanion by acids BH.[121] At pH < 5 the carbanion is believed to undergo a fast equilibrium protonation on the *p*-nitro group to form a nitronic acid.

Relief of steric strain on ionization of 9-alkylfluorenes contributes to their acidity and accounts for the failure of the acidity-oxidation potential (OAP) method to give a reasonable estimate of the radical-stabilizing effects of large alkyl substituents.[122] For fluorenide ions bearing 9-R, 9-RO, 9-RS and 9-R_2N substituents, where R is varied in size, the rate constants ($\log k_{obsd}$) for single-electron transfer to an electron acceptor are linearly related to the oxidation potentials and much less subject to steric hindrance than corresponding rate constants for S_N2 reaction with alkyl halides.

The 11.68 kcal mol^{-1} more favourable free energy of ionization of Meldrum's acid (**43**), relative to its acyclic analogue diethyl malonate, represents one of the largest organic stereoelectronic effects known and has been ascribed to an initial-state E/Z difference (6–8 kcal mol^{-1}) and a cyclization factor (3 kcal mol^{-1}).[123] The size of the bislactone ring has a marked effect, there being a rapid decrease in acidity between the six-membered and ten-membered ring; the thirteen-membered ring has the same pK_a as dimethyl malonate. Restricted rotation removes the stereoelectronic factor which normally makes esters less acidic than ketones and thereby permits an electronegative acid-strengthening effect of the now unconjugated ether oxygen.

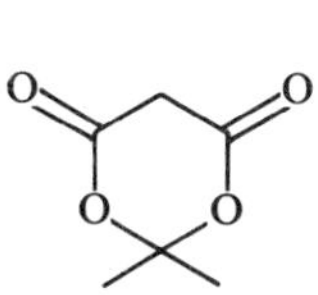

(43)

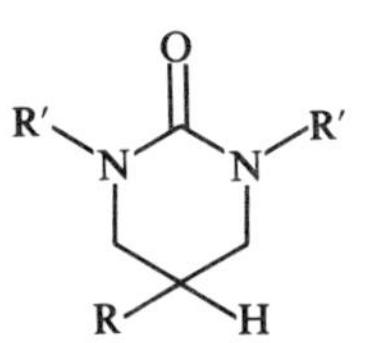

(44)

a; R = Bu^t, R′ = H; **b**; R = Bu^t, R′ = Me

c; R = H, R′ = Pr^i; **d**; R = Pr^i, R′R′ = Pr^i, H

A detailed study of equilibria and the slow rates of ionization and reprotonation of barbituric acids (**44a–d**) has been conducted.[124] For (**44b**) deprotonation is catalysed by general amine bases but a linear correlation with pK^{BH} is not observed since structural effects are important. In the case of (**44a**) ionization at C(5) is in competition with ionization at imide nitrogen. Correlations of carbon acidity with rates of C(5) deprotonation (by H_2O) and reprotonation (by H_3O^+), for which the respective Brønsted coefficients are 0.8 and 0.2, demonstrate that barbituric acid acidity is controlled primarily by substituent effects on the deprotonation reaction.

The Swain–Grunwald mechanism has been excluded for exchange of the acidic hydrogen of *t*-butylmalononitrile since the process is not inhibited by hydronium ions in aqueous solution.[125] Relative rates of bromination and reprotonation of the dicyanomethyl carbanion suggest that the latter process cannot be diffusion-controlled and that malononitrile is not therefore a 'normal' acid. It has been argued that the unusually large solvent kinetic isotope effects for cyanocarbon acid ionization are a consequence of proton transfer to a solvent cluster rather than a single water molecule.

The synthesis, properties (including CH acidity), and reactions of geminal sulphones have been reviewed,[126] and it has been established that α-deprotonation of sulphones is not inhibited by enforced pyramidalization of the resulting carbanion.[66] The regiochemistry of deprotonation/alkylation (MeI) of various 3-substituted 3-sulpholenes promoted, for example, by BuLi in THF-HMPA at −105°C, is as predicted.[127] Thus, anion-stabilizing 3-substituents (R) favour formation of anion (**45**) which alkylates at the 5-position, whereas anion-destabilizing groups promote 2-alkylation via anion (**46**).

(**45**)

(**46**)

Determination of the stereochemistry of H-D exchange of methyl phenylmethyl sulphoxide and of *t*-butyl phenylmethyl sulphoxide in THF, water, and methanol has revealed that deuteriation and methylation in THF occur with opposite stereochemistry for the former compound but with the same stereochemistry for the latter.[128] The stereochemistry of the products does not necessarily reflect the stable conformation of the free carbanion and may be influenced by kinetic acidity (governing the stereochemistry of the carbanion formed initially), thermodynamic acidity (which defines the stereochemistry of the conformation of the intermediate carbanion), and carbanion reactivity; each of which will be influenced by effects of cation association. In polar solvents, cation association is unimportant and the influence of kinetic acidity is dominant.

For benzylic carbon acids stabilized by adjacent cyano, carboalkoxy, or sulphonyl substituents, there is apparently little association of carbanion ion-pairs formed by transmetallation with R^-M^+ (M = Li or Cs) in THF since (as a consequence of anion charge delocalization) the electrostatic stabilization provided by aggregate formation is insufficient to overcome the unfavourable entropy of association.[129] However, cryptate effects suggest that ionic association influences the acidity of indene, fluorene, $PhCH_2CN$, Ph_2CHCN, and $PhCH_2CO_2Et$ in DMSO.[130]

The transition state for proton transfer from 4-nitrophenylphenylcyanomethane to tetramethylguanidine, $HN{=}C(NMe_2)_2$, is believed to occur early along the reaction path.[131] While the equilibrium constants show a good correlation with the solvent dielectric function $(D-1)/(2D+1)$ this is not the case for the rate constants, which are generally related to solvent parameters AN and DN. Rate constants for proton and deuteron transfer from 4-nitrophenylnitromethane to $XC_6H_4N{=}C(NMe_2)_2$ in acetonitrile show isotope effects which are dependent on the size and position of the substituent ($k_H/k_D \pm 0.3 = 9.7$, 12.5, and 15.6 for X = 4-Cl, 2-Cl, and 2-Br, respectively). That these figures may reflect increased degrees of proton tunnelling is suggested by the corresponding ratios of Arrhenius pre-exponential factors ($A_H/A_D \pm 0.1 = 0.98$, 0.87, and 0.77, respectively).[132]

Rate and activation parameters have been determined for iodination of acetophenone catalysed by carboxylate buffers in aqueous methanol.[133] The dependences on base strength, for each ketone, and on ketone aryl substitution, for each buffer base, have been quantified by application of the Brønsted and Hammett equations, respectively.

pK_a values (± 0.5) of 16.9 and 15.7 have been determined for acetaldehyde and isobutyraldehyde, from the kinetics of alkaline chlorination, by a method which was previously applied to ketones;[134] the conditions governing rate-limiting halogenation or enolization, respectively, have been discussed. The mechanism of deprotonation of acetaldehyde by lithium amide to form the lithium enolate-amine complex has been studied by MNDO and *ab initio* $(3-21+G)$ calculation.[135] Proton transfer occurs by a reactant-like transition state featuring a nearly colinear C—H—N arrangement; thermodynamics of the reversible process have been discussed.

Proton-transfer reactions of anions[136] and the formation of anions in the gas phase[137] have been reviewed. Theoretical studies of proton transfer have included: *ab initio* and MNDO calculations[138] on the reactions $XH_n \rightarrow XH_{n-1}^- + H^+$ and $XH_{n-1}^- + SiH_4 \rightarrow (SiH_4XH_{n-1})^-$ (X = C, N, O, F, Si, P, S, Cl); calculation of rates of proton transfer (including tunnelling) in $(H_3CH\cdots CH_3)^-$ by application of RRKM theory;[139] and *a priori* calculation of the ethane $pK_a = 50$ in water.[140]

A classical intersecting-state model of harmonic oscillators reveals that activation free-energy barriers for proton transfer are very dependent on the transition-state bond order $n^\dagger$, which ranges from 0.5 for carbon acids to 1.0 for HF; nitrogen and oxygen acids lie somewhere in between.[141] The maximum values for corresponding isotope effects decrease with increase in $n^\dagger$.

MNDO calculations have confirmed the feasibility of a base-catalysed 1,3-intramolecular proton shift which has been observed for cycloheptatriene derivatives; the transition-state structure is that of a 1,3-hydrogen-bond-bridged carbanion.[142] The ammonia-catalysed hydrogen migration in propene is also predicted to proceed by a direct 1,3-shift.

Factors affecting the stereochemistry and regiochemistry of products obtained upon protonation and alkylation of 1-arylpropenyllithium in ether have been studied.[143] Protonation occurs predominantly at C(3) whereas the preferred site for alkylation is the more sterically hindered C(1).

Anomalous temperature dependence of the kinetic isotope effect for DMSO/cumene hydrogen exchange, catalysed by *t*-BuOK,[144] and anomalous values of the Brønsted α-value for deprotonation of monoalkyl-pyridines, -biphenyls, and -benzenes by strong base[145] have been discussed.

It has been reported that the localization or delocalization of negative charge in the resulting carbanion does not alone determine the susceptibility of the proton-transfer reaction to the solvent basicity in H_2O/DMSO solutions.[146]

Brønsted relations of near unit slope previously reported for ionization of phenylacetylene and several other terminal monoalkynes support the view that the acetylenes exhibit the 'normal' acid-base behaviour found for oxygen and nitrogen systems whose basic electron pairs are also localized on a single atom. In contrast, studies of proton-exchange rates for deuteriated and tritiated cyanoacetylene and two conjugated diacetylenes have yielded Brønsted slopes (in the range 0.65–0.83) and sizeable primary isotope effects which are inconsistent with rate-determining separation of the products of fast proton transfer.[147]

By use of tetranitromethane as a carbanion trap it has been shown that oxidation of *n*-butylamine to *n*-butyraldehyde, catalysed by lysyl oxidase, proceeds by α-proton abstraction to give an intermediate enzyme-bound carbanion.[148]

For both 5,8-dihydroxy-1,4-naphthoquinone (pK_1^0 8.51, pK_2^0 11.49) and 1,4-dihydroxy-9,10-anthraquinone (pK_1^0 9.85, pK_2^0 12.28) the rate coefficient for the first and second acid dissociations, respectively, are *ca* 25-fold and 1000-fold lower than for normal proton transfer.[149] This is a consequence of a weak intramolecular hydrogen bond in the undissociated acids and a moderately strong hydrogen bond in the monoanions.

The reversible acid-base reactions of highly hindered diaminonaphthalenes (**47**) in aqueous sodium hydroxide have been found to occur slowly. A half-life of eight seconds has been reported for equilibration of (**47a**) with its protonated form in 60% v/v DMSO-H_2O containing 0.01 M NaOH; the 2- and 7-substituents markedly increase the basicity and decrease the rate of proton transfer.[150]

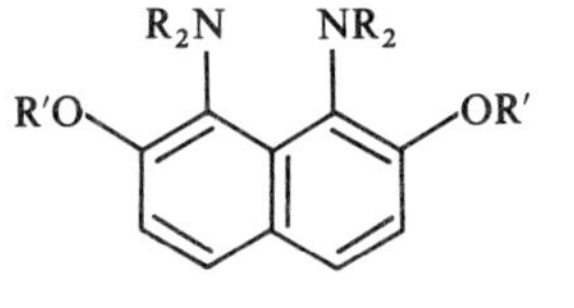

(**47**)

a; R = Me, R′ = Et (pK_a 16.1)
b; R = Et, R′ = Me (pK_a 15.9)
c; $R_2 = (CH_2CH_2OCH_2CH_2)$, R′ = Me ($pK_a$ 13.0)
d; $R_2 = (CH_2CH_2OCH_2CH_2)$, R′ = Et ($pK_a$ 12.5)

Dissociation constants for some β-keto amides and β-keto thioamides in non-aqueous media have been reported.[151]

The deuterium isotope exchange between *t*-butyl and isopropyl mercaptan in benzene is believed to occur by concerted transfer within a cyclic dimer held by two hydrogen bonds.[152]

Factors governing the rates of exchange of the amide NH protons of *d*-biotin sulphoxide stereoisomers with water, relative to results for biotin itself, have been discussed.[153] The geminate recombination of ion-pairs formed from amines and alcohols by laser-induced proton-concentration jump occurs on a picosecond time-scale and is governed by diffusion and coulombic attraction.[154]

Miscellaneous

The vinylic substitution of 9-(bromochloromethylene)fluorene and β,β-bis(*p*-nitrophenyl)-α-bromo-α-chloroethylene by *p*-toluenethiolate and *p*-cresolate ions gives monobromo, monochloro, and disubstitution products by addition-elimination via the corresponding carbanion.[155] The carbanion may be formed either directly or by single-electron-transfer followed by anion radical–radical combination. It has been concluded, from relative rates of expulsion of Br^- and Cl^-, that the halide is lost from the carbanion in an early transition state.

Vinylic substitution reactions of (**48**; Ar = *p*-MeC_6H_4 or *p*-$O_2NC_6H_4$) with MeO^- and substituted ArO^- and ArS^- ions have been shown, by NMR studies, to proceed with a high preference for retention of configuration.[156] *p*-Toluidine gives complete stereo-convergence and SCN^- induces *E*/*Z* isomerization of (**48**) before substitution. The degree of stereo-convergence (retention/inversion ratio) has been attributed to competition between rate-determining 60° and 120° rotations in the carbanion.

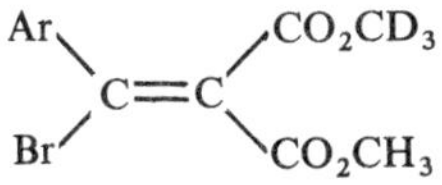

(**48**)

For allyl chloride the order of steric effects for reaction with 9-substituted fluorenide carbanions, 9-G-Fl^- (G = Me, PhS, Ph, *o*-tolyl, or *t*-Bu) is comparable to that found previously for S_N2 reaction with benzyl chloride.[157] The rate of reaction is increased by γ-methyl substitution, which causes the Brønsted β-value to decrease. The mechanism of the S_N2' reaction, which is induced by α-methyl substitution and favoured by increasing the size of the substituent G, has been discussed. For a series of fluorenide carbanions (9-G = Me, MeO, Me_2N) of nearly the same basicity, which undergo electron transfer to single-electron acceptors at progressively faster rates, a reverse order of S_N2 reactivity towards alkyl halides ($PhCH_2Cl$, *i*-BuBr, *i*-BuI) has been found.[158] While the S_N2 reaction is retarded 1400-fold by a change from 9-G = Me to *i*-Pr_2N, electron transfer to *c*-$C_6H_{10}(NO_2)Ts$ is eight times faster for the more sterically hindered fluorenide. The order of reactivity towards F_3CCH_2I, which is slow to react by the S_N2 process, suggests that electron transfer is rate-determining; the rates correlate with $E_{ox}(A^-)$ and are insensitive to steric effects in either donor or acceptor.

The sterically protected nitroarenes, 2,4,6-tri-*t*-butylnitrobenzene and 1,1,4,4,5,5,8,8-octamethyl-1,2,3,4,5,6,7,8-octahydro-9-nitroanthracene are resistant to attack by oxygen bases but oxidize 9-methoxyfluorenide and triphenylmethide ions; the resulting radical anions subsequently denitrate to the corresponding aryl radicals.[159]

Steric protection against bond formation allows electron transfer between Malachite Green [4,4′-bis(dimethylamino)trityl] cation and 4,4′,4″-trinitrotrityl anion in sulpholane with establishment of an equilibrium with the corresponding radicals.[160] Rate and equilibrium constants for cation-anion combination reactions have been reviewed.[161]

The oxidative cyclization of 1,3-dinitronate dianions with iodine in DMSO is believed to proceed by monoiodination followed by intramolecular coupling, perhaps involving single-electron transfer.[162] Photo-generated nitrobenzyl carbanion intermediates have been implicated in the photo-oxygenation of nitrobenzyl derivatives.[163] Transamination leading from ketimine to aldimine is believed to proceed via uncomplexed rather than metal-ion-complexed α,α'-carbanion.[164]

The Haller–Bauer cleavage of open-chain and cyclic non-enolizable α-phenyl ketones by $NaNH_2$ has been shown to occur with retention of configuration and not via racemic benzyl carbanions.[165] The stereochemistry of ligand coupling within the σ-sulphurane formed upon treatment of benzyl 2-pyridyl and related sulphoxides with Grignard reagents has been studied.[166] Medium-dependent competitive reactions of isomeric cyclohexadienyl carbanions have been discussed,[167] and generation of the dianion from tetrasilabicyclo[1.1.0]butane derivatives has been reported.[168]

Electrophilic Aliphatic Substitution

Benzyltrimethylsilane has been found to undergo electrophilic substitution at the benzylic position upon reaction with iodine monobromide in acetonitrile.[169] Substitution of optically active allylfluorosilanes (*S*)-(*Z*)-MeCH=CHCH(SiF_nMe_{3-n})Ph ($n = 1,2$) with $MeCOCl/AlCl_3$ and *t*-BuCl/$TiCl_4$ proceeds with *anti* stereochemistry to give S'_E products of *S* configuration.[170] *Anti* stereochemistry has been reported for the S'_E reaction of an optically active 2,4-dienylsilane.[171]

Halogenodemetallation reactions of mixed-functionality tetraorganotin compounds with iodine have been found to occur with reversal of the usually observed selectivity; breakage of alkyltin-carbon bonds in presence of aryltin- or vinyltin-carbon bonds occurs when intramolecular participation of a neighbouring hetero-atom at tin is possible.[172]

The rate-determining denitrosation step for hydrolysis of *N*-nitrosoamines has been discussed in the light of *ab initio* calculations,[173] and substituent effects on nitrosation of 1,3-diarylureas have been reported.[174]

References

[1] Moran, S., Ellis, H. B., DeFrees, D. J., McLean, A. D., and Ellison, G. B., *J. Am. Chem. Soc.*, **109**, 5996 (1987).

[2] Moran, S., Ellis, H. B., DeFrees, D. J., McLean, A. D., Paulson, S. E., and Ellison, G. B., *J. Am. Chem. Soc.*, **109**, 6004 (1987).
[3] Kalcher, J. and Janoschek, R., *Chem. Phys.*, **104**, 251, (1986); *Chem. Abs.*, **106**, 155589 (1987).
[4] Li, W. K., Nobes, R. H., and Radom, L., *THEOCHEM*, **34**, 67 (1987); *Chem. Abs.*, **107**, 58186 (1987).
[5] Hotokka, M. and Roos, B. O., *J. Chem. Soc., Perkin Trans. 2*, **1986**, 1979.
[6] Rajyaguru, I. and Rzepa, H. S., *J. Chem. Soc., Perkin Trans. 2*, **1987**, 359.
[7] Watts, J. D. and Stamper, J. G., *Tetrahedron*, **43**, 1019 (1987).
[8] Rajca, A., Streitwieser, A., and Tolbert, L. M., *J. Am. Chem. Soc.*, **109**, 1790 (1987).
[9] Lipkowitz, K. and Burkett, A., *J. Comput. Chem.*, **7**, 745 (1986); *Chem. Abs.*, **107**, 38795 (1987).
[10] Bank, S. and Dorr, R., *J. Org. Chem.*, **52**, 501 (1987).
[11] Kamlet, M. J. and Taft, R. W., *Acta Chem. Scand.*, **40B**, 619 (1986).
[12] Eliasson, B., Johnels, D., Wold, S., Edlund, U., and Sjostrom, M., *Acta Chem. Scand.*, **41B**, 291 (1987).
[13] Maercker, A., Graule, T., and Demuth, W., *Angew. Chem. Int. Edn*, **26**, 1032 (1987).
[14] Rajca, A. and Tolbert, L. M., *J. Am. Chem. Soc.*, **109**, 1782 (1987).
[15] Winchester, W. R., Bauer, W., and Schleyer, P. von R., *J. Chem. Soc., Chem. Commun.*, **1987**, 177.
[16] Christl, M. and Herzog, C., *Tetrahedron Lett.*, **28**, 187 (1987).
[17] Marsch, M., Massa, W., Harms, K., Baum, G., and Boche, G., *Angew. Chem. Int. Edn*, **25**, 1011 (1986).
[18] Gais, H.-J. and Dingerdissen, U., *J. Am. Chem. Soc.*, **109**, 3775 (1987).
[19] Sygula, A., Edlund, U., and Rabideau, P. W., *J. Chem. Res. (S)*, **1987**, 312.
[20] Sygula, A. and Rabideau, P. W., *J. Org. Chem.*, **52**, 3521 (1987).
[21] Knight, D. B. and Harrelson, J. A., *J. Chem. Soc., Chem. Commun.*, **1987**, 116.
[22] Pauli, A. and Meier, H., *J. Chem. Soc., Perkin Trans. 2*, **1987**, 1617.
[23] Karadakov, P. and Castano, O., *Theor. Chim. Acta*, **70**, 25 (1986); *Chem. Abs.*, **107**, 22722 (1987).
[24] Klabunde, K.-U., Mullen, K., and Vogler, H., *Tetrahedron*, **43**, 1183 (1987).
[25] Mullen, K., Meul, T., Schade, P., Schmickler, H., and Vogel, E., *J. Am. Chem. Soc.*, **109**, 4992 (1987).
[26] Cohen, Y., Roelofs, N. H., Reinhardt, G., Scott, L. T., and Rabinovitz, M., *J. Org. Chem.*, **52**, 4207 (1987).
[27] Frim, R., Rabinovitz, M., Hopf, H., and Hucker, J., *Angew. Chem. Int. Edn*, **26**, 232 (1987).
[28] Escher, A., Neuenschwander, M., and Engel, P., *Helv. Chim. Acta*, **70**, 1623 (1987).
[29] Hirayama, M., Suzuki, H., and Ohhata, H., *Chem. Lett.*, **1987**, 413.
[30] Zipperer, B., Fletschinger, M., Hunkler, D., and Prinzbach, H., *Tetrahedron Lett.*, **28**, 2513 (1987).
[31] Fletschinger, M., Zipperer, B., Fritz, H., and Prinzbach, H., *Tetrahedron Lett.*, **28**, 2517 (1987).
[32] Cohen, Y., Klein, J., and Rabinovitz, M., *J. Chem. Soc., Chem. Commun.*, **1987**, 1538.
[33] Trimitsis, G., Lin, F.-T., Eaton, R., Jones, S., Trimitsis, M., and Lane, S., *J. Chem. Soc., Chem. Commun.*, **1987**, 1704.
[34] Barchiesi, E., Bradamante, S., Ferraccioli, R., and Pagani, G. A., *J. Chem. Soc., Chem. Commun.*, **1987**, 1548.
[35] Yokoyama, Y. and Takahashi, K., *Chem. Lett.*, **87**, 589 (1987).
[36] Celina, M., Lazana, R. L. R., Franco, M. L. T. M. B., and Herold, B. J., *J. Chem. Soc., Perkin Trans. 2*, **1987**, 1399.
[37] Kawasaki, H., Tomioka, K., and Koga, K., *Kagaku (Kyoto)*, **41**, 361 (1986); *Chem. Abs.*, **106**, 118937 (1987).
[38] O'Neill, P. and Hegarty, A. F., *J. Org. Chem.*, **52**, 2113 (1987).
[39] Lee, T. V. and Toczek, J., *J. Chem. Soc., Perkin Trans. 1*, **1987**, 759.
[40] Hughes, D. L., Dolling, U.-H., Ryan, K. M., Schoenewaldt, E. F., and Grabowski, E. J. J., *J. Org. Chem.*, **52**, 4745 (1987).
[41] Ojima, I. and Qiu, X., *J. Am. Chem. Soc.*, **109**, 6537 (1987).
[42] Duhamel, P., Eddine, J. J., and Valnot, J.-Y., *Tetrahedron Lett.*, **28**, 3801 (1987).
[43] Narasaka, K., Ukaji, Y., and Watanabe, K., *Bull. Chem. Soc. Jpn*, **60**, 1457 (1987).
[44] Ferreira, J. T. B., Neeland, E. G., Ounsworth, J. P., and Weiler, L., *Can. J. Chem.*, **65**, 2314 (1987).
[45] Armesto, D., Ortiz, M. J., and Perez-Ossorio, R., *J. Chem. Soc., Perkin Trans. 1*, **1986**, 2021.
[46] Blackburn, G. M., Brown, D., Martin, S. J., and Parratt, M. J., *J. Chem. Soc., Perkin Trans. 1*, **1987**, 181.
[47] Molander, G. A. and Etter, J. B., *J. Am. Chem. Soc.*, **109**, 6556 (1987).
[48] Dewar, M. J. S. and Merz, K. M., *J. Am. Chem. Soc.*, **109**, 6553 (1987).
[49] Ballester, P., Costa, A., Garcia-Raso, A., and Mestres, R., *Tetrahedron Lett.*, **28**, 3853 (1987).
[50] Nguyen Trong Anh and Bui Tho Thanh, *Nouveau J. Chim.*, **10**, 681 (1986).
[51] Kaufman, M. J. and Streitwieser, A., *J. Am. Chem. Soc.*, **109**, 6092 (1987).
[52] Zimmerman, H. E., *Acc. Chem. Res.*, **20**, 263 (1987).

[53] Bouchoux, G. and Hoppilliard, Y., *Tetrahedron Lett.*, **28**, 4537 (1987).
[54] Bowie, J. H., Stringer, M. B., Hayes, R. N., Raferty, M. J., Currie, G. J., and Eichinger, P. C. H., *Spectroscopy (Ottawa)*, **4**, 277 (1985); *Chem. Abs.*, **107**, 133619 (1987).
[55] Raftery, M. J. and Bowie, J. H., *Aust. J. Chem.*, **40**, 711 (1987).
[56] Hayes, R. N., Sheldon, J. C., and Bowie, J. H., *Int. J. Mass Spectrom.*, **71**, 233 (1986); *Chem. Abs.*, **106**, 137811 (1987).
[57] Currie, G. J., Stringer, M. B., Bowie, J. H., and Holmes, J. L., *Aust. J. Chem.*, **40**, 1365 (1987).
[58] Ariga, M. and Matsumura, E., *Bull. Chem. Soc. Jpn*, **60**, 1198 (1987).
[59] Barluenga, J., Rubiera, C., Fernandez, J. R., and Yus, M., *J. Chem. Soc., Chem. Commun.*, **1987**, 425.
[60] Peiris, S., Ragauskas, A. J., and Stothers, J. B., *Can. J. Chem.*, **65**, 789 (1987).
[61] Assithianakis, P. and Stamm, H., *Chem. Ber.*, **120**, 855 (1987).
[62] Burgess, K., *J. Org. Chem.*, **52**, 2046 (1987).
[63] Gololobov, Y. G., Nesmeyanov, A. N., Lysenko, V. P., and Boldeskul, I. E., *Tetrahedron Lett.*, **43**, 2609 (1987).
[64] Sato, T., Itoh, T., and Fujisawa, T., *Tetrahedron Lett.*, **28**, 5677 (1987).
[65] Grossert, J. S., Hoyle, J., Cameron, T. S., Roe, S. P., and Vincent, B. R., *Can. J. Chem.*, **65**, 1407 (1987).
[66] Kirmse, W. and Mrotzeck, U., *J. Chem. Soc., Chem. Commun.*, **1987**, 709.
[67] Fuji, K., Usami, Y., Sumi, K., Ueda, M., and Kajiwara, K., *Chem. Lett.*, **1986**, 1655; *Chem. Abs.*, **107**, 58218 (1987).
[68] Hartman, G. D. and Halczenko, W., *Tetrahedron Lett.*, **28**, 3241 (1987).
[69] Simpkins, N. S., *Tetrahedron Lett.*, **28**, 989 (1987).
[70] Marji, D., *J. Chem. Soc., Chem. Commun.*, **1987**, 7.
[71] Johnson, C. R. and Dhanoa, D. S., *J. Org. Chem.*, **52**, 1885 (1987).
[72] Fava, A., Bongini, A., Cere, V., Paoulucci, C., Policcino, S., and Sandri, E., *Pure Appl. Chem.*, **59**, 955 (1987).
[73] Bernardi, F., Bottoni, A., Venturini, A., and Mangini, A., *J. Am. Chem. Soc.*, **108**, 8171 (1986).
[74] Haynes, R. K., Schober, P. A., and Binns, M. R., *Aust. J. Chem.*, **40**, 1223 (1987).
[75] Binns, M. R. and Haynes, R. K., *Aust. J. Chem.*, **40**, 937 (1987).
[76] Fang, J.-M., Hong, B.-C., and Liao, L.-F., *J. Org. Chem.*, **52**, 855 (1987).
[77] Okazaki, R., O-oka, M., Akiyama, T., Inamoto, N., Niwa, J., and Kato, S., *J. Am. Chem. Soc.*, **109**, 5413 (1987).
[78] Jonczyk, A., Owczarczyk, Z., Makosza, M., and Winiarski, J., *Bull. Soc. Chim. Belg.*, **96**, 303 (1987).
[79] Garner, P. and Ramakanth, S., *J. Org. Chem.*, **52**, 2629 (1987).
[80] Charrier, C. and Mathey, F., *Tetrahedron Lett.*, **28**, 5025 (1987).
[81] Bestman, H. J. and Schmidt, M., *Angew. Chem. Int. Edn*, **26**, 79 (1987).
[82] Snegirev, V. F., Delyagina, N. I., and Bakhmutov, V. I., *Izv. Akad. Nauk SSSR, Ser. Khim.*, **1986**, 1325; *Chem. Abs.*, **106**, 119170 (1987).
[83] Lochmann, L. and Trekoval, J., *J. Organomet. Chem.*, **326**, 1 (1987).
[84] Maercker, A. and Stotzel, R., *Chem. Ber.*, **120**, 1695 (1987).
[85] Beak, P., Meyers, A. I., *Acc. Chem. Res.*, **19**, 356 (1986).
[86] Meyers, A. I. and Dickman, D. A., *J. Am. Chem. Soc.*, **109**, 1263 (1987).
[87] Gawley, R. E., *J. Am. Chem. Soc.*, **109**, 1265 (1987).
[88] Padwa, A., Wannamaker, M. W., and Dyszlewski, A. D., *J. Org. Chem.*, **52**, 4760 (1987).
[89] Eaton, P. E., Daniels, R. G., Casucci, D., Cunkle, G. T., and Engel, P., *J. Org. Chem.*, **52**, 2100 (1987).
[90] Eaton, P. E. and Castaldi, G., *J. Am. Chem. Soc.*, **107**, 724 (1985).
[91] Jayasuriya, K., Alster, J., and Politzer, P., *J. Org. Chem.*, **52**, 2306 (1987).
[92] Beak, P., Hunter, J. E., Jun, Y. M., and Wallin, A. P., *J. Am. Chem. Soc.*, **109**, 5403 (1987).
[93] Schlicksupp, L. and Natale, N. R., *J. Heterocycl. Chem.*, **24**, 1345 (1987).
[94] Vos, M., de Kanter, F. J. J., Schakel, M., van Eikema Hommes, N. J. R., and Klumpp, G. W., *J. Am. Chem. Soc.*, **109**, 2187 (1987).
[95] Maercker, A., *Angew. Chem. Int. Edn*, **26**, 972 (1987).
[96] Schakel, M., Vrielink, J. J., and Klumpp, G. W., *Tetrahedron Lett.*, **28**, 5747 (1987).
[97] Stanssens, D., De Keukeleire, D., and Vandewalle, M., *Tetrahedron Lett.*, **28**, 4195 (1987).
[98] Feit, B.-A., Haag, B., and Schmidt, R. R., *J. Org. Chem.*, **52**, 3825 (1987).
[99] Mitchell, T. N. and Reimann, W., *J. Organomet. Chem.*, **322**, 141 (1987).
[100] Mitchell, T. N. and Reimann, W., *J. Organomet. Chem.*, **322**, 151 (1987).
[101] Fukuto, J. M., Newmann, D. A., and Jensen, F. R., *Organometallics*, **6**, 415 (1987).
[102] Ceccon, A., Gambaro, A., Venzo, A., Lucchini, V., Bitterwolf, T. E., and Shade, J., *J. Organomet. Chem.*, **327**, 55 (1987).

[103] Kundig, E. P., Grivet, C., and Spichiger, S., *J. Organomet. Chem.*, **332**, C13 (1987).
[104] Treichel, P. M. and Kirss, R. U., *Organometallics*, **6**, 249 (1987).
[105] Kirss, R. U., Treichel, P. M., and Haller, K. J., *Organometallics*, **6**, 242 (1987).
[106] Senechal-Tocquer, M.-C., Senechal, D., Le Bihan, J.-Y., Gentric, D., and Caro, B., *J. Organomet. Chem.*, **321**, 353 (1987).
[107] Brocard, J. and Lebibi, J., *J. Organomet. Chem.*, **320**, 295 (1987).
[108] Pearson, A. J. and Khetani, V. D., *J. Chem. Soc., Chem. Commun.*, **1986**, 1772.
[109] Davies, S. G. and Wills, M., *J. Organomet. Chem.*, **328**, C29 (1987).
[110] Semmelhack, M. F. and Fewkes, E. J., *Tetrahedron Lett.*, **28**, 1497 (1987).
[111] Ahlberg, P. and Davidsson, O., *J. Chem. Soc., Chem. Commun.*, **1987**, 623.
[112] Maccagnani, G., Schlegel, H. B., and Tonachini, G., *J. Org. Chem.*, **52**, 4961 (1987).
[113] Jonsall, G. and Ahlberg, P., *J. Chem. Soc., Perkin Trans. 2*, **1987**, 461.
[114] Beak, P. and Wilson, K. D., *J. Org. Chem.*, **51**, 4627 (1986).
[115] Trinks, R. and Mullen, K., *Chem. Ber.*, **120**, 1481 (1987).
[116] Yamataka, H., Matsuyama, T., and Hanafusa, T., *Chem. Lett.*, **87**, 647 (1987).
[117] Maruyama, K. and Katagiri, T., *Chem. Lett.*, **87**, 731 (1987).
[118] Maruyama, K. and Katagiri, T., *Chem. Lett.*, **87**, 735 (1987).
[119] Bernasconi, C. F., *Acc. Chem. Res.*, **20**, 301 (1987).
[120] Bernasconi, C. F. and Terrier, F., *J. Am. Chem. Soc.*, **109**, 7115 (1987).
[121] Terrier, F., Lelievre, J., Chatrousse, A.-P., Schaal, R., and Farrell, P. G., *Can. J. Chem.*, **65**, 1980 (1987).
[122] Bordwell, F. G., Bausch, M. J., and Wilson, C. A., *J. Am. Chem. Soc.*, **109**, 5465 (1987).
[123] Arnett, E. M. and Harrelson, J. A., *J. Am. Chem. Soc.*, **109**, 809 (1987).
[124] Buckingham, D. A., Clark, C. R., McKeown, R. H., and Wong, O., *J. Am. Chem. Soc.*, **109**, 466 (1987).
[125] Hojatti, M., Kresge, A. J., and Wang, W.-H., *J. Am. Chem. Soc.*, **109**, 4023 (1987).
[126] Neplyuev, V. M., Bazavova, I. M., and Lozinskii, M. O., *Usp. Khim.*, **55**, 1550 (1986); *Chem. Abs.*, **106**, 17575 (1987).
[127] Tao, Y.-T., Liu, C.-L., Lee, S.-J., and Chou, S.-S.P., *J. Org. Chem.*, **51**, 4718 (1986).
[128] Nakamura, K., Higaka, M., Adachi, S., Oka, S., and Ohno, A., *J. Org. Chem.*, **52**, 1414 (1987).
[129] Kaufman, M. J., Gromert, S., Bors, D. A., and Streitwieser, A., *J. Am. Chem. Soc.*, **109**, 602 (1987).
[130] Antipin, I. S., Vedernikov, A. N., and Konovalov, A. I., *Zh. Org. Khim.*, **22**, 1330 (1986); *Chem. Abs.*, **106**, 32272 (1987).
[131] Hojatti, M. and Leffek, K. T., *Can. J. Chem.*, **64**, 2365 (1986).
[132] Pruszynski, P., *Can. J. Chem.*, **65**, 2160 (1987).
[133] Satyanarayana, N. and Sundaram, E. V., *Z. Phys. Chem.* (*Leipzig*), **268**, 747 (1987).
[134] Guthrie, J. P. and Cossar, J., *Can. J. Chem.*, **64**, 2470 (1986).
[135] McKee, M. L., *J. Am. Chem. Soc.*, **109**, 559 (1987).
[136] DePuy, C. H. and Bierbaum, V. M., *NATO ASI Ser.*, **C193**, 293 (1987); *Chem. Abs.*, **107**, 133614 (1987).
[137] DePuy, C. H. and Bierbaum, V. M., *NATO ASI Ser.*, **C193**, 279 (1987); *Chem. Abs.*, **107**, 133613 (1987).
[138] Gordon, M. S., Davis, L. P., Burggraf, L. W., and Damrauer, R., *J. Am. Chem. Soc.*, **108**, 7889 (1986).
[139] Scheiner, S. and Latajka, Z., *J. Phys. Chem.*, **91**, 724 (1987).
[140] Jorgensen, W. L., Briggs, J. M., and Gao, J., *J. Am. Chem. Soc.*, **109**, 6857 (1987).
[141] Formosinho, S. J., *J. Chem. Soc., Perkin Trans. 2*, **1987**, 61.
[142] Takahashi, K., Suzuki, T., Toda, H., Takase, K., Koseki, S., and Nakajima, T., *J. Org. Chem.*, **52**, 2666 (1987).
[143] Tanaka, J., Nojima, M., Kusabayashi, S., and Nagase, S., *J. Chem. Soc., Perkin Trans. 2*, **1987**, 673.
[144] Shapiro, I. O., Ranneva, Y. I., and Shatenshtein, A. I., *Zh. Org. Khim.*, **22**, 1332 (1986); *Chem. Abs.*, **106**, 32290 (1987).
[145] Shapiro, I. O., Zharova, N. G., Ranneva, Y. I., Terekhova, M. I., and Shatenshtein, A. I., *Teor. Eksp. Khim.*, **22**, 443 (1986); *Chem. Abs.*, **106**, 119129 (1987).
[146] Kankaanpera, A., Lahti, M., and Scharlin, P., *Acta. Chem. Scand.*, **40A**, 387 (1986); *Chem. Abs.*, **107**, 6577 (1987).
[147] Aroella, T., Arrowsmith, C. H., Hojatti, M., Kresge, A. J., Powell, M. F., Tang, Y. S., and Wang, W.-H., *J. Am. Chem. Soc.*, **109**, 7198 (1987).
[148] Williamson, P. R. and Kagan, H. M., *J. Biol. Chem.*, **262**, 8196 (1987).
[149] Hibbert, F. and Spiers, K. J., *Chem. Ber.*, **120**, 1617 (1987).
[150] Hibbert, F. and Simpson, G. R., *J. Chem. Soc., Perkin Trans. 2*, **1987**, 243.

[151] Nashid, B., Salama, S. B., and Ibrahim, O., *Z. Phys. Chem. (Leipzig)*, **268**, 347 (1987).
[152] Wawer, A. and Wawer, I., *React. Kinet. Catal. Lett.*, **30**, 1 (1986); *Chem. Abs.*, **106**, 17804 (1987).
[153] Fox, T. L., Tipton, P. A., Cleland, W. W., and Mildvan, A. S., *J. Am. Chem. Soc.*, **109**, 2127 (1987).
[154] Huppert, D. and Pines, E., *NATO ASI Ser.*, **184**, 171 (1986); *Chem. Abs.*, **107**, 22717 (1987).
[155] Avramovitch, B., Weyerstahl, P., and Rappoport, Z., *J. Am. Chem. Soc.*, **109**, 6687 (1987).
[156] Rappoport, Z. and Gazit, A., *J. Am. Chem. Soc.*, **109**, 6698 (1987).
[157] Bordwell, F. G., Clemens, A. H., and Cheng, J.-P., *J. Am. Chem. Soc.*, **109**, 1773 (1987).
[158] Bordwell, F. G. and Wilson, C. A., *J. Am. Chem. Soc.*, **109**, 5470 (1987).
[159] Guthrie, R. D., Hartmann, C., Neill, R., and Nutter, D. E., *J. Org. Chem.*, **52**, 736 (1987).
[160] Arnett, E. M., Molter, K. E., Marchot, E. C., Donovan, W. H., and Smith, P., *J. Am. Chem. Soc.*, **109**, 3788 (1987).
[161] Ritchie, C. D., *Can. J. Chem.*, **64**, 2239 (1986).
[162] Wade, P. A., Dailey, W. P., and Carroll, P. J., *J. Am. Chem. Soc.*, **109**, 5452 (1987).
[163] Wan, P., Muralidharan, S., McAuley, I., and Babbage, C. A., *Can. J. Chem.*, **65**, 1775 (1987).
[164] Martin, R. B., *Inorg. Chem.*, **26**, 2197 (1987); *Chem. Abs.*, **107**, 96121 (1987).
[165] Paquette, L. A., Gilday, J. P., Ra, C. S., *J. Am. Chem. Soc.*, **109**, 6858 (1987).
[166] Oae, S., Kawai, T., Furukawa, N., and Iwasaki, F., *J. Chem. Soc., Perkin Trans. 2*, **1987**, 405.
[167] Venkatasubramanian, N., Hawkins, A., and Siegel, S., *J. Org. Chem.*, **52**, 1222 (1987).
[168] Kawase, T., Batcheller, S. A., and Masamune, S., *Chem. Lett.*, **1987**, 227.
[169] Bordeau, M., Villeneuve, P., Bennetau, B., and Dunogues, J., *J. Organomet. Chem.*, **331**, 169 (1987).
[170] Hayashi, T., Matsumoto, Y., and Ito, Y., *Organometallics*, **6**, 884 (1987).
[171] Hayashi, T., Matsumoto, Y., and Ito, Y., *Chem. Lett.*, **1987**, 2037.
[172] Jousseaume, B. and Villeneuve, P., *J. Chem. Soc., Chem. Commun.*, **1987**, 513.
[173] Nguyen, M. T. and Hegarty, A. F., *J. Chem. Soc., Perkin Trans. 2*, **1987**, 345.
[174] Miyahara, M., *Chem. Pharm. Bull.*, **34**, 1950 (1986); *Chem. Abs.*, **106**, 175664 (1987).

Organic Reaction Mechanisms 1987
Edited by A. C. Knipe and W. E. Watts

CHAPTER 12

Elimination Reactions

R. A. More O'Ferrall

Department of Chemistry, University College, Belfield, Dublin 4, Ireland

Ring-opening and Ring-closure

An extensive study of ring-opening reactions has shown that exocyclic eliminations forming both olefinic and carbonyl double bonds (Scheme 1) are consistently easier for fission of a cyclopropane than of the formally equally strained cyclobutane ring.[1–4] Leaving-groups (Y in Scheme 1), including alkoxide ions and nitro, cyano,

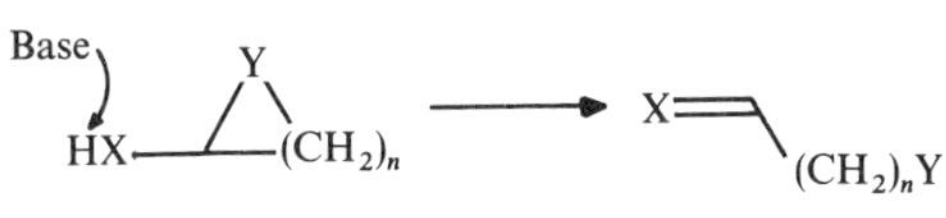

$XH = OH, CH_2SO_2R$

Scheme 1

carbomethoxy and sulphonyl carbanions, show an enhanced reactivity of up to 10^9 for fission of a 3-membered ring relative to a similar acyclic reaction. The low reactivity of the 4-membered ring is emphasized by the remarkable finding that hydrolysis of the amidinium ion (**1**) yields a mixture of β-lactam (**3**) and acyclic amide (**4**). These products imply that ring fission of the tetrahedral intermediate (**2**) competes on an equal footing with expulsion of the unconstrained exocyclic amine leaving-group.[5]

In discussing the discrepancy between 3- and 4-membered rings the different make-up of the strain energies has been emphasized. The smaller ring is dominated

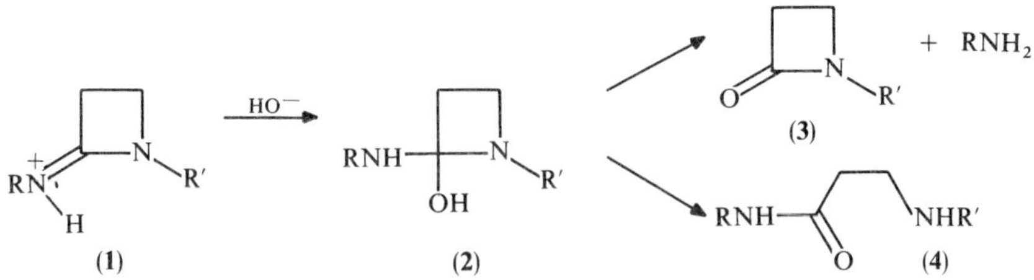

(1) (2) (3) (4)

by angle and torsional strain while the larger also includes cross-ring 1,3- and 2,4-carbon–carbon repulsions. MNDO calculations reproduce the differences in reactivity and are consistent with the view that 'strain energy declines less rapidly with bond extension' for the cyclobutyl than the cyclopropyl ring, leading to a less favourable compromise between product bond formation and strain release in balancing the energy of bond-breaking at the transition state.[4]

Endocyclic ring-opening (e.g. **5**→**6**) and inversion barriers of 3-membered ring anions (**5**) have been investigated by *ab initio* (STO-G) calculations.[6] As might be

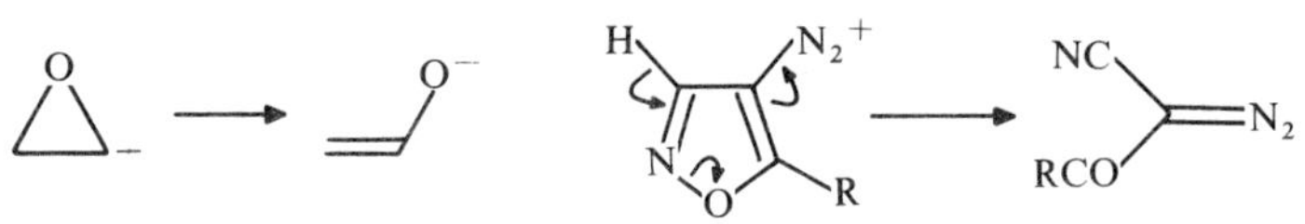

(**5**) (**6**) (**7**)

expected, carbanion inversion competes with conrotatory ring-opening of a cyclopropyl anion but not of the corresponding epoxide, aziridine or thiirane ions.[6] A probable example of endocyclic opening in the 5-membered ring is provided by nitrosation of aminoisoxazoles which yields diazo-ketones, presumably via the diazonium ion (**7**).[7] The reasons why thiiranium ions (**8**) show a tendency to undergo nucleophilic ring-opening by chloride ion at more and less alkylated carbons, respectively, in polar and non-polar media has been discussed:[8] it seems likely that the greater nucleophilicity of the substituting anion in the less polar solvent causes a switch from S_N1 (or S_N1-like) to S_N2 substitution, with a corresponding change in sensitivity to alkyl substitution.

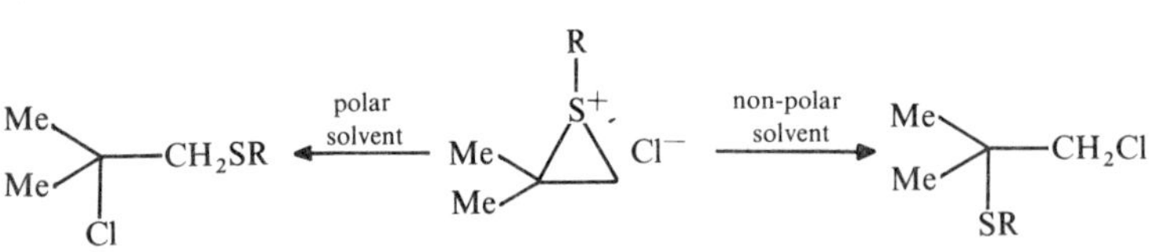

(**8**)

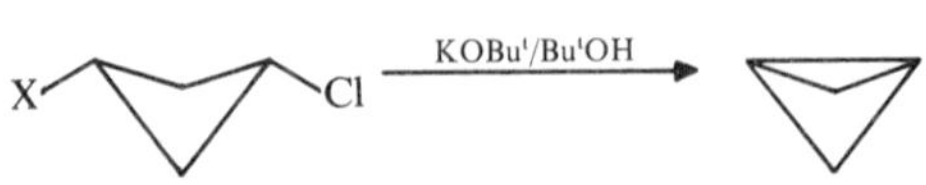

(**9**) (**10**)

X = CN, RCO

The reaction of episulphides with Wittig reagents has been found to yield olefins, in contrast to the corresponding reactions of epoxides which give cyclopropanes.[9]

Cyano- and keto-activated 1,3-dehydrohalogenations of cyclobutanes (**9**) to form bicyclobutanes (**10**) occur by a carbanion mechanism, with competing epimerization and hydrogen isotope exchange occurring in the case of the cyano substrate but not the keto. The rate of reaction shows little dependence on the stereochemistry of the substituents or upon the presence of an extra methyl or halo substituent α to the leaving-group. For the cyano substrate, leaving-group rate ratios as high as 70 indicate rate-determining loss of the leaving group ($ElcB_R$ mechanism), but for the keto substrate carbanion formation is probably rate-determining.[10] Cyclopropane ring formation accompanying dehydrochlorination within a cyclopentyl ring has been reported for the perchloro polycyclic insecticide derivatives photoaldrin and photodieldrin.[11]

Mechanism and Stereochemistry

Elimination from the diastereomeric substrates (**11**) containing β-halo and carbomethoxy activating substituents and leaving-groups OMe, OAc, F, Cl and Br appear to span the borderline between *ElcB* and *E*2 mechanisms. The *erythro* substrates show a larger rate for fluoro than chloro or bromo substrates (e.g. $k_{Br}/k_F = 0.1$) while for *threo* the reverse is true ($k_{Br}/k_F = 60$). For F, OAc, and OMe leaving-groups, rate ratios for *erythro* and *threo* substrates fall in the range 0.5–2.0 but for chlorides and bromides $k_{erythro}/k_{threo}$ lies between 30 and 400. This behaviour is consistent with operation of an irreversible *ElcB* mechanism in all cases except for the *threo* bromide and chloride for which the mechanism presumably is *E*2. Products are very predominantly ($\geqslant 98\%$) *Z*-olefins (**12**) except in the case of *erythro* bromides and chlorides. Here incursion of *anti*-elimination yielding *E*-olefin may reflect either competing *E*2 and *ElcB* mechanisms of elimination or loss of these

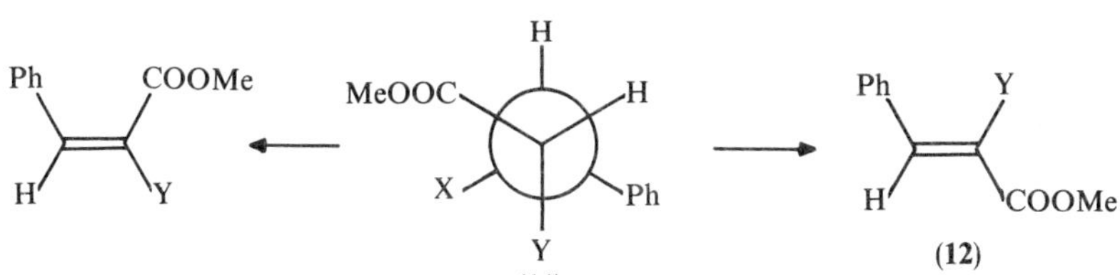

E-(*syn* product)

(11)
X = MeO, AcO, F, Cl, Br
Y = Cl or Br

(12)
Z-(*anti* product)

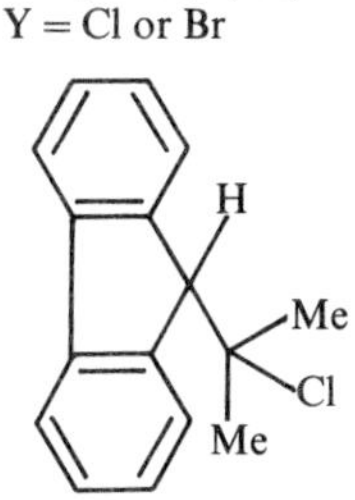

(13)

good leaving-groups from a carbanion intermediate that has not had time for rotational equilibration. The authors prefer the latter interpretation but in view of the undoubted operation of an $E2$ mechanism for the *threo* chloride and bromide, the likelihood of competing mechanisms seems high.[12]

Elimination from 9-(2-chloro-2-propyl)fluorene (**13**) in aqueous acetonitrile shows competing $E1$ and $E2$ or $E1cB$ mechanisms as well as substitution products. The fraction of elimination increases with the strength of added carboxylate anion bases ($\beta = 0.05$) and increase in basicity of the leaving-group, the latter indicating elimination within a carbenium ion-pair.[13] Solvolysis of benzyl azoxytosylate (**14**) in aqueous CF_3CH_2OH has also been studied and shows *partial* trapping by nucleophiles of a longer lived benzyl species than the cation or its ion-pair. Formation of benzaldehyde as a minor elimination product upon addition of acetate ion offers a clue that this species is a benzyloxydiazonium ion formed by rearrangement within an initially formed N_2O-separated benzyl/tosylate ion-pair (**15**) which, sensibly, is also the intermediate leading to the 'non-trappable' solvolysis products (Scheme 2).[14]

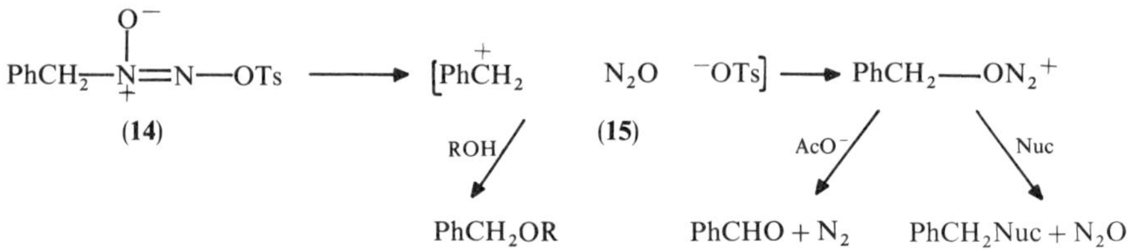

SCHEME 2

There have been several studies of the stereochemistry of elimination. In contrast to its reaction with alkoxide ions, (**16**) undergoes *syn*-elimination of HF in preference to *anti*-elimination of HCl in the base system $NaNH_2$—$NaOBu^t$—THF despite

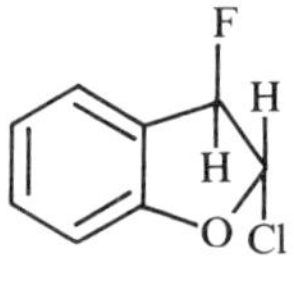

(**16**)

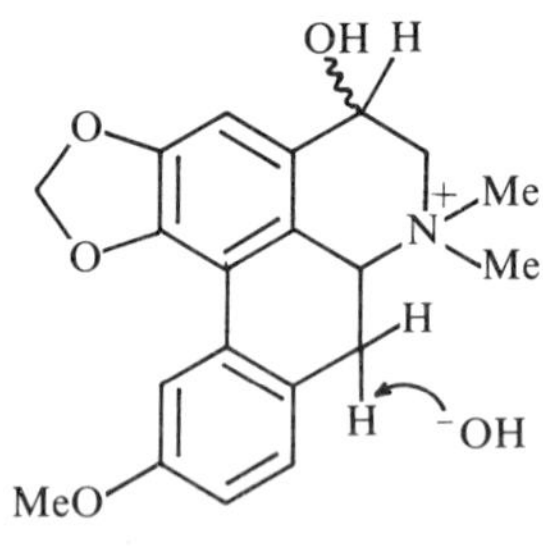

(**17**)

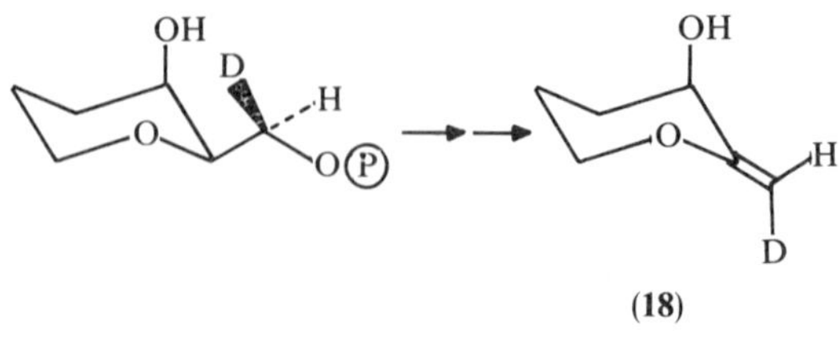

(**18**)

Cl^- being the better leaving-group and HCl elimination being activated by an aryl group. The behaviour is ascribed to coordination of Na^+ to fluorine in a cyclic *E*2 transition state[15] but the leaving ability of Cl will also be inhibited by a geminal interaction with oxygen. More predictably, a β-phenyl group does control the direction of elimination in the carbanion-like transition state for Hofmann elimination of the epimeric quarternized 4-hydroxyapoporphines (**16**); for neither epimer does elimination occur towards the β-hydroxy group.[16]

Elimination of phosphate from a 2-deoxy analogue (**18**) of 3-deoxy-D-arabinose heptulosonic acid 7-phosphate, which models an important step in the shikimate pathway for biosynthesis of aromatic acids, has been shown to proceed with *syn*-stereochemistry. Elimination is presumed to be preceded by oxidation of the 6-hydroxy to a keto group and followed by re-reduction to the alcohol.[17]

Studies of *E*-(1,4) and *E*-(1,6)-eliminations involving dehydration and decarboxylation, usually in the presence of DMF dineopentyl acetal, have shown the stereochemistry of these reactions to be predominantly *syn*, notwithstanding geometrical constraints on the eliminating groups imposed by incorporating the reacting carbon bonds into rings [e.g. as in (**19**)].[18] For the *syn*-thermal elimination of the dimethyl amine oxide $PhCH_2CD_2\overset{+}{N}Me_2O^-$ a rather large α-deuterium isotope effect (1.16 per D) has been measured, which contrasts with the small value

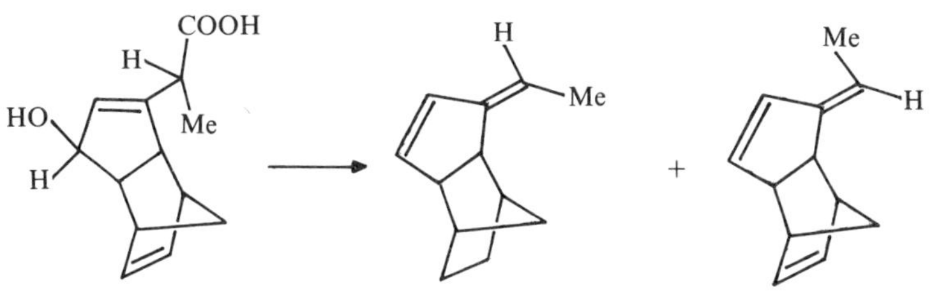

(**19**) *E*(99%) *Z*(1%)

(1.02 per D) found for base-promoted elimination of the structurally related phenethyltrimethylammonium ions: significant C—N bond-breaking but little double-bond development (despite carbanion character at the β-carbon atom) is suggested for the transition state.[19] There has also been further discussion of the interdependence of primary and secondary isotope effects (isotope effects upon isotope effects) which has been suggested as being a symptom of tunnelling in phenethyl elimination reactions.[20]

There have been several studies of dehydrohalogenation of α-halo-ketones[21–24] and -esters including isotope and Brønsted measurements.[21] The chloro- and bromo-lactones (**20**) have been shown to yield the corresponding butenolide (**21**) under typical *E2C* conditions,[22] and there has been an abortive but interesting attempt to effect the *E2C*-like demethylhalogenation (**22**)→(**23**).[25] Elimination reactions of β-halo-sulphones have been investigated as part of a search for neighbouring-group participation by the sulphone group,[26] and a Hammett correlation has been reported for elimination of substituted phenols activated by a β-cyano substituent.[27]

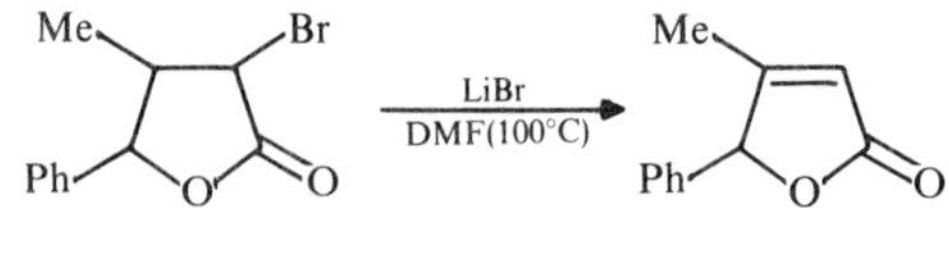

(20) (21)

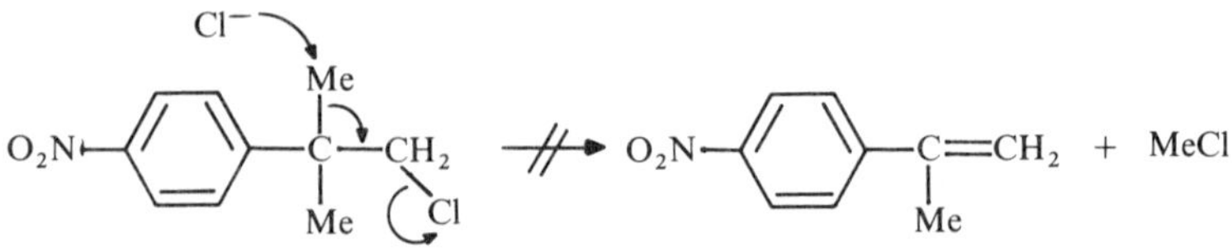

(22) (23)

Among elimination reactions yielding non-olefinic products, formation of imines has been most extensively investigated from a mechanistic point of view, and there has been a further study of the influence of base and leaving-group upon k_H/k_D and ρ for reaction of *N*-halogenobenzylamines ($ArCH_2NHHal$) with amine bases in acetonitrile.[28] An interesting study of molecular complexes of bromine with tertiary aliphatic amines has shown that for tribenzylamine the complex is not the *N*-bromonium bromide (**24**) but a mixture of benzylidinium and ammonium tribromides which, presumably, are formed from (**24**) by elimination.[29] Another imine-

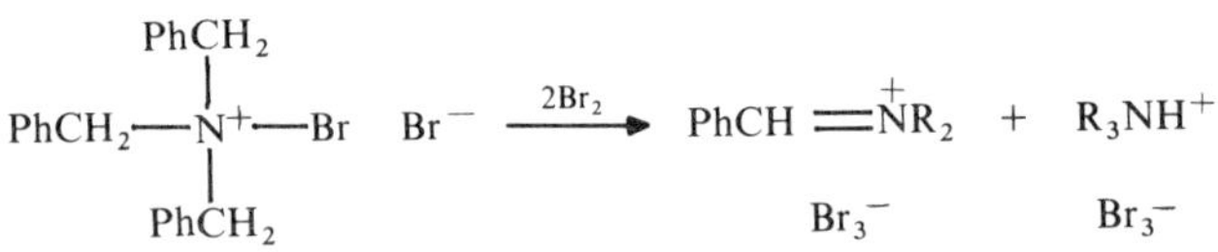

(24)

forming elimination provides the last step [(**26**)→(**27**)] of a base-catalysed conversion of dihydrofurooxazines (**25**) into acetylpyridine.[30]

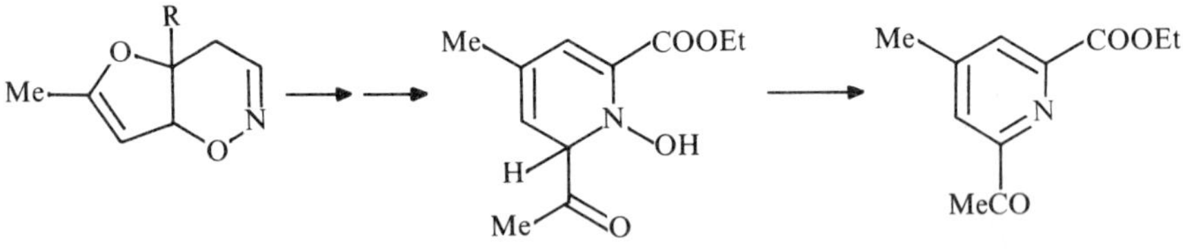

(25) (26) (27)

Sulphene-forming eliminations have been discussed in a short review[31] of mechanisms of sulphonyl transfer reactions: they have been invoked as intermediates in the hydrolysis of CH_3SO_2Cl ($CH_2{=}SO_2$) and β-styrenesulphonates ($PhCH{=}C{=}SO_2$).[32] An interesting study of eliminations of *E*- and *Z*-phosphaalkenes (**28**) and (**29**) has shown that thermolysis in the presence of oxygen yields the phospha-alkyne (**30**) with expulsion of $(Me_3Si)_2O$ in the case of the *Z*-isomer but not the *E*-, which decomposes, apparently with elimination of Bu^tSiMe_3.[33]

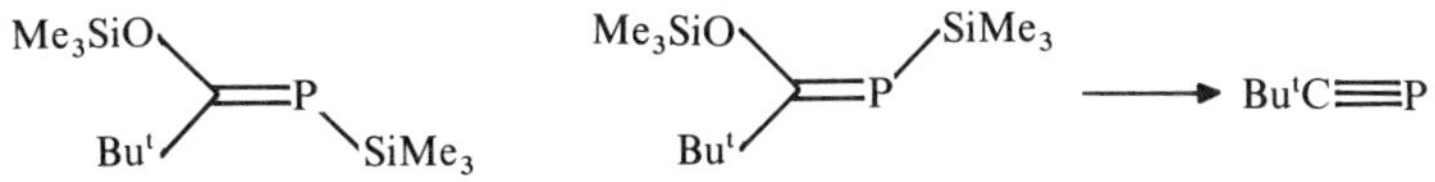

(**28**) *E* (**29**) *Z* (**30**)

Decarboxylation of allophanic acids ($ArNHCONHCOOH \rightarrow ArNHCONH_2 + CO_2$) has been studied as a model for the corresponding reactions of *N*-carboxybiotin. The reaction is general acid-catalysed and has a pH dependence consistent with spontaneous loss of CO_2 at high pH and an acid-catalysed loss at lower pH: Hammett ρ values are 0.84 and -0.72, respectively, in the two pH regions.[34] The possibility of a concerted mechanism for decarboxylation reactions under conditions where anion (usually carbanion) stabilizing substituents are absent has been discussed in the context of *ab initio* calculations (STO-3G and STO-31G) for the thermal decarboxylations of salicyclic and *p*-aminobenzoic acids in the presence of a chain of water molecules.[35]

Thermal Eliminations

Dehydrohalogenation, Decarboxylation

Neighbouring-group participation leading to cyclic products in competition with elimination has been studied in the gas phase for alkyl halides containing OH,[36] CH_3S,[37] and COOEt[38,39] substituents at the γ-carbon atom. The OH and SCH_3 groups yield tetrahydropyran and thiophene directly and the ester forms butyrolactone via pyrolysis to ethylene and 4-chlorobutyric acid and cyclization of the latter. A mild enhancement of the competing elimination reactions suggests that neighbouring groups provide some stabilization, probably by intramolecular solvation, of their transition states. In another study of substituent effects a value of $\rho = -0.46$ has been measured for the thermal 1,4-dehydrohalogenation of *O*-methylbenzyl chlorides (**31**) to benzocyclobutane.[40] Investigations have also been reported of elimination of CH_2CHF to give butyne at 400–570°C[41] and of CH_3CHF_2 to CH_2CHF at 300–400°C[42] (the latter in the presence of oxygen).

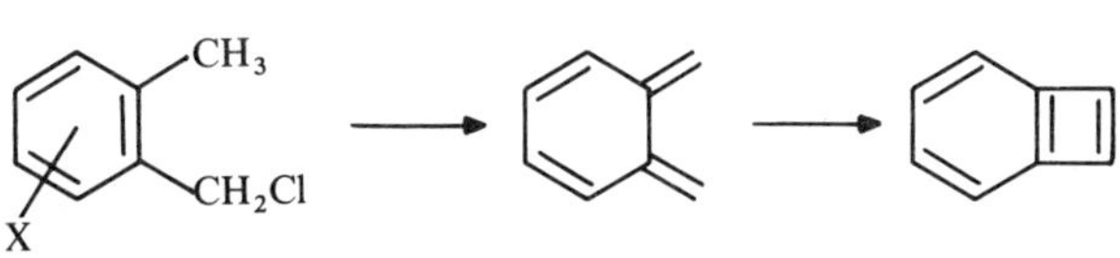

(**31**)

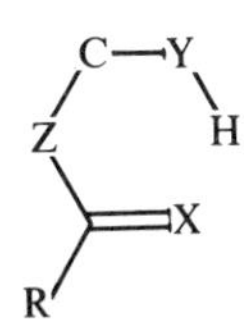

(**32**)

The thermal decomposition of pyruvic acid has been re-investigated and confirmed as being consistent with a unimolecular reaction, presumably via a

4-centre transition state (**32**). At 600 K the rate of reaction is 300 times less than that of oxalic acid and this is ascribed to the presence of an electron-rich carbonyl group inductively stabilized by OH in the latter[43] (although presumably an electron-deficient transition state stabilized by resonance with OH is not excluded). *Ab initio* calculations confirm that the preferred transition state for both decarboxylation and dehydration of acetic acid is a four-centre one.[44]

Ester Pyrolysis; Retroene and Related Reactions

The influence of structure upon reactivity of 6-centre thermal eliminations has been discussed in terms of an *ElcB-E2* spectrum of transition states for the concerted reactions. A distinction is drawn between substrates containing a relatively non-polar carbon-'leaving-group' bond (C—Z in **32**) and a nucleophilic double (C=X), such as in a ketone, in which the acidity of the bond to hydrogen (Y—H) is a significant influence upon reactivity, and substrates in which C—Z bond breaking is easier and the C=X bond less polar (e.g. an alkene), in which carbocation-stabilizing substituents at the α-carbon atom exert greater control over reactivity. The generalizations made are illustrated by reactions of β-keto-alcohols and β-hydroxy-alkenes together with literature data for vinyl ethers, 2-alkoxypyridines, esters, thioesters, and carbamates.[45] The effects of substituents on the thermolysis of carbamates have been examined by MINDO/3 calculations,[46] and data on the thermal elimination of bicycloalkyl tosylates have been reported.[47]

Substituent and solvent effects upon eliminations of sulphoxides have been investigated. For (**33**) values of $\rho_X = 0.32$ and $\rho_Y = 0.76$ are consistent with a strongly concerted process. However, the transition state is a flexible one, as illustrated by the 500-fold increase in rate caused by an α-phenyl substituent in $PhSOCH_2CH_3$ and 200-fold increase for a β-cyano group.[48] Alkyl substituent effects in the retroene

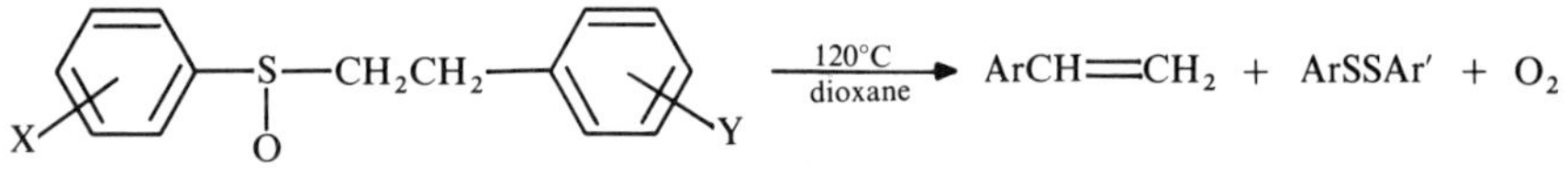

(**33**)

reaction of pyrazine ethanols (**34**) show a Thorpe–Ingold pattern which correlates with the extent of hydrogen-bonding between the OH group and pyrazine nitrogen atom in the reactants.[49] Retroene fragmentations of substituted 4-aza-1-phenyl-3-buten-1-ones have also been described[50] and it has been shown that at high temperatures thermal elimination of ethyl dicyanoacetate, $(CN)_2CHCOOEt$, switches from a pathway yielding ethylene and dicyanoketene to a retroene-like fragmentation forming HCN, cyanoketene, and formaldehyde.[51]

Cyclo-eliminations of the pyridine-*N*-oxides (**35**)[52] and of *N*-ethylpyrazoles[53] have been reported and presumably occur through 7-membered transition states by non-concerted mechanisms. There has been a further study of the complex pyrolytic eliminations of *N*-acylbenzothiazepinones (**36**).[54]

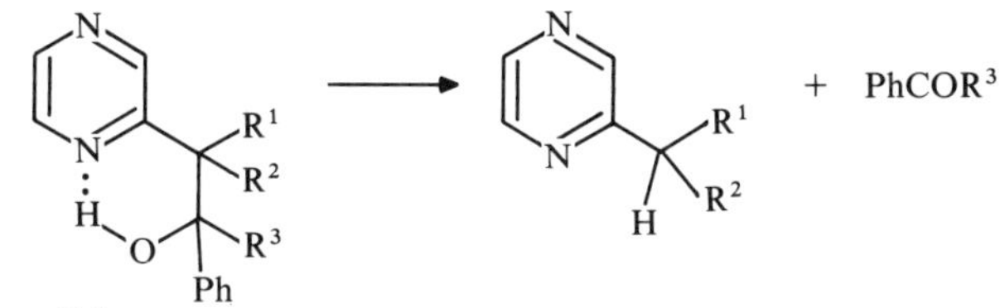

(34)

$R^1, R^2, R^3 = H$ or Me

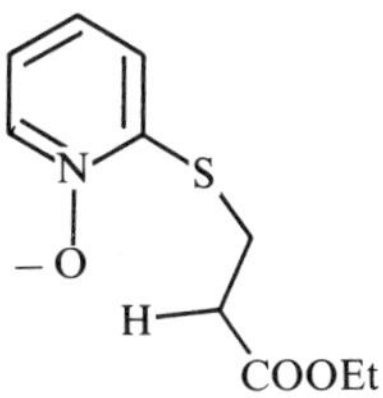

(35)

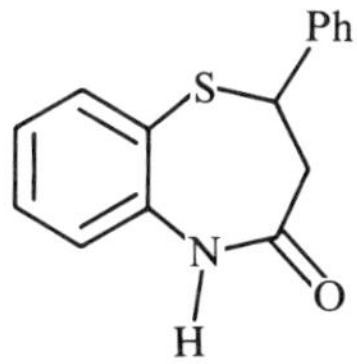

(36)

Loss of Nitrogen

Extrusion of nitrogen from the highly strained diazabenzvalene (**37**) occurs at −60°C with a half life of 20 minutes. No tetrahedrane is formed and the observed product is the tricyclooctadiene dimer of cyclobutadiene (**38**). The activation energy for this reaction is 16 kcal mol^{-1} less than for decomposition of the less strained (**39**).

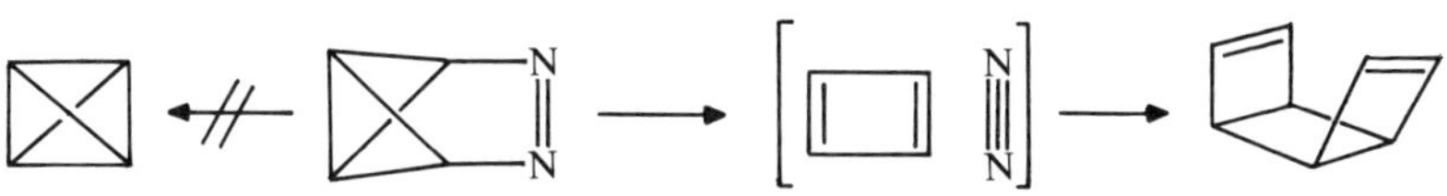

(37) **(38)**

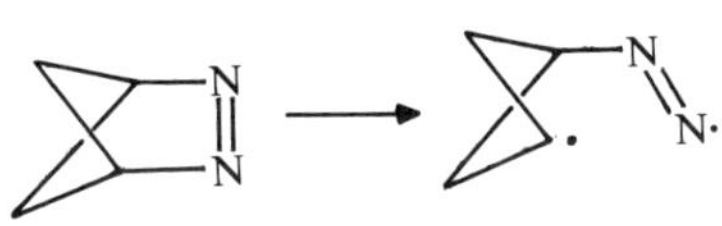

(39) **(40)**

There is good evidence that (**39**) reacts via a diazenyl diradical (**40**), and since the strain released in forming the corresponding diradical from (**37**) should hardly be greater than from (**39**) it is suggested that cyclobutadiene formation probably occurs via a concerted 6-electron transition state.[55]

Thermal or photochemical extrusion of nitrogen from (**41**) offers no evidence of products derived from the 'tetrayl' radical (**42**). Rather, successive eliminations of nitrogen occur, presumably via diyl radicals, to yield *exo*- and *endo*-bis-bicyclopentanes (**43**); the implications of the stereochemistry of these products are discussed.[56] Depending on the substituent, thermolysis of the tetrazoles (**44**) leads to

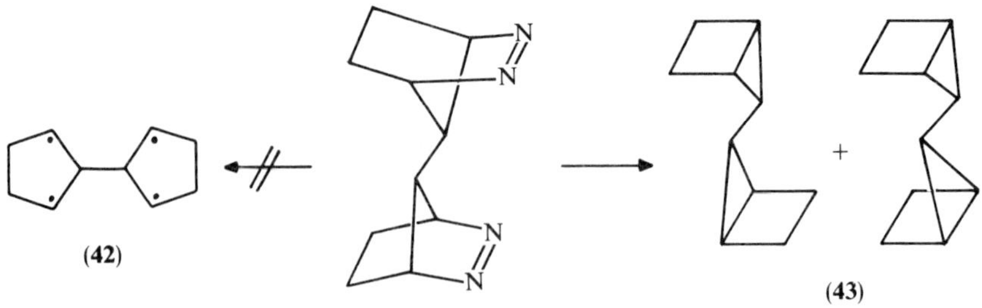

(42) **(41)** **(43)**

endo,endo *endo,exo*
(+*exo,exo* and *exo,endo*)

diaziridines (presumably also via loss of nitrogen and formation of a diyl radical) or, in a formal 4 + 2 cycloreversion, to an azide and imine (Scheme 3).[57]

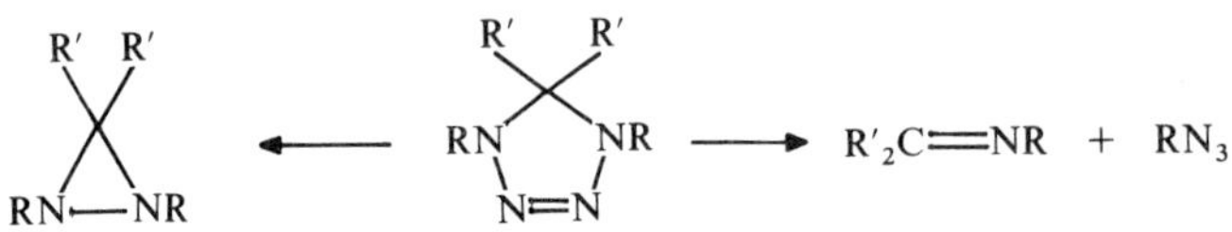

(44)

SCHEME 3

Cycloreversions

The interesting benzene dimer *cis,syn-o,o′*-dibenzene (**46**) is formed by mild thermal rearrangement of the photochemically prepared pentacyclic dimer of benzene (**45**). Interestingly, (**46**) itself is converted surprisingly easily (half life 6 hrs at 40°C) to benzene. The ease of this reaction seems inconsistent with the obvious forbidden pathway of 2 + 2 cycloreversion, and reaction via the intermediate (**47**) is suggested.[58]

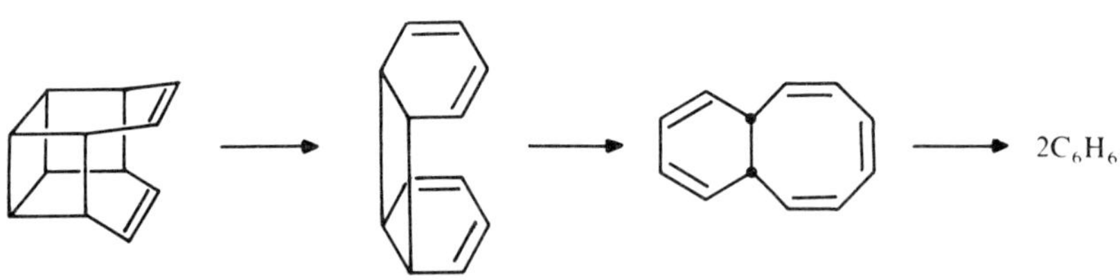

(45) **(46)** **(47)**

Another facile cycloreversion is the retro-Diels–Alder reaction of *N*-protonated azabicycloheptenes (**48**) which occurs in 2–5 hrs at 25–25°C depending on the substituent on the nitrogen atom.[59] By contrast cycloreversion of cyano-substituted bicyclooctenes (**49**) occurs between 520 and 630°C. On the basis of some *cis-trans* isomerization of the olefinic product from the latter a biradical mechanism is suggested.[60] The complex decomposition of silacyclobutane under flash vacuum

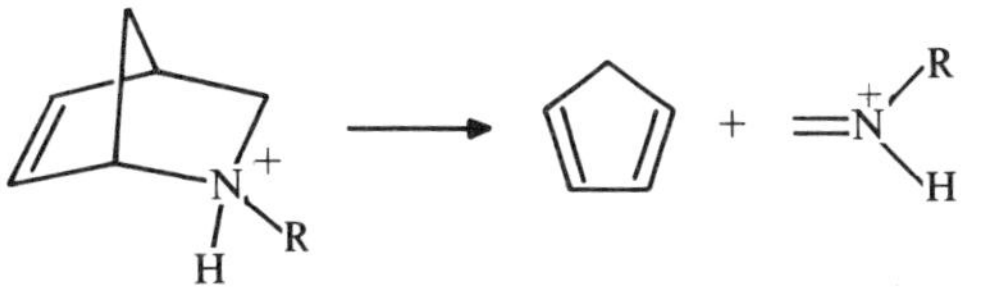

(48)

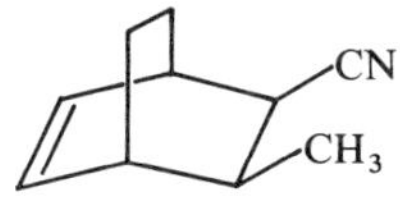

(49)

pyrolysis has been studied,[61] and also the reaction of tetramethylcyclobutane-1,3-dione to dimethylketene.[62]

Gas-phase Ionic Reactions

Isotope effects on elimination reactions of unsymmetrically deuteriated diethyl ether (**50**) promoted by H_2N^- ($k_H/k_D = 5.6$) and HO^- ($k_H/k_D = 3.2$) have been measured by intramolecular competition using FT ion cyclotron resonance, and their dependence on base strength interpreted in terms of an *E*2 mechanism with a variable transition state. Reaction paths yielding alkoxide leaving-groups solvated and unsolvated by the product water molecule are distinguished and evidence is adduced that these occur with *syn*- and *anti*-stereochemistry, respectively.[63] A

$$C_2H_5OC_2D_5 \xrightarrow{HO^-} CH_2{=}CH_2 + {}^-OC_2D_5 + H_2O$$

$$C_2H_5OC_2D_5 \xrightarrow{HO^-} CD_2{=}CD_2 + {}^-OC_2H_5 + HDO$$

(50)

related study of alcohol-assisted expulsion of F^- from the carbanion $(CF_3)_2CH^-$ shows that use of deuteriated alcohol leads to hydrogen exchange in the carbanion but not in the fluoride anion-alcohol cluster (F^-HOR) of the product. It is concluded that the alcohol can coordinate to both fluorine and carbon sites in the carbanion, and that expulsion of F^- occurs with an *anti*-stereochemistry and does not involve a carbon-bound alcohol molecule.[64] A mass spectrometric study of acetylacetonate ions shows that the predominant pathway for collision-activated reaction of the anion $MeCO\bar{C}HCOMe$ involves rearrangement to its isomer $\bar{C}H_2COCH_2COMe$ and fragmentation to ketene and acetone.[65]

Details have appeared of an isotope study of a multiphonon-induced fragmentation of vibrationally excited *tert*-butoxide ion (**51**) to acetone enolate anion and methane. Measurement of a large intramolecular secondary deuterium isotope effect (1.9 per H) strongly implies a stepwise reaction via acetone and a methyl anion with a transition state resembling these intermediates. The large isotope effect is imputed to the low 'out-of-plane' bending force constant of the methyl anion.[66]

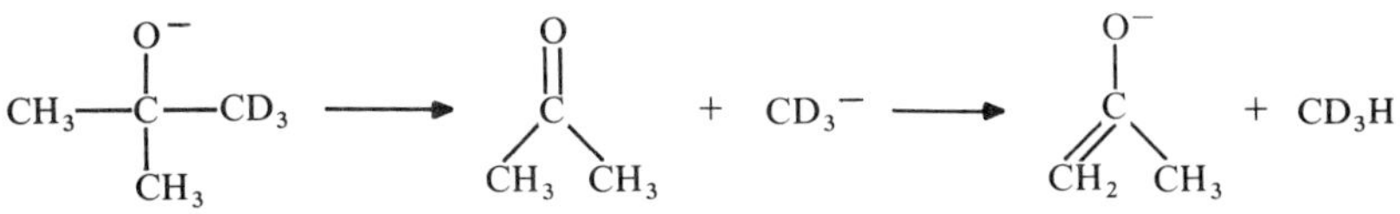

(51)

Elimination Reactions in Synthesis

Incorporation of elimination reactions into a computer-assisted synthesis programme that makes judgments between possible reaction paths has been discussed. Eliminations are dealt with in the general context of nucleophilic chemistry, as opposed to electrophilic, radical, carbenoid, pericyclic, and oxidation-reduction reactions, each of which requires separate analysis. In general the feasibility of the elimination is judged in terms of the pK_a value of the reacting base, the prospective C—H (or other) site of attack, and the leaving-group. Different types of elimination, such as 1,2-, 1,4-, fragmenting and annulene-forming (and their stereochemistry), are considered and compared with other nucleophilic pathways, notably addition and substitution.[67]

Dehydration is an important route to olefins and the advantages of different reagents, including copper(II) triflate,[68] in effecting mild and stereoselective reactions without competing rearrangement have been discussed. Use of Me_3SiCl has been recommended as producing *E*-olefin from diastereomeric alcohol precursors in the synthesis of alkenylfuranones such as the natural product geiparvarin (**52**), but the relative importance of kinetic and thermodynamic control appears not to have been fully clarified as the reagent also effects isomerization of *Z*- to *E*-olefins.[69] A surprisingly facile *syn*-stereospecific dehydration, to yield olefins (**53**)

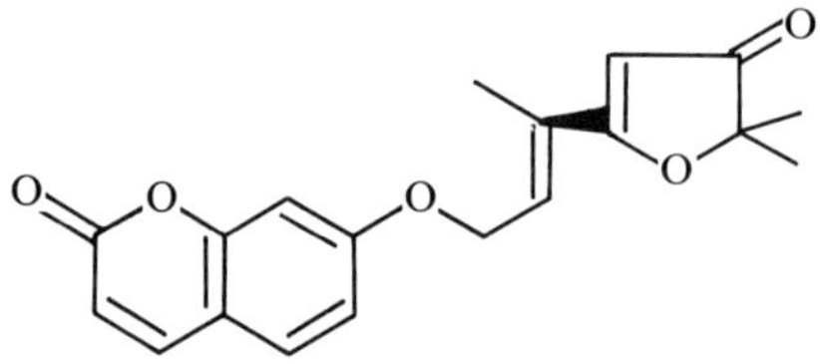

(52)

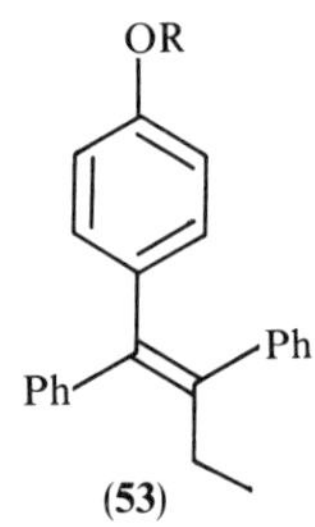

(53)

required in the synthesis of the anti-cancer agent tamoxifen (**53**; R = $CH_2CH_2NMe_2$), has been achieved in moderate yield by reacting the alkoxide anion of the alcohol with CS_2. That thermolysis of the dithiocarbonate anion intermediate ($ROCS_2^-$) occurs more readily than normal is ascribed to the activating effect of the several aryl substituents.[70]

There has been a number of applications of phenyl selenoxide eliminations to olefin synthesis. A general and regiospecific preparation of chloro-olefins involves treatment of a parent olefin with phenylselenyl chloride and chlorination of the product with sulphuryl chloride. A (chloro-alkyl)phenylselenyl dichloride (**54**) is obtained which could usually be separated from a small amount of regioisomer

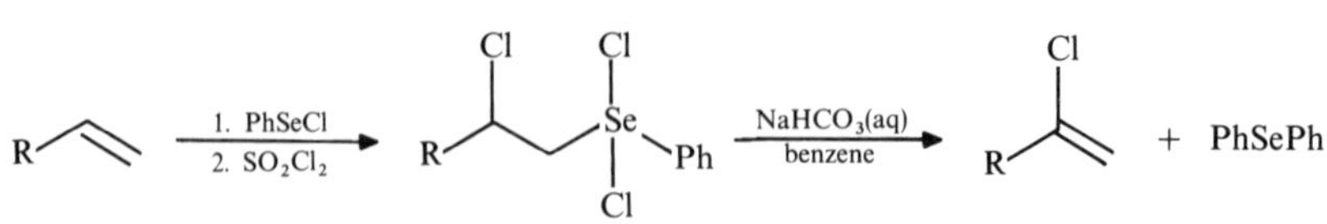

(54)

(formed by anti-Markownikoff addition of PhSeCl) by crystallization of the crude reaction mixture. A two-phase hydrolysis then gave the selenoxide which eliminated to the chloro-olefin. Where the original olefin contained a β-hydroxy or -acetoxy substituent anti-Markownikoff addition of PhSeCl gave the terminal chloro-olefin.[71,72] Phenyl selenoxide eliminations have also been used in a synthesis of dimethylhydropyrene[73] and in the isomerisation of *E*- to *Z*-nitro-alkanes.[74]

Cheleotropic extrusion of SO_2 has been exploited in the synthesis of a variety of dimethylenepyrrolidines (**56**) from the readily available precursors (**55**).[75] Stereo-

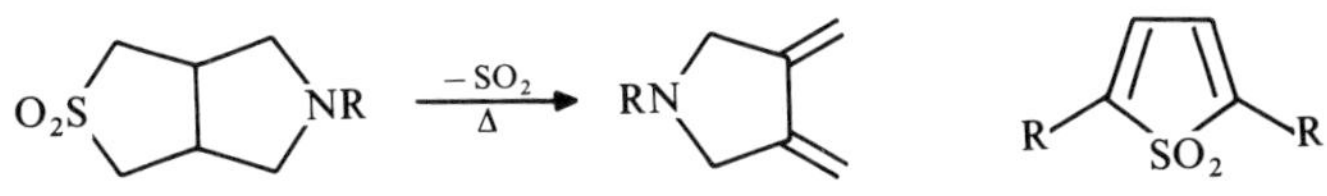

(**55**) (**56**) (**57**)

selective formation of (*E,Z*)-dienes from substituted sulpholenes (**57**) is reported to be promoted by ultrasonically dispersed metallic potassium.[76]

Enantioselective elimination from both alkyl halides and epoxides can be effected using lithium amides with a chiral auxiliary. For the epoxides an *ElcB syn*-elimination with coordination of metal ion to the epoxy oxygen is assumed.[77] Remarkably, dehydrohalogenation of the chlorocyclohexanecarboxylic acids (**58**) gives the cyclohexylideneacetic acid product (**59**) in up to 80% enantiomeric excess

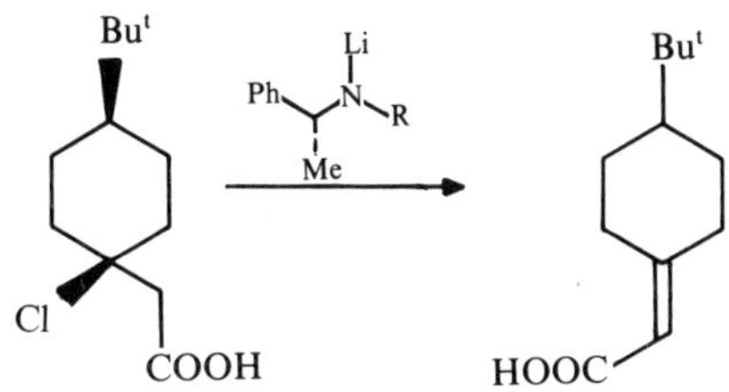

(**58**) (**59**)

of either configuration depending on the substituent in the lithium amide.[78] A diastereoselective deoxymercuration by iodide ion and base has been achieved in the presence of an adjacent carbonate function which is hydrolysed to the alcohol. Reaction via a cyclic carbonate (**60**) is assumed, but the origin of the selectivity is not altogether clear. The alternative configuration for the cyclic carbonate (**61**) would

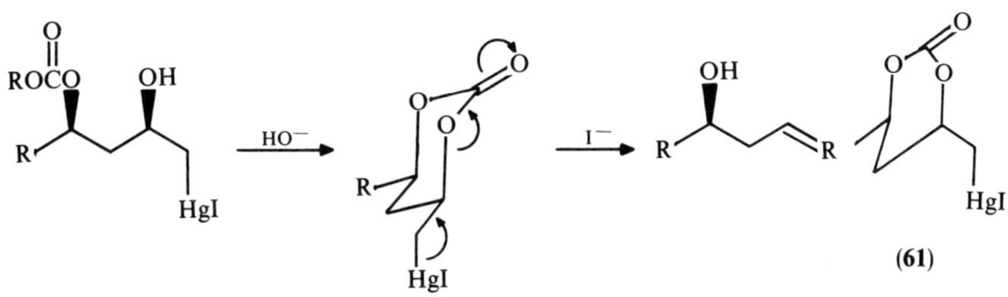

(**61**)

(**60**)

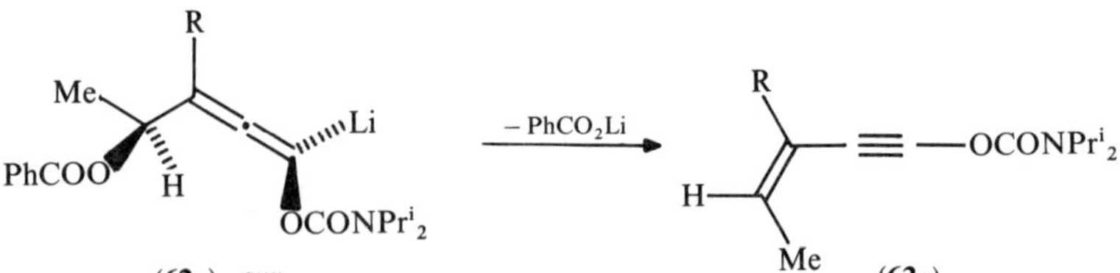

(62a) *syn* **(63a)**

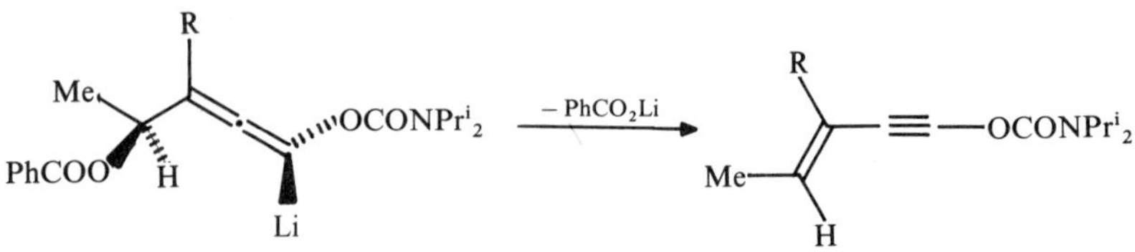

(62b) *anti* **(63b)**

place the CH_2HgI group in an axial position and either this is less stable or stereochemically less prone to fragmentation.[79]

A *syn*-stereospecific 1,4-elimination of the lithium allenyl benzoates (**62a**,**b**) at temperatures (−78°C) sufficiently low that they are configurationally stable has yielded the corresponding enyne carbamates (**63a**,**b**), possibly with a metal-(leaving-group) interaction influencing the stereochemistry.[80] The mechanism of the curious

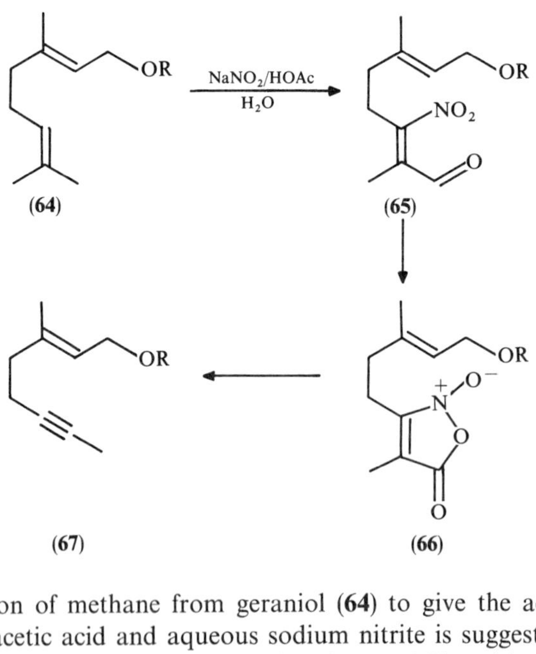

(64) **(65)** **(67)** **(66)**

formal elimination of methane from geraniol (**64**) to give the acetylene (**67**) on treatment with acetic acid and aqueous sodium nitrite is suggested as occurring through an oxidative nitrosation to the nitro-ketone (**65**) and isoxazalone (**66**).[81]

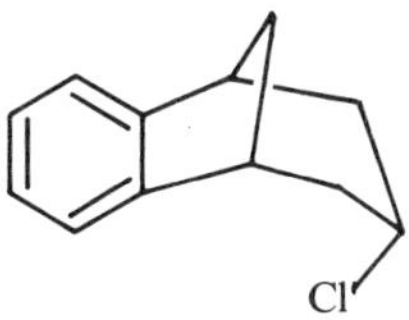

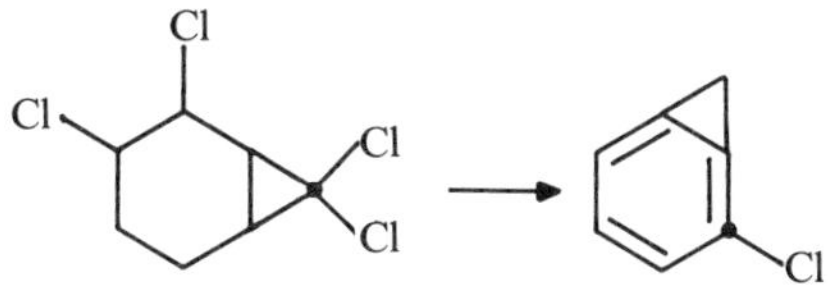

(68) **(69)** **(70)**

Further studies have been reported on the reaction of 1,4-dichlorobut-2-yne with dialkylboranes and methyllithium to yield 2-alkylbuta-1,3-dienes.[82] Arylation of *E*- and *Z*-alkenyltrimethylsilanes with arylpalladium acetates leads to alkenes with the stereochemistry of alkyl and $SiMe_3$ substituents reversed by successive *syn*-addition of the reagent and *syn*-elimination of Pd(0) and AcOH.[83]

Photolysis, Phase Transfer, Surface Reactions; Other Eliminations

Photolytically-induced solvolytic eliminations are possible at a site remote from an aromatic ring that provides the active chromophore for a reaction, and are illustrated by elimination from the chloride (**68**) in methanol on irradiation at 254 nm. It is thought that $\pi \rightarrow \pi^*$ excitation is followed by mixing of π,π^* and (C—Cl) σ,σ^* states. Significantly, the quantum efficiency for the *endo*-chloride (shown) is an order of magnitude greater than for the *exo*-chloride, where the C—Cl bond is more remote from the chromophore.[84]

Eliminations from dichloroethane[85] and tetrahexylammonium salts[86] under phase-transfer conditions have been reported. The latter reaction showed that increasing concentrations of sodium hydroxide in the aqueous phase (up to 50% NaOH) led to increased elimination rates but a *reduced* concentration of hydroxide ion in the chlorobenzene non-aqueous phase. Both factors, presumably, reflect the high basicity of hydroxide in the aqeuous phase arising from lack of solvating water molecules. It is interesting and reasonable that classical 'acidity-function' behaviour appears to be communicated to the non-aqueous phase.

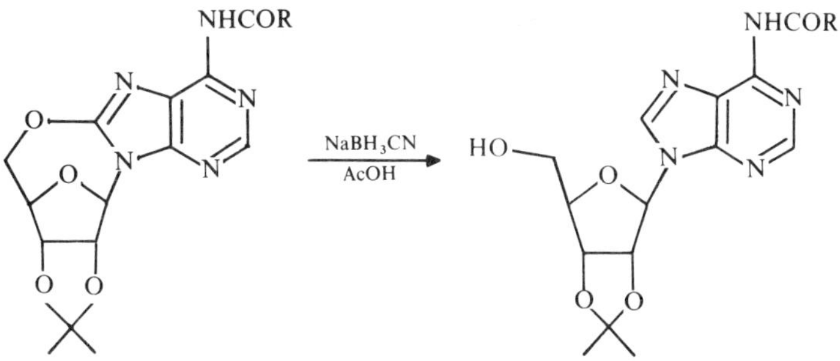

(71) **(72)**

In a study of elimination of perhalobicyclo[4.1.0]heptanes (**69**) to form cycloproparenes (**70**) it has been shown by ^{13}C labelling that partial rearrangement of the cyclopropane ring occurs; (**69**)→(**70**). Likely mechanisms for this reaction are discussed but the obvious possibility of a cycloheptatriene-norcaradiene rearrangement (which would presumably lead to a more random distribution of the ^{13}C label than observed) is rejected.[87]

A reductive ring-opening elimination from the cycloadenosine (**71**) to form (**72**) with $NaBH_3CN$ in acetic acid has been reported.[88] In an interesting paper it has been shown that the isomeric tetramethyloxolanes (**73**) and (**74**) on Al_2O_3 and other dehydration catalysts undergo ring-opening and dehydration to form the dienes (**75**) and (**76**), respectively. The products imply that the initial ring-opening occurs more readily to yield the Saytseff product from (**74**), in contrast to formation of the exocyclic Hofmann olefin from (**73**). The difference is ascribed to the operation of *E*1 and *E*2 mechanisms, respectively, but in this author's opinion more probably reflects poorer steric accessibility of the *β*-hydrogens in the ring of (**73**) than (**74**). It is curious that the dehydration of the alcohol intermediate formed from (**73**) also gives Hofmann orientation. Possibly the alcohol product of the first reaction dehydrates without desorption.[89] Conjugation of the two double bonds presumably controls

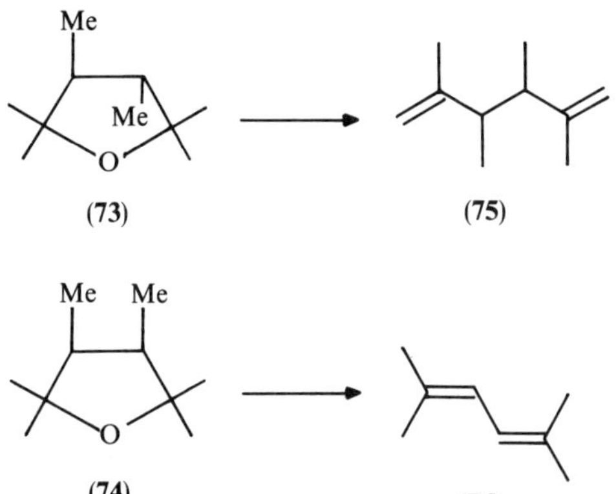

(**73**) (**75**)

(**74**) (**76**)

formation of (**76**) from the corresponding alcohol obtained from (**74**).

References

1 Hughes, S., Griffiths, G., and Stirling, C. J. M., *J. Chem. Soc., Perkin Trans. 2*, **1987**, 1253.
2 Piras, P. P. and Stirling, C. J. M., *J. Chem. Soc., Perkin Trans. 2*, **1987**, 1265.
3 Earl, H. A. and Stirling, C. J. M., *J. Chem. Soc., Perkin Trans. 2*, **1987**, 1273.
4 Bury, A., Earl, H. A., and Stirling, C. J. M., *J. Chem. Soc., Perkin Trans. 2*, **1987**, 1281.
5 Page, M. I., Webster, P., Bogdan, S., Tremerie, B., and Ghosez, L., *J. Chem. Soc., Chem. Commun.*, **1987**, 318.
6 Rajyaguru, I. and Rzepa, H. S., *J. Chem. Soc., Perkin Trans. 2*, **1987**, 359.
7 L'Abbé, G. and Godts, F., *Bull. Soc. Chim. Belg.*, **96**, 229 (1987).
8 Schmid, G. H., Strukelj, M., Dalipi, S., and Ryan, M. D., *J. Org. Chem.*, **52**, 2403 (1987).
9 Okuma, K., Yamasaki, Y., Komiya, T., Kodera, Y., and Ohta, H., *Chem. Lett.*, **1987**, 357.
10 Hoz, S. and Livneh, M., *J. Am. Chem. Soc.*, **109**, 7483 (1987).
11 Dureja, P., Walia, S., and Mukerjee, S. K., *Tetrahedron Lett.*, **28**, 895 (1987).
12 Cabaleiro, M. C. and Garay, R. O., *J. Chem. Soc., Perkin Trans. 2*, **1987**, 1473.
13 Thibblin, A., *J. Am. Chem. Soc.*, **109**, 2071 (1987).

[14] Maskill, H. and Jencks, W. P., *J. Am. Chem. Soc.*, **109**, 2062 (1987).
[15] Bartsch, R. A., Cho, B. R., and Pugia, M. J., *J. Org. Chem.*, **52**, 5494 (1987).
[16] Doshi, H., Cardis, A. B., Crelling, J. K., Miller, S. I., Dalton, D. R., Zacharias, D. E., and Glusker, J. P., *J. Org. Chem.*, **52**, 2604 (1987).
[17] Widlanski, T. S., Bender, S. C., and Knowles, J. R., *J. Am. Chem. Soc.*, **109**, 1873 (1987).
[18] Vogel, E., Caravatti, G., Franck, P., Aristoff, P., Moody, C., Becker, A.-M., Felix, D., and Eschenmoser, A., *Chem. Lett.*, **1987**, 219.
[19] Smith, P. J. and Westaway, K. C., *Can. J. Chem.*, **65**, 2149 (1987).
[20] Saunders, W. H., Subramanian, R., and Price, R. C., in *Synthetic Applications of Isotopically Labelled Compounds: Proceedings of Second International Symposium 1985*, ed. Muccino, R. R., Elsevier, Amsterdam, 1986, p. 95; *Chem. Abs.*, **107**, 225575, (1987).
[21] Popov, A. F., Matvienko, V. N., Piskunova, Zh. P., and Oleinik, N. M., *Zh. Org. Khim.*, **22**, 2142 (1986); *Chem. Abs.*, **107**, 153698 (1987).
[22] Gharbi-Bernarous, J., Essayegh, M. M., and Dara, G., *Can. J. Chem.*, **65**, 2031 (1987).
[23] Peli, E., Patonay, T., Zsuga, M., and Litkei, G., *Stud. Org. Chem.*, **23**, 75 (1986); *Chem. Abs.*, **106**, 175484 (1987).
[24] Popov, A. F., Piskunova, Zh. P., and Matvienko, V. N., *Zh. Org. Khim.*, **22**, 1442 (1986); *Chem. Abs.*, **106**, 155647 (1987).
[25] Eckert, T. S. and Rominger, R. L., *J. Org. Chem.*, **52**, 5474 (1987).
[26] Carretero, J. C., Ruano, J. L. G., Martinez, N. C., and Rodriguez, J. H., *Tetrahedron*, **43**, 4417 (1987).
[27] Vaitkus, R., Dieny, G., and Adomeniane, O., *Org. React. (Tartu)*, **22**, 454 (1985); *Chem. Abs.*, **106**, 118978 (1987).
[28] Cho, B. R., Namgoong, S. K., and Kim, T. R., *J. Chem. Soc., Perkin Trans. 2*, **1987**, 853.
[29] Bellucci, G., Bianchini, R., and Ambrosetti, R., *J. Chem. Soc., Perkin Trans. 2*, **1987**, 39.
[30] Gilchrist, T. L., Hughes, D., Stretch, W., and Chrystal, E. J. T., *J. Chem. Soc., Perkin Trans. 2*, **1987**, 2505.
[31] King, J. F., Skonieczny, S., Khemani, K. C., and Lock, J. D., *Adv. Chem. Ser.*, **215**, 385 (1987); *Chem. Abs.*, **107**, 153643 (1987).
[32] Vizgert, R. V., Enya, V. I., Tuchapskii, I. M., and Pekhtereva, T. M., *Zh. Obshch. Khim.*, **56**, 2618 (1986); *Chem. Abs.*, **107**, 96102 (1987).
[33] Barron, A. R., Cowley, A. H., and Hall, S. W., *J. Chem. Soc., Chem. Commun.*, **1987**, 980.
[34] Al Sabbagh, M. M., Calmon, M., and Calmon, J. P., *J. Chem. Soc., Perkin Trans. 2*, **1987**, 929.
[35] Ruelle, P., *J. Chem. Soc., Perkin Trans. 2*, **1986**, 1953.
[36] Chuchani, G. and Martin, I., *Int. J. Chem. Kinet.*, **19**, 183 (1987).
[37] Chuchani, G., Martin, I., and Dominguez, R. M., *Int. J. Chem. Kinet.*, **19**, 683 (1987).
[38] Chuchani, G. and Rotinov, A., *Int. J. Chem. Kinet.*, **19**, 789 (1987).
[39] Chuchani, G., Dominguez, R. M., and Martin, I., *React. Kinet. Catal. Lett.*, **30**, 77 (1986); *Chem. Abs.*, **106**, 17745 (1987).
[40] So, Y.-H., *J. Org. Chem.*, **52**, 1615 (1987).
[41] Cifuentes, L. O., Casas, F., and Aguilera, F. A., *Bol. Soc. Quim. Peru*, **51**, 224 (1985); *Chem. Abs.*, **107**, 38995 (1987).
[42] Kostyk, G. P., Kuchev, R. V., Lutsyk, N. V., and Pazderskii, Yu. A., *Ukr. Khim. Zh.*, **22**, 747 (1986); *Chem. Abs.*, **106**, 175504 (1987).
[43] Taylor, R., *Int. J. Chem. Kinet.*, **19**, 709 (1987).
[44] Ruelle, P., *Chem. Phys.*, **110**, 263 (1986); *Chem. Abs.*, **106**, 213166 (1987).
[45] August, R., McEwen, I., and Taylor, R., *J. Chem. Soc., Perkin Trans. 2*, **1987**, 1683.
[46] Dorofeeva, I. B., and Tarakanov, O. G., *Zh. Strukt. Khim.*, **27**, 42 (1986); *Chem. Abs.*, **106**, 155772 (1987).
[47] Qureshi, M. I., *Pak. J. Sci. Ind. Res.*, **29**, 91 (1986).
[48] Yoshimura, T., Yoshizawa, M., and Tsukurimichi, E., *Bull. Chem. Soc. Jpn*, **60**, 2491 (1987).
[49] Houminer, Y., Fennev, R. A., Secor, H. V., and Seeman, J. I., *J. Org. Chem.*, **52**, 3971 (1987).
[50] Armesto, D., Horspool, W. M., Perez-Ossorio, R., and Ramos, A., *J. Org. Chem.*, **52**, 3378 (1987).
[51] Gano, J. E. and Mahdi, F., *J. Org. Chem.*, **52**, 2102 (1987).
[52] Crich, D. and Lim, L. B. L., *J. Chem. Res. (S)*, **1987**, 353.
[53] Perez, J. D. and Phagouape, L. M., *Int. J. Chem. Kinet.*, **19**, 571 (1987).
[54] Kaupp, G. and Matthies, D., *Chem. Ber.*, **120**, 1741 (1987).
[55] Kaisaki, D. A. and Dougherty, D. A., *Tetrahedron Lett.*, **28**, 5263 (1987).
[56] Adam, W., Hannemann, K., Peters, E. M., Peters, K., von Schnering, H. G., and Wilson, R. M., *J. Am. Chem. Soc.*, **109**, 5250 (1987).
[57] Carboni, B., Tonnard, F., and Carrié, R., *Bull. Soc. Chim. Fr. II*, **1987**, 525.

[58] Yang, N. C., Hrujez, B. J., and Horner, M. G., *J. Am. Chem. Soc.*, **109**, 3158 (1987).
[59] Grieco, P. A., Parker, D. T., Fobare, W. F., and Ruckle, R., *J. Am. Chem. Soc.*, **109**, 5859 (1987).
[60] van Mele, B. and Huybrechts, G., *Int. J. Chem. Kinet.*, **19**, 885 (1987).
[61] Barton, T. J. and Tillman, N., *J. Am. Chem. Soc.*, **109**, 6711 (1987).
[62] Vala, M., Baiardo, J., Latham, D., Mukherjee, R., and Pascyz, S., *J. Indian. Chem. Soc.*, **63**, 16 (1986); *Chem. Abs.*, **106**, 101624 (1987).
[63] de Koning, L. J. and Nibbering, N. M. M., *J. Am. Chem. Soc.*, **109**, 1715 (1987).
[64] McDonald, R. N., McGhee, W. D., and Chowdhury, A. K., *J. Am. Chem. Soc.*, **109**, 7334 (1987).
[65] Hayes, R. N., Sheldon, J. C., and Bowie, J. H., *Int. J. Mass Spectrom. Ion Processes*, **71**, 233 (1986); *Chem. Abs.*, **106**, 137811 (1987).
[66] Tumas, W., Foster, R. F., Pellerite, M. J., and Brauman, J. I., *J. Am. Chem. Soc.*, **109**, 961 (1987).
[67] Metivier, P., Gushurst, A. J., and Jorgensen, W. L., *J. Org. Chem.*, **52**, 3724 (1987).
[68] Laali, K., Gerzina, R. J., Flajnik, C. M., Geric, C. M., and Dombroski, A. M., *Helv. Chim. Acta*, **70**, 607 (1987).
[69] Sakai, T., Kohda, K., Tsuboi, S., Utaka, M., and Takada, A., *Bull. Chem. Soc. Jpn*, **60**, 2911 (1987).
[70] McCague, R., *J. Chem. Soc., Perkin Trans. 1*, **1987**, 1011.
[71] Engman, L., *Tetrahedron Lett.*, **28**, 1463 (1987).
[72] Engman, L., *J. Org. Chem.*, **52**, 4086 (1987).
[73] Mitchell, R. H., Weerawarna, K. S., and Bushnell, G. W., *Tetrahedron Lett.*, **28**, 5119 (1987).
[74] Ono, N., Kamimura, A., Kawai, T., and Kaji, A., *J. Chem. Soc., Chem. Commun.*, **1987**, 1550.
[75] Ottenbrite, R. M., Chin, H., and Alston, P. V., *J. Heterocycl. Chem.*, **23**, 1725 (1986).
[76] Chou, T. and Yon, M.-L., *J. Org. Chem.*, **52**, 2224 (1987).
[77] Asami, M. and Kirihara, H., *Chem. Lett.*, **1987**, 389.
[78] Duhamel, L., Ravard, A., Plaquevent, J.-C., and Davoust, D., *Tetrahedron Lett.*, **28**, 5517 (1987).
[79] Lilitkarntakul, S., Hirama, M., and Itô, S., *Tetrahedron Lett.*, **28**, 1207 (1987).
[80] Hoppe, D. and Gonschorrek, C., *Tetrahedron Lett.*, **28**, 785 (1987).
[81] Corey, E. J., Seibel, W. L., and Kappos, J. C., *Tetrahedron Lett.*, **28**, 4921 (1987).
[82] Hoshi, M., Masuda, Y., and Arase, A., *J. Chem. Soc., Chem. Commun.*, 1629 (1987).
[83] Ikenaga, K., Kikukawa, K., and Matsuda, T., *J. Org. Chem.*, **52**, 1276 (1987).
[84] Morrison, H., Muthuramu, K., and Severance, D., *J. Org. Chem.*, **51**, 4681 (1986).
[85] Shavanov, S. S., Shutenkova, T. V., Ryabova, N. A., and Tolstikov, G. A., *Dokl. Akad. Nauk. SSSR*, **288**, 1379 (1986); *Chem. Abs.*, **106**, 32153 (1987).
[86] Landini, D., Maia, A., and Rampoldi, A., *J. Org. Chem.*, **51**, 5475 (1986).
[87] Halton, B., Randall, C. J., Gainsford, G. J., and Robinson, W. T., *Aust. J. Chem.*, **40**, 475 (1987).
[88] Sako, M., Saito, T., Kamegama, K., Hirota, K., and Maki, Y., *J. Chem. Soc., Chem. Commun.*, **1987**, 1299.
[89] Molnár, A. and Bartok, M., *Tetrahedron*, **43**, 131 (1987).

Organic Reaction Mechanisms 1987
Edited by A. C. Knipe and W. E. Watts

CHAPTER 13

Addition Reactions: Polar Addition

ARTHUR FRY

Department of Chemistry and Biochemistry, University of Arkansas, Fayetteville, AR 72701, USA

Reviews

The mechanistic aspects of polar addition reactions of cyclopropenes and methylenecyclopropenes, together with many other aspects of the chemistry of these interesting compounds, have been covered in a comprehensive review focusing on their use as multi-functional reagents in transition metal-catalysed reactions.[1] An important focus of a review on product stability in kinetically-controlled organic reactions is polar additions to carbon-carbon double bonds.[2] Electrophilic addition reactions have been classified into three groups, depending on the position in the Periodic Table of the atoms involved, and an orbital-domination rule has been proposed to account for mechanisms, stereospecificity, and regiospecificity.[3] In a general treatment of nucleophilic chemistry, an interactive computer program has been designed to predict the products of a variety of organic reactions, given starting materials and reaction conditions. Addition reactions constitute one of the major applications of the procedure.[4] Choices of Michael donors and acceptors and stereochemical aspects of conjugate additions are covered in a review of principles and methods in stereoselective syntheses.[5] Reference is made to reviews of more limited scope, or of a more specific nature, or which also include experimental work, in the various sections following.

Electrophilic Additions

Halogenation and Related Reactions

The use of cesium fluoroxysulphate in the presence of HF or methanol has proved to be an effective way to obtain vicinal difluorides and vicinal methoxy-fluorides from alkenes. Studies with (*E*)- and (*Z*)-stilbenes, acenaphthalene, and indene showed that *syn* addition predominated over *anti* addition, and that the regioselectivity was of the Markovnikov type leading to the suggestion that the reactions have mechanisms of a carbenium-ion nature.[6] With 1-phenylindene, where the carbocation would be expected to be stabilized, the regioselectivity remained high, but *anti* addition predominated.

The chlorination of 2-*tert*-butyl-4,6-dimethylphenol gave oxidation-substitution-addition products (**1**)–(**4**), presumably through the intermediacy of chlorodienone (**5**).[7] Addition of chlorine to (**5**) would result in the formation of (**1**) and (**2**); a straightforward combination of addition and elimination reactions would then lead to (**3**) and (**4**). The chlorination of 2,6-di-*tert*-butyl-4-methylphenol followed a similar, but somewhat more complex, path.

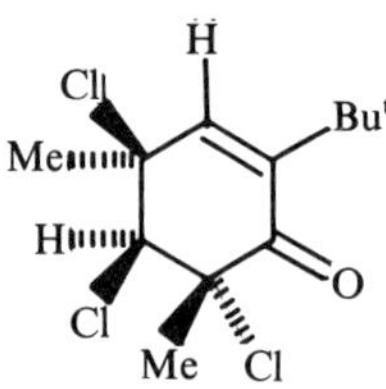

(**1**)

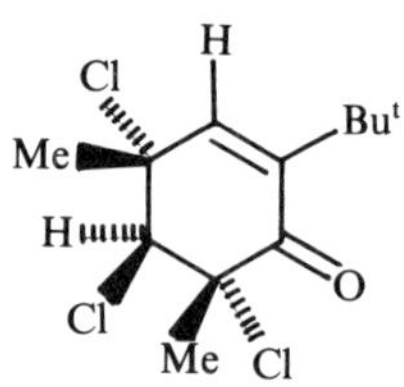

(**2**)

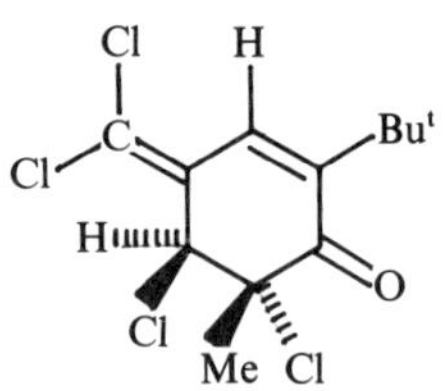

(**3**)

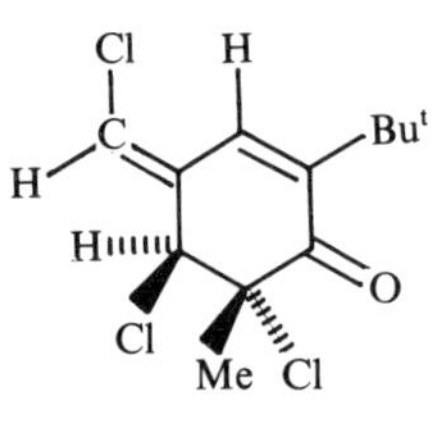

(**4**)

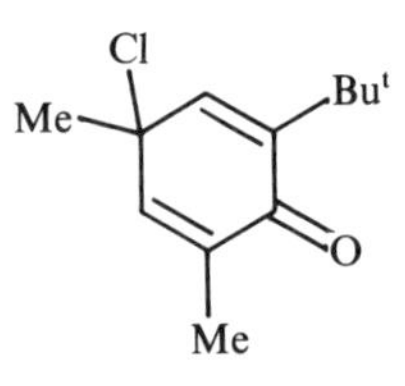

(**5**)

The critical step in the conversion of the aromatic hydrocarbon *syn*-1,6:8,13-bismethano[14]annulene (**6**) (and its olefinic *anti*-isomer) into dione (**8**) is the formation of bromine addition product (**7**);[8] the bromines are *syn* to each other and *syn* to the methylene bridges. The rationale for the stereochemistry is that the rehybridization at C(7) and C(14) from sp^2 to sp^3 substantially reduces the non-bonding interaction between the internal bridge hydrogen atoms by causing folding of the C_{14} perimeter. Thus this bromination 'may be regarded as an unusual example of reaction control exerted by strain relief.'

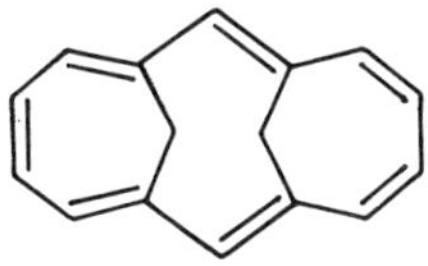

(6)

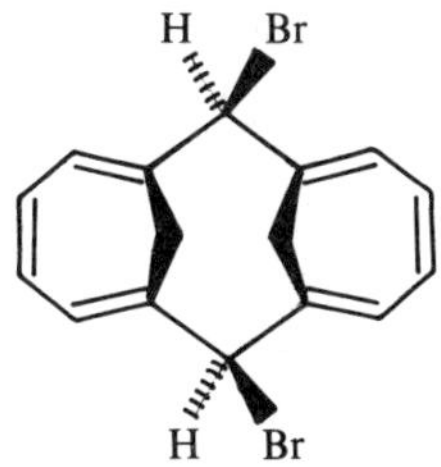

(7)

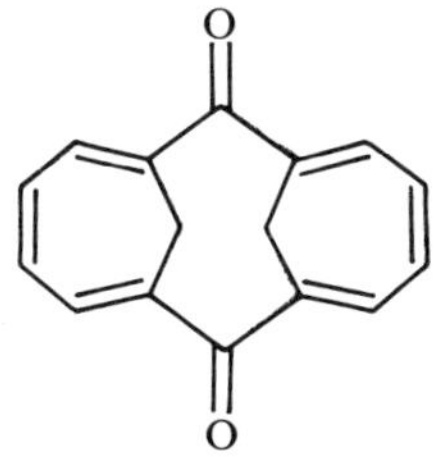

(8)

An extensive study of the reactions of the double bond and three-membered ring of a series of alkylidenecycloproparenes with halogens and a variety of other electrophilic reagents, has been carried out.[9] Several of the reactions are illustrated here with compound (**10**). Chlorination gives (**9**) by straightforward opening of the ring and addition to the double bond. With aqueous acid, water adds to the double bond, and the three-membered ring opens. Compound (**12**) is formed, probably via cyclopropenium ion (**11**). With bromine, (**14**) is formed, presumably by ring-expansion of carbenium ion (**13**). Compound (**15**) is formed by hydrolysis of (**14**), or directly by treatment of (**10**) with bromine water.

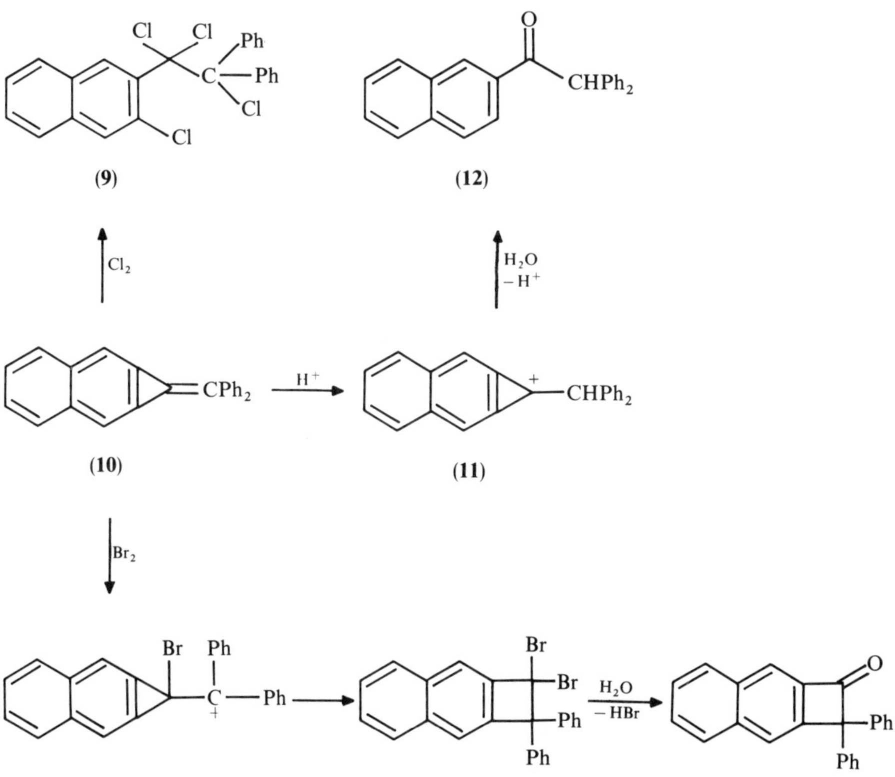

(9) (12) (10) (11) (13) (14) (15)

A detailed kinetic and product study of the mechanisms of bromination of *cis*- and *trans*-stilbenes and the related reactions of *erythro*- and *threo*-2-bromo-1,2-diphenylethanol with HBr has led to the conclusion that formation of bromonium-bromide or bromonium-tribromide ion-pair intermediates is reversible.[10] The addition of bromine to cyclohexen-1-yl benzoate, *p*-methyl benzoate, and *p*-methoxy benzoate gave, as expected, *anti* dibromo-addition products, (**16**) and (**17**), but also *syn* dibromo-addition products (**18**) and (**19**), through the intervention of intermediates of type (**20**).[11] The tetrafluoroborate salt (**20**; $X = BF_4^-$, Ar = *p*-anisyl) was isolated in crystalline form and characterized. The role of micellar charge in the bromination of various alkenes in several micelles has been investigated.[12] A quantum-chemical study of the addition of iodine to ethylene (and of several other reactions) has been reported.[13]

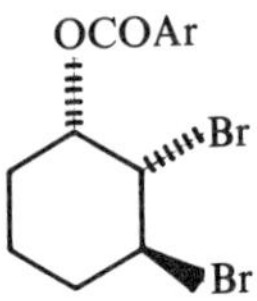

(**16**)

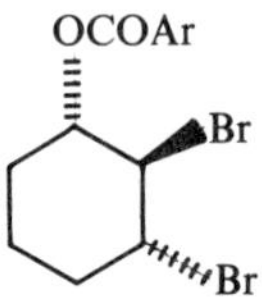

(**17**)

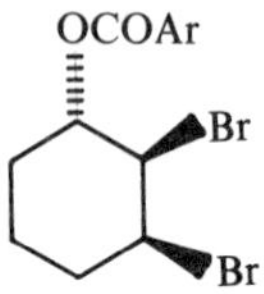

(**18**)

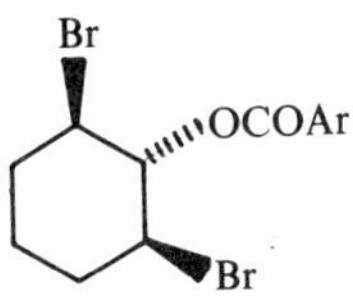

(**19**)

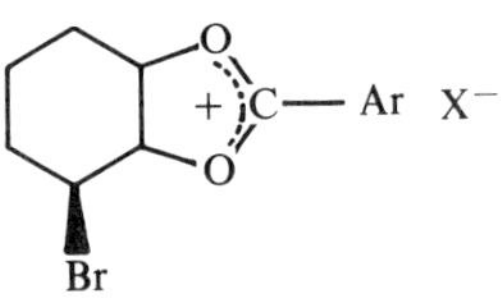

(**20**)

Several reports have appeared on the mechanisms of addition of inter-halogen compounds to alkenes[14–16] and alkynes.[17,18] The addition of BrCl to a series of eight substituted styrenes using tetrabutylammonium dichlorobromate was found to be highly regioselective (actually regiospecific in most cases in protic solvents) (product: $XC_6H_4CHClCH_2Br$).[14] In protic solvents, substantial quantities, about one-third of the total yield, of the solvent-derived adduct, $XC_6H_4CH(OS)CH_2Br$, were obtained. Attack of chloride ion or solvent on a bromonium-ion intermediate is the suggested mechanism. Some results are also reported for the addition of molecular BrCl to these styrenes.[15] The rate law for the addition of ICl to 22 alkenes (alkene in large excess) has been determined to be:

$$-\mathrm{d(ICl)}/\mathrm{d}t = k_{\mathrm{exp}}(\mathrm{alkene})_0(\mathrm{ICl})^3/[1 + \mathrm{C}_2(\mathrm{alkene})_0]^3$$

where C_2 is a measure of the formation constant for the molecular complex (charge-transfer complex) between the alkene and ICl.[16] The rate constants and C values for the alkenes are tabulated; it was found that the stability of the alkene-ICl complex increases as the hydrogens of ethene are successively replaced by alkyl groups; for the most part, the particular alkyl group had little effect.

The reactions of molecular BrCl and tetrabutylammonium dichlorobromate with several alkynes to give bromochloroalkenes were compared in terms of stereoselectivity and regioselectivity.[17] With the ammonium salt, all alkynes gave exclusively products of *anti* addition; with molecular BrCl, all alkynes except phenylethyne and phenylpropyne also gave stereospecific *anti*-addition products. The regioselectivity was generally high; for unsymmetrical alkynes, there was strong preference, ranging up to 100%, for the Cl to become attached to the carbon adjacent to the phenyl group (compared to H or alkyl) or adjacent to the alkyl group (compared to H) (addition in the Markovnikov sense). When there were two alkyl groups on the triple bond, the product ratio was near unity for both reagents. The results were interpreted in terms of an intermediate vinyl cation with BrCl and the phenylalkynes, and an intermediate bridged bromonium ion for all other BrCl cases and for all reactions of the ammonium salt. What seems to be a fairly general method for bifunctionalization (I-X addition) of alkynes utilizes an iodine-mercuric salt combination.[18] In all cases the yields are high and the products are exclusively those of *anti* addition. Unsymmetrical alkynes give Markovnikov regiospecificity. The mercury(II) salts used were chloride, acetate, methanesulphonate, toluene-*p*-sulphonate, toluene-*p*-sulphinate, and thiocyanate. Benzyl sulphide was also added. With thiocyanate ion, only thiocyanato derivatives were formed; no isothiocyanato ones were detected. Attack of the nucleophile on a bridged iodine-alkyne cation is the proposed mechanism for the reaction. Bifunctionalization of alkynes to give vinyl iodo-sulphones, $YC_6H_4SO_2CH{=}CIC_6H_4X$, can also be accomplished using substituted benzenesulphonyl iodides in benzene solution in visible light.[19] The reactions are regiospecific, and only *anti*-addition products were obtained. The effect on reaction rate of varying X and Y was studied; rate differences are not large; the reactions are interpreted in terms of a radical mechanism.

Although addition of X_2 is not strictly involved, several reactions involving 'positive halogen' are taken up at this point. A stereochemical study has been made[20] of the reactions of HOBr (from *N*-bromoacetamide and perchloric acid) with a number of steroidal alkenes containing various neighbouring substituents which subsequently react with the bromonium-ion adduct. The carbamoyloxy group was found to be a superior neighbouring-group. The relative importance of the electronic (Markovnikov) and stereoelectronic effects in these reactions has been evaluated. A pivotal point in the synthesis of some tricyclo[5.4.0.0^{2,8}]undecane derivatives is the highly stereo- and regio-selective addition of BrOX (generated from NBS in XOH) to an alkenone.[21] A study has been carried out[22] of the transannular *O*-cyclizations of the compounds formed by addition of MeOBr (formed from NBS in MeOH) to (*E*,*E*,*E*)-cyclododeca-1,5,9-triene. An extensive product and stereochemical study of the cyclization reactions of γ-hydroxy-alkenes bearing an alkoxy or alkyl group on the allylic carbon has been carried out.[23] Eight different electrophiles were used, including NBS, benzeneselenenyl chloride, I_2, and $Hg(OAc)_2$. The products are generally tetrahydrofurans with a great preference for *cis* stereochemistry of the 2- and 3-ring substituents. A mild, efficient, stereoselective synthesis of 3,4-disubstituted β-lactams has been devised which involves, as the critical step, closure of the 4-membered ring upon bromination of an unsaturated amide.[24] The regiochemistry of the iodo-cyclization (using *N*-iodosuccinimide) of allylic imidates to 4,5-dihydro-1,3-oxazoles or 4,5-dihydro-1,3-oxazines has been investigated.[25] The iodo-lactonization of 1,6-heptadiene-4-carboxylic acid proceeds with excellent group and face selectivity.[26]

Additions of Hydrogen Halides and Some Other Acids

The use of tetrabutylammonium and polymer-supported dihydrogentrifluoride has been shown to allow the addition of HF, under mild conditions, to carbon-carbon triple bonds activated by nitrile, ester, ketone, or aldehyde groups, affording the (*Z*)- and (*E*)-isomers of the fluoroalkenes.[27] A kinetic study has been carried out of the addition of gaseous HCl to gaseous (*E*)-2-butene, (*Z*)-2-butene, and (*Z*)-2-butene-2,3-D_2 in Vicor or Pyrex cells with NaCl or KBr windows (for following the course of the reaction by IR spectroscopy).[28] The follow-on elimination of HCl or DCl from the latter compound was also studied. The analysis of the kinetic data led to the conclusion that all the reactions were taking place between reactants absorbed on the NaCl or KBr windows. Laser excitation of the C—H overtones of 1,3-butadiene to well above the activation energy for addition of HCl failed to cause any detectable enhancement of product yields.[29]

The addition of HCl and a variety of other protic acids (CF_3CO_2H, CF_3SO_3H, TsOH, *m*-$NO_2C_6H_4SO_3H$) in methylene chloride at room temperature or below to alkynyl tosylates, RC≡COTs (R = Bu^t or Bu^s) gave quantitative yields of single products of regio- and stereo-specific *syn* addition, the previously unknown vinyl 1,1-(bis-esters) and related compounds.[30] The reactions are proposed to go through vinyl cation tight ion-pairs. Acid-catalysed hydrations of the alkynyl tosylates gave quantitative yields of the expected carboxylic acids. The alkenyl tosylates (and related mesylates), the first known examples of acetylenic esters, were prepared by an addition-elimination process using the alkylacetylenes and the (stable, crystalline) salts prepared by reaction of iodosobenzene diacetate with the monohydrates of TsOH or $MeSO_3H$.[31] The addition step of the synthesis reaction is proposed to proceed through a vinyl cation, $R\overset{+}{C}{=}CHI(OH)C_6H_5$, or a bridged iodonium ion.

Several studies[32–35] have appeared reporting details of the sulphonation of alkenes with sulphur trioxide. The mechanism of the reaction involves a concerted, stereospecific, *cis*-cycloaddition to give *β*-sultones (**21**) as the first products. In studies of the scope of the reaction, it was shown[32,33] that mono-, di-, tri-, and tetra-substituted ethylenes all proceeded through the *β*-sultones. With excess sulphur trioxide, the first formed *β*-sultones give a further stereospecific addition to the carbyl sulphates (**22**).[32,34]

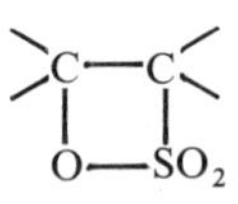

(**21**)

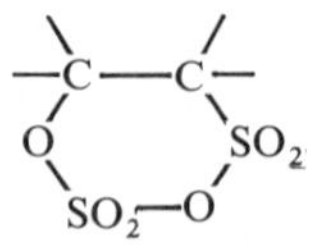

(**22**)

Reactions of alkenes with some other acids are covered below under hydration and miscellaneous electrophilic additions.

Additions of RX

The activation parameters and rate constants have been determined for addition of a series of *p*-substituted diarylmethyl chlorides to 2-methyl-2-butene.[36] The reactions involve the salts of diarylcarbenium ions with BCl_3. When di-*p*-tolylmethyl chloride is used, the rate of addition is not influenced by variation of temperature; the measured rate constant is termed 'isokinetic', and the di-*p*-tolylmethyl cation is termed the 'isokinetic substituent'. The same types of reactions have been used to refute the Reactivity-Selectivity Principle for these RX additions.[37] For the same group of diarylmethyl chlorides, the alkene pair, methylenecyclopentane/2-methyl-2-butene, showed decreasing selectivity with increasing reactivity; methylenecyclopentane/2-methylpentene and isoprene/2-methylpentene showed constant selectivity with increasing reactivity; 2-methyl-2-butene/2-methyl-2-pentene showed increasing selectivity with increasing reactivity.

A kinetic study[38] of the addition of tetrachloromethane to styrene in the presence of copper-amine complexes has shown that the reaction mechanism is not of the free-radical-chain type; rather, the rate-determining step involves the direct transfer of CCl_4 to the styrene from the $amine_2—CuCl—CCl_4$ complex.

Additions of ArSX, ArSeX, and Related Reactions

A point of contention in the addition of RSCl to unsymmetrical alkenes is whether (or under what circumstances) the position of chloride attack in the intermediate thiiranium ion is at the least-hindered or the most-hindered carbon. 'Stable' thiiranium ions may be different in this respect from the ones generated under the usual (non-polar solvent) addition-reaction conditions. To see if there are differences depending on the method of formation of the intermediate ion, other things being equal, thiiranium ion (**23**) was generated in the usual way by addition of RSCl to 2-methyl-2-butene, and also by addition of acid to allylic thioethers.[39] The addition-

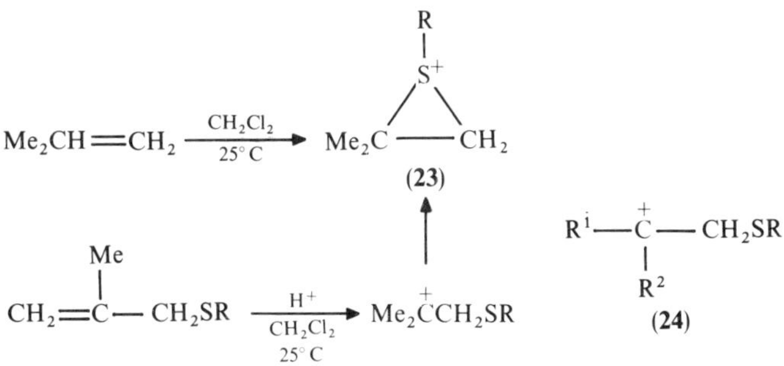

product mixture was the same for the two cases, involving preferential attack of chloride at the least-substituted position. Another question of significance in this reaction is whether open carbocations, e.g. (**24**), are involved in favourable cases, rather than bridged ions like (**23**). The stereochemistry of the reaction was studied[40] with a series of aliphatic (*E*)- and (*Z*)-alkenes which could generate tertiary carbocations. The products are those of *anti*-stereospecific addition, consistent with addition taking place by the usual thiiranium-ion mechanism.

Arylsulphenyl chlorides add to vinyl ethers in the normal way, and these adducts, e.g. (**25**), can serve as electrophilic alkylation agents for further reactions of vinyl ethers, presumably via a thiophanium salt, e.g. (**26**).[41] The scope of the reaction is quite broad. An interesting application is the formation of (**28**) from (**27**).

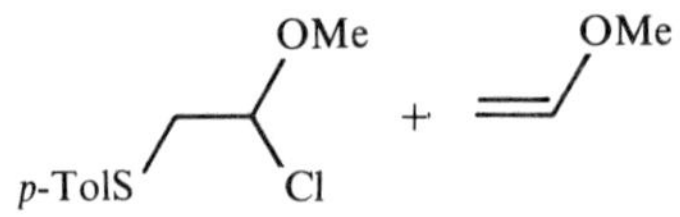

(**25**)

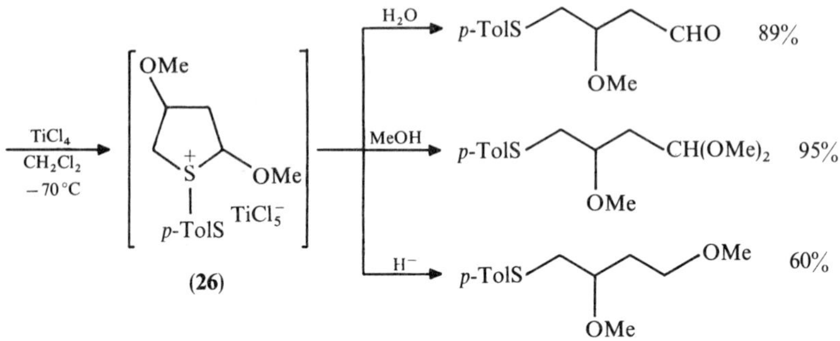

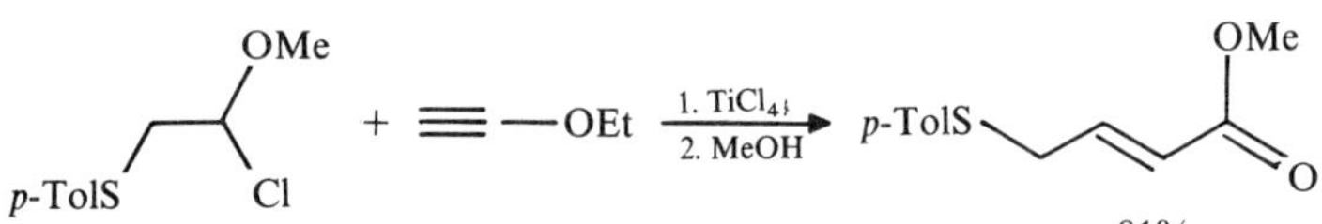

81%

(**27**) (**28**)

Treatment of spiro-alkenes (**29**), (**30**), (**33**), and (**35**) with arenesulphenyl chlorides afforded addition products (**31**), (**32**), (**34**), and (**36**) with regio- and stereo-specificity; only 1,2-*anti*-addition products were formed.[42] Not even traces of transannular ring-expansion by the spiro-cyclopropane group to homobrendanes or isotwistanes could be observed. The epoxidation of related spiro-alkenes was also studied.

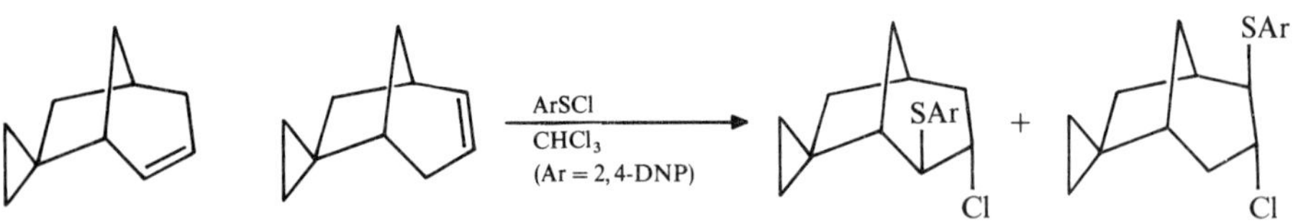

(**29**) (**30**) (**31**) (**32**)

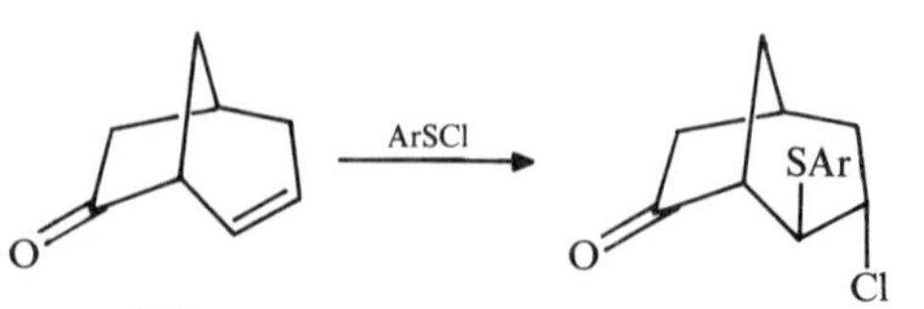

(**33**)

Ar = 2.4-DNP or Ph

(**34**)

(35) Ar = 2.4-DNP (36)

Use is made of the neighbouring-group properties of sulphur in the adducts of addition of arenesulphenyl chlorides to alkenes to effect the replacement (using dimethylzinc in the presence of $TiCl_4$, with retention of configuration) of the chlorine of the adduct with alkyl groups; the net result is the stereo- and regio-specific addition of the elements of ArSR to alkenes.[43]

The addition of PhSeCl and PhSCl to *exo*- and *endo*-alcohols (**37**) and (**40**) gave regio- and stereo-specific adducts (**38**), (**39**), (**41**), and (**42**).[44] It is noteworthy that the regiochemistry is opposite for the two substrates. The formation of the unexpected tricyclic oxetane (**42**) is also of special interest; this cyclization took place without

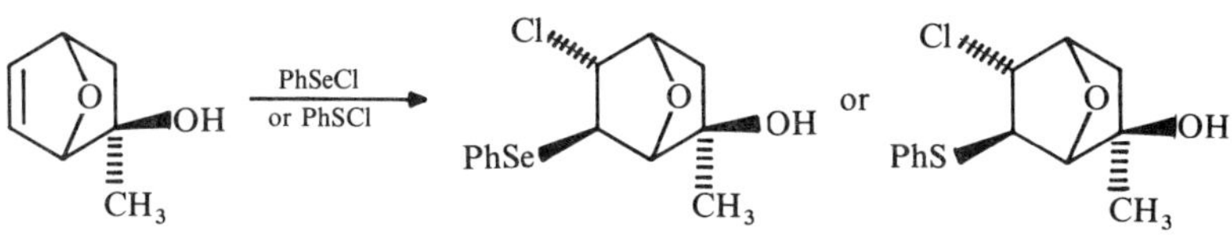

(37) (38) (39)

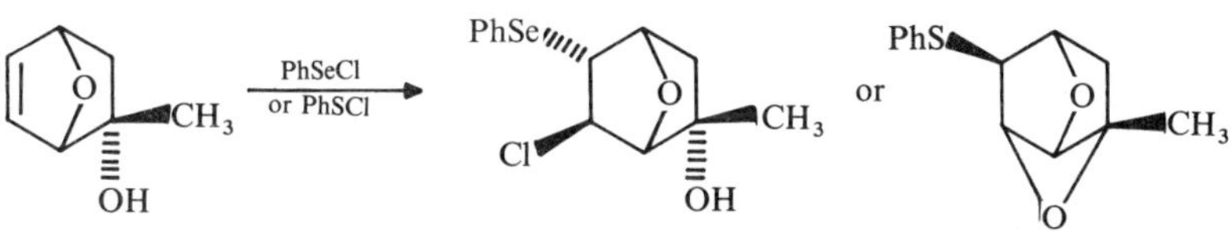

(40) (41) (42)

any added base. PhSeCl did not give a similar tricyclic product. Very similar results (regio- and stereo-specific *anti*-addition) were obtained in the addition of benzeneselenenyl chloride, 2-nitrobenzenesulphenyl, and 2,4-dinitrobenzenesulphenyl chloride to (**43**); adduct (**44**) was formed.[45] When the C(CN)OAc group of the substrate was hydrolysed to a carbonyl group, the same reagents gave the same type of adducts, but with opposite regiospecificity; (**45**) gave (**46**). Similar studies were carried out using substrates (**47**)–(**50**).[46] *N*-Phenylselenophthalimide has proved to be a very efficient *O*-cyclization reagent for the conversion of 1,5-hexadien-3-ols into tetrahydrofuran derivatives.[47]

When an electron-withdrawing group is present on an alkene, addition of benzeneselenenyl chloride, while regioselective, is not regiospecific; regioisomeric ratios range from 63:37 to 95:5, depending on the substituent.[48] The results are rationalized in terms of interaction between the substituent and an episelenonium ion.

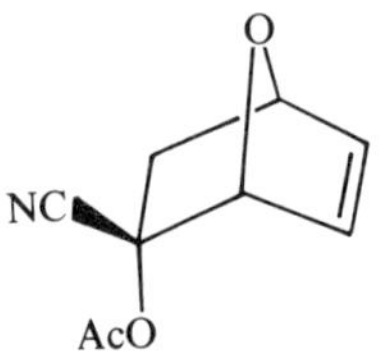

(43)

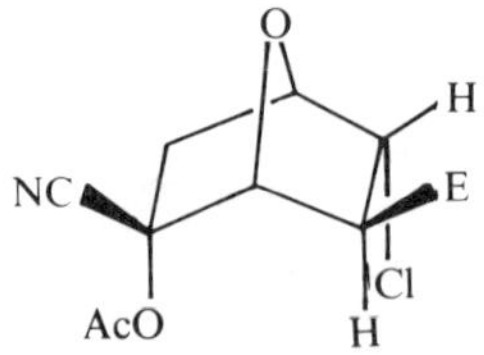

(44)

$E = C_6H_5Se$

or $2\text{-}NO_2C_6H_4S$

or $2,4\text{-}(NO_2)_2C_6H_3S$

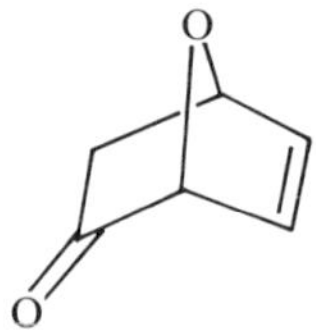

(45)

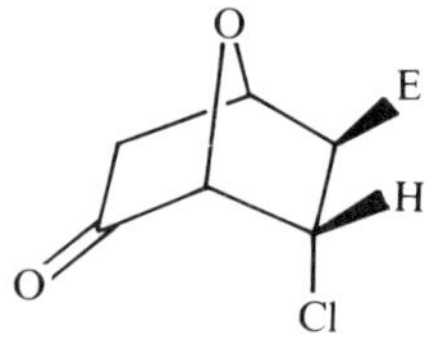

(46)

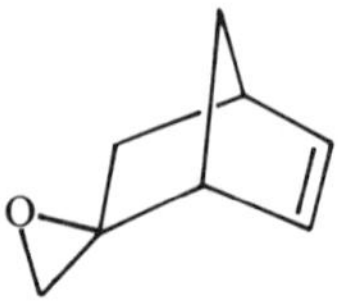

(47)

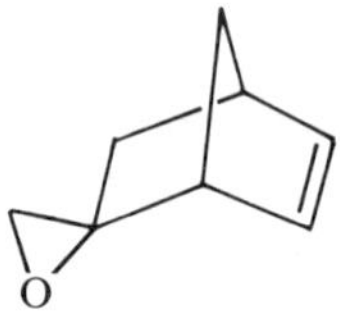

(48)

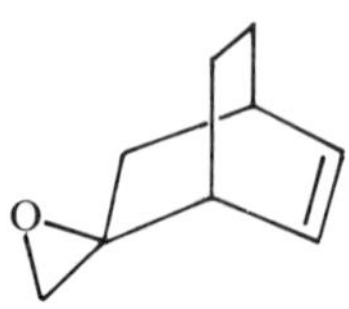

(49)

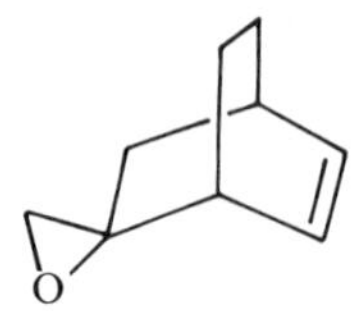

(50)

Treatment of γ,δ-unsaturated amides with PhSeCl or PhSeBr results in the formation of imino-lactones with the $PhSeCH_2$ group in the γ-position. The geometry of the molecule is favourable for backside attack of the carbonyl oxygen on the episelenonium ion formed by the addition to the double bond.[49] The scope of the reaction was studied. When γ,δ-unsaturated carboxylic acids are treated with benzeneselenenyl trifluoromethanesulphonate, the PhSe group adds to one end of the double bond, and a γ-lactone is formed at the other, the episelenonium ion being the presumed intermediate. Yields are better and there are fewer problems with side-reactions than when PhSeCl is used.[50]

The use of reactions of $PhSeN_3$ with a number of alkenes gives good yields of the azide/PhSe adducts.[51] The reaction proceeds stereospecifically but not regiospecifically. Nitrogen functions can also be introduced by additions to double and triple bonds using benzenesulphenylanilides, PhSNRAr, in the presence of BF_3-diethyl ether, presumably through thiirenium-ion intermediates.[52] The scope and mechanism of the additions of $PhSeCl_3$ and $PhSeBr_3$ have been investigated.[53] The addition of the elements $PhSeX_2$ and X proceeded smoothly and in good yields to give products of *anti* addition; regioselectivity (Markovnikov sense) was also high.

Hydroboration

In a most interesting new development, methylborane, the simplest of all the organoboranes, has been shown to be capable of adding to alkenes, hindered and unhindered, in THF, to give the simple methylalkylboranes $(RMeBH)_2$.[54] The yields are nearly quantitative; for the ten alkenes studied, only with 2-methyl-1-pentene could *any* (just 2%) of the R_2MeB compound be detected. The alkylmethylboranes show all the usual reactions of alkylboranes, including addition to another alkene.

This development makes it possible to make organoboranes with a methyl and two other (different) groups attached to boron.

MNDO calculations have shown a linear correlation between calculated and experimental regioselectivities in hydroboration of alkenes.[55] An acceptable linear correlation between calculated and experimental results was also found for asymmetric inductions of hydroboration with chiral hydroborating agents. The MNDO method has been extended to search for alternative hydroborating agents.

As part of a study of the chemistry of tricyclic hydrocarbons, it has been shown that diborane treatment, followed by the usual alkaline hydrogen peroxide oxidation, gives essentially pure 2-*exo*-hydroxy-*endo*-tricyclo[5.2.0^{2,6}]decane from tricyclo[5.2.0^{2,6}]dec-2(6)-ene.[56] Epoxidation of the alkene with *m*-$ClC_6H_4CO_3H$ also proceeded exclusively from the *exo* side. An alternative to the usual hydroboration addition/alkaline oxidation procedure for *anti*-Markovnikov hydration uses dichloroaluminium hydride and a catalytic amount of an organoboron compound, followed by dry-air oxidation.[57] A linear correlation has been found between rates of alkene hydroboration (8 alkenes) using 9-BBN and rates of oxymercuration using $Hg(OAc)_2$.[58] The conclusion is drawn that the transition states are similar in the two reactions.

Hydration, Addition of ROH, and Related Reactions

A kinetic and product study of the hydroxymercuration/demercuration reactions for a series of ring-substituted atropic acids, $XC_6H_4C(COOH){=}CH_2$, has been carried out in aqueous acetate buffers in the pH range 3.7–5.0.[59] The reactions are regiospecific, the demercuration products being the expected Markovnikov hydroxy-acids. The rates give a better correlation with σ^+ than with σ, with $\rho^+ = -2.14$. The results are consistent with the formation of an open α-mercurio-carbocation in the rate-determining step. A similar study reports on the kinetics of the methoxymercuration in methanol of a series of ring-substituted styrenes as a function of the amounts of co-solvents carbon tetrachloride, cyclohexane, and 18-crown-6.[60]

A noteworthy 'side reaction' was found to be the reason for the 'unsatisfactory' yields in the oxymercuration/demercuration of 1-phenyl-5-hexen-3-yl *tert*-butyl carbonate (**51**).[61] During the reduction with alkaline $NaBH_4$ the intermediate diastereomeric diol mercuric iodides, *syn*- and *anti*-(**52**), were found to decompose to alcohol (**53**) with a high degree of diastereoselectivity. It appears that a side-reaction hydrolysis of the *tert*-butyl carbonate group is involved, with decomposition through intermediate (**54**) taking place rapidly and decomposition via (**55**) proceeding only slowly.

When 1,2-epoxy-1-vinylcyclopentane is treated with mercuric acetate in methanol, followed by treatment with $NaBH_4$, the epoxy ring remains intact and normal Markovnikov addition of methanol to the double bond takes place.[62] Under the same conditions, when mercuric nitrate is used instead of mercuric acetate, the Markovnikov double-bond addition takes place, but in this case the epoxy ring is also opened (methanol addition). The epoxy-alkene is stable to methanol, so the

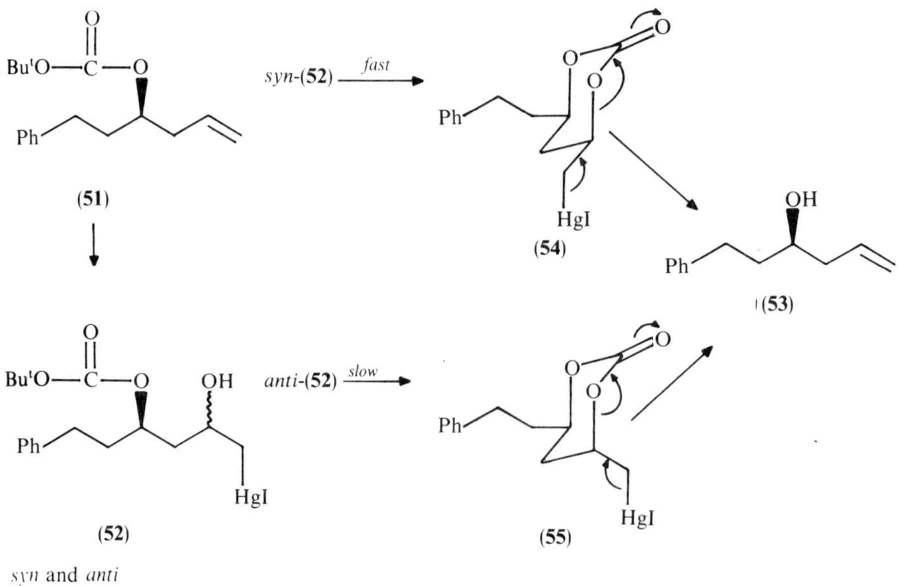

ring-opening mechanism must involve the mercuric salt. A series of 3-substituted methylenecyclohexanes has been utilized in a study of the effect of sterically remote substituents on π-facial stereoselectivity in addition reactions.[63] Oxymercuration, oxidation to the epoxides with *m*-chloroperbenzoic acid, and catalytic osmylation to mixtures of diastereomeric diols were the additions examined.

It has been demonstrated[64] that ether production can be enhanced in the alkoxymercuration/demercuration reaction by the use of sodium dodecyl sulphate (SDS) micelles. The course of the catalysis of hydration of a mixture of 2-methyl-1-butene and 2-methyl-2-butene by styrene-divinylbenzene ion-exchange resin has been investigated.[65] A comparison has been made of the effectiveness of *p*-toluenesulphonic acid and poly(styrenesulphonic acid) in catalysing the hydration of 2-methylpropene, in mixtures with water, 1,4-dioxane, and sulpholane.[66]

A detailed kinetic and deuterium-tracer study,[67] utilizing also the results of earlier carbon-14 rearrangement and oxygen-18 exchange studies, has led to the conclusion that there is not a common intermediate carbocation in the acid-catalysed hydration of 2-butene and the acid-catalysed reactions of 2-butanol. Rather, it is proposed, these reactions must proceed by parallel concerted mechanisms through transition states that have a large amount of carbocation character.

Comparisons of the equilibrium constants for the hydration of alkenes to form alcohols and of enols to form *gem*-diols have led to the conclusion that an oxygen substituent favours hydration as a result of the stabilizing geminal interaction of carbon-oxygen bonds in the diol.[68] Oxygen-substituent effects on other HX-addition equilibria were also investigated. The acid-catalysed photo-hydration of a series of ring-substituted styrenes and phenylacetylenes in aqueous buffer solutions has been studied.[69] General acid catalysis was detected, and Brønsted plots were approximately linear with α values in the range 0.14–0.18. In a study of the hydrolysis of 2-cinnamoylthiophene, at pH values below 10, the proposed mechanism involves rate-determining attack of water on the activated carbon-carbon bond.[70]

Hydrations and closely related reactions of ketenes, allenes, and alkynes have been the subject of considerable recent research.[71–78] The hydration reactivity of ketene itself and five alkyl- and dialkyl-ketenes has been studied in water and water/acetonitrile mixtures, as a function of acid and base catalysis.[71] For three of the compounds, solvent deuterium isotope effects were also measured (1.45–2.58). The reaction is strongly catalysed by both acid and base, with a large, flat, low reactivity well in the pH range 3–11. The acid-catalysed reaction is interpreted in terms of rate-determining proton transfer to the β-carbon perpendicular to the ketene plane, while the H_2O (uncatalysed) and HO^--induced reactions involve nucleophilic attack at the α-carbon in the ketene plane. The hydration reactivity of butyl-, phenyl-, and diphenyl-ketenes generated by flash photolysis has been studied as a function of acid and base concentration in aqueous solution.[72] The mechanistic interpretations parallel those mentioned immediately above.

An interesting intramolecular acid-catalysed alcohol addition (acetal formation) is involved in the conversion of allenic alkyne (**56**) into (**57**).[73] Other similar reactions involve the intramolecular additions of alcohols to ketene sulphoxides; e.g. (**58**) gives (**59**) upon treatment with NaH in THF.[74]

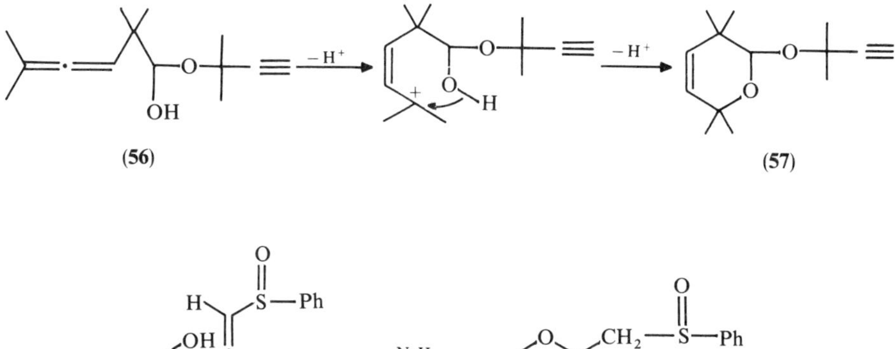

(**56**) (**57**)

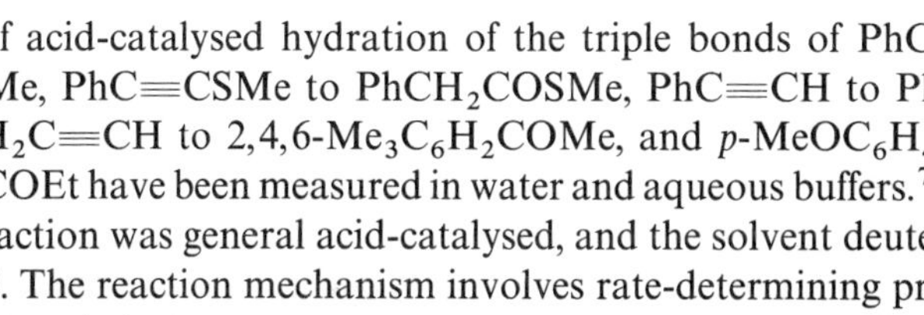

(**58**) (**59**)

The rates of acid-catalysed hydration of the triple bonds of PhC≡COMe to $PhCH_2COOMe$, PhC≡CSMe to $PhCH_2COSMe$, PhC≡CH to $PhCH_2COMe$, 2,4,6-$Me_3C_6H_2C$≡CH to 2,4,6-$Me_3C_6H_2COMe$, and *p*-$MeOC_6H_4C$≡CMe to *p*-$MeOC_6H_4COEt$ have been measured in water and aqueous buffers.[75] For PhC≡COMe, the reaction was general acid-catalysed, and the solvent deuterium isotope effect was 1.59. The reaction mechanism involves rate-determining proton transfer from catalyst to substrate.

Alkynoic acids (**60**) ($n = 1$, R = H, Me, and Ph; $n = 2$, R = H) were converted smoothly into the corresponding exocyclic enol lactones (**61**) using a newly developed Rh(I) complex.[76] Exclusive formation of the (*Z*)-alkene was observed; i.e. there was *anti* addition across the double bond. The 5-membered rings were formed

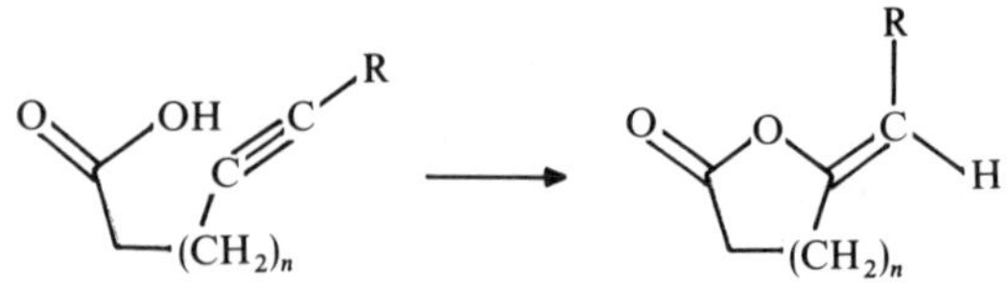

(60) **(61)**

in preference to the 6-membered rings. In the presence of a catalytic amount of bis (η^5-cyclopentadienyl)ruthenium/trialkylphosphine/maleic anhydride, a wide variety of carboxylic acids was found to add to several terminal alkynes (33 different combinations were examined) to give enol esters with a terminal methylene group.[77] The yields were reasonably good, and the regiospecificity was high, mostly above 90%. Deuterium scrambling was observed when MeCOOD was used. The reaction was first-order in carboxylic acid, alkyne, and catalyst. Hydration of methylacetylene in the presence of 3-pentanethiol gave a mixture of vinyl mercaptans, which underwent hydrolysis to acetone and propanal.[78]

As models for biological systems, some 2-furylethylenes were allowed to react in aqueous buffer solutions of pH 5–12 with the functions SH (thioglycolic acid), NH_2 (glycine), and OH (phenol or hydroxide ion).[79] The nucleophilic group attacks the carbon adjacent to the ring. The thiols are by far the most reactive. The kinetics and mechanism of addition of methanol to a series of 9 ring-substituted α-methoxystyrenes in acidic solution to give the dimethoxy ketals have been studied.[80] The reactions are interpreted in terms of rate-determining formation of an intermediate oxocarbenium ion. The compounds do not give a good linear Hammett plot, and the substituent effects are discussed in terms of the Yukawa–Tsuno–Sawada equation and the Young and Jencks treatment.

The mechanism of the conversion of norcamphene into 2-methylene-3-acetoxybicyclo[2.2.1]heptane by treatment with a Pd(II)/$CuCl_2$ catalyst probably proceeds via a (π-allylic)Pd(II) addition complex.[81] The addition (ring-opening) reactions of some small-ring compounds, while not really alkene-addition reactions, share many of the same mechanistic features as those of alkene addition. Among them are the acid-catalysed reaction of methanol with 3-X-bicyclobutanecarbonitriles,[82] and the protolytic cleavages of several tricyclohexanes, -heptanes, and -octanes.[83]

The addition of methanol to the carbon-silicon bond of both (*E*)- and (*Z*)-PhSi(Me)=C($SiMe_3$)CH_2CMe_3 proceeds stereospecifically to give separable diastereoisomers.[84] Additions to carbon-silicon double bonds have been little studied.

Miscellaneous Electrophilic Addition Reactions

A number of theoretical calculations have been carried out[85–88] on electrophilic addition reactions. The regio- and stereo-chemistry of electrophilic additions to chiral allylic double bonds were found[85] to be satisfactorily rationalized by a modelling procedure which focused on relative conformer abundance, relative reactivity of the various conformers, and the regio- and stereo-selectivities of the individual forms. Substrates examined included allylic alcohols, fluorides, and silanes; comparisons are made with experimental data from the literature. Calculated conformational-energy profiles for 3-buten-2-ol and 2-methoxy-3-butene are qualitatively similar, leading to the conclusion that conformational preferences are controlled mainly by minimization of (oxygen lone pair)-(π-bond) repulsion.[86] A preference is shown for electrophilic attack *anti* to OR or OH in the less abundant but more-reactive conformers. Calculations on allylic alcohols and related compounds with and without internal nucleophiles have shown[87] that there is a difference in mechanism for the two types of substrates. Specifically, where intramolecular nucleophilic attack is possible, cyclizations proceed via a transition state that resembles a reactant-like π-complex, whereas those reactions that involve intermolecular addition of the nucleophile are governed by one that resembles an 'onium' ion. Calculations show that these two extremes should exhibit opposite diastereofacial selectivity, in agreement with experimental data available in the literature. *Ab initio* calculations on the additions of water and ammonia to the nitrogen–carbon triple bond in isocyanides have been carried out.[88]

The critical step in the γ-phenylation (Heck Reaction via a Wacker-type intermediate) of chiral 3-methyl-3-but-3-en-2-ol in the presence of $PdCl_2$-$NaHCO_3$ appears to be the addition of a palladium-phenyl-iodide complex, oriented by coordination of the Pd with the alcohol OH.[89] This results in the formation of the product, 3-benzylbutanone, with up to 27% optical yield. Different stereoselectivities were observed for *cis*- and *trans*-pent-3-en-2-ols. In another study of the Heck Reaction[90] using butyl vinyl ether as a reference, it was established that methyl substitution at either the α- or β-olefinic carbon favours α-arylation, and that incorporation of the enol ether system in a ring completely inhibits β-arylation. Electron-rich aryl systems (4-methoxyphenyl) favour α-arylation whereas electron-poor aryl systems (4-nitrophenyl) favour β-arylation. Other reaction-system parameters were also examined.

Flash vacuum pyrolysis of aminomethylene derivatives of Meldrum's acid, e.g. (**62**), provides a simple way of making 1*H*-pyrrol-3(1*H*)-one derivatives. The reaction involves an intramolecular hydrogen transfer and ring-closure in an intermediate methyleneketene, e.g. (**63**), leading to product, e.g. (**65**).[91] From the fact that optically-active (**62**) leads to nearly racemic (**65**), it is now clear that the hydrogen-transferring closure reaction is not concerted, but that there must be an

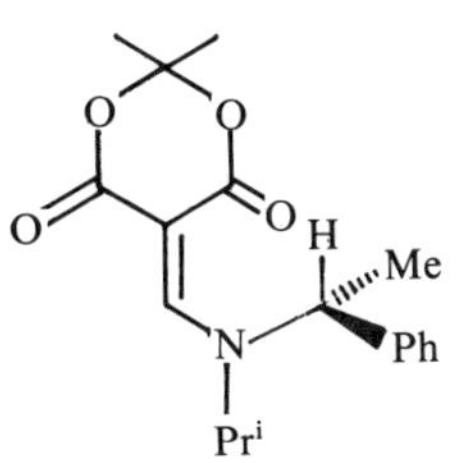

(*R*)-(**62**)

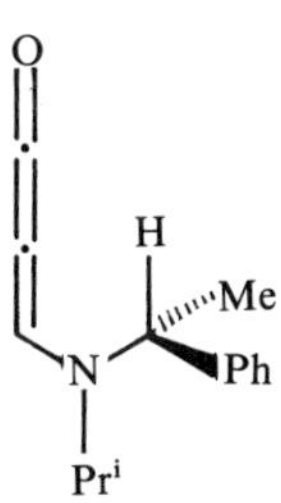

(*R*)-(**63**)

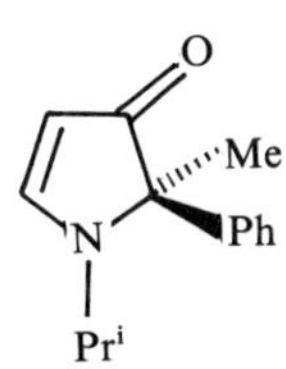

(*R*)-(**65**)

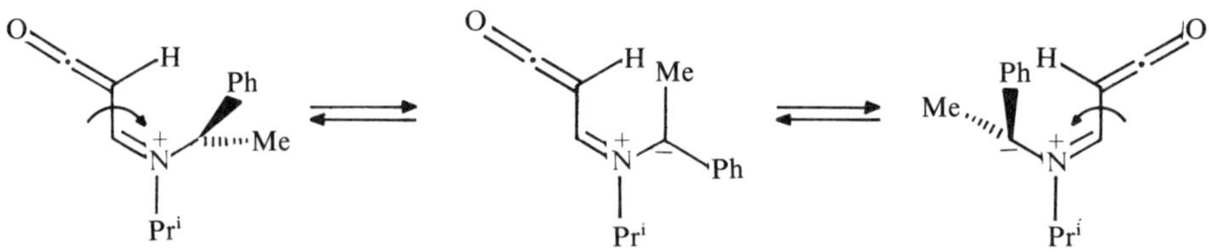

(*R*)-(**64**) (*S*)-(**64**)

additional intermediate, e.g. (**64**), capable of loss of chirality. The suggested intermediate (**64**) would be expected to exhibit rotation about the bond between the nitrogen and the original chiral centre. (In some other cases the racemization is complete.) This research project was extended to include vinylogous aminomethylene derivatives of Meldrum's acid.[92]

Quaternary ammonium chlorometallates have proved to be highly efficient bifunctional metal-complex and phase-transfer catalysts in addition reactions such as the hydrosilylation of phenylacetylene with triethyl- and (dimethyl)phenylsilane.[93] The formation of products of both α- and β-addition was observed in all cases, the latter being the *trans*-isomer. The effectiveness of several catalysts was evaluated. The follow-on addition of dichlorocarbene to the double bond to give cyclopropanes was also investigated using these phase-transfer catalysts.

Contrary to expectations, the hydrosilylation reaction of 2-methyl-1-buten-3-yne using a chloroplatinic acid catalyst was found to require a mixture of trichlorosilane and methyldichlorosilane in order to obtain satisfactory yields; neither of the silanes by itself was effective.[94] It was suggested that a radical mechanism was involved in this reaction. In a different study,[95] several *endo*-, *exo*- and *trans*-2,3-disubstituted bicyclo[2.2.1]hept-5-enes were found to react with dimethylchlorosilane in the presence of a platinum catalyst to form the corresponding 5-silyl derivatives in nearly quantitative yields. Silicon was found exclusively on the *exo* face of the bicycloheptyl skeleton, remote from substituents at the 2,3-positions. Purely steric arguments were not satisfactory for explaining the relative reactivity results; however, a field effect appears to provide an appropriate explanation.

Treatment of a series of 10 alkenes and cycloalkenes with trimethylchlorosilane-dimethyl sulphoxide resulted in the formation of the corresponding 2-chloroalkyl and 2-chlorocycloalkyl methyl sulphides.[96] The addition is both stereospecific and regioselective, affording *trans*-adducts from a preferential Markovnikov addition. Reaction between the DMSO and the chlorosilane probably gives a methanesulphenyl chloride, which adds as usual to the double bond. Use of trimethylbromosilane resulted in the formation of 1,2-dibromo derivatives. It is suggested that the stronger electrophilic nature of bromine than of sulphur gives a bromonium ion rather than an intermediate episulphonium ion.

A high degree of 1,4-diastereoselectivity [up to 26:1 for (**67**):(**68**)] was found in the addition of allyltrimethylsilane and three allylic tributylstannates in the presence of titanium tetrachloride to cyclohexenones, e.g. (**66**).[97] The role of $TiCl_4$ is proposed to be its simultaneous coordination with the two carbonyl groups as in (**69**); experiments with monodentate Lewis acids gave only low diastereoselectivity.

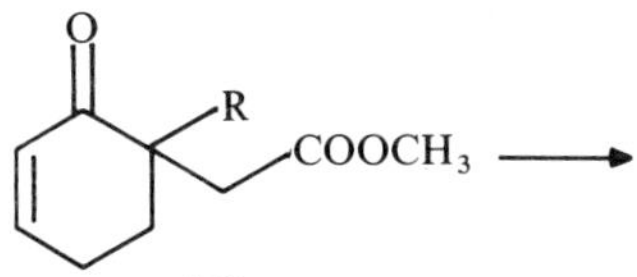

(66)

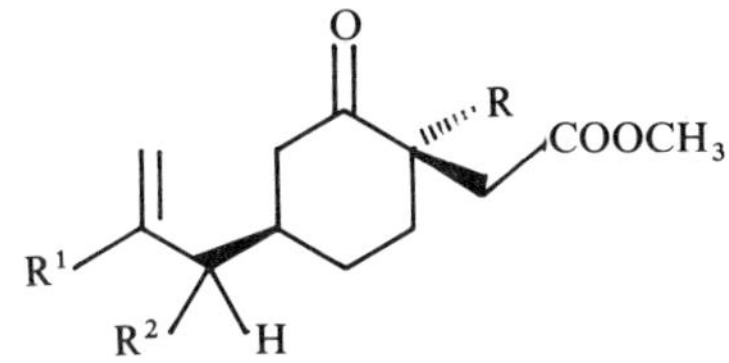

(67)

R = Me or Et, R^1, R^2 = H

R = Et, R^1 = H, R^2 = Me

R = Et, R^1 = Me, R^2 = H

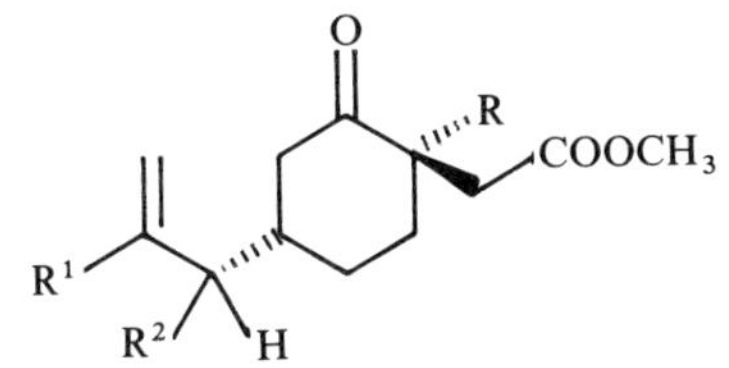

(68)

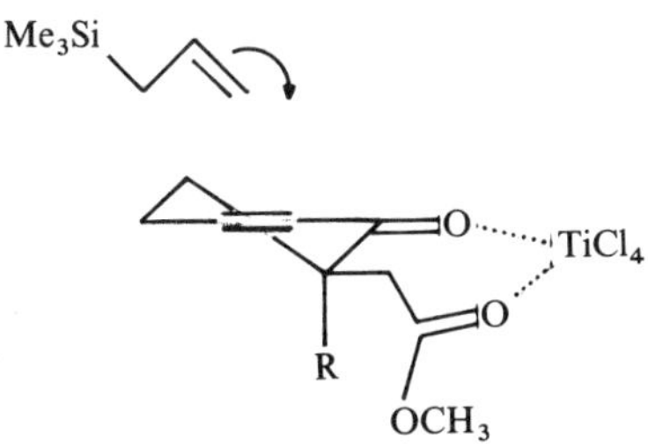

(69)

The stereoselectivity of additions to a series of three (R = Me, Pri, Ph) *trans*-(**70**) and *cis*-(**71**) pairs of phenyldimethylalkylsilanes has been studied using reactions with osmium tetraoxide (diol formation), *m*-chloroperbenzoic acid (epoxide formation), and Yamamoto's methylenation reagent.[98] With the single exception of the reaction of the *trans* compound (where R = Me) with osmium tetraoxide, the compounds of the (**72**) and (**74**) series are formed in preference to those of the (**73**) and (**75**) series. The preference is often quite large, ranging upwards of 20:1 in many cases. In all cases, the larger the R group, the more selective the reaction. The results are interpreted in terms of steric hindrance to attack near the large silyl group in the preferred conformation of the substrate.

Copper(I)-catalysed stannylalumination of 1-decyne proceeds via reversible stannylcupration.[99] The kinetically-favoured regioisomer is the 1-stannylalkene

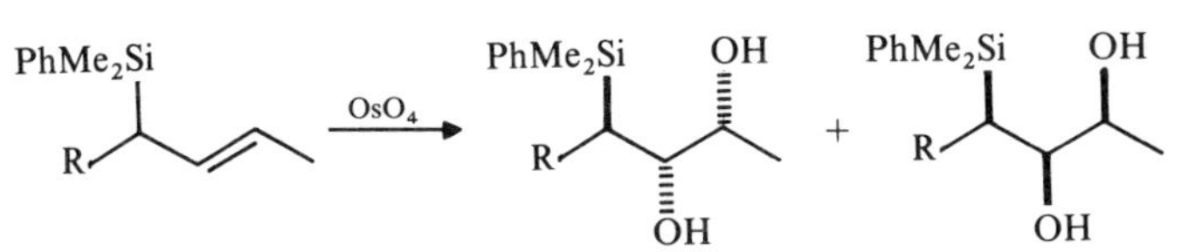

(70) (72) (73)

R	(72) : (73)
Me	34 : 66
Pr^i	67 : 33
Ph	92 : 8

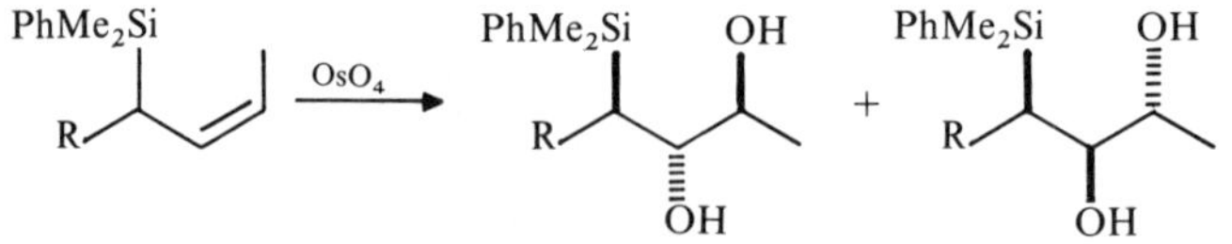

(71) **(74)** **(75)**

R	(74) : (75)
Me	78 : 22
Pr^i	85 : 15
Ph	83 : 17

whereas the thermodynamically-favoured regioisomer is the 2-stannyl derivative. Solvent and copper(I) counter-ion control the regiochemistry.

With all-aliphatic alkenes, hydrozirconation of the double bond leads to products in which the zirconium rapidly moves to the terminal position. It has now been found[100] that this migration is not the exclusive path for aromatic alkenes. In most cases the terminal Zr compound is mixed with one in which the Zr is in a benzylic position. Long-time heating increases the amount of the terminal Zr compound, but never to the exclusion of the isomer. Arenetellurinic anhydrides, $(ArTeO)_2O$, react with a variety of hydroxy-alkenes in acetic acid to give cyclic ethers bearing an aryltelluro group.[101] A mechanism via intramolecular oxytellurinylation with arenetellurinyl acetate is proposed.

The mechanism of the gas-phase reaction of iron(I) ions with neutral 4-octyne (a mass spectrometer experiment) has been studied using ^{2}H-labelled substrates.[102] No primary isotope effects are observed, so loss of ethylene to form the product $[Fe(1,2\text{-hexadiene})]^+$ does not involve a rate-determining hydrogen transfer as others have proposed. There is a secondary deuterium isotope effect of $k_H/k_D = 1.1$ per deuterium.

Terminal alkenes bearing electron-withdrawing groups are regioselectively acetalized at the terminal carbon by diols in the presence of $PdCl_2$ and CuCl.[103] Formation of a π-complex by addition of $PdCl_2$ to the double bond of an alkene is probably the critical step in the reaction.

A mathematical model has been developed[104] that can be used to estimate the effect of substrate and reagent concentration on the outcome of addition reactions that employ chiral non-racemic reagents with substrates that are equipped with unsaturated and enantiotopic ligands.

A study of the scope of the reaction of elemental sulphur with a variety of bicyclic, tricyclic, etc., ring systems has revealed that the main products are trithiolanes and pentathiepanes.[105] The reaction is favoured by polar solvents such as dimethyl formamide or dimethyl sulphoxide at temperatures as low as 25°C. As an example, benzonorbornadiene forms trithiolane **(76)** as a major product, and pentathiepane **(77)** as a minor product.

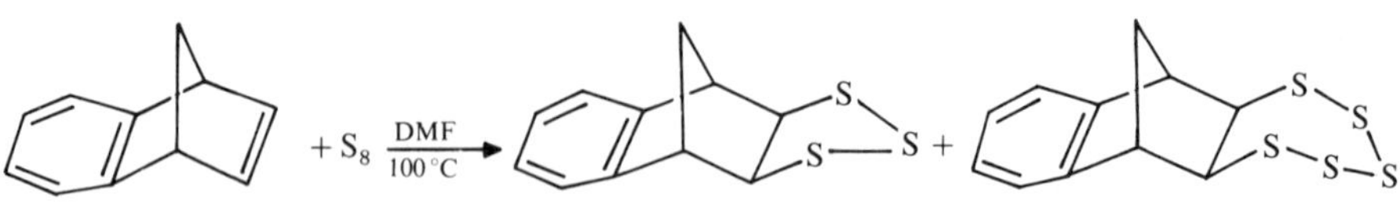

(76) **(77)**

The use of trityl salts to catalyse Michael additions and related reactions continues to provide interesting mechanistic information.[106–110] In the addition of silyl enol ethers of thioesters, e.g. (**78**), to α,β-unsaturated ketones in the presence of trityl hexachloroantimonate, it was found,[106] for instance, that the diastereoselectivity, (**79**):(**80**), inverted and then increased dramatically as the steric environment provided by the reagents became more crowded. Silyl enol ethers also add to α,β-unsaturated orthoesters regioselectively in the presence of catalytic amounts of trityl

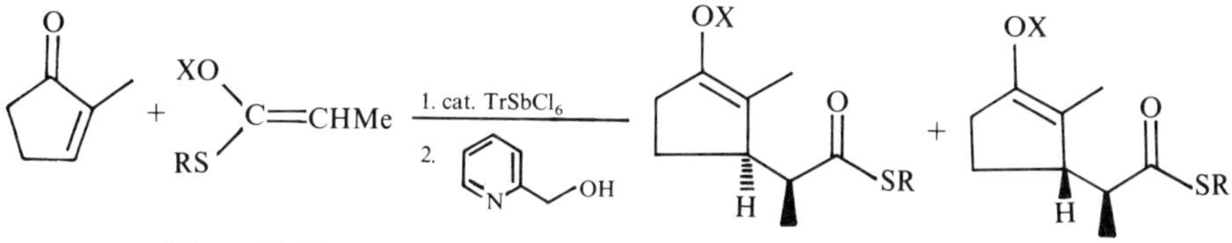

(*Z*)- or (*E*)-(**78**) (**79**) (**80**)

	(**79**) : (**80**)
Z-(**78**) R = Et, X = $PhMe_2Si$	67 : 33
E-(**78**) R = Et, X = $PhMe_2Si$	12 : 88
Z-(**78**) R = Bu^t, X = Bu^tMe_2Si	5 : 95
E-(**78**) R = Bu^t, X = Bu^tMe_2Si	5 : 95

perchlorate; no 1,2-addition products were formed.[107] The Michael addition of *O*-silylated ketene acetals with α,β-unsaturated esters in the presence of a catalytic amount of aluminium triflate proceeds smoothly under mild conditions to afford the corresponding glutarates in good yields.[108] Silyl enol ethers also react readily with ethylthioketene dithioacetals in the presence of trityl tetrafluoroborate to give the corresponding oxoketene dithioacetals in good yields.[109] The reaction provides an alternative to the Michael reaction between ketone enolates and α,β-unsaturated esters. These same ethylthioketene dithioacetals also react with allylstannanes and stannylated enolates in the presence of various trityl salts to give the corresponding adducts in good yields.[110]

The use of nitrosyl chloride gas as a means of simultaneous introduction of Cl and NO_2 has been explored for several unsaturated steroids.[111] For instance, 3,5-cholestadiene gave 3,6β-dinitro-5-chloro-5α-cholest-3-ene. With oximes, the nitro group (sometimes the nitrato group) appeared on the oxime nitrogen, and these compounds readily hydrolysed to the chloro-ketones.

Nucleophilic Additions

Additions to Multiple Bonds Conjugated with C=O, etc.

Michael (and related) addition reactions are routinely used in many syntheses, and no attempt is made here to cover all such reports; only papers offering some substantial mechanistic information are included. The choice is somewhat arbitrary.

The scope and generality of the addition of ketone enolates to α,β-unsaturated ketones to form 1,5-diketones has been studied.[112] For the (*Z*)-enolates, there was excellent *anti* diastereoselectivity; for the (*E*)-enolates there was some preference for the *syn*-addition product. For instance, (*Z*)-(**81**) reacts with (**82**) to form an 87% overall yield of 98% *anti*-addition product (**83**) and 2% of *syn*-addition product (**84**); for reaction of the (*E*)-enolate, the yield was 67% of (**83**):(**84**) in the ratio 17:83, respectively.

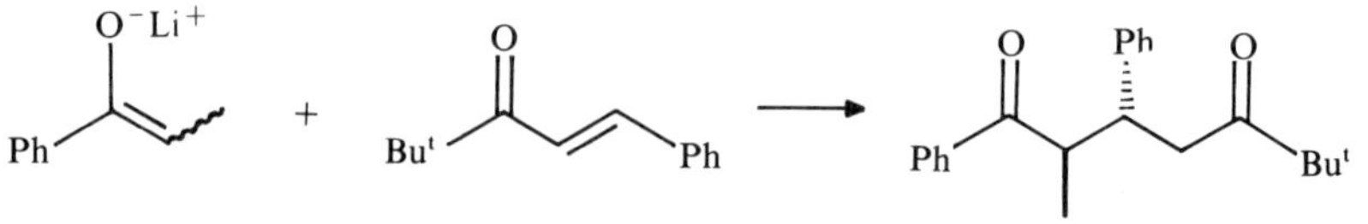

(**81**) (**82**) *anti*-(**83**)

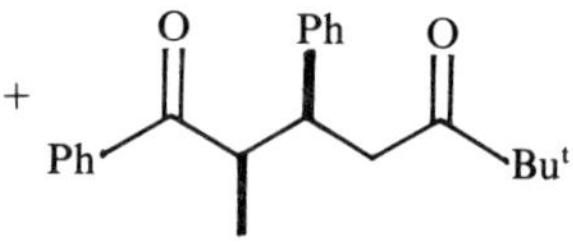

syn-(**84**)

Several ring-substituted 1,3-diaryl-2-propen-1-ones react with several ring-substituted arylacetonitriles in the presence of sodium ethoxide to give single diastereomers of the corresponding 2,3-diaryl-4-aroylbutyronitriles.[113] An intramolecular Michael addition (ring-closure) of a variety of aromatic enones (**85**) gave single isomers of nitro-tetralones (**86**).[114] Substituents R^1–R^4 were various combinations of H, Me, Ph, MeO, and EtO_2C.

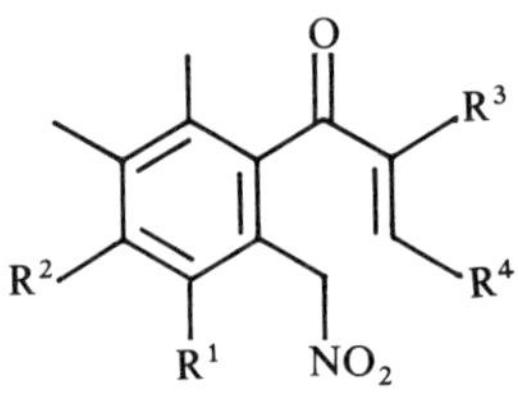

(**85**)

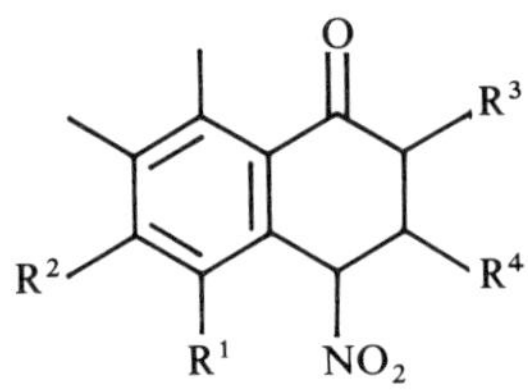

(**86**)

Chiral allylphosphonyl anions of (**87**) (and related compounds) undergo good enantioselective 1,4-additions with cyclopentene, cyclohexene, and cycloheptene to give 1,4-γ-adducts like (**88**).[115]

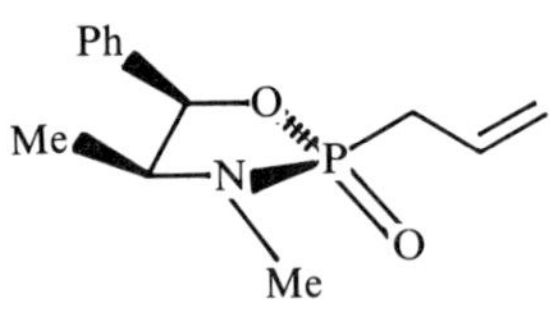

(**87**)

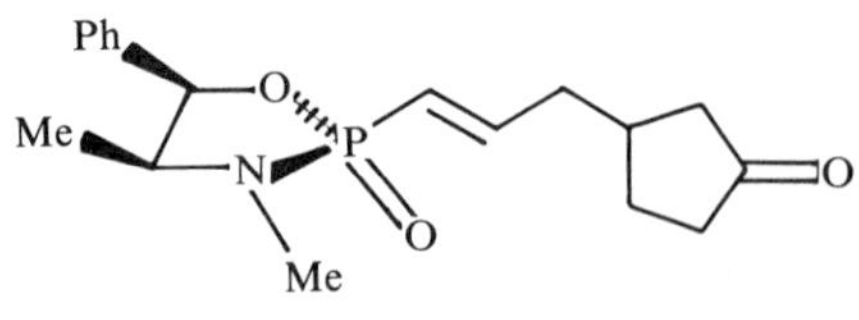

(**88**)

Some rearrangements that take place within some oxyanions of cyclopentene-substituted norbornenols seem to be best understood in terms of cleavage to a carbanion of an enone, followed by an intramolecular 1,4-Michael-addition ring-closure.[116] Michael addition of the lithium salt of methyl dithioacetate to a series of seven α,β-cyclic and -acyclic enones gives good yields of adducts with *syn* diastereoselectivities ranging from 76:24 to 95:5.[117] The synthesis of compounds containing adjacent tertiary and quaternary carbon centres, e.g. (**90**), in a highly stereoselective manner was accomplished by treatment of some α,β-unsaturated

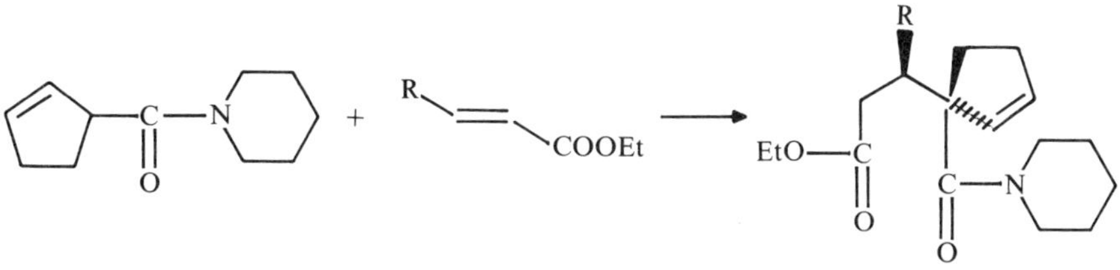

(**89**) (**90**)

esters with lithium salts of *N*-(2-cyclopentenylcarbonyl)piperidine (**89**).[118] Syntheses, with nearly complete enantio- and diastereo-selectivity, of similar systems with adjacent tertiary and quaternary carbon centres were also developed using chiral lithio-enamines of α-alkyl β-oxo esters with ethylidenemalonate.[119] 2-Phenyl-4-(1-pyrrolidinyl)thiophene reacts with methyl acetylenedicarboxylate by a Michael reaction to give (**91**).[120] This compound is formed by abstraction of H in the initially-formed 1,4-dipolar intermediate. Methyl 2-chlorocyclopropylideneacetate (**92**), when treated with lithium cyclohexadienolates, e.g. (**93**), undergoes highly regioselective iterative Michael additions with subsequent γ-elimination to give tricyclic γ-keto-esters (**94**).[121]

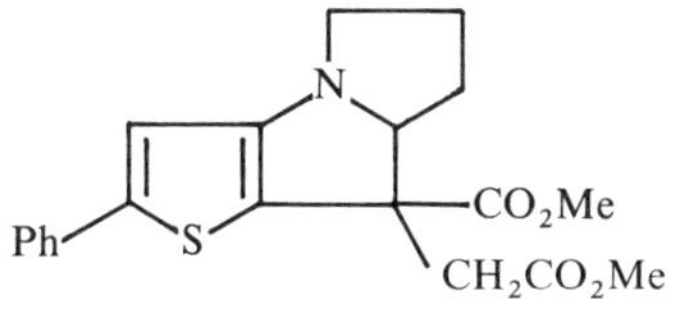

(**91**)

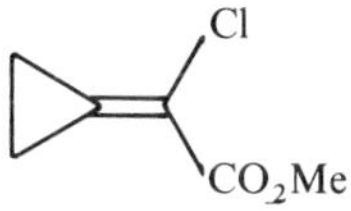

(**92**)

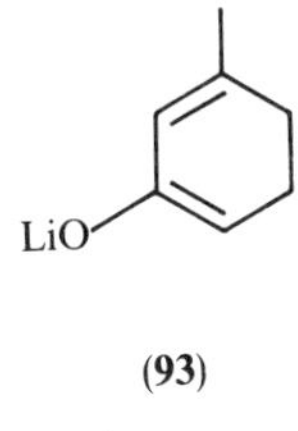

(**93**)

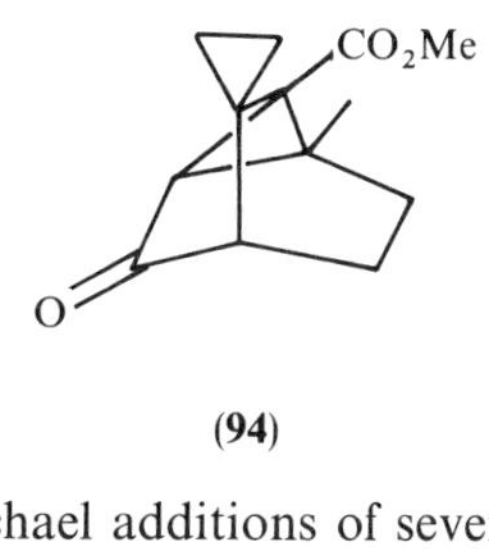

(**94**)

A study has been carried out of Michael additions of several active methylene compounds to chalcone on several $Ba(OH)_2$-based catalyst systems.[122] High absolute and relative stereochemical control is demonstrated for intramolecular addition of anions of chiral alcohols of β-keto-esters to α,β-unsaturated esters.[123] Michael addition of methyl cyanoacetate to benzylideneacetophenone and benzylideneacetone in THF under LiOBu^t or KF catalysis takes place only in the presence of complexing agents for the cation. Appropriate phase-transfer catalyst systems for these and related reactions are developed.[124]

Extensive studies have been carried out[125–127] of Michael-type additions of carbanions derived from some (*E*)- and (*Z*)-1-(phenylthio)oct-2-ene (**95**) and related compounds in the presence of hexamethylphosphoric triamide (HMPA). For instance, (*E*)-(**95**) adds to 4-*tert*-butoxycyclopent-2-enone (**96**) to give only (*E*)-(**97**), with equal quantities of *syn* and *anti* product. None of the vinyl sulphides were formed.[125] Conjugate addition of (**95**) and related compounds to γ-crotonolactone in the presence and absence of HMPA has also been studied extensively.[126,127] Some interesting follow-on chemistry is also discussed.

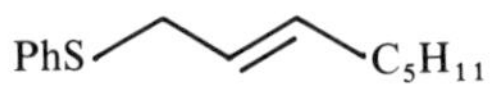

(*E*)-(**95**)

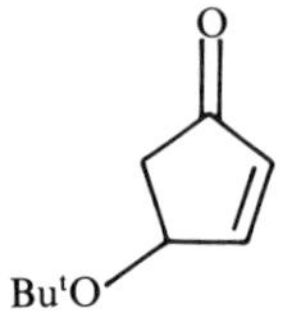

(**96**)

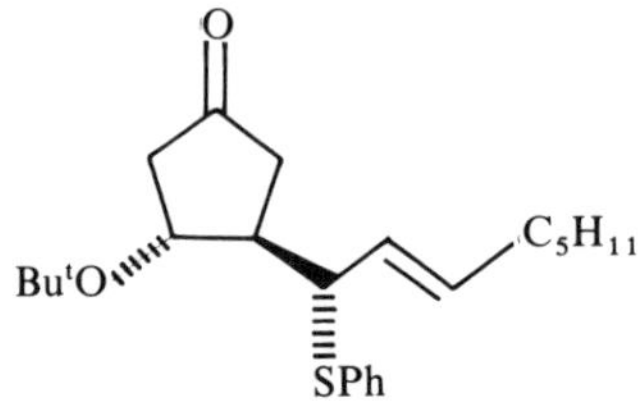

(**97**)

Asymmetric Michael additions of lithiated methyl ketone SAMP-hydrazones [SAMP is (*S*)-1-amino-2-methoxymethylpyrrolidine] to 2-benzylidenemalonates and dinitriles give adducts which are readily converted into δ-keto-diesters and -dinitriles in good overall yields and high enantiomeric excess.[128]

Mechanisms of addition of organocopper reagents to conjugated systems continue to receive attention.[129–136] One question of importance is whether the organocopper reagent adds 1,2 or 1,4. It appears to be 1,4-addition in at least one case. It has been found that treatment of methyl tiglate (**98**) and methyl angelate (**99**) with BuCu.BF_3 resulted, after protonation, in the same mixture of *syn*-(**100**) and *anti*-(**101**) adducts.[129] This is the result to be expected if the mechanism is 1,4-addition (enantiomeric pre-protonation intermediate enolates), but not if the mechanism were 1,2-addition (diastereomeric pre-protonation intermediates).

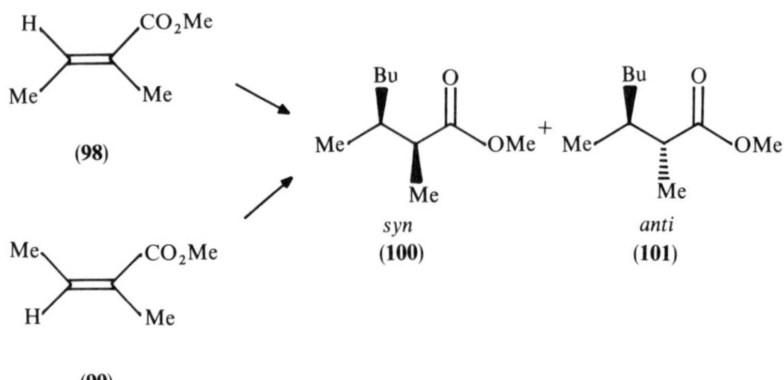

When alkoxy groups were present in the γ-position of α,β-unsaturated esters, addition of $RCu.BF_3$ to the *trans*-ester gave primarily the *anti*-adduct, whereas addition to the *cis*-ester gave mostly the *syn*-adduct.[130] Clearly the alkoxy oxygen plays an important role in establishing the reactive conformer. However, the situation becomes more complicated when γ-alkyl-α,β-unsaturated esters are used.[131] For *trans*-isomers, the *anti*-adduct preference holds for both the $BuCu.BF_3$ and $Bu_2CuLi.BF_3$ reagents, but for the *cis*-isomers, the *anti*/*syn* preference reverses for the two reagents. The results are rationalized through a conformational analysis.

Trimethylchlorosilane has a favourable influence on the conjugate addition of lithium methyl(2-thienyl)cuprate to methyl cinnamate and of 2-thienylcopper and butylcopper to 4-phenyl-3-propen-2-one.[132] An asymmetric conjugate addition of $Me_2CuI.BF_3$ is a key step in the synthesis of enantiomerically pure β-necrodol.[133] The reactions of 2-cyclohexenone, 2-cyclopentenone, (*E*)-3-penten-2-one, and (*E*)-3-octen-2-one with chiral organo(hetero)cuprates containing Me-, Bu-, and *tert*-Bu-transferable ligands gave the adducts in optical yields ranging up to 83%.[134] The conjugate addition of the phenyldimethylsilyl-cuprate reagent to cinnamate and crotonate esters and amides of known chiral auxiliaries is diastereoselective, making available β-silyl-esters of high enantiomeric excess.[135] Reaction of the same reagent with 4- and 5-substituted α,β-unsaturated lactones is highly stereoselective for the formation of the *trans*-adducts.[136]

An extensive kinetic and equilibrium study has been carried out on the addition of nitrogen nucleophiles to unsaturated systems of the type ArCH=CXY.[137–140] For amine nucleophiles, the order of decreasing intrinsic rate constants as a function of (X, Y) was found to be (CN, CN) > (CN, 4-$NO_2C_6H_4$) > (CN, 2,4-$(NO_2)_2C_6H_3$) > (H, NO_2) > (C_6H_5, NO_2).[137] An extensive study of the rate and equilibrium constants for piperidine and morpholine (R_2NH) additions to benzylidene Meldrum's acids, **(102)** ⇄ **(103)** ⇄ **(104)**, has contributed much to the understanding of this matter.[137] Data were measured for six different substituents (*Z*) in water and several water/DMSO solvent systems. The trends in the data were interpreted[137] in

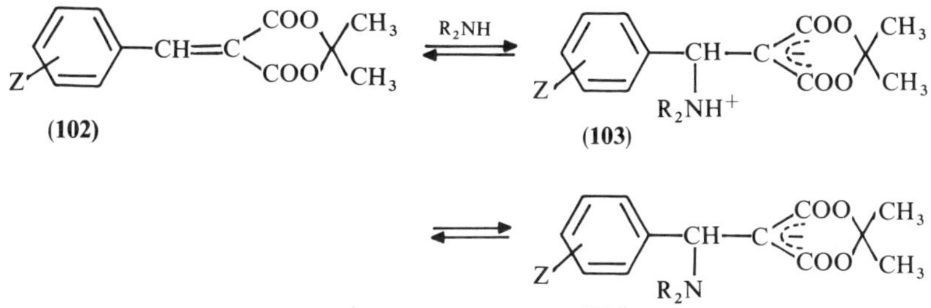

(102) **(103)** **(104)**

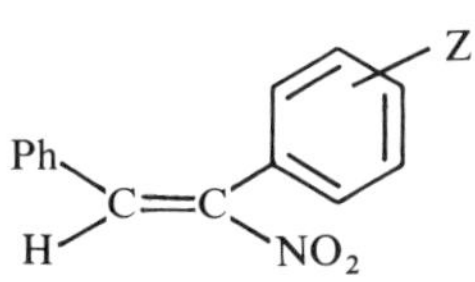

(105)

terms of the 'principle of nonperfect synchronization' (PNS),[139] by assuming that the development of carbanion resonance and solvation lags behind bond formation at the transition state. Closely related is the study of carbon protonation of amine adducts of benzylidene Meldrum's acid.[138] A similar study has used α-nitrostilbenes (**105**), again with piperidine and morpholine as the nucleophiles.[140] The results were again interpreted in terms of PNS.

Intramolecular Michael additions involving mechanistic stereochemical studies have been used in the syntheses of piperidenes,[141,142] pyrrolidines,[142] and indolizidines.[143] The major product from α,β-unsaturated ester (*Z*)-(**106**) was the *trans*-adducts (**107**), while the (*E*)-isomer gave mainly the *cis*-adduct (**108**).[141] There

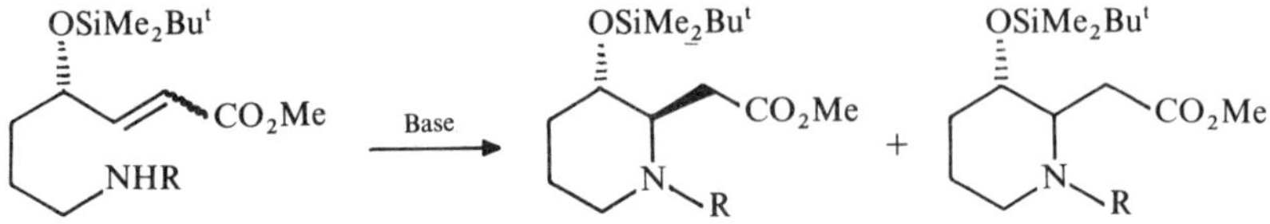

(**106**) (**107**) (**108**)

is sometimes difficulty in generating amino groups at remote places in a molecule which contains conjugated ester or other similar groups; this potential problem was solved[142] by reduction of azides with Ph_3P and water in THF to give the amino groups which, in turn, led to the intramolecular Michael cyclizations to pyrrolidines and piperidines with good diastereoselectivity.[142] This azide-reduction procedure was also a critical step in the formation of indolizidines, where the cyclization involved an intramolecular 1,6-Michael addition.[143]

Yields and enantioselectivities in the syntheses of β-amino-esters by Michael additions of amines to alkyl crotonates have been greatly improved by carrying out the reactions at high pressures and by the use of bulky chiral alkyl groups in the ester.[144]

A study of the $(Z)\rightleftarrows(E)$ interconversion of a series of aryl-ring-substituted 3-arylideneflavanones in pyridine has been carried out.[145] The proposed mechanism involves reversible 1,4-Michael addition of pyridine to the enone system. Except for the *m*- and *p*-nitro compounds, an excellent linear Hammett plot was obtained, using σ, with $\rho = 1.2$. The reasons for the anomalous results for the nitro compounds 'will form the subject of a future investigation'.

A study of the relative reactivities of the two nitrogens in phenylhydrazine in their 1,4-additions to chalcone has been carried out.[146] The two reactions have quite different mechanisms; base catalysis is observed for reaction at N(2) but not for reaction at N(1).

The rate constants for the additions of a variety of amines to substituted 1,3-diphenyl-1-propyn-3-ones, $XC_6H_4C{\equiv}CCOC_6H_4Y$, to give β-amino-α,β-unsaturated ketones in ethanol, acetonitrile, dimethyl formamide, and dioxane have been reported.[147,148] The Hammett plots are linear, with ρ values ranging from 1.14 to 2.21 as the solvent and amine are varied.

Michael addition to conjugated alkynes (**109**; R = Ph or Me_2COH) is also the critical step in the conversion of 2-benzimidazolethione (**110**), through intermediates (**111**), into 1,3-thiazinobenzimidazoles (**112**).[149]

(**109**)

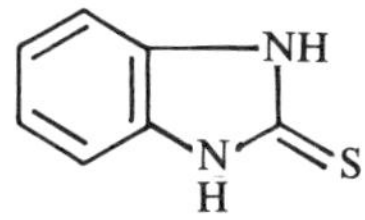

(**110**)

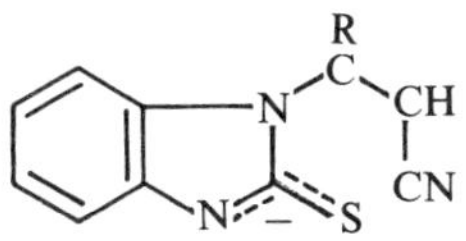

(**111**)

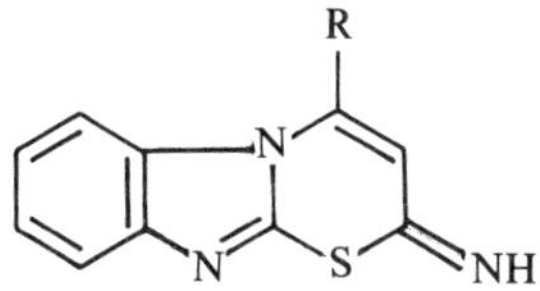

(**112**)

α-Chlorocyclopropylidine acetate (**113**) readily undergoes conjugate addition of sodium phenylselenide and sodium *p*-toluenesulphonate or benzenesulphonate to give (**114**) and (**115**), which are useful intermediates in a variety of further syntheses.[150]

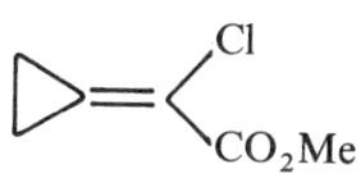

(**113**)

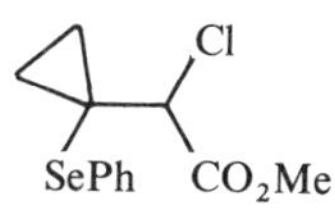

(**114**)

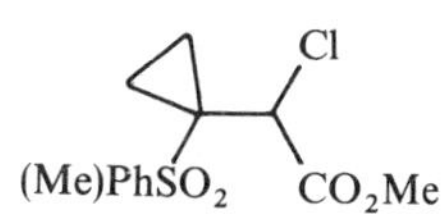

(**115**)

Asymmetric syntheses of α-amino-γ-nitro carboxylic esters have been carried out by the Michael addition of the titanium salt of the bislactim ether of cyclo (L-Val-Gly) and several nitroalkenes.[151] 1-Nitrocyclohexene readily undergoes Michael addition of lithium isopropyl sulphide, giving an anion which can be alkylated easily at either the α- or β-ring carbon, or can be converted into the synthetically useful silyl nitronate by reaction with *tert*-butyldimethylsilyl chloride.[152]

The optically-active vinyl sulphoxides, (*E*)- and (*Z*)-(**116**), readily undergo Michael additions with a variety of enolates to give nearly quantitative yields of adducts with a high degree of diastereoselectivity; e.g. acetophenone gives a 99% yield of nearly pure isomer (**117**).[153]

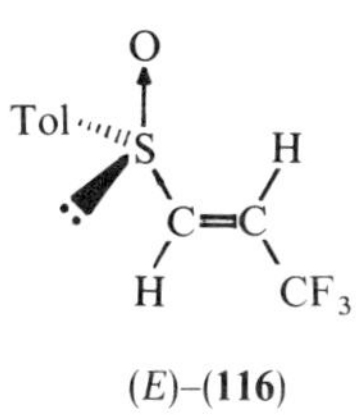

(*E*)-(**116**)

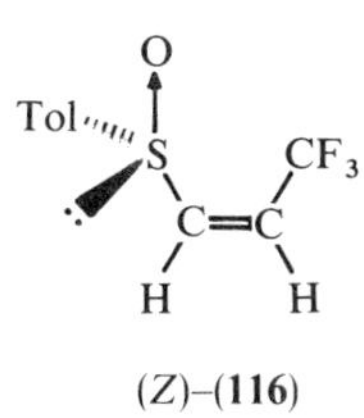

(*Z*)-(**116**)

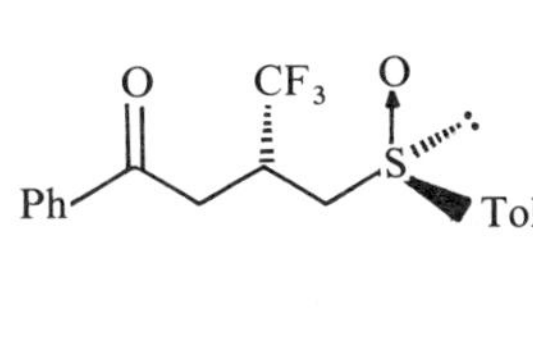

(**117**)

An extensive study has been carried out on the regioselectivity of the addition of thiols and amines to conjugated allenic ketones and esters; e.g. allene (**118**) added

PhSH in 75% overall yield, the product being almost 100% (**119**) with just a trace of (**120**).[154] This regioselectivity was the general pattern for neutral nucleophilic molecules, but salts of benzenethiols showed the opposite regioselectivity. For example, when the reaction shown was carried out in the presence of triethylamine, the (**119**):(**120**) ratio changed from ca. 100: trace to 17:83.

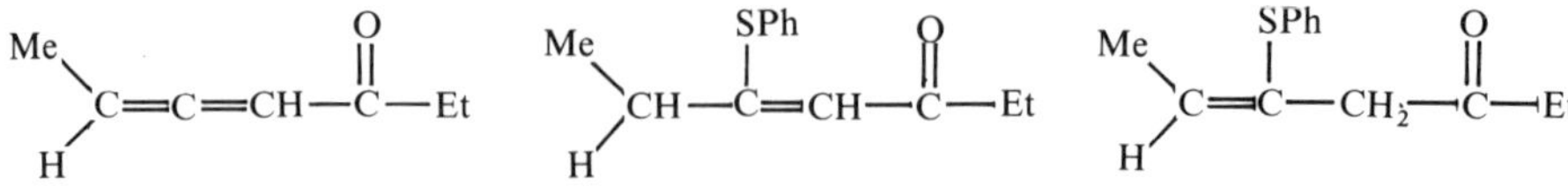

(**118**) (**119**) (**120**)

A study of the addition reactions of nucleophiles, in a chemical-ionization source, to 1,2,3-triaryl-propenones and -butenones reveals that stronger nucleophiles (e.g. MeO^-, EtO^-, and MeS^-) attack at the β- or carbonyl-carbon, while weaker nucleophiles such as Cl^- form the adduct by H-bonding with the β-hydrogen.[155]

The reversibility of conjugate-addition reactions provides a mechanism for vinylic substitution ('addition-elimination mechanism'). A few examples are reported here.[156–160] Methyl crotonate and several other α,β-unsaturated esters were treated with lithium amides [e.g. $LiNPr^i_2$ or $LiN(CH_2Ph)SiMe_3$], and the salts obtained were alkylated with several different alkyl halides. Several different procedures were utilized for the final elimination step.[156] The overall substitutions take place with high stereoselectivity. When a tertiary amine is used with a compound like acrylonitrile, enolates from aldehydes or ketones readily give α-substitution products; the amine adduct is not stable in such cases. That the rates of these reactions were greatly increased by increased pressure, led to the conclusion that the reaction of the adduct (zwitterion) with the carbonyl compound was rate-determining.[157] The rates of reaction of a series of (arylsulphonylvinyl)-trialkylammonium salts, $ArSO_2CH{=}CH\overset{+}{N}R^1R^2R^3$, with several primary and secondary amines, including piperidine, have been measured.[158,159] The results were interpreted in terms of an addition-elimination reaction The Hammett plots for the substituted-phenyl compounds studied were linear, with ρ in the 1.35–2.05 range for the various cases studied.[159] The effects of various salts on the rate of the reaction of *N*-phenylsulphonylbenzimidoyl chloride with 3-nitro-4-bromoaniline were studied.[160] The results were interpreted in terms of general base catalysis of an addition-elimination mechanism.

Miscellaneous Nucleophilic Addition Reactions

It is not unexpected that nucleophilic additions to non-conjugated alkenes are rare; complexing of the π-system with Hg(II), Pd(II), or with one of several transition metals can change the reactivity a great deal. This is reflected in the results of an *ab initio* molecular-orbital study on nucleophilic attack on free C_2H_4, and several of its complexes with metals.[161] Perturbational theory using extended Hückel molecular orbitals has been used to analyse the reactions of some nucleophiles with $(\eta^4\text{-diene})ML_n$ complexes.[162] The results accommodate well the available experimental observations.

Rates have been measured for the reactions between several bases and $(p\text{-}NO_2C_6H_4)_2C{=}CF_2$ to form the substitution products $(p\text{-}NO_2C_6H_4)_2C{=}CFNR^1R^2$.[163] The mechanism is considered to be one of addition-elimination, similar to those mentioned above for more obviously activated double bonds.

Arylcycloalkanes have been prepared by treatment of α-phenyl-ω-alkenyl methyl selenides (**121**) with butyllithium reagents to produce an intermediate (**122**) which undergoes intramolecular nucleophilic addition to the double bond. The final products are the *cis*- and *trans*-cyclopentane derivatives (**123**) and (**124**).[164]

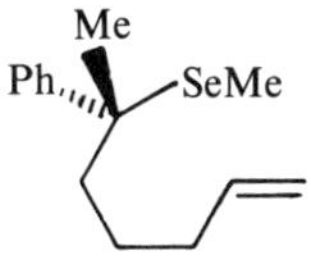

(**121**)

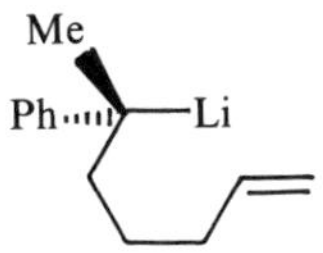

(**122**)

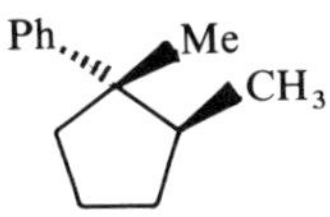

(**123**)

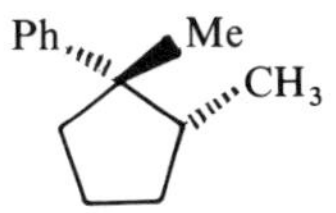

(**124**)

Complete control of the stereochemistry between the substituents of the cyclopentane derivatives can be achieved by proper choice of the experimental conditions. Along somewhat similar lines, cyclic 8-to-10-membered (*E*)-homoallylic sulphoxides undergo regiospecific BuLi-promoted transannular cyclization, yielding saturated bicyclic products. The reaction is interpreted as the addition of a 'carbanion' (an α-lithio-sulphoxide) to a non-activated double bond.[165]

A new route to useful organoboranes is provided by the additions of a variety of nucleophiles to the double bond in (trimethylsilylvinyl)dimesitylboranes (**125**) and related materials.[166] The adducts, e.g. (**126**), can easily be converted into the organoboranes (**127**) by treatment with fluoride ion.

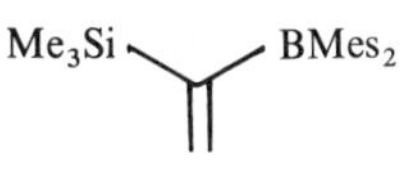

(**125**)

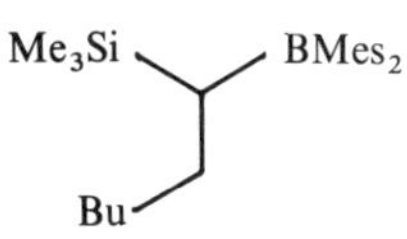

(**126**)

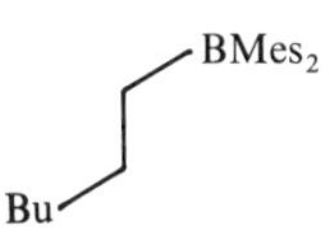

(**127**)

While not strictly alkene-addition reactions, some reactions of vinyl epoxides result in products of 1,4-addition of nucleophilic reagents to the alkene-oxirane system. As examples, epoxide (**128**) reacts with allyltrimethylstannate to form adduct (**129**),[167] and (**130**) reacts with Me_2CuLi to give an overall yield of 81% of an 84:16 mixture of (**131**) and (**132**).[168]

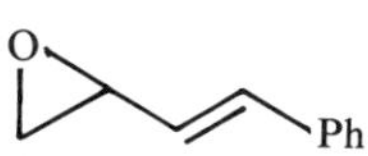

(**128**)

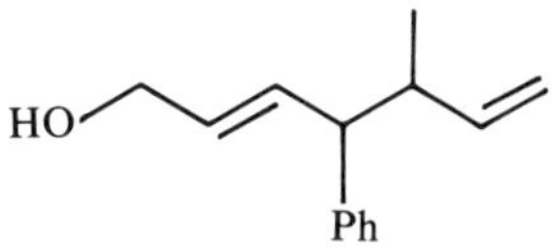

(**129**)

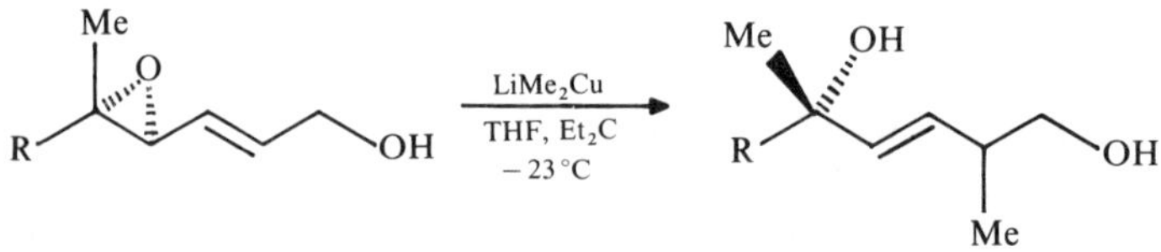

(130) **(131)**

R = $Me_2C{=}CHCH_2CH_2$

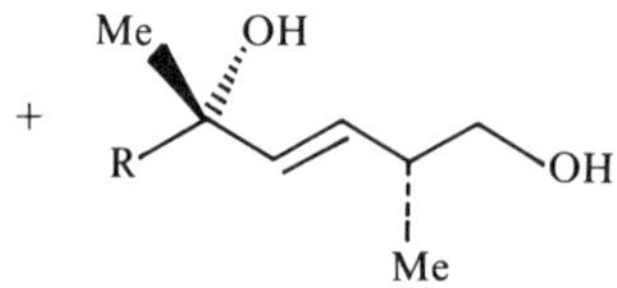

(132)

Acknowledgements

I wish to acknowledge the assistance of Miss Shirley Tsai, who did much of the library work for this chapter; of my wife, Lois, who typed the references and helped with the proof reading and in many other ways; and the hospitality of the Department of Chemistry of North Carolina State University, Raleigh, where the chapter was written during my study leave.

References

[1] Binger, P. and Buch, H. M., *Top. Curr. Chem.*, **135**, 77 (1987).
[2] Chandrasekhar, S., *Chem. Soc. Rev.*, **16**, 313 (1987).
[3] Dai, Q. and Schmid, G. H., *Beijing Gongye Daxue Xuebao*, **12**, 1 (1986); *Chem. Abs.*, **106**, 213211 (1987).
[4] Metivier, P., Gushurst, A. J., and Jorgensen, W. L., *J. Org. Chem.*, **52**, 3724 (1987).
[5] Winterfeldt, E., *Kontakte* (*Darmstadt*), **1987**, 20; *Chem. Abs.*, **107**, 153617 (1987).
[6] Stavber, S. and Zupan, M., *J. Org. Chem.*, **52**, 919 (1987).
[7] Hartshorn, M. P., Judd, M. C., and Robinson, W. T., *Aust. J. Chem.*, **39**, 2121 (1986).
[8] Vogel, E., Boehm, S., Hedwig, A., Hergarten, B. O., Lex, J., Uschmann, J., and Gleiter, R., *Angew. Chem. Int. Edn*, **25**, 1000 (1986).
[9] Buckland, S. J., Halton, B., Mei, Q., and Stang, P. J., *Aust. J. Chem.*, **40**, 1375 (1987).
[10] Bellucci, G., Chiappe, C., and Marioni, F., *J. Am. Chem. Soc.*, **109**, 515 (1987).
[11] Bellucci, G., Bianchini, R., and Vecchiani, S., *J. Org. Chem.*, **52**, 3355 (1987).
[12] Cerichelli, G., Grande, C., Luchetti, L., Mancini, G., and Bunton, C. A., *J. Org. Chem.*, **52**, 5167 (1987).
[13] Liu, R. Z., *Appl. Quantum Chem., Proc. Nobel Laureate Symp.*, 1984 (Publ. 1986), 85; *Chem. Abs.*, **106**, 175420 (1987).
[14] Negoro, T. and Ikeda, Y., *Bull. Chem. Soc. Jpn*, **59**, 3519 (1986).
[15] Negoro, T. and Ikeda, Y., *Wakayama Daigaku Kyoikugakubu Kiyo, Shizen. Kagaku*, **1986**, 15; *Chem. Abs.*, **107**, 58235 (1987).
[16] Schmid, G. H. and Gordon, J. W., *Can. J. Chem.*, **64**, 2171 (1986).
[17] Negoro, T. and Ikeda, Y., *Bull. Chem. Soc. Jpn*, **59**, 3515 (1986).
[18] Barluenga, J., Martinez-Gallo, J. M., Najera, C., and Yus, M., *J. Chem. Soc., Perkin Trans. 1*, **1987**, 1017.
[19] Correa, C. M. M. da S., and Fleming, M. D. C. M., *J. Chem. Soc., Perkin Trans. 2*, **1987**, 103.
[20] Kocovsky, P. and Stieborova, I., *J. Chem. Soc., Perkin Trans. 1*, **1987**, 1969.

[21] Gleiter, R., Muller, G., Huber-Patz, U., Rodewald, H., and Irngartinger, H., *Tetrahedron Lett.*, **28**, 1985 (1987).
[22] Haufe, G., *J. Chem. Res.* (*S*), **1987**, 100.
[23] Reitz, A. B., Nortey, S. O., Maryanoff, B. E., Liotta, D., and Monahan, R., *J. Org. Chem.*, **52**, 4191 (1987).
[24] Rajendra, G. and Miller, M. J., *J. Org. Chem.*, **52**, 4471 (1987).
[25] Bongini, A., Cardillo, G., Orena, M., Sandri, S., and Tomasini, C., *J. Org. Chem.*, **51**, 4905 (1986).
[26] Kurth, M. J. and Brown, E. G., *J. Am. Chem. Soc.*, **109**, 6844 (1987).
[27] Cousseau, J. and Albert, P., *Bull. Soc. Chim. Fr. II*, **1986**, 910.
[28] Tierney, J., Costello, F., and Dalton, D. R., *J. Org. Chem.*, **51**, 5191 (1986).
[29] Klenerman, D. and Zare, R. N., *Chem. Phys. Lett.*, **130**, 190 (1986); *Chem. Abs.*, **107**, 38852 (1987).
[30] Stang, P. J. and Roberts, K. A., *J. Org. Chem.*, **52**, 5213 (1987).
[31] Stang, P. J., Surber, B. W., Chen, Z. C., Roberts, K. A., and Anderson, A. G., *J. Am. Chem. Soc.*, **109**, 228 (1987).
[32] Roberts, D. W., Jackson, P. S., Saul, C. D., and Clemett, C. J., *Tetrahedron Lett.*, **28**, 3383 (1987).
[33] Bakker, B. H. and Cerfontain, H., *Tetrahedron Lett.*, **28**, 1699 (1987).
[34] Bakker, B. H. and Cerfontain, H., *Tetrahedron Lett.*, **28**, 1703 (1987).
[35] Boyer, J. L., Canselier, J. P., and Castro, V., *Commun. Jorn. Com. Esp. Deterg.*, **16**, 247 (1985); *Chem. Abs.*, **106**, 195634 (1987).
[36] Schneider, R. and Mayr, H., *Angew. Chem. Int. Edn*, **25**, 1016 (1986).
[37] Mayr, H., Schneider, R., and Grabis, U., *Angew. Chem.*, **98**, 1034 (1986); *Chem. Abs.*, **106**, 83759 (1987).
[38] Hun, L. G. and Nondek, L., *Collect. Czech. Chem. Commun.*, **52**, 1758 (1987).
[39] Schmid, G. H., Strukelj, M., Dalipi, S., and Ryan, M. D., *J. Org. Chem.*, **52**, 2403 (1987).
[40] Schmid, G. H., Garratt, D. G., and Dean, C. L., *Can. J. Chem.*, **65**, 1172 (1987).
[41] Smolyakova, I. P., Smit, V. A., and Lutsenko, A. I., *Izv. Akad. Nauk. SSSR, Ser. Khim.*, **1987**, 119; *Chem. Abs.*, **107**, 175155 (1987).
[42] Adam, W., Cramer, E., Lucchini, V., Peters, E.-M., Peters, K., von Schnering, H. G., and Staab, E., *Chem. Ber.*, **120**, 1719 (1987).
[43] Reetz, M. T. and Seitz, T., *Angew. Chem. Int. Edn*, **26**, 1028 (1987).
[44] Arjona, O., de la Pradilla, R. F., Perez, R. A., Plumet, J., and Viso, A., *Tetrahedron Lett.*, **28**, 5549 (1987).
[45] Black, K. A. and Vogel, P., *J. Org. Chem.*, **51**, 5341 (1986).
[46] Claret, F., Carrupt, P.-A., and Vogel, P., *Helv. Chim. Acta*, **70**, 1886 (1987).
[47] Magnol, E., Gore, J., and Malacria, M., *Bull. Soc. Chim. Fr. II*, **1987**, 455.
[48] Toshimitsu, A., Terao, K., and Uemura, S., *J. Chem. Soc., Perkin Trans. 1*, **1987**, 1059.
[49] Toshimitsu, A., Terao, K., and Uemura, S., *J. Org. Chem.*, **52**, 2018 (1987).
[50] Murata, S. and Suzuki, T., *Chem. Lett.*, **87**, 849 (1987).
[51] Hassner, A., and Amarasekara, A. S., *Tetrahedron Lett.*, **28**, 5185 (1987).
[52] Benati, L., Montevecchi, P. C., and Spagnolo, P., *J. Chem. Soc., Chem. Commun.*, **1987**, 1050.
[53] Engman, L., *J. Org. Chem.*, **52**, 4086 (1987).
[54] Srebnik, M., Cole, T. E., and Brown, H. C., *Tetrahedron Lett.*, **28**, 3771 (1987).
[55] Egger, M. and Keese, R., *Helv. Chim. Acta*, **70**, 1843 (1987).
[56] Takaishi, N., Takahashi, H., Yamashita, O., and Inamoto, Y., *J. Org. Chem.*, **51**, 4862 (1986).
[57] Maruoka, K., Sano, H., Shinoda, K., and Yamamoto, H., *Chem. Lett.*, **1987**, 73.
[58] Nelson, D. J., Cooper, P. J., and Coerver, J. M., *Tetrahedron Lett.*, **28**, 943 (1987).
[59] Lewis, A. and Carty, A., *J. Org. Chem.*, **52**, 1839 (1987).
[60] Lewis, A., *J. Org. Chem.*, **52**, 3099 (1987).
[61] Lilitkarntakul, S., Hirama, M., and Ito, S., *Tetrahedron Lett.*, **28**, 1207 (1987).
[62] Kagel, J. R. and Mertes, M. P., *J. Org. Chem.*, **52**, 2950 (1987).
[63] Johnson, C. R., Tait, B. D., and Cieplak, A. S., *J. Am. Chem. Soc.*, **109**, 5875 (1987).
[64] Livneh, M., Sutter, J. K., and Sukenik, C. N., *J. Org. Chem.*, **52**, 5039 (1987).
[65] Prokop, Z. and Setinek, K., *Collect. Czech. Chem. Commun.*, **52**, 1272 (1987).
[66] Meuldijk, J., Joosten, G. E. H., and Stamhuis, E. J., *J. Mol. Catal.*, **37**, 75 (1986); *Chem. Abs.*, **107**, 22762 (1987).
[67] Dietze, P. E. and Jencks, W. P., *J. Am. Chem. Soc.*, **109**, 2057 (1987).
[68] Harcourt, M. P. and More O'Ferrall, R. A., *J. Chem. Soc., Chem. Commun.*, **1987**, 823.
[69] McEwen, J. and Yates, K., *J. Am. Chem. Soc.*, **109**, 5800 (1987).
[70] Kwon, K. S., Mang, J. Y., and Park, C. K., *Chungnam Kwahak Yonguchi*, **12**, 107 (1985); *Chem. Abs.*, **107**, 77218 (1987).

[71] Allen, A. D. and Tidwell, T. T., *J. Am. Chem. Soc.*, **109**, 2774 (1987).
[72] Allen, A. D., Kresge, A. J., Schepp, N. P., and Tidwell, T. T., *Can. J. Chem.*, **65**, 1719 (1987).
[73] Nilsen, N. O., Skattebol, L., and Stenstrom, Y., *Acta Chem. Scand.*, **B41**, 459 (1987).
[74] Pairaudeau, G., Parsons, P. J., and Underwood, J. M., *J. Chem. Soc., Chem. Commun.*, **1987**, 1718.
[75] Banait, N., Hojatti, M., Findlay, P., and Kresge, A. J., *Can. J. Chem.*, **65**, 441 (1987).
[76] Chan, D. M. T., Marder, T. B., Milstein, D., and Taylor, N. J., *J. Am. Chem. Soc.*, **109**, 6385 (1987).
[77] Mitsudo, T., Hori, Y., Yamakawa, Y., and Watanabe, Y., *J. Org. Chem.*, **52**, 2230 (1987).
[78] Vsesvyatskaya, N. Yu., Shestakov, G. K., and Temkin, O. N., *Kinet. Katal.*, **27**, 1330 (1986); *Chem. Abs.*, **107**, 153706 (1987).
[79] Rosenberg, M., Balaz, S., Sturdik, E., and Kucher, A., *Collect. Czech. Chem. Commun.*, **52**, 425 (1987).
[80] Toullec, J., El-Alaoui, M., and Bertrand, R., *J. Chem. Soc., Perkin Trans. 2*, **1987**, 1517.
[81] Castanet, Y. and Petit, F., *J. Mol. Catal.*, **35**, 143 (1986); *Chem. Abs.*, **106**, 101510 (1987).
[82] Hoz, S., Livneh, M., and Cohen, D., *J. Org. Chem.*, **51**, 4537 (1986).
[83] Kirmse, W. and Streu, J., *J. Org. Chem.*, **52**, 515 (1987).
[84] Jones, P. R. and Bates, T. F., *J. Am. Chem. Soc.*, **109**, 913 (1987).
[85] Kahn, S. D., Pau, C. F., Chamberlin, A. R., and Hehre, W. J., *J. Am. Chem. Soc.*, **109**, 650 (1987).
[86] Kahn, S. D. and Hehre, W. J., *J. Am. Chem. Soc.*, **109**, 666 (1987).
[87] Chamberlin, A. R., Mulholland, R. L., Kahn, S. D., and Hehre, W. J., *J. Am. Chem. Soc.*, **109**, 672 (1987).
[88] Nguyen, M. T. and Hegarty, A. F., *J. Chem. Soc., Perkin Trans. 2*, **1987**, 1675.
[89] Smadja, W., Czernecki, S., Ville, G., and Georgoulis, C., *Organometallics*, **6**, 166 (1987).
[90] Andersson, C.-M. and Hallberg, A., *J. Org. Chem.*, **52**, 3529 (1987).
[91] McNab, H. and Monahan, L. C., *J. Chem. Soc., Chem. Commun.*, **1987**, 138.
[92] McNab, H., Monahan, L. C., and Gray, T., *J. Chem. Soc., Chem. Commun.*, **1987**, 140.
[93] Iovel, I. G., Goldberg, Y. S., Shymanska, M. V., and Lukevics, E., *J. Chem. Soc., Chem. Commun.*, **1987**, 31.
[94] Licchelli, M. and Greco, A., *Tetrahedron Lett.*, **28**, 3719 (1987).
[95] Eddy, V. J. and Hallgren, J. E., *J. Org. Chem.*, **52**, 1903 (1987).
[96] Bellesia, F., Ghelfi, F., Pagnoni, U. M., and Pinetti, A., *J. Chem. Res. (S)*, **1987**, 238.
[97] Mobilio, D. and De Lange, B., *Tetrahedron Lett.*, **28**, 1483 (1987).
[98] Fleming, I., Sarkar, A. K., and Thomas, A. P., *J. Chem. Soc., Chem. Commun.*, **1987**, 157.
[99] Sharma, S. and Oehlschlager, A. C., *Tetrahedron Lett.*, **27**, 6161 (1986).
[100] Gibson, T., *Organometallics*, **6**, 918 (1987).
[101] Hu, N. X., Aso, Y., Otsubo, T., and Ogura, F., *Tetrahedron Lett.*, **28**, 1281 (1987).
[102] Schulze, C., Weiske, T., and Schwarz, H., *Chimia*, **40**, 362 (1986); *Chem. Abs.*, **107**, 6588 (1987).
[103] Hosokawa, T., Ohta, T., Kanayama, S., and Murahashi, S.-I., *J. Org. Chem.*, **52**, 1758 (1987).
[104] Schreiber, S. L., Schreiber, T. S., and Smith, D. B., *J. Am. Chem. Soc.*, **109**, 1525 (1987).
[105] Bartlett, P. D. and Ghosh, T., *J. Org. Chem.*, **52**, 4937 (1987).
[106] Mukaiyama, T., Tamura, M., and Kobayashi, S., *Chem. Lett.*, **87**, 743 (1987).
[107] Kobayashi, S. and Mukaiyama, T., *Chem. Lett.*, **87**, 1183 (1987).
[108] Minowa, N. and Mukaiyama, T., *Chem. Lett.*, **87**, 1719 (1987).
[109] Hashimoto, Y., Sugumi, H., Okauchi, T., and Mukaiyama, T., *Chem. Lett.*, **87**, 1691 (1987).
[110] Hashimoto, Y., Sugumi, H., Okauchi, T., and Mukaiyama, T., *Chem. Lett.*, **87**, 1695 (1987).
[111] Shafiullah, Rafiuddin, A. M., and Husain, S., *Bull. Chem. Soc. Jpn*, **60**, 3411 (1987).
[112] Oare, D. A. and Heathcock, C. H., *Tetrahedron Lett.*, **27**, 6169 (1986).
[113] Al-Arab, M. M. and Issa, A. M., *Collect. Czech. Chem. Commun.*, **52**, 1021 (1987).
[114] Keumi, T., Inagaki, T., Nakayama, N., Morita, T., and Kitajima, H., *J. Chem. Soc., Chem. Commun.*, **1987**, 1091.
[115] Hua, D. H., Chan-Yu-King, R., McKie, J. A., and Myer, L., *J. Am. Chem. Soc.*, **109**, 5026 (1987).
[116] Paquette, L. A., Pierre, F., and Cottrell, C. E., *J. Am. Chem. Soc.*, **109**, 5731 (1987).
[117] Berrada, S. and Metzner, P., *Tetrahedron Lett.*, **28**, 409 (1987).
[118] Yamaguchi, M., Hamada, M., Nakashima, H., and Minami, T., *Tetrahedron Lett.*, **28**, 1785 (1987).
[119] Tomioka, K., Yasuda, K., and Koga, K., *J. Chem. Soc., Chem. Commun.*, **1987**, 1345.
[120] Okay, G., *Spektrosk. Derg.* 1985 (Publ. 1986), 98; *Chem. Abs.*, **106**, 175490 (1987).
[121] Spitzner, D., Engler, A., Wagner, P., de Meijere, A., Bengtson, G., Simon, A., Peters, K., and Peters, E.-M., *Tetrahedron*, **43**, 3213 (1987).
[122] Iglesias, M., Marinas, J. M., and Sinisterra, J. V., *Tetrahedron*, **43**, 2335 (1987).
[123] Stork, G. and Saccomano, N. A., *Nouveau J. Chim.*, **10**, 677 (1986).
[124] Cossentini, M., Strzalko, T., and Seyden-Penne, J., *Bull. Soc. Chim. Fr. II*, **1987**, 531.
[125] Binns, M. R. and Haynes, R. K., *Aust. J. Chem.*, **40**, 937 (1987).

[126] Haynes, R. K., Schober, P. A., and Binns, M. R., *Aust. J. Chem.*, **40**, 1223 (1987).
[127] Haynes, R. K. and Schober, P. A., *Aust. J. Chem.*, **40**, 1249 (1987).
[128] Enders, D., Demir, A. S., and Rendenbach, B. E. M., *Chem. Ber.*, **120**, 1731 (1987).
[129] Yamamoto, Y., Yamada, J., and Uyehara, T., *J. Am. Chem. Soc.*, **109**, 5820 (1987).
[130] Yamamoto, Y., Nishii, S., and Ibuka, T., *J. Chem. Soc., Chem. Commun.*, **1987**, 464.
[131] Yamamoto, Y., Nishii, S., and Ibuka, T., *J. Chem. Soc., Chem. Commun.*, **1987**, 1572.
[132] Lindstedt, E.-L., Nilsson, M., and Olsson, T., *J. Organomet. Chem.*, **334**, 255 (1987).
[133] Oppolzer, W. and Schneider, P., *Helv. Chim. Acta*, **69**, 1817 (1986).
[134] Dieter, R. K. and Tokles, M., *J. Am. Chem. Soc.*, **109**, 2040 (1987).
[135] Fleming, I. and Kindon, N. D., *J. Chem. Soc., Chem. Commun.*, **1987**, 1177.
[136] Fleming, I., Reddy, N. L., Takaki, K., and Ware, A. C., *J. Chem. Soc., Chem. Commun.*, **1987**, 1472.
[137] Bernasconi, C. F. and Panda, M., *J. Org. Chem.*, **52**, 3042 (1987).
[138] Bernasconi, C. F., Fairchild, D. E., and Murray, C. J., *J. Am. Chem. Soc.*, **109**, 3409 (1987).
[139] Bernasconi, C. F., *Adv. Chem. Ser.*, **215**, 115 (1987); *Chem. Abs.*, **107**, 175179 (1987).
[140] Bernasconi, C. F. and Renfrow, R. A., *J. Org. Chem.*, **52**, 3035 (1987).
[141] Hirama, M., Iwakuma, T., and Ito, S., *J. Chem. Soc., Chem. Commun.*, **1987**, 1523.
[142] Knouzi, N., Vaultier, M., Toupet, L., and Carrié, R., *Tetrahedron Lett.*, **28**, 1757 (1987).
[143] Boulaajaj, S., Le Gall, T., Vaultier, M., Gree, R., Toupet, L., and Carrié, R., *Tetrahedron Lett.*, **28**, 1761 (1987).
[144] d'Angelo, J. and Maddaluno, J., *J. Am. Chem. Soc.*, **108**, 8112 (1986).
[145] Dhavale, D. D., Joshi, P., and Marathe, K. G., *J. Chem. Soc., Perkin Trans. 2*, **1987**, 449.
[146] Kenny, P. W. and Robinson, M. J. T., *Tetrahedron*, **43**, 4043 (1987).
[147] Korshunov, S. P., Slyusareva, O. M., Korzhova, N. V., and Kazantseva, V. M., *Isv. Vyssh. Uchebn. Zaved., Khim. Khim. Tekhnol.*, **29**, 40 (1986); *Chem. Abs.*, **106**, 155662 (1987).
[148] Korshunov, S. P., Slyusareva, O. M., Korzhova, N. V., and Kazantseva, V. M., *Zh. Org. Khim.*, **22**, 913 (1986); *Chem. Abs.*, **106**, 4350 (1987).
[149] Andriyankova, L. V., Abramova, N. D., Pertsikov, B. Z., Afonin, A. V., and Pavlova, T. O., *Zh. Org. Khim.*, **22**, 2455, (1986); *Chem. Abs.*, **107**, 96129 (1987).
[150] Seyed-Mahdavi, F., Teichmann, S., and deMeijere, A., *Tetrahedron Lett.*, **27**, 6185 (1986).
[151] Schollkopf, U., Kuhnle, W., Egert, E., and Dyrbusch, M., *Angew. Chem. Int. Edn*, **26**, 480 (1987).
[152] Hwu, J. R. and Wang, N., *J. Chem. Soc., Chem. Commun.*, **1987**, 427.
[153] Yamazaki, T., Ishikawa, N., Iwatsubo, H., and Kitazume, T., *J. Chem. Soc., Chem. Commun.*, **1987**, 1410.
[154] Sugita, T., Eida, M., Ito, H., Komatsu, N., Abe, K., and Suama, M., *J. Org. Chem.*, **52**, 3789 (1987).
[155] Madhusudanan, K. P., Jain, R., Mittal, S., Durani, S., and Kapil, R. S., *Org. Mass Spectrom.*, **21**, 781 (1986); *Chem. Abs.*, **107**, 197206 (1987).
[156] Uyehara, T., Asao, N., and Yamamoto, Y., *J. Chem. Soc., Chem. Commun.*, **1987**, 1340.
[157] Isaacs, N. S., Hill, J., Castelijns, M., and Corvers, A., *Physica B + C (Amsterdam)*, **139–140**, 717 (1986); *Chem. Abs.*, **106**, 175477 (1987).
[158] Popov, A. F., Kravchenko, V. V., Kostenko, L. I., Kotenko, A. A., and Kovach, N. A., *Zh. Org. Khim.*, **22**, 2135 (1986); *Chem. Abs.*, **107**, 115054 (1987).
[159] Kravchenko, V. V., Kostenko, L. I., Popov, A. F., and Kotenko, A. A., *Zh. Org. Khim.*, **22**, 2322 (1986); *Chem. Abs.*, **107**, 197207 (1987).
[160] Mikhailov, V. A., Drizhd, L. P., and Savelova, V. A., *Zh. Org. Khim.*, **22**, 2319 (1986); *Chem. Abs.*, **107**, 38867 (1987).
[161] Sakaki, S., Maruta, K., and Ohkubo, K., *J. Chem. Soc., Dalton Trans.*, **1987**, 361.
[162] Sautet, P., Eisenstein, O., and Nicholas, K. M., *Organometallics*, **6**, 1845 (1987).
[163] Leffek, K. T. and Maciejewska, U., *Can. J. Chem.*, **64**, 2274 (1986).
[164] Krief, A. and Barbeaux, P., *J. Chem. Soc., Chem. Commun.*, **1987**, 1214.
[165] Cere, V., Paolucci, C., Pollicino, S., Sandri, E., and Fava, A., *J. Org. Chem.*, **51**, 4880 (1986).
[166] Cooke, M. P. and Widener, R. K., *J. Am. Chem. Soc.*, **109**, 931 (1987).
[167] Naruta, Y. and Maruyama, K., *Chem. Lett.*, **87**, 963 (1987).
[168] Marshall, J. A. and Trometer, J. D., *Tetrahedron Lett.*, **28**, 4985 (1987).

Organic Reaction Mechanisms 1987
Edited by A. C. Knipe and W. E. Watts

CHAPTER 14

Addition Reactions: Cycloaddition

R. M. PATON

Department of Chemistry, University of Edinburgh

Introduction

In recent years the proportion of the literature devoted to 2 + 2-, 2 + 3-, and miscellaneous cycloadditions has increased, reflecting their growing application in synthesis. This trend was maintained through 1987. Cation-radical pericyclic reactions[1] and capto-dative substituent effects on cycloadditions[2] have been reviewed as have photoadditions to aromatic compounds[3,4] and within organic crystals.[5] Cycloaddition reactions involving tetracyanoethylene,[6,7] chlorosulphonyl isocyanate,[8] *ortho*-quinodimethanes,[9] imines,[10] and tetrathiooxalates[11] have been discussed as parts of wider surveys of their chemistries. Reviews dealing with a single class of cycloaddition are dealt with in the appropriate section of this chapter.

2 + 2-Cycloaddition

There has been an increase in the literature[12–18] dealing with cycloaddition reactions in ordered systems such as organic crystals. Particular attention has been paid to the influence of environment on product selectivity and the subject has been reviewed.[5] It has also been observed[19] that stereoselectivity and rate of uracil photodimerization are enhanced in smectic liquid crystalline media.

In the photoreaction of electron-deficient alkenes with 5(*E*)-styryl-1,3-dimethyluracil cycloaddition takes place at C(5)=C(6) of the uracil rather than at the styryl unit.[20] Chloroethylenes add[21] to α-pyrones at both C(3)=C(4) and C(5)=C(6).

Acetylene reacts with carbohydrate-derived pyranone (**1**) face-selectively, the alkyne approaching *anti* to the 6-substituent.[22] Marked temperature and solvent dependence is observed for mixed 2 + 2-photocycloadditions with the chiral auxiliary in either the enone or alkene component.[23] The double stereo-differentiation for the photoreaction between chiral cyclohex-2-enone-3-carboxylates and optically active ketene acetals has been studied;[24] the induction is either 'matched' (increasing diastereoselectivity) or 'mismatched', thus paralleling experience in ground-state processes. Diastereoface selection has also been examined[25] for chiral α-enones such as (**2**) in which the auxiliary is rigidly attached. Photolysis of tetrahydronaphthalene derivative (**3**) affords, in addition to a 2 + 2-dimer, the triple 2 + 2-dimer (**4**), from which the parent hydrocarbon (**5**) can be prepared.[26] 1,4-Dioxene reacts with 3-methylcyclohex-2-enone to give a mixture of adducts.[27] Salem's intermolecular orbital theory applied[28] to the photoreaction of chromene and 4-quinoline with methyl vinyl ether indicates a concerted regioselective process. Significant diradical character is predicted[29] for the intermediate in the cycloaddition between hydroxyethene and acrolein.

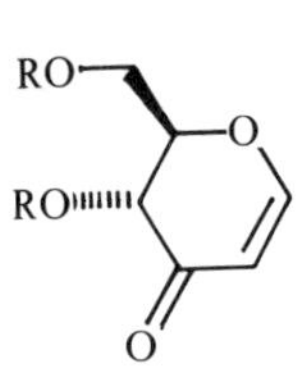

(**1**)

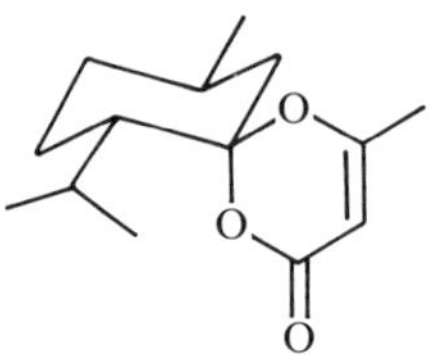

(**2**)

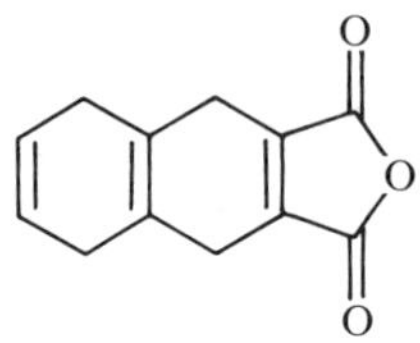

(**3**)

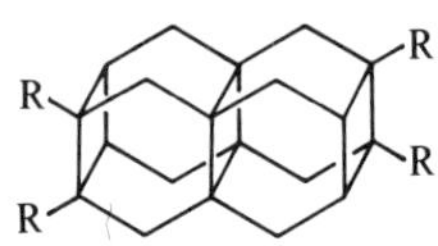

(**4**) RR = CO_2CO
(**5**) R = H

The cycloaddition reactions of ketenes and ketenimines continue to attract attention[30–33,41–46]. The balance between *cis*- and *trans*-products in the thermal dimerization of ketoketenes shows marked substituent dependence.[30] (Menthyloxy)-methylketene adds to *cis*-ethoxypropene with 3:1 diastereoselectivity.[31]

The expansion noted last year in the literature of *intra*molecular cycloadditions of α-enones and ketenes has been maintained.[34–46] A more detailed account has been presented[34] of the photolysis of 3-butenyloxycyclooct-2-enones. Steric effects play a key rôle in the corresponding reaction of 4-substituted-2-(pent-4-enyl)cyclohexenones.[35] For the dioxolenone system (**6**), exclusive formation[36,37] of 'inside-outside' adduct (**7**) is consistent with preferred transition state (**8**). Various 2-acyl-4-oxa- and 1-acyl-3-oxa-hexa-1,5-dienes afford predominantly 5-membered-ring cyclized products.[38,39] Chloroketenes undergo intramolecular 2 + 2-

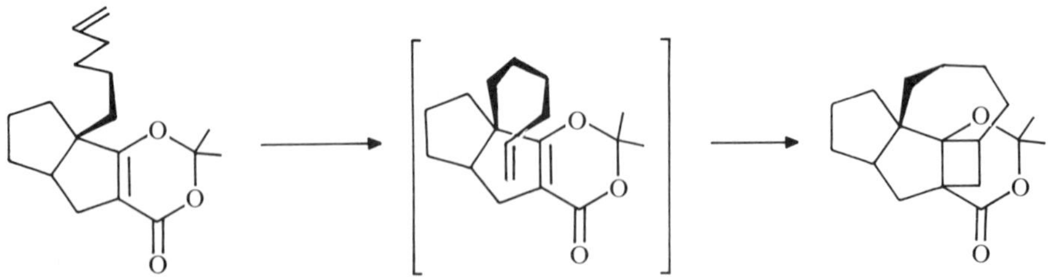

(**6**) (**8**) (**7**)

cycloaddition more readily than unsubstituted analogues;[41] further work on vinylketenes[42,43] and ketene-carbonyl cycloadditions[44] has also been reported. A comparison[45,46] of intramolecular cycloadditions to alkenes of ketenes and keteniminium salts shows that the former are more successful for ketoketenes whereas for aldoketenes the iminium salt method gives better yields.

The effect of pressure and viscosity on the regio- and stereo-chemical course of the reaction between styrenes and $F_2C{=}C{=}CH_2$ supports the previous hypothesis that the isomeric products are formed via two kinetically distinct conformations of the intermediate diradical.[47,48] Sensitized photolysis of aceanthrylene affords *syn* and *anti* head-to-head and head-to-tail dimers with slight *syn* preference.[49] Intramolecular photodimerization of α,ω-bis(*p*-vinylaryl)alkanes occurs readily yielding *cis*-cyclobutanes.[50] A stepwise diradical mechanism for the formation of 7-phenylbicyclo[4.2.0]oct-1-enes from styrene and cyclohexa-1,2-diene has been established[51] through deuterium-labelling experiments. Various intramolecular 2 + 2-cycloadditions of strained cage-dienes have been examined[52] as a preliminary to generating elusive prismanes. Further calculations have been performed[53] on the cross and parallel intramolecular 2 + 2-cycloaddition of α,ω-dienes. Butatriene (**9**; $n = 1$) spontaneously undergoes *syn* head-to-head dimerization at the double bond adjacent to the cyclopropane, presumably by a diradical mechanism;[54] in contrast, tetraene (**9**; $n = 2$) gives a *syn* head-to-head tetrayne (Scheme 1). Contrary to theoretical predictions, the newly isolated[55] *cis,syn-o,o'*-dibenzene proves to be more stable than the *anti*-isomer. *Z*-[6]Paracyclophen-3-ene reacts with tetracyanoethylene (TCNE) thermally, forming *ortho*-2 + 2-adduct (**10**), in contrast to the corresponding saturated [6]paracyclophane which gives a 4 + 2-product.[56] 2 + 2-Cyclodimerization of 3,3-dimethoxycyclopropenone can be accomplished in the presence of Pd(0) catalysts.[57] Kinetic and thermochemical measurements indicate[58] a formally forbidden $2\sigma + 2\pi$ pathway for the conversion of *endo*-(**11**) to (**12**). Benzocyclopropene reacts with furan and benzofuran to give $2\sigma + 2\pi$-cycloadducts with the furans acting as the 2π-components.[59]

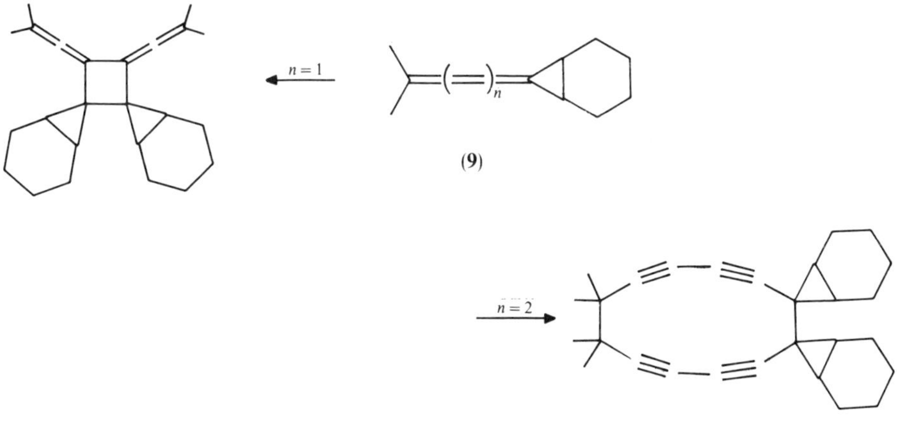

SCHEME 1

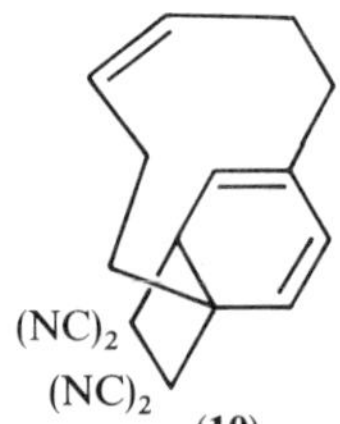

(10)

(11) **(12)**

Cycloadditions involving triazolinediones have been the focus of recent interest,[60,61] stimulated by mechanistic similarities to singlet oxygen reactions. Detailed examination of the addition of *N*-phenyltriazolinedione (PTAD) with substituted dienes reveals that, while some (e.g. *E,E*-hexa-2,4-diene) give the expected 4 + 2 products with high stereospecificity (> 200:1), others such as *Z,Z*-dienes (**13**) yield *cis* and *trans* adducts (**14**) and (**15**). A mechanism (Scheme 2) involving an intermediate diazetidine (**16**), detectable by NMR spectroscopy, is proposed; no direct evidence was obtained for aziridinium imides (**17**), the analogues of perepoxides in singlet oxygen reactions, but their involvement is assumed. The head-to-tail photodimerization of 2-arylbenzoxazoles represents a rare case involving C=N double bonds.[62] High pressure facilitates 2 + 2-cycloaddition of imines to electron-rich alkenes.[63] Lone pair-lone pair interactions are believed to make an important contribution to the high levels of stereoselectivity observed for the addition of azidoketenes to 3-imino-β-lactams.[64]

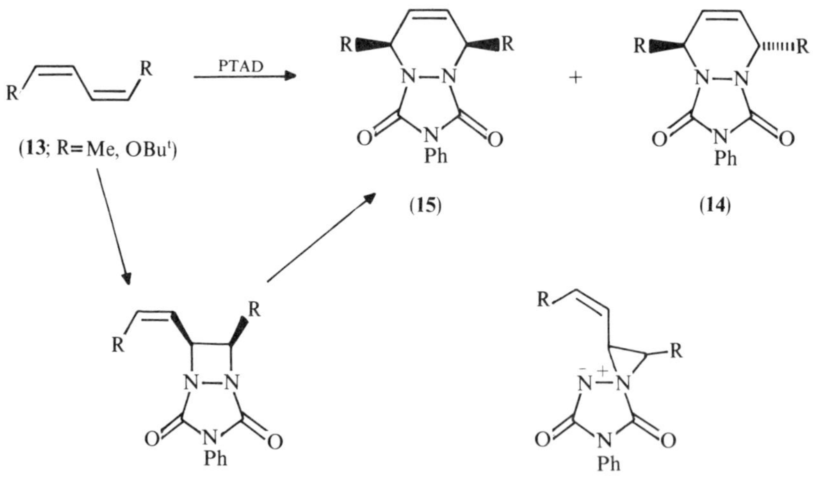

SCHEME 2

Hünig[65,66] and Prinzbach[67–69], with their co-workers, continue to present elegant examples of cycloaddition reactions involving conformationally rigid structures. A more detailed study[65] of the laticyclic [2 + 2 + 2]-conjugation in compound (**18**), using MO calculations and PE, UV, and X-ray techniques, reveals the origin of the prefered cyclization to (**19**) rather than (**20**) to be the smaller distance between the reacting centres and greater bond order in the excited state (Scheme 3). Photoconversions of (**21**; X = CH_2, NH, NCO_2Et, O) to corresponding (**22**) represent[66] the first examples of 2 + 2-photocycloadditions between 3-membered rings and azo groups (Scheme 4).

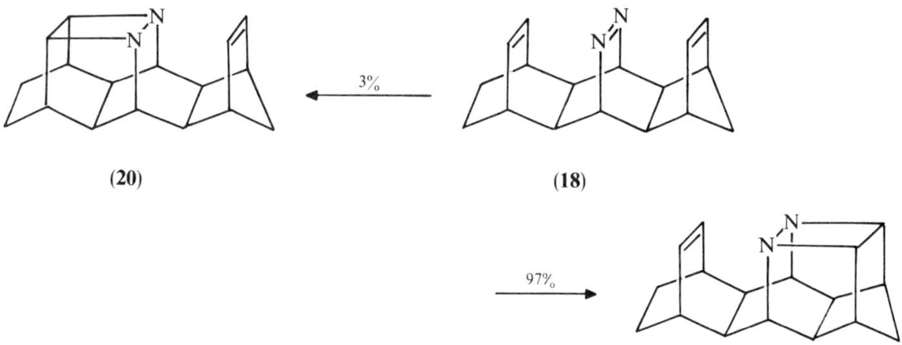

(**20**) (**18**) (**19**)

SCHEME 3

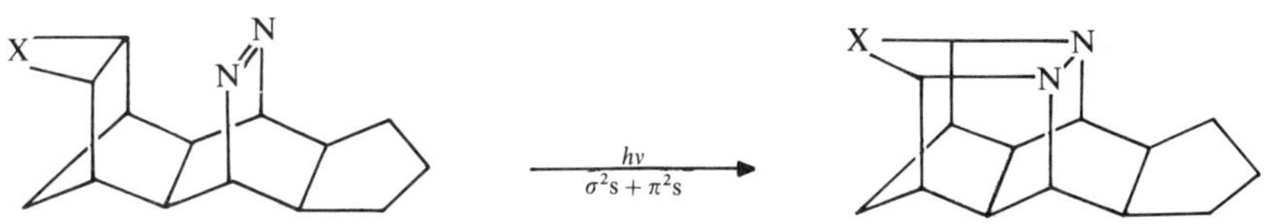

(**21**) (**22**)

SCHEME 4

Formation of oxetanes from the photoreaction of biacetyl and electron-rich alkenes has been discussed,[70] based on kinetic and ESR measurements, in terms of contact ion-pairs or at least an exciplex of extreme charge-transfer character. Photocycloaddition of methyl vinyl sulphides to benzophenone is more selective than the corresponding reaction of enol ethers, *trans*-4-substituted-3-methylthio-oxetanes predominating.[71] Kinetic analysis and trapping experiments for the addition of triplet aromatic thiones to allenes provide evidence, contrary to previous suggestions, for a 1,4-diradical intermediate which does *not* have an exciplex as its precursor[72,73]. Photocycloadditions of alkenes and alkynes to *N*-methyl-(thiophthalimide) have been examined;[74,75] in contrast to phthalimide itself, 2 + 2-products (e.g. thietanes) formed by a diradical mechanism predominate. The regiochemistry of the addition of η,π^*-excited thiocarbonyl compounds to ketene dimethyl acetals can also be rationalized[76] in terms of 1,4-diradical intermediates. Pyridine-4-thiones and acrydine-9-thiones react with electron-poor alkenes to give spirothietanes regioselectively.[77] 3-Allyldithiosuccinimides undergo intramolecular cyclization[78] yielding strained tricyclic adducts (**23**) and the vinylthiazolone (**24**) reacts similarly affording (**25**).[79] The preference for singlet oxygen, unlike TCNE or diphenylketene, to add to the more substituted double bond in *t*-butoxybutadienes can be rationalized[80] in terms of favourable frontier orbital interactions in the intermediate perepoxide. Matrix isolation of a tris-dioxetane from photooxidation of $Me_2C{=}C{=}C{=}CMe_2$ has been reported.[81] Low-temperature NMR studies[82] of the initial sultone adducts from SO_3 and octenes suggest that they are formed stereospecifically via a concerted *cis*-addition. *Ab initio* calculations have been performed[83] on the pathway of the Wittig reaction of $H_3P{=}CH_2$ and formaldehyde yielding H_3PO and ethene via an intermediate oxaphosphetane. Further examples of 1,2-disiloxetane formation from silenes with aldehydes and ketones have been described.[84,85]

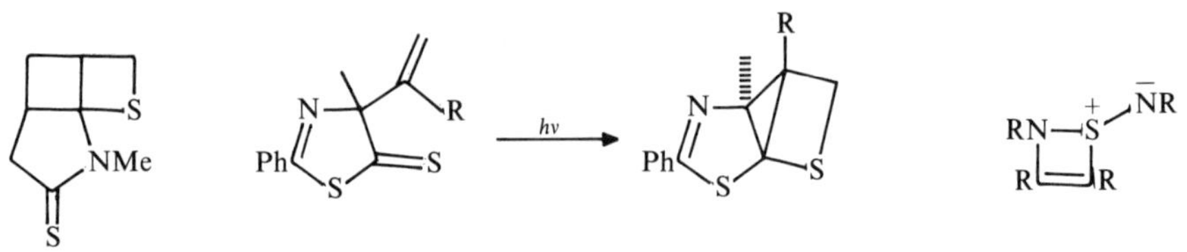

(23) (24) (25) (26)

Cycloaddition reactions of heterocumulenes have been the subject of several investigations.[86–96] MNDO calculations for the addition of $HOC{\equiv}COH$ (as a model for dialkoxyalkynes) with $HOCH{=}C{=}O$, $HN{=}C{=}O$, $O{=}CHN{=}C{=}O$, and $HN{=}C{=}S$ (as model hetero-cumulenes) predict in each case a stepwise mechanism proceeding via a short-lived diradical intermediate. Trichloroacetyl isocyanate adds stereospecifically *anti* with respect to the C(3)-substituent of glycals yielding both 2 + 2- and 4 + 2-products.[87,88] The cycloaddition-polymerization of unusual isocyanatoketenes of the form $O{=}C{=}N(CH_2)_n\,CH{=}C{=}O$, where $n = 0$–5, 9, and 10, have been examined;[89] for $n \geqslant 3$, reaction occurs solely between ketene groups, but for $n < 3$ both hetero-cumulene systems are involved. The regioselectivity of the reaction between *N*-alkyl-*N′*-arylcarbodiimides and isocyanates is controlled by interactions between the more nucleophilic nitrogen of the carbodiimide and the electrophilic carbon of the isocyanate;[90] the corresponding reaction with MeNCS yields a mixture of 1,3-diazetidinethione and 1,3-thiazetidinimines.[91] The products from the reaction of sulphur diimides with both electron-rich and -poor alkynes can be explained[92,93] by initial formation of 1,2-thiazete *S*-imide (**26**). A 2 + 2-cycloaddition mechanism has been postulated[94] for Rh-catalysed metathesis of CS_2-like hetero-allenes. Bis(trifluoromethyl)sulphene has been trapped[95] as 2 + 2-adducts with electron-rich alkenes. Cationic *N*-sulphinylamines react with ketenimines to form 3-imino-1,2-thiazetidinium salts via an initial $2^+ + 2$-cycloaddition[96] of $S{=}N^+$ with $C{=}N$ or $C{=}C$.

2 + 3-Cycloaddition

1,3-Dipolar cycloaddition reactions have been the subject of several reviews.[97–99] Topics covered include prototropic systems (X=Y—ZH),[97] nitrones,[98] and 3-oxopyrilium ylides.[99]

In a continuation of the extended debate about the concertedness of 1,3-dipolar cycloadditions, Firestone[100] claims that the exceptionally high stereospecificity observed in some cases does not necessarily rule out a stepwise diradical process; the basis of this postulate is (a) rotational barriers of free radicals > 0; (b) those for diradicals may exceed those for monoradicals; and (c) strain-free cyclizations to 5- and 6-membered rings generally enjoy a large rate advantage over their intermolecular counterparts. The use of Hartree–Fock theory to account for geometries of unsubstituted planar 1,3-dipoles has been examined.[101] For 24-electron (allyl type) systems, particularly those with oxygen as the central atom, open-shell (UHF) solutions consistent with some diradical character are preferred; calculated bond lengths and angles for 22-electron (propargyl-type) systems such as HCNO, CH_2N_2, HN_3, and N_2O suggest that each has an essentially fully formed double *and* triple bonds, and that they can therefore be described in terms of hypervalent structures in which the central atom is allocated more than its normal complement of electrons.

Allyl cations formed from 3-chlorocyclohexene in the presence of $ZnCl_2$ readily add to alkynes affording bicyclo[3.2.1]oct-6-enes.[102–104] Formation of cyclopentanes from trimethylenemethane complexes and allenes has been described.[105] Replacement of an α-carbon of cyclohexenones by oxygen (i.e. δ-oxaenones) increases reactivity and stereoselectivity in Pd-mediated cycloaddition.[106] Allylstannanes add to α,β-unsaturated acyliron complexes in the presence of $AlCl_3$ to give *cis*-1,3-disubstituted cyclopentane derivatives.[107] 1,3-Addition is also observed for the reactions of TCNE and PTAD with tricarbonyliron complexes of 7-methylenecycloheptatrienes[108] and tropone;[109] pressure and solvent effects are consistent with a concerted mechanism.[110] Anionic 3 + 2-cycloadditions of (2-carbamoyl)lithium reagents to electron-poor alkenes have also been reported.[111–112]

Although chloroformamidines of the form $(CF_3)_2CClN{=}CHNR_2$ react with alkynes, alkenes, and various hetero-cummulenes to give products formally derived from the corresponding nitrile ylide, rate measurements and regioselectivities differ markedly from those expected of the 1,3-dipole.[113] Silyl-substituted thioimidates have been examined[114] as nitrile ylide equivalents.

PE spectra of nitrile imines ($RC{\equiv}\overset{+}{N}{-}\overset{-}{N}{-}SiMe_3$), generated by flash vacuum pyrolysis of the corresponding tetrazoles, and assignment of ionization patterns establish the frontier orbital sequence and allow rationalization of regioselectivity of their cycloaddition reactions.[115] Formation of 5-alkoxypyrazoles from hydrazonoyl chlorides and triphenylphosphoranes in the presence of Et_3N can be explained[116] by a stepwise mechanism involving addition of nitrile imines to the Wittig reagent. Cinnamonitrile imines react with thio- and seleno-cyanate anions to give thia- and selena-diazolines.[117] 1:2-Adducts are formed with thieno[2,3-*c*]-pyridine by sequential reaction at the C=N and enaminic double bonds.[118] The regioselectivity of the reaction of benzonitrile *N*-phenylimide with 1,2-dihydronaphthalenes shows marked substituent dependence.[119] With alkyl isocyanides, 1-alkyl-2,3-diaryl-1,2,3-triazolium salts are formed presumably by cycloaddition and proton transfer.[120] Metal-carbonyl complexed 3,5-diphenyl-λ^3-

phosphinin (i.e. 3,5-diphenylphosphabenzene) react with nitrilium betaines at the C=P unit.[121]

MC-SCF gradient calculations and diabatic surface analysis for 'allowed' and 'forbidden' mechanisms for cycloadditions of fulminic acid to ethylene and acetylene indicate[122,123] a concerted rather than a diradical process. On the other hand, Sustmann *et al.* have used[124,125] the recently developed all-valence electron perturbation programme PERVAL based on MINDO/3 approximations to examine the transition states for cycloadditions involving HCNO and CH_2N_2. The former exhibits type II behaviour and it is concluded that regioselectivity can be explained better by non-covalent repulsion than by frontier orbital stabilizations; discussion in terms of the magnitude of eigenvector coefficients of the frontier orbitals does not give satisfactory results. Diazomethane shows type I behaviour, and reactivity can be explained by charge-transfer interactions with non-covalent interactions accounting for regioselectivity. Benzonitrile oxide and phenylimide react at both C=C and C=N groups in norbornene-fused dihydrooxazines.[126,127] A concerted mechanism is favoured[128] for the reaction of the same nitrile oxide with benzylideneanilines. The *exo*/*endo* ratio for the addition of CF_3CNO to norbornadienes is dependent on the electronic nature of substituents at the 2- and 3-positions,[129] electron-withdrawing groups increasing the proportion of *exo*-products. Nitrile oxides add exclusively to the less hindered face of 5-alkoxy-2(5*H*)-furanones[130] and to N(1)=C(7) in 5-methoxy-6*H*-1,4-diazepines;[131] 4-spirocyclopropylisoxazolines are the major products from methylenecyclopropanes.[132] Initial formation of dioxazolone (**27**) may explain[133] the isolation of oxazolone (**28**) from arenenitrile oxides and dimethylketene. Selectivity in the reaction of mesitonitrile oxide to the methylene group of (**29**) can be rationalized[134] on the basis of the stereochemical influences of the aziridine ring. Further examples of cycloaddition to seleno-aldehydes and -ketones have been described.[135,136] Keto or spiro groups at the 5-position have little influence on regioselectivity.[137] Various nitrile oxides add to chiral acrylates with π-facial stereoselectivity in accord with preferred *s-cis* conformation of the dipolarophile.[138] 3-Arylisothiazoles are formed from arenenitrile sulphides via a 1,3-dipolar cycloaddition/retro-Diels–Alder pathway.[139]

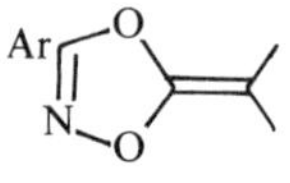

(**27**)

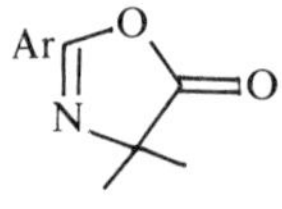

(**28**)

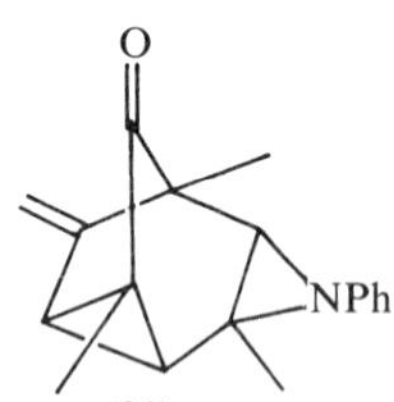

(**29**)

Considerable effort[140–152] has been devoted to the study of intramolecular 1,3-dipolar cycloadditions, particularly the stereochemical aspects of these reactions. Nitrile oxides, generated by conjugate addition of isocyanides to nitroalkenes, have been trapped by intramolecular alkene cycloaddition.[140] For nitrile oxides linked to the nitrogen of 2-vinylazetidines, stereoselectivity depends on the size of the newly formed ring.[141–143] Stereoselectivities for intramolecular nitrile oxide cycloadditions to chiral *Z*- and *E*-alkenes have been examined[144,145] in detail and models proposed on the basis of experiment and calculations (MM2): for *Z*-alkenes, the smallest substituent prefers to be inside, the medium-sized *anti*, and the largest outside, and the factors controlling selectivity are mainly steric; in contrast for *E*-alkenes, the small group is *outside*, the medium-sized inside, and the large *anti*. For the corresponding reaction[146] with chiral (*Z*)-allyl ethers, repulsive interaction between the oxygens of the incoming nitrile oxide and the alkoxy group results in transition state (**30**) being preferred to (**31**).

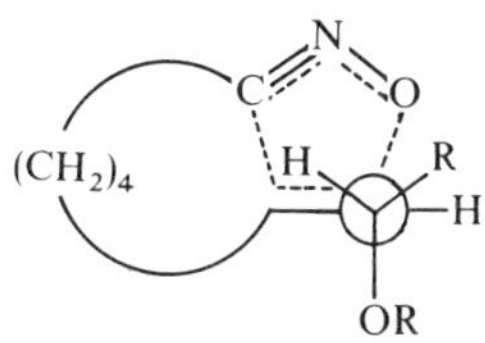

(**30**)

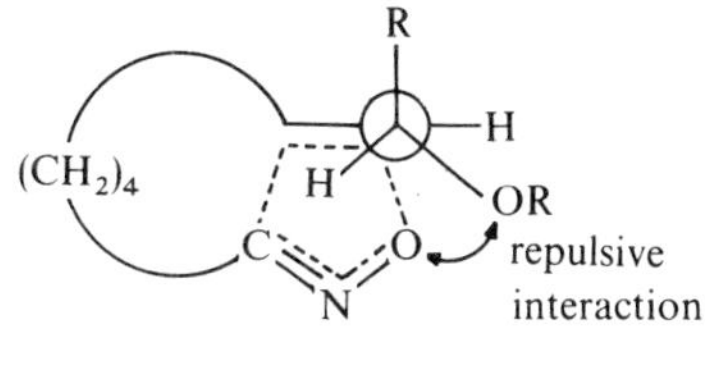

(**31**)

A fuller account has been published[149] of the formation of pyrroloindole (**32**) from azide (**33**); the reaction pathway involves intramolecular cycloaddition followed by cycloreversion with expulsion of diethyl diazomalonate (Scheme 5). Alkyl azides add to nitrilium ions yielding tetrazolium salts.[153]

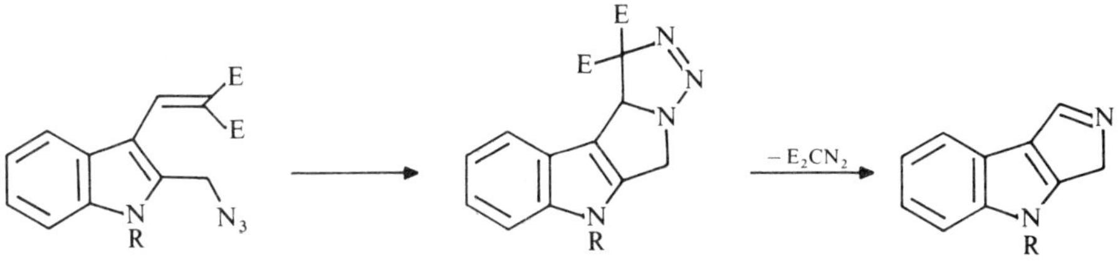

(**33**) $(E = CO_2Et)$ (**32**)

SCHEME 5

Non-frontier orbital charge-transfer interactions have been investigated[154] for the cycloaddition of diazoalkene to numerous mono- and di-substituted alkenes; they are believed to play a key rôle in determining both reaction rate and regiochemistry. The preference for 1-arylsulphonyl-2-trimethylsilylnorbornadiene to react with Me_2CN_2 at the unsubstituted double bond can be attributed[155] to steric rather than electronic factors. Whereas 3,4-dichlorocyclobutene reacts with diazoalkanes to give *syn*-adducts, 5,6-heteroatom-substituted bicyclo[2.2.0]hex-2-enes afford *anti*-products;[156] this effect is attributed, in part at least, to destabilizing influences of the heteroatoms and some pyramidalization of the olefinic carbon atoms. Contrary to previous reports, diazomethane *does* add to the alkene units of TCNE and ethenetetracarboxylates;[157] with dimethyl 2,3-dicyanofumarate, an array of pyrazoles and triazoles are formed,[158] all derived from an initial pyrazoline 1,3-dipolar cycloadduct. Diazoalkanes also react readily with 2,2-

difluoromethylenecyclopropane,[159] 5-methoxy-3-pyrrolin-2-one,[160] and the C(4)=C(5) link in isoxazoles[161] when activated by electron-withdrawing groups. The formation of annelated phosphiranes from Ph_2CN_2 and 1,2,3-diazaphospholes is believed[162] to involve formation of an initial 3 + 2-adduct in which the phosphorus is linked to the 1,3-dipole. Improved levels of asymmetric induction for the addition of diazofluorene to chiral acrylate and fumarate esters have been achieved[163] by using 8-phenylmenthyl as chiral auxiliary. The effect of solvent on the reaction of ethyl diazoacetate with 3-substituted propynals has been examined.[164] Diazomethylenephosphoranes ($R_3P{=}C{=}N_2$) undergo 3 + 2-cycloaddition with a variety of dipolarophiles.[165]

Various new dipolarophilic systems have been investigated. X-ray analysis of the monoadduct of *E,E*-cycloocta-1,5-diene with methyl *R*-2-azidopropanoate has provided evidence for a twist conformation for this diene.[166] The phosphaalkyne $Bu^tC{\equiv}P$ reacts readily[167–169] with diazoalkanes, azides, nitrile sulphides, and sydnones. The C≡N of aryl thiocyanates and selenocyanates can act as dipolarophile towards nitrilium betaines.[170] Various 2 + 2-, 2 + 3-, and ene-reactions of silanimines and germanimines ($Me_2X{=}NR$; X = Si, Ge) have been described.[171]

Grigg[172–176,180,181,183–185] and others[177–179,182] continue to explore the chemistry of prototropic 1,3-dipoles and the subject has been reviewed.[97] Further examples of ester-stabilized azomethine ylides, generated by decarboxylative condensation of α-amino-acids with carbonyl compounds have been described.[172–179] Ylide (**34**), formed from proline, yields ring-expanded products[172] with dimethyl acetylenedicarboxylate (DMAD) or methyl propiolate via a cycloaddition/Michael addition/rearrangement pathway (Scheme 6). Inter- and intra-molecular variants have been described for tandem addition-cycloaddition of oximes with Michael acceptors and dipolarophiles,[180,181] and also for the $3^+ + 2$-reaction of protonated hydrazones and alkenes.[182] Sulphinylaminomethanide, a novel 4π-electron species, is formed by sulphonyl transfer from *N*-sulphonylanilines to α-amino-acid esters and has been trapped by maleimides.[185]

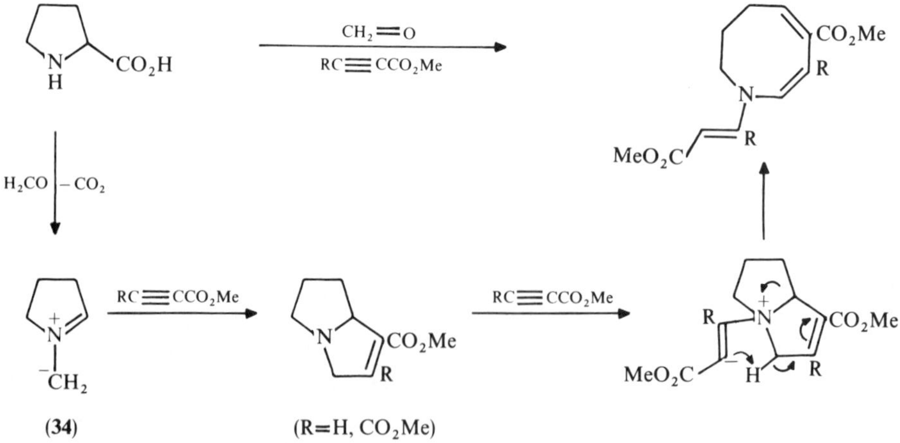

SCHEME 6

Pyridinium dicyanomethylide reacts with phenylsulphinylethene yielding, after extrusion of PhSOH and HCN, 3-cyanoindolizine;[186] with propiolate esters, a HOMO(ylide)-LUMO(dipolarophile) controlled process also yields indolizines.[187] Enhanced reactivity and selectivity are observed[188] for *N*-lithiated azomethine ylides. The adducts between olefins and pyridinium ylides bearing an anion-stabilizing substituent on the ylide carbon readily undergo alternative cycloreversion in the presence of silica gel forming alkylidenealkenes.[189] The formation of pyrrolidines from azomethine ylides, generated by base-induced elimination of PhSH from (phenylthiomethyl)-α-amino-acid esters, has been reported in fuller detail.[190] The stereochemistry of the reaction with nitroolefins has also been investigated.[191] 1,3-Dipolar character has been observed for pyrazolo[1,2-*a*]indazoles[192] and 1,5-dimethyl-3-oxidopyrazinium.[193] A note of caution has been raised[194] regarding the generation of azomethine ylides by thermolysis of 2-acylaziridines; the formation of rearranged product (**35**) from aziridine (**36**), rather than the previously reported azomethine ylide adduct (**37**), shows that the 1,3-dipole is not the only reactive species present (Scheme 7). Nitrosobenzene reacts with mesoionic Δ^2-oxazolin-5-ones to afford Δ^4-1,2,4-oxadiazoline-3-carboxylic acids.[195]

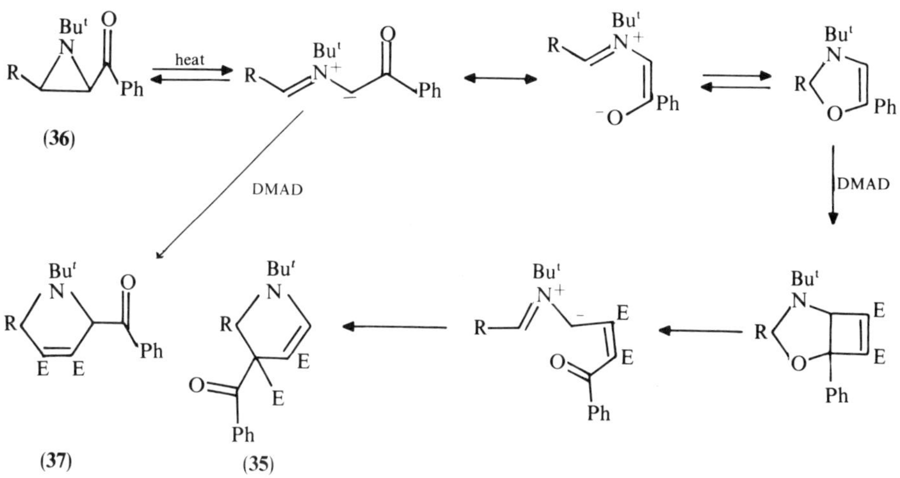

($E = CO_2Me$)

SCHEME 7

3-Hydroxy-5-methoxypyridinium *N*-imide can act as both azomethine imine and ylide.[196] FMO theory has been used[197] to explain the preferred formation of diazepin-5-ones from sydnones and isopropylidenecyclobutenone. A mechanism (Scheme 8) involving an initial nitrone-like cycloaddition to the $C{=}\overset{+}{N}{-}\overset{-}{O}$ unit of 1,2,3-triazole *N*-oxides has been postulated[198] to explain the formation of triazines on treatment with DMAD. An analagous cycloaddition ring-expansion pathway also accounts for the formation of 1,3,4,5-oxatriazines from 1,2,3-triazolium 1-imides and cinnamaldehyde.[199]

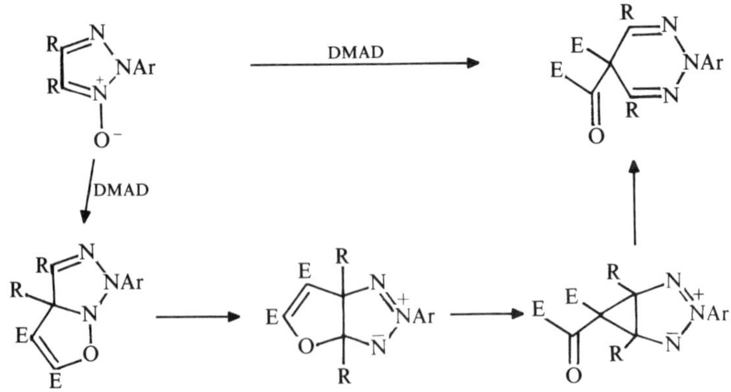

(R = H, CO_2Me)

SCHEME 8

Cycloaddition of cyclic nitrones to *E*- and *Z*-*β*-nitrostyrene has been examined[200] in detail; the latter reacts faster than the former with 3,4-dihydroisoquinoline *N*-oxide, and this is believed to be the first case of higher reactivity for a *cis*-rather than a *trans*-alkene in 1,3-dipolar cycloadditions. With styrene itself, tetrahydropyridine *N*-oxide shows very high stereoselectivity.[201] High levels of regio- and stereochemical control are also observed[202,203] for cycloadditions involving dihydropyrroline *N*-oxides 3,4-fused to cyclopropane and cyclobutane. The rôle of steric, electrostatic, and H-bonding effects as factors controlling face selectivity in reactions with norbornenes has also been investigated;[204] *syn*-attack is promoted by H-bonding in reactions of *endo*-5-hydroxy derivatives, whereas repulsive steric and electrostatic interactions account for the almost exclusive *anti*-selectivity observed for other substituents. Further kinetic and theoretical studies[205] of cycloadditions involving pyridine *N*-oxides have been carried out; charge-transfer complexes have been detected[206] in their reactions with phenyl isocyanate. *C*,*N*-Diphenylnitrone reacts with $PhSO_2C{\equiv}CR$ and $PhSO_2CH{=}C{=}CH_2$ to yield transient cycloadducts which rearrange to indoles with N—O bond scission.[207] With ethenesulphonic acid derivatives, 4-substituted isoxazolidines predominate in accord with FMO predictions.[206] Various nitrones react regiospecifically with geminal dinitriles,[209] a rare case of cycloaddition to C≡N. *C*-Phenyl-*N*-alkylnitrones add to electron-poor allenes in accord with FMO predictions yielding 5-methyleneisoxazolidines.[210] With fluoroallene, reaction takes place exclusively at the methylene (rather than fluoromethylene) unit and with a strong preference for attack *syn* to the fluorine.[211] The effect of allylic heteroatoms on regioselectivity[212] and the stereochemical course of intramolecular cycloaddition to chiral allyl ethers[152] have been investigated. In the presence of CsF, the benzaldoxime-derived iminium salt (**38**) can act as nitrone or azomethine ylide equivalents depending on the reactivity of the dipolarophile;[213] the proposed mechanism (Scheme 9) involves initial deprotonation to afford nitrone (**39**), which is either trapped by electron-deficient dipolarophiles, or rearranges via a 1,3-silicon shift to ylide (**40**) in the presence of less reactive dipolarophiles.

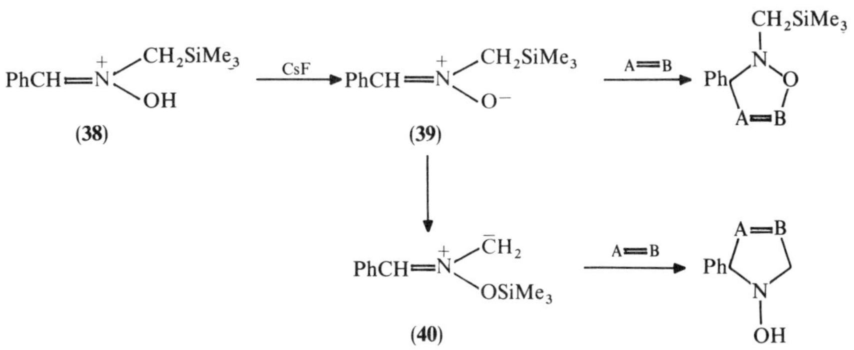

SCHEME 9

Moeller–Plesset perturbation theory has been used to study[214] the effect of correlation energy on mechanism of the parent carbonyl ylide-ethene reaction; its introduction appears to stabilize preferentially asymmetric structures. Further examples have been reported[215] of *endo*-selectivity for cycloaddition to alkenes of carbonyl ylides formed by photolysis of 2,3-epoxy-2,3-dihydronaphthoquinones. Acetone *O*-methylide, formed by photolysis of diazomethane in acetone, has been trapped by electron-deficient alkenes.[216] Kinetic studies[217] for the corresponding chloro-*p*-nitrophenylcarbene reaction show that formation of the 1,3-dipole is not reversible. 'Push-pull' carbonyl ylides of type $ROC_6H_4CH{=}\overset{+}{O}{-}\overset{-}{C}(CN)_2C_6H_4NO_2$, generated by thermal or photochemical ring-opening of the corresponding oxirane, show remarkably little regioselectivity towards unsymmetrical dipolarophiles.[218] 2-Benzopyrylium-4-olate, formed by thermal or base-catalysed elimination of acetic acid from 1-acetoxy-1*H*-2-benzopyran-4(3*H*)-one, undergoes cycloaddition reactions with a wide range of dipolarophiles with regiochemistry in accord with FMO predictions.[219] Formation of tetrahydrofuran carboxylates as well as oxetane 2 + 2-adducts from the photoreaction of benzophenones with haloketene acetals can be explained[220] in terms of 1,4-diradical/carbonyl ylide intermediates. The first example of an intramolecular ester-carbonyl oxide cycloaddition has been claimed.[150]

Thioketone *S*-methylides, generated from thiones and (trimethylsilyl)methyl triflate, have been trapped[221] by cycloaddition with thiones. Thioformylium methylide readily adds to carbonyl compounds affording 1,3-oxathiolanes,[222], apparently the first case of a thiocarbonyl ylide reacting with C=O. Mesoionic thiazolium- and thiolium-olates continue to attract attention;[223–225] 1,3λ^3-thiaphospholes are formed[225] from 1,4-dithiolium-4-olates and $ClP{=}CPhSiMe_3$, by elimination of COS and $ClSiMe_3$ from the initial 1,3-dipolar cycloadduct. The first examples of thioketene *S*-methylides have been generated[226] by release of disiloxane from $R_2C{=}C(SiMe_3)SOCH_2SiMe_3$ and their cycloaddition reactions investigated.

Although thiosulphines ($R_2C{=}S{=}S$) are mentioned frequently in the literature as transient intermediates, evidence for their existence is scanty. Recently, however, Huisgen *et al.*, regarding them as thiocarbonyl *S*-sulphides ($R_2C{=}\overset{+}{S}{-}\overset{-}{S}$) and thus a new class of 1,3-dipole, have reinvestigated[227] their formation. They can be generated by thermal cycloreversion of trithiolanes (**41**) or by sulphur-atom transfer to $R_2C{=}S$ from thiiranes and can be trapped by DMAD as thiole cycloadducts (Scheme 10).

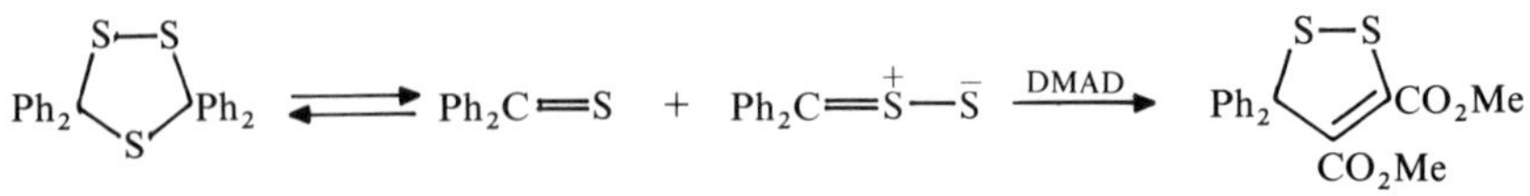

(**41**)

SCHEME 10

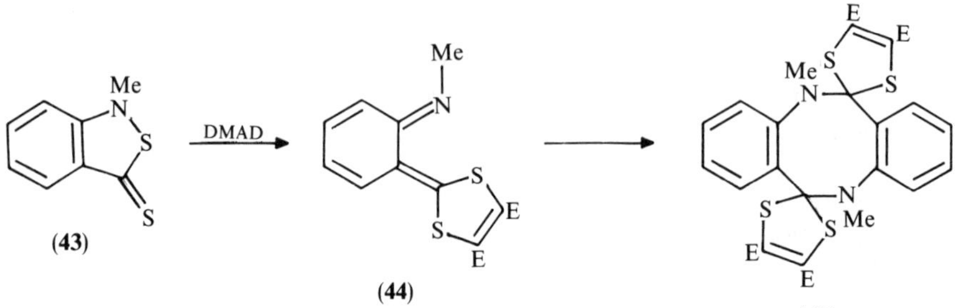

(**43**) (**44**) (**42**)

($E = CO_2Me$)

SCHEME 11

Nitroso oxides ($RN{=}\overset{+}{O}{-}\overset{-}{O} \leftrightarrow R\dot{N}{-}O{-}\dot{O}$), generated from nitrenes and oxygen, behave as electrophilic peroxy radicals rather than a 1,3-dipole, and isomerize unimolecularly to nitro-compounds.[228] Formation of diazocine (**42**) from benzoisothiazolinethione (**43**) and DMAD is explained[229] by a pathway (Scheme 11) involving dimerization of initial 1:1 adduct (**44**). Two distinct intermediates are responsible for the oxygenated and 3 + 2-products on irradiation of electron donor-acceptor complexes of 2,3-diarylmethylene-cyclopropanes, -oxiranes, and -aziridines with $TCNE/O_2$;[230] oxygenation occurs through solvent-separated radical cations diffused from photogenerated radical ion-pairs, whereas cage coupling is involved in the 3 + 2-process. For the reaction of TCNE with 2-methoxy-3-aryloxazoles, 3 + 2-adducts are formed via a zwitterionic mechanism.[231] Formation of 10π-electron trithiadiazepines from sulphur tetranitride and reactive alkynes ($RC{\equiv}CR$; $R = CN, CF_3$) can be explained[232–235] by a sequence (Scheme 12) involving 1,3-cycloaddition of the alkyne across nitrogen and 1,5-addition at sulphur to give 1:2 adducts, followed by cycloreversion. Theoretical calculations (MNDO and *ab initio*) performed[235] on this system indicate that 2:1 adduct formation is largely controlled by interaction between the alkyne LUMO and either the occupied $3a_2$ orbital of S_4N_4 favouring 2,4-addition across nitrogen, or the occupied $4b_1$ orbital resulting in 1,5-addition at sulphur; it is concluded that initial 1,5-attack is followed by 2,4-addition with elimination of the thiadiazole.

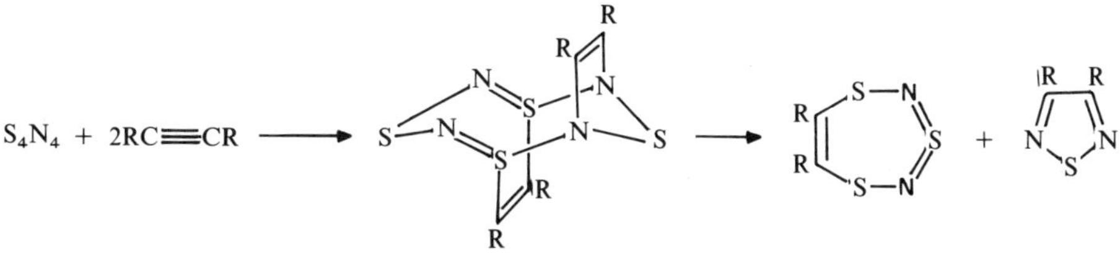

SCHEME 12

2 + 4-Cycloaddition

There have been several reviews dealing with Diels–Alder and related reactions.[9] Topics covered include cycloadditions of *ortho*-quinodimethanes,[9] 1,3-oxazin-6-ones[236] and 1,3,4-oxadiazin-6-ones,[237] and intra- and retro-Diels–Alder reactions including their application in natural-product synthesis.[238–240]

The relative importance of steric effects and secondary orbital overlap in Diels–Alder reactions has been investigated[241] with MNDO and AM 1 semi-empirical calculations using fulvene with cyclopentadiene and maleic anhydride as model systems; it is concluded that steric effects are at least as significant as secondary orbital interactions in controlling stereochemistry. Secondary deuterium kinetic isotope effects measured for the reaction of isoprene with various electron-deficient dienophiles indicate an almost synchronous, concerted pathway with an early transition state.[242] It has been noted[243] that the presence of a methyl group at the 2-position accelerates Diels–Alder reactions of 4-methylpenta-1,3-dienes; the effect is attributed to unfavourable steric interactions in the *s-cis*-conformation (**45**) being counterbalanced by interaction between the 1- and 3-methyl substituents in the *s-trans*-arrangement (**46**).

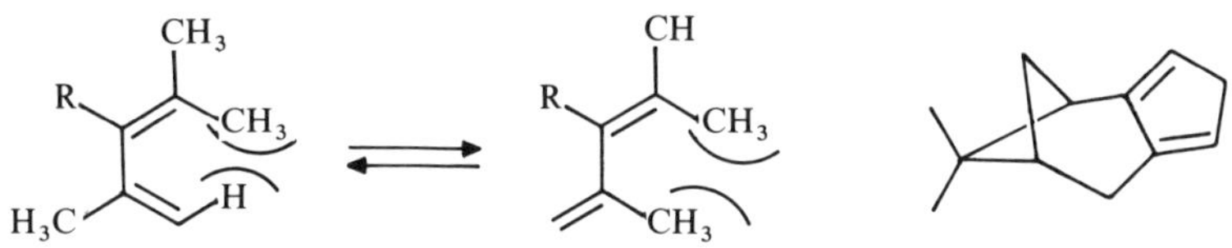

(45) **(46)** **(47)**

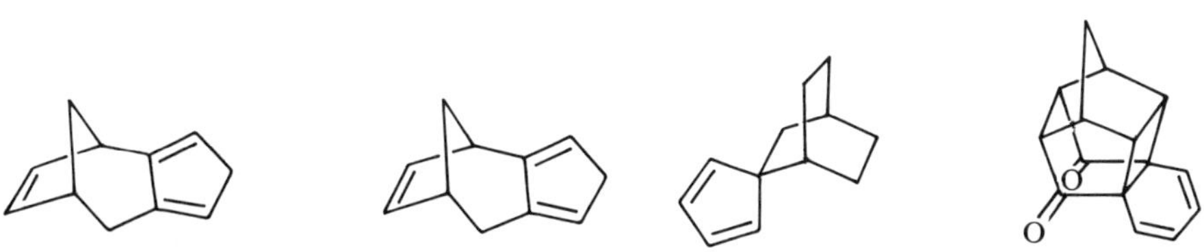

(48) **(49)** **(50)** **(51)**

Houk *et al.* have developed[244] their MM2 model approach to predict the stereochemical course of cycloadditions to unsymmetrically substituted cyclopentadienes and reasonable agreement with experimental results was obtained for pentamethyl and spiro-fused derivatives; the higher selectivity found for Lewis acid-catalysed reactions was attributed to an earlier rather than a tighter transition state. Studies of π-facial selectivity of cyclic dienes have been extended to include pinene-fused compound (**47**)[245] and homodehydroisocyclopentadiene derivatives (**48**)[246] and (**49**); in each case attack takes place at the less-hindered *exo*-face. In contrast little selectivity is observed[247] for spiro-compound (**50**). The rôles played by charge-transfer interactions, steric and polar effects, and orbital interactions have been assessed for 2,3-dimethylenenorbornane.[248,249] Control of stereo- and regio-selectivity by remote substituents in tetramethylenenorbornanes and related systems has been refined further,[250,251] and the 7-oxa-analogue has been utilized[252] for the synthesis of [12]cyclacene derivatives. Olefins add to hexacyclic diene (**51**) exclusively at the carbonyl-bearing face, but other dienophiles including benzyne, alkynes, and azo-compounds are less selective.[253] Various 1,3-dienic terpenes react with PTAD at the less-hindered face,[254] paralleling the behaviour of α-pinene but opposite to that of norbornane and [2.2.2]bicyclooctane analogues.

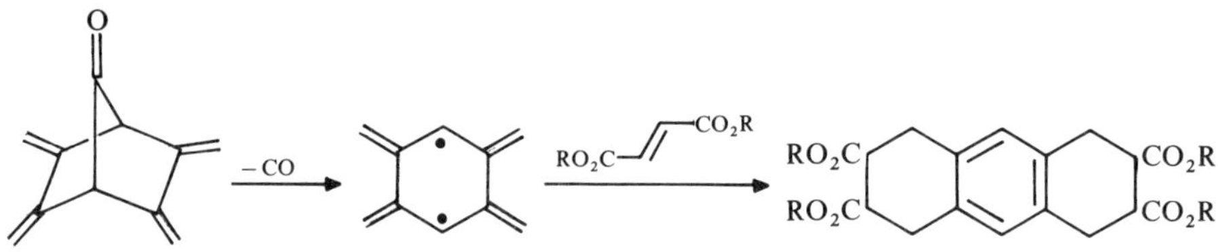

(**53**) (**52**)

SCHEME 13

Tetramethylene-1,4-cyclohexdiyl (**52**), a disjoint diradical with conjugated non-Kekulé structure, generated by extrusion of CO from ketone (**53**), forms 4 + 2-adducts with fumarate esters at both diene units (Scheme 13).[255] 2,3-Dimethylene-2,3-dihydrothiophene, the thiophene analogue of *ortho*-xylylene, has been generated and trapped as Diels–Alder adducts.[256] The corresponding 3,4-dimethylene-furan and -thiophene analogues, formed by thermal or photochemical expulsion of N_2 from the appropriate 1,2-diazines, can also be trapped by electron-deficient alkenes or oxygen;[257] in each case the intermediate diradical is believed to be in the singlet state.

A more detailed investigation[258] has been carried out of the synthesis and reactivity of 2-trialkylstannylbutadiene, and the 2-phenylseleno-analogue has been studied for the first time. 2-Phenylsulphonylbutadienes form 4 + 2-adducts with both electron-rich and -poor dienophiles.[259] Other acyclic diene systems examined include methyl 4-trimethylsilyloxy-2,4-pentadienoate,[260] dienylboronates,[261] 1-thiocyanato-[262] and 2,3-diazido-butadienes.[263] Cycloadditions involving alkenyl-substituted indoles,[264–267] pyrazoles,[268] and thiophenes[269,270] have also been reported.

Tricyclononatriene (**54**), a norcaradiene system influenced by antiaromaticity of the cyclobutadiene unit inherent in its cycloheptatriene tautomer (**55**), reacts with

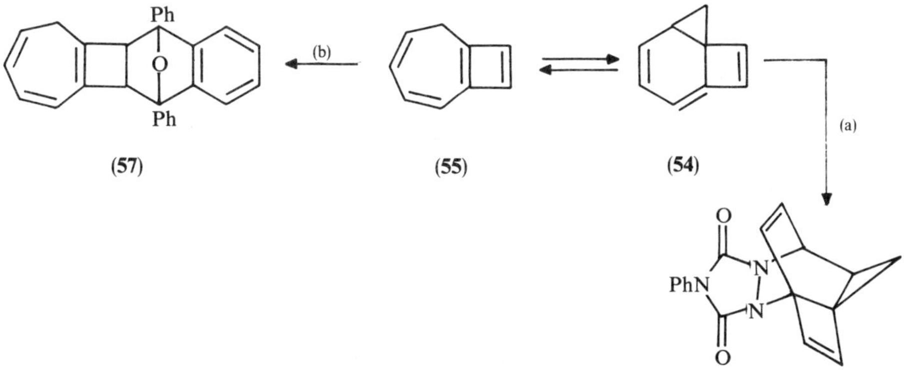

(57) (55) (54) (56)

(a) = PTAD; (b) = diphenylisobenzofuran

SCHEME 14

PTAD to give (**56**), whereas (**57**) is formed with an isobenzofuran presumably via tautomer (**55**) (Scheme 14).[271] Cycloheptatrienes fused to furanones react with PTAD to give 4 + 2- and 6 + 2-cycloadducts rather than norcaradiene-derived products.[272] Methyl tri-*t*-butylcyclobutadienecarboxylate, a kinetically stabilized cyclobutadiene, reacts with monoacceptor-substituted alkynes to afford regio-isomeric mixtures of benzene, Dewar-benzene, and prismanes;[273] its cycloaddition to benzophenone and fluorenone has also been reported.[274] A fuller account[275,276] has been published of the formation of bicyclo[2.2.2]oct-2-en-5-ones from phenols and *N*-arylmaleimides; the reaction is believed to involve cycloaddition to phenol itself rather than its keto-tautomer. Kinetic studies have shown[277] that the Diels–Alder reactions of dimethyl 3,4-phenylcyclopentadienone-2,5-dicarboxylate with cyanoethylenes are of diene donor-dienophile acceptor type, although those with *N*-arylmaleimides are close to neutral.

Ab initio MO calculations have been used[278] to evaluate the rôle of the Lewis acid in BF_3-catalysed reactions of acrolein. A Brønsted linear-free-energy relationship has been identified[279] for the acid-catalysed addition of diethyl maleate to cyclopentadiene. The large acceleration ($\times$ 10^4–10^5) of the reaction between maleic anhydride and anthracenes in the presence of $GaCl_3$ or $AlCl_3$ is attributed[280] to stabilization of the dienophile LUMO. The effect of varying acid strength[281] and the relationship between ΔH_f of *n,v*-complexes and the accelerating effect of Lewis acids[282] have been investigated. CdS-Photomediated dimerization of 1,1-dianisylethylene affords a mixture of 4 + 2-adducts.[283] The rate increases ($\times$ 70) and control of *exo*/*endo* ratio observed for β-cyclodextrin-catalysed reaction of cyclopentadiene with fumarate and maleate esters can be attributed[284,285] to ternary complex formation. Ultracentrifugation accelerates the reaction between maleic anhydride and furan,[286] the increase in rate depending on the centrifugal pressure exerted. The Diels–Alder adduct of 2-methylenecyclopentane-1,3-dione with anthracene readily undergoes acid-catalysed cycloreversion.[287] Further mechanistic studies involving hexachlorocyclopentadiene and related systems have been reported.[288–290] The greater reactivity towards PTAD of 5,5-dimethoxytetrabromocyclopentadiene compared with the hexabromo analogue is attributed[288] to a larger donor-acceptor interaction and smaller π-delocalization energies. 3,4-Dichlorothiophene *S,S*-dioxide readily undergoes inverse-electron-demand cycloaddition to the 5,6-double bond of 4-(2-oxoalkyl)-1,4-dihydronicotinates.[291] Maleic anhydride adds sequentially to pyran-2-ones yielding after spontaneous decarboxylation *syn,syn*-2:1 adducts;[292] with electron-poor alkynes the expected arenes are formed.[293] Competition experiments performed with a series of 1,3-substituted isobenzofurans indicate[294] that the 2-ethoxy derivative is the most reactive diene yet isolated. Four of the six possible isomeric dihydropentalenes have been trapped[295] using TCNE. 1,3-Disulphonyl-2-pyridones add regioselectively to vinyl ethers yielding unsaturated bridged bicyclic lactams.[296] The cyclopentadiene moiety of azulene acts as a diene component towards benzyne.[297]

Periselectivity is observed[298] for cycloadditions of vinylmethylketene and the isoelectronic vinylketenimines, the isomer distribution depending on the nature of the dienophile; with very electrophilic partners (e.g. TCNE, PTAD) only 4 + 2-adducts involving the C=C of the cumulene and the vinyl group are observed, whereas several 2 + 2- and/or 4 + 2-products are formed with less electrophilic or electron-rich substrates (e.g. thioketones, vinyl ethers). The reaction between 2-benzenesulphonyl-3-trimethylsilylnorbornadiene is unexpectedly selective, only one of eight possible products being formed;[299] *exo,anti*-addition takes place at the unsubstituted 5,6-double bond. Norbornadiene-2,3-dicarboxylates and their cyclopentadienylnickel complexes also react with tetrachlorodimethoxycyclopentadiene at the less hindered *exo*-face of the dienophile.[300]

X-Ray and NMR evidence for the Diels–Alder adducts from several cyclopropenes has necessitated reinterpretation of earlier results; in some cases the intrinsic *endo*-preference can be overridden by steric factors.[301] Formation of 3,4,5-trichlorotropone from tetrachlorocyclopropene and Danishefsky's diene can be rationalized[302] in terms of an initial 2 + 4-cycloaddition followed by elimination of Me_3SiOMe and ring-expansion with loss of HCl. The reactivities of 2-methyleneimidazolines[303,304] and cyclic ketone *N,O*-acetals[305,306] in inverse-electron-demand Diels–Alder reactions have been investigated. From kinetic studies of the gas-phase reaction of acrylonitrile, methacrylonitrile, and crotononitrile with cyclohexa-1,3-diene the accelerating effect of CN vs Me substituents in the dienophile fragment has been estimated[307] as *ca* 4 kcal mol^{-1}. Intermolecular perturbation theory has been used[308] to predict the regioselectivity of cycloaddition to quinones. The reactivity and regioselectivity of 1,4-naphthoquinones in increased by PhS, PhSO, and $PhSO_2$ groups at the 2-position.[309] The influence of *gem*-dimethyl groups on cycloaddition to cyclohex-2-enones has been examined.[310,311] Secondary orbital interactions are regarded[312] as the key factor in determining *exo*/*endo* ratios for the addition of cyclopentadiene to *N*-acyl-α,β-dehydroalaninates, and 1-acetylvinyl arenecarboxylates have been studied[313] as captodative dienophiles; the regioselectivity observed for the latter can be accounted for solely on the basis of HOMO(diene)/LUMO(dienophile) interaction. Cyclopentadiene affords *trans*- and *cis-endo* adducts with enones (**58**; X = O, NCO_2Et), whereas isobenzofurans yield only the thermodynamically-controlled *exo*-compounds, presumably via cycloreversion of the initial kinetic product.[314] The nitro group has been proposed[315] as a regiochemical control element for Diels–Alder cycloadditions; its presence activates olefinic dienophiles, enhances regioselectivity, and it can be removed from the adduct by treatment with Bu_3SnH. β-Phenylsulphinyl-nitroalkenes[316] act as nitroacetylene equivalents. Propargyl cations[317] and alkenylpyridinium salts[318] are highly reactive dienophiles; acrolein acetals[319] are active allyl cation equivalents. Some regioselectivity is observed[320] for the reaction of arynes and pyridynes with alkoxyisobenzofurans despite the low activation energies and early transition state. A single adduct (**59**) is obtained[321] from 1,4-benzadiyne equivalent (**60**) with hexamethylcyclohexa-2,4-dienone in the presence of $Pb(OAc)_4$. A fuller account has been presented[322] of the photoreaction of indene with cyclic dienes in the presence of 1,4-dicyanonaphthalene; the proposed mechanism involves combination of the diene with indene-dicyanoarene exciplex to form a ternary complex from which the products are formed. Transient alkenes trapped as Diels–Alder adducts include selenirene (**61**),[323] dicylopropa[*a*,*c*]naphthalene,[324] and 1-oxacyclohexa-3,4-diene.[325]

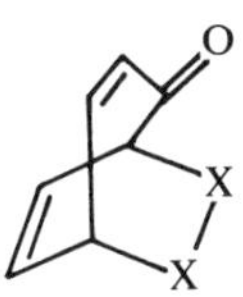

(**58**)

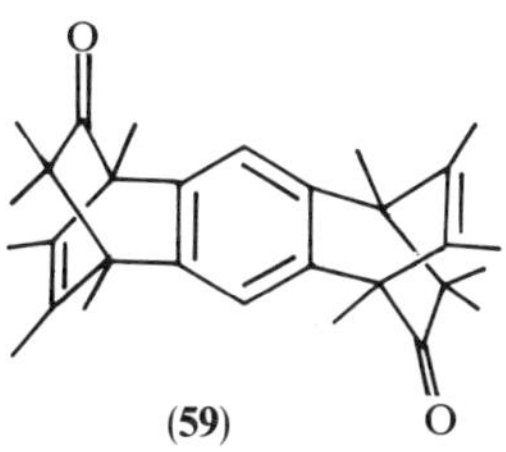

(**59**)

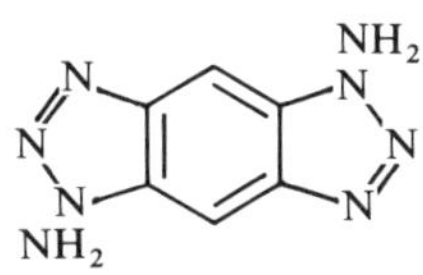

(**60**)

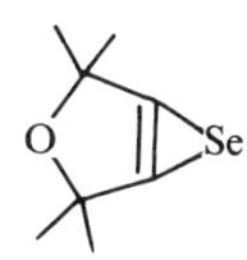

(**61**)

The increased interest in chiral Diels–Alder reactions has been maintained.[326–342] Results of modelling studies[326] involving chiral electron-rich dienes and electron-poor dienophiles are consistent with addition preferentially onto the diene face which is more reactive toward electrophiles and the nucleophilic face of the dienophile in accord with most available stereochemical evidence. MO calculations have been performed[327] on acrolein, acrylates, and on their complexes with Lewis acids as an aid to rationalizing observed stereoselectivities. Thermal additions of cyclopentadiene to chiral acrylates, in contrast to the corresponding catalysed reactions, show slight preference for transition states involving an *s-cis*-conformation.[138] High and opposite stereoselectivity is found[328] for Lewis-acid and uncatalysed cycloadditions of the fumarate of ethyl (*S*)-lactate (i.e. *trans*-$RO_2CCH{=}CHCO_2R$; R = $\overset{*}{C}HMeCO_2Et$). Further evidence, based on X-ray analysis and CD spectra, has been presented[329] in support of the earlier hypothesis that for chiral vinyl sulphoxides the most stable conformation (*s-trans* or *s-cis*) dictates diastereoselectivity. Asymmetric induction in excess of 97% is observed[330] for the photoaddition of chiral α-pyrrolidinoacrylonitrile to electronically excited 1-acetylnaphthalene. Various carbohydrates have also been utilized[331–334] as chiral auxiliaries. The use of chiral catalysts to promote asymmetric Diels–Alder reactions is attracting increased attention.[335–342].

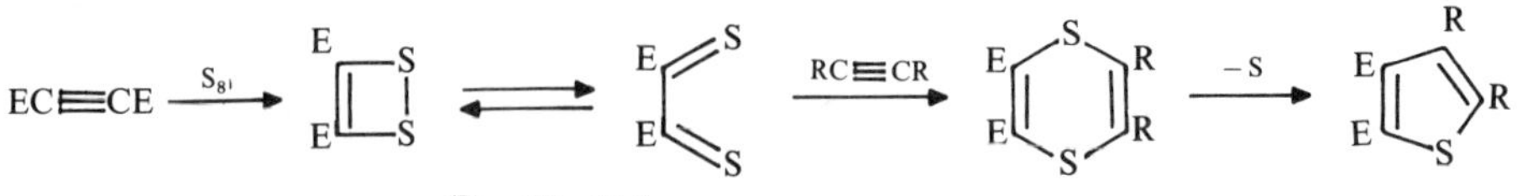

(E = CO_2Me)

SCHEME 15

A mechanism (Scheme 15) involving 1,2-dithione intermediates, which are trappable as Diels–Alder adducts, has been postulated[343] to account for the formation of dithiins and thiophenes on heating electron-poor alkynes with sulphur. A stepwise diradical mechanism accounts[344] for the *trans*-adducts from electron-rich alkenes and the *ortho*-quinoidal form of benzothiete. The periselectivity observed[345] in reactions of *N*-sulphinylaminoazines with 1,4-epoxy-1,4-dihydronaphthalenes has been rationalized from MO calculations. The recent availability of electron-deficient 2-azadienes has allowed their cycloaddition reactions to electron-rich alkenes to be explored.[346] Pentachloroazacyclopentadiene can act as a 2π-component at C=N yielding various cycloadducts.[347] Benzene and linearly annelated arenes can act as the dienophile component in inverse-electron-demand Diels–Alder reactions with 2,6-bis(trifluoromethyl)-1,2,4,5-tetrazine.[348] Steric factors are believed to account for the high regioselectivity found[349] for the reaction of tetrazines with cyclic ketene-*N*,*S*-acetals. The use of $SnCl_4$-catalysis to promote 4 + 2-cycloaddition of 1-nitroalkenes to unactivated alkenes has been extended[350] to include intermolecular examples; 1-nitrocyclohexene reacts with cycloalkenes to give predominantly *exo*-adducts. Glycals undergo highly diastereoselective 2 + 4-addition to the N=N—C=O unit of azodicarboxylates.[351] New heterodiene systems examined include the pyrrole portion of 2*H*-pyrrolo-[3,4-*b*]- and -[3,4-*c*]-pyridines[352] and 2,4,6-triaryl-1,3λ^3-azaphosphirines.[353] 1,2,2-Trichloronitrosoethene can react as heterodiene or heterodienophile depending on the co-reactant.[354]

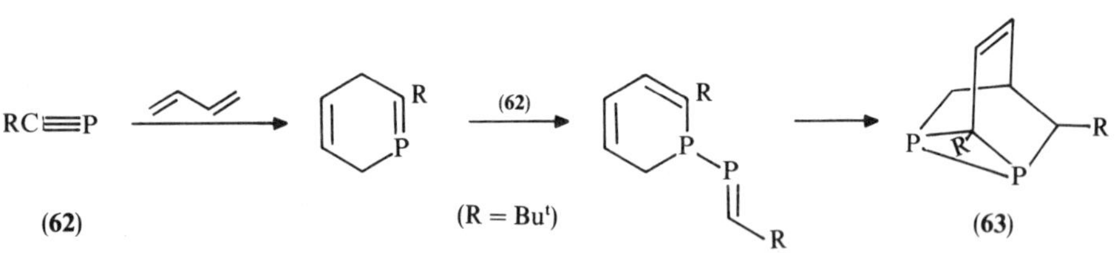

(62) (R = Bu^t) **(63)**

SCHEME 16

Whereas phosphaalkynes add to anthracene to give 2-phosphabarrelenes,[355] the reaction with acyclic dienes[356] is more complex; with butadiene and (**62**), a 2:1 adduct (**63**) is formed presumably by a pathway (Scheme 16) involving cycloaddition, ene, and intramolecular cycloaddition. Metal carbonyl-complexed 3,5-diphenyl-λ^3-phosphinines react with 1,3-dienes at the C=P unit.[121] Phosphaethenes $RP{=}CH_2$ (R = H, Me, Ph) have been trapped[357,358] as 2 + 4-adducts. The effect of pressure and catalyst on asymmetric induction for the addition of 1-methoxybutadiene to various sugar aldehydes has been studied.[359] Thermal cycloreversion of Diels–Alder adducts of $EtO_2CCH{=}S{=}O$ provides a convenient method of generation of the sulphine,[360] which may be trapped by various dienes. A fuller account has been presented[361] of trapping by cyclopentadiene of $CH_2{=}SO$ and $CH_2{=}S{=}O$. Silaethene $Me_2Si{=}C(SiMe_3)_2$ undergoes concerted 4 + 2-addition to *trans*-piperylene;[362] in contrast, the *cis*-isomer which cannot so readily achieve the *cisoid* conformation forms only a 2 + 2-adduct. Further examples of trapping of disulphur and various seleno-aldehydes and ketones have been reported.[135,136,363,364] The regiochemical control exerted by seleno-aldehydes is substituent dependent;[136] electron-withdrawing groups exhibit *ortho*/*para* selectivity, whereas electron-donating and conjugating substituents give rise to *meta-products*.

MINDO/3 calculations for ammonium ion-catalysed cycloaddition of acrolein to 1,3-dienes indicate[365] a two-step mechanism via a stable intermediate with total proton transfer to dienophile. A two-step pathway is also favoured[366] for the reaction of *N*-arylisoquinolinium salt with vinyl ethers. Protonation in conjunction with Lewis acid catalysis increases the reactivity of imines.[367] The $4^+ + 2$- and $4 + 2^+$-cycloadducts, (**64**) and (**65**), formed from protonated 4,4-dimethyl-4*H*-pyrazole and cyclopentadiene undergo proton-controlled equilibration; it has been established[368] that the process involves a [3.3] aza-Cope-type rearrangement rather than a cycloreversion-alternative cycloaddition sequence. Vinylnitrosonium cations, generated from α,β-epoxynitronates, act as reactive 4π-components in inter- and intra-molecular $4^+ + 2$-cycloadditions with unactivated alkenes.[369] Detailed studies[370,371] have been made of various aminium salt-initiated Diels–Alder reactions of diene/diene systems including dimerization and cross-addition; cyclodimerization of $CH_2{=}CMeCH{=}CMe_2$ is Brønsted acid-catalysed, but in every other case examined a cation-radical chain mechanism was confirmed. The first intramolecular examples have been reported;[372] triarylammonium ion-catalysed cyclization of deca-1,6,8-trienes affords *trans*-hydroindanes with high *endo*-selectivity. A mechanism involving the cation radical of the diene rather than dienophile is preferred[373] for electrochemically-induced cycloaddition of electron-rich dienes with electron-rich dienophiles.

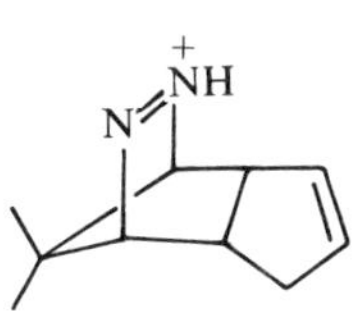

(**64**)

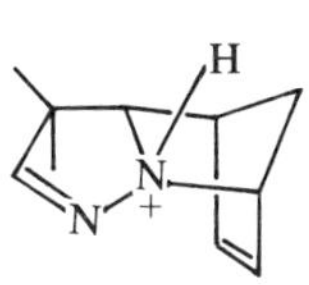

(**65**)

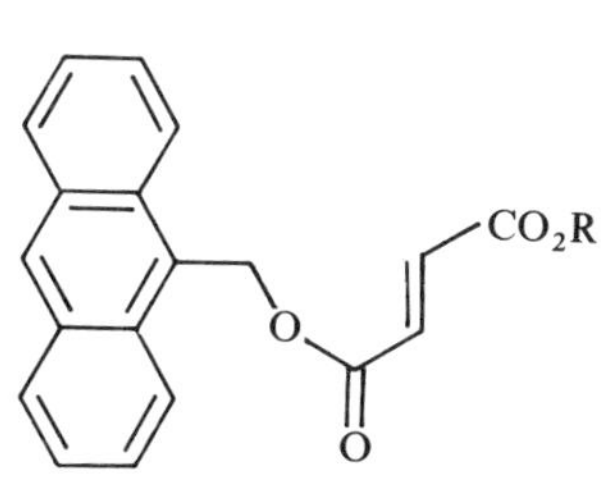

(**66**)

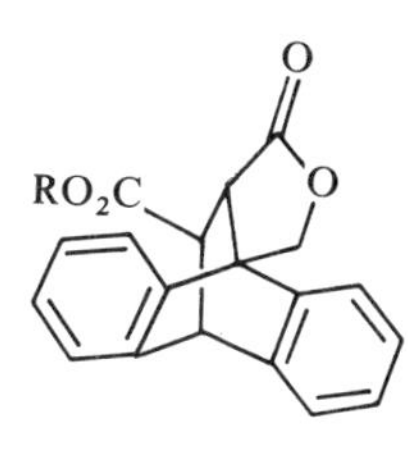

(**67**)

The literature dealing with intramolecular Diels–Alder reactions continues to grow, with an increasing proportion of the papers dealing with synthetic rather than mechanistic aspects. Fumarate esters (**66**) and the isomeric maleates on photolysis both afford 2 + 4-product (**67**), the latter arising from *cis*-to-*trans* isomerization of the dienophile prior to cyclization;[374] the cycloaddition is believed to proceed in the singlet state and the isomerization via the triplet. There has been further investigation[375–390] on the effect of substituents and the diene-dienophile link on the course of 1,3,8- and 1,3,9-alkatriene cyclizations. For trienes with the dienophile fragment terminally activated by electron-withdrawing groups (R), the extent of selectivity increases in the order R = $CONH_2 < CO_2Me < COMe < CHO$, consistent with predictions based on the previously postulated non-synchronous transition-state hypothesis.[375] Although conformational arguments have suggested that such intramolecular Diels–Alder cyclizations involve a chair-link transition state, triene (**68**) is converted into (**69**) by boat-conformation (**70**) (Scheme 17).[376] A series of transannular Diels–Alder reactions involving 13-membered macrocyclic trienes has been examined;[377–379] boat-like transition states are postulated for the cyclization of *cis,trans,trans*-compound (**71**) to (**72**)[378] and for *trans,cis,cis*-analogues.[377] Further examples of intramolecular allene-arene cycloadditions have

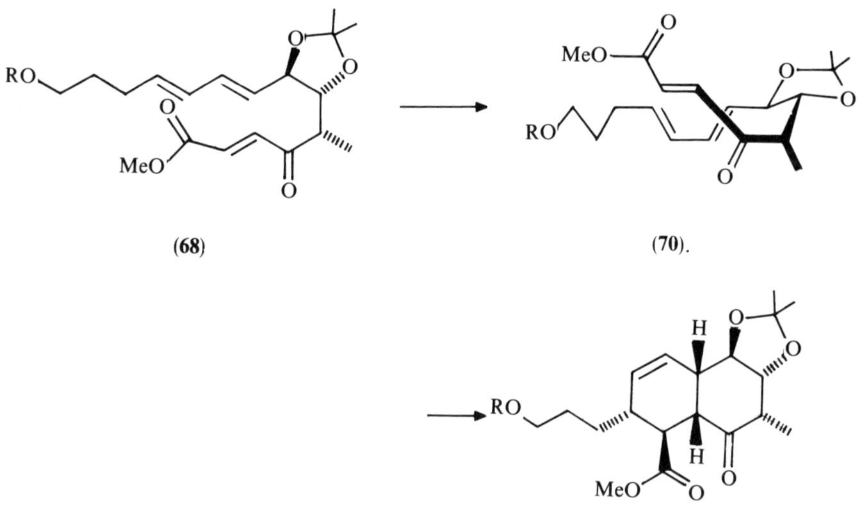

(**68**) (**70**) (**69**)

SCHEME 17

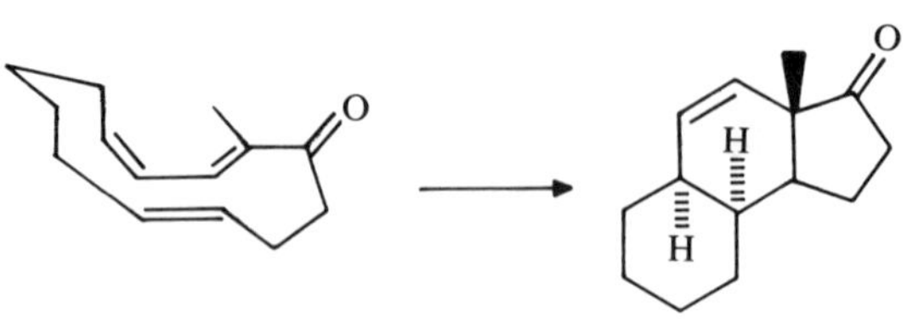

(**71**) (**72**)

been reported.[391,392] Intramolecular Diels–Alder reactions involving furan[393–398] and other 5-membered heterocyclic systems including isoxazoles[399] and thiazoles[400] have also been studied. Oxidation of 1,2-dihydro-3*H*-pyrazol-3-ones bearing a diene unit linked to C(5) affords cyclized product (**73**) via 3*H*-pyrazolone (**74**).[401] A more detailed study has been made[402,403] of the intramolecular cycloadditions of azoalkenes of structure $H_2C{=}CH(CH_2)_nCON{=}NCR{=}CHR$. ^{13}C NMR labelling experiments demonstrate[404] that the cyclization of 2-benzylidene-1,3-dicarbonyl compounds proceeds by a concerted pathway; asymmetric induction in such reactions has been attributed[405,406] to an *exo-Z-syn* transition state. Bicyclo[6.4.0]dodeca-4,9,11-triene readily cyclizes via conformation (**75**) to tetracyclodecene (**76**).[407] The highly strained bicyclo[3.3.1]nonadiene (**77**) has been generated by intramolecular cycloaddition of ynediene (**78**);[408] the product rapidly undergoes a homo-1,5-hydrogen shift to afford [3.3.1]propellane (**79**) (Scheme 18).

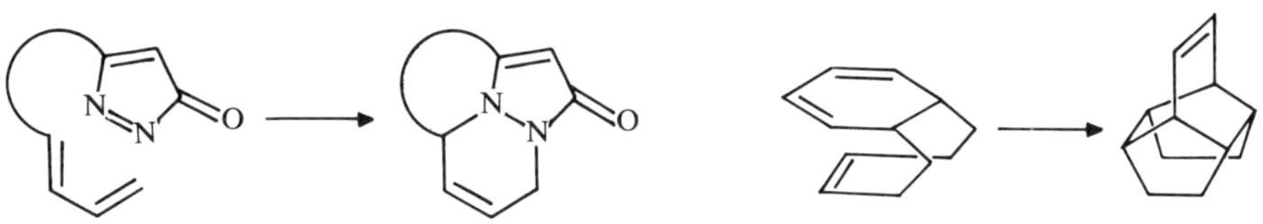

(**74**) (**73**) (**75**) (**76**)

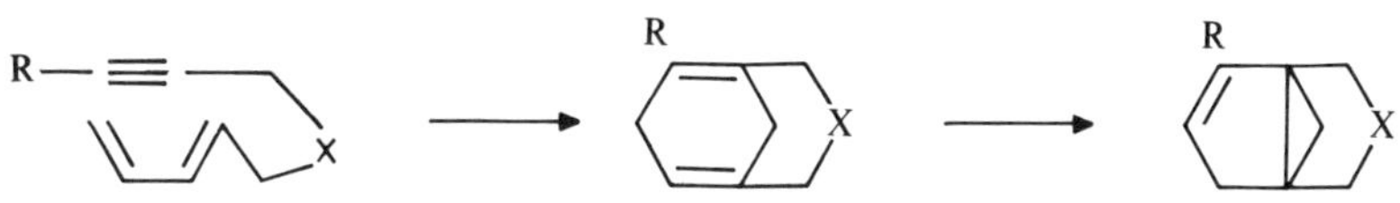

(**78**) (**77**) (**79**)

SCHEME 18

A more detailed study[409] has been made of the competition between homo-Diels–Alder reactivity vs insertion, rearrangement, and 2 + 2-cycloaddition for the reaction between PTAD and 2-substituted norbornadienes.

High-pressure-induced homo-Diels–Alder ($\pi^2 + \pi^2 + \pi^2$) and bishomo-diene ($\pi^2 + \sigma^2 + \sigma^2$) reactions of norbornadiene and quadricyclane with methyl propiolate have been combined into a sequential process leading to polycyclic bis-adducts.[410] Further examples of 1,4-dipolar cycloaddition of 6-oxopyrimidinium-4-olates have been reported[411–414] and a 6-thioxo derivative has also been examined.[413]

Miscellaneous Cycloadditions

MO calculations performed[415] on carbon monosulphide suggest a reaction pattern comparable with singlet carbenes; for cycloaddition to alkynes, the first step probably involves electrophilic approach of the vacant *p*-orbital to a π-bond of the alkyne. The formation of a disilacyclopropanimine from ArNC and $Ar_2Si{=}SiAr_2$ represents the first case of a 1 + 2-cycloaddition of an isonitrile to a disilene.[416] Di-*t*-butylsilylene, generated by photolysis of hexa-*t*-butylcyclotrisilane, is trapped by phosphaalkynes to give phosphasilirenes.[417] Both norbornadiene and quadricyclane react with phosphenium ions to give the same homo-1,4-cycloadduct.[418] Formation of ketones and lactones from photo-oxygenation of adamantylidenecyclopropanes has been rationalized[419] in terms of an initial perepoxide adduct. Immonium ions can undergo $2^+ + 1$-cycloaddition with nucleophilic carbenes.[420] 1 + 2-Addition of nitrenes to azo groups in conformationally rigid systems, analogous to the 2 + 2-reactions mentioned earlier (e.g. ref. 65), have been reported.[67,68] Substituent effects have been examined[421] for the reaction of Me_2Ge with 1,3-dienes; as well as the expected 1-germacyclopent-3-enes, 3,4-dialkenyl-1-germacyclopentanes are formed, presumably by 1,2-addition followed by insertion of a second diene. Transition metal carbonyl-mediated cycloadditions of tetrafluorodisilacyclobutadienes to 1,3-dienes, yielding e.g. 2-alkylidene-1,3-disilacyclopent-4-enes, have been studied in detail.[422–424]

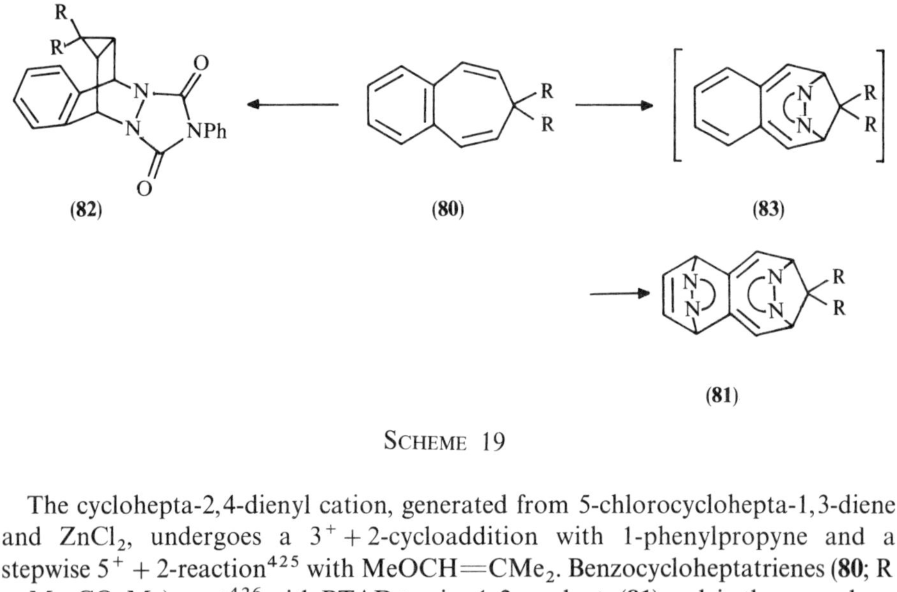

SCHEME 19

The cyclohepta-2,4-dienyl cation, generated from 5-chlorocyclohepta-1,3-diene and $ZnCl_2$, undergoes a $3^+ + 2$-cycloaddition with 1-phenylpropyne and a stepwise $5^+ + 2$-reaction[425] with $MeOCH{=}CMe_2$. Benzocycloheptatrienes (**80**; R = Me, CO_2Me) react[426] with PTAD to give 1:2 products (**81**) and, in the case where R = CO_2Me, a 1:1 adduct (**82**); the latter is presumably formed by 2 + 2 + 2-cycloaddition to the norcaradiene tautomer of (**80**), whereas the 1:2 product may result from 2 + 6- or 2 + 10-adduct (**83**) (Scheme 19). Photolysis of 2-(δ-alkynyl)tropones yields barbaralones via intramolecular 2 + 6-cycloaddition followed by rearrangement.[427] Various γ,δ-unsaturated-acylheptafulvenes undergo 10π-electrocyclization and intramolecular 2 + 8-cycloaddition,[428,429] the stereochemistry of the cycloadducts being dependent on the nature of tetraenophile, consistent with involvement of secondary orbital interactions. Tropone and thiotropone show divergent behaviour towards cyclopentadiene; the former undergoes the well-established 6 + 4-reaction whereas the latter gives an 8 + 2-product with cyclopentadiene acting as the 2π component; kinetic studies[430] confirm that the process is concerted, proceeding in a suprafacial-suprafacial manner with *endo*-approach and involving the highly oriented transition state of a 10π peripheral system. A stepwise mechanism involving initial Michael addition accounts for the products from the reaction of DMAD with 2-aryl-2*H*-cycloheptathiazoles.[431]

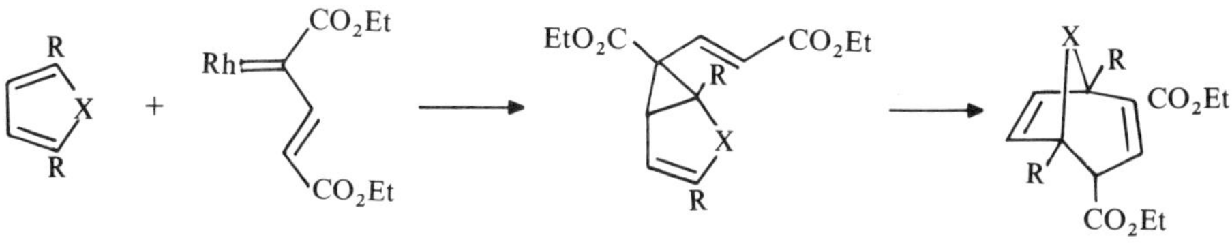

($X = CH_2,O$)

SCHEME 20

A 3 + 3-cycloaddition mechanism involving diazoalkane radical cations has been proposed[432] for the formation of olefinic dimers from one-electron oxidation of diazo compounds. Isolation of divinylcyclopropane intermediates from the 3 + 4-reaction between Rh(II)-stabilized vinylcarbenoids and cyclopentadiene (or furan) supports a tandem cyclopropanation/Cope rearrangement pathway rather than a concerted process (Scheme 20).[433,434] The proportion of 4 + 3- or 2 + 3-products from dienes and trimethylenemethane-metal complexes can be increased by geometrically restricting the diene to a *cisoid* conformation.[435] Cycloadditions of oxyallyl cations to various furans have been described.[436,437] Isolation of oxaazepine (**84**) in addition to normal 3 + 2-adducts from the reaction of *C,N*-diphenylnitrone (DPN) with diene (**85**) represents the first case of 3 + 4-cycloaddition of a nitrone to a 1,3-diene;[438] compound (**84**) is presumed to arise from ring-closure of intermediate diradical (**86**) (Scheme 21). 2-[(Trimethylsilyl)methyl]allyl carboxylates undergo Pd-catalysed 3 + 6-cycloaddition to tropones yielding 9-membered carbocycles, apparently the first case of such a 3 + 6-reaction.[439] Formation of trioxacycloheptene (**87**) from ozonolysis of tetraphenylcyclopentadiene is the first example of 3 + 4-cycloaddition of a carbonyl oxide to an α,β-unsaturated carbonyl compound (Scheme 22).[440]

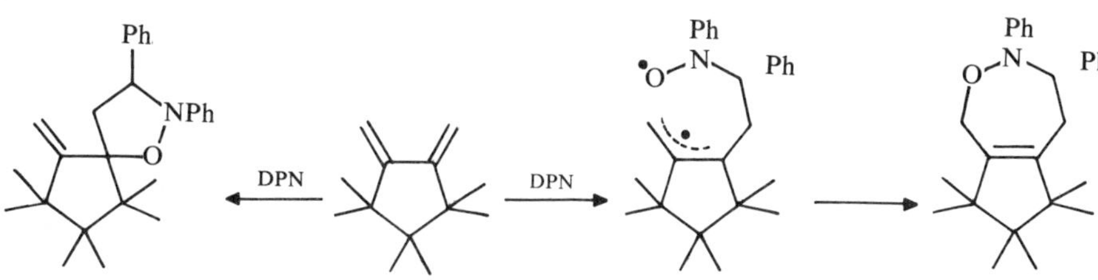

(**85**) (**86**) (**84**)

SCHEME 21

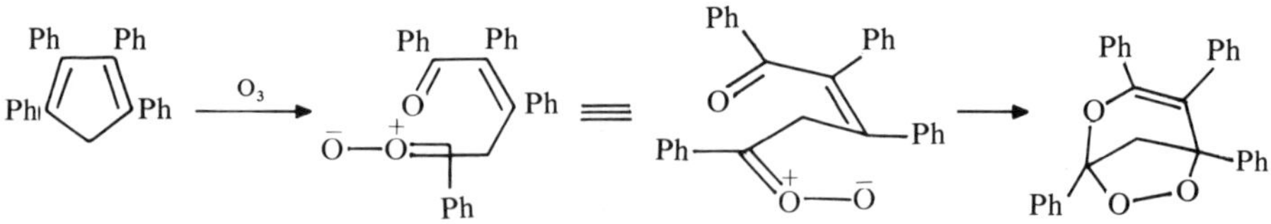

(87)

Scheme 22

Further examples of Ni-catalysed intramolecular 4 + 4-cycloadditions of bis-dienes have been reported;[441,442] model studies[441] show that the extent of allylic stereo-induction in the cyclization of 5-substituted-1,3,9,11-dodecatetraenes is highly dependent on the steric nature of the substituent. Intramolecular 4 + 4-dimerization of the anthryl units in 1,2-di(9-anthryl)ethene[443] and selective photodimerization of anthracene included in the cavity of cyclodextrins[444] have also been described. Cross-dimerization of anthracene-9-carbonitrile with naphthalenes and anthracene itself appears to take place via a significantly polar intermediate.[445] Photolysis of dione (**88**) affords hexacyclic ketone (**89**) and pentacyclic hydrocarbon (**90**) via extrusion of CO and intramolecular 4 + 2- or 4 + 4-cycloadditions (Scheme 23);[446] only (**89**) is formed under thermal conditions. The concerted nature of the allowed 6 + 4-addition of tropone to 2,3-dimethylbutadiene is confirmed by high-pressure kinetic analysis;[447] secondary orbital interactions, which predict the correct stereochemistry, exert no effect on the volume profile. Formation of pentacyclic compound (**91**) from 3,5-diphenylphosphinimine and diazomethane in aprotic solvents is attributed[448] to a remarkable intramolecular ionic $5_s + 5_s$ (6 + 4)-cycloaddition (Scheme 24).

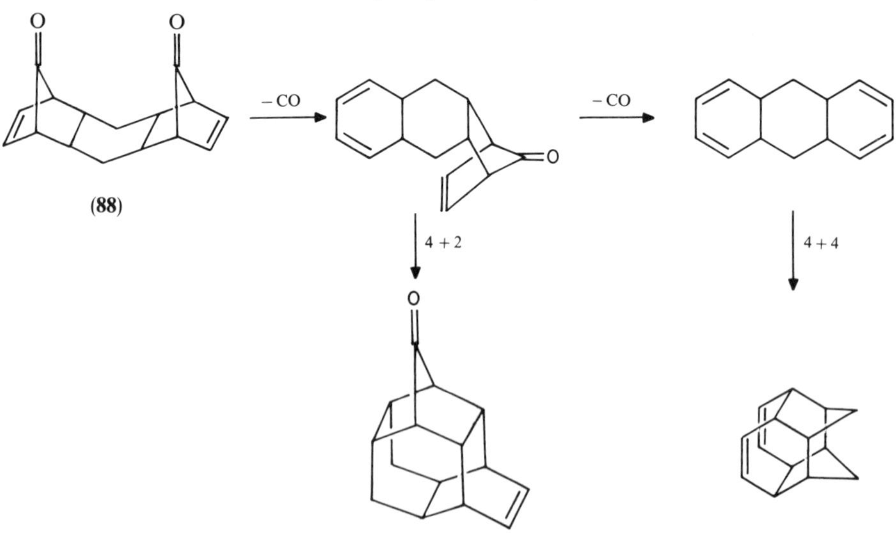

(88) **(89)** **(90)**

Scheme 23

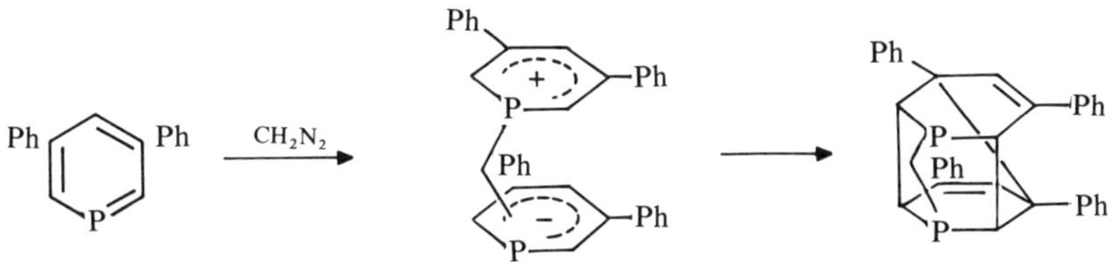

(91)

SCHEME 24

The combination of alkynes, allenes, and carbon monoxide in the presence of $Fe_2(CO)_9$ to form β-methylenecyclopentenones is an example of template-induced 2 + 2 + 1-cycloaddition via carbonyliron complexes as intermediates.[449] Further studies[450–453] have been made of Co-catalysed 2 + 2 + 2-reactions and the subject has been reviewed.[454] Formation of a 2 + 2 + 2-product from Pd-catalysed cyclization of 1,6-enynes in the presence of DMAD provides evidence in support of an intermediate palladacyclopentene.[455] A mechanism involving a 1,5-dipolar intermediate has been proposed[456] for the formation of 2:1 adducts between dichloroketene and 2,5-dimethoxy-2,5-dihydrofuran. Quaternary ammonium salt-catalysed trimerization of phenyl isocyanate[457] and metal-promoted trimerization of $Bu^tC{\equiv}P$[458,459] have also been reported. A radical-chain mechanism has been proposed[460] to account for the formation of a cyclic peroxide from the electrochemical oxidation of 2-methylcyclopentene-1,3-dione in the presence of molecular oxygen. The geometrical and stereoelectronic factors governing inter-/intra-cyclic thermal $(\pi^2 + \sigma^2 + \sigma^2)$ and photochemical $(\pi^2 + \pi^2)$ cycloadditions of sterically fixed norbornadienes and quadricyclanes of types (**92**) and (**93**) have been explored (Scheme 25).[461] Multiple intercyclic $[\pi^2 + (\sigma^2 + \sigma^2)_2]$ and $[\pi^2 + (\sigma^2 + \sigma^2)_3]$ bond formation occurs on thermolysis of the polycyclic system (**94**) which incorporates three adjacent quadricyclane units.[462] Norbornadiene dimerizes regiospecifically in the presence of $Mo(CO)_6$ yielding symmetrical heptacyclotetradecane (**95**).[463]

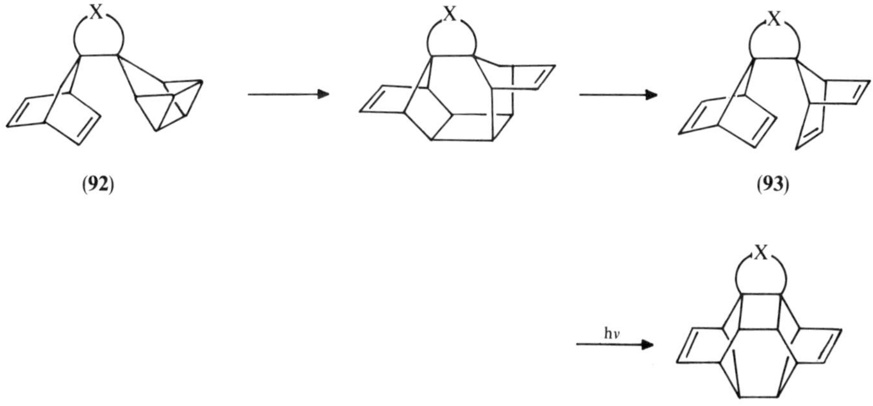

(92) **(93)**

SCHEME 25

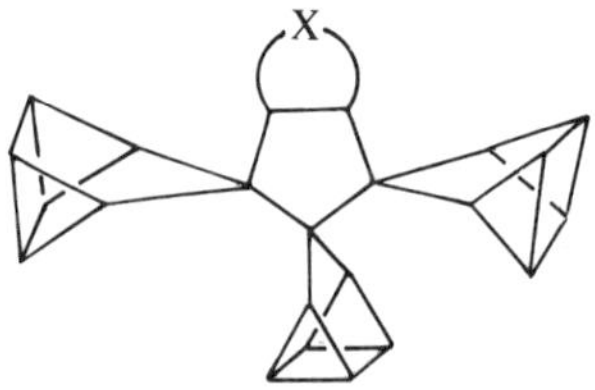

(94)

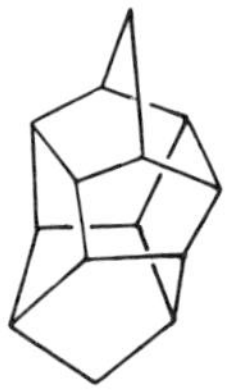

(95)

Photocycloadditions to arenes have been the subject of further investigation[464–470] and the topic has been reviewed.[3,4] 2 + 2-Adducts are the major products on photolysis of hexafluorobenzene with cycloalkenes.[464] Formation of *ortho*- and *meta*-adducts on irradiation of anisole in the presence of 1,3-dioxoles is rationalized[465] in terms of the exciplex-zwitterion mechanism; the high selectivities observed for *meta*-cycloaddition of *m*-$(CF_3)_2C_6H_4$ to alkenes also support the involvement of dipolar intermediates.[466] Some electron-rich alkenes add to the C≡N of benzonitrile-yielding azetidines rather than to the arene ring.[467] Intramolecular photocycloaddition of 5-arylalkynylsilanes yields cyclopentane-annulated cyclooctatetraenes with notable regioselectivity.[468] Photolysis of *N*-substituted phthalimides in the presence of cyclohexene yields *para*-cycloadducts.[469] Naphthols and their methyl ethers undergo photocycloaddition to acetylacetone only when the latter component is excited.[471]

MNDO calculations[472] for the ene reaction between ethene and propenol predict a stepwise mechanism involving initial H-atom transfer followed by C—C bond formation. A detailed analysis of regio-, diastereo-, and *cis/trans*-selectivity has been made[473] using ^{13}C NMR spectroscopy for the ene reaction between alkenes and maleic anhydride. Primary isotope effects for the corresponding reaction of PTAD with d_3-butenes indicate a mechanism involving an intermediate diaziridinium imide.[474]

References

1 Bauld, N. L., Bellville, D. J., Harirchian, B., Lorenz, K. T., Pabon, R. A., Reynolds, D. W., Wirth, D. D., Chiou, H.-S., and Marsh, B. K., *Acc. Chem. Res.*, **20**, 371 (1987).
2 Stella, L., *NATO ASI Ser.*, **189C**, 361 (1986); *Chem. Abs.*, **107**, 6542 (1987).
3 McCullough, J. J., *Chem. Rev.*, **87**, 811 (1987).
4 Wender, P. A. and Von Goeldern, T. W., *Spec. Publ. R. Soc. Chem.*, **57**, 226 (1986); *Chem. Abs.*, **106**, 195569 (1987).
5 Ramamurthy, V. and Venkatesan, K., *Chem. Rev.*, **87**, 433 (1987).
6 Fatiadi, A. J., *Synthesis*, **1987**, 749.
7 Fatiadi, A. J., *Synthesis*, **1987**, 959.
8 Kamal, A. and Sattur, P. B., *Heterocycles*, **26**, 1051 (1987).
9 Charlton, J. L. and Alauddin, M. M., *Tetrahedron*, **43**, 2873 (1987).
10 Sandhu, J. S. and Sain, B., *Heterocycles*, **26**, 777 (1987).
11 Jeroschewski, P. and Hansen, P., *Sulfur Reports*, **7**, 1 (1986).
12 Frey, H., Behmann, G., and Kaupp, G., *Chem. Ber.*, **120**, 387 (1987).
13 Cohen, M. D., *Tetrahedron*, **43**, 1211 (1987).
14 Murthy, G. S., Arjunan, P., Venkatesan, K., and Ramamurthy, V., *Tetrahedron*, **43**, 1225 (1987).

[15] Wang, W.-N. and Jones, W., *Tetrahedron*, **43**, 1273 (1987).
[16] Nalini, V. and Desiraju, G. R., *Tetrahedron*, **43**, 1313 (1987).
[17] Pfoertner, K. H., Englert, G., and Schoenholzer, P., *Tetrahedron*, **43**, 1321 (1987).
[18] Kaftory, M., *Tetrahedron*, **43**, 1503 (1987).
[19] Nagamatsu, T., Kawano, C., Orita, Y., and Kunieda, T., *Tetrahedron Lett.*, **28**, 3263 (1987).
[20] Shim, S. C. and Shin, E. J., *Chem. Lett.*, **1987**, 45.
[21] Shimo, T., Somekawa, K., Wakikawa, Y., Uemura, H., Tsuge, O., Imada, K., and Tanabe, K., *Bull. Chem. Soc. Jpn*, **60**, 621 (1987).
[22] Matsui, T., Morooka, T., and Nakayama, M., *Bull. Chem. Soc. Jpn*, **60**, 417 (1987).
[23] Lange, G. L., Decicco, C., and Lee, M., *Tetrahedron Lett.*, **28**, 2833 (1987).
[24] Herzog, H., Koch, H., Scharf, H.-D., and Runsink, J., *Chem. Ber.*, **120**, 1737 (1987).
[25] Demuth, M., Palomer, A., Sluma, H.-D., Dey, A. K., Krüger, C., and Tsay, Y.-H., *Angew. Chem. Int. Edn*, **25**, 1117 (1986).
[26] Hoffmann, V. T. and Musso, H., *Angew. Chem. Int. Edn*, **26**, 1006 (1987).
[27] Bernassau, J.-M., Bouillot, A., Fétizon, M., Hanna, I., Maia, E. R., and Prange, T., *J. Org. Chem.*, **52**, 1993 (1987).
[28] Chadha, R. and Ray, N. K., *Indian J. Chem.*, **25A**, 111 (1986).
[29] Balcells, M., Duran, M., Lledos, A., and Bertran, J., *THEOCHEM*, **34**, 153 (1987); *Chem. Abs.*, **107**, 153710 (1987).
[30] Dehmlow, E. V., Pickardt, J., Slopianka, M., Fastabend, U., Drechsler, K., and Soufi, J., *Liebigs Ann. Chem.*, **1987**, 377.
[31] Frater, G., Müller, U., and Günther, W., *Helv. Chim. Acta*, **69**, 1858 (1986).
[32] Greene, A. E., Charbonnier, F., Luche, M.-J., and Moyano, A., *J. Am. Chem. Soc.*, **109**, 4752 (1987).
[33] Johnston, B. D., Czyzewska, E., and Oehlschlager, A. C., *J. Org. Chem.*, **52**, 3693 (1987).
[34] Pirrung, M. C. and Webster, N. J. G., *J. Org. Chem.*, **52**, 3603 (1987).
[35] Becker, D. and Haddad, N., *Tetrahedron Lett.*, **27**, 6393 (1986).
[36] Winkler, J. D. and Heneger, K. E., *J. Am. Chem. Soc.*, **109**, 2850 (1987).
[37] Winkler, J. D., Hey, J. P., Hannon, F. J., and Willard, P. G., *Heterocycles*, **25**, 55 (1987).
[38] Matlin, A. R., Leckta, T. C., McGarvey, D. J., Jacob, P. W., and Picken, H. A., *Tetrahedron Lett.*, **28**, 5083 (1987).
[39] Matlin, A. R. and McGarvey, D. J., *Tetrahedron Lett.*, **28**, 5087 (1987).
[40] Crimmins, M. T. and Gould, L. D., *J. Am. Chem. Soc.*, **109**, 6199 (1987).
[41] Snider, B. B. and Kulkarni, Y. S., *J. Org. Chem.*, **52**, 307 (1987).
[42] Snider, B. B., Ron, E., and Burbaum, B. W., *J. Org. Chem.*, **52**, 5413 (1987).
[43] Kulkarni, Y. S., Niwa, M., Ron, E., and Snider, B. B., *J. Org. Chem.*, **52**, 1568 (1987).
[44] Brady, W. T., Giang, Y. F., Marchand, A. P., and Wu, A., *J. Org. Chem.*, **52**, 3457 (1987).
[45] Brady, W. T., Giang, Y. F., Weng, L., and Dad, M. M., *J. Org. Chem.*, **52**, 2216 (1987).
[46] Brady, W. T., Marchand, A. P., Giang, Y. F., and Wu, A., *Synthesis*, **1987**, 395.
[47] Dolbier, W. R. and Seabury, M. J., *Tetrahedron Lett.*, **28**, 1491 (1987).
[48] Dolbier, W. R. and Seabury, M. J., *J. Am. Chem. Soc.*, **109**, 4393 (1987).
[49] Plummer, B. F. and Singleton, S. F., *Tetrahedron Lett.*, **28**, 4801 (1987).
[50] Nishimura, J., Doi, H., Ueda, E., Ohbayashi, A., and Oku, A., *J. Am. Chem. Soc.*, **109**, 5293 (1987).
[51] Christl, M. and Schreck, M., *Chem. Ber.*, **120**, 915 (1987).
[52] Mehta, G., Padma, S., Osawa, E., Barbiric, D. A., and Mochizuki, Y., *Tetrahedron Lett.*, **28**, 1295 (1987).
[53] Ohsaku, M., *J. Chem. Soc., Perkin Trans. 2*, **1987**, 1027.
[54] Kaftory, M., Agmon, I., Ladika, M., and Stang, P. J., *J. Am. Chem. Soc.*, **109**, 782 (1987).
[55] Yang, N. C., Hrnjez, B. J., and Horner, M. G., *J. Am. Chem. Soc.*, **109**, 3158 (1987).
[56] Tobe, Y., Sorori, T., Kobiro, K., Kakiuchi, K., and Odaira, Y., *Tetrahedron Lett.*, **28**, 2861 (1987).
[57] Binger, P. and Biedenbach, B., *Chem. Ber.*, **120**, 601 (1987).
[58] Hassenrück, K., Martin, H.-D., Mayer, B., Urbanek, T., Zirwes, T., Walsh, R., and Beckhaus, H.-D., *Chem. Ber.*, **120**, 177 (1987).
[59] Saito, K., Ishihara, H., and Kagabu, S., *Bull. Chem. Soc. Jpn*, **60**, 4141 (1987).
[60] Jensen, F. and Foote, C. S., *J. Am. Chem. Soc.*, **109**, 6376 (1987).
[61] Clennan, E. L. and Earlywine, A. D., *J. Am. Chem. Soc.*, **109**, 7104 (1987).
[62] Paillous, N., Forgues, S. F., Jaud, J., and Devillers, J., *J. Chem. Soc., Chem. Commun.*, **1987**, 578.
[63] Aben, R. W. M., Smit, R., and Scheeren, J. W., *J. Org. Chem.*, **52**, 365 (1987).
[64] Ojima, I., Nakahashi, K., Brandstadter, S. M., and Hatanaka, N., *J. Am. Chem. Soc.*, **109**, 1798 (1987).
[65] Hünig, S., Martin, H.-D., Mayer, B., Peters, K., Prokschy, F., Schmitt, M., and von Schnering, H. G., *Chem. Ber.*, **120**, 195 (1987).

[66] Hünig, S. and Schmitt, M., *Tetrahedron Lett.*, **28**, 4521 (1987).
[67] Klingler, O. and Prinzbach, H., *Angew. Chem. Int. Edn*, **26**, 566 (1987).
[68] Marterer, W., Fritz, H., and Prinzbach, H., *Tetrahedron Lett.*, **28**, 5497 (1987).
[69] Fessner, W.-D., Sedelmeier, G., Knothe, L., Prinzbach, H., Rihs, G., Yang, Z., Kovac, G., and Heilbronner, E., *Helv. Chim. Acta*, **70**, 1816 (1987).
[70] Mattay, J., Gersdorf, J., and Buchkremer, K., *Chem. Ber.*, **120**, 307 (1987).
[71] Morris, T. H., Smith, E. H., and Walsh, R., *J. Chem. Soc., Chem. Commun.*, **1987**, 964.
[72] Kamphuis, J., Bos, H. J. T., Visser, R. J., Huizer, B. H., and Varma, C. A. G. O., *J. Chem. Soc., Perkin Trans. 2*, **1986**, 1867.
[73] Kamphuis, J., Ruijter, A. P., and Bos, H. J. T., *J. Chem. Soc., Perkin Trans. 2*, **1987**, 907.
[74] Coyle, J. D., Rapley, P. A., Kamphuis, J., and Bos, H. J. T., *J. Chem. Soc., Perkin Trans. 1*, **1986**, 2173.
[75] Coyle, J. D. and Rapley, P. A., *J. Chem. Soc., Perkin Trans. 1*, **1986**, 2273.
[76] Ooms, P. and Hartmann, W., *Tetrahedron Lett.*, **28**, 2701 (1987).
[77] Nishio, T., *J. Chem. Soc., Perkin Trans. 1*, **1987**, 1225.
[78] Oda, K., Machida, M., Aoe, K., Nishibata, Y., Sato, Y., and Kanaoka, Y., *Chem. Pharm. Bull.*, **34**, 1411 (1986).
[79] Wipf, P. and Heimgartner, H., *Helv. Chim. Acta*, **70**, 992 (1987).
[80] Clennan, E. L. and Lewis, K. K., *J. Am. Chem. Soc.*, **109**, 2475 (1987).
[81] Sander, W. and Patyk, A., *Angew. Chem. Int. Edn*, **26**, 475 (1987).
[82] Bakker, B. H. and Cerfontain, H., *Tetrahedron Lett.*, **28**, 1699 (1987).
[83] Volatron, F. and Eisenstein, O., *J. Am. Chem. Soc.*, **109**, 1 (1987).
[84] Brook, A. G., Chatterton, W. J., Sawyer, J. F., Hughes, D. W., and Vorspohl, K., *Organometallics*, **6**, 1246 (1987).
[85] Brook, A. G., Vorspohl, K., Ford, R. R., Hesse, M., and Chatterton, W. J., *Organometallics*, **6**, 2128 (1987).
[86] Pericas, M. A., Serratosa, F., and Valenti, E., *J. Chem. Soc., Perkin Trans. 2*, **1987**, 151.
[87] Chmielewski, M. and Kaluza, Z., *Carbohydr. Res.*, **167**, 143 (1987).
[88] Chmielewski, M., Kaluza, Z., Mostowicz, D., Belzecki, C., Baranowska, E., Jacobsen, J. P., Salanski, P., and Jurczak, J., *Tetrahedron*, **43**, 4555 (1987).
[89] Mormann, W. and Hoffmann, S., *Chem. Ber.*, **120**, 285 (1987).
[90] Ulrich, H., Richter, R., and Tucker, B., *J. Heterocycl. Chem.*, **24**, 1121 (1987).
[91] Ulrich, H., Richter, R., and Tucker, B., *Chem. Ber.*, **120**, 849 (1987).
[92] Gotthardt, H., Löhr, T., and Brauer, D. J., *Chem. Ber.*, **120**, 751 (1987).
[93] Gotthardt, H., Löhr, T., and Brauer, D. J., *Chem. Ber.*, **120**, 747 (1987).
[94] Bianchini, C., Meli, A., and Vizza, F., *Angew. Chem. Int. Edn*, **26**, 767 (1987).
[95] Smart, B. E. and Middleton, W. J., *J. Am. Chem. Soc.*, **109**, 4982 (1987).
[96] Lux, R. and Kresze, G., *Liebigs Ann. Chem.*, **1987**, 527.
[97] Grigg, R., *Chem. Soc. Rev.*, **16**, 89 (1987).
[98] Banerji, A. and Sahu, A., *J. Sci. Ind. Res.*, **45**, 355 (1986).
[99] Sammes, P. G., *Gazz. Chim. Ital.*, **116**, 109 (1986).
[100] Firestone, R. A., *Heterocycles*, **25**, 61 (1987).
[101] Kahn, S. D., Hehre, W. J., and Popte, J. A., *J. Am. Chem. Soc.*, **109**, 1871 (1987).
[102] Gray, B. D., McMillan, C. M., Miller, J. A., and Moore, M., *Tetrahedron Lett.*, **28**, 235 (1987).
[103] Gray, B. D. and Miller, J. A., *J. Chem. Res. (S)*, **1987**, 206.
[104] Gray, B. D., McMillan, C. M., Miller, J. A., and Ullah, G. M., *Tetrahedron Lett.*, **28**, 689 (1987).
[105] Aumann, R. and Uphoff, J., *Angew. Chem. Int. Edn*, **26**, 357 (1987).
[106] Trost, B. M., King, S. A., and Nanninga, T. N., *Chem. Lett.*, **1987**, 15.
[107] Herndon, J. W., *J. Am. Chem. Soc.*, **109**, 3165 (1987).
[108] Cunningham, D., Hallinan, N., Moran, G., and McArdle, P., *J. Organomet. Chem.*, **333**, 85 (1987).
[109] Goldschmidt, Z., Gottlieb, H. E., and Almadhoun, A., *J. Organomet. Chem.*, **326**, 405 (1987).
[110] Hallinan, N., McArdle, P., Burgess, J., and Guardado, P., *J. Organomet. Chem.*, **333**, 77 (1987).
[111] Beak, P. and Wilson, K. D., *J. Org. Chem.*, **51**, 4627 (1986).
[112] Beak, P. and Wilson, K. D., *J. Org. Chem.*, **52**, 218 (1987).
[113] Grützmacher, H. and Roesky, H. W., *Chem. Ber.*, **120**, 995 (1987).
[114] Padwa, A., Gasdaska, J. R., Haffmanns, G., and Rebello, H., *J. Org. Chem.*, **52**, 1027 (1987).
[115] Bock, H., Dammel, R., Fischer, S., and Wentrup, C., *Tetrahedron Lett.*, **28**, 617 (1987).
[116] Padwa, A. and MacDonald, J. G., *J. Heterocycl. Chem.*, **24**, 1225 (1987).
[117] Hassaneen, H. M., Farag, A. M., Shawali, A. S., and Algharib, M. S., *J. Heterocycl. Chem.*, **24**, 577 (1987).
[118] Fathi, T., An, N. D., Schmitt, G., Vebrel, J., and Laude, B., *Bull. Soc. Chim. Belg.*, **96**, 145 (1987).

[119] Tshiamala, K., Kitane, S., Vebrel, J., and Laude, B., *Bull. Soc. Chim. Belg.*, **95**, 1083 (1986).
[120] Moderhack, D. and Lorke, M., *Heterocycles*, **26**, 1751 (1987).
[121] Märkl, G. and Beckh, H. J., *Tetrahedron Lett.*, **28**, 3475 (1987).
[122] Bernardi, F., Olivucci, M., McDouall, J. J. W., and Robb, M. A., *J. Am. Chem. Soc.*, **109**, 544 (1987).
[123] McDouall, J. J. W., Robb, M. A., Niazi, U., Bernardi, F., and Schlegel, H. B., *J. Am. Chem. Soc.*, **109**, 4642 (1987).
[124] Sustmann, R. and Sicking, W., *Chem. Ber.*, **120**, 1471 (1987).
[125] Sustmann, R. and Sicking, W., *Chem. Ber.*, **120**, 1653 (1987).
[126] Stajer, G., Szabo, A. E., Bernath, G., and Sohar, P., *Tetrahedron*, **43**, 1931 (1987).
[127] Stajer, G., Bernath, G., Szabo, A. E., Sohar, P., Argay, G., and Kalman, A., *Tetrahedron*, **43**, 5461 (1987).
[128] Alcaide, B., Mardomingo, C. L., Plumet, J., Cativiela, C., and Mayoral, J. A., *Can. J. Chem.*, **65**, 2050 (1987).
[129] Tanaka, K., Masuda, H., and Mitsuhashi, K., *Bull. Chem. Soc. Jpn*, **59**, 3901 (1986).
[130] Fisera, L. and Oravec, P., *Collect. Czech. Chem. Commun.*, **52**, 1315 (1987).
[131] Ohba, Y., Matsukura, I., and Nishiwaki, T., *J. Chem. Res. (S)*, **1987**, 103.
[132] Brandi, A., Carli, S., Guarna, A., and De Sarlo, F., *Tetrahedron Lett.*, **28**, 3845 (1987).
[133] Evans, A. R. and Taylor, G. A., *J. Chem. Soc., Perkin Trans. 1*, **1987**, 567.
[134] Kanomata, N., Kobayashi, T., Uetake, T., Okada, S., and Nitta, M., *Bull. Chem. Soc. Jpn*, **60**, 683 (1987).
[135] Meinke, P. T., Krafft, G. A., and Spencer, J. T., *Tetrahedron Lett.*, **28**, 3887 (1987).
[136] Meinke, P. T. and Krafft, G. A., *Tetrahedron Lett.*, **28**, 5121 (1987).
[137] Adam, W., Carballeira, N., Crämer, E., Lucchini, V., Peters, E.-V., Peters, K., and von Schnering, H. G., *Chem. Ber.*, **120**, 695 (1987).
[138] Curran, D. P., Kim, B. H., Piyasena, H. P., Loncharich, R. J., and Houk, K. N., *J. Org. Chem.*, **52**, 2137 (1987).
[139] McKie, M. C. and Paton, R. M., *J. Chem. Res. (S)*, **1987**, 245.
[140] Knight, J. and Parsons, P. J., *J. Chem. Soc., Chem. Commun.*, **1987**, 189.
[141] Murthy, K. S. K. and Hassner, A., *Tetrahedron Lett.*, **28**, 97 (1987).
[142] Hassner, A. and Murthy, K. S. K., *Tetrahedron Lett.*, **28**, 4097 (1987).
[143] Hassner, A. and Murthy, K. S. K., *Tetrahedron Lett.*, **28**, 683 (1987).
[144] Annunziata, R., Cinquini, M., Cozzi, F., Dondio, G., and Raimondi, L., *Tetrahedron*, **43**, 2369 (1987).
[145] Annunziata, R., Cinquini, M., Cozzi, F., Gennari, C., and Raimondi, L., *J. Org. Chem.*, **52**, 4674 (1987).
[146] Annunziata, R., Cinquini, M., Cozzi, F., and Raimondi, L., *J. Chem. Soc., Chem. Commun.*, **1987**, 529.
[147] Buchanan, J. G., Edger, A. R., and Hewitt, B. D., *J. Chem. Soc., Perkin Trans. 1*, **1987**, 2371.
[148] Mohiuddin, G., Reddy, P. S., Ahmed, K., and Ratnam, C. V., *J. Chem. Res. (S)*, **1987**, 224.
[149] Sha, C.-K., Chuang, K.-S., and Wey, S.-J., *J. Chem. Soc., Perkin Trans. 1*, **1987**, 977.
[150] Bunnelle, W. H. and Schlemper, E. O., *J. Am. Chem. Soc.*, **109**, 612 (1987).
[151] Schultz, A. G., McMahon, W. G., and Kullnig, R. K., *J. Org. Chem.*, **52**, 3905 (1987).
[152] Annunziata, R., Cinquini, M., Cozzi, F., and Raimondi, L., *Tetrahedron*, **43**, 4051 (1987).
[153] Quast, H., Bieber, L., and Meichsner, G., *Liebigs Ann. Chem.*, **1987**, 469.
[154] Burdisso, M., Gandolfi, R., Quartieri, S., and Rastelli, A., *Tetrahedron*, **43**, 159 (1987).
[155] Padwa, A. and Wannamaker, M. W., *J. Chem. Soc., Chem. Commun.*, **1987**, 1742.
[156] Hassenrück, K., Höchstetter, H., Martin, H.-D., Steigel, A., and Wingen, H.-P., *Chem. Ber.*, **120**, 203 (1987).
[157] Huisgen, R., Eichenauer, U., Langhals, E., Mitra, A., and Moran, J. R., *Chem. Ber.*, **120**, 153 (1987).
[158] Huisgen, R., Mitra, A., and Moran, J. R., *Chem. Ber.*, **120**, 159 (1987).
[159] Dolbier, W. R. and Seabury, M. J., *Tetrahedron*, **43**, 2437 (1987).
[160] Farina, F., Martin, M. V., Paredes, M. C., and Tito, A., *J. Heterocycl. Chem.*, **24**, 1269 (1987).
[161] Nesi, R., Giomi, D., Quartara, L., Bracci, S., and Papaleo, S., *J. Chem. Soc., Chem. Commun.*, **1987**, 1077.
[162] Arbuzov, B. A., Dianova, E. N., Galiaskarova, R. T., and Schmidpeter, A., *Chem. Ber.*, **120**, 597 (1987).
[163] Okada, K., Samizo, F., and Oda, M., *Chem. Lett.*, **1987**, 93.
[164] Medvedeva, A. S., Margorskaya, O. I., Golovanova, N. I., Kalikhman, I. D., Demina, M. M., and Vyazankin, N. S., *Zh. Obshch. Khim.*, **56**, 2106 (1986); *Chem. Abs.*, **107**, 115050 (1987).
[165] Sotiropoulos, J.-M., Baceiredo, A., and Bertrand, G., *J. Am. Chem. Soc.*, **109**, 4711 (1987).
[166] Boeckh, D., Huisgen, R., and Nöth, H., *J. Am. Chem. Soc.*, **109**, 1248 (1987).
[167] Rösch, W., Hees, U., and Regitz, M., *Chem. Ber.*, **120**, 1645 (1987).

168 Rösch, W., Facklam, T., and Regitz, M., *Tetrahedron*, **43**, 3247 (1987).
169 Rösch, W., Richter, H., and Regitz, M., *Chem. Ber.*, **120**, 1809 (1987).
170 Greig, D. J., Hamilton, D. G., McPherson, M., Paton, R. M., and Crosby, J., *J. Chem. Soc., Perkin Trans. 1*, **1987**, 607.
171 Wiberg, N., Preiner, G., Karampatses, P., Kim, C.-K., and Schurz, K., *Chem. Ber.*, **120**, 1357 (1987).
172 Ardill, H., Grigg, R., Sridharan, V., and Malone, J., *J. Chem. Soc., Chem. Commun.*, **1987**, 1296.
173 Grigg, R., Surendrakumar, S., Thianpatanagul, S., and Vipond, D., *J. Chem. Soc., Chem. Commun.*, **1987**, 47.
174 Grigg, R., Idle, J., McMeekin, P., and Vipond, D., *J. Chem. Soc., Chem. Commun.*, **1987**, 49.
175 Grigg, R., Kemp, J., and Warnock, W. J., *J. Chem. Soc., Perkin Trans. 1*, **1987**, 2275.
176 Grigg, R., Gunaratne, N. Q. N., and Sridharan, V., *Tetrahedron*, **43**, 5887 (1987).
177 Tsuge, O., Kanemasa, S., Ohe, M., Yorozu, K., Takenaka, S., and Ueno, K., *Bull. Chem. Soc. Jpn*, **60**, 4067 (1987).
178 Tsuge, O., Kanemasa, S., Ohe, M., and Takenaka, S., *Bull. Chem. Soc. Jpn*, **60**, 4079 (1987).
179 Tsuge, O., Kanemasa, S., Yamada, T., and Matsuda, K., *J. Org. Chem.*, **52**, 2523 (1987).
180 Armstrong, P., Grigg, R., and Warnock, W. J., *J. Chem. Soc., Chem. Commun.*, **1987**, 1325.
181 Armstrong, P., Grigg, R., Surendrakumar, S., and Warnock, W. J., *J. Chem. Soc., Chem. Commun.*, **1987**, 1327.
182 Shimizu, T., Hayashi, Y., Miki, M., and Teramura, K., *J. Org. Chem.*, **52**, 2277 (1987).
183 Amornraksa, K., Grigg, R., Gunaratne, N. Q. N., Kemp, J., and Sridharan, V., *J. Chem. Soc., Perkin Trans. 1*, **1987**, 2285.
184 Grigg, R., Dowling, M., Jordan, M. W., Sridharan, V., and Thianpatanagul, S., *Tetrahedron*, **43**, 5873 (1987).
185 Grigg, R., Dowling, M., and Sridharan, V., *J. Chem. Soc., Chem. Commun.*, **1986**, 1777.
186 Matsumoto, K., Uchida, T., Ikemi, Y., Tanaka, T., Asahi, M., Kato, T., and Konishi, H., *Bull. Chem. Soc. Jpn.*, **60**, 3645 (1987).
187 Matsumoto, K., Uchida, T., Konishi, H., Watanabe, Y., Aoyama, K., and Asahi, M., *Chem. Lett.*, **87**, 807 (1987).
188 Tsuge, O., Kanemasa, S., Yorozu, K., and Ueno, K., *Bull. Chem. Soc. Jpn*, **60**, 3359 (1987).
189 Tsuge, O., Kanemasa, S., and Takenaka, S., *Bull. Chem. Soc. Jpn*, **60**, 1489 (1987).
190 Imai, N., Nemoto, M., Terao, Y., Achiwa, K., and Sekiya, M., *Chem. Pharm. Bull.*, **34**, 1080 (1986).
191 Benhaoua, H., Piet, J.-C., Danion-Bougot, R., Toupet, L., and Carrie, R., *Bull. Soc. Chim. Fr. II*, **1987**, 325.
192 Fujimura, Y., Nawata, Y., and Hamana, M., *Heterocycles*, **26**, 133 (1987).
193 Kiss, M., Russell-Maynard, J., and Joule, J. A., *Tetrahedron Lett.*, **28**, 2187 (1987).
194 Vedejs, E., Grissom, J. W., and Preston, J. K., *J. Org. Chem.*, **52**, 3488 (1987).
195 Pavez, H., Marquez, A., Navarette, P., Tzichinovsky, S., and Rodriquez, H., *Tetrahedron*, **43**, 2223 (1987).
196 Chen, L.-C. and Yang, S.-C., *Heterocycles*, **24**, 3411 (1986).
197 Mais, F.-J., Dickopp, H., Middelhauve, B., Martin, H.-D., Mootz, D., and Steigel, A., *Chem. Ber.*, **120**, 275 (1987).
198 Butler, R. N., Cunningham, D., Marren, E. G., and McArdle, P., *J. Chem. Soc., Chem. Commun.*, **1987**, 706.
199 Butler, R. N., Evans, A. M., McArdle, P., and Cunningham, D., *J. Chem. Soc., Chem. Commun.*, **1987**, 1090.
200 Burdisso, M., Gamba, A., Gandolfi, R., and Pevarello, P., *Tetrahedron*, **43**, 1835 (1987).
201 Hootelé, C., Ibebeke-Bomangwa, W., Driessens, F., and Sabil, S., *Bull. Soc. Chim. Belg.*, **96**, 57 (1987).
202 Tufariello, J. J., Milowsky, A. S., Al-Nuri, M., and Goldstein, S., *Tetrahedron Lett.*, **28**, 263 (1987).
203 Tufariello, J. J., Milowsky, A. S., Al-Nuri, M., and Goldstein, S., *Tetrahedron Lett.*, **28**, 267 (1987).
204 Burdisso, M., Gandolfi, R., Pevarello, P., and Raselli, A., *Tetrahedron Lett.*, **28**, 1225 (1987).
205 Matsuoka, T., Harano, K., Kubo, H., and Hisano, T., *Chem. Pharm. Bull.*, **34**, 572 (1986).
206 Harano, K., Kondo, R., Murase, M., Matsuoka, T., and Hisano, T., *Chem. Pharm. Bull.*, **34**, 966 (1986).
207 Parpani, P. and Zecchi, G., *J. Org. Chem.*, **52**, 1417 (1987).
208 Chanet-Ray, J., Vessière, R., and Zéroual, A., *Heterocycles*, **26**, 101 (1987).
209 Plate, R., Hermkens, P. H. H., Smits, J. M. M., Nivard, R. J. F., and Ottenheijm, H. C. J., *J. Org. Chem.*, **52**, 1047 (1987).
210 Padwa, A., Kline, D. N., Koehler, K. F., Matzinger, M., and Venkatramanan, M. K., *J. Org. Chem.*, **52**, 3909 (1987).
211 Dolbier, W. R., Wicks, G. E., and Burkholder, C. R., *J. Org. Chem.*, **52**, 2196 (1987).

[212] Ibouye, Y., Niwayama, S., Okada, T., and Kakisawa, H., *Heterocycles*, **25**, 109 (1987).
[213] Padwa, A., Dent, W., and Yeske, P. E., *J. Org. Chem.*, **52**, 3944 (1987).
[214] Ortega, M., Lluch, J. M., Oliva, A., and Bertran, J., *An. Quim.*, **82A**, 361 (1986); *Chem. Abs.*, **107**, 133687 (1987).
[215] Maruyama, K., Osuka, A., and Nakagawa, K., *Bull. Chem. Soc. Jpn*, **62**, 1021 (1987).
[216] Turro, N. J. and Cha, Y., *Tetrahedron Lett.*, **28**, 1723 (1987).
[217] Liu, M. T. H., Soundarajan, N., Anand, S. M., and Ibata, T., *Tetrahedron Lett.*, **28**, 1011 (1987).
[218] Clawson, P. and Whiting, D. A., *Tetrahedron Lett.*, **28**, 3155 (1987).
[219] Sammes, P. G. and Whitby, R. J., *J. Chem. Soc.*, *Perkin Trans. 1*, **1987**, 195.
[220] Ooms, P. and Hartmann, W., *Tetrahedron Lett.*, **28**, 4271 (1987).
[221] Aono, M., Terao, Y., and Achiwa, K., *Chem. Lett.*, **87**, 1557 (1987).
[222] Hosomi, A., Hayashi, S., Hoashi, K., Kohra, S., and Tominaga, Y., *J. Chem. Soc.*, *Chem. Commun.*, **1987**, 1442.
[223] Sheradsky, T., and Itzhak, N., *J. Chem. Soc.*, *Perkin Trans. 1*, **1987**, 1979.
[224] Gotthardt, H. and Pflaumbaum, W., *Chem. Ber.*, **120**, 1017 (1987).
[225] Märkl, G., Eckl, E., Jakobs, U., Ziegler, M. L., and Nuber, B., *Tetrahedron Lett.*, **28**, 2119 (1987).
[226] Terao, Y., Aono, M., Takahashi, I., and Achiwa, K., *Chem. Lett.*, **1986**, 2089.
[227] Huisgen, R. and Rapp, J., *J. Am. Chem. Soc.*, **109**, 902 (1987).
[228] Sawaki, Y., Ishikawa, S., and Iwamura, H., *J. Am. Chem. Soc.*, **109**, 584 (1987).
[229] McKinnon, D. M., Secco, A. S., and Duncan, K. A., *Can. J. Chem.*, **65**, 1247 (1987).
[230] Miyashi, T., Kamata, M., and Mukai, T., *J. Am. Chem. Soc.*, **109**, 2780 (1987).
[231] Ibata, T., Isogami, Y., Nakano, S., Nakawa, H., and Tamura, H., *J. Chem. Soc.*, *Chem. Commun.*, **1986**, 1692.
[232] Dunn, P. J. and Rees, C. W., *J. Chem. Soc.*, *Chem. Commun.*, **1987**, 59.
[233] Daley, S. T. A. K. and Rees, C. W., *J. Chem. Soc.*, *Perkin Trans. 1*, **1987**, 203.
[234] Dunn, P. J. and Rees, C. W., *J. Chem. Soc.*, *Perkin Trans. 1*, **1987**, 1579.
[235] Dunn, P. J. and Rzepa, H. S., *J. Chem. Soc.*, *Perkin Trans. 2*, **1987**, 1669.
[236] Sleglich, W., Jeschke, R., and Buschmann, E., *Gazz. Chim. Ital.*, **116**, 361 (1986).
[237] Christl, M., *Gazz. Chim. Ital.*, **116**, 1 (1986).
[238] Craig, D., *Chem. Soc. Rev.*, **18**, 187 (1987).
[239] Salakhov, M. S. and Ismailov, S. A., *Usp. Khim.*, **55**, 2008 (1986); *Chem. Abs.*, **106**, 118873 (1987).
[240] Ichihara, A., *Synthesis*, **1987**, 207.
[241] Fox, M. A., Cardona, R., and Kiwiet, N. J., *J. Org. Chem.*, **52**, 1469 (1987).
[242] Gajewski, J. J., Peterson, K. B., and Kagel, J. R., *J. Am. Chem. Soc.*, **109**, 5545 (1987).
[243] Bonnert, R. V. and Jenkins, P. R., *Tetrahedron Lett.*, **28**, 697 (1987).
[244] Brown, F. K., Houk, K. N., Burnell, D. J., and Valenta, Z., *J. Org. Chem.*, **52**, 3050 (1987).
[245] Paquette, L. A., Gugelchuk, M., and McLaughlin, M. L., *J. Org. Chem.*, **52**, 4732 (1987).
[246] Ipaktschi, J., Herber, J., Kalinowski, H.-O., and Boese, R., *Tetrahedron Lett.*, **28**, 3467 (1987).
[247] Burnell, D. J., Goodbrand, H. B., Kaiser, S. M., and Valenta, Z., *Can. J. Chem.*, **65**, 154 (1987).
[248] Sustmann, R. and Sicking, W., *Chem. Ber.*, **120**, 1323 (1987).
[249] Sustmann, R., Dern, M., Kasten, R., and Sicking, W., *Chem. Ber.*, **120**, 1315 (1987).
[250] Zwick, J.-C., Vogel, P., Mange, V., and Chapuis, G., *Helv. Chim. Acta*, **70**, 1231 (1987).
[251] Rubello, A., Vogel, P., and Chapuis, G., *Helv. Chim. Acta*, **70**, 1638 (1987).
[252] Kohnke, F. H., Slawin, A. M. Z., Stoddart, J. F., and Williams, D. J., *Angew. Chem. Int. Edn*, **26**, 892 (1987).
[253] Coxon, J. M., O'Connell, M. J., and Steel, P. J., *J. Org. Chem.*, **52**, 4726 (1987).
[254] Rienäcker, R., Ashkenazi, P., Kapon, M., Benn, R., and Ginsburg, D., *Tetrahedron*, **43**, 1161 (1987).
[255] Roth, W. R., Langer, R., Bartmann, M., Stevermann, B., Maier, G., Reisenauer, H. P., Sustmann, R., and Müller, W., *Angew. Chem. Int. Edn*, **26**, 256 (1987).
[256] Chadwick, D. J. and Plant, A., *Tetrahedron Lett.*, **28**, 6085 (1987).
[257] Stone, K. J., Greenberg, M. M., Goodman, J. L., Peters, K. S., and Berson, J. A., *J. Am. Chem. Soc.*, **108**, 8089 (1986).
[258] Bates, G. S., Fryzuk, M. D., and Stone, C., *Can. J. Chem.*, **65**, 2612 (1987).
[259] Bäckvall, J-E. and Juntunen, S. K., *J. Am. Chem. Soc.*, **109**, 6396 (1987).
[260] Oren, J., Demuth, M., and Fuchs, B., *Synthesis*, **1987**, 851.
[261] Vaultier, M., Truchet, F., Carboni, B., Hoffmann, R. W., and Denne, J., *Tetrahedron Lett.*, **28**, 4169 (1987).
[262] Huber, S., Stamouli, P., Jenny, T. A., and Neier, R., *Helv. Chim. Acta*, **69**, 1898 (1986).
[263] Banert, K., *Angew. Chem. Int. Edn*, **26**, 879 (1987).
[264] Pfeuffer, L. and Pindur, U., *Helv. Chim. Acta.*, **70**, 1419 (1987).

[265] Pindur, U. and Pfeuffer, L., *Tetrahedron Lett.*, **28**, 3079 (1987).
[266] Pindur, U. and Pfeuffer, L., *Heterocycles*, **26**, 325 (1987).
[267] Pfeuffer, L. and Pindur, U., *Monatsh. Chem.*, **118**, 1073 (1987).
[268] Simon, M. M. and Arques, J. S., *Tetrahedron*, **42**, 6683 (1986).
[269] Abarca, B., Ballesteros, R., Enriquez, E., and Jones, G., *Tetrahedron*, **43**, 269 (1987).
[270] Abarca, B., Ballesteros, R., and Soriano, C., *Tetrahedron*, **43**, 991 (1987).
[271] Kawase, T., Iyoda, M., and Oda, M., *Angew. Chem. Int. Edn*, **26**, 559 (1987).
[272] Saba, A. and De Lucchi, O., *Heterocycles*, **26**, 2339 (1987).
[273] Wingert, H., Irngartinger, H., Kallfass, D., and Regitz, M., *Chem. Ber.*, **120**, 825 (1987).
[274] Fink, J., Gümbel, H., Eisenbarth, P., and Regitz, M., *Chem. Ber.*, **120**, 1027 (1987).
[275] Bryce-Smith, D., Gilbert, A., McColl, I. S., Drew, M. G. B., and Yianni, P., *J. Chem. Soc., Perkin Trans. 1*, **1987**, 1147.
[276] Arriau, J., Fernandez, J., and Yianni, P., *J. Chem. Soc., Perkin Trans. 2*, **1986**, 2013.
[277] Samuilov, Y. D., Hiavi, J., Bukharov, S. V., and Konovalov, A. I., *Zh. Obshch. Khim.*, **56**, 2638 (1986); *Chem. Abs.*, **107**, 96130 (1987).
[278] Guner, O. F., Ottenbrite, R. M., Shillady, D. D., and Alston, P. V., *J. Org. Chem.*, **52**, 391 (1987).
[279] Bischoff, S. and Kasper, F., *J. Prakt. Chem.*, **328**, 449 (1986); *Chem. Abs.*, **107**, 38897 (1987).
[280] Kiselev, V. D., Shakirov, I. M., and Konovalov, A. I., *Zh. Org. Khim.*, **22**, 1034 (1986); *Chem. Abs.*, **106**, 17670 (1987).
[281] Branchadell, V., Oliva, A., and Bertran, J., *Afinidad*, **43**, 480 (1986); *Chem. Abs.*, **107**, 133661 (1987).
[282] Kiselev, V. D., Veisman, E. A., and Konovalov, A. I., *Zh. Obshch. Khim.*, **56**, 2378 (1986); *Chem. Abs.*, **106**, 175518 (1987).
[283] Al-Ekabi, H. and de Mayo, P., *J. Org. Chem.*, **52**, 4756 (1987).
[284] Schneider, H.-J. and Sangwan, N. K., *Angew. Chem. Int. Edn*, **26**, 896 (1987).
[285] Schneider, H.-J. and Sangwan, N. K., *J. Chem. Soc., Chem. Commun.*, **1986**, 1787.
[286] Dolata, D. P. and Bergman, R., *Tetrahedron Lett.*, **28**, 707 (1987).
[287] Bunnelle, W. H. and Shangraw, W. R., *Tetrahedron*, **43**, 2005 (1987).
[288] Mustafaev, A. M., Kasymov, K. I., and Guseinov, M. M., *Zh. Org. Khim.*, **22**, 1423 (1986); *Chem. Abs.*, **106**, 155678 (1987).
[289] Palm, V. and Musaeva, M. F., *Org. React.* (*Tartu*), **23**, 96 (1986); *Chem. Abs.*, **107**, 77043 (1987).
[290] Mustafaev, A. M., Novikov, S. N., Guseinov, M. M., and Adigezalov, N. R., *Org. React.* (*Tartu*), **23**, 3 (1986); *Chem. Abs.*, **107**, 6601 (1987).
[291] Akiba, K. Ohtani, A., and Yamamoto, Y., *J. Org. Chem.*, **51**, 5328 (1986).
[292] Effenberger, F. and Ziegler, T., *Chem. Ber.*, **120**, 1339 (1987).
[293] Ziegler, T., Layh, M., and Effenberger, F., *Chem. Ber.*, **120**, 1347 (1987).
[294] Tobia, D. and Rickborn, B., *J. Org. Chem.*, **52**, 2611 (1987).
[295] Pauli, A., Kolshorn, H., and Meier, H., *Chem. Ber.*, **120**, 1611 (1987).
[296] Posner, G. H. and Switzer, C., *J. Org. Chem.*, **52**, 1642 (1987).
[297] Cresp, T. M. and Wege, D., *Tetrahedron*, **42**, 6713 (1986).
[298] Barbaro, G., Battaglia, A., and Giorgianni, P., *J. Org. Chem.*, **52**, 3289 (1987).
[299] Williams, R. V. and Sung, C.-L. A., *J. Chem. Soc., Chem. Commun.*, **1987**, 590.
[300] Battiste, M. A., Griggs, B. G., and Sackett, D., *J. Organomet. Chem.*, **330**, 437 (1987).
[301] Apeloig, Y., Arad, D., Kapon, M., and Wallerstein, M., *Tetrahedron Lett.*, **28**, 5917 (1987).
[302] Banwell, M. G. and Knight, J. H., *J. Chem. Soc., Chem. Commun.*, **1987**, 1082.
[303] Gruseck, U. and Heuschmann, M., *Tetrahedron Lett.*, **28**, 2681 (1987).
[304] Gruseck, U. and Heuschmann, M., *Tetrahedron Lett.*, **28**, 6027 (1987).
[305] Möhrle, H. and Dwuletzki, H., *Chem. Ber.*, **119**, 3591 (1986).
[306] Möhrle, H. and Dwuletzki, H., *Chem. Ber.*, **119**, 3600 (1986).
[307] Van Mele, B. and Huybrechts, G., *Int. J. Chem. Kinet.*, **19**, 363 (1987).
[308] Fabian, W., *Monatsh. Chem.*, **117**, 1057 (1986).
[309] Iwao, M. and Kuraishi, T., *Bull. Chem. Soc. Jpn*, **60**, 4051 (1987).
[310] Angell, E. C., Fringuelli, F., Minuti, L., Pizzo, F., Taticchi, A., and Wenkert, E., *J. Org. Chem.*, **51**, 5177 (1986).
[311] Liu, H.-J. and Browne, E. N. C., *Can. J. Chem.*, **65**, 1262 (1987).
[312] Bueno, M. P., Cativiela, C., Finol, C., Mayoral, J. A., and Jaime, C., *Can. J. Chem.*, **65**, 2182 (1987).
[313] Aguilas, R., Reyes, A., Tamariz, J., and Birbaum, J.-L., *Tetrahedron Lett.*, **28**, 865 (1987).
[314] Takeshita, H., Mori, A., Sugiyama, S., and Tsuda, T., *Bull. Chem. Soc. Jpn*, **59**, 3988 (1986).
[315] Ono, N., Miyake, H., Kamimura, A., and Kaji, A., *J. Chem. Soc., Perkin Trans. 1*, **1987**, 1929.
[316] Jung, M. E. and Grove, D. D., *J. Chem. Soc., Chem. Commun.*, **1987**, 753.
[317] Gassman, P. G. and Singleton, D. A., *Tetrahedron Lett.*, **28**, 5969 (1987).

318 Jung, M. E. and Buszek, K. R., *Tetrahedron Lett.*, **27**, 6165 (1986).
319 Gassman, P. G., Singleton, D. A., Wilwerding, J. J., and Chavan, S. P., *J. Am. Chem. Soc.*, **109**, 2182 (1987).
320 Pollart, D. J. and Rickborn, B., *J. Org. Chem.*, **51**, 792 (1987).
321 Ok, D. and Hart, H., *J. Org. Chem.*, **52**, 3835 (1987).
322 Calhoun, G. C. And Schuster, G. B., *J. Am. Chem. Soc.*, **108**, 8021 (1986).
323 Ando, W., Kumamoto, Y., and Tokitoh, N., *Tetrahedron Lett.*, **28**, 2867 (1987).
324 Brinker, U. H., Wüster, H., and Maas, G., *Angew. Chem. Int. Edn*, **26**, 577 (1987).
325 Schreck, M. and Christl, M., *Angew. Chem. Int. Edn*, **26**, 690 (1987).
326 Kahn, S. D. and Hehre, W. J., *J. Am. Chem. Soc.*, **109**, 663 (1987).
327 Loncharich, R. J., Schwartz, T. R., and Houk, K. N., *J. Am. Chem. Soc.*, **109**, 14 (1987).
328 Hartmann, H., Hady, A. F. A., Sartor, K., Weetman, J., and Helmchen, G., *Angew. Chem. Int. Edn*, **26**, 1143 (1987).
329 Koizumi, T., Arai, Y., Takayama, H., Kuriyama, K., and Shiro, M., *Tetrahedron Lett.*, **28**, 3689 (1987).
330 Döpp, D. and Pies, M., *J. Chem. Soc., Chem. Commun.*, **1987**, 1734.
331 Lubineau, A. and Queneau, Y., *J. Org. Chem.*, **52**, 1001 (1987).
332 Shing, T. K. M. and Lloyd-Williams, P., *J. Chem. Soc., Chem. Commun.*, **1987**, 423.
333 Kunz, H., Müller, B., and Schanzenbach, D., *Angew. Chem. Int. Edn*, **26**, 267 (1987).
334 Chapleur, Y. and Euvrard, M.-N., *J. Chem. Soc., Chem. Commun.*, **1987**, 884.
335 Takemura, H., Komeshima, N., Takahashi, I., Hashimoto, S., Ikota, N., Tomioka, K., and Koga, K., *Tetrahedron Lett.*, **28**, 5687 (1987).
336 Bir, G. and Kaufmann, D., *Tetrahedron Lett.*, **28**, 777 (1987).
337 De Lucchi, O., Buso, M., and Modena, G., *Tetrahedron Lett.*, **28**, 107 (1987).
338 Northcott, C. J. and Valenta, Z., *Can. J. Chem.*, **65**, 1917 (1987).
339 Quimpère, M. and Jankowski, K., *J. Chem. Soc., Chem. Commun.*, **1987**, 676.
340 Chapuis, C. and Jurczak, J., *Helv. Chim. Acta*, **70**, 436 (1987).
341 Seebach, D., Beck, A. K., Imwinkelreid, R., Roggo, S., and Wonnacott, A., *Helv. Chim. Acta*, **70**, 954 (1987).
342 Reetz, M. T., Kyung, S. H., Bolm, C., and Zierke, T., *Chem. Ind. (London)*, **1986**, 824.
343 Nakayama, J., Yomoda, R., and Hoshino, M., *Heterocycles*, **26**, 2215 (1987).
344 Meier, H., Eckes, H.-L., Niedermann, H.-P., and Kolshorn, H., *Angew. Chem. Int. Edn*, **26**, 1046 (1987).
345 Hanson, P. and Wren, S. A. C. *J. Chem. Soc., Perkin Trans. 2*, **1987**, 197.
346 Barluenga, J., Tomas, M., Ballesteros, A., and Gotor, V., *J. Chem. Soc., Chem. Commun.*, **1987**, 1195.
347 Rammash, B. K., Atwood, J. L., Weeks, J. A., and Wong, J. L., *J. Org. Chem.*, **52**, 2712 (1987).
348 Seitz, G., Hoferichter, R., and Mohr, R., *Angew. Chem. Int. Edn*, **26**, 332 (1987).
349 Marcelis, A. T. M. and van der Plas, H. C., *J. Heterocycl. Chem.*, **24**, 545 (1987).
350 Denmark, S. E., Cramer, C. J., and Sternberg, J. A., *Helv. Chim. Acta*, **69**, 1971 (1986).
351 Fitzsimmons, B. J., Leblanc, Y., and Rokach, J., *J. Am. Chem. Soc.*, **109**, 285 (1987).
352 Tsai, C.-Y. and Sha, C.-K., *Tetrahedron Lett.*, **28**, 1419 (1987).
353 Märkl, G. and Dorfmeister, G., *Tetrahedron Lett.*, **28**, 1093 (1987).
354 Ascherl, B., Kresze, G., Vaerman, J. L., Vandenbulcke-Coyette, B., and Viehe, H. G., *Bull. Soc. Chim. Belg.*, **96**, 51 (1987).
355 Annen, U. and Regitz, M., *Tetrahedron Lett.*, **28**, 5141 (1987).
356 Fuchs, E. P. O., Rösch, W., and Regitz, M., *Angew. Chem. Int. Edn*, **26**, 1011 (1987).
357 Quin, L. D., Hughes, A. N., and Pete, B., *Tetrahedron Lett.*, **28**, 5783 (1987).
358 Pellerin, B., Guenot, P., and Denis, J.-M., *Tetrahedron Lett.*, **28**, 5811 (1987).
359 Jurczak, J., Bauer, T., and Jarosz, S., *Tetrahedron*, **42**, 6477 (1986).
360 Freer, A. A., Kirby, G. W., and Lewis, R. A., *J. Chem. Soc., Chem. Commun.*, **1987**, 718.
361 Block, E. and Wall, A., *J. Org. Chem.*, **52**, 809 (1987).
362 Wiberg, N., Fisher, G., and Schurz, K., *Chem. Ber.*, **120**, 1605 (1987).
363 Schmidt, M. and Görl, U., *Angew. Chem. Int. Edn*, **26**, 887 (1987).
364 Nakayama, J., Akimoto, K., Niijima, J., and Hoshino, M., *Tetrahedron Lett.*, **28**, 4423 (1987).
365 Branchadell, V., Oliva, A., and Bertran, J., *J. Mol. Catal.*, **35**, 39 (1986).
366 Gupta, R. B. and Franck, R. W., *J. Am. Chem. Soc.*, **109**, 5393 (1987).
367 Ryan, K. M., Reamer, R. A., Volante, R. P., and Shinkai, I., *Tetrahedron Lett.*, **28**, 2103 (1987).
368 Beck, K. and Hünig, S., *Angew. Chem. Int. Edn*, **26**, 670 (1987).
369 Denmark, S. E., Cramer, C. J., and Dappen, M. S., *J. Org. Chem.*, **52**, 877 (1987).
370 Lorenz, K. T. and Bauld, N. L., *J. Am. Chem. Soc.*, **109**, 1157 (1987).

[371] Raynolds, D. W., Lorenz, K. T., Chiou, H.-S., Bellville, D. J., Pabon, R. A., and Bauld, N. L., *J. Am. Chem. Soc.*, **109**, 4960 (1987).
[372] Harirchian, B. and Bauld, N. L., *Tetrahedron Lett.*, **28**, 927 (1987).
[373] Mlcoch, J. and Steckhan, E., *Tetrahedron Lett.*, **28**, 1081 (1987).
[374] Okada, K., Samizo, F., and Oda, M., *Tetrahedron Lett.*, **28**, 3819 (1987).
[375] Roush, W. R., Essenfeld, A. P., and Warmus, J. S., *Tetrahedron Lett.*, **28**, 2447 (1987).
[376] Roush, W. R. and Coe, J. W., *Tetrahedron Lett.*, **28**, 931 (1987).
[377] Baettig, K., Dallaire, C., Pitteloud, R., and Deslongchamps, P., *Tetrahedron Lett.*, **28**, 5249 (1987).
[378] Baettig, K., Marinier, A., Pitteloud, R., and Deslongchamps, P., *Tetrahedron Lett.*, **28**, 5253 (1987).
[379] Bérubé, G. and Deslongchamps, P., *Tetrahedron Lett.*, **28**, 5255 (1987).
[380] Zschiesche, R., Grimm, E. L., and Reissig, H.-U., *Angew. Chem. Int. Edn*, **25**, 1086 (1986).
[381] Shea, K. J., Fruscella, W. M., Carr, R. C., Burke, L. D., and Cooper, D. K., *J. Am. Chem. Soc.*, **109**, 447 (1987).
[382] Handa, S., Jones, K., and Newton, C. G., *J. Chem. Soc., Chem. Commun.*, **1986**, 1797.
[383] Hauser, A., Gruendemann, E., Preuschof, H., Schmitz, E., and Kuban, R. J., *J. Prakt. Chem.*, **328**, 488 (1986); *Chem. Abs.*, **107**, 77036 (1987).
[384] Saxton, H. M., Sutherland, J. K., and Whaley, C., *J. Chem. Soc., Chem. Commun.*, **1987**, 1449.
[385] Schreiber, S. L., Meyers, H. V., and Wiberg, K. B., *J. Am. Chem. Soc.*, **108**, 8274 (1986).
[386] Parker, K. A. and Iqbal, T., *J. Org. Chem.*, **52**, 4369 (1987).
[387] Parker, K. A. and Iqbal, T., *Tetrahedron Lett.*, **27**, 6291 (1986).
[388] Guy, A., Lemaire, M., Graillot, Y., Negre, M., and Guette, J. P., *Tetrahedron Lett.*, **28**, 2969 (1987).
[389] Marshall, J. A., Grote, J., and Audia, J. E., *J. Am. Chem. Soc.*, **109**, 1186 (1987).
[390] Marshall, J. A., Shearer, B. G., and Crooks, S. L., *J. Org. Chem.*, **52**, 1236 (1987).
[391] Trifonov, L. S., Simova, S. D., and Orahovats, A. S., *Tetrahedron Lett.*, **28**, 3391 (1987).
[392] Diehl, K. and Himbert, G., *Chem. Ber.*, **119**, 3812 (1986).
[393] Keay, B. A., *J. Chem. Soc., Chem. Commun.*, **1987**, 419.
[394] Wu, H.-J. and Pan, K., *J. Chem. Soc., Chem. Commun.*, **1987**, 898.
[395] Fischer, K. and Hünig, S., *Chem. Ber.*, **120**, 325 (1987).
[396] Fischer, K. and Hünig, S., *J. Org. Chem.*, **52**, 564 (1987).
[397] Yamaguchi, Y., Yamada, H., Hayakawa, K., and Kanematsu, K., *J. Org. Chem.*, **52**, 2040 (1987).
[398] Missiaen, P. and De Clercq, P. J., *Bull. Soc. Chim. Belg.*, **96**, 105 (1987).
[399] Brandi, A. and Nuti, P., *Heterocycles*, **26**, 47 (1987).
[400] Jacobi, P. A. and Frechette, R. F., *Tetrahedron Lett.*, **28**, 2937 (1987).
[401] Medina, J. C., Cadilla, R., and Kyler, K. S., *Tetrahedron Lett.*, **28**, 1059 (1987).
[402] Gilchrist, T. L., Wasson, R. C., King, F. D., and Wootton, G., *J. Chem. Soc., Perkin Trans. 1*, **1987**, 2511.
[403] Gilchrist, T. L., Wasson, R. C., King, F. D., and Wootton, G., *J. Chem. Soc., Perkin Trans. 1*, **1987**, 2517.
[404] Tietze, L. F., Bratz, M., Machinek, R., and Kiedrowski, G., *J. Org. Chem.*, **52**, 1638 (1987).
[405] Tietze, L. F., Brand, S., Pfeiffer, T., Antel, J., Harms, K., and Sheldrick, G. M., *J. Am. Chem. Soc.*, **109**, 921 (1987).
[406] Tietze, L.-F., Brumby, T., and Pretor, M., *Synthesis*, **1987**, 700.
[407] Grimme, W. and Wiechers, G., *Tetrahedron Lett.*, **28**, 6035 (1987).
[408] Shea, K. J. and Burke, L. D., *Tetrahedron Lett.*, **28**, 735 (1987).
[409] Adam, W., de Lucchi, O., Pasquato, L., and Will, B., *Chem. Ber.*, **120**, 531 (1987).
[410] Jenner, G., *Tetrahedron Lett.*, **28**, 3927 (1987).
[411] Potts, K. T., Dery, M. O., and Kullnig, R. K., *J. Chem. Soc., Chem. Commun.*, **1987**, 840.
[412] Gotthardt, H. and Blum, J., *Chem. Ber.*, **120**, 109 (1987).
[413] Gotthardt, H. and Blum, J., *Chem. Ber.*, **120**, 115 (1987).
[414] Gotthardt, H. and Riegels, M., *Chem. Ber.*, **120**, 445 (1987).
[415] Jorgensen, K. A., Moltzen, E. K., and Senning, A., *J. Org. Chem.*, **52**, 2505 (1987).
[416] Yokelson, H. B., Millevolte, A. J., Haller, K. J., and West, R., *J. Chem. Soc., Chem. Commun.*, **1987**, 1605.
[417] Schäfer, A., Weidenbruch, M., Saak, W., and Pohl, S., *Angew. Chem. Int. Edn*, **26**, 776 (1987).
[418] Weissman, S. A. and Baxter, S. G., *Tetrahedron Lett.*, **28**, 603 (1987).
[419] Akasaka, T. and Ando, W., *Tetrahedron Lett.*, **28**, 217 (1987).
[420] Möhrle, H. and Dwuletzki, H., *Arch. Pharm. (Weinheim, Ger.)*, **320**, 298 (1987).
[421] Neuman, W. P., Michels, E., and Köcher, J., *Tetrahedron Lett.*, **28**, 3783 (1987).
[422] Lin, C. H., Lee, C., and Liu, C., *Organometallics*, **6**, 1861 (1987).
[423] Lin, C. H., Lee, C., and Liu, C., *Organometallics*, **6**, 1869 (1987).

[424] Lee, C., Lin, C. H., and Liu, C., *Organometallics*, **6**, 1878 (1987).
[425] Heilmann, W., Koschinsky, R., and Mayr, H., *J. Org. Chem.*, **52**, 1989 (1987).
[426] Atasoy, B., Balci, M., and Buyukgungor, O., *Tetrahedron Lett.*, **28**, 555 (1987).
[427] Feldman, K. S., Come, J. H., Fegley, G. J., and Smith, B. D., and Parvez, M., *Tetrahedron Lett.*, **28**, 607 (1987).
[428] Liu, C.-Y. and Houk, K. N., *Tetrahedron Lett.*, **28**, 1367 (1987).
[429] Liu, C.-Y. and Houk, K. N., *Tetrahedron Lett.*, **28**, 1371 (1987).
[430] Machiguchi, T., Hasegawa, T., Otani, H., and Ishii, Y., *J. Chem. Soc., Chem. Commun.*, **1987**, 1375.
[431] Abe, N., Kondo, S., and Morita, K., *Bull. Chem. Soc. Jpn*, **60**, 1201 (1987).
[432] Ishiguro, K., Sawaki, Y., and Iwamura, H., *Chem. Lett.*, **1987**, 1853.
[433] Davies, H. M. L., Smith, H. D., and Korkor, O., *Tetrahedron Lett.*, **28**, 1853 (1987).
[434] Davies, H. M. L., Clark, D. M., Alligood, D. B., and Eiband, G. R., *Tetrahedron*, **43**, 4265 (1987).
[435] Trost, B. M. and MacPherson, D. T., *J. Am. Chem. Soc.*, **109**, 3483 (1987).
[436] Föhlisch, B., Gehrlach, E., and Gewitz, B., *Chem. Ber.*, **120**, 1815 (1987).
[437] Mann, J., Holland, H. J., and Lewis, T., *Tetrahedron*, **43**, 2533 (1987).
[438] Baran, J. and Mayr, H., *J. Am. Chem. Soc.*, **109**, 6519 (1987).
[439] Trost, B. M. and Seone, P. R., *J. Am. Chem. Soc.*, **109**, 615 (1987).
[440] Mori, M., Nojima, M., and Kusabayashi, S., *J. Am. Chem. Soc.*, **109**, 4407 (1987).
[441] Wender, P. A. and Snapper, M. L., *Tetrahedron Lett.*, **28**, 2221 (1987).
[442] Wender, P. A. and Ihle, N. C., *Tetrahedron Lett.*, **28**, 2451 (1987).
[443] Becker, H.-D. and Andersson, K., *Tetrahedron Lett.*, **28**, 1323 (1987).
[444] Tamaki, T., Kokubu, T., and Ichimura, K., *Tetrahedron*, **43**, 1485 (1987).
[445] Albini, A., Fasani, E., and Faiardi, D., *J. Org. Chem.*, **52**, 155 (1987).
[446] Srikrishna, A. and Sunderbabu, G., *J. Org. Chem.*, **52**, 5037 (1987).
[447] Sugiyama, S. and Takeshita, H., *Bull. Chem. Soc. Jpn*, **60**, 977 (1987).
[448] Märkl, G., Beckh, H. J., Mayer, K. K., Ziegler, M. L., and Zahn, T., *Angew. Chem. Int. Edn*, **26**, 236 (1987).
[449] Aumann, R. and Weidenhaupt, H.-J., *Chem. Ber.*, **120**, 23 (1987).
[450] Bönnemann, H. and Brijoux, W., *Nouv. J. Chim.*, **11**, 549 (1987).
[451] Vitulli, G., Bertozzi, S., Vignali, M., Lazzaroni, R., and Salvadori, P., *J. Organomet. Chem.*, **326**, C33 (1987).
[452] Boese, R., Knölker, H.-J., and Vollhardt, K. P. C., *Angew. Chem. Int. Edn*, **26**, 1035 (1987).
[453] Sheppard, G. S. and Vollhardt, K. P. C., *J. Org. Chem.*, **51**, 5496 (1986).
[454] Yamazaki, H. and Wakatsuki, Y., *Kagaku, Zokan* (*Tokyo*), **1987**, 87; *Chem. Abs.*, **107**, 153636 (1987).
[455] Trost, B. M. and Tanoury, G. J., *J. Am. Chem. Soc.*, **109**, 4753 (1987).
[456] Yang, C. H., Lin, H. C., and Whang, M. J., *J. Chin. Chem. Soc.* (*Taipei*), **33**, 139 (1986); *Chem. Abs.*, **107**, 115065 (1987).
[457] Matsunaga, K. and Yamashita, T., *Toyo Daigaku Kogakubu Kenkyu Hokoku*, **21**, 57 (1985); *Chem. Abs.*, **107**, 77225 (1987).
[458] Barron, A. R. and Cowley, A. H., *Angew Chem. Int. Edn*, **26**, 907 (1987).
[459] Milczarek, R., Rüsseler, W., Binger, P., Jonas, K., Angermund, K., Krüger, C., and Regitz, M., *Angew. Chem. Int. Edn*, **26**, 908 (1987).
[460] Yoshida, J., Sakaguchi, K., Isoe, S., and Hirotsu, K., *Tetrahedron Lett.*, **28**, 667 (1987).
[461] Trah, S., Weidmann, K., Fritz, H., and Prinzbach, H., *Tetrahedron Lett.*, **28**, 4399 (1987).
[462] Trah, S., Weidmann, K., Fritz, H., and Prinzbach, H., *Tetrahedron Lett.*, **28**, 4403 (1987).
[463] Chow, T. J., Chao, Y.-S., and Liu, L.-K., *J. Am. Chem. Soc.*, **109**, 797 (1987).
[464] Sket, B., Zupancic, N., and Zupan, M., *J. Chem. Soc., Perkin Trans. 1*, **1987**, 981.
[465] Mattay, J., Runsink, J., Piccirilli, J. A., Jans, A. W. H., and Cornelisse, J., *J. Chem. Soc., Perkin Trans. 1*, **1987**, 15.
[466] Weber, G., Runsink, J., and Mattay, J., *J. Chem. Soc., Perkin Trans. 1*, **1987**, 2333.
[467] Mattay, J., Runsink, J., Heckendom, R., and Winkler, T., *Tetrahedron*, **43**, 5781 (1987).
[468] Pirrung, M. C., *J. Org. Chem.*, **52**, 1635 (1987).
[469] Schwack, W., *Tetrahedron Lett.*, **28**, 1869 (1987).
[470] Gilbert, A. and Heath, P., *Tetrahedron Lett.*, **28**, 5909 (1987).
[471] Chow, Y. L., Buono-Core, G. E., Liu, X.-Y., and Qian, P., *J. Chem. Soc., Chem. Commun.*, **1987**, 913.
[472] Gad El Karim, I. and Rzepa, H. S., *J. Chem. Soc., Chem. Commun.*, **1987**, 193.
[473] Nahm, S. H. and Cheng, H. N., *J. Org. Chem.*, **51**, 5093 (1986).
[474] Orfanopoulos, M., Foote, C. S., and Smonov, I., *Tetrahedron Lett.*, **28**, 15 (1987).

Organic Reaction Mechanisms 1987
Edited by A. C. Knipe and W. E. Watts

CHAPTER 15

Molecular Rearrangements

A. W. MURRAY

Department of Chemistry, University of Dundee

Aromatic Rearrangements

Benzene Derivatives

Rearrangement reactions that involve N—N bond-cleavage have been reviewed;[1] as has the Wallach rearrangement of azoxyarenes,[2,3] while an account of the use of heavy-atom kinetic isotope effects in deducing the mechanism of the photo-Wallach rearrangement, and the acid-catalysed and thermal benzidine rearrangements, has appeared.[4] ^{15}N and ^{13}C kinetic isotope effects have been measured for both the acid-catalysed and the thermal rearrangement of 2,2′-hydrazonaphthalene.[5] The relatively small size of the ^{13}C effect and the large size of the ^{13}N effect in this study suggest that N—N bond-breaking and C—C bond-forming are not synchronous. Interestingly, in a recent study,[6] the nature of the products obtained in the base- and acid-catalysed reactions of phenylazo-1-naphthyl ether does not support the Kekulé diazo-coupling mechanism which suggests the intermediacy of a diazo-ether in the reaction. A new rearrangement of diacylamines containing N-protected amino-acids has been described,[7] and a recent study[8] of the *N,N*-diphenylsulphamide rearrangement has confirmed that 2-aminobenzenesulphonanilide is also a product of the rearrangement. Thermochemical and kinetic parameters have been determined[9] for the acid-base and stereoisomeric conversion of nitrobenzenediazonium compounds in aqueous media. The Fischer–Hepp rearrangement of *o*-nitrosodiphenylamines has been studied,[10] and a recent investigation has indicated that the rearrangement of *N*-(4-substituted phenyl)-piperidine and -morpholine *N*-oxides to the corresponding *N*-hydroxylamines can best be described by the intramolecular nucleophilic substitution presented in Scheme 1.[11,12] Unexpected migration of an amino group from P to N was observed to occur when *N*-(aminophosphinoyl)-*O*-sulphonylhydroxylamines were treated with base,[13] while a detailed study of the acid-catalysed rearrangement of *N*-(2,6-dimethylphenyl)hydroxylamine has established the intermediacy of a free nitrenium ion in the process.[14] Reaction of an initially formed *N-tert*-butylhydroxylamine, to give both the hydroxamic acid and the *O*-acylhydroxylamine, has been proposed[15] as the most satisfactory interpretation of the thermal conversion of *N-tert*-butylbenzohydroxamic acids to the corresponding *O*-benzoylhydroxylamines.

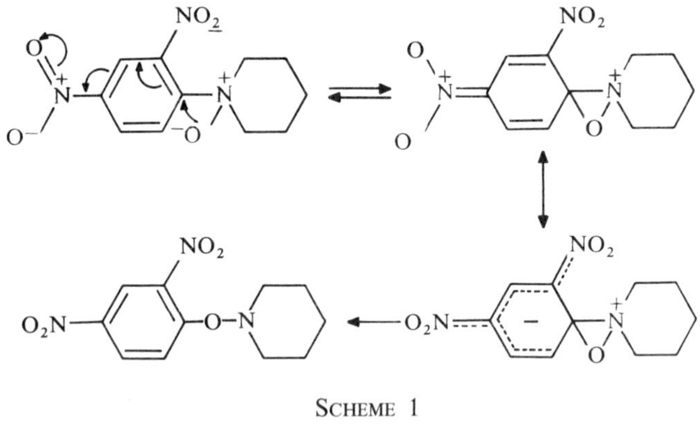

SCHEME 1

Equilibrium and rate constants and activation energies have been determined for the thermal rearrangement of phenyl- and alkylphenyl-ammonium hydrogen sulphates to aminobenzenesulphonic acids.[16] The process is thought to involve thermolysis of the reactant to the free base and sulphuric acid, followed by sulphonation of the amine. The postulation of an intramolecular radical rearrangement to account for the photo-rearrangement of sodium phenylsulphamic acid to the isomeric anilinesulphonic acids has gained credence,[17] while the same authors have proposed[18] that a series of σ-complexes give rise to the isomeric aniline sulphonic acids obtained on photolysis of the sodium salt of *m*-aminosulphonic acid.

The rearrangement of *O*-(*o*-acetylaryl)dimethylthiocarbamates to their corresponding *S*-aryl dimethylthiocarbamates has been achieved in reasonable yield.[19] The mechanism of the Chapman rearrangement of *N*-arylbenzimidates such as (**1**) which gives dibenz[*ch*]acridinone (**2**), has been interpreted on the basis of X-ray data.[20] In particular, the observation of an *N*...*C* interaction (2.648 Å) supports the mechanism shown in Scheme 2.

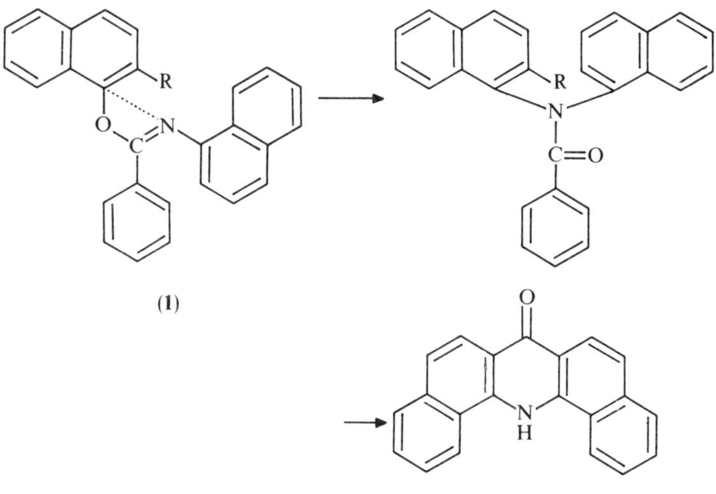

(**1**)

(**2**)
R = COOMe

SCHEME 2

The formation of the Meisenheimer spiro-adduct of *N*-(2,4,6-trinitrophenyl)-aniline methylamide, and its rearrangement to 2-amino-*N*-methyl-*N*-(2,4,6-trinitrophenyl)propanamide, has been reported,[21] and a aziridinone intermediate has been invoked[22] to account for the formation of 4,6-dinitro-2-nitrosoaniline from base treatment of the methyl and ethyl esters of *N*-(2,4,6-trinitrophenyl)-glycine and -alanine. An instance of the Smiles rearrangement being activated by fluorine atoms is exemplified by the conversion of acetaminotetrafluorodiphenyl ethers into acetyltrifluorophenoxazines.[23] Gas-phase rearrangement of 2-(arylamino)phenylaminyl radicals has been shown to lead to diphenylamine and phenazine products which apparently arise by equilibration of the initial aminyl radical via a spirodienyl radical.[24]

The reported butyllithium-promoted rearrangement of *o*-bromophenyl esters (**3**) to the corresponding *o*-hydroxy-ketones (**4**) after hydrolysis, represents an example of a metal-promoted Fries rearrangement.[25] Evidence for an intermolecular mechanism has been presented[26] for the Fries rearrangement of phenyl *m*-nitrobenzoates, and a mechanism invoking a Fries rearrangement has been considered[27] for the conversion of 2-hydroxybenzoylmethanes (**5**) into 2-benzoyl-2-methylchromones (**6**). 5,8-Dihydroxy-4,4-dimethyl-3,4-dihydro-1(2*H*)-naphthal-

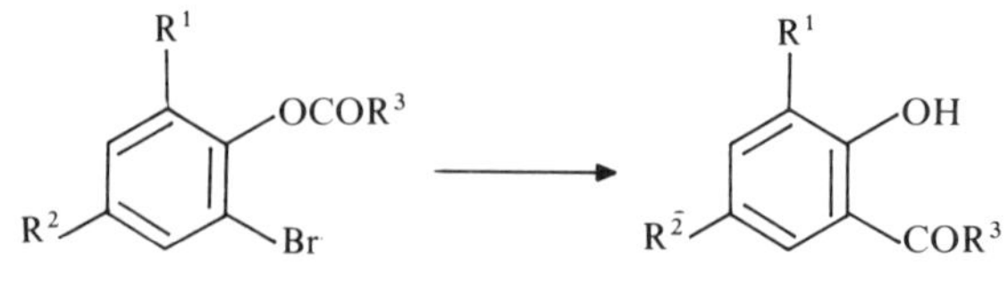

(**3**) (**4**)

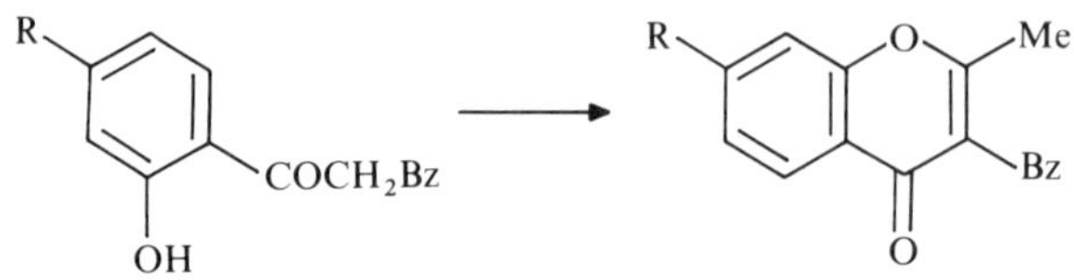

(**5**) (**6**)

enone has been produced unexpectedly during the Fries rearrangement of 1-methoxy-4-(4-methyl-2-and-3-pentenoyloxy)benzenes,[28] while irradiation of a series of aryl *trans*-2-butenoates has been found to produce the corresponding photo-Fries products together with isomerized starting material.[29] 2-Acetyl-4-methoxyphenol, the product of the photo-Fries rearrangement of 4-methoxyphenyl acetate, is considered to arise by recombination of a caged radical-pair originating from an excited singlet state,[30] and *p*-acetylcalix[4]arene has been synthesized via a Fries rearrangement route.[31] On nitration in acetic acid, 3-hydroxyacetophenones have been found to undergo substitution of the acyl group by a nitro group, followed by a 1,3-acyl shift reminiscent of a retro-Fries rearrangement, to yield phenyl esters.[32]

The novel photochemical 1,4-acyl migration that occurs during the photo-conversion of enol acetates (**7**) to 1,4-diketones (**8**) has been explained[33] by postulating a primary homolytic carbonyl-oxygen bond cleavage followed by a 1,2-hydrogen shift and recombination with the acetyl radical. The 1,3-oxygen-to-oxygen rearrangement of group IV esters of benzoic acid has been studied using ^{18}O isotopic substitution (see Scheme 3). The authors conclude[34] that when R = $SiMe_3$, $SiMe_2Bu^t$ or $SiMe_2Bu^s$, the rearrangement is intramolecular via a pentacoordinate silicon intermediate, or a transition state that does not involve significant charge separation. A new class of selenonium zwitterion (**9**) has been prepared by treating phenols with diphenylselenium bis(trifluoroacetate). These zwitterions have been observed to undergo a novel thermal rearrangement (see Scheme 4) to produce diaryl ethers (**10**).[35] Benzyne intermediates have been implicated[36] as being responsible for activation of aliphatic C—H bonds and the facile isomerization processes that occur on cyclometalation of 2,6-di-*tert*-butylphenoxide ligands by tantalum aryl groups.

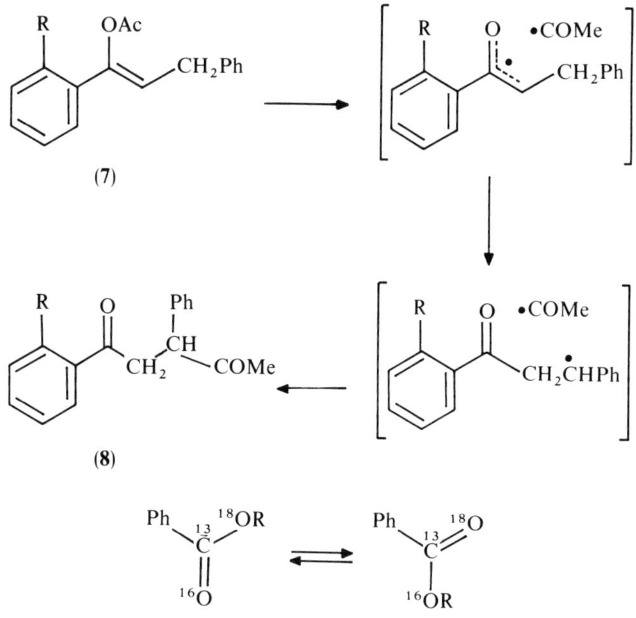

SCHEME 3

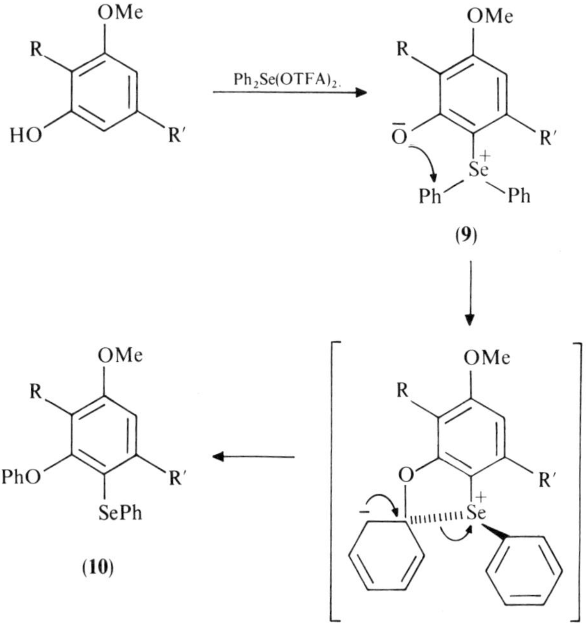

SCHEME 4

The photochemistry of 4-chlorobiphenyl in water has been studied in detail. The study established that arenium ions, arynes, and addition-with-elimination mechanisms are not involved in the formation of 2-hydroxybiphenyl. The plausible alternative put forward[37] is that 4-chlorobiphenyl isomerizes to the 3-isomer which then undergoes photo-hydrolysis. The *meta* migration of a chlorine atom has been observed during the photolysis of polychlorobenzenes,[38] while rearrangement of a *tert*-amyl group has been noted in the Friedel–Crafts reaction of *tert*-amylbenzene derivatives.[39] Zeolite HZSM-5 has been used to catalyse the equilibration of toluidines and ethylanilines by an intramolecular 1,2-shift mechanism,[40] and the same catalyst has been found to catalyse the equilibration of dihydroxybenzenes.[41] On heating in the presence of $AlCl_3$, 1-(1-phenanthryl)pyrene did not yield the product of dehydrogenative cyclization, but underwent complete rearrangement to isomeric biaryls.[42]

Differences in the regioselectivity of degenerate 1,2-shifts of nitro groups, and hydrogen and chlorine atoms in the 1-substituted 1,2:3,4:5,6-tributanobenzenonium carbocation have been studied by NMR spectroscopy.[43] Ion-pairing, ring-opening, and the rates of formation of aryl-bridged chlorinated intermediates have been investigated through reactions of mixed 1,2-dihalo-3-arylpropanes and 1,3-dihalo-2-arylpropanes. This has allowed an estimation of the free-energy difference between isomeric aryl-bridged (3-bromopropylene)benzenonium ions and bromine-bridged 3-arylpropylenebromonium ions.[44] 1,3-Migration of a phenyl group via a phenyl-bridged silicocation (**11**) has been observed.[45]

It has been demonstrated that at high temperatures in a quartz flow system, the atoms of benzene interchange positions intramolecularly. Several mechanistic possibilities that could explain this automerization of benzene have been considered,[46] but of these, the reversible formation of benzvalene (see Scheme 5) appears to be the most reasonable. Dicyanoprismanes have been suggested as intermediates

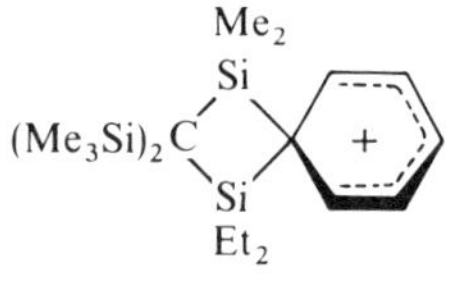

(**11**)

SCHEME 5

in the migration of a cyano group in dicyanobenzene in a radiofrequency plasma.[47] A thorough investigation of the roles of phenylmethylene, bicyclo[4.1.0]-hepta-2,4,6-triene, cycloheptatrienylidene, 1,2,4,6-cycloheptatetraene and bicyclo[3.2.0]hepta-1,3,6-triene in arylcarbene interconversions and related C_7H_6

rearrangements has been initiated.[48] Flash vacuum pyrolysis of 9-methyl-1,3-dihydronaphtho[1,2-*c*]furan-1,3-dione (**12**) gave exclusively 1*H*-cyclopent[*cd*]indene (**15**). These experimental findings have been interpreted[49] in terms of the following series of reactions: loss of CO_2 and CO from the anhydride, rearrangement of 8-methyl-1,2-didehydronaphthalene (**13**) to 7-methyl-1*H*-indenylidenecarbene (**14**), and insertion of the carbene into the CH bond of the

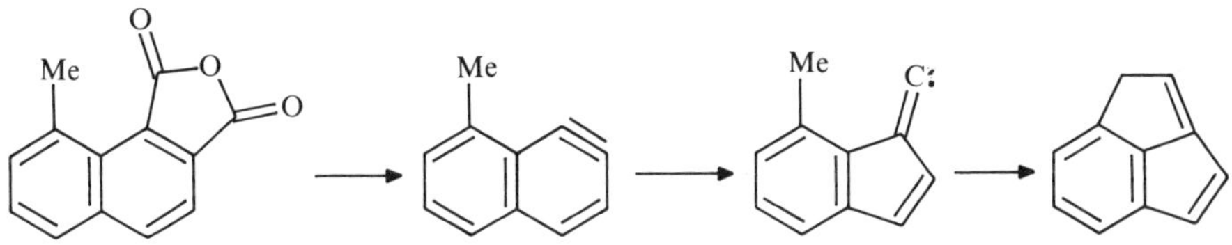

(**12**) (**13**) (**14**) (**15**)

adjacent methyl group. The predicted interconversions of aceanthrylene, acephenanthrylene, and fluoranthene at high temperatures have been verified,[50] and represent the first examples of thermal rearrangements of aromatic hydrocarbons in which one benzenoid skeleton is transformed into another benzenoid skeleton. The mechanism by which these rearrangements occur appears to involve a carbon 1,2-shift (benzene ring-contraction) and a hydrogen 1,2-shift, both across the same carbon–carbon bond but in opposite directions. The precise timing of these two shifts relative to each other remains uncertain. The observed formation of 9-ethylanthracene from thermolysis of 8*a*,9-dihydro-8*a*-methyl-9-methyleneanthracene represents the first example of a molecular rearrangement in which migration of a single substituent results in aromatization of two alicyclic rings. The most obvious mechanism for this reaction is a free-radical dissociation-recombination reaction occurring within a solvent cage.[51]

[4]Paracyclophane has been generated photochemically from 1,4-tetramethylene Dewar benzene,[52] and a new preparative route to [2.2.2](1,2,3)-cyclophanes has been described.[53]

Heterocyclic Derivatives

A review on the mechanism of S,N-, O,N-, and N,N-rearrangements has appeared,[54] and heteroatom rearrangements such as S,N, O,N, and N,N double rearrangements have been observed[55] when acrylonitriles substituted with two β-hetero-atom groups (MeS and SH, OH, or NHPh) are condensed with aliphatic carboxylic acids in the presence of toluene-*p*-sulphonic acid.

Experiments designed to prepare 3*a*-azaazulen-1-ones (**17**) from pyrolysis of enaminones (**16**) resulted[56] in formation of rearranged γ-pyridone derivatives (**18**) in a reaction which is formally analogous to the transformation of azulene into naphthalene. Kinetic studies[57,58] have indicated that the acid-catalysed rearrangement of 3,3-disubstituted indolenines to 2,3-disubstituted indoles takes place via a spiro(cyclopentane-3′-indolenine). Grignard reagents from *N*-unsubstituted indoles

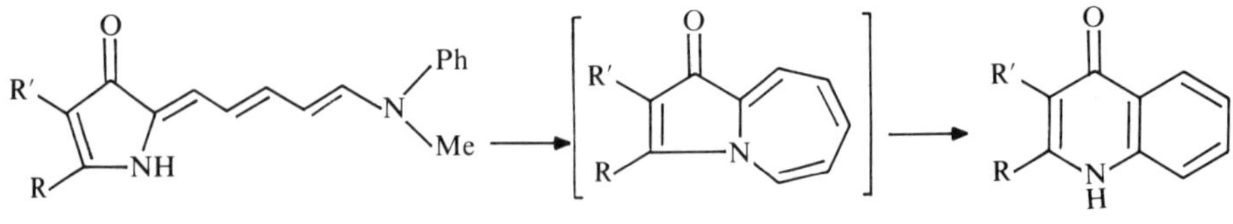

(16) **(17)** **(18)**

and benzenesulphonyl chloride have been found to undergo reaction, such as to produce unstable intermediates of β-chlorination and α-sulphonylation which are transformed into oxindoles or substituted indoles.[59] The relative quantities of chlorocarbazoles obtained[60] from acid treatment of *N*-chlorocarbazole strongly support the idea that they are formed in an acid-catalysed, intermolecular rearrangement analogous to that observed with *N*-chloropyrrole or with *N*-chloroanilines.

A review has appeared[61] on the use of ^{13}C- and ^{17}O-labelling to detect unusual novel ketoketene- and furandione–furandione rearrangements during pyrolysis or cycloaddition reactions of five-numbered 4-acyl substituted 2,3-dioxo heterocycles. A recent study[62] has shown that protonated 3-methylfuran isomerizes to protonated 2-methylfuran by a methyl migration, while the mechanism of the photoisomerization of 2-methylfuran to 3-methylfuran has been investigated[63] with a semiempirical MO method (SINDO 1). A diradicaloid was found to originate through ring-opening in this latter process. Acid-catalysed cationic rearrangements of 2,3-dihydrobenzofuran derivatives to xanthenes have been reported.[64] A novel photo-rearrangement of 3,3,5-triaryl-2(3*H*)-furanones to 3,4,5-triaryl-2(5*H*)-furanones has been described,[65] and the sensitized photo-rearrangement of 2,4-dioxofuro[2,3-*d*]pyrimidines has afforded 5,7-diazaspiro[2,5]octanes.[66] Thermolysis of 1*a*,1*b*,6*b*,6*c*-tetrahydro-azirino[2′,3′:3,4]cyclobuta[1,2-*b*][1]benzofurans has resulted in ring transformations to form benzo[2,3-*b*]pyrroles and their [3,2-*b*] isomers.[67]

The details of a new route to 1,2-diazetidinones (**20**) based on the photochemical ring-contraction of symmetrically substituted 4-diazopyrazolidine-3,5-diones (**19**) in the presence of nucleophiles has been reported[68,69] (see Scheme 6), and the rearrangement of 2-vinylpyrazolium salts (**21**) to 1,2-dihydropyrimidines (**22**) has been catalysed by alkoxides.[70] The thermal rearrangement of 2-aryl-1-cyanoindazol-3-ones to the corresponding benzimidazo[2,1-*b*]quinazolones has been postulated[71] to proceed as shown in Scheme 7. In the case of the 2-(2,6-

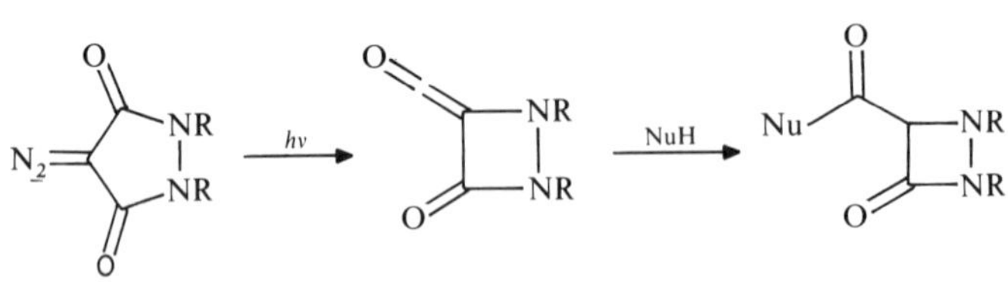

(19) **(20)**

SCHEME 6

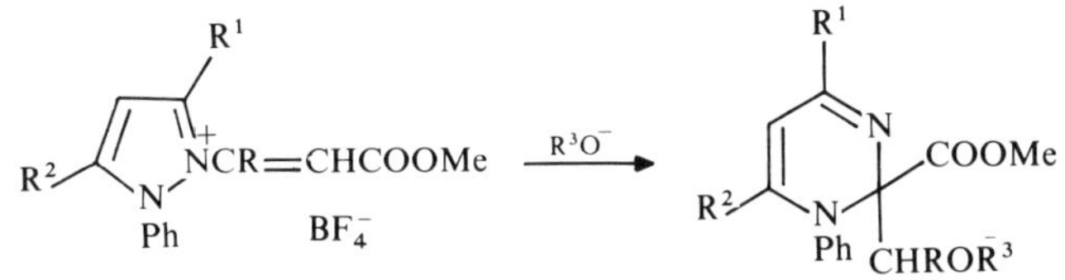

(21) (22)

dimethylphenyl)- and 2-(2,4,6-trimethylphenyl) compounds, rearrangement is accompanied by [1,9]-sigmatropic shifts of the obstructing methyl groups.

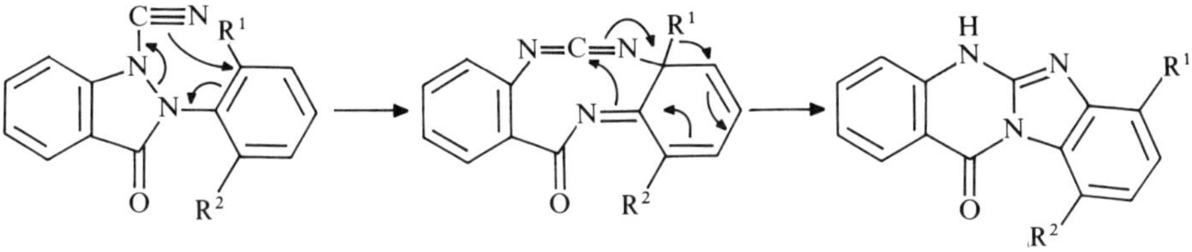

SCHEME 7

Oxadiazolecarboxamides have been obtained by photochemical rearrangement of nitroimidazoles in aqueous solution,[72] and in an analogous manner, UV irradiation of 1-hydroxy-2-methyl-5-nitroimidazole in oxygen-free aqueous solution has led to *N*-(2-hydroxyethyl)-5-methyl-1,2,4-oxadiazole-3-carboxamide.[73] Both these rearrangements possibly proceed by a nitro-to-nitrite rearrangement, illustrated for the second of these reactions in Scheme 8.

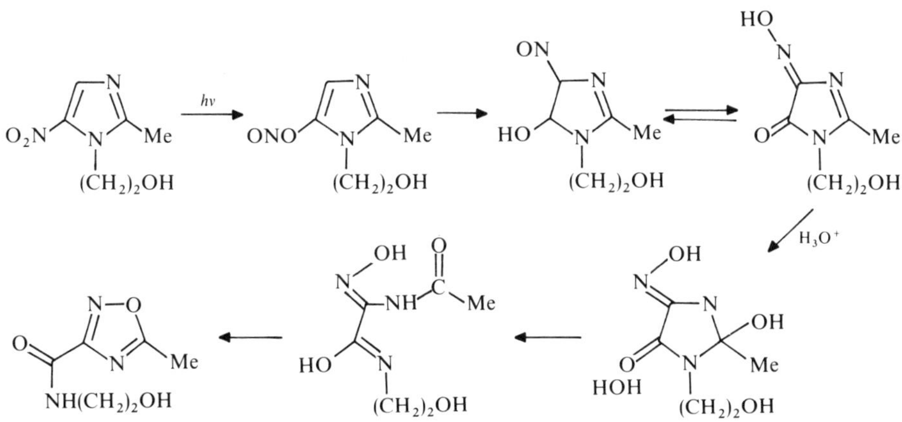

SCHEME 8

The rearrangement of 2-allyl-5(2*H*)-oxazolones to tetrahydro-3H-furo[3,4-*b*]pyrrole derivatives has been catalysed by Lewis acids,[74] and rearrangements of 2-amino-5-phenacyloxazoles (**23**) to imidazoles (**24**) and pyrimidines (**25**) on treatment with dimethylamine have been described.[75] The putative intermediates in these rearrangements are illustrated in Scheme 9. An electron-withdrawing group R in (**23**) would result in a favouring of the benzoyl carbonyl group toward nucleophilic attack, thus leading to a pyrimidine product. Photo-induced rearrangement of isoxazolines (**26**) has produced mixtures of enamino-aldehydes (**27**) and (**28**),[76] while photo-rearrangement of (**29**) afforded 2-substituted-6-phenyl-7-formyl-2, 4, 5, 8-tetrahydro-1, 3-dioxa-5-azocines (**30**).[77] The photo-conversion of 3-(aryloxy)anthra[1, 9-*e*, *d*]-6-isoxazolones to naphtho[2, 3-*a*]phenoxazine-8, 13-diones is thought[78] to proceed from the triplet and singlet excited states of the former. A new heterocyclic system, viz. derivatives of 2-aza-

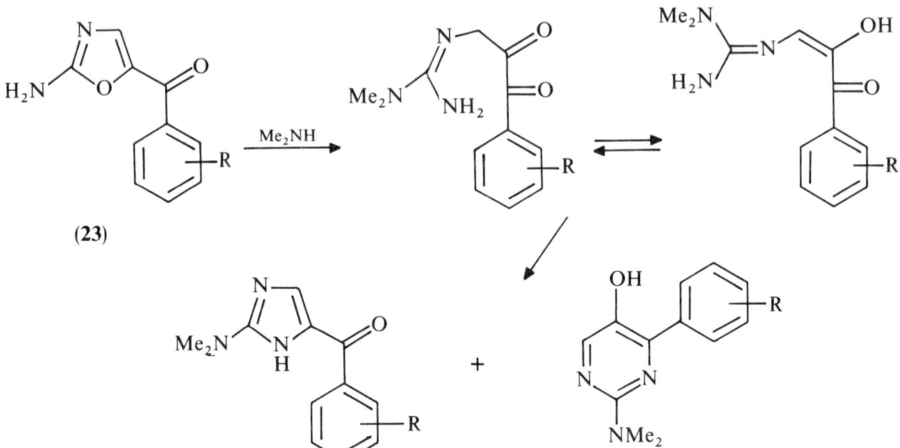

(**23**) (**24**) (**25**)

SCHEME 9

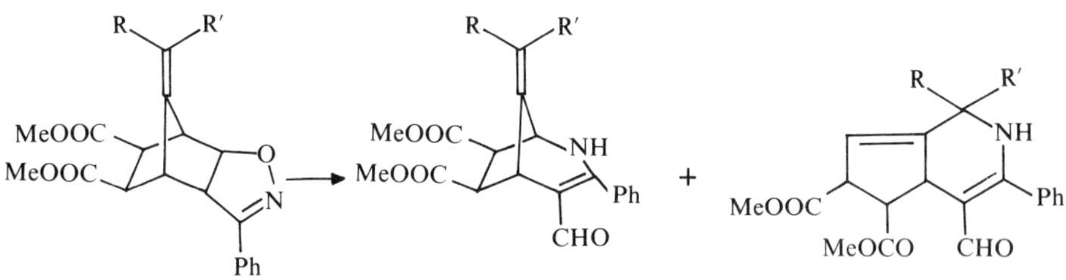

(**26**) (**27**) (**28**)

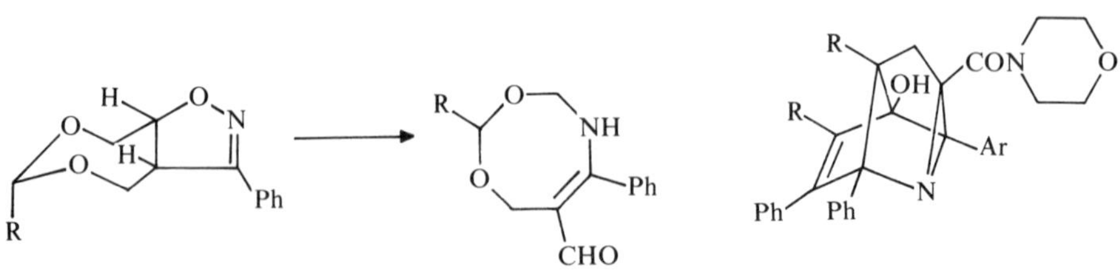

(**29**) (**30**) (**31**)

tetracyclo[4.3.0.0^{2,4}.0^{3,7}]nonane (**31**), has been produced[79] by thermal rearrangement of the cycloadducts obtained on reacting 5-amino-4,5-dihydro-4-methyleneisoxazoles and cyclopentadienones. A study has been made[80] of the 5-*exo*-methylene-isoxazolidine to 3-pyrrolidinone rearrangement. A novel transformation of 2,1-benzisoxazoles, involving N—O bond rupture with concomitant N—C bond formation by reaction with aryl-, methyl-, or 2-thienyl-zinc chlorides in the presence of a nickel catalyst, has been described.[81] The products of the reaction are *o*-(substituted-amino)-benzaldehydes and -benzophenones.

Ring transformations of thiazoles and imidazoles have been described.[82] A process involving homolytic cleavage of the S—N bond and intramolecular recombination of the biradical, followed by a hydrogen shift to regenerate the aromatic ring, has been proposed[83] to account for the unexpected photochemical transformation of naphthylbenzisothiazolone (**32**) into the isomeric benzonaphthothiazepinone (**33**). The S—C, S—N, and N—C bonds in the five-membered ring of (**34**) do not appear to undergo cleavage when it is irradiated. Instead, sulphinohydroxamic acid (**35**) is formed; (**35**) is stable in crystalline form but isomerizes back to (**34**) in solution.[84]

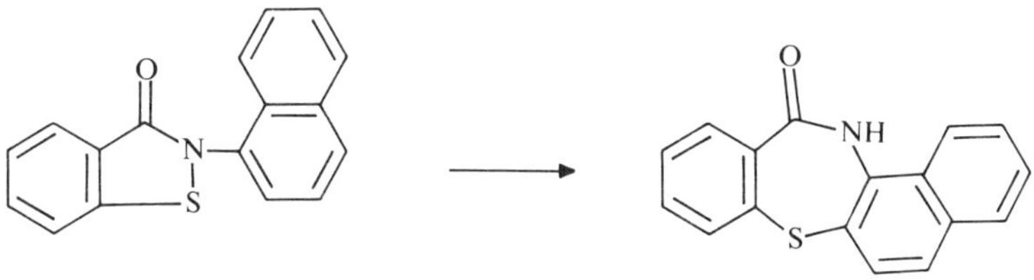

(32) (33)

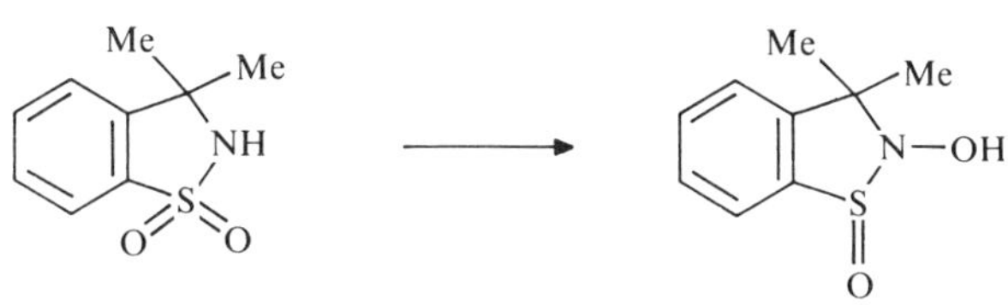

(34) (35)

The formation of indoles (**40**) and (**41**) on photolysis of 1-aryl-1,2,3-triazoles (**36**) with an electron-withdrawing group at C(5) has been explained[85] by a mechanism (see Scheme 10) in which the less stable carbene intermediate (**37**) rearranges to the more stable carbene (**39**), via the 1*H*-azirine (**38**). 2,5-Dihydro-1,2,3-triazine derivatives have arisen from attempted 1,3-dipolar cycloaddition of 1,2,3-triazole *N*-oxides with dimethyl acetylenedicarboxylate. This new ring-expansion is thought to proceed via the precursory intermediates (**42**) and (**43**) shown in Scheme 11.[86] The same group[87] has discovered that hexahydro-oxazolo[4,5-*d*]-1,2,3-triazoles undergo an analogous thermally-induced ring-expansion providing a simple, new route to the unusual 1,3,4,5-oxatriazine ring.

The thermal rearrangement of 2-aryl-5-methoxy-1,3,4-oxadiazoles to 2-aryl-4-methyl-1,3,4-oxadiazol-5(4*H*)-ones has been studied[88] by differential scanning

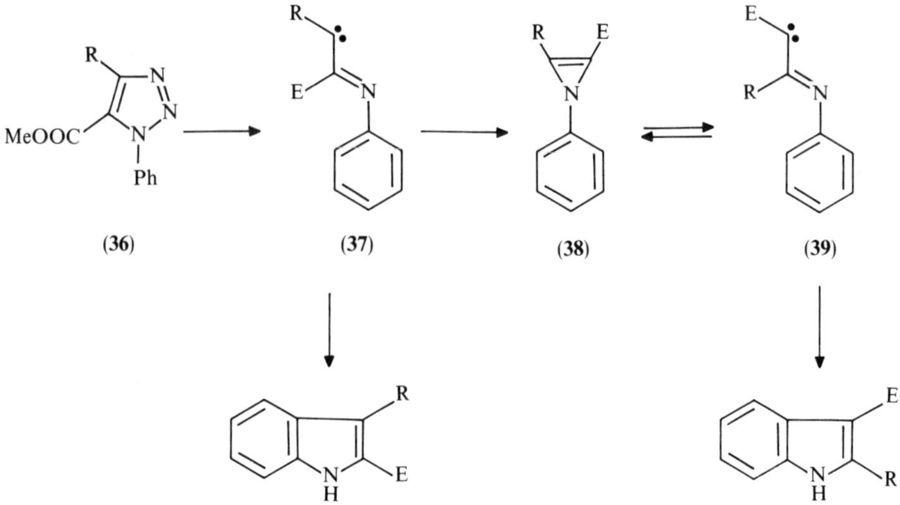

(36) (37) (38) (39)

E = COOMe (40) (41)

SCHEME 10

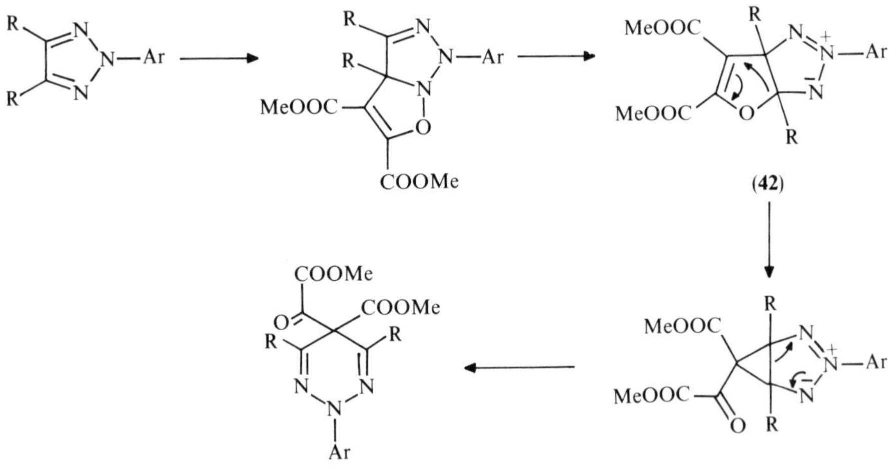

(42) (43)

SCHEME 11

calorimetry, and rearrangements of 1,2,4- and 1,2,5-oxadiazoles with carbethoxythiourea have been investigated.[89] In a number of cases, the corresponding thioureas were isolated, but these were found to rearrange easily to thiadiazoles. 1,2,5-Thiadiazole 1-oxides have been transformed into pyrazines, 1,2,4-thiadiazoles, and pyrroles using Huenig's base,[90] and a novel acid-catalysed rearrangement of 3-acyl-5-acylamino-2,3-dihydro-1,3,4-thiadiazole 1-oxides to 1,3,4-oxadiazoles has been reported.[91] Imidazolones have been produced by the $NaHCO_3$-induced rearrangement of *N*-phenacyltetrazolium salts.[92]

3,4-Dimethyl-1-oxa-6,6aλ^4-diselena-2-azapentalene (**44**) was found to react with tetraphosphorus decasulphide to give product (**45**) and various analogues by a new type of molecular rearrangement. The authors[93] have proposed that the transformation (**44**) to (**45**) occurs by the mechanism in Scheme 12. A new imidophosphonate-amidophosphite rearrangement, viz. (**46**)→(**47**), has been reported.[94]

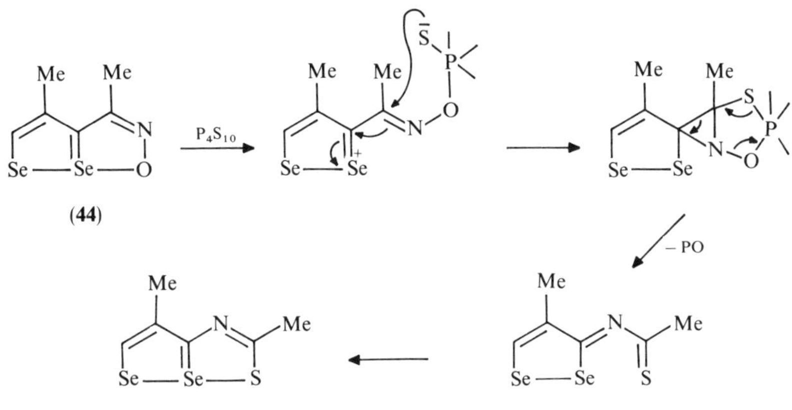

SCHEME 12

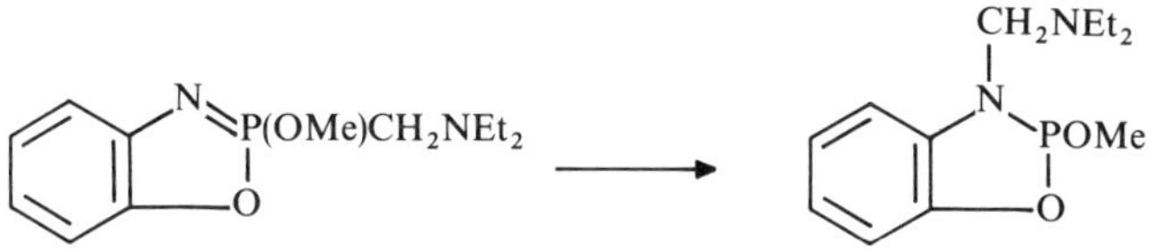

Hexahydro-1*H*-1-pyridines have been produced by the acid-catalysed rearrangement of 9-alkylidene-5-(*m*-methoxyphenyl)-2-methylmorphans,[95] and 4-(2-aminophenyl)-1,4-dihydro-2,6-dimethyl-3,5-pyridinedicarboxylic acid diethyl ester (**48**) in refluxing pyridine afforded 1,2,3,6-tetrahydro-2,4-dimethyl-2,6-methano-1,3-benzodiazocine-5,11-dicarboxylic acid diethyl ester and 2-methyl-3-quinolinecarboxylic acid ethyl ester (**49**). The formation of (**49**) has been rationalized[96] by the mechanistic scheme shown in Scheme 13.

An investigation[97] has been made of substituent effects on the photolytic rearrangement of nitraminopyridines in which the side-chain nitro group migrates to the vicinal β-position, while steric hindrance has been found to cause an unusual N-to-C migration, and hence the formation of (pyridylmethyl)pyridines (**50**) and (**51**) when 2-amino-6-methylpyridine was heated with CH_2I_2.[98] 1-Amino-4-methylpyridinium salts have been rearranged to 4-aminomethylpyridine by refluxing with base,[99] and an example of the conversion of a pyridinium salt into an

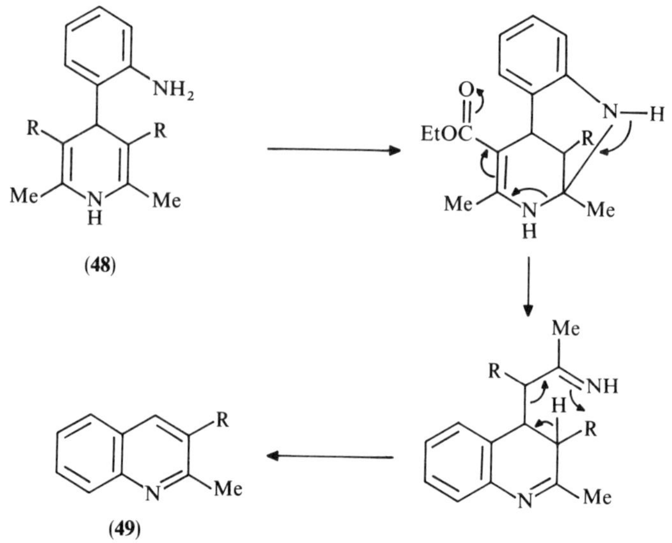

R = COOEt

SCHEME 13

azepine has been exemplified by the thermal, base-induced rearrangement of 1-benzyl-2, 4, 6-triphenylpyridinium tetrafluoroborate to 3*H*-2, 4, 6, 7-tetraphenylazepine.[100] 4-Acetylaminoisoquinoline-3(2*H*)-ones have been prepared[101] by treating 4-hydroxyimino-1,4-dihydroisoquinolin-3(2*H*)-ones with an HCl-AcOH-Ac_2O mixture.

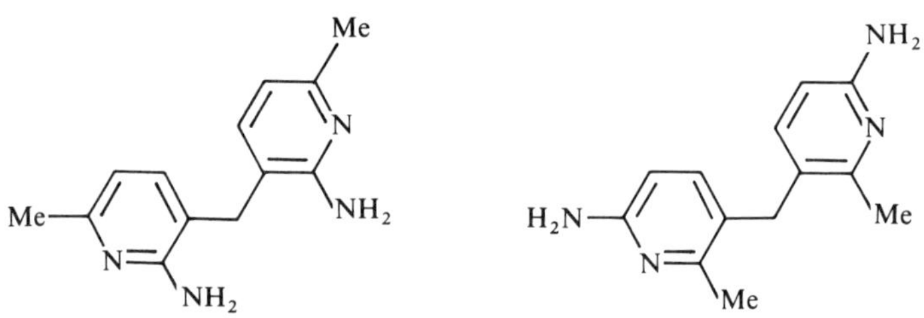

(50) **(51)**

A novel photochemical ring-contraction of some 4*H*-pyrans (**52**) to cyclobutenes (**53**) has been described,[102] the favoured route is shown in Scheme 14. α-Arylideneisochromenes have been converted into α-naphthols on heating,[103] μ-allylpalladium chloride-catalysed decomposition of 2(or 4)-(diazomethyl)-2(or 4)*H*-benzopyrans has afforded benzoxepines via ring-enlargement,[104] and an interesting and novel photo-rearrangement of benzopyrans to 2,3-pyranobenzopyrones has been described.[105]

Bipyrimidine (**55**) has been produced by the thermal transformation of azoxypyrimidines (**54**; X,Y,R,R′ = N,CH,Ph,Ph).[106] In the reaction of 5,6-

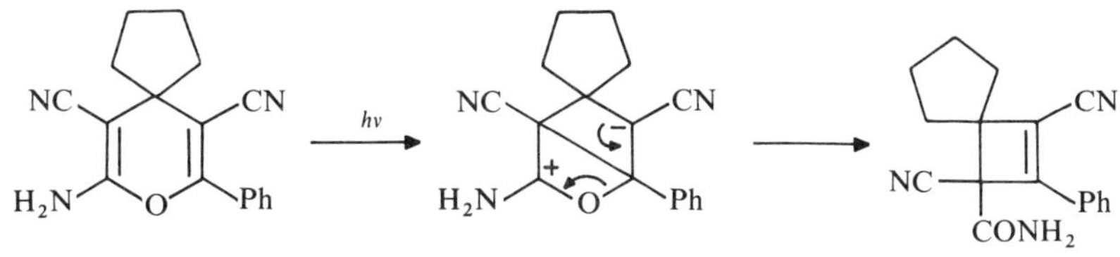

(52) (53)

SCHEME 14

dihydropyrimidine-4-(3*H*)-ones (**56**) with $POCl_3$,[107] a novel Pummerer-type rearrangement occurred with air oxidation of the dihydropyrimidine ring to afford the pyrimidine derivatives (**57**) quantitatively, while in a new type of Dimroth rearrangement, 1,2-dihydro-3*H*-quinazolone 4-oximes have been prepared from 4-amino-1,2-dihydroquinazoline 3-oxides.[108]

An interesting potassium amide-promoted ring-transformation of 1,2,4-triazines into 1,2,4-triazoles and 1,3,5-triazines has been reported.[109] The ring-contraction has been explained by invoking initial addition of the amide ion to C(6), ring-opening by fission of the N(1)—C(6) bond, and ring-closure. Transformation into the 1,3,5-triazine ring is thought to be initiated by addition of the amide ion to C(5), ring-opening between C(5) and C(6), a dehydrogenative rearrangement of the open-chain intermediate, 1-amino-2,4,5-triazahexatriene into 1-amino-4-cyano-2,4-diaza-1,3-butadiene, and ring-closure. Both pathways are outlined in Scheme 15.

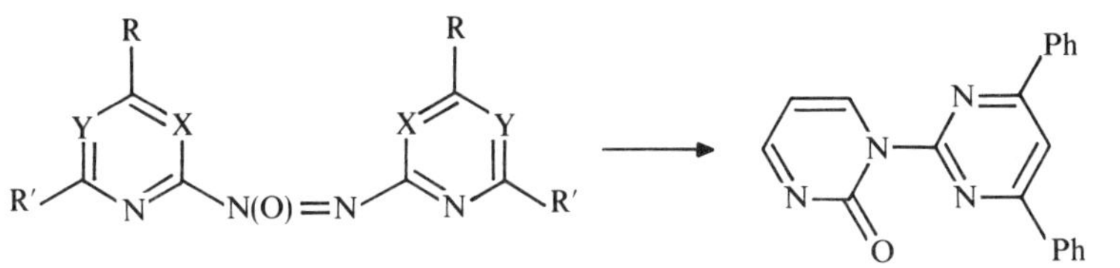

(54) (55)

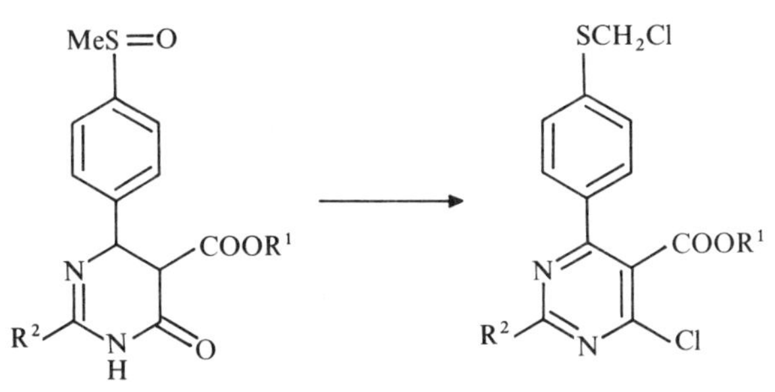

(56) (57)

6*H*-Pyrimido[1,2-*b*][1,2,4,5]tetrazin-6-ones have been shown to undergo a thermal, acid-catalysed rearrangement to 1,2,4-triazolo[4,3-*a*]pyrimidin-7(8*H*)-ones,[110] and a paper has appeared[111] which describes a novel ring-transformation

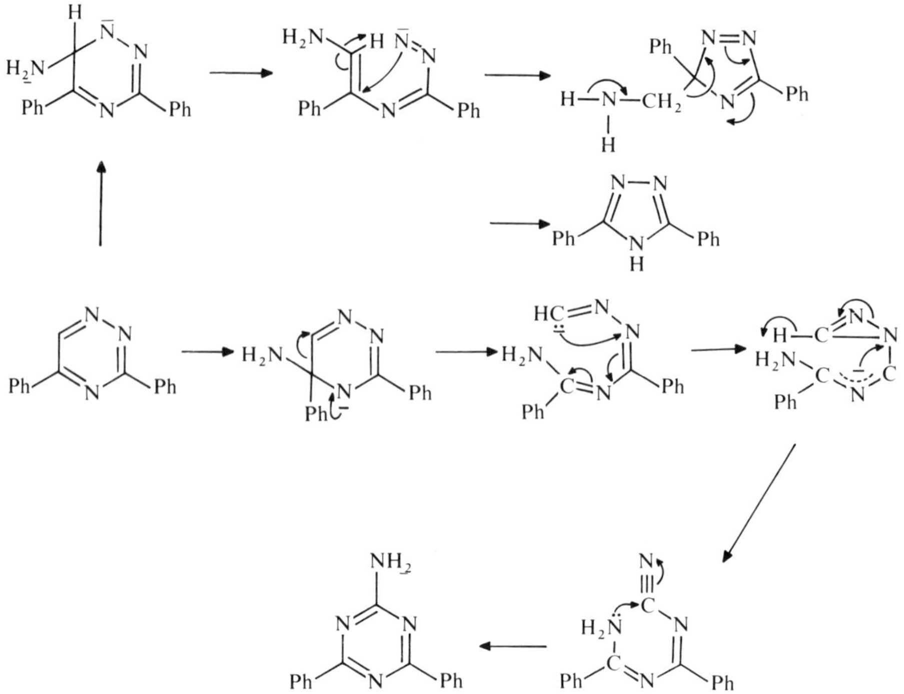

SCHEME 15

of 3,6-disubstituted 1,2,4,5-tetrazines to 4-amino-3*H*- or 1*H*-pyrazoles by cyanotrimethylsilane. The rearrangement appears to be induced by initial addition of the reagent to the C=N bond of the tetrazine.

Dioxodibenz[*bf*]azocine (**58**) has been found to react with ethylamine and with hydroxylamine to yield derivatives of 5,12-diaza- (**59**) and 5-oxa-12-aza-chrysene (**60**), respectively. The authors[112] envisage that the products arise by transannular interaction between a substituent at C(12) (formed after attack by the reagent) and the amide carbonyl group, resulting in cleavage of the amide, and followed immediately by a favoured 6-*endo-trig* ring-closure involving attack of the resulting amine on the β-carbon atom of the former α,β-unsaturated ketone. Treatment of the 2-(diethylamino)-3-(ethoxycarbonyl)-5-phenyl-3*H*-azepinyl anion with methyl iodide has led[113] to the formation of 2-(diethylamino)-7-(ethoxycarbonyl)-7-methyl-3,4-dihydro-3,4-methano-4-phenylpyridine. Photo-rearrangement of the first known representative of the spiro-2*H*-naphtho[1,8-*b*,*c*]oxepin system, viz. (**61**), has given (**62**).[114]

A novel ring-contraction of the 1,2-diazocine system (**63**) to pyrazoles (**65**) by way of [5.5]bicyclic intermediates (**64**) has been established[115] for the hydrolysis of 7-substituted 4-acetoxy-3,8-diphenyl-1,2-diazocines, and novel ring-transformations of 4-acylamino- and 4-dimethylamino-methyleneamino-1*H*-1,5-benzodiazepine-3-carbonitriles to pyrimidine-5-carbonitriles have been noted.[116] An unusual rear-

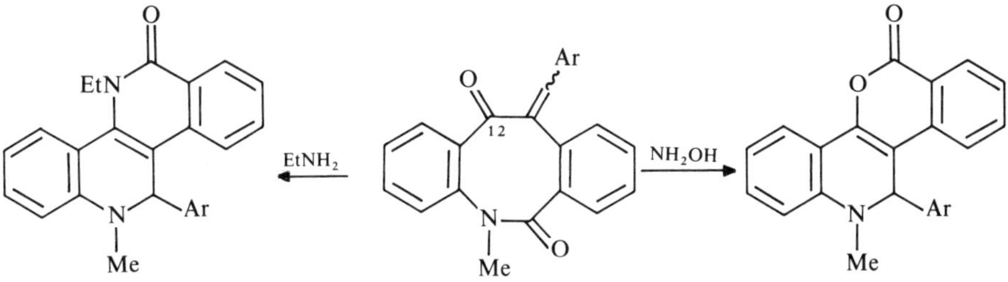

(59) (58) (60)

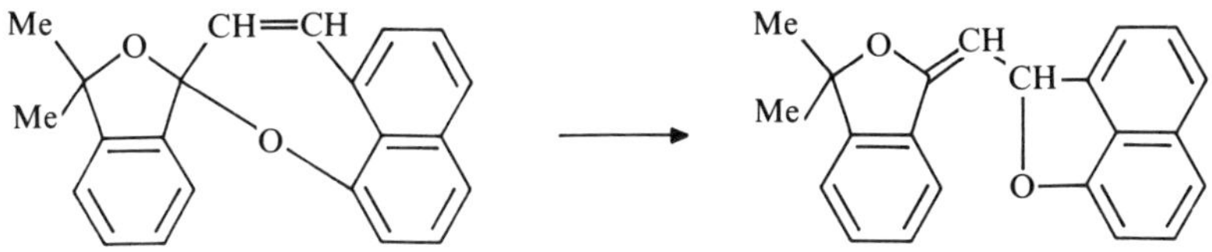

(61) (62)

rangement of a benzoxazepinone (**66**) to an indoline (**67**) has been observed[117] during $LiAlH_4$ reduction of the former. A possible rationale for the reaction is shown in Scheme 16. Photolysis of 2-methyl-1,5-benzothiazepines (**68**; R = Cl, OMe, OEt, SCN) was found to produce isomerization products (**69**); whereas under the same conditions (**68**; R = $ArCO_2$) gave Fries rearrangement products (**70**).[118] Finally, new ring-contraction reactions of some dihydro- and tetrahydro-1,4-thiazepine derivatives have been observed to occur under Pummerer conditions. These processes appear to involve a common bicyclic intermediate (**71**), a typical example of the rearrangement being represented in Scheme 17.[119]

Cyclohexadiene Derivatives

Photo-reactions involving enone and dienone rearrangements have been reviewed.[120] Photo-initiated dienone-phenol rearrangements of 4-methyl-4-phenyl-2,5-cyclohexadienones in the solid state have been reported,[121] and the 2,5-cyclohexadienone to bicyclo[3.1.0]hexenone photo-rearrangement has been developed for synthetic purposes.[122] Benzoic acid derivatives have been converted into 4-alkylphenols and 3-carbalkoxy-4-alkylphenols by a process[123] which combines recently reported methodology to convert benzoic acid derivatives into 2,5-cyclohexadienones with the familiar rearrangement of dienones to phenols. 6-(3-Azidopropyl)-2,4-cyclohexadienones have been readily converted into a variety of highly functionalized nitrogen-containing ring systems,[124] thus opening up a potential new route to pyrrolizidines and indolizidines. An extension of this reaction[125] to 2-(4-azidobutyl)-1,4-benzoquinone yielded the tricyclic lactone (**75**), which is thought to originate from the ring-contracted imine (**72**), which then undergoes a 1-aza-Cope rearrangement to the ketene enamide (**73**). Cyclization to zwitterion (**74**) followed by hydrogen-atom rearrangement and dehydrogenation by the starting quinone completes a possible mechanistic hypothesis; see Scheme 18.

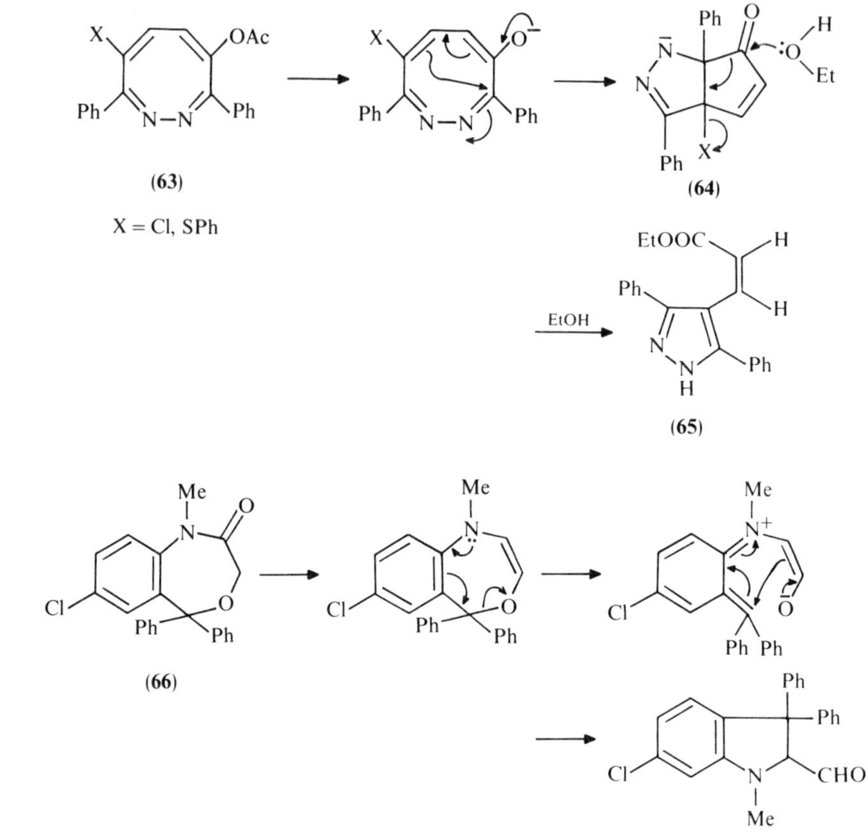

(63) X = Cl, SPh

(64)

(65)

(66)

(67)

SCHEME 16

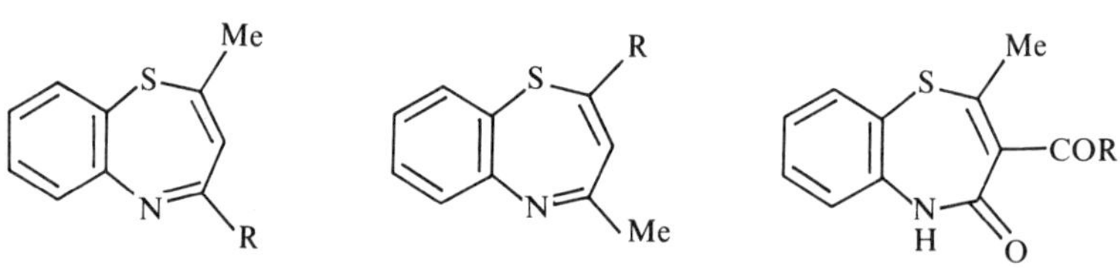

(68) (69) (70)

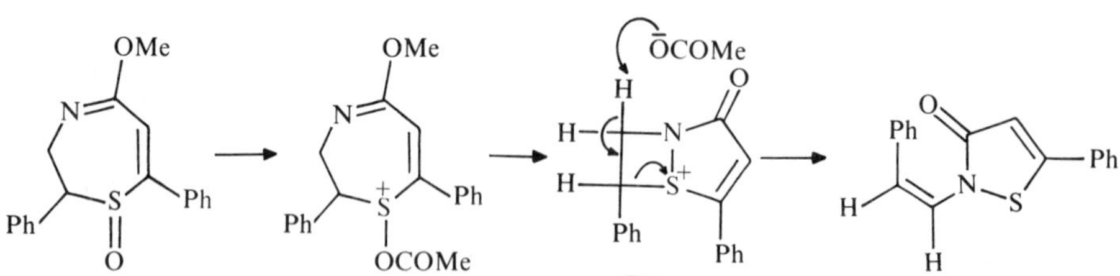

(71)

SCHEME 17

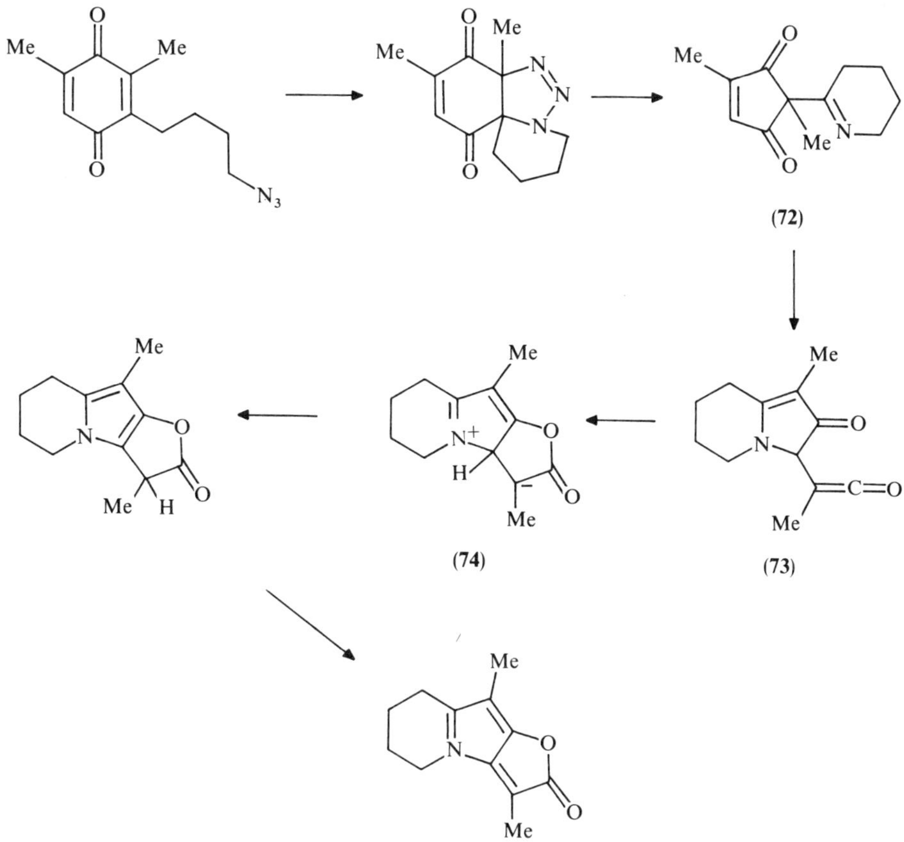

SCHEME 18

Pyrrolo[1,2-*a*]azepin-7-ones (**77**) have been prepared by heating spiro-compounds (**76**). The reaction is thought to proceed[126] by way of a tricyclic intermediate of a type already postulated in the azulene-naphthalene rearrangement. The *p*-methoxyphenoxyl group of 4-*p*-methoxyphenoxy-5-methoxy-*o*-benzoquinone (**78**) was found to be readily substituted in basic media by alcohols (ROH). The alkyl component (R) of the resulting 4-alkoxy-5-methoxy-*o*-benzoquinones (**79**) subsequently migrated spontaneously (or in acidic medium) to afford selectively, 3-alkyl-2-hydroxy-5-methoxy-*p*-benzoquinones (**80**) which have proved to be inacce ssible by other routes.[127] A novel and efficient route has been developed[128] for the preparation of 5,6,7,8-tetrahydrophenanthrene-1,4-diones from acylbenzoquinones through a four-step approach mediated by a dienone-phenol rearrangement.

It has been reported[129] that a dienone-phenol rearrangement occurs when the oxathiolane and dithiolane derivatives of 3-oxo-steroids are treated with copper(II) bromide, while the neutral form of the spiro-cation intermediate in the dienone-phenol rearrangement of a steroidal *p*-quinol has been isolated,[130] thus providing direct proof of the accepted mechanism for the process.

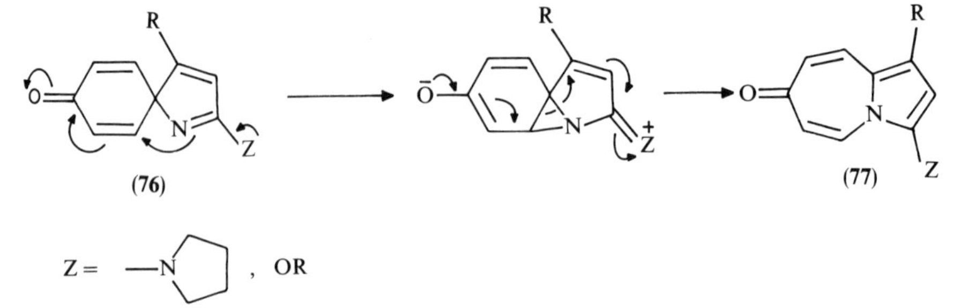

(76) (77)

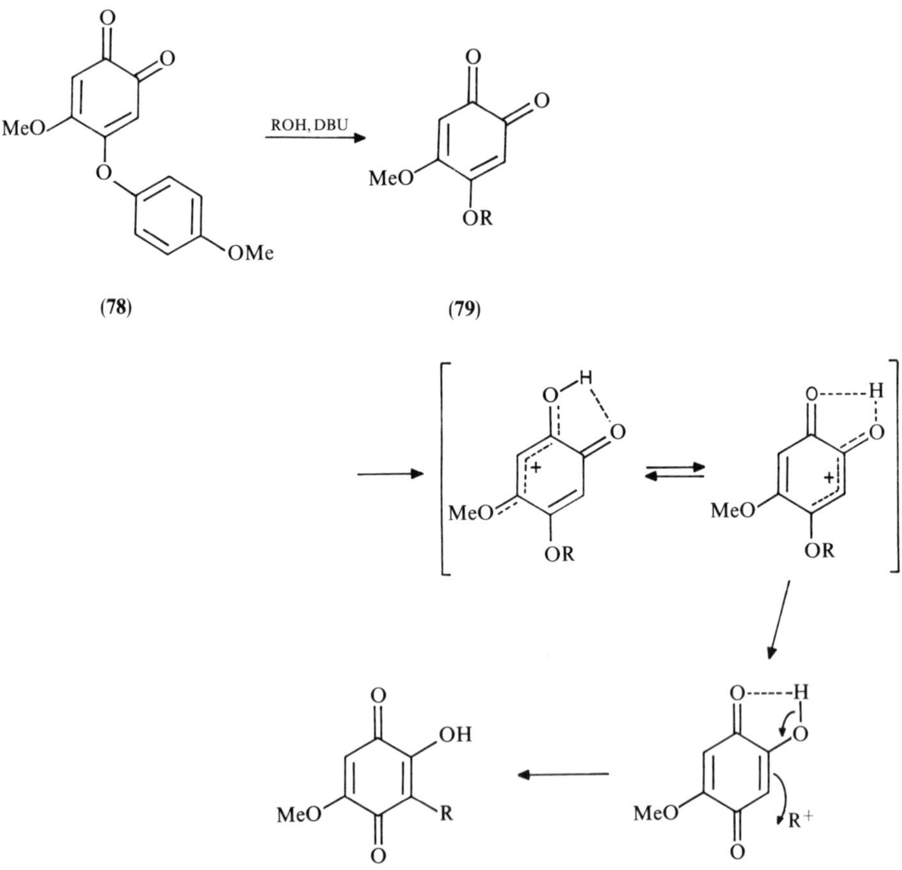

(78) (79) (80)

A study[131] of the photochemistry of 2,2-dimethyl- and 2-aryl-2-methyl-naphthalen-1(2*H*)-ones has shown that these compounds undergo an oxa-di-π-methane rearrangement to cyclopropindenones, while a recent investigation[132] of the photochemistry of 3,4-dimethyl-4-(trichloromethyl)-2,5-cyclohexadienone has shown that the rearrangement and fragmentation pathways available to this system are clearly far more numerous and complex than had been anticipated from earlier studies. Formation of the 1-naphthalenone (**83**) by dehydrobromination of 8*a*-benzyl-2-*tert*-butyl-6,7-dibromo-5,6,7,8-tetrahydro-1-(8*aH*)-naphthalenone (**81**) has been attributed[133] to a [1,5]-shift in the intermediate ketone (**82**). On heating in glacial acetic acid, anthracenetrione (**84**) also undergoes a [1,5]-shift of the acetoxy group while it reacts readily with nucleophiles to afford 3-substituted 1-acetylamino-4-hydroxy-9,10-anthraquinones (**85**).[134]

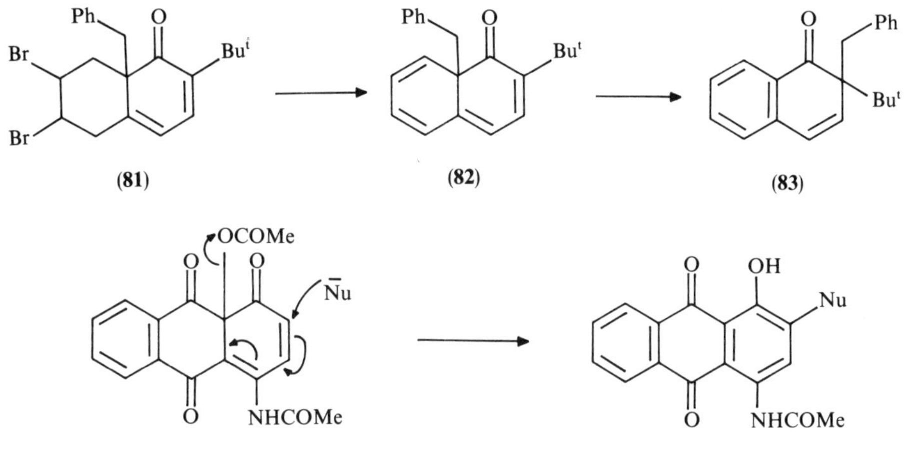

(**81**) (**82**) (**83**)

(**84**) (**85**)

Treatment of cyclohexadienone (**86**) with methanolic potassium hydroxide affords[135] the two isomeric cyclopentanecarboxylic acid derivatives (**88**) together

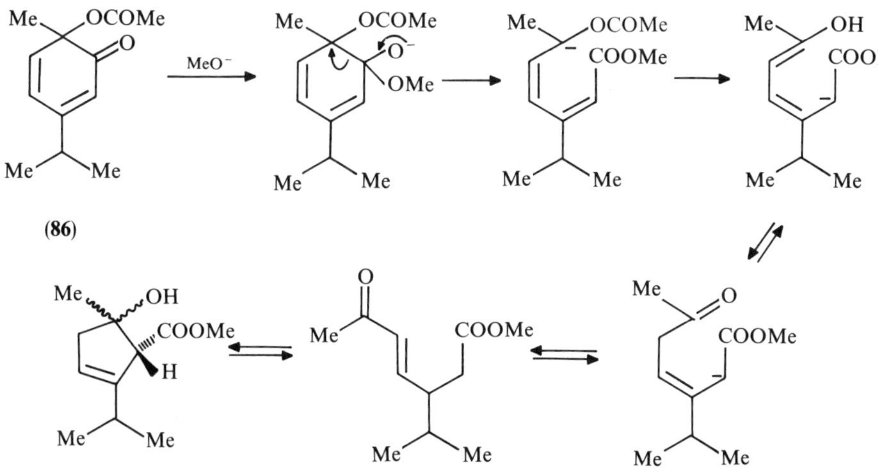

(**86**)

(**88**) (**87**)

SCHEME 19

with the acyclic ester (**87**). A plausible route for the reaction is presented in Scheme 19. The formation of 2-chloro-2-trifluoromethylcyclopentane-1,3-dione during the chlorous acid oxidation of 2- and 3-trifluoromethylphenol has been shown[136] to proceed via decarboxylation of 5-chloro-1-hydroxy-4-oxo-5-trifluoromethylcyclopent-2-enecarboxylic acid; see Scheme 20. Santonin (**89**) was found to undergo an unexpected type of photo-isomerization reaction in the crystalline state,[137] yielding the cyclopentadienone (**90**) instead of lumisantonin (**91**) which is recognized as the major photo-product of santonin in solution.

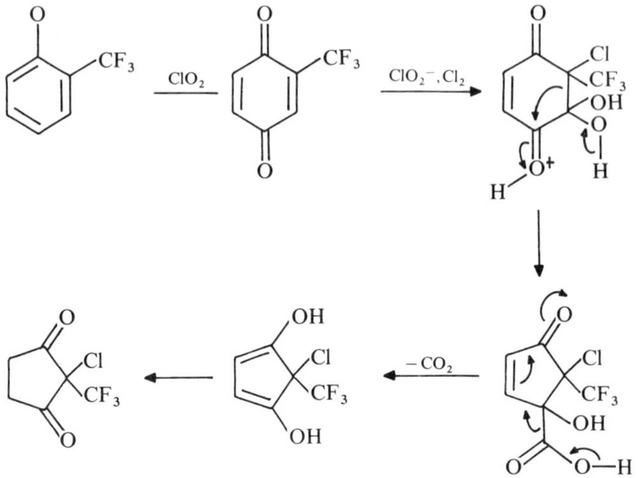

SCHEME 20

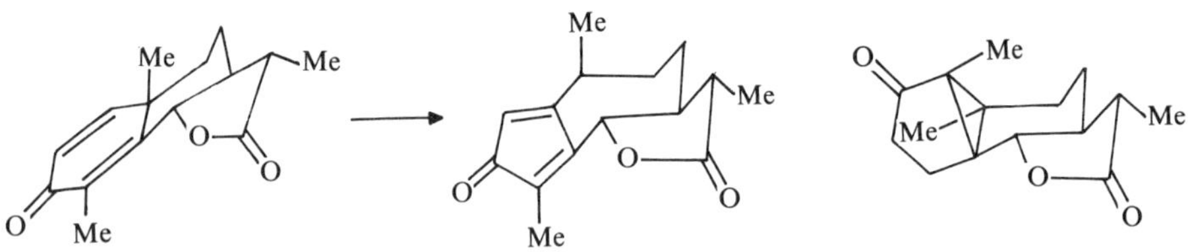

(**89**) (**90**) (**91**)

Sigmatropic Rearrangements

[3,3]-*Migrations*

Claisen and related rearrangements. A general formal-logical approach to sigmatropic rearrangements based on structural and graph theories has been developed,[138] and it has been shown that the free-energy profiles along and perpendicular to the reaction coordinate determine a two-dimensional energy for describing substituent effects in organic chemistry.[139] The various models have been compared with experimental results for free energy of activation of the Claisen rearrangement of allyl vinyl ether and the Cope rearrangement of substituted hexadienes. The importance of sigmatropic rearrangements in asymmetric synthesis has been reviewed,[140] and cyclization approaches to the synthesis of natural products, including the use of the Claisen rearrangement in the total synthesis of biflora-4,15,19-triene, have been described.[141] Difference-contour energy maps for the Claisen rearrangement of pent-4-enal have indicated[142] that the AMI and MNDO SCF-MO methods differ most significantly at C—C and C—O bond distances of approximately 2.3 and 2.05 Å, respectively.

A detailed study of the Claisen rearrangement of chorismic acid to prephenic acid has shown for the first time that the polar medium in which it generally occurs is a significant factor in the facility of the rearrangement.[143] Experiments designed to define the conformation of chorismate in solution, and to illuminate the nature of both the non-enzymic and the enzyme-catalysed transformations to prephenate, have indicated[144] that the transition state may be intrinsically dipolar with some of the character of a tight ion-pair between the enolpyruvate anion and the cyclohexadienyl cation. The accelerated rates and markedly enhanced solvent effects associated with the Claisen rearrangement of 4- and 6-alkoxyallyl enol ethers have provided strong evidence for a pronounced dipolar character of the pericyclic transition state.[145] Recent synthetic studies[146] on betaenone B and stemphyloxin 1 have provided yet another illustration of the potentially dominant role of steric effects in determining the stereochemical outcome of Claisen rearrangements which append an allylic unit to a cyclohexanone. The palladium(II)-catalysed Claisen rearrangement of allyl vinyl ethers has proved successful provided that the vinylic C=C bond is protected sterically from coordinating to the metal.[147] Through the judicious choice of either 2,6-dimethylphenol or $PdCl_2(RCN)_2$ as catalyst, Claisen rearrangements of enol ethers of cyclic ketones have been made to proceed with a high level of either *anti* or *syn* diastereoselection, respectively;[148] see Scheme 21. An

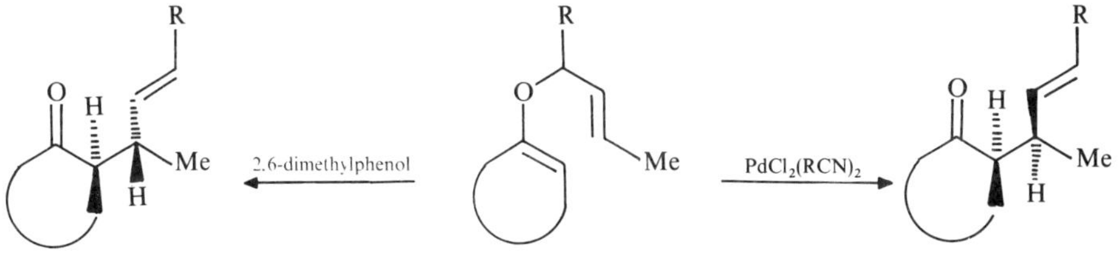

SCHEME 21

efficient and regiospecific transformation of 1,2,3-triols into fluoromethyl ketone enol ethers using the fluoro-olefination reaction with tetrabutylammonium fluoride and methanesulphonyl fluoride, and subsequent allylation by thermal [3,3]-sigmatropic rearrangement, has been described.[149] The 1,3-dioxolane ring system of the trichothecenes, sambucinol and sporol, has been constructed[150] by a stereoselective Claisen rearrangement, and α-methylenecyclopentanones have been produced by the photo-Claisen rearrangement of alkyl-substituted 2-(2-oxa-3-butenyl)cyclopent-2-enones.[151] The Claisen rearrangement of a series of (allyloxy)enamines (**92**) was found to afford 2-aminopent-4-enals (**93**) with high selectivity,[152] possibly a consequence of the reaction occurring via the more stable chair-like transition state. A different [1,3]-sigmatropic course was observed when the allylic double bond is part of an aromatic system. In these cases the tendency of the system to preserve its aromaticity seems to account for the [1,3]-sigmatropic shift which leads to 2-amino-2-methylpropanals.

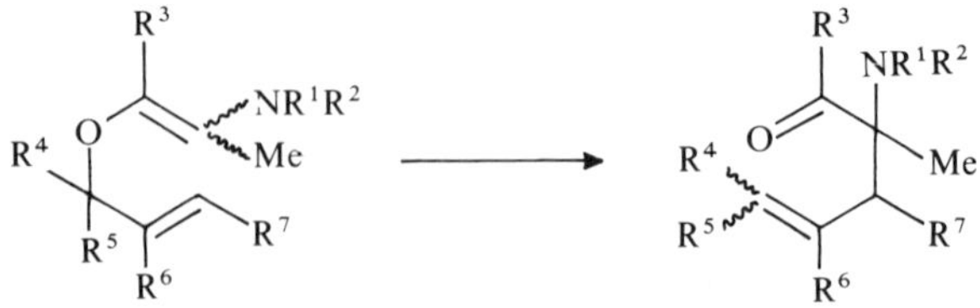

(**92**)) (**93**)

A reinvestigation[153] of the Claisen rearrangement of 2,5-bis(allyloxy)tropone has led to the isolation of three diallyl-5-hydroxytropolones along with an intermediary thermolysate, 3-allyl-5-allyloxytropolone. The thermal Claisen rearrangement of 4,6-diacetylresorcinol diallyl ether resulted[154] in a mixture of products some of which arose by either loss of, or [1,5]-sigmatropic migration of, the acetyl group, in addition to loss of the allyl group after [3,3]-migration. The Claisen rearrangement of allyl phenyl ether to 2-allylphenol has been investigated[155] by IR multiphoton excitation, while sodium and potassium oxido groups in the *ortho*-position of allyl phenyl ethers have been found to accelerate the rearrangement of these ethers, and at the same time produce higher ratios of *para*-to-*ortho*-rearranged products.[156] Several cationic Fe(II)(allyloxy)-carbene and related complexes have been shown to undergo (on deprotonation with KH and 18-crown-6) an *in situ* thermal Claisen rearrangement to yield the corresponding γ,δ-unsaturated acyliron complexes.[157] Labelling studies[158] have shown that the rearrangement of aryl pentadienyl ethers in the presence of $BF_3.OEt_2$ affords pentadienyl phenols without formation of the corresponding [3,3]-rearrangement products. Aryloxycyclopentenylphenols have been prepared by Claisen rearrangement of diaryloxycyclopentenes,[159] and 3-hydroxyphenyl allyl ethers have been shown to undergo [3,3]-sigmatropic rearrangement in refluxing aqueous methanolic KOH, in the presence of oxygen, whereas no reaction occurs in the absence of base and oxygen. Evidence has been presented[160] for the process being radical- rather than anion-mediated. A further investigation has been made of the [3,3]-sigmatropic rearrangement of 4-hydroxycoumarin allyl ethers.[161] Repeated sequences of [3,3]-sigmatropic rearrangements initiated by a Claisen rearrangement have permitted the synthesis of 3,6-disubstituted umbelliferone derivatives[162] from umbelliferone, thus demonstrating further the synthetic utility of multiple [3,3]-sigmatropic rearrangements for constructing prenyl-substituted coumarins. Several methyl γ-aryloxycrotonates have been prepared and successfully rearranged to a mixture of (*Z*)- and (*E*)-3-ethylidenebenzofuran-2-(3*H*)-ones in refluxing ethylene glycol,[163] while 3-(2-hydroxyphenylmethylene)-3,4-dihydro-2*H*-1-benzopyran-2-ones and 3-methoxycarbonyl-2*H*-1-benzopyrans are the products of the Claisen rearrangements of methyl-3-aryloxy-2-(aryloxymethyl)prop-2-enoates.[164]

The gas-phase rearrangement of deprotonated allyl phenylacetates has been shown to occur through six- and four-centre transition states with the six-centre (Claisen) rearrangement predominating.[165] The reported easy conversion of acetoacetic esters of tertiary allyl alcohols, viz. (**94**), into respective γ,δ-unsaturated ketones (**95**) on the surface of alumina[166] provides yet another example of the importance of adsorption in promoting organic reactions. The Claisen rearrangement of ester enolates with appropriate metal ions bearing bulky cyclopentadienyl

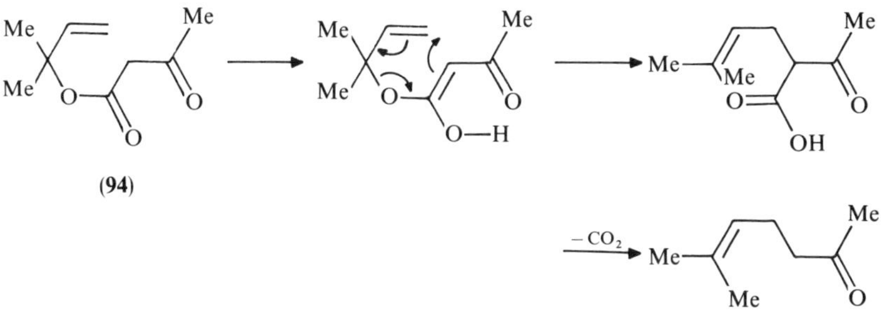

(**94**)

(**95**)

ligands has been found to proceed through a boat-like transition state,[167] thus producing products with a different configuration from those expected to result from a chair-like transition state. A conceptually new approach to the total synthesis of the antibiotic pseudomonic acid C has been described.[168] One of the key features of the approach includes the use of the glycolate ester enolate Claisen rearrangement to introduce the remote side-chain stereochemistry with good selectivity. The stereo-controlled synthesis of highly oxygenated acyclic systems has been achieved via the enolate Claisen rearrangement of *O*-protected allylic glycolates.[169] It has been shown[170] that the ester enolate Claisen rearrangement of (*E*)- and (*Z*)-allyl 2-hydroxyalkanoates diastereoselectively gives *erythro*- and *threo*-2-hydroxy-2,3-dialkyl-4-alkenoic acids, respectively. The first total synthesis of the Calabar bean alkaloid, geneserine, has been accomplished.[171] The key step in the synthetic strategy involves a tandem electrocyclic [3,3]-sigmatropic reaction of an *o*-quinodimethane (**97**) generated *in situ* by thermolysis of the benzocyclobutene (**96**); see Scheme 22. Combination of the Claisen rearrangement–nitrile oxide cycloaddition sequence has been used in a recent stereoselective synthesis of specionin,[172] while iterative applications of the dioxanone-to-dihydropyran Claisen rearrangement, a variant of the enolate Claisen rearrangement, have resulted in the efficient enantioselective production of the C(7)–C(13) fragments of erythronolides A and B.[173,174] Chiral trisubstituted pyrrolidines (**100**) have been obtained exclusively by the Ireland enolate Claisen rearrangement of the corresponding aza-lactones (**98**), the rearrangement proceeding exclusively via a boat-like transition state (**99**).[175] Using this methodology, an enantiospecific synthesis of (−)-α-kainic acid (**104**) from L-aspartic acid (**101**) has been achieved[176] in which the C(3)–C(4) bond is

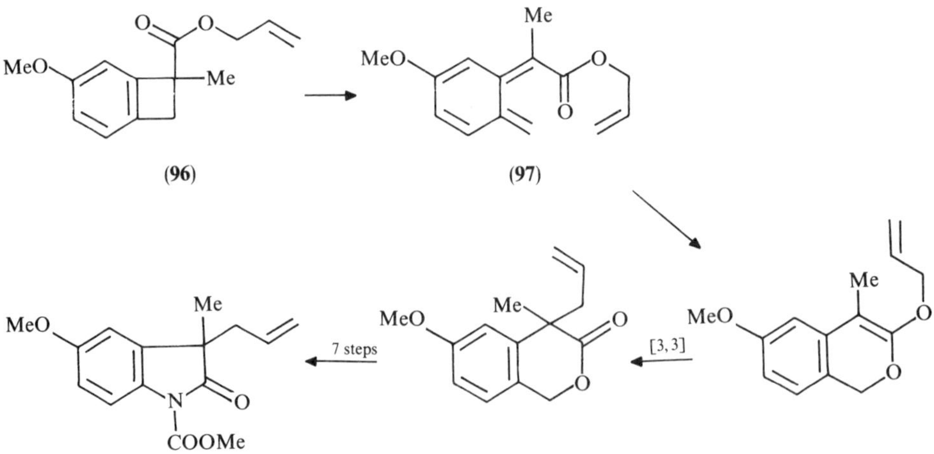

SCHEME 22

formed by a stereo-controlled enolate Claisen rearrangement, (**102**→(**103**), which delivers the correct geometry at the C(3) and C(4) asymmetric centres. A symmary of this process is illustrated in Scheme 23. Seleno-substituted γ-butyrolactones (**106**)

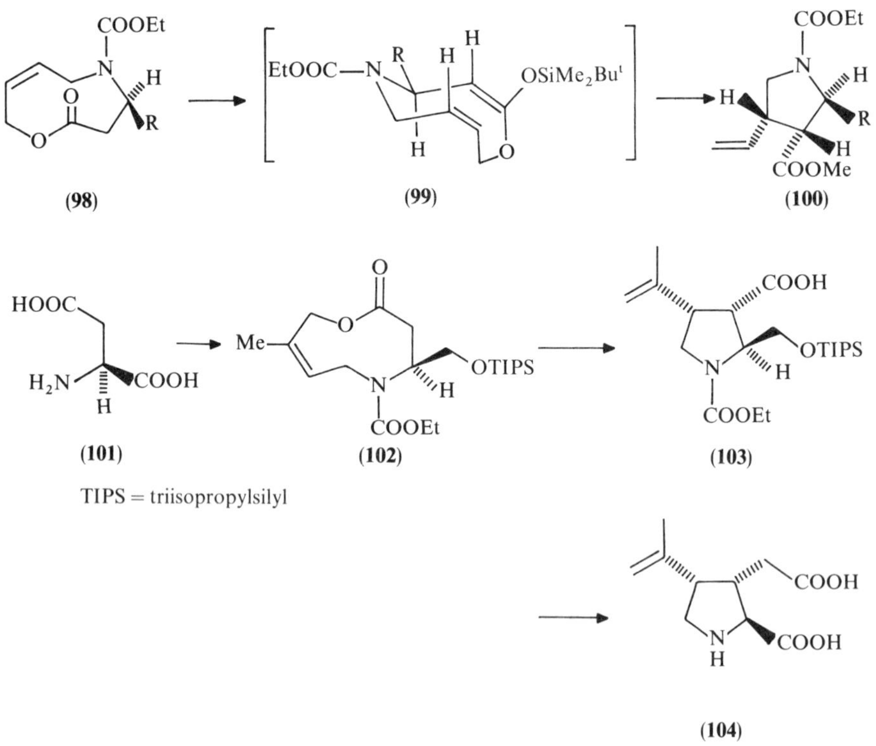

SCHEME 23

have been prepared by the [3,3]-sigmatropic rearrangement of enolate oxyselenonium species (**105**), themselves prepared by the reaction of selenoxides with chlorocyanoketene; see Scheme 24.[177]

A base-mediated alternative to the orthoester Claisen rearrangement has been described,[178] and stereoselectivity in the orthoester rearrangement of a series of diastereoisomeric tertiary allylic alcohols has been found to be dependent upon the chiral centre adjacent to the allylic alcohol.[179] An enantioselective synthesis of *trans*-octahydro 1,6-dioxoinden-4-ylpropionic acid has been achieved from (*R*)-2,3-*O*-isopropylideneglyceraldehyde by a tandem orthoester Claisen rearrangement.[180] Orthoester Claisen rearrangements of 3-C-(hydroxymethyl)methylene derivatives of hexafuranose have been used[181] for the stereoselective introduction of a

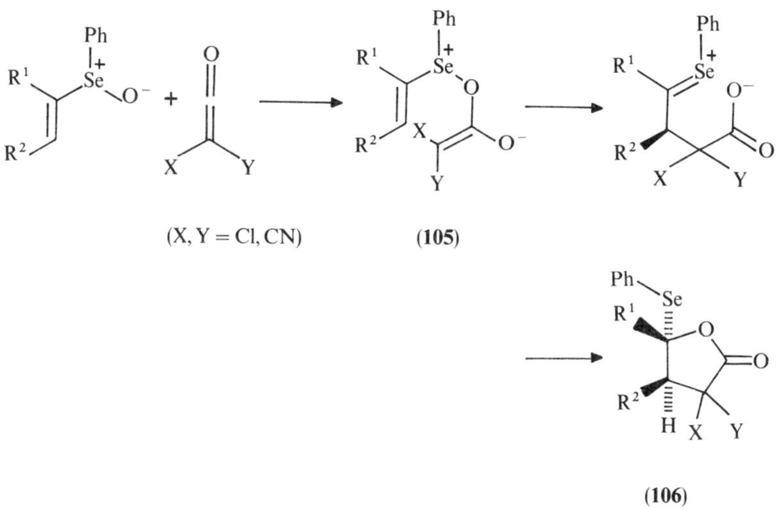

SCHEME 24

quaternary centre at C(3) of D-ribo, L-lyxo-, and D-arabino-hexofuranoses. The Claisen orthoester rearrangement has been used[182] as a key step in the synthesis of (4*E*,7*Z*)-4,7-tridecadienyl acetate and (4*E*,7*Z*,10*Z*)-4,7,10-tridecatrienyl acetate, the sex pheromones of the potato tubeworm moth. The total synthesis of dactylol and africanol have been achieved[183] from bicyclo[5.1.0]octane precursors by routes that involve two orthoester Claisen rearrangements.

Ketal Claisen rearrangements of the ketals of cyclic ketones have been shown to proceed with fair-to-good diastereoselectivity,[184] and ketal Claisen technology has been utilized for the synthesis of cyclic sequiterpenes containing quaternary centres,[185] and for the synthesis of ipsdienone (**107**);[186] see Scheme 25. A new and versatile synthesis of the tricyclic furan (**108**), a potential precursor of viridin and related compounds, has been described.[187] The method is based on a previously developed furan-transfer reaction and angular alkylation via the Claisen rearrangement.

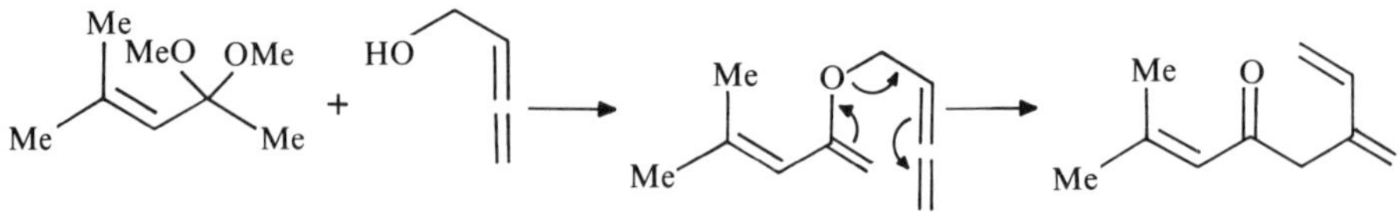

(107)

SCHEME 25

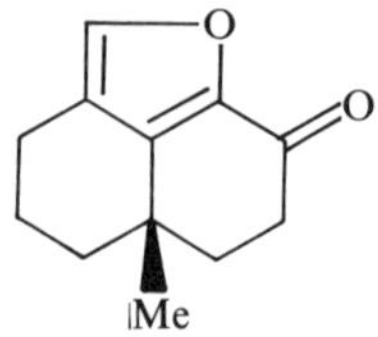

(108)

The pyridinium dichromate oxidative rearrangement of (*Z*)-substituted 1-ene-4-yne-3-ols (**109**) has been shown to be completely regiospecific, and has therefore been represented[188] (Scheme 26) as involving a [3,3]-sigmatropic rearrangement. *syn*-Allene carbinol (**110**), and its esters, have been transformed[189] to sterically loaded 1,3-alkadien-1-yl carbamates (**111**) by the Claisen rearrangement (see Scheme 27), and substituted buta-1,3-dienes have been conveniently prepared by a strategy[190] in which a prop-2-ynyl ester is converted, through the Claisen ester rearrangement, into the allenyl acetic acid, which is thermally decarboxylated; see Scheme 28. Substituted 2*H*-1-benzopyrans and benzofurans have been prepared from 2-methyl-2-phenoxy-propionyl-cyanomethylenetriphenylphosphorane derivatives in one step by a method[191] that merges the preparation of phenyl propargyl ethers by intramolecular Wittig reaction, and their subsequent Claisen rearrangement. A similar methodology[192] has been used to generate 2*H*-chromenes.

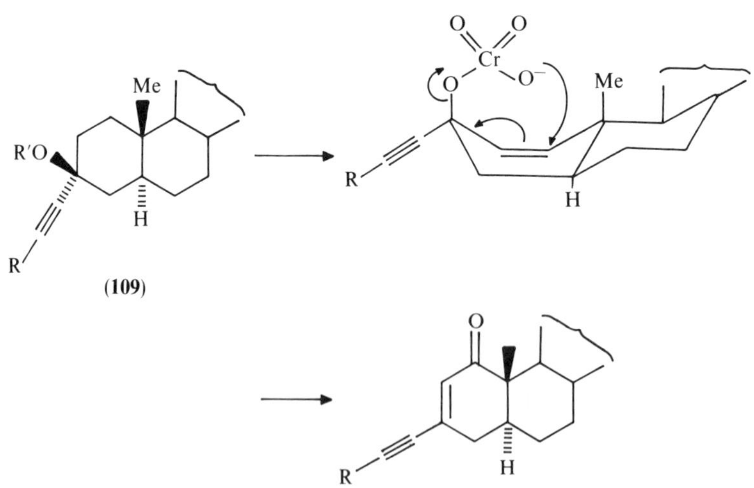

SCHEME 26

The Claisen rearrangement of propargyloxydienephosphonates has been investigated.[193]

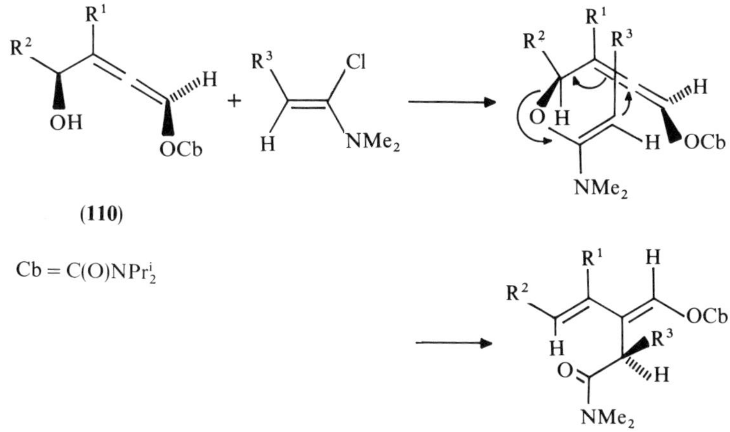

(110)

Cb = C(O)NPr$^i_2$

(111)

SCHEME 27

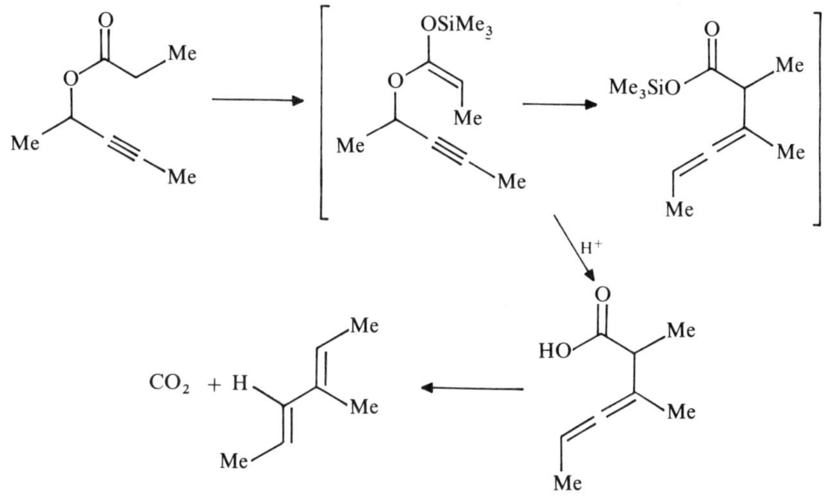

SCHEME 28

The amino-Claisen rearrangement has been observed to occur in the molecular ions of *N*-allylaniline,[194] and the kinetics of the thermal rearrangement of this compound and its derivatives have been determined.[195] The reaction of some tertiary allylamines (**112**) with dimethyl acetylenedicarboxylate has resulted[196] in the formation of 1:1 adducts (**113**) via a [3,3]-sigmatropic rearrangement of the allyl group from N to C; see Scheme 29. (+)-Dihydropallescensin-2 has been synthesized by application of a recently developed asymmetric aza-Claisen protocol which provides absolute stereo-control in the preparation of a C(β)-stereogenic centre which is a ring carbon; see (**114**)→(**115**).[197] *N*-Aryl-1-alkynesulphenamides have been reported[198] to undergo thermal [3,3]-sigmatropic rearrangement to yield indolinethiones, while *N*-(1-naphthyl)-ethenesulphenamides have likewise yielded 1*H*-benz[*g*]indoles via cyclization of intermediate amino-thioaldehydes.[199]

The thermal [3,3]-sigmatropic rearrangement of cinnamyl dithiocarbamic anhydrides to the corresponding β-(dithiocarbamyl)phenylpropionic acids has been described,[200] and an interesting asymmetric synthesis of thiotetronic acids has been developed[201] which is based upon an allyl xanthate-to-dithiocarbonate rearrangement. The first example of a synthetically promising Claisen-type rearrangement

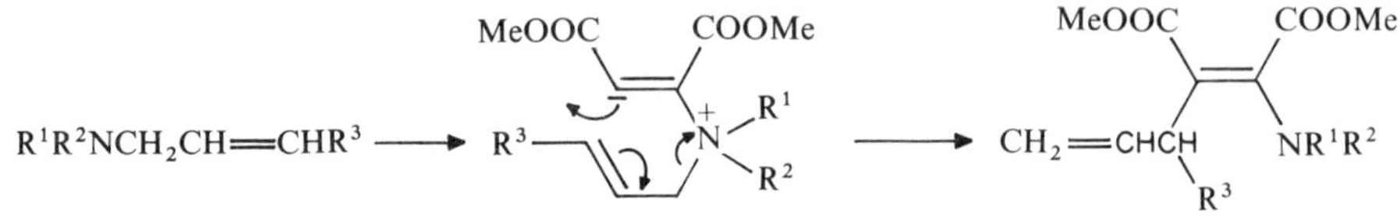

(**112**) (**113**)

SCHEME 29

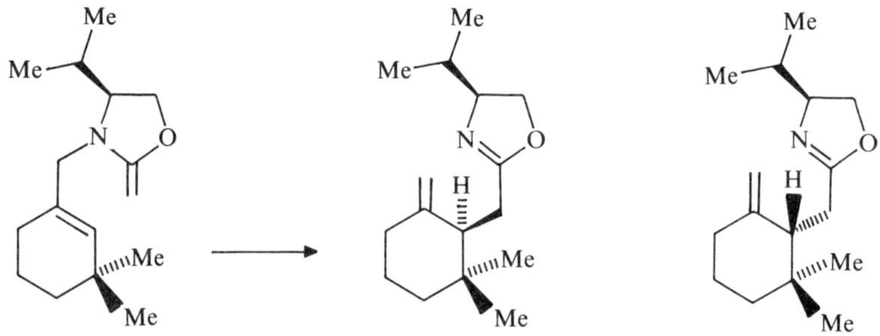

(**114**) 9 : 1

(**115**)

using γ-thionolactone substrates has been reported,[202] and a regioselective thio-Claisen rearrangement of *S*-allyl ketene-*S*,*N*-acetals generated from cyclic *S*-allylmonothiodicarboximide salts has been described.[203] The observed thio-Claisen rearrangement of sulphonium salt (**117**),[204] itself derived from the 2-(ethylthio)tryptophan derivative (**116**), to (**118**) supports the postulate that a thio-Claisen rearrangement is involved in indole alkaloid biosynthesis.

A skeletal rearrangement, observed during mass-spectral fragmentation of allylhydrazone molecular ions, has been classified as a [3,3]-sigmatropic process,[205] while a kinetic study of the thermal and acid-catalysed Fischer cyclizations of cyclohexanone arylhydrazones has indicated that a [3,3]-sigmatropic shift is involved during C—C bond formation.[206]

A detailed theoretical study of the Reformatsky reaction has shown that it involves a [3,3]-sigmatropic shift analogous to a Claisen or Cope rearrangement.

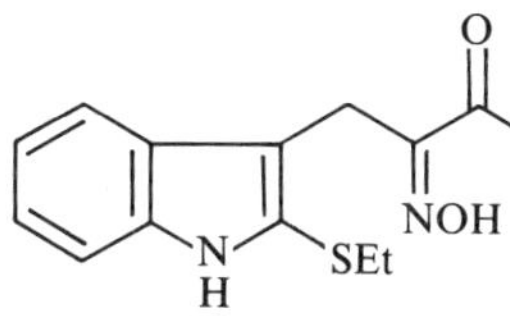

(116)

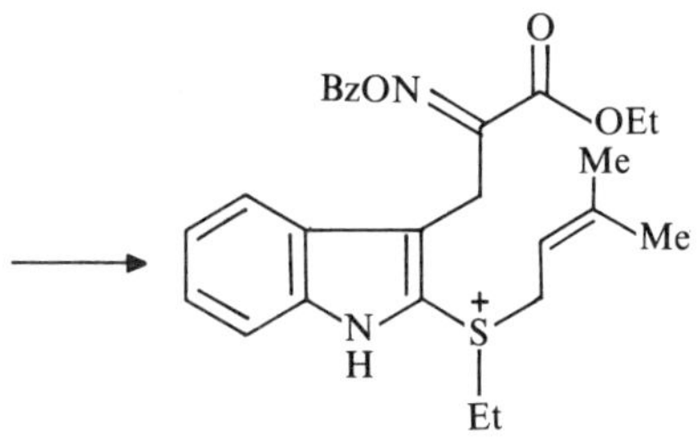

(117)

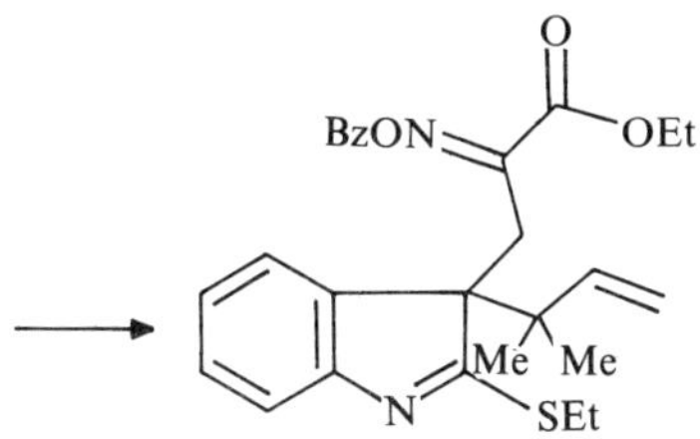

(118)

Consequently it has been classified[207] as a metallo-[3,3]-sigmatropic shift or a metallo-Claisen rearrangement.

Cope and related rearrangements. The problem of calculating dynamic quadrupole echo spectra in ordered systems in which exchanges take place between non-interacting nuclei with different quadrupole interactions and chemical shifts has been solved. The theory has been applied to the Cope rearrangement of bullvalenes in liquid crystal solution.[208] AM 1 calculations have shown that the boat and chair Cope rearrangements of hexa-1,5-diene can each proceed by two different paths, one involving an aromatic and the other a biradical-like transition state.[209,210] In order to obtain a complete characterization of the energy surface of a relatively simple hexa-1,5-diene system, the kinetics and stereochemistry of the thermal interconversion of racemic and optically active 4,5-dimethyl-2,6-octadienes have been studied,[211] while the kinetic and stereochemical effect of a fluorine substituent on the Cope rearrangement has been assessed.[212]

The degenerate Cope rearrangement of 4-vinylcyclopentene has been examined.[213] A Cope rearrangement has been utilized[214] in the synthesis of functionalized bicyclo[3.2.1]octa-2,6-dienes from substituted 6-*exo*-(1-alkenyl)-bicyclo[3.1.0]hex-2-enes, and a development of this methodology[215] has resulted in the synthesis of sinularene. The fact that divinylcyclopropane intermediates can be isolated from the [3 + 4]-cycloaddition reaction between rhodium(II) acetate-stabilized vinylcarbenoids and cyclopentadiene has indicated that this reaction proceeds through a tandem cyclopropanation-Cope rearrangement sequence.[216]

The titanium-induced cyclization of diketone (**119**) was found to yield diene (**121**), presumably by Cope rearrangement of tricycle[$8.2.^{1,10}0.^{1,7}0^{4,10}$]-1(7),4(10)-dodecadiene (**120**).[217] Molecular mechanic calculations support the belief that such a rearrangement would be rapid. It has been confirmed experimentally, for the first

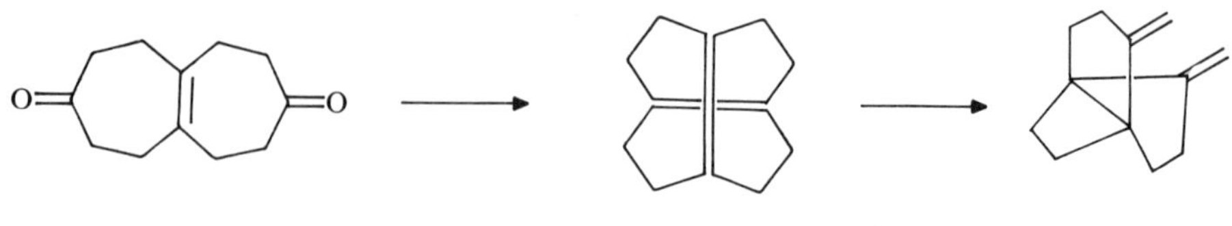

(**119**) (**120**) (**121**)

time,[218] that a Cope rearrangement from a medium-sized ring to a less strained product can be effected under mild conditions, while a recently observed rearrangement of the mecrolide latrunculin B to a new 16-membered macrolide has been presented as involving a Cope-type rearrangement.[219] The total synthesis of the germacrane sesquiterpene, (+)-dihydrocostunolide, has been accomplished[220] via a tandem Cope–Claisen rearrangement. A siloxy-Cope rearrangement has been used[221] as the key step in a recent synthesis of *N*-acetylactinobolamine, and an examination of the thermal and anionic rearrangements of *trans*-1-vinyl-3-cyclotridecen-1-ols, substituted in the α- and β-vinyl positions, has provided a useful contribution to an understanding of periselectivity in the oxy-Cope rearrangement.[222] The highly stereoselective Cope rearrangement of trienenitrile (**122**) has afforded cinnamonitrile (**123**)[223] having the correct relative stereochemistry at C(8) and C(14) for steroid synthesis, while the forskolin prototype (**126**) has been elaborated in nine steps from 2,4,4-trimethylcyclohexanone (**124**), the entire ring-system materializing during an oxyanionic Cope rearrangement of alcohol (**125**).[224] Deprotonated diallyl ether has been found to rearrange in the gas phase by an oxy-Cope mechanism;[225] see Scheme 30. The reactivity of diversely substituted 1,5-hexadien-3-ols in the presence of *N*-phenyl selenophthalimide and catalytic amounts of toluene-*p*-sulphonic acid has been studied.[226] This reagent mixture was found to exhibit a very poor trend in effecting the electrophilic-mediated oxy-Cope rearrangement, although it proved very efficient in effecting *O*-cyclization to tetrahydrofuran derivatives. (7*R*)-7-Methyl-7-vinylbicyclo[3.2.0]hept-2-en-6-one has been condensed with several chiral cyclopentenyllithium reagents for the purpose of assaying the diastereoselectivity of this coupling. However, due to ring strain, the alkoxides that were formed spontaneously underwent anionic oxy-Cope rearrangement. The initial diastereoselectivity was thereby incorporated into additional stereogenic centres; see Scheme 31.[227]

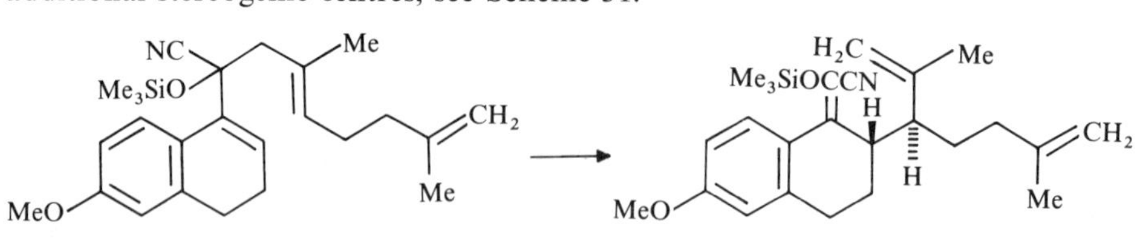

(**122**) (**123**)

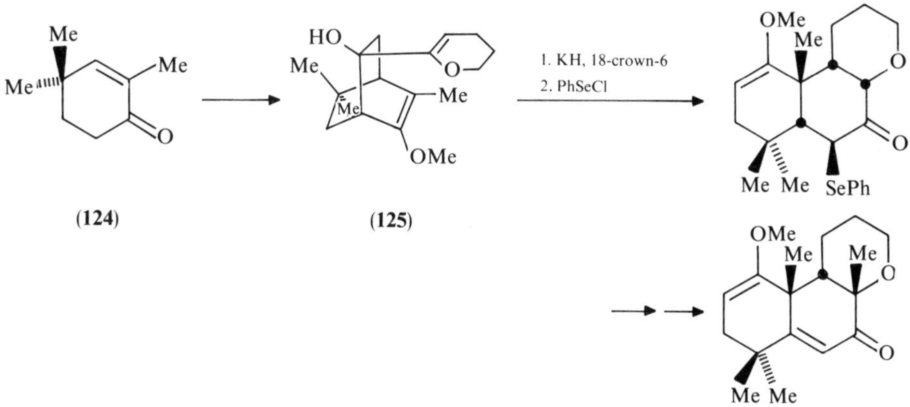

(124) (125) (126)

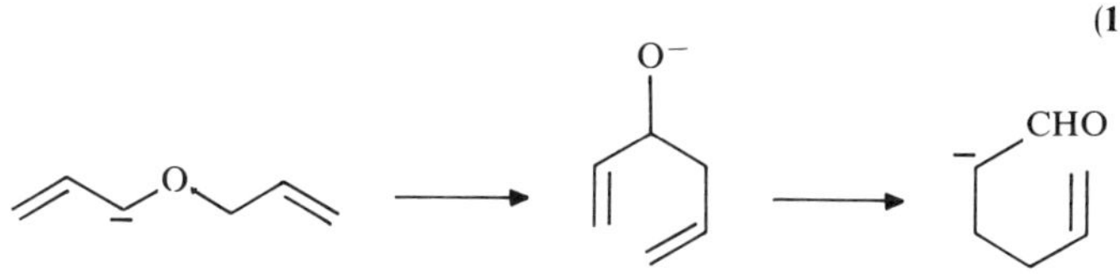

SCHEME 30

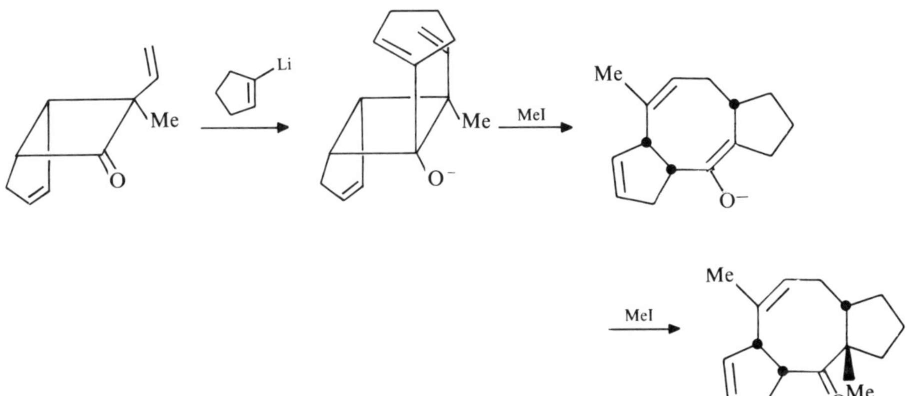

SCHEME 31

An efficient method for the first synthesis of two natural indoles produced by acorn worms, Enteropneusta, has been presented,[228] in which the key step is a hetero-Cope rearrangement of an intermediate *N*-phenyl-*O*-vinylhydroxylamine derivative. The production of 2-vinylindoles (**131**) from *N*-phenylhydroxylamine (**127**), aldehydes (**128**), and electron-deficient allenes (**129**) in a one-pot reaction, has been explained[229] by invoking a tandem hetero-Cope-retro-Michael addition of an intermediate *N*-phenylnitrone (**130**) followed by indolylation. The transformation of protonated azo-compounds of the type (**132**) to protonated (**133**) has been shown to proceed via an intramolecular [3,3]-Cope-type rearrangement, and not via the appreciably slower retro Diels–Alder reaction.[230,231]

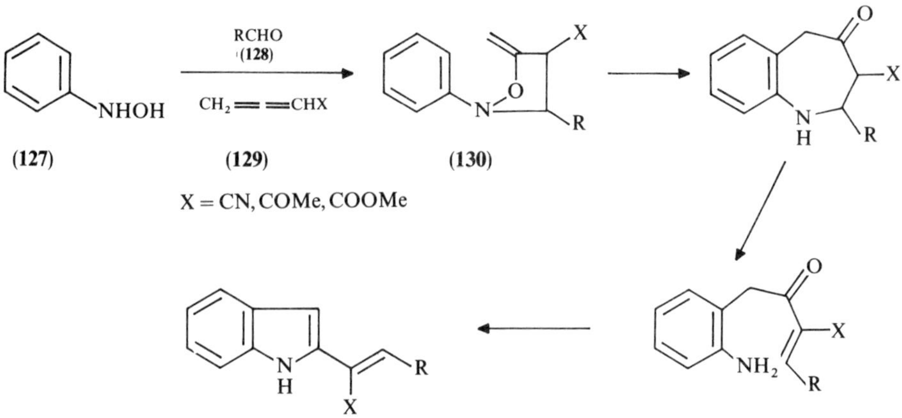

(127) (129) (130)

X = CN, COMe, COOMe

(131)

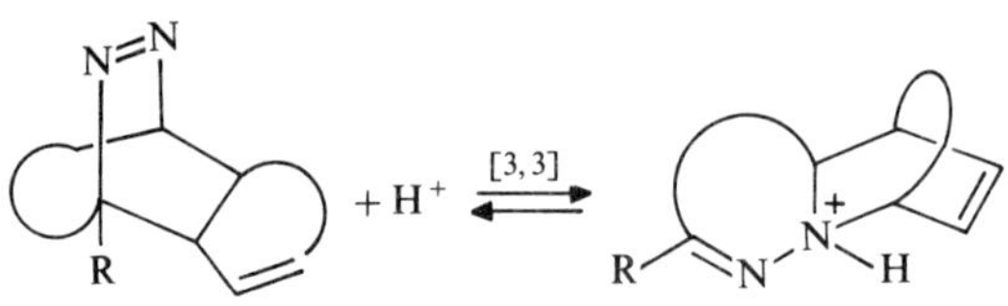

(132) (133)

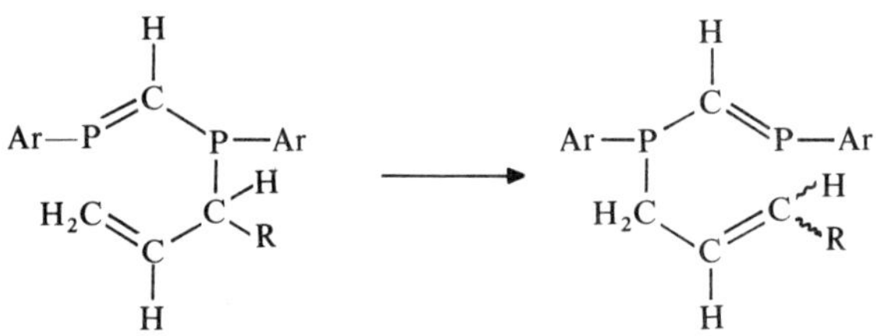

(134) (135)

Ar = 2, 4, 6-tri-tert-butylphenyl

The Cope rearrangement has been used to synthesize phosphaalkenes;[232] for example, 1,3-diphospha-1,5-hexadienes (**134**) have been converted into a mixture of (*E*,*E*)- and (*E*,*Z*)-(**135**).[233]

[2,3]-Migrations

Application of [2,3]-sigmatropic rearrangements[234] and applications of the highly stereoselective zirconium-mediated [2,3]-Wittig rearrangement,[235] to the synthesis of natural products, have been detailed. The asymmetric enolate [2,3]-Wittig rearrangement of chiral (*E*-2-alkenyloxy)acetic esters derived from (−)-8-phenylmenthol has provided a new and highly efficient method for diastereo- and enantio-selective synthesis of α-hydroxy-β-alkyl carboxylic acid derivatives.[236] Cembranoid natural products (**137**; $n = 17$) and bridged germacranolides (**138**; $n = 13$) have been synthesized by a novel variant of the [2,3]-Wittig rearrangement whereby cyclic allylic propargylic ethers (**136**) are found to undergo a remarkably facile ring-contraction to cycloalkynols.[237-239] Use of chiral bases in this process has allowed enantioselective [2,3]-Wittig ring-contraction.[240] The total syntheses of the germacrane sesquiterpenes, costunolide and haageanolide, have been achieved using transannular [2,3]-Wittig rearrangement of 13-membered diallylic ethers as the key reaction; see (**139**) → (**140**), Scheme 32.[241] An efficient route to the C(1)-to-C(10) fragment of the milbemycins and the avermectins has been accomplished[242] utilizing a selenium-initiated electrophilic cyclization in conjunction with a [2,3]-sigmatropic selenoxide rearrangement in the key transformation.

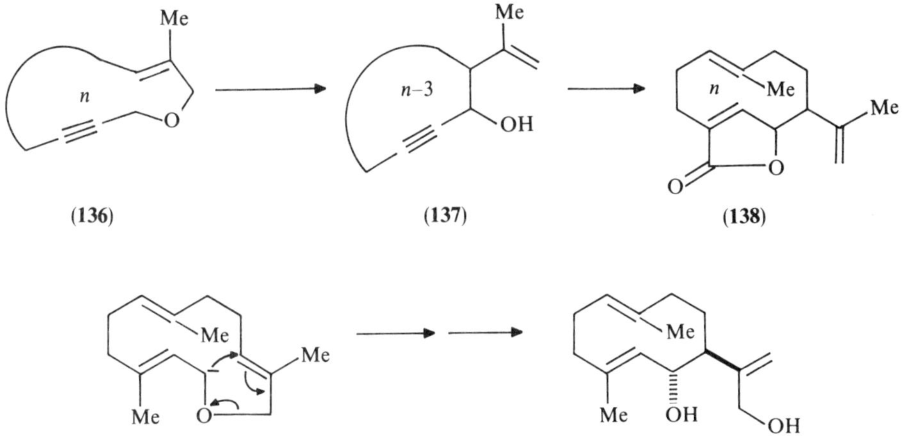

(**136**) (**137**) (**138**)

(**139**) (**140**)

SCHEME 32

Functionalized alkyl groups have been introduced regioselectively into the *ortho*-position of phenols via [2,3]-sigmatropic rearrangement of isopropylphenoxysulphonium alkylides.[243,244] The first examples have been reported[245] of [2,3]-sigmatropic rearrangements in polyfluoro-aromatic and -heteroaromatic systems in which an *ortho*-carbon bearing a fluorine, or an *ortho*-nitrogen, is the migration terminus. The alkylation of the difficulty accessible C(5) position of 4-hydroxy-6-methyl-2*H*-pyran-2-one has been achieved by transfer of functionality from position C(6) to position C(5) by a [2,3]-sigmatropic rearrangement of its 6-phenylthiomethyl ylide.[246] The catalytic effectiveness of rhodium(II) compounds for spontaneous ylide generation and [2,3]-sigmatropic rearrangement in reactions of diazo-compounds with allylic hetero-compounds has been further demonstrated[247] by the efficient preparation of γ,δ-unsaturated carbonyl compounds from allyl sulphides and α-diazo-carbonyls; see Scheme 33. A procedure employing a [2,3]-sulphoxide rearrangement has been used[248] as a general method for inverting the stereochemistry of bicyclic vinyl *endo*-alcohols (**141**) to afford the corresponding *exo*

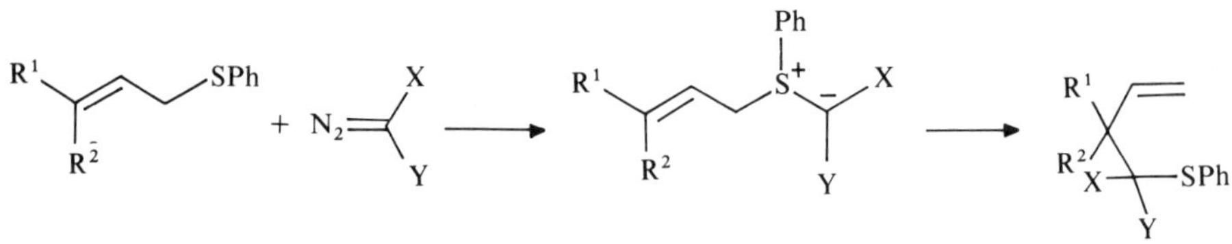

SCHEME 33

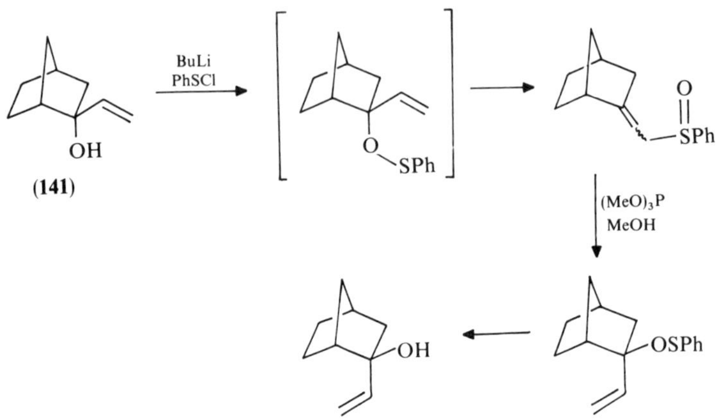

SCHEME 34

isomers; see Scheme 34. The mechanism of rearrangement of a number of allylic sulphinate esters to the corresponding allylic sulphones on heating in formamide has been investigated as a function of substrate structure.[249] Simple systems such as crotyl and α-methylallyl sulphinate appear to follow a [2,3]-sigmatropic shift pathway, while others, particularly cyclohex-2-enyl sulphinates, show the characteristics of an ion-pair mechanism. Thermolysis of cyclic α-vinyl sulphonium *p*-toluenesulphonylimides (**142**) resulted in [2,3]-sigmatropic rearrangement with ring-expansion to give thiazocines (**143**),[250] while 3,6-dihydrothiazine-1-imines (**144**) on reaction with Grignard reagent have yielded allylic sulphilimines which undergo rapid [2,3]-sigmatropic rearrangement to give sulphenamides.[251] These on desulphurization afford *threo* vicinal diamine derivatives (**145**). Treatment of *N*,*N*-dimethylhexahydrothienoazocinium or azoninium salts by sodium amide in liquid ammonia has afforded spiro-compounds which are considered to arise from ammonium ylides by [2,3]-sigmatropy.[252] The use of fluoride anion to cleave the Si—C bonds of benzyldimethyl[(trimethylsilyl)methyl]ammonium halides has been successfully accomplished[253] in the search for a new route to the Sommelet–Hauser rearrangement in non-basic media. 1,1-Captodative butadienes carrying amino, thioether, or methoxy groups combined with nitrile and ester substituents have been obtained from the corresponding propargylic ammonium or sulphonium

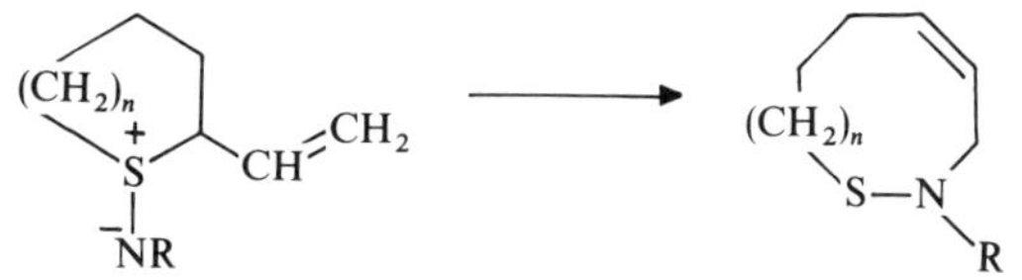

(142) **(143)**

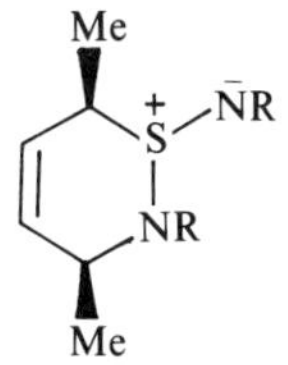

(144)

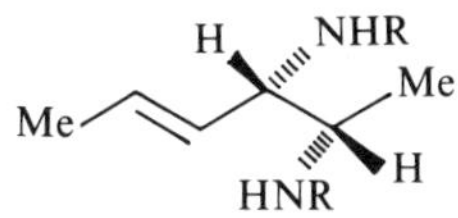

(145)

salts via the [2,3]-sigmatropic rearrangement.[254] The formation of 5,5′,6,6′-tetrahydro-2, 2′-diphenoxymethyl-1, 1′-methylenebis(4*H*-pyrrolo[3, 2, 1-*ij*]quinoline) (**147**) during the amine oxide rearrangement of 1-(4-phenoxybut-2-ynyl)tetrahydroquinoline *N*-oxide (**146**) has been explained by the pathway outlined in Scheme 35.[255] Highly regio- and stereo-selective oxidation sequences (see

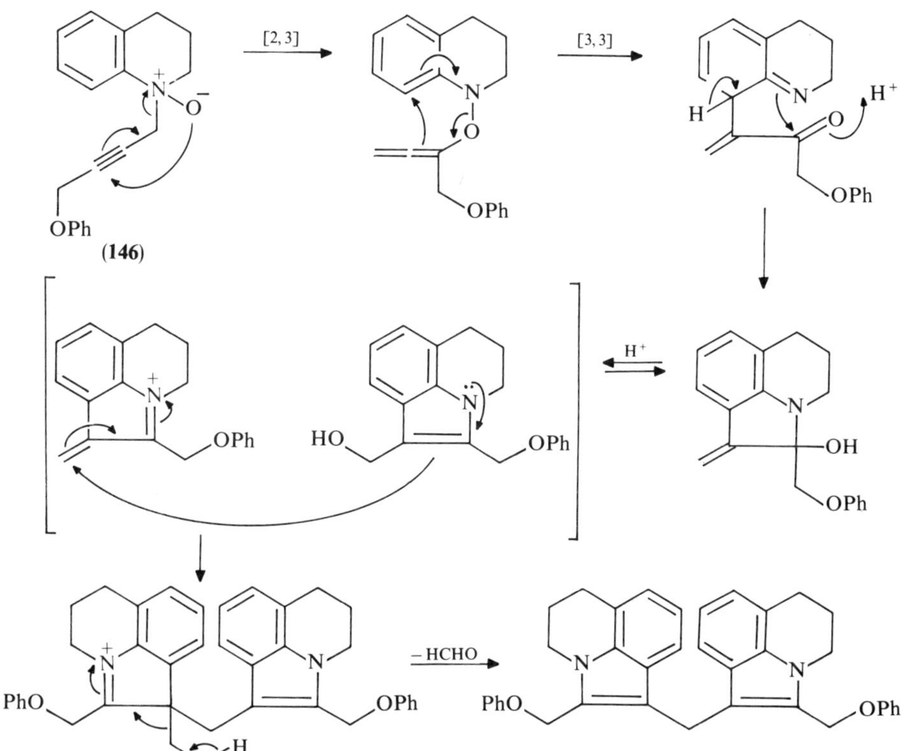

(146) **(147)**

SCHEME 35

Scheme 36) have been described[256] for the efficient conversion of the *gem*-dimethyl alkene terminus of acyclic terpenes into terminal *trans*-allylic alcohols and *trans*-α,β-unsaturated aldehydes.

Contrary to previous reports, the reaction of 2-butyne-1,4-diol with chlorodiphenylphosphine has been found to yield 2,3-bis(diphenylphosphinyl)-1,3-butadiene (**150**) instead of the 1,4-isomer. A possible mechanism for the formation of (**150**) is presented in Scheme 37[257] where the primary product, the bisphosphinous ester (**148**), is considered to undergo a [2,3]-sigmatropic rearrangement to the allenic intermediate (**149**) which subsequently is subjected to a second [2,3]-sigmatropic shift to afford (**150**).

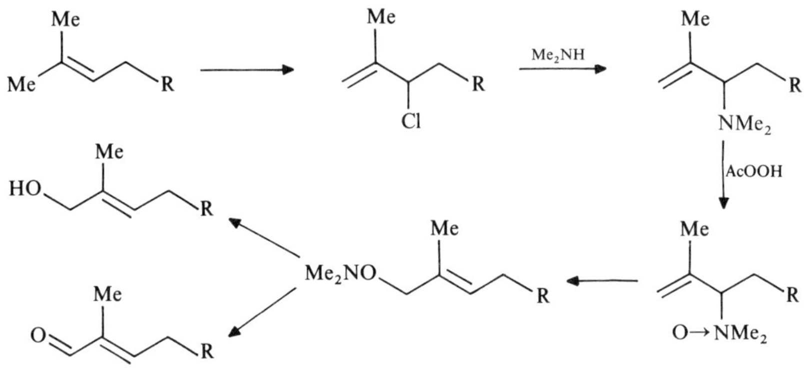

SCHEME 36

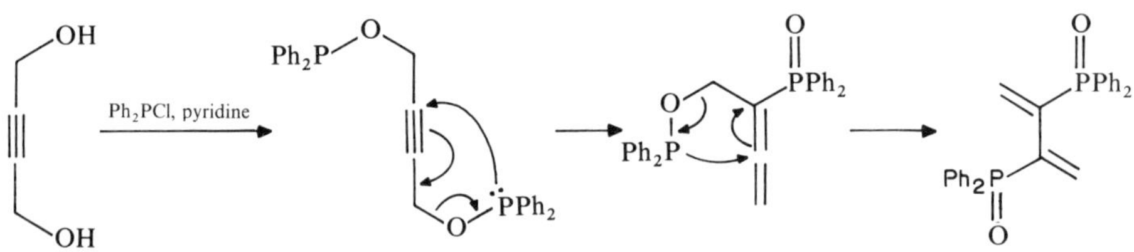

(**148**) (**149**) (**150**)

SCHEME 37

[*1,3*]-*Migrations*

Ab initio calculations have been carried out[258] for the [1,3]-shift in the isoelectronic series, HX—CH=Y ⇄ X=CH—YH (X,Y = CH_2, NH and O). An investigation of the base-catalysed intramolecular hydrogen transfer in 7-substituted cycloheptatrienes has shown[259] that the [1,3]-hydrogen shift occurs directly rather than via two successive [1,2]-shifts, and that the interaction between the migrating hydrogen atom and the central C(2) of the allyl system is repulsive, preventing the formation of a C(2) hydrogen-bonded intermediate. Allene carboxylates, and allene lactones, have been found to undergo thermally induced [1,3]-hydrogen shifts in the presence of octacarbonyldicobalt,[260] while the iron carbonyl-induced thermal conversion of 4-vinylcyclohexenes containing electron-donating and electron-withdrawing groups on the exocyclic double bond, into (η^4-cyclohexadiene)tricarbonyliron complexes, has been shown to involve η^2-intermediates formed by successive [1,3]-hydrogen shifts.[261] A study of CdS-photo-induced reactions of cyclic derivatives of diphenylcyclobutene in the absence of oxygen has established the first example of a suprafacial [1,3]-sigmatropic hydrogen shift via a radical cation.[262] Such a suprafacial pathway is symmetry-forbidden in closed-shell systems.

A report appeared[263] which considers transition structures for a variety of [1,3] N→O shifts that occur in *N*-nitrosamines. The relevance of these shifts to the formation and decomposition of carcinogenic nitrosamines is discussed. A new photochemical N→O acyl migration has been reported.[264] The unimolecular reactions of methyl acetate, methyl monothioacetates, and methyl dithioacetate, in terms of their relationship to ketene and thioketene formation in the gas phase, have been investigated[265] using the semi-empirical MNDO method. The most facile reaction for all but the *S*-methyl monothioacetate was found to be [1,3]-methyl migration. A new type of [1,3]-alkyl shift has been observed[266] during thermolysis of 4-alkyl-3,5-diphenyl-1,2,4-4*H*-triazoles. A kinetic and molecular mechanics study[267] of the thermal rearrangement of 9-carbomethoxy-*trans*-5,6-trimethylene-2-norbornene (**151**) to the tricyclodecene (**152**) has indicated that the concerted [1,3]-shift mechanism predicted for the process by orbital-symmetry rules would seem to be the simplest explanation of the data. Oxyanion formation within several *anti*-7-norbornenols has been shown to foster rearrangement predominantly via formal antarafacial-retention [1,3]-sigmatropic bridgehead carbon migration.[268] Oxy-Cope bond reorganization takes place to a lesser extent. The favoured

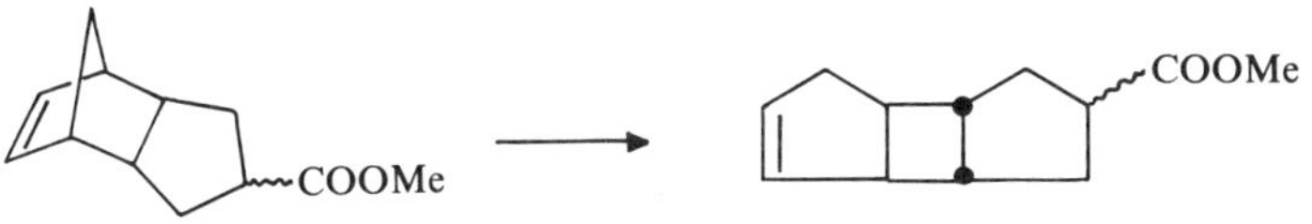

(**151**) (**152**)

mechanism for the [1,3]-shift pathway is thought to involve heterolysis of a norbornene bridgehead/apical bond, proper conformational alignment within the resulting intermediate 3-cyclohexenyl anion, and intramolecular Michael addition. The observed formation of 3,3-dicyano-1,2,3,8*b*-tetrahydrocyclopent[*a*]indene on irradiation of 9-dicyanomethylene-1,2,3,4-tetrahydro-1,4-methanonaphthalene has been explained[269] in terms of the [1,3]-carbon shift of an ethano bridge.

By analogy with the known base-promoted [1,3]-sigmatropic rearrangements of sterically crowded α-phenylselenenyl ketones, it has been postulated[270] that 3-(phenylsulphenyl)bicyclo[4.3.0]-1(6)-nonen-2-one (**154**) is formed from base treatment of *cis*-7*a*-hydroxy-3*a*-(phenylsulphenyl)-3*a*,4,5,6,7,7*a*-hexahydro-4-indanone (**153**) by the pathway shown in Scheme 38. [1,3]-Rearrangements of phenylselenenyl ketones which lack steric bulk at the α-carbon have been achieved[271] by using their sodium enolates instead of their lithium enolates. Photolysis of the 2,6-

dithioadamantane-2,2,6,6-tetraoxide (**155**), in contrast to its pyrolysis which results in loss of both SO_2 groups, has been shown to proceed by loss of only one SO_2 group, a double [1,3]-shift of the second sulphonyl group, and formation of (**156**).[272]

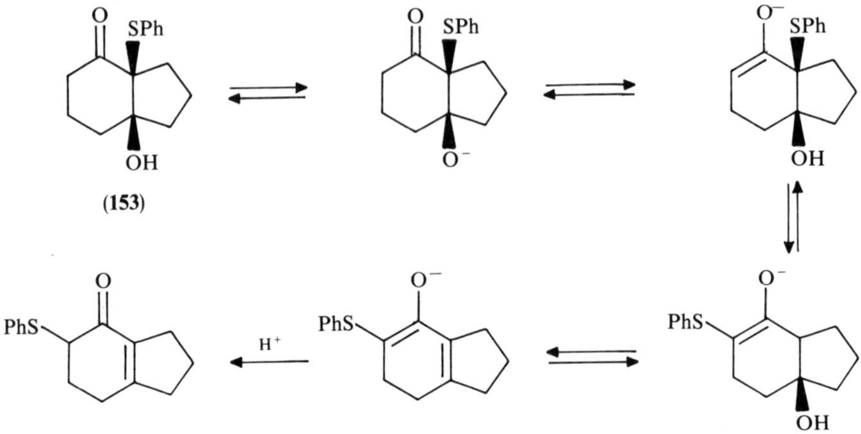

SCHEME 38

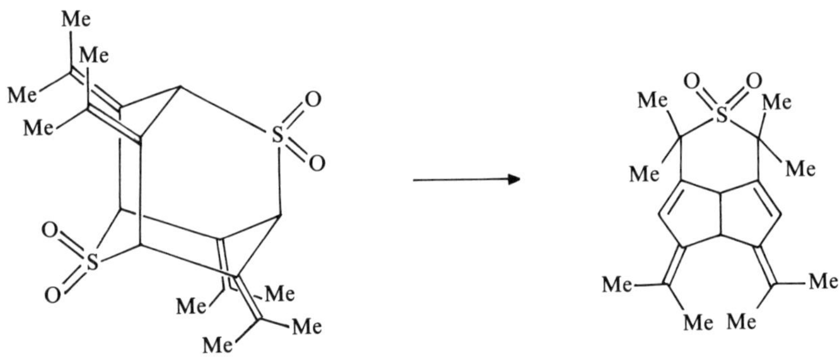

It has been shown[273] that the reaction of bis(trimethylsilyl)telluride with equimolar amounts of pivaloyl chloride affords trimethylsilyl telluropivalate which exists as a mixture of C=O and C=Te bonded isomers related by a [1,3] silyl-shift equilibrium. An MO study[274] of the mechanism of acyl migration in mixed carboxylic-phosphoric anhydride systems has supported the hypothesis that these reactions are concerted with no detectable intermediate intervening. A new synthesis of β-keto-phosphonates, which involves a [1,3]-phosphorus migration in vinyl phosphates, has been described.[275] Oxidative photolysis of *N*-aroylcarbazoles has been shown to lead to [1,3]- or [1,5]-aroyl migration from nitrogen to the ring, the product ratio being wavelength-dependent.[276]

[1,5]-Migrations

A potential-surface study has been made of the thermal rearrangement of *trans*-4A,10*a*-dihydrobenzocyclooctene into 7,8-dihydrobenzocyclooctene.[277] This rearrangement proceeds via two consecutive [1,5]-hydrogen shifts. MO theoretical studies have been carried out on [1,5]-sigmatropic hydrogen shifts in systems containing terminal oxygen and sulphur atoms,[278] and extensive calculations have appeared[279] which show that the [1,5]-sigmatropic hydrogen shift in 1,3-pentadiene does not occur via a 'planar nonfacial' C_{2v} geometry as has been recently proposed.

Pyrolysis of *N*,*N*-dimethyl-*N*-(2-methylphenyl)acetamidines has been found to produce mainly quinolines in a reaction which reveals a novel high-temperature rearrangement of 3,4-dihydroquinolines. Thus (**157**) gives not only the expected quinoline (**158**) but also the isomeric quinoline (**159**) by a rearrangement which the authors[280] have explained in terms of a series of allowed [1,5]-hydrogen shifts and six-electron electrocyclic reactions, as outlined in Scheme 39. The thermal [1,5]-

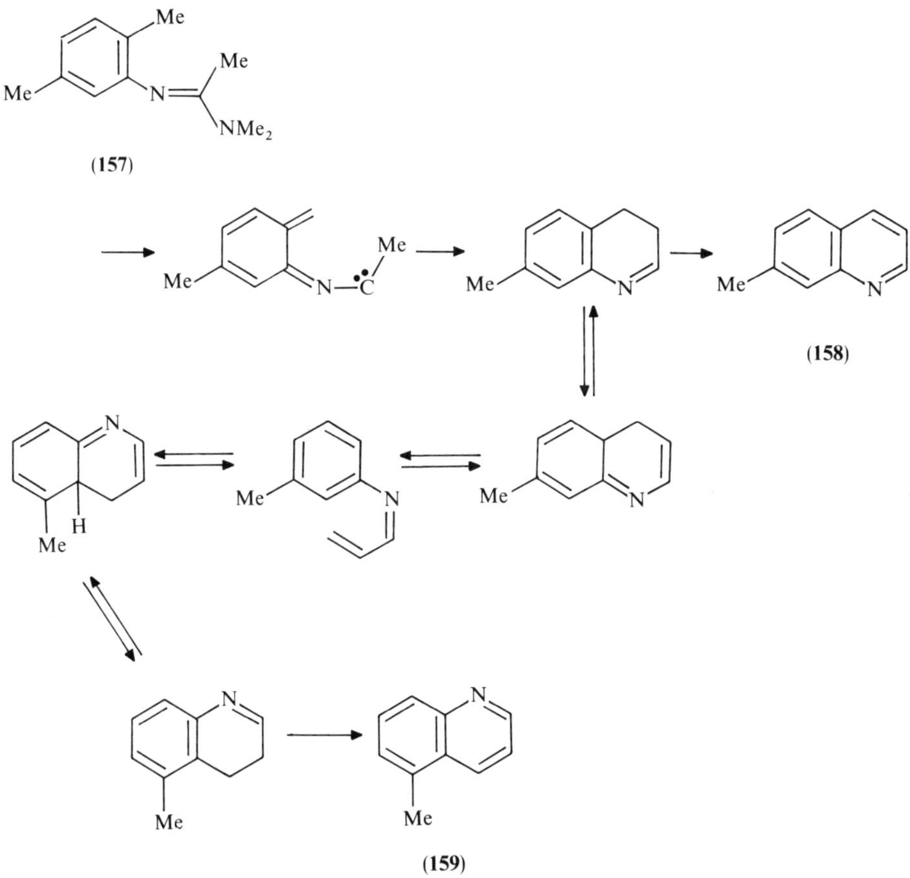

SCHEME 39

sigmatropic rearrangement of *tert*-butyl substituted 9,10- and 11,12-allenic retinoids has allowed easy access[281] to several difficult-to-prepare hindered 11-*cis*-retinoid analogues. A stereoselective method for the preparation of 6β-vitamin D-type vinylallenes has been developed,[282] and their thermal [1,5]-sigmatropic hydrogen shifts studied. Substituent-effect studies on the thermal [1,5]-sigmatropic hydrogen shifts of vinylallenes have been carried out.[283] The observed accelerating effect caused by substituents of increasing electron-withdrawing ability has been explained on the basis of a hydride-like (as opposed to a proton-like) [1,5]-sigmatropic hydrogen shift. The first example of photo-induced cyclization of spiro-indoles to carbazoles has been reported.[284] Thus, 3*H*-indole(spiro)cyclopropanes (**160**) have been transformed into (**163**), presumably by way of the dihydro intermediate (**162**) which could arise by an irreversible [1,5]-hydrogen shift in the intermediate (**161**). Self-reproduction of chirality in the thermal isomerization of

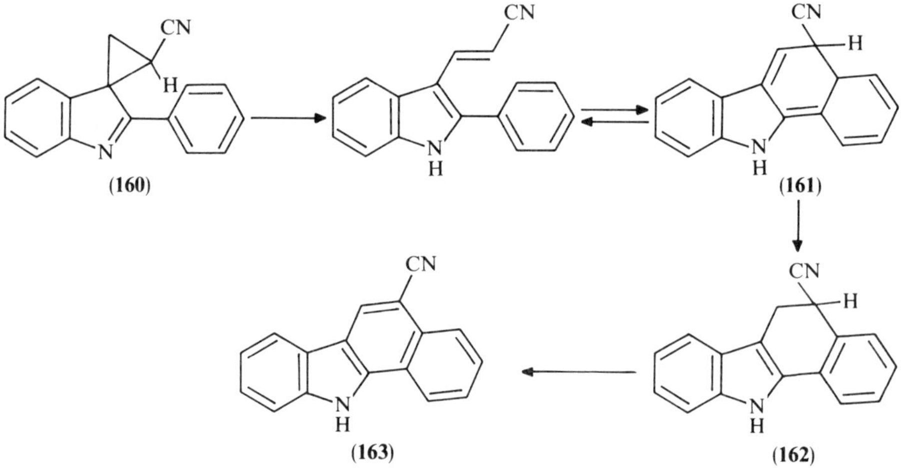

{[2-(1-pyrrolidinyl)-phenyl]methylene}propanedinitriles to pyrroloquinoline derivatives has been observed.[285] During this conversion, which is considered to take place via a [1,5]-hydrogen shift and subsequent cyclization, the centre of chirality in the starting material is lost in the corresponding dipolar intermediate, but reproduced with retention of configuration upon cyclization. In addition, the suprafacial [1,5]-hydrogen shift offers the possibility to introduce a second novel chiral centre with $> 98\%$ enantioselectivity; for an example, see (**164**)→(**165**). The highly strained, bridged *Z*,*Z*-bicyclo[3.3.1]nona-1(1),4(5)-diene system has been

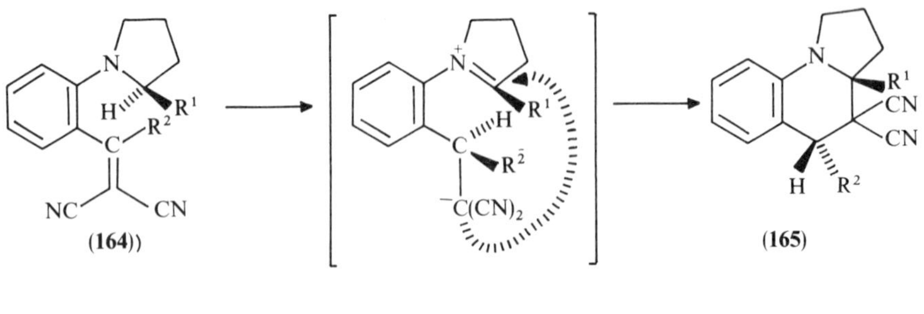

found to undergo a homo-[1,5]-hydrogen shift to yield a [3.3.1]propellane product.[286]

Thermal [1,5]-sigmatropic rearrangements of ethyl 2-cyano-3-(3,4-dimethylphenyl)-5-(4-methoxyphenyl)-2*H*-pyrrole-2-carboxylate have been studied[287] and found to yield mixtures of substituted pyridines, and 2-ethoxycarbonyl-3*H*-pyrroles have been converted via thermal [1,5]-ester shifts to the isomeric 1*H*-pyrrole-4- and -*N*-esters. Isolable intermediate 2*H*-pyrroles have been converted into the same products and also into 3*H*-pyrroles, thus demonstrating conclusively the reversibility of the 3*H*- to 2*H*-pyrrole interconversion.[288] In connection with investigations of the unique enzymic rearrangement involved in the formation of uroporphyrinogen III, a study of the chemistry of 2,2-disubstituted 2*H*-pyrroles has shown[289] that these compounds only rearrange easily by [1,5]-sigmatropic shifts if they do not have further substituents on C(3) and C(4). Such results indicate that, if the biosynthesis of uroporphyrinogen III occurs via a 2*H*-pyrrole, then it is much more likely that the rearrangement proceeds by a fragmentation-recombination mechanism.

The first complete kinetic study[290] has been made of the pyridine-catalysed reaction of the [1,5]-acyl migration of benzoquinone adducts (**166**) in order to establish the mechanism of formation of the analogue hydroquinones (**168**). A plausible mechanism for the [1,5]-acyl migration step involving a concerted process via formation of the transition state (**167**) has been presented. The Michael-type reaction of quinizarinquinone and 1-ethoxy-1-(trimethylsiloxy)buta-1,3-diene, followed by a Lewis acid-catalysed [1,5]-sigmatropic rearrangement and Friedel–Crafts cyclization, has provided a new route to variously functionalized tetracycles.[291] [1,5]-Benzoyl migration in azaenones has been reported.[292]

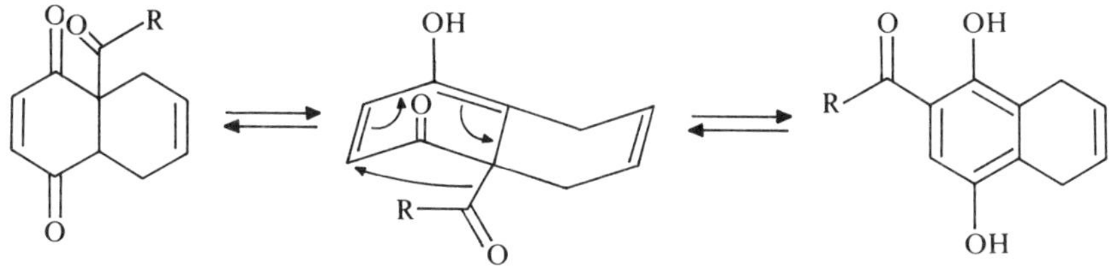

(**166**) (**167**) (**168**)

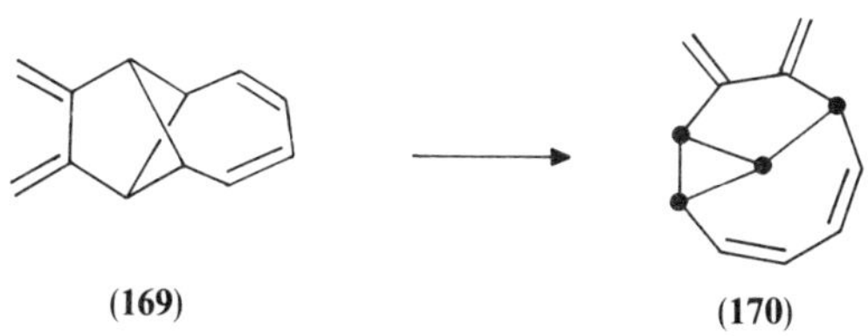

(**169**) (**170**)

9,10-Dimethylenetricyclo[5.3.0.0^{2,8}]deca-3,5-diene (**169**) has been prepared and its thermal stability assessed.[293] Heating (**169**) to 80°C induced rearrangement to the isomeric hydrocarbon (**170**) by formal [1,5]-sigmatropic migration of a central cyclobutane bond across the flanking 1,4-bridged butadiene π-system.

A rapid intramolecular suprafacial metallotropic rearrangement proceeding as a [1,5]-sigmatropic shift of an $SnMe_3$ group has been observed[294] in *trans*-(*E*)- and *cis*-(*Z*)-bis(trimethylstannyl)dihydropentalene.

Miscellaneous

It has been proposed[295] that the selection rules for [1,*n*]-sigmatropic rearrangements with an odd number of *n* can readily be deduced by the theory of alternance, but not by symmetry.

The thermal isomerization of 1-(2,4-dimethylpent-1-en-3-ylidene)-2,2,3,3-tetramethylcyclopropane (**171**) to 4-isopropyl-3-isopropylidene-2,5-dimethylhexa-1,4-diene (**173**), which involves a vinyl-trimethylenemethane intermediate (**172**), has been shown by deuterium labelling to involve a [1,6]-sigmatropic hydrogen shift.[296] The authors suggest that such sigmatropic rearrangements of non-Kekulé polyenes may involve orbital-symmetry control. The antarafacial stereochemistry of the

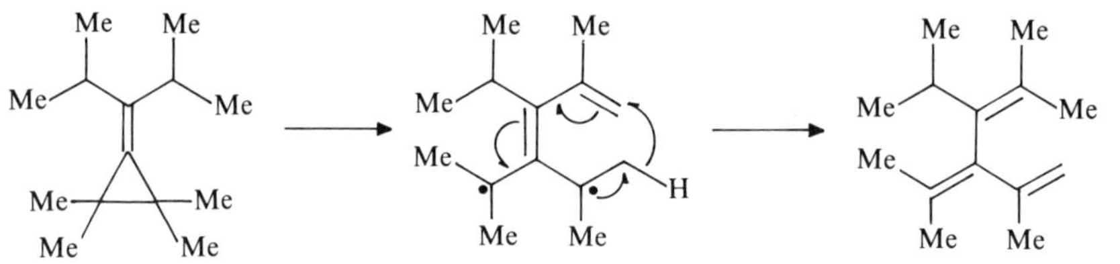

(**171**) (**172**) (**173**)

thermal [1,7]-sigmatropic hydrogen shift has been demonstrated for the first time,[297] while the thermal rearrangement of 2-(benzyloxy)-3-bromotropone to 2-(benzyloxy)-7-bromotropone, regarded to be the first reported [1,9]-sigmatropic rearrangement, has been analysed by high-pressure kinetics.[298] The data confirmed the concerted nature of the rearrangement. [3,5]-Sigmatropy has been favoured[299] as the process responsible for the formation of dihydroindenes (**175**) from the norcaradienes (**174**). The rearrangements of allylic hydroperoxides derived from oleic acid have been studied by labelling experiments.[300] The results appear to support a concerted 3,2-allylperoxyl pathway via a five-membered cyclic transition state.

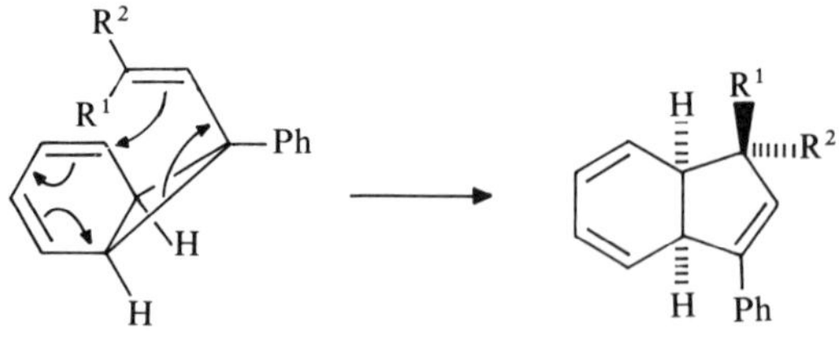

(**174**) (**175**)

Ab initio MO calculations have been applied in a study of the thermal walk rearrangement of bicyclo[2.1.0]pent-2-ene. The process has been observed to occur with inversion at the migrating centre and is thought to be a two-step reaction.[301]

The unexpectedly facile rearrangement of a range of arene oxides of polycyclic aromatic hydrocarbons to isolable oxepines via a photochemical oxygen-walk pathway has been reported.[302,303] Oxepine formation has been predicted from PMO calculations which indicate the associated loss of resonance energy during the oxygen-walk process.

A new, expedient, high-yield method for the preparation of α-substituted spiro-butyrolactones from cyclohexanecarboxaldehyde and acetic acid derivatives has been developed utilizing a dyotropic rearrangement as the pivotal step.[304] The three-step sequence is illustrated in Scheme 40. Finally, the observed rearrangement of pentacyclo[7.6.0.0^{2,13}.0^{3,8}.0^{10,14}]pentadeca-4,6,11-triene (**176**) to (**177**) represents a rare example of an intramolecular dyotropic hydrogen migration between two non-conjugated ethylene moieties.[305]

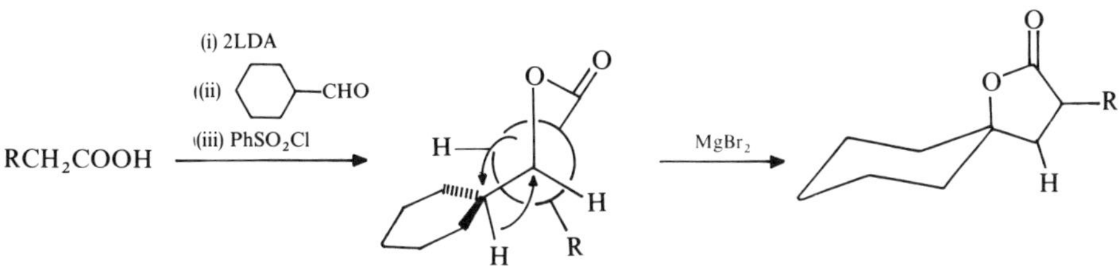

SCHEME 40

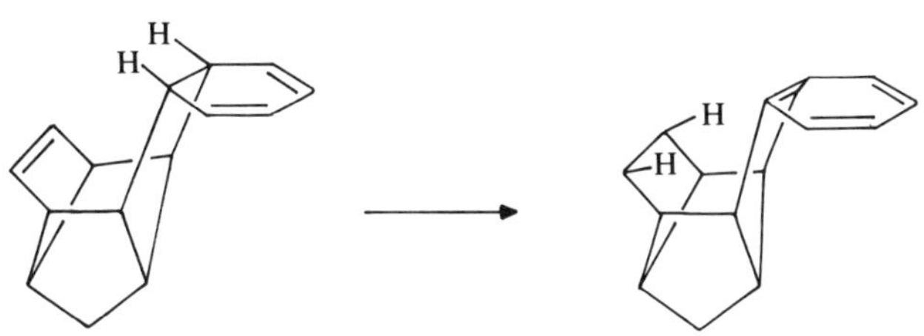

(**176**) (**177**)

Electrocyclic Reactions

Pericyclic reactions have been reviewed.[306] The nature of the dihydropentalenes prepared by the thermolysis of cyclooctatetraenes has been found to depend very much on reaction conditions and work-up.[307] The isomers are easily interconvertible by various hydrogen shifts, but thermodynamic control has been found to lead irreversibly from the equilibrium between the dihydropentalenes to styrene via 1-vinylfulvene.[308] Valence isomerization of nonapentafulvalene (**178**) into (**179**) has been found[309] to proceed even at low temperatures, while 8-(*trans*-3-ethoxycarbonylallylacetyl)heptafulvene (**180**) has been observed to undergo both an intramolecular [8 + 2]-cycloaddition and a 10π-electron electrocyclization followed by [1,5]-sigmatropic hydrogen shifts to give (**181**) and (**182**), respectively.[310]

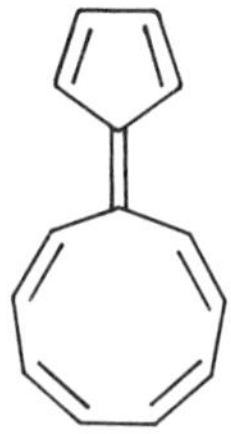

(178)

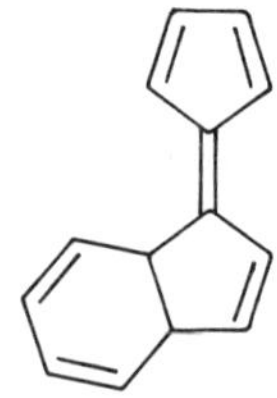

(179)

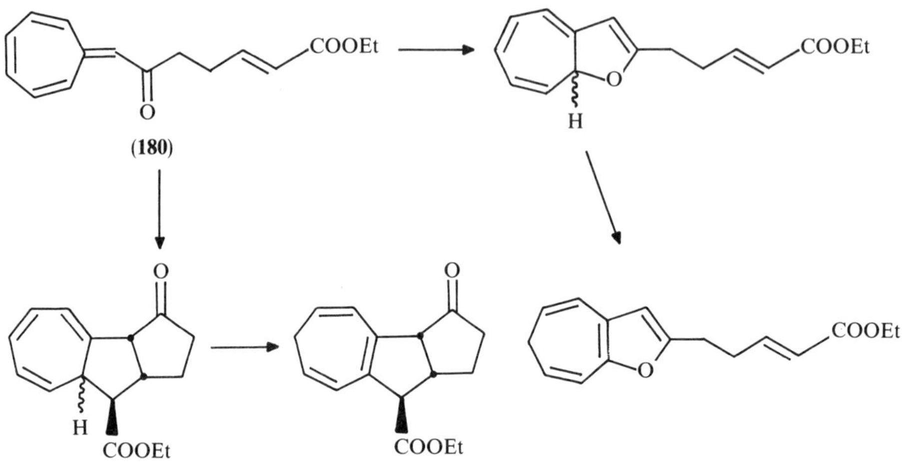

(180) **(181)** **(182)**

The electrocyclic ring-closure of conjugated trienes, in which the central double bond is part of a ring system, has been used as a viable method of constructing the C-ring of aromatic steroids.[311] An efficient and practical synthesis of the antipsoriatic drug lonapalene, involving the thermal rearrangement of 4-(4-chlorophenyl)-4-hydroxy-2,3-dimethoxycyclobutenone to 6-chloro-1,4-dihydroxy-2,3-dimethoxynaphthalene via the intermediacy of a vinylketene, has been reported.[312] The Diels–Alder reaction of donor-substituted 1,3-dienes with tetrachlorocyclopropene has led to novel bicyclo[4.1.0]heptenes which on acidic hydrolysis and oxidation yield a new series of chloro-substituted derivatives of tropone by way of elimination of hydrogen chloride concerted with an electrocyclic opening of the three-membered ring.[313] A novel thermal rearrangement of 1-chlorobicyclo[6.1.0]nona-2,4,6-triene (**183**) has resulted in the synthesis of the highly strained, unsubstituted hydrocarbon, tricyclo[5.2.0.0^{1,3}]nona-4,6,8-triene (**184**),[314] and the ring-closure of 2,3-dimethylene-2,3-dihydrofuran to dihydrocyclobutafuran has been accomplished photochemically.[315] The oxabicyclohexene (**187**), formed by photochemical addition of fluorenone (**185**) on to the cyclobutadiene (**186**), has been found to undergo spontaneous ring-opening to yield the 1,3-diene (**188**) which isomerizes by electrocyclic ring-closure to the cyclobutene (**189**).[316]

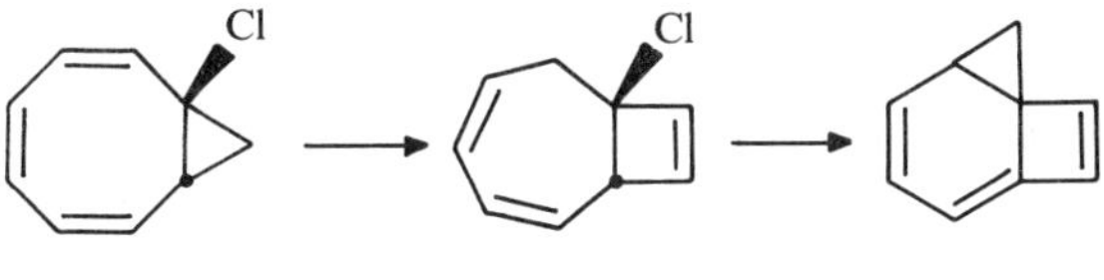

(183) (184)

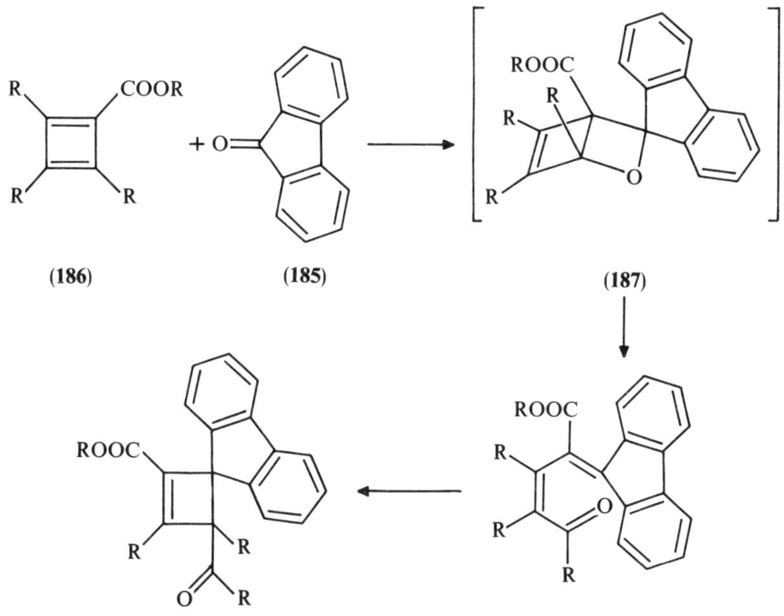

(186) (185) (187)

(189) (188)

R = Bu[t]

The electrocyclization reactions of *N*-substituted-1*H*-azepin-3(2*H*)-ones have been compared[317] with those of pyridin-2-one and azepin-2-one. Oxapyrazolinium ylides (**192**) and pyrazolinones (**193**) have been identified as the main products of thermolysis of enhydrazino-diesters (**190**). The formation of these products can be rationalized[318] by invoking the intermediacy of hydrazono-ketenes (**191**), their cyclization to (**192**), and a subsequent intramolecular [1,4]-sigmatropic alkyl shift to yield the pyrazolinones (**193**). 2-Diazomethylpyrans (**195**), readily obtained from pyrylium salts (**194**), have been found to isomerize to 1*H*-1,2-diazepines (**196**). This previously unknown rearrangement has been interpreted by postulating an initial electrocyclic ring-opening as illustrated in Scheme 41.[319] It has been shown[320] that, in the case of certain dimethylamino-nitrilimines (**197**), the 1,5-electrocyclization to 1,2,4-triazoles (**198**) is suppressed in favour of a 1,7-cyclization at the *N*-aryl group. The resulting benzotriazepines (**199**) represent a new example of a dipolar 1,7-electrocyclization at a benzene ring. The formation of enamines (**202**) by reaction of 4-*tert*-butyl-5-methylene-3,3-dimethyl-Δ^1-1,2,4-triazoline (**200**) with dimethyl acetylenedicarboxylate has been explained[321] by a series of electrocyclic-closures

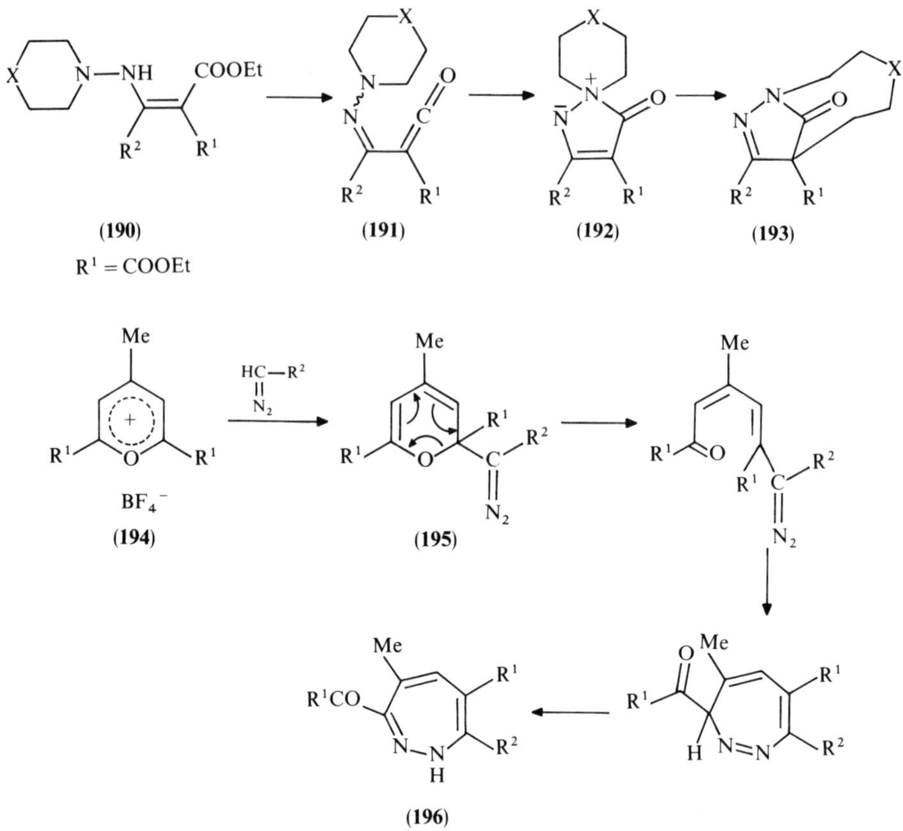

SCHEME 41

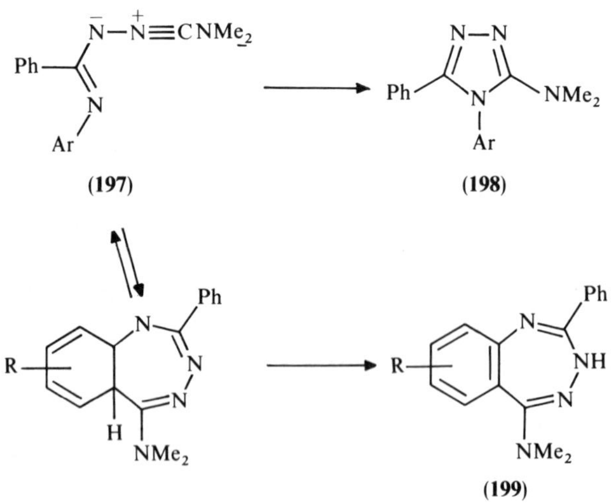

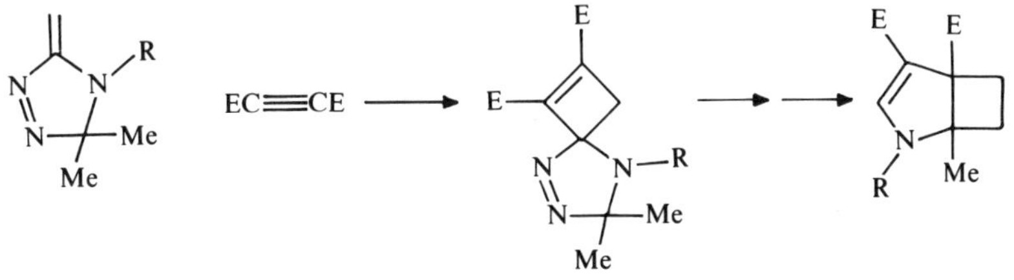

(200) (201) (202)

E = COOMe

R = Bu^t

and ring-openings proceeding by way of the azabicycloheptene (**201**) as key intermediate. Heating cyclic secondary α-amino-acids with an excess of paraformaldehyde and methyl propiolate has been shown to lead to a decarboxylative ring-expansion yielding 8- and 9-membered cyclic aza-dienes.[322] The mechanism of this remarkable ring-expansion is thought to involve formation of an azomethine ylide (**203**); see Scheme 42. 1,3-Dipolar cycloaddition of methyl propiolate can give rise to two regioisomers. Although the dipole-dipolarophile frontier-orbital interactions would be expected to be similar for both regioisomeric transition states, preference for (**204**) would be expected to arise from steric factors, with 2,3-bond formation being more advanced than 4,5-bond formation in the transition state leading to cycloadducts. Michael addition of the lone pair of the bridgehead nitrogen atom in (**204**) to methyl propiolate would then generate the zwitterion (**205**) which could fragment to product (**206**).

Apparently, only the product (**208**) of normal butadiene-cyclobutene cyclization has been observed[323] on photo-isomerization of the monosubstituted 1-benzothiepines (**207**). A unique photo-transformation of a 2,4′-bithiazole leading to a

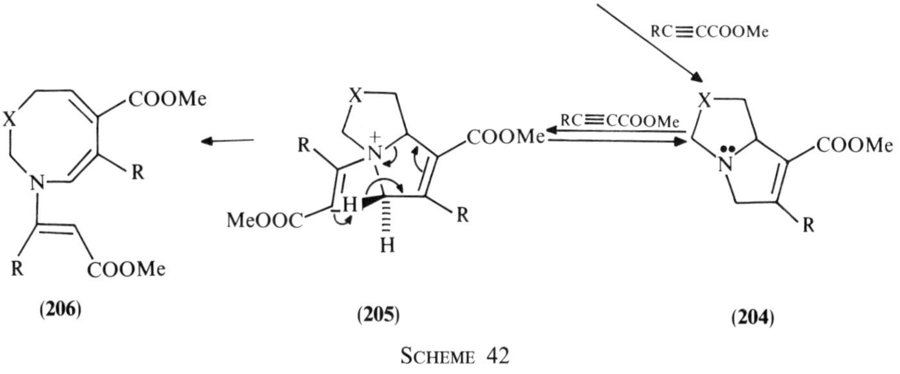

X = CH_2, S

(203)

(206) (205) (204)

SCHEME 42

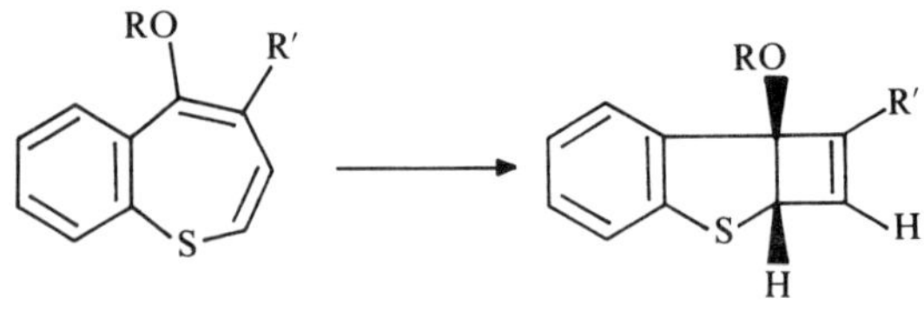

(207) (208)

selective ring-atom transformation at only one of the thiazole rings has been described;[324] see (**209**) → (**212**). It seems reasonable to assume that singlet excitation of (**209**) would result in an initial disrotatory closure of C(2)–C(5) to Dewar intermediate (**210**) which could rapidly undergo a [1,3]-shift of the sulphur atom to give (**211**) as possible precursor of (**212**). 3,6-Sulphur-bridged hexahydro-1,2,4-triazin-5-ones (**214**), prepared by reaction of thiazolium-4-olates (**213**) with dialkyl azidocarboxylates, have been shown (see Scheme 43) to rearrange to the blue thioxohydrazides (**215**) on heating in benzene.[325] Diazasilacycloheptadienes have been readily prepared by thermal rearrangement of diazasilabicyclo[3.2.0]hept-6-enes.[326]

The photolysis of the tropone endoperoxide, 6,7-dioxabicyclo[3.2.2]nona-3,8-dien-2-one, has produced a number of photolysates, which include new products, 2,7-dioxabicyclo[3.2.2]nona-3,8-dien-6-one and 7-*exo*-formyl-3-oxa-bicyclo[4.1.0]hept-4-en-2-one.[327] Scheme 44, which invokes the intermediacy of phosphoranyl 1,3-biradicals (**217**), is thought to correlate best the current facts concerning the recently reported photo-induced electrocyclic rearrangement of allyl

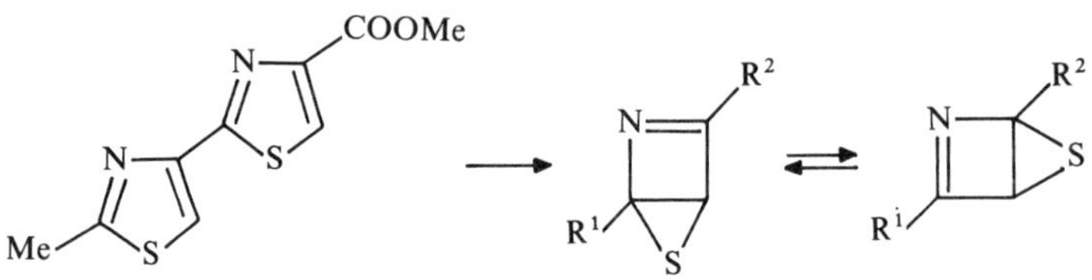

(209) (210) (211)

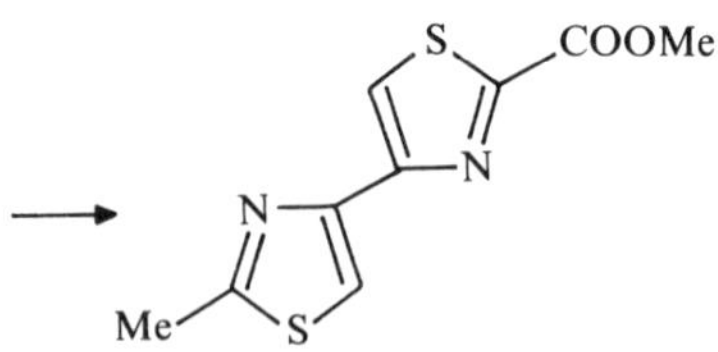

(212)

R¹ =

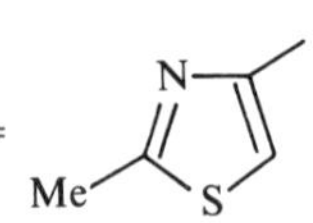

R² = COOMe

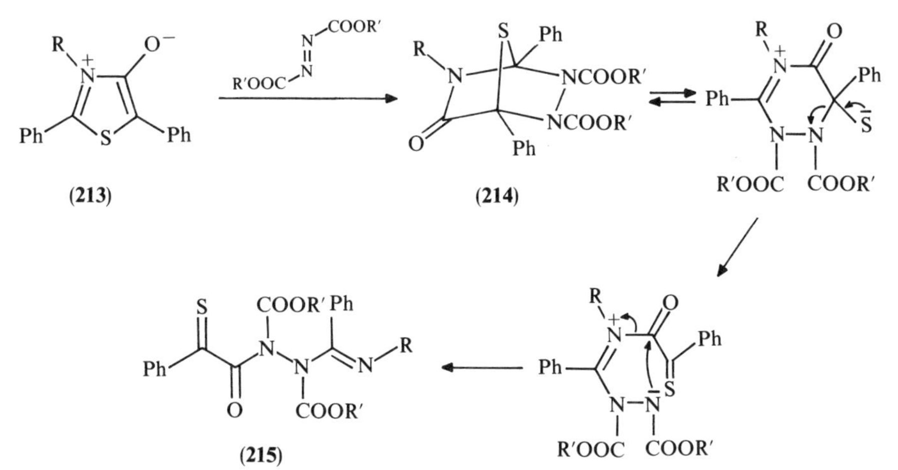

(213) (214) (215)

SCHEME 43

phosphites (**216**).[328] 1-Dewar-phosphinines, phosphaprismanes, and phosphabenzvalenes have been synthesized starting from 2-Dewar-phosphinines,[329] while the recently observed[330] disrotatory, electrocyclic valence isomerization (**218**)⇄(**219**) has demonstrated that the mutual conversion of $\lambda^3\sigma^3$- and $\lambda^3\sigma^2$-phosphorus in pericyclic reactions is possible.

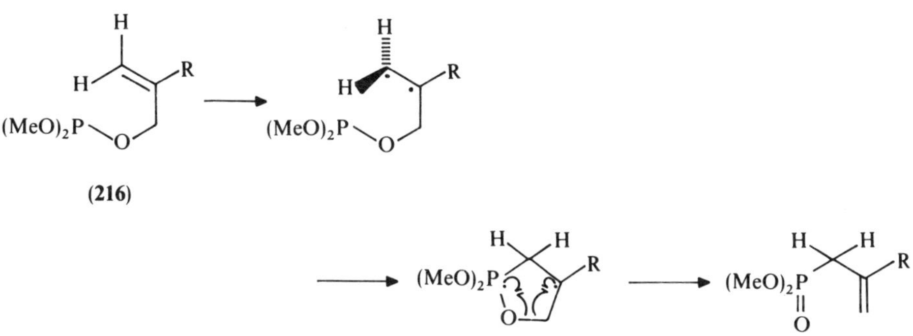

(216) (217)

SCHEME 44

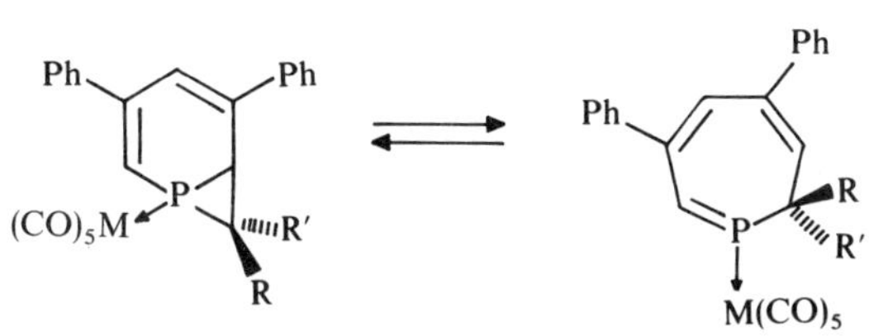

(218) (219)

M = Cr, Mo, W

Transition structures for the ene reactions of propene with ethylene to form 1-pentene, and propene with formaldehyde to form 3-buten-1-ol have been located[331] with the STO-3G and 3-21G basis sets. The ene reaction of 5-allyl-2ζ-hydroxy-5α-

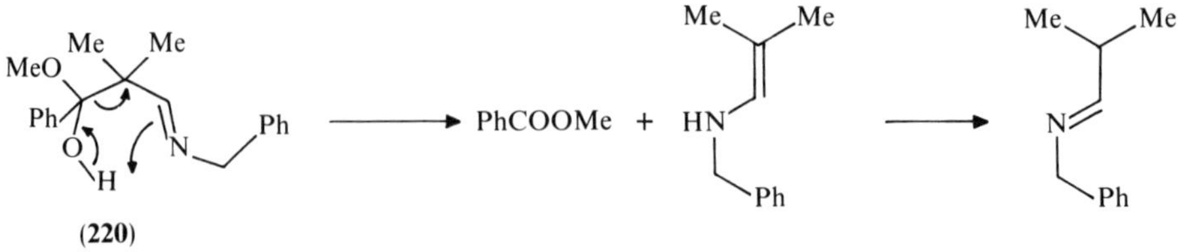

SCHEME 45

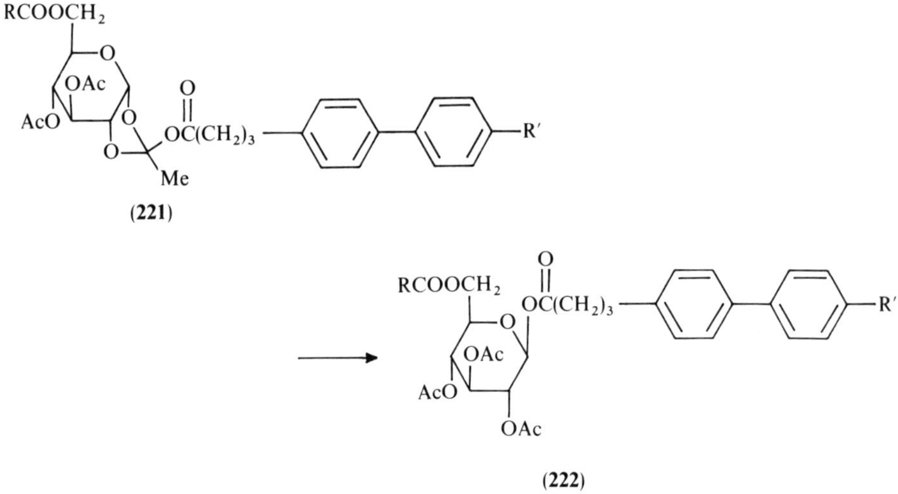

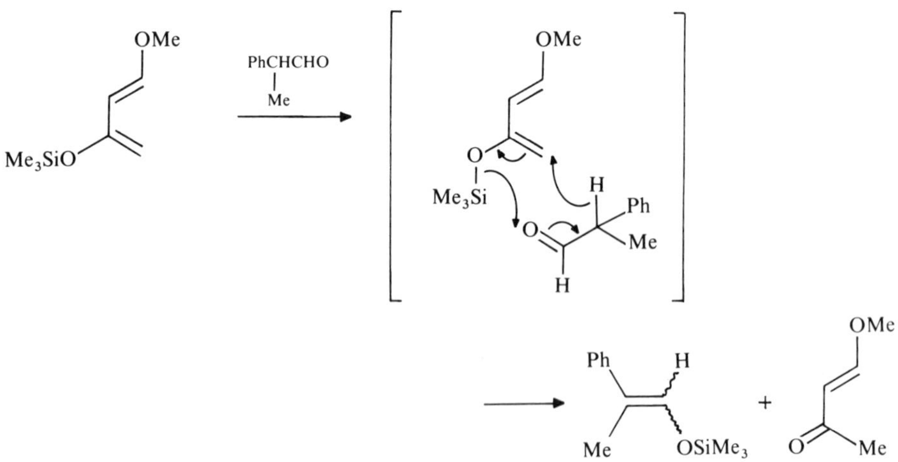

SCHEME 46

cholestane-3-one has been observed to proceed with remarkable facility,[332] while a unique hydroxyl-assisted intramolecular ene reaction has provided a key transformation in recent syntheses of tetracyclic diterpenes.[333] The retro-ene reaction of *N*-substituted derivatives of 4-aza-2,2-dimethyl-1-phenyl-3-butenone with methanol and benzylamine is considered to proceed via the hemiacetal (**220**), as shown in Scheme 45.[334] Acyl orthoesters (**221**) have been found to undergo a facile rearrangement to *trans*-diacyl derivatives (**222**) by an intramolecular pericyclic transition state similar to that of a retro-ene reaction.[335] The first example of the high-pressure-induced intermolecular silicon migration from oxygen to oxygen has been reported;[336] see Scheme 46.

Anionic Rearrangements

Localized orbital analysis has shown[337] that the 1,2-hydrogen shift reactions of neutral and anionic acetylene follow very different reorganization patterns from each other. Namely, the migrating hydrogen behaves as a hydride ion in the neutral C_2H_2 system and in the anion non-planar path, and as a hydrogen radical in the C_2H_2 anion planar reaction path. A theoretical study[338] of rearrangement processes for ethyl, ethenyl, and ethynyl anions has been initiated, and a study of the mechanism of 1,3-prototropic shifts in acetylene-allene isomerizations has demonstrated[339] that such shifts effected by alkali metal amides of diamines proceed with some loss of stereochemical integrity, thought to be caused by discrete anionic intermediates. Although a cyclic concerted bimolecular mechanism is considered to be operating in part, it cannot be the exclusive mode of proton transfer. On exposure to potassium *tert*-butoxide in dimethyl sulphoxide-d_6, optically active terminal-β,γ-acetylenic alcohol has been found to rearrange to β-trideuteriomethyl-α,β-acetylenic alcohol.[340] 4-Hetero-hepta-1,6-diynes have been rearranged by base catalysis to the corresponding 4-hetero-hepta-1,2,5,6-tetraenes. Thermal rearrangement of these tetraenes has yielded the corresponding dimers by way of diradical intermediates.[341] The ratio of 2-alkylthio-1,3-butadienes to 1-alkylthio-1,3-butadienes, formed from 1-alkylthio-2,3-butadienes under the influence of superbases, has been found to depend on the alkyl structure and on the nature of the alkali metal cation.[342] In order to probe the occurrence and relative ease of carbon-skeletal [1,2]-sigmatropic rearrangements of the free-radical, anionic, or radical-anionic type, derivatives of β,β-diphenyl-β-(2-pyridyl)- and β,β-diphenyl-β-(4-pyridyl)ethane systems have been treated with reagents expected to generate radical or anionic sites. The ensuing competitive [1,2]-shifts of the phenyl and/or pyridyl groups have been used as a diagnostic sign of the mechanism of the rearrangement.[343] The same authors[344] have treated a wide variety of benzylic amines with a selection of bases and interpreted the results of the ensuing rearrangements in terms of [1,2]-intramolecular shifts of aryl groups from N to benzylic carbon proceeding by way of a bridged aryl transition state or intermediate.

A study[345] aimed at resolving the controversy regarding the structure of the bicyclo[3.2.2]nona-3,6,8-trienyl anion and further elucidating the mechanism of its degenerate rearrangement has supported the existence of a symmetric ion in which the negative charge has been considerably delocalized into the vinyl bridges. The authors have proposed that the controversial stabilization of carbanions is not caused by delocalization and homoaromaticity, but is rather due to a previously ignored, important interaction with the metal ion. The radical anion of 1,5-methano[10]annulene and its rearrangement products have been studied recently by ESR and, in part, by ENDOR spectroscopy.[346]

A review[347] of ether cleavage with organo-alkali metal compounds and alkali metals has shown that, in the case of aliphatic ethers, the Wittig rearrangement occurs to a significant extent. The collisional activation mass spectra of benzylic carbanions derived from benzyl alkyl ethers and benzyl phenyl ethers have been interpreted[348] in terms of Wittig rearrangements. The [1,2]-Wittig rearrangement of *β*-alkoxyalkyl allyl ethers has been found to provide *syn*-1,3-diol derivatives with useful levels of diastereoselection.[349] Treatment of 7*H*,9*H*,16*H*,18*H*-dinaphtho[1,8-*cd*:1′,8′-*ij*][1,7]dithiacyclododecin with butyllithium and methyl iodide has been found to yield *anti*-7,8,15,26-tetrahydro-7,15-bis-(methylthio)cyclodeca[1,2,3-*de*:6,7,8-*d′e′*]dinaphthalene as the main product and 1′,2′,7,8-tetrahydro-1′,8-bis[(methylthio)methylene]-1,2′:7,8′-dimethano-dinaphthalene as a side-product. The origin of these products has been rationalized[350] by assuming a [1,2]- and a [1,4]-shift, respectively, in a Wittig thioether rearrangement. 2-(4-Cyano-, 2-nitro-, and 4-nitro-benzyloxy)tropones and 2-(2-oxo-2-phenethyloxy)tropone have been rearranged by base treatment to 2[(4-cyano-, 2-nitro-, and 4-nitro-phenyl)hydroxymethyl]tropones and 2(1-hydroxy-2-oxo-2-phenethyl)tropone, respectively,[351] in a reaction which has no precedent in troponoid chemistry; see Scheme 47.

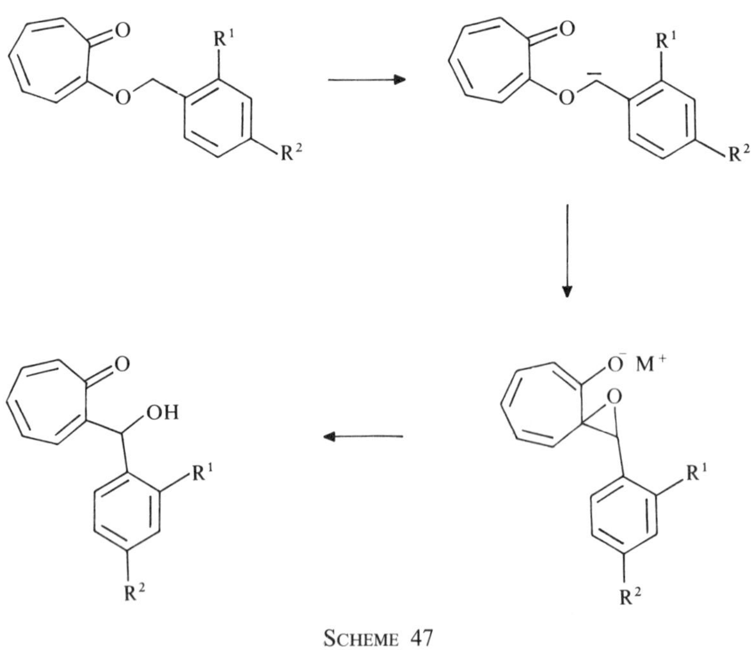

SCHEME 47

A combination of labelling studies and *ab initio* calculations has suggested[352] the pathway shown in Scheme 48 for the collision-induced loss of ketene from acetylacetone enolate anions. The first direct observation of an intermediate dienolate ion during the base-catalysed isomerization of a β,γ-unsaturated ketone to its conjugated isomer has been made.[353] A comprehensive investigation of the detailed mechanism of the methoxide-catalysed cyclization of benzylidene phthalides, methyl *o*-phenylacetylbenzoates,[354] methyl *o*-acylbenzoates, and 8-acyl-1-naphthoates[355] to indane-1,3-diones and related diones has been reported. The same authors have extended their study to the base-catalysed rearrangement and methoxide-catalysed methanolysis of a series of 2-acetylphenyl-3- or -4-substituted benzoates and related compounds.[356]

$$\mathrm{MeCO\bar{C}HCOMe} \longrightarrow \mathrm{MeCOCH_2CO\bar{C}H_2} \longrightarrow \mathrm{MeCO\bar{C}H_2} + \mathrm{CH_2CO}$$

SCHEME 48

Hydride transfer from triisobutylaluminium to the C(5) of 2-ethoxy-4-alkyl-2,3-dihydrofurans has been found to occur regioselectively, resulting in construction of the ring; see **(223)** → **(224)**.[357] An examination of the base-catalysed rearrangement of polymeric procyanidins has shown[358] that the interflavanoid bond of the system is more labile than the pyran ether to cleavage under alkaline conditions.

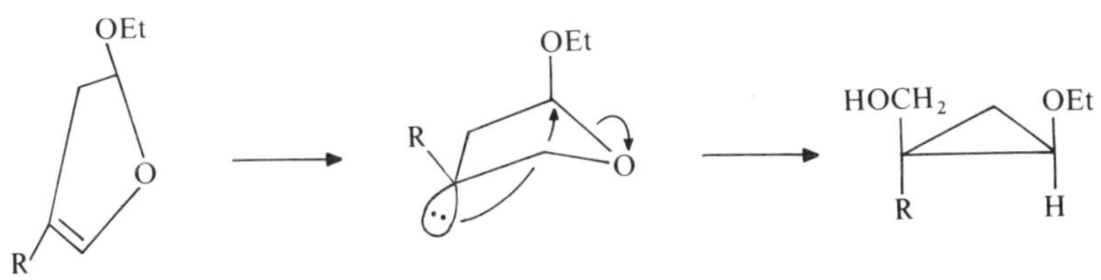

(223) **(224)**

The reaction of aromatic or hetero-aromatic aldehydes with 3-chloro-3-methyl-2-butanone in ethanolic KOH has been shown to afford 3,3-dimethyl-5-aryltetrahydro-2-furanone. The reaction has been tentatively explained[359] by invoking a combined Favorskii rearrangement and aldol reaction; see Scheme 49. Favorskii-type rearrangements of α,α'-dihalo-ketones induced by either the enolates of (diethoxyphosphinyl)acetic esters[360] or with sodiomalonates[361] and their application to the synthesis of α,α'-divinyl ketones and conjugated enones, have been described. The Favorskii rearrangement of 2-bromobicyclo[3.3.1]nonan-3-one has afforded the bicyclo[3.2.1]octane-6β-carboxylate.[362] A new type of Favorskii rearrangement which includes initiation by the Michael addition of alkoxide anion to 6-chloro-2-vinylidenecyclohexanone has been described,[363] while the Lewis acid-promoted Favorskii rearrangement of some cyclic α-bromo-ketones has been observed to furnish ring-contracted products containing a *gem* methyl carbomethoxy function.[364]

Selenoxy-ammonium salts, prepared from tertiary amine *N*-oxides and benzeneselenenyl triflate, have been noted to undergo a 'selenium Polonovski rearrangement' in the presence of base to yield α-selenoxy-amines (**225**),[365] which react *in situ* with an electrophile or a nucleophile to afford new *sec*- or *tert*-amine derivatives; see Scheme 50. The stereospecificity of the photo-rearrangement of nitronate anions to hydroxamic acids, and the utilization of this process for the stereospecific cleavage of cyclic compounds[366] is illustrated by the conversion in Scheme 51. When treated with potassium ferricyanide-sodium nitrite in the presence of base, methyl 3,10-dinitropentacyclo[5.3.0.0.^{2,5}0.^{3,9}0^{4,8}]decane-1-carboxylate was found to undergo an unusual skeletal rearrangement to afford methyl 3,9-dinitro-*exo*-10-methoxypentacyclo[5.3.0.0.^{2,5}0.^{3,9}0^{4,8}]decane-8-carboxylate.[367]

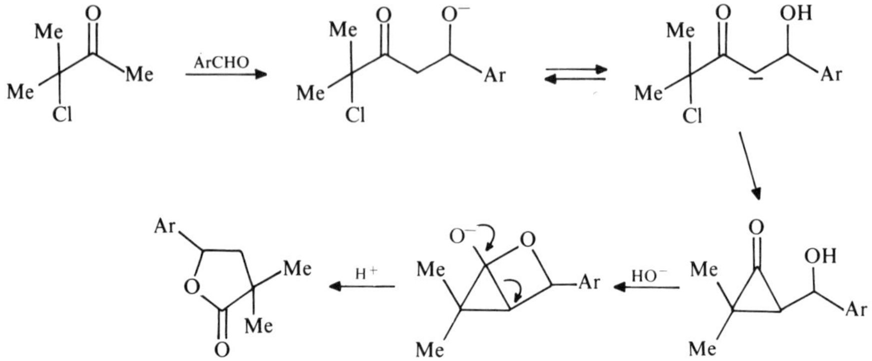

SCHEME 49

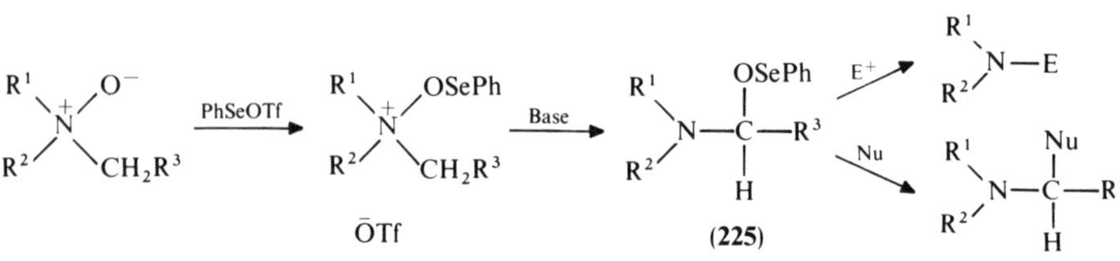

SCHEME 50

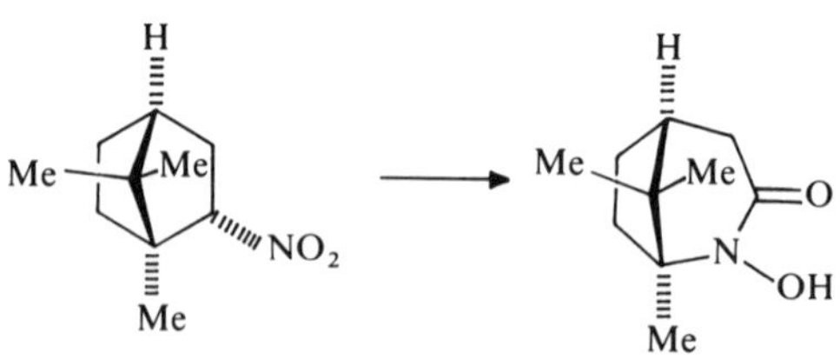

SCHEME 51

Unsaturated α-(dimethylamino)carboxylic acid amides have been prepared[368] by applying Stevens rearrangement methodology. Cyclic sulphonium ylides of the type (**226**) have been observed to undergo smooth rearrangement on heating. However, none of the substrates was found to form the usual products derived from a Stevens-type 1,2-rearrangement; instead they gave either fragmentation products or products of an overall 1,4-rearrangement.[369] Sulphone-stabilized anions have been utilized[370] in a regioselective ring-expansion that results in the formation of α-methoxy-ketones, and a carbanionic-induced ring-opening has been invoked[371] to explain the products of base-catalysed rearrangement of bis(ethylsulphonyl)(2,3-*O*-isopropylidene-α-D-lyxopyranosyl)methane.

Acylsilanes have been found to react with fluoride ion in the presence of water to give products arising either from intermediates equivalent to acyl anions, or from rearrangements involving alkyl-group migrations from silicon to carbon (see Scheme 52) depending upon the reaction conditions.[372] The nature of the solvent has also been found to affect the carbanion rearrangements of ω-phenyl-ω-(trimethylsilyl)alkyllithium compounds. Change from ether to tetrahydrofuran as solvent has the effect of accelerating silyl shifts with respect to proton shifts.[373] Numerous 3-[(trialkylsilyl)oxymethyl]-furans and -thiophenes have been found to undergo a 1,4-oxygen-to-carbon silyl migration when treated with butyllithium, to produce 2-(trialkylsilyl)-3-(hydroxymethyl)-furans and -thiophenes.[374]

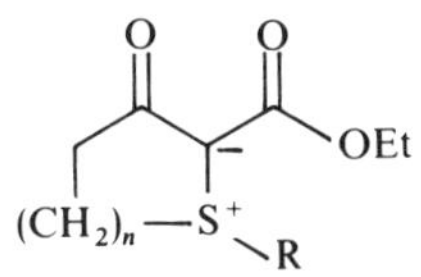

(226)

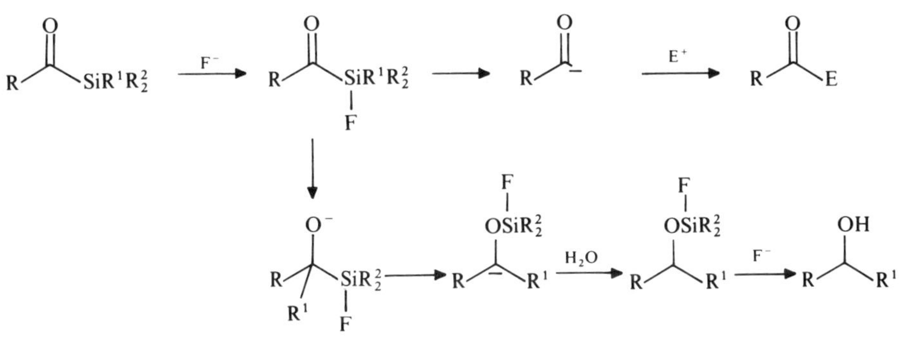

SCHEME 52

A new method for the preparation of β-keto-phosphonates has been developed,[375] involving rearrangement of vinyl phosphates upon treatment with strong base, while *o*-phosphinophenols have been prepared via a 1,3-carbanionic rearrangement of *o*-metalated (aryloxy)-phosphines.[376]

Cationic and Related Rearrangements

Ab initio MO calculations have been carried out for the various possible isomeric structures of the substituted methylene dications HCX^{2+} (X = F, OH, NH_2, etc.), and for the transition structures for their rearrangements.[377] The MINDO/3 method has been used to study the isomerizations of the tertiary 1-methylcyclohexylium ion to the 1,2-dimethylcyclopentylium ion[378] and to the tertiary 1-ethylcyclopentylium ion.[379] A reaction mechanism has been formulated for these rearrangements and the generality of the proposed mechanism has been discussed. The observed, facile acid-catalysed intramolecular allylic isomerization of 2-methoxyisopropylideneindan to 3-(1-methoxy-1-methyl)ethylindene[380] in aqueous solution represents the first demonstration of a carbocation-molecule pair as intermediate in a heterolytic reaction in a highly aqueous solvent.

A review which includes rearrangements in bimolecular nucleophilic substitutions at vinylic centres has appeared,[381] and the recently reported solvolytic rearrangement studies with (*E*)- and (*Z*)-2-anisyl-1,2-ditolyl[2-^{13}C] vinyl bromides[382] have completed the series of nine possible triarylvinyl cations with various combinations of phenyl, *p*-tolyl, and *p*-anisyl as the three aryl groups that can give rise to degenerate 1,2-aryl shifts across the double bond. Various cyclic dienynyl triflates (**227**) have been solvolysed into aryl ethers (**229**), presumably via intermediate aryl cations (**228**),[383] and a new propargylic rearrangement of acetylenic alcohols to thioketone derivatives has been induced by phosphorus pentasulphide.[384] As yet it is not clear whether this rearrangement occurs through a concerted pathway or is a stepwise process proceeding by way of an acetylenic thiol intermediate. A direct synthesis of methyl arylalkanoate via oxidative rearrangement of alkynes using [hydroxy(tosyloxy)iodo]benzene in methanol has been described;[385] see Scheme 53. CIDNP evidence for a conformationally excited or

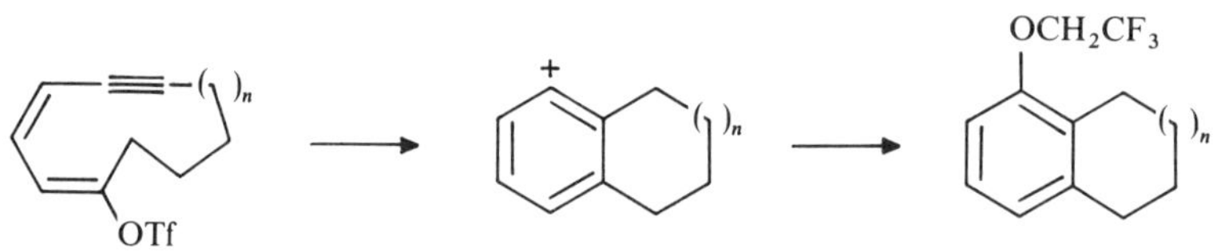

(**227**) (**228**) (**229**)

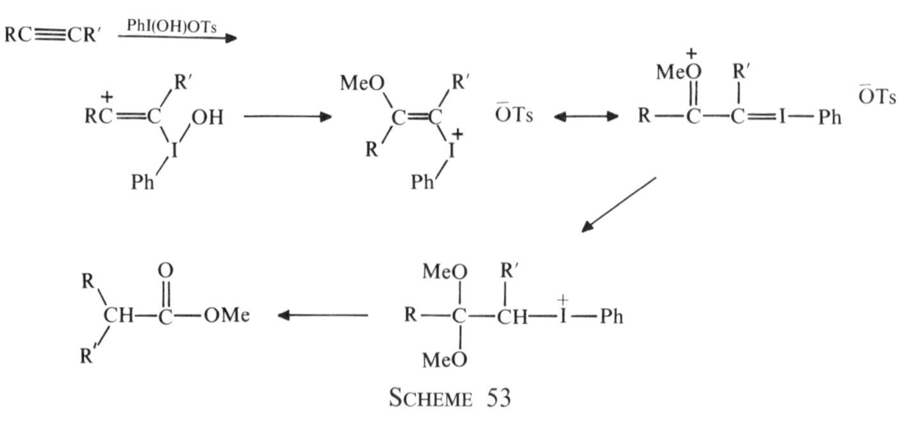

SCHEME 53

'hot' carbocation produced by electron transfer within a radical-pair has been presented.[386] The radical-pair precursor employed was β-phenyl[α,α-$^2H_2$]propionyl *m*-chlorobenzoyl peroxide. The rearrangement of 1-benzyloxymethyl-4,5-dimethoxycarbonyl-1,2,3-triazole and related compounds to the 2-benzyloxymethyl isomer has been shown to be consistent with a mechanism involving dissociation and formation of an essentially free carbocation which then recombines; thermodynamic control ultimately gives the less compressed isomer.[387] *p*-Cyanobenzylthianthrenium trifluoromethanesulphonate has been found to undergo facile photo- and thermal rearrangements in which the benzyl group migrates from sulphur to the aryl ring.[388] Product formation appears to be consistent with the development of a thianthrene cation-radical - *p*-cyanobenzyl radical pair which recombines either reversibly at sulphur or at the 1- or 2-position of the thianthrene cation radical.

The rearrangements of radical cations taking place in solution on photo-induced electron transfer have been reviewed.[389] An *ab initio* study has indicated[390] that suprafacial 1,3-hydrogen shifts in olefin radical cations should be facile processes. The study also hints at a general willingness of radical cations to undergo sigmatropic rearrangements that would be difficult in the neutral parent molecules. Upon irradiation, 3,3-diaryl-1-dideuteriomethylenecyclobutanes have been found to undergo a degenerate methylenecyclobutane rearrangement through allylically stabilized 1,4-cation radicals[391] which can be successfully intercepted by molecular oxygen. *Ab initio* MO theory using split-valence plus polarization basis sets and incorporating electron correlation, has been used to examine the barriers to intramolecular 1, *n*-hydrogen migration for simple ionized amines, including the radical cations of various alkylamines.[392] The calculated barriers and rate constants for hydrogen migration were found to be consistent with rearrangements taking place in the gas phase in long-chain amine radical cations.

It has been reported that the outcome of the Mannich-type cyclizations of amines of the type (**230**) is dramatically affected by the nature of the reaction solvent and the presence of external nucleophiles. The study[393] also presents clear evidence that the cyclization of iminium ions with simple alkenes proceeds via a cationic intermediate that is capable of partitioning between product formation and reversal to the starting iminium cation. The first total synthesis of (+)-streptazolin has been accomplished[394] in which controlled formation of the bicyclic system utilizes an iminium ion-vinylsilane cyclization coupled with an intramolecular acylation. Treatment of 3-substituted 3-(trimethylsilyl)methylcyclohexyl mesylates (**231**) with dimethylaluminium triflate has led to the formation of two types of rearrangement products,[395] a methylenecyclohexane (**232**) and a vinylcyclopentane (**233**). These results indicate that the silyl group has the effect of attracting a cationic site from the δ-position to the γ- and finally to the β-position, which induces successive hydrideion and alkyl-group shifts. The thermal rearrangements of (*S*,*S*)-1-naphthylphenylmethyl(1-chloroethyl)silane and (*S*)-(1-chloroethyl)-phenyldimethylsilane appear to proceed with inversion of configuration at carbon, a result which is inconsistent with the formation of an open carbocation intermediate. Instead, an intramolecular, simultaneous migration of chlorine and aryl groups is invoked.[396] The photochemical formation of bicyclooctenes (**237**) from 12,13-didehydrosqualenes (**234**) has been reported.[397] The authors propose that the electronically excited triene yields the zwitterionic intermediate (**235**) with twisting about the central bond of the excited triene until the two allylic systems are orthogonal. Trapping of the positively charged allylic system by an isolated double bond leads to the cyclohexenyl zwitterion (**236**). Ring-closure, producing a cyclobutane by bond formation between the allylic anion and the tertiary carbocation, gives the observed bicyclooctenes.

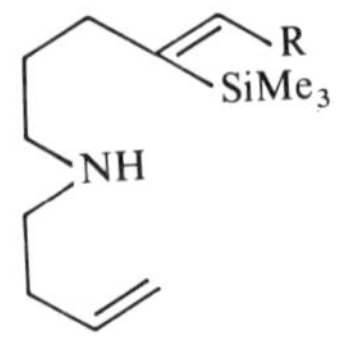

(**230**)

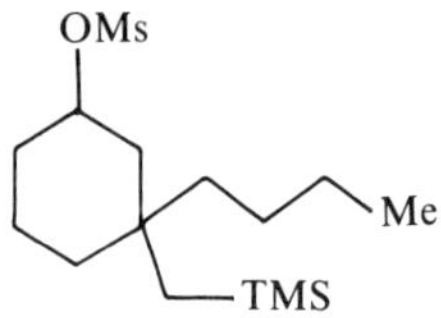

(**231**)

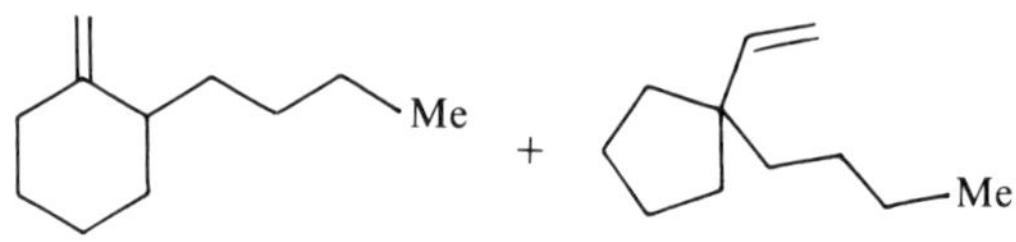

(**232**) (**233**)

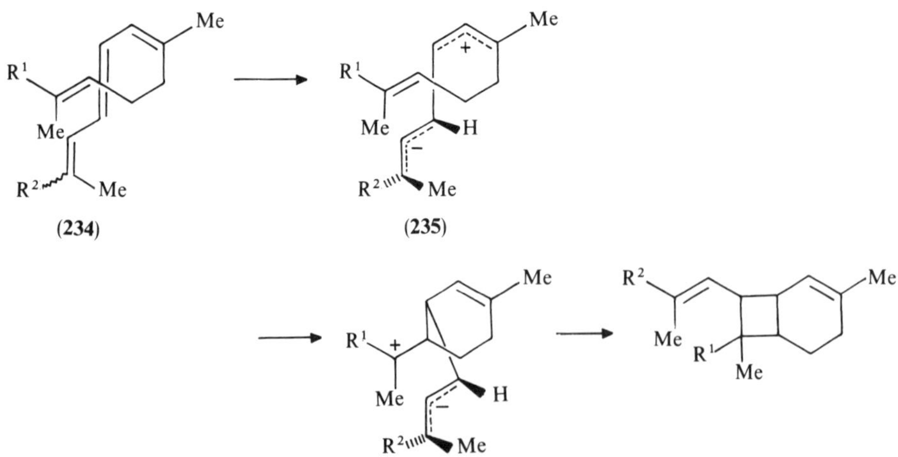

(**234**) (**235**) (**236**) (**237**)

The fast isomerization of trityl isocyanide to trityl cyanide in polar solvents has been attributed[398] to an ionic chain reaction. A detailed study by the same group of structure-reactivity relationships in the isonitrile-to-nitrile rearrangement has shown[399] that the rearrangement can be considered properly prototypical for a sigmatropic 1,2-rearrangement via a non-polarized, tight, hypervalent three-centred transition state. The reactivity data and the transition-state structures are discussed in context with other cationotropic 1,2-shifts. (*S*)-1-Arylalkanoic esters (**239**) have been prepared from (*S*)-1-aryl-2-sulphonyloxy-1-alkanone acetals (**238**). It has been assumed[400] that the hydrolytic rearrangement proceeds by a concerted process which involves concomitant attack by H_2O, 1,2-aryl shift, and elimination of the sulphonyloxy group; see Scheme 54. A similar methodology has been utilized[401] to prepare chiral β-aryl(or alkenyl) α-oxo nitrile acetals. Treatment of 1-halogenoethyl aryl ketones with thallium(III) nitrate and perchloric acid in trialkyl orthoformate has afforded[402] alkyl esters of 2-arylpropionic acid in excellent yields via 1,2-aryl migration, while in a similar manner, treatment of propiophenones (**240**) with lead(IV) acetate in trialkyl orthoformate in the presence of acid catalyst, has afforded alkyl esters of 2-arylpropionic acids (**241**). A reaction pathway involving the formation of a monoalkoxy lead(IV) species, and its decomposition accompanied with aryl migration (see Scheme 55) has been discussed for the process.[403] Optically

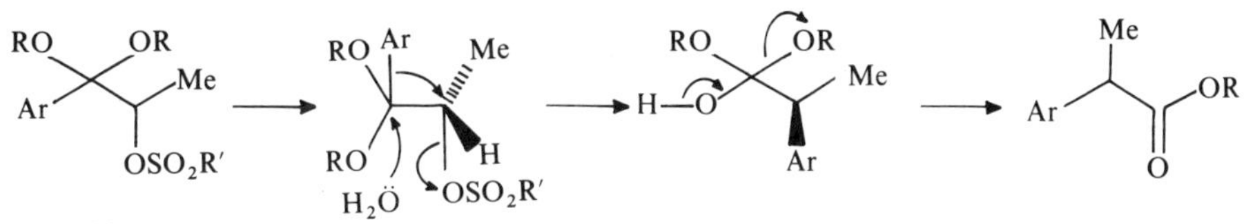

(**238**) (**239**)

SCHEME 54

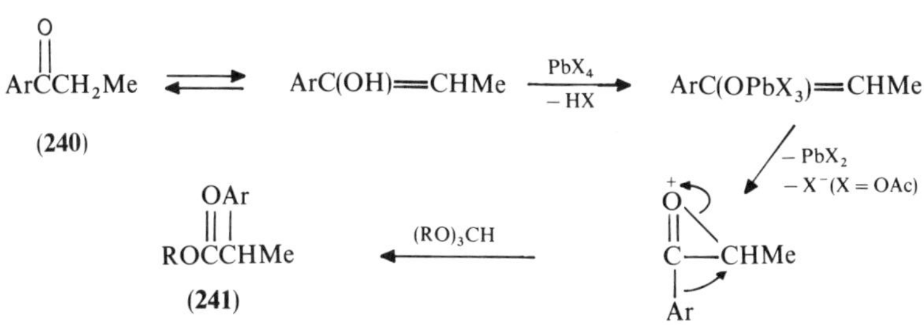

(**240**) (**241**)

SCHEME 55

active 2-arylpropionic acids and esters have also been prepared by the zinc salt-catalysed rearrangement of acetals of optically active aryl 1-chloroethyl ketones.[404] A recently devised method[405] for the synthesis of methyl indane-1-carboxylates by ring-contraction of tetralones, using lead(IV) acetate in the presence of boron trifluoride etherate and methanol is outlined in Scheme 56. A similar protocol has been extended to 5- and 7-membered cyclic aromatic ketones. 8-Hydroxyalkyl selenides derived from unsymmetrically substituted ketones have been rearranged to ring-expanded substituted ketones in the presence of soft electrophiles such as silver tetrafluoroborate and dichlorocarbenes.[406] Although the migration of the more substituted carbon often prevails, subtle variations have been observed depending upon the structure of the starting material and the experimental conditions utilized. A new ring-expansion route to olefinic lactones has been developed.[407] Oxidative 1,4-fragmentation of tributylstannyl lactols (**242**) utilizing (diacetoxyiodo)benzene gave unsaturated medium-ring lactones (**243**) and (**244**) stereospecifically, the relative stereochemistry of the tributylstannyl group of the lactols determining the double-bond geometry of the lactones. A variety of

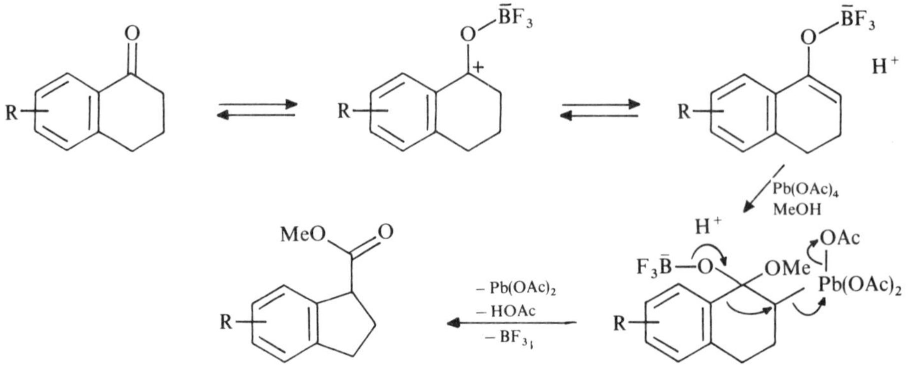

SCHEME 56

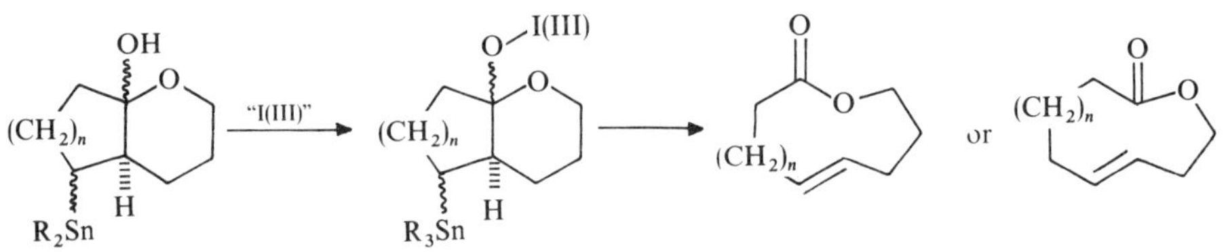

(**242**) (**243**) (**244**)

substituted *cis*-fused octahydrobenzofurans (**245**; $n = 1$) and cycloheptatetrahydrofurans (**245**; $n = 2$) have been prepared in stereo-controlled fashion by the acid-promoted rearrangement of acetals derived from 1-alkenyl-2-hydroxycyclohexanols and 1-alkenyl-2-hydroxycyclopentanols, respectively.[408] These transformations are considered to occur by a Prins cyclization-pinacol rearrangement sequence as outlined in Scheme 57. The same methodology has been used to prepare 3-acyltetrahydrofurans from allylic acetals.[409] Strategic use has been made of the stereoselective protonation of both cyclic and acyclic allenol ethers to generate oxonium ion intermediates that undergo tandem cyclization. In this way the novel 1,6-dioxaspiro[4.5]dec-3-ene system and the better known 1,7-dioxaspiro[5.5]undecene system have been prepared, respectively.[410] Electrophiles have been reported[411] to cause suitably substituted 1,2,4-trioxanes to rearrange to furans and pyrans via oxonium cations, while an O(1)→C(5) transannular interaction in an oxo-carbocation has been invoked[412] to explain the observed ring-contraction of 5-alkoxy-1,4-dioxepanes. In the presence of $Bz^+SbCl_6^-$, carbonyl compounds $RR^1R^2CCOR^3$ have been found[413,414] to undergo a retro-pinacol rearrangement to produce the dioxolanium salts (**246**) which can be hydrolysed to $HOCRR^1CR^2R^3OH$.

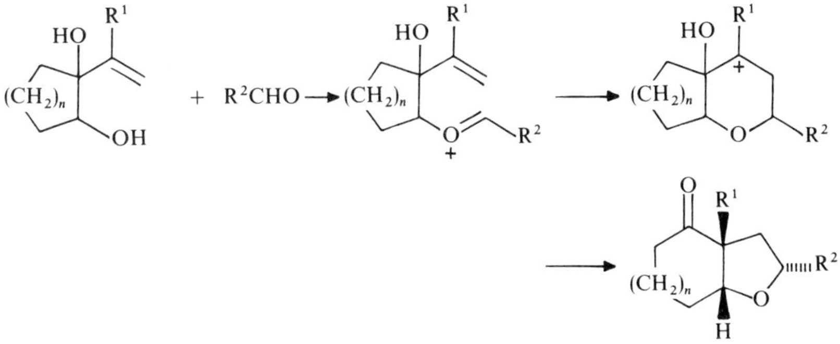

(**245**)

SCHEME 57

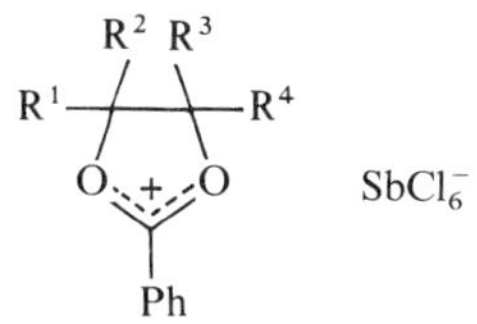

(**246**)

1-Methoxy-3-(arylthio)propenes have been constructed by acid-catalysed rearrangement of acrolein O,S-acetals,[415] while adducts of cyclic ketones with bis(phenylthio)methyllithium have been reacted with an alkyllithium to yield ring-expanded α-(phenylthio)ketones in a pinacol-type ring expansion.[416] The allylic 1,3-rearrangement of thiophenyl-substituted sulphones has been reported,[417] and *syn*- and *anti*-(*E*)-2,4-dimethylhex-3-ene-1,5-diols, compounds with 1,4-related chiral centres, have been synthesized by phenylthio migration protocol.[418] At low temperatures, sulphonyl isocyanates (**247**) have been found to react with ketene O, N-acetals (**248**) to give 1,4-dipoles (**249**). On warming, (**249**) rearranged further in a formal $O \rightarrow N$ methyl shift, to afford malonamides (**250**). Labelling experiments have confirmed the intramolecular nature of the methyl transfer.[419]

Di-*n*-butyltin oxide has been used to catalyse the rearrangement of 2-hydroxy-2-oxocarboxylic acid esters into the corresponding 2-hydroxy-3-oxy-esters,[420] and ribose-1,5-biphosphate has been observed[421] to undergo a sequence of elimination and benzilic acid rearrangement in a process with features that parallel those of the rearrangement occurring during the incorporation of pentose precursors into riboflavin. The Huang–Minlon reduction of 7α-hydroxy-methylbicyclo[3.3.1]nonan-3-one (**251**; $R^1 = R^2 = H$) has been found to yield 7β-methylbicyclo[3.3.1]nonan-3β-ol (**252**; $R^1 = R^2 = H$), formed as a result of a transannular 1,6-hydride shift enforced by relief of the steric constraint in the system.[422]

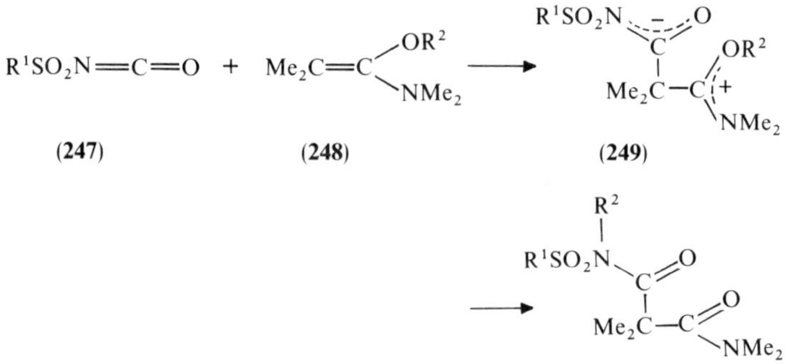

(250)

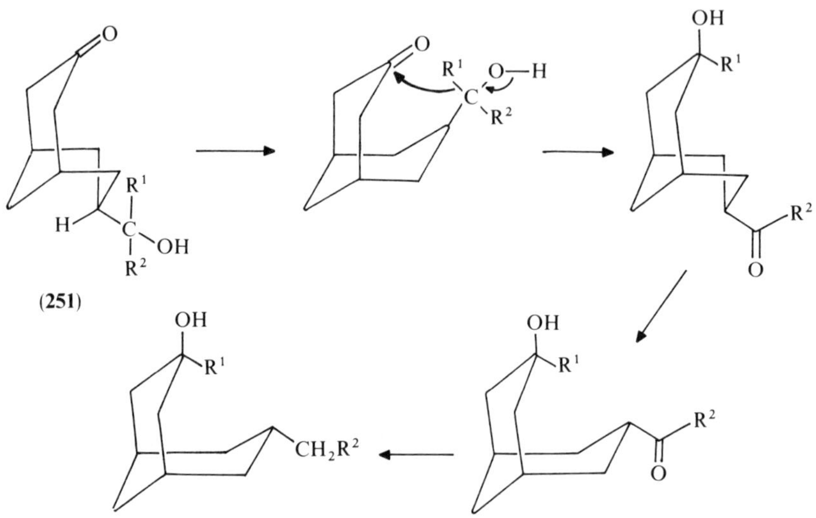

(252)

Ab initio and semi-empirical MO calculations of the gas-phase hydrolysis of protonated oxirane (**253**) have predicted[423] that the inductively destabilized 2-hydroxycarbenium ion (**254**) is a transition state for its rearrangement to the conjugate acid of acetaldehyde (**255**). An MINDO study[424] of the gas-phase reaction of chloride ion with epoxide (**256**) has shown that, although formyl migration is thermodynamically favoured over hydride migration, the latter predominates kinetically. Results of the spontaneous rearrangement of *p*-methoxystyrene oxide have been re-examined[425] in the light of known stereoselection observed in the rearrangement of C(1)-substituted epoxides, and a study of

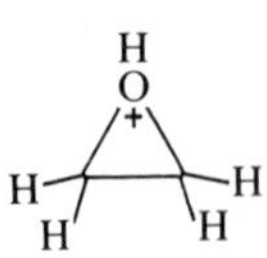

(**253**)

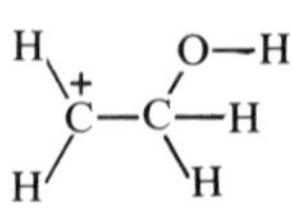

(**254**)

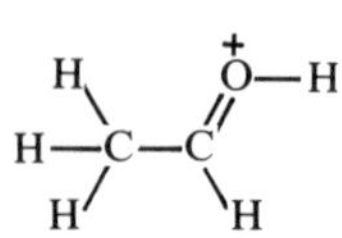

(**255**)

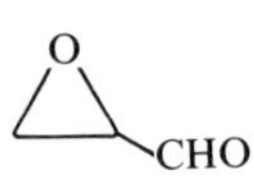

(**256**)

substituent effects on the stereoselectivity of the acid-catalysed solvolysis of rigid 1-arylcyclohexane oxides has provided further evidence for the process proceeding by different benzylic carbocationic species.[426] The transformations of 1,2,2,3,3,4,4,5-octamethyl-6-oxabicyclo[3.1.0]hexane into a variety of reaction products have been rationalized[427] by the steps outlined in Scheme 58. When a carbon-oxygen bond in the complex (**257**) breaks, two different methyl migrations are feasible. Methyl shift (*a*) yields the cyclopentanone complex (**258**) which would be thermodynamically favoured over the 1,5-dipole (**259**), the precursor of (**260**) and (**261**). However, in spite of that, path (*b*) is observed, presumably since in the early stage of the migration a perfectly eclipsed interaction between two carbon–carbon bonds is abandoned. The facile Lewis-acid catalysed acyl migration of several bicyclic α,β-epoxy ketones has afforded a practical route to both *cis*- and *trans*-8-formyl-1-hydrindanones and a variety of related perhydroindanone derivatives,[428] while a new method for the construction of quaternary carbon centres has been reported.[429] It is based on the Lewis acid-promoted 1,2-rearrangement of β,γ-epoxy alcohols which gives α,α-disubstituted aldols stereospecifically in high yields. A catalytic version of the epoxy silyl ether rearrangement using Me_3SiX (X = I, OTf) as the Lewis acid catalyst has been described.[430] This methodology, illustrated in Scheme 59, has been found to

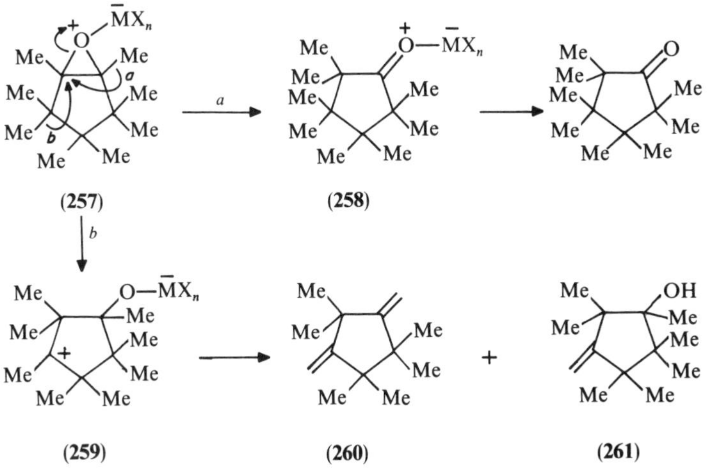

SCHEME 58

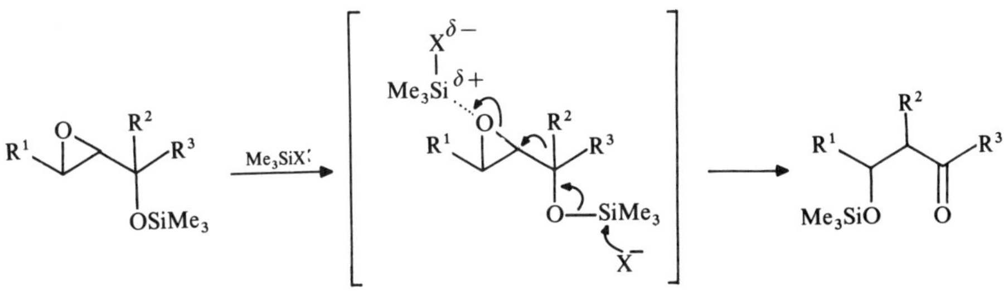

SCHEME 59

lead generally to higher yields of the aldol products than does the stoichiometric version. A new and efficient route to terminally-substituted (*E*)-allylsilanes has appeared in the literature.[431] It is based on the preference of a stereoselective *anti*-elimination of the hydroxy and selenyl moieties from various β-seleno-γ-silyl alcohols. Reactions involving migration of a phenylseleno group around the perimeter of a cyclopentadiene ring have been described.[432,433]

Various new 1-fluoro-1-alkyl (or aryl)-2-substituted cyclobutanes have been synthesized stereoselectively from 1-alkyl (or aryl)cyclopropyl carbinols by ring-expansion and fluorination, using an amine-metal fluoride-pyridinium poly(hydrogen fluoride)-complex.[434] A potentially valuable method has been described[435] for the enantioselective creation of quaternary carbons by thermal 1,2-asymmetric rearrangements in cyclopropane systems possessing a chiral sulphinyl group on the ring. The mechanistic pathway for this asymmetric induction is represented in Scheme 60. Clearly the degree of asymmetric induction depends on the thermodynamic stability of (**262**) and (**263**), that is, on the difference of the steric interference between R^1 or R^2 and the lone pair on the oxygen atom of the chiral sulphoxide. Upon protonation, dicyclopropenyl ketone has been found to undergo a tandem walk rearrangement via the symmetrical transition state (**264**), in the course of which the carbon atom bearing the hydroxyl group migrates around six 'equivalent' ring positions.[436] A stereoselective rearrangement of spiro[5.2]octan-4-ol in H_2O_2, namely (**265**) → (**267**), has been described.[437] In this rearrangement the ring-expansion occurs with overall inversion of configuration at the carbinyl centre, and the process has been exploited in a short, efficient synthesis of keto-butyrolactones.[438] Although multiple sites are offered by bis(bicyclo[1.1.0]butanes) (**267**) to electrophilic reagents, experimental evidence has been presented[439] to indicate that only one of the ten available strained carbon–carbon σ-bonds is cleaved by the attacking electrophile in remarkably regiospecific fashion, and that the initiation step proceeds with retention of configuration at the reaction centre. Moreover, the ensuing cascade of electronic rearrangements proceeds unidirectionally, irrespective of the length of the polymethylene ether in ring B. These observations are best reconciled with a mechanism in which only the central, and not the flanking bicyclobutane subunit, experiences initial electrophilic-induced cleavage, to generate carbocation (**268**).

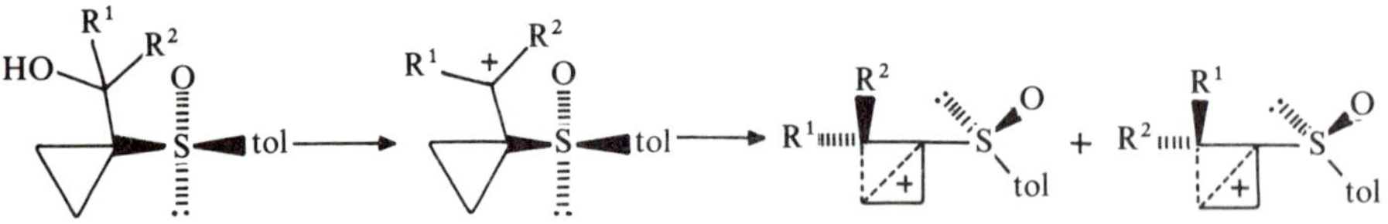

(**262**) (**263**)

SCHEME 60

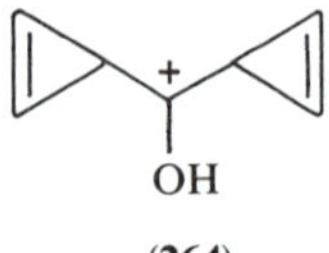

(**264**)

(265) (266)

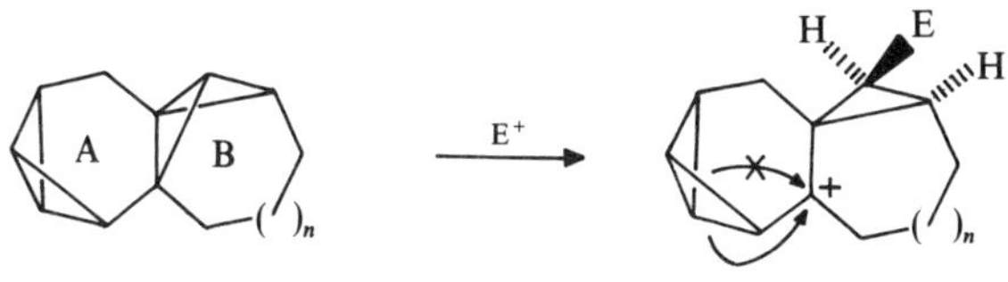

(267) (268)

Labelling and NMR studies have been used to establish the mechanism of the isomerization of methyl-substituted 5-oxabicyclo[2.1.0]pent-2-enes (**269**) (Dewar furans) to acetylcyclopropenes (**271**). A cyclobutenyl cation (**270**) has been proposed as intermediate.[440] It has been concluded[441] that the diverse rearrangement products that are produced on irradiation of 2,3,6-trialkyl-4-pyrimidinones arise by way of azetidinyl and acyl cations generated from intermediate Dewar 4-pyrimidinones. The formation of tetrahydronaphtho[2,3-*b*]thiophenes (**273**) during acid-catalysed dehydration of α-thienylcyclobutanols (**272**) has been rationalized[442] by the pathway outlined in Scheme 61. A convenient new preparation of 1,1′-

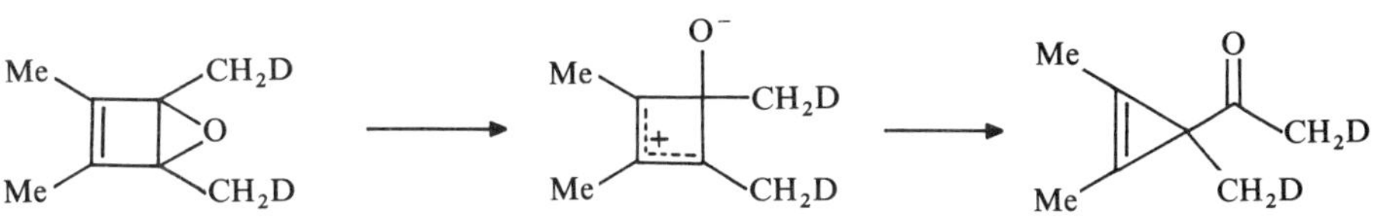

(269) (270) (271)

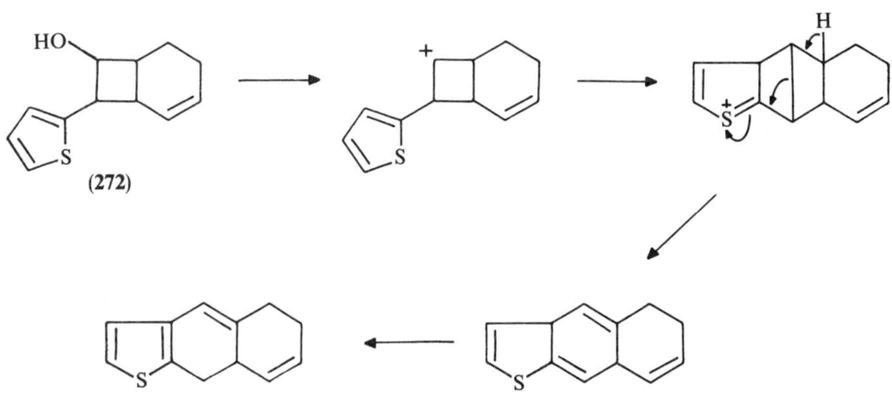

(272)

(273)

SCHEME 61

bi(benzocyclobutylidene) has been described. The hydrocarbon and its derivatives have been found to be susceptible to cationic rearrangement, and it is readily converted into 5,10-dihydroindeno[2,1-*a*]indene on protonation in Ac_2O.[443] A possible mechanism for the acid-catalysed rearrangement of 8-hydroxy-4*a*,5,8,8*a*-tetrahydrobiphenylene-5-carboxylic acid to the lactone of 6,9-dihydro-10-hydroxy-5,9-methano-5*H*-benzocycloheptene-6-carboxylic acid has been discussed,[444] the Lewis acid-catalysed rearrangement (**274**)→(**276**) has been explained[445] by a carbocationic 1,3-alkenyl shift via the cyclobutylcarbinyl cation (**275**), while *cis,trans*-tricyclo[6.3.0.0^{1,4}]undecan-5-one (**277**); R = H) has been found to rearrange under acidic conditions through unusual pathways to give *cis,cis*-tricyclo[6.3.0.0^{1,5}]undecan-4-one (**278**; R = H).[446] Application of this latter methodology to the α-methylene derivative (**277**; R = CH_2) afforded the angularly fused triquinane (**278**; R = CH_2) having an α-methylenecyclopentanone unit. The preparative value of rearrangement cascades on poly(spirocyclic) compounds has been expressed;[447] thus tetraspiroketone (**279**) has been observed to react with acids by undergoing a five-fold 1,2-shift via the β-hydroxycarbenium ion (**280**) to yield the ketone (**281**). Photochemical decarbonylation of (**281**) led to the [4,5]coronane (**282**).

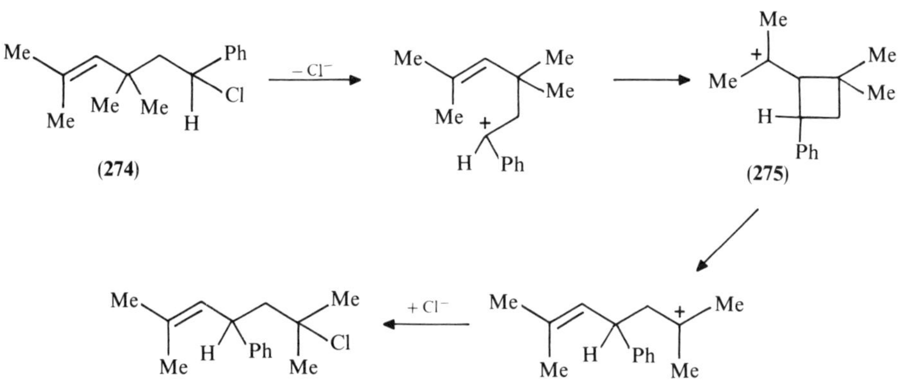

(**274**) (**275**) (**276**)

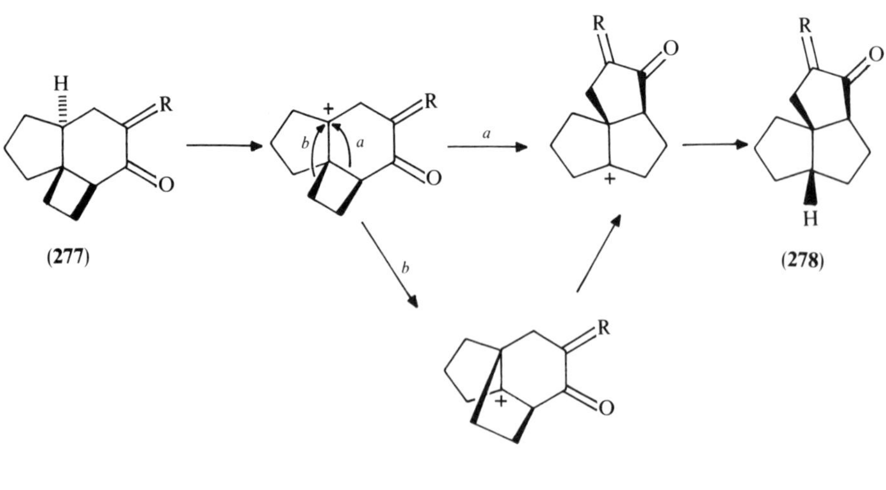

(**277**) (**278**)

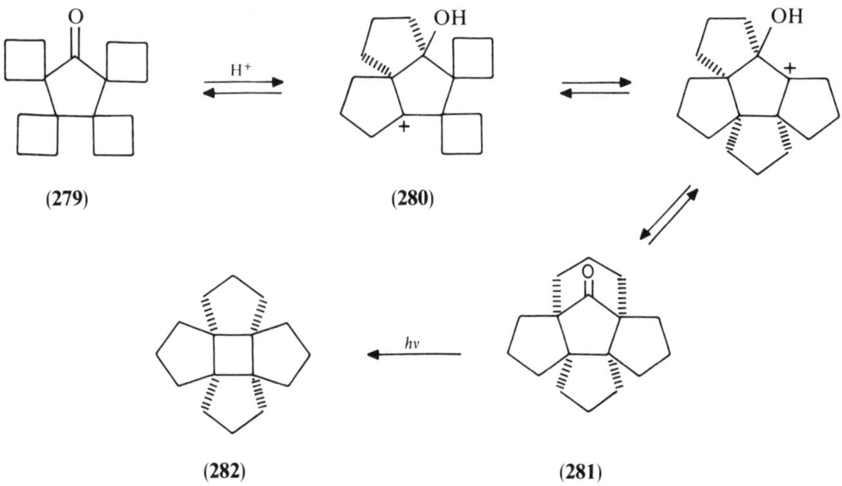

Pinacol-type rearrangements of bridged bicyclic ketones have been assessed[448] as a means of ring-expansion, and an examination of the potential-energy surface for the rearrangement of several bridgehead bicycloalkylium cations has been initiated.[449] [2,3-^{13}C]Norborn-2-yl chloride has been ionized (with SbF_5) to the 2-norbornyl cation which rapidly rearranges with complete scrambling of the ^{13}C labels,[450] while labelling and NMR studies of the stable carbocation formed from either 3-nortricyclyl or norbornenyl precursors have revealed[451] that the cation undergoes at least three different scrambling processes below 0°C. Scrambling appears to proceed by skeletal rearrangements of a bicyclo[3.1.1]heptenyl cation. The reaction of 2-*exo*-benzylbicyclo[2.2.1]heptan-2-*endo*-ol with HSO_3F at −78°C has been found to result in quantitative formation of the 6-phenylbicyclo[3.2.1]octan-6-yl cation by way of a mechanism (see Scheme 62) which has been unambiguously elucidated by deuterium-labelling experiments.[452] In superacid media, at −78°C, *anti*-sesquinorbornene (**283**) has been shown to yield the directly related carbocation (**284**); however, at higher temperatures this cation appears to rearrange irreversibly with ring-opening to the 1-(2-norbornyl)-2-cyclopentenyl cation (**285**). The transfer of charge from the norbornyl ring to the cyclopentenyl ring is considered to occur by successive 1,2-hydride shifts.[453] On treatment with excess aluminium isopropoxide in refluxing toluene, 2-*endo*-hydroxy-2-*exo*-phenyl-5,6-epoxybicyclo[2.2.1]heptene (**286**) has been converted into the monocyclic glycol (**287**). A likely mechanism for this novel rearrangement is outlined in Scheme 63.[454] In the presence of various electrophiles, spirocyclopropane-substituted norbornenes have been observed to undergo transannular ring-expansion of the spirocyclopropane moiety with formation of functionalized brendanes.[455] In a similar manner, isotwistanes and homobrendanes have been prepared from the corresponding spirocyclopropane-substituted bicyclo[2.2.2]oct-2-ene oxide.[456] General and stereospecific pinacolic-type rearrangement reactions of vicinal methoxy mesylates of bicyclo[2.2.2]octanes under neutral conditions have been developed[457] for the construction of functionalized

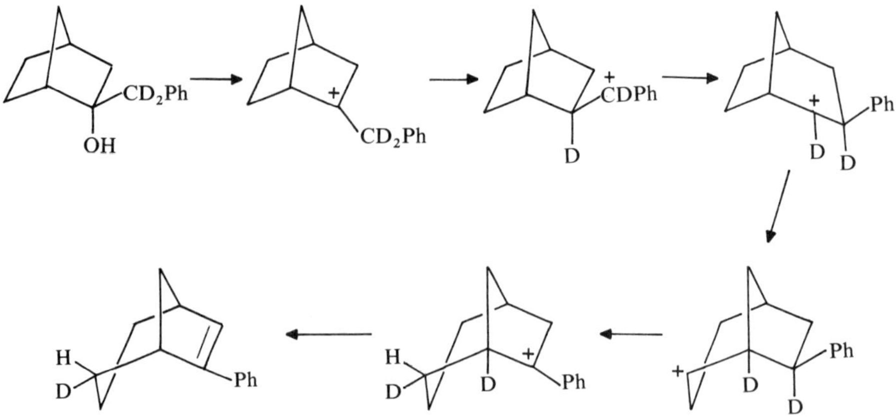

SCHEME 62

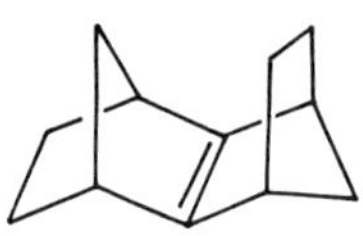

(**283**)

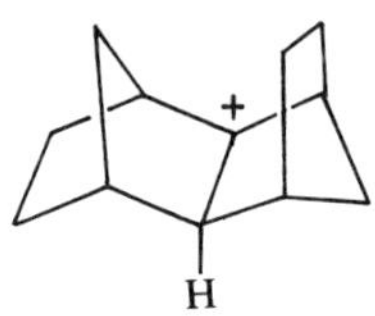

(**284**)

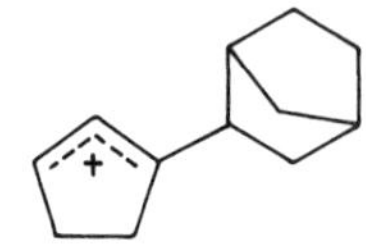

(**285**)

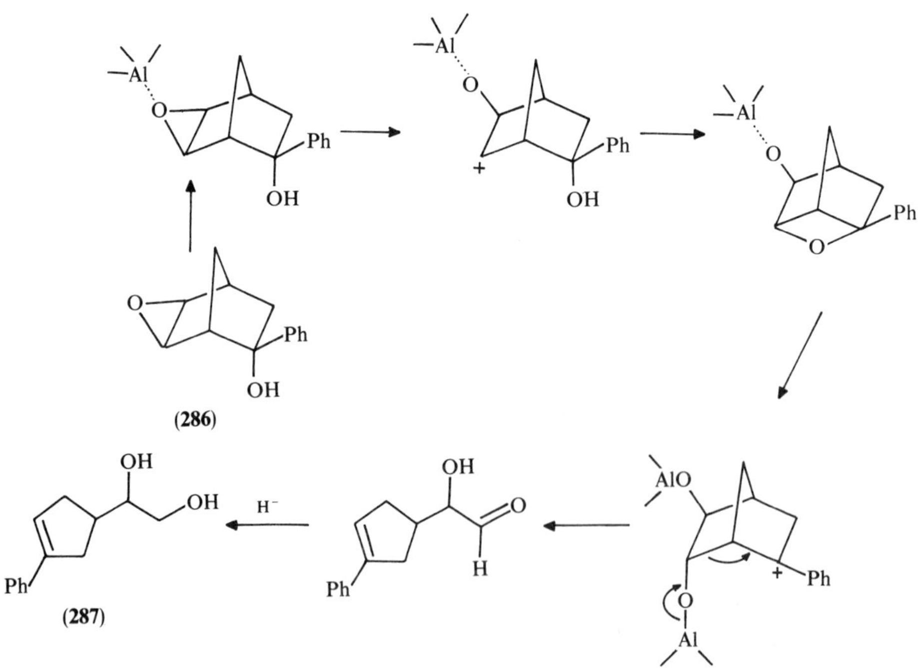

SCHEME 63

bicyclo[3.2.1]octanes, while trimethyl group-substituted bicyclo[2.2.2]octane derivatives have been synthesized and subjected to a trimethylsilyl group-directed Wagner–Meerwein rearrangement to afford bicyclo[3.2.1]octene derivatives in high yields.[458] Hydrolysis of *trans*-[tetracyclo(3.3.2.0^{2,4}.0^{6,8})doc-1-yl]methyl tosylate (**288**) has been shown to produce the rearranged alcohol (**289**).[459] Photochemical transformations of *m*-methoxy-substituted 2,3:6,7-dibenzobicyclo[2.2.2]octa-2,5-dienes[460] and the various isomers of 7-substituted and 7,8-disubstituted dibenzobicyclo[2.2.2]octa-2,5-dienes[461] have been studied. All observed products appear to arise from Wagner–Meerwein skeletal rearrangements with *syn*-migration predominating over *anti*-migration regardless of whether migration involves the benzo- or the anisolo-ring. The authors go on to point out that ideas regarding electron transfer as the key step in photo-solvolysis and Wagner–Meerwein photo-rearrangements in a variety of these bicyclo[2.2.2]octadienes are consistent with, and confirmed by, the results they have obtained. The pinacol-type rearrangement of 1-methoxybicyclo[3.2.2]nona-3,6-dien-2-ols has been shown to proceed preferentially to give bicyclo[3.2.2]nona-6,8-dien-2-ones. This transformation has been incorporated into a new synthesis of homobarrelenone and a large number of its derivatives,[462] while an analogous transformation of 1-methoxybicyclo[3.2.2]nona-6-en-2-one has been applied to the total synthesis of the sesquiterpene, widdrol.[463] Under acid conditions, the photoadducts of 1,4-dioxene and 3-methylcyclohex-2-en-1-one have been observed to undergo an unusual rearrangement leading to a number of products whose formation can be rationalized[464] in terms of an intermediate cation stabilized by oxygen, viz. (**290**).

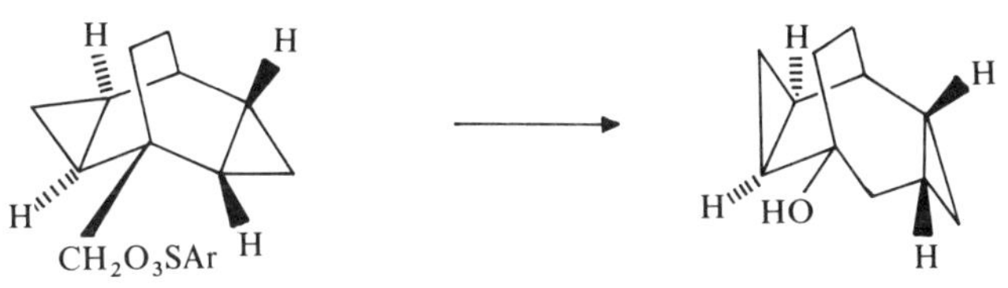

(288) **(289)**

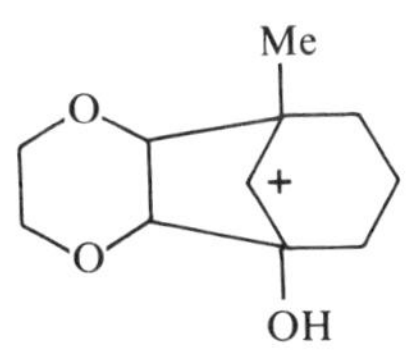

(290)

The rearrangement of 7-oxabicyclo[2.2.1]hept-2-enes to 2-oxabicyclo-[2.2.1]heptane derivatives upon treatment with iodine and silver acetate has provided further insight into the migratory preferences of alkyl groups involved in Wagner–Meerwein 1,2-bond shifts in unbiased situations.[465] Studies of acid-promoted rearrangements of 5,6-*exo*-epoxy-7-oxa-2-norbornyl derivatives have allowed the comparison of the true migratory aptitudes of acyl, alkyl, and α-substituted alkyl groups in Wagner–Meerwein rearrangements.[466] In addition, these studies have opened up new approaches to the stereoselective syntheses of polysubstituted cyclopentyl and 7-oxanorbornyl derivatives.[467]

Derivatives of both 2-azabicyclo[2.2.2]oct-5-ene and -octane, bearing electronegative substituents on nitrogen, undergo several types of rearrangement which appear to occur via ion-pairs of different types. The intermediacy of nitrenium ions in these processes has been excluded.[468] It has been reported[469] that the excited electron acceptor, triphenylpyrylium tetrafluoroborate, acquires an electron from ground-state azoalkanes as donors, resulting in the respective azoalkane radical cations. Loss of nitrogen from the latter then afforded rearranged photo-products which are probably derived from the intermediary radical cations. Diazenyl diradicals have been proposed as the immediate precursors to benzobicyclo[3.2.1]octatrienes in the thermal and photochemical denitrogenation of 8,9-benzo-4,5-diazatricyclo[4.3.1.0^{3,7}]deca-4,8-diene and its spirocyclopropane derivatives.[470] The acid-induced rearrangements of 2,8-diazabicyclo[3.2.1]octa-3,6-dienes and their C(6)–C(7) dihydro analogues to 1,4-dihydropyridines and pyrroles have been studied.[471] Both processes appear to proceed via azepinium cations.

A possible pathway for the conversion of the bullvalenyl cation into protonated naphthalene has been presented.[472] Both the *syn-* and *anti*-tricyclo-[4.2.1.1^{2,5}]decanes have been found to undergo well-defined $AlBr_3$-catalysed adamantane rearrangements.[473] The mechanism proposed[474] for the transannular iodination of 3,7-dimethylenebicyclo[3.3.1]nonanes to adamantanes involves sequential formation of 1:1 and 1:2 complexes of the starting material with iodine, followed by synchronous ring-closure with ion-pair formation and recombination. Carbonylation of the 1-adamantyl cation with CO in the presence of adamantane has afforded[475] 3-hydroxy-4-homoadamantyl-1-adamantane carboxylate. The reaction is considered to proceed via addition of the 1-adamantanecarbonyl cation to 1-adamantanecarbaldehyde, followed by Wagner–Meerwein rearrangement. In an analogous manner, acylation of bridgehead aldehydes with the 1-adamantanecarbonyl cation has been used as a novel route to bicyclic and tricyclic 1,2-diols which are not easily accessible by other methods.[476]

A review has appeared[477] on the use of the acid-catalysed rearrangement of tricycloundecanes to build up basic carbon skeletons of natural di- and triquinane sesquiterpenes, and a relatively simple, general route to the tricyclo[5.4.0.0^{2,8}]undecane ring system from the readily available 7-methoxybenzosuberone has been described.[478] Tetracyclic compound (**291**) has been converted via a series of carbocation rearrangements into (**292**),[479] while pentacyclo[5.4.0.0^{2,6}.0^{3,10}.0^{5,9}]undecane-4,7,11-trione and 4,4,7,7,11,11-hexanitro[6.3.0.0^{2,6}.0^{3,10}.0^{5,9}]undecane, members of a rare class of rigid, polycyclic organic molecules that belong to the chiral D_3 point group, have been synthesized by way of acid-catalysed rearrangement of 4,4-dimethoxypentacyclo[5.4.0.0^{2,6}.0^{3,10}.0^{5,9}]undecane-*exo,exo*-8,11-diol.[480] Hexacyclic propellanedione (**293**) has been observed to respond readily to Lewis acids with the production of novel products, (**294**), (**295**), and (**296**).[481]

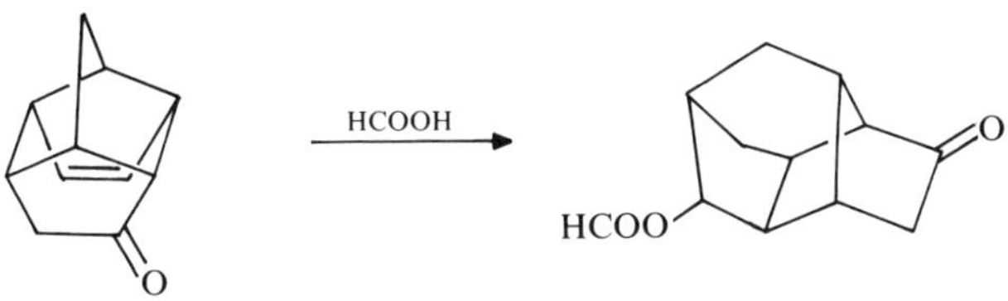

(291) (292)

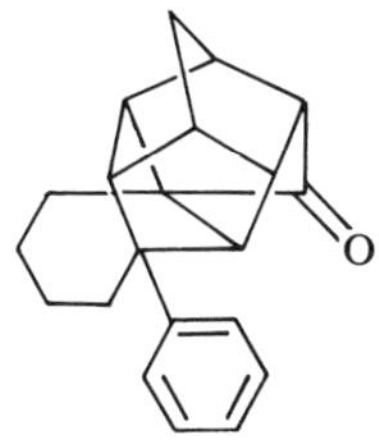

(295)

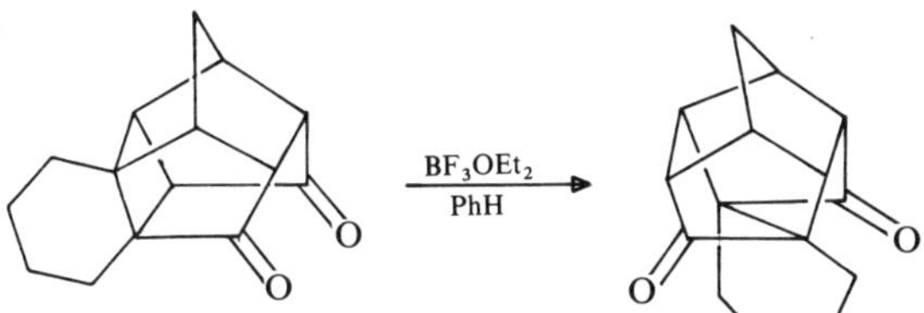

(293) (294)

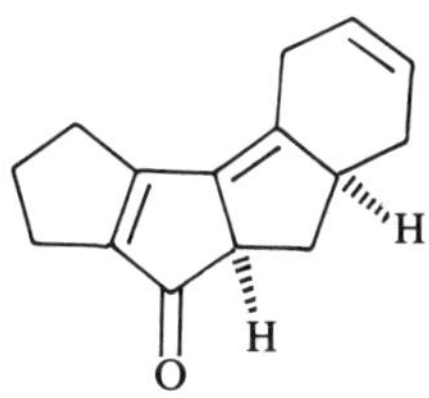

(296)

Strong evidence has been presented[482] to suggest that the elusive silicenium ion (**298**) can be generated via a 1,2-methyl migration of a solvolytically produced α-silyl carbenium ion (**297**). In addition, a study of the dissociative isomerization of p-$MeC_6H_4O(CH_2)_2SiMe_3$, $MeCH(Cl)SiMe_3$, $Me_2CHSiMe_2Cl$, and $Me(CH_2)_2SiMe_2Cl$ in the gas phase has revealed[483] a complex scheme of rearrangements that interconvert the resulting isomeric $C_5H_{13}Si^+$ ions with silicenium ions and α- and β-silyl-substituted carbocations. An *ab initio* MO study of the SiH_4^+ cation potential-energy surface has suggested,[484] in contrast with previous studies, that its most stable structure has C_s symmetry. The 1,3-silicon-to-silicon migration of the OMe group observed during the reaction of (bromodiphenylsilyl) (methoxydimethylsilyl)bis(trimethylsilyl)methane with $AgBF_4$, has been interpreted[485] in terms of rate-determining formation of an MeO-bridged cation formed by abstraction of Br^- by Ag^+ with anchimeric assistance by the OMe group. 1,3-Migration of a vinyl group in a similar situation has also been noted,[486] while 1,3-migration of a methyl group from a tin atom to a silicon atom within a cationic intermediate in which a methyl group bridges between these two atoms has been observed.[487] Treatment of α-chloro-α-silyl sulphides (**299**) with trisubstituted alkenes (**300**) in the presence of $SnCl_4$ has been found to yield vinyl sulphides (**301**), the formation of which presumably involves a carbocation rearrangement controlled by the silyl group.[488] A desilylative Pummerer-like rearrangement of α-trimethylsilyl-substituted sulphoxides and sulphides has been observed[489] when such compounds are treated respectively with N-chlorosuccinimide in the presence of trifluoroacetic acid; see Scheme 64. The first study of the sila-Pummerer rearrangement in a cyclopropane system has established[490] that several

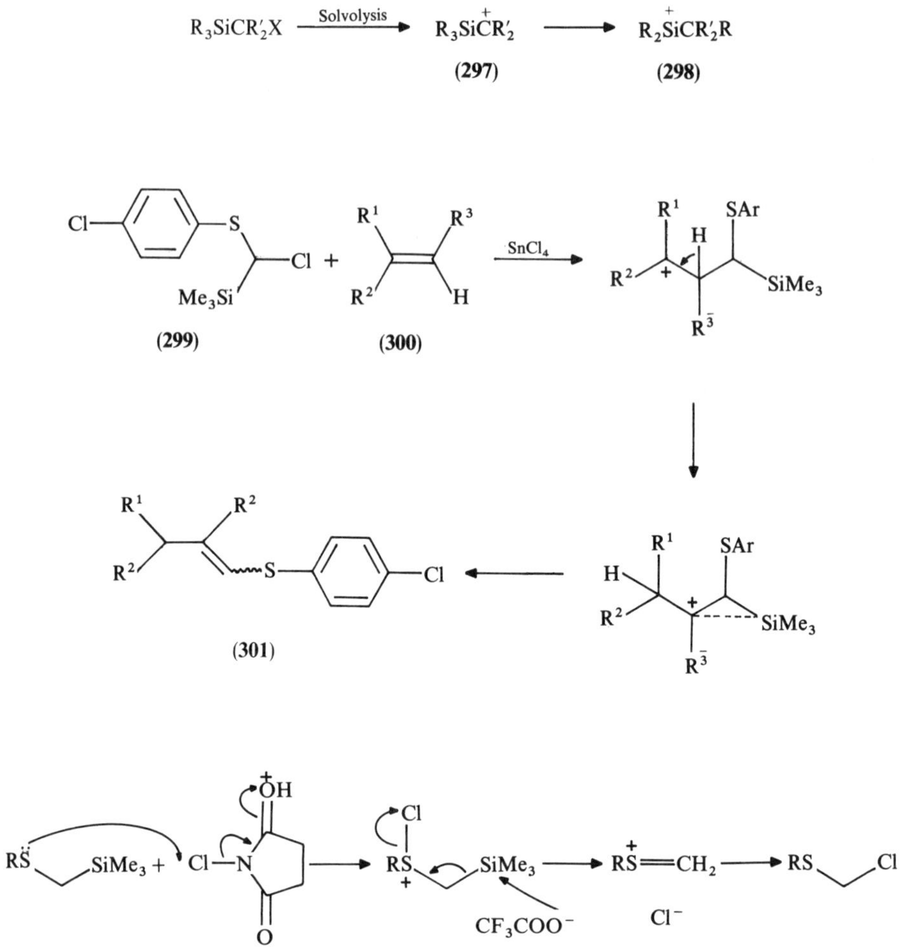

SCHEME 64

1-trimethylsiloxy-1-(phenylthio)cyclopropanes can be prepared stereoselectively through putative sulphur-stabilized carbocationic intermediates, while similar intermediates are invoked[491] in the Pummerer rearrangement of cyclopropyl(methoxy)phenylsulphonium salts to 1-methoxy-1-(phenylthio)-cyclopropanes. The Pummerer rearrangement of alkyl aryl sulphoxides has been studied,[492] and a two-step mechanism involving oxidative fluorination of sulphur and Pummerer-type rearrangement has been proposed[493] for the α-fluorination of sulphides with *N*-fluoropyridinium triflates. Alkylation of cyclic β-keto-ester (**302**) with racemic sulphonium salts containing optically active alkyl groups has been used to produce *C*-alkylated products with inversion of configuration at the asymmetric alkyl carbon atom. A reaction proceeding by enolate anion attack on the cationic sulphur atom to form an S—O sulphurane intermediate, and successive stereoselective intramolecular alkyl migration to the enolate, has been described[494] for the process; see Scheme 65.

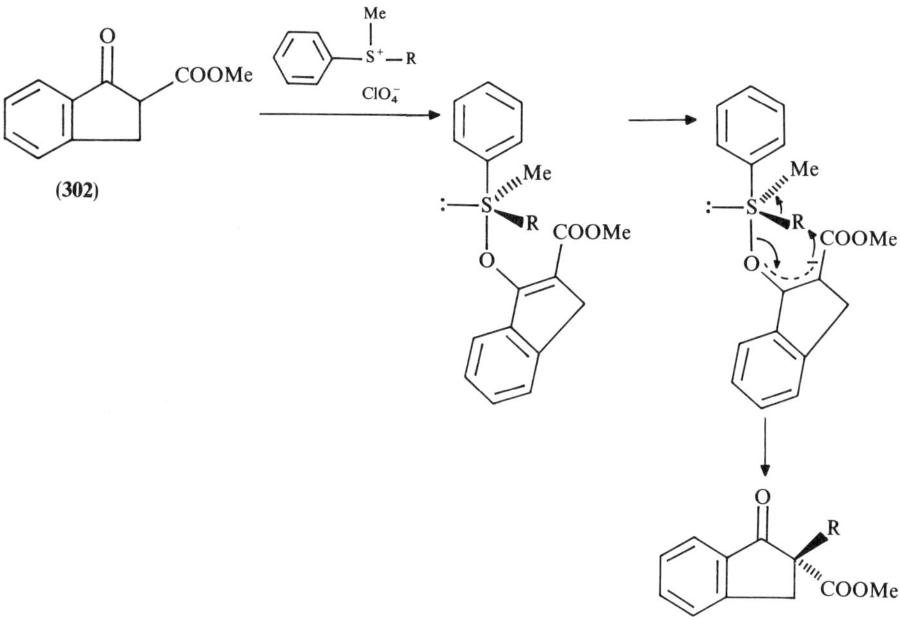

SCHEME 65

It has been proposed[495] that a dipolar structure is a reasonable part description for 9-phenyl-1(9)-homocubene, and this description has presented a possible way of rationalizing its observed chemistry. Thus a 'simple' Wagner–Meerwein shift can be invoked to account for the observed skeletal reorganization of **(303)** to **(304)**. An *ab initio* study of the cyclization mechanism of α-alkynones has indicated that a vinylidene can be regarded as a plausible intermediate for the process, and that its conversion to cyclopent-2-enones constitutes the rate-determining step for the overall reaction.[496] The formation of various dihydropentalenes on treatment of 8,8-dibromobicyclo[5.1.0]octa-2,4-diene **(305)** with methyllithium has been rationalized[497] by cleavage of the lateral bond '*a*' in **(306)**, which induces a carbene–carbene rearrangement with 1,3-carbon migration to yield carbene **(307)**. This species is stabilized by a 1,2-vinyl shift to afford 1,6*a*-dihydropentalene **(308)**; 1,4-, 1,5- and 1,6-dihydropentalenes are formed by subsequent 1,5-hydrogen shifts. Photolysis and pyrolysis of polysilylated diazomethanes has been shown to result in the formation of π-bonded silenes by the exclusive migration of a trimethylsilyl group to a carbene centre.[498] The dynamics of the Wolff rearrangement has been observed in detail by laser spectroscopy. The primary precursor to a ketene has been spectrally detected and assigned as an oxirene.[499]

(303) (304)

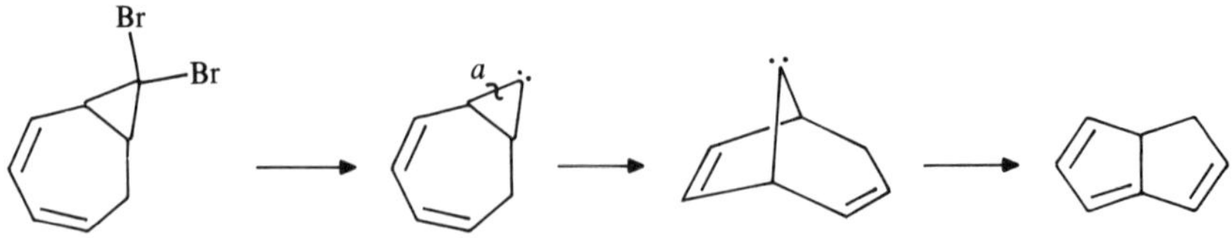

(305) (306) (307) (308)

Rearrangements in Natural-product Systems

Examples of stereospecific 1,2-rearrangement reactions in the synthesis of optically active natural products have been reviewed.[500]

Acetolysis of 5α-cholestane-5,19-diol 19-methanesulphonate (**309**) has been observed to afford the new spirocyclic compound (**310**) as a result of tandem migrations of two antiparallel skeletal carbon-carbon bonds.[501] This stereospecific, cascade rearrangement has been established as a general reaction of 5α-hydroxy and 5α-alkoxy steroids with an electron-deficient centre at C(19), and can be viewed as a double pinacol rearrangement; see Scheme 66. Solvolytic displacement of the 3-acetoxy and 19-*p*-toluenesulphonyloxy groups in 3-acetoxy-4β,5β-cyclopropancholestane-19-*p*-toluenesulphonates has led to products with rearranged bi- and tri-cyclic AB ring systems,[502] and thallic nitrite treatment of 5α-cholestan-3-one has been shown to produce a mixture of

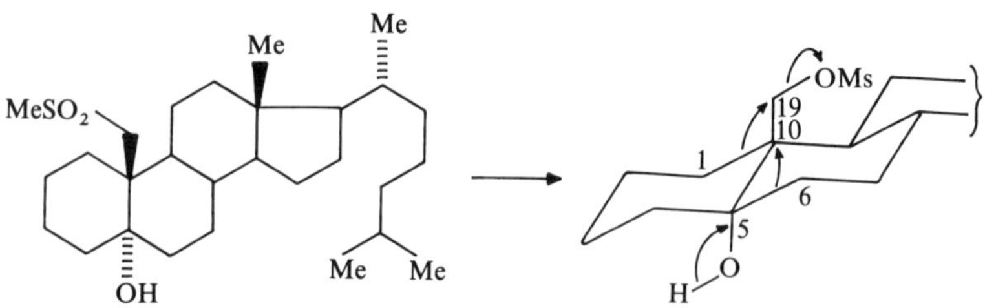

(309)

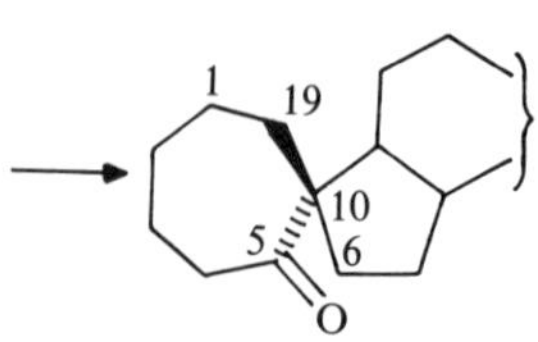

(310)

Scheme 66

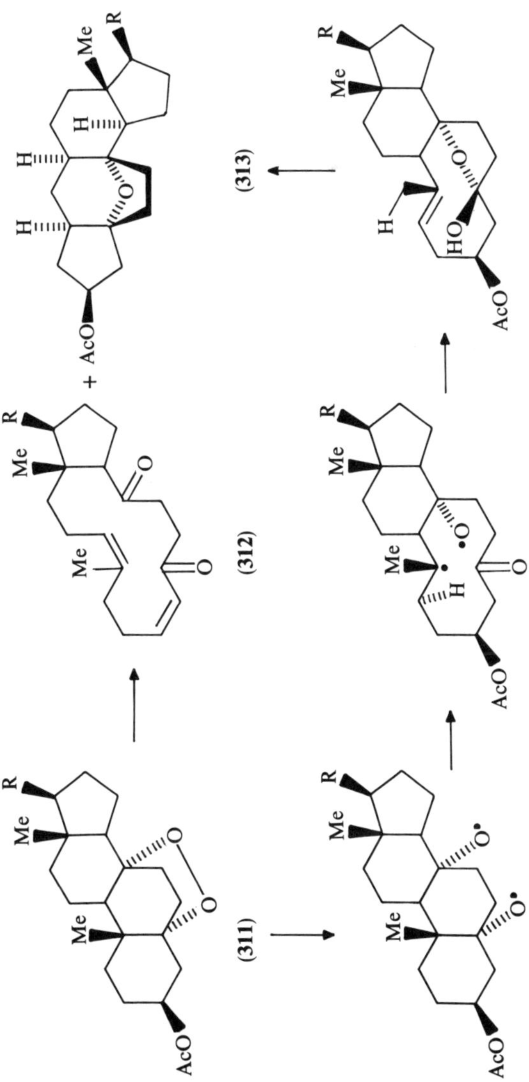

SCHEME 67

2α-carboxy-A-nor-5α-cholestane and its 2β epimer by an oxidative ring-contraction.[503] In addition to producing the expected 5,10:8,9-diseco derivative (**312**), thermolysis of 5,8α-epidioxy-5α-cholestan-3β-yl acetate (**311**) under acid conditions has been reported[504] to yield the rearranged 5α,8α-epoxide (**313**); see Scheme 67. Grignard reagents and lithium salts prepared from the epimers of 3β-chloro-7-ξ-methoxycholest-5-ene have been shown to undergo a stereoselective rearrangement-elimination process to yield 3α,5α-cyclocholest-6-ene as the major product.[505] An unprecedented stereospecific rearrangement of an excited steroidal α,β-unsaturated cyclic ketoxime (**314**) to a cholestane isoxazole (**315**) has been reported.[506] A pathway which accommodates the process is outlined in Scheme 68. A homollylic rearrangement has been observed during the peracid oxidation of a 19-iodo-5-androstene derivative,[507] while the reaction of 17- nor-5α,13α-androstane-16β-carbonyl chloride with *m*-chloroperbenzoic acid has been found to induce rearrangement to an *abeo* homobisnorsteroid.[508] A convenient and stereospecific synthesis of 5-hydroxy-9,10-methylene steroids by Simmons–Smith methylenation and subsequent rearrangement of the intermediate cyclopropyl carbenium ion has been described.[509]

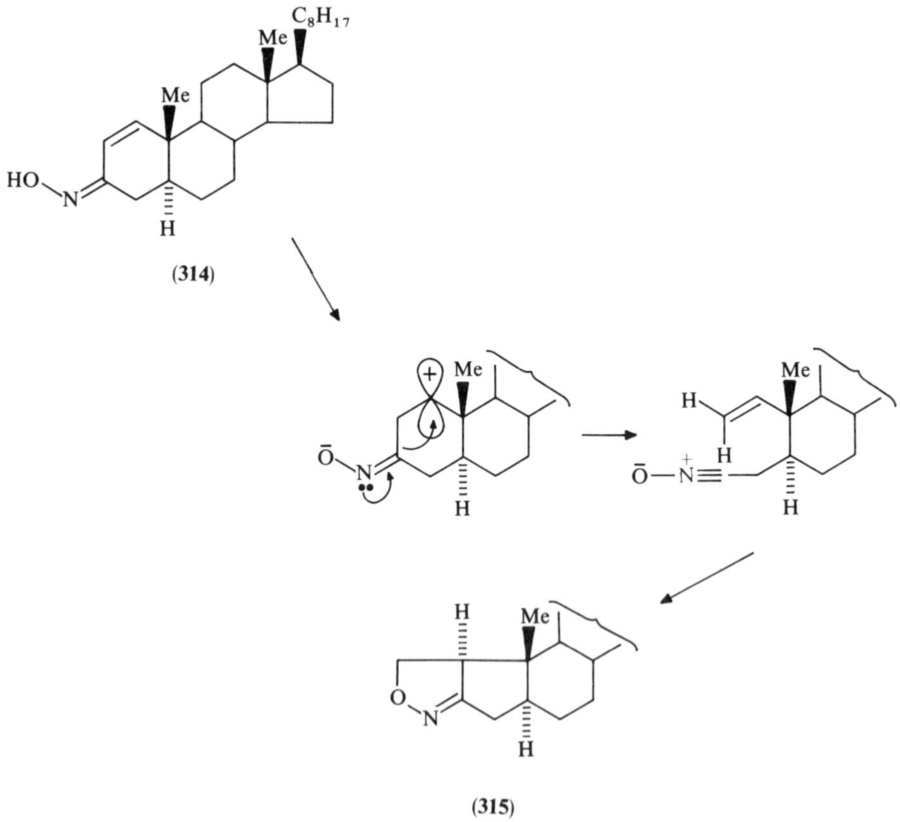

SCHEME 68

A number of monoterpenols have been synthesized by utilizing the cyclopropylcarbinyl rearrangement.[510] The Ritter reaction of isocamphanol (**316**) and nitriles has yielded amides (**317**) via a Wagner–Meerwein rearrangement,[511] while two consecutive Wagner–Meerwein rearrangements have been invoked[512] to explain the outcome of the acid-induced hydration of koumine. The first total synthesis of the macrolide, mycinolide IV, has been achieved by virtue of a new rearrangement of epoxyalcohol derivatives,[513] namely (**318**) → (**319**). The α,β-epoxyketone photo-rearrangement has been explored as a method for constructing the diterpenoid dolastane skeleton, and found to be well suited for this purpose.[514]

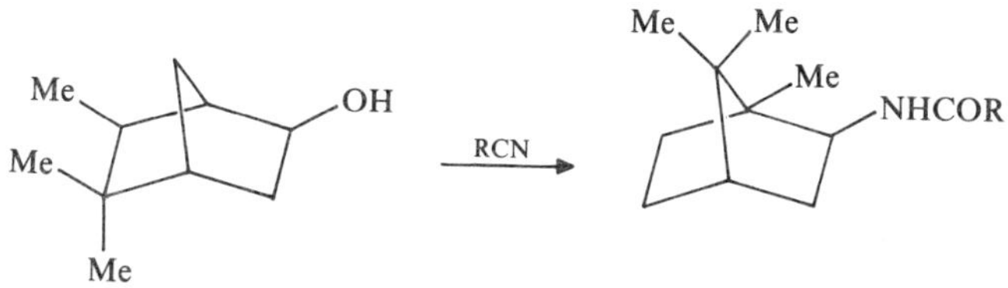

(**316**) (**317**)

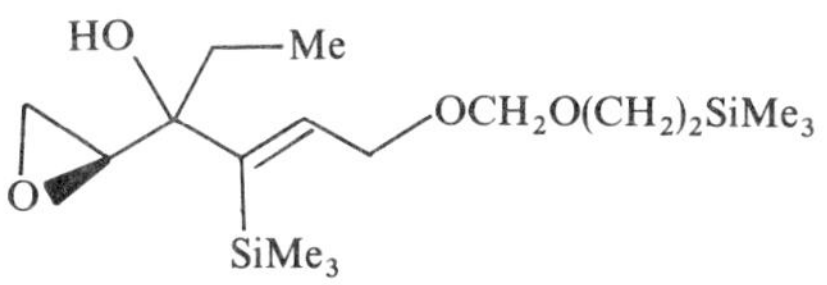

(**318**)

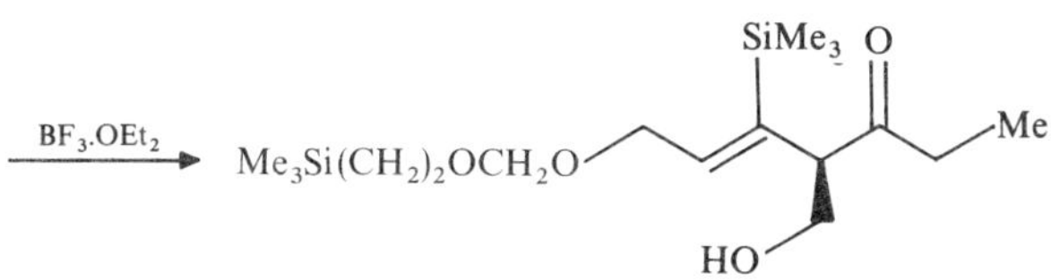

(**319**)

Thermal treatment of a series of natural neo-clerodanes having a 4α,18-oxirane ring and different functions at C(6), C(7), and C(19) has been utilized to produce good yields of different rearranged products which have proved useful intermediates for the synthesis of other natural neo-clerodane diterpenoids.[515] Direct evidence has been presented[516] to support the accepted hypothetical carbocation pathway for the biosynthesis of the diterpene, cryptotanshinone, while an interesting investigation[517] has shown that *exo*- and *endo*-D-norbeyerane-15-methyl tosylates undergo ring expansion and Wagner–Meerwein rearrangements in accord with a recently considered mechanism for diterpene biosynthesis. Biosynthetic studies on a novel metabolite, 7,12-dihydroxysterpurene, have shown[518] how farnesyl pyrophosphate (**320**) is rearranged in the cyclization reaction (see Scheme 69) to generate the tricyclic sterpurene carbon skeleton (**321**). Model experiments support the premise

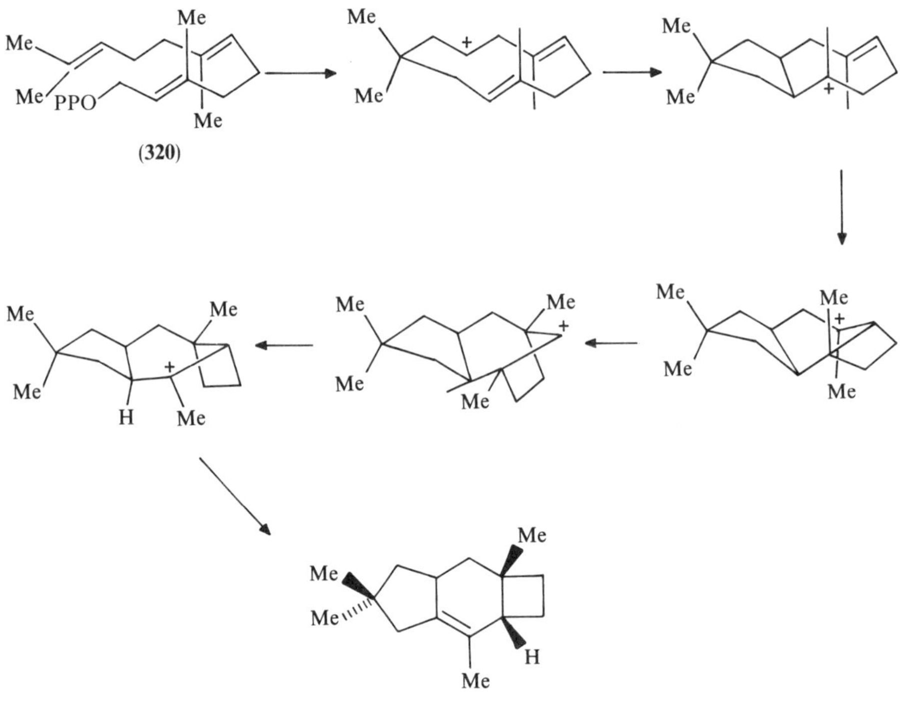

SCHEME 69

that heliangolides are the biogenetic precursors of cadinane and muurolane sesquiterpene acids.[519] Treatment of 9-methylenelongibornane with $BF_3.OEt_2$ in refluxing benzene has yielded 9-methylisolongifolene via hydride shift in a cationic intermediate,[520] while longibornane alcohols, when exposed to $HClO_4$, have been rearranged to isolongifolene through a series of oxonium ions.[521] A new type of tricyclic compound, viz. 12-hydroxy-10-methylenetricyclo[7.2.1.$0^{1,6}$] dodecan-11-one (**323**) has been synthesized by solvolytic rearrangement of *exo*-3-methylenetricyclo[6.4.0.$0^{1,4}$]dodecan-5-ol (**322**),[522] and an acid-catalysed rearrangement of a homobrendane to a perhydro-1,4-methanoindene skeleton has been described.[523] The intramolecular base-induced rearrangement of perhydronaphthalenes of type (**324**) has provided a feasible approach to 1,5-*cis*-fused 10-methyleneperhydroazulenes (**325**), and is assumed[524] to proceed in a concerted manner directed by selective intramolecular deprotonation of the angular methyl group by the proximate alkoxide; see Scheme 70. An intramolecular carbonyl-mediated electrochemical oxidation of the oleanene triterpene, methyl 3-*O*-acetylglycyrrhetate (**326**), has produced the skeletally rearranged triterpene (**327**) having an A-nor-B-homo ring system.[525]

A recently discovered rearrangement of carbohydrate triflates[526] has been applied to an efficient synthesis of a group of partially protected 2,6-dideoxy

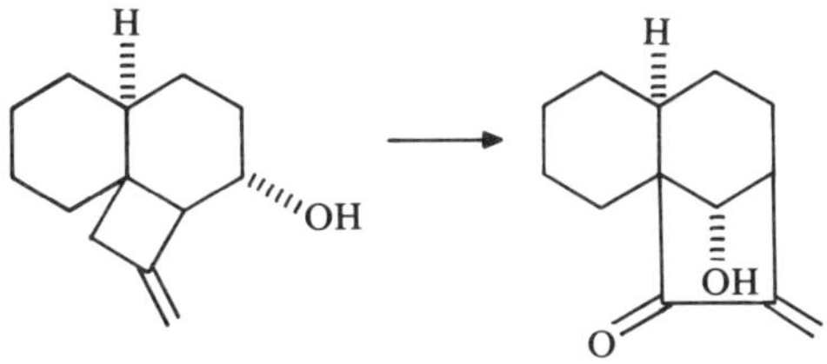

(322) (323)

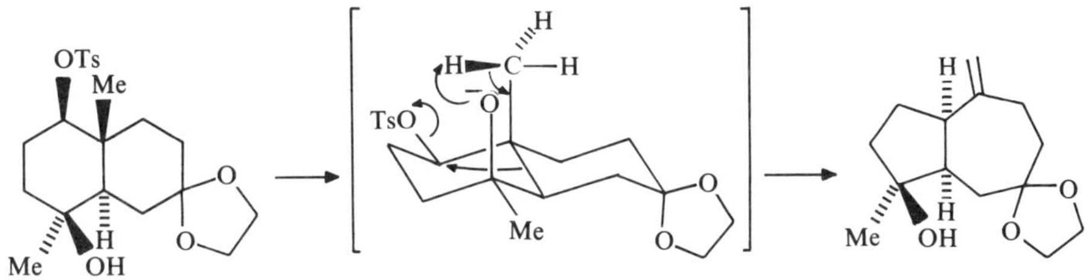

(324) (325)

SCHEME 70

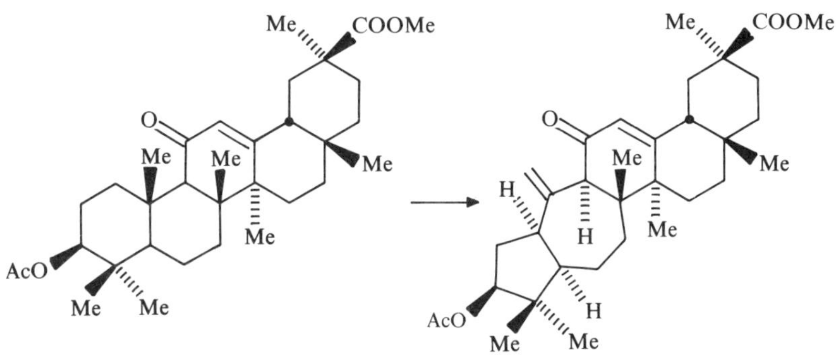

(326) (327)

sugars,[527] while a new method by which glycosyl halides can be converted directly into 2-deoxy sugars has been developed.[528] The procedure is based on the rapid and irreversible migration of the acyloxy group in the β-acyloxyalkyl radical (**328**) to give the acyloxyalkyl radical (**329**). On treatment with glacial acetic acid, the furanoid glycal (**330**) has been observed to undergo an unusual skeletal rearrangement, giving the septanose derivative (**331**). A tentative mechanism for the reaction is depicted in Scheme 71.[529] Migration of a C(4) benzoyl group to C(6) of an alkenylpyranoside has been activated by electrophilic attack on the C(6)—C(7) double bond. Subsequent treatment of the product with DBN causes migration of the benzoyl group back to C(4) with the formation of a C(6)—C(7) oxirane.[530] Terminal monosubstituted 5-membered ring *O*-benzylidene derivatives of aldose diethyl dithioacetals have been readily rearranged to structures containing terminal 6-membered *O*-benzylidene rings.[531] These rearrangements have been shown to be intramolecular by means of cross-over experiments.[532]

It has been observed[533] that the preferred mode of acid-catalysed rearrangement of the di(pyrrolylmethyl)-2*H*-pyrrole (**332**) corresponds to the specific direction of cleavage and rearrangement required for the conversion of the spiro system (**333**) into the uroporphyrinogen-III (**334**). An unusual rearrangement of a quebrachamine derivative by cyanide anion has been reported,[534] while the rearrangement of cinchona alkaloids to the corresponding toxine derivatives has been achieved using acetic acid.[535] Interestingly, the alkaloids (±)-xylopinine and (±)-tetrahydropalmatine have been synthesized by a photo-induced rearrangement of spirobenzylisoquinolines in the presence of vitamin C.[536]

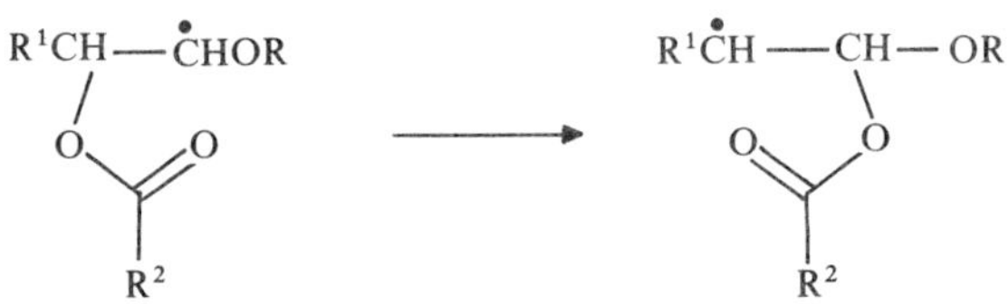

(**328**) (**329**)

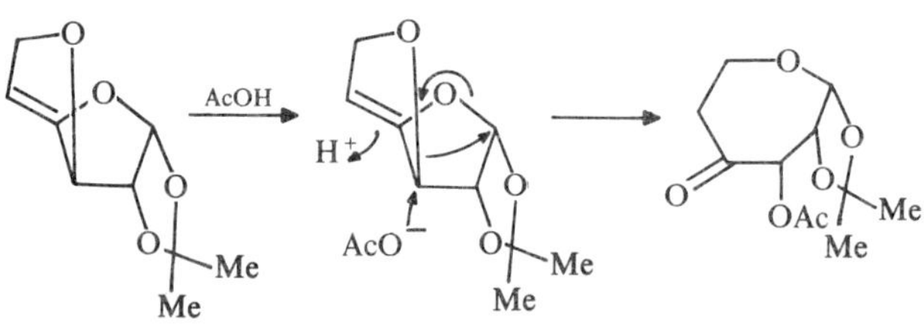

(**330**) (**331**)

SCHEME 71

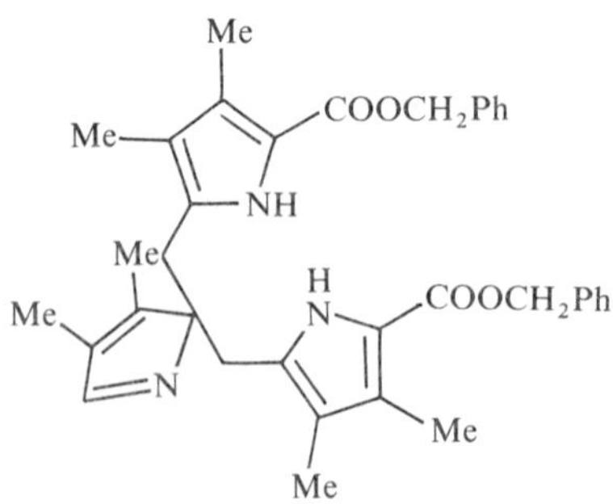

(**332**)

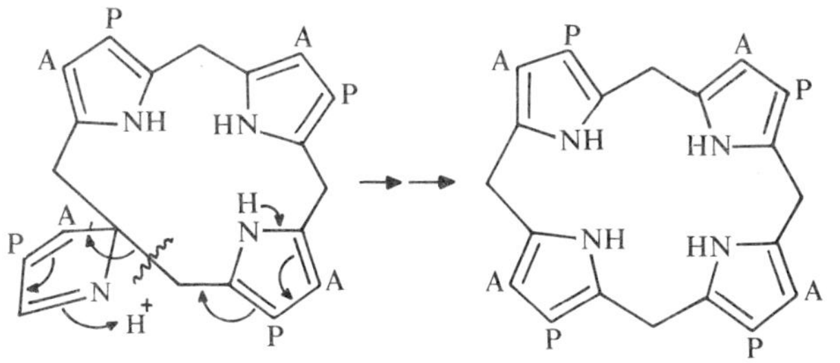

(333) (334)

A = CH_2COOH
P = CH_2CH_2COOH

Rearrangements Involving Electron-deficient Heteroatoms

Studies of the vapour-phase Beckmann rearrangement over alumina-supported boria catalysts have been described,[537,538] the Beckmann rearrangement of cyclohexanone oxime on HNaY zeolites has been studied,[539] and the effect of stirring on the Beckmann rearrangement and on its secondary reactions has been investigated.[540] The role of acids in the Beckmann rearrangement of (−)-(1*R*,*trans*)-*p*-menthan-3-one (*E*)-oxime has been examined[541] while, in sharp contrast with earlier results, the major products of the photolysis of camphor oxime have been found to be almost equal amounts of two isomeric lactams arising from the photo-Beckmann rearrangement and ring-opened amides; these results were also in line with those obtained for bicyclo[2.2.1]heptanone oxime and fenchone oxime.[542] It has been observed[543] that only the products of vinyl migration are obtained during the Beckmann rearrangement of Δ^4-6-hydroxyimino-steroids. A series of carbon-bridged dibenzazocines and dibenzodiazocines has been prepared, relying upon the Beckmann rearrangement for the key transformation. The results of the study have also clarified some factors controlling regiochemistry in the Beckmann rearrangement.[544]

An *ab initio* study[545] of the borylnitrene (H_2B—N) to iminoborane (HB═NH) isomerization, has indicated that the Curtius-type thermolysis of boron azide to iminoborane does not involve the intermediacy of borylnitrene. Similar calculations[546] on the silylnitrene-silanimine rearrangement have likewise confirmed that a silylnitrene is not an intermediate in the Curtius-type rearrangement that appears to occur during the thermolysis, or direct photolysis, of silyl azide. *N*-(Alkylphenylphosphinoyl)hydroxylamines (**335**) have been prepared and converted into their *O*-methylsulphonyl derivatives (**336**), which readily rearrange on treatment with base to give only the products (**338**, X = NHR or OMe) resulting from migration of the phenyl group.[547] A likely mechanism for the transformation involves base-induced Lossen-like rearrangement to a monomeric metaphosphonimidate (**337**) which then reacts rapidly with the nucleophile; see Scheme 72. A reinvestigation of the major product from the Schmidt reaction on camphor has led to the assignment of its structure as 1,9,9-trimethyl-3-oxo-2,8-diazabicyclo[3.3.1]nonane.[548] Reaction of silyl enol ethers (**339**) with *N*-chlorosuccinimide has been found to afford siloxycarbinyl cations (**340**) which have been trapped by nucleophiles (azide anion and methanol). Subsequent treatment of the azide product (**341**; Nu = N_3) under reflux in decalin has led to anilide derivatives (**342**) via a Schmidt-type rearrangement.[549]

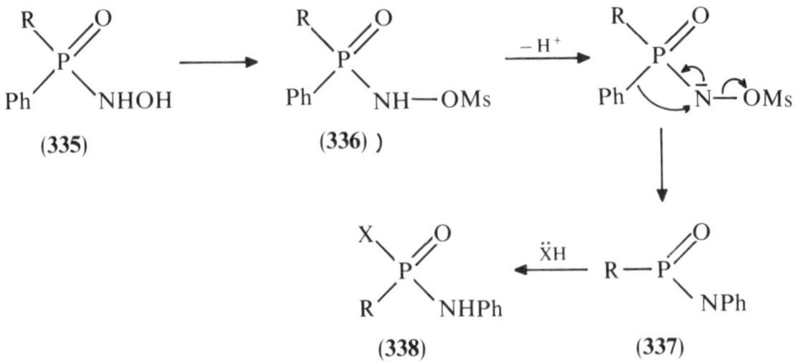

SCHEME 72

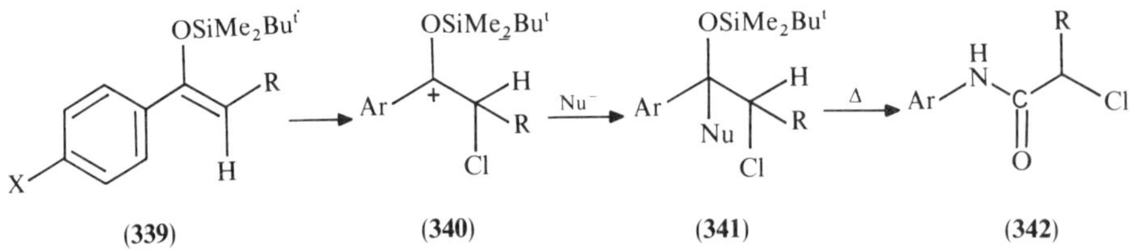

The benzylic hydroperoxide rearrangement has provided an effective complement to the Baeyer–Villiger rearrangement of acetophenones, especially suited for use on occasions when peracid treatment would prove to be expectantly poor.[550] A preformed reagent derived from $BF_3.OEt_2$ and 90% H_2O_2 has been used[551] as a useful reagent for promoting this rearrangement in readily oxidizable substrates bearing a free amino group, and such a rearrangement has been applied to the synthesis of selectively protected L-dopa derivatives.[552]

Metal-catalysed Rearrangements

Rearrangements, intramolecular exchanges, and isomerizations have been reviewed for a wide variety of mono-, dinuclear, and cluster organometallic compounds.[553]

The structure of pentadienyllithium has been examined by MO calculations which have shown that a U-shaped molecule with the lithium atom in the middle is the most stable arrangement. A mechanism for fast rearrangement of pentadienyllithium, involving lithium migration along the ligand's carbon skeleton, has been proposed.[554] Aldoses have been rapidly epimerized at C(2) by combination of alkaline earth or rare earth metal ions and monoamines, and by Ni(II) complexes.[555,556] ^{13}C NMR studies utilizing [1-^{13}C]-D-glucose have revealed[557] that the apparent C(2) epimerization of the molecule in fact proceeds via a molecular rearrangement in which the C(1) carbon label is found at the C(2) position of the product mannose.

A series of pentamethylcyclopentadienyl(acyl)titanium complexes has been observed to undergo acyl C—O bond fission with concomitant formation of benzene derivatives,[558] and the dichlorobis[π-cyclopentadienyl]titanium-catalysed reaction of α-methylalkylmagnesium bromides with isocyanates has been reported to yield only normal carboxanilides, with no trace of anilides containing the original alkyl group.[559] Recently observed molecular rearrangements of (η^5-pentamethylcyclopentadienyl)tris(*N,N*-dimethyldithiocarbamato)zirconium(IV)[560] are in accord with a ligand-centred C—N bond rotation mechanism and a metal-centred polytopal rearrangement. The insertion of one equivalent of CO into the 1-sila-3-zirconacyclobutane ring of metallacycles (**343**) was found to be followed by an intramolecular 1,2-silyl shift and the formation of an oligomeric derivative of (**344**).[561] Isomerization pathways of η^4-diene complexes of (*tert*-butylcyclopentadienyl)-zirconium and -hafnium have been discussed,[562] and the mechanism of the rearrangement of (η^5-C_5Me_5) (η^5,η^1-$C_5Me_4CH_2$)Hf$CH_2C_6H_5$ to its hafnabenzocyclobutene tautomer, (η^5-C_5Me_5)$_2$Hf(*o*-$CH_2C_6H_4$), has been explored.[563] A vanadium-organosiloxane catalyst has been used[564] to promote the allylic rearrangement of 1-alkenylethynylcarbinols.

Bicyclo[6.1.0]nona-2,4,6-triene complexes of chromium, molybdenum, and tungsten have been found to rearrange by a [1,7]-sigmatropic process in which the metal does not directly participate in cleavage of the carbon–carbon bond,[565] and the first examples of rearrangement of cyclopropenes to naphthols promoted by chromium and molybdenum carbonyls have been reported;[566] see Scheme 73. Allyl isocyanides (**345**) have been reacted with $(CO)_5M{=}C(OEt)C_6H_5$ (**346**; M = Cr, W) to give 3-aza-1,2,5-hexatriene complexes (**347**) which in the absence of suitable reactants isomerize spontaneously at 0° via a metal-induced N/C-migration of the allyl group to afford 4-pentenenitrile complexes (**348**).[567] Labelling experiments have proved that the rearrangement corresponds to the [3,3]-type. An MO study of the methyl migration observed in methylpentacarbonylmanganese has been undertaken.[568] Evidence has been forthcoming[569] for the occurrence of a $\sigma \rightarrow \eta^3$ rearrangement reaction in gaseous ions of allylic derivatives of manganese, rhenium, and iron and rhenium and ruthenium complexes have been used in the photo-isomerization of 1,1′-dibenzyl-1,1′,4,4′-tetrahydro-4,4′-binicotinamide and 1,1′-dibenzyl-1,1′,4,6′-tetrahydro-4,6′-binicotinamide.[570] Novel anionic rearrangements in hexacarbonyldiiron complexes of chelating organosulphur ligands have been recorded,[571] as has the photo-induced ring-expansion of cycloalkyl iron σ-complexes (**349**) to cyclic complexes (**350**);[572] see Scheme 74. Iron carbonyl-mediated isomerizations of 1,4-disilacyclohexa-2,5-dienes, by a mechanism involving metal-assisted cleavage of two carbon-silicon bonds and formation of an alkyne-iron-metallacycle intermediate, have been described.[573] The acid-catalysed rearrangement of [η^5-

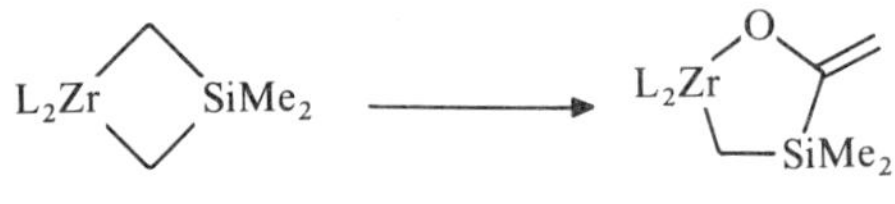

(**343**) (**344**)

$L = \eta^5\text{-}C_5H_5, \eta^5\text{-}C_5Me_5$

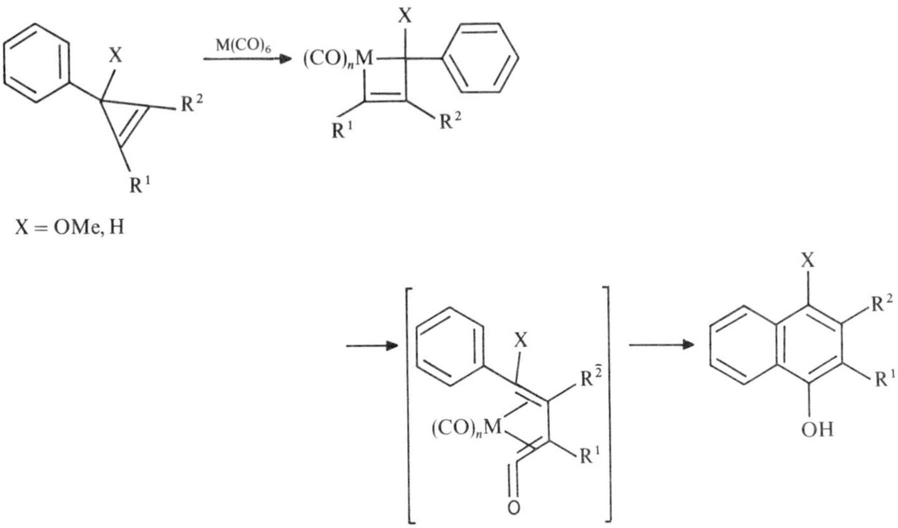

X = OMe, H

SCHEME 73

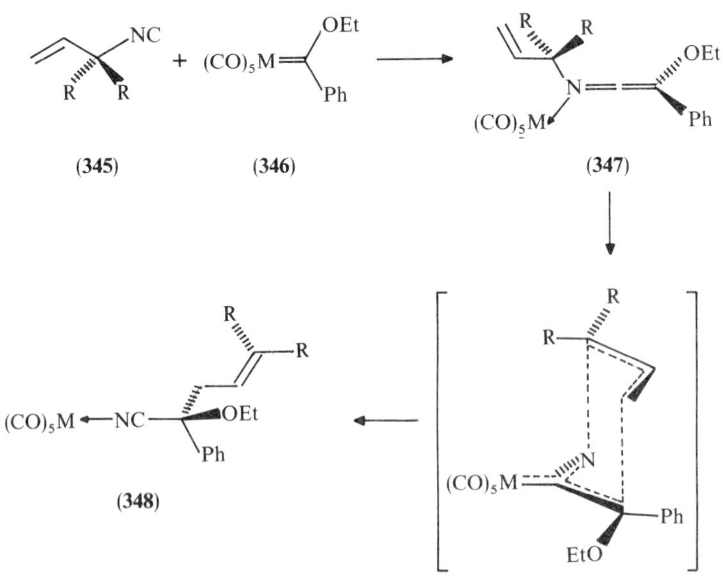

(**345**) (**346**) (**347**)

(**348**)

$(C_5H_5)Fe(CO)(PPh_3)COCH_2OCH_2Ph]$ to the ester $[\eta^5\text{-}(C_5H_5)Fe(CO)\text{-}(PPh_3)CH_2CO_2CH_2Ph]$ has been found to occur stereospecifically with inversion of configuration at carbon.[574] Two rearrangements of alkenes, known to proceed through the intermediacy of the alkene cation radical, have been observed to accompany the hemin-catalysed epoxidation of these alkenes.[575] These observations have been interpreted in terms of an electron transfer from alkene to the high-valent iron intermediate, leading to both rearrangement and epoxidation. Recent ruthenium-catalysed rearrangements of 15,16-epoxybeyerane diterpenes functionalized at C(14) have led to several derivatives[576] not reported in previous studies of such rearrangements.

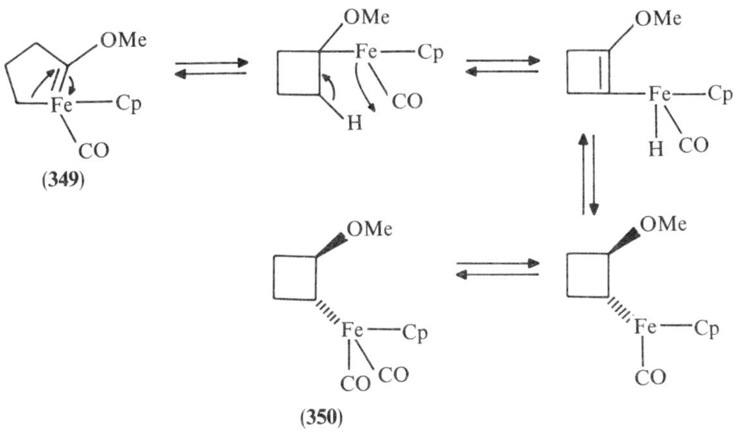

SCHEME 74

A review of free-radical mechanisms in bio-organometallic chemistry has included an account of coenzyme B_{12}-dependent rearrangements.[577] Thermolysis of cobalt-*meso*-tetraphenylporphyrin-catalysed reactions of endoperoxides derived from 2,3-dihydro-1(2*H*)-azulenone have been studied, and a new endoperoxide-endoperoxide rearrangement (see Scheme 75) has been discovered.[578] Carbon-skeleton rearrangements of 2,2-bis(ethoxycarbonyl)-1-bromopropane, a model substrate for methylmalonyl-CoA, catalysed by heptamethyl cobyrinate perchlorate[579] and by the cobalt complex of 2,10-diethyl-3,9-dipropyl-1,4,8,11-tetraazaundeca-1,3,8,10-tetraene-1,10-diol[580] have been investigated under electrochemical conditions. A novel reaction of cobaltacyclopentenedione (**351**) with terminal alkynes to provide 5-alkylidene cyclopent-2-ene-1,4-diones (**353**) has been described.[581] It is thought to proceed via the vinylidene tautomer (**352**).

In vitro experiments have been reported[582] showing that several transition metal ions (Co^{2+}, Ni^{2+}, Cu^{2+}) exert a profound influence on both the kinetics and chemical course of the rearrangement of dopachrome, a key step in the biosynthesis of melanins. Transition metal complexes have been used to catalyse the cyclization of alkynoic acids to alkylidene lactones,[583] and to promote ring-opening reactions of vinylcyclopropenes,[584] while evidence has been presented[585] for alkyl rearrangement during the iodide-promoted, rhodium-catalysed carbonylation of ethanol. The

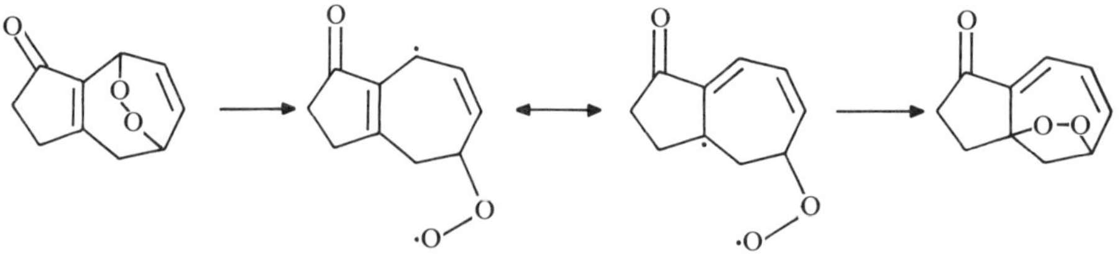

SCHEME 75

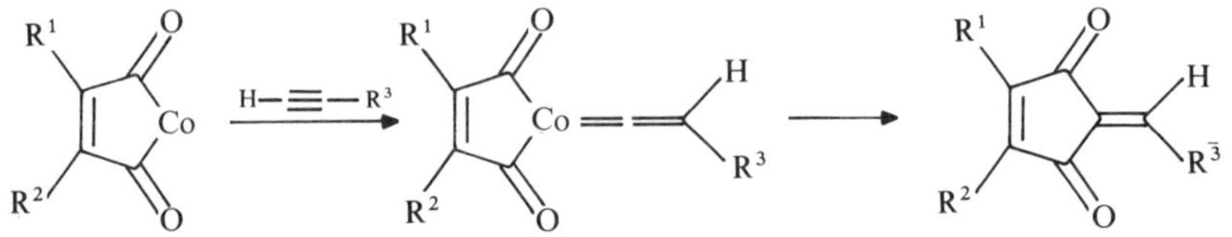

(351) (352) (353)

isomerization of *cis*-dihydridobromo(carbonyl)[bis(diphenylphosphino)ethane]-iridium (**354**) to its more stable isomer (**356**) has been studied kinetically and the reaction shown to proceed via two different mechanisms.[586] The first is a reductive elimination oxidative addition sequence which is first-order in cokplex, while the second is a bimolecular pathway involving dihydride transfer from (**354**) to the intermediate (**355**) to produce (**356**).

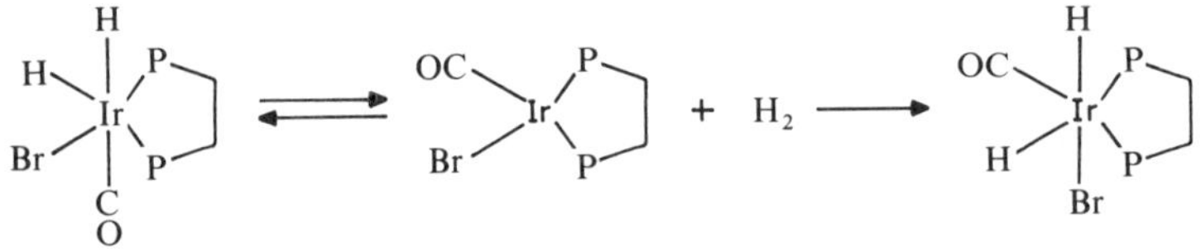

(354)) (355) (356)

Allylic phosphites have been rearranged to the corresponding allylic phosphonates using nickel chloride.[587] Methylenecyclopentanes have been prepared by Pd(0)-catalysed isomerization of 5-(bromozincmethyl)-3-methyleneoxacycloalkanes, the first reported case of Pd(0)-catalysed coupling between a saturated organozinc compound and an allylic ether.[588] Various rearrangements of platinacyclobutane complexes from *cis*-disubstituted cyclopropanes, such as those found in bicyclo[*n*.1.0]hydrocarbons ($n = 4$ or 6), have been described.[589]

The wavelength dependence of the photo-rearrangements of cuprous chloride complexes of 1,5,9-cyclododecatrienes has been established.[590] Di-2-pyridyl ketone complexes of copper(I) have proved efficient photo-catalysts for the norbornadiene-quadricyclane conversion,[591] and reports of various investigations of the silver(I)-catalysed isomerization of quadricyclanes to norbornadienes have appeared.[592–594] The valence isomerization of quadricyclane to norbornadiene on the surface of illuminated CdS and ZnO has been reported[595] and, contrary to a recent report, the process has been described as proceeding via a radical cation. A carbene→carbyne rearrangement of $ZnCH_2$ to HZnCH has been observed[596] during photolysis of the 1:1 adduct formed by co-condensing monoatomic zinc with diazomethane on to a cold surface.

Rearrangements Involving Ring-opening and Ring-closure

Three-membered Rings

Irradiation of 2,3-dihydro-2,3-methano-1,4-naphthoquinones has been observed to produce cyclobutanols which further rearrange to tricyclic β,γ-unsaturated ketones, presumably via homolytic fission of the cyclopropane ring followed by 1,2-alkyl migration.[597] A new and efficient method for introducing a 2-oxoalkyl group at the C(3) position of 2,3-dehydropiperidine, based on rearrangement of a (β-aminocyclopropyl)carbonyl intermediate, has been explored,[598] and 3-fluoroethyl-2-cyclopenten-1-ones have been prepared in a single step by rearrangement of cyclopropyl-methoxyallenyl-carbinols with hydrogen fluoride–pyridine-metal fluoride.[599] A new tri-*n*-butyltin hydride-based rearrangement of bromomethyl β-keto-esters (**357**) has provided[600] a synthetically useful ring expansion to γ-keto esters (**358**); see Scheme 76. A possible mechanism for the observed KF-induced dimerization of α-(sulphonio)ketone triflates is presented in Scheme 77.[601] Reactive cyclopropanone intermediates (**361**), generated via the sulphonium ylides (**360**) from the triflates (**359**), are considered to react rapidly with (**360**) to yield betaine intermediates (**362**) which rearrange to cyclobutanones (**363**).

Vinylcyclopropanes of type (**364**) have been converted into either cyclopentenes (**365**) or bicyclo[3.2.1]octenes (**366**) through carefully controlled rearrangement pathways.[602] Factors affecting these rearrangements and their utility in the stereo-controlled syntheses of substances such as pentalenic acid and pentalene have been described.[603] The facile photo-rearrangement of 1,1-bis(bromomethyl)-2-vinylcyclopropane to 6-bromo-2-(bromomethyl)-1,4-hexadiene has been reported,[604] and 8-methylenetricyclo[4.2.1.0^{1,3}]nonane (**368**; $n = 2$) has been identified

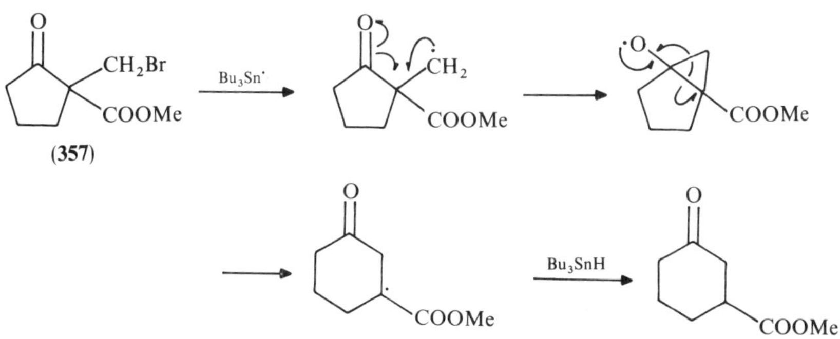

SCHEME 76

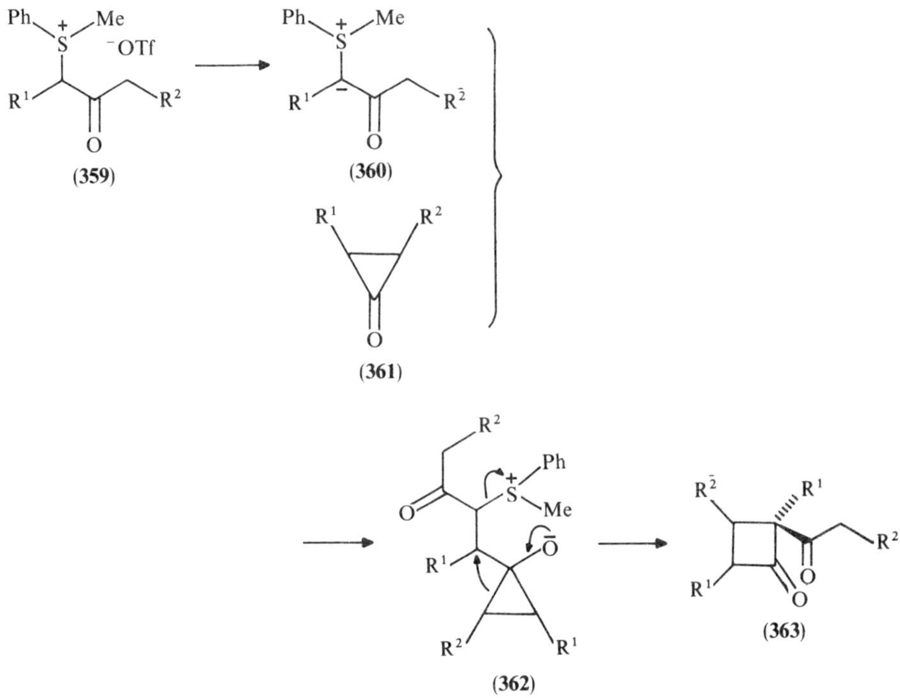

SCHEME 77

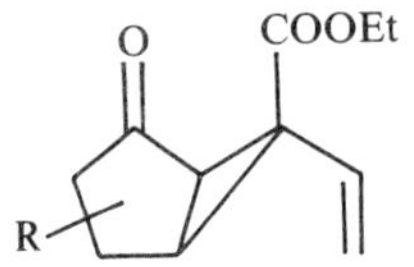
(364)

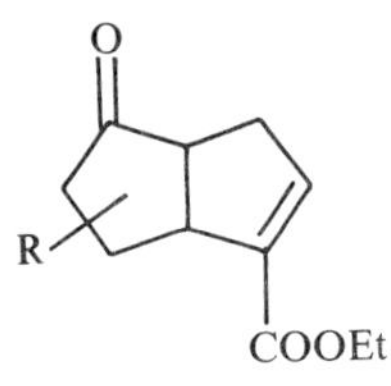
(365)

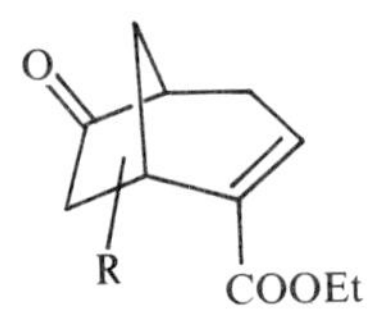
(366)

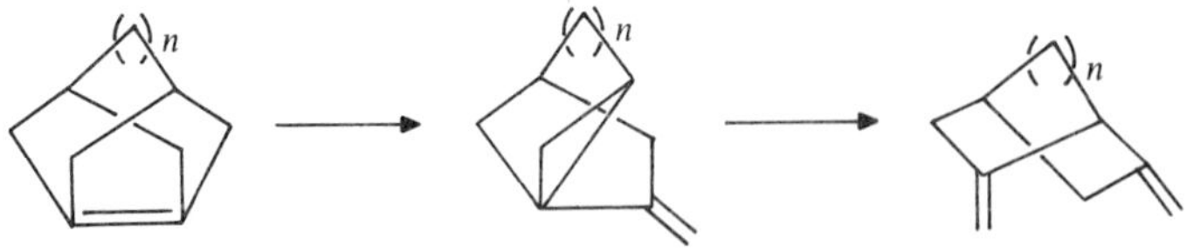
(367) (368) (369)

as an intermediate in the rearrangement of alkene (**367**; $n = 2$) to 2,6-dimethylenebicyclo[2.2.2]octane (**369**; $n = 2$). The reaction by which (**368**) is formed has been viewed[605] as the reverse of the well-known vinylcyclopropane rearrangement. The ring enlargement of a spiro 3-vinylcyclopropene (**370**) to a cyclopentadiene has been described. The process is considered to involve cyclopropene ring-opening as the key step and is formulated[606] as in Scheme 78. The thermal rearrangement of 1-functionalized-3,3-dimethylcyclopropene derivatives to allenes via vinylcarbene intermediates has been described,[607] and the allene ⇄ propyne isomerization has been studied behind shock waves and found to proceed in a series of successive reactions via cyclopropene.[608]

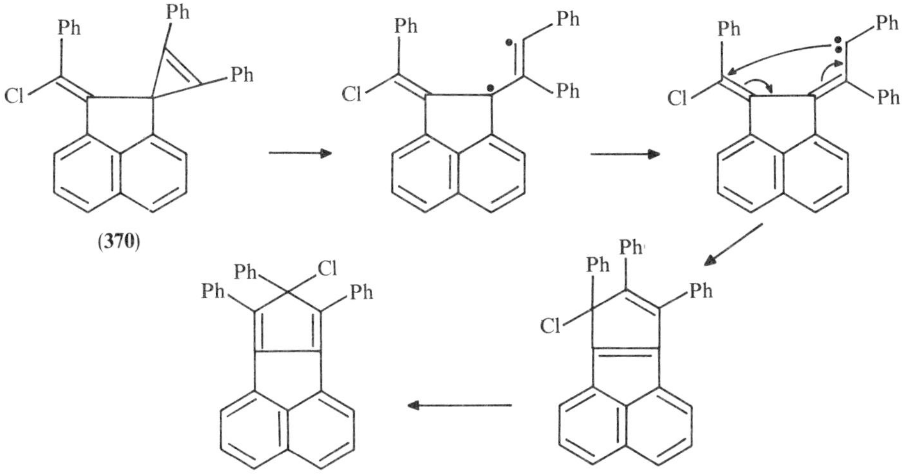

SCHEME 78

The effect of substituents on the thermal rearrangement of 3-aryl-2,2-dimethylmethylenecyclopropanes (**371**) to isopropylidenecyclopropanes (**372**) has been investigated,[609,610] and an X-ray crystallographic structure determination for methylenecyclopropane-2-carboxamide has revealed molecular distortions that correlate with known stereochemical preferences characteristic of the methylene-cyclopropane rearrangement.[611] Tricyclo[2.1.0.0^{1,3}]pentane has been thermally rearranged to cyclopentadiene,[612] and pentacyclo[4.3.1.0$^{1.6}$.0$^{7.9}$.0$^{8.10}$]decane (**373**) has been rearranged to the isomeric fulvene (**374**) in an unexpected thermal reaction.[613]

A facile synthesis of 3-chloro-2-oxoalkanoates vIa thermal rearrangement of 2-chloro-2,3-epoxyalkanoates promoted by alumina has been reported,[614] and *N*-benzoylcarbamates formed *in situ* from 2,3-epoxy-alcohols and PhCONCO have been observed to undergo base-catalysed N-to-C(2) cyclization followed by N-to-O

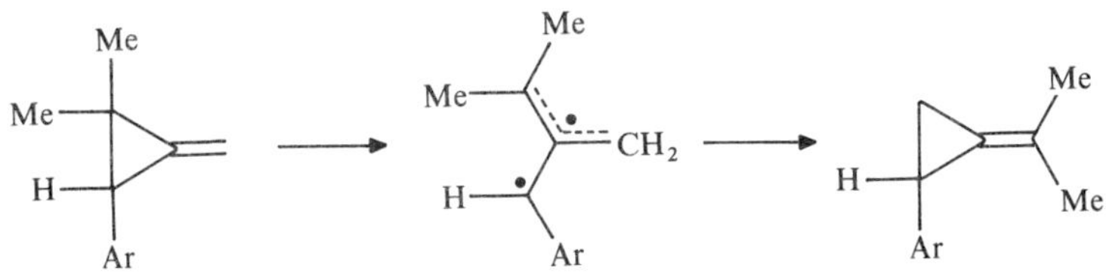

(**371**) (**372**)

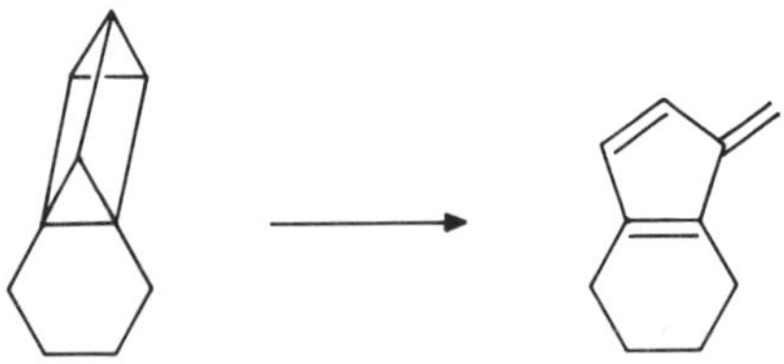

(373) (374)

acyl migration.[615] Subsequent benzoate cleavage was accompanied by equilibration of the *N*-unsubstituted oxazolidinones. The methodology has been applied to a diastereospecific conversion of methyl 4-bromophenyl sulphone into racemic thiamphenicol. The thermal conversion of (*Z*)-epoxyhexenynes into three pairs of isomeric vinylfuran derivatives has been explained on the basis of a multistep mechanism with butenynyl carbonyl ylides, cycloallenes, and vinylcarbenes as intermediates,[616] and products arising from a carbonyl ylide or a carbene intermediate have been identified during the photo-rearrangement of α,β-unsaturated γ,δ-epoxy-nitriles.[617] Thermolysis of the endoperoxide derived from 6,6-dimethylfulvene, viz. (**357**), has been observed to yield a mixture of (**378**), (**379**), and (**380**). An attractive rationalization of the formation of these compounds is shown in Scheme 79.[618] In particular, the formation of (**378**) presumes an intramolecular 1,3-dipolar cycloaddition of an allene oxide or a cyclopropanone, as in (**376**) and (**377**), to the aldehyde functionality. Conversion of oxaquadricyclanes (**381**) with a 2-oxopropyl group in the 2-position, to oxatricycloheptanes (**382**) and to the oxepine/benzene oxide systems (**383**)/(**384**) has been described and possible intermediates discussed.[619]

Allene episulphides (**385**) have been found to undergo thermal and acid-catalysed isomerization, and a desulphurization, which have been rationalized[620] by the tautomeric recyclization and sulphur elimination of a thioxallyl intermediate (**386**); see Scheme 80. Formation of an intermediate episulphonium ion appears to be the crucial step in the observed formation of various unexpected pyrazoles during the reaction of α-aroyl-α-bromoketene dithioacetals with hydrazine hydrate.[621]

Pyrolysis of *N*-*tert*-butylaziridines (**387**) in the presence of dimethyl acetylenedicarboxylate has yielded adducts of structure (**388**), which themselves are rearranged to (**389**), rather than to the expected ylide adducts (**390**).[622] In the presence of Lewis acid catalysts, spiroaziridinepenicillanates have been rearranged to pyrrolothiazoles.[623] Treatment of 7*a*-trichloromethyl-2,3,5,6,7,7*a*-hexahydro-1*H*-pyrrolizine and related compounds with pyridine has provided a new synthesis of 1-azabicyclo[3.3.1]nonanes, the formation of which has been envisaged as proceeding via a rearrangement involving an aziridinium salt.[624] An aziridinium intermediate has also been postulated[625] as intermediate in a novel skeletal rearrangement of the indolizidine alkaloid, swainsonine, while the phenoxide anion-induced ring expansion of 4-benzyl-3-chloromethylmorpholine to 4-benzyl-3-phenoxy-1,4-oxazepane has been explained[626] in terms of neighbouring-group participation leading to an intermediate ambident aziridinium cation. 1,2-Diaroyldiaziridines have been found to react with octacarbonyldicobalt to afford dihydro-oxazoles by an intramolecular displacement-rearrangement reaction.[627]

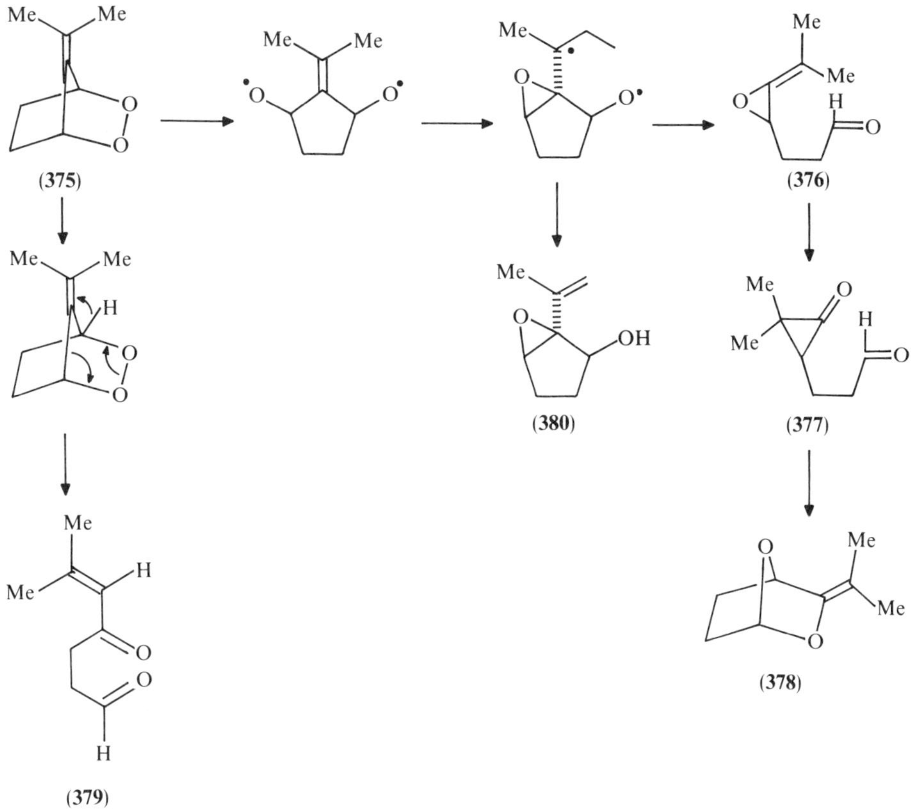

SCHEME 79

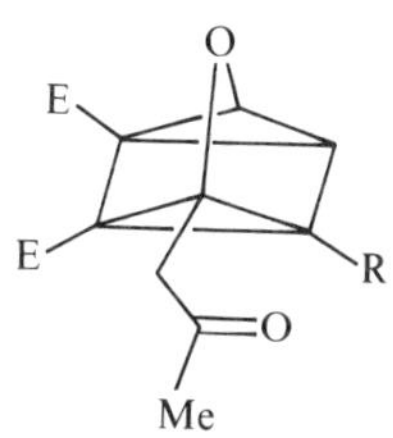

(381)

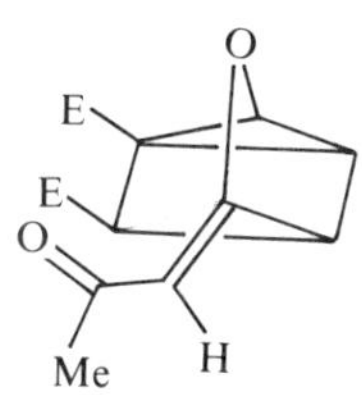

(382)

E = COOMe

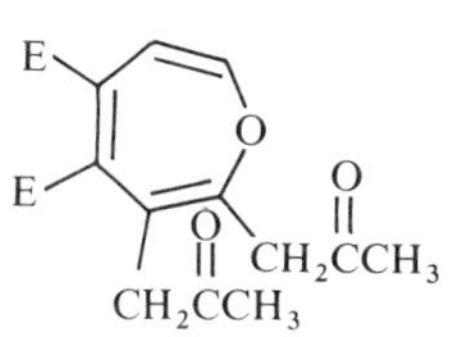

(383)

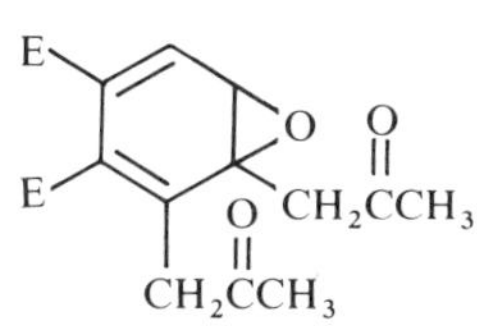

(384)

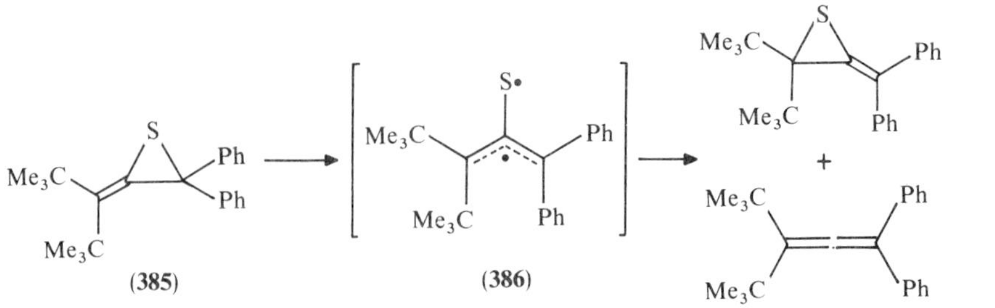

(385) (386)

SCHEME 80

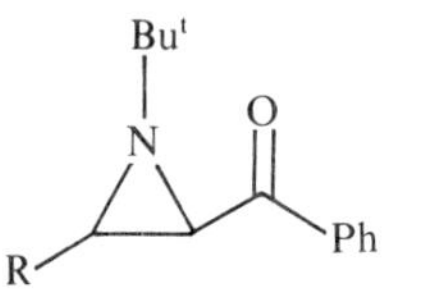

(387)

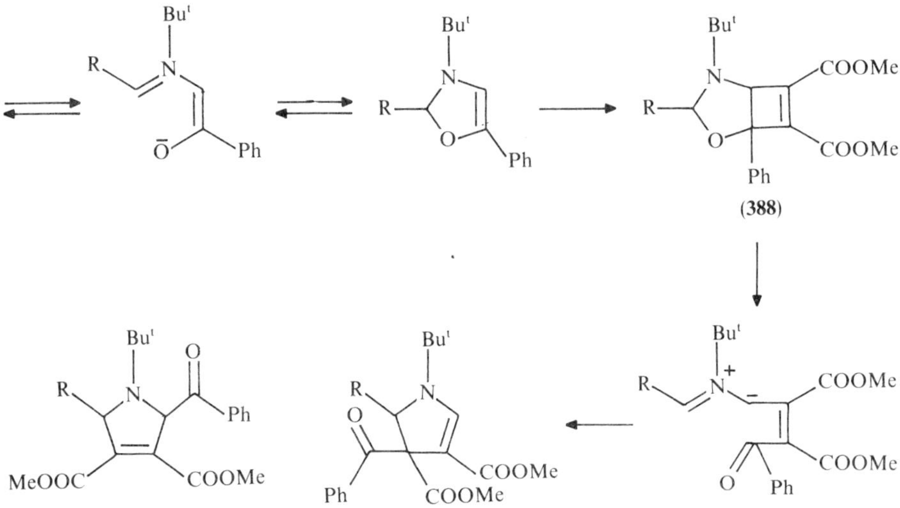

(388) (389) (390)

A formal synthesis of the dipeptide sweetener aspartame via the oxaziridine-amide rearrangement has been reported,[628] and a double rearrangement proceeding by opening of an oxaziridine ring has been invoked[629] to explain the acid-catalysed conversion of ethyl 6-oxa-2,2,4,4-tetramethyl-1-azabicyclo[3.1.0]hexane-3-carboxylate into ethyl 1,2,4,5-tetramethylpyrrole-3-carboxylate. The products from the photo-isomerization of simply substituted nitrile oxides (**391**) have been accounted for[630] in terms of a theoretically justifiable mechanism by assuming that they are converted via oxazirenes (**392**) into acylnitrenes (**393**). The latter can subsequently rearrange to isocyanates (**394**) or be cleaved in a second photo-step to give nitrene and CO.

The thermal decay of oxadiaziridines, photochemically generated from a series of bicyclic azoxy compounds, has been examined,[631] and isomerization of nitroso oxides (**395**) to nitro compounds (**396**) has been shown to proceed in an intramolecular manner by the pathway outlined in Scheme 81.[632]

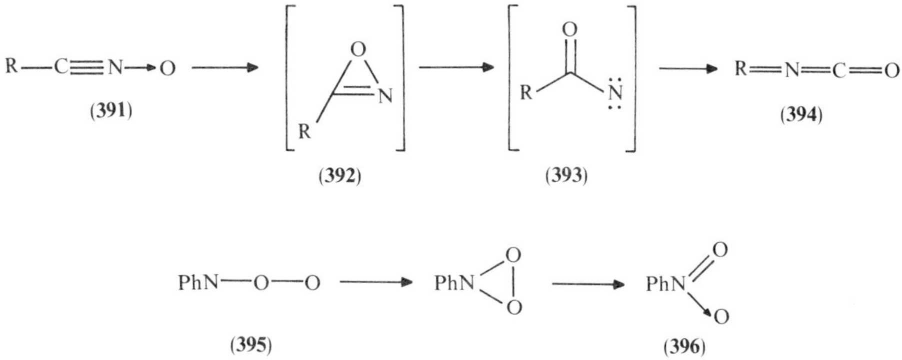

SCHEME 81

Four-membered Rings

A review has appeared on the chemistry of 6-diazopenicillanates and certain penam rearrangements,[633] while in the presence of $BF_3.OEt_2$, 6-diazopenicillanate (**397**) has been shown to react with ketones to yield mixtures of 6α-alkyl-6β-acylpenicillanates and thiazolo-oxazinones. The formation of the latter can be rationalized[634] by the pathway in Scheme 82.

In the opening of 3-formylcyclobutene, the formyl group has been predicted to rotate inward, and this non-intuitive course of events has been verified experimentally.[635] A study[636] of the thermal interconversion of bicyclo[4.2.0]octa-2,7-dienes of the type (**398**) and cyclooctatetraenes (**400**) has indicated that ring-opening of the cyclobutene is not a concerted, electrocyclic reaction, but probably the result of the weakened character of the middle carbon-carbon bond caused by donor-acceptor conjugation as illustrated in (**399**). A recent labelling study[637] of the thermal isomerization of benzocyclobutene has established for the process a mechanism which involves interconversion of the isomeric tolylmethylenes and the methylcycloheptatetraenes, and shock-tube kinetic techniques have been used to study the competitive fragmentation and geometric isomerization reactions of 1,2-dideuteriocyclobutane.[638] Difficultly accessible, or previously unknown, hexahydronaphthalenes (**403**) have been prepared via the [2 + 2]-cycloaddition of 1,3-butadiene to 1,2-cyclohexadiene. The initial products are 8-vinylbicyclo[4.2.0]oct-1-enes (**401**) which are considered to rearrange to (**403**) via the diradical (**402**).[639] Labelling experiments have established to stepwise nature of this process.[640] *cis*-Methyltetrahydroindene has been obtained by base treatment of 7-methyl-7-vinylbicyclo[3.2.0]hept-2-ene,[641] the thermal rearrangement of alkynyl-substituted benzocyclobutenones has provided a viable route to the

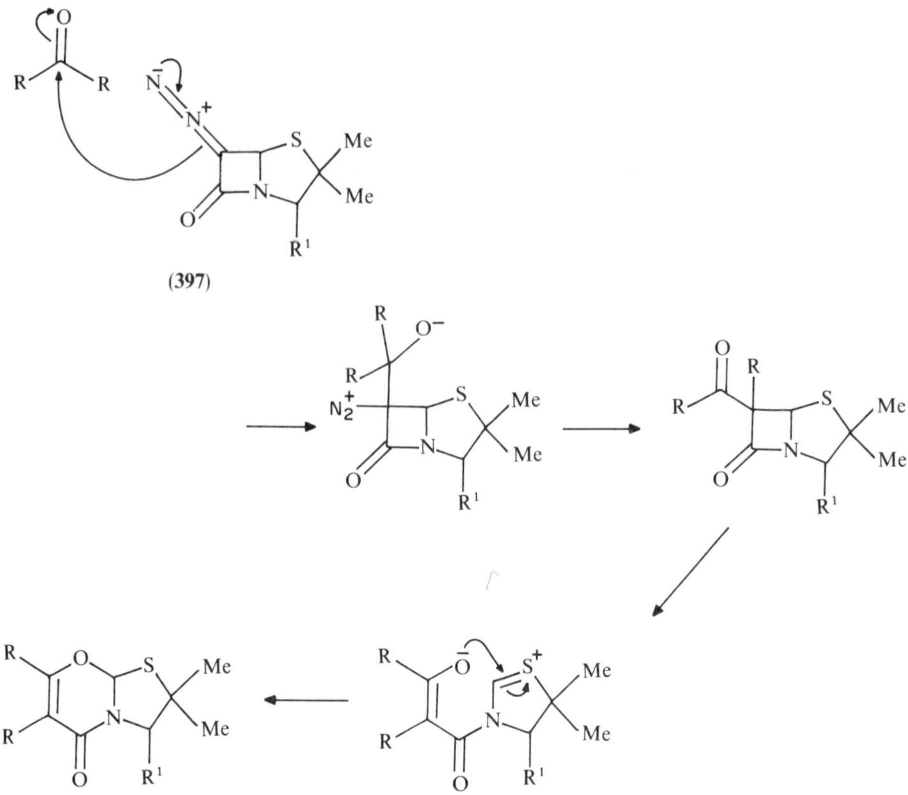

(397)

SCHEME 82

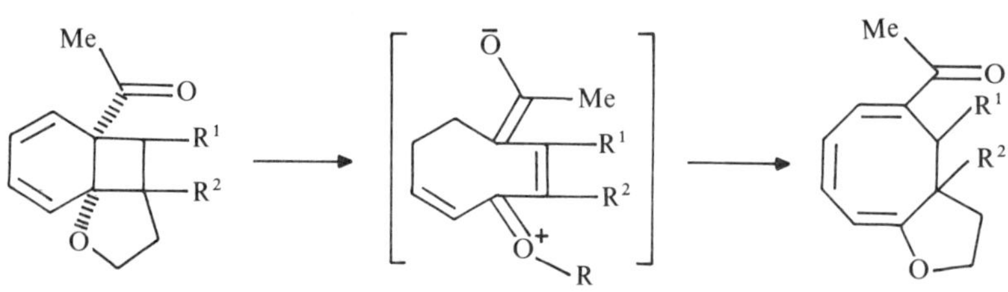

(398) **(399)** **(400)**

preparation of naphthoquinones of specific substitution patterns,[642] and the rearrangement of a furylcyclobutenone to a benzofuran has appeared as the key step in a short, efficient synthesis of khellinone from diethyl squarate.[643] The thermolysis of cyclobutanediones in hydrocarbon solvents has shown marked dependence on structure and appears to involve bis-decarbonylation to olefins, monodecarbonylation with rearrangement to cyclopentenone derivatives, or rearrangement without loss of carbon monoxide.[644]

Kinetic data have indicated that the observed rearrangement of dialkylideneureas to α-(alkylideneamino)alkyl isocyanates[645] involves an intramolecular process with a four-membered transition state. Finally, 1,2-siloxetanes (**404**) have been observed to undergo intramolecular rearrangement involving siloxy and aryl group migrations.[646]

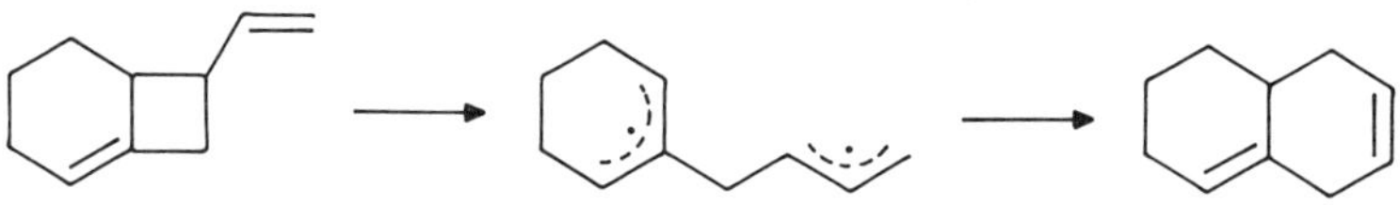

(**401**) (**402**) (**403**)

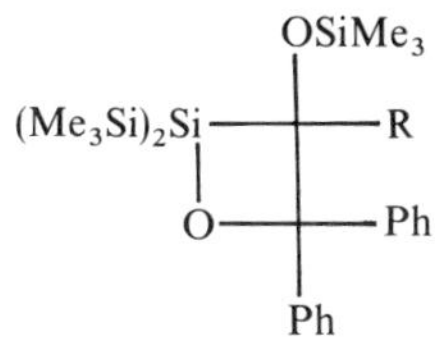

(**404**)

Five-membered and Larger Rings

Treatment of cyano-triketone (**405**) with base has been reported to yield tetracyclic lactam (**406**), the formation of which has been rationalized[647] by invoking a novel intramolecular rearrangement which is illustrated in Scheme 83. Acid treatment of dihydro-2,2,5,5-tetramethyl-3(2*H*)-furanone has been shown to yield products

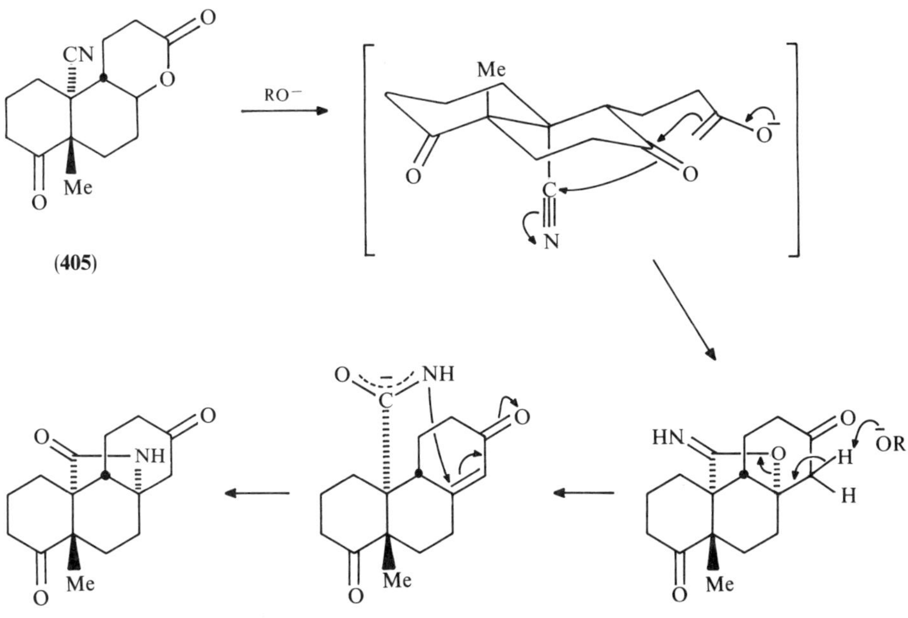

(**405**) (**406**)

SCHEME 83

resulting from ring-opening followed by reclosure, and by methyl migration from C(2) to C(3),[648] while a novel 3(2*H*)-furanone to 2(5*H*)-furanone rearrangement promoted by *N*,*N*-bis(2-oxo-3-oxazolidinyl)phosphorodiamidic chloride, has been used to generate a new class of γ-lactone amides.[649] The modes of thermal rearrangement of the 2,5-peroxides (**407**) of 2,5-unsubstituted furans to *cis*-diepoxides (**408**) and epoxyfuranones (**409**) and (**410**) have been reported and assumed to proceed via concerted processes in apolar solvents, and via ion-pairs in basic solvents.[650] Decomposition of a mixture of ^{34}S and deuterium-labelled thio-ozonides (**411**), prepared by photo-oxidation of dimethylthiophenes in $CDCl_3$, has shown[651] that the intramolecular rearrangement of these compounds to thioketone *S*-oxides (**412**) proceeds without sulphur-atom scrambling in the presence of sulphur allotropes. The most likely routes for formation of (**412**) are shown in Scheme 84; no information exists which discriminates between these pathways.

2-Aryl-4(3*H*)-quinazolinone-5-carboxylic acids have been prepared by base-catalysed rearrangement of (aroylamino)phthalimides,[652] and a mechanism involving the formation and opening of a five-membered ring (see Secheme 85) has been invoked[653] to explain the transfer of an ester group during the reaction of 3,5-diaryl-2*H*-pyrrole-2,2-dicarboxylic esters with dimethyl acetylenedicarboxylate. A ^{15}N NMR study of the transformation of methyl-sydnonimine hydrochloride to the (*E*)-and (*Z*)-isomers of *N*-methyl-*N*-(cyanomethyl)nitrosamine has been undertaken.[654]

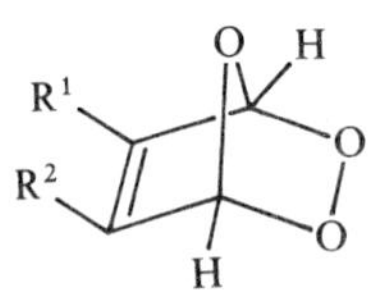

(**407**)

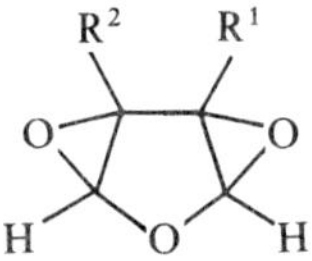

(**408**)

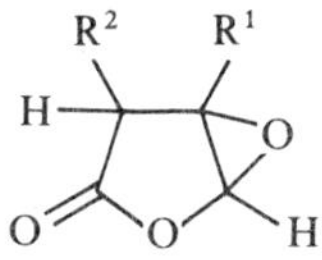

(**409**)

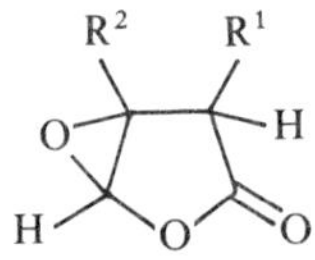

(**410**)

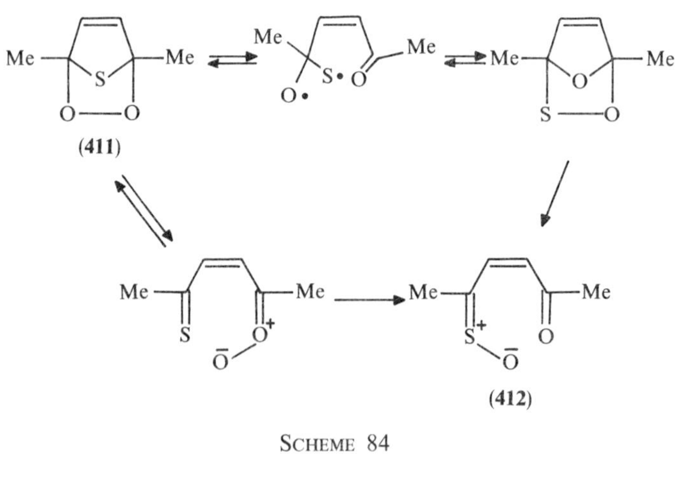

SCHEME 84

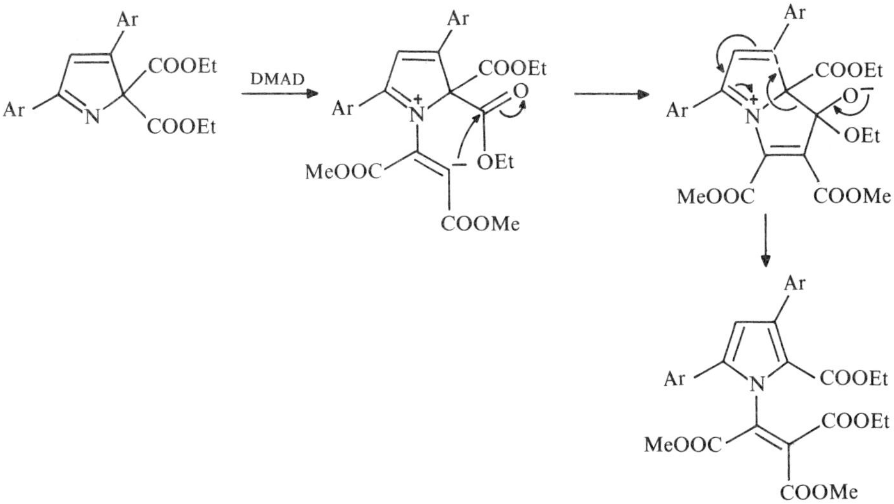

SCHEME 85

A five-membered ring intermediate (**414**) has been postulated[655] to account for the tributylstannane-induced migration of the cyano group in (**413**) to yield (**415**). Homolytic reductive debromination of 2*β*-bromomethylpenam by triphenyltin hydride has provided, via ring-expansion of the derived 2*β*-methyl radical, the corresponding cepham system.[656] A similar process has been invoked to explain the biosynthetic ring-expansion of penicillins to cephalosporins. An NMR study of the degenerate rearrangement of a bicyclic 1,3,2-oxazoniaphospholane salt, involving nitrogen exchange at tervalent phosphorus, has been described.[657]

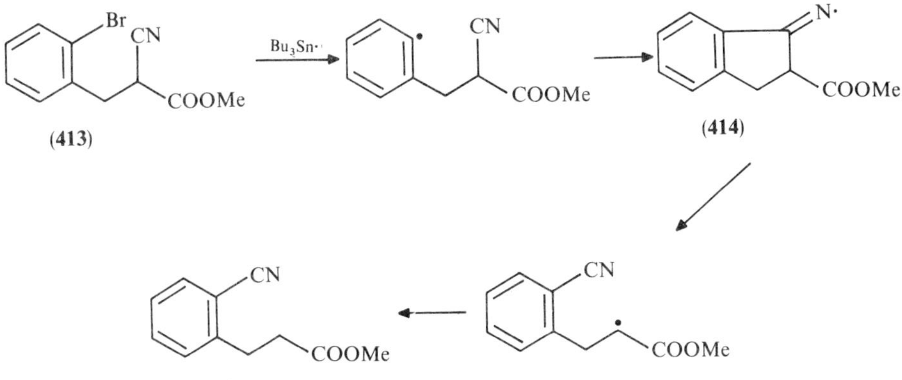

(**413**) (**414**) (**415**)

A kinetic study[658] has been carried out on the thermal conversion of 1′,3′,3′-trimethylspiro[2*H*-1-benzopyran-2,2′-indoline]-8-carboxylic acid from its spiro form (**416**) into its merocyanine form (**417**). Results of the study suggest that the structure of the transition state for the rearrangement is non-polar and indeed similar to that previously observed for the thermal conversion of 6-nitrospiropyran from its merocyanine form into its spiro form. Fragmentation and rearrangement of CO-bridged compounds prepared by carbonylation of alka-1,6-diynes, has been found to give rise to new spirocycles containing azetidine rings,[659] while the reaction of aromatic aldehydes with *N,N*-dimethyl-*O*-(3-lithioquinolyl)carbamate has been observed to produce 4-(dimethylaminomethyl)-3-hydroxyquinolines[660] by way of a new rearrangement; see Scheme 86.

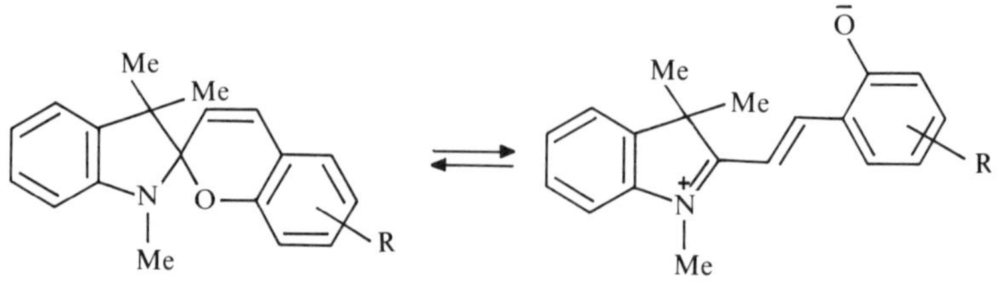

(416) **(417)**

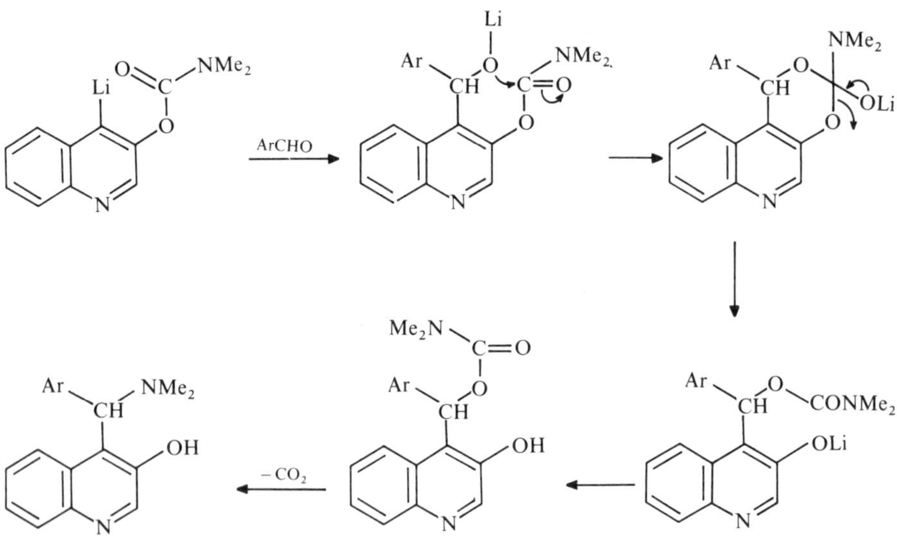

SCHEME 86

An interesting fragmentation reaction, **(418)**,[661,662] has been used to transform 2-aminomethyl-2-(3-hydroxypropyl)cyclododecanone into 12-methylene-15-pentadecanolide.[663] The complete sequence is shown in Scheme 87. A new route to medium-sized rings based upon a closely related free radical-promoted insertion of a three- or four-carbon chain into a cyclic *β*-keto-ester system has been described.[664]

Isomerizations

The isomerization of *cis*-2-pentene to a variety of other alkenes has been investigated,[665] and a quantitative study of the (*E*)/(*Z*) photo-isomerization of *trans*,

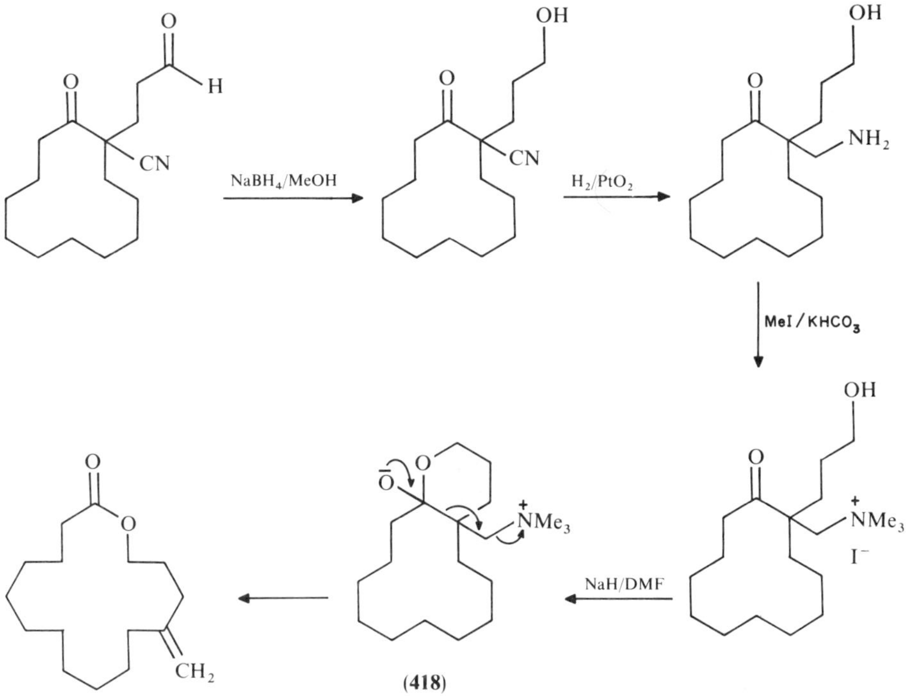

(**418**)

SCHEME 87

trans-p-styrylstilbenes has been undertaken.[666] The (*Z*)/(*E*) photoisomerization of α-unsubstituted pyrromethenones has been studied[667] using a direct NMR spectroscopic approach, while an NMR study has provided, for the first time, evidence of an acid-catalysed (*E*)/(*Z*) isomerization of substituted tetrathiafulvalenes prior to the selective precipitation of the (*E*)-isomer.[668] It has been shown[669] that α,β-substitution of dianthrylethylenes greatly enhances the efficiency of their photochemical reactivity, while the nature of the substituents has a marked effect on the course of the isomerization of these compounds. Photo-isomerization of tetrabenzoylethylene (**419**) to (**421**) is thought to proceed via formation of a ketene (**420**) and rotation about a single bond followed by intramolecular cyclization;[670] see Scheme 88. Upon irradiation, 5,6,8,10-tetramethylheptalene-1,2-dicarboxylic anhydride has been observed to rearrange to its double-bond-shift isomer in an equilibrium reaction.[671]

New kinetic evidence has been obtained[672] for an inversional mechanism in the thermal geometrical isomerization about carbon-nitrogen double bonds, while a reversed pressure effect has been observed in the thermal isomerization of 4-(dimethylamino)-4′-nitroazobenzene. The authors[673] conclude that the most probable reason for this reversed pressure effect is retardation of the intramolecular rotational movement by viscous interactions between the reactant and the solvent molecules.

2-Arylhexachloronorbornadiene has been found to undergo photo-isomerization to the corresponding quadricyclane which in turn reverts to the norbornadiene upon heating without a catalyst.[674] The observed effect of aryl substituents in the thermal reversion has been explained by suggesting that both spin delocalization and polar effect contribute equally to the stabilization of the transition state. The enthalpies of isomerization of norbornadiene and of several substituted norbornadienes to the corresponding quadricyclenes has been determined[675] by the method of photo-calorimetry, and a mechanistic approach to the sensitization process of aromatic ketones in the isomerization between norbornadiene and quadricyclane has been presented.[676] A simple electrochemical 'switch' that allows the conversion of quadricyclanes into norbornadienes has been developed.[677] The switch is based on the principle of a single-electron-transfer-promoted isomerization of quadricyclane to norbornadiene. Quadricyclene has been found to undergo valence isomerization to norbornadiene in the presence of photo-excited *n*-type semiconductor powders. The observation has been interpreted[678] in terms of the redox chemistry that results upon interaction of the photo-generated electron-hole pairs in the semiconductor with the surrounding organic medium. Sensitized irradiation of 1,7-cyclohexenonorbornadiene (**422**) has led to the quadricyclane (**423**).[679]

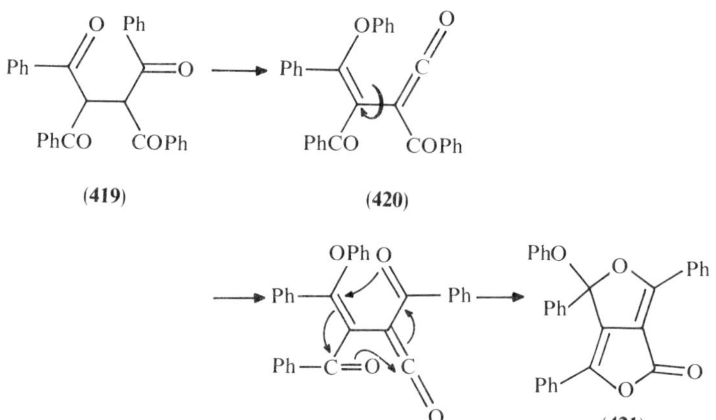

(**419**) (**420**) (**421**)

SCHEME 88

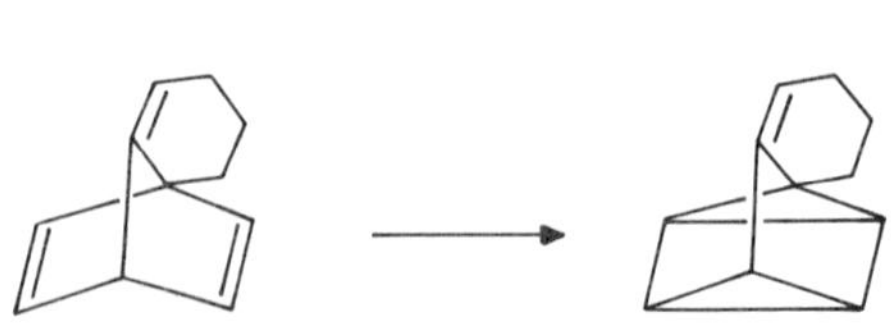

(**422**) (**423**)

The synthetic potential of the isocyanide-cyanide rearrangement has been realized and has been used[680] in the synthesis of a number of interesting optically active synthons. In this way, secondary and tertiary amino-acids have been converted into β-hydroxy-cyanides and γ-amino-alcohols. A novel reductive hydride shift has been proposed[681] to account for the isomerization of phospha-allene (**424**) to the phospha-alkene (**425**), and strong evidence has been presented[682] for the rearrangement of silylsilylene to disilene, i.e. $SiH_3SiH: \rightarrow SiH_2{=}SiH_2$. Finally, calculations have been reported that support the proposal of the diamond-square-diamond rearrangement mechanism for the 8-, 10-, and 11-atom boranes and carboranes.[683]

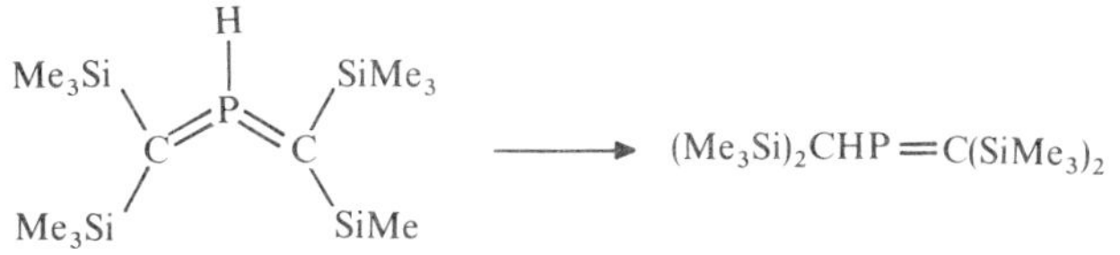

(**424**) (**425**)

Tautomerism

The quantum-mechanical prediction of tautomeric equilibria has been reviewed,[684] and a report which includes the tautomerism of dihydropyrimidines has appeared.[685]

Hydroxyethylidene (CH_3—C—OH), but not ethenol (CH_2=CH—OH), has been shown to tautomerize to ethanal.[686] The photo-induced proton-transfer tautomerization of benzanilide to its imidol form has been reported.[687]

Comparative studies have been carried out on the ring-chain tautomerism of 2-phenyl-substituted tetrahydro-1,3-oxazines[688] and of 1,3-oxazines condensed with a cyclopentane ring,[689] and ring-chain tautomerism in 2-substituted and 1,3-disubstituted diazolidines has been examined.[690] *Ab initio* calculations have been used to investigate the energetics of the tautomerism in alloxan,[691] and the kinetics of tautomerism in a range of 2-substituted 5,10,15,20-tetraphenylporphyrins have been studied[692] by NMR spectroscopy. α-Pyridylphenyldiazonium salts have been observed to undergo valence-bond isomerization to yield new pyrido[1,2*c*]benzo-*v*-triazinium salts.[693]

References

1 Marshalkin, M. F. and Yakhontov, L. N., *Usp. Khim.*, **55**, 1785 (1986); *Chem. Abs.*, **106**, 101423 (1987).

2 Kum, S. R., *Hwahak Kwa Kongop Ui Chinbo*, **25**, 774 (1985); *Chem. Abs.*, **106**, 83661 (1987).

3 Furin, G. G., *Usp. Khim.*, **56**, 911 (1987); *Chem. Abs.*, **107**, 217141 (1987).

4 Shine, H. J., Gruszecka, E., Kupczyk-Subotkowska, L., Rhee, E. S., and Subotkowski, W., *Synth. Appl. Isot. Labeled Compd., Proc. Int. Symp., 2nd* **1985**, 101; *Chem. Abs.*, **106**, 119032 (1987).

5 Shine, H. J., Kupczyk-Subotkowska, L., Subotkowski, W., and Gruszecka, E., *THEOCHEM*, **31**, 107 (1986); *Chem. Abs.*, **105**, 225563 (1987).

6 Tezuka, T., Sasaki, K., and Ando, S., *Tetrahedron Lett.*, **28**, 4427 (1987).

7 Gruszecki, W., Gruszecka, M., and Bradaczek, H., *Liebigs Ann. Chem.*, **1987**, 483.

8 Spillane, W. J., McHugh, F. A., Cunningham, D., and McArdle, P., *J. Chem. Res.* (*S*), **1987**, 252.

9 Luchkevich, E. R., Mileiko, V. E., Bagal, I. L., and El'tsov, A. V., *Zh. Obshch. Khim.*, **56**, 1574 (1986); *Chem. Abs.*, **107**, 115191 (1987).

10 Titova, S. P., Arinich, A. K., and Gorelik, M. V., *Zh. Org. Khim.*, **22**, 1562 (1986); *Chem. Abs.*, **107**, 39293 (1987).

[11] Khuthier, A.-H., Al-Mallah, K. Y., and Hanna, S. Y., *J. Chem. Soc., Perkin Trans. 2*, **1987**, 109.
[12] Khuthier, A.-H., Al-Mallah, K. Y., Hanna, S. Y., and Abdulla, N.-A. I., *J. Org. Chem.*, **52**, 1710 (1987).
[13] Harger, M. J. P. and Smith, A., *J. Chem. Soc., Perkin Trans. 1*, **1986**, 2169.
[14] Fishbein, J. C. and McClelland, R. A., *J. Am. Chem. Soc.*, **109**, 2824 (1987).
[15] Hussain, S. A. and Perkins, M. J., *J. Chem. Soc. Pak.*, **8**, 217 (1986); *Chem. Abs.*, **106**, 101832 (1987).
[16] Khelevin, R. N., *Zh. Org. Khim.*, **22**, 2126 (1986); *Chem. Abs.*, **107**, 153739 (1987).
[17] Lally, J. M. and Spillane, W. J., *J. Chem. Soc., Chem. Commun.*, **1987**, 8.
[18] Lallyl, J. M. and Spillane, W. J., *J. Chem. Soc., Chem. Commun.*, **1987**, 1571.
[19] Lau, C. K., Bélanger, P. C., Dufresne, C., and Scheigetz, J., *J. Org. Chem.*, **52**, 1670 (1987).
[20] Kimura, M., *J. Chem. Soc., Perkin Trans. 1*, **1987**, 205.
[21] Macháćek, V., Hassanien, M. M. M., Śtérba, V., and Lyćka, A., *Collect. Czech. Chem. Commun.*, **51**, 1972 (1986).
[22] Macháćek, V., Hassanien, M. M. M., and Śtérba, V., and Lyćka, A., *J. Chem. Soc., Perkin Trans. 2*, **1987**, 867.
[23] Kolchina, E. F., Kargapolova, I. Y., and Gerasimova, T. N., *Izv. Akad. Nauk SSSR, Ser. Khim.*, **1986**, 1855; *Chem. Abs.*, **107**, 39719 (1987).
[24] Cadogan, J. I. G., Hutchison, H. S., and McNab, H., *J. Chem. Soc., Perkin Trans. 1*, **1987**, 1407.
[25] Miller, J. A., *J. Org. Chem.*, **52**, 322 (1987).
[26] Maguer, P., *Bull. Soc. Chim. Fr. II*, **1987**, 199.
[27] Ali, S. M. and Ilyas, M., *J. Chem. Res. (S)*, **1987**, 52.
[28] Tanoue, Y. and Terada, A., *Bull. Chem. Soc. Jpn*, **60**, 3039 (1987).
[29] Miranda, M. A., Primo, J., and Tormos, R., *Tetrahedron*, **43**, 2323 (1987).
[30] Shine, H. J. and Subotkowski, W., *J. Org. Chem.*, **52**, 3815 (1987).
[31] No, K., Noh, Y., and Kim, Y., *Bull. Korean Chem. Soc.*, **7**, 442 (1986); *Chem. Abs.*, **107**, 96418 (1987).
[32] Benedikt, G. M. and Traynor, L., *Tetrahedron Lett.*, **28**, 763 (1987).
[33] Alvaro, M., Baldoví, V., García, H., Miranda, M. A., and Primo, J., *Tetrahedron Lett.*, **28**, 3613 (1987).
[34] Young, D. J. and Robinson, M. J. T., *Tetrahedron Lett.*, **28**, 3631 (1987).
[35] Marino, J. P. and Dax, S., *Tetrahedron Lett.*, **28**, 4007 (1987).
[36] Chamberlain, L. R., Kerschner, J. L., Rothwell, A. P., Rothwell, I. P., and Huffman, J. C., *J. Am. Chem. Soc.*, **109**, 6471 (1987).
[37] Moore, T. and Pagni, R. M., *J. Org. Chem.*, **52**, 770 (1987).
[38] Morisaki, K., Miura, Y., Abe, K., Hirota, M., and Nakada, M., *Chem. Lett.*, **1987**, 1589.
[39] Yamamoto, S., Nishiguchi, I., Hirashima, T., Kitano, A., Iwasaki, Y., and Yamamoto, K., *Kagaku to Kogyo (Osaka)*, **60**, 112 (1986); *Chem. Abs.*, **106**, 155694 (1987).
[40] Weigert, F. J., *J. Org. Chem.*, **52**, 3296 (1987).
[41] Weigert, F. J., *J. Org. Chem.*, **52**, 921 (1987).
[42] Bunte, R., Gundermann, K. D., Leitich, J., Polansky, O. E., and Zander, M., *Chem. Ber.*, **120**, 247 (1987).
[43] Loktev, V. F., Shakirov, M. M., and Shubin, V. G., *Zh. Org. Khim.*, **22**, 1110 (1986); *Chem. Abs.*, **105**, 225562 (1987).
[44] Faïn, D. and Dubois, J.-E., *J. Org. Chem.*, **52**, 260 (1987).
[45] Eaborn, C., Lickiss, P. D., Najim, S. T., and Stańczyk, W. A., *J. Chem. Soc., Chem. Commun.*, **1987**, 1461.
[46] Scott, L. T., Roelofs, N. H., and Tsang, T.-H., *J. Am. Chem. Soc.*, **109**, 5456 (1987).
[47] Yajima, T., Tsuchiya, A., and Tezuka, M., *J. Chem. Soc., Chem. Commun.*, **1987**, 1390.
[48] McMahon, R. J., Abelt, C. J., Chapman, O. L., Johnson, J. W., Kreil, C. L., LeRoux, J.-P., Mooring, A. M., and West, P. R., *J. Am. Chem. Soc.*, **109**, 2456 (1987).
[49] Brown, R. F. C., Coulston, K. J., Eastwood, F. W., and Saminathan, S., *Aust. J. Chem.*, **40**, 107 (1987).
[50] Scott, L. T. and Roelofs, N. H., *J. Am. Chem. Soc.*, **109**, 5461 (1987).
[51] Miller, B. and Baghdadchi, J., *J. Org. Chem.*, **52**, 2391 (1987).
[52] Tsuji, T. and Nishida, S., *J. Chem. Soc., Chem. Commun.*, **1987**, 1189.
[53] Yamato, T. and Tashiro, M., *J. Chem. Res. (S)*, **1987**, 198.
[54] Yokoyama, M., *Yuki Gosei Kagaku Kyokaishi*, **44**, 767 (1986); *Chem. Abs.*, **106**, 83669 (1987).
[55] Yokoyama, M., Sato, K., Tateno, H., and Hatanaka, H., *J. Chem. Soc., Perkin Trans. 1*, **1987**, 623.
[56] Flitsch, W., Jones, R. A., and Hohenhorst, M., *Tetrahedron Lett.*, **28**, 4397 (1987).
[57] Jackson, A. H. and Lynch, P. P., *J. Chem. Soc., Perkin Trans. 2*, **1987**, 1215.
[58] Ibaceta-Lizana, J. S. L., Jackson, A. H., Prasitpan, N., and Shannon, P. V. R., *J. Chem. Soc., Perkin Trans. 2*, **1987**, 1221.
[59] Wenkert, E., Moeller, P. D. R., Piettre, S. R., and McPhail, A. T., *J. Org. Chem.*, **52**, 3404 (1987).

[60] De Rosa, M., Quesada, P. A., and Dodsworth, D. J., *J. Org. Chem.*, **52**, 173 (1987).
[61] Wentrup, C. and Kollenz, G., *Synth. App. Isot. Labeled Compd., Proc. Int. Symp., 2nd*, 515 (1985); *Chem. Abs.*, **106**, 67012 (1987).
[62] Audier, H. E. and Robin, D., *Org. Mass Spectrom.*, **21**, 445 (1986).
[62] Buss, S. and Jug, K., *J. Am. Chem. Soc.*, **109**, 1044 (1987).
[64] Kozak, J. and Kozerski, L., *Bull. Pol. Acad. Sci., Chem.*, **33**, 421 (1985); *Chem. Abs.*, **106**, 213236 (1987).
[65] Gopidas, K. R., Lohray, B. B., Rajadurai, S., Das, P. K., and George, M. V., *J. Org. Chem.*, **52**, 2831 (1987).
[66] Ming, Y. and Wamhoff, H., *Chem. Ber.*, **120**, 1433 (1987).
[67] Kurita, J., Sakai, H., Yamada, S., and Tsuchiya, T., *J. Chem. Soc., Chem. Commun.*, **1987**, 285.
[68] Lawton, G., Moody, C. J., and Pearson, C. J., *J. Chem. Soc., Perkin Trans. 1*, **1987**, 877.
[69] Lawton, G., Moody, C. J., Pearson, C. J., and Williams, D. J., *J. Chem. Soc., Perkin Trans. 1*, **1987**, 885.
[70] Elguero, J., De la Hoz, A., and Pardo, C., *Croat. Chem. Acta*, **59**, 153 (1986); *Chem. Abs.*, **106**, 176310 (1987).
[71] Bird, C. W. and Kapili, M., *Tetrahedron*, **43**, 4621 (1987).
[72] Pfoertner, K. H. and Daly, J. J., *Helv. Chim. Acta*, **70**, 171 (1987).
[73] Wilkins, B. J., Gainsford, G. J., and Moore, D. E., *J. Chem. Soc., Perkin Trans. 1*, **1987**, 1817.
[74] Wild, H. and Steglich, W., *Liebigs Ann. Chem.*, **1986**, 1910.
[75] Walker, F. J. and Kraus, K. G., *J. Heterocycl. Chem.*, **24**, 1485 (1987).
[76] Zalupsky, P., Fisera, L., and Timpe, H. J., *Z. Chem.*, **26**, 441 (1986).
[77] Fisera, L., Oremus, V., Stibranyi, L., Zalupsky, P., Timpe, H. J., Dusi, A., and Dusi, R., *J. Prakt. Chem.*, **328**, 445 (1986); *Chem. Abs.*, **107**, 23299 (1987).
[78] El'tsov, A. V., Dmitriev, F. M., Gornostaev, L. M., and Rtishchev, N. I., *Zh. Org. Khim.*, **22**, 2361 (1986); *Chem. Abs.*, **107**, 58268 (1987).
[79] Ballabio, M., Destro, R., Franzoi, L., Gelmi, M. L., Pocar, D., and Trimarco, P., *Chem. Ber.*, **120**, 1797 (1987).
[80] Padwa, A., Tomioka, Y., and Venkatramanan, M. K., *Tetrahedron Lett.*, **28**, 755 (1987).
[81] Baum, J. S., Condon, M. E., and Shook, D. A., *J. Org. Chem.*, **52**, 2983 (1987).
[82] L'Abbe, G., Meutermans, W., and Bruynseels, M., *Bull. Soc. Chim. Belg.*, **95**, 1129 (1986).
[83] Kamigata, N., Iizuka, H., and Kobayashi, M., *Heterocycles*, **24**, 919 (1986).
[84] Döpp, D., Krüger, C., Lauterfeld, P., and Raabe, E., *Angew. Chem. Int. Edn*, **26**, 146 (1987).
[85] Mitchell, G. and Rees, C. W., *J. Chem. Soc., Perkin Trans. 1*, **1987**, 413.
[86] Butler, R. N., Cunningham, D., Marren, E. G., and McArdle, P., *J. Chem. Soc., Chem. Commun.*, **1987**, 706.
[87] Butler, R. N., Evans, A. M., McArdle, P., and Cunningham, D., *J. Chem. Soc., Chem. Commun.*, **1987**, 1090.
[88] Dessolin, M., Golfier, M., Gonthier-Vassal, A., and Szwarc, H., *Nouv. J. Chem.*, **10**, 753 (1986).
[89] Macaluso, G., Cusmano, G., Buscemi, S., Frenna, V., Vivona, N., and Ruccia, M., *Heterocycles*, **24**, 3433 (1986).
[90] Karady, S., Amato, J. S., Reamer, R. A., and Weinstock, L. M., *Heterocycles*, **24**, 1193 (1986).
[91] Kubota, S., Toyooka, K., Yamamoto, N., Kasai, T., and Shibuya, M., *Heterocycles*, **24**, 21 (1986).
[92] Moderhack, D. and Lembcke, A., *Chem.-Ztg.*, **109**, 432 (1985); *Chem. Abs.*, **106**, 18433 (1987).
[93] Allen, C., Boeyens, J. C. A., Briggs, A. G., Denner, L., Markwell, A. J., Reid, D. H., and Rose, B. G., *J. Chem. Soc., Chem. Commun.*, **1987**, 967.
[94] Terent'eva, S. A., Pudovik, M. A., and Pudovik, A. N., *Zh. Obshch. Khim.*, **56**, 712 (1986); *Chem. Abs.*, **106**, 18702 (1987).
[95] Awaya, H., May, E. L., Aceto, M. D., Harris, L. S., Silverton, J. V., Rice, K. C., Mattson, M. V., and Jacobson, A. E., *J. Med. Chem.*, **30**, 947 (1987).
[96] Kim, D. H., *J. Heterocycl. Chem.*, **23**, 1471 (1986).
[97] Sepiol, J. and Tomasik, P., *Acta Chim. Hung.*, **121**, 333 (1986); *Chem. Abs.*, **107**, 96556 (1987).
[98] Munavalli, S., Hsu, F. L., and Pozcomek, E. J., *Heterocycles*, **26**, 23 (1987).
[99] Pragst, F., Köpnick, T., and Boche, E., *Z. Chem.*, **27**, 289 (1987).
[100] Katritzky, A. R., Aurrecoechea, J. M., Quian, K., Koziol, A. E., and Palenik, G. J., *Heterocycles*, **25**, 387 (1987).
[101] Tikk, I., Deák, G., Sohár, P., and Tamás, J., *J. Chem. Res. (S)*, **1987**, 95.
[102] Armesto, D., Horspool, W. M., Martin, N., Ramos, A., and Seoane, C., *J. Chem. Soc., Chem. Commun.*, **1987**, 1231.
[103] Korobka, I. V. and Kuznetsov, E. V., *Khim. Geterotsikl. Soedin.*, **1986**, 276; *Chem. Abs.*, **106**, 32507 (1987).

[104] Facklam, T., Hoffmann, K.-L., and Regitz, M., *Chem. Ber.*, **120**, 1397 (1987).
[105] Yadav, N. S. and Gupta, S. C., *Tetrahedron Lett.*, **28**, 2049 (1987).
[106] Moskalenko, G. G., Sedova, V. F., and Manaev, V. P., *Izv. Akad. Nauk SSSR, Ser. Khim.*, **1987**, 701; *Chem. Abs.*, **107**, 217582 (1987).
[107] Cho, H., Ohnaka, Y., Hayashimatsu, M., Ueda, M., and Shima, K., *Tetrahedron Lett.*, **27**, 6377 (1986).
[108] Korbonits, D. and Kolonits, P., *J. Chem. Soc., Perkin Trans. 1*, **1986**, 2163.
[109] Rykowski, A. and van der Plas, H. C., *J. Org. Chem.*, **52**, 71 (1987).
[110] Bitha, P., Hlavka, J. J., and Lin, Y., *J. Org. Chem.*, **52**, 2220 (1987).
[111] Takahashi, M. and Kikuchi, H., *Tetrahedron Lett.*, **28**, 2139 (1987).
[112] Letcher, R. M., Ng, K.-W., and Cheung, K.-K., *J. Chem. Soc., Chem. Commun.*, **1987**, 1602.
[113] Streef, J. W. and van der Plas, H. C., *Heterocycles*, **26**, 685 (1987).
[114] Lokshin, V. A., Mezheritskii, V. V., Minkin, V. I., Aldoshin, S. M., Atovmyan, L. O., Knyazhanskii, M. I., Kozina, O. A., and Zhukovskaya, O. I., *Dokl. Akad. Nauk SSSR*, **291**, 639 (1986); *Chem. Abs.*, **107**, 175859 (1987).
[115] Yogi, S., Hokama, K., and Tsuge, O., *Chem. Lett.*, **1987**, 157.
[116] Takagi, K., Aotsuka, T., Morita, H., and Okamoto, Y., *Synthesis*, **1987**, 379.
[117] Szmuszkovicz, J., Chidester, C. G., Laurian, L. G., and Scahill, T. A., *J. Org. Chem.*, **51**, 5001 (1986).
[118] Wamhoff, H. and Ma, J., *Huaxue Xuebao*, **45**, 172 (1987); *Chem. Abs.*, **107**, 176009 (1987).
[119] Yamamoto, K., Yamazaki, S., Murata, I., and Fukazawa, Y., *J. Org. Chem.*, **52**, 5239 (1987).
[120] Carless, H. A. J., *Spec. Publ. - R. Soc. Chem.*, **57**, 118 (1986); *Chem. Abs.*, **106**, 195568 (1987).
[121] Meng, J.-B., Matsuura, T., and Ito, Y., *Youji Huaxue*, **1986**, 223; *Chem. Abs.*, **106**, 101561 (1987).
[122] Schultz, A. G., Lavieri, F. P., Macielag, M., and Plummer, M., *J. Am. Chem. Soc.*, **109**, 3991 (1987).
[123] Schultz, A. G., Harrington, R. E., Macielag, M., Mehta, P. G., and Taveras, A. G., *J. Org. Chem.*, **52**, 5482 (1987).
[124] Schultz, A. G., Staib, R. R., and Eng, K. K., *J. Org. Chem.*, **52**, 2968 (1987).
[125] Schultz, A. G., McMahon, W. G., Staib, R. R., and Kullnig, R. K., *Tetrahedron Lett.*, **28**, 4929 (1987).
[126] Barreau, M. and Ponsinet, G., *Synthesis*, **1987**, 262.
[127] Reinaud, O., Capdevielle, P., and Maumy, M., *Tetrahedron*, **43**, 4167 (1987).
[128] Cassis, R., Scholz, M., Tapia, R., and Valderrama, J. A., *J. Chem. Soc., Perkin Trans. 1*, **1987**, 2855.
[129] Satoh, J. Y., Haruta, A. M., Satoh, T., Satoh, K., and Takahashi, T. T., *J. Chem. Soc., Chem. Commun.*, **1986**, 1765.
[130] Planas, A., Tomás, J., and Bonet, J.-J., *Tetrahedron Lett.*, **28**, 471 (1987).
[131] Greenland, H., Kozyrod, R. P., and Pinhey, J. T., *J. Chem. Soc., Perkin Trans. 1*, **1986**, 2011.
[132] Schuster, D. I. and Brisimitzakis, A. C., *J. Org. Chem.*, **52**, 3644 (1987).
[133] Miller, B. and Baghdadchi, J., *J. Org. Chem.*, **52**, 3390 (1987).
[134] Fariña, F., Paredes, M. C., and Stefani, V., *Tetrahedron Lett.*, **28**, 4109 (1987).
[135] Carman, R. M. and Van Dongen, J. M. A. M., *Aust. J. Chem.*, **39**, 2171 (1986).
[136] Blazejewski, J.-C., Dorme, R., and Wakselman, C., *J. Chem. Soc., Perkin Trans. 1*, **1987**, 1861.
[137] Matsuura, T., Meng, J.-B., Ito, Y., Irie, M., and Fukuyama, K., *Tetrahedron*, **43**, 2451 (1987).
[138] Zefirov, N. S., *Acc. Chem. Res.*, **20**, 237 (1987).
[139] Agmon, N., *J. Org. Chem.*, **52**, 2192 (1987).
[140] Nakai, T., *Petrotech (Tokyo)*, **10**, 58 (1987); *Chem. Abs.*, **107**, 153612 (1987).
[141] Parker, K. A., *Nat. Prod. Chem., Proc. Int. Symp. Pak.-US Binatl. Workshop, 1st*, 307 (1984) (Pub. 1986); *Chem. Abs.*, **107**, 134539 (1987).
[142] Grierson, L., Perkins, M. J., and Rzepa, H. S., *J. Chem. Soc., Chem. Commun.*, **1987**, 1779.
[143] Gajewski, J. J., Jurayj, J., Kimbrough, D. R., Gande, M. E., Ganem, B., and Carpenter, B. K., *J. Am. Chem. Soc.*, **109**, 1170 (1987).
[144] Copley, S. D. and Knowles, J. R., *J. Am. Chem. Soc.*, **109**, 5008 (1987).
[145] Coates, R. M., Rogers, B. D., Hobbs, S. J., Peck, D. R., and Curran, D. P., *J. Am. Chem. Soc.*, **109**, 1160 (1987).
[146] Pratt, D. V. and Hopkins, P. B., *Tetrahedron Lett.*, **28**, 3065 (1987).
[147] Van der Baan, J. L. and Bickelhaupt, F., *Tetrahedron Lett.*, **27**, 6267 (1986).
[148] Mikami, K., Takahashi, K., and Nakai, T., *Tetrahedron Lett.*, **28**, 5879 (1987).
[149] Shimizu, M., Nakahara, Y., Kanemoto, S., and Yoshioka, H., *Tetrahedron Lett.*, **28**, 1677 (1987).
[150] Ziegler, F. E., Nangia, A., and Schulte, G., *J. Am. Chem. Soc.*, **109**, 3987 (1987).
[151] Matlin, A. R., Leckta, T. C., McGarvey, D. J., Jacob, P. W., and Picken, H. A., *Tetrahedron Lett.*, **28**, 5083 (1987).
[152] Barluenga, J., Aznar, F., Liz, R., and Bayod, M., *J. Org. Chem.*, **52**, 5190 (1987).

[153] Takeshita, H., Mori, A., Goto, Y., and Nagao, T., *Bull. Chem. Soc. Jpn*, **60**, 1747 (1987).
[154] Anjaneyulu, A. S. R. and Mallavadhani, U. V., *Indian J. Chem.*, **25B**, 515 (1986).
[155] Benmair, R. M. J. and Yogev, A., *Mol. Phys.*, **60**, 717 (1987).
[156] Goto, M., Hayashi, T., Goda, K., and Okada, Y., *Nippon Kagaku Kaishi*, **1986**, 1096; *Chem. Abs.*, **106**, 195671 (1987).
[157] Barrett, A. G. M. and Carpenter, N. E., *Organometallics*, **6**, 2249 (1987).
[158] Maruyama, K., Nagai, N., and Naruta, Y., *J. Org. Chem.*, **51**, 5083 (1986).
[159] Chattopadhyay, S. K., Sen, B. K., and Majumdar, K. C., *Indian J. Chem.*, **25B**, 56 (1986).
[160] Harwood, L. M., Oxford, A. J., and Thomson, C., *J. Chem. Soc., Chem. Commun.*, **1987**, 1615.
[161] Patra, A., Mukhopadhyay, A. K., and Mitra, A. K., *Indian J. Chem.*, **25B**, 1167 (1986).
[162] Cairns, N., Harwood, L. M., and Astles, D. P., *J. Chem. Soc., Chem. Commun.*, **1987**, 400.
[163] Sunitha, K. and Balasubramanian, K. K., *Tetrahedron*, **43**, 3269 (1987).
[164] Gopal, D. and Rajagopalan, K., *Tetrahedron Lett.*, **28**, 5327 (1987).
[165] Eichinger, P. C. H., Bowie, J. H., and Hayes, R. N., *J. Org. Chem.*, **52**, 5224 (1987).
[166] Pogrebnoi, S. I., Kalyan, Y. B., Krimer, M. Z., and Smit, W. A., *Tetrahedron Lett.*, **28**, 4893 (1987).
[167] Uchiyama, H., Kawano, M., Katsuki, T., and Yamaguchi, M., *Chem. Lett.*, **1987**, 351.
[168] Barrish, J. C., Lee, H. L., Baggiolini, E. G., and Uskokovic, M. R., *J. Org. Chem.*, **52**, 1372 (1987).
[169] Gould, T. J., Balestra, M., Wittman, M. D., Gary, J. A., Rossano, L. T., and Kallmerten, J., *J. Org. Chem.*, **52**, 3889 (1987).
[170] Sato, T., Tanaka, A., Tajima, K., and Fujisawa, T., *Chem. Lett.*, **1987**, 1979.
[171] Shishido, K., Hiroya, K., Komatsu, H., Fukumoto, K., and Kametani, T., *J. Chem. Soc., Perkin Trans. 1*, **1987**, 2491.
[172] Curran, D. P., Jacobs, P. B., Elliott, R. L., and Kim, B. H., *J. Am. Chem. Soc.*, **109**, 5280 (1987).
[173] Burke, S. D., Schoenen, F. J., and Nair, M. S., *Tetrahedron Lett.*, **28**, 4143 (1987).
[174] Burke, S. D., Chandler, A. C., Nair, M. S., and Campopiano, O., *Tetrahedron Lett.*, **28**, 4147 (1987).
[175] Cooper, J., Knight, D. W., and Gallagher, P. T., *Tetrahedron Lett.*, **28**, 3031 (1987).
[176] Cooper, J., Knight, D. W., and Gallagher, P. T., *J. Chem. Soc., Chem. Commun.*, **1987**, 1220.
[177] Marino, J. P. and Kim, M. W., *Tetrahedron Lett.*, **28**, 4925 (1987).
[178] Lidbetter, P. S. and Marples, B. A., *Synth. Commun.*, **16**, 1529 (1986).
[179] Davidson, A. H. and Wallace, I. H., *J. Chem. Soc., Chem. Commun.*, **1986**, 1759.
[180] Suzuki, T., Sato, E., Unno, K., and Kametani, T., *J. Chem. Soc., Perkin Trans. 1*, **1986**, 2263.
[181] Tadano, K., Idogaki, Y., Yamada, H., and Suami, T., *J. Org. Chem.*, **52**, 1201 (1987).
[182] Yadav, J. S., Kulkarni, A. D., and Reddy, P. S., *Indian J. Chem.*, **25B**, 1220 (1986).
[183] Paquette, L. A. and Ham, W. H., *J. Am. Chem. Soc.*, **109**, 3025 (1987).
[184] Daub, G. W. and Griffith, D. A., *Tetrahedron Lett.*, **27**, 6311 (1986).
[185] Schuda, P. F., Potlock, S. J., and Ziffer, H., *Tetrahedron*, **43**, 463 (1987).
[186] Baeckström, P., Stridh, K., Li, L., and Norin, T., *Acta Chem. Scand.*, **41B**, 442 (1987).
[187] Yamaguchi, Y., Hayakawa, K., and Kanematsu, K., *J. Chem. Soc., Chem. Commun.*, **1987**, 515.
[188] Liotta, D., Brown, D., Hoekstra, W., and Monahan, R., *Tetrahedron Lett.*, **28**, 1069 (1987).
[189] Egert, E., Beck, H., Schmidt, D., Gonschorrek, C., and Hoppe, D., *Tetrahedron Lett.*, **28**, 789 (1987).
[190] Baldwin, J. E., Bennett, P. A. R., and Forrest, A. K., *J. Chem. Soc., Chem. Commun.*, **1987**, 250.
[191] Rehman, H. and Rao, J. M., *Tetrahedron*, **43**, 5335 (1987).
[192] Yadla, R. and Rao, J. M., *Heterocycles*, **26**, 329 (1987).
[193] Dangyan, Y. M., Turakyan, M. R., Panosyan, G. A., and Badanyan, S. O., *Zh. Obshch. Khim.*, **56**, 1517 (1986); *Chem. Abs.*, **106**, 156579 (1987).
[194] Vendezonneville, A., Flammang, R., Maquestian, A., Kingston, E., and Beynon, J. H., *Org. Mass Spectrom.*, **21**, 351 (1986).
[195] Abdrakhmanov, I. B., Saraeva, Z. N., Nigmatullin, N. G., and Tolstikov, G. A., *Izv. Akad. Nauk SSSR, Ser. Khim.*, **1986**, 383; *Chem. Abs.*, **105**, 225561 (1987).
[196] Kandeel, K. A. and Vernon, J. M., *J. Chem. Soc., Perkin Trans. 1*, **1987**, 2023.
[197] Kurth, M. J. and Soares, C. J., *Tetrahedron Lett.*, **28**, 1031 (1987).
[198] Baudin, J.-B., Julia, S. A., and Lorne, R., *Bull. Soc. Chim. Fr. II*, **1987**, 181.
[199] Baudin, J.-B., Julia, S. A., and Ruel, O., *Tetrahedron*, **43**, 881 (1987).
[200] Shah, K. K. and Shah, A., *Indian J. Chem.*, **24B**, 1077 (1985).
[201] Chambers, M. S., Thomas, E. J., and Williams, D. J., *J. Chem. Soc., Chem. Commun.*, **1987**, 1228.
[202] Takano, S., Tomita, S., Takahashi, M., and Ogasawara, K., *Chem. Lett.*, **1987**, 1379.
[203] Takahata, H., Banba, Y., Mozumi, M., and Yamazaki, T., *Heterocycles*, **24**, 3347 (1986).
[204] Plate, R., Theunisse, A. W. G., and Ottenheijm, H. C. J., *J. Org. Chem.*, **52**, 370 (1987).
[205] Zenkevich, I. G., Isidorov, V. A., and Ioffe, B. V., *Zh. Org. Khim.*, **22**, 1797 (1986); *Chem. Abs.*, **107**, 115140 (1987).

[206] Vysotskii, Y. B., Przheval'skii, N. M., Zemskii, B. P., Grandberg, I. I., and Kostromina, L. Y., *Khim. Geterotsikl. Soedin.*, **1986**, 898; *Chem. Abs.*, **107**, 197253 (1987).
[207] Dewar, M. J. S. and Merz, K. M., *J. Am. Chem. Soc.*, **109**, 6553 (1987).
[208] Poupko, R., Luz, Z., Vega, A., and Zimmermann, H., *J. Chem. Phys.*, **86**, 5358 (1987).
[209] Dewar, M. J. S. and Jie, C., *J. Chem. Soc., Chem. Commun.*, **1987**, 1451.
[210] Dewar, M. J. S. and Jie, C., *J. Am. Chem. Soc.*, **109**, 5893 (1987).
[211] Gajewski, J. J., Benner, C. W., and Hawkins, C. M., *J. Org. Chem.*, **52**, 5198 (1987).
[212] Dolbier, W. R., Alty, A. C., and Phanstiel, O., *J. Am. Chem. Soc.*, **109**, 3046 (1987).
[213] Baldwin, J. E., Andrews, G. D., and Parker, D. W., *J. Org. Chem.*, **52**, 676 (1987).
[214] Piers, E., Jung, G. L., and Ruediger, E. H., *Can. J. Chem.*, **65**, 670 (1987).
[215] Piers, E. and Jung, G. L., *Can. J. Chem.*, **65**, 1668 (1987).
[216] Davies, H. M. L., Smith, H. D., and Korkor, O., *Tetrahedron Lett.*, **28**, 1853 (1987).
[217] McMurry, J. E. and Swenson, R., *Tetrahedron Lett.*, **28**, 3209 (1987).
[218] Eisen, N. and Vögtle, F., *Angew. Chem. Int. Edn*, **25**, 1026 (1986).
[219] Blasberyer, D., Green, D., Carmely, S., Spector, I., and Kashman, Y., *Tetrahedron Lett.*, **28**, 459 (1987).
[220] Raucher, S., Hwang, K.-W. C. K.-J., and Burks, J. E., *J. Org. Chem.*, **51** 5503 (1986).
[221] Askin, D., Angst, C., and Danishefsky, S., *J. Org. Chem.*, **52**, 622 (1987).
[222] Thies, R. W. and Daruwala, K. P., *J. Org. Chem.*, **52**, 3798 (1987).
[223] Wang, T. F. and Ziegler, F. E., *Bull. Inst. Chem., Acad. Sin.*, **33**, 19 (1986); *Chem. Abs.*, **106**, 196668 (1987).
[224] Oplinger, J. A. and Paquette, L. A., *Tetrahedron Lett.*, **28**, 5441 (1987).
[225] Eichinger, P. C. H. and Bowie, J. H., *J. Chem. Soc., Perkin Trans. 2*, **1987**, 1499.
[226] Magnol, E., Gore, J., and Malacria, M., *Bull. Soc. Chim. Fr. II*, **1987**, 455.
[227] Paquette, L. A., Learn, K. S., and Romine, J. L., *Tetrahedron*, **43**, 4989 (1987).
[228] Martin, P., *Tetrahedron Lett.*, **28**, 1645 (1987).
[229] Wilkens, J., Kuhling, A., and Blechert, S., *Tetrahedron*, **43**, 3237 (1987).
[230] Beck, K. and Hunig, S., *Angew. Chem. Int. Edn*, **26**, 670 (1987).
[231] Beck, K., Burghard, H., Fischer, G., Hunig, S., and Reinold, P., *Angew. Chem. Int. Edn*, **26**, 672 (1987).
[232] Appel, R., Winkhaus, V., and Knoch, F., *Chem. Ber.*, **120**, 125 (1987).
[233] Appel, R., Kochta, J., and Knoch, F., *Chem. Ber.*, **120**, 131 (1987).
[234] Li, J. S. and Sha, C.-K., *K'o Hsueh Fa Chan Yueh K'an*, **14**, 1253 (1986); *Chem. Abs.*, **107**, 23116 (1987).
[235] Uchikawa, M., Kuroda, S., Katsuki, T., and Yamaguchi, M., *Tennen Yuki Kagobutsu Toronkai Koen Yoshishu, 27th*, 62 (1985); *Chem. Abs.*, **106**, 18174 (1987).
[236] Takahashi, O., Mikami, K., and Nakai, T., *Chem. Lett.*, **1987**, 69.
[237] Marshall, J. A., Lebreton, J., DeHoff, B. S., and Jenson, T. M., *Tetrahedron Lett.*, **28**, 723 (1987).
[238] Marshall, J. A., Jenson, T. M., and DeHoff, B. S., *J. Org. Chem.*, **52**, 3860 (1987).
[239] Marshall, J. A., Lebreton, J., DeHoff, B. S., and Jenson, T. M., *J. Org. Chem.*, **52**, 3883 (1987).
[240] Marshall, J. A. and Lebreton, J., *Tetrahedron Lett.*, **28**, 3323 (1987).
[241] Takahashi, T., Nemoto, H., Kanda, Y., Tsuji, J., Fukazawa, Y., Okajima, T., and Fujise, Y., *Tetrahedron*, **43**, 5499 (1987).
[242] Crimmins, M. T., Hollis, W. G., and Lever, J. G., *Tetrahedron Lett.*, **28**, 3647 (1987).
[243] Sato, K., Inoue, S., Ozawa, K., Kobayashi, T., Ota, T., and Tazaki, M., *J. Chem. Soc., Perkin Trans. 1*, **1987**, 1753.
[244] Inoue, S., Ikeda, H., Sato, S., Horie, K., Ota, T., Miyamoto, O., and Sato, K., *J. Org. Chem.*, **52**, 5495 (1987).
[245] Brooke, G. M. and Ferguson, J. A. K. J., *J. Chem. Soc., Perkin Trans. 1*, **1987**, 2091.
[246] de March, P., Moreno-Mañas, M., and Ripoll, I., *Chem. Ber.*, **120**, 1413 (1987).
[247] Takano, S., Tomita, S., Takahashi, M., and Ogasawara, K., *Chem. Lett.*, **1987**, 1569.
[248] Brown, W. L. and Fallis, A. G., *Can. J. Chem.*, **65**, 1828 (1987).
[249] Knight, D. J., Whitham, G. H., and Williams, J. G., *J. Chem. Soc., Perkin Trans. 1*, **1987**, 2149.
[250] Sashida, H. and Tsuchiya, T., *Chem. Pharm. Bull.*, **34**, 3682 (1986).
[251] Natsugari, H., Turos, E., Weinreb, S. M., and Cvetovich, R. J., *Heterocycles*, **25**, 19 (1987).
[252] Cotelle, P., Hasiak, B., Barbry, D., and Couturier, D., *Chem. Lett.*, **1987**, 1005.
[253] Nakano, M. and Sato, Y., *J. Org. Chem.*, **52**, 1844 (1987).
[254] Stévenast-De Mesmaeker, N., Merényi, R., and Viehe, H. G., *Tetrahedron Lett.*, **28**, 2591 (1987).
[255] Majumdar, K. C. and Chattopadhyay, S. K., *J. Chem. Soc., Chem. Commun.*, **1987**, 524.
[256] Inoue, S., Iwase, N., Mujamoto, O., and Sato, K., *Chem. Lett.*, **1986**, 2035.
[257] Pollok, T. and Schmidbaur, H., *Tetrahedron Lett.*, **28**, 1085 (1987).

[258] Poirier, R. A., Majlessi, D., and Zielinski, T. J., *J. Comput. Chem.*, **7**, 464 (1986); *Chem. Abs.*, **106**, 83695 (1987).
[259] Takahashi, K., Suzuki, T., Toda, H., Takase, K., Koseki, S., and Nakajima, T., *J. Org. Chem.*, **52**, 2666 (1987).
[260] Trifonov, L. S., Orahovats, A. S., and Heimgartner, H., *Helv. Chim. Acta*, **70**, 1070 (1987).
[261] Rodriguez, J., Brun, P., and Waegell, B., *J. Organomet. Chem.*, **333**, C25 (1987).
[262] de Mayo, P. and Wenska, G., *Tetrahedron*, **43**, 1661 (1987).
[263] Reynolds, C. A. and Thomson, C., *J. Chem. Soc., Faraday Trans. 2*, **83**, 485 (1987).
[264] Henning, H.-G. and Hentschel, C., *Z. Chem.*, **27**, 339 (1987).
[265] Lebrilla, C. B., and Schwarz, H., *J. Chem. Soc., Perkin Trans. 2*, **1987**, 237.
[266] Carlsen, P. H. J., *Acta Chem. Scand.*, **41B**, 302 (1987).
[267] Clemans, G. B. and Blaho, J. K., *J. Org. Chem.*, **52**, 1621 (1987).
[268] Paquette, L. A., Pierre, F., and Cottrell, C. E., *J. Am. Chem. Soc.*, **109**, 5731 (1987).
[269] Tanaka, N., Kumagai, T., and Mukai, T., *Tetrahedron Lett.*, **27**, 6221 (1986).
[270] Caine, D., McCloskey, C. J., Atwood, J. L., Bott, S. G., Zhang, H.-M., and VanDerveer, D., *J. Org. Chem.*, **52**, 1280 (1987).
[271] Liotta, D., Saindane, M., Monahan, R., Brothers, D., and Fivush, A., *Synth. Commun.*, **16**, 1461 (1986).
[272] Braverman, S., Reisman, D., and Freund, M., *Tetrahedron Lett.*, **28**, 5189 (1987).
[273] Severengiz, T. and du Mont, W.-W., *J. Chem. Soc., Chem. Commun.*, **1987**, 820.
[274] Modro, T. A., Rzepa, H. S., and Symes, J., *J. Chem. Soc., Perkin Trans. 2*, **1987**, 767.
[275] Calogeropoulou, T., Hammond, G. B., and Wiemer, D. F., *J. Org. Chem.*, **52**, 4185 (1987).
[276] Ghosh, S., Das, T. K., Datta, D. B., and Mehta, S., *Tetrahedron Lett.*, **28**, 4611 (1987).
[277] Baumann, H., Cometta-Morini, C., and Oth, J. F. M., *THEOCHEM*, **31**, 229 (1986); *Chem. Abs.*, **105**, 225480 (1986).
[278] Kim, H. S. and Lee, I., *Taehan Hwahakhoe Chi*, **30**, 195 (1986); *Chem. Abs.*, **106**, 17613 (1987).
[279] Jensen, F. and Houk, K. N., *J. Am. Chem. Soc.*, **109**, 3139 (1987).
[280] Randles, K. R. and Storr, R. C., *Tetrahedron Lett.*, **28**, 5555 (1987).
[281] Shen, G.-Y., de Lera, A. R., Norman, T. C., Haces, A., and Okamura, W. H., *Tetrahedron Lett.*, **28**, 2917 (1987).
[282] Gibbs, R. A. and Okamura, W. H., *Tetrahedron Lett.*, **28**, 6021 (1987).
[283] Shen, G.-Y., Tapia, R., and Okamura, W. H., *J. Am. Chem. Soc.*, **109**, 7499 (1987).
[284] Galvez, C., Fernandez, I., Segura, R., and Vazquez, J., *J. Chem. Res. (S)*, **1987**, 16.
[285] Nijhuis, W. H. N., Verboom, W., Reinhoudt, D. N., and Harkema, S., *J. Am. Chem. Soc.*, **109**, 3136 (1987).
[286] Shea, K. J. and Burke, L. D., *Tetrahedron Lett.*, **28**, 735 (1987).
[287] Ip, S.-H. and Sammes, M. P., *J. Chem. Res. (S)*, **1987**, 330.
[288] Chiu, P.-K. and Sammes, M. P., *Tetrahedron Lett.*, **28**, 2775 (1987).
[289] Battersby, A. R., Baker, M. G., Broadbent, H. A., Fookes, C. J. R., and Leeper, F. J., *J. Chem. Soc., Perkin Trans. 1*, **1987**, 2027.
[290] Dabbagh, A. M. and Al-Hamdany, R., *Int. J. Chem. Kinet.*, **19**, 49 (1987).
[291] Hayakawa, K., Aso, M., and Kanematsu, K., *J. Chem. Soc., Chem. Commun.*, **1987**, 102.
[292] Armesto, D., Horspool, W. M., Langa, F., and Perez-Ossorio, R., *J. Chem. Soc., Perkin Trans. 2*, **1987**, 1039.
[293] Paquette, L. A., Dressel, J., and Pansegrau, P. D., *Tetrahedron Lett.*, **28**, 4965 (1987).
[294] Ustynyuk, Y. A., Shestakova, A. K., Chertkov, V. A., Zemlyansky, N. N., Borisova, I. V., Gusev, A. I., Tchuklanova, E. B., and Chernyshev, E. A., *J. Organomet. Chem.*, **335**, 43 (1987).
[295] Chen, Z. and Zhang, Y., *Wuli Huaxue Xuebao*, **3**, 4 (1987); *Chem. Abs.*, **107**, 133697 (1987).
[296] Bushby, R. J. and Mann, S., *J. Chem. Soc., Perkin Trans. 1*, **1987**, 2497.
[297] Hoeger, C. A., Johnston, A. D., and Okamura, W. H., *J. Am. Chem. Soc.*, **109**, 4690 (1987).
[298] Sugiyama, S., Mori, A., and Takeshita, H., *Chem. Lett.*, **1987**, 1247.
[299] Battye, P. J. and Jones, D. W., *J. Chem. Soc., Chem. Commun.*, **1986**, 1807.
[300] Porter, N. A. and Wujek, J. S., *J. Org. Chem.*, **52**, 5085 (1987).
[301] Skancke, P. N., Yamashita, K., and Morokuma, K., *J. Am. Chem. Soc.*, **109**, 4157 (1987).
[302] Boyd, D. R., Agarwal, S. K., Balani, S. K., Dunlop, R., Gadaginamath, G. S., O'Kane, G. A., Sharma, N. D., Jennings, W. B., Yagi, H., and Jerina, D. M., *J. Chem. Soc., Chem. Commun.*, **1987**, 1633.
[303] Boyd, D. R. and O'Kane, G. A., *Tetrahedron Lett.*, **28**, 6395 (1987).
[304] Black, T. H. and DuBay, W. J., *Tetrahedron Lett.*, **28**, 4787 (1987).
[305] Chow, T. J. and Ding, M.-F., *Angew. Chem. Int. Edn*, **25**, 1121 (1986).

306 Atkinson, R. S., *Annu. Rep. Prog. Chem., Sect. B, 1985* (Pub. 1986), **82**, 37; *Chem. Abs.*, **106**, 195577 (1987).
307 Meier, H., Pauli, A., Kolshorn, H., and Kochhan, P., *Chem. Ber.*, **120**, 1607 (1987).
308 Pauli, A., Kolshorn, H., and Meier, H., *Chem. Ber.*, **120**, 1611 (1987).
309 Escher, A. and Neuenschwander, M., *Helv. Chim. Acta*, **70**, 49 (1987).
310 Liu, C.-Y. and Houk, K. N., *Tetrahedron Lett.*, **28**, 1367 (1987).
311 Gilchrist, T. L. and Stanford, J. E., *J. Chem. Soc., Perkin Trans. 1*, **1987**, 225.
312 Perri, S. T. and Moore, H. W., *Tetrahedron Lett.*, **28**, 4507 (1987).
313 Seitz, G. and van Gemmern, R., *Synthesis*, **1987**, 953.
314 Kawase, T., Iyoda, M., and Oda, M., *Angew. Chem. Int. Edn*, **26**, 559 (1987).
315 Münzel, N. and Schweig, A., *Angew. Chem. Int. Edn*, **26**, 471 (1987).
316 Fink, J., Gümbel, H., Eisenbarth, P., and Regitz, M., *Chem. Ber.*, **120**, 1027 (1987).
317 McNab, H. and Monahan, L. C., *J. Chem. Soc., Chem. Commun.*, **1987**, 141.
318 Coqueret, X., Bourelle-Wargnier, F., Chuche, J., and Toubet, L., *Bull. Soc. Chim. Fr. II*, **1987**, 365.
319 Hoffmann, K.-L., Maas, G., and Regitz, M., *J. Org. Chem.*, **52**, 3851 (1987).
320 Boyd, G. V., Cobb, J., Lindley, P. F., Mitchell, J. C., and Nicolaou, G. A., *J. Chem. Soc., Chem. Commun.*, **1987**, 99.
321 Schwan, A. L. and Warkentin, J., *Can. J. Chem.*, **65**, 1200 (1987).
322 Ardill, H., Grigg, R., Sridharan, V., and Malone, J., *J. Chem. Soc., Chem. Commun.*, **1987**, 1296.
323 Hofmann, H. and Djafari, H., *Liebigs Ann. Chem.*, **1987**, 505.
324 Saito, I., Morii, T., Okumura, Y., Mori, S., Yamaguchi, K., and Matsuura, T., *Tetrahedron Lett.*, **27**, 6385 (1986).
325 Sheradsky, T. and Itzhak, N., *J. Chem. Soc., Perkin Trans. 1*, **1987**, 1979.
326 Heindirk, T. D. and Zettlitzer, M., *Chem. Ber.*, **120**, 795 (1987).
327 Mori, A. and Takeshita, H., *Bull. Chem. Soc. Jpn*, **60**, 3037 (1987).
328 Bentrude, W. G., Lee, S.-G., Akutagawa, K., Ye. C., and Charbonnel, Y., *J. Am. Chem. Soc.*, **109**, 1577 (1987).
329 Blatter, K., Rösch, W., Vogelbacher, U.-J., Fink, J., and Regitz, M., *Angew. Chem. Int. Edn*, **26**, 85 (1987).
330 Märkl, G., Beckh, H. J., Ziegler, M. L., and Nuber, B., *Angew. Chem. Int. Edn*, **26**, 1134 (1987).
331 Loncharich, R. J. and Houk, K. N., *J. Am. Chem. Soc.*, **109**, 6947 (1987).
332 Marples, B. A. and Spilling, C. D., *Tetrahedron Lett.*, **28**, 581 (1987).
333 White, J. D. and Somers, T. C., *J. Am. Chem. Soc.*, **109**, 4424 (1987).
334 Armesto, D., Horspool, W. M., Pérez-Ossorio, R., and Ramos, A., *J. Org. Chem.*, **52**, 3378 (1987).
335 Wulff, G., Patt, H., and Wichelhaus, J., *Nouv. J. Chim.*, **10**, 143 (1986).
336 Yamamoto, Y., *Chem. Lett.*, **1987**, 945.
337 Sakai, S. and Morokuma, K., *J. Phys. Chem.*, **91**, 3661 (1987).
338 Li, W.-K., Nobes, R. H., and Radom, L., *THEOCHEM*, **34**, 67 (1987); *Chem. Abs.*, **107**, 58186 (1987).
339 Abram, S. R. and Shaw, A. C., *J. Org. Chem.*, **52**, 1835 (1987).
340 Takano, S., Sekiguchi, Y., Shimazaki, Y., Takahashi, M., and Ogasawara, K., *Chem. Lett.*, **1987**, 1905.
341 Cheng, Y. S. P., Garratt, P. J., Neoh, S. B., and Rumjanek, V. M., *Isr. J. Chem.*, **26**, 101 (1985); *Chem. Abs.*, **106**, 49433 (1987).
342 Trofimov, B. A., Musorin, G. K., Amosova, S. V., and Keiko, V. V., *Sulfur Lett.*, **4**, 125 (1986); *Chem. Abs.*, **106**, 213237 (1987).
343 Eisch, J. J., Kovacs, C. A., Chobe, P., and Boleslawski, M. P., *J. Org. Chem.*, **52**, 4427 (1987).
344 Eisch, J. J., Dua, S. K., and Kovacs, C. A., *J. Org. Chem.*, **52**, 4437 (1987).
345 Jonsäll, G. and Ahlberg, P., *J. Chem. Soc., Perkin Trans. 2*, **1987**, 461.
346 Gerson, F., Knöbel, J., Metzger, A., Scott, L. T., Kirms, M. A., Oda, M., and Sumpter, C. A., *J. Am. Chem. Soc.*, **108**, 7920 (1986).
347 Maercker, A., *Angew. Chem. Int. Edn*, **26**, 972 (1987).
348 Eichinger, P. C. H., Bowie, J. H., and Bleementhal, T., *J. Org. Chem.*, **51**, 5078 (1986).
349 Schreiber, S. L. and Goulet, M. T., *Tetrahedron Lett.*, **28**, 1043 (1987).
350 Gleiter, R., Schaaff, H. P., Huber-Patz, U., Rodewald, H., Goetzmann, W., and Irngartinger, H., *J. Org. Chem.*, **52**, 3979 (1987).
351 Takeshita, H., Mori, A., and Suizu, H., *Bull. Chem. Soc. Jpn*, **60**, 1429 (1987).
352 Hayes, R. N., Sheldon, J. C., and Bowie, J. H., *Int. J. Mass Spectrom. Ion Processes*, **71**, 233 (1986); *Chem. Abs.*, **106**, 137811 (1987).
353 Pollack, R. M., Mack, J. P. G., and Eldin, S., *J. Am. Chem. Soc.*, **109**, 5048 (1987).
354 Bowden, K. and Chehel-Amiran, M., *J. Chem. Soc., Perkin Trans. 2*, **1986**, 2031.
355 Bowden, K. and Chehel-Amiran, M., *J. Chem. Soc., Perkin Trans. 2*, **1986**, 2035.

[356] Bowden, K. and Chehel-Amiran, M., *J. Chem. Soc., Perkin Trans. 2*, **1986**, 2039.
[357] Menicagli, R., Malanga, C., Guidi, M., and Lardicci, L., *Tetrahedron*, **43**, 171 (1987).
[358] Laks, P. E., Hemingway, R. W., and Conner, A. H., *J. Chem. Soc., Perkin Trans. 1*, **1987**, 1875.
[359] Sakai, T., Yamawaki, A., Katayama, T., Okada, H., Utaka, M., and Takeda, A., *Bull. Chem. Soc. Jpn*, **60**, 1067 (1987).
[360] Sakai, T., Amano, E., Miyata, K., Utaka, M., and Takeda, A., *Bull. Chem. Soc. Jpn*, **60**, 1945 (1987).
[361] Sakai, T., Ishikawa, M., Amano, E., Utaka, M., and Takeda, A., *Bull. Chem. Soc. Jpn*, **60**, 2295 (1987).
[362] Itooka, T., Matoba, K., Yamazaki, T., Muraoka, O., and Momose, T., *Chem. Pharm. Bull.*, **34**, 2391 (1986).
[363] Tsuboi, S., Nagae, H., Yamato, H., and Takeda, A., *Bull. Chem. Soc. Jpn*, **60**, 836 (1987).
[364] Maiti, S. B., Chaidhuri, S. R. R., and Chatterjee, A., *Synthesis*, **1987**, 806.
[365] Okazaki, R. and Itoh, Y., *Chem. Lett.*, **1987**, 1575.
[366] Yamada, K., Kishikawa, K., and Yamamoto, M., *J. Org. Chem.*, **52**, 2327 (1987).
[367] Marchand, A. P., Jin, P., Flippen-Anderson, J. L., Gilardi, R., and George, C., *J. Chem. Soc., Chem. Commun.*, **1987**, 1108.
[368] Kocharyan, S. T., Voskanyan, V. S., Razina, T. L., and Babayan, A. T., *Arm. Khim. Zh.*, **39**, 300 (1986); *Chem. Abs.*, **107**, 39336 (1987).
[369] Davies, H. M. L. and Crisco, L. Van T., *Tetrahedron Lett.*, **28**, 371 (1987).
[370] Trost, B. M. and Mikhail, G. K., *J. Am. Chem. Soc.*, **109**, 4124 (1987).
[371] Buchanan, J. G., Edgar, A. R., and Hutchison, R. J., *Carbohydr. Res.*, **164**, 403 (1987).
[372] Page, P. C. B., Rosenthal, S., and Williams, R. V., *Tetrahedron Lett.*, **28**, 4455 (1987).
[373] Maercker, A. and Stötzel, R., *Chem. Ber.*, **120**, 1695 (1987).
[374] Bures, E. J. and Keay, B. A., *Tetrahedron Lett.*, **28**, 5965 (1987).
[375] Calogeropoulou, T., Hammond, G. B., and Wiemer, D. F., *J. Org. Chem.*, **52**, 4185 (1987).
[376] Heinicke, J., Nietzschmann, E., and Tzschach, A., *J. Organomet. Chem.*, **310**, C17 (1986).
[377] Wong, M. W., Yates, B. F., Nobes, R. H., and Radom, L., *J. Am. Chem. Soc.*, **109**, 3181 (1987).
[378] Viruela-Martin, P. M., Nebot-Gil, I., Tomas, F., and Viruela-Martin, R., *J. Comput. Chem.*, **7**, 417 (1986); *Chem. Abs.*, **106**, 101564 (1987).
[379] Viruela-Martin, P. M., Nebot-Gil, I., Viruela-Martin, R., and Planelles, J., *J. Chem. Soc., Perkin Trans. 2*, **1987**, 307.
[380] Thibblin, A., *J. Chem. Soc., Perkin Trans. 2*, **1987**, 1629.
[381] Shainyan, B. A., *Usp. Khim.*, **55**, 942 (1986); *Chem. Abs.*, **106**, 49362 (1987).
[382] Lee, C. C. and Wanigasekera, D., *Can. J. Chem.*, **65**, 933 (1987).
[383] Hanack, M. and Rieth, R., *Chem. Ber.*, **120**, 1659 (1987).
[384] Toda, F. and Tokunaga, Y., *Chem. Lett.*, **1987**, 1299.
[385] Moriarty, R. M., Vaid, R. K., Duncan, M. P., and Vaid, B. K., *Tetrahedron Lett.*, **28**, 2845 (1987).
[386] Lee, S.-G., *J. Chem. Soc., Chem. Commun.*, **1987**, 1115.
[387] Jones, J. H. and Wyatt, P. B., *J. Chem. Res. (S)*, **1987**, 396.
[388] Saeva, F. D., *J. Chem. Soc., Chem. Commun.*, **1987**, 37.
[389] Roth, H. D., *Acc. Chem. Res.*, **20**, 343 (1987).
[390] Clark, T., *J. Am. Chem. Soc.*, **109**, 6838 (1987).
[391] Miyashi, T., Takahashi, Y., Yokogawa, K., and Mukai, T., *J. Chem. Soc., Chem. Commun.*, **1987**, 175.
[392] Yates, B. F. and Radom, L., *J. Am. Chem. Soc.*, **109**, 2910 (1987).
[393] McCann, S. F. and Overman, L. E., *J. Am. Chem. Soc.*, **109**, 6107 (1987).
[394] Flann, C. J. and Overman, L. E., *J. Am. Chem. Soc.*, **109**, 6115 (1987).
[395] Tanino, K., Hatanaka, Y., and Kuwajima, I., *Chem. Lett.*, **1987**, 385.
[396] Larson, G. L., Klesse, R., and Cartledge, F. K., *Organometallics*, **6**, 2250 (1987).
[397] McWhorter, W. W. and Jaun, B., *Helv. Chim. Acta*, **70**, 1095 (1987).
[398] Meier, M. and Rüchardt, C., *Chimia*, **40**, 238 (1986); *Chem. Abs.*, **106**, 32184 (1987).
[399] Meier, M., Müller, B., and Rüchardt, C., *J. Org. Chem.*, **52**, 648 (1987).
[400] Honda, Y., Ori, A., and Tsuchihashi, G., *Bull. Chem. Soc. Jpn*, **60**, 1027 (1987).
[401] Honda, Y. and Tsuchihashi, G., *Bull. Chem. Soc. Jpn*, **60**, 2698 (1987).
[402] Yamauchi, T., Nakao, K., and Fujii, K., *J. Chem. Soc., Perkin Trans. 1*, **1987**, 1255.
[403] Yamauchi, T., Nakao, K., and Fujii, K., *J. Chem. Soc., Perkin Trans. 1*, **1987**, 1433.
[404] Piccolo, O., Spreafico, F., Visentin, G., and Valoti, E., *J. Org. Chem.*, **52**, 10 (1987).
[405] Nongrum, F. M. and Myrboh, B., *Synthesis*, **1987**, 845.
[406] Krief, A. and Laboureur, J. L., *Tetrahedron Lett.*, **28**, 1545 (1987).
[407] Ochiai, M., Iwaki, S., Ukita, T., and Nagao, Y., *Chem. Lett.*, **1987**, 133.
[408] Herrinton, P. H., Hopkins, M. H., Mishra, P., Brown, M. J., and Overman, L. E., *J. Org. Chem.*, **52**, 3711 (1987).

409 Hopkins, M. H. and Overman, L. E., *J. Am. Chem. Soc.*, **109**, 4748 (1987).
410 Whitby, R. and Kocieński, P., *Tetrahedron Lett.*, **28**, 3619 (1987).
411 Jefford, C. W., Rossier, J.-C., and Boukouvalas, J., *J. Chem. Soc., Chem. Commun.*, **1987**, 713.
412 Espinosa, A., Gallo, M. A., Campos, J., and Entrena, A., *Bull. Soc. Chim. Fr. II*, **1987**, 379.
413 Zhdanov, Y. A., Luk'yanov, S. M., Borodaev, S. V., and Borodaeva, S. V., *Zh. Org. Khim.*, **22**, 1173 (1986); *Chem. Abs.*, **107**, 77663 (1987).
414 Borodaev, S. V., Luk'yanov, S. M., and Zhdanov, Y. A., *Dokl. Akad. Nauk SSSR*, **287**, 862 (1986); *Chem. Abs.*, **106**, 176217 (1987).
415 Hagen, J. P., Harris, J. J., and Lakin, D., *J. Org. Chem.*, **52**, 782 (1987).
416 Abraham, W. D., Bhupathy, M., and Cohen, T., *Tetrahedron Lett.*, **28**, 2203 (1987).
417 Padwa, A., Bullock, W. H., and Dyszlewski, A. D., *Tetrahedron Lett.*, **28**, 3193 (1987).
418 Aggarwal, V. K. and Warren, S., *Tetrahedron Lett.*, **28**, 1925 (1987).
419 Schaumann, E., Marr, T., Nimmesgern, H., and Sieveking, S., *Chem. Ber.*, **120**, 335 (1987).
420 Crout, D. H. G. and Rathbone, D. L., *J. Chem. Soc., Chem. Commun.*, **1987**, 290.
421 Crout, D. H. G. and Hadfield, J. A., *J. Chem. Soc., Chem. Commun.*, **1987**, 742.
422 Momose, T., Itooka, T., Nishi, T., Uchimoto, M., Ohnishi, K., and Muraoka, O., *Tetrahedron*, **43**, 3713 (1987).
423 Ford, G. P. and Smith, C. T., *J. Am. Chem. Soc.*, **109**, 1325 (1987).
424 Kim, W. K., Sohn, C. K., and Lee, I., *Bull. Korean Chem. Soc.*, **7**, 279 (1986); *Chem. Abs.*, **106**, 195663 (1987).
425 Coxon, J. M. and Hartshorn, M. P., *Tetrahedron Lett.*, **28**, 1333 (1987).
426 Crotti, P., Dell'Omodarme, G., Ferretti, M., and Macchia, F., *J. Am. Chem. Soc.*, **109**, 1463 (1987).
427 Mayr, H., Koschinsky, R., Will, E., and Bäuml, E., *J. Org. Chem.*, **52**, 1342 (1987).
428 Klix, R. C. and Bach, R. D., *J. Org. Chem.*, **52**, 580 (1987).
429 Shimazaki, M., Hara, H., Suzuki, K., and Tsuchihashi, G., *Tetrahedron Lett.*, **28**, 5891 (1987).
430 Suzuki, K., Miyazawa, M., and Tsuchihashi, G., *Tetrahedron Lett.*, **28**, 3515 (1987).
431 Sarkar, T. K. and Ghosh, S. K., *Tetrahedron Lett.*, **28**, 2061 (1987).
432 Mikhailov, I. E., Minkin, V. I., Klenkin, A. A., and Olekhnovich, L. P., *Zh. Org. Khim.*, **22**, 1331 (1986); *Chem. Abs.*, **106**, 49447 (1987).
433 Borodkin, G. I., Susharin, E. R., and Shubin, V. G., *Izv. Akad. Nauk SSSR, Ser. Khim.*, **1986**, 1687; *Chem. Abs.*, **106**, 155691 (1987).
434 Kanemoto, S., Shimizu, M., and Yoshioka, H., *Tetrahedron Lett.*, **28**, 6313 (1987).
435 Hiroi, K., Nakamura, H., and Anzai, T., *J. Am. Chem. Soc.*, **109**, 1249 (1987).
436 Maier, G., Bauer, I., Born, D., and Kalinowski, H.-O., *Angew. Chem. Int. Edn*, **25**, 1093 (1986).
437 Ronald, R. C., Ruder, S. M., and Lillie, T. S., *Tetrahedron Lett.*, **28**, 131 (1987).
438 Ruder, S. M. and Ronald, R. C., *Tetrahedron Lett.*, **28**, 135 (1987).
439 Paquette, L. A., Lau, C. J., Browne, A. R., and O'Brien, M. E., *J. Am. Chem. Soc.*, **108**, 8111 (1986).
440 Pitt, I. G., Russell, R. A., Waring, R. B., and Warrener, R. N., *J. Chem. Soc., Chem. Commun.*, **1987**, 1489.
441 Hirokami, S., Takahashi, T., Nagata, M., and Yamazaki, T., *J. Org. Chem.*, **52**, 2455 (1987).
442 Lee-Ruff, E. and Ablenas, F. J., *Can. J. Chem.*, **65**, 1663 (1987).
443 Barton, J. W. and Shepherd, M. K., *J. Chem. Soc., Perkin Trans. 1*, **1987**, 1561.
444 Barton, J. W., Howard, J. A. K., Shepherd, M. K., and Stringer, A. M., *J. Chem. Soc., Perkin Trans. 1*, **1987**, 2443.
445 Bäuml, E., Kolberg, G., and Mayr, H., *Tetrahedron Lett.*, **28**, 387 (1987).
446 Ue, M., Tsukahara, H., Kobiro, K., Kakiuchi, K., Tobe, Y., and Odaira, Y., *Tetrahedron Lett.*, **28**, 3979 (1987).
447 Fitjer, L. and Quabeck, U., *Angew. Chem. Int. Edn*, **26**, 1023 (1987).
448 Krow, G. R., *Tetrahedron*, **43**, 3 (1987).
449 Della, E. W. and Schlesser, C. H., *Tetrahedron Lett.*, **28**, 3869 (1987).
450 Jarret, R. M. and Saunders, M., *J. Am. Chem. Soc.*, **109**, 3366 (1987).
451 Saunders, M., Jarret, R. M., and Pramanik, P., *J. Am. Chem. Soc.*, **109**, 3735 (1987).
452 Barrow, C. J., Bright, S. T., Coxon, J. M., and Steel, P. J., *J. Org. Chem.*, **52**, 5300 (1987).
453 Bartlett, P. D., Ravenscroft, M. D., and Roof, A. A. M., *J. Org. Chem.*, **52**, 1847 (1987).
454 Waddell, T. G. and Ross, P. A., *J. Org. Chem.*, **52**, 4802 (1987).
455 Adam, W., Carballeira, N., Crämer, E., Peters, E.-M., Peters, K., and von Schnering, H. G., *Chem. Ber.*, **120**, 521 (1987).
456 Adam, W., Crämer, E., Peters, E.-M., Peters, K., and von Schnering, H. G., *Chem. Ber.*, **120**, 705 (1987).
457 Hua, D. H., Gung, W. Y., Ostrander, R. A., and Takusagawa, F., *J. Org. Chem.*, **52**, 2509 (1987).

[458] Asaoka, M. and Takei, H., *Tetrahedron Lett.*, **28**, 6343 (1987).
[459] Adams, B. R. and Chenier, P. J., *J. Org. Chem.*, **51**, 5016 (1986).
[460] Cristol, S. J., Aeling, E. O., and Heng, R., *J. Am. Chem. Soc.*, **109**, 830 (1987).
[461] Cristol, S. J., Aeling, E. O., Strickler, S. J., and Ito, R. D., *J. Am. Chem. Soc.*, **109**, 7101 (1987).
[462] Uyehara, T., Kawai, Y., Yamada, J., and Kato, T., *Chem. Lett.*, **1987**, 137.
[463] Uyehara, T., Yamada, J., Furuta, Y., Kato, T., and Yamamoto, Y., *Tetrahedron*, **43**, 5605 (1987).
[464] Bernassau, J.-M., Bouillot, A., Fétizon, M., Hann, I., Maia, E. R., and Prangé, T., *J. Org. Chem.*, **52**, 1993 (1987).
[465] Campbell, M. M., Sainsbury, M., and West, R., *Tetrahedron Lett.*, **28**, 3865 (1987).
[466] Le Drian, C. and Vogel, P., *Helv. Chim. Acta*, **70**, 1703 (1987).
[467] Le Drian, C. and Vogel, P., *Tetrahedron Lett.*, **28**, 1523 (1987).
[468] Bussmann, R. and Heesing, A., *Chem. Ber.*, **120**, 1767 (1987).
[469] Adam, W. and Dörr, M., *J. Am. Chem. Soc.*, **109**, 1570 (1987).
[470] Adam, W. and Dörr, M., Kron, J., and Rosenthal, R. J., *J. Am. Chem. Soc.*, **109**, 7074 (1987).
[471] Kumagai, T., Katayose, M., and Mukai, T., *Chem. Lett.*, **1987**, 899.
[472] Saunders, M. and Jarret, R. M., *J. Org. Chem.*, **51**, 5427 (1986).
[473] Brossi, M. and Ganter, C., *Helv. Chim. Acta*, **70**, 1963 (1987).
[474] Serguchev, Y. A., Krasutskii, P. A., Khotkevich, A. B., and Yurchenko, A. G., *Teor. Eksp. Khim.*, **22**, 743 (1986); *Chem. Abs.*, **107**, 77080 (1987).
[475] Takeuchi, K., Akiyama, F., Miyazaki, T., Kitagawa, I., and Okamoto, K., *Tetrahedron*, **43**, 701 (1987).
[476] Takeuchi, K., Kitagawa, I., Akiyama, F., Shibata, T., Kato, M., and Okamoto, K., *Synthesis*, **1987**, 612.
[477] Odaira, Y. and Kakiuchi, K., *Yuki Gosei Kagaku Kyokaishi*, **44**, 930 (1986); *Chem. Abs.*, **106**, 156684 (1987).
[478] Gleiter, R., Müller, G., Huber-Patz, U., Rodewald, H., Irngartinger, H., *Tetrahedron Lett.*, **28**, 1985 (1987).
[479] Ipaktschi, J., Herber, J., Kalinowski, H.-O., and Boese, R., *Tetrahedron Lett.*, **28**, 3467 (1987).
[480] Marchand, A. P., Sharma, G. V. M., Annapurna, G. S., and Pednekar, P. R., *J. Org. Chem.*, **52**, 4784 (1987).
[481] Mehta, G., Srikrishna, A., Rao, K. S., Reddy, K. R., Acharya, K. A., Puranik, V. G., Tavale, S. S., and Row, T. N. G., *J. Org. Chem.*, **52**, 457 (1987).
[482] Apeloig, Y. and Stanger, A., *J. Am. Chem. Soc.*, **109**, 272 (1987).
[483] Apeloig, Y., Karni, M., Stanger, A., Schwarz, H., Drewello, T., and Czekay, G., *J. Chem. Soc., Chem. Commun.*, **1987**, 989.
[484] Paddon-Row, M. N. and Wong, S. S., *J. Chem. Soc., Chem. Commun.*, **1987**, 1585.
[485] Buttrus, N. H., Eaborn, C., Hitchcock, P. B., Lickiss, P. D., and Najim, S. T., *J. Chem. Soc., Perkin Trans. 2*, **1987**, 891.
[486] Ayoko, G. A. and Eaborn, C., *J. Chem. Soc., Perkin Trans. 2*, **1987**, 1047.
[487] Dhaker, S. M., Eaborn, C., and Smith, J. D., *J. Chem. Soc., Chem. Commun.*, **1987**, 1183.
[488] Ishibashi, H., Nakatani, H., Sakashita, H., and Ikeda, M., *J. Chem. Soc., Chem. Commun.*, **1987**, 338.
[489] Ishibashi, H., Nakatani, H., Maruyama, K., Minami, K., and Ikeda, M., *J. Chem. Soc., Chem. Commun.*, **1987**, 1443.
[490] Bhupathy, M. and Cohen, T., *Tetrahedron Lett.*, **28**, 4793 (1987).
[491] Bhupathy, M. and Cohen, T., *Tetrahedron Lett.*, **28**, 4797 (1987).
[492] Fedorov, N. V., Anisimov, A. V., and Viktorova, E. A., *Vestn. Mosk. Univ., Ser. 2: Khim.*, **27**, 581 (1986); *Chem. Abs.*, **107**, 154005 (1987).
[493] Umemoto, T. and Tomizawa, G., *Bull. Chem. Soc. Jpn*, **59**, 3625 (1986).
[494] Kobayashi, M., Umemura, K., and Matsuyama, H., *Chem. Lett.*, **1987**, 327.
[495] Eaton, P. E. and Hoffmann, K.-L., *J. Am. Chem. Soc.*, **109**, 5285 (1987).
[496] Nguyen, M. T. and Hegarty, A. F., *J. Chem. Soc., Perkin Trans. 2*, **1987**, 55.
[497] Fleischhauer, I. and Brinker, U. H., *Chem. Ber.*, **120**, 501 (1987).
[498] Sekiguchi, A. and Ando, W., *Organometallics*, **6**, 1857 (1987).
[499] Tanigaki, K. and Ebbesen, T. W., *J. Am. Chem. Soc.*, **109**, 5883 (1987).
[500] Tsuchihashi, G. and Suzuki, K., *Kagaku, Zokan (Kyoto)*, **1986**, 119; *Chem. Abs.*, **107**, 58682 (1987).
[501] Koĉovský, P., Tureček, F., Langer, V., Podlahová, J., and Podlaha, J., *J. Org. Chem.*, **51**, 4888 (1986).
[502] Joska, J., Fajkoš, J., and Tureček, F., *Tetrahedron Lett.*, **28**, 2901 (1987).
[503] Balakrishnan, P. and Bhattacharyya, S. C., *Indian J. Chem.*, **25B**, 1050 (1986).
[504] Fuhrer, H., Rihs, G., Lorenc, L., Bondarenko, L., Kalvoda, J., and Mihailović, M. L., *Helv. Chim. Acta*, **70**, 37 (1987).

505 Fischer, P. M. and Howden, M. E. H., *J. Chem. Soc., Perkin Trans. 1*, **1987**, 475.
506 Suginome, H., Kaji, M., Ohtsuka, T., Yamada, S., and Furusaki, A., *J. Chem. Soc., Chem. Commun.*, **1987**, 283.
507 Sicinski, R. R. and Szczepek, W. J., *Tetrahedron Lett.*, **28**, 5729 (1987).
508 Suginome, H., Ohue, Y., and Orito, K., *J. Chem. Soc., Perkin Trans. 1*, **1987**, 1247.
509 Neef, G., Cleve, G., Ottow, E., Seeger, A., and Wiechert, R., *J. Org. Chem.*, **52**, 4143 (1987).
510 Moiseenkovs, A. M. and Ceskis, B., *Collect. Czech. Chem. Commun.*, **51**, 1316 (1986).
511 Kozlov, N. G., Popova, L. A., and Nesterov, G. V., *Zh. Obshch. Khim.*, **56**, 1562 (1986); *Chem. Abs.*, **107**, 40090 (1987).
512 Liu, Z., Wang, Q., Fan, Z., Wang, Q., and Pei, Y., *Huaxue Xuebao*, **44**, 239 (1986); *Chem. Abs.*, **106**, 18909 (1987).
513 Suzuki, K., Matsumoto, T., Tomooka, K., Matsumoto, K., and Tsuchihashi, G., *Chem. Lett.*, **1987**, 113.
514 Paquette, L. A., Lin, H.-S., Belmont, D. T., and Springer, J. P., *J. Org. Chem.*, **51**, 4807 (1986).
515 De La Torre, M. C., Fernandez, P., and Rodriguez, B., *Tetrahedron*, **43**, 4679 (1987).
516 Tomita, Y. and Ikeshiro, Y., *J. Chem. Soc., Chem. Commun.*, **1987**, 1311.
517 Coates, R. M. and Kang, H.-Y., *J. Org. Chem.*, **52**, 2065 (1987).
518 Abell, C. and Leech, A. P., *Tetrahedron Lett.*, **28**, 4887 (1987).
519 Teresa, J. de P., Gonzãlez, M. S., Caballero, M. C., Parra, T., and Bellido, I. S., *Tetrahedron Lett.*, **28**, 821 (1987).
520 Vaidya, S. P. and Nayak, U. R., *Indian J. Chem.*, **25B**, 40 (1986).
521 Reddy, R. T. and Nayak, U. R., *Indian J. Chem.*, **25B**, 793 (1986).
522 Ue, M., Kobiro, K., Kakiuchi, K., Tobe, Y., and Odaira, Y., *Chem. Lett.*, **1987**, 1903.
523 Kasturi, T. R., Chandra, R., Ramachandra, Reddy, S. M., Venkatesan, K., Srikrishnan, T., Row, T. N. G., Puranik, V. G., and Tavale, S. S., *J. Chem. Soc., Chem. Commun.*, **1987**, 75.
524 Wijnberg, J. B. P. A and de Groot, A., *Tetrahedron Lett.*, **28**, 3007 (1987).
525 Yoshikawa, M., Murakami, N., Taniyama, T., Hamamoto, Y., Nakae, T., and Kitagawa, I., *Tetrahedron Lett.*, **28**, 2029 (1987).
526 Binkley, R. W. and Sivik, M. R., *J. Org. Chem.*, **51**, 2619 (1986).
527 Binkley, R. W. and Abdulaziz, M. M., *J. Org. Chem.*, **52**, 4713 (1987).
528 Giese, B., Groeninger, K. S., Witzel, T., Korth, H.-G., and Sustmann, R., *Angew. Chem. Int. Edn*, **26**, 233 (1987).
529 Csuk, R., Glanzer, B. I., Furstner, A., Weidmann, H., and Formacek, V., *Carbohydr. Res.*, **157**, 235 (1986).
530 Danishefsky, S., DeNinno, M. P., and Schulte, G., *J. Org. Chem.*, **52**, 3171 (1987).
531 Grindley, T. B. and Kusuma, S., *Can. J. Chem.*, **64**, 2397 (1986).
532 Grindley, T. B. and Wickramage, C., *Carbohydr. Res.;* **167**, 105 (1987).
533 Hawker, C. J., Stark, W. M., and Battersby, A. R., *J. Chem. Soc., Chem. Commun.*, **1987**, 1313.
534 Yoshida, K., Nomura, S., Nishibata, Y., and Ban, Y., *Heterocycles*, **24**, 2239 (1986).
535 Yanuka, Y., Geryes, A., and Heller, M., *Tetrahedron*, **43**, 911 (1987).
536 Takeshita, K., Takeda, S., and Irie, H., *Chem. Pharm. Bull.*, **34**, 3919 (1986).
537 Sato, S., Urabe, K., and Izumi, Y., *J. Catal.*, **102**, 99 (1986).
538 Sato, S., Hasebe, S., Sakurai, H., Urabe, K., and Izumi, Y., *Appl. Catal.*, **29**, 107 (1987).
539 Aucejo, A., Burguet, M. C., Corma, A., and Fornes, V., *Appl. Catal.*, **22**, 187 (1986).
540 Hlidkova-Kadlecova, H., Orsag, J., Zdimal, V., and Skrivanek, J., *Chem. Prum.*, **36**, 580 (1986); *Chem. Abs.*, **107**, 77090 (1987).
541 Fernández, F. and Pérez, C., *J. Chem. Res. (S)*, **1987**, 340.
542 Suginome, H., Furukawa, K., and Orito, K., *J. Chem. Soc., Chem. Commun.*, **1987**, 1004.
543 Anastasia, M., Allevi, P., Ciuffreda, P., Fiecchi, A., and Scala, A., *J. Chem. Soc., Perkin Trans. 1*, **1986**, 2123.
544 Confalone, P. N. and Huie, E. M., *J. Org. Chem.*, **52**, 79 (1987).
545 Nguyen, M. T., *J. Chem. Soc., Chem. Commun.*, **1987**, 342.
546 Nguyen, M. T., Faul, M., and Fitzpatrick, N. J., *J. Chem. Soc., Perkin Trans. 2*, **1987**, 1289.
547 Harger, M. J. P. and Smith, A., *J. Chem. Soc., Perkin Trans. 1*, **1987**, 683.
548 Hunter, N. R., Khan, M. Z., Marat, K., El-Kabbani, O. A. L., and Delbaere, L. T. J., *Can. J. Chem.*, **65**, 137 (1987).
549 Ohkata, K., Mase, M., and Akiba, K., *J. Chem. Soc., Commun.*, **1987**, 1727.
550 Boger, D. L. and Coleman, R. S., *J. Org. Chem.*, **51**, 5436 (1986).
551 Boger, D. L. and Coleman, R. S., *Tetrahedron Lett.*, **28**, 1027 (1987).
552 Boger, D. L. abd Yohannes, D., *J. Org. Chem.*, **52**, 5283 (1987).

[553] Deeming, A. J., *Inorg. Organomet. React.*, **4**, 377 (1986).
[554] Raevskii, N. I. and Borisov, Y. A., *Izv. Akad. Nauk SSSR, Ser. Khim.*, **1986**, 2132; *Chem. Abs.*, **107**, 134342 (1987).
[555] Tanase, T., Shimizu, F., Kuse, M., Yano, S., Yoshikawa, S., and Hidai, M., *J. Chem. Soc., Chem. Commun.*, **1987**, 659.
[556] Tanase, T., Murata, T., Yano, S., Hidai, M., and Yoshikawa, S., *Chem. Lett.*, **1987**, 1409.
[557] London, R. E., *J. Chem. Soc., Chem. Commun.*, **1987**, 661.
[558] De Boer, E. J. M. and De With, J., *J. Organomet. Chem.*, **320**, 289 (1987).
[559] Zhang, Y., Jiang, J., and Chen, Y., *Tetrahedron Lett.*, **28**, 3815 (1987).
[560] Gau, H.-M. and Fay, R. C., *Inorg. Chem.*, **26**, 3701 (1987).
[561] Petersen, J. L. and Egan, J. W., *Organometallics*, **6**, 2007 (1987).
[562] Erker, G., Mühlenbernd, T., Rufińska, A., and Benn, R., *Chem. Ber.*, **120**, 507 (1987).
[563] Bulls, A. R., Schaefer, W. P., Serfas, M., and Bercaw, J. E., *Organometallics*, **6**, 1219 (1987).
[564] Gulyi, S. E., Novikov, N. A., and Erman, M. B., *Zh. Org. Khim.*, **22**, 1558 (1986); *Chem. Abs.*, **107**, 115220 (1987).
[565] Liotta, F. J., Van Duyne, G., and Carpenter, B. K., *Organometallics*, **6**, 1010 (1987).
[566] Semmelhack, M. F., Ho, S., Steigerwald, M., and Lee, M. C., *J. Am. Chem. Soc.*, **109**, 4397 (1987).
[567] Aumann, R., Kuckert, E., and Heinen, H., *Chem. Ber.*, **120**, 1293 (1987).
[568] Axe, F. U. and Marynick, D. S., *Organometallics*, **6**, 572 (1987).
[569] Innorta, G., Torroni, S., Foffani, A., and Perugini, D., *Inorg. Chim. Acta*, **112**, 183 (1986).
[570] Ishitani, O., Yanagida, S., Takamuku, S., and Pac, C., *Bull. Chem. Soc. Jpn*, **60**, 2419 (1987).
[571] Seyferth, D., Womack, G. B., Gallagher, M. K., Cowie, M., Hames, B. W., Fackler, J. P., and Mazany, A. M., *Organometallics*, **6**, 283 (1987).
[572] Stenstrøm, Y., Koziol, A. E., Palenik, G. J., and Jones, W. M., *Organometallics*, **6**, 2079 (1987).
[573] Chen, Y. C., Lin, C. H., Lee, C. Y., and Liu, C. S., *Organometallics*, **6**, 1882 (1987).
[574] Davies, S. G. and Wills, M., *J. Chem. Soc., Chem. Commun.*, **1987**, 1647.
[575] Traylor, T. G. and Miksztal, A. R., *J. Am. Chem. Soc.*, **109**, 2770 (1987).
[576] Garcia-Granados, A., Martinez, A., and Onorato, M. E., *J. Org. Chem.*, **52**, 606 (1987).
[577] Halpern, J., *Pure Appl. Chem.*, **58**, 575 (1987).
[578] Akbulut, N., Menzek, A., and Balci, M., *Tetrahedron Lett.*, **28**, 1689 (1987).
[579] Murakami, Y., Hisaeda, Y., Ozaki, T., Tashiro, T., Ono, T., Tani, Y., and Matsuda, Y., *Bull. Chem. Soc. Jpn*, **60**, 311 (1987).
[580] Murakami, Y., Hisaeda, Y., and Fan, S.-D., *Chem. Lett.*, **1987**, 655.
[581] Liebeskind, L. S. and Chidambaram, R., *J. Am. Chem. Soc.*, **109**, 5025 (1987).
[582] Palumbo, A., D'Ischia, M., Misuraca, G., and Prota, G., *Biochim. Biophys. Acta*, **925**, 203 (1987).
[583] Chan, D. M. T., Marder, T. B., Milstein, D., and Taylor, N. J., *J. Am. Chem. Soc.*, **109**, 6385 (1987).
[584] Egan, J. W., Hughes, R. P., and Rheingold, A. L., *Organometallics*, **6**, 1578 (1987).
[585] Dekleva, T. W. and Forster, D., *J. Mol. Catal.*, **33**, 269 (1985).
[586] Kunin, A. J., Johnson, C. E., Maguire, J. A., Jones, W. D., and Eisenberg, R., *J. Am. Chem. Soc.*, **109**, 2963 (1987).
[587] Lu, X. and Zhu, J., *J. Organomet. Chem.*, **304**, 239 (1986).
[588] van der Louw, J., van der Baan, J. L., Stieltjes, H., Bickelhaupt, F., and Klumpp, G. W., *Tetrahedron Lett.*, **28**, 5929 (1987).
[589] Parsons, E. J. and Jennings, P. W., *J. Am. Chem. Soc.*, **109**, 3973 (1987).
[590] Chow, Y. L., Chjeng, X. E., and Buono-Core, G. E., *Organometallics*, **6**, 1126 (1987).
[591] Basu, A., Saple, A. R., and Sapre, N. Y., *J. Chem. Soc., Dalton Trans.*, **1987**, 1797.
[592] Maruyama, K. and Tamiaki, H., *Chem. Lett.*, **1987**, 485.
[593] Maruyama, K. and Tamiaki, H., *Chem. Lett.*, **1987**, 683.
[594] Maruyama, K. and Tamiaki, H., *J. Org. Chem.*, **52**, 3967 (1987).
[595] Draper, A. M. and de Mayo, P., *Tetrahedron Lett.*, **27**, 6157 (1986).
[596] Chang, S.-C., Hauge, R. H., Kafafi, Z. H., Margrave, J. L., and Billups, W. E., *J. Chem. Soc., Chem. Commun.*, **1987**, 1682.
[597] Osuka, A., Shimizu, H., Suzuki, H., and Maruyama, K., *Chem. Lett.*, **1987**, 1061.
[598] Inokuchi, T., Takagishi, S., Akahoshi, F., and Torii, S., *Chem. Lett.*, **1987**, 1553.
[599] Shimizu, M. and Yoshioka, H., *Tetrahedron Lett.*, **28**, 3119 (1987).
[600] Dowd, P. and Choi, S.-C., *J. Am. Chem. Soc.*, **109**, 3493 (1987).
[601] Ito, Y., Nakajo, E., Shō, K., and Tamao, K., *Tetrahedron Lett.*, **28**, 2247 (1987).
[602] Fleming, A., Sinai-Zingde, G., Natchus, M. G., and Hudlicky, T., *Tetrahedron Lett.*, **28**, 167 (1987).
[603] Hudlicky, T., Sinai-Zingde, G., Natchus, M. G., Ranu, B. C., and Papadopolous, P., *Tetrahedron*, **43**, 5721 (1987).

604 Zefirov, N. S., Kozhushkov, S. I., Kuznetsova, T. S., and Sosonkin, I. M., *Zh. Org. Khim.*, **22**, 666 (1986); *Chem. Abs.*, **106**, 101752 (1987).
605 Yin, T.-K., Radziszewski, J. G., Renzoni, G. E., Downing, J. W., Michl, J., and Borden, W. T., *J. Am. Chem. Soc.*, **109**, 820 (1987).
606 Komatsu, K., Arai, M., Hattori, Y., Fukuyama, K., Katsube, Y., and Okamoto, K., *J. Org. Chem.*, **52**, 2183 (1987).
607 Kirms, M. A., Primke, H., Stohlemeier, M., and De Meijere, A., *Recl. Trav. Chim. Pays-Bas*, **105**, 462 (1986).
608 Kakumoto, T., Ushirogouchi, T., Saito, K., and Imamura, A., *J. Phys. Chem.*, **91**, 183 (1987).
609 Creary, X., *NATO ASI Ser., Ser. C.*, **189**, 245 (1986); *Chem. Abs.*, **107**, 6538 (1987).
610 Creary, X., Mehrsheikh-Mohammadi, M. E., and McDonald, S., *J. Org. Chem.*, **52**, 3254 (1987).
611 Van Derveer, D. G., Baldwin, J. E., and Parker, D. W., *J. Org. Chem.*, **52**, 1173 (1987).
612 Wiberg, K. B. and McClusky, J. V., *Tetrahedron Lett.*, **28**, 5411 (1987).
613 Kostermans, G. B. M., Hogenbirk, M., Turkenburg, L. A. M., de Wolf, W. H., and Bickelhaupt, F., *J. Am. Chem. Soc.*, **109**, 2855 (1987).
614 Tsuboi, S., Furutani, H., and Takeda, A., *Synthesis*, **1987**, 292.
615 McCombie, S. W. and Nagabhushan, T. L., *Tetrahedron Lett.*, **28**, 5395 (1987).
616 Eberbach, W. and Roser, J., *Tetrahedron Lett.*, **28**, 2685 (1987).
617 Ishii, K., Abe, M., and Sakamoto, M., *J. Chem. Soc., Perkin Trans. 1*, **1987**, 1937.
618 Erden, I. and Amputch, M. A., *Tetrahedron Lett.*, **28**, 3779 (1987).
619 Glombik, H., Wolff, C., and Tochtermann, W., *Chem. Ber.*, **120**, 775 (1987).
620 Ando, W., Itami, A., Furuhata, T., and Tokitoh, N., *Tetrahedron Lett.*, **28**, 1787 (1987).
621 Singh, G., Ila, H., and Junjappa, H., *J. Chem. Soc., Perkin Trans. 1*, **1987**, 1945.
622 Vedejs, E., Grissom, J. W., and Preston, J. K., *J. Org. Chem.*, **52**, 3487 (1987).
623 Jephcote, V. J., John, D. I., and Williams, D. J., *J. Chem. Soc., Perkin Trans. 1*, **1986**, 2195.
624 Miyano, S., Mibu, N., Irie, M., Fujii, S., Fujisaki, F., Abe, N., and Sumoto, K., *J. Chem. Soc., Perkin Trans. 1*, **1987**, 313.
625 Setoi, H., Takeno, H., and Hashimoto, M., *Heterocycles*, **24**, 1261 (1986).
626 Brown, G. R., Foubister, A. J., and Wright, B., *J. Chem. Soc., Perkin Trans. 1*, **1987**, 553.
627 Roberts, D. and Alper, H., *J. Chem. Soc., Chem. Commun.*, **1987**, 212.
628 Duhamel, P., Goument, B., and Plaquevent, J.-C., *Tetrahedron Lett.*, **28**, 2595 (1987).
629 Dehnel, A. and Kanabus-Kaminska, J. M., *Tetrahedron Lett.*, **28**, 3907 (1987).
630 Maier, G. and Teles, J. H., *Angew. Chem. Int. Edn*, **26**, 155 (1987).
631 Squillacote, M. E., Bergman, A., De Felippis, J., and West, E. M., *Tetrahedron Lett.*, **28**, 275 (1987).
632 Sawaki, Y., Ishikawa, S., and Iwamura, H., *J. Am. Chem. Soc.*, **109**, 584 (1987).
633 John, D. I., *Spec. Publ.–R. Soc. Chem.*, **52**, 193 (1985).
634 Jephcote, V. J. and John, D. I., *J. Chem. Soc., Perkin Trans. 1*, **1986**, 2203.
635 Rudolf, K., Spellmeyer, D. C., and Houk, K. N., *J. Org. Chem.*, **52**, 3708 (1987).
636 Wagner, P. J. and Nahm, K., *J. Am. Chem. Soc.*, **109**, 6528 (1987).
637 Chapman, O. L., Tsou, U.-P. E., and Johnson, J. W., *J. Am. Chem. Soc.*, **109**, 553 (1987).
638 Lewis, D. K., Glenar, D. A., Kalra, B. L., Baldwin, J. E., and Cianciosi, S. J., *J. Am. Chem. Soc.*, **109**, 7225 (1987).
639 Christl, M. and Schreck, M., *Angew. Chem. Int. Edn*, **26**, 449 (1987).
640 Christl, M. and Schreck, M., *Chem. Ber.*, **120**, 915 (1987).
641 Burkey, J. D., Leber, P. A., and Silverman, L. S., *Synth. Commun.*, **16**, 1363 (1986).
642 Decker, O. H. W. and Moore, H. W., *J. Org. Chem.*, **52**, 1174 (1987).
643 Reed, M. W. and Moore, H. W., *J. Org. Chem.*, **52**, 3491 (1987).
644 Rubin, M. B. and Harel, Y., *Tetrahedron Lett.*, **28**, 5373 (1987).
645 Gorbatenko, V. I., Mel'nichenko, N. V., and Samarai, L. I., *Zh. Org. Khim.*, **22**, 1184 (1986); *Chem. Abs.*, **106**, 17678 (1987).
646 Brook, A. G., Chatterton, W. J., Sawyer, J. F., Hughes, D. W., and Vorspohl, K., *Organometallics*, **6**, 1246 (1987).
647 Girard, M., ApSimon, J. W., and Huber, C. P., *Can. J. Chem.*, **65**, 191 (1987).
648 Yates, P. and Burke, P. M., *Can. J. Chem.*, **65**, 1695 (1987).
649 Mack, R. A. and Georgiev, V. St., *J. Org. Chem.*, **52**, 477 (1987).
650 Graziano, M. L., Iesce, M. R., Cinotti, A., and Scarpati, R., *J. Chem. Soc., Perkin Trans. 1*, **1987**, 1833.
651 Matturro, M. G. and Reynolds, R. P., *Tetrahedron Lett.*, **28**, 4981 (1987).
652 Caswell, L. R. and Chao, A. H. M., *J. Chem. Eng. Data*, **32**, 389 (1987).
653 Chung, M. W. L., Lai, T.-F., and Sammes, M. P., *J. Chem. Res. (S)*, **1987**, 366.

[654] Roberts, J. D., Stefaniak, L., Webb, G. A., and Witanowski, M., *Bull. Pol. Acad. Sci., Chem.*, **34**, 81 (1986); *Chem. Abs.*, **107**, 115089 (1987).
[655] Beckwith, A. L. J., O'Shea, D. M., Gerba, S., and Westwood, S. W., *J. Chem. Soc., Chem. Commun.*, **1987**, 666.
[656] Baldwin, J. E., Adlington, R. M., Kang, T. W., Lee, E., and Schofield, C. J., *J. Chem. Soc., Chem. Commun.*, **1987**, 104.
[657] Picavet, J. P. and Sliwa, H., *J. Chem. Res. (S)*, **1987**, 230.
[658] Sueishi, Y., Ohcho, M., Yamamoto, S., and Nishimura, N., *Bull. Chem. Soc. Jpn*, **59**, 3666 (1986).
[659] Bocelli, G., Chiusoli, G. P., Costa, M., and Fambri, L., *J. Chem. Soc., Chem. Commun.*, **1987**, 1182.
[660] Robin, Y., Godard, A., and Queguiner, G., *J. Heterocycl. Chem.*, **24**, 1487 (1987).
[661] Milenkov, B. and Hesse, M., *Helv. Chim. Acta*, **70**, 308 (1987).
[662] Milenkov, B., Guggisberg, A., and Hesse, H., *Helv. Chim. Acta*, **70**, 760 (1987).
[663] Milenkov, B., Guggisberg, A., and Hesse, H., *Tetrahedron Lett.*, **28**, 315 (1987).
[664] Dowd, P. and Choi, S.-C., *J. Am. Chem. Soc.*, **109**, 6548 (1987).
[665] Perrin, D., Richard, C., and Martin, R., *J. Chim. Phys. Phys.-Chim. Biol.*, **83**, 329 (1986); *Chem. Abs.*, **106**, 137750 (1987).
[666] Ito, Y., Uozu, Y., and Matsuura, T., *Tetrahedron Lett.*, **28**, 3493 (1987).
[667] Bonnett, R., Hamzetash, D., and Vallés, M. A., *J. Chem. Soc., Perkin Trans. 1*, **1987**, 1383.
[668] Souizi, A., Robert, A., Batail, P., and Oiahab, L., *J. Org. Chem.*, **52**, 1610 (1987).
[669] Becker, H.-D. and Andersson, K., *Tetrahedron Lett.*, **28**, 1323 (1987).
[670] Rubin, M. B. and Sander, W. W., *Tetrahedron Lett.*, **28**, 5137 (1987).
[671] Weber, R. H., Brügger, P., Jenny, T. A., and Hansen, H.-J., *Helv. Chim. Acta*, **70**, 742 (1987).
[672] Asano, T. and Okada, T., *Chem. Lett.*, **1987**, 691.
[673] Asano, T. and Okada, T., *Chem. Lett.*, **1987**, 695.
[674] Tomioka, H., Hamano, Y., and Izawa, Y., *Bull. Chem. Soc. Jpn*, **60**, 821 (1987).
[675] Harel, Y., Adamson, A. W., Kutal, C., Grutsch, P. A., and Yasufuku, K., *J. Phys. Chem.*, **91**, 901 (1987).
[676] Arai, T., Oguchi, T., Wakabayashi, T., Tsuchiya, M., Nishimura, Y., Oishi, S., Sakuragi, H., and Tokumaru, K., *Bull. Chem. Soc. Jpn*, **60**, 2937 (1987).
[677] Gassman, P. G. and Hershberger, J. W., *J. Org. Chem.*, **52**, 1337 (1987).
[678] Ikezawa, H. and Kutal, C., *J. Org. Chem.*, **52**, 3299 (1987).
[679] Paquette, L. A. and Racherla, U. S., *J. Org. Chem.*, **52**, 3250 (1987).
[680] Meier, M. and Ruechardt, C., *Chem. Ber.*, **120**, 1 (1987).
[681] Barron, A. R. and Cowley, A. H., *J. Chem. Soc., Chem. Commun.*, **1987**, 1092.
[682] Gaspar, P. P., Boo, B. H., and Svoboda, D. L., *J. Phys. Chem.*, **91**, 5011 (1987).
[683] Gimarc, B. H. and Ott, J. J., *J. Am. Chem. Soc.*, **109**, 1388 (1987).
[684] Kwiatkowski, J. S., Zielinski, T. J., and Rein, R., *Adv. Quantum Chem.*, **18**, 85 (1986); *Chem. Abs.*, **107**, 77008 (1987).
[685] Weis, A. and Van der Plas, H. C., *Heterocycles*, **24**, 1433 (1986).
[686] Wesdemiotis, C. and McLafferty, F. W., *J. Am. Chem. Soc.*, **109**, 4760 (1987).
[687] Tang, G.-Q., MacInnis, J., and Kasha, M., *J. Am. Chem. Soc.*, **109**, 2531 (1987).
[688] Fülöp, F., Pihlaja, K., Mattinen, J., and Bernáth, G., *J. Org. Chem.*, **52**, 3821 (1987).
[689] Fülöp, F., Pihlaja, K., Mattinen, J., and Bernáth, G., *Tetrahedron*, **43** 1863 (1987).
[690] Lambert, J. B., Wang, G., Huseland, D. E., and Takiff, L. C., *J. Org. Chem.*, **52**, 68 (1987).
[691] Millefiori, S. and Millefiori, A., *J. Heterocycl. Chem.*, **24**, 525 (1987).
[692] Crossley, M. J., Field, L. D., Harding, M. M., and Sternhell, S., *J. Am. Chem. Soc.*, **109**, 2335 (1987).
[693] Messmer, A., Hajós, G., Giber, J., and Holly, S., *J. Heterocycl. Chem.*, **24**, 1133 (1987).

Author Index

In this index bold figures relate to chapter numbers, roman figures are reference numbers.

Subject Index